ADVANCING MITIGATION TECHNOLOGIES AND DISASTER RESPONSE FOR LIFELINE SYSTEMS

PROCEEDINGS OF THE SIXTH U.S. CONFERENCE AND WORKSHOP ON
LIFELINE EARTHQUAKE ENGINEERING

August 10–13, 2003
Long Beach, California

EDITED BY
James E. Beavers

Technical Council on Lifeline Earthquake Engineering
Monograph No. 25

Published by
American Society of Civil Engineers

Library of Congress Cataloging-in-Publication Data

U.S. Conference on Lifeline Earthquake Engineering (6th : 2003 : Long Beach, Calif.)
  Advancing mitigation technologies and disaster response for lifeline systems : proceedings of the sixth U.S. Conference on Lifeline Earthquake Engineering, August 10-13, 2003, Long Beach, California / edited by James E. Beavers.
    p. cm. -- (Technical Council on Lifeline Earthquake Engineering monograph ; no. 25)
  Includes bibliographical references and index.
  ISBN 0-7844-0687-1
  1. Lifeline earthquake engineering--Congresses. I. Beavers, James E. II. Title. III. Monograph (American Society of Civil Engineers. Technical Council on Lifeline Earthquake Engineering) ; no. 25.

TA654.6 U15 2003
624.1'762--dc21   2003052065

American Society of Civil Engineers
1801 Alexander Bell Drive
Reston, Virginia, 20191-4400

www.pubs.asce.org

Any statements expressed in these materials are those of the individual authors and do not necessarily represent the views of ASCE, which takes no responsibility for any statement made herein. No reference made in this publication to any specific method, product, process, or service constitutes or implies an endorsement, recommendation, or warranty thereof by ASCE. The materials are for general information only and do not represent a standard of ASCE, nor are they intended as a reference in purchase specifications, contracts, regulations, statutes, or any other legal document. ASCE makes no representation or warranty of any kind, whether express or implied, concerning the accuracy, completeness, suitability, or utility of any information, apparatus, product, or process discussed in this publication, and assumes no liability therefore. This information should not be used without first securing competent advice with respect to its suitability for any general or specific application. Anyone utilizing this information assumes all liability arising from such use, including but not limited to infringement of any patent or patents.

ASCE and American Society of Civil Engineers—Registered in U.S. Patent and Trademark Office.

Photocopies: Authorization to photocopy material for internal or personal use under circumstances not falling within the fair use provisions of the Copyright Act is granted by ASCE to libraries and other users registered with the Copyright Clearance Center (CCC) Transactional Reporting Service, provided that the base fee of $18.00 per article is paid directly to CCC, 222 Rosewood Drive, Danvers, MA 01923. The identification for ASCE Books is 0-7844-0687-1/03/ $18.00. Requests for special permission or bulk copying should be addressed to Permissions & Copyright Dept., ASCE.

Copyright © 2003 by the American Society of Civil Engineers. All Rights Reserved.
Library of Congress Catalog Card No: 2003052065
ISBN 0-7844-0687-1
Manufactured in the United States of America.

Foreword

The Sixth U.S. Conference and Workshops on Lifeline Earthquake Engineering (TCLEE 2003) was held in Long Beach, California, on the 70th anniversary year of the 1933 Long Beach earthquake. The First U. S. Conference on Lifeline Earthquake Engineering was held in 1977 following the 1971 San Fernando earthquake. Unfortunately, the fragility of lifelines infrastructure and support systems continue to be a subject of concern as we move into the 21st Century. Recent earthquakes all over the world have shown the vulnerability of lifelines to earthquakes not properly designed for such severe earthquake ground motions. Since the fifth conference held in Seattle, Washington, in 1999, we have seen this during post-earthquake investigations by TCLEE members from earthquakes in El Salvador; India; Peru; Taiwan; Turkey (Izmit/Kocalei and Duzce); and in the United States (Napa Valley, Nisqually, and Hector Mine). In addition, new research has been conducted on the seismic performance of lifelines, new standards and codes have been developed, and major lifeline projects have been rehabilitated, designed, and constructed. We know that lifelines properly designed and constructed for severe earthquake ground motions can perform well as was demonstrated by the Trans-Alaska Pipeline during the November 3, 2002 Denali Fault earthquake.

Times have changed since the first lifelines conference. At that time we were primarily focused on the response of lifelines to only earthquakes, one of several natural hazards. Today we need to focus our efforts on all natural and technological hazards. As a result, the theme for TCLEE 2003 was **Advancing Mitigation Technologies and Disaster Response**. The conference goals were to provide the opportunity:

1. To bring together engineers, seismologist, geologist, social scientists, and mangers to exchange information about earthquakes and lifeline performance.
2. To discuss state-of-the-art lifeline earthquake engineering by individuals in their respective areas of expertise.
3. To explore similarities between response and vulnerability of lifelines to earthquakes and other types of catastrophic external loading events.
4. For ASCE members and nonmembers to participate in workshops in various specialty areas.
5. To introduce industry-standard systems and renowned market applications and products through vendor and organization exhibits.
6. For attendees to interface with each other and exchange information for improving seismic performance of lifelines, especially one-on-one discussions through poster sessions.

This Proceedings presents invited reports and results of studies by investigators who have analyzed and evaluated the effects of natural and technological hazards on lifelines. Unfortunately, we did not have as many papers as hoped on the subject of technological hazards, but this is a start for the future. Technical sections of this volume include: (1) Lifeline Management and Policy Issues, (2) Transportation System Preparedness, Response and Recovery, (3) Retrofit and Policy Issues with the Bay Area Rapid Transit System, (4) Post-Earthquake Investigations and Lifeline Performance During Earthquakes, (5) Lifeline Guidelines and Codes, (6) Performance of Ports and Wharves, (7) Water and Waste Water Systems and Components, (8) The Trans-Alaska Pipeline and the 2002 Denali Fault Earthquake, (9) Electric Power Guidelines and Performance of Equipment, (10) Design, Mitigation and Performance of Underground Pipelines, (11) Seismic Hazard and Risk Issues, (12) Seismic Risk Analysis of Highway Systems, (13), Geotechnical Lifeline Issues and (14) Innovative Mitigation of Bridges.

The conference was organized and convened by ASCE under the direction of the Technical Council on Lifeline Earthquake Engineering. The Conference Organizing Committee consisted of the following:

Executive Committee
Donald Ballantyne, *ABS Group International*
Ian G. Buckle, *University of Reno*
Anne S. Kiremidjian, *Stanford University*
Ronald T. Eguchi, *ImageCat, Inc.*
Thomas D. O'Rourke, *Cornell University*

Technical Program Committee
John Eidinger, *G&E Engineering Systems, Inc.*
Nesrin Basoz, *St. Paul Fire & Marine Fire Insurance Co.*
Curtis Edwards, *Pountney Consulting*
John A. Egan, *Geomatrix Consultants Inc.*
James D. Hart, *SSD, Inc.*
William F. Heubach, *Seattle Public Utilities*
Chih-Hung Lee, *Pacific Gas & Electric*
LeVal Lund, *Civil Engineer*
Anshel Schiff, *Stanford University*
Craig E. Taylor, *Natural Hazards Management Inc.*

The Organizing Committee extends special thanks for the financial support and cooperation of the United States Geological Survey, Mid-America Earthquake Center, Pacific Earthquake Engineering Research Center, Multidisciplinary Center for Earthquake Engineering, Port of Los Angeles, and Pacific Gas and Electric. We also thank the Port of Long Beach for the ports tour and our co-sponsors: Building Seismic Safety Council, Center for Earthquake Research and Information, Earthquake

Engineering Research Institute, Electric Power Research Institute, Federal Highway Administration, Global Alliance for Disaster Reduction, Multi-Hazard Mitigation Council, Seismological Society of America, and the Southern California Earthquake Center.

Each of the papers included in the *Proceedings* has been accepted for publication by the Proceedings Editor. All papers are eligible for discussion in the appropriate journals of ASCE. All papers are eligible for ASCE awards.

James E. Beavers
Editor and Conference Chair

TCLEE Monograph Series

These publications may be purchased from ASCE via telephone 1-800-548-ASCE (2723) or the world wide web www.asce.org. The TCLEE publications Web site is www.asce.org/disasterreduction/tclee_pubs.cfm

Kiremidjian, Ann, editor. *Recent Lifeline Seismic Risk Studies.* TCLEE Monograph No. 1, 1990.

Taylor, Craig, editor. *Seismic Loss Estimates for a Hypothetical Water System.* TCLEE Monograph No. 2, 1991.

Schiff, Anshel J., editor. *Guide to Post Earthquake Investigations of Lifelines.* TCLEE Monograph No. 3, 1991.

Cassaro, Michael, editor. *Lifeline Earthquake Engineering. Proceedings of the Third U.S. Conference.* TCLEE Monograph No. 4, August 1991.

Ballantyne, Donald, editor. *Lifeline Earthquake Engineering in the Central and Eastern U.S.* TCLEE Monograph No. 5, September 1992.

O'Rourke, Michael, editor. *Lifeline Earthquake Engineering. Proceedings of the Fourth U.S. Conference.* TCLEE Monograph No. 6, August 1995.

Schiff, Anshel J. and Buckle, Ian, editors. *Critical Issues and State of the Art on Lifeline Earthquake Engineering.* TCLEE Monograph No. 7, October 1995.

Schiff, Anshel J., editor. *Northridge Earthquake: Lifeline Performance and Post-Earthquake Response.* TCLEE Monograph No. 8, August 1995.

McDonough, Peter W., editor. *Seismic Design for Natural Gas Distributors.* TCLEE Monograph No. 9, August 1995.

Tang, Alex and Schiff, Anshel, editors. *Methods of Achieving Improved Seismic Performance of Communication Systems.* TCLEE Monograph No. 10, September 1996.

Schiff, Anshel, editor. *Guide to Post-Earthquake Investigation of Lifelines.* TCLEE Monograph No. 11, July 1997.

Werner, Stuart D., editor. *Seismic Guide to Ports.* TCLEE Monograph No. 12, March 1998.

Taylor, C, Mittler, E., Lund, L., editors. *Overcoming Barriers: Lifeline Seismic Improvement Programs*. TCLEE Monograph No. 13, September 1998.

Schiff, A.J., editor. *Hyogoken-Nanbu (Kobe) Earthquake of January 17, 1995—Lifeline Performance*. TCLEE Monograph No. 14, September 1998.

Eidinger, John M., and Avila, Ernesto A., editors. *Guidelines for the Seismic Upgrade of Water Transmission Facilities*. TCLEE Monograph No. 15, January 1999.

Elliott, William M. and McDonough, Peter, editors. *Optimizing Post Earthquake Lifeline System Reliability, Proceedings of the Fifth National Conference on Lifeline Earthquake Engineering*. TCLEE Monograph No. 16, September 1999.

Tang, Alex K., editor. *Izmit (Kocaeli) Earthquake of August 17, 1999 Including Duzce Earthquake of November 12, 1999—Lifeline Performance*. TCLEE Monograph No. 17, March 2000.

Schiff, Anshel J., and Tang, Alex K., editors. *Chi-Chi Taiwan Earthquake of September 21, 1999—Lifeline Performance*. TCLEE Monograph No. 18, 2000.

Eidenger, John M., editor. *Kujurat (Kutch) India M7.7 Earthquake of January 26, 2001 and Napa M5.2 Earthquake of September 3, 2000*. TCLEE Monograph No. 19, 2001.

McDonough Peter, editor. *The Nisqually, Washington Earthquake of February 28, 2001—Lifeline Performance*. TCLEE Monograph No. 20, 2002.

Taylor, Craig and VanMarcke, Erik, editors. *Acceptable Risk Processes: Lifelines and Natural Hazards*. TCLEE Monograph No. 21, 2002.

Heubach, William F., editor. *Seismic Screening Checklists for Water and Wastewater Facilities*. TCLEE Monograph No. 22, 2003

Edwards, Curtis L., editor. *Atico, Peru M_W 8.4 Earthquake of June 23, 2001*. TCLEE Monograph No. 23, 2003

Lund, Le Val and Sepponen, Carl. editors. *Lifeline Performance of El Salvador Earthquakes of January 13 and February 13, 2001*. TCLEE Monograph No. 24, 2003

Beavers, James E., editor. *Advancing Mitigation Technologies and Disaster Response for Lifeline Systems: Proceedings of the Sixth U.S. Conference and Workshop on Lifeline Earthquake Engineering*. TCLEE Monograph No. 25, 2003.

TCLEE Publications

Duke, C. Martin, editor. *The Current State of Knowledge of Lifeline Earthquake Engineering*, 1977.

Dowd, Munson, editor. *Annotated Bibliography on Lifeline Earthquake Engineering*, 1980.

Smith, D. J. Jr., editor. *Lifeline Earthquake Engineering: The Current State of Knowledge*, 1981.

Hall, William, editor. *Advisory Notes on Lifeline Earthquake Engineering*, 1983.

Nyman, Douglas, editor. *Guidelines for the Seismic Design of Oil and Gas Pipeline Systems, TCLEE Committee on Gas and Liquid Fuel Lifelines*, 1984.

Coopers, James, editor. *Lifeline Earthquake Engineering: Performance, Design and Construction*, 1984.

Eguchi, Ron and Crouse, C. B., editors. *Lifeline Seismic Risk Analysis—Case Studies*, 1986.

Cassaro, Michael and Martinez-Romero, E., editors. *The Mexico Earthquakes—1985 Factors Involved and Lessons Learned*, 1986.

Wang, Leon R. L. and Whitman, Robert, editors. *Seismic Evaluation of Lifeline Systems—Case Studies*, 1986.

Cassaro, Michael and Cooper, James, editors. *Seismic Design and Construction of Complex Civil Engineering Systems*, 1988.

Werner, Stuart D. and Dickenson, Stephen E., editors. *Hyogoken-Nanbu (Kobe) Earthquake of January 17, 1995: A Post-Earthquake Reconnaissance of Port Facilities, TCLEE Committee on Ports and Harbors Lifelines*, 1996.

Schiff, Anshel J., editor. *Guide to Improved Earthquake Performance of Electric Power Systems, ASCE Manual 96*, 1999.

Contents

Lifeline Management and Policy Issues

The Policy Legacies of California's Dam Act of 1929 1
    Robert A. Olson

Development of Disaster Management Equipment in India: A Way Towards Resurgence 9
    D.N. Singh and Vipin C. Shukla

Community Preparedness and Response Model for Protecting Infrastructures 19
    Nasim Uddin, Dennis Engi, and A. Haque

Integrating Lifelines Engineering with Emergency Management: The New Zealand Approach 29
    David Brunsdon and Noel Evans

Time and Space Restoration Process and Prediction of Recovery Period for Damaged Water Supply Systems Based on Geographic Information System (GIS) Data of the 1995 Kobe Earthquake 39
    Shiro Takada and Tatsuhiko Imanishi

Vulnerability of Energy Distribution Systems to an Earthquake in the Central and Eastern United States—An Update 49
    John M. Nichols and James E. Beavers

Planning and Mitigating for Local Tsunami Effects 59
    Craig Erdman, Jane Preuss, Elson T. Barnett, and Vivyan Murphy

Lifelines and Earthquakes in Switzerland 73
    Blaise Duvernay, Anne Eckhardt, and Kerstin Lang

Seismic Risk Assessment and Upgrade Strategy of Hospital-Lifeline Performance 82
    Yasuko Kuwata and Shiro Takada

Managing the Earthquake Risk: Research and Implementation Efforts on Utah's Highway Infrastructure 92
    Blaine D. Leonard

Development of Seismic Disaster Mitigation Master Plan for Asia-Pacific Regions Through Implementation of Risk Management Framework 103
    Hiromichi Higashihara

Transportation System Preparedness, Response, and Recovery

Development of a ShakeMap-Based, Earthquake Response System Within Caltrans 113
    David J. Wald, Philip A. Naecker, Cliff Roblee, and Loren Turner

Seismic Risk Assessment of Transportation System: Evaluation Immediately
After Earthquake 123
    Chin Hsiung Loh, Chun-Yu Lee, and Chin-Hsun Yeh

Development of a Handbook for Seismic Performance Testing of Bridge Piers 133
    J. Jerry Shen, W. Phillip Yen, and John O'Fallon

Operational Performance Seismic Design of Highway Bridges for 2500-Year
Earthquake Using Proposed National Cooperative Highway Research
Program (NCHRP) Provisions 143
    Bardia Emami and W.N. Marianos

Pushover Analysis of Bridge Intermediate Bents 153
    Jeffrey Ger and Phillip Yen

Retrofit and Policy Issues with the Bay Area Rapid Transit System (BART)

Bay Area Rapid Transit (BART) Seismic Retrofit Program: Characterization
of Design Ground Motion 163
    J. Litehiser, N. Gregor, J. Marrone, F. Ostadan, and R. Youngs

Seismic Risk Analysis of the Bay Area Rapid Transit (BART) System 173
    John Eidinger, Ed Matsuda, Tom Horton, and Ching Wu

Fragility Formulations for the Bay Area Rapid Transit (BART) System 183
    Mark Salmon, James Wang, David Jones, and Ching Wu

Seismic Retrofit Concepts for Bay Area Rapid Transit (BART) Aerial Structures 193
    Chip Mallare, Ed Matsuda, Bill Hughes, and Eric Fok

Seismic Assessment and Retrofit Concepts of the Bay Area Rapid Transit (BART)
Transbay Tube 203
    Ching Wu, Eric Fok, George Fotinos, Wen Tseng, and Gary Oberholtzer

*Post-Earthquake Investigations
and Lifeline Performance During Earthquakes*

Participating in International Post-Earthquake Lifelines Investigations 213
    Curt Edwards and Anshel J. Schiff

Lifelines Performance, Long Beach Earthquake, March 10, 1933: A Historical
Perspective 219
    Le Val Lund

Hospital Lifeline Response to the 1999 Izmit Turkey Earthquake 224
    Mark A. Pickett

Obtaining the Emergency Transportation Network for Rescue and Relief Activities
in Large Cities Based on the Life Loss Mitigation Criteria 231
    Afshin Shariat Mohaymany, Mahmood Hosseini, and Hossein Motevalli
    Habini

Performance of Yen-Feng Bridge During The 921 Taiwan Chi-Chi Earthquake 241
    Kuo-Chun Chang, Kung-Yuan Kuo, and Chih-Hung Lu

Post-Earthquake Lifeline Service Restoration Modeling..255
Z. Çagnan and R. Davidson

Lifeline Performance, El Salvador Earthquakes, January 13 and
February 13, 2001 265
Le Val Lund

Effects of Six Recent Earthquakes on Railroads 274
William G. Byers

Damage of Gas and Water Pipelines in Slope City, Kure, Due to the 2001 Geiyo
Earthquake...284
Junichi Ueno and Shiro Takada

Performance of Corrugated Metal Pipe (CMP) Culverts During Past Earthquakes....294
T. Leslie Youd and Chris J. Beckman

Lifeline Damage from the January 26, 2001 M7.7 Gujarat, India Earthquake 308
Curtis Edwards

Lifeline Guidelines and Codes

The American Lifelines Alliance Progress in the Development of National Consensus
Guidelines for the Design of Lifeline Systems for Natural and Man-Made Hazards....318
Douglas G. Honegger, Edward M. Laatsch, and Timothy D. Sheckler

Seismic Design Standards and Guidelines of Steel and Concrete Liquid
Storage Tanks 327
Lisa Yunxia Wang

Simplified Models for Flexibly Supported Liquid Storage Tanks and Its Application to
Eurocode 8, Part 4 339
J. Habenberger and J. Schwarz

Developing the First National Code for Gas Lifeline System in Iran: Possibilities and
Challenges..349
Mahmood Hosseini

The American Lifelines Alliance Approach: Four Years of Progress and Future
Directions...359
Edward M. Laatsch, Douglas G. Honegger, and Timothy D. Sheckler

Improving Natural Gas Safety in Earthquakes—California Recommendations 368
Fred Turner and Douglas Honegger

Seismic Design and Retrofit of Piping Systems: Overview of a Recent
American Lifelines Alliance (ALA) Report 378
George Antaki

Performance of Ports and Wharves

Analyzing the Seismic Performance of Wharves, Part 1: Structural-Engineering
Approach 385
W.H. Roth, E.M. Dawson, M. Mehrain, and A. Sayegh

Analyzing the Seismic Performance of Wharves, Part 2: Soil-Structure Interaction (SSI) Analysis with Non-Linear, Effective-Stress Soil Models 395
    W.H. Roth and E.M. Dawson

Seattle Alaskan Way Seawall Emergency Earthquake Evaluation and Repair 405
    Farhad Rowshanzamir, Dave Swanson, John Buswell, and Daniel Mageau

Container Wharf Upgrade and Seismic Strengthening Guidelines at the Port of Los Angeles 415
    Peter Yin, Stacey Jones, and Max Weismair

The Use of Simulation in Disaster Response Planning and Risk Management of Ports and Harbors 425
    Dimitris Pachakis and Anne S. Kiremidjian

Water and Waste Water Systems and Components

Economics of Seismic Retrofit of Water Transmission and Distribution Systems 435
    John Eidinger

Seismic Reliability of Urban Pipeline Network Systems 445
    Han Yang and Sun Shaoping

Guidelines for Defining Natural Hazards Performance Objectives for Water Systems 455
    W.P. Graf, C.E. Taylor, J.H. Wiggins, L. Lund, and T. Volz

Online Monitoring of Seismic Damage in a Water Delivery System 464
    Jianwen Liang

Evaluating Mitigation of Urban Infrastructure Systems: Application to the Los Angeles Department of Water and Power 474
    Stephanie E. Chang and Hope A. Seligson

URAMP (Utilities Regional Assessment of Mitigation Priorities)—A Benefit-Cost Analysis Tool for Water, Wastewater and Drainage Utilities: Software Development 484
    Charles K. Huyck, Ronald T. Eguchi, Reid M. Watkins, Hope A. Seligson, Stephen Bucknam, and Edward Bortugno

URAMP (Utilities Regional Assessment of Mitigation Priorities)—A Benefit-Cost Analysis Tool for Water, Wastewater and Drainage Utilities: Methodology Development 494
    Hope A. Seligson, Donald B. Ballantyne, Charles K. Huyck, Ronald T. Eguchi, Stephen Bucknam and Edward Bortugno

Standard Guidelines to Assess the Seismic Fragility of Water Transmission Systems 504
    Ronald T. Eguchi and Douglas G. Honegger

Comparison of Mitigation Alternatives for Water Distribution Pipelines Installed in Liquefiable Soils 512
    Donald Ballantyne and William Heubach

The Trans-Alaska Pipeline and the 2002 Denali Fault Earthquake

Performance of the Trans-Alaska Pipeline in the November 3, 2002 Denali Fault Earthquake..522
William J. Hall, Douglas J. Nyman, Elden R. Johnson, and J. David Norton

Seismic Hazard Exposure for the Trans-Alaska Pipeline..535
Lloyd S. Cluff, Robert A. Page, D. Burton Slemmons, and C.B. Crouse

Effect of the Denali Fault Rupture on the Trans-Alaska Pipeline..................................547
Steve P. Sorensen and Keith J. Meyer

Response of the Above-Ground Trans-Alaska Pipeline to the Magnitude 7.9 Denali Fault Earthquake..556
Steve P. Sorensen, Keith J. Meyer, Paul A. Carson, and William J. Hall

Assessment of the Below-Ground Trans-Alaska Pipeline Following the Magnitude 7.9 Denali Fault Earthquake..566
Elden R. Johnson, Michael C. Metz, and David A. Hackney

Trans-Alaska Pipeline Emergency Response and Recovery Following the November 3, 2002 Denali Fault Earthquake..576
Douglas J. Nyman, Elden R. Johnson, and Christopher H. Roach

Electric Power Guidelines and Performance of Equipment

System Earthquake Risk Assessment II (SERA II)..587
Dennis K. Ostrom

Seismic Displacement at Interconnection Points of Substation Equipment..................597
Jean-Bernard Dastous and André Filiatrault

Issues and Guidance for Institute of Electrical and Electronics Engineers (IEEE) 693 Equipment Qualification Tests..607
Anshel J. Schiff and Leon Kempner, Jr.

Interaction Between Electrical Substation Equipment Connected by Rigid Bus Slider..617
Junho Song, Armen Der Kiureghian, and Jerome L. Sackman

Seismic Evaluation of Hybrid Bus Connections...627
L. Kempner, Jr., W.H. Mueller III, and N.J. Hutson

Seismic Design of Secondary Systems..637
T.S. Aziz

Seismic Response of High Voltage Transformers..647
Howard Matt and André Filiatrault

Interpretation and Application of Hilbert-Huang Transformation for Seismic Performance Analyses..657
J. Jerry Shen, W. Phillip Yen, and John O'Fallon

An Experimental Study on the Seismic Response of Electrical Substation
Equipment Interconnected by Flexible Conductors 667
André Filiatrault and Christopher Stearns

Seismic Risk Management System for Electric Power Facilities 677
Yoshiharu Shumuta

A Comparison of Seismic (Dynamic) and Static Load Cases for Electric Transmission
Structures 687
Michael J. Riley, Leon Kempner, Jr., and Wendelin H. Mueller III

Simplified Seismic Calculation Method for Coupled System of Transmission Lines
and Their Supporting Tower 697
Hong-Nan Li, Wen-Long Shi, and Su-Yan Wang

Development of a Probabilistic Assessment Model for Post-Earthquake Residual
Capacity of Utility Lifeline Systems 707
Nobuoto Nojima and Masata Sugito

Design, Mitigation, and Performance of Underground Pipelines

Numerical Simulation of the Behaviour of Buried Jointed Pipelines Under Extremely
Large Fault Displacements 717
Radan Ivanov and Shiro Takada

Southern Loop Pipeline—Seismic Installation in Today's Urban Environment 727
Tom Shastid, Javier Prospero, and John Eidinger

Pipeline Seismic Mitigation Using Trenchless Technology 736
Le Val Lund

Considerations for the Design of Buried Natural Gas and Liquid Hydrocarbon
Pipeline Fault Crossings 744
Douglas J. Nyman, Douglas G. Honegger, and Paul C. Thenhaus

Centrifuge Modeling of Buried Pipelines 757
Michael O'Rourke, Vikram Gadicherla, and Tarek Abdoun

Seismic Hazard and Risk Issues

Modeling of Phase Spectra for Simulation of Near-Fault Design Earthquake
Motions 769
T. Sato, Y. Murono, and M. Murakami

The Role of Urban Planning and Design in Lifeline-Related Seismic Risk
Mitigation 779
Mahmood Hosseini and Leila Niazi Shemirani

Earthquake and Terrorism Risk Assessment: Similarities and Differences 789
Stephanie A. King, Hamid R. Adib, John Drobny, and James Buchanan

Prevention and Repair Measures for Infrastructure Natural Disaster Risk
Management 799
Jacob Greenstein

Regional Assessment of Earthquake Hazard in Japan (Part 1: Overall Framework) 809
  Takayuki Shimazu, Haruki Shimazu, and Naoki Shimazu

Regional Assessment of Earthquake Hazard in Japan (Part 2: Lifeline and Transportation) 819
  Takayuki Shimazu, Naoki Shimazu, and Haruki Shimazu

Quantitative Method for Developing Hazard-Consistent Earthquake Scenarios 829
  Kenneth W. Campbell and Hope A. Seligson

Seismic Hazard Analysis and Developing the Uniform Hazard Spectra for an Under-Construction Railroad Bridge 839
  Fariborz Yaghoobi Vayeghan, Maryam Firoozi Nezamabadi, and Mahmood Hosseini

Seismic Risk Analysis of Highway Systems

Current Developments and Future Directions for Seismic Risk Analysis of Highway Systems 849
  Stuart D. Werner

Earthquake Occurrence Modeling for Evaluating Seismic Risks to Roadway Systems 859
  David Perkins and Craig Taylor

Modeling Transportation Network Flows as a Simultaneous Function of Travel Demand, Earthquake Damage, and Network Level Service 868
  Sungbin Cho, Yue Yue Fan, and James E. Moore II

A Validation Study of the Risks from Earthquake Damage to Roadway Systems (REDARS) Earthquake Loss Estimation Software Program 878
  Sungbin Cho, Charles K. Huyck, Shubharoop Ghosh, and Ronald T. Eguchi

Application of Seismic Risk Assessment Procedures to the Performance-Based Design of Highway Systems 886
  Ian G. Buckle

The Pacific Earthquake Engineering Research (PEER) Highway Demonstration Project 896
  Anne Kiremidjian, James Moore, Yue Yue Fan, Ozgur Yazlali, Nesrin Basoz, and Meredith Williams

Fragility Curves for Concrete Bridges Retrofitted by Column Jacketing and Restrainers 906
  Sang-Hoon Kim and Masanobu Shinozuka

Post-Earthquake Road Unblocked Reliability Estimation Based on an Analysis of Randomicity of Traffic Demands and Road Capacities 916
  Yanyan Chen and Ronald T. Eguchi

A Geographic Information System (GIS)-Based Emergency Response System for Transportation Networks 926
  Nesrin Basoz, Meredith Williams, and Anne Kiremidjian

Geotechnical Lifeline Issues

Large-Displacement Soil-Structure Interaction Facility for Lifeline Systems 936
S.L. Jones, T.D. O'Rourke, H.E. Stewart, and S.L. Billington

Seismically Induced Lateral Earth Pressures on a Cantilever Retaining Wall 946
Russell A. Green, C. Guney Olgun, Robert M. Ebeling, and Wanda I. Cameron

A Simplified Two Dimensional Soil Model for the New Madrid Seismic Zone 956
Wei Zheng and Ronaldo Luna

Pipe-Soil Interaction During Transverse Permanent Ground Deformation 967
Moon Kyum Kim, Yunmook Lim, TaeWook Kim, and SungHee Chang

Modeling of Unbounded Domain in Seismic Soil-Pile-Structure Interaction 977
Dongmei Chu and Kevin Z. Truman

Ground Improvement Effectiveness for Liquefaction Mitigation at an Existing Highway Bridge 987
Harry G. Cooke and James K. Mitchell

Lateral Seismic Pressures for Design of Rigid Underground Lifeline Structures 1001
Craig A. Davis

Characterizing the Effects of Pile Foundations for Evaluation of Performance Based Seismic Design of Critical Lifeline Structures 1011
W.D. Liam Finn, N. Fujita, and T. Thavaraj

Liquefaction and Non-Liquefaction from 1999 Chi-Chi, Taiwan, Earthquake 1021
Jonathan P. Stewart, Daniel B. Chu, Shannon Lee, J.S. Tsai, P.S. Lin, B.L. Chu, Robb E.S. Moss, Raymond B. Seed, S.C. Hsu, M.S. Yu, and Mark C.H. Wang

Innovative Mitigation of Bridges

Performance of Viscous Damper and Its Acceptance Criteria 1031
Li-Hong Sheng and Don Lee

Impact of Friction Pendulum Bearings on the Seismic Retrofitting Cost of Typical Bridges with Wall Type Piers in the State of Illinois 1040
Murat Dicleli, Mouhamad Y. Mansour, Anoop Mokha, Victor Zayas, and Michael C. Constantinou

Development and Analysis of Composite Steel-Concrete Girder for Bridge 1050
Cui Yuping, Shi Zhongzhu, and Sun Shaoping

Seismic Fragility Curves for Bridges: A Tool for Retrofit Prioritization 1060
Bryant Nielson and Reginald DesRoches

Subject Index 1071

Author Index 1075

The Policy Legacies of California's Dam Act of 1929

Robert A. Olson[1]

Abstract

Following the sudden failure on March 28, 1928 of the St. Francis Dam in southern California, the legislature passed a law, the Dam Act of 1929, to greatly strengthen the state's role in supervising non-federal dams in California. This law meant that state government essentially would be directly administering a virtually new program that included "(1) examination and approval or repair of dams completed prior to the effective date of the statute, August 14, 1929; (2) approval of plans and specifications, and supervision of construction of new dams, and of the enlargement, alteration, repair, or removal of existing dams; and (3) supervision over maintenance and operation of all dams of jurisdictional size." This paper summarizes the legislative history of the Dam Act of 1929, presents data on how it was used following the 1933 Long Beach earthquake, and comments on how some of the Dam Act's original principles were incorporated into other seismic safety legislation—a classic example of "disaster learning."

Introduction

People were told that the newly constructed St. Francis Dam, 42 miles north of Los Angeles, was "the safest dam ever built" and "It was the last word in engineering." But "no one told those sleeping in the fertile valley below the dam [at 11:50 p.m. on March 12, 1928]. There wasn't enough time" as telegraph machines clicked "The dam has broken! The St. Francis is out!" (Hillinger, 1)

At least 500 people were killed by 36,000 acre feet of onrushing water and debris, including timber, bridges, roads, automobiles on the roads, railroad cars and tracks, boulders, animals, barns, homes, and people. The estimated maximum height of the water was 125 feet, and it came down a narrow canyon ultimately covering 65 miles from the dam's site to the coast near Ventura. Devastated communities included Piru, Fillmore, Bardsdale, Santa Paula, Saticoy, and Montalvo.

California's Governor C.C. Young arrived shortly after dawn on March 12, and President Coolidge "wired that the nation was stunned by the disaster and added that every effort would be made by the Federal government to aid State and local agencies in restoration projects." (Ibid.) St, Francis Dam's catastrophic failure was a classic

---

[1] President, Robert Olson Associates, Inc., 100 Egloff Circle, Folsom, CA 95630; phone 916-989-6201; roa1@attbi.com

"triggering event" that opened the also classic "window of opportunity" for public policy change.

Political Mobilization

The disaster set city, county, State of California, and Federal Government investigations in motion. A Coroners jury "returned the verdict of 'No evidence of criminal act or intent on the part of the Board of Water Works and Supply of the City of Los Angeles, or any engineers or employee in the construction or operation of the St. Francis Dam." (Ibid.)

Within a few months California's Legislature dealt with four draft legislative proposals to "prevent repetition of the St. Francis dam disaster" by placing about 600 non-Federal dams in California under control of state government. Interestingly, all four proposals were patterned after an existing State of Pennsylvania statute. One was drafted by State Engineer Edward Hyatt, Jr. on behalf of Governor Young's administration, a second by Senator Walter Duvall of Santa Paula, the third by Assemblyman Dan E. Williams of Jacksonville, and the fourth was "arranged by the farm bureau federation." (*Sacramento Bee*, January 15, 1929, 19)

Following an evening meeting on January 14, 1929 involving Hyatt, Duval, and Williams it was announced to the press that the four proposals would be consolidated into one, and that an identical version of each ("companion" bills) would be introduced into the Senate and the Assembly so both houses could consider the same legislation simultaneously. A *Sacramento Bee* article noted that:

> Jurisdiction over the dams is vested in the state engineering department, which is authorized to make an inventory of all non-federal dams and reservoirs. The engineer also is directed to supervise the operation and maintenance of the structures, and also the construction of new projects.
>
> Wide police power is conferred upon the engineer in the enforcement of the act, and obstruction of him or any of his agents is made a felony.
>
> For purposes of administering the act, a revolving fund of $250,000 is created, but fees will be charged for supervising the dams…(Ibid., 18)

Processing the Legislation

The formal introduction of Senate Bill 723 by Senators Duval and Mueller (who joined Duval as a coauthor) occurred on January 18, 1929. There is no further mention of a separate Assembly Bill, but it may have been unnecessary given the speed that SB 723 moved through the political process. The fact that several bills had been combined into one meant that strong consensus existed between Governor Young's administration, the legislative sponsors, and at least the farm bureau lobby.

The Senate's Committee on Governmental Efficiency clearly was where SB 723's substantive issues were managed. Between February 20 and March 28, 1929 the legislation was heard and amended three times. Finally, on March 28 it received a "Do pass as amended" recommendation whereupon SB 723 made its second committee stop—the Senate Committee on Finance.

Finance passed SB 723 out of committee "without recommendation" on April 2. It moved to the Senate floor where SB 723 was amended three more times while it was on the floor agenda. On May 1, 1929 the Senate approved the bill and sent it to the Assembly for consideration.

SB 723 moved very quickly through the Assembly without any apparent amendments. On May 1 the legislation was referred to the Assembly's Committee on Governmental Efficiency and Economy, and on May 6 it received a "Do pass" committee recommendation. Its second stop was the Assembly's Ways and Means Committee, which gave it a "Do pass" recommendation on May 10.

SB 723 passed the Assembly on May 14, and it was returned the same day to the Senate for "concurrence." Concurrence was achieved on the same day, and the legislature sent this landmark legislation to Governor Young on May 15, 1929. Governor Young signed the new Dam Act on June 10, 1929. (State of California, 248) Governor Young stated that:

> This bill provides for a centralized state supervision over dams in California, as to their design, construction and maintenance. In the past an unsatisfactory conditions has existed due to the fact that supervision was not thorough or complete [by the state railroad commission].
>
> The present bill concentrates authority in one office, that of the state engineer, who is given authority to do all things necessary to see that dams in California are safe. The bill makes possible the storage of water for irrigation and assures the public that this water is being stored behind safe structures. (*Sacramento Bee*, June 11,1929, 4)

The original Dam Act provided $200,000 so a survey could be made of all existing dams in California. A *Sacramento Bee* article noted that "If any of these structures are found to be faulty, the state engineer is empowered to compel the owners to make such changes and reinforcements as to assure their safety." (Ibid.)

The Dam Act's Enduring Legacies

The 1929 act institutionalized a strong regulatory role for state government over non-Federal dams in California. It not only addressed new dams but it called for the evaluation and upgrading or replacement of existing ones—always a difficult political challenge because of American society's general reluctance to "change the rules after the game has been played." In addition, the Dam Act required the state to greatly increase its technical competence so it would have the independent capabilities to review and approve design plans, actual construction, and the long-term maintenance of dams. These were important—landmark actually—principles that have found their way into several California seismic safety laws. Three examples are worth discussing.

The Safety of Design and Construction of Public School Buildings Act of 1933

On March 10, 1933 the City of Long Beach and much of the greater Los Angeles area was shaken by an estimated Richter Magnitude 6.3 moderate earthquake shortly after 5:00 p.m. Public schools suffered a disproportionate amount of damage, and it was fortunate that children were not present. This event, likened to a "climate change" by others (Geschwind, 1996), led to many seismic safety changes one of which was the enactment of what has become known as the "Field Act" for its author, Assemblyman Don C. Field from Glendale: The Safety of Design and Construction of Public School Buildings Act of 1933.

Assemblyman Field introduced Assembly Bill 2342 on March 23, 1933, and Governor James Rolph, Jr. signed the Field Act on April 10, 1933—30 days after the earthquake. It moved quickly through the legislative process.

The state was now responsible for independently reviewing and approving plans and supervising the construction of new public schools. What to do about existing schools (commonly known as "pre-Field Act schools") had to wait until the latter 1960s for policy attention, and private schools were not included. The justification was that state law required children to attend school and their construction was financed with public money.

Where did the Field Act's ideas come from so quickly? The 1929 Dam Act was used as a model. D.C. Willett, who in 1933 was an engineer serving as the chief assistant to the state architect, was tasked to draft legislation after a short meeting with Willett, Assemblyman Field, and the State Architect (a "Mr. McDougall"). In 1957 Willett reported how the Field Act started:

> [W]e were discussing...what Assemblyman Field had asked us and how it could be worked out—and Fred [Fred Green, a senior engineer with the division of architecture] took the paper and he said, "you know what you ought to have; you ought to have a law governing school construction." Well, gee, I [Willett] said, "Fred, you've given me an idea." So I immediately went upstairs to the Division of Water Resources; they had an act [Dam Act of 1929] regulating the design of dams...
>
> I went up and got the Dam Act and as soon as I finished lunch went over to the Assembly and got hold of Don Field. I said, "Don, I've got another idea. I

don't know whether it's worth a darn or whether you want it or not, but here it is…So I showed him the Dam Act; I showed him the pictures of the schools, and I said, "Now, listen," if you'll make a law to make school buildings safe, we can enforce it."

We took the Dam Act as the foundation, and applied it to schools. The structural features and the regulations that would control schools were added and put under architects and structural engineers. (Meehan and Jephcott, 5-12)

The Field Act contained several of the Dam Act's key provisions, and it institutionalized the state's role in supervising public school construction. Other policy issues such as retrofitting, replacing, or changing the uses of pre-Field Act public schools had to wait until later, but the principles were the same.

The Hospital Seismic Safety Act of 1973

Fifty-eight people died from the Richter Magnitude 6.4 moderate February 9, 1971 San Fernando earthquake, which also shook much of the Los Angeles area. While the great majority of the casualties occurred in an old Federal Veterans Administration hospital, it was the nearly complete collapse of the just-opened (and thankfully nearly unoccupied) $30 million nearby Olive View Medical Center that prompted questions about California's hospital design and construction requirements and practices.

The author, standing on Olive View's grounds with several prominent earthquake engineers a few days after the earthquake, asked "Can't we do better than this?" The engineers cautiously answered yes. That discussion set in motion work by those of us who were voluntarily working for the legislature's very recently formed (1969) Joint Committee on Seismic Safety, which had been charged with recommending how the state's risk could be reduced.

The state had nearly 30 years experience with the Dam Act's legacy—the Field Act— by now, and it provided a model for the hospital seismic safety legislation. The first legislative effort (Senate Bill 352 of 1971), sponsored by the Joint Committee's Chairman Senator Alfred Alquist and other members, failed largely because of the opposition to having to upgrade or replace existing hospitals. The committee's second effort, embodied in Senate Bill 519 of 1972, addressed only new hospitals. It was introduced on March 8, 1972, passed the legislative process, and was signed by Governor Ronald Reagan on November 21, 1972. The intent noted:

> It is the intent of the Legislature that hospitals, which house patients who have less than the capacity of normal healthy persons to protect themselves, and which must be reasonably capable of providing services to the public after a disaster, shall be designed and constructed to resist, insofar, as practical forces generated by earthquakes, gravity, and wind.

This law institutionalized the state's role in hospital plan and design review and construction supervision. Recently, an engineer summarized the new law' features:

- The law was patterned after the Field Act covering schools in California, specifying the same state review agency, and stipulating design by specially experienced and approved "Structural Engineers;"
- The law covered new buildings only;
- The law provided a "Building Safety Board" of industry design professionals and facility experts...to advise the state on implementation of the requirements; and
- The law and [implementing] regulations included four main considerations:
    - geologic hazard studies for sites [which was amended into the Field Act in the 1970s]
    - structural design forces in excess of those used for "normal" buildings...
    - specific design requirements for nonstructural elements [and]
    - strict review of design and supervision of construction (Holmes, 2-3)

The Hospital Seismic Safety Act of 1973 contained several of the Field and Dam Acts' key provisions, and it institutionalized the state's role in supervising new public and private hospital construction. Other policy issues such as retrofitting, replacing, or changing the uses of pre-Hospital Act facilities had to wait until later (Senate Bill 1953 of 1994), but the principles were the same.

The Essential Services Buildings Seismic Safety Act of 1986

With the Dam and Field Acts on the books and a hospital seismic safety act underway in the 1971-72 period, the Joint Committee's volunteer advisors started to think about a loosely defined broad category of "essential services buildings:" fire stations, police stations, emergency operations and communications centers, high occupancy buildings (e.g., apartments, high rise offices, auditoriums), and others. What should govern the design and construction of these, and who should be responsible?

Late in its existence (1973) the Joint Committee on Seismic Safety introduced three separate but closely linked measures: Senate Bills 1372, 1374, and 1375. SB 1372 would have required the State Office of Architecture and Construction (which administered the Field Act and did the plan checking for the Hospital Act) to prepare design and construction regulations for emergency services structures, with enforcement being left to local governments. SB 1374 would have required the seismic rehabilitation of existing emergency services facilities to an established building code, and authorized the issuance of state and local revenue bonds to finance the work. SB 1375 would have prohibited additions to, modifications of, or the installation of "new emergency communications or disaster equipment" in existing emergency services structures which did not meet prescribed earthquake standards. The three measures failed to pass, but some of the principles were carried forward by the Seismic Safety Commission after it came into existence in 1975.

On February 2, 1984 Senator Leroy Greene, an engineer who had played major roles in amending the Field Act to address pre-Field Act public schools, took up the cause of emergency services structures by introducing Senate Bill 1547. Referring to the 1971 San Fernando earthquake, the Seismic Safety Commission sponsored this legislation. Had it passed, SB 1547 would have substituted state standards and independent plan review and construction inspection for local control. The work would have been assigned to the Division of the State Architect, and the bill would have governed new construction and substantial modifications to existing buildings. A Commission staff analysis noted that "The ideas and concepts in SB 1547 are not new. Identical programs have been underway for some time in the design and construction of schools and hospitals in the state. These programs have worked well in upgrading the safety and performance of these facilities...(MacLeod, emphasis in original)

Senator Greene "voluntarily withdrew" this bill, apparently because of the issue of local versus state control. The legislation reemerged as Senate Bill 239 early (January 24, 1985) in the 1985 Regular Session, where it eventually would succeed. The new legislation accommodated local interests, and it was signed into law by Governor George Deukmejian on October 2, 1985, not long after the devastating major Mexico City earthquake of September 19, 1985.

The Essential Services Buildings Act of 1986 referred to the hospital seismic safety law and restated its intent almost verbatim for the defined essential services buildings. However, the new law governed new, reconstructed, altered, or substantial additions to essential services buildings defined as fire stations, police stations, emergency operations centers, California Highway Patrol offices, sheriff's office, or emergency communications dispatch center.

The law further required that the "enforcement agencies" (Office of the State Architect for state-owned and leased buildings and local building safety agencies for their structures) conform to state specified standards and regulations. The new law also carefully defined how independent plan review, design certifications, and construction inspection were to be independently accomplished at the local level so as to meet the same degree of scrutiny as though the state was performing these functions.

While not full state preemption of this class of buildings, SB 239 came very close. It drew on previous state legislation, established standards, and mandated independent review processes. It did not deal with existing buildings, except as they might in the future be governed by the provisions affecting alterations or other modifications.

Conclusion

The Dam Act of 1929 became the "spiritual centerpiece" of many of California's key state seismic safety laws, including the Field Act governing the construction of public schools, the Hospital Seismic Safety Act, and the Essential Services Buildings Act. To varying degrees the four share some common characteristics: state preemption, or nearly so of the issue in question; development of above-normal building standards and regulations; and processes to assure independent review by qualified people of design plans, specifications, construction drawings, and actual construction. In 1928 no one could have foretold the policy legacies of the St. Francis Dam's catastrophic failure. It also is clear that the 1933 Long Beach, 1971 San Fernando, and 1985 Mexico City earthquakes contributed to the passage of each of the laws discussed above.

References

Birkland, T.A. (1997) *After Disaster: Agenda Setting, Public Policy, and Focusing Events*, Georgetown University Press, Washington, DC.

Geschwind, C.H. (1996) *Earthquakes and Their Interpretation: The Campaign for Seismic Safety in California, 1906-1933*, UMI Dissertation Services, Ann Arbor, MI.

Hillinger, Charles (1952) Dam Break Greatest Southland Disaster, *Los Angeles Times*, March 25.

Holmes, William (2002) *Background and History of the Seismic Hospital Program in California*, Rutherford & Chekene, Oakland, CA.

MacLeod, John (undated memorandum) *Commission Issue Piece for Essential Services Building Seismic Safety Act (SB 1547 Introduced by Senator Greene)*, Seismic Safety Commission, Sacramento, CA.

Meehan, J.F. and D.K. Jephcott (1993) Appendix I: A transcript of a conversation between Mr. D. C. Willett and Mr. Frank Durkee, October 21, 1957 in *Task 4: The Review and Analysis of the Experience in Mitigating Earthquake Damage in California Public School Buildings* (unpublished), Building Technology, Inc., Silver Spring, MD.

Olson, R.A. (2003, pending) Legislative Politics and Seismic Safety: California's Early Years and the "Field Act," 1925-1933, *Earthquake Spectra*, Earthquake Engineering Research Institute, Oakland, CA.

Rogers, J. David (1995) A Man, A Dam and A Disaster: Mulholland and the St. Francis Dam, *Ventura County Historical Society Quarterly*, 40, Nos. 3-5 (Spring/Summer 1995).

Sacramento Bee (January 15, 1929) Dam Supervision Bill Being Drafted.

Sacramento Bee, (January 18, 1929) State Control of Dams Asked in Bills To-Day.

Sacramento Bee, (June 11,1929) Governor Signs Bill to Control Dams of State.

State of California (1929) *Final Calendar of Legislative Business*, Sacramento, CA.

Development of Disaster Management Equipments in India: A Way towards Resurgence

D.N. Singh[1], Vipin C. Shukla[2]

Introduction

Year after years devastating disasters ravage the land mass and leave behind countless victims. Pre-disaster planning can have a significant impact on minimizing the effect of tragedy and sufferings, and may result in a more efficient and coordinated (non-panicked) response in saving lives and properties. With seismic studies revealing sizable portions of the Indian Sub-continent prone to earthquake(s), it is desirable for India to be equipped with disaster management equipment at all times. The country is also prone to other calamities like cyclones, droughts, floods, fires etc. In a report by UN Office for the Coordination of Humanitarian Affairs (OCHA) on Gujarat earthquake, it is mentioned that in such situations for saving of lives and rehabilitation to be taken on a war footing, it is imperative to develop disaster management equipments on priority such as cutting concrete slabs, mobile communication equipment etc. which can be deployed for clearing debris (UN Office for the Coordination of Humanitarian Affairs OCHA, 2001).

After Gujarat earthquake disaster on 26th January 2001 in which thousands of people were killed and Lac rendered homeless, *Technology Information, Forecasting & Assessment Council (TIFAC), New Delhi (Under the Government of India)* has taken initiatives to help the nation in design and development of such equipments in the country. In this context, TIFAC has commissioned projects for the development of two different types of equipment namely Disaster Management Equipment and Radio Control System with Bharat Earth Movers Limited (BEML), a public sector in India, that has expertise in developing these types of equipment. In earthquake-hit areas, disaster management equipment will demolish the effected buildings prone to fall or big blocks required to be removed for saving the underlyings and then radio control system to remove the debris from remote distance. Presently these equipments are not manufactured in the country and unavailability of these causes huge loss to the country in such situations.

---

[1] Adviser, Technology Information, Forecasting & Assessment Council (TIFAC), Technology Bhawan, New Mehrauli Road, New Delhi-110 016; phone: 91-11-26857651

[2] Senior Scientific Officer, Technology Information, Forecasting & Assessment Council (TIFAC), Technology Bhawan, New Mehrauli Road, New Delhi-110 016; phone: 91-11-26527487, vipins@tifac.org

TIFAC & its Vision

The key role of Science and Technology in economic & social development has long been recognized by the visionary leaders of independent India. Since then, the national developmental plans put thrust towards creation of strong technical and scientific base. In late eighties of twentieth century, TIFAC was established as 'think tank' to keep watch on global technology development and assess the preferred option for India in various important sectors. Accordingly, Technology Information, Forecasting and Assessment Council (TIFAC), was established on 10th February 1988 as an autonomous society under Department of Science & Technology, Govt. of India with the prime objectives:

- To look ahead in technology development with a watch on global trends and formulate preferred options for India,
- To undertake technology assessment and forecasting studies in selected areas of national importance
- To promote key technologies
- Selected Information Services on technologies

Towards such objectives, TIFAC has undertaken the following activities and programmes:
- Generation of business linked Technology Reports
- Technology Information Services
- Home Grown Technology Programme
- Technopreneur Promotion Programme
- Implementation of Technology Projects in Mission Mode
- Patent Facilitating Centre
- International Linkages
- Technology Vision 2020-Follow Up Missions & Actions

Knowledge based network action with industries, industry associations, academia, government institutions, non-government organizations and national experts is the key of TIFAC's flexible management system in implementing above programs.

In 1993, TIFAC embarked on a first major long-term technology forecasting exercise - known as 'Technology Vision 2020' with the involvement of 500 user agencies and 5000 experts, encompassing major areas – Agro Food Processing, Advanced Sensors, Electric Power, Food & Agriculture, Road Transportation, Civil aviation, Waterways, Telecommunications, Engineering Industries, HealthCare, Life Sciences & Biotechnology, Materials & Processing, Services, Strategic Industries, Electronics & Communication, Chemicals Process Industries and Driving Forces-Impedances. The objectives of the above exercise has been as below:

- To generate necessary linkages and specific project proposals to realize the Vision into Mission(s).
- To make efforts to bring together project teams and suggest action packages.

The recommendations emerging out of technology Vision 2020 exercise have been disseminated through a spectrum of specific workshops/fora being held nationally and internationally. Many government departments, industries and other agencies have adopted the recommendations for implementation.

As a further follow up, in order to speed up the realization of the recommendations of the report(s), Government of India entrusted TIFAC for undertaking a few demonstration projects in six identified sectors including the one - Road Construction & Transportation Equipment.

It is well known that efficient transportation system is a pre-requisite for the development of every aspect of the society be it agriculture, economic development, processing industries, health care and education as well. This is an important segment in the engineering sector, which incidentally serves for road construction and transportation.

Initiatives by TIFAC for Disaster Mitigation

The post disaster management demands action in several areas like immediate relief measures in terms of saving life of injured/effected people, property, rehabilitation (quick) programs and also a mass awareness program to make the people understand what to do and not to do to reduce the effects of miseries as precaution. Some of the actions taken by TIFAC after Gujarat earthquake are as below:

In wake of disastrous damages by the earthquake in Gujarat, the Advanced Composites Mission of TIFAC has contributed to the national effort of re-building and rehabilitation. A wide array of innovative composite products suitable for low-cost building and construction sector and bio-medical appliances has been developed to help the people. Jute-coir composite boards, rice husk particle boards, fibre reinforced plastic (FRP) toilet blocks, multipurpose FRP handcarts and composite artificial limbs address the crucial need of the hour-post-disaster relief at the quickest possible time (TIFAC, 2002).

TIFAC is preparing a report on "Technology for the retrofitting existing buildings to make them earthquake resistant" with the help of Indian Institute of technology Roorkee which is supposed to be top engineering institute of India in earthquake engineering related fields. It is expected to be completed very soon.

For creating mass awareness among the people, TIFAC has made a documentary film on "Earthquake – Surviving Nature's Fury". It has beautifully demonstrated the various ways of preparedness for the earthquake. It was telecasted in the national channel of Doordarshan in India.

One of Mission program is Relevance and Excellence in Achieving new Heights in Educational Institutions (Mission REACH), which aims towards creating Centres of Relevane and Excellence (COREs) in higher technical educational institutes in partnership with industry and users. TIFAC has established a CORE for disaster mitigation at Indian Institute of Technology Roorkee (Uttaranchal State, India) that aims towards consolidate, augment and strengthen the facilities for research and extension work in the area of disaster mitigation through human resource development, advanced R&D activities, evolving strategies for mitigation and management of disasters and establishment of a national database for rapid dissemination of information and knowledge.

As mentioned earlier the two projects namely Development of Disaster management Equipment and Radio Control System under the 'Road Construction and Transportation Equipment' program with BEML relate to disaster management. A brief of the two projects are given below:

Disaster Management Equipment-Excavator with attachments (BEML, 2001)

It is a 30-ton class excavator with requisite attachments like Combi-cutter and Hydraulic Hammer and Scrap Grappler. In the earthquake-hit areas, this will be deployed with its all attachments. During normal periods, the standard equipment could be used for infrastructure development like road construction, construction purposes, irrigation, canal works, public works etc. The photographs are attached in the **Gallery 1 as Photograph-1**.

The objective of the equipment is to demolish buildings at a faster rate, which are hanging precariously so that new buildings are built, with which people are rehabilitated at the earliest.

Key Features.
A) Basic Equipment: 30-ton class hydraulic excavator fitted with 1.57 cubic meter backhoe bucket. It has 360° swing.

B) Combi-cutter: It serves the purpose of both the shearing and the crushing. Combi-cutter is not only cost saving but also considerably reduces the time for replacement of individual shearer and crusher attachments. In other words, it is a combination of shearer and crusher both.

Crusher: Crusher is ideally suited for crushing large-scale foundations, buildings, bridges, columns etc. in earthquake hit areas. The jumbo crusher having two crushing claws is actuated by two hydraulic cylinders respectively. So the crusher can catch at the center of the target to be crushed without giving any excessive force to the boom and the arm. It has working pressure of 320 kg/cm^2 and Jaw opening of 850 mm.

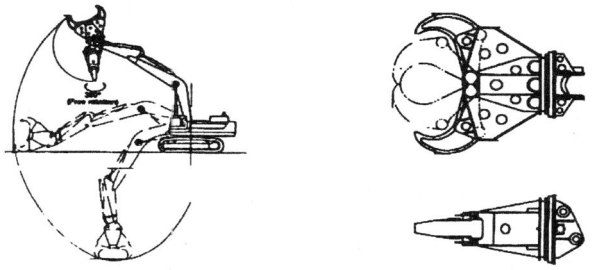

Figure-1. General arrangement of a crusher

Shearer: Shearer is a heavy duty jaw type shearing attachment, which is capable of cutting any steel frames, steel girders, reinforcing rods etc.(which are part of the damaged buildings) at disaster sites. Its blade edge is well shaped to catch an object easily but firmly. It can effectively hold any steel section whether it is square, round or of any irregular shape. It has working pressure of 320 kg/cm^2 and Jaw opening of 400 mm.

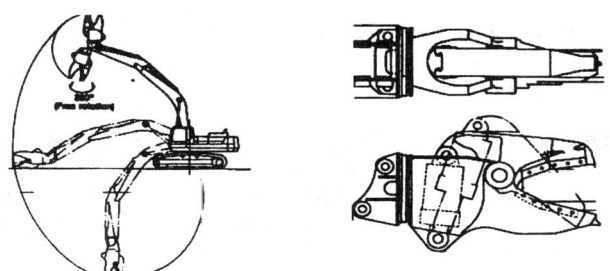

Figure-2. General arrangement of shearer

C) Hydraulic Hammer: It is best suited for breaking rocks, heavy concrete blocks etc into small pieces and can be handled easily and cleared from the quake hit sites. It has working pressure of 160-180 kg/cm^2, service weight 1700 kg and impact rate of 320-600 blow per minute.

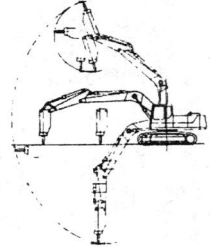 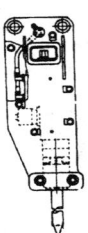

Figure-3. General arrangement of a breaker

D) Scrap Grappler: It is best suited for safe handling of massive and irregularly shaped rocks, debris, scraps etc. It has got four claws which open or close by independent hydraulic cylinders. The grapple rotates freely to make readjustments safe and easy. It has working pressure of 320 kg/cm^2, jaw opening approximately 1.42-2.24 meter and capacity 0.46 – 2.16 cubic meter per minute.

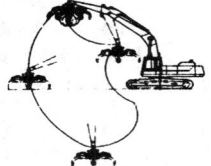 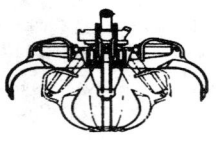

Figure-4. General arrangement of a grappler

Radio Control System (BEML, 2001)

This is an unmanned dozer that can be used for removing the debris in dangerous areas such as damaged multi-storied buildings, bridges etc. The photographs are attached in the **Gallery 1 as Photograph-2**. Objective is to provide a method of control machine without physical connection between the operator and the machine. It can also be used for handling toxic materials. It enables the operator to perform all vehicle operations from a distance at any time from any position. Hence, radio control technology when integrated with vehicle makes it possible for deployment in hazardous environment with no risk to the operator.

The operating system mainly comprises of handheld transmitter and vehicle mounted receiver. Typical block diagram of the system i.e. how the radio control function is achieved is shown in **Gallery 2 as Figure-5**. The block schematic of pendant function is also shown in **Gallery 2 as Figure-6**.

Key Features:

- Operating range is 60 m under normal operating conditions.
- Press to Transmit (PTT) switch interlock: Without actuating this switch no remote operation except 'Brake' & 'Stop' can be performed. This will take care of any accidental fall of transmitter.
- Starting safety has been incorporated to ensure Neutral starting and avoid repeated cranking to safe guard starting motor.
- Stopping delay time: upon actuation of stop control, pre determined time-out is ensured for complete shut down of engine.
- Visual indication on equipment and at command panel for radio link (RL) and equipment fault has been provided.
- In case of Radio Link failure, Brake and Stop command is automatically actuated.
- Compact control box using Quick pressing connector and miniaturized valves have been used for easy field servicing.
- Enables operator to move to most advantageous position in potentially hazardous areas and operate the vehicle.
- Automatic self-test on start-up.
- Uses of security codes in order to check integrity of Tele command.
- Pendant operation in case of RL Failure / Servicing / emergency is possible from the operator seat.

Conclusion

The indigenous development of these equipments has been welcomed by several user agencies like Central/State/District level governmental agencies, industries and Non-Government organizations. During a major international exhibition and conference on construction equipment and technologies EXCON 2002 held at Bangalore (India) on December 16-19, 2002, BEML displayed these equipments. It proved to be one of the most crowd puller in the event (The Hindu, 2002). Many press media wrote about these two equipment and appreciated the effort of TIFAC and BEML. The interest shown by various armed personnel, scientists, engineers, industrialists and media

(Business Line, 2002) has been overwhelming and the indigenous development plans of these equipments has widely been appreciated.

Subsequently several queries in this regard is indicative of its future development / utilities by several agencies dealing with disaster management. BEML has capability and developed expertise to design such equipments of various specifications required by users to meet their specific demands. This is vital for preparedness by the concerned departments (like Irrigation/ Agriculture of Central/State governments or by the Municipal councils/City corporations of the respective regions) to meet such situations, as and when required. *It is being considered very useful, certainly in wiping off tears from the victim's eyes with a dream of a brighter tomorrow.*

Acknowledgement

The authors are thankful to M/s Bharat Earth Movers Limited for providing the technical details mentioned in the paper and sharing their experiences related to the development of these equipments. In particular: Dr. K. Aprameyan, Ex-Chairman & Managing Director; Mr. V.Rs. Natarajan, Chairman; Mr. Venkatanathan, Director-R&D; Mr. V.A. Ashok Kumar, Dy General Manager-R&D; Mr. V. Palanisamy, Dy General Manager-R&D and Mr. H.B. Ramnath, Dy. General Manager-R&D (all from BEML)

The authors also thank their colleagues in TIFAC who have contributed to the stimulating environment that has led to the work reported in this paper. Special thanks to Mr. Sanjay Singh, Mr. T. Chandrasekhar, and Mr. T. Selvan for sharing their experiences and helping the authors in preparing this paper.

TIFAC's encouragement in sharing the experience is thankfully acknowledged.

References

BEML, (2001) *Report on Disaster Management Equipment & Radio Control System*

Business Line, (2002) *BEML dozer, big draw at Excon*, (Dec. 17, 2002)

The Hindu, (2002) *BEML's radio control dozer evokes interest*, (Dec. 18, 2002)

TIFAC, (2002) *Annual Report 2001-2002*

UN Office for the Coordination of Humanitarian Affairs (OCHA). (2001). " India-Earthquake OCHA Situation Report No.4". http://www.reliefweb.int

Gallery 1

Photograph-1 : Disaster Management Equipment

Photograph-2 : Radio Control System

Gallery 2

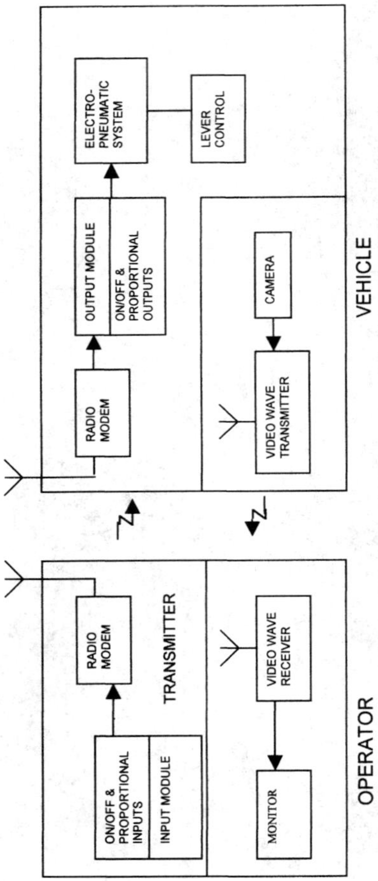

Figure-5. Schematic Diagram of Radio Control Dozer Functions

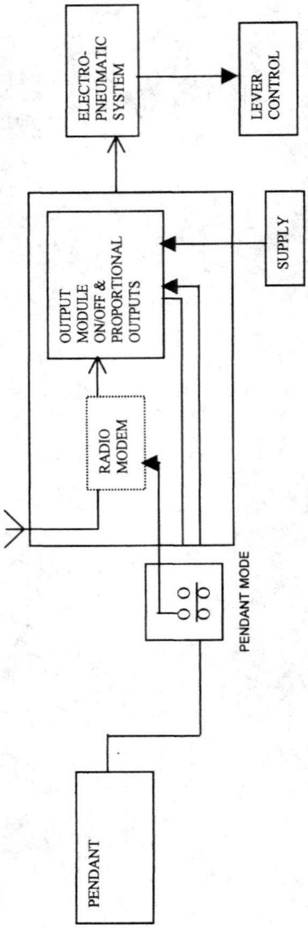

Figure-6. Schematic Diagram of Pendant Mode operation

Community Preparedness and Response Model for Protecting Infrastructures

Nasim Uddin, Ph.D., P.E[1]., Dennis Engi, Ph.D.[2], and A. Haque, Ph.D.[3]

Abstract:
This paper investigates a community preparedness and response model for protecting infrastructure that will significantly lessen loss of human life and lower the cost of disaster recovery in a region.

Introduction

In the aftermath of 9/11 the potential ramifications of critical infrastructure failures have never been so greater, and the scope never so comprehensive. Because of immediate threat and limited resources, government policy has been more towards security, robustness and reliability of infrastructures, including communication networks at the federal level. However, all disasters are first local and the first people on the scene in a crisis are local emergency response teams. Indeed the community represents an essential focal point for building a foundation for any national initiative that seeks to influence the public to make them proactive for facing unintended consequences (US Department of Commerce, 2002). Therefore, to comprehend a disaster into a manageable preparedness-and-response mechanism, it must be localized to an extent where it is possible to understand, monitor and utilize available information, infrastructure and human resource to the best possible comprehensive operation with least possible human and physical cost to the community.

According to the National Strategy for the Physical Protection of Critical Infrastructures and Key Assets (2003), "currently there is no central coordinating mechanism to assess the impact of sensitive information" coupled with the "lack of technical communications systems to enable the secure transmittal of classified threat information to owners and operators of concern" (p. 26). Furthermore, the report re-affirms that there is no clear understanding of the risks associated with vulnerabilities of the critical infrastructures that demands modeling simulation and analysis capabilities to enable decision support and planning activities related to national defense and intelligence missions (p. 33).

The primary objective of the paper is to explore the understanding of the information technology required to build systems addressing multidisciplinary interacting information. This computational framework can be the key to scientific decision-making involving large-scale multidisciplinary modeling, simulation, and computation for the 21st century. Large-scale integration of measurements/observations with modeling and simulation is becoming increasingly crucial to ensuring prosperity and well-being for the society at large. The notion that

---

[1] Department of Civil and Environmental Engineering, University of Alabama at Birmingham, AL 35294
[2] School of Industrial Engineering, Purdue University, West Lafayette, IN 47907
[3] Department of Governments, University of Alabama at Birmingham, AL 35294

the information revolution favors and strengthens networked forms of organizations over bureaucratic hierarchies has, by now, become widely accepted. While most governmental agencies were created to respond to particular crisis, new policy approaches are quite complex and require coordinated efforts across multiple agency jurisdictions. Information residing in one agency could be critical for another agency's decision-making --a fact that has been greatly acknowledged in recent times after the terrorist attacks in New York and Washington D.C. Effective governance requires the capacity of the public service to leverage information, knowledge and technology across jurisdictions to provide integrated responses to complex social problems (Dawes, 1996). Moreover, successful policy execution must incorporate joint or shared decision making among formally independent organizational units. Information sharing (IS) avoids duplicate data collection, processing and storage, and thereby increases productivity and reduces the overall operating cost. On an organizational level, cooperative IS arrangements improve the quality of decision-making and increases the quantity and availability of data (Dawes et al, 1999). IS procedures depend upon the establishment of a central data repository which functions as a traditional clearinghouse and a catalyst in attracting agencies and citizens into one single IT arena where they can converse with one another through the use of a common and integrated language. Unlike more traditional areas of public activity, IS does not constitute a formal public policy enacted by elected officials and implemented by third-party bureaucrats. To the contrary, IS is an inter-agency agreement over procedures and IT processes. The agreement may be "legalistic," such as when a regional council of governments (COG) chooses to routinize procedure. Or it may also be "behavioral," such as when agency directors agree unofficially to share data so that each may accomplish disparate tasks. The agreement may also acknowledge potential differences in agency mandates and professional values regarding the sharing of information. All this makes the implementation of IS somewhat unique yet increasingly common in the 21st century.

Moreover, modeling, predictions, integrative observation, and acquired knowledge are each potentially important components in the complex process of developing and using knowledge about extreme events to reduce vulnerability and respond to impacts. However the appropriate design, role, and application of these integrating mechanisms including their relations to each other-are far from well understood, and are themselves inherently contextual and dynamic. This means that constant communication between researchers and decision makers is an absolute essential component of addressing the rising threat of disasters and other extreme events.

Community Preparedness and Response (CPR) Model

In the following a Community Preparedness and Response (CPR) model is proposed by using available state-of-the-art technology to understand and answer key questions of interoperability across agency jurisdictions ("stovepipes") and community risks related to vulnerabilities of critical infrastructures. The overall aim of the idea is to support the existing regional Department of Homeland Security (DHS) under the guidelines of the *Homeland Security Act of 2002*. The proposed CPR is designed to provide benefits pre, during, and after a disaster or emergency event.

Ideally, the CPR should provide value to the region even in the absence of a disaster or emergency. During a disaster and for the purposes of this discussion the first 72 hours after a disaster, the proposed CPR should facilitate reductions in human losses by allowing faster and more effective response actions, more rapid identification of locations with the most injuries and casualties, optimal movement of injured to both functional and capable medical care facilities, and improved performance of the complete emergency medical care system in region. With respect to achieving reductions in economic losses, many aspects of the CPR are specifically designed to inventory the status of the critical infrastructures in the region during and immediately after a disaster. This capability should allow prioritized response procedures and better coordinated multiple responses bases on accurate assessments of the infrastructure damage as well as actions to minimize the cascading effect of infrastructure failures due to interdependencies, which in turn should reduce both infrastructure losses and the time necessary to restore infrastructure operations. This result should translate directly into reduced economic losses in the region.

After a disaster or during the disaster recovery period, the proposed CPR will contribute mostly to reductions in economic losses. These reductions should be obtainable through an accelerated return to normalcy of infrastructure and business operations in the region. In addition, reductions in human losses and more rapid restoration of critical services attributable to the CPR should have a direct positive effect on the psychological environment in which recovery operations must proceed, thereby facilitating a faster overall economic recovery in the region. In the absence of a disaster, the presence of the proposed CPR should improve the competitiveness of the region's economy. Minor, and possibility major, emergencies should be much less disruptive because of the increased level of disaster resistance of the community achieved through the development, implementation, and operation of the CPR as well as the state-of-the-art interagency communications systems installed as part of the CPR. Insurance premiums may be lowered. Moreover, the region will as a result of implementing the CPR have a unique economic, as well as quality-of-life, asset that should contribute to the success of ay business recruitment efforts.

Modeling Approach: The proposed approach is to apply CPR model to design a self-sustaining region called "cell" (Figure 1) which will enable adaptive and robust critical infrastructure via robust backup contingency system, such that each civil critical infrastructure system in the "cell" survive during the event, capable of sustaining itself.

Defining the concept of a "Cell": The "cell" is a critical zone defined by the affected critical decision apparatus in the aftermath of an event. The "cell" zone (geography) is identified automatically depending on the affected administrative apparatus or Critical Decision Apparatus (See Fig 1). Relational database (Table 1) will be the key in defining the "cell" and its size depending on the linkages of the affected infrastructure to other infrastructures (i.e., cascading affect) and their decision personnel. A backup critical decision zone will also "trigger" giving support to the "cell" (See Fig 1). Therefore, the "cell" takes its form from the network of decision-makers identified by the relational database and the CPR triggers giving "life" to the "cell".

The CPR includes two major components: (i) improved community awareness and readiness for disasters via Digital Community Model (DCM), and (ii) disaster simulation, planning, and mitigation via Disaster Management System (DMS). The latter is primarily addressed through the physical infrastructure support application (Engineering and Technology); the former requires social, managerial, and administrative planning and coordination (Social and Behavioral Science and Technology). The overall purpose of the CPR is to support a proactive administrative apparatus DCM of a regional community in making target-based decisions during crisis based on continuous information feed from DMS.

Digital Community Model (DCM): The purpose of the Digital Community Model (DCM) is to build upon the existing social, political and institutional infrastructure of a regional community to maintain a proactive digital administrative decision-making apparatus for target-based decision-support during critical infrastructure failures. The outcome of the DCM will be empowering the key decision-makers by up-to-date information of critical infrastructure vulnerability and provide readiness through the CPR backbone. The focus of this model will be upon developing a robust information technology infrastructure to support a region-wide information system under the CPR schema. In doing so, the "cell" will overcome hurdles that have historically blocked intergovernmental cooperation during times of crisis. The role of the sophisticated electronic interface

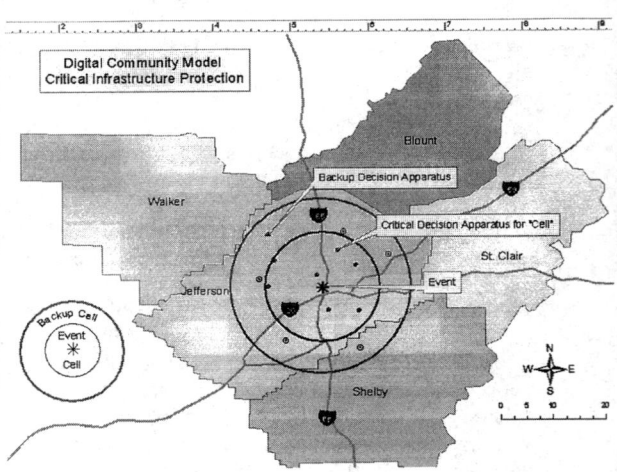

Fig.1: DCM Model Concept: A snapshot of a "cell" in the Birmingham

Table 1: An example of Cascading Linkage

ID	CRITICAL INFRASTRUCTURE			ADMINISTRATIVE PERSONNEL (CRITICAL)			ADMINISTRATIVE PERSONNEL (BACKUP)		
	Name	Address	XY Coordinate	Name	Address	XY Coord.	Name	Address	XY Coord

will boost efficiency while allowing interdependent jurisdictions in the region to retain control over future catastrophes. The outcome of such efforts will serve a long-desired predictive function for critical decision-making bodies in a region.

Disaster Management System (DMS): The proposed architecture for the Disaster Management System (DMS) is depicted in Figure 2. The five vital issues identify the information needs for the system, which in turn drive the system architecture. The bottom layer is the information communication infrastructure, labeled "Common Communication Info-structure." All data, control signals, and interaction will flow in this layer. Examples of information using this structure are sensor outputs, accumulated data from surveying, and voice interactions via Telephone. The next layer in the system model is the shared information repository. Here, all data will be stored in either object-oriented or relational schemas. The "Decision Support Tool-kit" component will include applications that use the data in the repository to promote better disaster management. This layer will include a *large-scale disaster simulator*, along with certain automated applications regarding enhanced community awareness related to disaster planning, preparedness and mitigation.

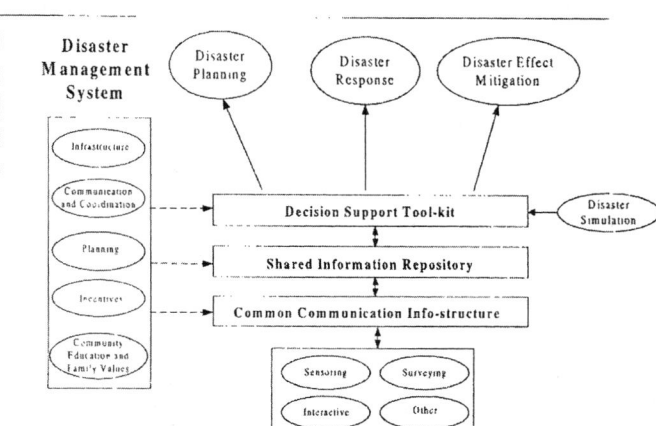

Fig.2: Architecture for the DMS

The DMS will support disaster analysis in the pre-disaster phase (i.e., planning), trans-disaster phase (i.e., disaster response), and post-disaster phase (i.e., disaster effect mitigation). The disaster simulator element is a very important component of the disaster management package. This package will support a wide range of decision-making activities. The design of the program will be flexible and will

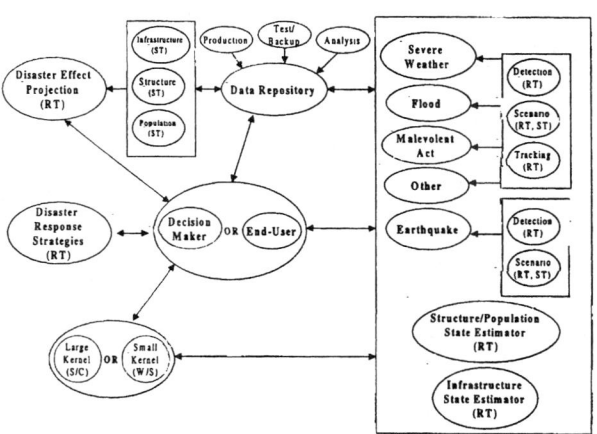

Fig. 3: Disaster Simulator

include a server/module approach. Simulations will operate in simulated time (ST) or real time (RT). Figure 3 depicts the event, simulations, and disaster effect projection modules that will be available for use by decision makers to develop and review disaster response strategies. These modules will interface with a shared data repository, which could evolve into a data warehouse environment. Decision makers to further refine disaster response strategies can use validated information from the production data repository.

The following describes two conceptual dynamic simulation models primarily intended to show the approach and to serve as demonstrations of the potential of the proposed DMS.

Infrastructure Modeling Approach

The conceptual Infrastructure Modeling tool is developed using AweSim simulation software (Pritsker et al 1997), a window simulation product that supports multiple concurrent animations from a single model. AweSim was designed as a discrete-event simulator, originally used in industrial process control and optimization applications. For modeling purposes, the software components for AweSim include Entities (Basic unit of flow in the models), Activities (Control the direction and duration of entity flow), Nodes (define starting and stopping points for activities), Resources (used by entities during activities), and Events (Control the flow of the simulation program, define by entry or exit of any entity at a node). For the demonstration purposes, the approach is to develop a tool with basic modeling techniques that demonstrated only three main features of the proposed model: (1) effects of disasters and outages on infrastructures, (2) dependencies and interdependencies of infrastructures, and (3) variable detail/zoom features. The modeling approach is to develop three animations with the Infrastructure Modeling tool. Each scenario represents different view of the five-county region, thus simulating to zoom features of a true multi-perspective tool. Each scenario also demonstrates different disaster events and interactions. The three scenarios are focused on the Evansville, Indiana area and illustrate the ability of simulation to analyze disaster response strategies at various level of detail. The following describes the animation of each scenario in brief:

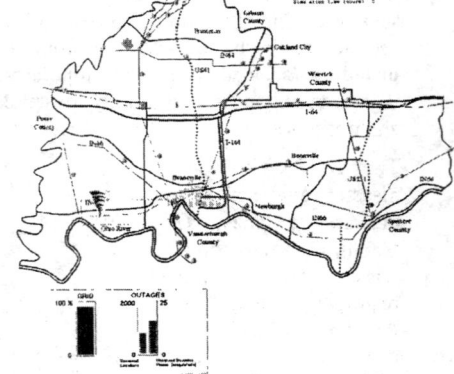

Fig. 4: A Tornado Passes South to North

Scenario 1 (Fig. 4): A tornado passes south to north on the west side of Evansville destroying a segment of electrical grid. As the simulation begins, various power icons move across the screen, and the tornado moves from south to northwest of the city. The bar charts shown on the bottom of the screen indicate the effects of the tornado as the segment of the grid is taken out of service, repaired, and brought back online.

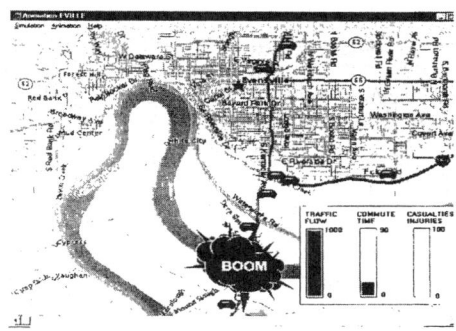

Scenario 2 (Fig. 5): The Evansville/Henderson Bridge is blown up in a terrorist attack, causing injuries and traffic disruption. In this mid-level perspective, simulated traffic flows normally on highway 41 north to south and also east to west on highway 164. At the bottom of the screen, several bar charts indicate current performance of the system. These charts show traffic flow through this area, commute time between Henderson and Evansville, and casualties and injuries from traffic accidents. As the simulation starts, traffic is flowing normally on highways 41 and 164. The bar charts at the bottom of the screen reflect the current performance of the system in terms of the traffic flow; commute time, and casualties or injuries.

Fig. 5: The Evansville/Henderson Bridge is blown up

Policy Portfolio Analysis Tool

A conceptual Policy Portfolio Analysis tool is developed to illustrate how such a simulation tool could be used to compare the effects of implementing various disaster response policies in protecting infrastructures. The purpose of the tool is to allow users to gain a

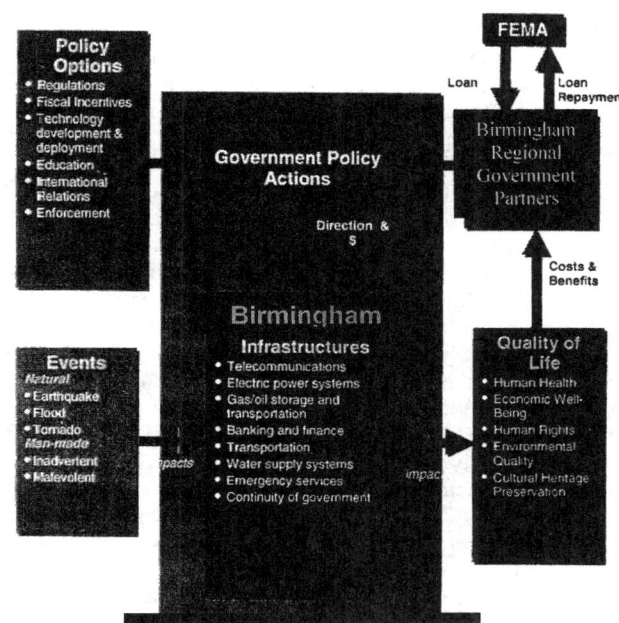

Figure 6: Policy portfolio analysis Tool

measure of understanding about how various policies might influence disaster response before the users commit time and resources to policy implementation.
The prototype tool currently provides a rough approximation of the effect of the policies, permitting decision makers to get a sense of each policy's importance to the system. For example, a user might choose to use this tool to investigate the effect upon quality of life during a Tornado with and without the policy of conducting forced evacuations. In addition, project stakeholders used the current tool during the initial analysis to visualize how a fully developed Policy Portfolio Analysis tool could be integrated into the disaster management system. The value of the Policy Portfolio Analysis tool is that it permits the user to experiment with various options and conditions before time and resources have been committed. As an example the prototype illustrates the potential of the tool and provides rough approximations of the effects of specific policies. The initial model is tested using largely hypothetical data inputs, and results verified the structure of the model. The fully developed tool would incorporate realistic data gathered from experts in the region, making it a powerful technique for establishing priorities and cost/benefit analysis.

As illustrated in Figure 6, the conceptual model, which provides the foundation for the actual computer simulation, depicts the interrelationships between the major systems and subsystems that comprise the disaster management system for the region. In this model, the effects of policy options on critical infrastructures during disaster events are examined, with results shown as impacts on quality of life indicators (health, economic well-being, human rights, environmental quality, and cultural heritage preservation).

The simulation modeling approach for the Policy Portfolio analysis prototype is to develop a tool that provided users an opportunity to experiment with the effects of implementing various policies under certain conditions. The primary emphasis was on demonstrating the concept of using the tool to prioritize policies.

CONCLUSION

A concept of CPR model to protect critical infrastructure is presented here to support a proactive administrative apparatus Digital Community Model (DCM) of a regional community in making target-based decisions during crisis based on continuous information feed from a disaster management system (DMS). A disaster management system (DMS) is defined that will integrate sensor technologies, modeling and simulation tools, telemetry systems, and computing platforms, in addition to non-automated elements including increased community education and involvement. The system is expected to provide information (pre-event, during event, and event recovery) to community leaders that will significantly enhance the ability of the community to manage disaster response. The value of the system will be manifest in both reducing loss of human life as well as staying the economic well being of the community. To aid in system definition during the process, two dynamic simulation models were also developed: the Infrastructure Modeling tool and the Policy Portfolio Analysis tool.

We showed in this preliminary study that the conceptual infrastructure model, which provides the foundation for the actual computer simulation, depicts the interrelationships between the major systems and subsystems that comprise the disaster management system for region. The infrastructure model developed during the analysis illustrates the ideal structure of the production model. The prototype, which was developed using AweSim simulation software, presents animations that convey the tool's potential to display disaster effects and responses with various levels of detail. The animations demonstrate infrastructure interdependencies during disaster events and also simulate a zooming feature for the graphical display, both of which could be included in the fully developed tool. In the policy analysis model, the effects of policy options on critical infrastructures during disaster events are examined, with results shown as impacts on quality of life indicators (health, economic well-being, human rights, environmental quality, and cultural heritage preservation).

REFERENCES

Dawes, Sharon S., (1996). Interagency Information Sharing: Expected Benefits, Manageable Risks Journal of Policy Analysis & Management, Vol. 15 Issue 3.

Dawes, Sharon S., Bloniarz, p., Kelly k., and Fletcher, P. (1999). Some Assembly Required for the 21st Century, Center for Technology in Government, National Science Foundation,WashingtonD.C.,p.31.

http://www.ctg.albany.edu/research/workshop/dgfinalreport.pdf

Haque, Akhlaque, (2001). "GIS, Public Service and the Issue of Democratic Governance," Public Administration Review, 61 (3), 259-65.

Powersim Corporation, *Powersim Construction 2.5 User Manual Set*, Powersim Press: reston, Virginia. 1999.

Pritsker, A. lan B., O'Reilly, Jean J., and LaVal, David K., (1997) *Simulation with Visual SLAM and AweSim*, Systems Publishing Corporation: West Lafayette, Indiana, 1997.

Southwest Indiana Disaster resistant Community Initiative (2000), *Proposal for the Design and Implementation of a Disaster Management System for a Five County Region in Southwest Indiana*, Sandia National Laboratories, February 2000.

Southwest Indiana Disaster resistant Community Initiative (2000), *Report of the Disaster Resistant Community Initiative Stakeholder Panel Meetings*, Sandia National Laboratories, February 2000.

Southwest Indiana Disaster resistant Community Initiative (2000), *Dynamic Simulation Models for the Disaster Resistant Community Initiative*, Sandia National Laboratories, February 2000.

Uddin, N., 2002. "Lessons Learned: The Failure of a Hydroelectric Power Project Dam", *J. Performance of Constructed Facilities*, ASCE, (in press).

Uddin, N., and D. Engi. "Disaster Management System for South Western Indiana", *J. Natural Hazards Review,* ASCE, 3(1), 2002: pp. 19-31.

Integrating Lifelines Engineering with Emergency Management: The New Zealand Approach

David Brunsdon[1] and Noel Evans[2]

Abstract

There have been significant developments in Lifelines engineering in New Zealand over the past few years. The regionally-based multi-disciplinary Lifelines engineering model has expanded to involve Lifelines Projects in each of the 16 regions of New Zealand. This model is based on the application of the principles of the international Risk Management Standard AS/NZS 4360: 1999 via a collaborative process.

One important factor influencing the growing activity in this field is the recent passage of new Civil Defence Emergency Management legislation. This internationally unique legislation requires Lifeline utilities to become actively involved in regional and national emergency management planning, and to be able to ensure service continuity to the fullest possible extent following emergency events, including earthquake.

This paper outlines the current developments in Lifelines engineering in New Zealand, and the influences that are shaping the future direction of this work. Recent and current projects of interest are also summarised, including:

- The recently completed Hawke's Bay Engineering Lifelines Project
- A regional risk management perspective of Fire Following Earthquake
- Priority restoration of utility services to key community facilities

1. Introduction

Lifelines engineering in New Zealand was initiated in 1989, and is now an established technical discipline. There are now Lifelines Projects and Groups established or being planned in each of the 16 regions of New Zealand. The Lifelines engineering process represents a very effective regional scale collaborative model which is being viewed by other sectors as a model framework for integrating technical processes with community considerations.

---

[1] National Lifelines Co-ordinator, New Zealand   drb@spencerholmes.co.nz
[2] Project Manager, Hawke's Bay Engineering Lifelines Project   n.evans@opus.co.nz

This paper provides an update on New Zealand presentations at previous TCLEE conferences (Hopkins et al, 1995), and outlines the current developments in Lifelines engineering in New Zealand, and the influences that are shaping the future direction of this work. Of particular significance is the recently enacted Civil Defence Emergency Management legislation that requires utilities to meet certain obligations with regard to emergency planning and performance.

2. The New Zealand Lifelines Engineering Process

The New Zealand Lifelines Engineering process is based around the following risk management steps:

- Identifying the *hazards* which could affect each lifelines network
- Compiling *inventories* of the various utility and transportation networks
- Assessing the *vulnerability* of each lifeline network to those hazards (the *potential damage to* and *consequences for* each network)
- Identifying and implementing practical *mitigation* measures
- Facilitating the preparation of comprehensive *emergency response* plans

This process is based on risk management methodology encapsulated in AS/NZS 4360:1999 (SA & SNZ, 1999), and is described more fully elsewhere (Brunsdon, 2001). It is applied on a regional basis, rather than on an organisation-by-organisation basis. The responsibility for taking appropriate mitigation and preparedness steps however remains with the individual organisations.

With respect to hazards, the focus of lifelines work in New Zealand is on **regional scale events** that are beyond the ability of individual organisations to respond to and control. While Lifelines Engineering in New Zealand had its origins with earthquake, subsequent work has extended to address all hazards. Earthquake hazard has however been found to present the most significant hazard to utility systems in virtually all of the regions studied to date.

The five key Lifelines steps typically take from 3 to 5 years to work through for each region, and result in a major report. Reports have been completed by separate Lifelines Projects in the major metropolitan centers of Wellington (CAE, 1991), Christchurch (CAE, 1998), Dunedin (DELP, 1999) and Auckland (ARC, 1999). The recently completed Hawke's Bay Engineering Lifelines Project (HBRC, 2001) was the first to adapt the metropolitan methodologies to suit the smaller and more dispersed centres with much less dense and/ or widely spread utility networks.

The Lifelines process is however an ongoing one, reflecting the iterative nature of risk management generally. Each of the completed Lifelines Projects have metamorphosed into ongoing *Lifelines Groups*. Lifelines Groups are responsible for continuing the task of communicating the findings, outcomes and recommended mitigation and response preparedness measures to the wider community, and ensuring that the individual utilities address the regional level recommendations within their own asset management plans. A review of mitigation and preparedness progress and achievements across all utility organisations in each region is typically conducted on an annual basis. This important step maintains the momentum and information exchange achieved by the earlier work.

There has been a range of physical mitigation undertaken by the various utility sectors over the past decade. While some of this work was or would have been initiated by the respective individual utility asset management plans, the lifelines process has provided a sharper focus and often a greater sense of urgency in the 'toughening' of networks. The extent of physical work undertaken to date has however varied between organisations depending on the level of priority assigned by senior management.

3. Organisational Vulnerabilities

While the physical vulnerability of utility networks and facilities to earthquake is generally well appreciated by Lifelines practitioners, organisational vulnerabilities are of equal concern in New Zealand. Lifeline utilities in New Zealand have undergone considerable transformation over the past decade, as in other countries. The restructuring in most sectors has generated a number of newer utility organisations and a greater commercial focus, particularly for those in sectors with revenue directly at risk. While this has led to significant advances in short- and medium-term financial risk management, many of the newer utilities have not given the same level of attention to mitigation and preparedness for longer return period hazard events. The same can also be said for some from the category of utilities that do not have revenue directly at risk.

The dividing up of some utility sectors into component pieces (eg. *generation*, *transmission* and *distribution* for electricity) and highly competitive sectors working within anti-competition legislative frameworks (telecommunications and energy) has led directly to a 'silo' approach for emergency response. This clearly has an adverse influence on the ability of these sectors to develop integrated plans to respond to a major event such as earthquake. Studies of the 1995 Kobe earthquake highlighted that the effective response of utilities such as Kansai Electric Power Company was strongly influenced by their integrated nature, covering generation, transmission and distribution (WELG, 1995).

Restructuring has also led to extensive outsourcing for design and maintenance, with a resulting heavy dependence of many utility organisations on contractors, some of whom are shared with other organisations. While maintenance contracts place considerable emphasis on 24 hours/ 7 days a week response as part of 'business as usual', they need to be subjected to more careful scrutiny to ensure that the procedures will also be effective for extreme events such as earthquake. For example, the ability of external contractors to carry out the critical initial impact assessment immediately after a significant earthquake is open to question.

4. The New Legislative Context in New Zealand

On 1 December 2002, new Civil Defence Emergency Management (CDEM) legislation was enacted. Under this Act, CDEM Groups will form across each region to integrate and co-ordinate hazard risk management planning and emergency management activity. CDEM Groups are essentially a consortia of local authorities, emergency services, lifeline utilities and others delivering emergency management within regional boundaries.

The Act defines lifeline utilities as key agencies, including airport authorities, port companies, gas production, supply and distribution companies, electricity generators and distributors, providers of networks for water and waste-water, telecommunications, rail and road, and producers or distributors of bulk petroleum energy products. The Act does not impose new business requirements or alter responsibility for risk, asset and emergency management. The emphasis is on providing continuity of operation, particularly where their service supports essential CDEM activity (e.g. a hospital).

Given their importance to the nation's disaster resilience, utilities within the definition are required under section 60 of the Act to:

a) ensure that it is able to function to the fullest possible extent, even though this may be at a reduced level, during and after an emergency

b) have plans for such continuity, that can be made available to the Director if requested

c) participate in the development of the National Strategy and CDEM Plans (where requested)

d) provide technical advice on CDEM issues, where reasonably requested by CDEM Groups or the Director

These duties have been interpreted within a set of Director's Guidelines (MCDEM, 2002) as expectations for lifeline utilities which include:

- Planning for and being able to implement procedures to ensure continuity of business and service to customers – it is not an option to be unprepared. This includes understanding the full range of hazards that could impact an operation and considering external risks such as dependence on utilities from other sectors and outsourcing/contractor arrangements.

- Establishing planning and operational relationships with CDEM Groups (local government and emergency services – Police, Fire, Health) and other external agencies.

- Communicating and planning across their sector to optimise service during emergencies.

To meet the requirements of the Act, there will clearly need to be greater emphasis by lifeline utilities on and commitment to the key elements of earthquake preparedness across the 4Rs of *reduction, readiness, response* and *recovery*.

5. Case Study: The Hawke's Bay Lifelines Engineering Project

The Hawke's Bay region is located on the east coast of the North Island of New Zealand, and covers an area of approximately 190 km x 75 km. Its 150,000 population includes two cities, Napier and Hastings each of about 55,000 people. In 1931 the region suffered NZ's most recent devastating urban earthquake. This magnitude 7.8 event resulted in the loss of 243 lives, a major fire in the Napier Business District and also caused an extensive area of land to rise up to 1.5 metres.

The Hawke's Bay Engineering Lifelines Project, which started in 1998, had by 2002 completed Stage One that involved defining the risk posed to lifelines from natural hazards and elements of highest risk. Further stages will involve monitoring and encouraging mitigation of the identified risks and using the information gathered within emergency management and response planning.

The Project
The project has experienced a number of issues that are common to regional (non-metropolitan) projects. These include:

- The major networks are often sections of national networks and so can be affected by, and can affect, events outside the region.
- Some of the networks consist of single strands running the full length of the region.
- There are many individual sites along each network that can be subject to risk from one or more natural hazard.
- Most of the project participants are the local network managers and operators who are not specialists in risk assessment or risk mitigation.

The Hawke's Bay project followed the now well-established methodologies conforming with AS/NZS 4360:1999.

Natural Hazards
This project adopted an "all natural hazards" approach that included:

- Earthquake
- Meteorological Hazards
- Flooding
- Volcanic Impacts
- Landslides
- Tsunami

Earthquake
Studies on the seismicity of Hawke's Bay indicate that the region is one of the most earthquake prone areas of New Zealand. The region has at least 22 known active faults and folds that are capable of producing strong earthquake shaking in the future.

At least 5 of these are capable of producing levels of earthquake shaking similar to those experienced in 1931.

Large subduction thrust earthquakes on the interface between the Australian and Pacific plates underlying Hawke's Bay occur frequently and are capable of producing high levels of shaking over a large part of the region. The hazard associated with subduction faults is difficult to quantify but is a very important element of seismic assessment for the Hawke's Bay region.

The subduction zone in the North Island is unusual in two ways. Firstly, much of the convergent margin is above sea level, meaning that subduction events that are normally offshore in Japan, for example, occur much closer to populated areas. Secondly the direction of convergence is not perpendicular to the general NE-SW strike direction. The component of motion perpendicular to the coast is assumed to be taken up by combination of slip on the subduction interface and slip during thrust type earthquakes on shallow faults.

Reyners (2000) has estimated that large subduction zone thrust events with a magnitude of 7.7 have a recurrence of about 550 years in Hawke's Bay and involve an average slip of about 3 metres of a segment of tectonic plate with an area of 120 km long by 45 km wide (see Figure 1). Such an event would be expected to produce both uplift and subsidence.

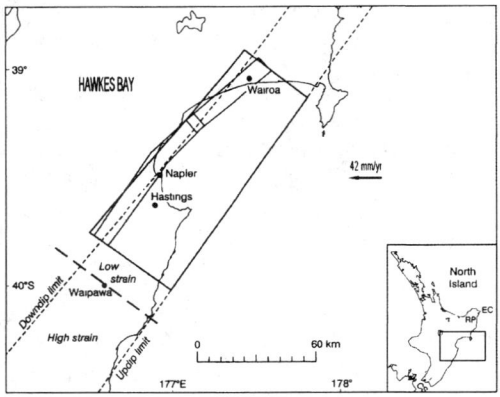

Figure 1: The heavy-lined rectangle indicates the portion of the seismogenic zone which might rupture in a large subduction thrust earthquake

Tsunami
The low-lying parts of Hawke's Bay are the most densely populated and also the most prone to damage from tsunami. This combined with the fact that Hawke's Bay is very close to an active tectonic plate boundary, makes tsunami a real and

significant hazard for the region. Further research is needed to better understand the tsunami hazard.

Major Findings of the Project
The seismic hazard poses the greatest risk to all networks especially to structures such as bridges and wharves, control centres and studios. Landslips and flooding are potentially the next most serious hazards.

The supply of electric power to Hawke's Bay is limited by the capacity of the single 220 kW line from outside the region to just north of Napier. If this supply were lost, local sources of electricity would not be capable of maintaining full economic production in the region.

Hawke's Bay needs a well-designed and constructed regional civil defence emergency operating centre.

Mitigation Measures
Already some networks have set about mitigation measures based on assessments made during the project. Where the measures are appropriate they can often be more economically and easily incorporated as part of upgrade or redevelopment programmes.

6. Fire Following Earthquake

The Wellington Lifelines Group has recently completed a national research project to identify the key issues for New Zealand with regard to fire following earthquake. This project was financially supported by the NZ Fire Service Research Fund (WeLG, 2002), and took a systematic look at this issue by applying the steps of the risk management process (SA & SNZ, 1999). While fire following earthquake has been intensively studied and modelled at the local level internationally, one of the objectives of the study was to provide a framework for establishing whether fire following earthquake is sufficient a risk in an overall regional context to warrant specific detailed studies. Whilst regional seismic hazard is clearly a significant influence, other factors such as (i) the presence and extent of reticulated gas, (ii) the age and nature of buildings (predominant construction type and separation), (iii) proximity to high fuel load vegetation and (iv) prevailing wind conditions are arguably more significant factors in considering the potential for individual post-earthquake fires developing into an escalating conflagration.

The other side of the equation is of course the ability of the Fire Service to extinguish individual fires after a major event. The general planning assumption for fire following earthquake in most of New Zealand's major cities is that the majority of fires will not be extinguished following a major earthquake, due to the combination of *lack of access* due to extensive damage to many of the buildings of earlier construction coupled with the *likely lack of water* due to poor performance of key water mains.

The project report contains recommendations under the headings of (i) *Short-term mitigation actions*, (ii) *Medium-term planning processes* and (iii) *Further research activities*. Examples of short-term mitigation actions include emphasising the seismic restraint of heavy and vulnerable items such as boilers, heating units, stoves and tanks in commercial and apartment buildings, and developing integrated procedures for post-earthquake shutting off and restoration/ reconnection of electricity and gas services. Recommended areas for further research include a more detailed review of international earthquakes to better understand why no conflagrations have occurred in the more recent events such as the Chi-chi, Taiwan earthquake. The post-earthquake fire spread potential in high-rise CBD buildings also requires further modelling, given the three significantly differing fire resistance design eras of pre-1965, 1965-1991 and post-1991. It is anticipated that such research would include a review of the pro-active building regulatory approach for fire following earthquake adopted by cities such as Vancouver.

7. Priority Restoration of Utility Services to Key Community Facilities

The Auckland Engineering Lifelines Group in conjunction with the Ministry of Civil Defence & Emergency Management is currently undertaking a project to identify the community facilities where utilities need to focus the restoration of their services after a major emergency such as earthquake. While many aspects of the response and recovery for both utilities and the community will be significantly influenced by the actual physical impacts, there are certain facilities that need to be accorded high priority for restoration. These include hospital and medical centres, Civil Defence and Emergency Services headquarters and control centres for energy, communications, water and wastewater processes. While these and other essential facilities should have robust alternative energy supplies, backup communications and stored water, the extent of these will be finite, and having "normal" services restored at an early stage will clearly benefit the effectiveness of the functions undertaken in those facilities.

For critical community facilities such as hospitals, an indication of the utility restoration priorities in general terms along with the anticipated reduced levels of utility service in an emergency situation will enable those organisations to review their business continuity planning preparations, and ensure that their emergency supply arrangement are adequate. The expectations that utilities will act for the "public good" during emergencies must be balanced against commercial or performance imperatives.

This project will generate a national template, and illustrates how the Lifelines Engineering process and emergency management practice is intended to interface under the new legislation.

8. Summary

There have been significant developments in Lifelines engineering in New Zealand over the past few years. The regionally-based multi-disciplinary Lifelines engineering model has expanded to involve Lifelines Projects in each of the 16 regions of New Zealand. This model is based on the application of the principles of the international Risk Management Standard AS/NZS 4360: 1999 via a collaborative process. Although an all-hazards approach is followed, earthquake remains the principal hazard event for utilities to plan for.

Notwithstanding this good progress at regional and national levels, however, much remains to be done by individual utilities in order to significantly reduce the vulnerability of major utilities to earthquake. Relevant observations include:

- The extent of physical work undertaken to date has varied between organisations depending on the level of priority assigned by senior management
- Many utilities have not made use of the information available from regional Lifelines Projects
- The degree of action taken by individual utilities is still governed by their internal economic perspective, rather than the assessed regional level of risk
- Many utilities are still planning in isolation, without much regard to their dependence on other lifeline sectors

The backdrop to the limited progress by some utilities is that no major New Zealand urban area has experienced a damaging earthquake since the devastating 1931 Napier earthquake. It is this fact that makes the credibility of earthquake risk so hard to convince the decision makers (senior management, elected councillors, board members, etc) to commit funds for mitigation and ongoing work.

The new Civil Defence Emergency Management legislation however provides a regulatory framework that sets out much more clearly the obligations and expectations of lifeline utilities in planning for major natural and technological hazard events. To meet the requirements of the Act, there will need to be greater emphasis by lifeline utilities on and commitment to the key elements of earthquake preparedness across the 4Rs of *reduction*, *readiness*, *response* and *recovery*.

With its emphasis on the much-needed external perspective by highlighting the *interdependencies* involved, the Lifelines Engineering process will continue to be an important interaction mechanism for the utility and emergency management sectors in the near future.

References

Auckland Regional Council. (1999). Auckland Engineering Lifelines Project: Final Report – Stage 1 *ARC Technical Publication* No. 112, ISSN No. 1172 6415, Auckland, New Zealand.

Brunsdon D. R. (2001). Lifelines Engineering in New Zealand: A Decade of Collaboration and Achievement, *Proc. Wellington Lifelines Group International Workshop*, Wellington Regional Council, ISBN 0909016771, Wellington, New Zealand.

Centre for Advanced Engineering. (1991). Lifelines in Engineering - Wellington Case Study, *Project Summary and Project Report, Centre for Advanced Engineering*, University of Canterbury, Christchurch, New Zealand.

Centre for Advanced Engineering. (1998). Risks & Realities: A Multi-disciplinary Approach to the Vulnerability of Lifelines to Natural Hazards, *Christchurch Engineering Lifelines Group Report,* University of Canterbury, Christchurch, New Zealand.

Dunedin City Lifelines Project. (1999). *Dunedin City Lifelines Project Report*, ISBN No. 0-9597722-2-7, Dunedin, New Zealand.

Hawke's Bay Regional Council. (2001). Facing the Risks, *Hawke's Bay Engineering Lifelines Project Report,* Napier.

Hopkins D.C., Norton J.A. and Brunsdon D. R. (1995). A Comprehensive Framework for Response Planning, *Proc 4th Conference of the Technical Council for Lifeline Earthquake Engineering*.

Ministry of Civil Defence and Emergency Management (2002). Working together: Lifeline Utilities and Emergency Management *Director's Guidelines DGL 3/02* www.civildefence.govt.nz

Reyners, M. (2000). Quantifying the Hazard of Large Subduction Thrust Earthquakes in Hawke's Bay, Institute of Geological and Nuclear Sciences, *Bulletin of the New Zealand Society for Earthquake Engineering* Vol 33, No. 4 December 2000.

Standards Australia and Standards New Zealand. (1999). *Risk Management AS/ NZS 4360*, Sydney and Wellington.

Wellington Lifelines Group (2002). Fire Following Earthquake: Identifying Issues for New Zealand. *Report for the NZ Fire Service Contestable Research Fund,* Wellington

Time and Space Restoration Process and Prediction of Recovery Period for Damaged Water Supply Systems Based on GIS Data of The 1995 Kobe Earthquake

Shiro Takada[1] and Tatsuhiko Imanishi[2]

Abstract

Present study addresses time and space restoration process for damaged water supply systems in Kobe Water Bureau and proposes a formula to predict recovery periods of water supply based on the GIS analyses for space and time recovery process of the systems. Present GIS involves the database of damage sites of transmission and distribution pipelines giving the classes of damage mode, pipe material and joint type, diameter as well as various kinds of data related to recovery works. From the view point of performance based design, by employing the proposed equation, a target level of strengthening the water supply system corresponding to the target recovery periods could be decided, which is also respondent to the level of earthquake ground motions related with the pipe damage ratio.

Introduction

Among flow materials given by lifeline functions, water is vital for keeping the human life safety and no any other alternative materials in lieu of it . During the 1995 Kobe Earthquake, emergency water supply and restoration works were so heavy under the situation of the confused water control center without grasping total damage states all over the city during 2-3 days after the Earthquake.

The existing disaster prevention plan was mainly focused on the countermeasures to strengthen facilities of pipelines, pump stations and so on. The emergency action plan was not necessarily enough and unpractical in existing plans and it was impossible to predict the required periods to restore the damaged systems in spite of the existence of some theories predicting the restoration periods based on complicated network theories and/or water flow analyses.

---

[1] Professor, Kobe University, Japan, +81-78-803-6037, takada@kobe-u.ac.jp
[2] Graduate student, Japan, Kobe University, +81-78-803-6037

At the end of January, 1995, Kobe Water Bureau reported the final recovery dates would be within 2-3 weeks in respond to the demand by citizens and then, on February, the Bureau counted the recovery dates based on the past data on gas supply systems which could not give satisfied answers to the verdict of the public. The predict of water recovery period with the sufficient accuracy is extremely important to the society after the earthquake though earthquake preparedness to mitigate the damage of water supply systems is required before the earthquake. It takes time and cost to make up the water supply system having no damage under any level of earthquake motions. From the view point of performance based design, we have to have taking consideration of the states of malfunction in water supply lifelines under high level of earthquakes.

Malfunction of Water Supply Systems During The Kobe Earthquake

1.3 million customers in Hyogo Prefecture and 0.3 million customers in Osaka Prefecture lost a drinking water just after the 1995 Kobe Earthquake. The malfunction areas spread to north-east direction from the epicenter in Awaji Island corresponding to the direction of the fault rapture. The main cause of the malfunction of the water supply in Hyogo Prefecture was the break of water transmission trunk pipelines in Hanshin Water Industry supplying water to Water Business Sectors in 4 cities from Yodo River and also huge numbers of break of water distribution pipelines in each Sector.

Figure 1 shows water supply recovery process in space and time in Kobe Water Bureau during 90days after the occurrence of the earthquake. Figure. 1 (a) is showing the date in each block the recovery works started. East and west areas and north mountain areas in the City where there did not occur house damage so heavily were given the priority to start the recovery work and the center areas in the City so called "Heavy Disaster Belt" needed more than 60 days to start the work. Figure. 1 (b) is required periods of recovery works after the start in each block. The west and mountain areas did not need so many days for the work, however the east area starting earlier needed more than 60 days in some blocks. The center area did not spend many days in spite of the late start of the work. Those recovery work was supported by a lot of workers from Water Supply Sectors all over Japan.

Figure 2 is the change of malfunction rate of water supply in 10 cities and 7 blocks in Hyogo Prefecture. The maximum period for the recovery work was 91days in Kobe Water Bureau. Water supply was recovered so quickly in Awaji Island due to relatively rural areas. The definition of the end of recovery work was different Water Sector by Sector. Some means the date customers could use water in residential houses and some the date when distribution pipelines were repaired and water came to the connection point to houses. When water supplied dates are compared for the data by Kobe Water Bureau and a questionnaire investigation by author, there exists 1-2 weeks difference for 50% recovery rate, which means it took 1-2 weeks in average to repair water supply pipelines inside house yards.

Figure 1 (a) Starting date for recovery work in each Block

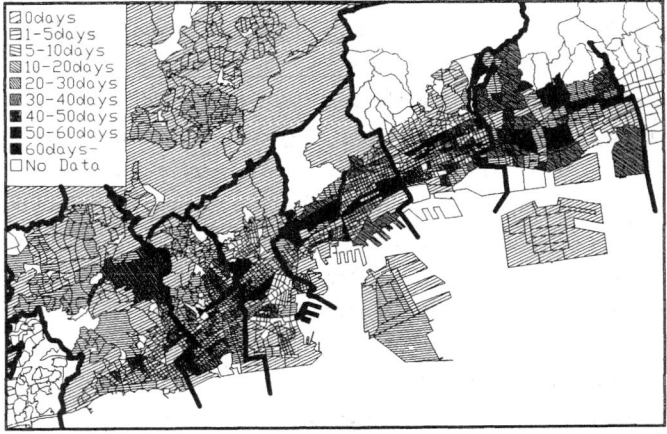

Figure 1 (b) Periods required for recovery works

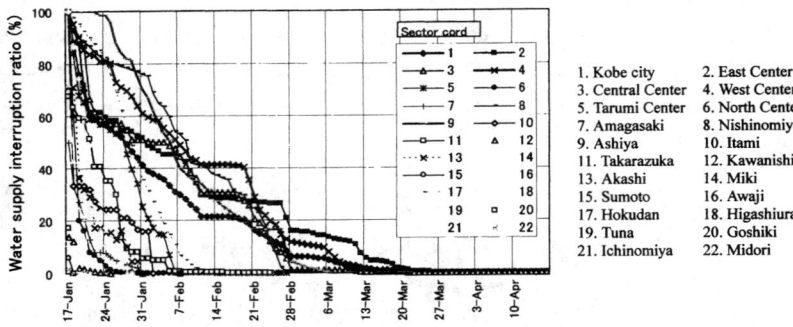

Figure 2 Water supply recovery process

Macroscopic Analyses For Water Supply Recovery Periods

This chapter analyses a macroscopic recovery process for the unit of each Water Supply Sector corresponding to Cities or Blocks.

(1) Recovery rate related with pipe damage ratio

Figsure 3 and 4 depicts area-maps for a 90% recovery period and pipe damage ratio respectively. The recovery rate is defined by the number of water-supplied households divided by all households in the City or Block, and the damage ratio of pipelines gives the number of leakage points divided by total length of distribution pipelines in the City or Block. These figures indicate that the recovery rate is not necessarily proportional to the pipe damage ratio. The disaster belt areas needs the more recovery periods in spite of not so high pipe damage ratio. Figure 5 shows these relations in terms of the recovery rates for 90% related with damage ratio of distribution pipelines. A densely populated areas of Hanshin district (Osaka and Hyogo Prefectures) needs the more recovery periods compared with Awaji district of rural areas.

(2) Factor analyses by quantity theory

A quantity theory I is applied to investigate the effects of various factors to the recovery periods. The quantity Yi associated to the recovery dates is given by a next equation when a category number a_{ij}, which is defined by the number for a category k in an item i, is optimized by the least square method to fit the data base of the recovery dates.

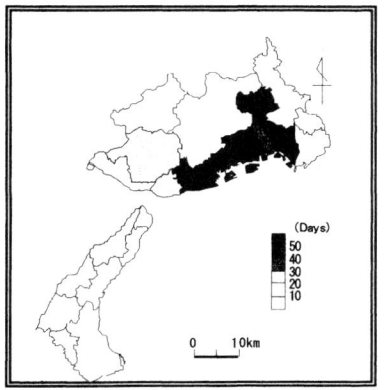
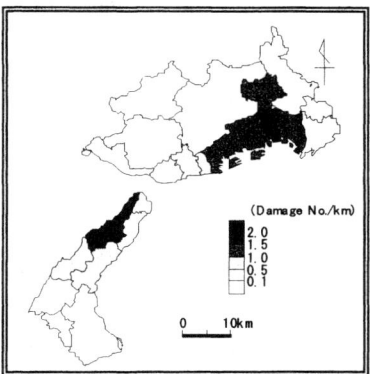

Figure 3 Working days for 90% recovery days in each city

Figure 4 Pipe damage ratio in each city

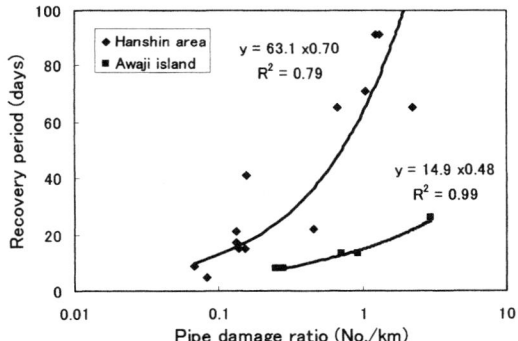

Figure 5 Recovery days related with pipe damage ratio

$$Y_i = \Sigma \Sigma \ a_{ij} \ \delta_j(jk) \qquad (i = 1,2,\cdots,n) \qquad (1)$$

$$\delta_j(jk) = 1 \ or \ 0$$

The obtained category number a_{ij} means the degree of affection to the recovery periods by each category. Analytical results are shown in Table 1 and Figure 6. Pipe damage ratio, population density and residential house damage ratio are selected for the items in the analyses. Table 1 gives the results that the recovery periods increase in case of the higher pipe damage ratio more than 0.5 No./km and densely populated areas need the more recovery periods. Figure 6 shows the good accuracy for prediction and actual data values of the recovery periods of 90%. On the contrary, only a damage ratio of residential houses does not necessarily cause the longer period for the recovery.

Table1 Result of 90% recovery periods by quantity theory I

Item	Category	Category Num.	Extent	Correlation
Pipe damage ratio(No./km)	~0.15	-16.5	40.50	0.93
	0.15~0.5	-15.3		
	0.5~1	13.0		
	1~	24.0		
Population density (Person/km²)	~2000	-5.6	11.22	0.67
	2000~7000	1.2		
	7000~	5.6		
Damage ratio for houses	~0.15	5.5	20.99	0.79
	0.15~0.3	6.1		
	0.3~	-14.9		
Constant Value		19.4	$R^2=0.90$	

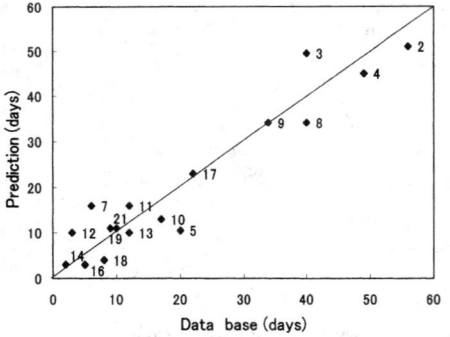

1. Kobe city
2. East Center
3. Central Center
4. West Center
5. Tarumi Center
6. North Center
7. Amagasaki
8. Nishinomiya
9. Ashiya
10. Itami
11. Takarazuka
12. Kawanishi
13. Akashi
14. Miki
15. Sumoto
16. Awaji
17. Hokudan
18. Higashiura
19. Tuna
20. Goshiki
21. Ichinomiya
22. Midori

Figure 6 Comparison of recovery periods between Prediction and database

(3) Prediction formula for recovery period

In general, the recovery process of damages water supply systems are affected by factors like as a pipe damage number, a length of pipelines, system characteristics of water supply networks, a mutual correlation with other lifelines, a damage ratio of residential houses, a damage states of infrastructures related to recovery works and also an urbanized factors. Present paper introduces a new factor of the urbanized degree adding to the pipe damage ratio to predict the recovery periods for damaged water supply systems.

Introducing a number of customers per 1 km distribution pipeline (we call a customer density) as the new factor, the relation of the 90% recovery period related with the customer density is shown in Figure 7, which shows roughly proportional relation. Figure 8 depicts the distribution of the customer density in each city in an area map. The figure indicates that the customer density is less than 100 in rural areas, less than 150 in satellite cities and more than 150 in highly urbanized cities. By using the customer density, we could characterized the water supply scale in each Water Supply Sector.

By applying a multi-regression analyses, the 90% recovery period can be expressed in terms of the customer density and pipe damage ratio as follows:

$$Day_{90} = 0.38 \cdot x_1^{0.86} \cdot x_2^{0.59} \quad (R^2 = 0.79) \tag{2}$$

Day_{90}: 90% recovery period (Day)
x_1: Customer density(Customer number/1 km distribution pipeline)
x_2: Pipe damage ratio(Damage number /1 km distribution pipeline)

Above equation depicts as shown in Figure 9 which could give an information that a less than 0.5 pipe damage ratio in the city of customers density 150 when the recovery works should end within 3 weeks.

By employing the proposed equation, a target level of strengthening the water supply system corresponding to the target recovery periods could be decided, which is also respondent to the level of earthquake ground motions related with the pipe damage ratio.

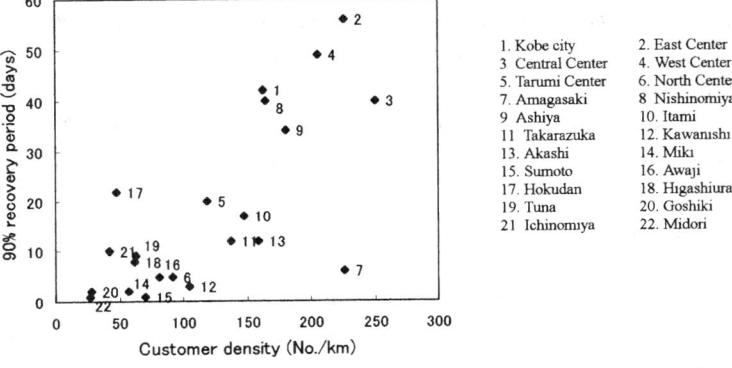

Figure 7   90% recovery period related with customer density

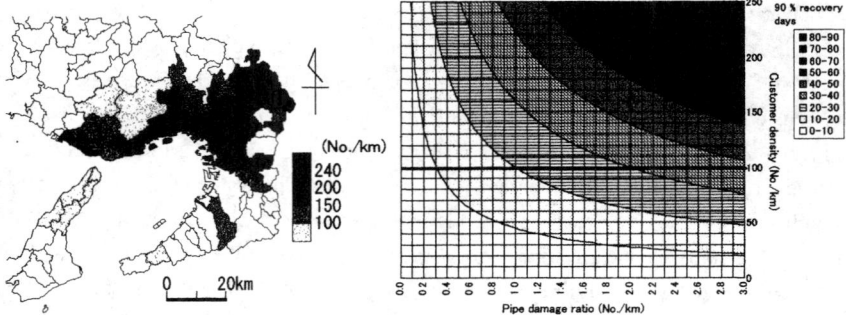

Figure 8 Map of customer density **Figure 9** Prediction chart for recovery

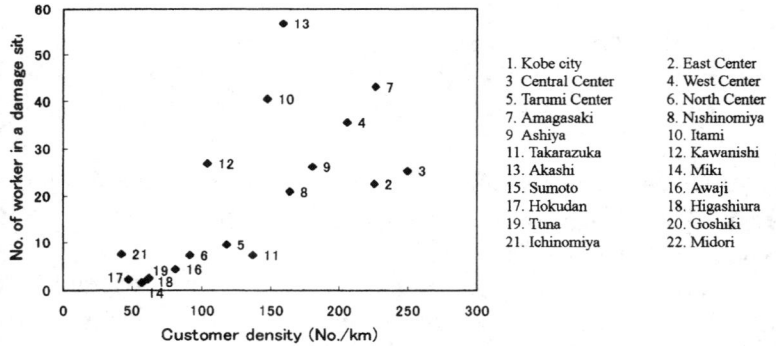

1. Kobe city   2. East Center
3. Central Center   4. West Center
5. Tarumi Center   6. North Center
7. Amagasaki   8. Nishinomiya
9. Ashiya   10. Itami
11. Takarazuka   12. Kawanishi
13. Akashi   14. Miki
15. Sumoto   16. Awaji
17. Hokudan   18. Higashiura
19. Tuna   20. Goshiki
21. Ichinomiya   22. Midori

Figure 10 No. of workers related with customer density

(4) Prediction formula for number of workers necessary for recovery process

Figure 10 shows the relation between the customer density and number of workers required to repair one site damage of pipelines in each city. The number of workers is almost proportional to the customer density because of the lower working efficiency in highly urbanized cities. A multi-regressive analyses give the following formula:

$$Y = 34.3x_1 + 18.2 x_3 - 3488 \quad (R^2 = 0.92) \tag{3}$$

Y is the number of workers, x_3 is a number of damaged pipes and x_1 is the same in equation (2). By introducing the customer density, we can estimate the necessary numbers of workers in high accuracy with correlation coefficient 0.92.

(5) Malfunction impact by water supply interruption

Malfunction impact due to water supply interruption is dependent on the process of recovery rate as shown in Figure 11. Case 2 in the figure would give the more impact to the society than Case 1. Here the malfunction impact is defined by the product of water interruption rate {1- (water supply rate)} by recovery period in each time-dependent recovery process. Figure 12 is the malfunction impact in each city. The scale of impact differs from the recovery period and the impact is much more large scale in the center areas in Kobe City. Table 2 shows analytical results of the quantitative theory I as the factors giving the serious effects on the malfunction impact. The impact increases rapidly for more than 0.4 pipeline damage ratio in densely population areas. Combining the results of the malfunction impact with the previous results of recovery period, we can say pipe damage ratio should be less than 0.4~0.5 to mitigate the secondary spread of disasters.

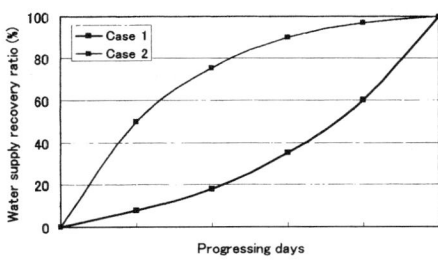

Table 2 Results of guantitative analysis

Item	Category	Category No.	Extent
Pipe damage ratio (No./km)	~0.135	-642.9	1591.40
	~0.4	-501.2	
	~1.0	83.7	
	1.0~	948.6	
Length of distribution pipelines(km)	~80	-132.9	251.50
	~500	-68.2	
	~660	69.9	
	660~	118.6	
Customer density	~100	-254.5	416.40
	100~	161.9	
Constant value		894.0	R2=90

Figure 14 Examples of restoration curve

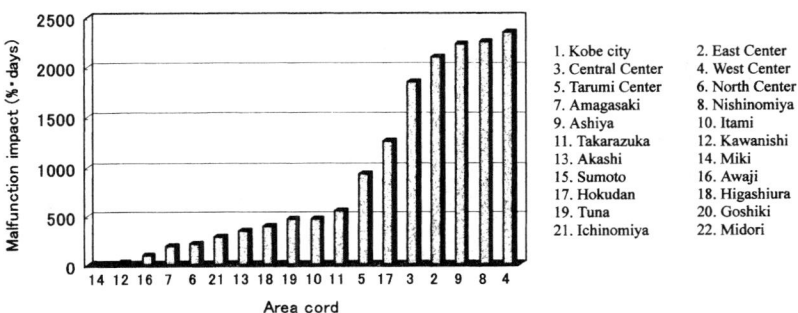

1. Kobe city 2. East Center
3. Central Center 4. West Center
5. Tarumi Center 6. North Center
7. Amagasaki 8. Nishinomiya
9. Ashiya 10. Itami
11. Takarazuka 12. Kawanishi
13. Akashi 14. Miki
15. Sumoto 16. Awaji
17. Hokudan 18. Higashiura
19. Tuna 20. Goshiki
21. Ichinomiya 22. Midori

Figure 15 Malfunction impact in each water supply sector

Conclusions

Based on GIS database, space and time recovery process for damaged water supply systems are analyzed by using a quantitative theory I and a multi-regressive analyses. Followings are the results obtained by the present research.

1. Only the pipe damage ratio does not necessarily govern the recovery period for the damaged water supply systems.
2. A damage ratio of residential houses is closely related with the speed of recovery works, and the date on which customers receive water does not necessarily coincide with the end day of recovery works reported by Water Supply Sectors.
3. By introducing a new factor of the customer density, a predicting formula for the recovery period is proposed with a sufficient accuracy. The applicability of the proposed formula should be investigated for other type of earthquake disasters besides the Kobe Earthquake.
4. A predicting formula for the number of workers necessary for recovery works is proposed also by introducing the customer density.
5. From the view point of performance based design, by employing the proposed equation, a target level of strengthening the water supply system corresponding to the target recovery periods could be decided, which is also respondent to the level of earthquake ground motions related with the pipe damage ratio.

References

Kobe Water Bureau (1997). "Record on Restoration of Kobe Water after the Hanshin-Awaji Earthquake Disasters."
Amagasaki Water Bureau (1996). "Record on Restoration of Amagasaki Water after the Hanshin-Awaji Earthquake Disasters."
Nishinomiya Water Bureau (1995). "Hanshin-Awaji Earthquake Disasters, Restoration Record on Water.
Takarazuka Water Bureau (1996). "Damage and Restoration of Takarazuka Water after the Hanshin^-Awaji Earthquake Disasters."
Ashiya Water Bureau (1997). "Report on Restoration of Ashiya Water after Hanshin-Awaji Earthquake Disasters."

Vulnerability of Energy Distribution Systems to an Earthquake in the Central and Eastern United States – An Update

John M Nichols[1] and James E Beavers[2]

Abstract

This paper updates previous research on the vulnerability of energy transmission systems in the central and eastern United States to earthquake hazard. The paper will present and summarize the recent research completed on these systems with reference to a range of earthquake scenarios commencing at a M6 earthquake. An assessment will be provided of the vulnerability of the transmission systems to earthquake loading, and a summary provided of future research that may reduce the hazard to energy transmission systems. This region represents 3 percent of the world's population, and it has been seismically quiet for 100 years. The probability of a major earthquake in this century will be considered as part of the assessment.

Introduction

The United States energy distribution systems are vulnerable to failure from a variety of sources both natural and from human intervention. The AAES (1986) report reviews the vulnerability of the energy distribution system to an earthquake for the central and eastern United States with the data coverage for the previous decade to the AAES report. The energy distribution[3] systems across the U.S. consist primarily of pipelines, railroads, water carriers, motor carriers, and electrical transmission. While pipelines, water carriers and railroads are typically interstate distribution systems, electrical and motor carrier distribution systems are primarily intrastate. This paper reviews the vulnerability of these energy distribution systems to earthquake damage.

The oil distribution system has changed significantly from the movement of barrels by wagon and train common in the late 1800s to modern pipelines and the consumption of energy products has increased in the USA since the 1800s. This increased consumption continues a trend from the time of the 40 gallon barrels[4] of kerosene in Pennsylvania that anecdotal evidence recounts were filled to 42 gallons if 40 gallons was purchased, creating the defined barrel of oil product [abbrev: bbl(US, petrol) equivalent to ~ 0.1589 m^3](Cardarelli 1997). The total US energy consumption at the beginning of the twenty-first century is about 105 exajoules (~ 99 quadrillion BTU, abbreviated Quad). As the complexity of the urban landscapes and human economy changes, so there are changes and increasing complexity in the

---

[1] Assistant Professor, Department of Construction Science, TAMU, College Station, TX, 77843-5700; phone 979-846-6541; nicholsj@tamu.edu
[2] Visiting Associate Director, External Affairs, Department of Civil Engineering, UIUC, Urbana, IL
[3] The term "distribution" in this paper is used in a broad sense representing the movement of energy.
[4] The non SI term is used here to reflect the story not the quantity which is irrelevant for this work.

distribution system for all energy forms, which can result in some regional areas being dependent on a particular energy delivery system. Disruption of the energy delivery system can lead to disruption of the economy and the safety of the inhabitants of such a regional area. The supply of petroleum to Maine is one example of reliance on shipping. Shipping can become icebound, disrupting supply. The vulnerability aspect of interest in this study is the damage to the distribution system from an earthquake loading. The area of significant previous seismic activity is a regional band stretching from Midland, Texas to Eastport, Maine which has been the subject of significant research work by the NSF earthquake research centers in Illinois and New York in the last score years.

Literature Review

The second author in preparing the research for the AAES report provides a through summary of the issues and data related to the earthquake threat, the recent earthquake research, the threat issues between the east and western coasts, extent of the energy distribution systems within the USA and the volume of material being transported in these systems. The critical finding from the AAES research is the vulnerability of the interstate pipeline distribution systems to an earthquake, with particular emphasis on the threat posed by an event originating within the New Madrid Seismic Zone. The AAES report shows the location of the major and moderate earthquakes in the lower 48 states up to 1986, Figure 1 shows earthquakes from 1568 to 2003 for the USA.

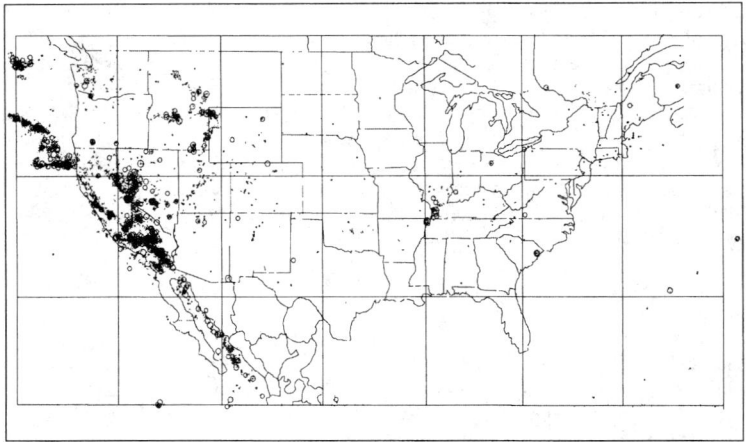

Figure 1. Major and Moderate Earthquakes in the USA up to 2003

Richter (1958) provides a convenient starting point for considering the developments in the field of seismology in the last two score and five years. There are still a number of significant issues to be resolved in the field of seismology particularly for intraplate regions. Rather than diverging into these issues, the objective of this literature review is to consider the current level of knowledge for those research areas that impact on an assessment of the vulnerability of the distribution systems to an earthquake threat.

Johnson and Kanter (1990) outline the problems of determining the seismicity of intraplate regions. The New Madrid Seismic zone (NMSZ) represents a distinct intraplate earthquake hazard (Nuttli 1986). Richter discusses the problem of great earthquakes with a magnitude greater than M8. There is no credible research, since the time of Nuttli and Richter, disputing the accepted findings that the 1811-1812 earthquakes were in the great category, as clearly defined in Richter's text. The critical issue is in determining the hazard from this time-limited dataset and from recent geological research. The past frequency of great events in the NMZS is 400 to 500 years from sand boil data (Tuttle et al. 2002).

The 2001 Gujarat earthquake and the NMZS sand boil data highlight the issue of making a Poissonian or a characteristic assumptions in determining the likely frequency of a great future earthquakes (MAE Center 2002). An earthquake with a probable recurrence interval of 1,400 to 2,600 years struck the Abruzzo region of Italy on January 13, 1915 and killed 32,000 (Ward and Valensise 1989). Even a near complete written history for a fault near Rome can be insufficient to warn of silent active faults, as Kafka and Levin (2000) discuss in relation to several tectonic areas. The earthquake hazard in the CEUS is significant and real and corresponds in large part to the population centres (Figure 2).

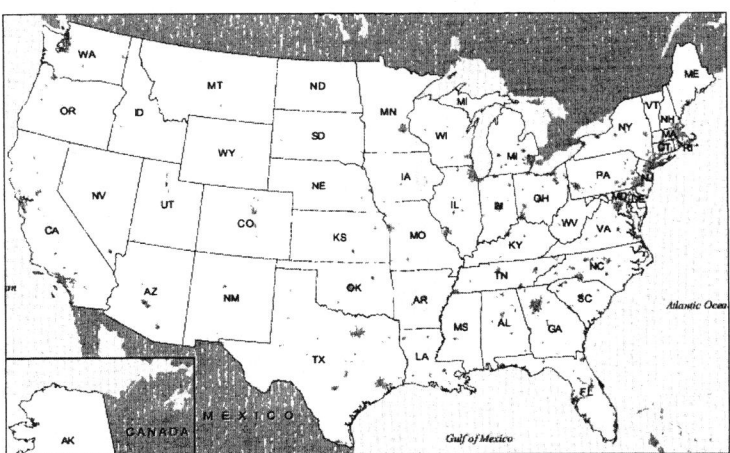

Figure 2. Population Centers (after USDOT - OPS)

Urban development in the south and southwest as well as the other more established regions of the CEUS in the last two decades has required a commensurate development for the natural gas pipeline delivery system. The AAES report provides a map of the major pipeline routes for natural gas in 1986 and provides the sizes and capacities of the main pipelines (Figure 3). Natural gas accounts for about 20 percent of the US energy supply. The potential for increased supply in the US is in the Alaskan fields holding 1 Tm^3. The development of system of pipelines to move this energy to California and perhaps the eastern States will require a substantial national investment (NEPDG 2001) traversing a region of significant interplate seismic risk.

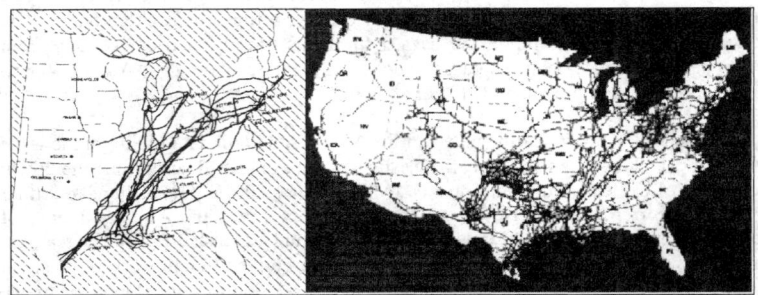

Figure 3. Major Natural Gas Pipeline Routes 1986 and 2002
(after AAES and NEPDG Reports)

Sixty six per cent of the petroleum products movements or transport occurs in pipelines. A significant number of these pipelines traverse the seismic corridor from the NMSZ to New York (Figure 4).

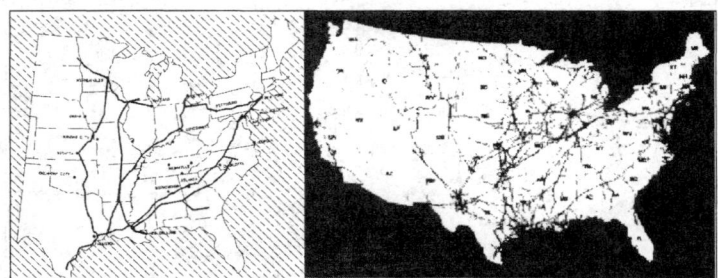

Figure 4. Major Petroleum and Crude Products (2002) Pipelines
(after AAES and NEPDG Reports)

AAES provides details of the specific pipes and the capacities, and the crude oil pipeline routes. The critical crude oil pipelines, including a 1000, 560 and 510 mm set pass through the Mississippi floodplain and meet at Patoka, Illinois. These three pipelines can deliver (approx.) 1.18, 0.37, and 0.29 m^3/s respectively, assuming a velocity of flow of 1.5 m/s. Crude oil (~ 1.8 Mm^3/day) is imported to the US. The CEUS has major refineries accepting imported oil in Pennsylvania and New Jersey.

Coal production rose from 644 in 1985 to 974 teragrams (Tg) in 2000. The production of coal has risen in the states west of the Mississippi River in this period so that the split of coal production is approximately 50 percent in each region compared to 70 percent in the eastern region in 1985. The comments in the AAES report are still applicable to the vulnerability of the coal system to earthquake damage. The damage is still likely to be limited because of the diversity of the location of the coalmines and the use of trains as the primary transport method. The rail system has significant built in redundancy compared to a pipeline system.

The NEPDG Report identifies a number of constraint regions in the electrical transmission system (Figure 5). Demand for a stable electricity supply is exceeding the US economy and political systems capacity to deliver this increased transmission capacity. The constraint point in southern Missouri is close to the NMZS which was a significant problem identified in the AAES report.

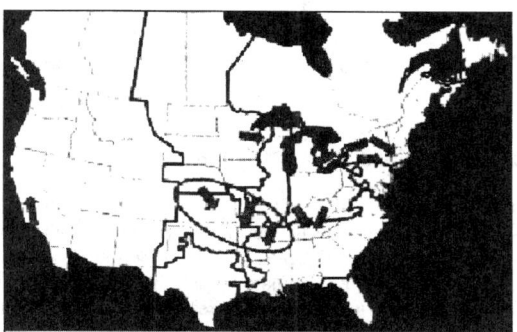

Figure 5. Electrical Transmission Constraint Regions
(after NEPDG Report)

Energy Usage in the USA

Energy usage continues to increase in the USA in line with the increase in population (Figure 6). This sixth chart shows the annual energy usage (in exajoules ~1.055 Quads), the population of the United States (millions of persons), the average energy consumption per person (GJ per annum) and the annual usage of energy in the generation of electricity (EJ) (AAES 1986; Flexible Energy Inc 2001; NEPDG 2001; U.S. Census Bureau 2000)

The interesting feature of this graph is the relative stability of the annual energy consumption per person in the US from 1970 to the present time, at 350 GJ.

This statistical observation suggests that the energy usage will grow in direct proportion to the population increase and is reasonable independent of the economic growth . The annual rate of population increase in the last 30-year period is 1.0 ± 0.1%.

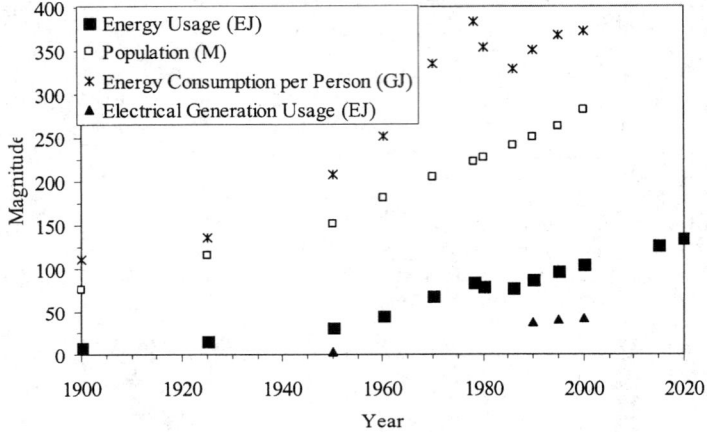

Figure 6. Population and Energy Consumption Data USA 1900 to 2000

US National Energy Policy Development Group (2001) confirms the reasonably constant rate of energy usage per person in the USA since the 1970s. The projected population using a one percent rate for the period to 2020 is 330 million persons, with a upper and lower estimate of 355 to 320 million, based on the variation in the growth rate in the last 50 years. The estimated annual energy consumption in the year 2020 ranges from 114 to 132 EJ for the range of the population estimate and the average and peak values for the annual energy consumption per person. The National Energy Policy Development Group estimate of the energy consumption in 2020 suggests a population used for this estimate of 422 million representing a 2 per cent population increase for the 20-year period. The population growth rate peaked at 2 percent in 1950 for a brief period and was at that sustained level prior to the First World War.

The conclusion remains that the increase in energy consumption will require the development of new energy sources or places an increased reliance on imported energy. The other element that has not been explored is the issue of increased cost for energy as the remainder of the world moves towards the level of energy usage in the developed world. This type of problem would tend to reinforce the development of efficient energy sources in the US over imported product. This significant national issue is barely addressed in the recent NEPDG Report.

Seismic Risk

USGS (2002) provide the estimated hazard from earthquake in the CEUS (Figure 7). The critical hazard areas are the NMZS, Charleston SC, and the upper reaches of NY and into Canada. AAES (1986, Figure 7) show the principal fault areas in the CEUS. Johnson and Kanter (1990) provide an excellent summary of the issues in developing an understanding of the problems associated with determining the hazard from intraplate earthquakes. Richter's comments on the issue of "seismically quiet areas of the world" reinforce the views of Johnson and Kanter.

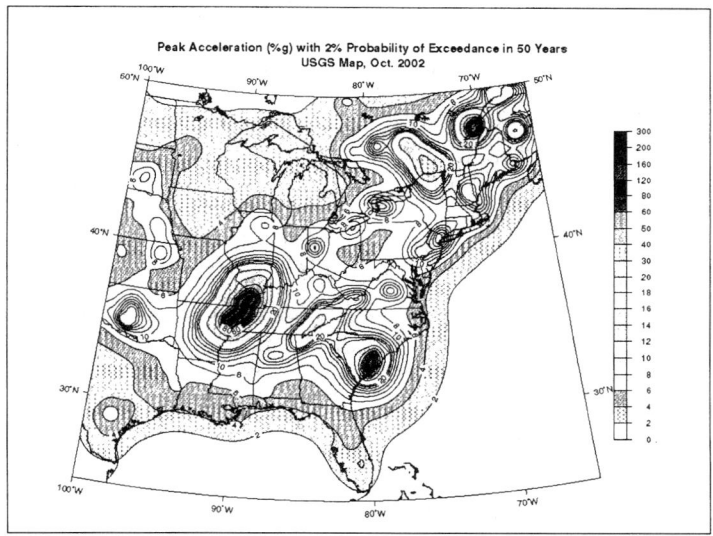

Figure 7. CEUS Earthquake Hazard 2500-year return period

There are two significant issues to answer reliably the larger question of the vulnerability of the energy distribution systems in the Central and Eastern United States. The AAES report adequately addresses the issues in the western US. The problem of 0.5 to 2 Hz pulses with a peak at 1 Hz in large intraplate events represents a significant pulse loading to structures such as nuclear power stations and transmission towers. Newmark and Hall (Newmark and Hall 1978)developed a method to consider this problem, although for a lack of data on intraplate events maintained the constant velocity region for the typical nuclear power station spectra. Nichols in the MAE report (2002) considered the mathematical properties of the 0.5 to 3 Hertz region of the frequency spectra in intraplate events after discussions with Hall[5]. The critical point relates to the non-conservative physical interpretation of the

---

[5] Hall, W. J., (2001) personal conversation dated 26 April 2001 at UIUC.

"constant velocity region" of all current design earthquake spectra. Newmark and Hall demonstrate clearly the superposition of waves that create the apparent constant velocity region in smaller events. Nevertheless, from basic thermodynamics and wave mechanics considerations, and confirmed from actual event data such as the Fast Fourier transform of the 1985 Nahanni event the hypothesis "<u>that the constant velocity region of the spectra exists in all events and forms part of any design spectra</u>" cannot be supported for all intraplate events. This is the most significant issue in the CEUS requiring research.

Silent active faults occur. A "fundamental buckling" of the intraplate surface in the CEUS is suggested in Figure 7 as noted by Denham *et. al.*, (1981) for the Australian continental region. The USGS attempts with the estimated seismic hazard to determine a reasonable bounding function expressed in terms of a 475 and 2500-year hazard at a frequency of 1 and 5 Hz for the design of structures. The assumption is that smaller event delineate where larger events will occur and that we have sufficient historical records to accurately locate the hazardous faults and intrusions such as the NMSZ. There are insufficient historical data and subsurface geological data to determine if all silent active faults in the CEUS and the associated hazard were identified by the USGS. A simple consideration of the mathematical probability of earthquakes faults with a 2500-year return period within a written history that is really no longer than 400 to 500 years suggests that further peaks will develop or existing peaks be enlarged on the CEUS earthquake hazard map. The 1915 Abruzzo earthquake clearly shows this problem, and highlights that a scientific solution does not currently exist beyond the work of the USGS. In the simplest terms, there is a non-zero probability that the bounding function developed for the CEUS by the USS underestimates the hazard, but we lack the data to prove or disprove this hypothesis.

Transmission System Vulnerability

The second author in preparing the AAES report demonstrated the earthquake vulnerability of the petroleum, electrical, and gas systems in the CEUS. These systems are lifelines that support the human population and are critical to maintaining the health and economic welfare of the nation. The vulnerability of each system is greatest in the NMZS to earthquakes. The largest earthquakes would probably occur within the epicentral area for the 1811 – 1812 events, but there is significant activity in Southern Illinois and towards St Louis that could result in smaller events with a magnitude of M5 to M7.

The oil distribution system if breached in the Mississippi River Valley and discharging the equivalent of 20 kilometres of pipeline oil would release about 35,000 m^3 of product that is equivalent to 9 million barrels or 125 Olympic sized swimming pools. This represents a spill as large as the EXXON Valdez loss. The secondary loss is in the time taken to repair the pipeline and the time to divert oil from the national reserve to overcome the likely shortfalls in the East Coast. This system vulnerability is high to such a loss and it is recommended that no more oil distribution pipelines be routed on this corridor. It is recommended that any additional pipelines be routed further south well outside the NMZS. A similar comment applies to the natural gas and petroleum products pipelines. The National Security issues here are significant, and the design of all pipelines should include shut off systems to limit discharge to the Mississippi River or other rivers in the event of a sustained break or terrorist attack or earthquake. The response of authorities to a spill becomes problematic in an earthquake with anecdotal evidence from recent major international events suggesting a non-local response time of 24 hours. The spill could spread 150+ kilometres in one day at a river velocity of two metres per second.

The bottleneck in the electrical distribution system identified in the NEPDG Report coincides with the NMZS. The likely damage from earthquakes is to the secondary components of the system including the switchyards, substations, and converter stations, with the most probably damage from falling elements that have not been tied down to limit earthquake damage. The area without power in any event will probably be at least ten times the size of a similar interplate earthquake and significant in any are where the felt intensity is VII+. The issue of coal transport, and power station vulnerability has not changed since the release of the AAES report. The issue of pulse loading of power stations represents a risk that will need to be addressed in the renewal of licenses of power stations, where any building or element has a natural frequency in the range of 0.5 to 2 Hz.

Conclusions

If a destructive earthquake occurs in the cental or eastern United States then the likely damage to the energy distribution systems is dependent on the magnitude and location of the event. Three scenarios cover the likely range of events, a M5 to M6, M6 to M7, and M7+. The impact of an M5 to M6 would be felt within the meisoseismal area with the likely failure of the electrical distribution systems. The secondary impact on oil delivery that relies on electricity for pumping stations could cause temporary shortages in the eastern part of the US. The probability of an oil spill increases as the epicenter gets close to the pipeline and the magnitude of the event increases. The impact of an M7+ event would be significant, with the probable location for this size of event in the area of the New Madrid Seismic zone. The damage to the electrical distribution systems would be extensive in the meisoseismal area particularly for any felt intensity VII+ region. The damage to the pipeline distribution has two potential effects on the human population, the first is the disruption of the petrol-oil supply, and the second is the damage to the environment with an oil spill. A spill of the magnitude of the Exxon Valdez spill in the Mississippi River Valley or another river system would cause significant long-term pollutant problems. The threat of an earthquake in the CEUS is real. The likely consequences depend on the magnitude and location of the event, however there are a number of steps that can be taken to plan for this occurrence. The response of the authorities in the first day is critical to contain the damage and disruption to lifelines. The planning for the recovery is at this stage the critical element in any planning work for a loss scenario that could extend to a loss of lives and release of pollutants, particularly as there is limited opportunity to retrofit this massive human infrastructure, and the length of time since the last events provides a false sense of safety.

References

AAES. (1986). *Vulnerability of Energy Distribution Systems to an Earthquake in the Eastern United States - An Overview*, American Association of Engineering Societies, Washington.

Cardarelli, F. (1997). *Scientific Unit Conversion: A Practical Guide to Metrication*, M. J. Shields, translator, Springer-Verlag, London.

Flexible Energy Inc. (2001). "Energy Review - Fall 2001." FE Inc., Oak View.

Johnston, A. C., and Kanter, L. R. (1990). "Earthquakes in Stable Continental Crusts." *Scientific American*, 262(3), 42 - 9.

Kafka, A. L., and Levin, S. Z. (2000). "Does the Spatial Distribution of Smaller Earthquakes Delineate Areas where Larger Earthquakes are Likely to Occur?" *Bulletin of the Seismological Society of America*, 90(3), 724 - 38.

MAE Center. (2002). *Gujarat Report*, Mid America Earthquake Center, Urbana.

NEPDG. (2001). "National Energy Policy." National Energy Policy Development Group, Washington.

Newmark, N. M., and Hall, W. J. (1978). "Development of Criteria for Seismic Review of Selected Nuclear Power Plants." *NUREG/GR-0098*, NRC, Urbana.

Nuttli, O. W. (1986). "The Current and Projected State of the Knowledge on Earthquake Hazard." *FEMA / USGS Workshop*, Nashville.

Richter, C. F. (1958). *Elementary Seismology*, Freeman, San Francisco.

Tuttle, M. P., Schweig, E. S., Sims, J. D., Lafferty, R. H., Wolf, L. W., and Haynes, M. L. (2002). "The earthquake potential of the New Madrid seismic zone." *Bulletin of the Seismological Society of America*, 92, 2080-2089.

U.S. Census Bureau. (2000). "Historical National Population Estimates: July 1, 1900 to July 1, 1999", Population Estimates Program, Population Division, U.S. Census Bureau, >(2 January 2002).

USGS. (2002). "Peak Acceleration (%g) with 2% Probability of Occurrence in 50 Years", USGS Earthquake Hazards Program National Seismic Mapping Program, http://geohazards.cr.usgs.gov/eq/2002October/CEUS/CEUSpga2500v3.pdf, >(26 January 2003).

Ward, S. N., and Valensise, G. R. (1989). "Fault parameters and slip distribution of the 1915 Avezzano, Italy, Earthquake derived from geodetic observations." *Bulletin of the Seismological Society of America*, 79(3), 690 - 710.

PLANNING AND MITIGATING FOR LOCAL TSUNAMI EFFECTS
Craig Erdman[1], Jane Preuss[2], Elson T. Barnett[3] and Vivyan Murphy[4]

INTRODUCTION

Over the last two decades, earth scientists have recognized that earthquakes recur on the Cascadia Subduction Zone (see Figure 1) off of the Pacific Northwest coast on an average of every 300 to 500 years (Atwater and Hemphill-Haley, 1996). According to geologic evidence, such earthquakes tend to generate tsunamis with both near-source and distal impacts. Tectonic displacement, and/or subaerial or submarine landslides can generate the attendant tsunamis.

Despite awareness of the potential for tsunamis it is not yet possible to predict their precise location or the characteristics of their focusing with respect to land based developments. It is impractical and even undesirable to remove many or most structures and facilities out of the zone of hazard (as one might for river or other types of flooding) because coastal areas accommodate vital economic resources. They also constitute a unique concentration of critical lifelines including ports, various transportation facilities such as roadways and bridges as well as utilities and communication networks.

The essential first step to reduce the impacts of tsunamis is to analyze damage characteristics that have been documented from recent events. The authors used a case study approach to analyze the tsunami impacts to Crescent City, California (see Figure 1 for location) from the March 28th, 1964 (Good Friday) Alaskan Earthquake. Impacts have been organized into two categories: primary impacts and secondary or interactive impacts. Primary impacts include from flooding of structures, displacement of structures from foundations, structural damage from fluid pressure and erosion. Secondary impacts include structural damage from debris impact, fire debris accumulation across the impacted area such as roadways. Note that the majority of impacts to lifelines are the result of secondary or interactive effects. It should also be noted however that data sources primarily documented damage to structures, and did not systematically document disruption to lifelines such as from debris that obstructed access, especially for response vehicles.

Tsunamis are rare. They are also extremely destructive, and their effects are interactive. Thus it is critical that engineers and planner integrate the heightened awareness of the attributes of tsunamis caused damage to improve public awareness and tsunami resistant planning in routing and design of structures and infrastructure. The output of the analysis is a set of factors that can be useful in development of performance based standards for coastal facilities located in probable tsunami risk zones.

---

[1] *Affiliate Member*,Certified Engineering Geologist, GeoEngineers, 8410-154th Avenue NE, Redmond, WA 98052; phone (425)-861-6000; cerdman@geoengineers.com
[2] American Institute of Certified Planners, GeoEngineers, 8410-154th Avenue NE, Redmond, WA; phone (425)-861-6000; jpreuss@geoengineers.com
[3] Registered Geologist, GeoEngineers, 8410-154th Avenue NE, Redmond, WA 98052; phone (425)-861-6000; cbarnett@geoengineers.com
[4] Murphy & Associates, 3120 - 17th Street, Eureka, California 95501

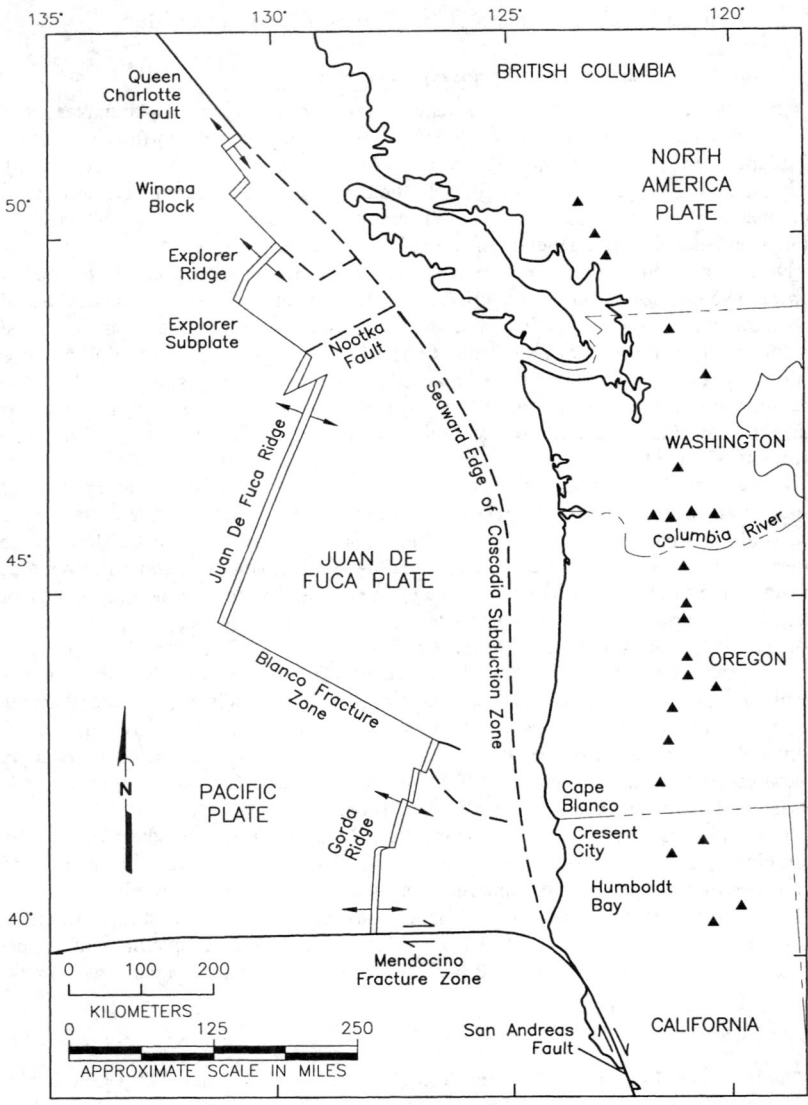

Figure 1. Vicinity Map of the Northwest Coast of Conterminous United States Showing Majority Tectonic Elements Along Northern California, Oregon and Washington (After Rogers et al., 1996).

TSUNAMI OCCURRENCE ALONG THE WESTERN COAST OF THE CONTERMINOUS UNITED STATES

The record for historical tsunami along the west coast is limited to approximately the last two hundred years, during which time over 100 events have either been observed or recorded (Lander et al., 1993).

As our understanding of the processes forming the earth's surface have evolved over the last thirty years, earth scientists have better recognized the sources and causes of earthquakes. Even though plate tectonics was broadly accepted by the scientific community by the early to mid-1970s, the potential that large, subduction earthquakes could be generated along the Cascadia Subduction Zone off of northern California, Oregon and Washington (see Figure 1) has only been accepted for about the last fifteen years. The potential that west coast areas of the mainland United States could also be affected by transoceanic earthquakes was dramatically demonstrated by the experience of many communities along the west coast, including Crescent City, as documented in the National Academy of Sciences report on the Good Friday Alaskan Earthquake (Cox, 1972).

Studies of coastal areas over the last twenty years have shown that the Pacific Northwest is a seismically active area (Atwater, 1987; Peterson and Darienzo, 1990; Jacoby et al., 1995). Many studies have also reported paleotsunamis that are believed to correspond to past seismic activity in this region.

Buried forests and wetland deposits have been observed extensively along the Cascadia margin in Vancouver, Canada, Washington, Oregon and northernmost California (Atwater, 1987, Clague and Bobrowsky, 1994, Darienzo and Peterson, 1990, and Jacoby et al., 1995). Locally, a clean sand overlying the buried wetland deposit is interpreted as an indicator of tsunami inundation. Radiocarbon dating of the most recent buried wetland and forest deposits indicates a strong possibility that the entire length of the Cascadia Subduction Zone ruptured about 300 years ago.

These studies have helped increase the awareness for the potential of locally generated tsunamis. The potential for locally generated tsunamis has been affirmed in the last decade by small tsunamis generated by the 1992 and 1994 earthquakes offshore the northern coast of California (Dengler and Moley, 1999).

Indications of tsunami occurrence from a Cascadia Subuction Zone earthquake about 300 years ago consist of written records by observers in Japan (Satake et al., 1996) and local Native American oral history (e.g. Heaton and Snavely, 1985). Based on the study by Satake et al. (1996), a date of January 26, 1700 is estimated for the most recent great Cascadia Subduction Zone earthquake.

A great Cascadia earthquake locally provides a complex condition along the margin as the North America plate would likely have an elastic response similar to that observed in great earthquakes in Chile in 1960 and Alaska in 1964 (Vita-Finzi and Mann, 1994; and Plafker, 1972). The response includes areas of subsidence, no vertical displacement and uplift dependent on the distance from the subducting (Juan de Fuca) plate. Because of this complexity and the potential effects to the runup of a tsunami it is useful to study a setting where there was no vertical displacement from an earthquake, but tsunami impacts were carefully documented.

Over seventeen tsunami have been recorded at tide gauges or reported at Crescent City, California since about 1885 (Lander et al., 1993). Along the west coast, only San Diego and San Francisco have greater number of events observed or measured during this same period. At Crescent City, most of the observed runups or amplitudes ranged from <0.1 to 0.9m as measured above mean low low water (mllw). Two events, generated by the May 22, 1960 Chilean earthquake and the March 28, 1964 Good Friday Alaskan Earthquake, resulted in damages to Crescent City and had recorded run ups of 1.7m and 6.3m above mllw, respectively.

CASE STUDY: CRESCENT CITY, CALIFORNIA-DAMAGE FOR 1964 ALASKA EARTHQUAKE GENERATED TSUNAMI

Crescent City, California was selected as a case study for analysis of tsunami impacts because 1) damages from the 1964 Good Friday Alaskan Earthquake are well-documented (e.g. Magoon, 1965; Wilson and Torem, 1968, Griffin, 1984, 2) some structures remain that survived the tsunami, 3) Crescent City represents an opportunity to analyze damages caused solely by the tsunami without complexities of damage caused by an earthquake and 4) the structures within the zone of inundation represent a variety of construction methods and materials and a range of structure ages.

1964 GOOD FRIDAY ALASKAN EARTHQUAKE

The tsunami generated by the March 28, 1964 Alaskan Earthquake devastated the harbor district and the southern portions of the city. It was characterized by five waves of which the second and third waves were successively higher than the preceding wave. Following the third wave, the water reportedly receded from the harbor, exposing the bottom. The fourth wave was the highest (approximately 6.3m above mllw) and the flow was turbulent. Severe damage was observed in areas where the tsunami exceeded 4 to 6 feet above the ground surface (Magoon 1965).

The tsunami resulted in 11 deaths and reportedly 289 businesses, residences or structures were damaged or destroyed (Griffin, 1984). Many of the deaths occurred because people re-entered the area affected by the first few waves, thinking that the worst had come to pass (Griffin, 1984). Five people drowned in the Elk River when their boat overturned.

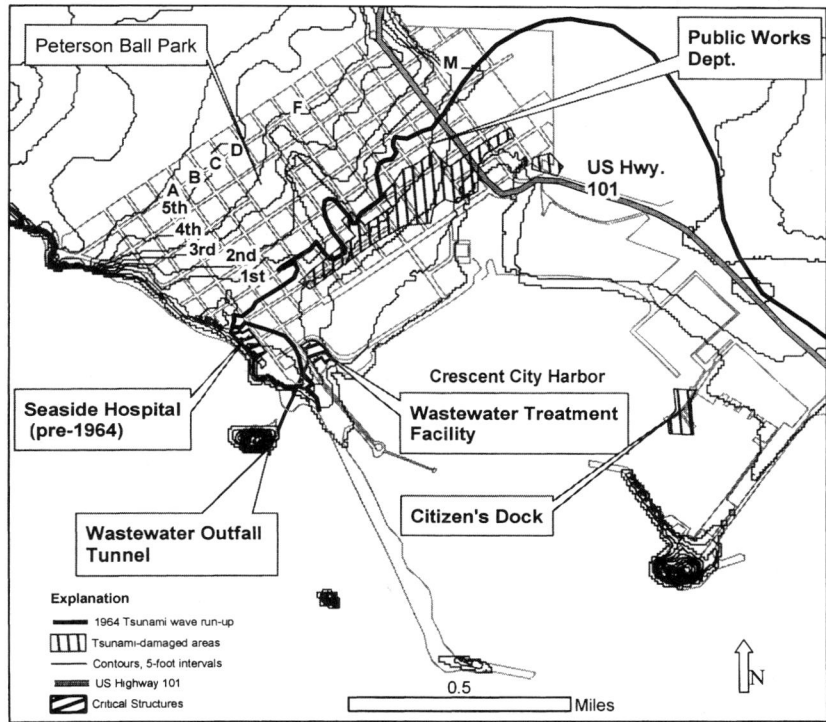

Figure 2. Map of Crescent City, California showing the location of roads, public facilities and critical lifelines and areas affected by the 1964 tsunami.

Building Damages

The types of damage resulting from a tsunami can be divided into two categories, primary effects and secondary effects. Primary effects are the damages directly resulting from the wave, such as inundation, displacement by floating, displacement from the force of the wave and damage from wave impact. Secondary effects are damages that occur indirectly from the tsunami and include debris impact, access impediment and fire.

From our review of existing maps, documents and literature, we compiled the principal cause of damage for over 100 structures in Crescent City, California (see Table 1 and Figure 2). Of these, more than 40 were principally affected by high water. About 28 structures were floated or 'swept' off of their foundations. In many cases, it was not possible to tell whether structures had floated or were displaced by the force of the waves. While most structures that were structurally tied to foundations did not move, at least one building is reported to have 'torn loose' from its foundation. Many structures were moved out into streets, affecting access. Perhaps the most dramatic example of a building displacement is the Coast Guard station, which is reported to have floated off its foundation and was found three miles offshore.

Table 1 Damage Summary

Cause	Structure / Facility Type						
	Houses	Business	Electrical	Fuel	Harbor-Marine	Waste-water	Roads
Swept-Off Foundation	5	21			2		Houses landed on roads
High Water (over 4 ft.) & Erosion	2	38			Docks damaged ?	1 – Outfall Tunnel	X
Walls Buckling / Other Unknown Cause		9					
Fire	2	2		3			
Impact from Debris	1	10	?		3 Docks impacted ?	Wastewater Treatment Plant?	X
Site or Cause Not Specified			Power outages; Sparking downed wires	Leaking Natural Gas Pipes			X

Sources: Griffin (1984), Wilson and Torem (1968); Magoon (1965).

In some cases where structural damage occurred, it wasn't possible to tell if the damage was from debris impacts or from the force of the waves. About nine of the structures fall into this category. It is likely that at least some of these were from the impact forces of the wave.

Figure 3. This propane tank is unanchored and could be carried away by a tsunami, potentially striking other objects. In addition, the leaking propane provides a source of fuel.

The principal damage for at least fourteen of the structures was caused by debris impact. In Crescent City, debris included automobiles, propane tanks, boats, sawn logs, driftwood, and timbers broken loose from abandoned piers remaining along the waterfront. In at least one case, one building was struck by another building that had floated or been swept off its foundation. Drums or other unsecured tanks also represent potential hazards. Figure 3 shows an unanchored propane tank that could become a floating projectile.

A small tank truck was shoved into an automobile dealership building where it broke loose an electrical junction box, igniting a fire. The fire burnt the building down, and spread the fire to an aboveground storage tank (AST) farm. The tank farm burnt for three days, spreading to and damaging two other fuel storage facilities. Loss of communication by emergency personnel and access interruption by high water prevented access by the fire trucks to stop the spread of the fire.

Sparks were also reported along the suspended electrical power lines. Residents also reported smelling gas, likely from pipes ruptured when buildings were displaced or damaged.

Figure 4. Tanks like these within the zone of tsunami inundation are subject to debris impacts, fire and may be weakened by the effects from a seismic event preceding the tsunami. Lack of containment berms increases the potential for spills and fire.

IMPACTS TO CRITICAL LIFELINES

Virtually all aspects of critical lifelines can potentially be affected across an area as a result of a tsunami. These include transportation, utility lifelines, and critical buildings and structures.

Transportation

In terms of transportation lifelines, highway and marine facilities were primarily impacted. Both highways and local roads were temporarily impacted by high water, which hampered the ability for emergency access to fight fires and evacuate residents. Access was disrupted within the downtown area as well as along highway 101 at the south end of town where the fire raged. Longer-term disruption of access occurred during the response stage as a result of the debris carried by the tsunami, including logs, vehicles, and structures that were floated or swept from foundations.

The most critically impacted transportation lifeline facilities affected by the 1964 tsunami include the port and harbor facilities. Twenty-one commercial boats were lost, affecting the economic recovery of the area. Both the new harbor office and the Coast Guard station buildings floated out to sea. The new jetty was seriously damaged. Extensive damage also occurred to old and new pier facilities, resulting in loss of decking and displacement of pile supports. An unmaintained, abandoned dock provided the source for debris that impacted other structures inboard.

Utility Lifelines

Electrical outages were reported in the vicinity of the area affected by the tsunami. The outages likely resulted from toppling of power poles from debris impact. Although no major power corridors were in the zone of inundation, disruption of electrical power service has the potential to affect other lifelines, such as water supply and various elements of wastewater treatment. Communications were also affected by the damages incurred, slowing emergency response time.

There were several anecdotal reports of leaking gas following the tsunami (Griffin, 1984). No fires caused by leaks were reported, perhaps only because the requisite ignition source was not present.

There were no reported impacts to the water supply. However, since many pipes were broken as the result of damage and displacement of structures, it is likely, similar to other flooding scenarios, that water supply could have been contaminated.

The only indication of impacts to the wastewater system is a report that a grant was obtained to repair the wastewater outfall, located along the shoreline near Battery Point (Griffin, 1984). The current wastewater plant is located along the western edge of the zone of inundation. Potential impacts to the wastewater plant include inundation by water and debris and damage from debris impact.

Fuel storage tanks were located near the harbor area and were affected by conflagration that started nearby. Fuel lines were present along the piers extending out into the harbor, as they do in many harbor facilities.

Communications

Telephone lines went down throughout town. Radio stations that had been giving advice during the first wave went off line, leaving the community with no form of communication. Even emergency personnel had difficulty communicating with each other (Griffin, personal interview, 2002).

Critical Facilities

The main critical structures affected by the 1964 tsunami in Crescent City have largely been discussed above. The existing wastewater treatment plant will remain within the zone of inundation identified by Toppozada et al. (1995) and may be impacted. The Seaside Hospital was also located near the edge of the 1964 inundation zone. Present-day hospitals are now outside the expected zone of inundation for a tsunami from a Cascadia Subduction Zone earthquake. Many of the city and county buildings including city hall, the city fire station, the county sheriff's office and the police station are within the expected inundation zone for a Cascadia event.

CONSIDERATIONS FOR COASTAL AREAS PRONE TO TSUNAMI HAZARDS LESSONS

The west coast of the United States has experienced a limited number of tsunami. The case study of the impacts to Crescent City, California therefore provides an opportunity to 1) qualitatively assess the impacts on structures and associated lifelines, 2) evaluate damages from a tsunami without local effects earthquake effects such as strong ground motion, local or regional subsidence, uplift, liquefaction or lateral spreading and 3) infer from descriptions additional mitigation issues to reduce secondary impacts and life loss.

TSUNAMI AND EARTHQUAKE INTERACTIONS

The primary, or initial, impacts from a tsunami are inundation and wave impacts. As noted above, the vast majority of damaged buildings floated off of their foundations. These buildings tended to be post and pier foundations or slab or grade. They were constructed prior to requirements for anchoring. Many of the structure that floated off of their foundations were essentially undamaged. It is unclear whether anchoring the structures would have prevented damage—or made them more susceptible to fluid pressures.

Harbor facilities as well as other buildings, such as docks and piers are subject to wave and debris impacts that can displace pilings or rip timbers loose. It was reported that while the majority of the port-related buildings on the harbor were destroyed, one that had been constructed with lateral bracing did survive. These anecdotal descriptions attest to the importance of performance-based design. They also attest to the value of enforcing current seismic based design in tsunami prone areas.

In the event of an earthquake, the vulnerability to ground failure impacts such as liquefaction and lateral spreading are particularly acute. Loss of soil support may be amplified by erosion and scour caused by the tsunami (e.g. Yeh et al., 2001). Therefore, structures weakened from seismic shaking may be more susceptible to tsunami impacts.

REDUCING INTERACTIVE EFFECTS: FUEL AND UTILITY LINES

A particularly problematic aspect of tsunami research is that descriptions of damage have focused on analysis of particular buildings and structures without consideration of the interactions of the two. Thus roadways blocked by debris hamper response and the ability to suppress fire before it becomes a conflagration.

These interactions have not been well-defined in the literature—and thus the context for mitigation has not been fully explored.

Pipelines located along the docks, piers and greater port area can be subject to the same forces as structures. Rupture of fuel and water lines may occur from failure of the support dock or pier. Though no significant spills were reported in the Crescent City case, the potential exists, increasing the potential for fire. Loss of fuel will also impede local recovery efforts.

Fuel tanks may also be susceptible to wave impacts, but are more likely susceptible to debris impacts. Power outages can also occur in areas of suspended or subsurface powerlines, since poles can be toppled and transformers can become non-functioning if flooded and could catch fire. Depending upon the distribution network, power outages may extend outside the zone of inundation. These damages will be magnified in the case of a locally generated tsunami, since liquefaction, lateral spreading and settlement may have already adversely affected coastal facilities.

Secondary impacts leading to structure damage include debris impact, debris inundation and fire. Shoreline areas often include not only boats and ships, but often include cargo that may consist of logs, shipping containers or other items that may readily float or become entrained. These elements can then damage docks, piers, bridges, wastewater facilities, storage tanks and pipelines. The displaced materials may end up on typically dry land areas where they obstruct access. High water can move equipment, vehicles and other debris into roadways and waterways, blocking access. Modern building codes for seismic considerations include anchoring of structures to foundations, so this is likely less of a concern except in areas where older buildings exist. Provisions must be made to have access to equipment that can deal with the debris clearing necessary.

Fire can occur wherever fuel, oxygen and ignition exist. Spilled petroleum products from tanks, vessels or vehicles, leaking natural gas or petroleum fuel lines can supply the fuel necessary for fires. Ignition sources can be provided by downed electrical wires or wiring ripped loose by impact, as was seen in Crescent City.

IMPLICATIONS FOR MITIGATION

Many communities in coastal areas are small. Ports and small boat harbors are administered by a staff that generally consists of one or two people. Many of the facilities do not comply with current codes e.g. tanks are not anchored and many do not have containment diking. These conditions significantly contribute to the hazards, yet there is no program to assist such communities with very limited funds to comply with national standards. A program needs to be in place to assist in reduction rupture potential during a seismic event and susceptibility to fire. Alternatively, other methods of mitigation could be used. Elevating containment dikes around tanks may be suitable, but they will need to be designed and constructed to address seismic considerations such as liquefaction, settlement and spreading, that may lead to dike or tank failure. Permeable barriers that allow flow of water but catch debris may be another method that could be implemented to reduce debris damages. Some mitigation measures cannot be completed until the hazard is better characterized.

HAZARD MITIGATION AND EVACUATION

In the mean time, emergency response plans can be developed that can be integrated with other mitigation and emergency preparedness efforts. Warning systems deployed in offshore areas recognize the impending arrival of trans-oceanic tsunami. Strong ground shaking in coastal areas will provide the best indication that a tsunami may occur.

Many jurisdictions, including Crescent City, have developed public education materials and, in concert with earth scientists and planners, have developed appropriate evacuation routes. Evacuation can be horizontal routes to higher ground, but vertical evacuation up buildings or other features can save lives. Critical facility workers (and residents) in coastal areas must be trained to use proper evacuation routes and awareness that an event causing a tsunami can result in more than one wave is imperative. They must also understand that previous events do not necessarily provide a clear indication of how another tsunami may affect an area. Many residents in Crescent City re-entered affected areas after arrival of the first three waves since the waves were similar to what they had seen before, only to be caught by the fourth wave (Griffin, 1984). In addition, agreements must be made to obtain equipment needed in the response and recovery period, especially for clearing debris.

The majority of fatalities from tsunamis occur when there is insufficient time for evacuation (i.e. less than 5 to 10 minutes). While experience has shown that the majority of people can self-evacuate (Chung 1995) there are inevitably people with limited mobility. It is therefore important that communities have some substantial structures which can be designated as vertical evacuation sites. Such structures must be built to withstand the tsunami. Current studies using computer and physical modeling are underway to evaluate and better understand the forces acting on structures from tsunamis to evaluate potential structural mitigation strategies. Once these forces are better understood, it may become possible to develop design criteria for critical structures that cannot be located to avoid the hazard. Orientation and siting of structures can have significant effects on adjacent or nearby structures. Structures located landward of other structures may focus return-wave forces on the back face of outboard structures (Preuss et al., 1999). Broad-faced buildings present a much greater surface area for wave impacts (as well as debris impacts). Preliminary results of physical modeling (Petroff and Arnason, 2002) shows that cylindrical structures and structures oriented with a corner facing the on-coming wave experience lower pressures than a flat face. Physical modeling will need to include multiple elements to better understand the cumulative effects of numerous structures and the interaction with the tsunami.

Acknowledgements:
Research for this project was funded by NSF grant Number CMS-9907945CM.

REFERENCES

Atwater, B.F., and Hemphill-Haley E., (1996). *Preliminary estimates of recurrence intervals for great earthquakes of the past 3500 years at northwestern Willapa Bay, Washington.* USGS. Open-file Report 96-001.

Atwater, B.F., (1987). "Evidence for great Holocene earthquakes along the outer coast of Washington state." *Science*, 236, 942-944.

Chung, Riley M., (1995). *Hokkaido-Nansei-Oki Earthquake and Tsunami of July 12, 1993 Reconnaissance Report.* EERI Earthquake Spectra, Publication 95-01.

Clague, J. J. and Bobrowsky, P. T. (1994). "Tsunami deposits beneath tidal marshes on Vancouver Island, British Columbia." *Geological Society of America Bulletin*, 106, 1293-1303.

Cox, D.C., (1972). National Academy of Sciences (NAS), "Oceanography and Coastal Engineering" in *The Great Alaska Earthquake of 1964*. National Academy of Sciences, Washington D.C.

Darienzo, M. E. and Peterson, C. D. (1990). "Episodic tectonic subsidence of Late Holocene salt marsh sequences in Netarts Bay, Oregon, central Cascadia margin, USA." *Tectonics*, 9, 1-22.

Dengler, Lori, and Moley, Kathy, (1999) "Living On Shaky ground, How to Survive Earthquakes and Tsunamis on the North Coast." Humboldt Earthquake Education Center, Humboldt State University, Arcata, CA.

Griffin, Wallace, (1984). *Crescent City's Dark Disaster*. The Crescent City Publishing Company, Crescent City, California.

Heaton, T. H., and Snavely, P. D., Jr. (1985). "Possible tsunami along the northwestern coast of the United States inferred from Indian traditions." *Bull. Seism. Soc. Am.*, 75, 1455-1460.

Jacoby, G., Carver, G., and Wagner, W. (1995) "Tree and herbs killed by an earthquake approximately 300 years ago at Humboldt Bay, California." *Geology*, 23, 77-80.

Lander,J.F., Lockridge,P., and Kozuch, M. (1993). *Tsunamis Affecting the West Coast of the United States, 1806 – 1992.* U.S. Dept. of Commerce, NGDC Key to Geophysical Records Documentation No. 29.

Magoon, Orville (1965), "Structural Damage by Tsunamis" in *Coastal Engineering Conference Proceedings October 1965.* American Society of Civil Engineers.

Petroff, Catherine, and Arnason, Halgor, (2002) personal communication, University of Washington College of Engineering, August 27.

Plafker, G., (1972). "Alaskan earthquake of 1964 and Chilean earthquake of 1960: implications for arc tectonics." *Journal of Geophysical Research*, 77, 901-925.

Preuss, Jane, Radd, Peter, and Bidoae, Razwan (1999). "Coastal Earthquake Effects: Tsunami." *TsuInfo Alert*, Washington State Department of Natural Resources, 1(6), 6-17.

Rogers, Albert M., Walsh, Timothy J., Kockelman and Priest, George, R, (1996), *Assessing Earthquake Hazards in the Pacific Northwest, Volume I*, U.S. Geological Survey, Professional Paper 1560.

Satake, K., Shimazaki, K., Tsuji, Y., and Ueda, K. (1996). "Time and size of a giant earthquake in Cascadia inferred from Japanese tsunami records of January 1700." *Nature*, 379, 246-249.

Toppozada, Tousson, Borchardt, Glenn, Haydon, Wayne, and Petersen, Mark (1995). *Planning scenario in Humboldt County and Del Norte County, California for a great earthquake on the Cascadia Subduction zone*, California Department of Conservation, Division of Mines and Geology, Special Publication 115.

Wilson, B.W. and Torum, A., (1968) *The tsunami of the Alaskan Earthquake, 1964: Engineering Evaluation*, U.S. Army Corp of Engineers, Coastal Research Center.

Vita-Finzi, C., and Mann, C. D., (1994). "Seismic folding in coastal south central Chile." *Journal of Geophysical Research*, 99, 12,289-12,299.

Yeh, Harry, Fuminori Kato, Shinji Sato (2001) "Tsunami Scour Mechanisms around a Cylinder" in *Tsunami Research at the End of a Critical Decade.* Kluwer Academic Publishers, Norwell, Massachussetts, 33-46.

Lifelines and Earthquakes in Switzerland

Blaise Duvernay[1], Anne Eckhardt[2], Kerstin Lang[3]

Abstract

Switzerland is a country with a low to moderate seismicity on a world scale. Seismic provisions in building codes first appeared in 1970 and were largely insufficient until the revision of the building codes in 1989. This means that about 90% of the buildings have an unknown and potentially insufficient seismic safety and the same can be said about lifelines and essential facilities.

The potential losses due to earthquakes are high and specialists view earthquake risk as the predominant natural risk in Switzerland, a fact still largely ignored by the general public and by many construction professionals.

The Swiss Federal government initiated an earthquake risk mitigation program by the end of the year 2000 and one of the consequences was the creation of a working group "Lifelines and Earthquakes" with representatives of all federal offices concerned.

The main tasks of this working group are to define which lifelines and essential facilities are critical after an earthquake and provide a methodology for their inventory, their vulnerability assessment and their protection objectives.

A demonstration project was initiated in the canton Nidwald in Central Switzerland to test the developed concepts. Nidwald is located in a zone were several important historical earthquakes have been recorded and is very active in the field of hazard assessment and mitigation programs for natural hazards in general.

The inventory of critical lifelines elements and essential facilities has been performed using the Delphi-method based on individual discussion with experts in charge of the different lifelines and a consensus discussion with all experts.

---

[1] Lead Civil Engineer, Coordination Center for Earthquake Risk Mitigation, Federal Office for Water and Geology, 20 Ländtestrasse, CH-2501 Biel, Switzerland; phone ++41 (21) 328 87 48; blaise.duvernay@bwg.admin.ch

[2] Head of section "Technology and Society", Basler & Hofmann, Consulting Engineers, 395 Forchstrasse, CH-8029 Zurich, Switzerland; phone ++41 (1) 387 11 22; aeckhardt@bhz.ch

[3] Senior Scientist, Basler & Hofmann, Consulting Engineers, 395 Forchstrasse, CH-8029 Zurich, Switzerland; phone ++41 (1) 387 11 22; klang@bhz.ch

The critical lifeline elements and the essential facilities have then been integrated in databases to be used in an adaptation of the US Hazus®99-SR2 software for Switzerland. This adaptation will serve to test the usefulness of such a tool for earthquake loss estimation and lifeline damage assessment in Switzerland.

The lessons learned from this project will help in establishing guidelines and tools in the domain of lifeline earthquake engineering and vulnerability assessment in Switzerland.

Introduction - Seismic risk and mitigation programs in Switzerland

Hazard studies in Switzerland were mostly based on the analysis of an historical earthquake catalogue of the Swiss Seismological Service [SED]. The 1356 Basel earthquake is the strongest reported earthquake with an epicentral intensity of IX and an estimated magnitude Mw of 6.5 to 7.0. For a 475 years return period, earthquake intensities ranging from VI+ to VIII are expected [Grünthal et al, 1998] (Figure 1, left). Based on hazard studies, 4 seismic zones are defined in the Swiss Engineers and Architects Society building codes [SIA261, 2003]. The maximum horizontal accelerations for a 475 years return period are 0.06g, 0.10g, 0.13g and 0.16g respectively for zones 1, 2, 3a and 3b (Figure 1, right). Globally the seismicity in Switzerland can be described as low to moderate. Nidwald (rectangle in figure 1), the canton in which the lifelines demonstration project is carried out, is situated in seismic zone 2.

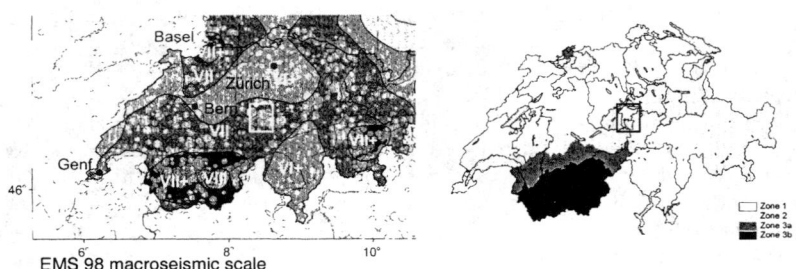

Figure 1. Left: seismic hazard map of Switzerland for a 475 years return period [Grünthal et al., 1998]; Right: seismic zones according to the SIA 261 building code.

The first generation of building codes with relatively adequate earthquake design provisions was issued in 1989 [SIA160, 1989]. The codes have been updated in 2003 and the seismic provisions are now largely inspired by the Eurocode 8. Ninety (90) percent of the building stock in Switzerland has been built before 1989. These buildings were designed using insufficient lateral loads and construction provisions to adequately take into account earthquake loading. Their seismic safety is largely unknown. The same can be said about most lifelines and essential facilities.

A low to moderate seismicity and a relatively high vulnerability leads to a high-risk situation. Studies performed by the Federal Office for Civil Protection

[BZS, 2003] have shown that earthquake risk is the predominant risk in Switzerland (Table 1), a fact largely ignored by the population and construction professionals that still frequently disregard earthquake provisions of the building codes.

Hazard	Percentage of risk [%] (with aversion factors[1])
Earthquake	37%
Geological instabilities	1%
Floods	13%
Storms / Hail	4%
Tempests / Hurricanes	7%
Avalanches	2%
Cold waves	7%
Heat waves / drought	7%
Dam breaks	1%
Nuclear accidents (civil)	9%
Migration	2%
Epidemics	9%
Others	< 1%

Table 1. Risk distribution with aversion factors for regional importance and national importance events in Switzerland according to the KATARISK study [BZS, 2003].

In December 2000, the Federal Council has decided an initial earthquake risk mitigation program running from 2001 to 2004 (Table 2). To implement and manage the program at the federal level, it has created the Coordination Center for Earthquake Risk Mitigation (CCERM) at the Federal Office for Water and Geology.

	Theme
1	Enforcement of the most recent building codes for new federal constructions or constructions that receive federal subsidies
2	Seismic safety control of all existing federal buildings to be modified. Implementation of upgrading measures if needed
3	Seismic safety control for all the important federal buildings in seismic zones 2 to 3. Prioritization of upgrading measures
4	Report on seismic safety concepts for cultural heritage
5	Legislation enhancements to include earthquakes as a natural hazard in the Constitution in order to develop legislative prerogatives for the federal government
6	Report on concepts for the financing of earthquake losses
7	Development of intervention procedures in case of earthquakes

Table 2. Federal Earthquake Risk Mitigation Program (2001-2004)

---

[1] With aversion factors societal risk perception and exceptional challenges for society through catastrophic events are considered.

At the end of 2004 a general report will be presented at the Federal Council which will decide of a continuing program.

The working group " Lifelines and Earthquakes" is one of the working groups that has been initiated for the mitigation program. Its activities have implications in themes 1, 3 and 7 described in table 2.

It is to be mentioned that due to a gap in the Constitution, the mitigation measures of the federal program can nowadays only be enforced on the federal level and not on the state or private level. Nevertheless, synergies are developed with state authorities in order to promote the voluntary application of federal mitigation measures. The Lifelines and Earthquake demonstration project in Nidwald is one of these initiatives that imply a proactive participation of the canton.

Working group "Lifelines and Earthquakes"

The working group " Lifelines and Earthquakes" was founded in 2001 and is entrusted with
- defining the notion of lifelines
- providing a concept for the identification of lifelines
- elaborating design criteria and protection objectives for lifelines
- pointing out, how substantial flaws can be identified and eliminated

The group is constituted by representatives of various federal departments and federal offices under the responsibility of the Federal Office for Water and Geology. Having started work in November 2001, it has so far presented propositions for the definition and identification of lifelines as well as for protection objectives. To evaluate the practicability of its results, in December 2002 it initiated the lifelines demonstration project in Nidwald.

The tentative definition of lifelines developed by the working group is the following:

"Lifelines are buildings and installations that are essential for rescue, recovery and reconstruction after an earthquake"

This definition corresponds more or less to the US-definition of essential facilities and critical lifelines, but is very different from the broad US-definition of lifelines in general that considers all systems contributing to society's sustainability.

Lifelines as defined by the working group should have protection objectives according to their importance in the rescue, recovery and reconstruction phases respectively. As an example, buildings classified as highly important in the rescue phase, such as hospitals, have to be designed for the 475-year return period earthquake loads multiplied by 1.4 and should guarantee serviceability for a 475-years return period earthquake. Protection objectives for the other combinations of importance versus post earthquake phase still have to be defined. The final result will be a matrix of protection objectives.

The working group defined the following list of systems, from which elements should be considered as potential lifelines.

- **Essential facilities:** Alarm system, commanding posts, civil protection facilities, fire protection facilities, police facilities, health care facilities, radio and telecommunication facilities
- **Utilities:** power distribution, potable water distribution, waste water system, fuel and combustible distribution, waste disposal
- **Transport:** road network, rail network, air transport facilities, ferry facilities

Lifelines inventory procedure: concept

So far, no systematic method of network analysis for the identification of lifelines in a defined region is available. The working group therefore decided to apply the Delphi Method. It started from the assumption that experts working daily with a certain type of facility or infrastructure are most competent to identify the elements that are essential for survival. The objective of the Delphi applications in general is to furnish reliable information for decision making. The method is based on a structured process for collecting and distilling knowledge from a group of experts by means of questionnaires or interviews interspersed with controlled opinion feedback. In recent years, the Delphi Method has been widely used to generate among others forecasts in technology, for instance in the studies "Nanotechnologie in der Medizin" of the Swiss Centre for Technology Assessment in Berne (TA-Swiss, 2003) or "Innovations for our future" of the Fraunhofer Institute for Systems and Innovation Research (ISI) in Karlsruhe (Cuhls et al., 2002).

Lifelines inventory procedure: demonstration project in the canton Nidwald

Nidwald is one of the cantons of Central Switzerland, bounded in the north by Lake Lucerne, and in all other directions by prominent chains of mountains. The canton comprises 38'000 inhabitants, living in 11 municipalities. The economy is characterized by a multitude of small and medium-sized companies of different industrial branches, building trade and agriculture. Nidwald was chosen for the demonstration project because of its size, which is easy to survey, its well developed and manifold infrastructure, the availability of all relevant data on geographical information systems and its intense activities in the fields of hazard assessment and mitigation programs for natural hazards in general.

The demonstration project was initiated and financed by the Federal Office for Water and Geology and the cantonal Property Insurance of Nidwald. The experts consulted in the Delphi were representatives of the cantonal environmental protection, civil engineering and civil protection authorities, the cantons' police department, the fire protection inspectorate, the Lucerne-Stans-Engelberg railway, the electricity supply Nidwald, the Swiss National Emergency Operations Centre and of Swisscom, the company, which ensures telecommunication in emergency situations. The Delphi Method was applied in the form of structured interviews with all experts, evaluation of these interviews, back-reporting to the experts and a final consensus discussion with all experts, surveyed by an independent professional in the field of civil infrastructure.

The course of the project showed, that the Delphi is an appropriate method for the identification of lifelines. The experts appreciated the personal interviews and the opportunity for a final discussion. For future Delphis, they suggested that an introductory information on seismic scenarios and their possible consequences should be given to all consulted experts.

For the identification of lifelines the differentiation between rescue (first days after a seismic event), recovery (weeks after a seismic event) and reconstruction (months after a seismic event) proved to be important .A good co-ordination with experts responsible for intervention in case of catastrophes also proved to be essential as they have a good overview of available emergency systems that can temporarily replace lifeline elements.

Table 3 gives an overview of the results of the Delphi Method in Nidwald. This table shows for each potential lifeline system where some of its elements are considered essential. Important to notice is its dependency on the particularities of the situation in Nidwald. Up to this point the expert discussion tended to over-focus on the rescue phase which is the easiest to apprehend in terms of what elements are needed. Further work has to be devoted to the choice of lifeline elements that are essential for the recovery and reconstruction phases.

System	**Rescue**
Alarm	All protected Sirens with emergency power supply
Command Posts (CP)	1 cantonal CP and 11 communal CPs
Civil Protection	17 big communal installations
Fire protection	1 central fire house and 11 communal fire houses
Police	1 central police station with 2 locations
Health care	1 cantonal hospital, 1 first aid center, 8 patient centers
Radio	1 emergency broadcast station
Telecommunication	3 broadcast stations of the cantonal police
	3 antennas of the fire departments
	1 protected emergency network in telematics
Potable water distribution	No essential elements because of numerous sources and emergency back up systems for the distribution of water
Waste water disposal	No essential elements
Waste disposal	No essential elements
Power distribution	No essential elements because of well-established emergency power supply
Fuel distribution	No essential elements
Road network	2 highways, 1 main road
Rail network	No essential elements
Air transport	No essential elements, transport provided by helicopters
Transport by ships	No essential elements, transport provided by ships on Lake Lucerne

Table 3. Provisory selection of lifeline elements selected by the experts in the demonstration project of Nidwald

For the rescue phase, emergency backup systems for the distribution of water and electricity can be put into use very rapidly. Wastewater can be evacuated without treatment on a temporary basis. Fuel is stocked in numerous highly decentralized small tanks and can be brought from out of the canton as soon as roads are reestablished.

For the recovery and reconstruction phases, essential needs will have to be primarily defined for water distribution, waste and waste water treatment, power distribution, transport and telecommunications.

HAZUS® adaptation for Switzerland

The Software Hazus®99-SR2 developed and distributed by the US Federal Emergency Management Agency (FEMA) through agreement with the National Institute of Building Sciences (NIBS) has been adapted to the Swiss context in order to serve as a demonstration tool for the calculation of losses in case of earthquakes. The adaptation has been performed by the University of Geneva without access to the source code. It has several limitations compared to the original version but the most important features remain operational. Attenuation functions for Switzerland could not be integrated and shake maps have to be produced independently from the software.

At the time at which this paper is written, the data specific to the canton Nidwald is being prepared for its integration in Hazus®. In this demonstration project, the first objective is to evaluate the damage to specific lifelines and essential facilities based on different earthquake scenarios. The second objective is to evaluate the potential losses for the whole building stock of Nidwald.

The first series of results will be produced without evaluating the correctness of Hazus® fragility functions for the construction typology of Switzerland. Those results will only serve to demonstrate the capabilities of such a tool and evaluate its potential use and development at a larger scale in Switzerland.

If the results are judged to be useful, a research project for the adaptation of Hazus® fragility functions to local construction typology will be initiated. At the same time the development of a tool for generating deterministic shake maps based on the Swiss Seismological Service attenuation functions [Bay, 2002] will be launched.

HAZUS® application in Nidwald: data preparation

The lifelines elements inventoried by the Delphi Method are in the process of being integrated in the specific databases of Hazus®. The databases for the general building stock are also being prepared based on very accurate information of the cantonal Property Insurance of Nidwald, a state organization that has the monopoly for building insurance in Nidwald. This process is particularly cumbersome although Nidwald is probably the best-organized canton in terms of building stock database.

Four (4) return periods are considered for the earthquake scenarios (100, 500, 2'500 and 10'000 years). The shake maps have been produced by the Swiss

Seismological Service (SED) and correspond to probabilistic hazard maps and not to discrete scenarios.

An indicative microzonation map has been produced by the Federal Office for Water and Geology and will be used for the hazard calculations in Hazus®. Furthermore, as earthquake triggered landslides can play a very important role in Nidwald, particularly in conjunction transportation and distribution lifelines, a landslide susceptibility map has also been produced for integration in Hazus®.

Submarine landslides triggered by historical earthquakes have produced seiches up to 3 m high in the nearby Lake Luzern [Schnellmann et al., 2002]. An inundation map would therefore also be meaningful for the loss estimation calculations of this project. Such a map will be integrated at a later stage.

Discussion and conclusion

The demonstration project permitted to test the concepts developed by the federal working group "Lifeline and Earthquakes" in terms of definition and identification of lifelines essential for the rescue, recovery and reconstruction phases after an earthquake.

The application of the Delphi Method in Nidwald, one the 26 cantons of Switzerland demonstrated that this method based on local expert opinion is suitable for the identification of lifelines. A tendency to over-focus on the rescue phase was observed and further work to identify lifelines essential for the recovery and reconstruction phase is needed.

It has also been observed that induced hazards could play a major role in damage caused to lifelines in Nidwald. Access roads and railways to the canton are very few and threatened by landslides prone slopes. Furthermore seiches in the lake Lucerne could reduce the capacity of transport on water by damaging ships and port installations.

Once all lifelines will be identified, their integration in an adaptation of Hazus® for Switzerland will serve to judge the potential use of such a tool for damage assessment to lifelines.

Acknowledgments

The authors thank the federal working group " Lifelines and Earthquakes" as well as the cantonal Property Insurance of Nidwald for the initiation and support of this study. Thanks are extended to all local lifeline experts for discussions and input into the project.

References

Bay, F. (2002), "Ground Motion Scaling in Switzerland: Implications for Hazard Assessment", PhD-thesis ETH No. 14567, Swiss Federal Institute of Technology, Zürich

BZS - Federal Office for Civil Protection (2003), "KATARISK, Katastrophen und Notlagen in der Schweiz", Bundesamt für Zivilschutz, Bern

Cuhls, K. et al. (2002), "Innovations for our future: Delphi 98 – new foresight on science and technology", Physica-Verlag, Heidelberg 2002

Grünthal, G., Mayer-Rosa, D., Lenhardt, W. (1998). "Abschätzung der Erdbebengefährdung für die D-A-CH Staaten Deutschland, Österreich, Schweiz." *Bautechnik*, Verlag Ernst & Sohn, Berlin, 75 (10), 753-767.

SED - Swiss Seismological Service, "Earthquake Catalogue of Switzerland", http://histserver.ethz.ch/intro_e.html>(Jun. 22, 2002).

Schnellmann, M., Anselmetti, F.S., Giardini, D., McKenzie, J.A., Ward, S.N. (2002), "Prehistoric earthquake history revealed by lacustrine slump deposits", *Geology*, Geological Society of America, December 2002, v. 30 (12), 1131-1134

SIA – Swiss Society of Civil Engineers and Architects, (1989), "SIA 160 (Norm): Einwirkungen auf Tragwärke", Schweizerische Ingenieur- und Architekten- Verein, Zürich, 1989

SIA – Swiss Society of Civil Engineers and Architects, (2003), "SIA 261 (Norm): Einwirkungen auf Tragwärke", Schweizerische Ingenieur- und Architekten- Verein, Zürich, 2003

TA-Swiss – Swiss Centre for Technology Assessment, (2003), "Nanotechnologie in der Medizin", Berne, 2003 (in preparation)

Seismic Risk Assessment and Upgrade Strategy of Hospital-lifeline Performance

Yasuko Kuwata [1] and Shiro Takada [2]

Abstract

Medical facility needs to sustain lifeline functions in order to receive and take care of a large number of injured people whenever the catastrophic earthquake occurs. Malfunction of hospital-lifeline due to damage to lifelines of outside hospital has caused significant impact to the injured people. This paper addresses risks of physical damage and malfunction impact due to water outages to hospital. Social benefit is analyzed when some strategies are taken to sustain hospital-lifeline function.

Introduction

Hospital in emergency situation, especially immediate after catastrophic earthquakes, should play a very essential role of managing mass injured people. It is task for hospitals to give emergency medical care and to rehabilitate their lives even though they had more or less damaged. In actual cases of disasters and crises, emergency responses of medical sector have been mostly characterized by their organization and coordination along their emergency plans. That follows emergency response for ordinal accident such as traffic accidents and mass food poisoning. In ordinary cases medical facility would run as normal and the hospital would accept as many people as possible. On the contrary, in earthquake disasters, medical facilities may also be damaged. Prerequisites of both conditions are quite different. Emergency coordinators in medical sectors are specialists in disaster medical management. They can deal with particular demands of injured people and its relevant medical operations, but have little knowledge in seismic reliability of medical equipments and facilities exposed to seismic hazards. Earthquake engineering contribution needs to medical emergency planning and preparedness.

----

[1]Graduate student, Graduate School of Science and Technology, Kobe University, Rokkodai 1, Nada, Kobe 6578501, JAPAN; Tel +81-78-803-6047; e-mail 994d838n@y01.kobe-u.ac.jp, [2]Professor, Department of Architecture and Civil Engineering, Kobe University, Rokkodai 1, Nada, Kobe 6578501, JAPAN; Tel +81-78-803-6037; e-mail takada@kobe-u.ac.jp

This issue can be argued from functional reliability of medical facilities during recent earthquakes. These reports show that medical facilities, which were out of function or little functioning due to physical damage, hampered to receive and to take care of injured people, and sometime forced to evacuate them to outside facilities (e.g., Pickett, 1995; Takada and Kuwata, 2002). More recent report says that 6,433th person who lost his life due to stop of respiratory during the 1995 Kobe earthquake is counted as the earthquake-related casualty. It was caused due to outages of electric power at the hospital. That means emergency response of hospitals depends on not only coordination of personnel and medical resources but also functional reliability of hospital facilities, especially lifeline systems. Hospital buildings are generally constructed better seismic design than other general ones, while hospital-lifelines also should be taken into more consideration, because this reliability is determined by a product of inside and outside medical facilities. Risk of lifeline system at the hospital therefore should be assessed from more widespread point of view.

Studies on risk assessment of medical facilities have been investigated so far, that considers structural and nonstructural performances inside facilities (Porter, et al. 1993; Cruz and Castillo, 2000). Moreover, lifeline companies have taken into consideration of functional performance of medical sectors in emergency (East Bay Municipal Utility District, 1994). Advanced scheme to bridge between medical and engineering communities by sharing information on vulnerable factors as well as tasks to implement earthquake mitigation program at each system is required now.

The focus of present paper is on seismic performance of hospital-lifeline related to saving human lives, rather than physical reliability itself. On the risk assessment of hospital-lifelines, all the lifelines should be taken into account because of intra-dependences on other lifelines. At the first stage of this study, water supply system was only dealt with as an independent system. The reason is that water supply system is the most critical among hospital-lifelines. It takes comparatively long time to restore, and outages of water are highly dependent on facilities and equipments related to medical operations (Kuwata et al, 2003). Besides highlighting the vulnerable water supply system quantitatively, social benefits are given attention to when some strategies for strengthening lifelines are taken. This does not cover all the vulnerability factors of medical facility, but would be possible to contribute decision-makers to give good information as one of criteria in physical capability.

Risk assessment of water supply system to hospital

Methodology. Structures of water supply system can be regarded to be composed of three elements; water distributing reservoir, pipelines and inside building systems at hospital. During the 1995 Kobe earthquake transmission pipes were little damaged, the main damage concentrated on distribution pipes. Source supplying water to the system is therefore counted for the distributing water reservoir. In the water supply network, there are several routes linking reservoir to hospital. Since switching route channels is impossible to carry out in emergencies after the earthquake, the route of water network actually follows the ordinary main line. These elements are connected in series from reservoir to hospital.

When considering physical system, vulnerability factors of each element include geological and structural components. Seismic risk of each component can be expressed by the probability of these vulnerabilities provoked by the ground motion. Functional performance of water supply system means a product of these risks.

$$F_W(V_R, V_P, V_{IN} | M_j) = F_R(A_R) * F_P(A_P) * F_{IN}(A_{IN}) \tag{1}$$

Where, $F_W(V_R, V_P, V_{IN} | M_j)$ is the functional probability of water system when given vulnerability sets (V_R, V_P, V_{IN}) of the water distributing reservoir, pipelines and inside building systems provoked by an earthquake M_j ($0 \le F_W \le 1$), and $F_R(A_R), F_P(A_P)$ and $F_{IN}(A_{IN})$ are the functional probabilities of water distributing reservoir, pipelines and inside building system when the ground motions A_R, A_P and A_{IN} are given respectively. When each component works functionally, the every functional probability F equals 1.0.

This probability lets us understand the most vulnerable water supply system to the hospitals. Another advantage of this probability is to use another risk assessment of hospital performance as well as to know reliability of water supply system itself. Water outages provoke the limitation of medical care operation for patients, lack of drinking water, malfunction of the other hospital lifeline systems. In this study, malfunction impact related to saving human lives is tackled. The degree of impact depends on occupied patients in that hospital. When assuming the hospital is full of patients, the people those who are exposed to water outages are defined in terms of the number of beds B multiplied by the malfunction probability ($1-F_W$). The malfunction impact I_m can be explained as follows.

$$I_m = B * (1 - F_W) \tag{2}$$

Application. Case study of quantitative analysis is applied to 10 hospitals in Kobe City, which are relatively large and used for emergency hospitals. Pipelines from reservoirs to hospitals were charted in Figure 1. Ground motion of this application is estimated from the seismic intensity on the JMA scale observed during the Kobe earthquake. The PGA is applied to the reservoir and inside building system, while the PGV is applied to the pipelines.

Reservoir. Functional probability of the reservoir is estimated by using the procedure of HAZUS99 (FEMA, 1999). The functional probability can be expressed

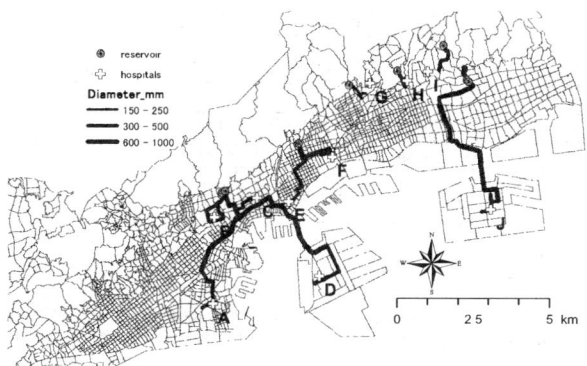

Figure 1. Water pipelines from reservoir to hospital

with its damage probability using the standard normal cumulative distribution function as shown in Eq.(4).

$$F_R(A_R) = 1 - P_R(A_R) \tag{3}$$

$$P_R(A_R) = \Phi\left[\frac{1}{\beta}\ln(\frac{a_R}{\overline{a_{R,ds}}})\right] \tag{4}$$

Where, P_R is the damage probability, a_R is the PGA (g), $\overline{a_{R,ds}}$ is the median value of PGA (g) at which the tank reaches the threshold of damage state ds, β is the standard deviation of the natural logarithm of the PGA for damage state ds. All the water reservoir tanks in this case are built of concrete anchored on the ground. When referring to parameters of median value and standard deviation for the moderate damage, $\overline{a_{R,ds}}$ is 0.52, and β is 0.72.

Pipelines. Under the condition that pipeline topology is one-way series without pump station, the functional probability of pipelines can be expressed by the product of damage probability of each pipes as shown in Eq.(5).

$$F_P(A_P) = \prod_{i(i \in P)} (1 - p_i(A_{pi})) \tag{5}$$

Where, $p_i(A_{pi})$ is damage probability of pipe i when given the ground motion A_{pi}.

Damage probability can be referred from the formula proposed by Takada et al. (2001) based on the damage data during the Kobe earthquake. When given a peak ground motion (either PGA or PGV), damage ratio S_i (damage numbers/km) can be calculated by the standard damage ratio S_{di} (damage numbers/km) with several correction coefficients: type of pipe, diameter and liquefaction condition. These coefficients are listed in Table 1.

Table 1. Correction coefficients

Type of pipes	C_{Pi}	Diameter (mm)	C_{di}	Liquefaction condition	C_{li}
DIP(A,K,T)	0.3	$\phi 75$	1.6	No (0<PL<5)	1.0
DIP(S, SII)	0.0	$\phi 100$- $\phi 150$	1.0	Partially (5<PL<15)	2.0
CIP	1.0	$\phi 200$- $\phi 250$	0.9		
SP	0.3	$\phi 300$- $\phi 450$	0.7	Totally (15<PL)	2.4
VP	1.0	$\phi 500$-	0.5		
SGP	4.0				
ACP	2.5				
Others	-				

Note: DIP is the ductile cast iron pipe, CIP is the cast iron pipe, SP is the welded steel pipe, VP is the polyvinyl chloride pipe, SGP is the steel gas pipe with screw joint, and ACP is the asbestos concrete pipe.

$$S_i = S_{di} * C_{Pi} * C_{di} * C_{li} \quad (6)$$

where, the standard damage ratio is estimated when PGV is given as v_i (kine).

$$S_{di} = 6.33 * 10^{-5} * v_i^{2.10} \qquad (v_i \leq 110 \text{ kine}) \quad (7)$$

When assuming that pipeline consists of around 5m-unit of pipes and pipe damage occurs at most one for each pipe unit, damage probability of pipe unit p_{ui} (damage number/ pipe unit) can be expressed as follows.

$$p_{ui} \cong S_i / 200 \quad (8)$$

As the result of the assumptions, Eq.(5) can be expressed as shown in Eq.(9).

$$F_P(A_P) = \prod_{i(i \in P)} (1 - p_{ui}(A_{pi}))^{\frac{L_{pi}}{\Delta l}} \quad (9)$$

where, L_{pi} is the pipe length (m) and Δl is the length of pipe unit (m).

Inside building system. Inside building system consists of pipelines in building and several facilities, such as water receiving tank, elevated reservoir tank that is sometimes divided into two tanks for drinking and miscellaneous uses. It needs to be treated as the more complicated system. In the analysis addressed here, vulnerability of the inside building system is regarded as that of hospital building. More detail analysis on that system is under investigation.

The functional probability of the hospital building can be expressed with its damage probability using the normal cumulative distribution function as shown in Eq.(11).

$$F_{IN}(A_{IN}) = 1 - P_{IN}(A_{IN}) \quad (10)$$

$$P_{IN}(A_{IN}) = \frac{1}{\sqrt{2\pi}\sigma} \int_{-\infty}^{K} e^{-\frac{(K-K_0)^2}{2\sigma^2}} dK \tag{11}$$

Where, P_{IN} is the damage probability, K is the engineering seismic intensity, which equals the ratio of PGA a_{IN} (gal) to gravitate acceleration g (gal) as expressed a_{IN}/g, K_0 and σ are the median value and its standard deviation of engineering seismic intensity respectively. When referring to a study on the Kobe earthquake damage, as far as reinforced concrete buildings, K_0 is 0.72 and σ is 0.085.

Results. Table 2 shows the functional probabilities of reservoir, pipelines, inside building system and water supply system among 10 hospitals. It can be seen that probability of pipelines dominates those of the water supply systems because they are the most fragile among components of the system. Water supply system at hospital should be considered the more on the reliability of outside pipeline system.

Pipelines of the hospitals D and J are crossing the Kobe Bay through bridges up to the manmade islands. The pipelines are relatively long and easy to be influenced by liquefactions. Functional probability of pipes of the hospital F is quite lower. The reason is that most of pipes have very small diameters and located at high seismic intensity area. The longer pipeline increases vulnerability due to series morphology of pipelines, while intensity of the ground motion, properties of pipe and local ground condition are also influenced.

Table 2 also shows malfunction impact on occupied patients. Whatever there was high probability of water supply system, malfunction impact is more critical according to the occupancy of hospital; in particular larger scale of hospitals B and D. Priority of countermeasures depends on the type of exposures. In this case, the risk of

Table 2. Functional probability of 10 hospitals

Hospital	Reservoir F_R	Pipelines F_P	Inside building system F_{IN}	Water system F_W	Numbers of beds B	Malfunction impacts $B*(1-F_W)$
A	0.97	0.81	1.00	0.79	242	51.9
B	0.97	0.76	1.00	0.73	920	246.0
C	0.97	0.90	1.00	0.88	126	15.5
D	0.97	0.77	1.00	0.75	1,000	251.2
E	0.90	0.83	0.99	0.74	151	39.0
F	0.90	0.64	0.99	0.57	325	139.1
G	0.97	0.94	1.00	0.91	222	20.1
H	0.97	0.71	1.00	0.69	178	55.8
I	0.90	0.94	1.00	0.84	400	62.9
J	0.90	0.56	1.00	0.50	307	152.8
Total					3,871	1,034.2

hospital water system was dealt with an indicator in terms of malfunction impacts. The probability obtained here can be applied to another kind of exposures in various purposes.

Effects of new pipe installing strategies

Social benefit. Cost-benefit analysis has been widely done in risk assessment programs. A conceptual problem seems to exist when treating human lives in cost-benefit analysis. Risks in loss of human life are often impossible to quantify well in the economic sense. The cost-benefit analysis is at least reasonable scheme while the decision should not be always determined only by its monetary result. To reduce expected malfunction impact proposed in this paper is dealt with as a measure to explain social benefit in a sense. This indicator is not easy to compare with another monetary benefit, but would show a new criterion when making a decision in earthquake countermeasures for critical places, for example, disaster headquarters, hospital, and civil protection services.

Pipe installing management strategy. Under the same condition of hazard factors, cost to implement several strategies for pipeline and effect to malfunction impact are examined. Two strategies are taken into account.

Strategy1. Following the Kobe earthquake, Kobe Water Bureau planed a big project that large capacity (ϕ 2,400mm) transmission pipelines run parallel to the existing transmission pipes in Kobe area as charted in Figure 2. In the year of 2010, the construction project will be completed. The new transmission pipeline will ensure that, in case of such disasters, the redundancy increases and an effective dispatch of repair manpower could shorten restoration period. Lessens from the earthquake taught that repair site was located at the top of network tree in the first stage of post-earthquake even though the large number of repair workers came. The new pipeline provides some emergency water supply stations in lieu of existing

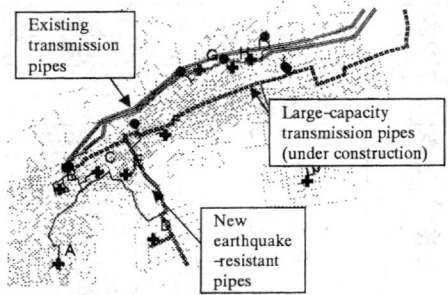

Figure 2. Plan of large capacity transmission pipes

Table 3. Replacement cost

Pipe Diameter (mm)	Repair cost (Thousand USD)
$-\phi$ 150	2.99
ϕ 200 $-\phi$ 300	4.27
ϕ 300 $-$	9.40

analysis it is considered that the hospital D has the other route from new transmission pipelines when completed this project. New earthquake-resistant pipelines run inside tunnel beneath the Kobe Bay.

Strategy2. Significant damage to pipes mostly concentrated on CIP during the earthquake. This strategy is to replace pipe units of CIP to those of DIP. Two procedures of replacing pipe units are introduced as follows;

Case1: choose a water supply route which has the lowest functional probability, and then replace the most vulnerable pipe unit along the route, and continue to replace the next vulnerable one until the functional probability of the water supply system exceeds that of the other route to hospitals. When it exceeds, find another route to hospital with the lowest probability.

Case2: choose a hospital which has the highest malfunction impact, and then replace the most vulnerable pipe unit along the pipelines to that hospital, and continue to replace the next vulnerable one until the malfunction impact of the hospital is less than that of the other hospitals. When it is, find another hospital with the highest impact.

The former is paid a priority to increase the probability of each system, while the latter is to increase the total of impact. The replacement cost for pipe unit (5m) is listed in Table 3 (Takada, 1998). 500 pipe units (2,500m) correspond to 23 % of all CIP along analyzed pipelines.

Costs and effects. Table 4 shows the result of costs and total impact for 10 hospitals when these strategies are adopted. In case of strategy1, the strong effect is appeared thanks to the functioning of pipeline to the hospital D. If another pipeline to the other hospitals is installed, the expected malfunction impact may reduce the more. Replacement cost of the strategy 2 is far less than strategy 1. When comparing these strategies as shown in Figure 3, Case 1 lets all the functional probability increase,

Table 4. Replacement cost and expected malfunction impact

Strategy #			Notes	Installing or replacement cost (million US$)	Total malfunction impact (persons)
0			Before strategy programs	-	1034.2
1			Install large capacity transmission pipelines	427.35*	783.0
2	Case 1	500p	Replace 500 or 1,000 pipe units of CIP to those of DIP paying priority of Case1 procedure	2.56	961.7
		1,000p		7.03	925.0
	Case 2	500p	Replace 500 or 1,000 pipe units of CIP to those of DIP paying priority of Case2 procedure	2.13	836.1
		1,000p		6.22	798.1

*this is the project expenses of transmission pipeline in Figure 2

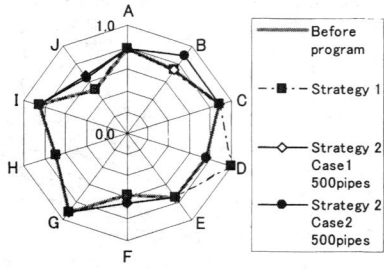 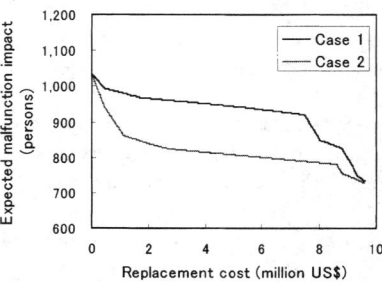

Figure 3. Functional probability of water supply system

Figure 4. Expected malfunction impact related to replacement cost

while Case 2 does the functional probability of the hospital B. The reason is that the hospital B has the largest number of patients. If the policy of earthquake countermeasures focuses on that each hospital's capability is emphasized, Case 1 is useful. On the other hand, if it focuses on capability of the whole area and transportation system is well functioning after the event, Case 2 has an advantage.

Figure 4 explains the relationship between sum of replacement cost and effect according to process of placing pipes. The expected malfunction impact is possible to decrease by around 1 million dollars if Case 2 is adopted. Even though it takes 2 to 8 million dollars for replacement cost, not so much effects cannot be expected. When taking over 9 million dollars of cost, there is little difference in effects between both replacement procedures, and the effect will provide same as strategy 1.

Conclusive remarks

This paper addressed the seismic risks of water outages at hospitals. Follows can be concluded.
- Risk assessment methodology of water supply system from reservoir to hospital was proposed and applied to 10 hospitals in the Kobe area. Functional probability of buried water pipelines was lower than that of other components.
- Proposed malfunction impacts due to water outages were calculated in terms of patient number and functional probability. Risk of malfunction impact is higher at large hospital with many patients even though functional probability of physical system is not so low.
- In case of installing and replacing to new pipes, cost for new pipes and effect to the expected malfunction impact are examined. The large capacity transmission pipeline provides strong effect on the hospital D. According to the procedures to replace pipes, there is difference in effects. However, when the investing cost exceeds a certain level, expected effects reached similar values.

Acknowledgements

Authors would like to thank Water Bureau of Kobe City, Mr. Makoto Matsushita for giving the information on water system in the Kobe City. First author acknowledgments the Japanese Society for the Promotion of Science (Grant-in-Aid for Scientific Research, DC1, No.09368) for its support of this research.

References

Cruz, M. and Castillo, R. (2000). "Disaster mitigation in health facilities: Structural issues", Emergency Preparedness and Disaster Relief Coordination Program, Osorio, C (Ed.) PAHO/WHO, Pan American Health Organization.

Federal Emergency Management Agency (1999). HUZUS 99, Technical manual.

Kuwata, Y., Jin, Y-H, Takada, S. (2003). "Risk assessment of hospital-lifelines in emergency situation", Proc. of Joint Workshop on US-Japan Cooperative Research in Urban Earthquake Disaster Mitigation, Los Angels, Jan. 2003.

Pickett, M. A. (1995). "The effects of the 17 January 1994 Northridge earthquake on hospital lifelines", Technical Council on Lifeline Earthquake Engineering Monograph No. 6, Lifeline Earthquake Engineering, Proceedings of the 4th U.S. Conference, ASCE, New York, Aug. 795-802.

Porter, K., Johnson, G.S., Zadeh, M.M., Scawthorn, C.R., and Eder, S.J. (1993). "Seismic vulnerability of equipment in critical facilities: Life-Safety and Operational Consequences", National Center for Earthquake Engineering Research, State University of New York at Buffalo, Technical Report NCEER-93-0022, November, 1993.

Takada, S. (1998). "Direct and indirect economic losses of Kobe water systems during the 1995 Hyogoken-Nanbu earthquake", Proc. of Third China-Japan-US Trilateral Symposium on Lifeline Earthquake Engineering, Kunming, Aug. 1998, 291-300.

Takada, S., Fujiwara, M., Miyajima, M., Suzuki, Y., Yoda, M. and Toda, T. (2001). "Study on the methodology of prediction for pipe damage based on characteristics of urban earthquake", Journal of Japan Water Works Association 71, 3, 21-37 (in Japanese).

Takada, S. and Kuwata, Y. (2002). "Casualty and life-saving lifeline function in the Chi-Chi village during the 921 Taiwan earthquake", Proc. of the 2nd Japan –Taiwan Workshop on Lifeline Performance and Disaster Mitigation, Kobe, May, 150-157.

Managing the Earthquake Risk:
Research and Implementation Efforts on Utah's Highway Infrastructure

Blaine D. Leonard[1], P.E., M.ASCE

Abstract

Most of the population of Utah lives along the Wasatch Front, the western flank of the Wasatch Mountain Range. This region is situated in the Intermountain Seismic Belt, a zone of historical seismicity containing several major normal faults, including the Wasatch Fault. The Wasatch Fault is considered capable of producing seismic events with magnitudes of up to 7.5. During the recent reconstruction of I-15 through the Salt Lake Valley, the Utah Department of Transportation embarked on a series of unique research projects aimed, in part, at identifying and quantifying the earthquake risk to Utah's highways and bridges and developing or modifying design and maintenance procedures to improve the survivability and response of the critical transportation lifelines. Some of these projects included: 1) Preparation of deterministic maximum ground motion maps for the entire state; 2) Installation of strong motion instrumentation on a high, reinforced concrete bridge structure, at the ground surface, and several hundred feet deep within the soft soil to measure the dynamic response during seismic events; 3) Development of surface response spectra specifically suitable for soft soil sites found in Northern Utah, and determination of the influence of soil softening and liquefaction on this spectra; 4) Full-scale, destructive, lateral testing of existing reinforced concrete bridges, including testing of post-failure bridges which were reinforced with carbon fiber reinforced polymer (FRP) composites; 5) Implementation of a carbon FRP composite repair on an in-service reinforced concrete bridge and a long term evaluation of the behavior of the retrofit and the materials; 6) Full scale evaluation of the shear capacities of column-bent connections and the improvements gained by retrofitting with composite materials; 7) Characterization of the dynamic properties of full-scale bridges by imposing vibrations on the structures in various states of damage with the intent of developing methods to evaluate bridge damage from dynamic signatures; 8) Prioritization of critical highway and bridge lifelines so that appropriate design standards and resources can be applied during new construction, maintenance, and disaster response; 9) Full scale static and dynamic lateral load testing on groups of steel pipe piles to determine their capacities in a seismic event; and 10) Evaluation of the capacities and responses of Geopier foundation systems. Each of these initiatives will be briefly described and references given for more detailed understanding of the projects.

---

[1] Senior Research Project Manager, Utah Department of Transportation, P.O. Box 148410, Salt Lake City, UT 84114-8410, bleonard@utah.gov

Introduction

As a Department of Transportation, we are committed to saving lives and resources by providing a quality transportation system that is safe, effectively designed, well preserved, well maintained, and efficient. Toward that goal, we are dedicated to the use of new technologies, and are continually seeking information to help us develop the tools and implement the new technologies for better transportation tomorrow.

Like many areas in the world, part of our challenge is to design, build and maintain a transportation system that will withstand and survive, to some degree, a large magnitude earthquake. Most of the population of Utah lives along the Wasatch Front, the western flank of the Wasatch Mountain Range that runs north and south through Northern Utah. This region is situated in the Intermountain Seismic Belt, a zone of historical seismicity stretching from Northern Arizona to Western Montana. The Intermountain Seismic Belt contains several major normal faults, including the Wasatch Fault. The Wasatch Fault is considered capable of producing seismic events with magnitudes of up to 7.5 and peak ground accelerations of 0.8g or higher. Since the faults found in Utah can produce large earthquake events but on a relatively infrequent timetable, our experience with these ground motions is limited and the resources needed to understand and prepare for these events are often inadequate. Considering, however, that the recurrence intervals for large events on the entire Wasatch Fault is on the order of 350 years, and the most recent documented surface rupture likely occurred more than 600 years ago (Dames & Moore 1996), our risk is imminent, at least from a seismic prospective.

In conjunction with the recent reconstruction of 27 km (17 miles) of I-15 through the Salt Lake Valley, the Utah Department of Transportation (UDOT) embarked on a series of unique research projects aimed, in part, at identifying and quantifying the earthquake risk to our highways and bridges and developing or modifying design and maintenance procedures to improve the survivability and response of our critical transportation lifeline. This effort was funded through a combination of a special Federal appropriation, federal and state highway research funds, pooled research funds contributed by other states, and contributions from private industry. The research was, and still is, conceived, managed, performed, and implemented by the Utah Department of Transportation and our many partners in the transportation industry, including the Federal Highway Administration Turner-Fairbank Highway Research Center, Brigham Young University (BYU), the University of Utah (U of U), Utah State University (USU), and several consultants, with cooperation from the contractors and suppliers who constructed the I-15 project.

Our research efforts have been focused on better understanding the nature of the earthquake risk, more fully characterizing the soils which underlie the highway system and their response to earthquake motions, evaluating how the bridge structures will respond to earthquake events, analyzing the capacities and responses of the foundation systems, and improving our ability to prepare for and respond to seismic events.

This paper will describe some of the varied efforts UDOT has undertaken relative to local seismic conditions and risk, and provide some insight into what we have learned and still expect to learn from this research. None of the work described herein is the work of the author; rather, this paper is a summary of the completed and on-going work of numerous capable researchers working with UDOT. The technical details of these various projects can be found in the individual research project reports and associated papers.

Understanding the Earthquake Risk

Seismic hazard maps have been available to the engineering community for design purposes for over half a century. One of the more widely used maps is the "Seismic Zone Map of the United States" published in various editions of the Uniform Building Code (ICBO 1997). This map places most of the populated Wasatch Front within a Zone III area, indicating the probability of major damage associated with significant seismic events. Recent efforts by a number of government agencies and private organizations to provide more detailed, up to date, and consistent information on ground shaking hazards have resulted in a series of probabilistic earthquake ground motion maps attached to the *Recommended Provisions for Seismic Regulations* (NEHRP 1997a, 1997b). Because of the uncertainties associated with the infrequent earthquake events in Utah, researchers at Utah State University (USU) prepared a deterministic map showing peak ground accelerations without coupling those accelerations with a probability of occurrence (Halling et al. 2002). In this effort, the researchers identified and characterized the faults which would contribute to earthquake ground motions, assigned maximum considered earthquake (MCE) magnitudes to each of these faults, calculated the peak horizontal and vertical accelerations at a series of grid points throughout the state resulting from each seismic source (on a 1-km grid), and created contours of the maximum accelerations calculated for each of these grid points. This systematic assessment of the maximum peak bedrock accelerations was the first such effort in Utah and provides a context for the existing probabilistic ground motion maps and another tool to the engineer designing critical infrastructure. The resulting maps were produced using Geographic Information System (GIS) technology so the user would have maximum flexibility in the use of the data, and are distributed on a CD-ROM (Halling et al. 2002).

A notable deficiency in our attempt to understand the earthquake hazard in Northern Utah is our lack of site specific strong motion data. Although the Wasatch Fault is believed capable of producing earthquakes on the order of magnitude 7.5, none larger than about magnitude 6.6 have been experienced by modern inhabitants in the area and little strong motion data is available. In 2000, researchers from USU, in conjunction with the University of Utah Seismograph Station (UUSS) and the US Geologic Survey (USGS), installed permanent strong motion accelerometers on several spans of a newly constructed freeway bridge on the I-15 corridor (Halling and

Petty 2001). The 18-channel installation included uni-axial, bi-axial, and tri-axial forced balance accelerometers located on the decks and bents of three bridge spans. In addition, a three-channel free field site, consisting of one tri-axial instrument, was installed adjacent to this bridge. Under a separate contract with Brigham Young University (BYU), we are in the process of augmenting this site with an array of four downhole accelerometers at depths of 8 to 120 meters (25 to 400 feet) below the ground surface.

The complete installation at this site includes data acquisition, triggered data recording, and remote access capabilities. This instrumentation will provide long-term information on natural vibrations and earthquake events, and allow engineers and seismologists to evaluate the seismic motions beneath the ground and at the surface resulting from local earthquakes, study the amplification of seismic waves through soft soils, and evaluate the resulting bridge response during seismic events. This is the first complete strong motion installation in Utah, comprising above ground, ground surface, and below-ground instruments.

Characterizing the Behavior of the Soils

Most of the Wasatch Front in Utah is underlain by thick silt, clay and sand soils deposited by ancient Lake Bonneville. In addition, the groundwater table is generally shallow. While the efforts described above are useful to define the bedrock accelerations that can be expected from a large earthquake, it is also essential to understand how those accelerations will amplify or deamplify as they travel through hundreds of meters of these unconsolidated, soft soils. Current methods for modifying strong ground motions for soil sites involve the use of site coefficients which are dependant upon the soil conditions and the level of ground shaking (NEHRP 1997a, 1997b). Particularly for the soft soil conditions found in Northern Utah, these factors represent simplifications and extrapolations that deserve further attention. UDOT has contracted with the University of Utah (U of U) to develop systematic guidance for performing site-specific analyses on soft soil sites and developing an appropriate response spectra for these sites. Their work, which is on-going, includes developing a detailed subsurface profile using existing field and laboratory data from two locations along the I-15 corridor, development of earthquake time histories appropriate for the seismic setting, a deconvolution and convolution analysis of the ground shaking spectrum to transform the spectrum from a surface rock site to a spectrum at depth and back to a spectrum at a soft soil surface, and calculation of site-specific amplification and deamplification factors (Bartlett 2002). The analysis will be performed using the equivalent linear ground response analysis with the SHAKE computer program.

In addition to the work being performed at the U of U, UDOT has contracted with BYU to develop similar guidance for sites that are subject to soil softening and liquefaction. This softening tends to amplify the long period seismic ground motions as they propagate through the soil, but no accepted procedure or theoretical basis exists for accurately accounting for the influence of this soil softening. During the I-

15 reconstruction, engineering judgment was applied to the design ground motions to account for these amplifications, but a rigorous analysis was determined to be unavailable. This recent work by researchers at BYU began with the evaluation of records from liquefiable sites associated with past earthquakes (Youd and Carter 2003a). Using this background information, they then developed design guidance for the modification of the response spectra in areas along the Wasatch Front where soil softening and liquefaction is likely (Youd and Carter 2003b). This work supplements the ground response work being done by the U of U.

Evaluating the Structural Response

One of the more unique opportunities for research in conjunction with the I-15 reconstruction project was a series of full-scale, destructive bridge tests. Two reinforced concrete bridge structures, originally built in 1963, were made available for testing prior to their complete removal and replacement. Like much of the Interstate in Utah, these structures were designed and built before the advent of current structural codes. Since the lateral capacity of reinforced concrete bridge structures is of critical importance during a seismic event, considerable attention was focused on this series of tests. Research teams from BYU, USU and the U of U combined efforts to maximize the benefits of this opportunity.

The bridge elements tested consisted of a stand-alone bent structure and two single span bridge segments, each consisting of two bents and a slab connecting the bents. To facilitate the testing, a load frame was constructed next to the bridge bent to be tested, and a hydraulic ram was placed on the load frame. The load frame and actuator system had the capability to push or pull the bent with a force of about 1780 kN (400 kips), up to a total deflection of 0.76 meters (30 inches). The bents being tested consisted of three square columns with a beam cap, all supported on steel pipe piles. One bent was tested free standing, the others were tested with a slab spanning to an adjacent bent. (Pantelides et al. 2000) End caps were placed on either end of the bent cap and tied together with pre-stressing tendons to allow the loading ram to pull the bent structure, as well as push it. These repetitive cycles caused the column-bent connection to yield. For obvious reasons, we refer to these tests as the "push over tests".

Three general scenarios were tested in the push-over tests: 1) bridge bents in an existing condition, 2) failed bents (having yielded during other phases of the push-over tests) which were repaired using a Fiber Reinforced Polymer (FRP) composite material, and 3) existing condition bents upgraded with FRP prior to loading. The intent of the testing was to simulate the kind of motion and damage which would result from credible earthquake activity, to measure the strength and ductility of the as-is structure, to determine whether the damage could be repaired sufficiently to allow it to be returned to regular service after a seismic event, and to determine if strength and ductility gain could be realized on intact, but aging, structures. Additional objectives were aimed at understanding the failure modes involved and assessing the adequacy of existing codes. (Pantelides et al. 2001)

Similar tests have been performed previously by other researchers on columns, but we believe this to be the first time the beam-column structural systems were reinforced with FRP materials and tested to failure in a full scale test.

Much was learned from these tests. Results demonstrated that capacity and ductility of older reinforced concrete bridges could be improved by using FRP composites. Wrapping the un-damaged bents with FRP increased their strength by up to 20% and their ductility by up to 100%. For damaged structures, wrapping the joints restored about 90% of the strength and essentially all of the ductility (Pantelides et al. 2001).

Encouraged by the results of the push-over tests and other similar tests elsewhere in the country, the Department moved ahead with a seismic upgrade of an existing structure using these advanced composite materials. The I-80 bridge crossing State Street, at approximately 2400 South in Salt Lake City, was designed in 1965 according to the specifications and standards prevalent at that time. As such, this bridge had not been designed to resist any earthquake-induced forces or displacements. When the structure was identified for rehabilitation, because of the normal deterioration from age, it was decided that a seismic upgrade would be also be undertaken. Previous to this project, the Department had not included seismic upgrades in routine rehabilitation projects.

The goal of the seismic retrofit was to improve the lateral displacement ductility of the bridge bent and to reduce the earthquake-induced forces on the bridge. Researchers at the U of U, in conjunction with consultants and suppliers with expertise in composite fabric applications, analyzed the reinforced concrete structure and found that it had confinement and shear deficiencies. Based on this detailed analysis, structural rehabilitation was undertaken using FRP composites by wrapping the columns, wrapping and underlaying the beam cap, and diagonally wrapping the beam - cap joints. This work was accomplished while traffic continued to flow over and beneath the structure, and at a significantly lower cost and time impact than for a complete replacement. The analysis indicates that this retrofit will result in less bridge damage from a small or moderate earthquake event than an un-retrofitted bridge, and the prevention of total failure for the 10% in 250 year design earthquake (Pantelides et al. 1999a).

The researchers at the U of U are continuing to monitor the performance of this structure and the composites used for the retrofit, to assess the long-term impacts and functionality of these methods and materials. Additional research at both the U of U and BYU is evaluating the shear capacities of FRP-wrapped column-bent connections and the mechanisms involved in FRP-wrapped square and rectangular columns.

Another component of the push-over tests was a project undertaken by USU researchers to measure the response of the bridge structures to various frequencies of excitation after each different state of damage. These "forced vibration" tests measured the modal shapes, natural frequencies, and damping characteristics of the structure due to vibrations. The vibrations were caused by simple impact loads or were forced by an eccentric mass shaker with offset flywheels capable of producing 89 kN (20,000 lb) force in any horizontal direction. The vibrations were measured by accelerometers, placed on the bridge deck, bents, and foundations.

Tests were performed on the single-span push-over structures and on several multi-span structures which were scheduled for demolition during the I-15 reconstruction. A total of seven different structures were tested. Since the structures were being taken out of service, the researchers were able to experimentally determine the dynamic characteristics of an intact structure, inflict a known damage condition, and then reassess the dynamic characteristics. Some of the structures were tested in up to seven different damage states. Since the dynamic characteristics of the bridge would change with changes in mass, stiffness, and damping coefficients, a correlation can be drawn between the damage and the measured characteristics. Early results demonstrate that damage inflicted on full-scale structures can be detected through changes in modal frequencies, but that experimental error and testing variability make it difficult to directly correlate small changes in measured response quantities to actual damage (Womack et al. 2001, Halling et al. 2001, Halling et al. 2003).

Further studies are being conducted on new structures along the I-15 corridor in an attempt to refine the measurement and evaluation techniques. It is hoped that, with further development of this technology, the function and health of our transportation structures can be more intelligently monitored in the future, and this important asset can be better managed through more informed decision making. In addition, post-earthquake damage assessments will be facilitated by a more complete knowledge of the structural characteristics of our bridges, and scientific techniques that will allow for more complete evaluations.

Building on the knowledge we have gained from our seismic risks to our bridges, we are now considering the critical lifelines along our transportation routes and determining which bridges are most essential to a viable post-earthquake transportation system. Our intent with this study is to prioritize resources during new construction, maintenance, and disaster response efforts so that critical bridges have the highest chance of survivability. This effort involves a multi-disciplined, multi-agency approach. One of the background components of this work is a 1998 study of highway bridge sites throughout Utah that prioritized the relative risk of liquefaction hazards at each of these bridges, and proposed a screening procedure to evaluate them (Youd et al. 1998). In addition to considering the earthquake shaking, liquefaction risks and structural response issues, we are also incorporating traffic demands, traffic patterns, emergency response, and security planning, and are coordinating with other state and local agencies to insure that all applicable factors are taken into consideration.

Two possible outcomes from this study might be an increased program of retrofitting seismically deficient bridges and a two-tiered approach to the design of new bridges. At present, UDOT designs interstate bridges to a strong ground motion hazard corresponding to a 2 percent probability of exceedance in 50 years (i.e., 2500 year return period). Bridges of lower criticality might be designed for lower hazards, thus freeing up resources for higher priority structures.

Analyzing the Foundation Support

Most of the significant bridge structures in Northern Utah are supported by pile foundations. The lateral load capacity of these groups of piles during an earthquake event is of critical importance to the viability of the structure. It is well known that individual piles within a closely spaced pile group will undergo significantly more displacement and higher bending moments for a given load than an isolated single pile will. (Rollins et al. 2003). This group interaction effect has been evaluated through numerical analysis, model studies, and a few full-scale field tests. In an attempt to gain better understanding of this phenomenon, we have participated in several additional full-scale lateral load tests on pile groups in recent years. These studies have included tests in saturated sands, some of which have been caused to liquefy by setting off explosive charges (Ashford and Rollins, 2000), and in saturated clays (Rollins et al. 2000), with varying numbers of piles, pile diameter, and spacing. Lateral loads have included static and dynamic loads. Piles have been tested in free-head and fixed head (reinforced pile caps) conditions, and the backfill behind the pile caps have been varied.

Data from these lateral load tests, and others that are still being performed, are providing valuable insights into the design of piles for our bridge structures. Studies in liquefied soils indicate that those soils provide more lateral resistance to the pile systems than had previously been estimated (Ashford and Rollins, 2000). Results of static tests in sands and clays indicate that p-multipliers used to reduce the carrying capacity of piles in a group, particularly those in back rows, are lower than might be predicted by some common computer analysis techniques. Further, gaps formed in clay soils adjacent to piles in a lateral load scenario can also reduce the pile group capacity, and must be considered in design (Rollins et al. 2003). Dynamic loads were applied to the pile groups with a Statnamic device, capable of producing large, quick loads such as would be experienced during an earthquake. The piles reacted differently under dynamic loads, providing greater stiffness during the first load, then decreasing significantly during subsequent cycles (Rollins et al. 2003).

Additional static, lateral load studies were performed on another foundation system, Geopiers. These short aggregate piers, a proprietary product of the Geopier Foundation Company, have the capability to withstand significant lateral loads, with some deformation, without significant damage, and still provide foundation support. In addition to compression and lateral loads, a steel plate can be inserted at the bottom of the Geopier, and attached to vertical uplift bars, to provide resistance to uplift forces. In some of the push-over tests described above, the load frame constructed for the tests was placed on foundations consisting of ten 0.9-meter (36-inch) diameter, 4.6-meter (1.5 foot) high Geopiers (Lawton, 2000b). During the load testing, the Geopiers we e subjected to compression, uplift, and lateral loads. The capacity of these Geopiers was evaluated in each of these modes, and the soil structure interactions studied to compare actual behavior to theoretical models (Lawton, 2000a). Comparisons were made to the conventional steel pipe piles also used during the testing, with the intent of determining how Geopiers might be used to withstand and survive earthquake forces. Further studies on these systems are being performed to compare them to the compressive and uplift capacities of drilled piers.

Summary

In summary, our efforts at understanding and preparing for the earthquake risk in Utah begin deep under the ground, and incorporate studies of the seismic conditions and behavior of the soil we are building upon, innovations in foundation systems and designs, evaluations of structural system responses and new construction and retrofitting techniques, and consideration of the uses and functions we have come to expect from our transportation network. We do not know when the large earthquake event will strike, but we are going to great length to be ready when it does.

The projects described in this paper were performed by a variety of researchers. Detailed descriptions of their work can be found in the papers referenced herein. In many cases, complete reports can be obtained from the Utah Department of Transportation Research Division.

References

Ashford, S.A., and Rollins, K.M. (2000) "Full-Scale Behavior of Laterally Loaded Deep Foundations in Liquefied Sand: Test Results", University of California San Diego, Structural Systems Research Project Report No. SSRP-2000/09.

Bartlett, S. (2002) "Development of Site-Specific Response Spectra for UDOT Bridge Design: Modelling Approach Paper", University of Utah unpublished manuscript.

Halling, M.W., Muhammad, I. and Womack, K.C. (2001) "Dynamic Field Testing for Condition Assessment of Bridge Bents", ASCE Journal of Structural Engineering, Vol. 127, No. 2, pp. 161-167.

Halling, M.W., and Petty, T. (2001) "Strong Motion Instrumentation of I-15 Bridge C-846", Utah Department of Transportation Report No. UT-01.12.

Halling, M.W., Keaton, J.R., Anderson, L.R., and Kohler, W. (2002) "Deterministic Maximum Peak Bedrock Acceleration Maps for Utah", Utah Geologic Survey Miscellaneous Publication 02-11, (UDOT Report UT-99.07, joint publication).

Halling, M.W., Bay, J.A., Womack, K.C., Achter, J., Robinson, M., Gottipati, A., Huber, M., and Christensen, C. (2003) "Structural Condition Assessment Using Dynamic Testing: I-15 Testbed Research, Phase II", Utah Department of Transportation Report No. UT-02.18.

ICBO. (1997). *Uniform Building Code*, International Conference of Building Officials, Whittier, California.

Lawton, E.C. (2000a) "Performance of Geopier Reinforced Soil Foundations During Simulated Seismic Tests on I-15 Bridge Bents", Transportation Research Record 1736, Paper No. 00-1307, Transportation Research Board, Washington, D.C.

Lawton, E.C. (2000b) "Performance of Geopier-Supported Foundations During Simulated Seismic Tests on Northbound Interstate 15 Bridge over South Temple, Salt Lake City, Utah", University of Utah Report No. UUCVEEN 00-03.

NEHRP (1997a) *NEHRP Recommended Provisions for Seismic Regulations for New Buildings and Other Structures, Part I: Provisions* (FEMA 302), Building Seismic Safety Council.

NEHRP (1997b) *NEHRP Recommended Provisions for Seismic Regulations for New Buildings and Other Structures," Part II: Commentary* (FEMA 303), Building Seismic Safety Council.

Pantelides, C.P., Okahashi, Y. and Moran, D.A. (1999a) "Seismic Rehabilitation of State Street Bridge", University of Utah Department of Civil and Environmental Engineering Report No. UUCVEEN 99-02

Pantelides, C.P., Reaveley, L.D. and Gergely, J. (1999b) "In-Situ Tests at South Temple Bridge on Interstate 15 - Construction Report", University of Utah Department of Civil and Environmental Engineering Report No. UUCVEEN 99-01.

Pantelides, C.P., Gergely, J., Marriott, N. and Reaveley, L.D. (2000) "Seismic Rehabilitation of Concrete Bridges: Verification Using In-Situ Tests at South Temple Bridge on Interstate 15", University of Utah Department of Civil and Environmental Engineering Report No. UUCVEEN 00/1.

Pantelides, C.P., Duffin, J.B., Ward, J., Delahanty, C. and Reaveley, L.D. (2001) "In-Situ Tests of Three Bridge Bents at South Temple Bridge on Interstate 15 - Interim Report", University of Utah Department of Civil and Environmental Engineering Report No. UUCVEEN 02/01.

Rollins, K.M., Sparks, A.E. and Peterson, K.T. (2000) "Lateral load Capacity and Passive Resistance of Full-Scale Pile Group and Cap", Transportation Research Record 1736, Paper No. 00-1411, Transportation Research Board, Washington, D.C.

Rollins, K.M. , Olsen, R.J., Egbert, J.J., Olsen, K.G., Jensen, D.H. and Garrett, B.H. (2003) "Response, Analysis, and Design of Pile Groups Subjected to Static & Dynamic Lateral Loads", Utah Department of Transportation Research Report No. UT-03.03.

Womack, K.C., Halling, M.W., Ghasemi, H. and Bay, J.A. (2001) "Modal Analysis Research on the I-15 Corridor, Salt Lake City, Utah", *Transportation Research Board* preprint, Washington, D.C.

Youd, T.L., Willey, P.S., Gilstrap, S.G. and Peterson, C.R. (1998) "Liquefaction Hazard Evaluation of Interstate, Federal, and State Highway Bridge Sites in Utah", unpublished report to the Utah Department of Transportation, Brigham Young University, Provo, Utah

Youd, T.L. and Carter, B. (2003a) "Influence of Liquefaction on Response Spectra at Wildlife, California and Port Island, Japan Instrument Sites", Proceedings, 8th US Japan Workshop on Earthquake Resistant Design of Lifeline Facilities and Countermeasures against Liquefaction.

Youd, T.L. and Carter, B. (2003b) "Influence of Soil Softening and Liquefaction on Response Spectra for Bridge Design", Utah Department of Transportation Research Report No. UT-03.07.

Development of Seismic Disaster Mitigation Master Plan for Asia-Pacific Regions through Implementation of Risk Management Framework

Hiromichi Higashihara[1]

Abstract

Some theoretical aspects of development of seismic disaster mitigation plan are presented. Over these years, a multi-lateral collaborative research has been carried out toward Master Plan for Asia-Pacific Regions funded by Japanese government. In this paper, application of risk management framework to the development of EqTAP Master Plan is discussed. This attempt can afford a basic methodological basis to every external review or oversight of societal human activities like public administration or projects management and therefore informative to many Asian countries that are aiming at economical reconstruction through establishment of regulation system.

Introduction

Since 1998, Japanese Government has funded a multi-lateral collaborative research project entitled:" Development of Earthquake and Tsunami Disaster Mitigation Technologies and Their Integration for the Asia-Pacific Region"(abbreviated as EqTAP). The Earthquake Disaster Mitigation Research Center (abbreviated as EDM), National Research Institute for Earth Science and Disaster Prevention, Ministry of Education, Culture, Sports, Science and Technology, headed by Prof. Hiroyuki Kameda of Kyoto University has been in charge of management of the project (EDM,2003) and researchers from a lot of universities or research institutes joined across the APEC region.

This project started as a loose consortium of Japanese research groups, who then introduced their research partners from Asia-Pacific region. Consequently, it is an extensive package of individual researches including hazard assessment of built structures and cities, hazard estimation of active faults, seismic design of urban transportation systems or composite masonry buildings, retrofitting of brick/block masonry buildings, tsunami disaster mitigation etc. It even includes sophisticated researches like multi-national parallel on-line experiments linked on the Internet. Urban planning approach to disaster risk assessment of mega cities or urban disaster management is also included.

International Advisory Panel of experts of disaster management had, however, long been pointing out the lack of integration of the entire project when it carried out the mid-term review of EqTAP in autumn of 2001. Thus, the review criticized the

---

[1]Member ASCE, Dr. Eng., Prof., Earthquake Research Institute, University of Tokyo, Yayoi 1-1-1, Bunkyo-ku, Tokyo, 113-0032, JAPAN, e-mail: higashi@eri.u-tokyo.ac.jp, telephone: +81-3-5841-8268, fax: +81-3-5841-5784

major aspects of Phase I activities of EqTAP as "individual technology developments limited to activities by researchers", and requested to characterize Phase II (2002 and 2003 fiscal years) by integration of information, implementation of technology and involvement of stakeholders. The present paper describes the theory and practice of integrating researches done in the project. This affords a theoretical basis to the EqTAP Master Plan that the Japanese Government has called for, intending to use it as one of guidelines for its international cooperation.

At the same time, the Panel recommended the latest version of *Risk Management Standard of Australia and New Zealand* (Standards Australia, 1999) so that the project can adopt a more structured approach. Japanese members accepted it because of its enhanced capability of stakeholder involvement and high accountability realization. EqTAP Master Plan has thus been so far developed as a customization of this Standard.

EqTAP project

According to resolutions of several workshops, EqTAP is going to achieve its goals, safety and sustainability of Asia-Pacific region, through reduction of preventable deaths and injuries, social and economic disruption, psychological impacts, and environmental damage caused by earthquakes and tsunamis.

Its all outputs are to be compiled into the Master Plan, which are expected to consist of risk management framework, the Metro Manila Case Study, the Toolbox, and the Digital City. The Metro Manila Case Study, the Toolbox and the Digital City are described elsewhere (Britton, 2002). The present paper raises and discusses the potential issues to be anticipated for the introduction of risk management framework.

Close relation exists among these outputs. First, the Metro Manila Case Study is expected to provide materials, through conducting a complete cycle of risk management process, to the risk management framework. The Toolbox and the Digital City have been devised and constructed on the Internet as a means to show the result of researches. They are supposed to be portal site on earthquake and tsunami disaster alleviation for the Asia-Pacific region, through which stakeholders can access on their demand to relevant information.

The outputs of individual researches of EqTAP will be eventually arranged and loaded in the Toolbox. For this purpose, however, we need to first establish the architecture of the Toolbox. But its blueprint requires full integration of individual researches and established context of the problem and the risk management framework must lead this.

On the other hand, EqTAP's individual researches are being encouraged to consider operating principles including regional perspective, risk management framework, inter-disciplinarity, implementation, and integration. As is seen above, the risk management framework appears twice as an *output* and an *operating principle*. In order to distinguish them in this paper, we will tentatively call the risk management framework as *output* the EqTAP risk management scheme. Genealogy of the EqTAP Master Plan is outlined in Fig.1.

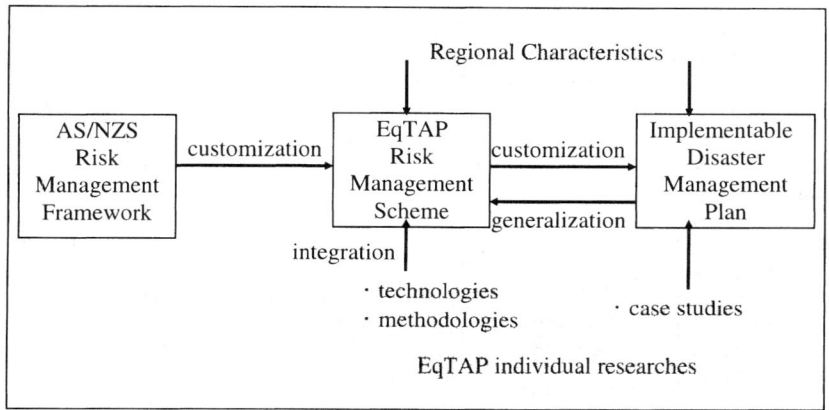

Fig.1 Pedigree of EqTAP Risk Management Scheme

The risk management framework is carrying two fold missions in EqTAP. First, it is used for management of the EqTAP project itself. Second, it should afford the basis to the EqTAP risk management scheme. This paper does not treat the management of EqTAP but solely focuses on the customization of the basic risk management framework to get its executable form. Management of EqTAP is also summarized in EDM documents (Britton, 2002).

General risk management frameworks

According to the Australian and New Zealand Risk Management Standard, risk management is conducted iteratively through a cycle shown in Fig.2.

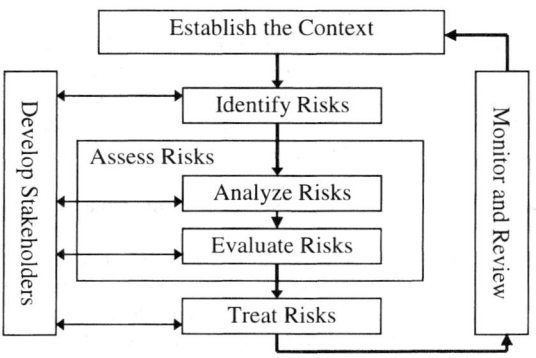

Fig.2 AS/NZS risk management cycle

Unique to this Standard was explicit treatment of context and stakeholders. In 2001, Japanese Standards Association, after two years committee activity, released Guidelines for development and implementation of risk management system (Japanese Standards Association, 2001). In contrast with AS/NZS, this guideline had two significant defects: shortage of recognition of the value of the context exploration and apparent indifference about stakeholders. This appeared enough to make the practical value of this guideline suspicious and EqTAP team leaders did not adopt it.

Emphasis of the context is not new in Japan; researches about public administration have recently argued a variety of context related matters. But as to the adherence to stakeholders, AS/NZS seemed superior to Japanese researches. This may have been indicating the delay of dissemination of accountability-based public administration in Japan. From the very reason, AS/NZS's stance will be valuable to many Asian countries because they are currently aiming at globally admissible standards of accountability of governmental behavior.

AS/NZS appears, however, to have some noticeable implicit premises, which may have reflected the regional characteristics of both countries. First, it assumes that users are well-established organizations where top management enjoys legitimacy and power under mature legal control. For example, the Standard is optimistic about attainment of consensus about risk management criteria, and this could be one expression of the premise.

Similarly, the reference to the "context" in AS/NZS sounds to reflect the needs of well-defined body like established corporations of industrialized countries where the context is easy to see. Unfortunately this is not the case in EqTAP; many Asian disaster problems face almost all aspects of human activity and, accordingly, the Master Plan needs, *e.g.*, ethnological consideration of regional perspective like cultures, religions etc.

Another implicit premise is that of moderate disaster. As a matter of fact, the total monetary loss of the disaster that was exemplified in a lecture about AS/NZS was only ten million dollars (Jull; 2002). This is only 0.01% of the purely monetary loss of the Kobe disaster. Confidence of AS/NZS in the possibility of risk assessment in such a high precision as to suffice detailed judgment could be an indicator of that premise, while many Asian disasters are rather similar to that of Kobe; enormous damage with hardly tractable uncertainty. Because of these, we need some modification of AS/NZS. In the present study, therefore, we need to seek for the "context" in some extended manner.

Logical aspects of implementation of risk management frameworks

The risk management framework is completely universal; *i.e.*, it can be applicable to from nation-wide public administration to business management of private companies. But just this universality disables its applicability to given specific objects automatically and we must customize the basic AS/NZS risk management framework. This is a kind of unfreeze. As a matter of course, any frameworks are ordinarily a condensed or crystallized essence of huge amount of real experiences and learning. It is to inherit the wisdom of predecessors and can be an origin of versatile know-how.

Generic vs. specific. Risk management framework itself is generic; it can be implemented only through complete account of characteristic feature that is specific to the object. Let us see one example that is popular to us. Differential equations have general solutions that can represent any particular solution. Such generality is guaranteed by its abundant arbitrariness that is expressible in terms of a set of many indefinite parameters included. By choosing these parameters appropriately, the solution can satisfy any well-posed boundary conditions or initial conditions. And thus obtained specific solution affords thorough information about the phenomenon under concern. On the contrary, general solutions have little practical meaning because of its arbitrariness so long as they remain unspecified.

A similar relation exists between thermodynamics and statistical mechanics. Thermodynamics is generic; it is valid for *any* thermal system. But just because it is generic, it can say little about any particular system. It only says about the laws that *any* system obeys. Therefore if we want to analyze a given actual system, we have to develop a model that is specific to the system. But this requires a different paradigm, *e.g.*, statistical mechanics. The risk management framework corresponds to the general solution or thermodynamics here.

Theory vs. practice. Another important distinction exists between theory and practice. A well-known example might be the relation between economics and political economy but let's compare here structural mechanics and structural design. Although the latter is an application of the former, logical structure is totally different. The mechanics can determine the stress or deflection of the structure *if* the details of the structure and the external loads are given. But who determines the dimension first, and how? Designers. Mathematically, the designers' work belongs to the so-called inverse problem, while structural mechanicians treat forward problems. Inverse analysis requires full knowledge about corresponding forward ones. Consequently, designers must carry comprehensive knowledge. On the other hand, researchers do not necessarily need complete coverage. However narrow their knowledge may be, it will not prevent them from being capable researchers.

Inverse problems are generally more difficult than forward ones and demand different type of intellectual ability. At the EqTAP's Manila workshop of 2000, the author attempted to express this ability with "imagination" and "insight". These are well-known as the two pillars of ancient Hebrew wisdom, which showed a famous example of how a strict religion could bear practical worldly ethos. This perspective is undoubtedly important in the study of Asian ethos, too.

Doctrine vs. catechism. Revolutions in the history like religious revolution of Moses, Muhammad or Protestants or communist revolution of Lenin or Mao Ze-dong provide us, in the present context, a useful reference. They used to have a condensed and keen doctrine. But they had, at the same time, to resolve countless affairs or troubles from daily lives of people, based on their very compact doctrine (or dogma). This is the catechism and gives profound insight about the nature of implementation.

Integration of component researches

Significance of integration. As was pointed out by the International Advisory Panel, EqTAP Phase I was a simple aggregate of disconnected researches that had been proposed at the very beginning of EqTAP by individual participants. Many researches of Phase II still are the prolongation of Phase I. It is therefore necessary to integrate them. All the worse, because EqTAP started as a simple assembly, the coverage of researches is not complete and this injures the usefulness of the Toolbox. We must therefore identify and fill up the lacking items.

Integration itself has a wide applicability; it is the core of any oversight and external review. Recently in Japan, external or third-party review has become popular as a measure to oversee activities of public organizations. The key point here is that the overseers must get an integrated knowledge about the object because only such knowledge and not a simple assembly of fragmentary information enables right judgment. Development of the methodology of integration is therefore urgent in Japan as well as in EqTAP. All the more, the situation is quite similar in many Asian countries, where development of effective regulatory methods of various economic activities has growing needs since the economic crisis of 1997. Development of methodology of integration of researches is thus timely.

Basic tools required for integration in EqTAP. It is a matter of course for integrators to fully understand every material. But this is only the beginning; *i.e.*, any integration has as its core a mechanism of synthesis. This belongs to the world of "imagination" and "insight" and preparation of some tools is vital.

In general, a common platform is desirable that enables plotting the individual researches. In EqTAP's Bangkok workshop, the author pointed the value of development of a glossary and flow charts of disaster mitigation activities (Higashihara; 2002). Glossaries of disaster-related technical terms have been published in Japan. Some of them are multi-language (Japan National Committee for IDNDR, 1996). But they appear irrelevant for EqTAP use for several reasons and looks like EqTAP needs its own.

Integration of many individual risk assessment and treatment techniques becomes much easier if we can prepare flow charts that describe the dynamic sequence of decision-making of multiple subjects. This is truly challenging because the process has vast feedback mechanism and quite a lot of links traversing the society.

Establish the context and strategy

Exploration of the context. As is shown in Fig.2, the risk management cycle starts from risk identification. But in case of complex cases, it is likely to miss important issues. Here is the *raison d'être* of context exploration. But neither

AS/NZS nor JIS are telling anything about how to do it and we have to work out the methodology of context exploration. There are deductive approach and inductive ones. The latter is generalization of concrete specimens. Actually, study must almost always start with some samples; *i.e.*, inductive approach is indispensable to trigger the research while deductive work is needed to expand and deepen the understanding.

The deductive approach consists of centripetal approach and centrifugal one. Centripetal approach draws first the whole entity in *a priori* manner and then assigns individual researches from overall perspectives. On the contrary, the centrifugal approach starts from existing individual researches and integrates them into one body. We must employ these two methods simultaneously but, as is mentioned above, because EqTAP's individual researches existed in advance, a centrifugal approach plays a primary role. On the other hand, a lot of materials for the inductive approach have been developed in numerous case studies of EqTAP: not only in the large-scale Metro Manila Case Study but also in many individual researches, big cities like Bangkok, Beijing, Hanoi, Kobe, Shanghai and Singapore etc.

Some basic aspects of the context in EqTAP. In many EqTAP countries, their rapid economic growth through drastic admission of investment from oversea countries has attained a certain success but then abrupt economic crisis occurred since 1997 (World Bank, 1993, 1998). Many western criticisms pointed out (*e.g.*, Krugman 1994) low productivity of industries, deceptive competitiveness depending on external economy or excessive risk exposure, and lack of governance.

Although some controversy remains about these statements, Asian countries rather promptly initiated drastic reforms that appeared to consider these criticisms. In effect, their challenge is very similar: augmentation of productivity through innovation, improved governance and effective regulations. Obviously they are aiming at another economic success but in a more sustainable manner this time. This is harmonic with EqTAP Goals (see Chapter 2).

Productivity enhancement of the society is expected to attain through human development, establishment of governance, and generation of organized markets. It is worth noticing that the target of human development has been changing from workers to entrepreneurs. In other words, acquisition of business mind or entrepreneur's mind that requires management skills and/or even investment preference is aimed at; i.e., they have to learn to take risks of investment for higher return through realization of qualitative discrimination of products and services. This could be one promising response to globalization that realizes sustainable growth of Asia.

In EqTAP a noteworthy approach exists that is along this line. One of them is improvement of earthquake resistant ability of structures being done by negotiation between users and suppliers. National or other official codes are adjusted or modified in a *de facto* manner through consultation and persuasion. It is also pointed out the opportunity for insurance companies to play some role. Also noticeable is replacement of former governmental activity like vulnerability assessment of buildings and other facilities through private firms' profitable business. These are a kind of nongovernmental, market-oriented approach.

Functional market demands effective regulation and oversight. Recently, a series of scandals of Japanese corporations repeated: nuclear power, meat, pharmacy, bank etc. This has been basically attributed to regulatory inability of government. For example, *Nature* criticized Japanese practice of public safety administration just after the criticality accident of a nuclear fuel factory (Oct.1999) saying "the Japanese government seems unable to set up competent regulatory bodies with sufficient staff and expertise....it must commit funding, manpower, expertise and accountability."

On the other hand, it is true that Japanese legislature toward transparent administration has attained drastic progress. One of the Japanese authorities of administrative law asserted that enactment of the Japanese law of freedom of information based on the accountability of national government afforded a revolution rather than an evolution to Japanese administration since Meiji era.

Accommodation of disaster policy into National development plan. Under the current trends where Asian national governments have good appetite for further development under traditional top-down method, their national development plan is important; EqTAP Master Plan must consider and be harmonic with these plans.

Their integral part includes poverty reduction and disaster mitigation. This is reasonable because reinforcement of the weakest points of the reproduction ring of the society is of primary importance for sustainability. In most countries, there is a serious environmental degradation and elevated disaster risks and correspondingly there is on-going concern for sustaining development that can improve the quality of environment, amenity and traffic, and cope with disasters.

But the most important is the balance among quite a many political needs and, as a matter of course, economical efficiency seems the most important issues; it may be unrealistic to expect that disaster management policy will enjoy priorities. Above all, energy problem and urban infrastructure/transportation problem will be crucial. For example, poor transportation facilities and inappropriate price policy of fuels are allowing incessant traffic congestion and waste of resources. Further invention is therefore necessary to effectuate the earthquake disaster policy.

Comprehensive counter-disaster plan. Experiences of disasters are telling that it is ineffective to try to switch to some special organizational arrangements for disaster purposes. EqTAP countries have suffered from a variety of natural hazards including cyclones, drought, floods, landslides, volcanic eruptions, wildfires as well as earthquakes and tsunamis. But the feasibility of organized response to disasters depends strongly upon their frequency and magnitude.

From this perspective, seismic hazards are sudden, unexpected and widespread while the needed interval of management cycle is extremely long. This hinders systematic planning. On the contrary, flood is, if any, much more frequent and has wide stricken area. In flood-prone region, therefore, earthquake disaster mitigation plan should utilize existing resources against flood. This incorporated plan should eventually be built in the development plan.

Public Involvement

Several EqTAP documents have consistently been urging researchers toward understanding the requirements of local stakeholders. As is mentioned in Chapter 2, AS/NZS framework recommends, more direct involvement of stakeholders and dialog with them. EqTAP's effort of public involvement is in harmony with the current effort of Asian countries toward decentralized anti-disaster administration.

The documents even claim, "EqTAP will *directly* help meet the *stated* needs of the stakeholders" or "Procuring information can only come from talking with stakeholders". But here we must recall the serious question of how we can identify the stakeholders' needs. Reliability of direct interview or questionnaire has long been questioned seriously. Even the validity of the concept "public needs" itself has been controversial. Stakeholders have diversity of needs and their opinion could be volatile; *i.e.*, our results are not free from dependency upon the choice of stakeholders. On the other hand, it is well recognized that human factors play primary role in emergency response operations. Although the so-called crisis pressure or disaster impact is extremely difficult to forecast in advance, experiences are even telling the special difficulty in identifying assistance needs of traumatized recipients. Stakeholders development thus requires further investigation.

In this context, we should pay attention to the monitoring ability of local administrative organization and non-governmental organization. Non-governmental organizations are known to have strong out-reach and contact capability. Another important thing in public involvement is about leaders. Leadership is most critical in emergency response. Even apart from the disaster context, overall promotion of leaders in the society has immeasurable value for Asian countries because leadership in the market economy is the true driving force of innovation and high productivity.

Confronting intensive uncertainty and tremendous damage induced by dense population, seismic disaster mitigation planning always suffers from the divergence of scenarios. Just after the Kobe earthquake, a French journalist asserted, based on his long life in Japan since before the World War II, that the well known resignation of life of Japanese typically expressed in a famous essay of medieval times was primarily formed by the terror of earthquakes that strike them suddenly. It is undeniable that strong uncertainty included in disasters can alter the socio-cultural style of risk acceptance. Similarly, how to accept the pain of life has long been the center of concern of Indian thinking that underlay deeply the EqTAP region. These are not past thing; *e.g.*, a document published by the Asian Development Bank is mentioning the existence of longstanding acceptance of disaster risks by governments and communities, who may feel that traditional measures are adequate (Carter, 1991). Risk perception is thus of practical importance.

Understanding Asian mega-cities - substance of regional perspective

Many recent researches are reporting that, in Asia, the globalization has changed the structure of employment while destroying traditional societal safety nets. This is a challenge to the EqTAP goals and consistent study about Asian big cities is needed.

The so-called "over-urbanization" model has long been used in the analysis of Asian big cities including Bangkok, Manila, etc. (Kingsley, D. and Golden, H., 1955). It discussed the explosive increase of slums and squatters, permanent unemployment, enlargement of informal sectors, collapse or absence of public services and overcrowded population, where all were regarded negative and to be overcome. But in these decades, antitheses have been developed that illuminated positive features of the problem. In concert with these perspectives, governmental practice has also changed at the same time, *e.g.*, from enforced emigration from slums to support of livelihood in slum. According to these researches, there are quite a many functions of slum dwellers; they are contributing to the city by carrying out a diversity of inhabitant-supporting services. Other researches revealed how the nationwide migration had desirable functions on welfare of poorest people.

More than thirty years ago, a Japanese anthropologist criticized the western approach to South-Eastern Asia saying "cut the objects into pieces without shedding their own blood" and proposed a kind of sympathetic, participatory approach. This sounds like questioning the EqTAP's fundamental attitude. We admit, *e.g.*, that high population density is one of the key features of the disaster-prone regions. But at the same time it can be an expression of the fertility of rice crop culture in contrast with the poor land of Western Europe. A Japanese myth reports a male God's opposition with daily production of thousand men against the curse of a Goddess of the underworld, his former wife, to daily kill thousand men.

It is true that habits and attitude that create vulnerability are prevailing in EqTAP region, but these habits and attitude usually have deep roots, long history and many justifications. This is just the regional perspective that anthropology can serve. All these are suggesting a sort of sympathetic model that EqTAP Master Plan needs.

References

Britton, N. (2002), EqTAP Protocol
Carter, W.N. (1991), *Disaster Management*, Asian Development Bank
EDM (2003), http://www.edm.bosai.go.jp/english.htm
Higashihara, H. (2002), Implementation of the risk management framework and the regional perspective of East Asia
Japanese Standards Association (2001), *Guidelines for development and implementation of risk management system*, JIS Q 2001
Japan National Committee for IDNDR (1996), *Multi-language Glossary on Natural Disasters*
Jull, J. (2002), Case Study: Application of the AS/NZS4360 for Hazard Management in New Zealand
Kingsley, D. and Golden, H. (1955), Urbanization and the Development of Pre-Industrial Areas, in *Economic Development and Cultural Change*
Krugman, P. (1994), The Myth of Asia's Miracle, *Foreign Affairs*
Standards Australia (1999), Risk Management, AS/NZS 4360:1999
World Bank (1993), *The East Asian Miracle: Economic Growth and Public Policy*
World Bank (1998), *East Asia: The Road to Recovery*

DEVELOPMENT OF A SHAKEMAP-BASED, EARTHQUAKE RESPONSE SYSTEM WITHIN CALTRANS

David J. Wald[1], Philip A. Naecker[2], Cliff Roblee[3], and Loren Turner[4]

Abstract

ShakeMap is a system for automatically generating maps of ground motion and intensity in the minutes immediately following an earthquake. Caltrans is beginning to take advantage of the information provided by ShakeMap and currently is test deploying a simplified protocol for its direct use for post-earthquake prioritization of bridge inspection. The current process involves manually retrieving the GIS-formatted, response-spectral acceleration ShakeMaps from a website and employing a GIS spatial analysis to identify the bridges that were most strongly shaken. The simplified protocol sets threshold values for ShakeMap 1-second spectral acceleration that defines GIS map zones that correspond to possible damage states for bridges built during pre- and post-ductile design eras. Output from this operation includes lists of bridges within each zone grouped by route and ordered by postmile for ease of inspection. These lists can be distributed to local maintenance crews to facilitate rapid inspection and response to the most severely shaken areas.

To advance and fully automate this process, we introduce "ShakeCast" (for "ShakeMap Broadcast"), which will allow Caltrans, and others, to automatically and reliably receive desired ShakeMaps and trigger post-processing tools to initiate an established response protocol. The system will initiate software applications and automatically generate alarms in response to predefined shaking conditions. Currently, USGS "pushes" ShakeMap electronically (using ftp) to utilities and other critical users, but ShakeCast will allow this to be replaced with a subscriber service, providing more robust delivery from redundant ShakeMap generation sites and distributed ShakeCast servers. ShakeCast will also allow agencies such as Caltrans to receive and process ShakeMap at multiple divisions within the agency that require different post-earthquake actions, from bridge inspection and repair to traffic management. Caltrans plans to further develop damage estimation tools including incorporation of more sophisticated bridge fragility relationships along with more detailed bridge information. We expect that the lessons learned and tools developed during the implementation of this prototype with Caltrans will facilitate use by other organizations in the lifeline arena.

---

[1]Seismologist, U.S. Geological Survey, Golden, CO, wald@usgs.gov; [2]Chief Scientist, Gatekeeper Systems, Pasadena, CA, pan@gatekeeper.com; [3]Senior Research Engineer, Caltrans, Sacramento, CA, cliff_roblee@dot.ca.gov; and [4]Senior Transportation Engineer, Caltrans, Sacramento, CA, loren.turner@dot.ca.gov

ShakeMap Background

ShakeMap was designed as a rapid response tool to portray the extent and variation of ground shaking throughout the affected region immediately following significant earthquakes. ShakeMaps summarize the shaking intensity over the entire affected area by interpolating the ground motions measured at seismic stations with geologically based site correction factors and by empirically estimating ground motions in sparse areas of the network or in areas for which data have not yet been received. Hence, ShakeMaps are constrained both directly from seismological data, and by tools that developed from both seismological research and empirical observations. Figure 1 shows an example of an Instrumental Intensity ShakeMap for the 1994 Northridge earthquake.

As a rapid response tool, the ShakeMap ground motion values are used for emergency response, loss estimation, assessment of damage to the lifeline and utility networks, and for providing information to the general public through the internet and through the media. Ground motion values portrayed with ShakeMap are distributed to the California Office of Emergency Services and the Federal Emergency Management Agency for input into the HAZUS loss estimation software, replacing simpler but approximate estimates based on location and magnitude. The ground motion values are also distributed to regional and state utility providers to enable them to determine areas of their networks that may have sustained damage. ShakeMap is also now used as a planning tool by generating ground motion estimates for a suite of earthquake scenarios. Estimates based on scenarios can provide a firm basis for loss estimation on a regional scale as well as providing utilities and other users a means of exercising their emergency response capabilities.

ShakeMap was originally developed under the TriNet project in southern California, which began in the years following the 1994, magnitude 6.7 Northridge earthquake. Ongoing development of ShakeMap is under the auspices of the U.S. Geological Survey's Advanced National Seismic System (ANSS). Under this program, ShakeMap now runs in southern and northern California, as well as the Seattle and Salt Lake City areas. It will be available in other seismically active regions of the country if sufficient numbers of real-time strong motion stations are installed as outlined in the ANSS strategic plan.

The ground motion from the Hector Mine event was widely felt in urban Los Angeles and, based on past experience, responders, the media and public had legitimate concerns regarding its source and potential damage. For this event, the ShakeMap provided rapid evidence that large-scale emergency response mobilization was unnecessary.

Ground motion metrics from the seismic stations summarized with ShakeMap include peak ground acceleration (PGA), peak ground velocity (PGV), and peak response spectral amplitudes (at 0.3, 1.0, and 3.0 sec). At the same time, maps of instrumental intensity are generated through newly developed relationships between recorded ground-motion parameters and expected shaking intensity values (Wald *et* al., 1999a) as shown in Figure 1. Production of the maps is automatic, triggered by any significant earthquake in California (for more details see Wald et al., 1999b).

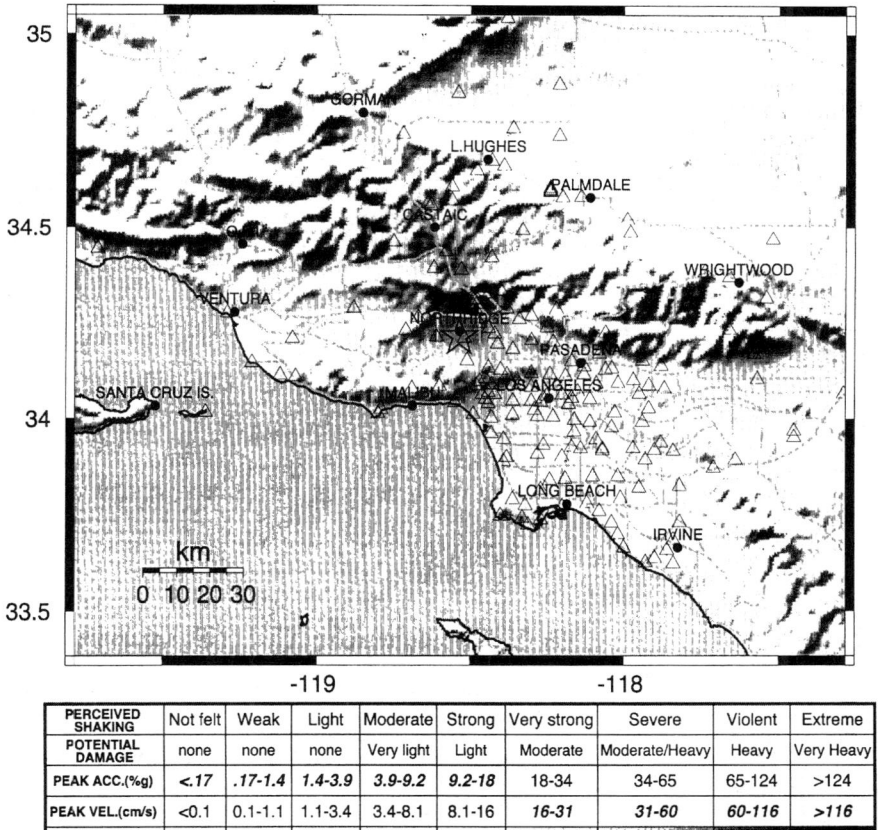

Figure 1. Instrumental Intensity ShakeMap for the 1994, Northridge, California, earthquake. Stations are indicated by triangles; the epicenter is shown with a star. ShakeMap was not in operation at the time; analogue strong motion data were processed with the ShakeMap software in hindsight to show the capacity of the system for a damaging earthquake.

Although ShakeMap was initially designed to be primarily a web-based real-time display (see http://earthquake.usgs.gov/shakemap), ShakeMap products have evolved to include high-resolution graphics files, maps made specifically for television, GIS files for direct input into the HAZUS loss estimation software, as well as gridded ASCII data files, all of which are also now automatically generated. Once individuals or organizations know that an earthquake has occurred, they can visit the ShakeMap web sites to obtain maps and data. However, few organizations have the technical means to use this data other than by "just looking at it". Critical users, including emergency responders and utilities, for example, prefer to receive ShakeMap products automatically and without the potential competition from other internet users. Automatic delivery is fundamental for some, since the arrival of ShakeMap files can also trigger user-designed software applications.

Currently, ShakeMap products are automatically distributed to the California (Governor's) Office of Emergency Services (OES), state agencies and utility providers to enable them to determine the extent of damaging shaking in their districts. In addition, a number of media news organizations and private companies, including several engineering and financial institutions, also receive the automatic electronic delivery of ShakeMap.

ShakeCast: Improved ShakeMap Access and User Tools

To date, the delivery mechanism is a standard File Transfer Protocol (FTP) "push", requiring access through the user's internet firewall. This is awkward for some users, and it is now impossible for other potential clients given the more rigorous approach to computer security in recent years. It is often difficult to setup the initial "push" delivery, since this requires substantial coordination with IT security personnel in addition to the communications with the direct ShakeMap users within an organization. In addition, our daily diagnostic tests reveal various failure modes, making long-term maintenance problematic for ShakeMap operators.

Disseminating earthquake shaking information is a difficult problem, for a variety of technical reasons:

o Wide adoption of such a system will depend upon reliable transmission of data, even if communications networks have been damaged

o There are a wide variety of data products that must be delivered for different users, including different metrics of shaking intensity (e.g., acceleration, velocity, instrumental intensity, spectral acceleration etc.).

o The spatial variability of shaking intensity is critical, and most consumers of the data are likely interested only in data for specific locations where they own or manage facilities that might be damaged by earthquake shaking.

o It is important to keep track of updated versions of each data product so that as new information arrives it can be seamlessly related back to the original event and previous version of the data and to decisions that may have been made based on that data.

o There is wide variance in the types, sophistication, and purposes of the organizations and computer systems that will consume the data.

A second limitation with the current set of ShakeMap products is that substantial expertise is required to automatically process downloaded or delivered maps. The rapid and reliable dissemination of detailed earthquake information is of great importance for public safety and emergency response. This information is needed by all kinds of facility owners, such as municipalities, utilities, building managers, schools, and many others. Such organizations would like to be "consumers" of earthquake information, but currently have no suitable technology that they can use to readily access and make use of earthquake information.

To address these problems, the ShakeCast System is designed to be a simple, reliable, and widely deployable software tool that any modestly capable computer user can install on their computer to receive and make use of customized and personalized earthquake information. We call the system ShakeCast (short for "ShakeMap Broadcast") because its purpose is to broadcast ShakeMaps. ShakeCast consists of a receiver component (client) and a transmitter component (server). The information to be disseminated via ShakeCast is the output of the ShakeMap system, which provides early estimates of the severity of shaking during an earthquake and thus is a good tool for estimating the likelihood of damage to structures.

The ShakeCast software will also:

o Automatically download and display maps of the areas affected by an earthquake.
o Automatically receive and process notifications of earthquakes
o Let users define locations (representing structures and facilities) of interest to, and set shaking thresholds that will trigger automatic notification
o Provide users with options for electronic notification (pager, email, personal web pages, etc.) of events and projected shaking intensity at specified facilities
o Reliably manage the receipt of updated shaking data from multiple ShakeCast servers distributed around the internet, providing an excellent chance of receiving an uninterrupted and authenticated data feed even after a major event
o Easily integrated with in-house GIS systems, control systems, utility outage management systems, and other business systems in organizations
o Provides a mechanism for continual end-to-end testing of the system, assuring that the system is working properly when it is eventually needed

An overview of the main features of the ShakeCast system being developed is shown in Table 1. ShakeCast allows individuals and facility owners to make widespread and immediate use of the beneficial information already produced by ShakeMap. It takes advantage of the very substantial investment already made in ShakeMap and in the very large seismic monitoring infrastructure behind it. It also provides quantitative metrics on the use of ShakeMaps both before and after an earthquake. These data will then be available for policy decisions on the future direction of the ShakeMap and ShakeCast systems. Finally, ShakeCast should help engage and involve managers and policy makers at a wide variety of institutions (e.g., state transportation departments, municipal governments, emergency responders, utilities, etc.) who are concerned about timely receipt of earthquake shaking data.

Feature	Description
Client (Receiver) Software Features	
Multi-platform	Available on PCs and Unix systems
Easy installation and configuration	Installation and basic configuration in less than an hour in most cases
Automated registration	Automatic software registration with ShakeCast broadcast systems, including registration with servers in multiple regions
Integrated quality assurance and testing	The client software will participate in the ShakeCast system's comprehensive end-to-end testing procedures to provide high confidence in proper system function during an earthquake. Broadcast data will be checked for authenticity, correctness and completeness.
Automated notification	The client software will notify a list of people of earthquake-related events via email, pager, and other mechanisms. Notification can be based on shaking intensity (e.g., "peak ground acceleration at Mom's house greater than 0.3g") using any of the shaking metrics of the current or future ShakeMap system. Users can "sign up" for notification via a Web page on their local ShakeCast system.
Personal web pages	Provide local ShakeCast users the ability to view shaking data (including maps, events, and alarms) on personalized web pages served from their local ShakeCast server without each user needing to access the main USGS ShakeMap systems.
Data version support	Revise and re-issue notifications as new data arrives. Maintain permanent record of the sequence of notifications issued.
Locations and thresholds database	Maintain local list of locations of interest and notification thresholds.
External program integration	ShakeCast can trigger the execution of external programs for further event and data processing.
Basic GIS tools	Tools for working with GIS format ShakeMap data. Display users own facilities and ShakeMap data in a Web-based map generated locally on the client system.
Simple administration	Web-based configuration and administration interfaces
High quality documentation	Professionally developed documentation and support materials

Table 1. Overview of ShakeCast system features for the client.

The design of ShakeCast takes full advantage of the software tools that are widely available and commonly used on the internet. The ShakeCast system is intended not as a static software solution to be used identically by every site, but as a reference implementation: that is working software that others can either use or extend as they wish. The reference implementation will be freely downloadable and will perform all of the basic functions needed by a consumer of earthquake shaking data.

The design goal for the ShakeCast reference implementation is that a modestly knowledgeable personal computer user should be able to install the entire system and begin receiving useful, reliable, authenticated, location-specific reports of earthquake shaking with less than an hour of effort. A further goal is to create a community of software developers who work for enterprises who are consumers of ShakeMap data. This community of developers will then be able to continue enhancement and extension of ShakeCast so that ShakeMap information can be made more readily usable by their organizations.

ShakeCast is intended for a wide variety of user organizations. Public Utilities may use ShakeCast to generate alarms indicating possible impacted facilities and to direct initial response and post-earthquake inspection efforts. Public Safety Agencies may use ShakeCast to help plan the deployment of emergency resources. Civil Engineers and Property Mangers will use ShakeCast to prioritize the dispatch of inspection and repair personnel to their properties and individuals will use ShakeCast to ascertain the likelihood of damage or possible injury to loved ones, homes, or property.

ShakeCast requires the client's software to periodically "poll" one of several redundant ShakeCast servers. In such poll-and-request interactions, a ShakeCast client periodically checks for new data on a server. When the server has no new ShakeMap data available, the client waits and tries again. If the server has new data, the client requests the products as preconfigured and, optionally, begins post-processing software, loss estimation, and alarming.

The ShakeCast System uses Extensible Markup Language (XML) to communicate information between ShakeCast servers. This information includes:

o Data about ShakeCast Servers and the ShakeCast software itself
o Data about events (earthquakes) and products (data files) available on the network
o Status information that helps the administrators of ShakeCast systems tell if their network is running smoothly

ShakeCast servers store much of the data used by the system in a relational database. This database contains information needed to interact with other ShakeCast Servers, data that will be presented to users, configuration information needed to perform notifications, and various other kinds of information. On the client side, the user can define a list of facilities, where a "facility" is a specific structure (e.g., bridge, school, pumping station, etc.) at a specific location. The location may be defined by a latitude/longitude for "point" facilities, or by a bounding box for non-point facilities. Upon downloading the ShakeMap data, the list is processed automatically to display (and optionally alarm) a summary list in order of either the shaking level or the likelihood of damage. In this model, having the facility data under the control of the user is fundamental since these data tend to change frequently, some data is proprietary, and keeping track of multiple users' inventories would be cumbersome on the ShakeMap production side. These data are best kept within the individual user's ShakeCast client configuration.

Use of ShakeMap and ShakeCast Within Caltrans

Following a major earthquake, Caltrans faces an array of decision-making challenges. Perhaps no other agency has a comparable earthquake exposure in the State of California. Caltrans recognizes that ShakeMap, in conjunction with ShakeCast, will be extremely beneficial in the decision-making process and Caltrans is currently working with existing ShakeMap technologies and earthquake responders within the Department to more effectively respond to earthquakes. In addition, Caltrans is serving as the implementation test bed for the ShakeCast prototype.

One of several critical tasks facing Caltrans after an earthquake is to rapidly assess the condition of all bridges and roadway corridors in the State highway system. Timely response is important to ensure public safety, aid routing of emergency vehicle traffic, and (re-) establish critical lifeline routes.

Figure 2. Caltrans Oakland, California, Traffic Management Center (TMC). Photo courtesy of Kane Wong, Caltrans.

The primary method for bridge and roadway damage and functionality assessments is a thorough onsite inspection by trained personnel from Caltrans' maintenance and structures design units. However, procedures used in the past for establishing inspection priorities were relatively unfocused due to lack of precise information about the distribution of damaging levels of shaking. In the absence of such information, the practice had been to use the epicenter location (available from the REDI/CUBE system immediately after the earthquake), find the closest mapped fault, and develop a list of bridges within a specified buffer zone surrounding that fault. Maintenance crews were dispersed widely within that region to perform initial reconnaissance.

The problem with epicenter-based or whole-fault based buffer zones is that earthquake shaking levels vary dramatically within the buffer zone. Even if the epicenter is aligned with a mapped fault, an earthquake rarely ruptures over the entire mapped length. Furthermore, ground shaking at the same distance from a rupture zone varies by nearly a factor of 10 due to a variety of seismological and geotechnical effects including fault rupture directivity, deep basin effects, and local site response. Buffer zones large enough to account for all areas that *could* be strongly shaken will include wide swaths with no damage, thus diverting inspection resources away from critical needs.

Given the scale of Caltrans inventory, a focused post-earthquake response is essential. There are nearly 24,000 bridges and overpasses under Caltrans authority. Caltrans also has 14 Traffic Management Centers (TMC), which act as nerve centers for this critical lifeline in California (Fig. 2). Caltrans recognizes that a more effective method for prioritizing earthquake response is possible using ShakeMaps and GIS data and technologies already in place.

For initial deployment, a preliminary rating scale was devised to simply categorize bridges based upon the year of construction, which resulted in a crude translation of susceptibility to damage from ground shaking. This proposed rating scale, combined with ShakeMap shaking intensity, was used to create a map of the impacted zone. Red regions denote "*severe damage possible (all bridges)*," yellow regions indicate "*severe damage possible (all bridges constructed before 1975)*," green regions correspond to "*some damage possible*," and clear regions indicate "*damage unlikely.*" A map using this rating scale for the 1994 Northridge earthquake is shown in Figure 3. Generally, these zones correspond reasonably well with observed bridge damage from that event

A simple GIS routine was also developed that allows the map and bridge data to be analyzed and quickly converted into a priority list, several entries of which are in Table 2. This table lists bridges within the affected region sorted by damage potential, route, and postmile to aid in the dispatch of inspection personnel. Other parameters on the list include geodetic coordinates and ShakeMap values of shaking intensity at that location. The GIS tools developed for this application are now functional and are currently being tested by clients in Caltrans structures and maintenance divisions. Additional functions have been requested within Caltrans including enhanced automation, internet dissemination, electronic field data collection and exchange for reconnaissance. Additional applied research using data developed through the bridge retrofit program is underway to refine the preliminary bridge rating scale and thereby yield a more accurate and focused inspection prioritization scheme. Further functional enhancements are also needed to make it easier to use with less GIS training, and more readily accessible using internet, and possibly wireless technologies. It is anticipated that ShakeCast combined with research and development within Caltrans will result in a greatly enhanced capacity to respond to the next earthquake disaster in California.

We anticipate that this prototype system will lead to similar uses with other lifeline systems.

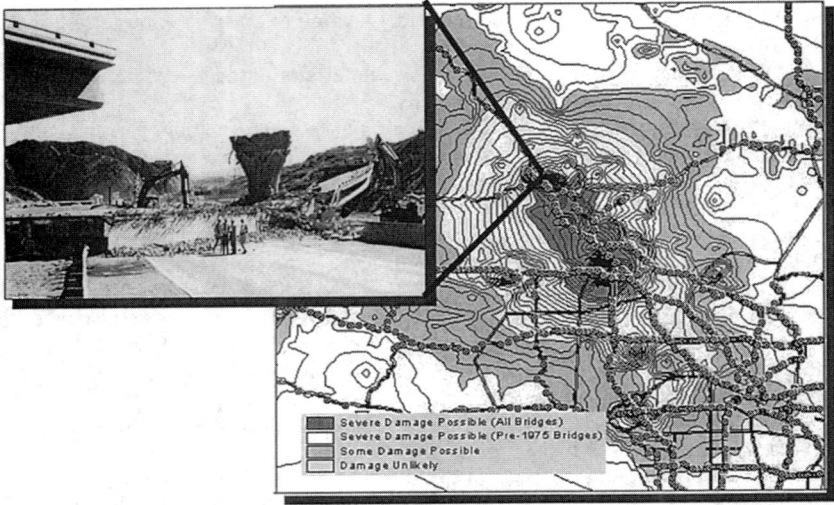

Figure 3. Caltrans bridge potential damage assessment using ShakeMap data (1.0 sec spectral acceleration) for the Northridge earthquake. Actual damage shown is from the Northridge earthquake at the I5-SR14 Interchange.

Bridge Name	PM	Lat.	Long	Dist	Rte	SA (1s.)	Potential Damage Status
Aliso Cr Culvt	R6.38	34.268	−118.518	7	118	1.04	Severe Dam. Possible (All Bridges)
Chimineas Av OC	R6.23	34.268	−118.517	7	118	1.04	Severe Dam. Possible (All Bridges)
Etiwanda Ave OC	R6.03	34.270	−118.527	7	118	1.00	Severe Dam. Possible (All Bridges)
Reseda Blvd OC	R5.81	34.277	−118.530	7	118	1.00	Severe Dam. Possible (All Bridges)
Wilbur Ave OC	R5.20	34.275	−118.545	7	118	0.96	Severe Dam. Possible (All Bridges)
Tampa Ave OC	R4.64	34.273	−118.547	7	118	0.88	Severe Dam. Poss. (Pre-'75 Bridges)
Limekiln Cany WA	R4.60	34.267	−118.548	7	118	0.88	Severe Dam. Poss. (Pre-'75 Bridges)

Table 2. Portion of priority list based upon GIS analysis.

References

Wald, D. J., V. Quitoriano, T. Heaton, and H. Kanamori (1999a). "Relationships between Peak Ground Acceleration, Peak Ground Velocity and Modified Mercalli Intensity in California" *Earthquake Spectra*, **15**, 557-564.

Wald, D. J., V. Quitoriano T. Heaton, H. Kanamori, C. W. Scrivner, and C. B. Worden (1999b). "TriNet 'ShakeMaps': Rapid Generation of Instrumental Ground Motion and Intensity Maps for Earthquakes in Southern California" *Earthquake Spectra*, **15**, 537-556.

Seismic Risk Assessment of Transportation System: Evaluation Immediately After Earthquake

Chin-Hsiung Loh[1], Chun-Yu Lee[2], Chin-Hsun Yeh[3]

Abstract

The seismic reliability assessment of transportation network requires consideration of several items including: (a) evaluation of regional seismicity, (b) microzonation with respect to local geologic and soil conditions at each site, (c) structural response and system analysis, (d) network analysis. The objective of this research is to perform the evaluation of transportation system immediately after earthquake hazard using limited number of recorded free-filed ground motion data. In this study two types of earthquake hazard are examined for highway transportation system: ground motion, and liquefaction.

Introduction

As modern urban functions rely on lifeline systems, and earthquake protection of lifeline systems is now a critical issue for the enhancement of seismic reliability of urban region. Particularly, highway network system plays a vital role in post-earthquake emergency, recover, and reconstruction stage, as well as in a normal situation. Generally, transportation system is a network system that is composed by three parts, which are road, railway, and bridge. To secure the functionality of highway networks, two types of countermeasures are considered, that is, the pre-earthquake structural measures such as retrofitting and the post-earthquake nonstructural measures such as traffic control. Several approaches to examining the reliability and serviceability of systems subjected to an earthquake have been developed. Sato and Der Kiureghian [1982] studied the seismic hazard analysis of lifeline incorporating soil and geological effects. Besides the ground motion seismic reliability analysis of lifeline system, the lifeline system analysis should also include methods for computing lifeline system as a network of interconnected nodes and links. Studies such as those of Taleb-Agha [1977], Shinozuka, Koike and Kameda [1988] developed a reliability method for system connectivity recognizing the hierarchical feature of lifeline system.

---

[1] Professor, Department of Civil Engineering, National Taiwan University, Taipei, Taiwan. E-mail: lohc@ncree.gov.tw

[2, 3] Assistant Engineer and Associate Research Fellow, Respectively National Center for Research on Earthquake Engineering, Taipei, Taiwan.

Various components of the transportation risk analysis methodology have been developed previously. For example, Hobeika and Ardekani [1989] developed a transportation decision support system to provide an efficient update of the physical and operational condition of the highway network, and traffic management strategies immediately after an earthquake. Sato and Toki [1994] studied the seismic reliability of lifelines with some variations with respect to seismic hazard and network analysis models. Hwang et al. [2000] presented a procedure for the evaluation of the expected seismic damage to bridges and highway system. In this study, susceptibility of transportation system for the seismic demand including liquefaction and ground motion is examined. Incorporate with the fragility curves for urban road and bridges the damage probability of transportation system is estimated. Finally, a scenario earthquake was studied to examine the reliability of the transportation network immediately after the earthquake. A flow chart of this study is shown in *Fig 1*.

Risk Analysis of Transportation Network with Mutually Dependent Tie-Sets

The main components of a risk analysis methodology include (a) hazard exposure, (b) highway component and network system configuration, (c) consequence analysis. A large scale transportation system is often hierarchical functionally and physically, consisting of high-traffic flow transmission and low-traffic flow network systems. Besides the traffic flow any network can be transformed to series system in parallel (SSP) and each series consist of links and nodes. Consider a mesh for a highway transportation system, as shown in *Fig. 2*, with "n" nodes and "m" connections. The topology of the mesh is summarized by the connection matrix **M** (a symmetric matrix) where rows and columns correspond to nodes, and in the location M_{ij} corresponding the link "k" connecting nodes i and j. The "node" indicated the intersection of two highway transportation lines. As mentioned above, the distribution of each link of the urban road network is "line type". Each link may cross several grids with different seismic demand (PGA, Sa, PGD) on contour map and bridges may also located along this link. Different procedures for developing SSP network, through "tree diagram" Taleb-Agha (1975), can also be used for evaluating the system failure probability.

To perform the failure analysis of transportation system, the physical network system is transformed into its corresponding series system in parallel (SSP) by adopting the connection matrix. Between each input and output a tie-set can be identified. Each tie-set consists of not only links connected in a series but also nodes at both ends of each link. In each link bridge may located in it. For the analysis of a specific bridge the fragility curves of a bridge system must be developed. It is defined that $F_N(\cdot / EQ)$ = the probability value obtained from the fragility curve. It is assumed three types of failure modes are defined: Failure (F: Severe damage), M (moderate damage), and S (slight damage). The probability of damage of a bridge subjected to an earthquake is defined as:

Severe Damage: $P[N_i(F)/EQ] = F_N(F/EQ)$
Moderate damage: $P[N_i(M)/EQ] = F_N(M/EQ) - F_N(F/EQ)$ (1)
Slight damage: $P[N_i(S)/EQ] = 1 - P[N_i(F)/EQ] - P[N_i(N_i(M)/EQ)]$

The probability of failure of a tie-set can be expressed as follows:

$$P[T_k(F)/EQ] = 1 - \prod_{i=1}^{TK_K}\{1 - P[L_i(F)/EQ]\} \cdot \prod_{i=1}^{TN_K}\{1 - P[N_i(F)/EQ]\} \quad (2)$$

With the aid of the occurrence rate for node failure described in the preceding section, the conditional of major, moderate, and minor damage of the j-th sub-link in the local network given an earthquake can be written as:

$$P[L_j(F)/EQ] = P[SL_j(F)/EQ[exp\{-k_j l_j\}] \quad (3)$$

where $SL_j(F)$ is the event that the j-th sub-link (urban road) is under the state of major damage. l_j is denotes the average distance of a sub-link in on the major link, and k_j denotes damage rate representing expected number of failure per unit length. Because the actual damage state is inevitably unknown before the occurrence of earthquake disaster, a large number of random damage patterns can be generated from previous disaster experiences.

Consider a highway transportation line, for example, as shown in *Fig.3*, to evaluate the probability for survival (P_{sl}) of a tie-set AG it is assumed that there are seven links in this tie-set and five nodes are defined between input and output nodes A and G. It also assumed that node D and node E are bridge structures. Six sub-links are assumed based on the distribution of seismic demand. The probability of survival along the specified link AG can be expressed as:

$$P_{s,AG} = (1 - P_{f,AD})(1 - P_{f,D})(1 - P_{f,DF})(1 - P_{f,F})(1 - P_{FG}) \quad (4)$$

where $P_{s,AD}$ is the survival probability for sub-link from A to D and $P_{f,AD}$ is the failure probability for link from A to D. The failure probability of segment AD is expressed as:

$$P_{f,AD} = \frac{P_{f,AB} \times l_{AB} + P_{f,BC} \times l_{BC} + P_{f,CD} \times l_{CD}}{l_{AD}} \quad (5)$$

l_{ij} is the length between two point i and j. Combining all the sub-links to become one link (or tie- set), the probability of survival, P_{sl}, of each link can be evaluated as:

$$P_{sl} = \coprod_{i=1}^{n}(1 - P_{f,i}) \quad (6)$$

where $P_{f,i}$ is the failure probability of each sub-link. For a network system it is possible to develop the series system in parallel (SSP network). In a SSP network, each path containing several links and is treated as a tie-set.

Spatial Distribution of Earthquake Ground Motion

Transportation system usually is spread over vast areas with varying geologic and soil conditions. To check the earthquake performance of a designed structural component or any node at a transportation system, it is necessary to estimate the seismic demand at any route of the transportation system. Assessment the potential earth science hazard at any site or any node in the transportation system, including ground motion and ground failure, is a key issue in seismic risk analysis of transportation network.

For ground motion estimation from attenuation model the framework of HAZUS-99 can provide the methodology (NIBS 1999). With modification, a HAZ-Taiwan risk assessment methodology developed by National Science Council of Taiwan, can also be used for ground motion estimation [Loh et al. 2000]. Based on the free field strong motion data collected by the Seismology Center of the Central Weather Bureau from the past earthquake, the scenario earthquake induced intensity measure (such as PGA or spectral intensity) around the island is calculated. The empirical attenuation form can be expressed as:

$$IM = Y_r = f(M,R) = b_1 e^{b_2 M} [R + b_4 \exp(b_5 M)]^{-b_3} \qquad (8)$$

where *IM* is the intensity measure of seismic demand (it can be expressed as peak ground acceleration, or spectral acceleration at any specified period); M and R are the earthquake magnitude and the site-to-source distance, respectively; b_1 through b_5 are constants given earthquake magnitude, focal depth, epicenter and/or surface fault rupture. The seismic demands in terms of peak ground acceleration and response spectrum are calculated for user-specified scenario earthquake. ***Table 1*** shows the estimated model parameters from Taiwan earthquake data. Normally the attenuation relationships of ground motion parameters predict the intensity in rock site. For other site conditions, the ground motion intensity parameters should be modified by some factors, which depend on the local soil condition and intensity level. This modification factor for different site condition can be developed in advance by using ground motion data collected from the dense strong motion array (Taiwan strong motion instrumentation program) with same soil conditions. For each specific site the revised intensity measure Y_s can be expressed as follows (Chang et al. 2002):

$$Ln(Y_S) = C_0 + C_1 Ln(Y_r) \qquad (9)$$

where C_0 and C_1 are coefficients of the regression line. ***Fig.4*** shows the flow chart of the ground motion estimation procedures. The real-time free-field strong motion data (from Taiwan rapid Information Release System, TRIRS) may also be used to

Table 1: Model parameters of attenuation model shown in Eq.(1)

Case	b_1	b_2	B_3	b_4	b_5	$\sigma_{\ln(Err)}$
PGA	0.0036944	1.7537666	2.0564446	0.1221955	0.7831508	0.68-0.75
S_{as}	0.0097360	1.7348416	2.0857212	0.1136533	0.8003162	0.67-0.75
S_{al}	0.0027914	1.7730463	2.0419005	0.1154175	0.7713924	0.85

upgrade the ground motion estimation. To be more realistic, the scenario earthquake should be selected based on probabilistic analysis in order to identify the important source areas that have large contribution factor to each specific site for a prescribed probability.

Vulnerability of Highway Transportation System

The highway transportation system consists of roadways and bridges (neglect the tunnels in this study). Roads located on soft soil or which cross a surface rupture can experience failure resulting in loss of functionality. Bridges may fail usually result in significant disruption to the transportation network. Damage functions or fragility curves for all highway system components mentioned above (bridge and road) must be developed first. Generally, the fragility curves are modeled as lognormal distributed functions that give the probability of reaching or exceeding different damage states for a given level of ground motion or ground failure.

Bridge Damage Caused by Ground Shaking Bridge structure classification scheme incorporates various parameters that affect damage in fragility analysis. A total of sixteen bridge classifications were selected. To define the damage states a total of five damage states are defined for highway system components. These are none (ds1), slight/minor (ds2), moderate (ds3), extensive (ds4) and complete (ds5). Medians of these damage functions, in terms of Sa(T=1.0 sec) were given (T-Y. Lin 2002). The fragility curves are modeled as lognormal-distributed functions that give the probability of reaching or exceeding different damage states for a given level of ground motion or ground failure. ***Fig. 5*** shows the fragility curves for TYPE-7 highway bridges considering the structure with seismic design. Based on this study the damage estimation algorithm for bridges can be summarized into several steps: (1) Identify bridge location and its classification, (2) Based on the estimation of spatial distribution of ground motion, evaluate the soil-amplified shaking at the bridge site (including PGA, Sa[0.3 sec] and Sa[1.0 sec]), (3) Evaluate the bridge system modification factor (including skew effect, shape and 3-dimensional effect), (4) Modify the "standard" fragility curves if necessary, (5) Use revised fragility curves to evaluate the ground shaking-related damage probability (Sa[1.0 sec] value was used as intensity measure in this study).

Bridge Damage Caused by Liquefaction Induced Settlement Both severe ground shaking and permanent ground deformation can cause damages to highway transportation systems. To evaluate damage state probabilities of various components in the highway transportation system due to permanent ground deformation, the effects of soil liquefaction are studied. In order to estimate soil liquefaction potential and the induced amount of permanent ground deformation, the soil types are classified into six liquefaction susceptibility categories, i.e., very high, high, moderate, low, very low and none (Yeh et al 2002a). The liquefaction susceptibility category of the site can be determined by borehole data. Earthquake magnitude, peak ground acceleration and ground water depth are factors in estimating liquefaction probability and the induced settlement as briefly explained below. Based on the idea proposed by Ishihara (1993) the maximum shear strain and volumetric strain due to consolidation following liquefaction and the relationship of maximum shear strain and safety factor to resist liquefaction were developed. With constant earthquake magnitude (7.5) and ground water depth (1.5 meters), *Fig. 6* shows the relationships between the liquefaction induced settlement and the PGA for different susceptibility categories (Yeh et al 2002b). It can be seen that the amount of settlement approaches a limiting value, which depends only on the liquefaction susceptibility categories. A lognormal distribution function is assumed between the relationship of settlement versus PGA, and only two parameters, i.e., median and deviation, are required to describe the relationship. Thus, the relationship between liquefaction settlement (S) and PGA (A) can be expressed as

$$S = \bar{S}_i \cdot \int_0^A \frac{1}{\sqrt{2\pi}\sigma_i x} e^{-\frac{[\ln(x/m_i)]^2}{\sigma_i^2}} dx = \bar{S}_i \cdot \Phi\left[\frac{\ln(A/m_i)}{\sigma_i}\right] \qquad (9)$$

where $\bar{S}_i$ is the limiting value for susceptibility category i; $\Phi(\cdot)$ is the standard normal distribution function; and m_i and σ_i are the median and deviation of the lognormal distribution function, respectively. It is noted that m_i and σ_i are functions of the earthquake magnitude and the ground water depth. Incorporate with the fragility curves for major road and urban road, as shown in *Fig. 7*, the damage probability of transportation system due to liquefaction induced settlement can be estimated.

Example

In this study a highway transportation system of southern Taiwan is selected for the analysis of network reliability. *Fig. 8* shows the transportation system of the study area. The transportation system includes: highway (90 km/hr), expressway (70 km/hr), provincial road (50km/hr) and county road (40km/hr). In the beginning, transform the transportation system to network system, as shown in *Fig.9*. It is assumed that "Node 2" is the input node and "Node 0" is the output node. Through SSP network analysis over thousands of tie-set were identified (a total of 35420 tie-sets between input and output). To extract the reasonable tie-set from the SSP network for reliability analysis the following two weighing schemes were adapted: (1) The shortest travel time along a tie-set route, (2) The ratio of highway road length to the total tie-set route length. Larger ratio of highway road length to the total tie-set length will have a higher rank. The identified top five tie-sets of the transportation system with higher rank between input and output were selected. Through scenario earthquake risk analysis along each tie-set was conducted to study the reliability of these five tie-sets.

It is assumed that an earthquake with magnitude 7.6 and depth 10 km occurred at Chu-Kuo fault near the highway transportation system. Ground motion intensity measures across the network system can be estimated from the ground motion attenuation model. To estimate the damage probability along the tie-set the fragility curves of bridge structure (in terms of Sa-value) and urban road (in terms of the permanent ground deformation caused by the liquefaction induced settlement, as shown in *Figs. 8 & 9*) are used. *Table 2* shows the estimated probability of survival in each sub-link of the identified first rank tie-set subjected to different earthquake magnitude excitation. The probability of survival for the specified tie-set (or path) can be determined using equation (6).

Discussion and Conclusions

The present study showed how a seismic reliability analysis of transportation system can actually be conducted under a scenario earthquake. Techniques for micro-zonation and simulation of spatially distribution earthquake motions have been developed. The method proposed herein requires simple information on network properties to determine important ranking: (1) network configuration of network systems, (2) a set of component reliability of all links and nodes. Through network analysis probabilistic evaluation of potential earth science hazard of the transportation system was examined. Using the fragility functions, the damage states for each bridge and urban road segment was estimated. The method proposed herein enables one to find valuable Original-Destination pairs and to understand systematic relations between physical performance of transportation facilities and functional performance of the transportation network. Through this study the following discussions are drawn:

1. In this study for each tie-set the seismic demand and the fragility curves of urban road and bridge are considered. The length of each sub-tie-set (or link) from one tie-set (or urban road) was selected to match with the same intensity measure. It means that each sub-tie-set may have the same seismic demand without considering the length effect. The vulnerability of urban road may higher if the length of the road is longer. It is necessary to consider the effect of urban road length on the estimation of link reliability in this study. The link reliability can be computed as $p_k = e^{-\lambda_k d_k}$, where λ_k denotes damage rate representing expected number of failure per unit length, and d_k denotes the link length.

2. The role of each link as a network component had not been examined using traffic flows assigned to selected links. Only the travel time for different grade of urban road is considered. As the network performance deteriorates, correlation among traffic volume at different route becomes weak.

References

Hwang, H., Jernigan, J.B., Lin, Y.W. (2000), "Evaluation of Seismic Damage to Memphis Bridges and Highway Systems," J. of Bridge Engineering, ASCE, Vol.3, No. 4, 322-330.

Ishikawa, Y. and Kameda, H. (1991) "Probability-Based Determination of Specific Scenario Earthquakes," *Proceedings of the 4th International Conference on Seismic Zonation*, Vol.II , Stanford, CA., August 25-29, 3-10.

Loh, C. H., S. R. Lawson and W. M. Dong (2000), "Development of A National Earthquake Risk Assessment Model for Taiwan," Proceedings of 12 WCEE, paper No. 380, Auckland, New Zealand.

National Institute of Building Sciences (1999), "Earthquake Loss Estimation Methodology – HAZUS97 Technical Mannual," *National Institute of building Sciences*, Washington, D.C.

Sato, T. and Der Kiureghian, A. (1982), "Seismic Hazard Analysis of Lifeline Incorporating Soil and Geologic Effects," Proceedings of 3rd Microzonation Conf., Seattle, July.

Sato, T., Toki, T., Sekiya, T. (1988) "Critical Components for Upgrading Seismic Reliability of Large Lifeline Networks," *Proceedings of 9WCEE*, Vol.II, 135-140.

Shinozuka, M., Koike, T. and Kameda, H., "Seismic Reliability of Hierarchical Lifeline Systems," Proceedings of ASCE Symposium on "Seismic Design and Construction of Complex Civil Engineering Systems," St. Louis, Missouri, October, 1988, 27-29.

Taleb-Agha, Ghiah,(1977). "Seismic risk analysis of life line network," *Bulletin of the Seismological Society of America*, Vol.67, No.6, pp.1625-1645.

Yeh, C. H., Hsieh, M. Y., and Loh, C. H. (2002a), "Classification and Parametric Study on Soil Liquefaction Potential", Proceedings of the Second Japan-Taiwan Workshop on Lifeline Performance and Disaster Mitigation, Kobe, Japan, May 13-15.

Yeh, C. H., Hsieh, M. Y., and Loh, C. H. (2002b), "Estimations of Soil Liquefaction Potential and Settlement in Scenario Earthquakes", Proceedings of the Canada-Taiwan National Hazards Mitigation Workshop, July 17-19, Ottawa, Canada.

T. Y. Lin International Consultant (Taiwan) (2002), "Seismic Analysis, Assessment, and Retrofit of Highway Bridges in Taiwan Freeway No.1," Technical Report (in Chinese), June.

Alessandro Baratta and Giulio Zuccaro, "Modeling and Seismic Risk Analysis of Networks," Proceedings of ASCE Lifeline Earthquake Engineering Conference, 1995.

Table 2: Estimated probability of survival for each link with different magnitude.

LinkID	M=5.5	M=6.0	M=6.5	M=7.6
201	1	1	0.64	0
206	1	1	0.89	0.23
212	1	0.94	0.42	0
221	0.99	0.37	0.01	0
227	0.93	0.36	0.03	0
239	1	0.99	0.89	0.45
251	1	1	1	1
261	1	1	1	1
269	1	1	1	1
273	1	1	1	1

DISASTER RESPONSE FOR LIFELINE SYSTEMS   131

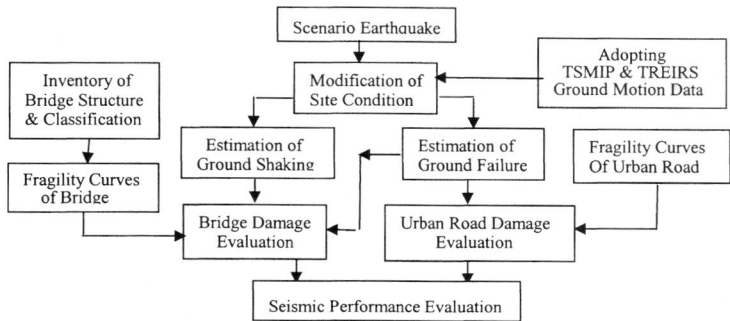

Figure 1: A flowchart to describe the modules for seismic performance of transportation system.

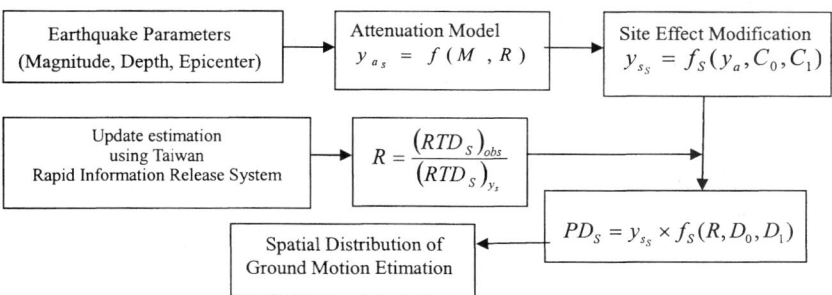

Figure 3: Procedures for the estimation of spatial distribution of ground motion.

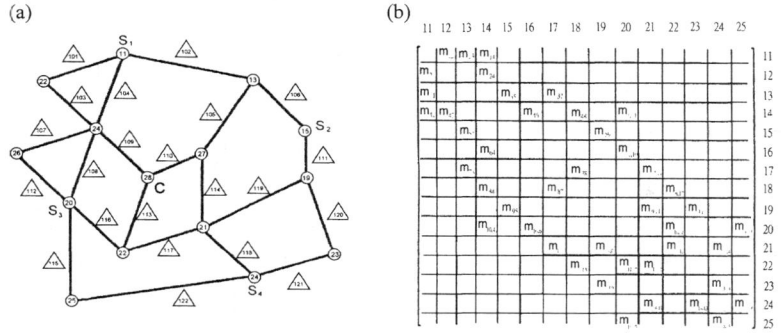

Figure 2: A network example and its corresponding connection matrix for network analysis. (a) S1 denotes input and S4 denotes output, (a) the column and row of the matrix indicate node number (Baratta & Zuccaro, 1995).

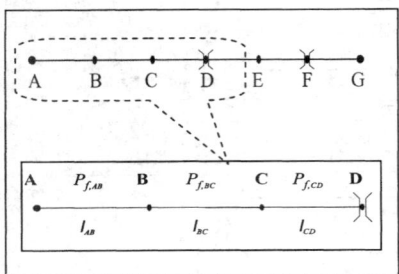

Figure 4: A schematic diagram shows the implementation of Eqs.(4) and (5).

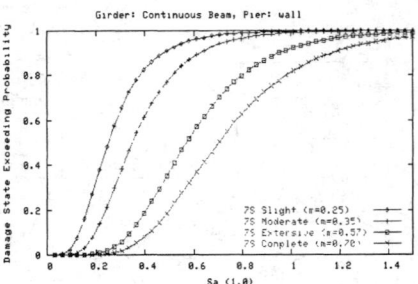

Figure 5: Fragility curves for bridge TYPE-7 : case of seismic design.

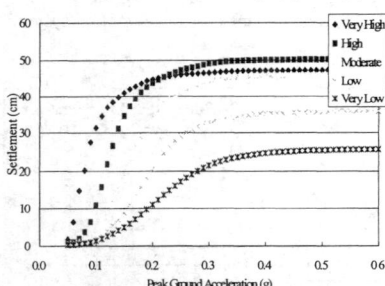

Figure 6: liquefaction settlement for different susceptibility categories.

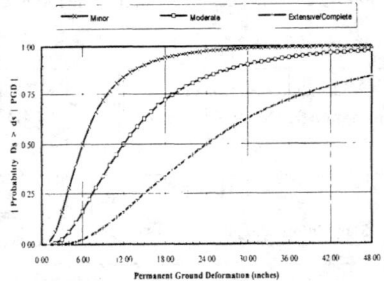

Figure 7: Fragility curves for urban road (HAZUS99).

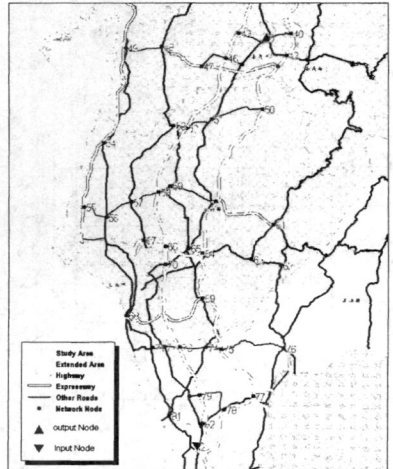

Figure 8: Major traffic network in the southern part of Taiwan as a study region.

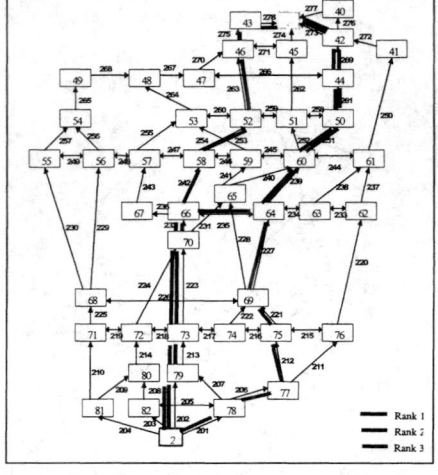

Figure 9: Transform the network system shown in figure 8 into node and link system.

Development of a Handbook for Seismic Performance Testing of Bridge Piers

J. Jerry Shen[1], W. Phillip Yen[2], and John O'Fallon[3]

Abstract

As the bridge design criteria gradually move towards the performance-based approach, there is an increasing need of reliable and comparable results from seismic performance tests. The Federal Highway Administration (FHWA) has conducted a study on the techniques and procedures of experimental bridge column tests in order to develop a document that can assist the experimentalists to conduct seismic performance tests that produce consistent, cost-efficient, and accessible results. The experiences of the major earthquake laboratories such as those in the national earthquake engineering research centers, which are capable of conducting various types of bridge tests, serve as the basic samples for studying typical and advanced practice of tests. The majority of the experiment issues including specimen constructions, loading procedures, and documentation are studied and summarized. An expert panel has been formed to advise the development of a handbook for Bridge pier testing. A task force of a smaller group is continuously searching and processing information on test methods to develop recommendations for general test procedures. This paper presents the progress of this project and gives a brief review on the first draft of the handbook.

Objective

This study conducted by Federal Highway Administration (FHWA) was targeted on the techniques and procedures of conducting experimental bridge tests. Information of testing methods collected in the past few years are summarized and a handbook for seismic performance testing on bridges is being produced. In production of this handbook, unnecessary variations of test procedures are synthesized. Important issues that dominate test quality are emphasized. Balanced coverage on academic research and practical engineering study will be maintained. The format of the handbook is designed to encourage rigorous and creative planning for experiments based on adequate rationale, rather than to regulate or forcefully unify entire test procedure.

Because of the absence of commonly accepted testing criteria in the past, a significant amount of time resource may have been spent on designing and verifying testing procedures while the results of tests are often recorded and written down in the format for only single use. After the single purpose is served, the data and most critical test condition become virtually useless and are often deserted. This phenomenon jeopardizes the credibility of the tests because others cannot reexamine details of the tests in a later

---

[1] LENDIS/FHWA, 6300 Georgetown Pike, McLean, VA 22101, Tel:202-493-3098; E-mail: Jerry.Shen@fhwa.dot.gov (through LENDIS Corporation)
[2] Office of Infrastructure, R&D Federal Highway Administration, 6300 Georgetown Pike, McLean, VA 22101, Tel:202-493-3056; E-mail: Wen-huei.Yen@fhwa.dot.gov
[3] Office of Infrastructure, R&D Federal Highway Administration, 6300 Georgetown Pike, McLean, VA 22101, Tel:202-493-3051; E-mail: John.O'Fallon@fhwa.dot.gov

time. The adequacy and accuracy of the test procedure cannot be verified. In the meantime, due to the absence of some critical measurements, analytical or empirical model developers cannot use these test results obtained by various experimental researchers. This greatly limits the available amount of verification on theories. These problems will be addressed in the handbook and resolutions will be provided.

An expert panel was formed to advise the development of this handbook. Members include academic researchers, laboratory experts, and state highway engineers. The diversified member composition of this panel ensures the quality and practicality of this handbook.

Scope and format

Bridge pier tests can reveal unknown properties of pier elements or verify theoretical anticipations. In order to provide assistance to both practical engineering tests and academic researches, this handbook maintains a balance between clearly specified details, which provides quick reference on standardized performance verification, and freedom in selecting and creating alternative test procedures, which allows unconventional test setup for explorative experiment.

Illustrations and examples are given in each of the three stages of experimental researches set forth in this handbook. Sensible combination can be selected to address the requirement of each project. The tested subject is presumed to be made of steel and/or concrete. Innovative material types (such as fiber reinforced polymer) are not excluded. Due to the unknown pattern of material behavior, less specific principles of testing are given for pier specimen containing material beyond steel and concrete.

Categories

A completely realistic seismic performance test on a full-scale true replica of bridge is not always practical due to the high cost and technical difficulties. There are various simplifications in each test, depending on the limitation of the research project and facility. Types of bridge pier tests are presented in Table 1. The types of tests listed in each block are commonly executed with the associated speed and loading options. Deliberate consideration on test plan should be made if tests need to be done using options beyond the table. Due to the difficulties in adequate scaling and test result interpretation, fast displacement-prescribed tests (category B) are not recommended unless found necessary for academic reasons.

Complete test procedures are not provided in the handbook. Instead, sensible methods for each stage of tests are illustrated. Users can select adequate pieces of procedures and construct a test plan that meets all requirements of a testing project.

	Displacement prescribed	Inertia-dependent loading	
		Distributed load	Point load
Slow	**(A)** Monotonic loading, cyclic loading	N/A	**(D)** Pseudodynamic
Fast	**(B)** Monotonic loading, cyclic loading. Not recommended for non-academic testing	**(C)** Shaking table tests	**(E)** Real-time pseudodynamic tests, hybrid tests

Table 1 Testing options and common names

Specimen

The general specimen design procedure is illustrated in Fig. 1 and Fig. 2. A prototype is designed to fulfill the research requirements. A scaled model is produced, if necessary, in accordance with the facility and funding limitations.

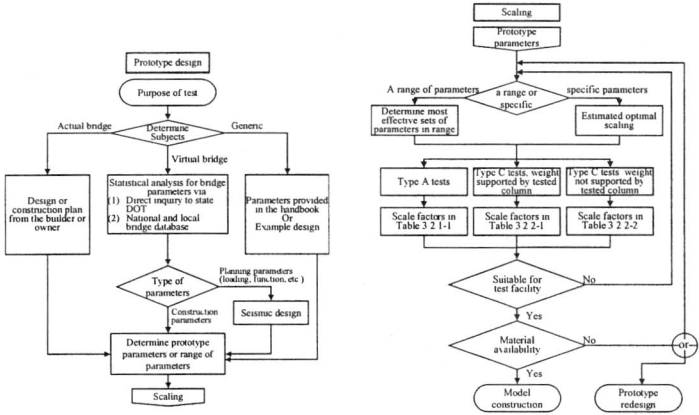

Fig. 1 Prototype design   Fig. 2 Scaling

There are three common sources for the design parameters of a testing prototype:
1. Actual bridge: The subject is a real bridge targeted by the project. The design parameters are unique and can be obtained from designer or from physical inspection of the bridges.
2. Generic bridge: A designated bridge design deemed most representative in a wide domain.
3. Virtual bridge: A nonexistent bridge specially designed for the research. The design can be an average or a maximum/minimum of a group of bridges selected by category, age, or geographic location, etc. These parameters can be identified through statistical investigations (Lowes and Moehle, 1995, Abo-Shadi et al, 2000). A

common approach to obtain the input data for the statistical analysis is through inquiries to state departments of transportation. They can be also retrieved from local or nationwide database (Lampe and Azizinamini, 2000).

The prototype determination is fundamentally based on the research objectives. However, some modification may be necessary if technical difficulties are found in the scaling process.

A dimensional analysis is used to find the proper scaling of each parameter (Moncarz, 1981; Bertero, 1984). Common scaling factors in seismic performance testing are provided in the handbook (Table 2). In the quasi-static bridge column tests (category A), all time-dependent parameters, i.e. those relate to velocity and acceleration, are ignored. The axial load and lateral load from the superstructure are reduced to two external forces applied vertically and horizontally to one point (normally at the top) on the column. The material properties, such as elastic modulus, yielding stress, and specific weight, are kept to the same as the prototype (scale factors = 1). The parameter that needs to be changed freely is the geometrical dimensions. The resultant scaling factors for quasi-static tests are listed in the 1st column of Table 2.

Time and time-dependent variables need to be considered in the scaling for inertia-activated tests (category C, D, and E) such as shaking table tests. The dynamic scaling system is shown in the 2nd column of Table 2.

Variable	Quasi-static	Dynamic
Geometric size l	S_L	S_L
Time t	N/A	$S_L^{0.5}$
Stress σ	1	1
Strain ε	1	1
Elastic modulus E	1	1
Density ρ	N/A	1
Force P	S_L^2	S_L^2
Bending moment M	S_L^3	S_L^3
Displacement U	S_L	S_L
Acceleration	N/A	1
Mass	N/A	S_L^2
Weight	N/A	S_L^2

Table 2 Scaling factors for different tests

In some dynamic tests, the axial load of the column is provided or augmented by a separated mechanism, such as hydraulic jacks (Fig. 3). In this case, the scale of horizontal acceleration can be different from that of gravitational acceleration. The investigator has the freedom of selecting the most convenient scale factor for earthquake acceleration. This additional freedom is reflected in the scale factors in Table 3.

Density is not scaled in all scaling system. The self-weight of the specimen is proportional to S_L^3 rather than S_L^2 or $S_L^2 S_a^{-1}$ listed as the mass scale. This effect is negligible when the mass (and weight) of the substructure is much smaller than that of the superstructure. For the subject with massive substructure, additional mass attachment on the substructure is needed to reduce the distortion.

Scaled material properties are difficult to control. It is intended in this handbook to keep them unscaled so that original construction material can be used. However, keeping material properties unchanged in scaled model is still a difficult, if not impossible, task. Detailed documentation on the material tests is encouraged in this handbook to allow distortion error analysis after the experiments.

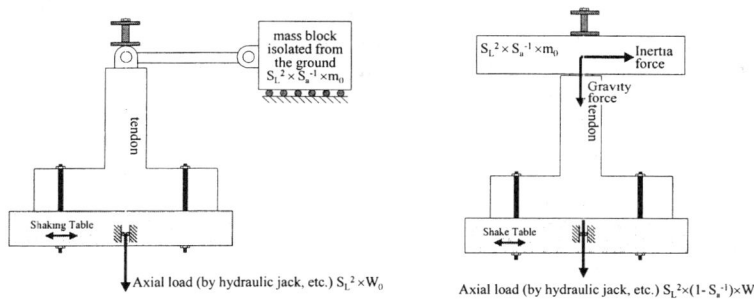

Fig. 3 Independent earthquake acceleration and gravitational acceleration

Variable	Dynamic (alternative)
Geometric size l	S_L
Time t	$S_L^{0.5} S_a^{-0.5}$
Stress σ	1
Strain ε	1
Elastic modulus E	1
Density ρ	1
Force P	S_L^2
Bending moment M	S_L^3
Displacement U	S_L
Acceleration	S_a
Mass	$S_L^2 S_a^{-1}$
Weight	S_L^2

Table 3 Scale factors for vertically supported mass

Loading Program

Category A: Monotonic loading program can provide useful information on the inelastic behavior that can assist conducting following tests. It is recommended before cyclic loading tests in order to identify various limit states that are used to define amplitude of each cycle (see Fig. 4). Some limit states has not been clearly defined in the past. They are intended to be made clear in this handbook for convenience in documentation. Monotonic test is not mandatory because of affordability issues. It is especially beneficial when a yielding point is not very clear for specific pier tests (Krawinkler et al, 2001). The upper and lower limits of testing speed are provided to reduce the possible bias caused by creep or rate-dependent characteristics.

Cyclic loading is the most general loading program to test seismic performance. It can be carried out unidirectionally or bidirectionally. Although not resembling the real earthquake load, results from a well-designed cyclic loading program can be a very non-specific indication of inelastic seismic performance for a structural component. Most existing unidirectional cyclic loading programs share similar incremental features but vary in details. The most representative ones are adopted and explained in the handbook. One example is shown in Fig. 5. The balance of demonstrating strength degradation and avoiding undesired early fatigue fracture need to be considered when determining the number of cycles repeated at one displacement amplitude (Lowes and Moehle, 1995).

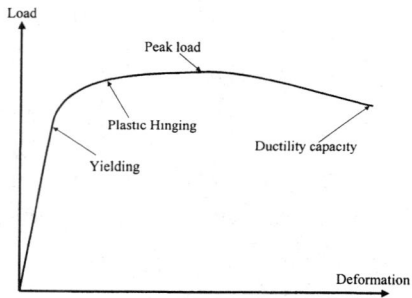

Fig. 4 Limit states

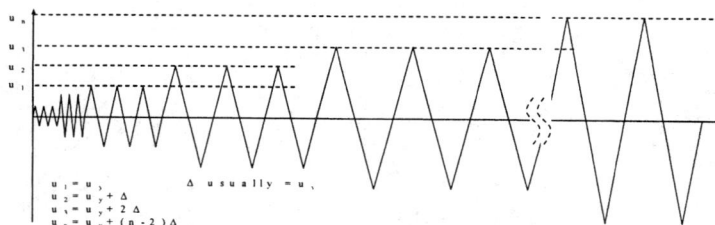

Fig. 5 Typical cyclic loading program

When specific earthquake record is to be tested on the specimen, a preliminary analysis based on assumed structural dynamic characteristics is required to estimate structural response, and consequently the displacement history of the specimen (El-Bahy et al, 1999). This procedure also applies to the artificially generate ground motions such as design spectrum compatible records.

There are various types of bidirectional loading program in practice. Each has advantages and disadvantages in terms of simulating bidirectional earthquake response. Some often-used programs are collected in the handbook to be used individually or in combination. Fig. 6 illustrates some of these loading programs.

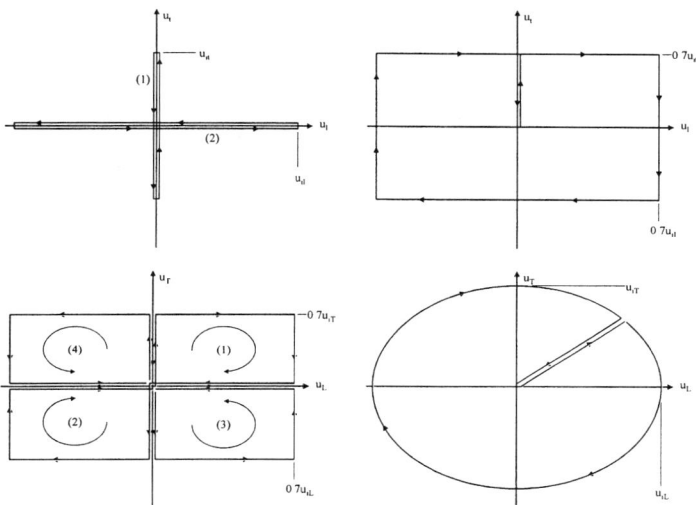

Fig. 6 Examples of bidirectional cyclic loading programs

Category C: Despite being the most realistic loading program, the shaking table tests do not provide credible information if not carried out with care. The calibration record of each table should be complete and accessible by all potential users of the test result. The combined dynamic characteristics of the specimen and the table need to be documented prior to large ground motion tests.

The shaking table tests are earthquake- and structure-specific. More than one tests for the same subject are needed to provide eligible general conclusions. Ground motions can be either real earthquake records or spectrum-compatible synthetic records. Filtering may be required for some earthquake records that are difficult to be reproduced by the table. Reliable records can be obtained from the websites of the earthquake engineering research centers.

Axial load effect: There are two parts of the axial load effect. One part comes from the horizontal offset of the loading point with respect to the element support point, which is referred as P-Δ effect. The other part comes from the swaying of the axial loading mechanism. The P-Δ effect is a realistic condition that occurs in bridges when excessive horizontal displacement occurs. The resulted bending moment distribution is not triangular and can only be simulated by the axial loading machine or weight that keeps the same loading direction (vertical) throughout the entire test. The swaying of the loading mechanism generates an additional shear force (positive or negative) varying with the amount of swaying, which does not exist in real seismic event. In the result of the test, the lateral force introduced by the swaying should be combined with the applied lateral load while the P-Δ effect should be reported separately. Fig. 7 shows the different bending moment diagram caused by the P-Δ effect and the swaying.

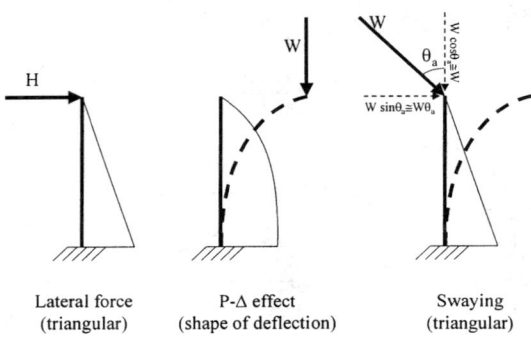

Fig. 7 Bending moment diagrams for lateral and axial loading

Measurement and Documentation

The technologies used in measurement systems evolve quickly. New issues emerge while old issues are resolved. It is not easy to setup general guidelines for all tests. However, well-configured measurement and acquisition systems are based on some similar principles. It is important to layout the principles to be followed.

A measurement plan should be created during the test design. Sensor and data-acquisition requirement should be carefully established. Some fundament measurements should be taken regardless the necessity in the specific project (Fig. 8). These fundamental measurements include the most demanded measurements such as applied force or displacement, curvature in the plastic hinge area, column base rocking, variation of axial load and displacement, etc. Other helpful measurements are those assisting error analyses, such as all slip-critical surface, boundary conditions, and redundant measurements (e.g. displacement and acceleration at same location and orientation).

A complete and accurate documentation is the key to maximize the usage of the test results. Many efforts are spent on this subject to handle the demand on seismic performance databases, which can support performance-based design in the near future. One of the most extensive projects for this purpose is from the Network for Earthquake Engineering Simulation (NEES) program funded by National Science Foundation (NSF). In order to integrate the network of experimental facilities and to establish scalable data depositories, a standard for test data reporting and storage will be established. In the

grand plot of the future earthquake engineering research, the role of the documentation provision in this handbook will be:
1. Before the general standard is established, the documentation standard set forth in this handbook can be practiced in the bridge seismic performance studies.
2. The practical provision can provide the developers of the future data standard a reference of the demanded data and documentation format in the bridge seismic performance studies.
3. The documentation standard in this handbook will be made compatible to the new standard developed by NEES program when the new standard is completed.

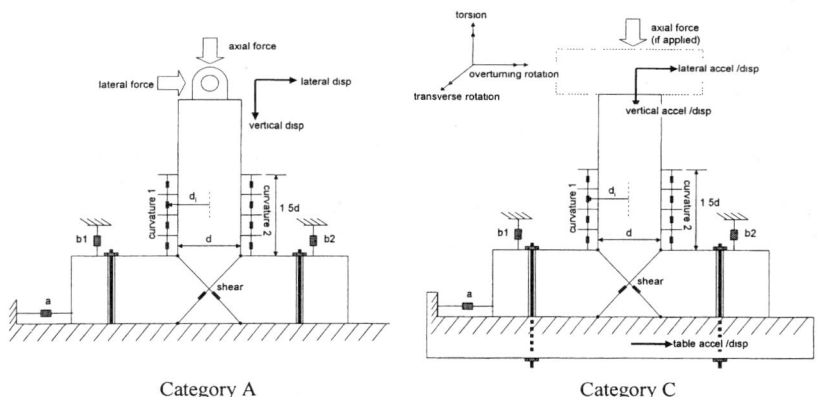

Fig. 8 Some fundamental measurements

Summary

1. The anticipated large demand on performance testing in the near future makes it necessary to produce a commonly accepted guiding document to increase credibility, consistency, and accessibility of test data.

2. This handbook is currently limited to seismic performance testing of bridge piers. The format of the handbook is designed to accommodate provisions for other component tests as well as bridge system tests.

3. Issues in specimen, loading, and documentation are examined and discussed with testing experts. Most sensible methods are recommended while other reasonable options are provided as reference.

4. Documentation is critical from many aspects. It is emphasized in a number of sections in this handbook. Completeness and compatibility are important for future integration to seismic performance database.

Reference

Abo-Shadi, N.; Saiidi, M.; Sanders, D. (2000). *Seismic Response of Reinforced Concrete Bridge Pier Walls in the Weak Direction*, Technical Report MCEER-00-0006.

Bertero, V. V.; Aktan, A. E.; Charney, F. A.; Sause, R. (1984). *Earthquake Simulation Tests and Associated Studies of a 1/5th-scale Model of a 7-story R/C Frame-wall Test Structures*, Report No. UCB/EERC-84/05, University of California, Berkeley.

El-Bahy, A.; Kunnath, S.; Stone, W.; Taylor, A. (1999). "Cumulative Seismic Damage of Circular Bridge Columns: Variable Amplitude Tests," *ACI Structural Journal*, v. 96 no. 5, 711-719.

Krawinkler, H., Parisi, F., and Ayoub, A.S. (2001). *Development of a testing protocol for wood frame structures*, Final report W-02, CUREe-Caltech Woodframe Project.

Lampe, N.; Azizinamini, A. (2000). "Steel Bridge System, Simple for Dead Load and Continuous for Live Load," *Proc. Conference of High Performance Steel Bridge*, Nov..

Moncarz, P. D. (1981). *Theory and Application of Experimental Model Analysis in Earthquake Engineering*, Dissertation, Stanford University.

Operational Performance Seismic Design of Highway Bridges for 2500-Year Earthquake Using Proposed NCHRP Provisions

Bardia Emami[1] and W. N. Marianos[2]

Abstract

The Applied Technology Council (ATC), in a joint venture with the Multidisciplinary Center for Earthquake Engineering Research (MCEER), has recently developed Comprehensive Specifications for the Seismic Design of Highway Bridges (National Cooperative Highway Research Program, NCHRP Project 12-49). The American Association of State Highway and Transportation Officials (AASHTO) is considering these recommended specifications for possible incorporation into the future AASHTO Load and Resistance Factor Design specifications. The primary objective of the NCHRP Project 12-49 was to develop seismic design provisions that reflect the latest research findings, design philosophies, and design approaches. Henceforth, implementation of the newly developed provisions will ensure enhanced seismic performance of highway bridges.

In this study, the highway bridge seismic design procedures of the current AASHTO and proposed NCHRP provisions are compared for the substructure and superstructure of the bridge remain the same for both provisions. An existing bridge located in a high seismic zone of Missouri with Seismic Performance Category "D" is redesigned according to the proposed NCHRP provisions. In this project, this bridge has been redesigned using the NCHRP provisions by Operational Performance Level for the Maximum Considered Earthquake (MCE) with a 3% of probability of exceedance in 75 years (with an approximate return period of 2500 years). By the new design the bridge will remain functional after the MCE as a critical bridge with lifeline performance.

This research indicates that changing from the current AASHTO Standard to the new provisions requires 49% increase of reinforcing steel for this bridge. Transverse reinforcement of the piles increases significantly for the requirements of the new provisions (2500% increase). By using the new provisions, concrete increases 72% at wing-walls and the piles.

---

[1] Senior Design Engineer, Optimum Engineering Solutions, Inc., #3 Country Club Executive Park Suite 200, Glen Carbon, IL 62034; phone 618-288-3131; bemami@openso.com
[2] Senior Associate, Modjeski and Masters Inc., 804 North 1st Street, St. Louis, MO 63102; phone 314-588-115; wnmarianos@modjeski.com, adjunct faculty member at Southern Illinois Univ. Edwardsville, Edwardsville, IL 62026-18000

Introduction

Bridges are key elements in transportation systems because of their high cost and importance (bridge failure may cause system failure). Bridges have had less than satisfactory performance in recent earthquakes. Many full and partial failures have occurred in different countries with significant economic losses. For example, $1.7 billion in the 1989 Loma Prieta, CA earthquake, $300 million in the 1994 Northridge, CA earthquake and $6.5 billion in the 1995 Kobe, Japan earthquake were only the loss due to bridge damage (not including the indirect economic losses to business and the transportation system). Significant damage to these bridges shows that it is important to improve the methods and specifications of bridge seismic design. The latest bridge seismic design specifications (AASHTO LRFD Bridge Design Specifications) are based on Division I-A of the AASHTO Specifications and Division I-A is originally based on 1983 research results (which were incorporated with little modification into the AASHTO Specifications in 1992). Damage in recent earthquakes shows that the current provisions can be improved. To address these weaknesses, in 1998, AASHTO sponsored NCHRP Project 12-49: Comprehensive Specifications for the Seismic Design of Bridges. This project was completed in November 2001 with partnership of Applied Technology Council (ATC) and the Multidisciplinary Center for Earthquake Engineering Research (MCEER), and it is being considered for adoption in the future AASHTO LRFD Bridge Design Standard Specification.

These provisions contain the New U.S. Geological Survey (USGS) maps (1996). This document contains provisions for two different earthquakes. The event with 50% probability of exceedance in 75 years is termed the Frequent Earthquake (FE). The event with 3% probability of exceedance in 75 years is termed the Maximum Considered Earthquake (MCE). By using a higher level of analysis (nonlinear static displacement capacity verification (pushover) analysis), the designer will be able to use a higher value of Response Modification Factor. For low seismic hazard areas and regular bridges, a no analysis option has been introduced in these guidelines. The effect of the fault distance zone has been considered by the vertical acceleration effect. Earthquake Resisting Systems and Elements (ERS and ERE) have been discussed with more detail in this document to aid the designer in choosing a better seismic resisting system for the bridge. Two performance levels of design have been introduced in these provisions. For Life Safety performance level, a bridge will have minimal damage and immediate service after the frequent earthquake. Significant damage may occur after the Maximum Considered Earthquake with significant disruption in service. For Operational performance level, the bridge will have minimal damage and immediate service after the Frequent and Maximum Considered Earthquake. Soil factors, spectral shapes and load combinations have been modified in this document and a new concept of spring constant for foundation modeling has been introduced. Liquefaction effects (one of the main problems in failure of bridges in past earthquakes) have been discussed with more details. Seismic isolation systems have been discussed with more details in these provisions. Figure 1 shows the general procedure of seismic design in this document.

DISASTER RESPONSE FOR LIFELINE SYSTEMS 145

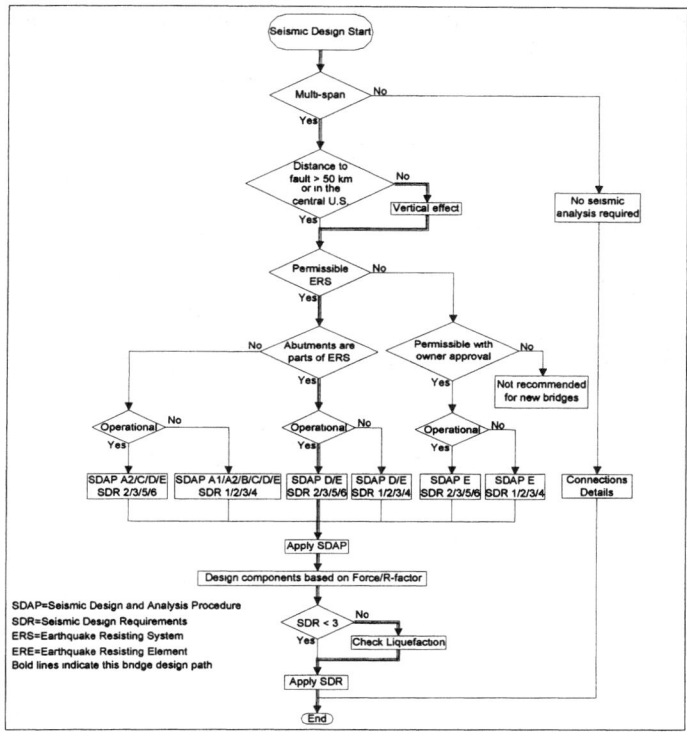

Figure 1. Design procedure in new provisions

Overview of Structure

The existing three-span bridge is comprised of a concrete deck supported by pre-stressed reinforced concrete girders on integral concrete abutments. Cast-In-Place piles support the integral abutments. Figure 2 shows the plan and elevation of the bridge. The interior substructure units (piers) are Cast-In-Place pile piers. The bridge has a 20-degree skew. This bridge is located in Pemiscot County, Missouri, USA and it has been designed in 2000 by AASHTO Standard Specifications (16th edition 1996). This bridge is located in seismic performance category "D" and according to current specifications a 0.36g acceleration should use for the seismic design of the bridge. Figure 2 shows the plan and geometry of the bridge.

Site Class and Soil Properties

As Figure 3 shows, the soil profile for this bridge contains cohesive soil (stiff clay) and sand at the bottom. The soil profile has potential for liquefaction. In this project, two separate analyses will be done for the liquefied and non-liquefied cases. For the liquefied case the minimum soil properties of the soil will be used. Site profile "III" has been used for the current design and for the proposed guideline site class "D" will be used for design.

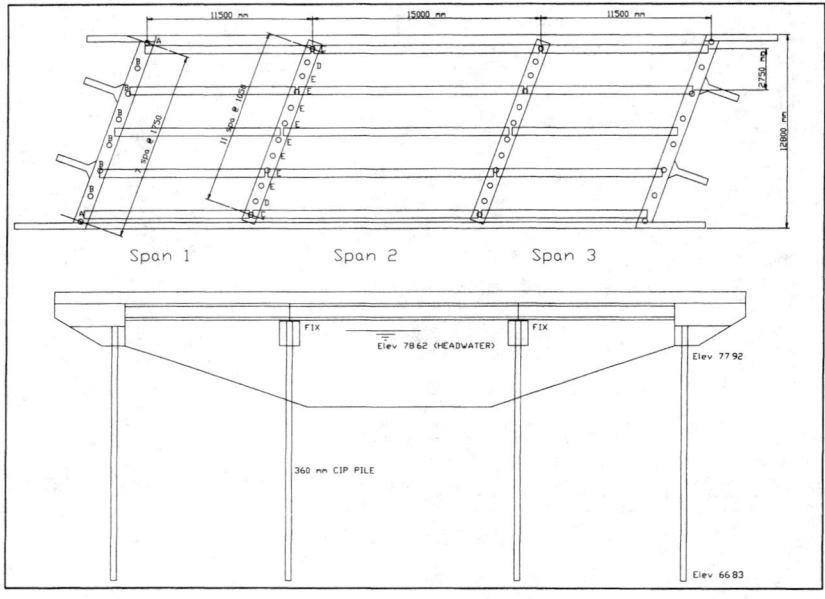

Figure 2. Plan and elevation

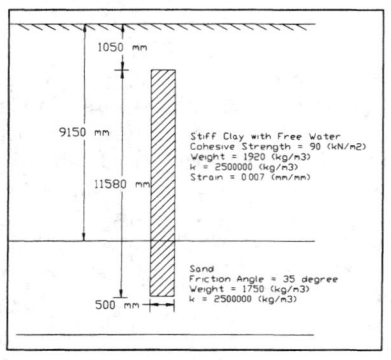

(a) At abutments

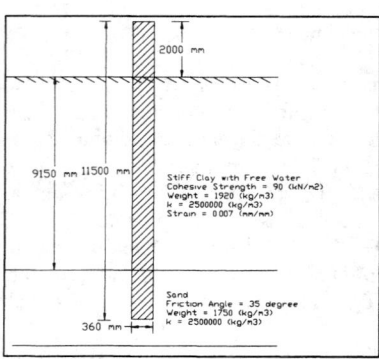

(b) At intermediate bent

Figure 3. Soil properties

Design and Analysis Requirements

In this project, the bridge will be designed for Operational Performance. According to the NCHRP provisions (based on the location of the bridge), seismic hazard level "IV" is permissible for this project. Because of the significant contribution of abutments in lateral resisting system Seismic Design and Analysis Procedure (SDAP) of this bridge is "E". Response spectrum analysis method will be used for this project. Seismic Design Requirement (SDR) six is appropriate for this bridge for the Operational Performance and the structure has an approach slab. Figure 4 presents the response spectrums for this bridge.

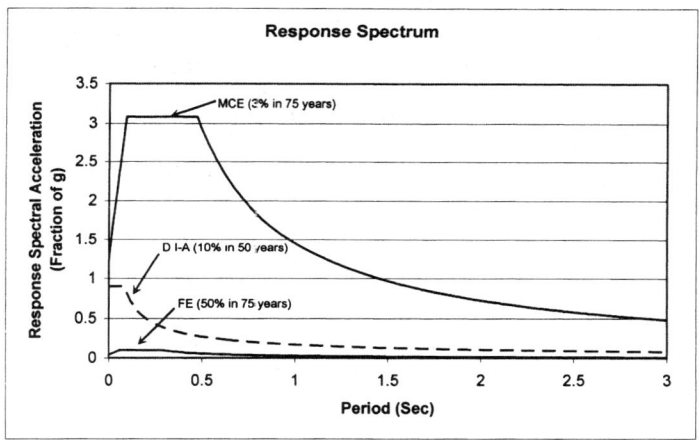

Figure 4. Design response spectrums

Seismic Load Path

The significant part of the dead load of the superstructure is carried to the integral abutments. A part of the lateral load of superstructure due to earthquake forces will transfer as the shear to the intermediate bent. At the intermediate bents, the piles will transfer the lateral load to the soil. At the abutments, the back wall, the wing walls and the Cast-In-Place piles will contribute to transfer the lateral load of superstructure to the soil. The exterior wing walls are not structural element in the lateral resisting system of the bridge and only the interior wing walls will contribute in the resisting system.

Finite Element Modeling

The superstructure of this bridge has two types of pre-stressed reinforced concrete girders and it has been modeled as a single frame member in the SAP2000 program. The equivalent section properties have been calculated for this member. The actual bridge superstructure is connected to intermediate bent the cap beam and they act as a rigid diaphragm. To address the rigid behavior of superstructure and cap beam, all of the joints in the cap beam in SAP2000 model are constrained for uniform distribution of the superstructure vertical load between the piles. For the real mass distribution of the finite element model, the cap beam and superstructure will be located at the center of gravity of the superstructure. A rigid zone area is provided at the top of the piles to capture the effect of girders' height. In the longitudinal direction, moment resistance at the top of the piles is released because the connection between the pier cap beam and superstructure is not able to transfer the moment.

At abutments, a vertical short and rigid element contains the mass of the abutment. At the bottom of this element, an equivalent spring is provided to represent the stiffness of the back wall, wing walls and the piles. The height of this element transfers the spring constant from the superstructure's center of gravity to the top of the piles. For effect of the skew, the local axis of the joints and the short element at the abutment has been rotated for 20 degrees. At the intermediate bent, according to the NCHRP Provisions, Foundation Modeling Method (FMM) "II" should be used for SDAP "E". Depth of fixity method is used to model the foundation of intermediate bents of bridge in the finite element model. The bottom of the pile at the depth of fixity is fixed.

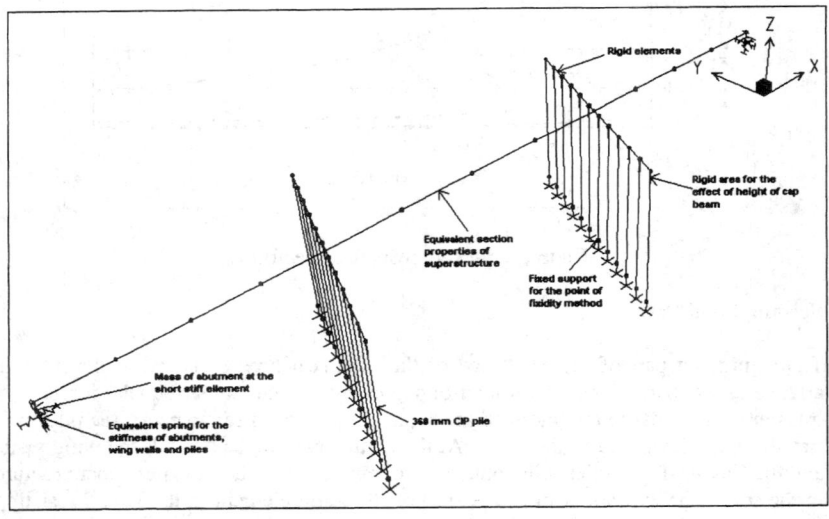

Figure 5. Finite element model

For finding the depth of fixity for the piles, the soil profile of bridge and the current diameter of the piles have been modeled in the L-pile program. For modeling the Cast-In-Place piles, a non-linear analysis has been used with a section with steel shell and without core. The analysis was repeated for liquefied and non-liquefied cases and the effect of liquefaction is not significant on depth of fixity for this bridge.

The average result of different cases of liquefied and non-liquefied, and fixed head and free head analysis of the pile gives a depth of fixity of 5640 mm. On other hand, 1980 mm of the top of the pile is above the ground so in the SAP2000 finite element model, a length of 7620 mm will be used as the depth of fixity of piles at intermediate bents. Figure 5 shows the finite element model of the bridge in the SAP 2000 program. Figure 6 presents the main mode shapes of the bridge in the SAP 2000 program.

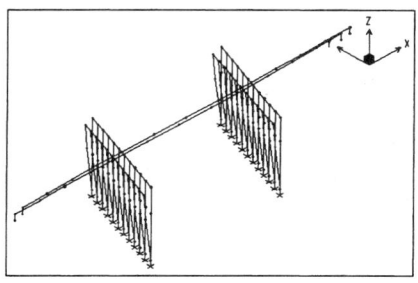 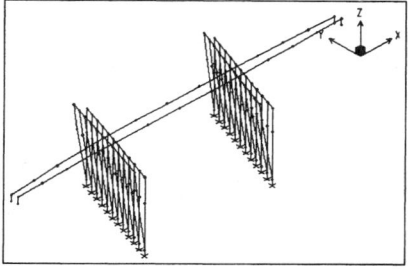

(a) Longitudinal direction (T=0.180 sec)   (b) Transverse direction (T=0.138 sec)

Figure 6. Main natural period

Reinforcement design of the piles

In the next step of the design of the piles at the abutments, the maximum displacement (includes the effect of 100% plus 40% combination from the SAP2000 model) will be applied to the L-pile model and the maximum moment at the top of the pile will be compared to the yield moment of the pile. The potential of plastic hinging will be studied for the piles due to the MCE forces. To avoid the potential for plastic hinging, the piles will be designed elastically and adequate reinforcing steel will be provided for the piles to guarantee that the maximum moment because of the MCE will be smaller than the yield moment of the piles. This will insure that no plastic hinge will occur in the pile due to the MCE. For this purpose and according to the displacement due to the MCE forces, the diameter of some of the piles has been increased at the abutment and the intermediate bents. To address the group pile effect of the piles, the forces are distributed between the piles by the P-multiplier method (recommended by FHWA). In this method the leading row of the piles in each direction is subjected to higher forces than other rows.

The NCHRP Provisions requires the designer to perform the displacement capacity verifications (pushover analysis). Elastic design of the piles and nonlinear analysis of the piles by the L-pile program indicates adequate displacement capacity of the piles for the displacement of the bridge due to the MCE forces.

In the provisions, two methods are introduced for the pile to cap connection design. Method 2 is applicable for the SDAP "E". In this method tensile and compression stress at connections will be calculated and compared with the allowable stress. Extra reinforcement will be provided at connections as necessary. Figure 7 shows the final results of the pile design.

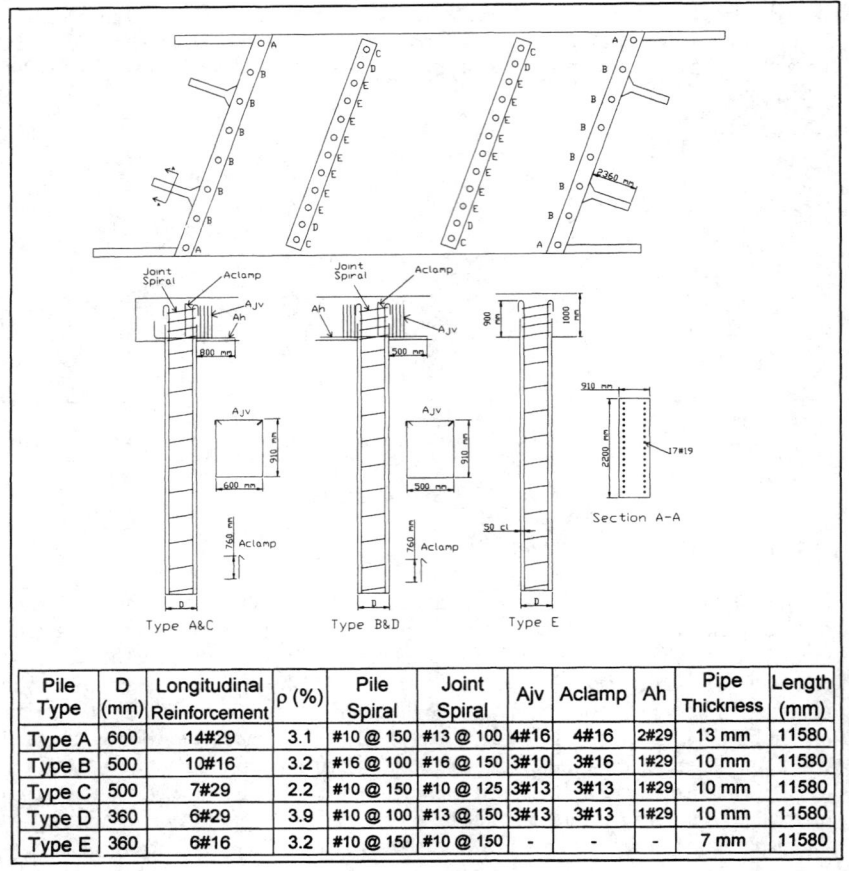

Pile Type	D (mm)	Longitudinal Reinforcement	ρ (%)	Pile Spiral	Joint Spiral	Ajv	Aclamp	Ah	Pipe Thickness	Length (mm)
Type A	600	14#29	3.1	#10 @ 150	#13 @ 100	4#16	4#16	2#29	13 mm	11580
Type B	500	10#16	3.2	#16 @ 100	#16 @ 150	3#10	3#16	1#29	10 mm	11580
Type C	500	7#29	2.2	#10 @ 150	#10 @ 125	3#13	3#13	1#29	10 mm	11580
Type D	360	6#29	3.9	#10 @ 100	#13 @ 150	3#13	3#13	1#29	10 mm	11580
Type E	360	6#16	3.2	#10 @ 150	#10 @ 150	-	-	-	7 mm	11580

Figure 7. Summary of substructure design

Conclusion

These provisions provide more guidelines and flexibility in the design of the substructure members and connections. Plastic hinge behavior at connections and members are treated directly in these provisions, hence, the designer is able to provide adequate and rational reinforcement for these critical bridge elements. More varieties of analysis are introduced in these provisions. For more complex bridges in high seismic zones or bridges with significant contributions from abutments in the resisting system, the bridge should be checked by pushover analysis (displacement capacity verification), which requires more computation effort comparison to current specifications. Defining the load path for the lateral forces providing a permissible earthquake resisting system and elements will help the designer to formulate better resisting seismic systems for the bridge from the beginning steps of design.

Figure 8 shows the significant impact on substructure quantities of using the new provision for this bridge. As Figure 4 shows, the maximum part of the design response spectrum is more than 3g. In many cases, the main period of vibration of the bridge will be in this range and it will cause large forces in the members and have a significant cost impact in comparison with the current specifications. We recommend providing a maximum limit for the design response spectrum in regions like the New Madrid zone with large value of spectral acceleration or adding new design option with smaller forces for non-critical bridges.

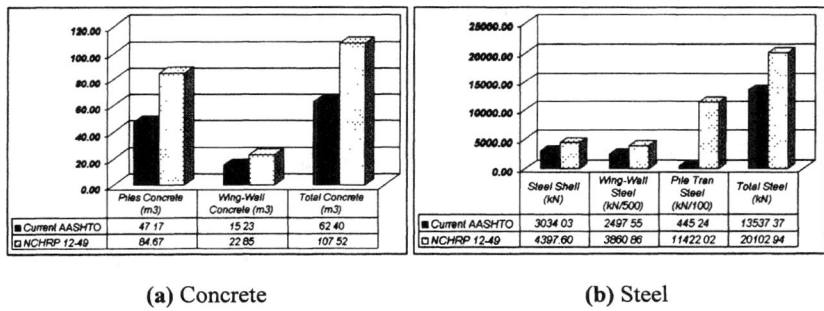

(a) Concrete (b) Steel

Figure 8. Substructure quantities

Comparison of the FE and MCE forces for this bridge shows that the MCE forces are governed in all cases. Even by using the larger value of R-factor (for more flexible structure) for the high seismic zone, the design forces for the MCE will be five times larger than the FE forces. It seems because of the significant different between the MCE and FE in Mid-West (small Frequent Earthquake and large Maximum Credible Earthquake), the MCE design forces will be governed in all cases. It is recommend providing guidelines (or sets of map) for this kind of regions to only check the MCE.

Acknowledgement

The authors sincerely thank Horner & Shifrin, Inc. (MO, USA) for providing the original drawings and calculations of the existing bridge based on current specifications and Mr. Jacques Darvish (Vice President and Director of Structural Engineering Department) for his assistance.

The authors also thank Dr. Nader Panahshahi (Chair Person of the Civil Engineering Department of Southern Illinois University Edwardsville) as the advisor of the project.

References

American Association for State Highway and Transportation Officials (AASHTO). (1998). *AASHTO LRFD Bridge Design Specifications*, 2nd edition, AASHTO, Washington.

American Association for State Highway and Transportation Officials (AASHTO). (1996). *AASHTO Standard Specifications for Highway Bridges*, 16th edition, AASHTO, Washington.

Applied Technology Council (ATC) and Multidisciplinary Center for Earthquake Engineering Research (MCEER). (2002). *Recommended LRFD Guidelines for the Seismic Design of highway Bridges Part I: Specifications*, MCEER, Buffalo.

Applied Technology Council (ATC) and Multidisciplinary Center for Earthquake Engineering Research (MCEER). (2002). *Recommended LRFD Guidelines for the Seismic Design of highway Bridges Part II: Commentary and Appendices*, MCEER, Buffalo.

Barker, R., and Puckett, J. A. (1997) *Design of Highway Bridges*, John Wiley, New York.

Emami, B., Siegfried, D. L. and Panahshahi, N. (2002). " A Comparison Study of Seismic Design of Highway Bridges Using AASHTO Standard Specifications and Proposed NCHRP Provisions." *6th International Conference on Civil Engineering*, ICCE2003, Isfehan, Iran.

Emami, B. (2002). "Development of Bridge Seismic Design in the United States." *12th Symposium on Earthquake Engineering*, 12SEE, Roorkee, India.

Friedland, I. M., Mayes, R. L., and Bruneau, M. (2001). "Recommended Changes to the AASHTO Specifications for the Seismic Design of Highway Bridges (NCHRP Project 12-49)." *MCEER Research Progress and Accomplishments*, Buffalo.

Walsh, K. D., Frechette, D. N., Houston, W. N, and Houston, S. L. (2000). "State of the Practice for Design of Group of Laterally Loaded Drilled Shafts", *Transportation Research Record 1736*, Paper No. 00.1306, Washington.

Pushover Analysis of Bridge Intermediate Bents

Jeffrey Ger[1] and Phillip Yen[2]

Abstract

Five methods for the push over analysis of multiple-column bridge bents are presented in this paper. They are Finite Segment–Finite String (FSFS), Finite Segment–Moment Curvature (FSMC), Interaction Axial load-Moment (PM), Constant Moment Ratio (CMR), and Moment-Rotation Hinge (MRH) methods. The lateral force-lateral displacement curves of a two-column bent generated by these methods are compared. The numerical results indicate that the post-yielding behavior of the two-column bent is sensitive to the p-δ effect. All the methods can predict the descending lateral force-lateral displacement (i.e. unstable) behavior of the bent. The results from these methods are in favorable agreement with each other if the p-δ effect is included in the analyses. Further study is needed to verify the numerical results with available test data from large-scale multiple-column bents.

Introduction

The nonlinear static analysis or pushover analysis in FEMA-273 and ATC-40 has become a standard procedure in current structural engineering practices. The recent NCHRP 12-49 study also recommends that AASHTO adopt the pushover analysis as one alternative in the seismic analysis and design of highway bridges (ATC/MCEER, 2001).

The computer program INRESS (Inelastic Analysis of Reinforced-Concrete and Steel Structures) is being developed for the nonlinear time history analysis and nonlinear static (pushover) analysis of 3-dimentional steel and reinforced concrete structures. Either force control or displacement control can be used in the pushover analysis. Currently five methods are incorporated into the program for the pushover analysis. They are Finite Segment–Finite String (FSFS), Finite Segment–Moment Curvature

---

[1] Structure Engineer, Federal Highway Administration, Florida Division, 227 N. Bronough Street, Suite 2015, Tallahassee, FL 32301; phone 850-942-9650; jeffrey.ger@fhwa.dot.gov
[2] Structural Research Engineer, Office of Infrastructure R&D, FHWA, 6300 Georgetown Pike, McLean, VA 22101.

(FSMC), Interaction Axial load-Moment (PM), Constant Moment Ratio (CMR), and Moment-Rotation Hinge (MRH) methods. The moment-curvature relationship of reinforced-concrete circular sections can also be generated by the FSFS method. This paper briefly describes the methodologies used in each individual method. The lateral force-lateral displacement curves of a two-column bent were calculated by the above-mentioned methods and the results herein compared.

Finite Segment–Finite String (FSFS) Method

A structural member is divided into several segments. Each segment has 12 degrees-of-freedom and its cross section is divided into many small elements (called strings) along the segment's longitudinal direction. The structural member can be steel or reinforced concrete. For a reinforced concrete member, the stress distribution throughout the cross section can be calculated by the concrete and steel stress-strain relationships (Mander, etc. 1988). Currently a bilinear steel stress-strain relationship is utilized in the developed computer program. A more refined steel stress-strain relationship will be added to the program in the future. For each small element on the cross section, the strain increment can be expressed as

$$\Delta \varepsilon_c^{ij} = \Delta \varepsilon_c^j + V_i \Delta \varphi_u^j - U_i \Delta \varphi_v^j \tag{1}$$

in which

$$\Delta \varepsilon_c^j = (\Delta W_b^j - \Delta W_a^j)/L \tag{2}$$

$$\Delta \varphi_u^j = (\Delta \theta_{ub}^j - \Delta \theta_{ua}^j)/L \tag{3}$$

$$\Delta \varphi_v^j = (\Delta \theta_{vb}^j - \Delta \theta_{va}^j)/L \tag{4}$$

where i = ith cross-sectional element; j = jth segment; $\Delta \varepsilon_c^{ij}$ = the strain increment of element i in jth segment; $\Delta \varepsilon_c^j$ = the normal strain at the centroid of jth segment; $\Delta \varphi_u$ and $\Delta \varphi_v$ = the bending curvature increments about the U and V axes, respectively; and U_i, V_i = the location of the ith cross-sectional element in the segmental coordinates U and V, respectively; W = the segmental coordinate along the longitudinal direction of the segment. Subscripts a and b represent the two ends of the segment. The current total strain for element i is

$$\varepsilon^{ij} = \varepsilon_p^{ij} + \Delta \varepsilon^{ij} \tag{5}$$

where ε_p^{ij} is the ith element total strain in the previous deformation state. Based on plastification of the cross section, the current principal axes, sectional properties, and the stiffness matrix of individual segments can be calculated (Ger, etc. 1993). The member's stiffness matrix is established by stacking up the segmental stiffness matrices for which a rotation matrix $[R]_{12 \times 12}$ is required for each segment by transferring the segment stiffness matrix from the segment coordinate system (U, V, W) to the member's coordinate system (i.e. small deflection and P-δ considerations). The moment-curvature relationship of a column can also be generated by the FSFS method. A comparison of moment-curvature relationships of a reinforced-concrete circular section generated by the computer program developed and SEQMC (<u>Seq</u>ad <u>M</u>oment <u>C</u>urvature Analysis Tools) software (SEQMC, 1998) is shown in Figure 1.

The cross sectional details from FHWA Seismic Design Example No. 4 (FHWA, 1996) were used herein, and these are: column diameter =48"; 34 - #11 longitudinal bars; f_c' =4 ksi; f_y =60 ksi; spiral = #5 @3.5"; concrete cover = 2.625" from the surface of longitudinal rebar to the surface of column; and the applied axial load = 660 kips. The post yield modulus of the reinforcing steel stress-strain curve is assumed to be 1% of the elastic modulus. The computation is terminated when the confined concrete strain reaches the ultimate concrete compression strain of 0.0153 (in/in). It can be seen that the curve generated by the developed program is almost identical to that generated by SEQMC except that the ultimate curvature calculated is larger than the SEQMC result.

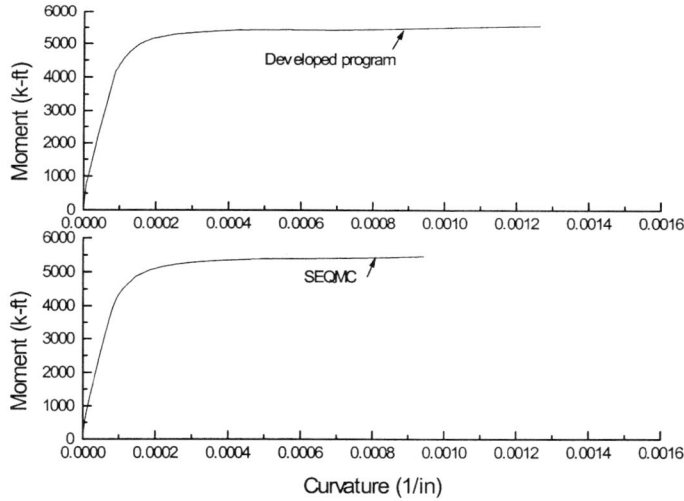

Figure 1. Moment-Curvature Comparison

Finite Segment–Moment Curvature (FSMC) Method

This method is similar to the FSFS method except that the cross section of each segment is not divided into many elements. The segment stiffness matrix at each incremental step is calculated based on the cross-sectional axial load-moment-curvature family of curves from which the flexural property, EI, can be obtained (from the slope of moment-curvature relationship for a given axial load). The total curvature at each step is the accumulation of the incremental curvatures from the previous steps based on equations (3) or (4). Similar to the FSFS method, the member stiffness matrix is established by stacking up the segment stiffness matrix with consideration of the segmental rotation matrix $[R]_{12 \times 12}$ and P-δ effect.

Interaction Axial Load–Moment (PM) Method

The bilinear moment-curvature relationship with certain axial loads are used to generate the corresponding bilinear moment-rotation curve for the member shown in Figure 2(a). The moment-rotation curve can be obtained by conjugate beam theory, and it is composed of two imaginary components as shown in Figures 2(b). In these figures, the initial slopes of the linear and elasto-plastic components are a1 = pa, a2 = qa, and p+q=1, where p is the fraction of stiffness apportioned to the linear component and q is the fraction of stiffness apportioned to the elasto-plastic component.

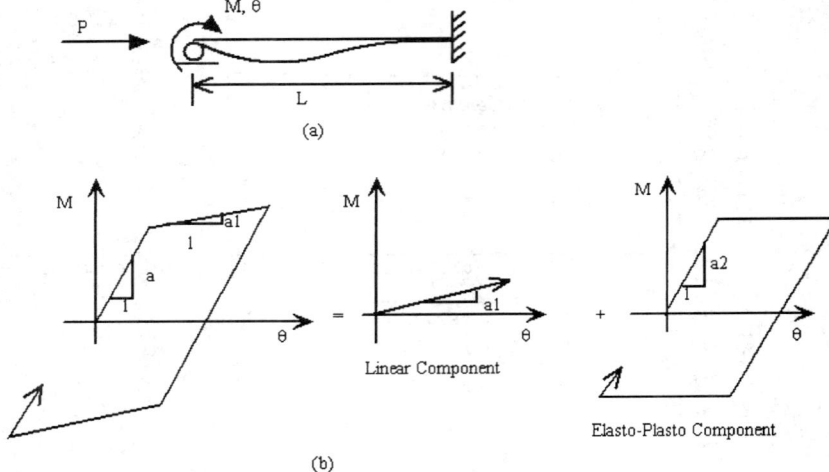

Figure 2. Bilinear moment-rotation model

Based on Figure 2(b), member stiffness matrix at any incremental step can be formulated according to the state of yield. The state of yield may be one of the following four conditions: (a) both ends linear, (b) i end nonlinear and j end linear, (c) i end linear and j end nonlinear, and (d) both ends nonlinear. For example, the stiffness matrix for condition (b) is

$$\begin{bmatrix} \Delta M_i \\ \Delta M_j \\ \Delta V_i \\ \Delta V_j \end{bmatrix} = \begin{bmatrix} pa & pb & -c & -c \\ pb & pa+qe & -pc-qf & -pc-qf \\ -c & -pc-qf & pd+qg & pd+qg \\ -c & -pc-qf & pd+qg & pd+qg \end{bmatrix} \begin{bmatrix} \Delta \theta_i \\ \Delta \theta_j \\ \Delta Y_i \\ \Delta Y_j \end{bmatrix} \quad (6)$$

in which a=4EI/L, b=2EI/L, c=6EI/L^2, d=12EI/L^3, e=3EI/L, f=3EI/L^2, and g=3EI/L^3.

Constant Moment Ratio (CMR) Method

Given a member of length L, the end moments M_i, M_j, and the moment-curvature relationship, the end moment-rotation relationship at each end can be derived by the conjugate beam theory as shown in Figure 3.

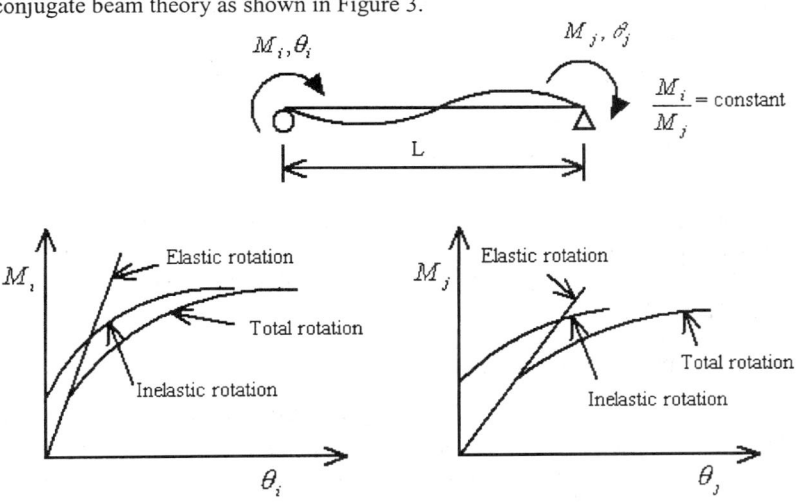

Figure 3. Moment-rotation relationship

The nonlinear bending stiffness matrix is formulated based on the assumption that the moment ratio, M_i/M_j is a constant and the flexibility of the inelastic rotation can be lumped at the member ends. Therefore the total rotation, elastic rotation, and the inelastic rotation expressed in terms of incremental forms are

$$\begin{bmatrix} \Delta\theta_i \\ \Delta\theta_j \end{bmatrix} = \begin{bmatrix} \Delta\theta_{iE} \\ \Delta\theta_{jE} \end{bmatrix} + \begin{bmatrix} \Delta\theta_{iIE} \\ \Delta\theta_{jIE} \end{bmatrix} \qquad (7)$$

$$\begin{bmatrix} \Delta\theta_{iE} \\ \Delta\theta_{jE} \end{bmatrix} = \frac{L}{EI} \begin{bmatrix} 1/3 & -1/6 \\ -1/6 & 1/3 \end{bmatrix} \begin{bmatrix} \Delta M_i \\ \Delta M_j \end{bmatrix} \qquad (8)$$

$$\begin{bmatrix} \Delta\theta_{iIE} \\ \Delta\theta_{jIE} \end{bmatrix} = \begin{bmatrix} f_i & 0 \\ 0 & f_j \end{bmatrix} \begin{bmatrix} \Delta M_i \\ \Delta M_j \end{bmatrix} \qquad (9)$$

where f_i and f_j are the flexibilities of inelastic rotations at member ends i and j, respectively. From equations (7), (8), and (9), the member stiffness matrix can be expressed as

$$\begin{bmatrix} \Delta M_i \\ \Delta M_j \\ \Delta V_i \\ \Delta V_j \end{bmatrix} = \frac{1}{D} \begin{bmatrix} \frac{L}{3EI}+f_i & \frac{L}{6EI} & -(\frac{1}{2EI}+\frac{f_i}{L}) & -(\frac{1}{2EI}+\frac{f_j}{L}) \\ & \frac{L}{3EI}+f_i & -(\frac{1}{2EI}+\frac{f_j}{L}) & -(\frac{1}{2EI}+\frac{f_j}{L}) \\ & & \frac{1}{EIL}+\frac{f_i+f_j}{L^2} & \frac{1}{EIL}+\frac{f_i+f_j}{L^2} \\ & \text{Symm.} & & \frac{1}{EIL}+\frac{f_i+f_j}{L^2} \end{bmatrix} \begin{bmatrix} \Delta\theta_i \\ \Delta\theta_j \\ \Delta Y_i \\ \Delta Y_j \end{bmatrix} \quad (10)$$

in which $D = \dfrac{L^2}{12(EI)^2} + \dfrac{L}{3EI}(f_i+f_j) + f_i f_j$ (11)

For a multiple-column bent subjected to earthquake, if a column deforms in a double-curvature shape, it may be assumed that the point of contraflexure is at the middle point of the column. In this case, $M_i/M_j \cong 1$ and $f_i = f_j$ can be used for the stiffness matrix formulation.

Moment-Rotation Hinge (MRH) Method

The stiffness matrix of a column is formulated by the combination of an elastic column element and a non-linear rotational spring connected at each end of that element. The rotational spring and the elastic bending stiffness of the column element behave as two springs in series. The stiffness of the rotational spring is governed by the moment-rotational curve of a hinge with length Lp, where Lp is the plastic hinge length (Priestly and Calvi, 1996) defined as the distance from the critical section of the column plastic hinge to the point of countraflexure of the column. The moment-rotation curve of Lp can be calculated by the FSFS method.

Numerical Example and Findings

A two-column bent used in the NCHRP 12-49 Design Example No. 8 (NCHRP 12-49, 2001) was chosen for the pushover analysis (see Figure 4). The details of the columns are: diameter =48"; 20-#10 longitudinal bars; $f_c^{'}$ =4 ksi; f_y =60 ksi; spiral = #5 @3.25"; concrete cover = 2.6"; and applied column axial dead loads = 765 kips. The post yield modulus of the steel stress-strain curve is assumed to be 1% of the elastic modulus. For the FSFS method, the computation stops when the confined concrete strain reaches the ultimate concrete compression strain. The ultimate concrete compression strain was calculated and found to equal 0.016 (in/in). For simplicity, the foundation of the bent structure has been assumed fixed in this example. Also, displacement control has been utilized. For the PM, CMR and MRH methods, the maximum required displacement of 31" (obtained from NCHRP 12-49 Design example 8) was used, and the calculated plastic rotations corresponding to the maximum required displacement were compared with the plastic rotational capacity of the columns. The plastic rotational capacity of the columns (from NCHRP 12-49

Example No. 8) is 0.055 radians. This is calculated based on the Recommended LRFD Guidelines (ATC/MCEER, 2001) Article 8.8.6.

The moment-curvature plots corresponding to a column axial dead load of 765 kips are shown in Figure 5. The moment-rotation relationships shown in Figure 6 were used in the PM and CMR methods, and these were calculated based on the idealized bi-linear moment-curvature relationship from Figure 5 and conjugate beam theory. In Figure 6, the slopes of the post-yield lines for both PM and CMR methods are equal to 4.3% of the initial slopes.

Figure 7 shows the lateral force-lateral displacement relationships generated from different methods with and without consideration of p-δ effects. The maximum lateral displacement from the FSFS method is 28.27", which is less than the required lateral displacement of 31". The maximum plastic rotations of the columns corresponding to the required displacement of 31" are 0.0483, 0.0483, and 0.0498 radians extracted from the outputs based on PM, CMR, and MRH methods, respectively. These values are less than the plastic rotational capacity of 0.055 radians. The numerical results indicate that the post-yielding behavior of the example bridge bent is sensitive to the p-δ effect. All the examined pushover analysis methods are capable of predicting the descending lateral force-lateral displacement (i.e. unstable) behavior of the bent, if geometric stiffness matrices are included in the analyses.

The recommended LRFD guidelines (ATC/MCEER, 2001) define the lateral displacement capacity as the displacement at which the first structural component reaches its inelastic deformation capacity (for example, column plastic rotation capacity). However these guidelines do not limit the displacement capacity to a point beyond which overall instability of a bent occurs. If the bridge owner desires to prevent possible occurrence of overall bent instability in design, the realistic moment-curvature relationship (i.e. not bi-linear models) should be used in the CMR and MRH methods in order to predict the critical point beyond which overall instability occurs. As shown in Figure 7, the FSFS and FSMC methods can adequately predict the displacement at which strength degradation begins. Figure 7 also shows that the PM and CMR methods produce almost identical results.

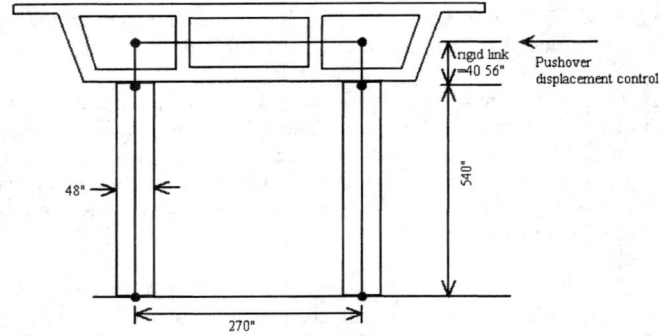

Figure 4. Two-column bent

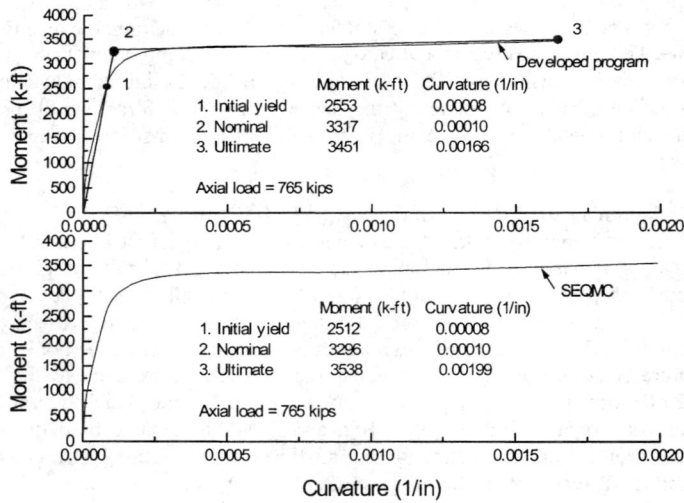

Figure 5. Moment-curvature relationships

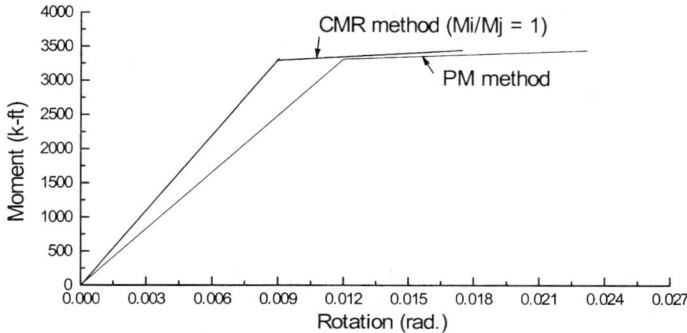

Figure 6. Moment-rotation relationships used in PM and CMR methods

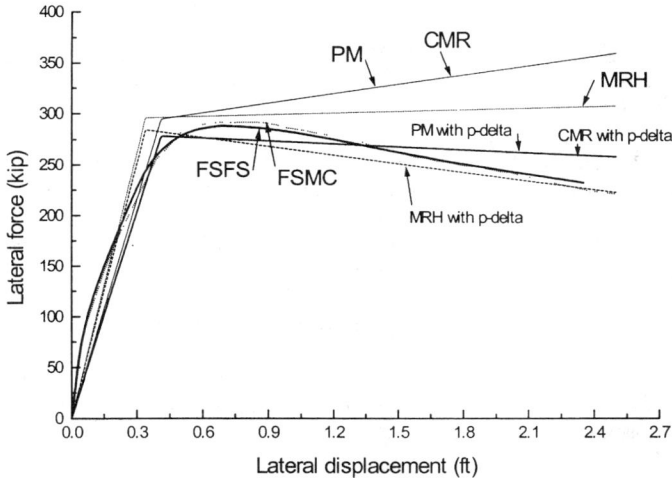

Figure 7. Lateral force-later displacement curves

Future Studies Needed

1. Further study is needed to verify the numerical results from individual methods with the available test data, especially the pushover data from the large-scale multiple-column bents.

2. The advantage of using the FSMC method is that the moment-curvature model can easily be adjusted to account for the column splice failure, shear failure, or joint degradation. The modified moment-curvature model will be included in the FSMC method.
3. Develop a user-friendly interface for the program.

References

ATC/MCEER joint venture (2001). *Recommended LRFD guidelines for the seismic design of highway bridges*, based on NCHRP 12-49.

FHWA (1996). *Seismic Design of Bridges, Design Example No. 4*, FHWA-SA-97-009.

Ger, J.F. and Cheng, F.Y. (1993). "Post-buckling and hysteresis models of open-web girders." *Journal of Structural Engineering*, ASCE, 119(3), 831-851.

Ger, J.F., Cheng, F.Y., and Lu, L.W. (1993). "Collapse behavior of Pino-Suarez building during 1985 Mexico earthquake." *Journal of Structural Engineering*, ASCE, 119(3), 852-870.

Mander, J.B., Priestly, J.N., and Park, R. (1988). "Theoretical stress-strain model for confined concrete." *Journal of Structural Engineering*, ASCE, 114(8), 1804-1826.

NCHRP Project 12-49 (2001). *Seismic Design of Bridges: Design example No. 8*, Prepared by Berger/Abam Engineers, Inc.

Priestly, J.N., Calvi, G.M. (1996). *Seismic design and retrofit of bridges*, Wiley, New York.

SEQMC, Demo version 1.00.06 (1998). Moment-curvature analysis package for symmetric sections.

BART Seismic Retrofit Program: Characterization of Design Ground Motion

J. Litehiser[1], N. Gregor[1], J. Marrone[1], F. Ostadan[1], R. Youngs[2].

Abstract

The original San Francisco Bay Area Rapid Transit (BART) system comprises over 70 miles of rapid transit line and 34 stations (underground, at grade, and aerial) and was completed in 1976 to 1968 state-of-the-art seismic design standards. These seismic criteria considered only the San Andreas and Hayward faults and concluded that a single "deterministic" design response spectrum throughout the BART system anchored at 0.33g or 0.5g for "shallow" and "deep" soil sites, respectively, was adequate.

Modern procedures to develop ground motions for the design of structures are significantly more complex. The seismic criteria for the BART East Bay, Colma, and San Francisco Airport (SFO) extensions, for example, began with a much more detailed model of potential earthquake sources, a much more complex way to model the effect of attenuation with distance of motions caused by earthquakes on these sources, a more comprehensive way to model the effect of site foundations, and some explicit consideration of uncertainties in all these model parameters.

Recognizing the evolution of seismic design criteria development over the past 30 years, both for BART, for other critical Bay Area structures (such as the Caltrans Toll Bridges), and for almost any modern engineered structure, ground motions for the BART Retrofit Program were developed through analysis of both "deterministic" and "probabilistic" characterizations.

Acceptable levels of conservatism for engineered structures have also evolved and become more demanding, as indicated by using deterministic ground motion estimates greater than the expected median motions or probabilistic ground motion estimates with longer return periods. The impetus for this additional conservatism arises from higher-than-expected recorded ground motions and greater-than-expected damage to some engineered structures since the 1960s.

The BART Program seismic design ground motion criteria incorporate the latest model available for Bay Area earthquake sources, fault dynamics, and soil column effects, concepts of deterministic motions from selected "scenario" earthquakes, concepts of motions with nonexceedance probabilities during a specified design life, and of acceptable consequences for alternative earthquake loads. The criteria were all guided by current practice, by practice established for the design of the BART extensions, and by judgment. Criteria were developed for two levels of ground motion (the Lower-level Design Basis Earthquake (LDBE) and the Design Basis Earthquake (DBE)), five zones plus several station-specific and three Transbay

---

[1] Bechtel Corporation, P. O. Box 193965, San Francisco, CA 94119; phone 415-768-1234

[2] Geomatrix Consultants, 2101 Webster Street 12th Floor, Oakland, CA 94612 phone 510-663-4100

Tube sites, five foundation conditions, and horizontal and vertical spectra for frequencies from tens of cycles per second to several seconds. High-frequency anchors for these spectra range from 0.43g to 0.72g (for LDBEs on firm foundations), 0.42g to 0.61g (for LDBEs on soft soils), 0.65g to 0.99g (for DBEs on firm foundations), and 0.58g to 0.89g (for DBEs on soft soils).

The seismic source characterization and ground motion modeling used to develop the design ground motion criteria were also used for other specific facets of the BART Retrofit Program, as detailed in companion papers within these Proceedings. The range of criteria ground motions was considered in the development of fragility curves (Salmon et al.). Criteria spectra were used to develop input time histories to assess current vulnerability and retrofit concepts of aerial structures (Mallare et al.). Deterministic ground motions from scenario earthquakes were developed for use in risk analyses of the BART system (Eidinger et al.). Both deterministic ground motions from scenario earthquakes and complete probabilistic ground motion analysis using all Bay Area seismic sources were integrated with assessment of conditional probability of liquefaction to derive estimates of scenario and annual probability of liquefaction along the BART Tube (Wu et al.).

Background to Development Ground Motion Criteria

Recent seismic event statistics have indicated that there is a 70% probability that a major earthquake will hit the Bay Area within the next 30 years (Michael et al., 1999). Although the BART original system performed well through the 1989 Ms=7.1 (Mw=6.9) Loma Prieta earthquake, it was built to seismic standards of the 1960s. The extensions, on the other hand, were built in the 1990s, to considerably updated and more stringent standards. In addition, retrofit designs for Bay Area Caltrans toll bridges developed in the 1990s represent recent acceptable standards.

The mission of the BART Seismic Retrofit Program is to seismically retrofit the entire original system, using state-of-the-art standards and methodologies, to enable the BART system to return to operation reasonably soon after an earthquake.

In developing ground motions to accomplish this mission a number of guidelines were established, among them: 1) Design ground motions should consider both deterministic and probabilistic evaluations of seismic hazard, consistent with other Bay Area lifeline projects. 2) Associated life safety hazard implied by the resultant seismic design ground motion criteria should not be less than that considered in 1997 UBC – that is, design ground motions should not, in general, be associated with a hazard level less than a 10% probability of exceedance in 50 years. 3) Development of the seismic retrofit design motions at SFO should not be greatly different than those used for the design of the SFO Extension, which should be fairly consistent with state-of-the-knowledge of seismic hazard. 4) The seismic retrofit design criteria should result in a suite of rock (NEHRP S_B/S_C) design spectra that in application encompass the entire BART alignment. 5) In developing the suite of design spectra, the period range of 0.5 to 2.0 seconds was considered most crucial.

As a result of these guidelines, the basic BART retrofit design ground motions for checking life-safety performance were generated based on the spectral values being the greater of the deterministic median + ½ σ or the probabilistic 500-year return period values (DBE). Lower values, based on median deterministic spectra, were generated for checking functionality performance (LDBE). Higher values, up to the greater of the median deterministic + σ or the probabilistic 1,000-year return period values for the DBE, were developed to evaluate the critical Transbay Tube.

Basic Elements for the Characterization of Design Ground Motions

Both deterministic and probabilistic require models of where future earthquakes might occur and how large they might be; how ground motion varies with earthquake magnitude and distance; and how ground motion varies for different foundation conditions. In addition, probabilistic estimates require a model of how often future earthquakes might occur as a function of their magnitudes.

The earthquake hazard in the Bay Area is dominated by active faults of the San Andreas system although smaller events are widely distributed (Figure 1). Details of the earthquake sources used in criteria development are given in WGCEP (1999) and Bechtel Infrastructure Corporation (2002).

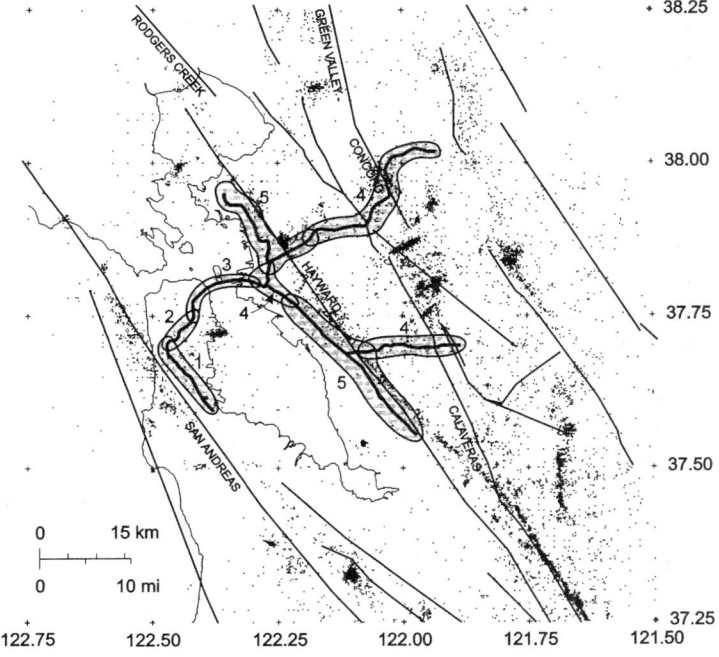

Figure 1 BART alignment, historical earthquakes (magnitude ≥ 1.5), and schematic traces of faults. Also shown are the five zones identified to group BART facilities for purposes of criteria development (see discussion below).

There are advantages and disadvantages to both the deterministic and probabilistic hazard methods. Deterministic estimates of appropriate earthquake design provide a clear and traceable method to compute seismic hazard (model elements are easily seen and directly examined); they also provide engineers and others with easily understandable scenarios. They do not, however, provide an open and formal way to account for uncertainty, and this can lead to the mistaken idea that there is no uncertainty. The deterministic estimates provided in this study were developed simply for models of significant regional earthquake sources and a median plus $x\sigma$ estimate of what the ground motions from the largest earthquakes associated with these sources would attenuate to selected BART facilities. This procedure, with use of engineering judgment, has historically been adopted for all BART facilities.

On the other hand, probabilistic estimates can incorporate a wide range of information and judgment, can handle uncertainty formally and explicitly, and can reach conclusions that are not easily upset by new data or hypotheses. However, their highly integrative nature can obscure those model elements that drive the results and their highly quantitative nature can lead to false impression of precision.

The deterministic method can be thought of as a special case of the probabilistic formulation. That is, let $v(z)$ be the average annual frequency for which the level of ground motion parameter Z exceeds value z at the site from all earthquakes on all sources in the region. Then,

$$v(z) = \sum \alpha_n(m^0) \int_{m^0}^{m_n^u} f_n(m) \left[\int_{r=0}^{\infty} f_n(r|m) \cdot P(Z > z|m,r) \cdot dr \right] \cdot dm \quad (1)$$

where $\alpha_n(m^0)$ is the annual frequency of earthquakes on source n above a minimum magnitude of engineering significance, m^0; $f_n(m)$ is the probability density of earthquake size between m^0 and a maximum earthquake the source can produce, m_n^u; $f_n(r|m)$ is the probability density function for distance to an earthquake of magnitude m occurring on source n; and $P(Z>z|m,r)$ is the probability that, given an earthquake of magnitude m at distance r from the site, the peak ground motion will exceed level z. These parameters are determined from the distribution in time and within the earth's crust of regional earthquakes and from models of the magnitude and distance dependance of Z. For the BART analysis, Z is the expected horizontal ground motion spectral acceleration assuming NEHRP (1997) site classification S_B/S_C (seismic shear-wave velocity = 760 m/s) and neglecting near-field and fault propagation effects. To incorporate these effects (potentially important near major faults), to develop vertical design ground motions, and/or to modify these base-case motions for other site classifications, simple scalar multipliers were found to be adequate.

It is assumed that the occurrence of damaging earthquakes can be represented as a Poisson process (where t is time in years):

$$P(Z > z|t) = 1 - e^{-v(z) \cdot t} \quad (2)$$

Deterministic estimates of "maximum credible" median or median-plus-some-percentile design ground motions neglect the second equation and simplify the first by eliminating the integrals (formally, $f_n(m)$ becomes $\delta_n(m_{max})$ and $f_n(r|m)$ becomes $\delta_n(r_{min}|m_{max})$ where $\delta(x)$ is the Dirac delta function), noting the value of z for which $v(z)$ jumps from 0 to 1 for each source, and keeping the maximum z from all sources.

Scenario earthquake ground motions assume deterministic-type magnitudes and distances but may incorporate uncertainty in the attenuation relations to develop a distributions of results, in the current case to provide a range of inputs to cost-benefit analyses or liquefaction potential evaluations (see below).

Modifications of probabilistic or deterministic estimates of horizontal rock design ground motions to include effects of alternative soil types, orientation relative to a near-by fault, or direction of fault rupture propagation relative to the site, or to estimate vertical motions, were made with scalar coefficients.

Development of Horizontal Spectra for Alternative Soil Types

Site amplification factors were used to develop soil motions from rock motions consistent with NEHRP site classification following the recommendations in the 1997 and 2000 NEHRP Provisions with adjustments described below.

The design rock response spectra for BART were developed using widely used rock attenuation relations (see Bechtel Infrastructure Corporation, 2002, for details). Examination of the ground motion databases used in developing these relationships indicates that some of the data are from NEHRP Site Class B (S_B) and some from Site Class C (S_C). Therefore, when using the NEHRP site factor tables, which are referenced or normalized to Site Class B (for which all site factors are 1.0 in the NEHRP table), the NEHRP site factors was renormalized to the predominant condition represented by the rock attenuation relationships. It was decided to renormalize the site factors to the S_B/S_C boundary.

Other than renormalization of the site factors to an appropriate reference condition, there is no current consensus for changes to the NEHRP site factors. A number of studies that have been carried out since the site factors were developed in 1992 and published in the 1994 edition of the NEHRP Provisions (see Bechtel Infrastructure Corporation, 2002, for details). Based on an evaluation of these studies an adjustment was made a slight increase was made to the NEHRP 1.0-second period site factor for Site Classes C and D.

Some of the recent site factor studies have examined how site factors change for periods longer than 1.0 second. In addition, soil attenuation relationships extended to periods longer than 1.0 second were examined for their predicted amplifications relative to rock attenuation relationships. On the basis of these examinations, a set of site factors for motion $\geq$ 2.0-second period was developed that is a factor of 1.2 times higher than site factors at 1.0-second for Site Class C and D.

The resulting complete set of site factors (for 0.2-second, 1.0-second, and $\geq$ 2.0-seconds is shown in Table 1. In this table, S_S is the normalized short period (0.2 s) response spectral acceleration for "rock" (the B/C boundary condition) and S_1 is the normalized long period (1.0 s) response spectral acceleration for rock.

The site amplification factors were applied to the horizontal rock spectral acceleration for each seismic zone to develop horizontal soil motions.

Table 1 Adjusted Site Amplification Factors

(a) Values of Short Period Site Factor: Fa (0.2 second period)

Site Class	Ss ≤ 0.25	Ss = 0.50	Ss = 0.75	Ss = 1.00	Ss ≥ 1.25
A	0.72	0.72	0.76	0.80	0.80
B	0.90	0.90	0.95	1.00	1.00
B/C	1.00	1.00	1.00	1.00	1.00
C	1.10	1.10	1.05	1.00	1.00
D	1.45	1.27	1.14	1.10	1.00
E	2.27	1.55	1.14	0.90	0.90

(b) Values of Long Period Site Factor: Fv (1.0 second period)

Site Class	$S_1 \leq 0.1$	$S_1 = 0.2$	$S_1 = 0.3$	$S_1 = 0.4$	$S_1 \geq 0.5$
A	0.59	0.62	0.64	0.66	0.70
B	0.75	0.77	0.80	0.83	0.87
B/C	1.00	1.00	1.00	1.00	1.00
C	1.30	1.30	1.25	1.25	1.20
D	1.80	1.60	1.50	1.45	1.40
E	2.60	2.45	2.25	2.00	2.00

(c) Values of Long Period Site Factor: Fv' (≥ 2.0 second period)

Site Class	$S_1 \leq 0.1$	$S_1 = 0.2$	$S_1 = 0.3$	$S_1 = 0.4$	$S_1 \geq 0.5$
A	0.59	0.62	0.64	0.66	0.70
B	0.74	0.77	0.80	0.83	0.87
B/C	1.00	1.00	1.00	1.00	1.00
C	1.55	1.55	1.50	1.50	1.45
D	2.10	1.90	1.80	1.75	1.70
E	2.60	2.45	2.25	2.00	2.00

Implementation of the Seismic Hazard Characterization

Retrofit Ground Motions

For the purpose of practical seismic retrofit criteria, it was recognized that a limited suite of criteria design spectra should be recommended for application to current BART structures and facilities. Preliminary DBE and LDBE design response spectral values were developed for about 20 sites distributed along the BART facilities alignment. It was determined that these spectra could be reasonably grouped into five zones for each of which a single horizontal rock design spectrum could be specified. These zones (shown in Figure 1) are: 1 – Near San Andreas Fault, 2 – San Francisco County, 3 – Transbay Crossing, 4 – Downtown Oakland and East of Hayward Fault, and 5 – Near Hayward Fault. The design spectra for each particular zone are those spectra representing strongest motion at any trial site within the zone so that the design spectrum for a given zone is intended to be a conservative estimate for all points within that zone.

Horizontal rock design response spectra were developed for each zone for 14 spectral periods from PGA (assumed to correspond to 0.0303 seconds) to 10.0 seconds. A high-order polynomial function was fit through these 14 points to produce a smooth spectrum defined at any desired period within this interval.

The modified NEHRP site amplification factors of Table 1 are specified only at three periods. To use the factors in scaling the response spectra at other site conditions, a smooth continuous curve covering the entire period range for the amplification factors was generated for each distinct soil condition. For each site condition, amplification factors were selected from Table 1 for periods at 0.2 second, 1.0 second, and 2.0 seconds and factors obtained at intermediate periods by semi-log-linear interpolation. The curve was then smoothed at the corners so that the spectrum curves at all five zones were smooth and continuous, with no sharp corners.

The final results of the horizontal design spectra for checking life-safety performance (and for all site conditions B, B/C, C, D, and E) are shown in Figure 2 for seismic zone 1. Similar figures were developed for all zones and the Transbay Tube.

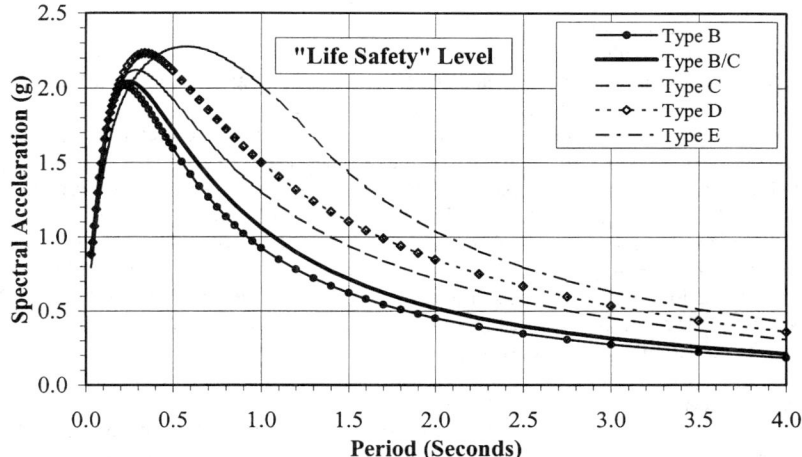

Figure 2 Horizontal Acceleration Design Response Spectra with 5% damping for Zone 1: "Near San Andreas Fault." (The B/C spectrum is based on the greater of deterministic median + 0.5σ or probabilistic 500-year return period values from all significant regional earthquake sources).

Vulnerability Analysis Ground Motions

The characterization of ground motions is a key initial input parameter in the seismic vulnerability assessment of BART. Seismic vulnerability studies can be performed using either deterministic or probabilistic estimates of the ground motions based on the formulation presented earlier in this paper. For the BART seismic vulnerability study (Eidinger et al., these Proceedings), the characterization of the ground motion was developed based on deterministic ground motion predictions for four scenario earthquakes: San Andreas (M8.0), Hayward (M7.0), Calaveras (M6.8) and Concord-Green Valley (M6.8). The sizes of these four earthquakes, which are not necessarily the largest earthquake magnitudes possible for a given fault or fault system in the Bay Area, were selected based on their likely occurrence in the next 30 years and the likelihood of significant damage to the BART system. The initiation point for each of the four scenarios was chosen to maximize the effects of rupture directivity on the ground motions over the entire BART system.

Given an earthquake scenario, the deterministic ground motions for approximately 15,000 sites along the BART system were estimated as the average ground motions from the suite of empirical attenuation relationships for rock site conditions used in developing the BART retrofit ground motion criteria. To account for the variability in ground motions, a Monte Carlo technique was used to randomly generate ground motions from 100 different realizations of a given scenario earthquake (i.e, fault source, rupture direction, and magnitude) based on the uncertainties given in the empirical attenuation relationships.

Ground motion values for soil sites were estimated by applying scalar amplification factors, as discussed earlier, to the rock ground motion values. Additional scalar factors were applied to incorporate the effects of rupture directivity for an extended fault source. These factors were based on a site-specific directivity scalar value for a given scenario and site location rather than on a global system-wide value developed for the retrofit ground motion criteria.

An additional uncertainty associated with the spatial correlation of ground motions from a given earthquake which can be important for extended systems (e.g., BART, power distribution systems, or large span bridges) was also considered in the Monte Carlo simulation of ground motions for the vulnerability study. This variability was based on the observed variation in ground motions from recently recorded large earthquakes. Finally, the ground motions from each of the 100 simulations for each of the four scenarios were convolved with the set of BART system-specific fragility curves (Salmon et al., these Proceedings) to estimate the seismic vulnerability of the BART system.

Liquefaction Potential Ground Motions

As the Transbay Tube is one of the more critical elements of the BART alignment, a detailed evaluation of the liquefaction hazard of the Tube was undertaken. An initial evaluation that considered the specific geotechnical properties of the underlying and overlying fill material indicated that the fills would reach liquefaction early in the progress of a major earthquake – magnitude 7 to 8 – along the Hayward or San Andreas faults. In concert with the quantification of seismic ground motion hazard that considered both scenario and cumulative (i.e., all sources) probabilistic hazards, similar assessment of the probabilistic hazard of liquefaction was undertaken.

Early methodologies of liquefaction hazard assessment, such as Seed and Idriss (1982), consider peak ground acceleration, the duration of shaking, and the geotechnical parameters characterizing the site-specific soil of interest to assess the likelihood of liquefaction in a "yes/no" or binary fashion. A relative quantitative measure of the likelihood or *factor of safety* can be obtained by determining the "distance" from the empirical curve that delineated the binary assessment of liquefaction potential. Starting in the late 1980s, a number of relations have been published that have statistically formalized the factors of safety to develop relations that assess the *conditional* probability of liquefaction (see discussion in Bechtel Infrastructure Corporation, 2002), such as shown in Figure 3. These relations are *conditional* in that they assume that the earthquake has occurred with the given ground motion intensity and duration. As part of BART's evaluation of its vulnerability to earthquake hazards these relations were considered to assess the probability of liquefaction given the median ground motions from the occurrence of various size events along the San Andreas or Hayward faults. Figure 4 shows an estimate of the conditional probability of liquefaction from various size events along these faults for a location along the Tube.

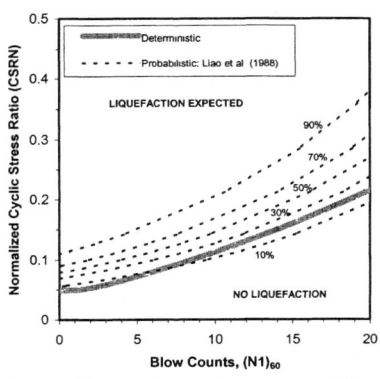
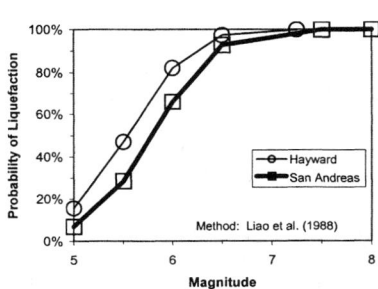

Figure 3. Comparison of binary ("deterministic") assessment of liquefaction potential from Seed and Idriss (1982) and conditional probability of liquefaction from Liao et al. (1988).

Figure 4. Conditional probability of liquefaction for various size scenario earthquakes for a site along the BART Transbay Tube, using conditional probability relation of Liao et al. (1988).

Given an estimate of the annual recurrence rate of these scenario events and their rupture proximity to the Tube location, it was assessed that the annual probability of liquefaction for any scenario event along the San Andreas or Hayward faults was less than 1.5%.

Four published relations of conditional probability of liquefaction were utilized further to assess the *joint* annual probability of liquefaction from all seismic sources and all magnitudes. The robust quantification of this hazard can be represented by an adaptation of Eq. 1:

$$v_L = \int\limits_{z=0}^{\infty} \left\{ \sum \alpha_n(m^0) \int\limits_{m^0}^{m_n^u} f_n(m) \left[\int\limits_{r=0}^{\infty} f_n(r|m) \cdot P(Z=z|m,r) \cdot P_L(L|z,m,CSRN,(N_1)_{60}) \cdot dr \right] \cdot dm \right\} \cdot dz \quad (3)$$

where v_L is the annual number of liquefaction events and $P_L(L|\ z,\ .\ .\ .)$ is the conditional probability of liquefaction. Note that since $P_L(L|\ z,\ .\ .\ .)$ is, among other dependences, a function of a given ground motion z, then the hazard has to be integrated over all possible ground motions.

The integration in Eq. 3 was performed as an equivalent post-processing routine on the ground motion hazard curve Eq. 2, where the magnitude and distance deaggregation matrices as a function of z were developed for several points along the ground motion hazard curve, and the integration over z in Eq. 3 was performed as a discrete summation. Uncertainty in the values of $(N_1)_{60}$ was incorporated by considering a weighted distribution of $(N_1)_{60}$ values. The final results indicated an annual probability of liquefaction of 5% to 7% at any location along the Transbay Tube. Since the DBE ground motion for the Tube is at least the median + σ deterministic motion from earthquakes greater than magnitude 7-1/4 on the Hayward or 8 for the San Andreas faults, Figure 4 indicates that there is greater than a 90% probability of liquefaction along the BART Transbay Tube in the event of the DBE.

References

Bechtel Infrastructure Corporation (2002). *BART Seismic Retrofit Program – Development of BART Seismic Retrofit Design Ground Motion Criteria*, Report submitted to San Francisco Bay Area Rapid Transit District.

Liao, S. S. C., D. Veneziano, and R. V. Whitman (1988). *Regression models for evaluating liquefaction probability.* J. Geotech. Engrg., ASCE, v. 114, n. 4, 389-411.

Michael, A. J., Ross, S. L., Schwartz, D. P., Hendley, J. W., II, and Stauffer, P. H. (1999). *Major quake likely to strike between 2000 and 2030—Understanding Earthquake Hazards in the San Francisco Bay Region*, U. S. Geological Survey Fact Sheet 152-99, 4 p.

National Earthquake Hazard Reduction Program NEHRP (1997). *Recommended Provisions for the Development of Seismic Regulations for New Buildings and Other Structures*, Vol. I (Provisions), Vol. 2 (Commentary).

Seed, H. B. and I. M. Idriss (1982). *Ground motions and Soil Liquefaction During Earthquakes.* Earthquake Engineering Research Institute, Monograph Series, 134p.

Working Group on California Earthquake Probabilities (WGCEP), 1999 (WG99), *Earthquake probabilities in the San Francisco Bay Region, 2000 to 2030 – A summary of findings*: U.S. Geological Survey Open File Report 99-517, 36 p. plus figures and tables.

Seismic Risk Analysis of the Bay Area Rapid Transit System

John Eidinger[1]
Ed Matsuda, Tom Horton[2]
Ching Wu[3]

1 Introduction

This paper presents a Seismic Risk Analysis of the Bay Area Rapid Transit (BART) system (Figure 1).

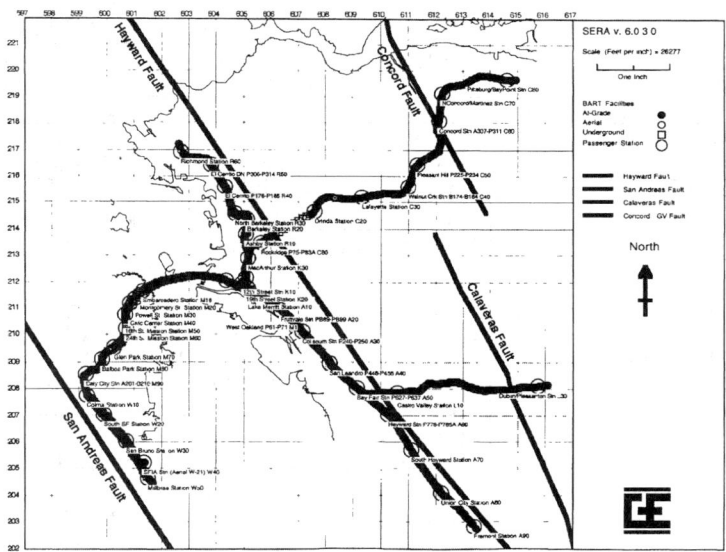

Figure 1. BART System, Showing Location of Nearby Faults

The Seismic Risk Analysis includes assessments of:

1. How the existing BART system might perform after large earthquakes on the San Andreas, Hayward, Calaveras and Concord faults.

2. How a seismically-upgraded BART system might perform after these same earthquakes, assuming implementation of any of five different retrofit alternatives (called Packages 1, 2, 3, 4, 5). The capital cost of these five packages ranges from $729,000,000 to $1,118,000,000. (All $ in this paper are in $2002).

---

[1] G&E Engineering Systems Inc., 6315 Swainland Rd, Oakland, CA 94611; eidinger@earthlink.net
[2] BART, 1330 Broadway, 12th Floor, Oakland, CA 94612; ematsud@bart.gov; thorton@bart.gov
[3] Bechtel, 1330 Broadway, 12th Floor, Oakland, CA 94612 ; cwu1@bart.gov

3. The benefits and costs of the six alternatives: do nothing, or implement any of the five retrofit packages. This assessment is done using benefit-cost analyses.

BART initiated the Seismic Risk Analysis in order to assess whether a retrofit program was needed, and if needed, demonstrate the need in a credible manner. BART also wanted the Seismic Risk Analysis to develop information that could be used to choose and justify a prudent and cost effective level of retrofit.

Component	Number	Description
At grade facilities	10,106	5,809 track segments, either bolted or ballasted; 542 embankments ranging from 5 to 35 feet in height; soil retaining walls; 3,755 aerial girders.
Aerial columns	1,983	1,983 aerial columns of 10 major types. A typical column carries 2 girders and 2 tracks
Underground facilities	112	27 bored tunnels, 28 cut-and-cover tunnels, 57 segments of the Transbay tube. In 363 segments.
Passenger stations	43	34 1968-vintage stations, 5-1990 vintage stations, 4-2002 vintage stations. Stations are either aerial, at-grade or underground. Underground stations include ventilation equipment. All stations include train control, communication and other low voltage equipment. Includes 1,964 pieces of equipment.
Ventilation facilities	90	Horizontal, vertical and centrifugal fans and controllers; communication cabinets; low voltage circuit breakers and transformers
Electric substations	68	Substations and switching substations. These convert 34,500 (or higher) volt PG&E power to 1,000 volt DC power used for traction power, and provide circuit protection between the PG&E and BART systems. Includes 351 individual electrical components (rectifiers, etc.)
Breaker stations	40	These provide circuit protection between adjacent contact (third) rails of the BART system
Administrative and maintenance yard buildings, parking structures	41	21 Yard buildings (towers, car washes, S&I buildings, blowpits, storerooms, traction motor repair facilities, training centers); 7 multi-story parking garages; office buildings; control facilities; 17 non-occupied equipment buildings

Table 1. Inventory for Evaluation of BART System

2 Inventory

The Seismic Risk Analysis was performed for essentially every structure and piece of equipment in the BART system. The analyses were performed using the System Earthquake Risk Assessment (SERA) program. The SERA program is a specialized geographical information system for the seismic evaluation of complex and geographically distributed lifeline systems (G&E, 2002).

The analysis includes a total of 15,078 individual structures and pieces of equipment located at 3,089 sites within the BART system. The year 2002 replacement value for these components is $10.85 Billion, excluding the value of land. These components are listed in Table 1.

3 Scenario Earthquakes and Ground Motion Parameters

The BART system was analyzed for four scenario earthquakes (Table 2). The scenario earthquakes represent the most likely earthquakes in the Bay Area that would cause considerable damage to the BART system.

Earthquake Source Fault	Magnitude M_w	Where (Approximately)
Hayward	7.0	From Richmond to Fremont
San Andreas	8.0	From North of Fort Bragg to south of Palo Alto
Calaveras	6.8	From Danville to south of Pleasanton
Concord	6.8	From Concord to north of Fairfield

Table 2. Scenario Earthquakes for Evaluation of BART System

Ground motions were calculated at each of the 3,089 locations of BART structures and equipment. Attenuation models are based on an average of five attenuation relationships (Abrahamson and Silva, Boore, Joyner and Fumal, Campbell, Idriss and Sadigh et al). The project report (G&E 2002) and Litehiser et al (2003) provide comprehensive descriptions.

The ground motions were applied to the model with allowance for:

- Randomness due to the attenuation relationships.

- Fault rupture directivity and fault normal / fault parallel characteristics.

- Local soil conditions, which may amplify or de-amplify the response spectral shape.

- Liquefaction susceptibility, which could create permanent ground deformations. Settlements could also occur at embankments.

- Fault offset. The BART tracks cross the Hayward fault at grade at two locations; the Calaveras fault at grade at one location, and the Concord fault aerially at one location.

- Inter-event variability. This is the variability of the median ground motion between different simulations of the same scenario earthquake. It is due to the variability of the average features of the seismic source. For example, it could be thought of as the effect of the variability of stress-drops of earthquakes. If a particular simulation of a scenario earthquake has a high stress-drop, then the average ground motion for that simulation will be above the median predicted by the attenuation relationship.

- Intra-site correlation. As each individual aerial column is modeled separately, and since we wished to examine the potential of no-damage versus any-damage between any two passenger stations, we included the correlation of ground motions between any two sites. For example, if one column experiences a "median plus one standard deviation" motion, it would make sense that the adjacent column, located about 70 feet away, would also likely experience close to a "median plus one standard deviation" motion.
- The detailed procedure to account for inter-event and intra-site correlations is based on work by Abrahamson, relying on observations from recent earthquakes such as Chi-Chi 1999. Analyses with consideration of intra-site and inter-event correlations tended to produce wider scatter for performance of the BART system as a whole than when these variabilities were not considered.

Retrofit Alternative	Description	Total Cost ($1,000,000)
Package 1. Safety Improvements and Transbay Tube	Upgrade aerial structures, passenger stations, occupied buildings, the Transbay Tube and equipment	$729
Package 2. Operability Improvements from Rockridge to Daly City Yard	All Package 1 retrofits, plus Operability Improvements from Rockridge Station to Daly City Yard, additional upgrade to the Lake Merritt Administration building, plus additional upgrades to equipment	$828
Package 3. Operability Improvements from North Berkeley to Coliseum	All Package 2 retrofits, plus Operability Improvements from MacArthur Station to North Berkeley Station and from the Oakland Wye to Coliseum Station	$882
Package 4. Operability Improvements to South Hayward	All Package 3 retrofits, plus Operability Improvements from Coliseum Station to South Hayward Station	$972
Package 5. Full Operability Improvements to Richmond, Fremont, and Concord	All Package 4 retrofits, plus Operability Improvements from South Hayward Station to Fremont Station, North Berkeley Station to Richmond Station, Orinda Station to Pittsburg / Bay Pointe Station	$1,118

Table 3. Retrofit Cost Summary

4 Retrofit Alternatives

The BART system was evaluated in its Status Quo condition, and for five possible retrofit alternatives. Table 3 describes the five retrofit alternatives.

5 Results of Seismic Risk Analysis

Each of the 15,078 structures and pieces of equipment were evaluated for 100 simulations for each of the four scenario earthquakes. The impacts for each scenario earthquake are presented in several ways:

- The cost to repair damage to BART structures and equipment. This is calculated using fragility curves. Each set of fragility curves for each component includes a series of damage states for inertial (or where applicable) permanent ground deformation movements; cost and time needed to make emergency repairs (like temporary bracing); cost and time needed to make permanent repairs; life-safety potential. Fault trees are used to consider the overall impact of a facility should a series of individual damage states occur.

- The number of riders that will leave the BART system while the damage is being repaired. This includes a BART-system wide model showing the number of riders making a trip from station A to B under normal (non-earthquake) conditions. After the earthquake, some of these station pairs may will remain out of service until sufficient repairs are made to all significant damage between the stations. The model allocates crews to repair components between the highest ridership stations, thereby tracking the restoration of ridership for each day after the earthquake until the entire system is functional.

- The number of sites in the BART system that sustain enough damage to cause a potential serious chance of fatality or injury to BART riders or BART employees.

- All results are tracked in terms of the range of outcomes for 100 simulations for each scenario earthquake. For example, it might take 894 days (average) to fully restore service after a Hayward M 7 earthquake; or 746 days (16th percentile) or 1,041 days (84th percentile).

Tables 4 and 5 summarize these three impacts for two of the scenario earthquakes for the BART system in its Status Quo (as-is) condition, or in each alternative retrofit condition.

Results are mapped using the SERA geographical information software. Each component is plotted in one of four colors: red (significantly damaged and trains cannot operate until at least temporary repairs are made); yellow (damaged and trains can operate at reduced speed); blue (lightly damaged, trains can operate at full speed while repairs are made); green (undamaged). For a typical simulation of the Hayward M 7 scenario earthquake, the dominant damage is as follows:

- Transbay tube. Suffers relative settlements due to liquefaction, possible dislocation of expansion joints. Consequences include expensive and long duration retrofit before the tube can be re-opened for general use and significant life safety risk. Figure 2 shows a two year outage time for the tube should it suffer heavy damage. Field investigations are ongoing (early 2003) to refine the level and consequences of damage and restoration time for the tube; current findings suggest that restoration time could significantly exceed two years, even if regulatory requirements dealing with construction in the San Francisco Bay are waived.

- Aerial columns. BART has some 1,983 individual aerial column, bent and abutment structures, most of which support two to four girders. Many of these might suffer significant damage to their base mat foundations, leading to some amount of permanent leaning of the column; a few could even outright collapse.

BART System Condition	Direct Damage ($ Millions)	Lost Ridership (Millions of Trips)	Sites with Significant Life Safety Potential
Status Quo	$1,097	82.4	286
Package 1	341	41.2	5
Package 2	310	21.2	5
Package 3	295	20.1	5
Package 4	269	18.5	5
Package 5	198	17.0	4

Table 4. Summary Impacts – Hayward M 7 Scenario Earthquake

BART System Condition	Direct Damage ($ Millions)	Lost Ridership (Millions of Trips)	Sites with Significant Life Safety Potential
Status Quo	$860	77.5	111
Package 1	174	19.6	2
Package 2	163	11.8	2
Package 3	162	11.7	2
Package 4	162	11.5	2
Package 5	145	11.3	2

Table 5. Summary Impacts – San Andreas M 8 Scenario Earthquake

Ridership losses are calculated using a BART system model. Figure 2 shows a typical result from the model, for the system in its as-is "status quo" and upgraded "package 1" conditions. Figure 2 highlights a few of the key issues: there are normally (as of mid-2002) about 250,000 average daily riders (average over 7 days – ridership on weekdays is higher); in the "status quo" condition, the undamaged part of BART can carry about 70,000 rides per day within a couple of days; then increasing to 125,000 rides per day upon completion of repair to aerial structures; then increasing to 240,000 rides per day once the Trans Bay Tube is put back in service; and finally increasing to 250,000 rides per day once the Berkeley Hills Tunnel is returned to normal service in about 2.3 years.

Figure 3 shows pie charts with the cumulative impacts caused by damage to the BART system for each of the four scenario earthquakes. Each pie chart is divided into six parts. The meaning of the pie charts is as follows:

- Repair costs. These are the out-of-pocket costs to BART to make the repairs following the earthquake. This is the same as the "direct damage" data in Tables 4 and 5.

- Fare box revenue losses. This is the lost revenue to BART from the impact of loss of ridership due to damage to the system.

- Bus bridge cost. This is the cost to BART to operate bus bridges following the earthquake.

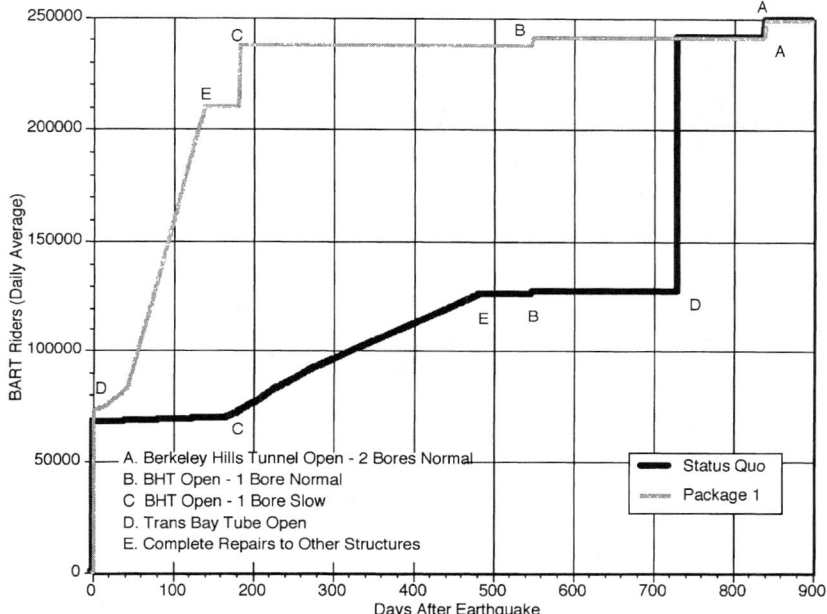

Figure 2. Example Restoration of Service After Hayward M 7 Scenario Earthquake

- Economic impact to BART riders. This is the cost to BART riders who take alternate transportation to reach their destinations, given that BART service is not available due to earthquake-caused damage to the BART system.

- Economic impact to Bay Area commuters. This is the cost to Bay Area commuters who are impacted by the increased congestion caused by the BART riders who take alternate transportation to reach their destination.

- Monetized casualty loss. This is the economic value of injuries and fatalities caused by earthquake-caused damage to the BART system.

- Number directly below each pie chart: the total scenario loss (year 2002 dollars). This is the sum of all the "wedges" of each pie chart. For example, the total scenario loss for the Hayward M 7.0 earthquake for BART in its Status Quo condition is $4,841,000,000.

- The order of the legend corresponds to a clockwise depiction of each "wedge" in the pie chart, beginning with the 12 o'clock position.

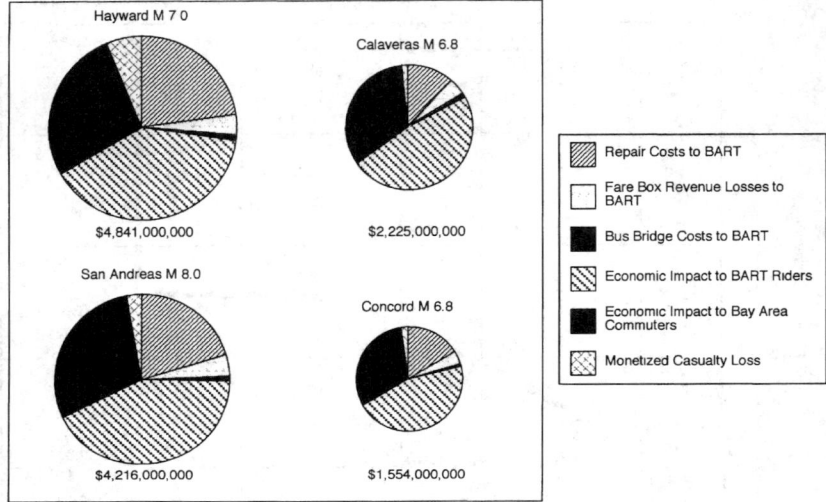

Figure 3. Total Impact of Scenario Earthquakes – BART System in its Status Quo Condition

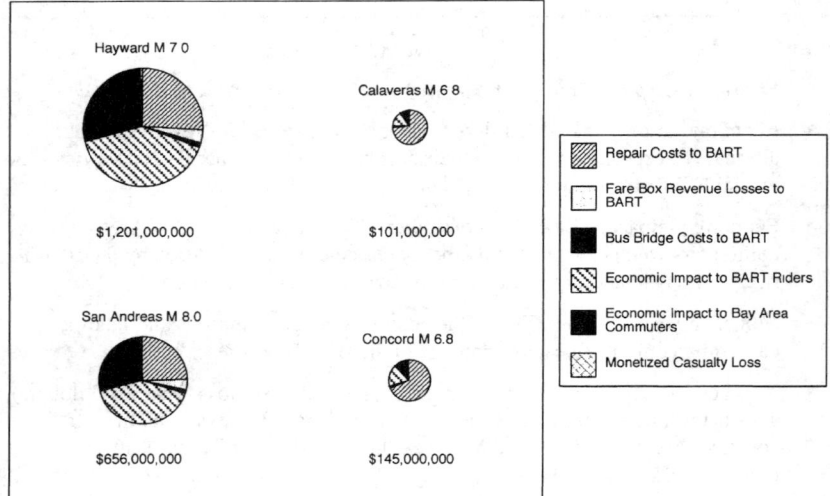

Figure 4. Total Impact of Scenario Earthquakes – BART System After Implementation of Seismic Retrofit Package 2

For example, Figure 3 shows that the expected economic impacts to the Bay Area are $4.841 Billion (year 2002 dollars) should a Hayward Magnitude 7 earthquake occur before BART implements any seismic retrofit program. About one quarter of the losses are "out of pocket" costs to BART (the sum of the cost to make repairs, lost fare box revenue and the cost to operate bus bridges while repairs are made). The remainder of the losses accrue to BART riders and Bay Area commuters, mostly due to the disruption of commute patterns, and some due to the potential for casualties.

Figure 4 presents similar pie charts as Figure 3, but with the assumption that BART implements retrofit Package 2, costing $828 million. The scenario losses are reduced from the Status Quo case by an average of 80%.

6 Benefits and Costs of Retrofit Alternatives

A Benefit Cost Analysis was performed to establish the relative cost effectiveness of each of the six retrofit alternatives. A Benefit Cost Ratio (BCR) of 1 or higher shows that the cost for the retrofits is less than the net present value of the benefits (benefits = reduction in future losses) from earthquakes, in consideration of the time value of money, the probabilities of different sized earthquakes, and their impacts. The four faults listed in Table 2 plus various magnitude earthquakes on additional faults were considered for purposes of benefit cost analysis. The three main impacts (direct damage, loss of ridership, life safety) were monetized for purposes of the Benefit Cost Analysis. The BCR values are based on the incremental benefit of each more expensive retrofit alternative. Table 7 summarizes the findings.

BART Condition	Incremental Costs ($millions)	Incremental Benefits ($millions)	Benefit-Cost Ratio
Status Quo	n/a	n/a	n/a
Package 1	$729	$2,394	3.28
Package 2	$98	$225	2.29
Package 3	$55	$23	0.42
Package 4	$90	$38	0.42
Package 5	$146	$67	0.46

Table 7. Costs and Benefits of Each Retrofit Alternative

Based on these findings, seismic retrofit of the BART system is economically sound, for either retrofit alternatives Package 1 or 2, costing $729 or $828 million respectively. Seismic retrofit Packages 3, 4 and 5, costing from $882 million to $1.1 billion, do not appear to be justifiable on an economic basis.

Besides the benefit-cost ratios given above, many other quantifiable and less tangible factors need and have been considered in recommending a retrofit package. One factor was the consequences of using lower deterministic acceptance criteria (like stresses, strains, allowable demand to capacity ratios, etc.), and allowing more damage and longer restoration/repair time in selected segments of the system versus incremental cost. Another factor was the level of uncertainty associated with specific fragility relationships, and associated risk due to the uncertainty. Broader economic impacts to the Bay Area from the extended loss of a vital, established, public transportation system, which might include loss of business, lowered real estate values, reduced consumer spending, etc., were not directly included in the benefit cost analysis. Since the "Package 5" option returns more of this vital transportation system to operation sooner, these broader impacts would be reduced, compared to the "Package 2" option. Although Caltrans and many cities are conducting retrofit programs of their own, the possibility of road closures and disruptions to other forms of transportation after a large earthquake remains. Experience following the Loma Prieta earthquake showed that BART system ridership could increase as a result. In addition, the impact of post-earthquake repairs on local communities near the BART alignment will be less if the "Package 5" option is implemented, since there will be fewer repairs required.

7 Implementation

Based on the findings of the seismic risk analysis, it was apparent that seismic retrofits included with Package 1 or 2 were the most cost effective. Additional retrofits beyond the package 2 level appear to provide decreasing benefits. In November 2002, a BART Retrofit Measure (bonds that would be repaid via property tax assessments) was put on the ballot in three counties: San Francisco, Alameda and Contra Costa. This bond measure would have provided funds to implement package 2 (with allowance for inflation). In San Francisco, Alameda and Contra Costa Counties, the bond measure had a 73.4%, 66.6% and 54.4% "yes" vote respectively. To pass the bond, the required cumulative "yes" vote was two-thirds majority, while the actual cumulative "yes" vote was 64.2%.

The vulnerability study uncovered unanticipated safety risks that led to the decision to place a Bond Measure on the November 2002 ballot. Although the BART Board was concerned about the short time available for campaigning, experts on both the Bechtel Design Review Panel and the Independent Peer Review Panel recommended that the vulnerabilities be addressed and mitigated as soon as practical. Despite a shortened campaign with very little publicity, the Bond almost passed.

After the defeat of the Bond Measure, BART has had to significantly curtail activities on its seismic retrofit project. Due to the critical nature of the program, BART is planning to bring the issue to the voters again in March or November of 2004 with an enhanced information effort.

8 References

G&E Engineering Systems Inc., BART Seismic Risk Analysis, Report 55.01.05, prepared for the Bay Area Rapid Transit District, May 17, 2002.

Litehiser, J., Gregor, N., Marrone, J., Ostadan F., and Youngs R., Characterization of Design Ground Motion for the BART Retrofit Project, 6th US Conference on Lifeline Earthquake Engineering, TCLEE, ASCE, Long Beach, CA August 2003 (in press).

SERA User and Theoretical Manual, G&E Engineering Systems Inc., v. 6.0.3.1, June 2002.

Fragility Formulations for the BART System

By Mark Salmon[1], James Wang[2], David Jones[3], Ching Wu[4]

Abstract

This paper presents fragilities developed for the seismic risk analysis of the San Francisco Bay Area Rapid Transit (BART) system, including treatment of dependent and independent damage states. Categories of fragilities include aerial guideway structures, passenger stations, tunnels and the Transbay Tube, with fragilities developed based on a combination of analytical and experiential data.

Introduction

The BART system provides rail transit in the San Francisco Bay area, carrying approximately 300,000 passengers daily. Constructed approximately 30 years ago, the facilities have a number of seismic deficiencies by today's standards, and a major earthquake in the region may result in significant damages and losses. Accordingly, two types of vulnerability studies were undertaken to determine the vulnerabilities and potential losses: a deterministic seismic evaluation of the structures, and a probabilistic seismic risk analysis to determine the likely consequences to the BART system resulting from a number of scenario earthquakes, and to compare the benefits of various retrofit packages.

The inventory of the BART system consisted of over 15,000 components, including aerial structures, tunnels, the Transbay Tube, stations, system equipment and buildings. Each inventory item was assigned fragility curves, which express the conditional probability of reaching various damage states as a function of a seismic demand parameter, together with impacts on life safety, outage times, and repair costs. A number of different types of fragilities were used, derived using different techniques and with varying levels of detail, depending on the type of component, its impact on the system, and the type of data available. These included experience data, test data, analytical data, and published fragility data such as in HAZUS (1999).

The fragility curves in the BART model were expressed as a truncated lognormal distribution, which is defined in terms of a median seismic level parameter at which the damage is expected to occur 50% of the time, a lognormal standard deviation beta, and a lower bound truncation value representing the value of the seismic level parameter below which the damage state cannot occur. The seismic level parameter is either spectral acceleration, peak ground acceleration, peak ground velocity or peak ground deformation, depending on the type of component.

Three sets of fragilities were defined: the as-built condition, a "Life Safety" retrofit, and an "Operability" retrofit.

---

[1]MGE Engineering Inc. (Mark Salmon, Corresponding Author, 1330 Broadway, Suite 1410, Oakland, CA 94612, phone 510-208-4320). [2]Bay Area Rapid Transit District, [3]HNTB Corporation, [4]Bechtel Infrastructure Corporation.

Analytical Fragilities

A number of different techniques were employed to determine fragility curves, depending on the analytical data available and the importance and vulnerability of the component. The deterministic evaluation of BART structures provided tabulated Demand/Capacity (D/C) ratios of critical components, for a Design Basis Earthquake (DBE), according to a set of design rules or criteria. The "Initial SA" was defined as the spectral acceleration at which the component D/C equals unity, based on 5% damped response spectra and the design criteria. These design criteria generally contain considerable conservatism, and the Initial SA was factored to obtain the median spectral acceleration, using the margin factors in Table 1.

Table 1 - Margin Factors

Margin Factor	Description	To Take Account of:	Methodologies and Parameters Used	Typical Values
F1	Material Strength	Expected strength > design strength	Recalculate D/C's using expected strengths	1.0 - 1.3
F2	Supplemental Damping	Energy absorption due to sliding, rocking, radiation damping, etc.	Capacity spectrum method; Newmark & Hall spectral reduction factors for estimated equivalent viscous damping.	1.0 – 1.5
F3	Ductility, Hysteretic Response	Hysteretic damping due to ductile mechanisms	Capacity spectrum method; Newmark & Hall spectral reduction factors for effective viscous damping; equal energy and equal displacement principles using ductility ratios.	1.0 – 3.0
F4	Higher Modes	Higher modes not included in analysis	Mass and modal participation factors	≤ 1.0
F5	Vertical Earthquake	Damage due to concurrent vertical motions not included in analysis	Comparison with analyses including vertical motions	≤ 1.0
F6	Near Fault	Damage due to near field effects (velocity pulses) not included in analysis	Comparison with non-linear time-history analyses	≤ 1.0
F7	Modeling Conservatism	Conservative assumptions, lateral resisting or energy absorbing elements not included in analysis model	Comparisons with more refined models and historical performance of similar structures; judgment.	1.0 – 1.5
Total			Product of F1 to F7	1.0 – 5.0

The Initial SAs and margin factors were tabulated for each damage state, and the product of these calculated as the "Final SA," the median spectral acceleration at which the damage state occurs. These tabulations were reviewed by an Expert Panel, for consistency, reasonableness and completeness.

Figure 1 shows the capacity spectrum method of obtaining the F3 margin factor. The Capacity Curve and the DBE 5% damped response spectrum (DBE, 5%) are plotted on the same axes. In step 1, the DBE spectrum is reduced (DBE, β_{eff}) to account for hysteretic damping, in accordance with HAZUS (1999) and ATC-40 (1996). In step 2, this reduced spectrum is factored up by X in order that it intersects the Capacity

Curve at the damage state threshold point. In step 3 the 5% damped DBE spectrum is factored up by the same amount: the spectral acceleration SA_{final} at a given reference period represents the median seismic level parameter for the damage state (assuming all other margin factors are unity). F3 is the ratio of SA_{final} to $SA_{initial}$.

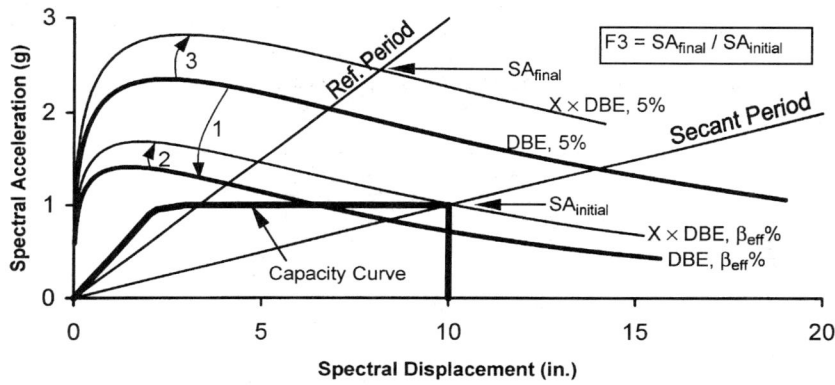

Figure 1 - Capacity Spectrum Method

Beta Values

The probabilistic uncertainty is represented by a Beta value, the log-normal standard deviation. The Beta values were, in general, judgmentally assigned, with published values, such as in HAZUS, used as reference. The damage state uncertainty consists of the structure (capacity) uncertainty combined with the ground motion (demand) uncertainty. The ground motion uncertainty is taken into account separately by the risk analysis software, thus the Beta values associated with the fragilities represent the structure uncertainty only. The following observations were made:

- Betas are greater for more severe damage states, due to greater uncertainty associated with the threshold of higher damage states.
- Betas are higher for failure modes where test results have shown a high degree of scatter or where there was a high degree of uncertainty in the analytical predictions.
- Higher Beta values tend to increase the total damage estimate, since the high "tails" have a disproportionately greater affect on total damage predicted by the simulations. In order not to generate over-estimates of the damage losses, a relatively low default Beta value of 0.3 was used, with most Beta values ranging from 0.2 to 0.5.
- The fragilities were truncated with a lower-bound "No Damage" value for the ground motion parameter, below which the damage state was considered as having essentially zero probability of occurring.

Independent and Dependent Damage States

The most common and simplest way to represent the damage to a structure is where the structure can only exist in one of several progressively severe damage states. This applies for a structure where there is only one single sequence of damage events, and would be represented by a single pushover curve, with a progression of yielding or failure events, as shown in Figure 2. An example would be a frame with several locations yielding in turn, or a structure with successive failures of lateral resisting elements. The damage states are considered dependent in that only one can occur at a time, and each damage state depends on the previous damage state having occurred first.

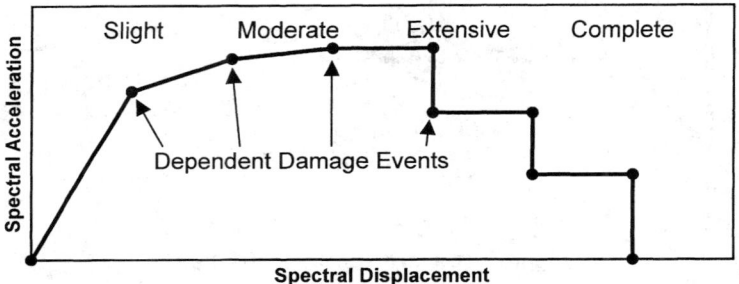

Figure 2 - Dependent Damage States

For more complex structures, there is a possibility that multiple damage states may occur simultaneously and independently. An example would be a building with two independent lateral systems in each orthogonal direction, or a structure with structurally independent portions, which would need to be represented by independent pushover curves, each with their own sequence, or group, of damage states, as shown in Figure 3. Each group would be considered independently, and the damage state and losses determined for each.

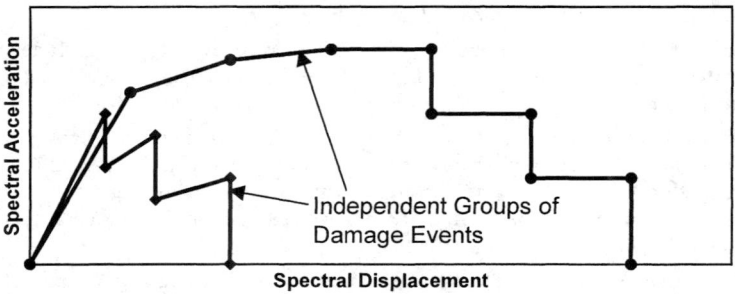

Figure 3 - Independent Damage States

A third case is where there are multiple possible modes of failure, but the modes cannot occur simultaneously and yet there is uncertainty about which mode occurs.

This is the case where a structure has a number of potential weak links in series, but it is uncertain which link is weakest. This may be represented by a pushover curve with multiple diverging branches or paths of failure, each path with its own sequence of damage states, but only one path is possible at a time, as shown in Figure 4. An example is an aerial column with three possible modes of failure: column hinging, pile failure, and pile cap failure. Only one of these modes can occur at a time, but it is uncertain which one is the weak link, and the consequences are greatly different. If the column hinges before the pile cap breaks, then the mode is ductile, whereas if the pile cap breaks first, it is brittle. For the column hinging mode, the "complete" damage state would occur at a much higher spectral displacement than for the pile cap failure mode.

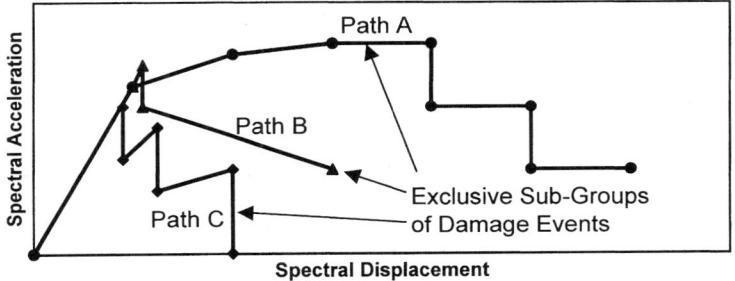

Figure 4 - Mutually Exclusive Damage Paths

The series of damage states along each of these paths is considered as a "sub-group." The mode of failure that occurs is governed by the relative probabilities (as defined by the fragility curves) of the initiating damage states for each sub-group. Once a particular path is selected (by Monte Carlo simulation), the other paths are excluded. This fault-tree logic for the sub-group failure modes is particularly applicable to structures such as the aerial columns, which have a number of possible paths in the pushover curves, with a high degree of uncertainty as to which path will actually occur, but very different consequences.

An alternative method of dealing with these multiple paths of failure is to combine, for example, the fragility curves for the "complete" damage states for each path into a single fragility curve that represents the "complete" damage state. This fragility curve would have a very high Beta value, due to the uncertainty regarding the mode of failure. Although the overall loss estimation may be correctly predicted using this approach, considerable detail information would be lost regarding the distribution and type of damage expected throughout the system.

Examples of Fragilities

Aerial Structures

The typical BART aerial structure consists of simply supported precast girders supported on single-column hammerhead piers on spread or pile foundations. Major damage mechanisms were flexural and joint shear failure of pile caps, pile connection failure, shear key failure, and column hinging. An example of fragility data for an aerial structure is given in Table 2.

Table 2 - West Oakland Aerial Guideway Pier Type IB Fragility Analysis

Damage Mechanism	Damage State	Median SA (g)	Beta	Ref. Freq (Hz)	SA No Damage	Function Factor	Life Safety	Group No.	Sub-Group	Short Term Cost	Long Term Cost
Pile cap Failure	Major	1.60	0.50	2.0	0.70	1.00	4	1	1		100%
	Minor	1.06	0.20	2.0	0.77	0.00	1	1	1	2%	20%
Pile Failure	Major	2.15	0.40	2.0	1.11	1.00	4	1	2	5%	60%
	Minor	1.44	0.20	2.0	1.04	0.00	1	1	2	2%	20%
Shear Key Failure	Major	2.08	0.30	2.0	1.26	0.50	3	1	3	2%	10%
	Minor	1.04	0.20	2.0	0.75	0.00	1	1	3		3%
Column Hinging	Major	3.93	0.20	2.0	2.81	0.50	2	1	4	2%	5%
	Minor	1.31	0.20	2.0	0.94	0.00	1	1	4		2%

In the above example, 4 paths ("Damage Mechanisms") are possible, represented by the 4 sub-groups. The mechanism that occurs is governed by relative probabilities of the initiating "minor damage" states. The loss of functionality was represented by the Functional Factor (1= no operation until long term repairs, 0.5= full operation after short term repairs, 0= no impact on operations), and the damage level and impact on life safety by the Life Safety Factor (1=slight, 2=moderate, 3=extensive, 4=complete/collapsed). The short and long term repair costs were represented as percentages of replacement cost.

At-Grade Stations

There are six at-grade stations in the original BART system. The at-grade stations generally consist of a combination of single-story concrete shear wall building-type structures and retaining wall structures, partially buried on one or more sides by adjacent embankments. Damage states were defined as:

- *Slight* - damage resulting in less than 1 inch of movement or minor cracking, so that operations are unaffected.
- *Moderate* - operations are impacted, but short-term repairs (e.g. track realignment) can return the station to full operation. Movements were taken as limited to about 3 inches, which corresponds to the track tolerances permissible for slow speed operation.
- *Extensive* - damage is where there is possibility of local failures, major structural damage, or movements greater than 6 inches. Normal operations are not possible, but there is no collapse or serious risk to life safety.
- *Complete* - total or partial collapse, high risk to life safety, and damage that may not be repairable.

An example of at-grade station fragilities is given in Table 3.

Table 3 - Example of At-Grade Station: MacArthur Station Fragility Analysis

Damage State	Median SA	Beta	Ref. Freq. (Hz)	Function Factor	Life Safety	Group No.	Short Duration (days)	Short Cost	Long Duration (days)	Long Cost
Complete	4.20	0.40	5.00	1.00	4	1			720	100%
Extensive	2.30	0.40	5.00	1.00	3	1			360	30%
Moderate	1.20	0.40	5.00	0.50	2	1	30	1%	90	5%
Slight	0.70	0.40	5.00	0.00	1	1	3	0.1%	30	1%

Cut-and-Cover Tunnels

Cut-and-cover tunnels were categorized as either one-cell, two-cell, or three-cell box structures. The major damage states were determined to be exterior wall bending, exterior wall cracking, and exterior wall shear, as shown in Table 4.

Table 4 - Example: Double Cell Box Fragility Analysis

Damage State	Median PGA	Beta	PGA No Damage	Function Factor	Life Safety	Group No.	Short Cost	Long Cost
Exterior Wall, Flexural Failure	5.42	0.3	2.71	1.00	4	1	N/A	50%
Exterior Wall, Significant Cracking	1.30	0.3	0.63	0.10	2	1	0.1%	1%
Exterior Wall, Shear Failure	1.81	0.3	0.91	1.00	4	2	N/A	50%

Bored Tunnels

The steel liner's longitudinal deformation and cross-sectional ovaling capacities control the seismic capacities of the bored tunnels. The nonlinear curves of liner axial forces and circumferential moments, as functions of their corresponding axial deformations and cross-section diameter changes, were developed for the different types of steel liner in the BART system. Using these capacity curves, the fragility data for the bored tunnel liners were obtained by extrapolating the seismic input levels such that the seismic D/C ratios computed for different tunnel locations in the BART system under the retrofit design seismic input condition were equal to 1.0. An example of bored tunnel fragility data is given in Table 5, characterized in terms of the peak rock acceleration (PRA) at the rock out-crop elevation of a site.

Table 5 - Bored Tunnel at San Francisco Approach Fragility Analysis

Damage State	Median PRA	Beta	Life Safety	Function Factor	PRA no Damage	Group No.	Short Cost	Long Cost
Failure in axial connection of ring plate	3.1	0.3	3	0.5	0.5	1	N/A	30%
Failure in circumferential connection of end plate	6.8	0.3	4	1.0	0.5	2	N/A	30%

Underground Stations

There are 14 underground stations in the original BART system. The analysis included soil column SHAKE analysis and SASSI soil-structure interaction analysis or racking analysis of cross-sections of the stations, together with SAP2000 pushover analyses, to determine component D/Cs under the DBE PGA levels at each location. The median fragility for a each component was expressed in terms of site PGA, derived from the D/Cs using margins analysis. An example is shown in Table 6.

Table 6 - Example Underground Station: 12th Street Station Fragility Analysis

Damage State	Median PGA	Beta	PGA No Damage	Function Factor	Life Safety	Group No.	Short Cost	Long Cost
Main Column Damage	1.13	0.30	0.57	1.00	4	1	N/A	50%
Roof Damage	1.09	0.30	0.55	1.00	4	2	N/A	50%
Mezzanine Floor Damage	1.38	0.30	0.69	1.00	4	3	N/A	30%
Upper Train Floor Damage	1.75	0.30	0.88	1.00	3	4	N/A	30%
Exterior Wall Damage Shear Failure	1.23	0.30	0.62	1.00	3	5	N/A	30%
Exterior Wall Damage Significant Cracking	1.50	0.30	0.75	0.00	2	5	0.1%	1%

Retaining and U-walls

Each height and type of wall was analyzed to determine the wall D/C's for stability and strength, based on a seismic earth pressure corresponding to a PGA of 0.7g. The PGAs corresponding to D/C=1 were derived, and multiplied by margin factors to obtain the median PGAs for three levels of damage:

- Major Stability Damage: large sliding or overturning movements resulting in short term loss of use of the trackway, with possible life safety hazard to adjacent retained areas.
- Major Wall Damage: major structural damage to walls, possibly resulting in short-term partial loss of functionality of trackway, and possible life safety risk.
- Minor Wall Damage: minor structural damage to walls such as cracking. No impact on functionality or life safety.

An example of retaining wall fragilities is given in Table 7.

Table 7 - Example Retaining Wall Fragility Analysis

Damage State	Median PGA	Beta	Function Factor	Life Safety	Group No.	Short Duration	Short Cost	Long Duration	Long Cost
Major stability damage	2.62	0.40	1.00	4	1	60 Days	5%	180 Days	50%
Major wall damage	1.11	0.40	0.50	3	1	30 Days	5%	180 Days	30%
Minor wall damage	0.55	0.40	0.00	2	1	3 Days	0.1%	60 Days	5%

DISASTER RESPONSE FOR LIFELINE SYSTEMS 191

Transbay Tube

The fragility data for the Transbay Tube were developed for the 57 tube segments and associated seismic joints at both ends of the tube in terms of peak ground acceleration (PGA) at rock outcrop based on extrapolation of the seismic demands obtained from the seismic vulnerability analysis to the structural capacities. The minimum level of PGA at rock outcrop that could cause damage to the structure (i.e. the no-damage PGA value) was taken as the cut-off value for the tail end of the fragility curve, which was assumed to follow the lognormal distribution. The damage-state modes for the Transbay Tube were determined to be as shown in Tables 8 and 9.

Table 8 - Damage States of Transbay Tube

Mode	No Damage	Damage
Steel shell tensile yield and concrete liner cracking	Tensile strain in steel shell < yield strain of 0.00124 in/in	When it reaches 2X yield strain with a beta (β) value of 0.4
Steel shell tensile rupture and concrete liner tensile failure	Tensile strain in steel shell is 6 times yield strain	When it reaches strain hardening of 0.0192 in/in with a beta (β) 0.5
Liquefaction beneath the tube	Soil beneath the tube is not liquefied	When it is liquefied assuming a beta (β) value of 0.4

Table 9 - Transbay Tube Segment Fragility Analysis

Damage State	Median PRA	Beta	Life Safety	Function Factor	PRA No Damage	Group No.	Short Cost	Long Cost
Steel shell tensile yield and concrete liner cracking	0.50	0.4	0	0	0.25	1	N/A	1%
Steel shell tensile rupture and concrete liner tensile failure	3.40	0.5	4	1	1.50	1	10%	50%
Liquefaction - major damage	0.30	0.4	4	1	0.15	1	N/A	70%
Liquefaction - minor damage	0.15	0.4	2	0	0.10	1	N/A	15%

Berkeley Hills Tunnel

Fragility data for the Berkeley Hills Tunnel was expressed as the consequences to the tunnel due to fault-offset on the nearby Hayward fault, as shown in Table 10.

Table 10 - Berkeley Hills Tunnel Fragility Analysis

	Damage State	Fault-offset displ.	(Beta)	Life Safety Impact	Function Factor	Group No.	Repair Cost (millions)
1	Hayward fault-offset exceeding 80 in.	>80 in.	0.1	4	1	1	$60
2	Hayward fault-offset 40 in. to 80 in.	>40 in.	0.1	4	1	1	$30
3	Hayward fault-offset 12 in. to 40 in.	>12 in.	0.1	3	1	1	$12
4	Hayward fault-offset 3 in. to 12 in.	>3 in.	0.1	2	0.5	1	$6

Conclusions

A variety of techniques have been used to derive component fragilities from deterministic analyses and evaluations, including margins analysis, capacity spectrum analysis based on pushover curves, and judgment. Expert Panel opinion and review was employed to validate assumptions and ensure consistency and completeness.

Explicit descriptions of structural damage states and logical treatment of dependent, independent, and mutually exclusive damage paths is useful for assessing the type and distribution of damage, in addition to overall loss estimation.

Further refinement of the fragilities and damage states can be recommended for:

- Components with high frequency of damage and losses, such as aerial structures.
- Components where more detailed evaluations, especially non-linear analyses, are performed.
- Correlating with test results and experience data, as they become available.
- Performing sensitivity analyses of the effect of variances in fragility parameters on the loss estimates.

References

Bechtel Infrastructure Corporation Bechtel/HNTB Team (2002), "BART Seismic Risk Analysis."

Applied Technology Council (1996), "ATC-40 – Seismic Evaluation and Retrofit of Concrete Buildings."

FEMA (1999), "HAZUS 99 Advanced Engineering Building Module – Technical and User's Manual."

Acknowledgements

The authors wish to thank the many contributors and reviewers for the BART Seismic Risk Analysis, including Mr. John Eidinger, G&E Engineering Systems Inc., the Bechtel/HNTB Team, BART Seismic Program personnel, the Seismic Program's Design Review Board and Expert Panel, and the Peer Review Panel.

Seismic Retrofit Concepts for BART Aerial Structures

By Chip Mallare[1], Ed Matsuda[2], Bill Hughes[1], Eric Fok[2]

Abstract

The California Legislature created the San Francisco Bay Area Rapid Transit District (BART) in 1957 to provide rapid transit facilities to the Bay Area. Today BART is a critical component in the economic success of the entire Bay Area, serving four counties and carrying about 300,000 passenger trips a day during weekdays and about 45 percent of the commuters crossing the Bay. In light of its criticality to the Bay Area, BART has begun the Seismic Retrofit Program to enhance the seismic performance of the system in the event of a future major earthquake.

The intent of this paper is to discuss the seismic retrofit of the BART aerial guideway structures. The structural details of the typical BART aerial guideway will be presented and insights into the seismic performance of as-built structures will be provided. Results of physical testing on a typical guideway pier, performed at the University of California at San Diego, will also be presented. This paper will discuss the seismic performance criteria initially established for the program by BART, and how information gained from the preliminary retrofit analyses, physical testing, and risk management plan could impact those original performance goals. This paper will also present typical structural details to retrofit the aerial guideways to meet BART's goal of providing a safe and reliable system.

Introduction

BART's critical contribution to Bay Area transportation was highlighted following the 1989 Loma Prieta earthquake, which damaged major roadways around the region and left numerous others unusable. BART was able to resume limited passenger service in the East Bay within 12 hours after the quake and full service within 2 days. In fact, BART carried an additional 100,000 passengers each day during this period, increasing patron loads by 40 percent over pre-earthquake service. However, future major earthquakes are likely to occur much closer to the BART system, with the potential for greater damage to its aging structures.

Aerial guideways are a vital component of the overall BART system and retrofit program, therefore a large portion of the total program expenditures will go toward retrofitting these structures. The original BART system included more than 25 miles of aerial guideway structures, comprised of approximately 1,918 piers and abutments.

---

[1]Bechtel/HNTB Team (Chip Mallare, Corresponding Author, 1330 Broadway, Sui 1200, Oakland, CA 94612, phone 510-874-7459). [2]Bay Area Rapid Transit District.

As-Built Aerial Guideway Structures

The typical BART aerial guideway structures consist of simply supported precast prestressed concrete box girders with spans between 80 and 100 feet (see Figures 1 and 2). In a few locations, composite steel girders, steel through girders, or cast-in-place prestressed concrete box girders are used. The prestressed girders are typically single-cell trapezoidal boxes with girder ends dapped, or notched, to allow pier caps to be partly contained in girder depths.

Superstructure spans are typically seated on elastomeric bearing pads. A closure pour is provided on top of pier caps between the ends of superstructure girders. One end of a superstructure girder is pinned with horizontal dowels into the closure pour, which in turn is restrained laterally by vertical concrete-filled steel pipe shear keys. The other end of the girder is free to slide longitudinally and restrained laterally by one horizontal concrete-filled steel pipe shear key connecting the girder end to the closure pour.

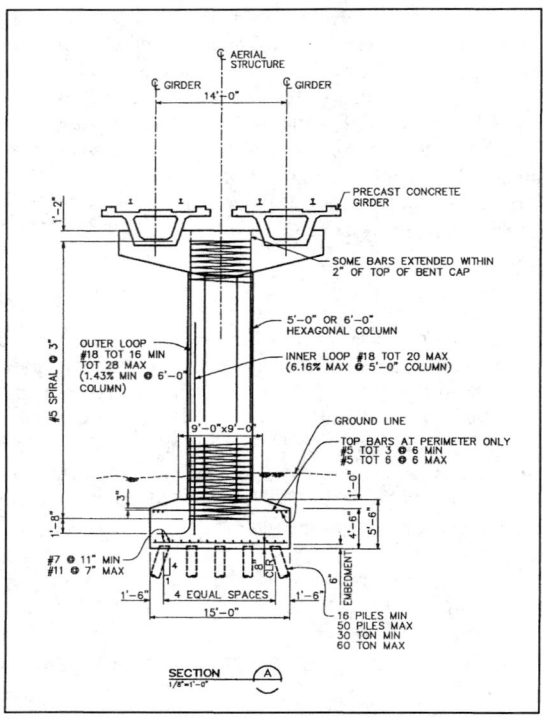

Figure 1. Typical As-Built Aerial Guideway Pier

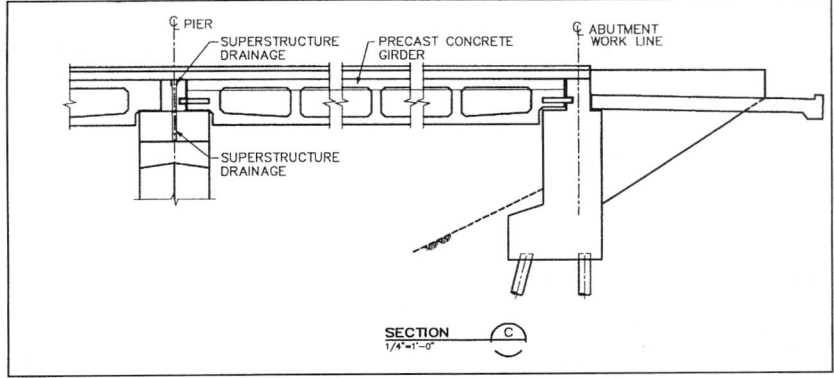

Figure 2. Typical As-Built Aerial Guideway

Substructures for aerial guideways consist of concrete columns, piers, and abutments. The majority of bents are single concrete T-bents, or hammerhead bents, that support two parallel track girders. Columns in the hammerhead bents are typically highly reinforced, with reinforcement ratios up to 6%, but have relatively good confinement (#5 @ 3") for columns built in this period. The columns are supported on either spread footings or pile foundations. Other bent types include single rectangular concrete columns supporting a single-track girder with no bent cap, single or multiple columns with in-fill walls that support multiple track girders, and multicolumn bents with bent caps.

Challenges and Uniqueness of the Program

Upon embarking on its seismic retrofit program, BART managers recognized that program challenges stem primarily from two sources – the uniqueness of the BART system and BART's rigorous performance requirements.

Unique Features of the BART System

One of the primary drivers behind the development of BART's initial performance criteria is the uniqueness of the BART system. Unlike most highway systems, BART has little redundancy. If a particular section of track, tunnel or station goes out of service, there are no alternate routes around it, and areas on opposite sides of the nonoperational section will be isolated from each other. Since BART does not have a loop system, failure at a critical location can essentially split the system in half, greatly reducing its effectiveness. This characteristic was a prime factor in BART requirements for maintaining operability.

The particulars of rail system vulnerability also drive criteria development. Unlike highway systems, railroads cannot tolerate large relative displacements. Large temporary displacements during an earthquake can cause train derailments or even cause trains to fall off aerial structures. Moreover, excessive permanent displacements cause difficulties in realigning tracks and may interfere with resumption of train service.

Performance Criteria

Because of these unique features, BART initially established Seismic Performance Categories (SPCs) to guide the seismic analysis and evaluation of the BART system.

SPC-1 required that primary revenue structures be capable of returning to revenue service within 72 hours after a "Maximum Credible Earthquake" (MCE). The 72-hour standard was designed to allow time to inspect the system and declare it safe for operation before resuming service. This criterion essentially meant that SPC-1 structures should be operational immediately after a major earthquake. Structures in this category included aerial guideways, stations, trackways, subways, and the Transbay Tube. For the aerial guideway structures, this meant allowing only minor damage to the structural elements of the guideways. To this end, BART's original direction was to keep the aerial guideway foundations essentially elastic and to force any plastic action into the columns, which could be accessed more easily for inspection and cased for additional ductility capacity.

SPC-2 dictated secondary facilities (including train control and power systems, mechanical and electrical systems, track, and the Central Operations Center) return to service within 60 days after an MCE. Restricted levels of service based on temporary repairs were allowed as part of this criterion. It was envisioned that the SPC-2 system components would be brought back on line gradually as mechanical, electrical, and track systems are restored, with the entire system back in operation by the 60-day mark.

SPC-3 applied to noncritical facilities such as shop buildings, training facilities, office structures, and parking structures. These facilities were required to meet life safety or non-collapse guidelines.

BART established the SPC classifications at the inception of the Seismic Retrofit Program. As a result of information learned from work performed on the program to date (including preliminary seismic analysis of the aerial guideways, physical testing, and the Risk Management Plan) BART is currently re-evaluating their original performance goals and criteria. For the aerial guideways this could mean increased allowable seismic damage and possibly longer periods needed to return to service following a major seismic event.

Aerial Guideway Retrofit Concepts

Although the original BART system was designed and constructed in the 1960s, BART had the foresight to include seismic demands in their original design criteria. As a result, many of the as-built aerial guideway structures include details that allow them to withstand seismic forces much better than other bridges designed and built during the same period. These features include well-anchored column reinforcement, closely spaced column horizontal reinforcement, moderate bearing seats with restrainers, and an absence of column lap splicing. However, the magnitude of seismic forces used in the original design criteria was well below that used today, and recent seismic analysis of BART elevated structures reveals a number of seismic vulnerabilities. Many of the seismic deficiencies found in BART aerial guideway structures are similar to those found in California highway bridges and therefore some of the retrofit approaches planned for the elevated structures are similar to those taken for California Department of Transportation (Caltrans) highway bridges.

The retrofit alternatives and details described in the following section reflect the SPC-1 performance criteria initially set by BART. They represent, more or less, upper bound solutions, which are meant to allow train operations to resume within 72 hours of a major seismic event. As stated earlier, BART is currently re-evaluating their original performance goals and may revise the acceptable performance level based on results from the physical testing and Risk Management Plan. This in turn would change the retrofit approach taken by the design engineers.

Foundations

Aerial guideway as-built pile foundations were found to have a number of seismic deficiencies that include inadequate pile capacities, inadequate pile-to-pile cap connections, inadequate pile cap flexural reinforcement, inadequate footing shear strength, and excessive joint shear stress. Retrofit details proposed for pile foundations include increasing the footing size to provide for additional piles and adding a footing overlay that incorporates a top mat of reinforcing. (See Figure 3.)

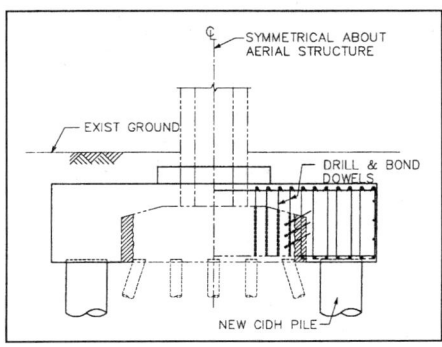

Figure 3. Foundation Retrofit Alternative

Because of low overhead and limited access, two pile alternatives are currently being investigated: large cast-in-drilled-hole (CIDH) piles and micro- or pin-piles. Micropiles may be battered, as are many of the as-built piles, to increase their lateral resistance and to limit lateral displacement. Footing pedestals are also being considered to increase joint shear performance. Pedestals will reduce the joint shear stresses in footings without further increasing the depth of the entire footing.

Another unique detail being considered is extending vertical drill and bond dowels to the bottom of footings. Extending dowels to the bottom of footings will increase footings' shear capacity and provide better joint shear performance.

Aerial guideway as-built spread footing foundations were found to have seismic deficiencies similar to the pile foundations and include inadequate footing flexural reinforcement, inadequate footing shear strength, and excessive joint shear stress. Retrofit details proposed for the spread footing foundations include increasing the footing size to resist the column plastic hinge forces and adding a footing overlay that incorporates a top mat of reinforcing.

Columns

As-built columns have generally good seismic performance, with the columns having sufficient strength, stiffness, and ductility to resist expected seismic demands. However, many columns are either deficient in shear capacity or close to their shear capacity as compared to their plastic shear. Therefore, column casings are proposed where needed to increase the overall shear capacity of columns. Common steel casing retrofit procedures would involve elliptical steel shells at rectangular columns and circular casings at hexagonal columns. However, since shear strength enhancement is the primary goal, fabricating casings to closely match the shape of the columns is a possible option. This would offer aesthetic advantages as well as potential savings in material costs. Furthermore, shaped casings would minimize column flexural strength increases that could require further costly strengthening at the footings. Where additional column ductility is needed, circular or elliptical steel casings, (or perhaps composite material jacketing) can be used. The composite material jacketing offers some advantages over steel jackets. For example, column flexural strength enhancement is negligible using composite jackets, provided unidirectional composites are used. Also, composite casing can readily conform to the rectangular or hexagonal shapes of BART columns while still providing greater ductility enhancement over the shaped steel jackets, and installation may be simpler in highly congested areas. Further consideration is needed in some areas on the appropriateness of the composite casing.

Bent Caps

Study reveals that as-built shear keys typically have insufficient strength to resist the plastic shear force of the columns. To resist transverse seismic forces, external shear keys are proposed to be attached to the bent caps at the exterior surfaces of the girders. Longitudinal forces will be resisted by placing bearing material, or "bumpers" such as steel plates or elastomeric pads, between the ends of the girders and the face of the bent caps. The existing shear keys also impart large torsional moments to the bent caps since the longitudinal seismic force is currently transferred through the shear keys at the top of the bent cap. The bumpers would decrease these torsional demands on the bent caps and thereby eliminate the need to retrofit the bent cap for torsional forces.

Abutments

The typical BART aerial guideway structure has short seat type abutments on piles, with the superstructure supported on low-profile elastomeric bearing pads. As-built abutments typically have a 1-inch expansion joint between the ends of the girders and the abutment backwall. The gap between the dapped face of the girder and the front face of the abutment is typically 3 to 4 inches. Concrete shear keys at both sides of the girders secure them from transverse movement. (See Figure 4.)

Longitudinal Response. In a typical Caltrans bridge retrofit, the abutment backwall is allowed to fail. The longitudinal restraining force is then supplied by the soil behind the backwall. However, the typical BART aerial guideway structure has a mildly reinforced, 2½-foot-thick backwall. Because of this large backwall, piles are expected to fail prior to failure of the backwall.

To satisfy BART's original requirement of preventing foundation damage, three possible retrofit schemes are being investigated for abutments. The first is to open abutment gaps to provide sufficient distance to either eliminate abutment engagement or reduce displacement and force demands on the abutments to an acceptable level. However, this approach is difficult to implement without affecting BART operations. The second approach is to weaken the abutment backwall so that it will fail prior to damaging the piles. This could be done by coring a hole transversely near the front face of the backwall, thereby removing the front face of reinforcing. If this approach is chosen, BART would have to accept some damage to the abutments after a major (or even moderate) seismic event. Although this backwall damage would eventually need to be repaired, it is not likely to affect train operations. The third approach would be to strengthen the abutment foundation sufficiently to prevent failure by providing additional piles. This is the most costly alternative, but it would limit the amount of damage to the existing abutment and abutment foundation.

Transverse Response. In normal Caltrans bridge design, transverse abutment shear keys are sized so that they fail prior to damaging pile foundations. However, shear keys on the BART aerial guideway structures are generally much stronger than the pile foundations. Two retrofit alternatives are currently being investigated for transverse response. The first is to replace existing shear keys with weaker keys or other sacrificial elements that will fail before damaging the piles. However, analyses have shown that the transverse movement of the girders using this type of system would be quite large, on the order of 1½ to 2½ feet. This amount of movement would severely

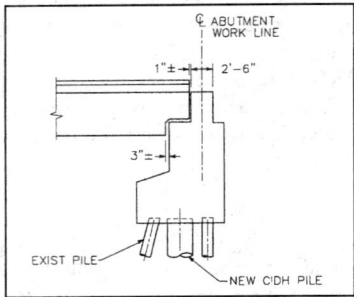

Figure 4. Abutment Retrofit Alternative

damage the rails and existing elastomeric bearing pads and would likely prevent train operation. Replacement of the rails could be done quickly, but replacement of the bearing pads would require jacking superstructure girders and could take longer than the 60-day guideline. The second alternative would be to increase the foundation's capacity by using either large diameter CIDH or battered micropiles. These piles could be used for both the transverse and longitudinal seismic forces.

Physical Testing

To better evaluate the performance of the as-built structures and possible retrofit details, BART commissioned the University of California, San Diego to conduct scale model tests of a typical BART aerial guideway pier. A 1:2 scale model of the West Oakland Viaduct Guideway Bent P 16 was built and then tested in two phases. The objective of Phase I was to analyze the response of the structure at the pile/pilecap connection. In Phase II, the objective was to examine the behavior of the joint at the column/pilecap connection in order to assess the column/footing joint performance after retrofitting the pier with a footing overlay and piles. In both phases, the model pier was partially retrofitted to include a footing overlay that provided a top mat of footing reinforcement and full length drill and bond dowels (the dowels extended as close to the bottom of the footing as practical). In Phase II, the footing was restrained to simulate retrofitting the structures with the addition of new piles to prevent pile failure.

The results from the Phase I tests showed that the pile/pilecap connection was the weak link in the partially retrofitted configuration (as noted above, the footings were not tested without a footing overlay). The piles failed due to fracture of the reinforcing in tension and spalling of the concrete in compression. The failure mode was desirable because the pilecap remained intact and allowed for a rocking mode of response to occur following the pile failure.

In Phase II, the damage occurred in the column's plastic hinge region, including spalling of the concrete, fracture of the transverse reinforcement, and buckling of the column's longitudinal reinforcement. A maximum drift of 7.9% at $\mu_\Delta = 4.0$ showed a satisfactory amount of ductility in the as-built columns. The nominal joint shear stresses reached 8.6 $(f'c)^{1/2}$. The footing overlay retrofit performed as intended and was sufficient to keep the pilecap joint shear distortion low and minimize pilecap cracking.

Risk Management Plan

As part of the overall Seismic Retrofit Program, BART performed a risk management plan, which had three main objectives. The first objective was to demonstrate the need for the BART system to be seismically retrofitted. The second objective was to assist BART in establishing acceptable seismic performance goals for the BART system. The final objective was to determine the level of funding needed to carry out such a program and to facilitate funding procurement. Included in the risk management plan was addressing the affordability and cost-effectiveness of the retrofit solutions proposed to meet the original SPC-1 classification for the aerial guideways.

Development of risk management-directed retrofit alternatives began with assessing the impact of high-likelihood, high-magnitude deterministic earthquake scenarios on the system in its present condition (to establish the *Status Quo* alternative). Retrofits required to mitigate life safety risks were then established and assessed, under various earthquake scenarios, for their impact on operability (assuming life safety mitigation retrofits are in place). This determined the *Life Safety* alternative. The *Operability* alternatives were established by assessing whether and where retrofit beyond life safety should be considered to improve operability, and developing retrofit schemes to achieve various levels of enhanced operability. Cost/benefit analysis of the various retrofit alternatives were then performed. The alternatives were evaluated based on acceptable operability/system recovery and cost benefit justification.

As a result of the initial analyses, physical testing, and risk management plan, BART determined that it may be prudent to allow more damage to the aerial guideways than was originally accepted under the SPC-1 classification. This includes more damage to abutments and possibly allowing rocking to occur in pier foundations. BART is currently evaluating these modified operability requirements, weighing the benefits of lower initial retrofit costs against possible increases in post-earthquake costs and decreases in operability within portions of the system.

The relaxed operability requirements would change the retrofit solutions previously proposed under the "upper bound" SPC-1 criteria. The retrofits would still be required to reduce the possibility of brittle type failures such as shear in the columns and footings through use of column encasement and footing overlays with shear dowels. However, providing new retrofit piles may not be needed depending upon the soil conditions at each pier site. Moreover, pier rocking would decrease the seismic load demands on the structures, possibly eliminating the need for shear key retrofits and limiting the expected ductility demands on the columns. Finally, costly abutment retrofit could be avoided by allowing damage to the abutments and retrofitting only to accommodate the expected structure movements through use of seat extenders.

Conclusion

BART has always maintained twin goals of protecting the community's investment in the system while planning for the future. Given the public's need for additional transportation capacity, the system has regularly expanded, with new extensions and capacity enhancements in various stages of conceptual planning, final design, and construction. However, portions of the current system are aging and in need of maintenance and upgrading. BART will use the information developed as a part of the Seismic Retrofit Program, and in particular the seismic risk management plan, to guide decision-making relative to seismic retrofit options.

Seismic upgrading of transit systems is a complex issue. Major factors that influence results include identifying appropriate ground motions, reasonably evaluating the integrity and performance of the structures, and developing the appropriate, affordable upgrades to protect and benefit the public who use and depend on the system. The initial seismic data developed suggest that some measures of seismic upgrading will be prudent. This important fact allows BART to begin implementation planning for the retrofit program internally, and with the public and interested external agencies. In particular, Caltrans and FHWA have become important partners assisting BART in defining and implementing the retrofit program.

The remaining challenges for the program will be to specifically define the level of retrofit, procure the balance of funding necessary, comply with environmental guidelines, and properly design and construct the improvements in an efficient manner with the least disruption to BART's patrons and neighbors. BART is committed to operating a safe and reliable system, and the seismic retrofit program is an expression of that commitment.

References

Horton, T., Fok, E., Matsuda, M., Hughes, W., Tseng, W., Mallare, C., and Chen, K. (2002). "Ground Motions, Design Criteria, and Seismic Retrofit Strategies for the Bay Area Rapid Transit District Bridge and Other Structures." Proceedings from the Third National Seismic Conference and Workshop on Bridges and Highways.

Schoettler, M., Restrepo, J.I., and Seible, F. (2002). "Report BART Unit 1, A 1:2 Scale Model Test of the Proposed Retrofit for Bent P16 of BART's Aerial Guideway." Department of Structural Engineering, University of California, San Diego.

Seismic Assessment and Retrofit Concepts of the BART Transbay Tube

By Ching Wu[1], Eric Fok[2], George Fotinos[1], Wen Tseng[3], Gary Oberholtzer[1]

Introduction

The Bay Area Rapid Transit (BART) system's Transbay Tube is 3.6 miles long and lies at the bottom of the San Francisco Bay, at a maximum depth of 132 feet below mean sea level (see Figure 1). The tube is constructed of 57 segments with an average segment length of 330 feet. A ventilation structure is located at each end of the Transbay Tube – the San Francisco Ventilation Structure and the Oakland Ventilation Structure – and each is connected to the tube by a seismic joint.

The Transbay Tube is a critical link in the BART system and a key system element analyzed as part of the BART Seismic Retrofit Program. Since the tube is submerged, it is important to ensure the structural integrity of the tube and the associated ventilation structures and seismic joints for the safety of passengers and BART personnel. Furthermore, for BART to provide an adequate level of post-earthquake transportation service to the public, satisfactory seismic performance of the tube is critical during any future strong earthquake in the Bay Area.

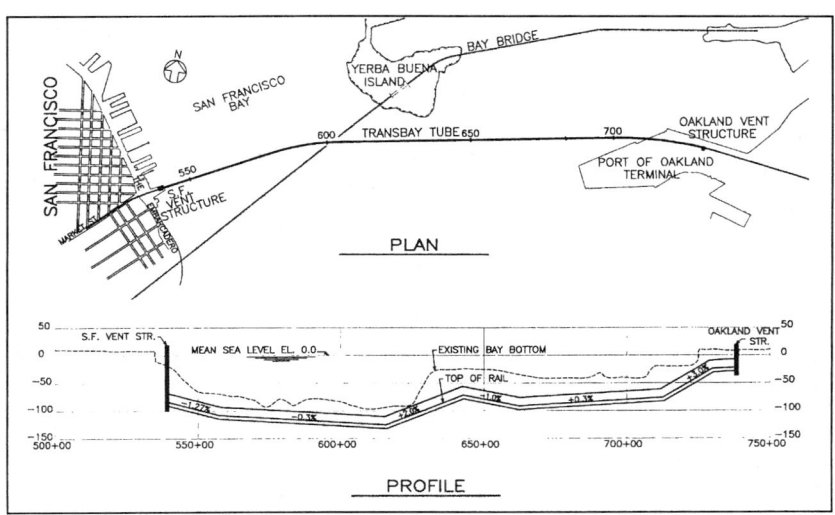

Figure 1. General Plan and Profile of BART Transbay Tube

[1]Bechtel Infrastructure (Ching Wu, Corresponding Author, 1330 Broadway, Suite 1200, Oakland, CA 94612, phone 510-287-4836). [2]Bay Area Rapid Transit District. [3]International Civil Engineering Consultants.

The overall BART system – and especially the Transbay Tube, ventilation structures, and seismic joints – performed well in the 1989 Loma Prieta earthquake. Small movements at the seismic joints were measured, however, no structural damage was observed in these structures. Although the ground-shaking intensity in the general area of the Transbay Tube during Loma Prieta was severe, future major earthquakes could produce much stronger ground-shaking intensity.

Analyses for Assessing Seismic Vulnerabilities

For assessing the potential seismic vulnerabilities of the Transbay Tube System (i.e., the Transbay Tube, the San Francisco and Oakland ventilation structures, and the seismic joints), seismic analyses were performed to provide response data required for determining the seismic demands of the system under the design basis seismic event. The potential major vulnerabilities that needed to be assessed include (a) potential for liquefaction of the granular backfill materials surrounding the tube that may impact stability of the tube and cause uplift (b) San Francisco shoreline slope movement that may impact the San Francisco Ventilation Structure and the seismic joint, (c) seismic movements at both ends of the tube relative to surrounding free-field soil caused by travelling seismic waves that may exceed the movement capacities of the seismic joints, and (d) effects of dynamic soil-structure interaction of the San Francisco Ventilation Structure relative to its surrounding soft soils that may further contribute to the movement demands at the seismic joint. Analyses performed for assessing each of the potential vulnerabilities mentioned above are briefly described below.

For this critical structure linking both sides of the Bay, two levels of design earthquakes have been established for analysis and retrofit design: (1) the **Design Basis Earthquake (DBE)** for ensuring life safety (an earthquake with 1,000 year return interval by probabilistic method or a median plus one standard deviation by deterministic approach, whichever governs) and (2) the Lower-level Design Basis Earthquake (LDBE) for maintaining operation (an earthquake with 500 year return interval by probabilistic method or a median earthquake by deterministic approach, whichever governs).

A. Liquefaction Potential of Tube Backfill Materials. All backfill materials for the tube are granular in nature and were placed underwater without any subsequent compaction. These materials are expected to be relatively loose and prone to liquefaction during a major earthquake. Evaluation of liquefaction potential of the backfill beneath and around the tube during the DBE involved three methods: (1) An investigation of case histories where liquefaction was known to have occurred in gravel materials with gradations and in-situ densities similar to the backfill materials around the tube was made. Fifteen cases were investigated. Cyclic stress resistance and cyclic stress demand of the backfill materials along the tube under the DBE were estimated and compared with each other. The demand was found to be several times higher than the estimated resistance, leading to the conclusion that there is high potential for liquefaction. (2) Probabilistic assessment of the liquefaction potential of the backfill materials was completed using four different methodologies, in which

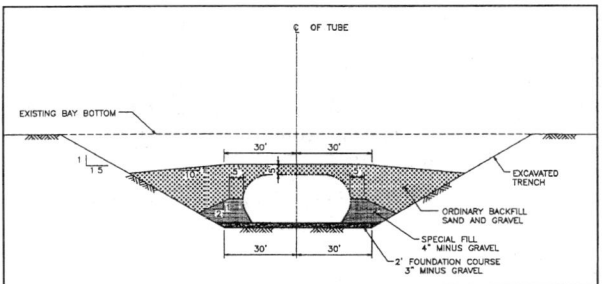

Figure 2. Cross Section of Transbay Tube and Backfill Materials

variations in peak ground acceleration and material properties were considered in the evaluation. The results indicated that, for the DBE, the estimated probability of liquefaction occurring in the backfill around the tube is greater than 90%. Furthermore, the annual probability of liquefaction from all seismic sources and all magnitudes was determined to be 5% to 7% at any location along the Transbay Tube. (3) The simplified Makdisi-Seed method and the computer program GADFLEA were used to determine pore-pressure generation and dissipation during cyclic earthquake loading condition to evaluate potential for liquefaction and its consequences. All three sets of analysis point to a high potential for liquefaction of the backfill materials surrounding the tube.

B. San Francisco Shoreline Slope Stability. The San Francisco Ventilation Structure is located off the shoreline and situated at the foot of the shoreline slope. It is embedded in loose sand fill and supported on relatively weak Bay Mud below. A deep-seated soil slope failure and associated down-slope movement of the slope during a major earthquake could result in large lateral soil pressure acting on the ventilation structure that will impact the structure as well as the seismic joint. To evaluate the slope stability and movement under the design seismic event, both simplified analyses using the Newmark method and the computer program SLOPE/W and more rigorous, nonlinear finite-difference, large-strain analyses utilizing the computer program FLAC were made. Results of these preliminary analyses indicate that the DBE can result in potentially significant down slope movement and lateral soil pressure. This will lead to potentially large sliding movement of the ventilation structure toward the Bay.

C. Tube Response Due to Travelling Seismic Waves. Because of the long length of the structurally continuous tube and the free-end condition at both ends, spatial variation of seismic ground motions along the tube length will cause deformations of the tube and movement of the ends of the tube relative to the surrounding free-field soil. Such movements are especially critical in the tube's longitudinal direction due to high axial rigidity. To evaluate the tube response due to travelling waves, spatially varying soil motions at the tube's centerline elevation were generated. The motions resulted from seismic waves propagating along the tube axis with an apparent horizontal wave propagation speed of 2.5 km/sec. Near-fault large velocity pulses and local site soil response effects were included in the generated motions. Simplified analyses using the so-called "wave propagation method" of the ASCE Standards and more refined nonlinear finite-element analyses using the nonlinear ADINA computer program were carried out. The analyses took into account the nonlinear behaviors of the tube structure and the soils surrounding the tube. Results of the analyses show that large relative longitudinal movements (up to 6 inches) of the tube at the San Francisco end can result under the DBE.

D. Soil-Structure Interaction Response of San Francisco Ventilation Structure. The San Francisco Ventilation Structure is a massive, relatively rigid, steel-plate-wrapped, reinforced concrete caisson embedded in deep, soft, backfill soils and supported on Bay Mud. Because of the soft and weak soil condition, significant soil-structure interaction (SSI) is expected to take place during earthquakes. To evaluate the complex seismic response behavior, SSI analyses were conducted using the computer program SASSI. The 3-D finite-element model included an equivalent linear soil-structure model of the irregular backfill soils around the structure. This analysis determined the global seismic response of the soil-structure system. Newmark's sliding block method and the energy balance method were employed to estimate the displacement of the ventilation structure due to nonlinear base sliding and uplifting. Additionally, nonlinear dynamic response analysis was performed using the nonlinear computer program ADINA. Local nonlinear behavior of the backfill soils surrounding the structure was explicitly modeled and included in the ADINA analysis. This analysis determined the nonlinear dynamic response, including the nonlinear sliding and rocking response, of the structure. Due to the soft and weak soil condition, the results of the analyses show that relatively large dynamic (including nonlinear sliding and rocking) displacements of the structure (up to 9 inches) relative to surrounding soil can take place during the DBE. These displacements contribute further to the seismic movement demands at the seismic joint, making the total demand 15 inches. This demand exceeds the existing ultimate capacities of the seismic joint by a large margin.

Seismic Joints

Specially designed flexible seismic joints connect the Transbay Tube to ventilation structures at each end of the tube. These joints allow 6 degree-of-freedom movements (3 translations and 3 rotations) of the tube relative to the ventilation structures. At each seismic joint, the tube can slide in the longitudinal (push/pull), lateral (right/left) and vertical (up/down) directions. The design capacities for joint-movement were 1.5 inches in push/pull motion, and 6 inches in the right/left and/or up/down motions that allowed for vertical joint-movement of 2 inches for long-term differential settlement and 4 inches for transient seismic motion. As originally designed, the joint had maximum design movement capacities of 4.25 inches for push/pull motion and 6.75 inches for right/left and up/down motions. Over time, the seismic joints have moved due to overloading during construction, settlement, earthquakes, and other reasons. As indicated in Table 1, the smallest remaining ultimate capacities (at which the restraining cables reach ultimate strength) of the seismic joint for the San Francisco

Table 1. Seismic Joint Capacities – San Francisco Ventilation Structure

Joint Location		Original Design Capacity (in)	Ultimate Design Capacity (in)	Ultimate Capacity (in)	Present Condition (in)	Remaining Ultimate Design Capacity (in)	Remaining Ultimate Capacity (in)
Longi-tudinal	Push	1.5	4.25	7.4	2.75	1.50	4.65
	Pull	1.5	4.25	7.4	-	7.00	10.15
Transverse		6	6.75	9.4	.25	6.50	9.15
Vertical	Up	6	6.75	9.4	-	12.25	14.9
	Down	6	6.75	9.4	5.50	1.25	3.9

Ventilation Structure on the Bay side are 4.7 inches in the push direction and 3.9 inches in the downward direction.

Retrofit Concepts for the Transbay Tube

The backfill surrounding the tube is prone to liquefaction. The uncertainty of the consequences of liquefaction and the criticality of the tube require that the worst-case scenario be considered. Two retrofit alternatives are currently under consideration.

A. Micropile Tube Tiedowns. The micropile tube tiedown concept would involve the installation of small diameter tension piles through the floor of the tube from within the gallery area, extending into underlying competent soils. This retrofit concept would be applied along the full length of the tube. By anchoring the tube in this manner, the upward buoyant force would be resisted in the event of liquefaction. This concept would require a minimum of 38, 400-kip micropile tie-downs for each 330-foot tube segment drilled to approximately 100 feet below the tube. The micropile casings would house an embedded #20 rod with a pressure-grouted concrete bulb deadman at the tip, and would be filled with grout. Existing utilities and BART equipment and emergency egress requirements in the gallery could be significant limitations to the installation of micropiles.

B. Vibro-Replacement. An alternative retrofit concept which could stabilize the granular backfill materials along the full length of the tube is vibro-replacement. As shown in Figure 3, vibro-replacement will consist of compaction of the granular backfill and placement of stone columns in a grid pattern about six feet by six feet on both sides of the tube to densify it to a relative density of 60-70 percent.

Field soil investigation and testing and a vibro-replacement demonstration on a test section of the existing tube backfill is recommended to verify the feasibility of this retrofit concept. This investigation and demonstration program would be performed on land at the Port of Oakland. Depending on the results of this initial testing, some additional testing of the tube backfill material may be conducted on the San Francisco side near the San Francisco Ventilation Structure.

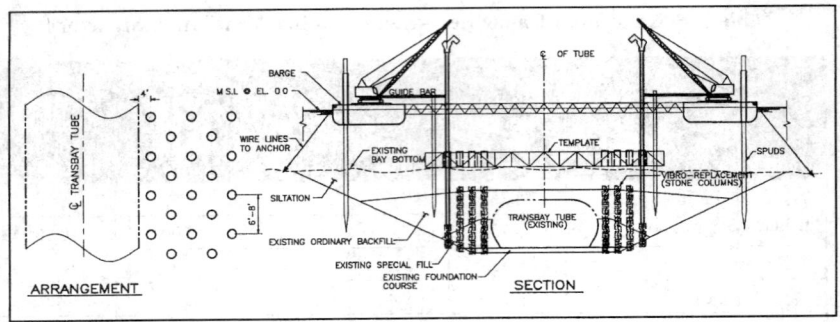

Figure 3. Vibro-Replacement Retrofit Concept

C. Alternatives Considered. Other alternative retrofit measures considered include exterior tube tie-downs, heavy riprap over the existing fills, and chemical or jet grouting of the backfill. These alternatives would be potentially costly and difficult to confirm effectiveness, and would cause greater environmental concerns than the micropile and vibro-replacement concepts.

Retrofit Concepts for the San Francisco End

The overall retrofit strategy for the San Francisco end of the Transbay Tube is to reduce the seismic motion demands on the existing seismic joint as much as practicable and, at the same time, increase the protection of the joint from potential water leakage where reduction of demands alone may not totally alleviate the vulnerability of the joint.

A. Reduction of Impact Due to Overall Shoreline Slope Movement

Permanent displacement of the shoreline embankment is predicted under the ground shaking condition of the DBE. Since the ventilation structure is founded on approximately 20 feet of soft Bay Mud, it too would undergo permanent displacement should shoreline movement occur.

A1. Array of Large-Diameter Piles. To reduce the impact of the overall shoreline embankment movement on the ventilation structure, an array of large-diameter steel piles (approximately 8 feet in diameter), independent of the Ferry Plaza Platform and placed to the approximate elevation of –200 feet, is proposed as shown in Figure 4. Their locations between the ventilation structure and the shoreline would reduce movement of the lower soils.

A2. Reinforced Shear Walls. Another method under consideration for stabilization of the shoreline slope is the installation of shear walls in the soils by means of deep soil mixing or slurry wall construction. Rows of parallel walls, extending like fingers from the ventilation structure toward the shoreline, would utilize shearing resistance between the walls and the surrounding soils to prevent slope movement. The walls would be reinforced and anchored by vertical structural steel members extending down into the underlying competent soils.

A3. Alternatives Considered. Several alternative concepts were considered, including a sheet pile barrier wall, a soil cement mix barrier wall, and adding additional larger-diameter piles to the platform. These concepts were rejected due to a variety of reasons, including inadequate load resistance or interference with the existing platform.

B. Reduction of Impact Due to Dynamic Movement from Soil-Structure Interaction

Dynamic interaction of the ventilation structure with its surrounding soils resulting from a seismic event could result in high shearing and bearing pressures under the structure. These pressures could exceed the shear and bearing capacities of the soft Bay Mud under the structure, resulting in large dynamic movement of the structure including base uplifting and sliding.

B1. Large Diameter Piles with Concrete Frame. The preferred retrofit measure to reduce seismic motion of the ventilation structure relative to soil (which could cause rocking, sliding and base-uplifting), as well as the shoreline movement described above, is to install approximately eight large-diameter steel pipe piles (10 to 12 feet in diameter) with concrete frames around the structure, as shown in Figure 4. This retrofit concept is expected to resist the downslope forces and dynamic movements acting on the structure from its surrounding soils, and also to prevent differential settlement due to movement of the Bay Mud under the ventilation structure.

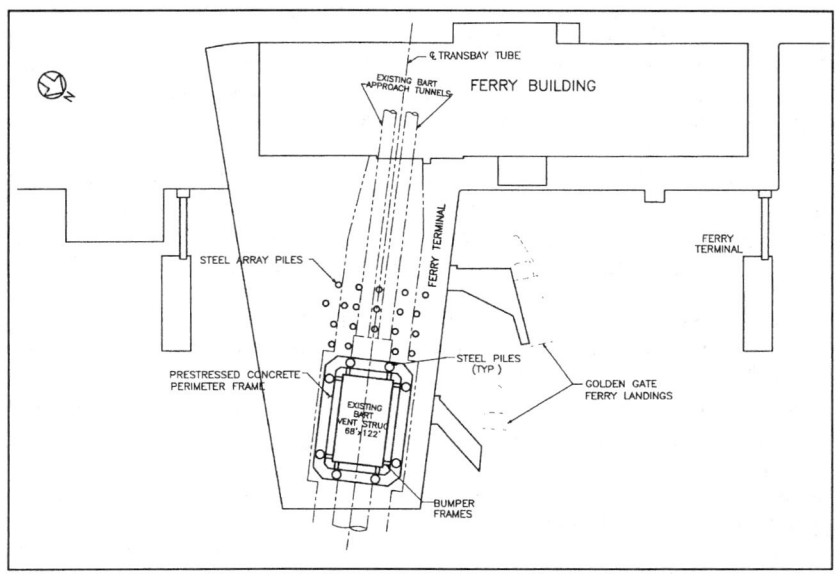

Figure 4. Array of Piles and Collar Retrofit Concepts

The large-diameter steel pipe piles would need to extend into underlying competent soils (to an approximate elevation of –200 feet) to develop adequate lateral resistance to the imposed lateral loads. The concrete frames, once assembled, would be lowered, adjacent to the ventilation structure walls, to an appropriate elevation under water near the Bay bottom.

B2. Alternatives Considered. As an alternative to installing large-diameter pilings and frames, jet or chemical grouting of the 20-foot-thick soft Bay Mud layer under the base of the ventilation structure was considered to improve the soil shearing and bearing capacity and to prevent bearing and sliding failures of the soil.

C. Reduction of Impact Due to Dynamic Movement of End Sections of the Tube

Relatively large axial or longitudinal movement between the end sections of the Transbay Tube and the free-field soil is predicted under the DBE. This relative movement would result, partially, from the loss of the longitudinal friction between the tube and the surrounding backfilled soils due to liquefaction. There are two preferred retrofit solutions, acting together, to mitigate this axial movement on the San Francisco side. These are pile tie-downs (or pile "stitching") and vibro-replacement. In addition, a tunnel liner sleeve at the seismic joint on the Bay side of the ventilation structure is planned as a backup to protect the joint from potential water leakage.

C1. Stitching Piles at Tube Ends. Anchoring the end of the tube using stitching piles will restrain the tube from longitudinal movement. This would be accomplished by installing large-diameter piles into lower, stronger soils. Clusters of four piles would be connected to the tube through precast concrete pile caps and tremie concrete around the tube dam plates. A cluster would be placed every 330 feet along the Transbay Tube for approximately 2,000 feet from the ventilation structure, as shown in Figure 5.

C2. Vibro-Replacement. The retrofit concept to supplement "stitching" piles is vibro-replacement of the granular backfill around the tube. This concept consists of compaction to a relative density of 60-70 percent by means of placement of stone columns in a grid pattern about six feet by six feet on each side of the tube, as shown in Figure 3. This retrofit concept would be the most compatible with the original tube installation design. A field verification investigation and testing program would be necessary to confirm the effectiveness and to refine the requirements for this alternative.

C3. Alternatives Considered. Chemical or jet grouting was considered for anchoring the tube's end to improve the friction between the tube and the soil. This alternative was determined to be less reliable and more expensive than micropiles or vibro-replacement. Also, installing a new seismic joint in the first section of the tube approximately 200 feet from the present seismic joint was considered as an alternative to accommodate the large movements at the seismic joint. The new joint would be constructed to have sufficiently large seismic movement capacity to accommodate the large predicted seismic motion demand at the end segment of the tube. This alternative was found not to be viable due to high costs and risks to BART operation during construction.

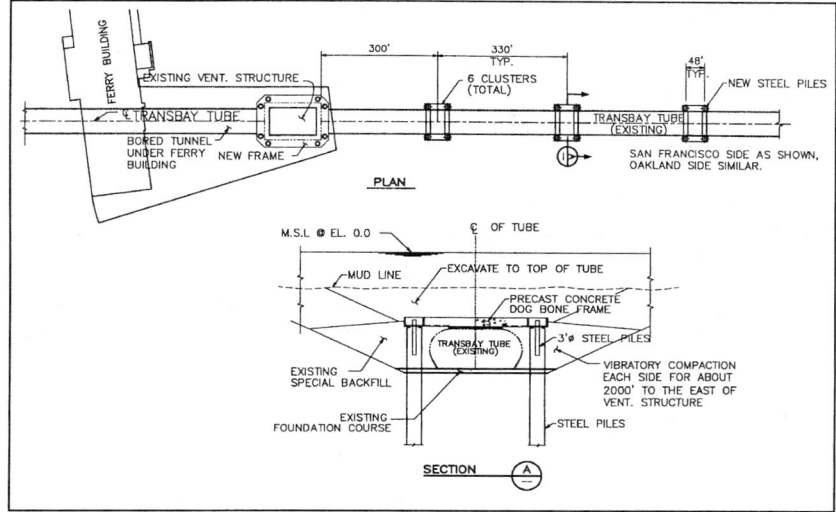

Figure 5. Stitching Piles Retrofit Concept

D. Secondary Barrier at Seismic Joint

D1. Tunnel Liner Sleeve at the Seismic Joint. To provide additional leakage protection at the seismic joint on the Bay side of the San Francisco Ventilation Structure, steel tunnel liner plate panels would be installed to line the interior of the two trackway tubes and the gallery between. The liner panels would be bolted together, covering all tube interior concrete surfaces (including under the tracks) from just east of the seismic joint through the exterior wall of the ventilation structure. A flexible sealing compound or a series of rubber "O" rings would be installed between the liner plate panels and the interior concrete surfaces to provide a positive seal if water leakage were to develop through excessive movement of or damage to the seismic joint. Removable access panels would be included in the sleeves at the location of the seismic joint couplings for future joint maintenance.

D.2 Alternatives Considered. The installation of a permanent cofferdam structure was considered as an alternative, interim safety measure prior to installation of all seismic retrofit measures and as a long-term redundant protection of the tube. The cofferdam would surround the ventilation structure and existing seismic joint, and would minimize the volume of Bay water entering the tube if water leaks developed at the seismic joint following excessive joint movement. This concept was not favorable because sealing the cofferdam as it crossed the tube on the Bay side would be very difficult to accomplish, and there would be a potential for damage to the tube and adjacent structures.

Retrofit Concepts for the Oakland End

As with the San Francisco end of the Transbay Tube, the retrofit strategy is to increase resistance between the section of the tube near the Oakland Ventilation Structure and the surrounding soils. Retrofit solutions similar to those proposed for the San Francisco end of the tube can also work for the tube at the Oakland end. The primary retrofit concept is densification of backfill by vibro-replacement.

The above-grade portion of the Oakland Vent Structure, which is located on land, requires strengthening of its steel frame. This will be accomplished by changing the existing lateral resisting system, which consists of steel bracing, by adding new reinforced concrete shear walls. While the existing steel bracing will remain in place, the more stiff shear walls will act as the primary lateral resisting system during a seismic event. The walls would be attached to the precast concrete panel through a grid of newly installed anchors.

Implementation

Before proceeding with design of the retrofit concepts presented in this paper, additional geotechnical investigations and further evaluation of the concepts are needed. Exploratory soil borings and cone penetrometer tests at both ends of the tube are being conducted and borings along the tube will be performed to better define subsurface conditions and confirm vulnerabilities. Also underway is a hydrographic survey along the entire tube alignment across the Bay to define the level of the existing Bay bottom. A future pilot program demonstrating the vibro-replacement concept at the Oakland end of the tube will test success in stabilizing the tube's granular backfill materials. Furthermore, as the Transbay Tube is a complex system with a variety of structures and vulnerabilities, use of soil data to model the global seismic response of the tube will identify potential interaction of the various retrofit concepts. Completion of these items will provide the data needed by Section Designers to design appropriate retrofit measures.

Acknowledgements

The authors wish to thank the members of the Seismic Retrofit Program's Design Review Board, including Prof. Ben Gerwick, Prof. Leslie Youd, Prof. Ray Seed, Dr. Wayne Clough, Dr. Ignacio Arango, Prof. Joseph Penzien, Prof. Thomas O'Rourke, Prof. Bruce Bolt, and Mr. Thomas Kuesel; and the members of the Peer Review Panel, including Prof. James Mitchell, Prof. I.M. Idriss, and Dr. Ignatius Po Lam.

Participating In International Post-Earthquake Lifelines Investigations

Curt Edwards, MB-ASCE[1] and Anshel J. Schiff, MB-ASCE[2]

Abstract

This paper reviews the procedures for organizing an international post-earthquake investigation. It reviews how individuals can participate in earthquake investigations and discusses training conducted by the Earthquake Investigation Committee to enhance investigations and the data that is collected.

Introduction

Information gained from post-earthquake investigations has been one of the main forces that has driven advances in building codes and other guidelines. Since the 1971 San Fernando, California, earthquake, and the creation of the Technical Council on Lifeline Engineering (TCLEE) by ASCE, additional effort in post-earthquake investigations has been devoted to lifelines. The Earthquake Investigation Committee of TCLEE has evolved from a small group of devoted utility engineers to a diverse group of members from utilities, universities, and consulting engineers. This group has developed an extensive guide to post-earthquake investigation of lifelines (see Reference 1), conducts annual training sessions to enhance the quality of data that is collected, and increases the diversity of lifelines that are investigated. While many investigators typically participate in investigation of significant earthquakes in the United States, it has been more difficult to field large teams when the earthquake is outside of the United States.

Should an International Investigation be Mounted

More often than not, the first indication that a lifeline-earthquake investigation will be mounted is shortly after the earthquake pictures appear on CNN. There are many earthquakes in which it is clear almost from the onset that there has been significant lifeline damage and that valuable lessons can be learned from the investigation. It is interesting to note that even in countries that have traditionally used different building materials, designs, and codes, lifeline facilities are similar to those found in the United States. Lifeline facility construction methods, their configuration, and the equipment that is used are amazingly uniform throughout the world. Thus, if lifelines are damaged or if they have survived significant shaking without damage, valuable insights can be gained from an investigation. It is more problematic to determine if an investigation should be mounted for earthquakes in less developed countries or centered in remote areas where the presence of modern lifelines is not known.

---

[1] Curtis Edwards, Chairman, Earthquake Investigation Committee; Vice President, Pountney Consulting Group, Inc., 4455 Murphy Canyon Road, Suite 200, San Diego, CA 92123, cedwards@pountney.com
[2] Anshel Schiff, PMI, 22750 Edgerton Rd., Los Altos Hills, CA. 94022; schiff@ce.stanford.edu.

Typically the chair of the Earthquake Investigations Committee, or someone he/she designates, takes the lead in determining if there has been lifeline damage or if lifeline facilities have been subjected to significant ground shaking. In addition, several other members of the committee will start to network with professional contacts to seek out information on their lifeline of interest. There is also coordination with the Earthquake Engineering Research Institute (EERI), which has a professional staff and standing contract to support investigation. There have been earthquakes where other investigative groups have not mounted an investigation because of the character of the earthquake and where it was located, yet TCLEE has found that the lifeline investigation was very rewarding.

In addition to the live television pictures that are usually available within a matter of hours after a major earthquake, the World Wide Web is an invaluable source of up to date information. News organizations also maintain web sites; CNN, BBC, Google, and others keep reports up to date. These typically need to be searched immediately after the event because "yesterday's news" will not show up the following day. These news sites often have links to other contacts/sources that can be used to evaluate lifeline damage. In most cases, however, these news sources mostly deal with the loss of life and major building destruction and, as a result, it is hard to determine the extent of lifeline damages. There are also several specialized organizations, such as the American Red Cross and ReliefWeb, that have detailed estimates of injuries, damage to infrastructure, and damage to housing stock.

The next step is to try to get direct information on lifeline damage. The earthquake investigation committee membership roster will be reviewed for local members. ASCE, EERI, and other groups have worldwide membership that can be used as a resource. Next, local universities typically have engineering departments that are very interested in learning about earthquake damage. In many cases, the professors and students can all speak some English, have knowledge of local damage, and could provide translation services while aiding the team in the field.

Many countries have their own earthquake organizations that monitor and document earthquake damage. Although they will be very busy immediately after an event, they usually can be contacted for information, advice, and assistance. Also, some countries have engineering organizations similar to ASCE or EERI. These and other groups can be found using search engines for various internet providers.

Lastly, many international consulting engineering firms maintain offices in affected areas. Typically, they can be contacted by talking to local, English speaking personnel for a reference to the international office. More details on the process used in deciding to field a team are given below.

Forming a Team for an International Investigation

As soon as it appears that an investigation will be mounted, an email asking for volunteers is sent to all committee members. An investigation involves a major commitment of time and other resources. Typically an investigation will take from 6 to 10 days, including travel time. Many organizations understand the benefits gained by having one of their members (or employees) participate in the investigation, so they may provide salary support while investigator is gone in addition to covering some of the expenses. TCLEE investigations are typically partially self-supported with limited funding from ASCE. Once the investigation has been performed, additional time must be spent by each team member to organize the information collected and contribute to the technical paper prepared to document the observations and relate them to existing standards or practices in the United States. The report is published as a TLCEE monograph and is usually 200 to 300 pages in length.

The selection of the team not only depends on the availability of the participant on short notice for a week's excursion, but on the technical specialties and language skills needed. From the pre-departure information that is gathered, the need for certain expertise may be clear and these special skills may be sought out, even from people not on the committee. For international investigations, the size of the team is usually limited by the availability of investigators and financial support. As a result, specialists for each lifeline may not be on the team. In addition, when in the field, investigators are usually broken into groups of two or three. In the course of a day's investigation a group may come upon valuable information on a lifeline for which a specialist is not available. This brings up the important issue of investigator training that is discussed below.

Individuals interested in participating in a lifeline post-earthquake investigation will learn that there are advantages to joining with a group such as TCLEE. TCLEE has participated in over 14 international earthquake investigations. It has, through its members and over time, developed a large network of contacts that can facilitate gathering information about facility damage and gaining access to restricted facilities. Members of the TCLEE team may have special expertise that allows the impact of the earthquake to be better understood and to seek out particulars of the damage and its impact that are of special interest to the technical community. There are also advantages to the profession for an investigator to join the TCLEE team. Observations are published and made available to the technical community so that many in the profession can benefit from the lessons that are learned. By joining with a group, the number of visits to a given facility will be reduced. This reduces the drain on the facility's resources while having to meet with many investigators. There have been cases where facilities have been closed to an investigation team because of the large number of earlier visits.

Training of Lifeline Investigators

As noted above, because of the limited size of an international investigation team, and the need to gather data that is encountered when a specific lifeline specialist is not present, team members need a broad knowledge of lifelines. The Earthquake Investigations Committee achieves this by conducting training sessions as an integral part of their annual meeting. The training takes three forms. First, the results of recent earthquake investigations are reviewed. Information can be exchanged between those who have recently participated in earthquake investigations and novice investigators who can learn the day-to-day procedure and get questions answered. Second, key parts of the Committee's investigation guide are reviewed. The third, and the most important part of the training, is provided through site visits to actual lifeline facilities located near the site of the annual meeting. At these facilities, a committee specialist will identify equipment that has been shown to be vulnerable to earthquake damage, identify failure modes, and identify the impact of damage on the facility and the overall lifeline. The special safety issues associated with facilities are also reviewed so that some of the hazards associated with an investigation can be reduced. In addition to personnel training, regular meetings also update team members on organizational changes, personnel changes, and review capabilities for members who are not regularly attending these meetings and training sessions.

Experience has shown that engineers who work for a utility that participate in earthquake investigation will have experience with facilities directly related to their job. For example, a civil engineer working for a power utility will be familiar with substation equipment support structures, control houses, and foundations. The engineer will generally be less familiar with switchyard equipment, control center functions, and network operation. Thus, there is often a need for training in lifeline system operations for investigators.

It should be noted that investigators need not be members of ASCE or the Earthquake Investigations Committee to participate in an investigation. Participants would be expected to follow committee practices and participate in report preparation.

Domestic/Foreign Contacts

It is important that investigation organizations maintain a database of key lifeline and assistance contacts in potential earthquake hazard areas. This will allow easy access for team leaders to gather information on lifeline damage and to solicit local assistance for hotel recommendations, translating, driving, and other local logistics. This database can be maintained by the group leader, secretary or lead organization. Since earthquakes frequently occur in the same locations, the easiest way to maintain a historical listing of contacts is to build upon contact information from previous reconnaissance reports for the same area.

Of course, the more historical data the group has on earthquake areas, the easier it is to contact local resources and make preparations for the investigations team. Please be aware that even with adequate resources, international contacts after an event can take a great deal of time due to communications disruptions, the fact that local personnel are very busy, and their lack of desire to call one of may unknown solicitors of information.

The Mechanics of Fielding a Team - Leaders Responsibilities

There are many activities that the person responsible for fielding a post-earthquake investigation needs to undertake. The order in which they occur is dependent on the availability of contact information and the database discussed previously. The following discussions are intended as a guideline, but in most cases, the specific order of events can vary.

Of course, the leader must determine if the event is worthy of fielding an investigation team. This can be the most difficult task, in that a mistake either way could mean valuable data is not observed and documented or, team time and resources are wasted on very little damage. If there were significant ground motions, this is not necessarily bad, as there are benefits to documenting both good as well as poor performance. In general, the USGS Real Time Earthquake site should be reviewed to determine the location and magnitude of the event (http://neic.usgs.gov/neis/bulletin/).

Earthquakes with a Moment Magnitude (Mm) of < 5.0 typically have very little lifeline damage and probably are not worth sending a team, unless direct reports identify damage. Events with Mm ranging from 5.0 to 5.9 are problematic and will require evidence of damage prior to sending a team. Those with a Mm range of 6.0 to 6.9 almost always have some lifeline damage, especially in those above 6.5 Mm. Any event 7.0 Mm or higher will always have lifeline damage. Of course, these criteria are dependent on whether or not there are any (or significant) lifeline facilities in the area. Review of internet map services can provide immediate information as to the locations of major cities near the epicenter of the earthquake. Earthquakes centered in rural, remote regions could cause damage to residential structures (such as adobe, stone, etc.), but not much lifeline damage; however, important lifeline facilities may still be present, such as a power plant or hydroelectric dam. Many of these rural areas have no electricity, water, sewer, telephone, etc. As a result, it would not be beneficial to send a team. As a guide, the affected area should include cities with populations greater than 10,000 people.

Organizations such as EERI typically send teams to the area within 7 days of the earthquake. Quite often, they can provide input as to the extent of lifeline damage. However, this is usually limited to major damage such as collapsed bridges, landslides, collapsed water tanks, etc., and quite often does not give an accurate representation on total lifeline damage or damage that would be of interest to the lifeline community.

When it is determined there is enough damage to send a team, the leader should send out a participation request to all group members. This request should include availability to go (available and unavailable dates), if they have funding, their particular area of interest, and their willingness to provide a written documentation of their findings.

The maintenance of a pool of diverse, qualified potential investigation team members is critical to the success of the organization and the quality of the investigative reports. However, turnover, waning interest (this effort requires a great deal of volunteer time), over commitments, competitive pressures to reduce overhead charges to companies, and age can reduce the number of people available to go on an international investigation and donate the many hours required to complete the documentation of the earthquake event.

Accordingly, the Earthquake Investigations Committee is always searching for new members who:

- Have a strong interest in one of the lifeline disciplines
- Want to improve the overall performance of lifelines and reduce the loss of life, improve emergency response times and procedures, and reduce the cost of lifeline repairs.
- Have the time and desire to actively participate in committee activities including:
  - Attending the annual meeting and lifeline training sessions
  - Participating in domestic and international earthquake investigations
  - Participating in the preparation of post-earthquake investigations reports and other special lifeline projects
- Is willing to donate or seek some financial resource to support their earthquake investigations

Interested individuals should contact the Earthquake Investigation Committee chair for additional information and membership requirements. For further information please contact the current chairman, whose contact information can be found at www.asce.org/inside/tclee_eqinvest.cfm.

References

1. "Guide to Post-Earthquake Investigation of Lifelines," Ed. A. J. Schiff, Technical Council on Lifeline Earthquake Engineering, ASCE, Monograph 11, July 1997.

Lifelines Performance, Long Beach Earthquake, March 10, 1933
A Historical Perspective

Le Val Lund, P.E., M., ASCE-TCLEE [1]

Abstract

This article, a historical perspective, was prepared on lifeline utilities for the Sixth United States Conference on Lifeline Earthquake Engineering, TCLEE 2003, August 10 to 13, 2003 in Long Beach, CA. The March 10, 1933 earthquake in Long Beach initiated early development of emergency response planning. The report discusses the lifeline performance to electric power, gas, harbor, petroleum, telephone and water lifelines for this historic earthquake.

Introduction

In the number of lives lost and the amount of property damage, the Long Beach earthquake, some times is referred to as the Southern California earthquake, of March 10, 1933 was one of the most disastrous of any that occurred in the United States in the first quarter of the Twentieth century. The number of deaths was reported as 120, of which 52 were in Long Beach and 17 in Compton. Probably two thirds of the loss of life was by persons being struck by falling debris from buildings.

The M 6.3 earthquake struck at 5:54 pm, Pacific Standard Time, on March 10, 1933 on the Newport-Inglewood fault, the epicenter was located 3.2 kilometers (2-miles) west of Newport Beach in the Pacific Ocean. The area seriously shaken was a north south line extending from Long Beach to Vernon with a number of small cities in between and estimated population of 300,000. The most pronounced damage occurred in Compton and vicinity. The total property loss in the shaken area was early on estimated to be $141,000,000.

Emergency Response

An early development of a emergency response organization was developed in 1926 prior to the Long Beach earthquake by Chief R. J. Scott, Los Angeles Fire City Department, various city officials and executives of power, gas and oil companies

---

Civil Engineer, 3245 Lowry Road, Los Angeles, CA 90027, telephone 323-664-4432, lundasan@earthlink.net

and others had the foresight in preparing a plan in anticipation of a major disaster, particularly designed for use in the event of an earthquake. Briefly this included safely stored maps, diagrams and data concerning water mains and valves, gas, electric and oil lines, supplies of explosives and a list of outside mutual aid agencies available. Consideration was given to the use of motorcycles, airplanes and radio communication, establishment of temporary headquarters, shutting off gas, electric, ammonia and oil lines. Detailing men for rescue work and using tank wagons, emergency water storage and wells for water supply.

Electric Power

The Southern California Edison Company (SCE) supplied electric power in Long Beach from an interconnected system of to large steam electric generating plants and a series of hydroelectric generating stations. Power interruption ranged from momentary to 7-hours in duration. Power lines failed due to broken guy wires, shorting of swinging crossed wires and damaged pole top transformers. Following failure of the supply transmission lines a standby steam electric generating station quickly picked up the Long Beach load.

The Los Angeles electric supply was provide by the Los Angeles Bureau of Power and Light (LABP&L) and the LAG&E and to a lesser extent by SCE, who also sold power to LABP&L. Minor damage occurred to the LABP&L system at its connection to the SCE system and anchorage of batteries used to operate switches. LAG&E had two steam electric generating stations, one in Los Angeles and the other in Seal Beach. The Los Angeles plant was able to operate fully, while the Seal Beach plant was not able to because of the arching of the switches shaken open while carrying current were destroyed. No power could be supplied to the system until temporary repairs were made. The overloaded condition of the Los Angeles plant prevented synchronizing with the Southern Sierra Power company supply.

Gas

The Long Beach Municipal Gas Department purchased gas from the Southern California Gas Company (SCG) and distributed gas to the citizens of the city. SCG supplied gas at 207-kilopascal (30-pounds per Square inch) pressure through nine service connections; Long Beach through 24 regulating stations reduced the pressure to 28-kPa (4-psi) for distribution. The system included two large gasholders with a total capacity of 350,000-cubic meters (12,500,000-cubic feet). As a result of the first shock numerous breaks in the high and low-pressure systems reduced the gas pressure service quickly. After field inspection closing valves isolated breaks. In areas outside of Long Beach gas service was provided by SCG and the LAG&E were similar breaks occurred and the closing of valves prior to restoring the system isolated the breaks.

Harbor

At Los Angeles Harbor, damage was confined mainly to the breakage of some oil and water lines and moderate damage to brick buildings, which included some fire sprinkler systems.. Long Beach Harbor was very small at this time and had a minimum of damage.

Petroleum

Los Angeles Basin was the largest concentration of oil storage in the world, about 12,000,000-cubic meters (75,000,000 barrels) and 60 percent of this was in the area severely shaken. Some of the oil tank farms and refineries were close to the Newport-Inglewood fault and were on damp swampy ground on which maximum damage would normally be expected. In spite of these unfavorable conditions, the loss to oil properties was a very slight percentage of the total investment in the industry.

Oil is stored in large earthen reservoirs or steel tanks. No damage was reported to the earthen reservoirs, but it is possible cracks may have occurred in the concrete lining. Tanks shifted on their foundations with breakage of some pipe connections. Retaining walls or dykes proved of value, leakage through tank connections largely being retained by such encloses. At Signal hill oil field, many small field tanks were damaged, mainly through broken connections. Three wooden oil well rigs were destroyed, when boilers set fire to crude oil escaping from the damaged field tanks.

Telephone

In Long Beach, the Associated Telephone Company served about 30,000 stations through four automatically operated exchanges. Two exchanges in buildings of modern construction passed through the earthquake without damage to the equipment; the other two buildings suffered considerable damage. One of these was removed from service for 72 hours. Within a few minutes after the event automatic equipment became dangerously overloaded, therefore all telephone service was shutdown, except the Long Beach Fire Department private system. Telephone service was restored on a priority basis as the system was restored.

Water

Prior to the earthquake the National Board of Fire Underwriters (NBFU) issued reports on the potential for earthquake performance on nine cities in the impact area. Probably because of the NBFU initiative the water systems supplying Santa Monica, Glendale, Pasadena, Alhambra, Pomona and a major part of Los Angeles were unharmed. Systems supplying Huntington Park and Santa Ana received only slight damage. Long Beach and the harbor district of Los Angeles were kept in water with some difficulty.

Long Beach most serious damage was from 127 breaks in cast iron distribution mains, ranging in diameter from 100- to 300-milimeters (4- to 12-inches). Three distribution pumping stations were placed out of service leaving the Alamitos Reservoir to supply the demand by gravity. The reservoir was replaced in 1931 with six steel tanks to withstand earthquakes with a total capacity of 75,700,000-liters (20,000,000 gallons). They were undamaged, although there were some cracks on the ground surrounding the tanks. The pumping plants had structural damage in the earthquake. Groundwater well pumps were available as soon as power was restored. The Los Angeles Bureau of Waterworks and Supply supplied mutual aid. The Long Beach Public Utilities Building a 3-story reinforced concrete structure performed very well. The structure with gypsum block partitions, suffered only a few plaster cracks.

In the harbor district of Los Angeles, which includes San Pedro, Terminal Island and Willington water supply was maintained with difficulty because of a broken 300-mm (12-in) submarine pipeline from Wilmington to Terminal Island. The main pulled 80- to 100-mm (3 1/2 to 4-in) from the rigid concrete abutments and had 21 leaks in the under water portion. The other 400-mm (16-in) main was found intact to the island. Since the 1929 NBFU report a 2,108,000-L (557,000-gal) elevated steel tank has been placed on the island, the tank was full at the time of the earthquake and was only slightly damaged. Number of the automatic sprinkler connections to the warehouses sheared off and the tank drained in 1 1/2 hours. When these connections were shut off, low-pressure water service was restored to the island system from the 400-mm (16-in) main.

There was no reported well casing damaged or pump shaft misalignments in the area. Virtually all well pumping stations were available to operate when electric power was restored. The event was particularly hard on large brick pumping stations housing steam-operated pumps. There was no material damage to concrete ground level reservoirs. Significant damaged occurred to a 28.4 million liters (7 1/2 million gallon) riveted steel tank in southwest Los Angeles, due to wave action on the tank shell. Twenty-five elevated steel tanks on steel legs were used by the cities in the area, two collapsed and two others were placed out of service because of broken risers. About one-third of the 25 elevated wooden tanks were destroyed. He limited number of concrete water mains perform well, but wood stave pipe and riveted steel pipe performance depended on the quality, age, banding, etc, of the pipe.

Lifelines

Some of the lessons learned from the Long Beach earthquake have been remedied by modern design, materials and construction. Some of the lessons learned are still prominent because the older lifeline infrastructure that still exists in some areas. The Long Beach event was certainly before C. Martin Duke, Professor of Civil Engineering, UCLA, brought forth the importance of lifeline concept in earthquake engineering following the 1971 San Fernando earthquake and he was the founder of the ASCE Technical Council on Lifeline Earthquake Engineering (TCLEE). Lifelines (water, power, communication, gas, liquid fuels and transportation systems) are necessary for the existence of a community and the restoration of a community after a disaster such as an earthquake.

Related References

Bryant, E. S. (1933), Memo "The Work Done by the (Long Beach) City Gas Department in Restoring Service from March 11 to 26th, 1933, Inclusive", Long Beach, CA, March 27, 1933.

Board of Harbor Commissioners, Long Beach, CA, (1933), Letter report R.G. McGlone to J. F. Collins, Re: "Earthquake Damage of March 10, 1933", Long Beach, CA, October 18, 1933.

Engineering News Record, (1933), New York, NY, March 16, 1933.

Los Angeles Department of Water and Power, "Thirty-Second Annual Report, Board of Water and Power Commissioners of the City of Los Angeles, California, Fiscal Year Ending June 30, 1933", Los Angeles, CA.

Long Beach Historical Society, Historical photos and references.

Metropolitan Water District of Southern California (MWD), (1933), "Preliminary Report on the Long Beach Earthquake of March 10, 1933 with Special Reference to the Aqueduct Distribution System, Proposed MWD Lower Feeder alignment", MWD, Los Angeles, CA, May 16, 1933.

National Board of Fire Underwriters (NBFU), Committee on Fire Prevention and Engineering Standards, "Southern California Earthquake, March 10, 1933".

O'Rourke, T. D. and Palmer, M. C., (1994) "Feasibility Study of Replacement Procedures and Earthquake Performance Related to Gas Transmission Pipelines", National Center for Earthquake Engineering Research, Buffalo, NY, Technical Report NCEER-94-0012, May 25, 1994.

Scott, J. R., (1933), Letter to A. F. Bridge, (Brief resume of experiences), Southern Counties Gas Co., Los Angeles, CA. April 4 1933.

ASCE-TCLEE website, www.asce.org

LL 012703 6USCLEE

Hospital Lifeline Response to the 1999 Izmit Turkey Earthquake

Mark A. Pickett[1]

Abstract

This paper describes the details of earthquake related problems of hospitals in the epicentral area, caused by the 17 August 1999 Izmit Turkey earthquake. The paper summaries the structural and lifeline damage to hospitals, and gives the damage details for several hospitals.

Data is presented regarding the number of patients treated for earthquake related injuries. The location of this treatment (inside vs. outside the facilities) is discussed. Numerical data is also presented regarding earthquake-injured patients served by hospitals and evacuated from damaged hospitals. The reasons patients were evacuated from various hospitals are detailed.

Comparisons are also made to hospital damage caused by the 1994 Northridge earthquake.

Introduction

The internal lifelines of hospitals received a limited amount of damage from the 17 August, 1999, Izmit earthquake. However, approximately 8000 patients with earthquake related injuries were evacuated from hospitals in the epicentral area. These evacuations were necessary primarily because of facility structural damage. Some hospitals suffered loss of commercial electrical power, or loss of interior communications capability. Many elevators systems were damaged and remained inoperable, long after the event.

---

[1] Associate Professor, Department of Civil Engineering, University of Toledo, 2801 W. Bancroft Street, Toledo, OH 43606; mark.pickett@utoledo.edu; phone (419) 530-8120, fax (419) 530-8116.

In Turkey, there are four types of hospitals; state hospitals, faculty and military hospitals, Social Insurance Association (SSK) hospitals, and private hospitals. In the Istanbul area in August 1999, there were approximately 80 hospitals, total from these four categories. As a result of the earthquake, several hospitals in the earthquake-affected zone were damaged. These hospitals evacuated patients (who were admitted before the earthquake) to undamaged hospitals located in cities on the eastern side of Istanbul or to hospitals in Bursa or Ankara. Additionally, these damaged hospitals sent patients with earthquake related injuries to undamaged hospitals on the outskirts of Istanbul. Table 1 presents a compilation of data on the patients who were sent from damaged hospitals in the affected zone to undamaged hospitals (Istanbul Il Saglik).

Table 1. Data regarding patients with earthquake related injuries and hospital status.

	State Hospital	Faculty Military Hospital	SSK Hospital	Private Hospital	Morgue	Total
Patients with earthquake related injuries treated in outpatient facilities	2899	858	307	3251		7315
Patients with Earthquake Related Injuries Admitted To Hospital	445	477	281	421		1624
Beds Available in Those Hospitals that Received Earthquake Patients	1390	1500	270	536		3696
Patients Discharged from Those Hospitals that Received Earthquake Patients	826	588	24	542		1980
Patients with Earthquake Related Injuries Who Died In Hospital or were received dead	205	78	32	140	124	579

Affect of Lifeline Performance on Hospitals

The affect of lifeline performance on specific hospitals was investigated by the author during on-site inspections and on-site interviews. Table 2 summaries the lifeline performance in five specific hospitals.

Table 2. Lifeline performance and affects on hospitals.

Notes: RC = reinforced concrete   NDRC = non-ductile reinforced concrete
       URM = unreinforced masonry

	Kartal Research Hospital	Izmit SSK Hospital	Izmit State Hospital	Adapazari SSK Hospital	Adapazari State Hospital
Distance from epicenter, estimated acceleration	85 km 0.21 g	10 km 0.225g	5 km 0.225 g	40 km 0.4 g	45 km 0.4 g
Buildings' dates of construction, Size in terms of number of beds, Structural type, Structural damage	1973, 750 beds, RC, no damage.	1938, 144 beds, NDRC, URM, patients were evacuated due to wall and column damage. 1978, 300 beds, RC, URM, Not damaged.	1989, clinics, NDRC, URM, No damage. 1939, 350 beds, NDRC, URM, Wall and column damage, patients evacuated.	1996, clinics, RC, URM, No damage. 1985, 300 beds, NDRC, URM, patients were evacuated due to wall and column damage.	1970 420 beds, NDRC, URM, patients were evacuated due to wall and column damage.
Earthquake related patients	625 treated inside	500 treated in tents	1000 treated in tents	400 treated, 160 evacuated.	3600 treated in tents
Lifelines:					
Power		Lost commercial power for 2 days, Emergency generator sufficient.	Lost commercial power for 2 days, Emergency generator sufficient.	Lost commercial power for 11 hours, Emergency generator sufficient.	Lost commercial power for 11 hours, Emergency generator sufficient.

	Kartal Research Hospital	Izmit SSK Hospital	Izmit State Hospital	Adapazari SSK Hospital	Adapazari State Hospital
Waste water				Several internal ruptures	
Communications			Lost exterior for > 2 days, including cellular phones; interior lost for > 25 days.	Lost exterior for > 2 days, including cellular phones; interior lost for > 25 days.	Lost exterior for > 25 days, cellular phones ok; interior lost for > 25 days
Medical gases		Gas bottles overturned		Gas bottles overturned	Gas bottles overturned
Transportation	Elevators inoperable > 25 days	Elevators inoperable > 25 days. Expansion plates buckled preventing interior patient transportation.	Expansion plates buckled preventing interior patient transportation.		
Stationary medical equipment			Coronary monitors fell	Shelf-mounted monitors fell, Nurses' stations overturned.	Shelf-mounted monitors fell.

Comparisons to Hospital Performance during the 1994 Northridge Earthquake

During on-site inspections, interviews and phone interviews, the performance of lifelines in specific hospitals affected by the 1994 Northridge was investigated. Table 3 summaries the findings.

Table 3. Lifeline performance and affects on hospitals during the 1994 Northridge earthquake.

Notes: RC = reinforced concrete   RM = reinforced masonry
NDRC = non-ductile reinforced concrete   URM = unreinforced masonry

Hospital	Performance	Lifeline Water	Lifeline Waste Water	Lifeline Power	Lifeline Fire Suppression
Los Angeles County - Psychiatric	NDRC column damage, Evacuated 100 patients, Red tagged for 15 months.	Roof tank rupture.			
Los Angeles County - Pediatric	NDRC column damage, Evacuated 67 patients, Red tagged for 15 months.				
Los Angeles County - Womens' Health	Not evacuated	Roof tank rupture.			
Univ. of Southern California	Base isolated steel frame, No damage.				
Olive View	Evacuated 300 patients, treated patients in parking lot	Lines ruptured due to wall failure.		Lost external power, Lost emergency power.	
Northridge	All functions interrupted. Triage in parking lot.		External lines ruptured.	Lost external power,	Fire main ruptured.

Hospital	Performance	Lifeline Water	Lifeline Waste Water	Lifeline Power	Lifeline Fire Suppression
Veterans Administration Sepulveda	RC frame, RM walls, Pounding of wings. Evacuated 331 patients	Lines ruptured due to wall failure. Water caused grounds in electrical system.		Lost external power, Lost emergency power due to grounds and start-up batteries overturning	
Holy Cross	Evacuated 50 patients, Closed for 3 weeks	External lines ruptured. Internal lines ruptured due to wall failure.		Patient died due to loss of emergency power due to grounds caused by water line ruptures,	
Granada Hills	Treated patients in parking lot	External lines ruptured. Roof tank rupture.			

Conclusions and Summary

In both the Izmit and Northridge earthquakes, very few hospital functionality problems were directly caused by lifelines failures, alone. In both earthquakes, structural damage occurred first and the structural damage caused lifeline problems or lifeline failures that affected the functionality of the hospital.

As a result of the Izmit earthquake, approximately 8000 patients with earthquake related injuries were evacuated from damaged hospitals. These patients were sent to hospitals in Istanbul, Bursa and Ankara. These evacuations were necessary due to structural damage to hospitals in the epicentral area. The evacuations were not due to damage to lifelines in the hospitals or exterior to the hospitals.

All of the investigated Turkish hospitals lost commercial electrical power for at least two days, but none reported problems with their emergency power generation system. However, there were problems with the emergency power generation systems of several hospitals in the aftermath of the Northridge earthquake.

The Turkish hospitals reported no problems with water systems, and only one facility reported breakage of waste water lines. During the Northridge earthquake, internal water line ruptures caused many electrical grounds and several electrical fires.

Most Turkish hospitals lost commercial telephone communications for at least 2 days. In at least one hospital, the telephone system was inoperable for more than 25 days. Some facilities had problems with interior communications. Cellular phones and walkie-talkies were used to solve these communications problems. However, some cellular systems were inoperable during the immediate aftermath. This was due to collapsed towers and the resultant system saturation.

The Turkish facilities were heated by fuel oil, and there were no reported leaks or fires in these systems. There were no reported problems with hazardous materials, or the fire suppression systems. Many unrestrained oxygen cylinders toppled over, but there were no reported leakage or fires from this problem, since most facilities shut down the central oxygen system. However, many unrestrained monitors fell from their shelf or mounting. Most Turkish hospitals reported problems with elevators. Some of these problems were still unresolved by the 25th day after the earthquake, because elevator repair technicians were unavailable.

General Recommendations

Hospital facilities need to ensure that portable equipment, monitors, and medical gas cylinders are properly restrained to their mounts and that tall desks (nurses' stations) are properly anchored to the floor.

Emergency response plans must be formulated to include actions that specified staff members must take in the event that some or all of the hospital functions are disrupted by earthquake damage. These plans must designate responsibilities, chain of command, and actions to be taken until an Emergency Operations Center is operational. These emergency response plans should be tested at least twice a year.

References

iSTANBUL iL SAGLIK MUDURLUGU KAYITLARI-TUM HASTANELER HASTA DURUMU, fax dated 11/09/1999 / 10:15, received from Istanbul Crisis Management Center.

Obtaining the Emergency Transportation Network for Rescue and Relief Activities in Large Cities Based on the Life Loss Mitigation Criteria

Afshin Shariat Mohaymany[1], Mahmood Hosseini[2], and Hossein Motevalli Habibi[3]

Abstract

In this paper a method is proposed for obtaining the "emergency paths" for the use of rescue and relief teams, to be kept available by the help of police forces in the aftermath of a great earthquake in a large city. The proposed method is based on two "life loss mitigation criteria", which are minimization of travel time between rescue and relief centers and help-needing population, and maximization of rescue and relief forces service capability to all help-needing population. The main elements considered in the process of emergency paths selection include the highly populated areas, and vulnerable and hazardous areas on the one hand, and rescue and relief centers on the other. The proposed path selection process is performed by using some developed computer programs in the GIS environment based on various information layers, in which all factors affecting the functionality of the transportation network are included. The proposed process, which is basically useful for the short-term period after the earthquake, can be also used for the long-term economic restoration time after the earthquake.

Introduction

Past earthquakes have shown that the rescue and relief activities in large cities can not be successful without the help of police forces for controlling the emergency situation, and keeping some vital paths open for rescue and relief teams. On the other hand, usually the number of police forces is not sufficient to keep control on all of the required paths in a big city. Therefore, having a set of pre-selected paths,

---

[1] Assistant Professor, Transportation Engineering Group, Civil Engineering Department, Iran University of Science and Technology, Tehran, Iran, Email: shariat@iust.ac.ir

[2] Associate Professor and Head of Lifeline Engineering Department, International Institute of Earthquake Engineering and Seismology (IIEES), P.O.Box 19395/3913, Tehran, Iran, Email: hosseini@dena.iiees.ac.ir

[3] Research Assistant, Planning and Training Bureau, Transportation and Terminals Organization, Ministry of Transportation, Tehran, Iran

to be controlled by police forces for the use of rescue and relief people, can be very effective is achieving a successful emergency management accomplishment in the aftermath of an earthquake in a large and populated city. In this paper after a brief review on previous studies on various methods for analyzing and evaluating the service reliability of the road or highway network in urban areas, or their emergency response, a method is presented for obtaining an "emergency network" for rescue and relief teams. At the end, application of the proposed method to Tehran metropolis is discussed.

Previous Studies on Emergency Response of Urban Transportation Systems

Several studies have been performed with regard to performance evaluation or functionality assessment of transportation systems in the aftermath of earthquakes in large cities. One of the early works on performance evaluation method for highway transportation systems during post-earthquake period has been proposed in late 80s (Igarashi et al., 1989). Mentioning the important role of highway transportation system in various post earthquake activities, such as fire-fighting, relief works, and repair activities, Igarashi and his colleagues have proposed a method for evaluating the performance of functionally disordered transportation systems after an earthquake. They have claimed that their results can be used for practical seismic design purposes and for establishing pre-disaster plans to aid in determining whether a highway transportation system can fulfill its function after an earthquake. As another early work the "transportation emergency management of post-disaster operations" (TEMPO) user manual can be mentioned (Ardekani and Jabri, 1992).

Many studies on this issue have been performed by using Geographic Information Systems (GIS). For example regional evaluation of transportation lines in New York State has been done with the aid of GIS technology (Shinozuka et al., 1992). Notifying the great concern with regard to transportation lifeline for purposes of emergency planning and response, and for establishing priorities for maintenance and retrofit if an unacceptable risk is found to exist, they have done an interactive regional risk analysis for bridges to illustrate the procedures involved. They have utilized Erie County, New York, as a demonstration area for regional analysis, which has been considered a representative of Eastern US conditions.

Many researchers have worked on various methods for analyzing and evaluating the reliability of urban transportation systems in emergency situations. Particularly in recent years, risk analysis has been paid more attention than before. Huang, Zhu, and Zhong (1996) have worked on the application of theory of graphs to reliability analysis of lifeline networks. They have introduced the concept of the "road matrix", and have presented two methods for calculating this matrix: the Warshall algorithm and the coloring method, and have discussed the advantages of these two methods.

Transportation systems analysis methods have been also used for bridge retrofit prioritization and emergency response (Kiremidjian and Basoz, 1997). They have embedded some network analysis methods in a general seismic risk analysis methodology to develop an approach for prioritizing the seismic retrofit of a large number of bridges and for an emergency response system. The prioritization and emergency response methods have been applied to the Palo Alto, California area. It has been discussed that the prioritization of the bridges is consistent with their seismic exposure and their importance within the transportation system. They have tried to identify "critical routes" within the system are and to delineate minimal paths for hypothetical earthquake events affecting the study region.

In late 90s a so-called seismic safety estimation method has been developed for emergency transportation road networks (Nagamatsu et al., 1998). They have claimed that since the 1995 Hyogoken-Nanbu earthquake, pre-earthquake assessments have been conducted to examine disaster prevention techniques. They have nominated emergency transportation road networks, have checked bridges and slopes, and have developed a database. Mentioning that earthquake damage can be reduced by adequate pre-earthquake countermeasures using these data, in that study a database has been constructed about bridges and road networks using GIS. That paper aims to determine the prioritization of ranks for seismic retrofitting, including bridge seismic safety estimations.

A network analysis based-GIS has been also introduced for urban road system in the earthquake emergency response (Shuai and Cheng, 1998). In that paper certain situations have been considered in the network analysis, including the effects of collapsed buildings and damaged bridges. The method has been applied in the case study of earthquake disaster estimation and developing of mitigation measures for the city of Urumchi to develop a computer program called "UrumEDIS" - Earthquake Damage Estimation and Disaster Mitigation Information System for Urumchi. Shuai and Cheng have claimed that by combining the function with the other measures of analysis, comprehensive decisions would be made rapidly for mitigation and rescue action.

Kiremidjian and Basoz (1998) have also worked on transportation system network analysis for earthquake emergency response and long term economic recovery. In that paper, a system has been presented that utilizes transportation risk analysis with network analysis, which identifies the shortest path available from single origins to single destinations, and from multiple origins to multiple destinations. Mentioning that most currently available network analysis methods provide information on critical paths for a single origin and single destination, they have stated that following a major earthquake, however, there are typically many origins (e.g., hospitals, fire stations, and police stations) and many destinations (e.g., locations of collapsed buildings scattered throughout the affected region). So they have developed network analysis capabilities for multiple origin and multiple destinations that can be particularly helpful in resource allocation decisions both for emergency response and for long-term economic recovery purposes

Nojima and Sugito (2000) have also worked on simulation and evaluation of post-earthquake functional performance of transportation network. Their model combines the Monte Carlo simulation method and the modified incremental assignment method (MIAM). The former generates damage states of the transportation network, and the latter simulates traffic behavior in each damage state, producing a set of link flows. On this basis, various performance measures for links, O-D (origin-destination) pairs, centroids, cross sections and the total network have been defined to evaluate aggregate and non-aggregate conditions of the network function. Proposed measures reflect mixed effects attributed to decrease in O-D trips due to facility damage and overload, increase in trip length due to detouring actions, increase in travel time due to detouring and congestion, and so forth.

A reliability importance analysis has been also performed for highway network systems for successful future highway construction, prioritization of a retrofit program, and efficient disaster traffic management (Wakabayashi, 2000). The first stage of that study has been carried out from the traffic engineering aspect. The second stage has been an integrated study combining traffic engineering and earthquake engineering. That research, however, have been involved studies of traffic aspects and functional damage only on the basis that the highway physical structure was perfectly sustained. Highway network reliability has been, however, achieved both from traffic/functional and from structural aspects. Wakabayashi has stated that a combination of traffic engineering and earthquake engineering is required and has tried to show the framework for such a study.

Proposed Method for Obtaining the "Emergency Network"

Based on the review of previous studies it is seen that in spite of several researches on the behavior of urban transportation systems after a major earthquake, the concept of "critical routes", namely the routes for the use of rescue and relief teams in emergency situation after a major earthquake has been paid little attention. In this section of the paper, based on "life loss mitigation criteria", a method is proposed for obtaining the best paths for the use of rescue and relief teams, to be kept available by the help of police forces. These paths are called the "emergency paths" or "emergency network". Accordingly, after a major earthquake the repair teams of various utilities in the city should be sent at first to the emergency paths, as the paths with the highest priority, to check and to make sure on their passibility, and do repairs, if required, to keep them functional.

The proposed path selection process is performed by using some developed computer programs in the GIS environment based on various information layers. Once the emergency paths are selected as explained, some shortcoming may be realized in some parts of the selected paths. This realization can lead to making decision for resolving those shortcomings by upgrading of the weak component of the system as the pre-earthquake period is passing, namely before the occurrence of the probable earthquake. The proposed process, which is basically useful for the short term period immediately after the earthquake, can be also used for the long term period after the earthquake for economic restoration of the society. The employed criteria for identifying and various stages for obtaining the emergency network are explained in the following parts of this section.

Life Loss Mitigation Criteria - As stated before, in this study the "emergency transportation network" for rescue and relief activities has been obtained based on the "life loss mitigation criteria". These criteria are:
- Minimization of travel time between "rescue and relief centers" (RRCs) and help-needing population

- Maximization of "rescue and relief forces" (RRFs) service capability to as much help-needing population as possible.

The first criterion is very well-known, since the minimum travel time between rescue and relief centers and help-needing population is obviously desired for saving the stricken people. In fact, the travel time has been used as an important factor in evaluation of the post-earthquake performance of transportation networks in large and populated cities (Nojima and Sugito, 2000). The average travel time concept has been also used in defining the "accessibility index" in a network (Shariat Mohaymany, 2002; Hosseini et al., 2002).

The second criterion is somehow a new concept, which is more emphasized in this study. By this criterion it is meant that relying only on the first criterion is not enough to make sure on having an "emergency network" with efficient functionality. This is because the existence of unpredictable problems which may occur in the aftermath of a major earthquake in a large city, mainly due to the public reaction (Hosseini and Mirza Hessabi, 1999). Therefore, it is emphasized in this paper to consider the use of police forces for controlling and keeping open some specific routes for RRFs. Noting that usually enough number of police forces is not existing to keep several paths open in all districts of the city at the same time, this criterion is itself a confirmation on the validity of the idea of having a predetermined network of routes for the use of RRFs, namely the "emergency network".

Elements Considered in the Path Selection Process - The main criteria for selecting the emergency paths are: a) the population density of the areas to which the paths serve, and b) the required travel time in those paths to reach the corresponding RRCs from the highly populated stricken areas or vice versa. Other factors which should be considered in the path selection process, is the existence of potential hazardous city components which threaten the path passibility such as bridges, adjacent high rise buildings, and other lifeline components which interact with the transportation network, such as power transmission towers, water mains, and main gas pipes in the city, and finally the number and the locations of other potentially hazardous points, such as fuel stations, chemical factories and plants, warehouses of flammable materials, which may need help of the RRFs. The help-needing population centers and other help-needing points in the city are called here the help-needing centers or briefly the HNCs.

The Analysis Procedure - For obtaining the emergency routes a GIS-based program has been used. The procedure has three main steps, each composed of some sub steps as follow:
- **The first step**
  Constructing the basic network
  Locating RRCs and HNCs
- **The second step**
  Prioritization of HNCs based on the population density or hazardousness
  Obtaining the shortest paths between RRCs and HNCs
  Prioritization of selected paths between RRCs and HNCs
  Obtaining the basic "emergency network"

- **The third step**
  Modification of basic "emergency network" based on path selection criteria
  Introducing the final "emergency network"

There are some modification criteria to be applied to the basic constructed network to improve it as the final network is sought. These are as follow:
- The selected paths should have appropriate width.
- Major routes are preferred, as RRFs and police drivers are more familiar with major routes.
- Highways and particularly freeways are preferred as they have less likelihood of interruption by other routes.
- Selected paths should be preferably of those routes which have the less cases of service level "E" (the service level in which the route is used by its maximum capacity) in normal conditions.
- The number of components which have the potential of making the route instable, such as co-level intersections, bridges, high trenches and embankments, tunnels, etc. should be as little as possible.
- Straight paths are better than paths with turns, as the quick maneuver possibility for both RRFs and police forces is much more in straight paths.

Application of the Proposed Method to Tehran Metropolis

Tehran, the capital and the largest city in Iran, is a metropolis with near 10 million population distributed in an area of almost 1200 square kilometers. This city is located in the highly seismic zone of the country with the average PGA value of 0.40g, as shown in figure 1.

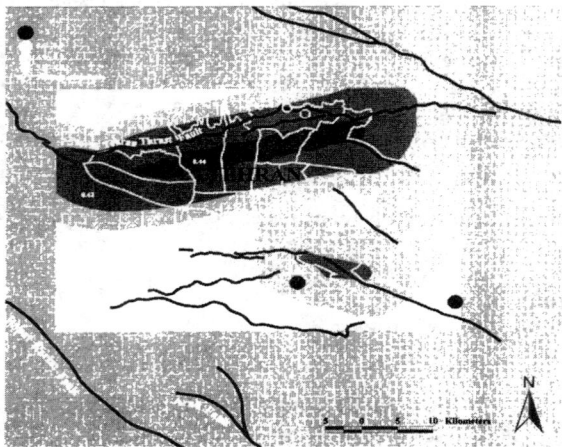

Figure 1. PGA microzonation map of Tehran for the 10% of exceedence in 50 years ground motion (IIEES)

The population density in some parts of the city exceeds 350 people per *ha*. Almost 50% of buildings in this city are more than 25 years old, and in many parts of the city more than 70% of buildings can not resist the probable earthquake (Ghafory-Ashtiany et al., 1992; JICA, 2000). On this basis, the predicted number of casualties and life losses due to the probable earthquake are respectively around 1,600,000 and 400,000. This is while in some parts of the city because of the very poor building conditions and lack of access of rescue and relief teams, the life loss can reach 20% of the corresponding population. As a sample of casualties' distribution in the city, the case of South Rey fault (shown in figure 1) night scenario is shown in figure 2.

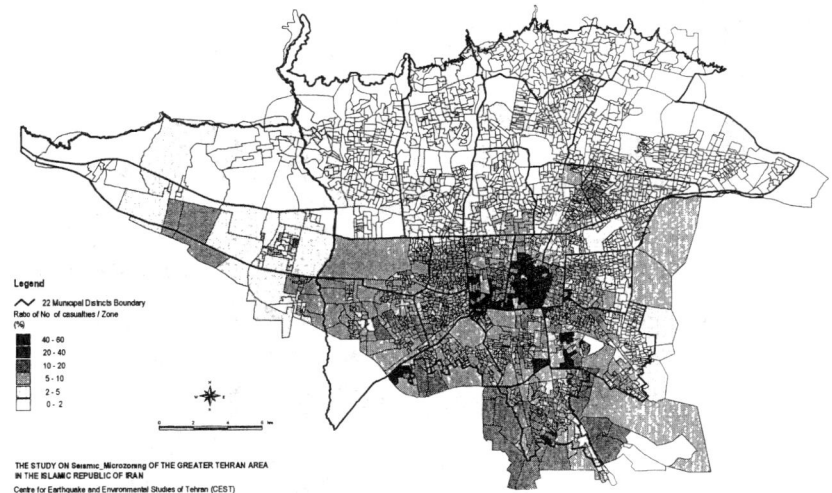

Figure 2. The casualties' distribution for South Rey fault night scenario

As it is seen in figure 2, the casualties' distribution is quite non-uniform. This is while in some of the city districts with high casualties' density, there are only few RRCs, and furthermore, 3 meters- and 6 meters-wide alleys comprise a great percent of the city routes in those districts, and open areas are very rare. There are totally 180 hospitals, 55 fire stations, 4 Red Crescent centers, and 109 police stations in the city. The public utilities centers, such as electric power, water and wastewater, as well as communication, should be added to these centers to consist the set of all centers to which the "emergency network" should provide reliable access.

The Analyzed Network - Tehran transportation network, which covers an area of almost 800 square kilometer comprise of almost 2356 *km* paths, including 209 *km* freeways and highways, 110 *km* ramps and access loops to highways, 454 *km* first degree main roads, 402 *km* second degree main roads, and 1181 *km* collectors and local roads. Highways and freeways consist almost 10% of the whole network. This network serves daily to more than 11,500,000 people just in its entrances and exits. The pick of travels in the rush hour is around 1,120,000. In the analyzed network all the paths having width of 3 meters or more have been included. On this basis, in the GIS environment the analyzed network is consisted of 141,856 members, which have been categorized in 19 different groups.

As the contribution of fire departments in the network analyses is essential, it is necessary to assume a travel time upper limit for the fire departments' cars to reach the help-needing points in the whole city. To show one of the shortcomings in the existing Tehran transportation system and distribution of fire stations, the coverage areas of fire stations by assuming a travel time upper limit of 5 minutes, which is far from most of national and international standards, is shown in figure 3, which indicates that with this high travel time upper limit still all the city area can not be covered by fire stations.

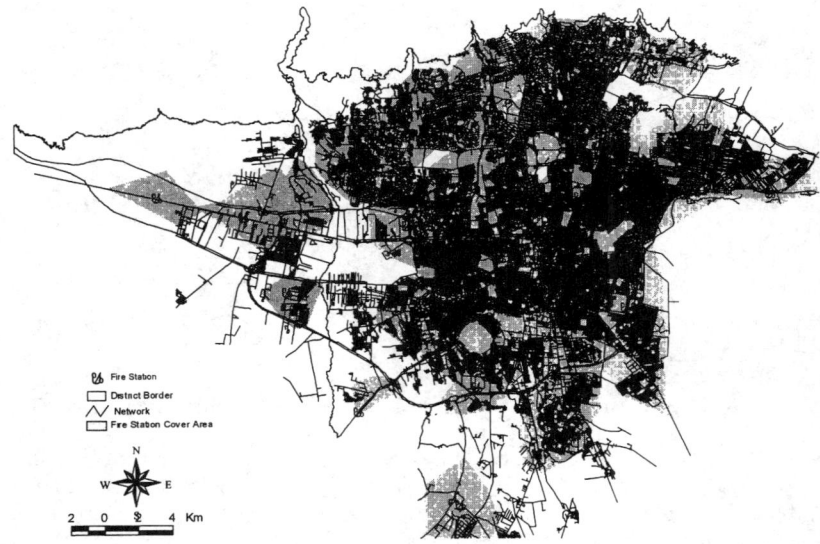

Figure 3. The total coverage area of all 55 fire stations in Tehran metropolis

Various Scenarios for "Emergency Network" Assessment - As the occurrence time of the earthquake is the key parameter to which the number of casualties and their distribution as well as the traffic condition are dependent, various scenarios should be defined based on the earthquake occurrence time, and accordingly various "emergency networks" should be assessed. At least, five different time periods can be considered for earthquake occurrence, each having its own characteristics. These are: (Hosseini et al., 2002)
1. Morning rush hour time

2. Mid-day (working hours) time
3. Afternoon rush hour time
4. Evening (after-work hours) time
5. Night (sleep) time

Each of these time periods has its own traffic pattern, people distribution, people preparedness for safety reactions, people's reaction with respect to their relatives, casualty likelihood, situation and preparedness of public utilities authorities, and finally the situation of rescue and relief authorities, who are responsible for disaster management actions. The results of Tehran "emergency networks" have been obtained for some of the aforementioned scenarios, which is now under finalization, and the study is presently ongoing for other scenarios.

Conclusions

This study shows that the life loss mitigation criteria are quite reasonable for making decision on the "emergency network" for rescue and relief activities. This will result in the optimum use of rescue and relief as well as police forces in the process of disaster management in the aftermath of a major earthquake in large cities. Once the emergency paths are selected, some shortcomings may be realized in some parts of the selected paths, so decision should be made for resolving those shortcomings by upgrading the weak component(s) of the system before the probable earthquake. The proposed process, which is basically useful for the short term period immediately after the earthquake, can be also used for the long term period after the earthquake for economic restoration of the society. This is possible by introducing the new situation after the earthquake as the basic information to the GIS environment.

References

1. Ardekani, Siamak A. and Jabri, M. Iyad, *"Transportation emergency management of post-disaster operations (TEMPO): user manual"*, Department of Civil Engineering, University of Texas, Arlington, Texas, 1992, 35 pages.
2. Ghafory-Ashtiany, M. and Hosseini, M., Jafari, M. K., Eshghi, S., Qureishi, M., and Shaditalab, J., Tehran Vulnerability Analysis, *Proceedings of the 10th World Conference on Earthquake Engineering*, Madrid, SPAIN, July 1992.
3. Hosseini, M., Mansour-Khaki A., and Shariat, A., "Functionality Assessment of Urban Transportation Systems for Rescue and Relief Activities in the Aftermath of Earthquake in Large Populated Cities", *Proceedings of the 7th US National Conference on Earthquake Engineering,* Boston, USA, July 2002
4. Hosseini, Mahmood and Mirza-Hessabi, Ali, "Lifelines interaction effects on the earthquake emergency response of fire departments in Tehran metropolis", *Technical Council on Lifeline Earthquake Engineering Monograph No. 16*, Proceedings of the 5th U.S. Conference on Lifeline Earthquake Engineering, American Society of Civil Eng., Reston, Virginia, Aug. 1999, pages 731-740.
5. Huang, S.; Zhu, J.; and Zhong, G., Application of theory of graphs to reliability analysis of network of lifeline engineering, *Proceedings of the Eleventh World*

Conference on Earthquake Engineering, Pergamon, Elsevier Science Ltd., 1996, Disc 3, Paper No. 1464.
6. Igarashi, A.; Yamada, Y.; and Noda, S., "Performance evaluation method of highway transportation systems during post-earthquake period", *Proceedings of the 9th World Conference on Earthquake Engineering*, Japan Assn. for Earthquake Disaster Prevention, Tokyo, Vol. VII, 1989, pages 153-158, Paper 11-3-6.
7. JICA, "*The study on seismic microzoning of the greater Tehran Area in the Islamic republic of Iran*", Japan International Cooperation Agency, 2000.
8. Kiremidjian, A. S.; Basoz, N. I., "Use of transportation network systems for bridge retrofit prioritization and emergency response", *Technical Memorandum of PWRI 3481*, Proceedings of the Third U.S.-Japan Workshop on Seismic Retrofit of Bridges, December 10 and 11, 1996, Public Works Research Institute, Tsukuba-shi, Japan, 1997, pages 51-63.
9. Kiremidjian, Anne S. and Basoz, Nesrin, "Transportation system network analysis for earthquake emergency response and long term economic recovery", *Proceedings of Third China-Japan-US Trilateral Symposium on Lifeline Earthquake Engineering*, Kunming, People's Republic of China, Aug. 1998, pages 215-222.
10. Nagamatsu, Yoshitaka; et al., "Development of seismic safety estimation method for emergency transportation road networks" (in Japanese), *Proceedings of the 10th Earthquake Engineering Symposium*, Architectural Inst. of Japan, Tokyo, 1998, pages 3407-3412, Vol. 3, Paper No. L1-10.
11. Nojima, Nobuoto and Sugito, Masata, "Simulation and evaluation of post-earthquake functional performance of transportation network", *Proceedings of the 12th World Conference on Earthquake Engineering* [computer file], New Zealand Society for Earthquake Engineering, Upper Hutt, New Zealand, 2000, Paper No. 1927.
12. Shariat Mohaymany, Afshin, "*Evaluation of Transportation Lifelines in the Aftermath of Earthquake*", Ph.D. Dissertation submitted to Transportation Engineering Group, Civil Engineering Department, Iran University of Science and Technology, Tehran, Iran, February 2002.
13. Shinozuka, Masanobu, et al., "Regional evaluation of transportation lines in New York State with the aid of GIS technology", *Technical Council on Lifeline Earthquake Engineering Monograph 5*, Lifeline Earthquake Engineering in the Central and Eastern U.S.: Proceedings of three sessions sponsored by the Technical Council on Lifeline Earthquake Engineering in conjunction with the ASCE National Convention in New York, September 1992, pages 102-109.
14. Shuai, Xianghua and Cheng, Xiaoping," Network analysis based-GIS for urban road system in the earthquake emergency response", *Proceedings of Third China-Japan-US Trilateral Symposium on Lifeline Earthquake Engineering*, Kunming, People's Republic of China, Aug. 1998, pages 315-322.
15. Wakabayashi, Hiroshi, "Reliability analysis and importance of highway network systems", *Confronting Urban Earthquakes: Report of Fundamental Research on the Mitigation of Urban Disasters Caused by Near-Field Earthquakes*, Kyoto University, Kyoto, Japan, Mar. 2000, pages 177-180.

Performance of Yen-Feng Bridge during the 921 Taiwan Chi-Chi Earthquake

Kuo-Chun Chang[1]   Kung-Yuan Kuo[2]   Chih-Hung Lu[3]

ABSTRACT

This paper is concerned with the performance of Yen-Feng Bridge during the September 21 Taiwan Chi-Chi Earthquake. In this earthquake, very high peak ground accelerations, near fault velocity pulses as well as permanent ground displacements were reorded. However, the extent of bridge damage is relatively minor when compared to those observed in the 1994 Northridge earthquake and 1995 Kobe earthquake. Most of the bridge damage appeared to be the movement of superstructure and separation of thermal expansion joints due to sliding or failure at the bearings, with the exception of seven bridges collapsed due to large fault displacements directly acrossed the bridges. It was also observed that the number of bridge column failure was surprisingly small. Since most of the bridges in the damaged area of Chi-Chi earthquake were designed without ductile detailing, it may be in contrast to the current seismic design concept emphasing the steel detailing to control the location of the plastic hinge zone in the bridge columns. Yen-Feng bridge is near the epicenter of the earthquake with only minor damage. A non-linear finite element model of this bridge is first established to simulate the type of damage and verified by the observed conditions. Parametric studies are then carried out to discuss why most bridges experienced with no or only minor damage during this earthquake.

INTRODUCTION

The 921 Taiwan Chi-Chi Earthquake with the magnitude of $M_L=7.3$ incurred tremendous disaster to the central region of the island, particularly to Taichung and Nantou countries. In this earthquake, very high peak ground accelerations, near fault velocity pulses as well as permanent ground displacements were reorded. However, the extent of bridge damage is relatively minor when compared to those observed in the 1994 Northridge earthquake and 1995 Kobe earthquake. Most of the bridge damage appeared to be the

---

[1] Professor, Department of Civil Engineering, National Taiwan University, Taipei, Taiwan, R.O.C.
[2] PhD Candidate, Department of Civil Engineering, National Taiwan University, Taipei, Taiwan, R.O.C.
[3] M. S., Department of Civil Engineering, National Taiwan University, Taipei, Taiwan, R.O.C.

movement of superstructure and separation of thermal expansion joints due to sliding or failure at the bearings, with the exception of seven bridges collapsed due to large fault displacements directly acrossed the bridges. It was also observed that the number of bridge column failure was surprisingly small.

The Yeng-Fang Bridge is located in between the Chu-Lon-Pu and Shuang Tung fault, as shown in Figure 1. It is near the epicenter of the earthquake with only minor damage. In this study, a non-linear finite element model was established, which can predict the dynamic characteristics of the Yen-Feng Bridge under ambient vibration condition as well as the seismic performance during the 921 Taiwan Chi-Chi Earthquake. And the results by using a detailed nonlinear time history analysis are compared with the actual damage of the bridge under this earthquake.

THE YENG-FENG BRIDGE

The Yen-Feng Bridge, located at the milepost of 26km+937m on Provincial Route 14, is one of the main links connecting Tsaotun and Poli town in Nantou country. The bridge is in between the Chu-Lon-Pu and Shung-Tung fault (approximately 5 km from Chu-Lon-Pu fault and 7km from Shung-Tung fault), as shown in Figure 1. It is actually 13-span bridges with a uniform span length of 35m and 16m wide. All spans are supported by neroprene bearings at each cap beam. The superstructure is composed of six simply supported, prestressed reinforced concrete girders and deck slab with additional 5 cm A.C. on top. The substructure comprises multicolumn bents (fram-type piers) on spread footing. At both ends, the bridge girders are supported on abutment. The schematic drawing and cross section of the whole bridge system are shown in Figure 2 and 3, respectively. Two shear keys are placed on each cap beam to prevent the girders from unseating. The Yeng-Feng Bridge is typical type of ordinary highway bridges in Taiwan.

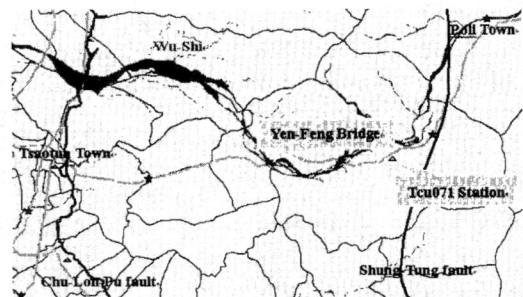

Figure 1. Geographical position of Yen-Feng Bridge

DISASTER RESPONSE FOR LIFELINE SYSTEMS 243

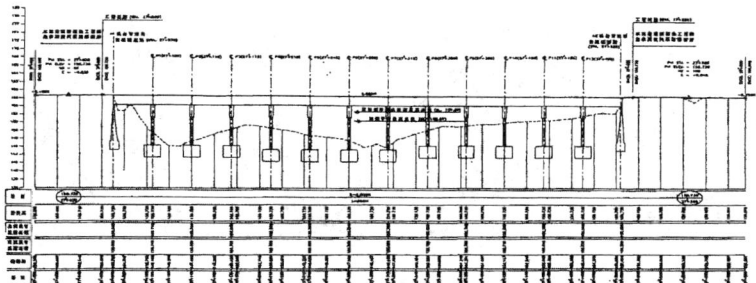

Figure 2. Schematic drawing of the bridge

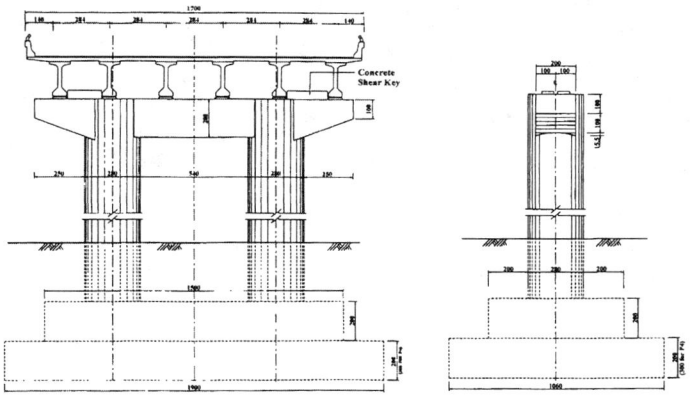

(a) Multicolumn bent

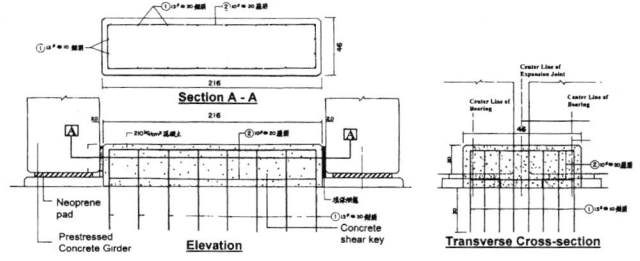

(b) Concrete Shear key

Figure 3. Elevation and sectional drawing

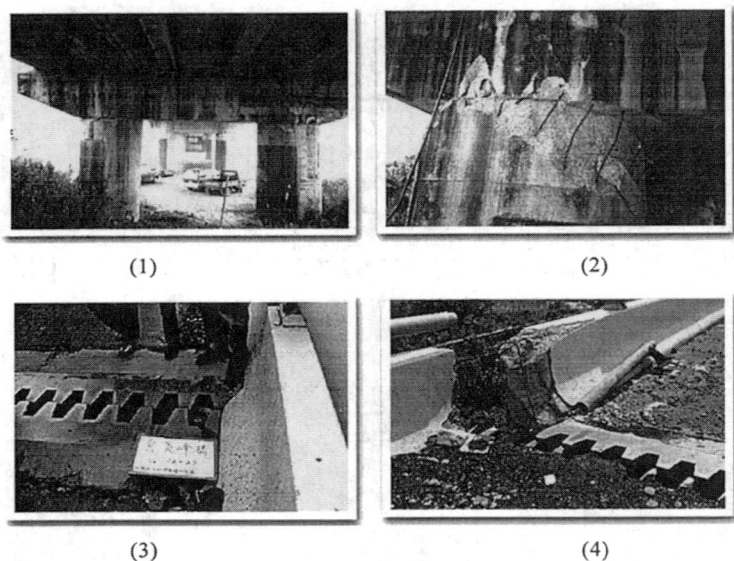

Picture. (1) Concrete spalling and cracking at pier top; (2) Relative displacement of cap beams at pier top; (3) Residual displacement at expansion joint; (4) Residual movement between two adjacent spans (at the fourth span)

BRIDGE DAMAGE

The Chu-Lon-Pu and Shuang Tung fault did not directly across the Yeng-Feng bridge, but evoked strong ground motion in vicinity of the bridge with a horizontal PGA of 500 gal in primary north-sourth, 400 gal in primary east-west and vertical PGA of 300 gal. During such large ground excitation, concrete spalling and cracking occured at the column top, and several concrete shear keys on the cap beam were crushed. According to the bridge-disaster reconnaissance reports, the damages of Yeng-Feng Bridge are summarized as followings:

1. Concrete spalling and cracking on the weak construction joint between cap beam and column top at bents P1 to P12, as shown in Picture 1.

2. Relative movement of cap beams at bents P9, P10 and P11 in transverse direction, as shown in Picture 2.

3. Several concrete shear keys were cracked.

4. Residual relative displacement occurred at several expansion joint. The bridge deck at the fourth span moved about 40 cm in transverse direction, as shown in Picture 3 and 4.
5. Parts of neroprene bearings had permanent displacement.

DESCRIPTION OF ANALTICAL MODELS

Analtical models

The Feng-Feng Bridge features a superstructure consisting of the daiprams, prestressed reinforced concrete girders, decks and bearings, as shown in Figure 3. In this study, the superstructure is expected to remain essentially elastic, limiting nonlinear modeling considerations to joint between superstructure segments, connections with cap beam under seismic force. The detail describtion of analytical model can be summarized as followings:

(1) Superstructure

The superstructure, consists of the prestressed reinforced concrete girder, diaphragm and deck slab, is modeled as a series of elastic beam column elememts.

(2) Substructure

In the analytical model, the multi-column bent can be modeled as a planar frame along the bent axis, consisting of beam and column elements with effective member properties. Except for the properly modeling of column, cap beam and footing, the effect of soil-structure interation at the abutment and footing was modeled by equivalent sets of linear springs. Furthermore, considering changing riverbed has an effect on the length changing of the pier, the soil-covered pier is also modeled as spring elements.

The construction joint between cap beam and column top is a discontinuous weak face, which has lower shear strength, and such damages occured during this earthquake. In order to understand this phenomenon, the weak construction joint is modeled as a bi-linear element and the shear strength is determined according to the shear friction approach, as shown in Figure 5.

The gap length between two adjacent spans is 10cm and the allowable movement for the expansion joint is 5cm. Differential displacements between bridge segments separated by movement joints can result in impact under differential response to seismic excitation. Therefore, a gap element (as shown in Figure 5) is adopted to estimate the expected differential displacement across expansion joints.

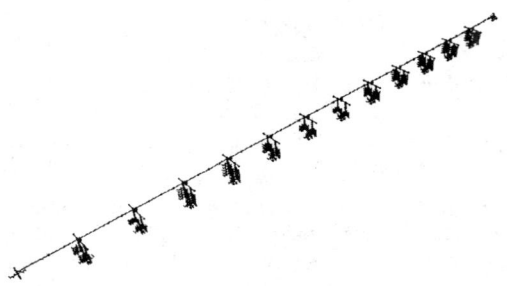

Figure 4. Analytical model of the bridge system

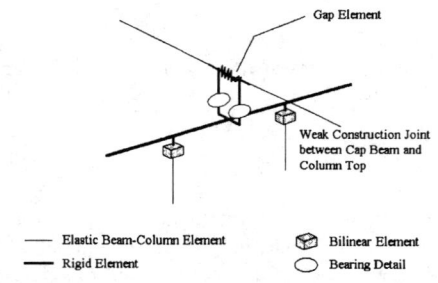

Figure 5. Non-linear elements of sub-structure

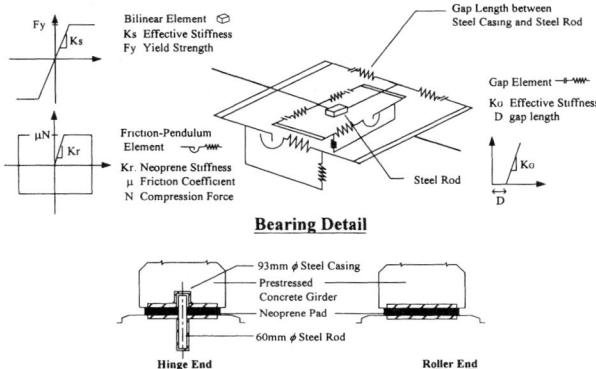

Figure 6. Non-linear elements of super-structure

(3) Bearing system

The analytical model of bearing system, including neoprene stiffness, friction and steel rod, can be shown in figure 6. Due to the friction between girder and neoprene pad, the bearings are modeled as friction elements with an initial stiffness equal to that of neroprene bearing and zero after slipping occurs. The function of the steel rod is to limit the displacement of superstructure in general use or during low to middle-earthquake. A bi-linear element can be used to model the steel rod and gap elements are considered as the gaps between PCI girder and steel rod.

Concrete shear keys between the superstructure and cap beam critically affect the seismic response bridge, especially the distribution of inertia forces and displacement between bents in transverse direction. In this study, shear key is modeled as a bi-linear element. Gap elements are used to consider the gaps between PCI girders and shear keys.

Characteristics of ground acceleration

The ground acceleration used in this study was selected to represent as closely as possible the actual excitation at the bridge site under the 921 Taiwan Chi-Chi Earthquake. The nearest strong motion accelerometer, TCU 071 is located approximately 4 km away from this site. Figures 7 are the acceleration recordings in primary east-west, primary north-south and vertical directions, respectively. The peak values of ground acceleration (PGA) measured in those recordings are 518 gal, 640gal and 416 gal related to primary east-west, primary north-south and vertical direction. After normalizing the PGA to 1.0g, the response spectrum shows that the design response spectrum is not conservative in few points, such as for short period of 0.16 ~ 0.4 sec and median period of 0.5~0.82 sec in primary east-west, and for short period of 0.15 ~ 0.35 sec in primary north-south. This indicates that the design seismic force is inadequate in short period.

NON-LINEAR TIME HISTORY ANALYSIS

In order to evaluate the dynamic response history of the bridge during the 921 Taiwan Chi-Chi Earthquake, all recordings in primary east-west, primary north-south and vertical direction are assumed to be the estimated ground motion at this site. The overall responses included a detailed study of bearings, expansion joints, and pier forces are summarized as followings:

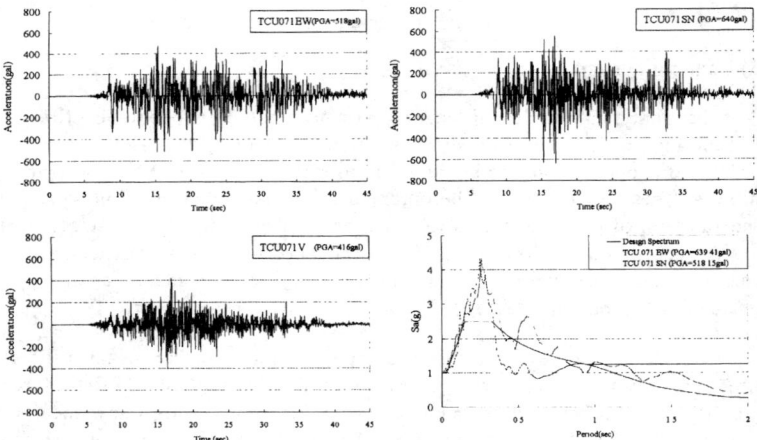

Figure 7. Ground Motion Recordings and Response Spectrum (TCU071)

1. Weak construction joint between the cap beam and pier top

Figure 8 shows the maximum shear force of the construction joint between the cap beam and column top. The weak joints at piers P8 ~ P11 had been yielded, and ones at piers P2, P5 and P8 are almost yielded. As the superstructure moved forward to the transverse in 2 centimeters and pounded the concrete shear keys, it transferred forces to the bridge column in this direction. In the longitudinal direction, the shear force was minor without shear keys. The results of numerical analysis almost represent the actual damage of the bridge. The dotted line means the analytical results of the bearing strength being reinforced. Apparently the shear force of the construction joint between the cap beam and column top increased and led to critical damage.

2. Flexural strength

In order to evaluate whether the flexure failure of column occurs or not, the axial load-moment interaction diagram is calculated. Comparing the axial force-moment time history response with axial load-moment interaction diagram, as shown in Figure 10(a), the maximum moment and axial force for P6 did not excess the axial load-moment interaction diagram that indicates the pier did not have flexure failure. Virtually only damage occurred at the weak construction joint between the cap beam and pier top, no flexure failure is found. Figure10 (b) shows the results of the bearing strength being reinforced. Apparently the column moment increases and caused damage.

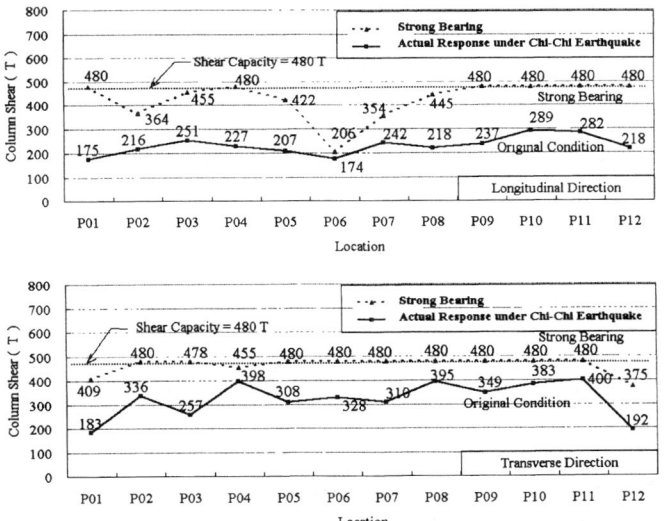

Figure 8. Maximum shear force of the construction joints between the cap beam and column top

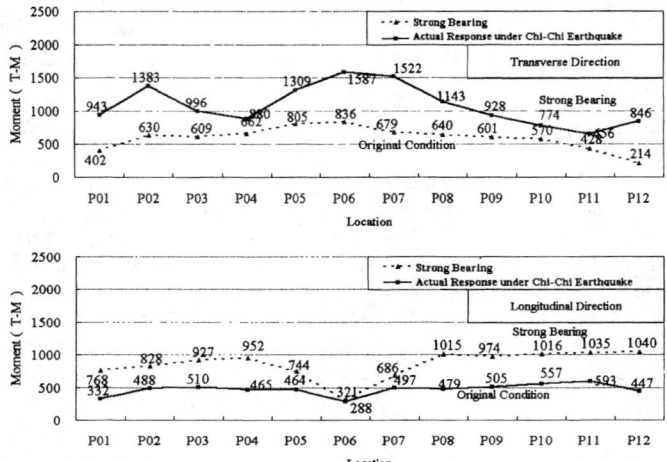

Figure 9. Maximum bending moment of the construction joints between the cap beam and column top

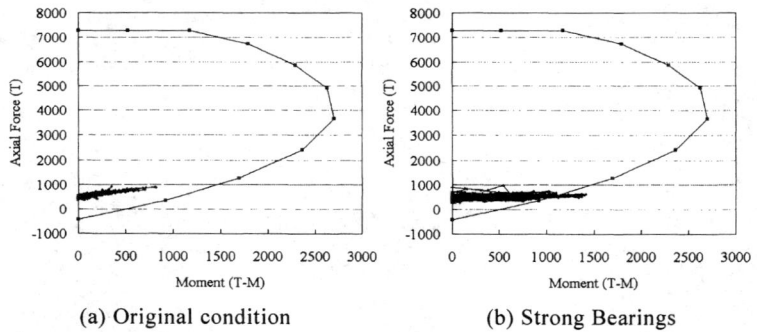

(a) Original condition (b) Strong Bearings

Figure 10. Comparison the axial force-moment time history response with axial load-moment interaction diagram

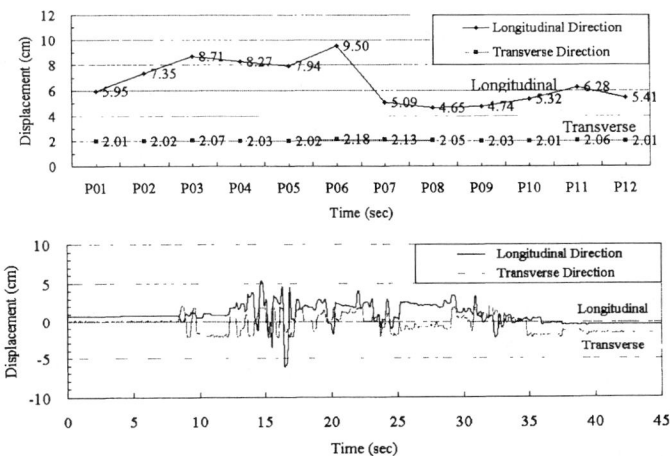

Figure 11. Maximum relative displacement distribution at bearings and relative displacement time history at bent P1

3. Bearing system

Figure 13 shows the maximun pounding force distribution of the concrete shear keys and pounding force time history at bent P3. When the PCI girders pounded the concrete shear keys, the displacements of superstructure were limited to 2 centimeters, and the forces were transferred to the substructure. The results of numerical analysis represent that all of the steel rods had been yielded during this earthquake.

4. Expansion joint

The maximum longitudinal displacement for the expansion joint can be obtained using the nonlinear time history analysis, as shown in Figure 14. Except for the span S1 ~ S2, S12 ~ S13 deck slabs had more than 5cm relative displacement, the rest parts of bridge have less relative movement; therefore, no pounding occured during this earthquake.

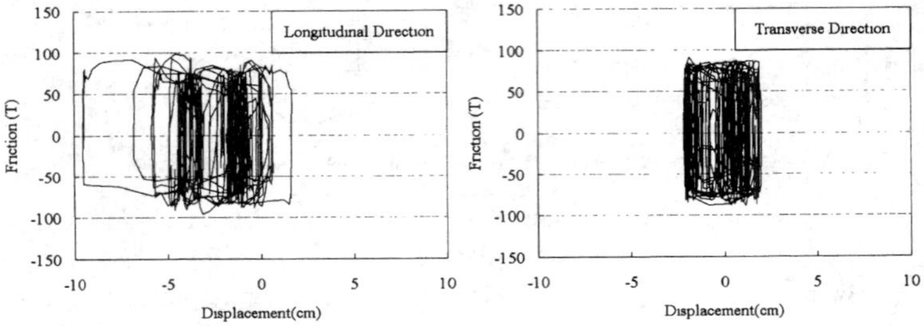

(a) Longitudinal direction (b) Transverse direction
Figure 12. Hysteresis loop of bearing friction at bent P6

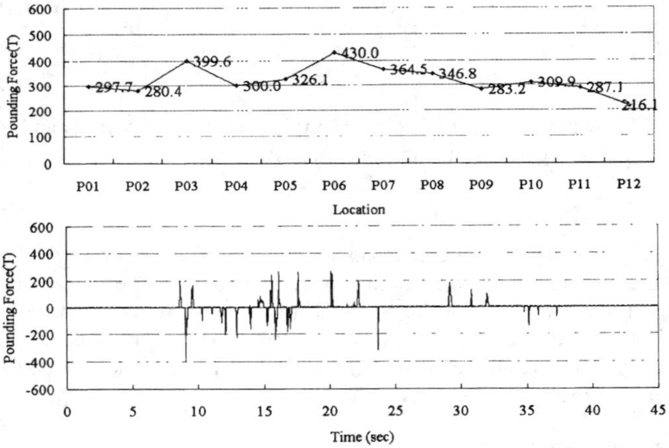

Figure 13. Maximun pounding force distribution of the concrete shear keys and pounding force time history at bent P3.

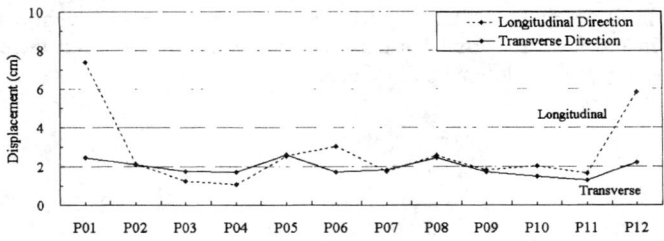

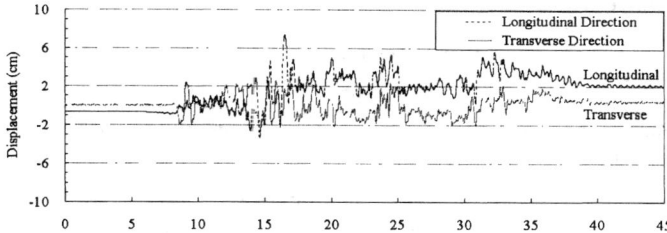

Figure 14. Maximum deformation distribution of the expansion joints and deformation time history at bent P1

CONCLUSION

Based on the analytical results presented in this study, the following conclusions can be made:

1. The analytical model, consists of bearings, concrete shear keys and other hystertic behaviors, predicted the actual seismic performance of the bridge and lead to results that were generally in agreement with the bridge-disaster reconnaissance reports.

2. The bridge column failure was small as a result of the movement of superstructure and separation of thermal expansion joints due to sliding at the bearings.

3. As the superstructure moved forward to the transverse in 2 centimeters and pounded the concrete shear keys, it transferred forces to the bridge column and led the substructure damage to a serious extent in this direction. In the longitudinal direction, the substructure damage is minor without shear keys.

REFERENCE

1. K.C. Chang, Jerry Shen, M.H. Tsai, and George C. Lee, 2000, "Performance of A Isolated Bridge under Near Fault Earthquake Ground Motions ", The Second International Workshop on Mitigation of Seismic Effects on Transportation Structures September 13-15, 2000, Taipei, Taiwan, R.O.C.

2. S. F. Pan, R. T. Chu, J. Y. Chang, 2000, "The bridge-disaster reconnaissance report of the Yeng-Feng Bridge on Provincial Route 14 during the 921 Taiwan Chi-Chi Earthquake", Taiwan Highway

Engineering, 2000.

3. Ministry of Transportation, Taiwan, 1987, "Design Code for Highway Bridges ".

4. Ministry of Transportation, Taiwan, 1995, "Seismic Resistant Design Code for Highway Bridges ".

5. M.J.N. Priestley, F. Seible, G.M. Calvi, 1995, "Seismic Design and Retrofit of Bridges ", John Wiley & Sons, Inc., New York.

6. National Center for Research on Earthquake Engineering, 1999, "Qreconnqissance report of the 921 Chi-Chi earthquake for bridges and transportation facilities" Report No. NCREE -99-055, Taipei, Taiwan, November.

7. Robert K. Dowell, Frieder Seible, and Edward L. Wilson, 1998, "Pivot Hysteresis Model for Reinforced Concrete Member", ACI Structural Journal/ September-October.

8. Roy A. Imbsen, Richard V. Nutt, and Joseph Penzien, "Seismic Response of Bridges – Case Studies," Report No. UCB/EERC-78/14, 1978.

POST-EARTHQUAKE LIFELINE SERVICE RESTORATION MODELING

Z. Çağnan[1] and R. Davidson[2]

Abstract

This paper describes a simulation-based methodology being developed to model post earthquake restoration processes as part of the MCEER research on the Los Angeles Department of Water and Power's (LADWP's) electric and water supply systems. Post-earthquake restoration models play an important role in estimating the economic impact of earthquake damage to lifeline systems. The focus of this study is on electric power system but in the future the proposed methodology will be extended to develop a multi-lifeline restoration model that can take into account the effects of infrastructure interdependencies on the restoration processes. This paper begins with a discussion of the available restoration modeling approaches, listing advantages and disadvantages of each. The new simulation-based methodology is described and key innovations that distinguish it from previous approaches explained. Then details of the preliminary model and plans for future work are discussed.

Introduction

Quantifying direct and indirect economic losses caused by earthquake damage has been one of the most active research areas in the recent years (Chang et al. 1999). In case of an earthquake, normal economic activity is disturbed by loss of infrastructural function as well as direct physical damage. Duration of functional loss is a critical determinant of the magnitude of economic disruption. Hence models of post-earthquake restoration processes are very important in evaluating economic losses.

The objective of this study is to develop an improved model of the post-earthquake restoration process of the electric power system. The model uses estimates of physical damage to the system and an understanding of the repair and recovery operations to estimate expected restoration time, as well as the uncertainty surrounding this estimate. This study is part of the MCEER research on the Los Angeles Department of Water and Power's (LADWP's) electric and water supply systems, aim of which is to assess the economic and societal resilience of a community. The proposed restoration model will serve between the damage estimation models and the models that aim to measure economic resilience.

---

1 and 2 - PhD Candidate (zc32@cornell.edu) and Assistant Professor (rad24@cornell.edu) respectively, School of Civil and Environmental Engineering, Cornell University, Ithaca, NY 14853-3501

Available Lifeline Restoration Modeling Approaches

Generally three different approaches have been used in order to model the restoration of lifelines. The first method is based on statistical restoration curves. One other approach is based on repair resource constraints. These approaches have been considered with the restoration of individual lifelines. There also exists a third approach, which considers the urban system as a collection of several subsystems, each competing for the limited rescue resource with the others.

Statistical Restoration Curves

Restoration curves typically represent the restored service capacity as a function of elapsed time since the earthquake under consideration. These curves are generally based on expert opinion data. In this approach, system restoration time is directly estimated from these statistical curves without modeling the actual restoration process. Restoration time estimates primarily depend on ground shaking intensity hence this approach considerably oversimplifies the restoration process.

Statistical restoration curves approach is explained by the ATC-25 (1992) document and the related reports ATC-25 (1991) and ATC-13 (1985). The ATC-25 procedure for estimation of restoration curves depends on data excerpted from ATC-13. Based on regression analysis of this expert opinion data, which was obtained through an iterative questionnaire process, time to complete restoration is formulated as function of MMI (Modified Mercalli Intensity) for different types of facilities and various regions of U.S. These "time to complete restoration curves" are then synthesized to restoration curves by performing connectivity and serviceability analyses, which enable taking into account the network system aspects (Nojima et al., 2001). Statistical restoration curves approach can not address the temporal and spatial variability within the restoration processes as well as the effects of organizational factors such as repair prioritization plans, mutual aid agreement etc.

Chang et al. (1996) compared opinion based outage estimates by ATC-25 with the historic performance of lifelines in several California earthquakes. It is concluded that the results obtained by ATC-25 procedure can result in unrealistic restoration time estimates mainly due to experts' subjective judgments, modeling errors in regression analysis and due to the aforementioned limitations of the statistical restoration curves approach.

In Chang et al. (1996), the restoration time estimates for the Memphis Light, Gas and Water Division's (MLGW) water delivery system were obtained at the census tract level from an empirical log-linear relationship with break density, that was developed by Seligson et al. (1991) based on data from the San Fernando earthquake. This procedure provided an improvement to the statistical restoration curves approach since it allowed depictions of restoration progress across both time and space (Chang et al., 1999).

Resource Constraint Approach

Resource constraint approach models the restoration time as function of available resources and damage state. This approach involves modeling the actual restoration process and is based on unit repair time and available repair crew estimates that are generally obtained from past earthquake and expert opinion data.

Ballantyne (1990) developed estimates of unit repair times in a study concerning the vulnerability assessment of Seattle water supply system. A fixed number of repair crews were assumed to be available following the disaster. It was also assumed that in more severe events, resource constraints would be more, because of the loss of electric power, telecommunication and worse transportation conditions. In the Ballantyne study the central business district was given higher priority and assumed that will be restored faster. This prioritization rule was established based on the observations made in past earthquakes.

HAZUS uses a similar approach toward modeling restoration of water delivery system (NIBS, 1997). In this procedure, the number of available workers is estimated as a fixed percentage of the population of the study area. Again, estimates of unit repair times are provided. It is assumed that larger diameter pipes would be repaired first. Then, the number of leaks and breaks in pipes would determine the total time required for the restoration of the system.

Isumi et al. (1985) expressed the restoration process by differential equations, each one of which is applicable for a certain service area. A simulation of the restoration process was carried out by using step by step calculations and Monte Carlo method. The restoration process of a lifeline system is described by the decrease in the structural damage and by the functional recovery following the structural recovery. Number of workers for each division is determined by the degree of urgency of restoration in the service area under consideration. The degree of urgency is determined on the basis of prioritization rule being employed. The time necessary for workers to move depends on the road conditions and is considered as a phase delay in the computation. In this study, number of workers varied with time throughout the restoration process.

Chang et al. (1999) underlined the need for improving the resource constraint approach in a variety of new directions than what has been done in previous studies. The assumptions of the resource constraint approach were tested using past earthquake data especially from Kobe earthquake and it was found that restoration rate changes with time, the number of available workers is not likely to be proportional to the resident population and it changes with time throughout the restoration period. It was also observed that relationship between restoration times and damage densities is not a simple linear relationship, especially in areas with high damage density. Hence the experience with the Kobe earthquake indicated that some of the basic assumptions of the resource constraint approach should be relaxed, especially mutual aid agreements should be considered.

Chang et al. (2000) improved the methodology used for modeling earthquake impact on Memphis and Shelby County water delivery system by employing resource constraint approach instead of statistical restoration curves approach for restoration modeling and by integrating economic loss and restoration models within the Monte

Carlo simulation process. In this study, the restoration model parameters are based on a survey of current lifeline restoration models and data from the Kobe earthquake. In addition, the restoration model specifies a sequence of restoration based on engineering priorities and observations in Northridge and Kobe earthquakes. This model is capable of evaluating the effects of post–disaster mitigation measures such as restoring damaged facilities according to an optimal restoration plan but does not explore the effects of mutual aid agreements.

Linked Subsystems Approach

In the third approach, the evolutionary restoration process of each of the subsystems is modeled as Markov chain. The model proposed by Zhang (1992) is capable of taking into account the interactions between different subsystems. Zhang's procedure can be useful only when various parameters and probability values of the model (e.g. state probability vectors, transition probability matrix) are in agreement with the characteristics of the real urban system. Extensive research is required to establish databases from the real life restoration processes and to interpret them into parameters that can be employed in this method. The other shortcoming of this approach is that an overall subsystem state is considered in the analysis; hence the spatial variability within the restoration process can not be addressed.

LADWP Project Methodology

As mentioned above, this study is part of the MCEER research on LADWP's electric and water supply systems. A similar study was conducted for Memphis area with the aim of estimating earthquake induced direct and indirect economic loss. Details of the Memphis study can be found in Shinozuka et al. (1998), Chang et al. (1999 and 2002). In this study as in Chang et al. (1999), for each scenario earthquake the entire loss estimation process from damage to economic loss is simulated multiple times within a Monte Carlo framework. The outline of this approach is illustrated in Figure 1. Direct and indirect economic loss is estimated based on damage and outage results which are obtained using a Monte Carlo simulation approach (Shinozuka et al. 1994) as in the Memphis project. For each simulated damage pattern, power flow and connectivity analyses are performed with the aim of identifying nodes at which power imbalance, abnormal voltage and isolation from rest of the system occur. Damage patterns are the input of the restoration model. Updated damage patterns at certain instants of the restoration process are the output of the restoration model. For each updated damage pattern, power flow and connectivity analyses are carried out giving updated outage pattern. These results are then input into the economic loss model of business interruption. This process allows the construction of economic fragility curves that reflect potential economic loss from earthquake-induced loss of lifeline system serviceability. When these are combined with probabilistic hazard data, expected annual loss can be estimated (Chang et. al. 1999).

Restoration Methodology

The proposed methodological approach for modeling post-earthquake restoration processes of lifelines includes a number of improvements and expansions to the previously developed approaches especially to the aforementioned Memphis project restoration process modeling approach.

Modeling Approach

In this study, the post-earthquake restoration process is modeled by a methodology that includes resource constraint approach and discrete event simulation. As mentioned above, resource constraint approach allows spatial and time wise depiction of the restoration process as well as consideration of effects of repair prioritization plans, mutual aid agreements etc. In this approach, restoration process does not only depend on the damage state but also on the available repair resources. Discrete event simulation (DES) has been used especially in Operational Research for decades now. This simulation technique bases simulations on the events that take place in the simulated system and then recognizes the effects that these events have on the state of the system. In DES, state changes occur instantaneously at specific points in time. The main shortcoming of the resource constraint approach is that it has been mainly applied to individual lifelines ignoring effects of interactions between lifelines. The aim of including DES into the proposed methodology is to compensate for this shortcoming of the resource constraint approach. The authors believe that interaction of lifelines has considerable effect on post earthquake restoration processes and aim to develop a multi-lifeline restoration model in the future as part of the LADWP study.

The fundamental problem in simulation is one of modeling e.g., using the abstractions built into the simulation language to describe the system being analyzed (Joines and Roberts, 1998). All simulation models represent an abstraction of reality. The construction of a simulation model using a particular language is limited by the structure of the language. The difficulty in simulation modeling of complex systems is the lack of an adequate description or reference model for the abstraction upon which the language or simulator is built. Object oriented programming which provides the possibility of having a one to one mapping between objects in the complex system being modeled and their abstractions in the simulation model, directly deals with this modeling problem. A good simulation tool must be able to take into account the interactions and co-operations between components of complex systems. Traditional procedural simulation techniques limit interactions and co-operations between entities for achieving system wide objectives since scheduling is distributed throughout the code (Davidsson, 2000). The other main limitation of the procedural approaches is their lack of extensibility. Hence by using OOP concepts, the functionality of DES softwares can be extended and more flexible, more realistic models compared to traditional procedural DES softwares can be developed. The simulations performed by these DES packages that are developed with OOP are known as Object Oriented Simulations (OOS). In this study since our main objective is to be able to include effects of interactions into restoration models, OOS are carried out. Flexibility of DES packages developed with OOP enables studying various prioritization rules, since making changes to the restoration model is relatively easy with these tools. Also, more realistic models of the restoration process can be developed with these tools because of the aforementioned characteristics of OOP.

In the proposed methodology, statistical variations in and uncertainties associated with key factors such as post earthquake damage inspection duration, start and finish time for repair of each damaged component, time needed for replacement of components that can not be repaired and resource allocations are taken into account by defining these durations and amount of available resources as probabilistic distributions rather than single deterministic values. For each damage pattern, the restoration process is simulated n times hence uncertainty related with restoration time estimates is quantified as well.

Improvements Provided

Key innovations that distinguish the proposed methodology from previously developed methodologies include:

- Temporal and spatial disaggregation of the restoration process which enables incorporation of the temporal and spatial dimension of loss.

- Explicit consideration of the decision variables such as number of repair crew, amount of repair resources and repair prioritization plans. This enables exploration of how post-event mitigation strategies such as mutual aid agreements and spatially prioritized restoration can reduce total economic loss.

- Incorporation of statistical variations and uncertainties associated with key factors such as post earthquake damage inspection duration, time needed for repair or replacement of damaged components etc. This allows quantification of uncertainty associated with restoration time estimates.

- Development of multi-lifeline restoration model. OOS enable inclusion of interaction effects into restoration models.

Preliminary Model

Work on collecting information about details of restoration processes is still in progress. The developed model is flexible enough to accommodate necessary changes as more data is obtained and the model details given below are that of the preliminary model, not the final model.

In the proposed restoration model, each substation is considered as a group of components such as circuit breakers, buses, disconnect switches and transformers. Breaking down each substation into several components enables developing more realistic model of the restoration process. Each component has a probabilistic distribution of repair and replacement time assigned to it. These probabilistic distributions are based on post-earthquake data and opinion of LADWP experts. A priority number is assigned to each substation depending on the prioritization rule under consideration. Among various possibilities, prioritizing according to damage state, number of customers in each service area and power utilization level per worker in the service area exists. Number of available repair crews changes with time depending on mutual aid agreements in effect. Repair crews visit substations according to the priority numbers assigned to them. Each repair crew stays at the substation until inspection, repair and or replacement are completed, and then it returns to the nearest service center to obtain more repair resources. Amount of repair resources at each substation is probabilistic as well. Substations and service centers are connected to each other with a path network. Repair crews travel on this network. Delays due to damaged roads are included into the restoration model by changing travel speed of repair crews on certain paths. Each simulation continues until restoration process is complete. For each damage pattern, restoration simulation is repeated n times hence as mentioned above uncertainty associated with restoration time estimates can be quantified. The output of each simulation is time history of the restoration process. This includes information both substation vise and repair crew vise, for example at which instant of the repair process, repair of each substation is completed, what is the utility level of each repair crew etc.

Plans for Future Work

After collecting all the required data about details of restoration process such as repair/replacement durations for different components, mutual aid agreements, amount of available restoration resources etc, the restoration model will be finalized. This model will then be extended for water supply systems and then interdependencies between different lifeline systems will be incorporated to be able to study effects of infrastructure interdependencies on restoration processes.

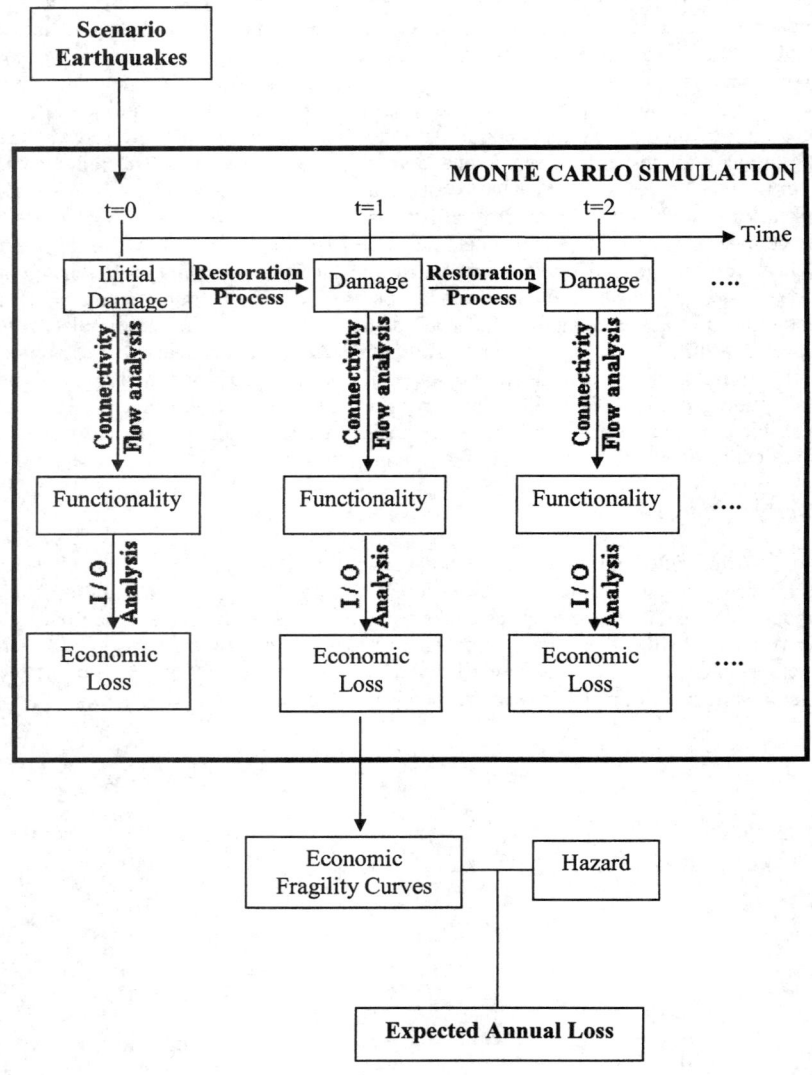

Figure 1: Flowchart of MCEER Loss Estimation Model

References

Applied Technology Council (ATC). (1985). *Earthquake Damage Evaluation Data for California*. Report No: ATC – 13, Redwood City, California.

ATC. (1991). *Seismic Vulnerability and Impact of Disruption of Lifelines in the Conterminous United States*. Report No: ATC – 25, Redwood City, California.

ATC. (1992). *A Model Methodology for Assessment of Seismic Vulnerability and Impact Distribution of Water Supply Systems*. Report No: ATC – 25 – 1, Redwood City, California.

Ballantyne, D.B. (1990). *Earthquake Loss Estimation Modeling of the Seattle Water System*. Report to the U.S. Geological Survey, Federal Way, Washington.

Chang, S.E., Seligson, H.A., and Eguchi, R.T. (1996). *Estimation of the Economic Impact of Multiple Lifeline Disruption: Memphis Light, Gas and Water Division Caste study*. Technical Report No. NCEER-96-0011, Buffalo, NY.

Chang, S.E., Shinozuka, M., and Svekla, W. (1999). "Modeling Post – Disaster Urban Lifeline Restoration." in W.M. Elliott and P. McDonough, eds. *Optimizing Post Earthquake Lifeline System Reliability: Proceedings of the 5th U.S. Conference on Lifeline Earthquake Engineering*. ASCE Technical Council on Lifeline Earthquake Engineering Monograph No. 16, pp. 602 – 611.

Chang, S.E., Rose, A., Shinozuka, M., Svekla, W., and Tierney, K.J. (2000). *Modeling Earthquake Impact on Urban Lifeline Systems: Advances and Integration*. Research Progress and Accomplishments 1999-2000, MCEER.

Chang S. E., Svelka W.D., and Shinozuka M. (2002). "Linking infrastructure and urban economy: simulation of water-disruption impacts in earthquakes." *Environment and Planning B: Planning and Design 2002*, volume 29, pages 281 – 301.

Davidsson, P. (2000). "Multi Agent Based Simulation: Beyond Social Simulation." http://www.ide.bth.se/~pdv/Papers/MABS2000.pdf.

Isumi, M. and Shibuya, T. (1985). "Simulation of Post Earthquake Restoration for Lifeline Systems." *International Journal of Mass Emergencies and Disasters*.

Joines, J.A., and Roberts, S.D. (1998). "Fundamentals of Object Oriented Simulation." http://www.informs-cs.org/wsc98papers/018.PDF.

National Institute of Building Sciences (NIBS). (1997). *HAZUS Technical Manual*. Washington, DC: NIBS.

Nojima, N., Ishikawa, Y., Okumura, T., and Sugito, M. (2001). "Empirical Estimation of Lifeline Outage Time in Seismic Disaster." *Proc. of U.S.-Japan Joint Workshop and Third Grantee Meeting*, U.S.-Japan Cooperative Research on Urban Earthquake Disaster Mitigation, Seattle, WA, USA, August, pp.516-517.

Seligson, H.A., Eguchi, R.T., Lund, L., and Taylor, C.E. (1991). *Survey of 15 Utility Agencies Serving the Areas Affected by the 1971 San Fernando and the 1987 Whittier Narrows Earthquakes*. Report Prepared for the NSF.

Shinozuka, M., Tanaka, S., and Koiwa, H. (1994). "Interaction of Lifeline Systems under Earthquake Conditions." *Proceedings of the 2nd China-U.S.-Japan Trilateral Symposium on Lifeline Earthquake Engineering*, pp.43-52.

Shinozuka, M., Rose, A., and Eguchi, R.T. (1998). *Engineering and Socioeconomic Impacts of Earthquakes: An Analysis of Electricity Lifeline Disruptions in the New Madrid Area*. Monograph No.2, MCEER, Buffalo NY.

Zhang, R. H. (1992). "Lifeline Interaction and Post Earthquake Urban System Reconstruction." *Proceedings of 10th World Conference on Earthquake Engineering*, Balkema, Rotterdam.

Lifeline Performance, El Salvador Earthquakes, January 13 and February 13, 2001

Le Val Lund, P.E., M. ASCE-TCLEE[1]

Abstract

The two earthquakes in El Salvador caused significant damage to lifelines, housing and buildings, especially in the rural areas of the country. The earthquakes resulted in an extraordinary number of landslides that disrupted roads and highways, damaged lifelines, and structures in the cities and villages. In general lifelines performed well in the capital city of San Salvador, good portion was due the lessons learned from the 1986 earthquake and the country's need to adapt during the 1977-1992 civil war.

Introduction

Two damaging earthquakes hit El Salvador, the smallest country in Central America, in January and February 2001. On January 13, 2001, 11:33 AM, the country was struck by 7.6 moment magnitude (Mw)(USGS) earthquake with the epicenter located about 110- kilometers (68-miles) south-southeast, off the coast, from the capital city of San Salvador. The earthquake had a focal depth 39-km (24-mi) and was located on the Caribbean Plate above the subducting Cocos Plate.

Exactly one month later on February 13, 8:22 AM, a second earthquake occurred, with a Mw 6.6 (USGS) and had an epicenter about 30-km (19-mi) east of San Salvador. The focal depth of this earthquake was 13-km (8-mi). It has been characterized as a strong shallow intraplate earthquake, which likely occurred in response to complicated stresses in the Caribbean Plate as it overrides the Cocos Plate. El Salvador continues to feel the affects of many aftershocks, some over Mw 5, to these events.

A team from the American Society of Civil Engineers-Technical Council on Lifeline Earthquake Engineering (ASCE-TCLEE) Earthquake Investigation Committee performed a preliminary reconnaissance survey of the performance of lifelines following these two earthquakes and the many aftershocks, during the period from February 25 to March 3, 2001. Lifelines are defined as network systems

---

[1]Civil Engineer, 3245 Lowry Road, Los Angeles, CA 90027, telephone 323-664-4432, lundasan@earthlink.

necessary for the existence of an urban community and also important for the emergency response and restoration of a community after a disaster, such as an earthquake. Examples of lifelines are power, communication, gas, liquid fuel, transportation, water and wastewater systems.

The field investigation team consisted of Le Val Lund, Civil Engineer, Los Angeles, CA, Field Leader, Rossana D'Antonio, Los Angeles County Department of Public Works, Alhambra, CA; Robert Lo, Klohn Crippen, Vancouver, B.C., Canada; Michael Salmon, Los Alamos National Laboratory, NM and Mario Velado, Caltrans, Sacramento CA. Carl Sepponen, Nolte Associates, San Diego, CA was the Administrative Leader.

The 2001 earthquakes in El Salvador resulted in an extraordinary number of landslides (preliminary number exceeds 500), that disrupted roads and highways, and destroyed lifelines, housing and buildings in the cities and villages. In general, lifelines performed well in the capital city of San Salvador. As reported by some of the lifeline managers, the lifelines performed well, a good portion was due to the lessons learned from the 1986 earthquake and the country's need to adapt to the hardships endured during the 1977-1992 civil war.

The performance of lifelines and housing in particular, in the rural areas and villages was very poor. In these areas much of the housing and shops are constructed of unreinforced adobe and "bajareque". The bajareque construction is common, in the rural villages, consists of mud walls reinforced with bamboo and did not perform well seismically. The bamboo is stems of plants called "vara de castilla" or "bamboo stems". Roofs on the bajareque constructed houses are typically light gage corrugated steel.

The total estimated cost of damage (In US dollars as reported by El Diario de Hoy, March 1, 2001 edition) was $1.6 billion for both earthquakes. It was estimated that the total cost of damage for the January 13 event was $1.255 billion and for the February 13 event was $348.5 million. The estimated cost, in millions of dollars, of the damage to lifelines as of March 1, 2001 is as follows:

Transportation	$432.8
Water	23.1
Electric Power	16.4

Of the lifelines contacted there seem to be no formal emergency response and recovery plans, although generally they respond to emergencies on a day to day basis. They indicated that in future they might develop these plans. The availability of system maps handicapped the investigation.

Communications

Three privatized companies provide the telecommunications (telephone, cellular phone and pager) service in El Salvador. In general, the telephone systems performed well following the earthquakes and general telephone service in the undamaged areas around San Salvador was restored within one day. The companies reported heavy congestion because of system overload following the earthquakes, which is common following large earthquakes. Telecom (800,000 customers) reported that three central switching offices were damaged; however, there was no loss of function. Out of a total of 250 business offices, Telecom had ten business offices that were so badly damaged, they were demolished; and 43 others were damaged and will be repaired. Ten civil/structural engineers from Telecom were evaluating their facilities. It was reported that wire, fiber optic, repeater, and satellite facilities worked well in both earthquakes. They reported the good performance of their systems may be due to the redundancy of their system.

Gas and Liquid Fuel

Gasoline and diesel fuel for operation of vehicles and equipment are produced at the Refinería Petrolera de Acajulta (RASA). The refinery is located on the coast adjacent to the Puerto de Acajulta. An offshore marine terminal supplies the refinery. There was no reported loss of refinery production due to the earthquakes. Texaco imports the remainder fossil fuel demand as "clean" fuels through the port of Acajutla.

There is no natural gas distribution system in El Salvador. Propane for cooking is provided in bottles distributed by trucks from central distribution facilities. The product is distributed mostly by Tropigas and is imported through a facility near the port of El Cutuco at the southerly end of the country. However, the RASA Refinery produces a small portion. In the rural areas firewood is used for cooking. Electricity is also used for cooking, but almost never used for heating.

By far, RASA is the most prepared entity in El Salvador to react to an emergency. It has an emergency response plan that includes an automated response for their refinery and an evacuation plan for their personnel. This emergency plan is rehearsed a least twice a year, according to their top managers.

Power

It was reported that the electric power service performed well with outages from a few hours to a day or two, except to those areas with significant structural damage. Previously, the electric power system in El Salvador was nationalized. However, several years ago the system was restructured and partially privatized and is divided into generation, transmission and distribution systems. The total generating capacity in 1998 was 943.5 megawatts (mw).

The major hydroelectric generation is provided by the Comision Ejecutiva Hidroeléctrica del Rio Lempa (CEL) from four plants (414 mw) located at reservoirs on the Rio Lempa, the largest river in the country. There are other private fossil fuel, hydroelectric and geothermal generating plants in the country, some of the plants are owned and operated by US companies. The transmission network (785 km)(487-miles) is operated by Empresa Transmisora de El Salvador. The transmission lines normally carry 115 kilovolts (kV), with substations, transmitting the electricity from the generating stations to the five privatized distribution systems serving the country. There was no reported major damage to the generating and transmission systems.

Five privatized companies serving five areas of the country distribute electricity in the country (3,375 gigawatt-hours)(GWh). Some of these companies are owned by US companies and use US construction standards. Team members contacted two of these companies Compañia de Alumbrado Eléctrico de San Salvador (CAESS) and Compañia de Luz Eléctrica de Santa Ana y Cia (CLESA), representing the urban areas of the country. They reported their systems were under control at the time of the investigation. However, they reported a substantial loss of customer base due to the large amount of housing damage. They reported significant overhead distribution line and transformer damage from landslides, falling structures and trees and foundation failures for their concrete poles. CAESS estimated the loss of 100,000 customers due to collapsed houses. There was damage to a vacuum circuit breaker and rigid buss connections. As a long-term seismic mitigation measure, they were in the process of replacing their live tank circuit breakers (high profile equipment) with vacuum and sulfur hexifloride (SF6) circuit breakers in their system.

Their operating facilities did have emergency generators, along with generators at their yards and offices, which operated in a satisfactory manner following the earthquakes. It was noted the emergency batteries associated with these generators were not completely anchored. However, there was no reported toppling of the batteries.

Transportation

Roads and Highways. The Pan American Highway (Carretera Panamericana-CA1) is the main highway, which runs the length of El Salvador. It transports and distributes the majority of the country's commercial goods and passenger traffic. Landslides blocked this highway in a number of locations requiring detours and complete closure of the highway.

In the Los Chorros Canyons, where CA1 and CA8 converged carries most of the commercial goods and passengers from the west into San Salvador. A series of landslides blocked, a 7-km (4.3-mi) stretch of highway and as of March 3, a control road was still in operation requiring one way traffic between designated hours. The highway itself incurred minor damages, with the exception of a blocked culvert, but the instability of the slopes adjacent to the roadway prevents the normal operation of the highway.

At La Leona, east of San Salvador, the highway continues to be closed. At the time of the investigation, excavation to remove the slide has been halted due to instability of the mountainside, with the continuous sizeable after shocks, which already has slid twice. Effective slope stabilization measures have to be completed before cleanup and rehabilitation of the highway can be continued.

The Comalapa Highway, a divided highway, from San Salvador to the El Salvador International Airport, suffered circumferential cracks at five sites on the outer side of the roadway along a hillside location. Core samples had been taken and observation wells installed to monitor the roadway. At the time of the investigation it required the shifting of traffic to two-way traffic on the inner side of the highway.

The five rock tunnels on the coastal road from La Libertad to Acajulta experienced minor damages to their portal reinforced concrete arches. There was no reported highway bridge damage, other than the usual settlement at the abutments and the expected shear key damage. The surprising good behavior of highway bridges is worth mentioning given the lack of maintenance and rehabilitation they endure.

Airport. In El Salvador, the administration of airport, port and harbor, and railroad are under the jurisdiction of Comision Ejecutiva Portuaria Autonoma (CEPA).

The El Salvador International Airport (Aeropuerto Internacional El Salvador) was opened in 1980 and has 14 gates. It was closed for one day after the earthquake to permit inspection of the runways and the terminals. At the east end of the runway a cold joint separated and other transverse cracks appeared in the runway. A contractor repaired the cracks and the runway was returned to service in one day.

Airport terminal damage included the popping out of control tower windows, cracking of plaster in the walls, falling of suspended ceilings, cracking of reinforced concrete walls and to columns and beams. The airport was returned to service after cleanup of debris, even though all repairs had not been made as of March 3. The airport performed a major role in providing international emergency response and materials following the earthquakes.

Railroad. The railroad, Ferrocarriles Nacionales de El Salvador (FENADSAL) was created in 1975 from the merger of two railroads. There are three railroad lines in the country, one from the west, one from the east and one from the north. These railroad lines are mainly used for freight, but have some limited passenger service; all converge in San Salvador. The cargoes transported include petroleum products, steel, milk products, cement and containers. It has a total of 565.5-km (351-mi) of track. At the time of the January 13 earthquake, only two lines were operating. The north line transports most of the cement produced in the country, and the west line connects the port of Acajutla with San Salvador. The east line was not in operation, but a railroad bridge, a steel arch truss, south of San Salvador, across the Rio Lempa collapsed, otherwise there was no reported major damage to the railroad.

Port. The major port facility is the Puerto de Acajulta consisting of eight berths, cranes, warehouses and storage areas. A conveyor transports bulk (wheat, corn and soybean) cargo from the ship to shore. There was some minor damage to one of the supports for this system. It was reported that there was no ground liquefaction in the port area. The port maintains a separate electric system, including its own generators.

The port is approximately 130-km (81-mi) from the epicenter of the January 13 event and 100-km (62-mi) from the February 13 event. Intensity at the port was apparently low due to the fact that many adobe homes and unreinforced brick walls showed little signs of damage in the area.

Transit. A variety of privately owned busses and taxis provide passenger transportation in the country. A government agency controls and administers the bus routes.

Water

The operation of the water and wastewater systems in El Salvador is under the jurisdiction of Administración Nacional de Acueductos y Alcantarillados (ANDA). ANDA-Water has 530,000 services and ANDA-wastewater (sewer) has more than 400,000 connections serving 2.6 million people, about 60% of the population. Extensive use of bottled water is used in the homes, hotels, restaurants and even the offices of the lifeline agencies. The reported water demand on the system is 8.3 cubic meters per second (189.4 million gallons per day). The major water supply is from wells and the Rio Lempa (Lempa River). Minor supplies are from springs and streams. The well supply is disinfected by chlorination and the river supply receives conventional treatment (sedimentation, filtration and chlorine disinfection). The Cacahuatal Treatment Plant received major damage. Since the wells and pumping plants did not have emergency generators, loss of commercial power affected their operation for approximately a day. Yards and offices did have emergency generators that performed well. Limited anchorage of batteries was also noted, without any problems.

The Cacahuatal Treatment Plant, treated water from Rio Lempa, has capacity of 100,000 cubic meters per day (26.4 mgd) and suffered damage from the February 13 earthquake. The President of ANDA-Water reported that it will take two to three months to repair the destroyed piping, chlorination facilities, retaining walls, basins, bridge and electrical facilities at the plant. Also, the damaged access road and associated pumping plant require repair. The plant services 22,000 families, in five communities and as of February 25 the citizens were relying on tanker trucks for potable water.

The treatment plant is located in San Vicente and is approximately 30-km (19-mi) from the epicenter of the February 13 event. The intensity in San Vicente was estimated to be Modified Mercalli Intensity (MMI) IX to X as most of the houses in San Vicente (approximately 90%) were destroyed. Damage to masonry structures of good quality consisted of cracks to partial collapse.

In San Salvador, ANDA-Water had no maps of their system available; however, they were beginning to document their system-using computer aided drafting (CAD) system. In the future they plan to implement a supervisory control and data acquisition (SCADA) system. They reported the loss of electric power for their wells and pumping plants for about 24-hours. ANDA-Water has 140 wells pumping water from the groundwater basin. Even though their operating facilities did not have emergency generators, their yards and offices did have generators, which they reported operated in a satisfactory manner following the earthquakes. It was noted, as in the electric power system, the emergency batteries associated with these generators were not completely anchored. However, there was no reported toppling of the batteries.

In the San Salvador area, ANDA-Water reported only three pipeline repairs on their Zona Norte (north zone) supply line. This line is a well water supply line, installed in 1999, and was a 1,215-millimeters (48-inch) bell and spigot, cement mortar lined, ductile iron pipe. A French manufactured mechanical joint secured the joints. It was reported that the bolts became loose on this joint and caused the failure.

A parallel, 1977 installed 1,215-mm (48-in.) spiral welded steel pipeline, undamaged, was used as a bypass to facilitate repairs to the 1999 pipeline. Another pipeline, an inlet/outlet to a pair of tanks, located on a very steep slope above a new subdivision pulled apart. This was a 610-mm (24-in.) wrapped coated steel pipeline with a rubber gasket bell and spigot joint. Welding the joint had repaired the pipeline; however, above this location cracks were noted in the ground and a concrete drainage ditch, which could be a problem in a future earthquake.

A limited inspection was made of the ANDA-Water reservoirs; it was reported there was no damage to reinforced concrete post tensioned tanks. It was reported that several brick and concrete block tanks were severely damaged and probably will be replaced. It is noted some of the tanks were partially full, because of the lack of water supply and may not have been subjected a normal full tank seismic loading.

Wastewater

ANDA-Alcantarillados (Wastewater) collects the wastewater in the urban areas. In the rural areas waste disposal is by private systems. The storm water and wastewater systems are separated in San Salvador. Flow in the wastewater system is $4 - m^3/s$ (91-mgd). In San Salvador the wastewater is collected in three systems, Rio Urbiña, Rio Cañas and the Rio Lempa which ultimately discharge, without treatment, into the Rio Lempa. The discharge point is well downstream from the ANDA-Water diversion point. Major collecting pipelines are concrete, ranging in size from 455-mm (18-in.) to 2,430-mm (96-in.). Visual inspection has been made at critical points in the system and no damage has been reported. No internal inspection has been made.

Conclusions

1. The majority of the deaths, injuries and financial loss can be attributed to earthquake induced landslides in both earthquakes and collapse of poorly constructed houses in the epicentral area of the February 13 earthquake.

2. Lifeline facilities in San Salvador performed generally well, with only minor repairable damage and outages.

3. In the San Vicente area, and in rural areas of high Modified Mercalli intensity, lifelines suffered a great deal of damage. Water distribution was particularly damaged and was still unavailable at the time of the investigation.

4. The 1986 earthquake provided the emphasis to seismically design and strengthen structures and anchor equipment.

5. The 1977-92 civil War experiences provided the country a more efficient emergency response.

6. A major impact on lifelines was due to landslides, which closed important highways required for emergency and recovery.

7. Lifeline agencies and citizens had emergency generators because of the need during the civil war.

8. Anchorage of equipment is important and it was noted at some of the emergency generator facilities that emergency batteries were not completely anchored.

9. There is a need for system maps delineating the lifeline facilities.

10. Supervisory Control and Data Acquisition (SCADA) system is being started in some of the organizations and needs to be completed.

11. Civil/geotechnical investigation needs to be made on the cut, fill and natural slopes of the highways and hillside pipeline locations to determine safety of the sites. Cracking at the tops of the slopes was observed adjacent to critical lifeline facilities.

12. A slope stabilization program needs to be implemented for all major. Critical highways in the country.

13. Although lifeline agencies have in the past responded to emergency situations, there is a need to develop and practice formal Emergency Response and Recovery Plans. Such plans should include government and the private sector as partners.

Acknowledgements

ASCE-TCLEE wishes to thank the Asociacion Salvadoreña de Ingenieros y Arquitectos (ASIA), Comision Ejecutiva Portuaria Autonoma (CEPA) and Ministeria de Obras Publicas (MOP) for providing personnel, equipment and a headquarters to coordinate the TCLEE team lifeline investigation activities. Special appreciation goes to ASIA General Manager, Lic. Jóse L. Pérez Sánchez, ASIA members CPA. Rene Amaya and Ing. Hernan Moz, CEPA Engineering Manager Edna Elizabeth Escobar and MOP Executive Director Rene Gomez. Also, TCLEE thanks the many lifeline managers and their staff for providing technical information and field investigations of their facilities. Special thanks to the Salvadorian people who allowed us into their country, homes and shared their experiences.

ASCE-TCLEE Monograph

A detailed report on the El Salvador Earthquakes can be found in the ASCE-TCLEE Monograph (163 pages) "Lifeline Performance El Salvador Earthquakes, of January

17 and February 13, 2001". Available for purchase from ASCE, Reston, VA by telephone

1-800-548-ASCE (2723) or

from the ASCE world wide web.

www.asce.org

LL 01-27-03 6USCLEE

EFFECTS OF SIX RECENT EARTHQUAKES ON RAILROADS

William G. Byers[1], F. ASCE

Introduction
The six earthquakes with magnitudes of 6.8 to 8.4 that damaged railroads in the 23 month period from August of 1999 through June of 2001 illustrate the effects of strong-to-great earthquakes. The locations of earthquake epicenters, indicated by stars, and railroads in the affected areas are shown in Figures 1 and 2.

Characteristics of Affected Railroads
Approximately 110 km of the Turkish State Railways line between the Istanbul suburb of Haydarpasa and Ankara, plus the 8 km line from Arifiye to Adapazari, lie within 16 km of the fault rupture of the August 17, 1999 Kocaeli earthquake. The rupture crossed the tracks at three locations. About 270 km of these electrified lines lie within 80 km of the epicenter. The affected portion of the railroad and the fault rupture zone are shown in Fig.1. Additional details are supplied in *Byers* (2000).

The portion of the Taiwan Railway affected by the September 21, 1999 Chi-Chi earthquake is described by *Abe et al*. A double track electrified line runs south from Keelung, through Taipei, in the west of Taiwan to near the southern tip of the island and connects to a single track line that follows the east coast back to Keelung. For part of this distance, as shown in Fig. 1, there are two double track lines, one along the coast and one through the mountains. There are also several branch lines.

Railroads in the vicinity of the October 16, 1999 Hector Mine, CA earthquake are shown in Fig. 1. The Burlington Northern and Santa Fe Railway (BNSF) owns a double track line from Los Angeles and a single track line from the north that join at Barstow and continue east through Albuquerque, NM. The Union Pacific Railroad (UP) owns a double track line running east from Daggett through Las Vegas, NV and operates over the BNSF from Daggett to its lines in the Los Angeles Basin.

The January 26, 2001 Gujarat, India earthquake affected four of eight divisions of the Western Railways. Tracks in the damaged area, shown in Fig. 2, are broad gage (1676 mm) or meter gage. Some of the heavier traffic meter gage lines were being converted to broad gage at the time of the earthquake.

The primary railroads in the Puget Sound region, which was affected by the February 28, 2001 Nisqually earthquake, are the BNSF and the UP. They have a joint operation over the double track line between Vancouver, WA and Tacoma as well as separately operated lines. Amtrak operates 10 passenger trains between Portland and Seattle over the Vancouver-Tacoma line. There are also short lines operating in the area. Some lines in the Seattle area handle commuter traffic.

---

[1] Burlington Northern and Santa Fe Railway, Kansas City, Kansas

Two railroads were affected by the June 23, 2001 Atico earthquake. PeruRail operates over 1000 km of lines in southern Peru. One line runs inland from the ports of Mollendo and Matarani, through Arequipa and Juliaca, to Puno. A connecting line runs northward from Juliaca. The Southern Peru Copper Corporation operates a 215 km industrial railroad from its smelter and refinery on the coast at Ilo to mines at Toquepala and Cuajone. Both the Mollendo-Matarani-Arequipa-Puno line of PeruRail and the Southern Peru Copper Corporation (SPCC) railroad sustained extensive damage in the areas indicated in Fig. 2. A model of the fault rupture by *Kiuchi and Yamanaka* indicates that a portion of the rupture surface extends under the western part of the PeruRail line at a depth of 30 to 40 km.

Effects on Operations

All six of the earthquakes caused significant interruptions in train operation while the safety of facilities was being determined. The amount of time required varied with the distribution of available inspection personnel and the accessibility of the railroad. At one extreme, appreciable numbers and strategic location of appropriate employees allowed restoration of service on lines without major damage within about 3 hours after the Gujarat earthquake. At the other, inspection of the SPCC railroad after the Atico earthquake required about 48 hours as large segments of the railroad are not accessible by road and could only be reached on foot because the track was blocked by slides. With the exception of the SPCC railroad, which experienced damage over nearly all of its length, the areas where operation of trains was stopped or restricted pending inspection considerably exceeded the damaged area and operations were adversely affected beyond the regions inspected.

Forty four hours after the Kocaeli earthquake, one of the two main tracks of the line between Haydarpasa and Arifiye was repaired and returned to service with a 30 km/hr speed restriction. Normal track speed is 110 km/hr. The second track was returned to service 5.7 days after the event when the line between Haydarpasa and Izmit, which had been closed due to a fire at the Tupras refinery adjacent to the right-of-way near KM 74, was reopened. Speed was still restricted in an area where the fault crossed the track six weeks after the earthquake due to inability to maintain track surface across the rupture. A power company's substation at Arifiye, which supplied power for locomotives, was inoperative as a result of the earthquake. However, power for locomotive operation was supplied through adjacent substations.

Following the Chi-Chi earthquake, the line along the west coast of Taiwan, which was not damaged, could not be reliably operated during the next 10 days because its power supply was interrupted by earthquake damage to power plants. The main line through the mountains was out of service for 17 days and branch lines for from one to six months depending on the extent of damage and importance of restoration.

The Hector Mine earthquake derailed a passenger train running on the north track of the BNSF, partially blocking the south track. The south track was restored to service 12 hours after the earthquake and the north track 20.5 hours after the event, both with speed limited to 25 mph (40.2 km/hr). Speed was increased to 40 mph (64.4 km/hr) after 5 days and all speed restrictions removed after 20 days.

After the Gujarat earthquake, all lines, except the line from Dahinsara to the port of Navlakhi, which was out of service for 9 weeks, were restored to service within 3 days with most lines in service within 2 days. Recovery was accomplished in stages starting with operation of trains with speed restrictions. At locations where track or bridges were damaged, service was restored with speed restrictions appreciably before permanent repairs were complete and disturbed track had stabilized to the extent required for normal operating speeds. Damage to signal and communication systems necessitated the use of alternative methods for dispatching and controlling train movements. While the Dahinsara-Navlakhi line was out of service, coal was trucked from the port.

BNSF lines north of Seattle were returned to service about 5 hours after the Nisqually earthquake. Lines used for commuter service and other main lines were returned to service within 7 hours, except for a 7 mile segment near East Olympia. On this segment, Main Track No. 1 was returned to service with a 25 mph (40.2 km/hr) speed restriction after 8.5 hours and Main Track No. 2 was returned to service with the same speed restriction after 12.5 hours. Passenger trains were not operated on this segment the following day. Speed restrictions were removed after enough traffic to consolidate disturbed ballast had run over the segments. A minor branch line in Olympia was returned to service after 6 days, completing the recovery.

Following the Atico earthquake, 3 days were required to restore service on PeruRail between Arequipa and Puno and 5 days were required to restore service between Arequipa and the ports. The SPCC railroad was out of service 7 days. Speed was restricted to 10 mph (16 km/hr) during the following week. Within a month 65 percent of the line was being operated at normal speed.

Derailments
An Amtrak passenger train, running at about 60 mph on the BNSF at the time of the Hector Mine earthquake, had 22 of the 24 cars and one wheel set each on two other cars derailed. The engineer's observation of the track moving ahead of the train and the fact that none of the three locomotive units was derailed strongly indicate that the derailment was caused by ground movement under the train, not by earthquake damage to the track. The derailment is discussed in detail by *Byers* (2001).

Although a number of trains, including 8 scheduled passenger trains, were operating in the area of the Gujarat earthquake, none were derailed. Two tank cars loaded with water and standing on tracks at Bhachau were derailed and a third was overturned.

No derailments were reported in the other four earthquakes. Four passenger trains were operating within 80 km of the epicenter of the Kocaeli earthquake at the locations shown in Figure 1. A number of train and switching movements were in progress in the area affected by the Nisqually earthquake. Two trains operating on the portion of the SPCC railroad damaged by the Atico earthquake were able to proceed to a location where the crews could be removed and transported by road.

Damage to Bridges

No railway bridges were damaged by the Kocaeli earthquake. There are 167 bridges with steel or concrete spans with a combined length of over 13 km and 14 concrete or stone arch bridges within 80 km of the epicenter. Nearly half of these are within 15 km of the fault rupture.

Damage from the Chi-Chi earthquake to a major railway bridge and to a bridge on an abandoned branch line is reported by *Abe et al.* Twelve hammer-head, concrete piers of a 29 span, 725 meter long, double track, concrete girder bridge across the Dai-jia River were damaged. Track buckling indicated possible relative movement between piers. Damage was readily repaired. A steel girder bridge on the abandoned line had anchor bolt damage and spalling at the base of some concrete pier columns.

Bridge damage in the Hector Mine earthquake was not extensive enough to require removal of any bridges from service but six bridges suffered minor damage, some of which required repair. The most prevalent damage was to unreinforced or lightly reinforced wing walls, primarily separations at cold joints, preexisting cracks or inadequately reinforced corners. One bridge with prestressed concrete slab spans and substructure built from precast concrete elements had damage concentrated at locations where the substructure elements were joined together. In the segment between the first and last damaged bridges, 13 bridges with independent spans under each track were undamaged. Both damaged and undamaged bridges are ballast deck.

In the Gujarat earthquake, bridges were damaged on the 135 km of broad gage line between Bridge 161 across the Little Rann of Kachchh and Bhuj and the 50 km of meter gage line between Samakhiyali and Gandhidham, which contain about 240 bridges and culverts, including 22 major bridges. Six of 24 arch bridges, 5 of 75 girder bridges and one of 41 reinforced concrete slab bridges sustained significant damage. Arch bridge damage included failures caused by outward earth pressure against parapets and spandrel walls, cracking in mortar joints of arch rings and displacement of stones in arch rings due to outward movement of abutments. Girder bridge damage included movement of girders relative to piers and/or pier displacement resulting in unacceptable track geometry and, in some cases bearing and/or anchor bolt damage. Cracking of horizontal mortar joints of stone substructure units occurred in both girder and concrete slab bridges. Other bridge damage included separation of wing walls from abutments and other damage to wing walls. One reinforced concrete box had wing wall damage. Three arch bridges required installation of temporary falsework to carry traffic before they could be returned to service. Two of these bridges lost major portions of their parapet and spandrel walls with a resulting loss of fill supporting the track. The third had been extended on both sides to allow conversion from meter gage to broad gage track. Longitudinal joints near the boundary between the original arches and the extensions separated. A fourth arch bridge suffered collapse of a portion of the parapet wall and about 50 mm downward displacement of stones in both end spans due to 6 to 10 mm outward movement of the abutments. Partial repairs were required before it was returned to service. One reinforced concrete slab bridge and two girder bridges with severely cracked masonry joints in substructure units were placed in service with restrictions. The restrictions included monitoring of cracks and a requirement that trains stop before crossing the bridge at a speed not exceeding 10 km/hr.

A bascule bridge in Seattle, about 50 km northeast of the epicenter, was open at the time of the Nisqually earthquake. Following the earthquake, it was impossible to close the bridge as the pivot pier and rest pier had moved together about 150 mm. Cracks from lateral spreading due to liquefaction were observed behind the pivot pier. A framed end bent of a trestle 15 km from the epicenter, was rotated toward the channel about the end of the deck, probably due to liquefaction and lateral spreading.

Only three railroad bridges were in the area affected by the Atico earthquake. None of the bridges were damaged although two were in areas with heavy track damage.

Damage to Tunnels
Eleven tunnels are located within 15 km of the fault rupture and a total of 29 are within 80 km of the epicenter of the Kocaeli earthquake. Five tunnels, located near KM 35, KM 146, KM 147, KM 241 and KM 242, had minor damage, primarily opening of preexisting cracks. Cracks in one portal and the adjacent lining of the tunnel at KM 240+869 were widened. A crack in its left wall near the spring line was the only tunnel damage that required repair. The left side of the tunnel is close to the surface of the ground as it slopes into a canyon, resulting in very limited available passive pressure. The tunnel originally had structural connections between the walls below the track which had been removed in lowering the track to improve clearance.

Minor damage from the Chi-Chi earthquake in the 7 km double track San-yi tunnel, including spalled concrete lining, detachment of the catenary and horizontal displacement of about 20 mm with associated track buckling is reported by *Abe et al.*

No tunnels were in the areas affected by the Hector Mine and Gujarat earthquakes.
There was no damage to tunnels in the Nisqually earthquake although several relatively long, concrete lined tunnels in soft ground are in the affected area.

There was minor damage to tunnels in the Atico earthquake consisting of fallen rocks in unlined PeruRail tunnels near the coast and spalled shotcrete lining in SPCC tunnels near the end of the line farthest from the coast.

Damage to Track and Roadbed
Significant track damage from the Kocaeli earthquake was all in an area that is located less than 8 km from the fault. Most damage was in the segment with two main tracks that is essentially parallel to the fault between KM 104+758 and KM 129+600. At KM 104+950, both tracks were offset over 3 meters where they were crossed by the fault. At KM 128+600, both tracks pulled apart at insulated joints.

This appears to be the result of a second crossing by the fault rupture The fault rupture crossed the single track line between Arifiye and Ankara a short distance south of Arifiye causing an offset in the track of nearly 2 meters.

Track damage caused by the Chi-Chi earthquake is reported by *Abe et al.* Track buckling and differential settlement due to liquefaction occurred on the Taichunkang branch line. Where the fault rupture crossed the Chi-Chi line, a vertical offset of nearly 1 m and track buckling from local compression occurred. Liquefaction and track buckling occurred at a number of other locations on this line. Landslides and ground failure caused extensive damage on the Mt. Ali line.

Track damage from the Hector Mine earthquake was all within 20 km of the epicenter and included buckled track, other disturbances to track alignment and surface and displacement of ballast from cribs. Track at the end of one bridge was virtually skeletonized by displacement of ballast and/or vertical movement of the track.

The most significant damage from the Gujarat earthquake was to about 1 km of embankment and track that shifted and settled into the Gulf of Kachchh near Navlakhi as a result of liquefaction and lateral spreading. The line was out of service slightly over 2 months. Additional roadbed damage and track displacement occurred from slides and cracking due to lateral spreading. These locations were primarily northeast of Gandhidam, in an 18 km stretch around Bhachau. Settlement at bridge ends occurred as far east as the vicinity of Maliya Miyana.

In an 11 km segment of double track main line about 20 km south of the epicenter of the Nisqually earthquake, track settled at bridge ends, embankment settled and shifted and ballast was lost under ties. Most damage was to Track 2. The embankment for Track 2 was constructed against the embankment for Track 1 after Track 1 had been in service for a number of years. Location is about 15 km from the epicenter. There was considerable liquefaction affecting track with settlement and sand ejection near Capitol Lake in Olympia, about 20 km southwest of the epicenter. Track was out of line near Ravensdale, about 60 km east of the epicenter. A landslide partially blocked the track ditch near Tacoma.

Segments of PeruRail between the coast and Guerreros (KM 39) are located about 150 km from the epicenter of the Atico earthquake and, probably, over the deeper portion of the fault rupture. Damage at 81 locations in this area affected an aggregate length of 11.56 km in a total of less than 60 km. Between Guerreros and KM 130, damage was limited. Beyond KM 130, there were 75 locations with damage affecting an aggregate length of 11.95 km in a total length of 113 km. at 150 to 180 km from the epicenter. From KM 24 to KM 73 of the SPCC railroad, 275 to 295 km from the epicenter, damage was significant. From KM 85 to KM 103, at 290 to 300 km from the epicenter, damage was severe, extreme between KM 85 and KM 93. From KM 103 to KM 142 damage was limited. From KM 142 to KM 155, at 300 to 310 km from the epicenter, damage was significant. Damage included slides and settlement in high fills, shifting of roadbed in side-hill cuts, slides in cuts, rock falls, some of which broke rails and/or crossties, and disturbed ballast and track geometry.

Damage to Buildings and Other Facilities
The most severe damage to railroad facilities in the Kocaeli earthquake was the nearly total destruction of a major part of a passenger car shop at Adapazari. There was also minor to severe damage in station buildings from Haydarpasa to near KM 200. There was minor damage to retaining walls. There was some damage to the catenary system where the fault rupture crossed the tracks.

There were no railroad buildings near the Hector Mine earthquake. Minor signal system damage affected train operation for several hours after the earthquake.

The Gujarat earthquake damaged thousands of railway owned buildings, with about one third not repairable. These included many control cabins, employee housing and other building types not usually owned by railroads in North America. Most stations between Samakhiyali and Bhuj were not repairable. Stations were also damaged over a much wider area. A foot bridge over the tracks at Samakhiyali had to be supported on falsework due to shear failures in the supporting concrete piers. There was severe damage to signal and communication systems.

Railroad buildings in Seattle and their contents had varying degrees of damage from the Nisqually earthquake. A 24 stall roundhouse with back-shop had significant damage. A small shop building at another location had major damage. There was minor signal system damage.

The Atico earthquake caused minor damage to railroad buildings in Arequipa.

Service and Social Considerations
Following the Gujarat earthquake, 57 trains for evacuating earthquake victims were run from affected cities. Thirty six special freight trains were run to carry relief supplies and equipment to the affected areas. The broad gage section between Gandhidham and Bhuj, which had recently been converted from meter gage and was due for pre-operation safety inspection, was rehabilitated and opened for moving relief material by February 4. Relief passenger trains were run between Gandhidham and Bhuj. Five railway owned accident-relief medical equipment units, including one each from Central, Northern and Southern Railways were sent to the area. The Railways mobile hospital, a six car train, was sent to the area.

The possibility of stranding a large number of commuters who had traveled to downtown Seattle by train was a major concern following the Nisqually earthquake. After considering the risks involved, evening trains were operated at restricted speed, preceded by on-track maintenance-of-way vehicles. Commuters were transported without incident. Amtrak trains were not run in the area on the following day because of a concern with possible aftershocks, which proved to be unwarranted.

The SPCC Railroad was out of service for 7 days following the Atico earthquake. Continued operation of the smelter and refinery depended on stockpiled concentrate and the 15,143 metric tons of concentrate, approximately one third of the quantity normally moved by rail, which was trucked to the smelter during this period.

Conclusions

The earthquakes, with magnitudes from 6.8 to 8.4, affecting both electrified and non-electrified rail lines traversing a variety of topography, provide a good sample of potential effects of earthquakes on railroads. The time required for post-earthquake inspection and restoration of service on operable lines depends on the availability and distribution of inspection personnel and accessibility of damaged portions of the railroad. Permanent ground movements, involving fault offset, landslides and rockfalls, embankment settlement and displacement of bridge piers, are likely to be the predominant cause of earthquake damage to rail lines. Displaced ballast and damage to bridges, particularly to substructures and earth-retaining elements, resulting from shaking also cause damage. Derailments of rolling stock resulted directly from earthquake accelerations, not from running over damaged track. Service resumption on electrified lines was dependent on power from outside sources. Damage to major shops and other support facilities, while not directly affecting the operation of trains, can have an important financial effect on railroads.

Acknowledgement

Material related to the Kocaeli, Gujarat and Atico earthquakes is the result of courtesies extended by personnel of the involved railroads who provided information and access to damaged facilities.

References

Abe, Masato, *et al.*, "Damage to transportation facilities", *The 1999 Ji-Ji earthquake, Taiwan – investigation into damage to civil engineering structures*, Earthquake Engineering Committee, Japan Soc. of Civil Engineers, 1999, 4-1 to 4-3

Byers, William G. (2000), Effects of the August 17, 1999 Izmit earthquake on the Turkish State Railways, *Proceedings of the 2000 Annual Conferences, American Railway Engineering and Maintenance of Way Association,* American Railway Engineering and Maintenance of Way Association, Landover, MD, 2000, 33-54

Byers, William G. (2001), Railroad Damage from the October 16, 1999 Hector Mine Earthquake, *Proceedings of the 2001 Annual Conferences, American Railway Engineering and Maintenance of Way Association,* American Railway Engineering and Maintenance of Way Association, Landover, MD, 2001, 1061-1081

Kiuchi, M. & Yamanaka, Y., EIC seismological note: No. 105, Earthquake Information Center, University of Tokyo, 2001

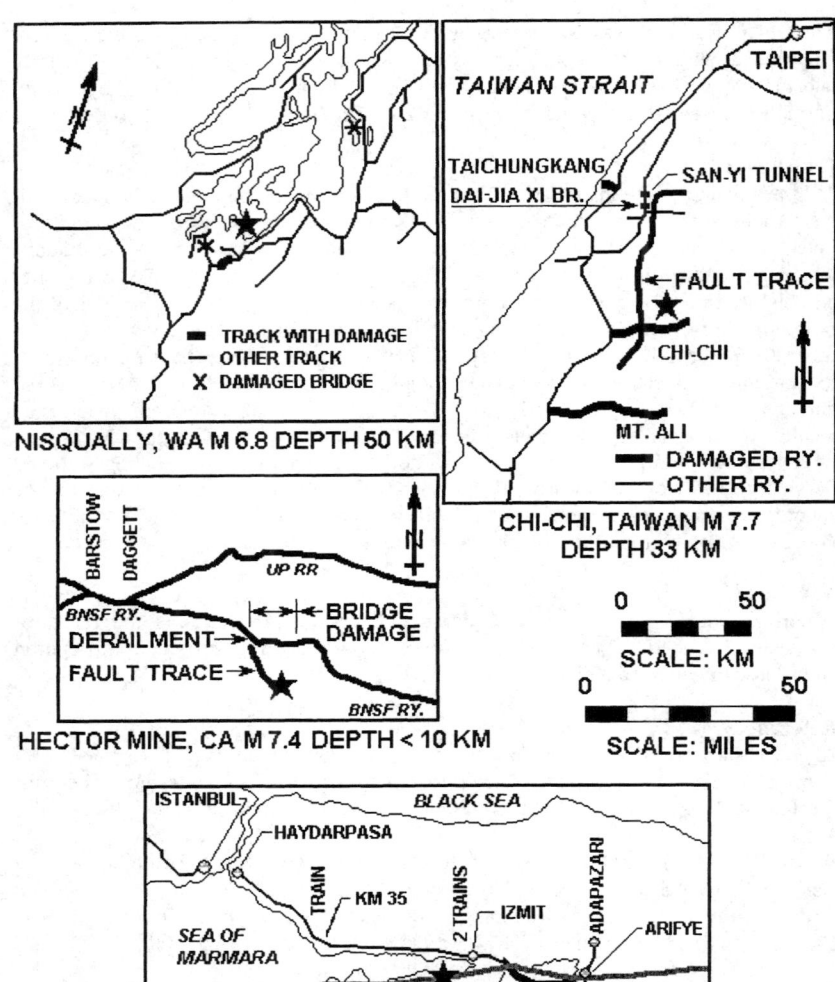

Figure 1

DISASTER RESPONSE FOR LIFELINE SYSTEMS 283

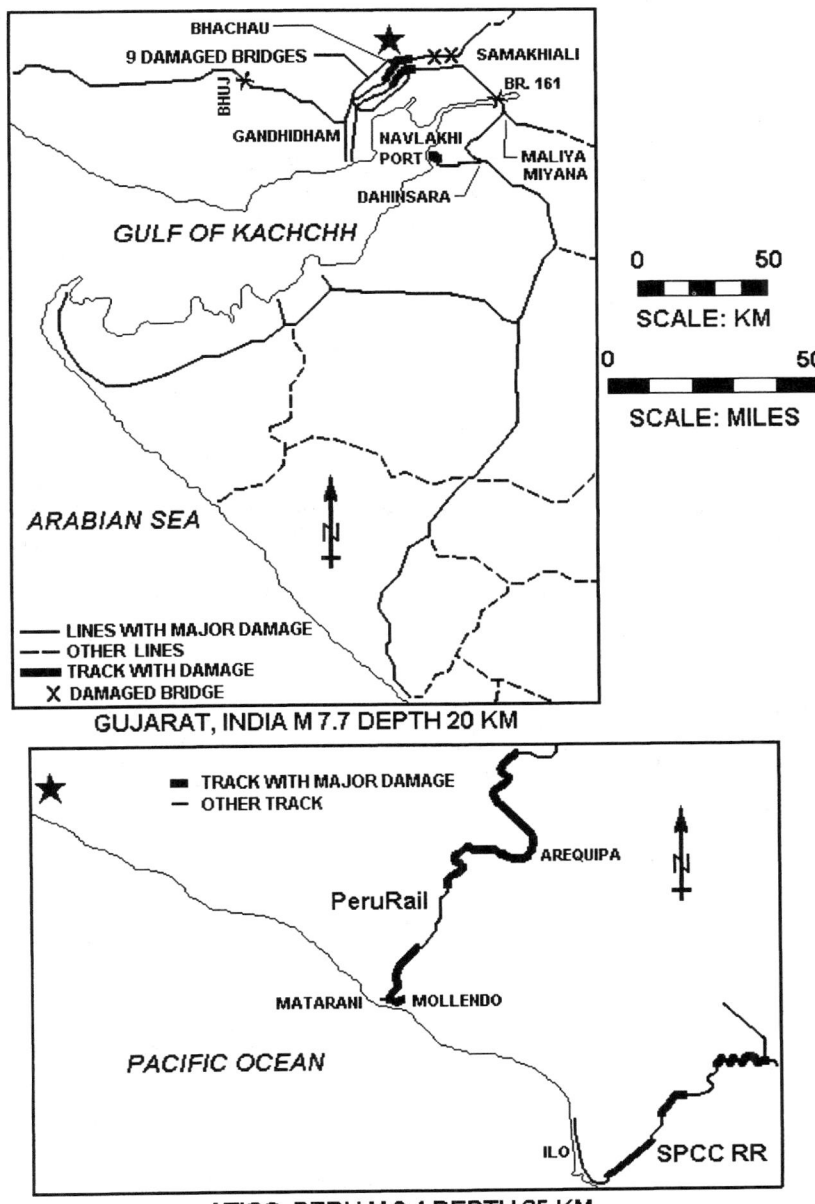

Figure 2

DAMAGE OF GAS AND WATER PIPELINES IN SLOPE CITY, KURE, DUE TO THE 2001 GEIYO EARTHQUAKE

Junichi Ueno[1] and Shiro Takada[2]

Abstract

At 3:27PM, March 24th, 2001, an earthquake of magnitude 6.4, with its epicenter at Aki islands, occurred. Its moment magnitude was 6.9 as same as Hyogoken Nanbu earthquake (1995), but as the depth of hypocenter was deep, damage was not so serious. Although damages for house/building and lifeline facilities had spread to broad area of Chugoku region and Shikoku region. Also two people died. This paper will describe the result of statistical investigation and analysis of the ground for lifeline facilities of gas and water in a slope city, KURE, based on the field survey and data collection.

Introduction

Figure 1 (NIED 2001) shows the strong EW ground motions in Kure city recorded by K-NET of NIED. Figure 2 shows its Fourier spectra. Maximum values of acceleration were 425.3gal (EW), 311.9gal (NS) and 203.1gal (UD). Maximum value was large but maximum velocity was not large as about 20 kine.

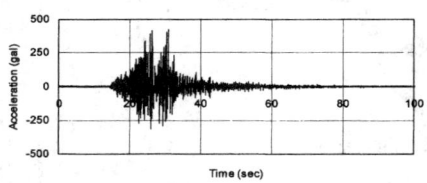
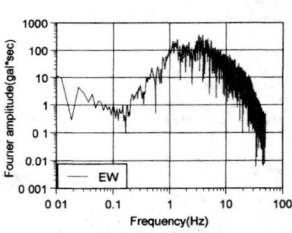

Figure 1. Strong ground motion (E-W, Kure city)   Figure 2. Fourier spectra

---

[1]Researcher, Department of Architecture and Civil Engineering, Kobe University, Rokkodai 1, Nada, Kobe 657-8501, JAPAN; phone +81-78-803-6037; ju55@columbia.edu
[2]Professor, ditto, takada@kobe-u.ac.jp,

Damage for lifelines in Kure City

Gas lifelines

Hiroshima Gas Co. supplies 240,033,000(m^3(m^3/10,000Kcal)) for 414,000 customers with pipelines of 3,629km(1999)(Gas Energy News 2001). Whole supplied area consists of 3 areas. Table 1 shows length of pipeline and number of customers (households) at each area.

Table 1. Length of gas pipeline and number of customer (Hiroshima city 2001)

Area	Length of gas pipeline (km)					Number of customers
	Total	High pressure	Middle pressure A-type	Middle pressure B-type	Low pressure	
Hiroshima	2,682	3	170	241	2,268	341,767
Kure	602	0	36	42	524	53,632
Bingo	345	0	36	38	271	18,614

Table 2. Number of leakage by area and type of pipe

Type of pipe \ Area	Hiroshima	Kure	Bingo	Total
Main pipe	24	9	9	42
Service pipe	13	17	16	46
house pipe (supplier's)	66	42	11	119
house pipe (customer's)	52	25	22	99
Total	155	93	58	306

(unit : number of damage)

Table 3. Relation between type of pipe and damage mode

Damage mode \ Type of pipe	Main	Supply	House (supplier's)	House (customer's)	Total
Cracked/broken without corrosion	10	7	19	12	48
Cracked/broken with corrosion	10	14	14	15	53
Corrosion	2	8	12	5	27
Slip out of joint	2	1	1	0	4
Loosen joint	13	5	19	8	45
Grease exhausted	1	0	8	6	15
Other	0	0	4	7	11
Unknown	4	11	42	46	103
Total	42	46	119	99	306

(unit : number of damage)

Table 2 shows the number of leakages by area and type of pipe. Largest number of leakages occurred in Hiroshima area, but by considering total length of pipe and number of customers shown in Table 1, it is assumed that the extent of damage was relatively moderate than other areas. Table 3 shows the relation between type of pipe and damage mode. In past earthquakes, damage for low pressure line was concentrated on the part of screw joint. But as shown in Table 3, number of damage due to corrosion recorded the maximum.

Water lifelines

Water supply for 47,000 households had to be cut off. Main feature of damages for water system is that damage was concentrated on specific area such as Kure city. Total number of households in Kure is about 84,000. Ratio of number of households between western Kure/Chuo area and eastern Hiro/Aga area partitioned by Yasumi Mountain is 2:1. Miyahara purification plant, the largest plant in Kure city, consists of 3 facilities. Water volume treated by each facility is 31,000m^3/day, 54,000m^3/day and 15,000m^3/day respectively (Kure city 2001). Immediately after the earthquake, volume of transmission line had increased to 3,000m^3/hour which is twice as mach as usual volume. As the water of the plant is about to run dry in 5 hours with this volume, water bureau suspended water supply (as of 17:45 on 25th) to save water for secondary disaster. By this measure, supply for Hiro/Aga area where has 21,000 households (24% of whole customer) and Geiyo islands had stopped. Water bureau had become aware of 7 damages for distribution pipe and 33 for service pipe under public roads just after earthquake. Increase of water volume was due to the damage of distribution main of $\phi 400$(CIP) at Hiromachida 2-chome. Other 6 damages didn't contribute for volume increase. After repairing above damage for distribution line, suspension was settled 15:00 on 25th. It was cleared finally that total number of damage was 192 (Water Industry News 2001) and its content was 12 for conveyance pipe, 1(ϕ150) for transmission pipe, 18 for distribution pipe and 161 for service pipe under public road. Total length of distribution line in Kure city was 903km and following damage rate was 0.02 damage/km. Its rate was one order larger than that of Hiroshima city. With this earthquake, suspension of supply occurred in the eastern part of Kure city, but damage for water pipe was not concentrated in the eastern part. It is supposed that the critical factor for pipe damage was not local seismic property but the aging effect. Above mentioned damaged distribution pipe caused supply-stop was buried before WWII. As for material, all the damage had occurred with CIP and VP. There was no damage for DCIP pipe. Since 1996, water bureau had started renovation program to replace old distribution main (over ϕ300) into earthquake-resistant pipe (DCIP with S-type joint). But old pipe buried 70 to 90 years ago, which is not yet replaced, still exists. Comparison between Kure and Hiroshima about damage for water system is shown in Table 4. From Table 4, it is assumed that the intensity of damage for water facility in Kure was more severe than Hiroshima, similarly as the gas facility.

Table 4. Damage for water system in Hiroshima and Kure

City	Number of damage for distribution pipe (as of Mar. 26th)	Number of repair works for service pipe (as of Mar. 26th)	Ability to supply (x 1000 m^3/day, 1996)	Number of persons served (x 1000, 1996)
Hiroshima	16	673	650	1,184
Kure	7	324	124	205

Damage distribution for lifelines in Kure city

Extent of damage for gas and water in Kure city were greater than those of Hiroshima city. Distribution of damage for gas pipe and water pipe are shown in Figure 3 and Figure 4.

DISASTER RESPONSE FOR LIFELINE SYSTEMS 287

Distribution of damage for low pressure main pipe and distribution main water pipe are shown in Figure 5 and Figure 6.

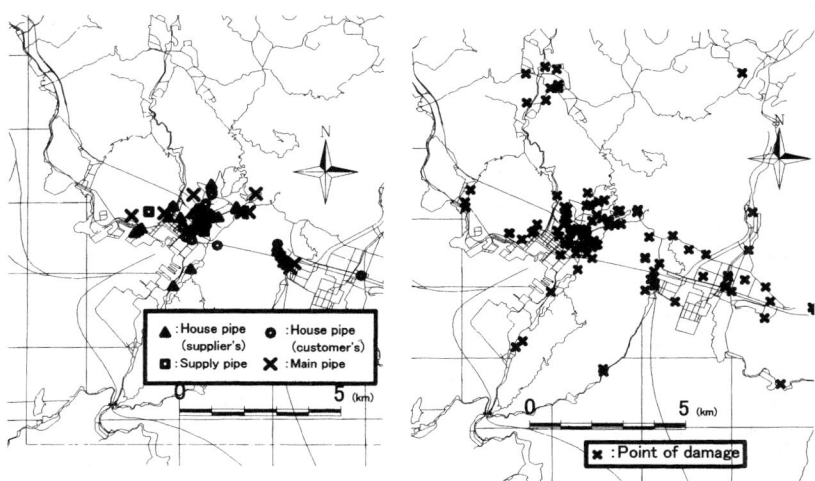

Figure 3. Distribution of damage for gas pipe in Kure city

Figure 4. Distribution of damage for water pipe in Kure city

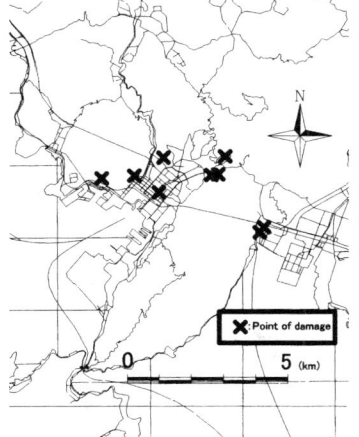

Figure 5. Distribution of damage for low pressure main gas pipe in Kure city

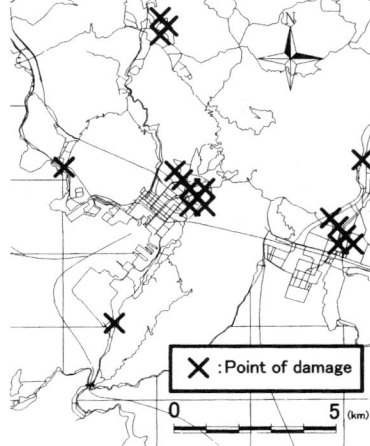

Figure 6. Distribution of damage for distribution main water pipe in Kure city

It was remarkable that damage for house located on the steep slope in the foothills of a mountain around western Kure/Chuo area, where is surrounded by mountains on three sides. 500 places were specified as danger zones of collapse due to steep slope by Hiroshima prefecture before Geiyo earthquake. The collapse of slope and retaining wall had occurred at 218 places with this earthquake (JSCE 2001). It is a logical result that the location of damage for gas house pipe and water service pipe occurred in sloped area as they are closely related to structural damage for houses as shown in Figure 3 and Figure 4. But the damage for gas house pipe and water service pipe were spread out into whole supplied area including lower flat area. Pipes located in this flat midtown area might be older than the surrounding residential area. It is assumed that the damage for low pressure gas main and water distribution main are buried under public roads so they are hardly to be affected by the damage of housing structure. But they occurred at the sloped zone close to completely collapsed houses in Kure/Chuo area as shown in Figure 5 and Figure 6. As a reason, it is assumed that the area for the public road was made by land development similarly to the space for houses in this sloped area, so they suffered from permanent displacement such as displacement of filled land or collapse of slope. It is impossible to deny the possibility that the house on a slope also affected by the amplification effect of acceleration due to topography.

Effect of irregular layering ground on damage of lifeline facilities in Kure city

Kure/Chuo area where is sedimentary land surrounded by mountains on three sides had suffered from this earthquake. In the edge of surface layer where the base rock has risen up to surface, magnitude of response is amplified due to refraction and reflection of wave propagated in surface layer. In this chapter, two dimensional equivalent linear and non-linear seismic response analysis were performed to study the relation between ground structure and damage for buried pipe by modeling the section in Kure/Chuo area

Model for analysis
The wave profile on bedrock surface (138.19gal, 9.15kine) shown in Figure 7 was calculated by deconvolution using equivalent linear analysis code from the wave observed in municipal office.

Figure 8 and Figure 9 show the plan of Kure/Chuo area and its cross section (Kure city 1971) along the line A-B. This time two models of surface layer were made as shown in Figure 10. Municipal office is located at the point where is 1,500m apart from left boundary. Tatsukawa elementary school is located at right boundary of Case1. This time, soil profile of the area to the north of this school was not available. But there are houses on the area and two houses had collapsed in this area. So, model of Case1 was supposed to have the geometry where the base rock had

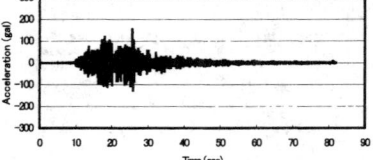

Figure 7. Wave profile on bedrock surface for analysis

risen up to surface around Tatsukawa elementary school and in Case2, the part to the north (right) side of this school, about 300m width, was added. Aspect ratio of scale in Figure 10 is 1:5.

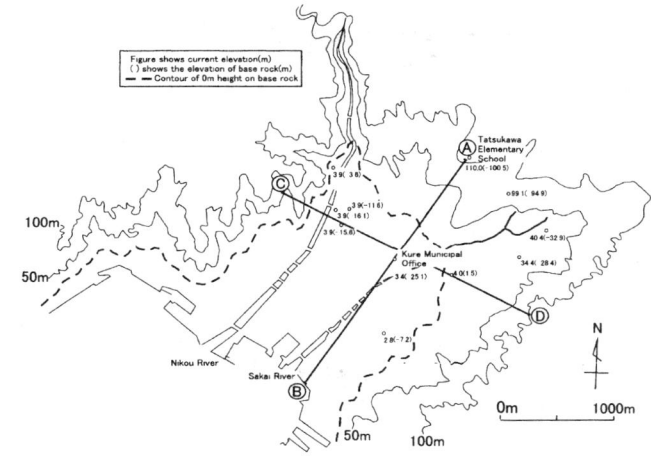

Figure 8. Plan of Kure/Chuo area

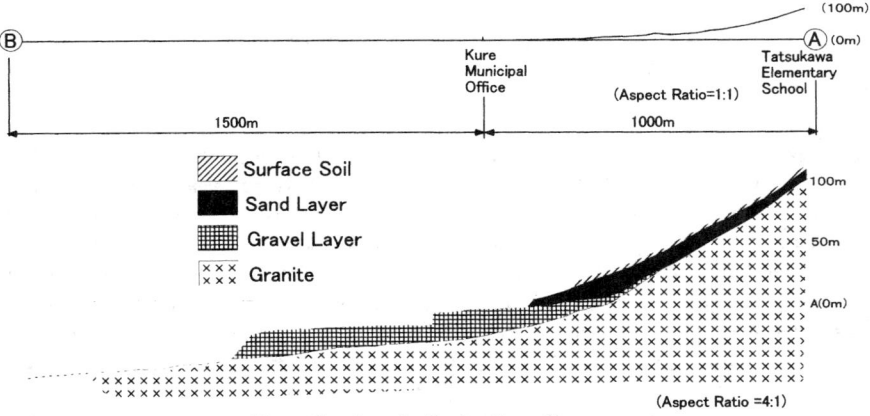

Figure 9. Longitudinal soil profile

As for boundary condition, south(left) side has viscous boundary and nodes on north(right) side coincide with the base rock surface. It is assumed that lower boundary is on rigid base rock. So waves is refracted or reflected at lower and right side boundary. Non-linier property of elastic shear coefficient due to strain magnitude of the ground was given by Iwasaki model (Iwasaki et al 1978).

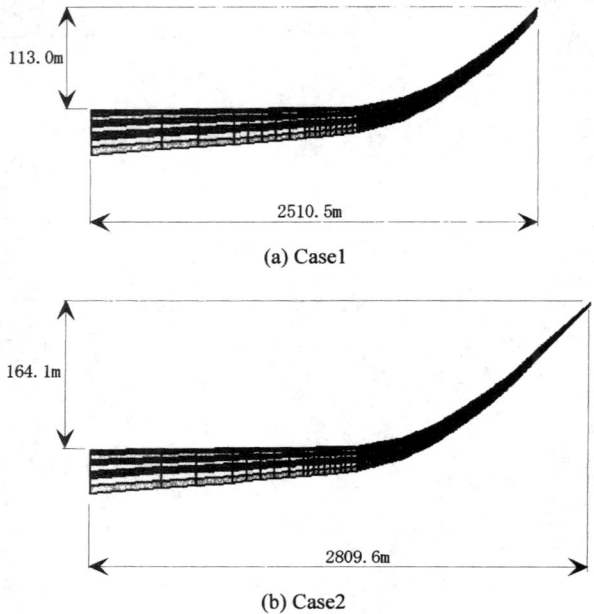

Figure 10. Shape of two models

Result

Figure 11(a) shows the time-history response of acceleration in Kure municipal office. Figure 11(b) shows that in Tatsukawa elementary school. Maximum value in these point was (a)262.01gal and (b)155.42gal. Maximum value, 262.01gal, is larger than that of observed value, 251.48gal and calculated value, 251.21gal, by the one dimension model used for deconvolution. But time-history of displacement at municipal office (max. 2.336cm) given by adding the time-history at base rock (max. 1.061cm) given by integrating input motion to time-history of relative displacement (max. 1.458cm) given by analysis has become identical with the time-history (max. 2.336cm) given by integrating observed motion. Therefore it is assumed that the response at municipal office is not affected by the change of thickness of surface layer and reflection at right end due two dimension modeling. Figure 12 shows the distribution of maximum value of acceleration response in each point along surface. This figure shows that response of 592.76gal was recorded at a point in an added area in Case2. Time-history response at the point (x-coordinates ; 2718.88m) is shown in Figure 13. In this area where is added to north of the school, there are houses along small river. 2 houses had collapsed in this area when the earthquake. Therefore it is assumed that the surface layer was extending to the north of Tatsukawa elementary school as Case2 and amplification of acceleration made its peak in this zone. Amplification where is 80m away from edge of surface layer, was very rapid. In this analysis, reason is assumed as the interference of refracted wave at boundary and reflected wave within surface layer because surface wave of lateral direction couldn't propagate back-and-forth due to viscous boundary of left side. Figure 14 shows the frequency response functions at municipal office, Tatsukawa elementary school and the point where maximum acceleration was observed as shown in Figure 13. Figure 14 shows that hi-frequency component had predominated in the edge of surface layer. As for other damages reported around Tatsukawa elementary school, 6 houses on the west side had collapsed. They were 100m to 200m away from the school. As a reason of these damages, the local amplification effect

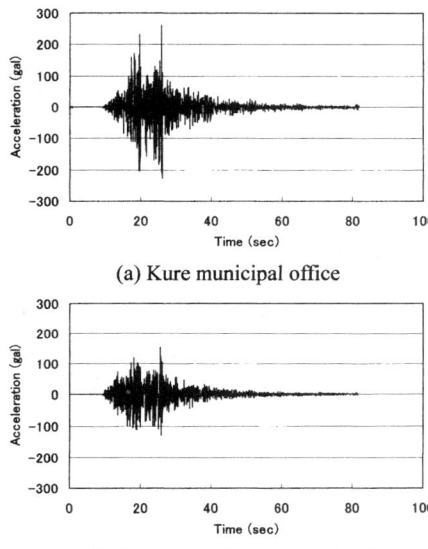

(a) Kure municipal office

(b) Tatsukawa elementary school

Figure 11. Acceleration response

due to small upheaval behind them is supposed other than the amplification effect along the N-S direction in parallel with line A-B. It is generally considered that damage of buried pipe is closely related with ground strain. This time, amplification on acceleration was clear but strain was not so amplified as large as allowable strain of a pipe in the edge of surface layer near a foothill of a mountain. But the damage for buried pipe was mainly occurred at this kind of sloped area in Kure city. Therefore, permanent displacement of filled land, landslide or a relative displacement at cut and fill are supposed as the main reason of the damage for buried pipe, especially main pipe. On the other hand, it is impossible to ignore the possibility that above mentioned amplification effect had contributed to house collapse. In Kure city, number of house damage occurred at the slope in west side was larger than that in east side. As a reason of this difference, inclination of ground surface/base rock and property of E-W input motion are supposed. Next, nonlinear analysis was done with the model of Case2. As nonlinear characteristics of stress-strain relation, Hardin-Drnevich model was adapted. Its parameters were set as the skeleton curve will conform to that of previous equivalent linear analysis. Other conditions such as input or boundary conditions are same as equivalent linear analysis of Case2. Figure 15 shows the distribution of maximum value of acceleration response along surface. Amplification up to 400gal occurred at a point similar to the case of equivalent linear analysis. Comparing with the Figure12, shape of distribution is similar to that of Figure 12, but value of response is smaller over the entire section of the model. Figure 16 shows the acceleration response at municipal office. Comparing with Figure 11(a), wave shape is similar but peak acceleration magnitude dropped lower. Damped hi-frequency component due to nonlinear behavior of ground is assumed as a factor of these differences.

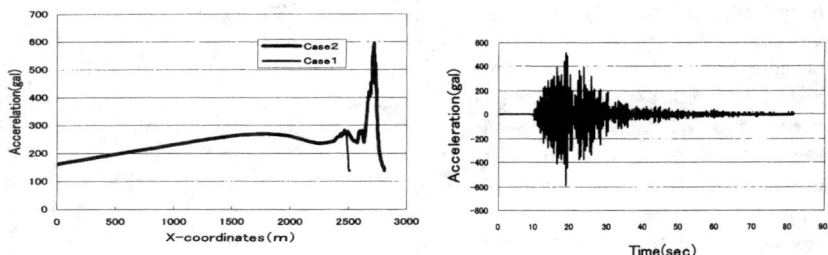

Figure 12. Maximum acceleration along surface

Figure 13. Response at the point showing max. acceleration in Case2

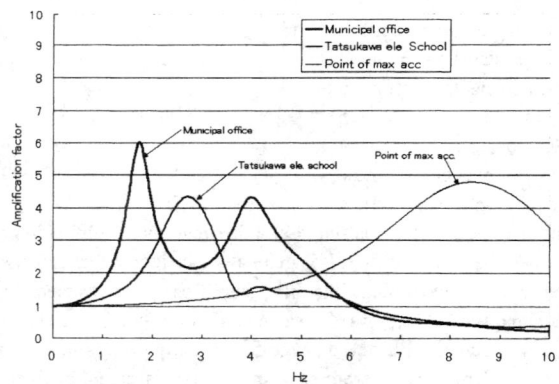

Figure 14 Frequency response functions

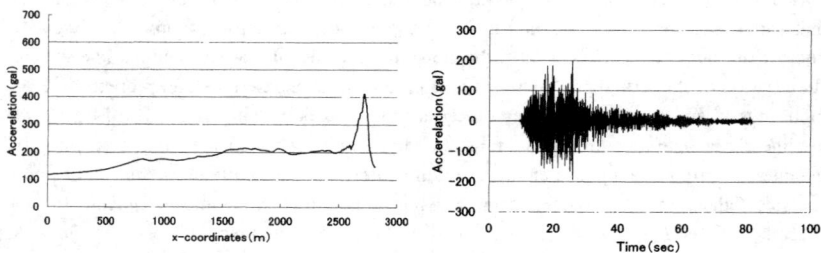

Figure 14. Maximum acceleration along surface (nonlinear)

Figure 13. Acceleration response at municipal office (nonlinear)

Conculision

(1) Geiyo earthquake had caused damage for lifeline of gas and water in Hiroshima, Kure and Bingo area. Judging from damage rate or other index for pipeline, it was cleared that the extent of damage in each area was very varied and that in Hiroshima area was lighter than Kure and Bing area. And furthermore, extent of damage in local site in a same area was very varied such as east side and west side in Kure city. Therefore, it is assumed that distribution of damages for lifeline was reflecting local site condition and local effect of seismic property.

(2) Two dimensional equivalent linear and nonlinear seismic response analyses were performed to investigate the reason of damage at Kure/Chuo area. According to the result, the point showing large response of acceleration in the edge of surface layer around foothills of a mountain was observed. The difference of natural frequency between lower flatlands and upper sloped area was also found. There is a possibility that these response or transition of property had contributed to the damage of house on a sloped zone. On the other hand, amplification for ground strain had not become so large as to cause damage for buried pipe.

(3) Generally, the damage for house/service gas pipe and water service pipe is supposed to be closely relating with the house damage. But this kind of relation was not found with Geiyo earthquake. Contrary to this supposition, distribution of damage for house and main pipe (low pressure gas main and water distribution main) are similar and their damage mainly occurred at steep sloped area. In this area, main pipe was buried under the road where is made by cut and fill similarly as the house. As the ground failure such as landslide and collapse of retaining/stone wall or relative displacement between cut and fill had occurred in this area, both houses and main pipes are supposed to be suffered.

Reference

Gas Energy News (2001), City gas suppliers in Japan '01, 2001

Hiroshima city(2001), Material prepared by Department of waterworks, March 27th, 2001

Iwasaki, T. and Tatsuoka, F. and Takagi, Y. (1978), Shear moduli of sands under cyclic torsional shear loading, Soils and Foundations Vol. 18, No.1, pp.39-56, 1978

JSCE (2001), March 24th, 2001 Geiyo Earthquake Damage Survey Report, pp.12-13, http://www.jsce.or.jp/report/13/01/report.pdf

Kure city (2001), Material prepared by Department of waterworks, March 27th, 2001

Kure city (1971), Guidance Material of Kure Board of Education 127, Earth Science in Kure, 1971

NIED (National Research Institute for Earth Science and Disaster Prevention) (2001), Strong-motion seismograph network (K-NET), Record at Kure, HRS0190103241528.tar.gz, http://www.k-net.bosai.go.jp/k-net

Water Industry News (2001), Water Industry News, Vol. 3824, July 2nd ,2001

PERFORMANCE OF CORRUGATED METAL PIPE (CMP) CULVERTS DURING PAST EARTHQUAKES

T. Leslie Youd* and Chris J. Beckman**

*Department of Civil and Environmental Engineering, Brigham Young University
Provo, Utah 84602-4081; PH 801-422-6327; tyoud@byu.edu
**Project Engineer, Applied Geotechnical Engineering Consultants, 600 Sandy Parkway, Sandy, UT 84070; PH 801-566-6399; beckman@agecinc.com

Abstract

To evaluate culvert performance during earthquakes, we reviewed reconnaissance reports from six earthquakes and conducted field investigations in areas shaken by three of those earthquakes. Hundreds of CMP culverts were in place in strongly shaken areas. Lack of reported or observed damage to all but ten of these structures indicate that CMP culverts generally perform very well during strong earthquake shaking. Of the ten damaged culverts, all but one were in areas of ground failure caused by liquefaction or slope instability. The one culvert not in an area of ground failure suffered minor damage because of increased lateral pressures cracking a head wall and slightly deforming the pipe inlet. Damaging ground failures included embankment penetration into softened or liquefied foundation soils, lateral spread, ground oscillation and slope instability. Diameters of the examined culverts ranged from 0.45 m to 3.6 m.

Introduction

The performance of culverts is important to the safety and reliability of highways during and after large earthquakes. Culverts, however, have received little attention in the earthquake engineering literature. To provide information on performance of these structures, we reviewed reconnaissance reports from several earthquakes and conducted field investigations to inspect both damaged and undamaged culverts shaken by recent earthquakes. Where damage occurred, we studied both the culvert and the surrounding ground to determine the cause of the poor performance. Because of space limitations, only corrugated metal pipe (CMP) culverts are considered in this paper. A more in-depth treatment of CMP and the performance of other types of culverts is given in Youd and Beckman (1996).

Following a literature review, we contacted highway officials in California and Idaho to obtain reports and locations of culverts damaged during recent earthquakes. In some instances, the highway officials had inspected both damaged and undamaged culverts and could confirm either good or poor performance. In other instances, we interpreted the lack of reported damage as an indicator of good performance. During the field reconnaissance, we inspected a number of culverts in these areas to verify that no damage occurred.

The types of culverts commonly used in highway construction include corrugated metal pipe, synthetic pipe, concrete pipe, and concrete boxes. The most common use of culverts is for drainage beneath roadway embankments. In these instances, culverts are placed well below the base courses to allow excess storm runoff to harmlessly flow beneath the roadway. Larger culverts may also be placed to allow pedestrian or vehicular traffic to pass beneath the roadway. Most drainage culverts are constructed of corrugated metal pipe

TABLE 1 Corrugated metal pipe (CMP) culverts specifically investigated during this study

No.	Location	Ground Effects	Diam.	Damage
1	Portage Highway, 1.6 km south of Portage, Alaska (1964 earthquake)	Penetration of fill into foundation	0.91 m	Ends deflected upward 2 m
2	Mile Marker 40.4, Seward Highway, Alaska (1964 earthquake)	Penetration of fill into foundation	not known	Ends deflected upward 0.5 m
3	Mile Marker 46.6, Seward Highway, Alaska (1964 earthquake)	Penetration of fill into foundation	not known	Ends deflected upward
4	Route 5/210 intersection, Sylmar, Calif. (1971 earthquake)	Lateral spread	1.07 m	Severe joint separation
5	Route 5, south of Roxford St., Sylmar, Calif. (1971 earthquake)	Increased lateral earth pressure	3.6 m	Fractured headwall
6-25	20 culverts, Summit Road, Santa Cruz County (1989 earthquake)	Area of diffuse tectonic deformation	various	No damage
26	Jetty Road north of Moss Landing, Calif. (1989 earthquake)	Penetration and lateral spread of fill	0.76 m	Split and ends deflected up
27	Elkhorn Road east of Moss Landing, Calif. (1989 earthquake)	Penetration of fill into foundation	0.76-0.91 m	Ends deflected upward 0.5 m
28	Spanish Ranch Road, Santa Cruz County (1989 earthquake)	Slump of hillside and roadway	0.91 m	Pipes tilted and separated
29	Port of Mori, Hokkaido, Japan (1993 earthquake)	Lateral Spread	0.46 m	Separated from headwall
30-31	Route 405, San Fernando, Calif. (1994 earthquake)	No visible ground disturbance	0.38-0.46 m	No damage
32	Lower San Fernando Dam, Calif. (1994 earthquake)	Increased lateral pressure	2.4 m	Lateral collapse

(CMP). The corrugations, parallel ridges and furrows around the circumference of the pipe, provide added pipe strength. CMP is usually galvanized for corrosion resistance. CMP is widely used because of its strength, versatility, ease of placement, and low cost.

Corrugated Metal Pipe (CMP) Culvert Performance

Table 1 lists 32 CMP culverts that we specifically studied either through field inspection or from critical review of descriptions in earthquake reconnaissance reports. Of these 32 culverts, one (no. 32) collapsed laterally due to increased horizontal pressures, six (nos. 1-3, 26-28) buckled or bent, one (no. 4) was pulled apart at joints, two (nos. 5 and 29) were slightly damaged due to fracture of head walls (easily repairable), and 21 were undamaged. Many more CMP culverts were located in the strongly shaken areas, the total of which we did not determine. None of these additional culverts were reported as damaged, either by local highway officials whom we interviewed, or in published reports. During our investigation, we drove over or by many of these culverts, briefly stopping to examine a sampling of these structures. We saw no evidence of culvert damage or deformed or patched pavement that could be indicative repairs to underlying culverts. Thus, we infer that culverts with no reported damage performed well and were not adversely affected by earthquake shaking.

As noted in Table 1, the seven most severely damaged culverts were at sites disturbed by ground failure. The following sections describe the types ground failures involved and the damage they induced.

Penetration of Embankments into Softened Foundation Soils

The most common mode of ground failure causing damage to CMP culverts was penetration of highway embankments into softened foundation materials. This softening was usually caused by increased pore water pressures or by liquefaction. At some localities, lateral spread as well as embankment penetration occurred. A schematic representation of embankment penetration and resulting damage is illustrated in Fig. 1. The following case histories illustrate this type of ground failure and culvert damage.

Alaskan Culverts. Prior to the 1964 Alaska earthquake, a 0.9-m-diameter CMP culvert was placed beneath a 1-m- to 2-m-high highway embankment that crosses a marshy area about 1.6 km south of Portage, Alaska. The highway provided access to the Portage Glacier area. Liquefaction was widespread in the marsh sediments as evidenced by sand boils, fissures, ground settlement, and lateral spreads in the area (McCulloch and Bonilla, 1970). These phenomena also led to severe damage or collapse of several nearby highway and railway bridges. As diagrammed in Fig. 1 and photographed in Fig. 2, penetration of the highway fill pressed the mid part of the culvert downward into the softened marsh deposit, buckling the pipe, most likely at a joint, and lifting the ends of the culvert well above the ground surface. The left end of the culvert shown in Fig. 2 rose about 2 m (McCulloch and Bonilla, 1970). Liquefaction of foundation sediment most likely caused this damage.

Similar damage occurred to two additional culverts beneath the Seward Highway, one at highway mile marker 40.4 and the other near 46.6. These localities lie 50 to 55 km southwest of Portage. Both culverts were placed in marshy areas to allow local seepage and runoff to pass beneath the roadway. Kachadoorian (1968) reported that the ends of the culvert at mile-marker 40.4 deflected upward by about 0.5 m, but did not note the amount of upward movement at mile-marker 46.6.

Jetty Road Culvert. During the 1989 Loma Prieta, Calif. earthquake, a 100-m-long segment of Jetty Road subsided as much as 1 m and spread laterally as much as 6 m. This locality is a few kilometers north of Moss Landing. The subsidence and spreading left open fissures as wide and as deep as a few meters. This ground disturbance blocked the egress of several vehicles and their occupants, leaving them stranded at a nearby beach. A single 0.76-m-diameter CMP culvert had been placed beneath the embankment to accommodate flow of drainage and tidal waters. At the time of the earthquake, that culvert was partially plugged with sediment, greatly restricting tidal flow. As a consequence, a freshwater marsh had formed inland from the culvert (Kenneth Gray, Calif. State Parks, Monterey, Calif., oral communication, June, 1994).

During the earthquake, the roadway embankment penetrated about 1 m into the foundation soil and spread laterally about 6 m. This action fractured and pulled apart the culvert and deflected the ends upward out of the marsh (Fig. 3). Eye witnesses reported that the ends of the culvert were about 6 m farther apart after the earthquake than before.

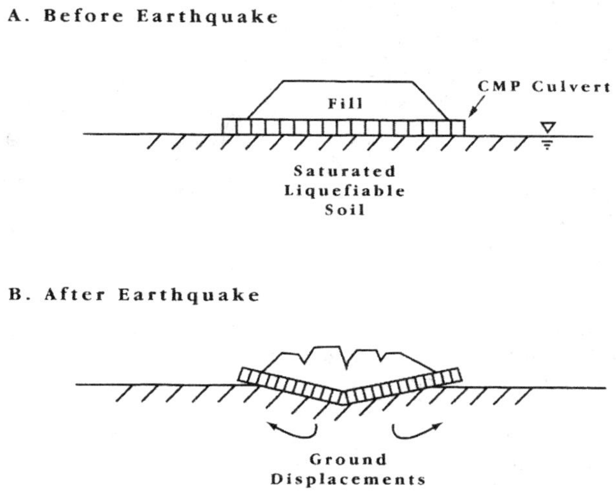

FIGURE. 1. Schematic diagram of embankment penetration into foundation and consequent culvert damage (Youd and Beckman, 1996)

FIGURE 2. End of 0.9-m-diameter CMP culvert lifted about 2 m above ground surface by embankment penetration during 1964 Alaska earthquake; culvert located south of Portage (McCulloch and Bonilla, 1970)

FIGURE 3. Damage to Jetty Road and culvert north of Moss Landing, Calif.; view southeastward of disturbed embankment and raised end of CMP culvert (Photograph by Ken Gray, Dept. of Parks and Recreation, Monterey County, Calif.)

By the time of our visit, nearly 5 years after the earthquake, the roadway had been rebuilt, with six new 0.76-m-diameter CMP culverts placed to carry the flow of Elkhorn Slough. Because of the increased capacity of these culverts, a considerable volume of tidal water could then flow through the culverts; this flow was rapidly converting the former freshwater marsh inland from the culvert into a nearly barren tidal marsh.

Fig. 1 shows the general mechanism of failure for the Jetty Road culvert. In this instance, liquefaction-induced lateral spread of the embankment fill also fractured the CMP and pulled it apart.

Elkhorn Road Culverts. About 5 km east of Moss Landing, Elkhorn Road crosses over Elkhorn Slough. Drainage and tidal water pass beneath the roadway through seven 0.9-m diameter CMP culverts constructed through the base of an approximately 2-m-high embankment (pre-earthquake). The embankment is founded on soft alluvial sediment. During the earthquake, the roadway subsided 0.5 m to 1 m over a distance of few hundred meters. Penetration of the fill into the foundation sediment, apparently as a consequence of soil softening, caused the roadway subsidence. The embankment penetration pressed all seven culverts downward beneath the roadway, causing the ends to deflect upward as much as 0.6 m as shown in Fig. 4. The culverts continued to function with near capacity quantities of water freely flowing through most pipes. Although the road surface had been smoothed and the pavement patched, no regrading had occurred prior to our visit. The damaged pipes and the embankment were in much the same condition as they were immediately after the earthquake. Considerable tidal flow passes through the damaged culverts which are mostly submerged, even at low tide.

Ground Displacement Due to Slumping of Fill or Foundation Materials

The slumping of fill and foundation materials is capable of damaging roadways and culverts placed on such unstable ground. This type of failure usually occurs along steep slopes, where weaker materials tend to slide during earthquake shaking. Our survey identified only one CMP culvert damaged by slumping.

FIGURE 4. CMP culverts deflected upward due to penetration of Elkhorn Road into softened marsh sediment

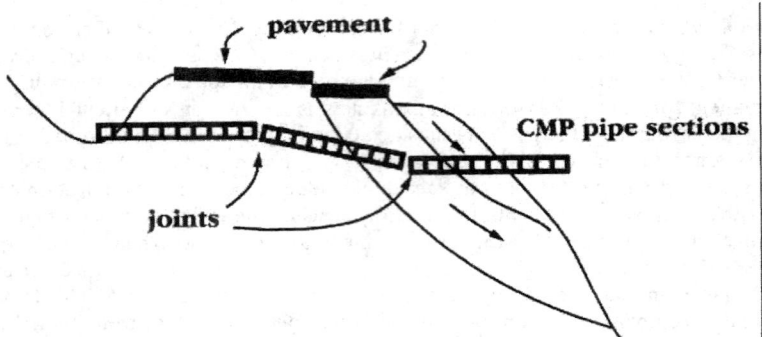

FIG. 5. Diagram of slump and culvert damage, Spanish Ranch Road, Santa Cruz County, Calif. (Youd and Beckman, 1996)

Spanish Ranch Road Culvert. A 0.9-m-diameter CMP culvert was placed beneath Spanish Ranch Road in the Santa Cruz Mountains prior to the 1989 Loma Prieta earthquake. Spanish Ranch road winds precipitously along the mountain slopes east of Santa Cruz, Calif. The culvert consisted of three 6-m long segments of CMP with 100-mm to 200-mm long over-wraps of corrugated metal over each joint. The culvert was covered by about 1 m of compacted fill.

Strong ground shaking during the 1989 Loma Prieta earthquake generated a slump that intersected the outer third of the roadway and underlying culvert. The slumped ground tilted and slipped downward about 1 m. The ground movement did not crush nor tear the CMP culvert, but rather rotated the pipes while stretching the joints (Fig. 5). Joints connecting the pipe sections were pulled apart as much as 150 mm on the top and the bottom of the first and second joints from the inlet, respectively. The CMP pipes were undamaged, but were rotated in a vertical plane as much as five degrees. The wrap of corrugated metal over the joints remained in place, covering the openings and preventing highway fill from entering the culvert. The culvert continued to function without interruption by allowing drainage to flow freely beneath the roadway. The highway was temporarily repaired by paving the roadway shoulder next to the mountain side.

Strong Ground Shaking

Summit Road CMP Culverts. As noted in Table 1, many culverts are in service beneath Summit Road which traverses the crest of the Santa Cruz Mountains east of Santa Cruz, Calif. Being in the epicentral region of the 1989 Loma Prieta earthquake, this area was strongly shaken (estimated peak accelerations in excess of 0.4 g), and disturbed by a diffuse zone of tectonic deformation. The zone of deformation was characterized by a broad band of disperse and discontinuous small surface ruptures (U.S.G.S. staff, 1990; Aydin et al.,

1992). We inspected three culverts at random and slowly drove over the remaining culverts searching for pavement damage or repairs indicative of damage or repairs to the underlying pipe. We also matched locations of culverts with fissures mapped by the U.S.G.S. staff (1990), and found that no single fracture intersected any of the culverts. No pavement damage or repairs were observed, which indicates that these culverts performed well. From this good behavior, we conclude that strong ground shaking in the absence of ground failure or ground deformation is not detrimental to well-constructed CMP culverts.

Increased Lateral Earth Pressures

Earthquake shaking may transiently or permanently increase lateral earth pressures acting against buried structures such as culverts. Visual inspection of the interior of the three CMP culverts located on Summit Road, the culverts beneath the San Diego Freeway and the damaged culvert beneath Spanish Ranch Road (noted above) revealed no signs of deformation due to increased lateral pressures from the 1989 Loma Prieta and Northridge earthquakes. Also, we did not find any notes in highway agency reports of deformation of CMP culverts beneath highways due to lateral pressures generated by earthquake shaking.

San Fernando Valley Culverts. During our investigation, we inspected two culverts beneath the San Diego Freeway (Interstate Highway 405) in the area strongly shaken by the 1994 Northridge earthquake. We detected no damage to either of these culverts.

Lower San Fernando Dam CMP Culvert. During the 1994 Northridge earthquake an approximately 75-m long segment of 2.4-m-diameter CMP collapsed (Figs 6 and 7). That culvert had been placed beneath the reshaped Lower San Fernando dam to allow water collected in the old reservoir to flow through the reshaped embankment. The reason for the reshaping of the dam and the placement of the CMP culvert was failure of the original dam during the 1971 San Fernando earthquake. During the 1971 earthquake, hydraulically a placed fill liquefied leading to a massive flow failure. That failure carried a large segment of embankment, including the crest of the dam, into the reservoir (Seed et al., 1975). After the earthquake, a replacement dam was constructed about one mile upstream from the failed dam. The old embankment was then reshaped to act as a flood retention basin. For added stability, the fore part of the reshaped dam was gently sloped, forming a berm.

While reshaping the old dam, a 2.4-m-diameter CMP storm drain was placed beneath the reconstructed fill. The drain was then connected to an old reinforced concrete pipe outlet conduit that had been placed beneath the undamaged part of the old embankment. As noted above, the culvert drains water from a pond that has formed in front of the reshaped dam. Water from the pond has also saturated parts of the foundation and lower layers of fill.

The Lower San Fernando dam was strongly shaken by the 1994 Northridge earthquake with estimated peak ground accelerations ranging from 0.4 g to 0. g (Davis and Bardet, 1998). This shaking liquefied saturated materials beneath the berm, generating numerous fissures and sand boils on or near the lower part of the berm (Fig. 6). The zone affected by fissures and sand boils was about 130 m wide (north to south), and 420 m long (east to west). Most of the fissures paralleled the toe of the berm and were characterized by open cracks as wide as 300 mm. These fissures were apparently caused by a fraction of a meter of down-slope lateral spread of the berm. Other sets of fissures, mostly near the abutments and over the buried storm drain, were oriented upslope and generally parallel to the buried CMP culvert. These fissures were likely generated by transient ground oscillations across the face of the berm and by the collapse of the CMP culvert.

Approximately 100 m of the new CMP storm drain passed through the zone of liquefied effects. In May 1994, the storm drain was excavated, revealing a 75-m-long section of CMP culvert that had collapsed laterally inward (Fig. 7). The northern terminus of the collapsed culvert was approximately 18 m inland from the toe of the fill; the southern terminus of collapsed CMP was near the connection with a pre-1971 outlet conduit composed of reinforced-concrete pipe (Davis and Bardet, 1998). The thickness of compacted fill over the collapsed CMP ranged from about 4 m at the northern end of the collapse to more than 12 m over the southern end. The ground water level at the time of failure varied from roughly mid height of the pipe to 1 m above the top of the pipe (Davis and Bardet, 1998).

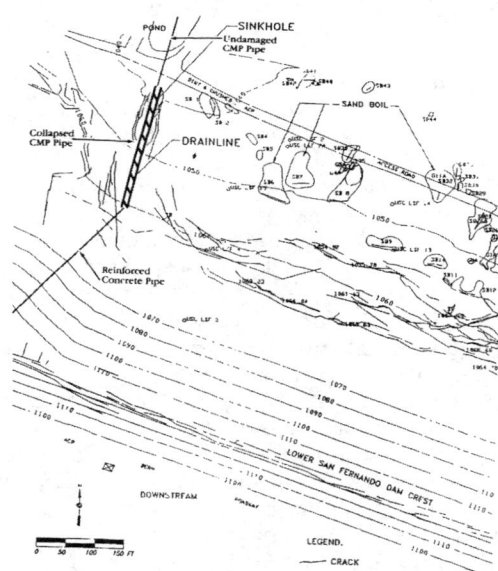

FIG. 6. Upstream berm of Lower San Fernando Dam with alignment of buried collapsed drain, and localities of ground fissures and sand boils (Bardet and Davis, 1995 (used with permission of ASCE))

FIG. 7. View southward of laterally collapsed 2.4-m-diameter CMP culvert beneath Lower San Fernando Dam (Davis and Bardet, 1998, (used with permission of ASCE))

The laterally inward collapse indicates that failure was induced either by increases of lateral earth pressure acting against the pipe or loss of foundation and side support. Prior to the earthquake, vertical earth pressures were likely somewhat greater than lateral pressures as a consequence of fill placement and compaction. Thus, a collapse due to increased lateral stress, requires a rather large increase of lateral earth pressures. We suggest that such a large increase could have been generated by liquefaction induced ground oscillation. Oscillations in an east-west direction, across the face of the berm, could have caused opening and closing of fissures, including those near and above the pipe, that created large lateral forces upon impact. Evidences for liquefaction ground oscillation include sand boils and north-south oriented ground fissures near the abutments and over the collapsed culvert. The magnitude of the dynamic lateral earth pressures, caused by the fissure closures, would be a function of depth of burial. By this mechanism, collapse would occur only at burial depths sufficient to cause lateral stresses greater than the collapse strength of the pipe.

Loss of foundation and side wall soil resistance could have allowed the pipe to collapse downward and inward under high overburden pressures. In such instances, the base of the culvert would tend to expand outward while the upper part of the pipe collapsed inward (Davis and Bardet, 1998).

Lateral Spread

A CMP culvert south of the intersection of Highways 5 and 210 was pulled apart in extension during the 1971 San Fernando earthquake. The Foundation Section of the California Department of Transportation (1973) gave the following brief report of damage to this culvert: "A 42-inch (1-m) corrugated steel pipe also suffered severe joint separation in this area. It was believed that this pipe could be repaired without replacement."

Although the location of this culvert is not clearly specified, the culvert likely was located in a lowland south of the interchange which was displaced by the Juvenile Hall lateral spread during the 1971 earthquake (Youd, 1971; 1973). During our 1995 field investigation, we searched unsuccessfully for this culvert. This culvert apparently was removed or replaced with a much stronger culvert that now connects a small flood control basin east of Interstate Highway 5 with a drainage canal west of the roadway. In that area, lateral displacements as large as 1.8 m generated by the Juvenile Hall lateral spread could have caused the separations in the culvert joints noted above.

Increased lateral pressures on headwalls

Fig. 8 is a photograph of a fractured head wall and distorted inlet to a 3.6-m-diameter CMP culvert beneath Route 5 about 600 m south of the Roxford Street underpass. The culvert was constructed from bolted curved plates of corrugated steel. More than 6 m of fill cover the central part of this culvert. During the 1971 San Fernando earthquake, a 200-mm-thick segment of head wall above the culvert crown at the inlet fractured and broke away from the wall. As shown in the photograph, the culvert was also locally deformed at the inlet. In this instance, the pipe deformation did not extend beyond 1 m inward from the inlet. A retaining wall that extended northward from the inlet was also fractured and bulged outward due to increased lateral earth pressures generated during the 1971 earthquake.

This culvert is within the area strongly shaken by both the 1971 San Fernando and 1994 Northridge earthquakes. A re-inspection of the culvert about a year after the 1994 earthquake revealed that no repairs had been made and no additional detectable damage occurred to the pipe and headwall during the 1994 earthquake. Ground shaking during both the 1971 and 1994 earthquakes was intense at this locality with peak accelerations exceeding 0.5 g. The reasons for the lack of significant additional damage during the 1994 earthquake are not clear. One possible explanation is that the general water table in the area likely dropped significantly after the 1971 event. In 1971, the culvert outlet was near the shoreline of lower Van Norman Lake. The lake was drained after the 1971 earthquake which likely lowered the water table beneath the freeway embankment. Although there were no visible signs of liquefaction or ground failure at the locality in 1971, the foundation soils may have been saturated and softer than in 1994, allowing minor permanent and transient ground displacements and increase in lateral pressure against the head wall.

FIG. 8. Fractured headwall and deformed inlet of culvert beneath Route 5 about 600 m south of Roxford Street, Sylmar California; damage occurred during 1971 San Fernando earthquake, but photograph taken after January 1995 Northridge earthquake.

Conclusions

To evaluate culvert performance during earthquakes, we reviewed reconnaissance reports from six earthquakes and conducted field investigations in areas shaken by three of those earthquakes. Hundreds of CMP culverts were in place in areas strongly shaken by these earthquakes. Lack of reported or observed damage to all but ten of these structures indicates that CMP culverts generally perform very well during earthquake shaking. Thirty-two CMP culverts, including the ten damaged structures, were specifically examined through field inspection or critical literature review. Diameters of these culverts ranged from 0.45 m to 3.6 m. Of the ten damaged culverts, eight required repairs or replacement. Only two of those culverts were nonfunctional after earthquake shaking. Based on this performance, we draw the following conclusions:

1. CMP culverts have performed very well during earthquakes, except in areas of ground failure or ground disturbance. Of the ten damaged culverts identified, all but one were in areas of ground failure caused by liquefaction or slope instability. The one culvert not in an area of ground failure suffered minor damage because of increased lateral pressures cracking a head wall and slightly deforming the pipe inlet.

2. Penetration of roadway embankments into softened or weakened foundation soils has been the most common cause of damage to CMP culverts. This damage was

usually a consequence of soil softening or liquefaction. Embankment penetration pushes mid-sections of culverts downward, bending or buckling the pipe and deflecting end-sections upward. In most instances the pavement and overlying roadway were also severely damaged.

3. Liquefaction-induced lateral spread pulled apart three CMP culverts primarily through failure at the joints. Differential axial and lateral ground displacements generated extensional or flexural forces within the culverts causing separations at the joints or tearing of the pipe.

4. Increased lateral earth pressures or softened foundation and supporting soils laterally collapsed one 2.4-m-diameter CMP culvert over a length of about 75 m. This culvert was buried beneath 4.2 m to 12 m of compacted fill overlying liquefiable deposits. Transient pulses of lateral stresses or support failure generated by liquefaction caused the collapse of this CMP culvert. The same pipe was undamaged where the depth of cover was less than 3.5 m.

5. Several smaller-diameter CMP culverts (<1.2 m) were inspected and found to be undeformed even though they were located in areas of intense seismic shaking or in areas where liquefaction occurred. These seismic effects should have transiently or permanently increased lateral earth pressures, but did not perceptibly deform the culvert sections. Depth of cover for these culverts ranged up to 6 m. This performance indicates that well constructed and uncorroded small-diameter (<1.2 m) culverts are sufficiently strong to resist lateral deformation due to seismic effects.

6. Slumping of fill and foundation soils down a steep slope deformed and pulled apart one CMP culvert. In this instance, about 1 m of differential vertical displacement across a 18-m-long culvert did not rupture or deform individual pipe sections, but rotated the pipes by as much as 5 degrees. This disturbance caused pull-apart of as much as 200 mm at the joints. An over-wrap of corrugated metal remained in place, however, and prevented fill or other debris from infiltrating though the extended joints. This culvert remained in service for several years after the earthquake.

Acknowledgements

This project is part of the Highway Project being conducted in behalf of the Federal Highway Administration (FHWA) by the National Center for Earthquake Engineering Research (NCEER) under FHWA Contract DTFH61-92-C-00106 Task 106-B(I). Support, including funding, from FHWA and NCEER is gratefully acknowledged. Information on damaged and undamaged culverts in their jurisdictions was provided by highway officials from the California Department of Transportation and the Idaho Department of Transportation, and Custer County, Idaho, and public works officials in Los Angeles, Monterey, and Santa Cruz Counties, California,

References

Aydin, Atilla, Johnson, A.M., and Flemming, R.W. (1992) "Right-lateral reverse surface rupture along the San Andreas and Sargent faults associated with the October 17, 1989, Loma Prieta, California earthquake," Geology 20, 1063-1067.

Bardet, J.P. and C.A. Davis (1995). "Lower San Fernando Corrugated Metal Pipe Failure," *Proc. 4th U.S. Conf. on Lifeline Earthquake Engineering*, ASCE, San Francisco, Aug., pp. 644-651.

Davis, C.A., and Bardet, J.P. (1998). "Seismic analysis of large-diameter flexible underground pipes." *J. Geotech. and Geoenviron. Engrg.*, 124 (10), 1005-1015.

California Division of Highways, Materials and Research Department, Foundation Section (1973). "Earthquake damage to California highways," *The San Fernando, California, earthquake of February 9, 1971*, U.S. Dept. of Commerce, 2, 235-246.

Kachadoorian, Reuben (1968). "Effects of the great earthquake of March 27, 1964, on the Alaska highway system," *U..Geological Survey Prof. Paper* 545-C, 1-66.

McCulloch, D.S. and Bonilla, M.G. (1970). "Effects of the Earthquake Of March 27, 1964, on The Alaska Railroad," *U.S. Geological Survey Professional Paper* 545-D, 1-161.

Seed, H.B., Lee, K.L., Idriss, I.M., and Makdisi, F.I. (1975). "The slides in the San Fernando Dams during the earthquake of February 9, 1971," *J. of the Geotechnical Enrg. Di.v*, ASCE, 101 (GT7), 651-688.

U.S.G.S. Staff (1990). *Fracture Map of the Santa Cruz Mountains, Loma Prieta Earthquake*, U.S. Geological Survey, Menlo Park, Calif.

Youd, T.L. (1971). "Landsliding in the vicinity of the Van Norman Lakes, *The San Fernando, California, Earthquake of February 9, l971, U.S. Geological Survey Professional Paper* 733, 105-109.

Youd, T.L. (1973). "Ground Movements in Van Norman Lake Vicinity During San Fernando Earthquake," in, *The San Fernando, California, earthquake of February 9, 1971*, U.S. Dept. of Commerce, 3, 197-206.

Youd, T.L., and Beckman, C.J. (1996). "Highway culvert performance during past earthquakes," Natnl Cent. for Earthquake Engrg. Resrch Technical Report NCEER-96-0015, 1-60.

Lifeline Damage from the January 26, 2001 M7.7 Gujarat, India Earthquake

Curtis Edwards, P.E.; M.ASCE[1]

ABSTRACT

On January 26, 2001 at 8:46 local time, there was a magnitude 7.9 (Mw) earthquake in the seismically active area of the Kachchh (Kutch) District of the Gujarat State of India about 600 km northwest of Mubai (Bombay). It is estimated there were nearly 20,000 deaths, 150,000 injuries and 15,700,000 people affected by the earthquake in the State of Gujarat (1991 population – 37,800,000). In the Kutch District (population 1,200,000) about 600,000 people live in the four largest cities (Gandhiham – 400,000, Bhuj – 110,000, Anjar – 60,000, and Bachau – 40,000). The remainder lives in 884 villages. In Kutch, 52.6% of the housing stock (257,500 houses) was destroyed and a total of [1]3,563,000 houses (7.5%) were damaged/destroyed in the Gujarat State. Most of these buildings were non-engineered, unreinforced stone masonry buildings. However, there were a few instances of engineered reinforced concrete structures collapsing, which would indicate poor design, material and/or construction.

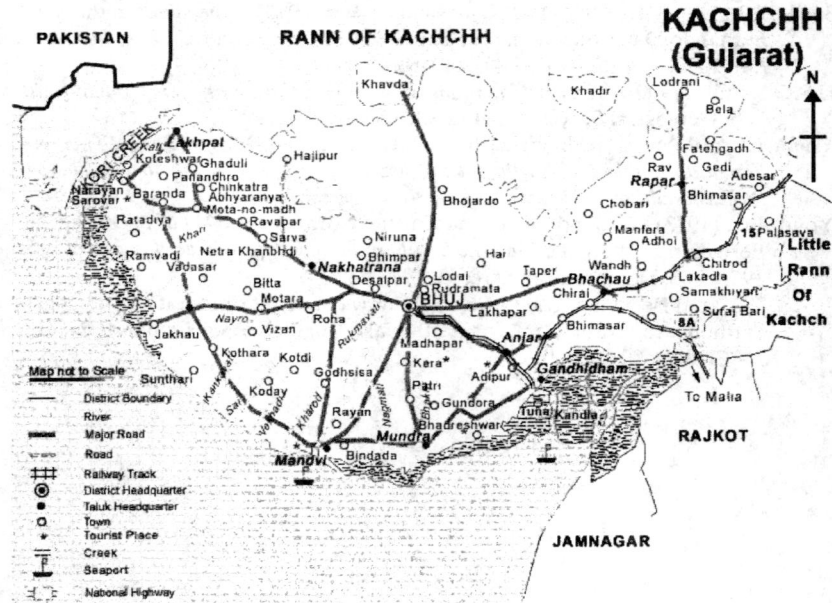

Figure 1- Map of Kutch District

---

[1] Chairman ASCE TCLEE Earthquake Investigation Committee; Vice President, Pountney Consulting Group, Inc., 4455 Murphy Canyon Rd. #200 San Diego, CA 92123; cedwards@pountney.com; Phone - 858-576-9200; Fax – 858-565-1738

Gujarat is a highly earthquake-prone region. Devastating earthquakes have rattled nearly all parts of the state. There have been 12 significant earthquakes that have shook Gujarat since 1819. Gujarat primarily lies in three different seismic zone regions, as defined by Indian Standard 1893. Implicit within the Indian Standard 1893 for earthquake design forces in that the Zone V regions of India, including the Kutch region, could expect to experience a maximum peak ground acceleration (PGA) in the range of 0.24g, within the reasonable planning horizon for many engineered structures. For Indian Zone V, seismic design forces are about 60% that commonly specified by the UBC (1994 version) for Zone 4 regions of California.

There was extensive surface cracking in the area northwest of Bhachau, which caused damage to the main highway, bridges, water pipelines and two railroad track bridges. Liquefaction caused minor damage to facilities in the Kandla Port south of Gandhidham and major damage to the Navlakhi port southeast of Gandidham on the southern shore of the Gulf of Kutch in the Rajkot District. At Navlakhi, there was extensive lateral spreading and subsidence, which caused the main access road and railroad track to drop below sea level interrupting the transport of coal and other goods from the port. In addition, lateral spreading caused a new reinforced concrete wharf to collapse into the sea.

WATER SUPPLY AND DISTRIBUTION

A large groundwater aquifer supplies water to 80-90% of the Kutch District. Surface water from the Tapar Dam and water treatment plant supplies the City of Gandhiham. Because of the great number of towns and villages in the Kutch district, it has not been economical to construct just a single water transmission and distribution network. Instead, a total of around 130 mostly hydraulically separate water systems have been developed to serve 10 towns and 884 villages in the Kutch district.

Most well fields are connected to the supported town (s) and village(s) by a single transmission pipeline. Typical transmission pipeline diameters are 12 to 16-inch (300 – 400 mm) nominal. For smaller villages, the transmission pipelines could be as small as 8-inch (200-mm) nominal diameter. There are distribution pipeline networks for the largest towns, with 4-inch to 6-inch (400 – 150 mm) diameter pipes being most typical. Typical pipeline materials are concrete (6-foot (2 m) long segments, joints using jute and cement); asbestos cement pipe (rubber gasket joints), PVC pipe (push on joints with no apparent gaskets, unlined). In a few cases, mild steel (unlined mostly) or ductile iron pipe was used (not typical).

There was extensive damage to the transmission system, due mostly to liquefaction at stream crossings. The following information is based on information collected between February 12 – 27, 2001. Table 1 provides a summary of the water service restoration process as of February 12, 2001.

Table 1 - Water Restoration as of February 12, 2001

District	Kutch	Rajkot	Jamnagar	Ahmneabad	Surend-ranagar	Total
Number of towns / cities affected	10	2	2	1	3	18
Number of villages affected	884	125	92	0	239	1340
Number of towns / cities where water supply has been restored via pipeline	3	1	1	1	3	9
Number of towns / cities where water supply has been restored via tanker truck	7	1	1	0	0	9
Number of villages where water supply has been restored via pipeline	539	86	41		200	866
Number of villages where water supply has been restored via tanker truck	311	32	51		10	404
Number of villages where water supply has been restored via other means	34	7			29	70

Figure 2 – Damaged Pump Station w/ Cistern

Figure 3 – Damaged Well House

Most of the observed damage at pump stations (approximately 300 locations) was due to pump buildings that were constructed out of unreinforced stone (Figure 2). The stone walls and roofs collapsed damaging electrical controls and emergency generators. The pumps and piping were usually not damaged in that they are sturdy and not affected by the falling stones. System restoration usually involved providing a new, generic control panel for the pumps that was located outside of the old building perimeter. Severely damaged generators were not restored at the time of the inspection. As a result, the restored stations were vulnerable to subsequent power failures.

In the Kutch district there are approximately 200 or 300 at grade or partially buried reinforced concrete tanks, most of which are located at wells (Figure 3). These range in capacity from 10,000 liters to 1,000,000 liters. In addition, there are approximately 75 elevated reinforced concrete tanks (Figure 4). There was no reported damage to the at-grade tanks. Most elevated reinforced concrete tanks performed very well with a reported two collapses and three or four with minor leaks.

In areas without water service, water tanker trucks (35 total) provided water delivery during the system outages. Approximately 500 water department personnel assisted in the restoration of the system. This included clearing rubble from around collapsed pump and well buildings, drilling 125 new wells, replacing pump control panels, and setting up temporary water treatment plants.

Figure 4 – Damaged Concrete Tank

ROADS AND BRIDGES

Roads in the Kutch district primarily consist of one or two lane asphalt or dirt improved roads. National Highway No. 8 runs east west across India and crosses through the Kutch region. Many of the Highway 8 bridges were under construction at the time of earthquake. The road, highway and bridge system performed very well, even in the area with severe ground shaking. There was a least one collapse of a reinforced concrete slab culvert due to failure of the unreinforced stone abutment (Figure 5). Bridges performed very well and with one exception, were useable after the earthquake. Of the 7,000 bridges in Gujarat, almost 200 had some sort of damage and about 20 were seriously damaged (Table 2). Most bridge damage occurred at the bridge approaches. In one case, the bridge crossing the Little Rann of Kutch (tidal area) had severe settlement and was restricted to one way (signalized) traffic. Its new replacement was slightly damaged, but was opened within four months after the earthquake. Because of the extended drought in the region, temporary bridge bypass roads could be constructed to divert traffic while the bridges were repaired (Figure 6). This highway network played a critical role in the transportation of emergency supplies into the damaged regions.

Table 2- Summary of Bridge Damage

Structures	Length	Number Damaged
Major Bridges	> 60 m	8
Minor Bridges	30-60 m	9
Minor Bridges	6-30 m	42
Culverts	arch / slab	73
Culverts	pipe	55
TOTAL		187

Figure 5 – Collapsed Slab Culvert

Figure 6 – Damaged Bridge Approach/Temporary Bypass Road

ELECTRICAL POWER SYSTEM

There was no reported damage to the electrical transmission network, which consists of 220-kV and 132 kV systems. There also was no reported damage to the fossil fuel and nuclear power plants that are all located over 150 km from the epicenter. There was extensive damage to substations in the high shaking areas causing immediate power outages (Figure 7 & 8). Most of the substation control buildings were either severely damaged or collapsed, heavily damaging and/or destroying control equipment (Figure 9). Fragile equipment in the high voltage substations was severely damaged. Porcelain insulators broke, transformers slid, and aluminum fittings broke at the three 220-kV stations, the two 132-kV stations and the 40, 66-kV stations. Temporary controls were in place and power restored to all 220-kV systems by January 30, 132-kV systems by February 4, and all 66-kV systems by February 12, 2001. Falling poles and buildings heavily damaged local distribution power systems; however, restoration was not a high priority due to limited demand in those areas. Pole mounted transformers were adequately braced and were not reported to have failed. Where control buildings completely collapsed, new control buildings were being constructed with metal frames. Damaged control buildings were braced with steel beams until such time as permanent repairs could be made.

Figure 7 – Typical Switchyard Damage

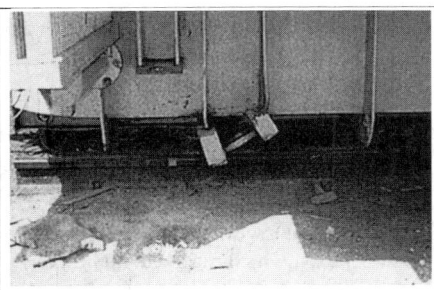

Figure 8 – Broken Transformer Mounts

Figure 9 – Damaged Control Building, Anjar

TELEPHONE FACILITIES

In the Bhuj, there are 17,000 services covering a city with a population of over 200,000 people. This indicates that many home do not have individual phone services. Most local service is provided by the STD/ISD call booths, which provide direct local, interstate and international metered dialing. Telephone facilities suffered damage similar to that of the electrical systems with the exception that the central office buildings were constructed of reinforced concrete frames with stone in-fill walls. In Bhuj, 50% of the infill walls collapsed damaging controls and switches, but the reinforced concrete frame remained. Electrical controls were damaged, but within one week of the earthquake, phone service was available to anyone that requested it. Emergency power provided adequate power to the exchanges until grid power was restored. Fiber Optic cables were severed causing significant interruptions in service. Cellular service, which was limited before the earthquake, was mostly restored when electrical power was restored.

OIL AND LIQUID FUELS

There is major oil refinery (Reliance Oil) in Jamnagar on the south shore of the Gulf of Kutch. There was no reported damage to the refinery or the 800-mm pipeline crossing the Gulf to the Kutch District. Most liquid petroleum is transported by truck to regional distribution centers and filling stations. There were temporary interruptions in deliveries due to bridge and road damage. However, these were minor and there were no major shortages of liquid fuels reported.

The Indian Oil Company (IOC) reported that it had restored adequate supplies of petrol (gasoline), diesel and kerosene for Bhuj, Kandla and other effected areas. They also said that operations at the Koyali Refinery had been unaffected by the earthquake and that the Koyali-Ahmedabad pipeline was operational.

The CRL tank farm is located at the south (new) Kandla port. The TCLEE team visited this tank farm. There are a total of 30 tanks at this tank farm constructed from 1994 to 2000. All tanks are made from mild steel plates with steel floors. All roofs are conical. All tanks are unanchored, resting on asphalt pads. The pads rest on the hydraulic fill of the port. Surrounding each tank pad is a concrete ring, which is independent of the tank structure. There was only minor damage to these tanks.

PORTS

Port of Kandla

The Kandla Port is India's largest port and was severely damaged by the earthquake. There are 10 cargo berths, which were all damaged. There are 8 berths with gantry cranes and no container cranes. Primary cargo includes wood, chemicals, coal, sulfur, lead ore, etc. By February 16, 2001, 5 berths were fully operational with the other five operating at a reduced capacity. There are 4 oil jetties of which only three were operational as of March 1. The port also has two hospitals, which received only minor damage. The Kandla Port Trust operates the port. The pile supported signal tower building was listing as a result of liquefaction.

There were many factories in the port that sustained damage. A fertilizer plant was reduced to rubble and two tanks containing dangerous chemicals leaked onto port grounds. Clean up of the spill was completed within one week. A significant number of warehouses suffered structural damage due to short column effect and flexing of end walls pulling out longitudinal bracing perpendicular to, and along the bottom of roof trusses.

There was one minor pipeline leak that was quickly plugged. Two private tank farms containing acryl nitrate had significant leakage. The owners of the tanks quickly neutralized the spills.

There were 7 tanks at various tank farms, of which 3 settled due to liquefaction and 3 ruptured due to tears at the tank bottom just inside the connection with the tank wall. Elephants foot buckling was not observed.

Shipping operations resumed by Saturday, January 27, 2001. The port engineer indicated that the biggest issue after the earthquake was that ship owners were demanding a certification that the berths were safe. Until the port provided such a certification, some ship's captains refused to dock.

Port of Navlakhi

The Port of Navlakhi is on the southern shore of the Gulf of Kutch. It has one main dock constructed on reinforced concrete piers. A relatively new wharf was completely damaged. Coal for power plant use is one of the main cargoes handled at this port.

Liquefaction and lateral spreading heavily damaged this port. The main access road and railroad line subsided into the gulf preventing the movement of goods. A temporary road was constructed allowing interim movement. A permanent road surface was under construction at the time of the investigation. Other than removal of damaged rails, no railroad repair appeared to be underway. It is likely; the railroad replacement (located on the bay side of the access corridor) will be replaced after the permanent road is in place. In lieu of train transport, coal was being loaded on trucks for transport out of the port.

Figure 10 – Damaged Wharf, Navlakhi

RAIL SYSTEMS

A number of separate railways were constructed in India beginning in 1853. In 1951, they were combined into national systems consisting of 9 geographical zones. The Gujarat region is in the Western Zone. There are three different track gages used in the Western Zone. However, only the meter gage (1000-mm) was in use in the damage area. Much of the system is electrified, but the Kutch region is not. Crossties are primarily concrete or steel as use of wood is essentially prohibited.

A number of trains were operating at the time of the earthquake including 8 passenger trains. None of the operating freight or passenger trains derailed. In Bhachau, a water tank car derailed and one tank body overturned. Inspection was completed very quickly and operation on most lines resumed within 3 hours after the earthquake. All lines except the line from Dahinsara to the port of Navlakhi were restored to service by January 29.

Where there was significant damage to signal and interlocking systems, operation was resumed with manual operation of switches and reduced speed for facing point moves. As repairs were made, speed restrictions were removed until normal operation was restored.

About 1 km of embankment and track shifted and settled into the Gulf of Kutch near Navlakhi as a result of liquefaction and lateral spreading. This line was fully returned to service on March 31, 2001. Other track damage occurred near Gandhidam and Bhachau primarily from settlement, slides and cracking. Of the 240 bridges and crossings in the region, six of 24 arch bridges, five of 75 girder bridges and one of 41 reinforced concrete bridges sustained significant damage. Temporary repairs included epoxy injection of displaced stone arches and temporary shoring of threatened spans.

EMERGENCY RESPONSE

Hospitals in Bhuj and Anjar were totally destroyed. The Indian military provided immediate emergency support, which was soon followed by international relief efforts. The International Federation of the Red Cross and Red Crescent Societies provided a mobile hospital, medical doctors, relief workers and relief supplies to the stricken areas of Kutch. The temporary hospital will remain in Bhuj until the replacement hospital is constructed. The Red Cross distributed food, water and tents to the stricken areas. In addition, they worked with local officials to restore a clean water supply and maintain sanitary conditions. In addition, they provided the TCLEE investigation team with valuable assistance during our stay.

The Gujarat Bureau of Roads and Buildings (R&B) is responsible for both the restoration of the roads and the removal of building debris. There was limited heavy equipment available for the removal of the rubble. R&B enlisted private truckers and animal drawn carts to remove the debris from the cities. Rubble was either stockpiled along roads or in floodplains.

LESSONS LEARNED

- There was less damage to lifelines as a result of the drought caused deep water table and the fact that dams/reservoirs were nearly empty.
- Damage to housing stock caused massive relocations causing a need for new emergency lifeline services rather that restoration of old ones.
- Unreinforced stone buildings and infill walls caused a great deal of damage to delicate control systems for water, telephone and electrical systems. This could have been avoided by using engineered buildings to house these facilities.
- There was only limited damage to utility structures that had been designed and constructed to seismic standards.
- There was a relatively high rate of damage to various bridge structures. It was fortunate that the drought allowed for rapid construction of temporary bypasses.
- There was no fire following this earthquake. This could be due to the fact that there are no natural gas systems, wood is not a typical construction material and there were lengthy outages of electrical power.

INVESTIGATION TEAM

The Post Earthquake Investigation Committee of the Technical Council on Lifeline Earthquake Engineering (TCLEE), a technical council of the American Society of Civil Engineers (ASCE) organized a team of four TCLEE members with support from ASCE and PEER to perform a reconnaissance of the lifeline in the earthquake areaThe investigation team consisted of the following ASCE members:

Mr. Curtis Edwards, Pountney & Associates, Inc. San Diego, CA

Mr. John Eidinger, G&E Engineering Systems, Oakland, CA

Mr. Mark Yashinsky, Caltrans, Sacramento, CA

Mr. William Byers, BNSFRR, Kansas City, KA

REFERENCES

Eidinger, John, editor. *Gujarat (Kutch) India M7.7 Earthquake of January 26, 2001 and Napa M5.2 Earthquake of September 3, 2000.* ASCE - TCLEE Monograph No. 19, June 2001. Material reproduced by permission of ASCE.

The American Lifelines Alliance Progress in the Development of National Consensus Guidelines for the Design of Lifeline Systems for Natural and Man-Made Hazards

Douglas G. Honegger*, Edward M. Laatsch**, and Timothy D. Sheckler

* ALA Principal Investigator, D.G. Honegger Consulting, 2690 Shetland Place, Arroyo Grande, CA 93420; PH 805-473-0856; dghconsult@aol.com
** Federal Emergency Management Agency, Federal Insurance Administration and Mitigation Directorate, 500 C Street S.W., Washington, D.C. 20472; PH 202-646-3885; Edward.laatsch@fema.gov
** Federal Emergency Management Agency, Federal Insurance Administration and Mitigation Directorate, 500 C Street S.W., Washington, D.C. 20472; PH 202-646-2834; tim.sheckler@fema.gov

Abstract

The American Lifelines Alliance (ALA) was initiated with financial support from the Federal Emergency Management Agency (FEMA) in 1998 with a primary goal of facilitating the development and improvement of the design of key utility and transportation systems (electric power, telecommunication, water, waste water, oil, natural gas, rail, and shipping ports) to achieve the desired level of performance in natural hazards. ALA has not attempted to focus efforts on all lifeline systems but has instead chosen to place priority on specific topics relevant to selected lifeline systems where a need for improved hazard mitigation practices is identified. To assist in identifying gaps in existing knowledge or practice, ALA prepared a matrix of existing guidelines and standards related to the design of lifeline systems for natural hazards. The natural hazards matrix is revised annually based on feedback from ALA corresponding members and comments solicited from standard developing organizations. Based on this matrix, ALA has prioritized a suite of potential projects related to natural hazards to be carried out over the next several years. The approach to developing the matrix of existing lifeline system guidelines and standards for natural hazards and the general parameters considered by ALA in prioritizing potential projects are discussed in the paper. In addition, ALA has recently developed a similar matrix for man-made hazards. The need for a man-made hazards matrix is a direct result of FEMA's desire to expand the goals of ALA to include man-made hazards. This expanded role for ALA is in part related to the fact that some measures, taken to improve lifeline system performance in natural hazards, also have benefits in reducing the susceptibility of prolonged system interruption from certain types of man-made hazards. As of the end of 2002, ALA has sponsored nine projects that affect three existing guidelines or standards and will result in five new guidelines under the direction of ANSI-accredited standards organizations or industry-sponsored organizations. This paper provides a summary of the key technical issues that have been investigated in these past ALA projects and identifies near-term needs that have been identified by the ALA project team.

Background on ALA

The engineering community has long worked to build safe and reliable lifeline systems -- that is, those systems necessary to provide electric power, natural gas, water and wastewater, and transportation facilities and services that are essential to the well being of the community served by these systems. Providing lifeline system function is especially important in assisting rapid recovery following natural hazards. Engineering approaches to limiting damage to lifeline systems from natural hazards have developed specifically for individual natural hazards and individual types of lifeline systems. Thus, the design of electric power transmission systems focused on loads from high wind and ice storms, while the design of natural gas transmission systems focused on landslides and fault crossings.

Using a system-based approach to assessing expected lifeline function is a relatively recent development that was largely driven by the need to prioritize efforts to improve system performance for large earthquakes capable of generating multiple hazards throughout the system.

However, identifying lifeline system risks and implementing measures to improve earthquake performance have not been uniformly carried out, even in regions where the frequency and severity of natural hazards is high (e.g. earthquake mitigation practices in California). In lower-hazard parts of the United States, implementation of natural hazard risk management has been very sporadic. A good example is the limited progress in developing and implementing consensus guidelines and standards to provide consistent improvements in the performance of new and existing utility and transportation systems in large, but rare, earthquake events that may occur outside of California such as the Pacific Northwest, Midwest, and Southeastern United States.

In 1998, the Federal Emergency Management Agency (FEMA) and the American Society of Civil Engineers entered into a cooperative agreement to establish the American Lifelines Alliance (ALA) to facilitate the "creation, adoption and implementation of design and retrofit guidelines and other national consensus documents that, when implemented by lifeline owners and operators, will systematically improve the performance of utility and transportation systems to acceptable levels in natural hazard events, including earthquakes." Inclusion of all natural hazards in the scope of ALA recognizes the benefit of managing the risks from multiple potentially damaging natural hazards events in a balanced fashion that meets the objectives of the lifeline owner or operator for lifeline functionality and the needs of their customers who rely on the lifeline services. Decisions on where to devote resources for improving lifeline system performance should be prioritized by considering the likelihood of experiencing natural hazard events, the impact of the natural hazards on the system, and the value of improving system performance to the owners of the system and their customers. Accordingly, the need to implement system improvements that address specific natural hazards will vary depending on the likelihood and severity of natural hazards and on the operational characteristics of the system.

Following the terrorist attacks on September 11, 2001, the scope of ALA was expanded to include man-made hazards. This change is consistent with the broader goals of FEMA and recognizes that actions that minimize the effects of natural hazards also can improve resistance of structures and systems to man-made hazards. Since, late 2002, ALA activities have been conducted under the Multihazard Mitigation Council through a contract between FEMA and the National Institute of Building Sciences.

The primary function of ALA is to bridge the gap between hazard mitigation practices for buildings and lifeline systems through the sponsorship of national consensus guidelines and standards specific to lifeline systems. ALA activities are focused on converting well-established practices within the utility and transportation industries into nationally applicable consensus guidelines. ALA projects take existing non-consensus guidelines, standards, and industry practices, improve upon them as appropriate, and shepherd the resulting documents through a formal consensus process. In some cases, practices are not sufficiently developed to carry out ALA's process. In these situations, the role of ALA may include funding or co-funding well-defined studies in order to improve or extend existing practices to the point where national consensus is achievable. ALA has sponsored eleven projects to date as summarized in Table 1. The consensus guidelines that result from ALA activities are being prepared to be used on a voluntary basis by owners and operators of lifelines; but they could also be adopted by lifeline regulators to provide suggested or required guidance for achieving acceptable performance in response to natural hazards or human threats. Products of ALA projects are displayed and downloadable on the ALA web site (www.americanlifelinesalliance.org) while being refined in the SDO process.

The following discussion summarizes the projects that have been taken or are underway. These projects can be placed into two categories, improvement of current industry and national design guidelines and standards and improvement of risk-management decisions for lifeline systems.

ALA Projects that Improve Existing Design Practices

ALA has sponsored five projects directly related to improving current design practices for components of lifeline systems. These projects have direct application to pipeline, electric power, telecommunication, and railway systems.

Updating Seismic Design Standards for Aboveground Steel Storage Tanks ALA is working with the American Petroleum Institute (API) and the American Water Works Association (AWWA) to facilitate revisions to seismic design standards for aboveground steel storage tanks. This project was initiated in part due to several significant changes have been incorporated into the 1997 and 2000 *NEHRP*

Recommended Provisions for Seismic Regulations for New Buildings and Other Structures (FEMA, 2001), ASCE 7-02 (ASCE, 2003), and the International Building Code (ICC, 2000). These changes have a significant impact on present tank design and the revised standards will influence all new tanks and many existing tanks during retrofit or modifications. Unchecked, these rules may increase the seismic design loads by more than 30% over levels in the present standards. It is unclear whether these load increases, which were initially developed for buildings, are either appropriate or inappropriate for tanks and vessels. This project is also important to ALA because it highlights the ALA approach of working directly with the affected standard developing groups to make change (in this case API and AWWA) as opposed to trying to impose change from the outside. The primary objective of the project, being implemented by Tank Industry Consultants Inc., is to facilitate revisions to existing API and AWWA tank standards and provide the basis for continual updating of seismic design requirements for aboveground steel storage tanks directly by API and AWWA. The primary objectives of the ALA project with API and AWWA are as follows:

1. Evaluate the impact of the new rules for a spectrum of welded steel storage tanks and locations. Summarize the impact in a "white paper" to the API CRE Committee.
2. Draft the proposed revisions to applicable API and AWWA codes based on guidelines contained in the *NEHRP Recommended Provisions for Seismic Regulations for New Buildings and Other Structures*.
3. Develop plausible recommendations on how the new building code rules can be revised or augmented to better suit the needs of the liquid storage tank stakeholders using the latest knowledge and documented historic performance.
4. Develop a performance basis concept to steer future code development efforts in the proper direction.
5. Develop proposed changes to API and other tank standards that incorporate newer technology and methods.
6. Publish a comprehensive recommended practice to serve as a design guideline on seismic design of ground-supported liquid storage tanks.

Guidelines for Assessing Railway Storm Scour ALA is working with the American Railway Engineering and Maintenance of Way Association (AREMA) to improve the ability of America's railroads to withstand scour and erosion at railroad embankments and bridges that can accompany major storms. The project with AREMA is a product of the identification of scour as a common hazard to many lifeline systems in the ALA guidelines matrix. The partnership with AREMA is the result of discussions over a several months between ALA team members and individuals representing the interests of a variety of lifelines systems. The AREMA project involves a series of 2-day seminars around the US conducted by Ayres Associates of Fort Collins, Colorado that present case studies of past bridge failures due to storm scour, stream stablility concepts and analysis, the mechanics of bridge scour analysis, countermeasure design, and inspection procedures focused on recognition of erosion hazards. This project will lead to an update of AREMA's Manual for Railway Engineering, a document sold worldwide as a reference for the railway engineering profession.

Ice Load Mapping The ALA has contracted with the Army Cold Regions Research and Engineering Laboratory to create new hazard maps of atmospheric ice loads for multiple return periods between 50 and 400 years. The ice loads mapping project resulted from responses to a request for possible ALA projects sent to our Corresponding Advisors. The ice loads mapping project is a three-year effort initiated in 2000 that built upon previous work for parts of the United States east of the Mississippi River. The project was conducted in two phases. The first year of the project covered the United States east of the Rocky Mountains. The second year of the project focused on parts of the United States west of the Rocky Mountains. It is intended that the revised icing hazard maps will be incorporated into national standards such as ASCE 7 and the National Electrical Safety Code.

Guidelines for the Design of Buried Pressure Pipelines ALA supported a task committee of members from the American Society of Civil Engineers and the American Society of Mechanical Engineers (ASME) charged with developing design guidelines for buried pressure pipelines. The task committee was formed because of a recognized need to bridge the gap in knowledge between pipeline design practices, typically governed by ASME codes and consideration of soil-pipeline interaction, an area typically viewed as being within the expertise of civil engineers. While the task committee had been working on the guidelines for several years on a voluntary basis, progress had been slow due to the difficulties in bringing the task group members together and constraints that limited the time members could volunteer to the project. ALA support consisted of travel funds and honorarium for selected committee members to prepare draft materials for review by the committee. The guideline addresses a variety of load conditions common to buried pipelines including loads arising from soil overburden, permanent ground displacement, pipeline lowering, and seismic wave propagation. An appendix to the guideline also recommends acceptance criteria applicable to different load conditions.

Guidelines for the Seismic Design and Retrofit of Aboveground Piping While many piping systems in manufacturing and industrial facilities are typically designed using codes maintained by the American Society of Mechanical Engineers (ASME), these codes do not contain explicit procedures for addressing earthquake loads on the piping systems and several do not reference current seismic hazard definitions. Attempts to provide guidance for the seismic design of aboveground piping within the *NEHRP Recommended Provisions for Seismic Regulations for New Buildings and Other Structures* and explicit requirements contained in ASCE-7, *Minimum Design Loads for Buildings and Other Structures,* are often viewed as inadequate in that they do not recognize the design practices within the ASME codes. ALA sponsored a project with Mr. George Antaki to develop guidelines for the seismic design and retrofit of aboveground piping systems that addressed these two deficiencies. The guidelines developed from this project are currently being balloted within the ASME

B31 standards committee and are the topic of a separate paper in this conference (Antaki, 2003).

ALA Projects that Improve Risk Management Decisions for Lifeline Systems

ALA has sponsored five projects directly related to improving risk management decisions for lifeline systems. These projects have direct application to pipeline (water, wastewater, oil and gas) and electric power systems.

Seismic Vulnerability Assessment of Water Conveyance Systems A fundamental requirement for assessing the seismic performance of a water utility is the ability to quantify the potential for component damage as a function of the level of seismic hazards. The term vulnerability relationship is used to refer to a general deterministic, statistical, or probabilistic relationship between the component's damage state, functionality, economic losses, etc., given some measure of the intensity of the earthquake hazard. In 2001, ALA completed a project with G&E Engineering Systems, Inc. to develop detailed procedures that can be applied to any water transmission system in order to evaluate the probability of damage from earthquake hazards to various components of the system.

While there exists considerable project experience from implementing vulnerability assessments for water systems within industry, consulting and academic communities, specific procedures vary considerably. A consequence of this lack of guidance is the inability to directly compare the potential earthquake damage for water transmission systems among a diverse population of system owners and users. Lack of uniformity in risk assessment impedes the prioritization of what actions should be taken to reduce damage and where resources should be focused to improve performance.

The products of this project include the fragility curves for each type of component and appendices containing the data used in the analyses, comparisons of the fragility curves with those prepared by other researchers in the past, examples of application of the methods, and a description of the statistical analysis methods used in developing the fragility curves.

The vulnerability assessment guidelines are currently being considered as part of an American Society of Testing and Materials (ASTM) activity entitled "Seismic Fragility of Water Transmission Systems," This activity, F36.30, falls under ASTM Committee F36 on Technology and Underground Utilities.

Reliability Guidelines ALA has initiated four projects to develop guidelines on acceptable methods to develop information that can be used to define the level of performance reliability for water, waste water, electric power, oil and natural gas systems. The need for these projects was made apparent in discussions between ALA and individuals representing the interests of those promoting revisions of seismic design guidelines and codes for buildings, primarily the *NEHRP Recommended Provisions for Seismic Regulations for New Buildings and Other Structures* and

ASCE-7, *Minimum Design Loads for Buildings and Other Structures*. A particularly important issue that arose from these discussions was the basis for establishing performance requirements for lifeline systems.

The ability of a lifeline system to maintain an acceptable level of service after an earthquake will depend, not only on the seismic performance of its various spatially dispersed components, but also on the redundancy and service capacity of these components (e.g., number of lanes within roadway elements). To the extent that a lifeline system is comprised of redundant components of sufficient service capacity, it can maintain an acceptable level of service to a community even if some of the redundant components are damaged during the earthquake. Except for certain transportation structures (e.g., bridges and tunnels), earthquake damage to the lifeline system components generally rarely results in direct life-safety consequences. Therefore, acceptable seismic performance for a lifeline system is typically based on: (a) whether the system provides an adequate level of service to its users after an earthquake; (b) whether economic losses related to direct damage, lost revenue from an inoperable system, and liability exposure are within tolerable limits; and (c) whether any adverse political, legal, social, administrative, or environmental consequences are experienced. For these reasons, acceptable seismic performance requirements for lifeline systems are best established through interaction with the appropriate stakeholders, including the lifeline agency, its customers or users, and appropriate regulatory interests.

A project directed at water systems exposed to natural hazards was completed with Natural Hazards Management, Inc. in 2002. Three similar projects were initiated in 2003 for wastewater, oil and gas, and electric power systems that address both natural and man-made hazards. These guidelines will provide clear, concise guidance on specifying the procedures to follow and information to consider in performing a standardized evaluation of electric system performance during and after potential natural hazard and human threat events. The goal is to assist lifeline system owners and operators in defining what approaches are necessary to evaluate the anticipated performance of their systems and provide a defensible basis for risk management decisions. Implementation of the approaches recommended in the Guideline will allow owners and operators to define the scope of activities necessary to determine appropriate risk-management actions to reach acceptable performance levels for impacts from natural hazards and human threats to the lifeline system.

Conclusions

The projects that ALA has completed or initiated will have a significant impact on the practice of designing and managing the risks to lifeline systems. Within the next two to five years, it is expected that many of the ALA products will find their way into practice via adoption of national consensus guidelines and standards published by standards developing organizations within the US.

Future ALA projects have been identified based on a prioritized assessment of apparent "gaps" in generally accepted practice identified by ALA and noted on guidelines matrices that can be viewed on the ALA website. ALA will continue to seek opportunities to co-sponsor mutually beneficial projects with standards developing organizations, industry groups, and individuals. Those wishing to keep abreast of ALA activities are encouraged to become an ALA Corresponding Advisor by sending an e-mail request to jsteller@nibs.org. ALA views its Corresponding Advisors as a valuable resource, advising them of new ALA activities, requesting feedback from them on the value of projects being considered by ALA, and soliciting their ideas for potential future ALA projects.

References

1. American Lifelines Alliance, www.americanlifelinesalliance.org

2. American Society of Civil Engineers, 2003. "Minimum Design Loads for Buildings and Other Structures," SEI/ASCE 7-02.

3. Federal Emergency Management Agency, 2001. "NEHRP Recommended Provisions for Seismic Regulations for Buildings and Other Structures," 2000 Edition, FEMA 368.

4. International Code Council, 2001. "2000 International Building Code"

Table 1. Summary of Completed and Ongoing ALA Projects

Project Topic	Description	Status
Atmospheric Icing	Develop national maps of atmospheric icing thickness and concurrent wind speed for average return periods between 50 and 400 years	Complete Being submitted to standard developing organizations
Water Transmission Vulnerability	Develop state-of-practice relationships to relate seismic hazards to likely damage states for components of water transmission systems	Complete Submitted to ASTM F36 standards committee
Buried Pipelines	Develop guidelines for the design of buried pipelines for a variety of load conditions	Complete SDO being sought for national consensus processing
Aboveground Steel Storage Tanks	Facilitate modification of existing API and AWWA seismic standards for aboveground steel storage tanks to comply with current seismic hazard definitions	Ongoing
Aboveground Pressure Piping	Prepare guidelines for the seismic design of aboveground pressure piping to be used in conjunction with ASME piping codes	Complete Being balloted in ASME B31 committee
Railroad Bridge Scour	Assist AREMA in updating their design manuals to reflect current state-of-practice in designing scour-resistant bridge structures	Ongoing
Man-Made Hazards Matrix	Develop a matrix of existing guidelines and standards to be used to identify potential ALA projects that address the design and operation of lifeline systems to improve resistance to man-made hazards	Complete Posted on ALA web site
Water System Reliability	Prepare guidelines to assist water utilities in deciding what actions need to be taken to provide a defensible level of system reliability appropriate for natural hazards	Complete SDO being sought for national consensus processing
Wastewater System Reliability	Prepare guidelines to assist wastewater utilities in deciding what actions need to be taken to provide a defensible level of system reliability appropriate for natural and man-made hazards	Ongoing
Oil and Gas System Reliability	Prepare guidelines to assist oil and gas pipeline owners in deciding what actions need to be taken to provide a defensible level of system reliability appropriate for natural and man-made hazards	Ongoing
Electric Power Reliability	Prepare guidelines to assist electric power utilities in deciding what actions need to be taken to provide a defensible level of system reliability appropriate for natural and man-made hazards	Ongoing

Seismic Design Standards and Guidelines of Steel and Concrete Liquid Storage Tanks

Lisa Yunxia Wang, Ph.D., P.E.

Abstract

Practicing engineers face many issues and challenges on the design and seismic evaluation of liquid storage tanks. These challenges are generally either in the application of the current design codes and standards, or in choosing an appropriate design method. This paper addresses the design issues on the liquid storage tanks especially on the steel tanks, and the application of the ANSI/AWWA D-100 standard on the design of steel tanks as well as the ACI 350 code on the design of concrete tanks.

Introduction

Both Long Beach and the greater Los Angeles area have experienced lifeline damage due to earthquakes. Collapses of water tanks during the 1933 Long Beach and the 1971 San Fernando earthquakes had serious consequences. Liquid storage tanks are critical lifeline structures that are geographically dispersed over broad areas. These tanks are exposed to a wide range of seismic hazards, community users, and interactions with other sectors of the built environment. Problems associated with the seismic behavior of liquid storage tanks involve the analysis of three systems: the tank, the soil and the liquid, as well as the interaction between them along their boundaries. To achieve accurate results, a 3-D nonlinear finite element analysis of tank system shall be implemented considering (1) the interaction between the tank and the liquid (2) the dynamic soil-structure interaction. Such 3-D nonlinear finite element analysis, including the contained fluid as well as the foundation soil in the system, is complex and extremely time consuming. Figure 1 illustrates the technical complexity on the analysis of such systems, and such approach is generally not practical for engineers to pursue.

Currently practicing engineers apply generally two methods for seismic analysis/design of liquid storage tanks. One is the Standard Method, and the other one is the Response Spectrum Method. The Standard Method is used when the site response spectrum is not available and the procedure is included in the ANSI/AWWA D-100 standard. This method is based on the earlier simplified analyses of liquid storage tanks. The Response Spectrum method, on the other hand, shall be used if the site response spectrum is available. The present available codes and standards, such as AWWA D100, API 650 standards, and the New Zealand recommendations,

---

Assistant Professor, Civil Engineering Department, California State Polytechnic University, Pomona, CA 91768; phone 909-869-4641; fax 909-869-4342; ylwang@csupomona.edu

suggest the use of a response spectrum to evaluate the overturning moment exerted on the tank wall. For large or high-importance-factor tanks, it is essential that the site-specific earthquake response spectrum curves be provided.

Part I. Design and Evaluation of Steel Liquid Storage Tanks

1. Lateral Earthquake Load:

After the horizontal ground acceleration is determined from the response spectrum based on the fundamental natural period of a tank, a lateral push analysis is performed to calculate the overturning moment at various height of tank wall. Pseudo-dynamic load in a parabolic distribution is used to apply on the tank wall for hydrodynamic pressure. This pressure pushes the tank in the lateral direction, and the tank uplifts from its foundation and develops a similar uplifting mechanism to that occurred under earthquake excitation. As shown in Figure 3, a hydrodynamic pressure distribution is assumed on the tank wall as

$$p = p_0(1 - \frac{y^2}{H^2})\cos\theta$$

where y is the elevation of a point on the shell measured from the base, H is the fluid depth, θ is the angle measured from the axis of excitation and p_o is the pressure amplitude at the tank base at $\theta = 0°$.

Based on the hydrodynamic pressure distribution, the overturning moment at any height of tank shell can be derived through integration. Figure 4 shows how the variation of bending moment along tank wall. The overturning moment about the center of the base is $M = 0.25 \pi H^2 p_o R$. The base shear of the tank can be determined in a similar way for shear design, and the base shear $Q = 0.67 \pi H p_o R$. The response quantities evaluated include the maximum values of the hydrodynamic circumferential, shearing and axial stresses in the tank wall.

To determine the minimum shell thickness required for each ring, it is necessary to evaluate the shell buckling status for the design level of the seismic ground motions. Checking on the buckling of tank wall is very critical because most of the unanchored tanks failed in a buckling mechanism: the elephant foot buckling in broad wide tanks while tall tanks suffered a diamond shaped buckling spreading around the circumference.

2. Effects of Vertical Component of Ground Shaking:

The vertical component of the ground acceleration induces a hydrodynamic wall pressure in addition to that induced by the horizontal component. The pressure is uniformly distributed in the circumferential direction and varies in the axial direction as well as with respect to time. If the peak response of the vertical component of an earthquake occurred simultaneously and in the same direction with the peak response of the horizontal component, it may significantly increase the exerted hydrodynamic forces on the tank wall. Therefore vertical acceleration shall be taken into consideration on tank design. Seismic load case must include the horizontal and vertical components of earthquakes, and the load combinations could be (100% Horizontal + 40% Vertical) and (40% Horizontal + 100% Vertical) to be safe.

3. Uplift of Unanchored Tanks:

Nonlinear finite element analysis should be performed for unanchored tanks so that uplift displacement can be determined. This is especially critical for the design of tank attachments. Equivalent springs shall be used for formulation of ground flexibility. The response of the unanchored tank was governed primarily by a rocking motion. The contact characteristics of the unanchored tank with its foundation are important factors in evaluating the response of such tanks. The uplift displacement on the tension side is much higher than the penetration displacement on the compression side, which is demonstrated in Figure 5. Such a behavior is expected due to the tensionless nature of the foundation.

4. Fundamental Frequency/Period of Tank-Liquid System:

The fundamental natural periods are determined based on the method that was described in the publication of American Society of Civil Engineers, Guidelines for the Seismic Design of Oil and Gas Pipeline Systems. In this method, both the Veletsos-Yang and Haroun-Housner procedures are used to consider the effects of damping, roof mass, and vertical component of ground shaking.

5. Wave height/water-surface displacement:

After the sloshing wave period is determined, the horizontal seismic wave acceleration Ac can be obtained from the provided response spectrum. The sloshing wave height is calculated based on the horizontal seismic wave acceleration as 7.53 D Ac/18=0.418 D Ac, where D is the diameter of tank. The sloshing wave height needs to be determined for the fixed-roof tanks so that enough freeboard is specified for the design of such tanks.

6. Hoop Stress in Tank Wall:

The hydrodynamic hoop stress shall be combined with the hydrostatic stress in determining the total hoop stress based on the approach shown below--square root of sum of squares. The calculated stress in the tank wall shall not exceed the allowable design stress. The allowable stress must be reduced by the applicable joint efficiency, then increased by one-third for seismic allowable stress.

$$\sigma_s = \frac{\sqrt{N_i^2 + N_c^2 + (N_h a_v)^2}}{t}$$

Where:

σ_s = hydrodynamic hoop stress
Ni = impulsive hoop force
Nc = convective hoop force
Nh = hydrostatic force
t = thickness of the shell ring under consideration

In order to make the design process more efficient, a program was developed based on the AWWA D-100 standards, with my research work incorporated in it. The program is both for design of new tanks and for seismic evaluation of existing tanks.

7. Program development to facilitate tank design:

A program was developed for both tank design and seismic evaluation of existing tanks. Both the AWWA D-100 Standard Method and the Response Spectrum Method are included in the program. It is used to determine the wave height, overturning moment, anchorage requirement, and shell buckling status especially near the bottom of unanchored tanks during earthquakes. For the Response Spectrum approach, the means of calculating the fundamental horizontal period, vertical period and the liquid sloshing period are built in the program. Below is a selected portion of the program output on a 26-m (85-ft) diameter, 7.3-m (24-ft) high steel tank.

Seismic Analysis			Analysis Type:	Response Spectrum Method
AWWA Seismic Design Parameters				
Ratio D/H =	3.65			
Circular Frequency w =	1.03			
T_w =	6.08			
C_1 =	0.020			
Water Mass W_1/W_w =	0.315	W_1 =	11574	KN
Water Mass $W2/W_w$ =	0.635	W_2 =	23309	KN
Centroid Height X_1/H =	0.375	X_1 =	2.7	m
Centroid Height X_2/H =	0.539	X_2 =	3.8	m
Response Spectrum Method				
Tank Natural Period:				
Horiz., Tank and Water Impulse, T_i:	0.154	sec	(See next sheet for detail calculation)	
Vert., Tank and Water, T_v:	0.119	sec	(See next sheet for detail calculation)	
Horiz., Water Wave, T_c:	6.08	sec		
Horizontal and Vertical Accelerations:				
Horiz. Seismic Acceleration, A_i in g:	2.10			
Vert. Seismic Acceleration, A_v in g:	1.97			
Horiz. Seismic wave Accel., A_c in g:	0.03			

Site Reduction Factor, R_f	4.5	Use 2.5 per 96 AWWA for ground motion with a mean recurrence interval of 10,000 years

Standard Method

Zone Factor, Z	0.4
Importance Factor, I	1.25
Soil Factor, S	1.2
Force Reduction Factor, R_w	3.5 — When Unanchored (Use 4.5 When Anchored)

Accelerations to be used:

	Response Spectrum	Standard Method	Accel. to use
Horiz. Seismic Acceleration, in g	0.47	0.36	0.47
Vert. Seismic Acceleration, in g	0.44	0.24	0.44
Horiz. Seismic wave Accel., in g	0.01	0.06	0.06

Fundamental Frequency/Period of Tank-Liquid System

Horizontal Impulsive Period of Tank:

Weighted-average Tank Shell Thick, t:	0.313	in
Ratio t/R =	0.00061	
Ratio H/R =	0.55	
Coefficient C_L =	0.0577	
Fund. Nat. Frequency for Roofless Tank, f_0:	6.53	cps (For Roofless Tank)
Frequency from Roof Mass Effect, f_F:	283.17	cps (On Cantilever Flexural Beam)
Frequency from Roof Mass Effect, f_S:	57.26	cps (On Cantilever Shear Beam)
Freq. of Tank-Liquid System w/ Roof, f_0':	6.48	cps
Fund. Impulsive Natural Period, T_i:	**0.154**	**sec**

Vertical Period:

Coefficient, C_V:	0.0740	
Vert. Fund. Frequency, f_V:	8.37	cps
Vert Fund. Period, T_V:	**0.119**	**sec**

Note: This method is as described in the publication of American Society of Civil Engineers, *Guidelines for the Seismic Design of Oil and Gas Pipeline Systems*, 1984, Sections 7.5 and 7.7.

Part II. Concrete Storage Tanks

The design of conventional rectangular and circular concrete tanks is based on ACI 350 for environmental engineering concrete structure, and the PCA publications "Rectangular Concrete Tanks" and "Circular Concrete Tanks without Prestressing". Based on ACI 350, reinforced concrete tank walls that are in contact with liquids should have a minimum thickness of 305 mm (12 inch). For crack control, it is preferable to use a large number of small-diameter bars for min reinforcement rather than an equal area of larger bars. Temperature and shrinkage reinforcement should be spaced not greater than 305 mm (12 inch) on center, divided equally between the two surfaces of the concrete section.

For conventional circular concrete tanks (without prestressing or post-tensioning), the wall thickness should be sufficient to keep the concrete from cracking, but if the concrete does crack, the ring steel must be able to carry all the tension alone. On the analysis of a tank, it is assumed that the tank is either fixed or hinged at base. The assumption is not generally in conformity with the actual conditions of restraint. Adjustments need to be made in accordance with the judgment of the engineer.

The actual condition of restraint at a wall footing is between fixed and hinged, but generally closer to hinged. Usually it is difficult to predict the behavior of the subgrade and its effect upon the restraint at the base. Therefore it is more reasonable to assume that the base is hinged than fixed, and the hinged-base assumption gives a conservative although not wasteful design. Therefore it is recommended that the tank wall have a hinged base in the analysis.

For liquid storage tanks, the pressure on the tank wall is a triangular load. For tanks used for storage of gasoline, the pressure on the tank wall is a combination of the pressure due to the weight of the liquid plus a uniformly distributed loading due to the vapor pressure. The combined pressure has a trapezoidal distribution with a triangular element due to liquid weight and a rectangular element due to vapor pressure.

For example, there is a 12.2 m (40 ft) diameter, 8.5 m (28 ft) high digester tank with a thickness of 305 mm (12 inch). The concrete strength is 28 MPa (4000 psi), the weight of liquid is 9806 N/m^3 (62.5 lb/ft^3), and the vapor pressure is 165 KPa (288 psf). Figure 6 shows the variation of hoop tension force along the height of tank. The bending moment curve is plotted along the height of tank wall (Figure 7) so that flexural design may be performed for the maximum controlling values.

Conclusions

A lateral push analysis shall be implemented with a pseudo-dynamic hydrodynamic pressure. The response quantities include the maximum values of the hydrodynamic circumferential, shearing and axial stresses in the tank wall, as well as the buckling status of tank shell at any height. Such approach has advantage over the Standard Method in the AWWA D-100 standard. The hoop tension force, overturning moment and shear force at any height of the shell can be determined through analysis, thus it is more systematic and consistent on the design. In addition, sloshing wave height needs to be determined for the fixed-roof tanks so that enough freeboard is specified for the design of such tanks. Furthermore, both horizontal and vertical components of ground motions shall be considered for the seismic design of new tanks or evaluation of existing tanks.

For large or high-importance-factor tanks, it is essential that the site-specific earthquake response spectrum curves be provided. Thus the earthquake horizontal acceleration, the vertical acceleration, and the horizontal seismic wave acceleration can be determined more accurately based on the fundamental periods of the tank system. Furthermore, a 3-D nonlinear finite element analysis is recommended for the tank system including the contained fluid and the foundation soil, when the resources are possible. By doing this type of analysis, the fluid-structure interaction and the soil-structure interaction can be considered in the nonlinear analysis. Therefore the results are accurate enough to assure a ultimate design of critical lifeline structures-- liquid storage tanks.

References

American Society of Civil Engineers (1984) *Guidelines for the Seismic Design of Oil and Gas Pipeline System*, 313-318, 339-342.

M. A. Haroun, Y. Wang, W. Sbou-Izzeddine and N. Mode (1995). "*Large Amplitude Settlements of Oil Storage Tanks.*" Proceedings of the Third International Conference on Recent Advances in Geotechnical Earthquake Engineering and Soil Dynamics, Vol. III, St. Louis, Missouri, April 1995, pp. 1223-1226.

M.A. Haroun, Y. Wang and M. Martinez (1996). "*Computational Analysis of Large-Amplitude Settlement and Shell Ovalization of Oil Tanks.*" Advances in Computational Techniques for Structural Engineering, Third International Conference on Computational Structures Technology, Budapest, Hungary, August 1996, pp. 165-170.

American Water Works Association (AWWA). (1996). *AWWA Standards for Welded Steel Tanks for Water Storage*, ANSI/AWWA D100-96

American Concrete Institute, *Environmental Engineering Concrete Structures*, ACI 350.

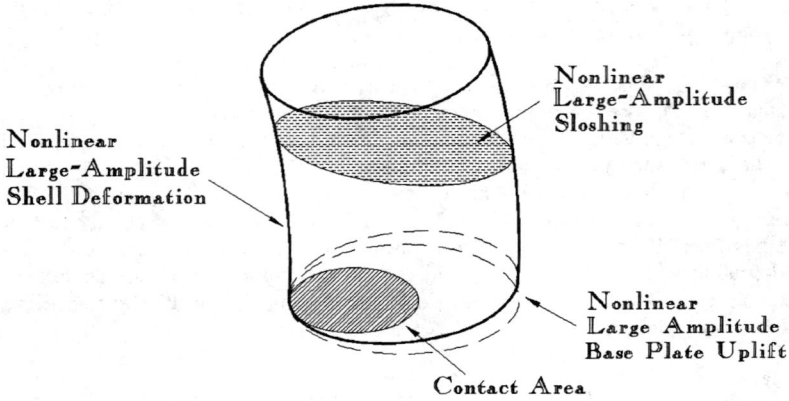

Figure 1. Nonlinear Complex Behavior of Liquid Storage Tanks

Column Supported Cone Roof Tank

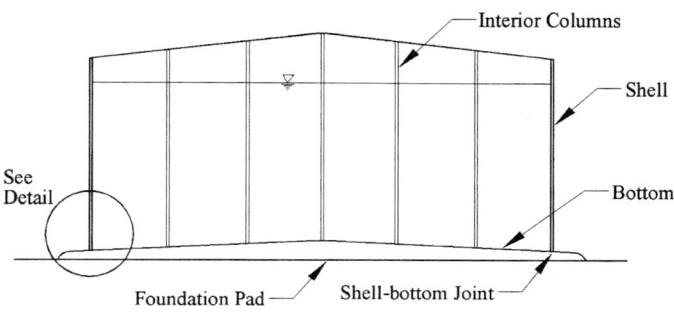

Floating Roof Tank

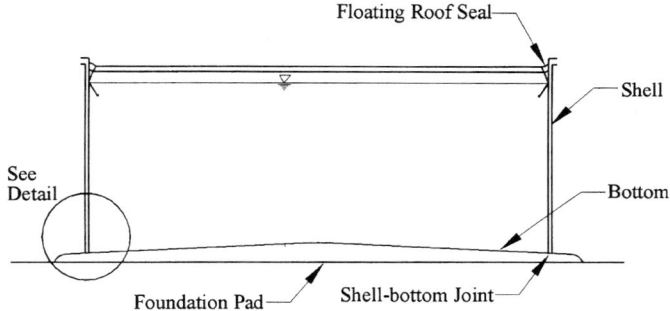

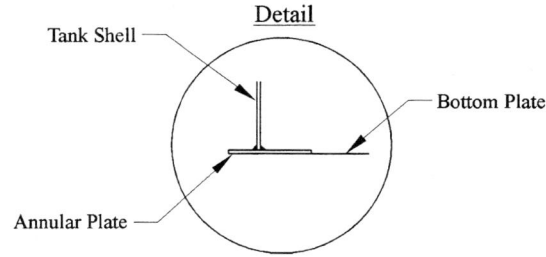

Figure 2. Fixed-roof and Floating-roof Tanks

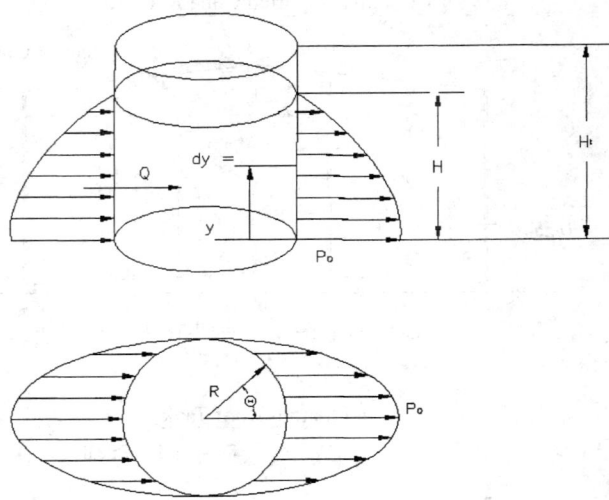

Figure 3. Pseudo-Dynamic Loads Applied on Tank Wall

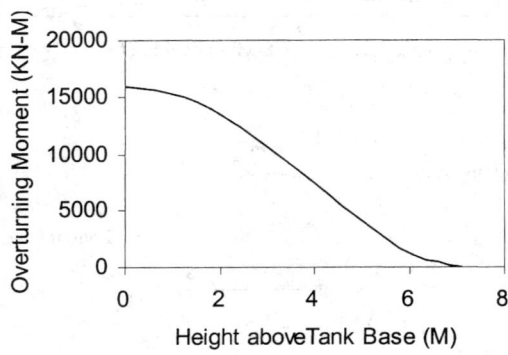

Figure 4. Variation of Overturning Moment along Tank Wall due to Lateral Earthquake Load

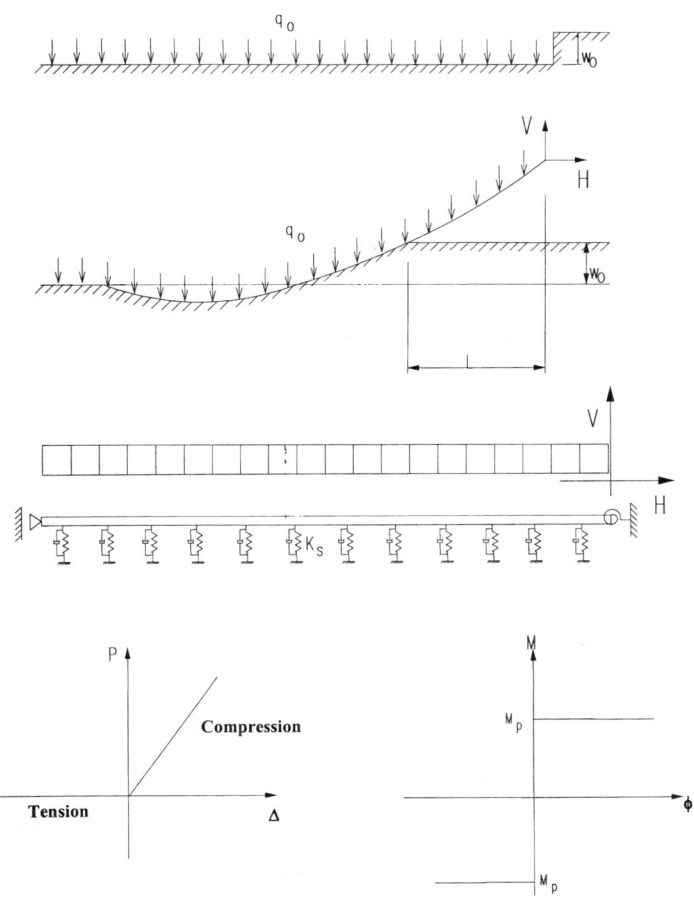

Figure 5. Simplified Nonlinear Computer Model of Tank System

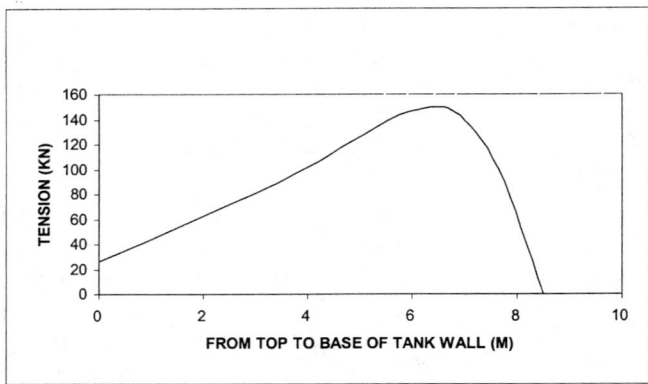

Figure 6. Variation of Ring Tension of Tank Wall with Hinged Base (with vapor pressure)

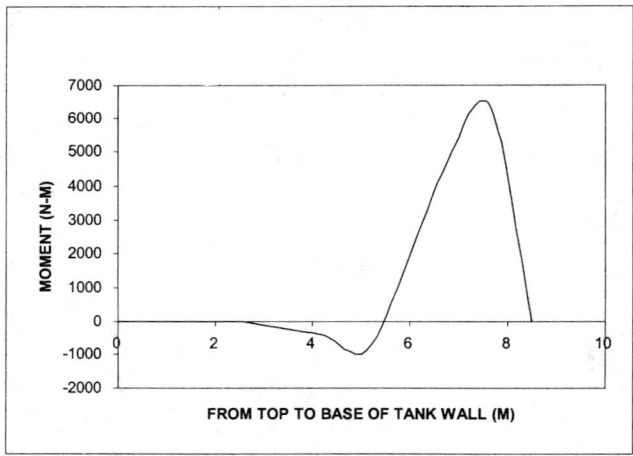

Figure 7. Variation of Bending Moment of Tank Wall

Simplified models for flexibly supported liquid storage tanks and its application to Eurocode 8, Part 4

J. Habenberger[1] and J. Schwarz[2]

Abstract

The following paper treats the influence of the foundation flexibility on the soil-structure-interaction of liquid storage tanks. Impedance functions are derived for a circular plate resting on the elastic half-space impacted with the seimic loading caused by the liquid and the tank wall. The impedance functions are implemented into a lumped mass model representing the soil-tank-liquid-system. The reliability and capability of an improved engineering approach are verified with more refined finite element calculations using special boundary conditions to incorporate the energy absorbing properties of the infinite soil. From the frequency dependend behaviour of the lumped mass model dynamic parameters are derived which might be incorporated into seismic codes.

Introduction

The dynamic behaviour of liquid storage tanks is mainly affected by the interaction of the flexible tank wall with the contained liquid and by the interaction of the tank with the supporting soil. Considering damages and failures of tanks during recent earthquakes the question arises whether design practice and guidelines suggested in seismic codes are appropriate.
The influence of the flexibility of the tank wall on its interaction with the contained liquid was the purpose of comprehensive investigations in previous works (Habenberger et al., 2001; Schwarz et al., 1998; Wunderlich et al., 2000b). Veletsos and Tang proposed in (Veletsos and Tang, 1992) a simplified method to examine the soil-tank-interaction. They assumed that the tank rests on the half-space with a rigid circular base mat. However, the tank base is often a plate with low bending stiffness. These flexibility can influence the dynamic behaviour of the soil-tank-liquid-system significantly.

Method

To investigate the influence of the foundation flexibility on the seismic response of tanks the mechanical model developed by (Veletsos and Tang, 1992) is used. This model is based on the substructure method. Subsequently, the tank-liquid-soil-system is divided into two independent subsystems: the foundation-soil-system and the tank-liquid-system. The foundation-soil-system can be modeled by spring-dashpot-elements (Lysmer, 1965). According to (Haroun and Housner, 1981) it is possible to idealize the tank-liquid-system by single-degree-of-freedom oscillators (SDOF-systems). The coupling between the two

[1] Seismotec GmbH, Weimar, Germany, email: habenber@bauing.uni-weimar.de
[2] Bauhaus-Universität Weimar, Germany, email: jochen.schwarz@uni-weimar.de

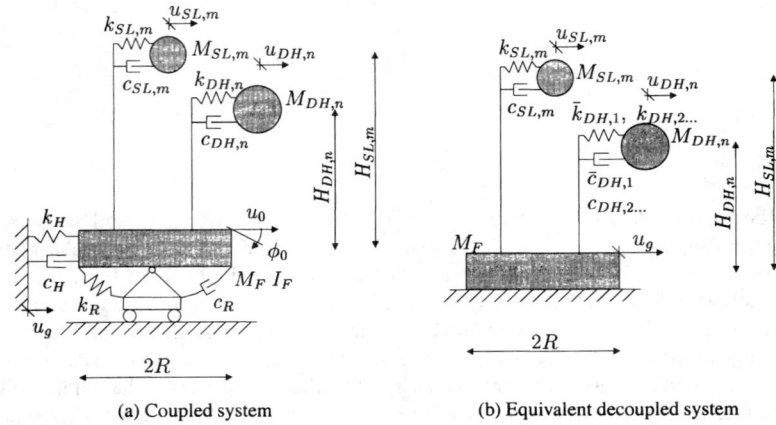

Figure 1. Mechanical models representing the soil-tank-liquid-system.

subsystems is provided by interactive forces having the equal amplitudes and opposite direction of action. This leeds to a mechanical model of the coupled system which is illustrated in Figure 1(a). The spring-dashpot-elements (k_H, c_H, k_R, c_R) are frequency dependent. Therefore, it may be appropriate to perform the analysis in the frequency domain.

Tank-liquid-subsystem. The hydrodynamic pressure acting on the tank wall and the tank base can be discribed by a model of two components:

- the impulsive pressure p_D due to rigid motion of the tank and vibration of the tank wall (denoted by index "DH")

- the convective pressure p_{SL} due to the sloshing of the free liquid surface (denoted by the index "SL")

For each mode of the pressure components a SDOF-system can be specified (see Figure 1(a)). The masses $M_{DH,n}$ and $M_{SL,m}$ of the modes n and m subjected to the corresponding spectral accelerations are aquivalent expressions of the resultants of the activated hydrodynamic pressure components. In addition to the lateral forces overturning moments $MM_{DH,n}$ and $MM_{SL,m}$ resulting from pressures on the tank wall and the tank base can be determined. They are represented by their heights $H_{DH,n}$ and $H_{SL,m}$ of the accelerated liquid masses in the equivalent SDOF-system (Figure 1(a)).
The dynamic behaviour of the SDOF-system depends on the natural circular frequencies ($\omega_{DH,n} = \sqrt{\frac{k_{DH,n}}{M_{DH,n}}}$, $\omega_{SL,m} = \sqrt{\frac{k_{SL,m}}{M_{SL,m}}}$) and the damping values $c_{DH,n}$ and $c_{SL,m}$ for the impulsive and convective components. Assuming an ideal fluid with small surface

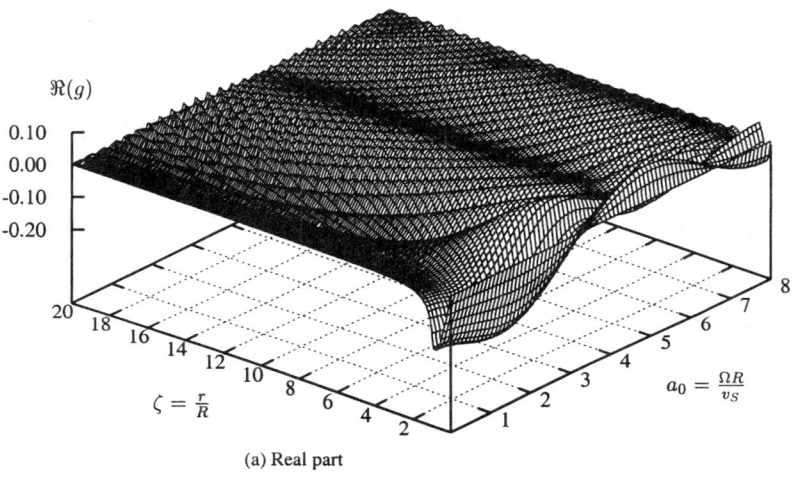

(a) Real part

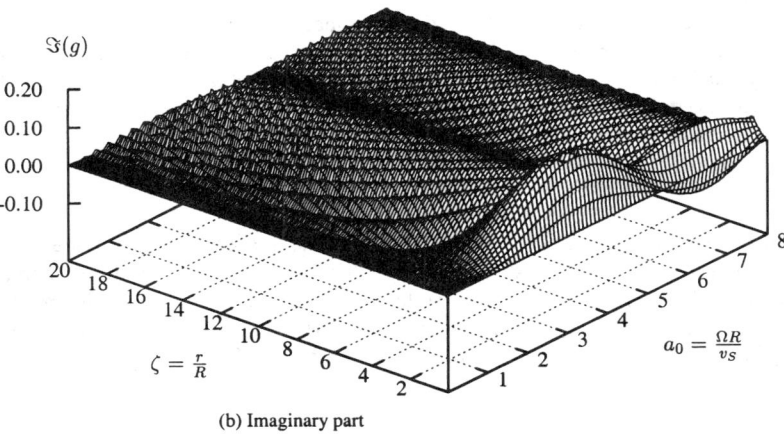

(b) Imaginary part

Figure 2. Flexibility function of elastic half-space under an antimetric harmonic circular load.

displacements and uncoupled convective and impulsive components an analytical solution for the impulsive and convective components can be taken from (Schwarz et al., 1998; Habenberger et al., 2001). The equivalent masses and heights of the tank-liquid-system are given by (Habenberger et al., 2001; Schwarz et al., 1998; Wunderlich et al., 2000b).

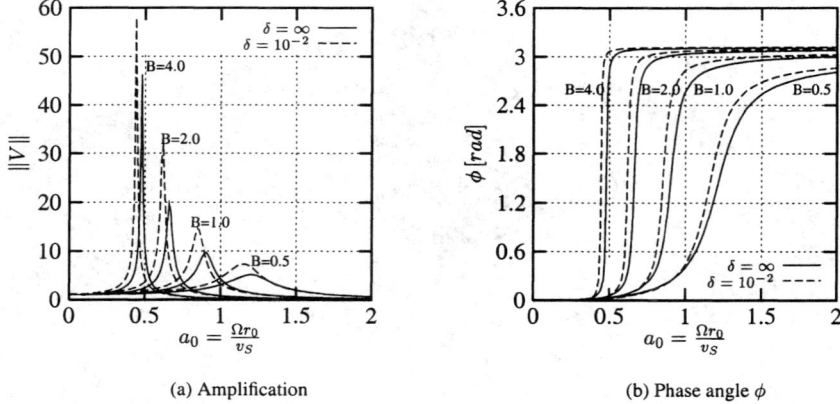

(a) Amplification (b) Phase angle ϕ

Figure 3. Transfer function of a circular plate subjected to an antimetric harmonic ring load supported by the elastic half-space.

Soil-foundation-subsystem. The impedance function discribes the dynamic behaviour of the soil-foundation-system subjected to a harmonic loading. According to (Liou and Huang, 1974) the impedance function of the horizontal force-displacement-relation is nearly independent off the foundation flexibility. Therefore, only the rocking impedance function has to be considered. The tank foundation is assumed to be a circular plate with the normalized stiffness parameter δ:

$$\delta = \frac{K}{GR^3} \qquad (1)$$

with K-bending stiffness; R-radius of the base plate; G-shear modulus of the soil. The loading of the plate can be assumed as a ring load caused by the overturning moment of the tank (MM_{wall}) and an antimetric vertical area load caused by the liquid pressure on the tank bottom (MM_{bottom}). The ratio of the moments (ϵ) on the circular plate from the ring load and the area load depend on the slenderness (α) of the tank:

$$\epsilon(\alpha) = \frac{MM_{wall}}{MM_{bottom}} \qquad (2)$$

To determine the rocking impedance function of a flexible circular plate the ring method developed by (Lysmer, 1965) is used. The inertia mass of the plate is neglected. For the application of the ring method the flexibility function of the unconstraint half-space under an antimetric harmonic circular load has to be known. The real and the imaginary part of this function are shown in Figure 2. For the rotation of the base plate ψ the vertical displacement $w(R)$ below the bottom of the tank wall is used. This is assumed to be decisively for the response of the whole tank-liquid-system. The displacements are small so that the rotation can be determined by Eq. 3:

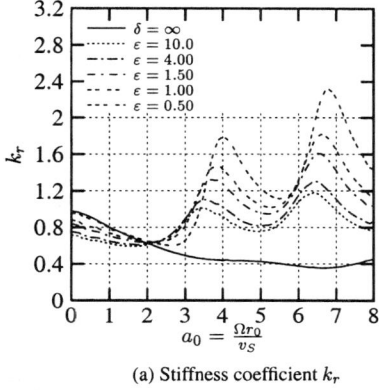 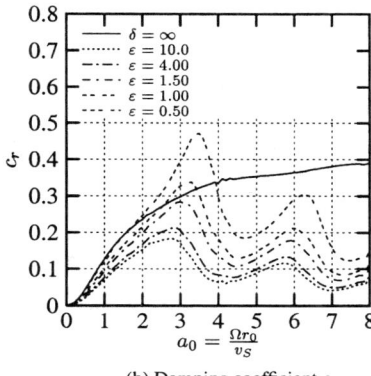

(a) Stiffness coefficient k_r (b) Damping coefficient c_r

Figure 4. Impedance function for storage tank foundation with $\delta = 10^{-6}$.

$$\tan(\psi) \approx \psi = \frac{w(R)}{R} \tag{3}$$

The relationship between rotation and rocking moment (the impedance function) can be given with the following equation:

$$MM = K_R \left[k_R + i a_0 c_r \right] \psi \tag{4}$$

with MM-total overturning moment; K_R-static stiffness of the rigid plate on the half-space; k_r, c_r-stiffness and damping parameters of the impedance function; $a_0 = (\Omega R)/v_s$-normalized frequency with the annular exciting frequency Ω; v_s-shear wave velocity of the soil.

If the inertia moment I_F of the plate is considered, a resulting transfer function of the foundation-soil-system can be elaborated. In this case the Eq. 4 takes the following form:

$$I_F \ddot{\psi} + c_r a_0 K_R \dot{\psi} k_r K_R \psi = MM e^{i\Omega t} \tag{5}$$

Fig. 3 shows the amplitude A and the phase angle ϕ of the transfer function for different normalized inertia moments:

$$B = \frac{3(1-\nu_s)}{8} \frac{I_F}{\varrho_s R^5} \tag{6}$$

with ν_s - Poisson ratio of the soil, ϱ_s - density of the soil and a normalized stiffness of $\delta = 10^{-2}$ under a ring load on the edge of the plate.

The increase of the flexibility of the plate leads to a decrease of the resonance frequency and of the damping of the soil-foundation-system. Impedance functions for different ratios ϵ and a normalized plate stiffness of $\delta = 10^{-2}$ are given by Figure 4.

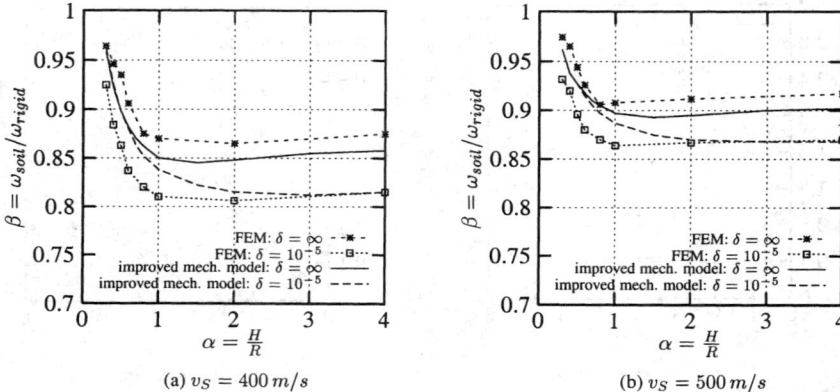

Figure 5. Ratio of resonance frequencies: $\omega_{soil}/\omega_{rigid}$ (*soil*=flexible system, *rigid*=rigid system).

Coupled soil-tank-liquid-system. Having determined the impedance function the transfer function of the masses of the coupled system according Figure 1(a) can be calculated. From the transfer functions the resonance frequencies and the damping can be evaluated. In Figure 5 the ratios of the resonance frequency of the flexibly supported system to the resonance frequency of the rigidly supported system are shown for a shear wave velocity of the soil $v_S = 400\ m/s$. The results of the simplified approach are compared with those of finite-element-calculations (FEM). From Figure 5 it can be concluded that a flexible foundation leads to a decrease of the resonance frequency of slender tanks ($\alpha \geq 0.8$) [Note: The decrease of the damping is less pronounced and occures only for broad tank ($\alpha \leq 0.5$)].

Equivalent SDOF-system. For engineering application the use of the response spectra method is convenient. Because of the strong damping of the mechanical system according to Figure 1(a) the modal decoupling is complicated. Veletsos and Tang suggested an approximate decoupling using the transfer function of the coupled system (Veletsos and Tang, 1992). From the transfer function the adapted damping value $\bar{c}_{DH}$ and frequency value $\bar{\omega}_{DH}$ for system of Figure 1(b) can be determined.

$$\bar{c}_{DH,1} = \xi_{R,1} 2\omega_{DH,1,rigid} M_{DH} \qquad (7)$$
$$\bar{\omega}_{DH} = \beta \omega_{DH} \qquad (8)$$

Figure 6 gives damping values $\xi_{R,1}$ and frequency ratios β depending on the tank slenderness $\alpha = H/R$ for different shear velocities v_S of the subsoil. The stiffness of the foundation is taken as $\delta = 10^{-6}$.

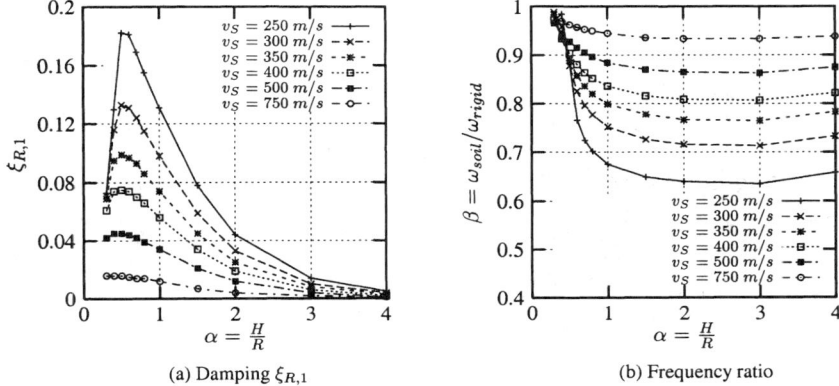

Figure 6. Dynamic coefficients for the 1th mode of the impulsive horizontal vibration of the equivalent system with flexible foundation ($\delta = 10^{-6}$).

Application to Eurocode 8, Part 4. The described SDOF-system can be applied to the current version of the European prenorm (EC 8, Part 4) (prEN 1998-4:200x, 1998) to consider the flexibility of the liquid tank basemat. The overall damping behaviour of the replacement system $\xi_{DH,1}$ has to be determined according to Eq. 9.

$$\xi_{DH,1} = \xi_{R,1} + \xi_S \beta^3 \qquad (9)$$

with the structural damping of the shell ξ_S; $\xi_{R,1}$ and β according to Figure 6. The natural frequency of the replacement system has to be calculated with Eq. 8. These modifications concerns only the appendix 1 to Eurocode 8, Part 4.

Example

To verify the reliability and capability of the developed method comparative calculations with the finite-element-method are carried out (Wunderlich et al., 2000a). The finite-element-model uses special boundary conditions to consider the radiation damping properties of the infinite half-space (Schäpertöns, 1996; Wunderlich et al., 1989). Figure 7 shows the principle finite-element-model. The parameters of the tank, the liquid and the soil used for the example are given by Figure 8. The examined tank was excited by the acceleration time history shown in Figure 9. The finite-element-calculation were carried out in the time domain, whereas the calculation with the simplified approach uses the FFT and DFT techniques in the frequency domain. The time histories of normalized activated liquid masses induced by pressure on the tank bottom and the tank wall indicate only minor differences between FEM and the improved mechanical model. Figure 10 gives the liquid mass M_V resulting from the hydrodynamic pressure acting on the tank bottom

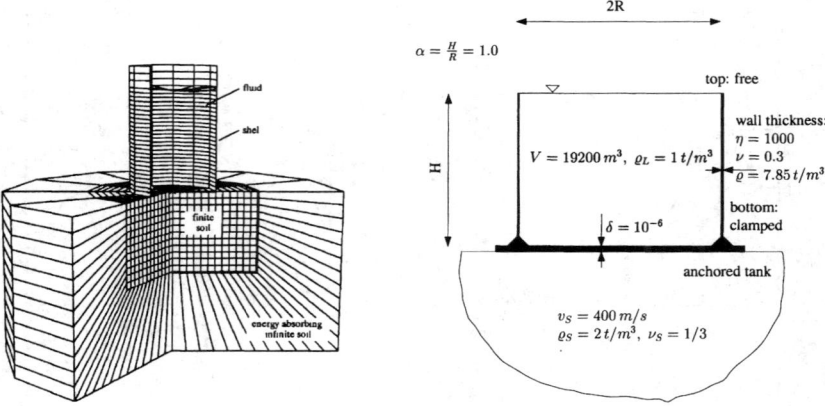

Figure 7. Finite element model (from Seiler 2000).

Figure 8. Geometrie and system of tank example.

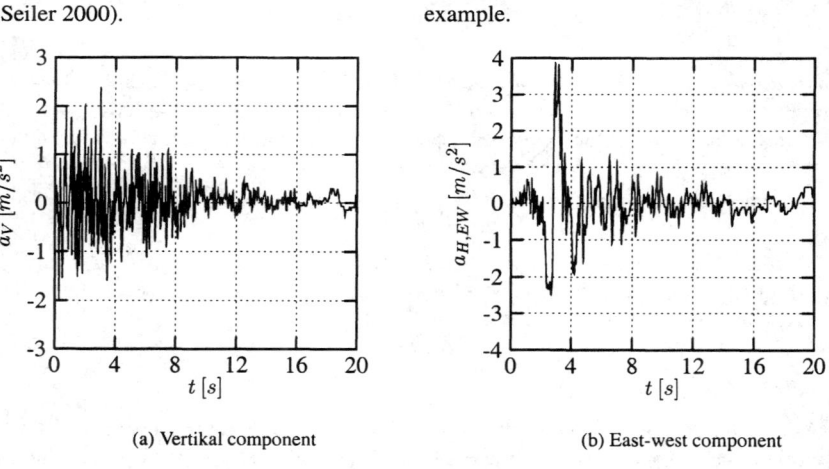

(a) Vertikal component

(b) East-west component

Figure 9. Time histories of the Erzincan earthquake 1992, mainshock (Meteorological Station).

and normalized by the weight of the tank liquid M_L. In Figure 11 the time history of the normalized liquid mass M_H resulting from the pressure on the tank wall is shown.

Conclusions

The ring method suggested by (Lysmer, 1965) was extended to antimetric loading and flexible circular foundations. Special impedance functions for tank foundations were de-

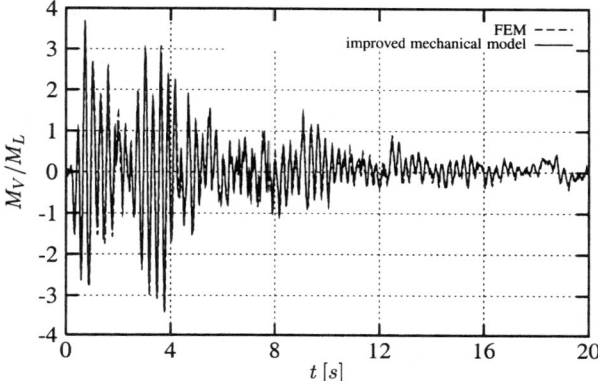

Figure 10. Time history of normalized activated liquid masses induced by pressure on the tank bottom.

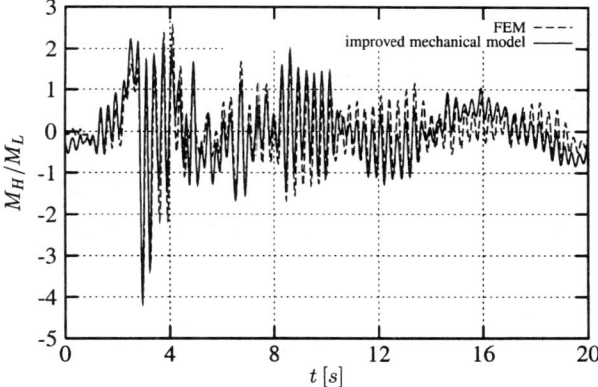

Figure 11. Time histories of normalized activated liquid masses induced by pressure on the tank wall.

rived and implemented into the simplified model proposed by (Veletsos and Tang, 1992). Comprehensive parameter studies were performed. They shows that the consideration of the foundation flexibility leads to a decrease of the resonance frequency of tall tanks ($\alpha \geq 0.8$) and to a decrease of the radiation damping of broad tanks ($\alpha \leq 0.5$). For engineering use damping and frequency values of a decoupled model were derived providing the base for response spectra calculations. The reliability and capability of the suggested engineering approach could be proved by comparison with more refined finite element calculations.

References

Habenberger, J., Schwarz, J., and Trabert, J. (2001). Zur Berechnung von flüssigkeitsgefüllten Tankbauwerken unter seismischen Einwirkungen. *Thesis Wissenschaftliche Zeitschrift der Bauhaus-Universität Weimar*, 47(1).

Haroun, M. and Housner, G. (1981). Earthquake response of deformable liquid storage tanks. *Journal of Applied Mechanics*, 48:411–418.

Liou, G.-S. and Huang, P.-H. (1974). Effect of flexibility on impedance functions for circular foundations. *Journal of Engineering Mechanics*, 120:1429–1446.

Lysmer, J. (1965). *Vertical Motion of Rigid Footings*. PhD thesis, University of Michigan.

prEN 1998-4:200x (1998). Eurocode 8, part 4: Design of structures for earthquake resistance: Silos, tanks and pipelines, Project Team Draft No. 1, June 2002.

Schäpertöns, B. (1996). *Über die Wellenausbreitung im Baugrund und deren Einfluß auf das Tragverhalten von flüssigkeitsgefüllten Behältern*. PhD thesis, TU München.

Schwarz, J., Habenberger, J., Wunderlich, W., and Seiler, C. (1998). Critical loading conditions of anchored liquid storage tanks under earthquake excitation. In A. Pecker, P. P., editor, *11th European Conference on Earthquake Engineering*, Paris, France. Balkema Rotterdam, 1998.

Seiler, C. (2000). *Näherungsmethoden zur Stabilitätsuntersuchung von erdbebenerregten verankerten Flüssigkeitsbehältern*. PhD thesis, TU München.

Veletsos, A. and Tang, Y. (1992). Dynamic response of flexibly supported liquid storage tanks. *Journal of Structural Engineering*, 118(1):264–283.

Wunderlich, W., Seiler, C., Schwarz, J., and Habenberger, J. (2000a). Berechnung von Behälter- und Silotragwerken unter seismischen Einwirkungen. Technical report, TU München, Bauhaus-Universität Weimar. Abschlußbericht zum DFG Forschungsvorhaben Wu 67/24-4 und Ha 1839/1-3.

Wunderlich, W., Seiler, C., Schwarz, J., and Habenberger, J. (2000b). Seismic response and failure mechanism of flexibly supported liquid storage tanks. In *12th World Conference on Earthquake Engineering*, volume 1, Auckland, New Zealand.

Wunderlich, W., Springer, H., and Goebel, W. (1989). Discretization and Solution Techniques for Liquid Filled Shells of Revolution under Dynamic Loading. In Kuhn, G. and Mang, H., editors, *Discretization Methods in Structural Mechanics*, pages 145–155. Springer Verlag.

Developing the First National Code for Gas Lifeline System in Iran: Possibilities and Challenges

Mahmood Hosseini[1]

Abstract

This paper deals with the various aspects of developing the first National Code for Seismic Design of Gas Transmission and Distribution System to be used by the corresponding people of one of the major lifeline systems in a developing country, the National Iranian Gas Company. At first a short history of the code development project and a brief description of the different parts of the developed code, as well as resources and knowledge bases used for developing the code are given. Then the problems to which seismologists and particularly earthquake engineering specialists have encountered during their attempts for developing the code are discussed, and the experienced challenges are highlighted. Finally, it is tried to make a comparison between the process of code developing in developed and developing countries.

Introduction

In many countries, including Iran, natural gas is the main source of heating energy for domestic consumption as well as a large part of the industry. In spite of the wide use of gas on the one hand, and the high seismicity of the country on the other, the major part of gas transmission, purification, and distribution systems in Iran have not been designed thoroughly for earthquake effects. There are three main reasons for this serious shortcoming. The first reason has been the limited knowledge of the authorities in the initial stage of the project. The second reason, which has convinced the National Iranian Gas Company (NIGC) to accept the conventional design and construction methods and procedures, has been the availability of some foreign design and construction standards, for example Japanese ones. Finally, the third and the most important reason has been the lack of a national standard seismic design code applicable to all components of the gas systems.
It should be pointed out that despite the existence of many codes and standards, including seismic design ones in several countries, based on the existing publications, there is no specific code for seismic design of gas systems, even in

---

[1] Associate Professor and Head of Lifeline Engineering Department, International Institute of Earthquake Engineering and Seismology (IIEES), P.O.Box 19395/3913, Tehran, Iran, Phone: +98 21 283 1116-9 and 283 3634, Fax: +98 21 229 9479, Email: hosseini@dena.iiees.ac.ir

developed countries, or if it exists it has not been released internationally. But at least, developed countries possess some codes and standards in various engineering fields, which when combined can provide the design engineers with the required knowledge bases for their needs for almost any specific purpose. The reasons behind making the decision to develop a specific code for seismic design of gas systems in Iran was firstly the occurrence of some destructive earthquakes in recent decades, particularly 1990 Manjil earthquake in northern part of the country, and secondly the establishment of the International Institute of Earthquake Engineering and Seismology (IIEES) in 1989, and specially the founding of IIEES Lifeline Engineering Department in 1991.

A Short History of the Code Development Project

IIEES authorities based on their responsibilities started to develop the knowledge in various branches of earthquake engineering and seismology, including lifeline engineering, and at the same time they tried to drag the attention of the government and officials more and more to the high seismic risk of the county in general, and more specifically to lifeline systems. Following IIEES's warning to the country authorities and decision makers, after few years, in 1994, the NIGC responded with a good and surprising reaction. This company asked IIEES to develop the detailed seismic immunization or security plan for all the existing and future facilities of the company, including transmission, refinement, and distribution systems. Obviously, it was not possible in that time to fulfill all of the NIGC demands, since the definition which the company authorities had in their mind for "Seismic Security" was beyond the seismic safety plan, namely a state in which none of the system components including buildings and facilities get even slight damage in the case of probable earthquake in Tehran. IIEES people, particularly the author of this paper as the Head of Lifeline Engineering Department, tried to convince the NIGC authorities that what they can expect with regard to earthquake is not "seismic security", but it is some provisions for seismic design of future facilities as well as some remedies for upgrading the existing facilities. Finally, NIGC and IIEES came to this agreement that IIEES prepares a seismic design code for gas systems and at the same time provides the possible remedies for the existing facilities.

The author of the paper was nominated as the manager of the "code development project", and started the task in early 1995. In this way, beside the possibilities found out for performing the job, many difficulties showed up which postponed some stages of the project. The main objective of this paper is to investigate the causes of the problems, and especially the difficulties, in which a group of seismology and earthquake engineering specialists have been involved for developing the desired code as the first National Code for a lifeline system in a developing country as Iran. The project was divided to various phases, including:

- Studying the existing codes and guidelines with regard to seismic design of gas systems and other similar facilities
- Considering an appropriate format for the code and its sections
- Dividing the project to some main parts based on the code organization and assigning the required teams for performing the corresponding task

- Gathering and classifying the existing materials for completion of the code body
- Selecting the proper materials to be included in the code
- Modifying the selected materials for the code format and adjusting them to conditions of the country
- Preparing new materials for the parts on which there has not been any published material

In the final stages of the project, some other issues were realized to be useful as parts of the code, which are related basically to the maintenance of the gas systems for keeping them prepared for future earthquakes. In the following sections of the paper various stages of code development project and related issues are discussed. At first, various aforementioned phases of the code development process are briefly explained, the specialties of the people who worked on various sections, the time they put on the job, and the resources they used are mentioned. Then, the difficulties which the responsible people working on every section encountered, and the way they overcame those difficulties or the reasons of their failure are explained. Finally, a comparison is made between the case at hand and those in the developed countries.

Studying the Existing Codes and Guidelines

The history of code development for Gas systems goes back to early 80s. One of the first publications in this regard was concentrated on "high pressure gas manufacturing facilities" (High Pressure Gas and Explosive Safety Council, 1981). That publication was in fact the draft of anti-earthquake design code prepared by a sub-committee, and the final version of that code seems not to be released internationally. However, that document was not realized to be helpful for the desired code development.

Another available document in this regard was a useful book published by ASCE (1984), whose purpose was to provide guidance for the design of most major components of oil and gas pipeline systems. The guidelines have been developed over a three-year period by the Technical Council on Lifeline Earthquake Engineering Gas and Liquid Fuel Lifelines Committee. That book was studied thoroughly, and was evaluated as a very well organized guidelines book for oil systems, particularly pipelines and oil tanks, but some shortcomings were realized in it, especially for gas facilities. Therefore, some selected parts of the book were modified to make them usable for developing the code for gas lifeline systems.

More than one decade passed till a specific monograph was published by ASCE on gas systems (McDonough, 1995), but it was only with regard to gas distributors and not the whole system. Furthermore, it has been mentioned in the monograph that the contents of that publication are not intended to be and should not be construed to be a standard of the American Society of Civil Engineers and are not intended for use as a reference in purchase specifications, contracts, regulations, statutes, or any other legal document. Although that publication could be a very good base for an important part of the desired code, regarding the content of the monograph and the aforementioned statement using that document also needed several additions and modifications.

The aforementioned documents were used to make some preliminary format or structure for the desired code, but there were no more code, guidelines book, monograph, or standard available to be used as a complete basis for the desired code; therefore, to have a more appropriate and applicable text, and also to make it as much updated as possible it was necessary to get more materials, and prepare some new materials in some cases.

Considering an Appropriate Format

As the second phase of the code development it was necessary to prepare a format or structure for the code and its sections. By modifying and completing the format of the aforementioned documents the following structure was finally considered:
Section 1- General Considerations
    1-1-   Introduction
    1-2-   Definitions
    1-3-   The scope of the code
    1-4-   General provisions
Section 2- Seismic Hazard Studies
    2-1-   Introduction
    2-2-   Faulting hazard
    2-3-   Strong ground motion parameters
Section 3- Geotechnical Studies
    3-1-   Introduction
    3-2-   Site effects
    3-3-   Landslide hazard estimation
    3-4-   Liquefaction hazard estimation
    3-5-   Subsidence hazard estimation
Section 4- Seismic Loading and Analyses
    4-1-   Introduction
    4-2-   Analyses of buried pipes
    4-3-   Analyses of aboveground pipes
    4-4-   Seismic analyses of concentrated buildings and facilities
    4-5-   Seismic analyses of supporting structures
Section 5- Seismic Design of Pipes, Buildings, and Concentrated Facilities
    5-1-   Design of buried pipes for wave propagation effects
    5-2-   Design of buried pipes for fault crossing
    5-3-   Design of buried pipes for large ground deformation
    5-4-   Design of aboveground pipes
    5-5-   Design of buildings and concentrated facilities
    5-6-   Design of distribution network
    5-7-   Design of supportive structures
    5-8-   Monitoring and maintenance provisions
Section 6- Disaster Management Planning
    6-1-   Introduction
    6-2-   Prediction of emergency conditions
    6-3-   Training specific teams for emergency reaction
    6-4-   Detailed program for disaster management
    6-5-   Planning for repair and restoration after earthquake

Assigning the Code Development Teams

The third phase of the project was making decision on the specialists who should prepare various parts of the code based on the accepted format. The available expertises in the country in this case include:
- Seismology and engineering seismology for section 2
- Geotechnical earthquake engineering for section 3
- Lifeline earthquake engineering for sections 1, 4 and 5, and
- Crisis management for section 6

The number of specialists in above categories was respectively two, five, three, and one. The time spent by each of the above categories to prepare the corresponding section(s), was respectively one and a half, three, four, and one year(s).

Gathering and Classifying the Existing Materials

As mentioned before, the obtained materials from the available codes and guidelines were not quite enough, so to have a more appropriate and applicable provisions, and to make the code as much updated as possible it was necessary to get more materials, and prepare some new materials in some cases. These include:
a) Seismic hazard analysis for lifeline systems particularly gas system facilities,
b) Defining the seismic analyses inputs considering the specific configurations and characteristics of the gas system,
c) The methods for seismic analysis of various components of the gas system, and
d) The seismic design criteria for gas systems and its components.

The number of publications on gas systems is not so large except for the LNG tanks. In fact, there are more than fifty papers or reports with regard to LNG tanks, while there are a little more than twenty publications with regard to all other parts of the gas system. The main publications with regard to the above subjects, excluding LNG related ones, are listed hereinafter, and just one paper is mentioned with regard to LNG tanks as a sample. The selected publications are sorted here based on their date of publication:
- Seismic design of LNG storage facilities--current Canadian practice (Charlwood et al., 1983)
- Recommended practice for earthquake resistant design of medium and low pressure gas pipelines (Saito et al., 1983)
- Recommended practice for earthquake resistant design of high pressure gas pipelines (Toki, K.; et al., 1983)
- Application of Risk Analysis to Offshore Oil and Gas Operations--Proceedings of an International Workshop (Yokel, F. Y.; Simiu, 1985)
- Development of seismic anchorage guidelines for nuclear plant equipment (Czarnecki, R. M.; Sliter, 1988)
- Recommended practice for earthquake resistant design of gas pipelines considering liquefaction (Nakane, 1989)
- Computerized databases for earthquake hazard analysis of gas and electric systems (McLaren et al., 1992)

- Seismic analysis of a regulator in Memphis gas distribution systems (Lu et al., 1993)
- Guidelines for disaster information systems for important infra-structures (Kawashima, 1993)
- Systems analysis for Memphis Light, Gas and Water (Shinozuka, 1994)
- Seismic design methods for oil and gas transmission pipelines: a comparative study (Zarea et al., 1995)
- Seismic design for oil and gas pipeline in north-west of China (Chen, 1998)
- Seismic risk analysis of liquid fuel systems: a conceptual and procedural framework for guidelines development (Shinozuka, Masanobu; Eguchi, 1998)
- Summary report on new regulatory codes and criteria and guidelines: IAEA guidelines for WWER reactor coolant system integrity assessment (Havel, 2000)
- Seismic code development for civil infrastructures after the 1995 Hyogoken-nanbu (Kobe) earthquake (Hamada, 2000)

Selecting and modifying the materials

For a code to be followed and used effectively by the NIGC design engineers, it was necessary to provide easy-to-use formulas or procedures, which need minimum computer usage and calculation time. So, it was realized that the materials should be selected and modified to develop some simplified methods and techniques which cover the following topics based on the considered format of the code:

- **Section 1**: Definition of technical terms
- **Section 2**: Identifying the existing faults locations, their faulting zone territories dimensions, and their geometry and movements values; Predicting the values of acceleration, velocity and displacement on concentrated and extended sites bedrock
- **Section 3**: Path selection for pipelines; Bed rock motion; Alluvia classification; Soil identification logging program; Slope stability analyses; Soil stabilizing techniques; Liquefaction potential prediction; Soil modification techniques; Land subsidence evaluation: Identification of karstic areas and site modification
- **Section 4**: Buried pipes response to wave propagation (equivalent static, and dynamic methods); Buried pipes response to fault movement, landslide, and liquefaction; Aboveground pipes response to wave propagation (equivalent static, and dynamic methods); Aboveground pipes response to fault movement, landslide, and liquefaction; Response analyses of LNG tanks
- **Section 5**: Design for wave propagation: straight segment, bents, and tees or forks; Design for fault movement, landslide, and liquefaction; Design of buildings which house the crucial facilities; Design of LNG tanks; Complementary design of distribution network (location of network discretization valves and shut-off valves)
- **Section 6**: Emergency situation prediction (disaster scenario); Training the specific emergency teams; Disaster management planning; Repair and restoration planning.

Materials were selected form the publications reviewed in the previous sections and then the required modifications were performed.

Preparing new materials

In spite of the existence of several publications, still considering the specific situation and conditions of the country, there were some areas in which more investigations were required to make the code more usable and practical for NIGC engineers. The main subjects which were realized to need more studies include:
- Seismic hazard estimation for lifeline systems including gas facilities based on near-field earthquakes
- Prediction of faulting zones dimensions
- Dynamic analysis of pipes subjected to wave propagation
- Design of bent and tees for wave propagation effects
- Design of pipes for landslide
- Disaster management plan for gas system authorities

Several research projects were started by IIEES faculties in response to the above needs, of which some are still ongoing.

Encountered Problems in Developing the Code

As IIEES experts were trying to prepare the required materials for the code, they encountered several problems, which needed to be solved to achieve the final goal of the project. These challenges can be listed as follow:

• Very low number of experts: As mentioned before, number of specialist who worked on the code was totally twelve. Obviously this number of experts is too low for developing a national code. In fact the lack of enough specialists in this field has been one of the main difficulties of the country. Presently, there are almost 70 experts in seismology and earthquake engineering in the whole country, which is really few for a seismic country as Iran. Out of this number twenty-five belong to IIEES, however there are just less than ten lifeline earthquake engineering specialists in the country, of them three belong to IIEES.

• Lack of effective contribution of NIGC experts in developing the code: In fact, because of the low knowledge of earthquake engineering in NIGC technical staff, it was not possible for them to have a remarkable role in the project. This lack of knowledge relates somehow to the fact that they have not experienced any major earthquake in the gas system of the country. The other reason could be the little attention which has been generally paid to research works in the country. Fortunately, this shortcoming is being resolved these years, as the sustainable development has become the major goal of the government.

• Lack of technical knowledge on some subjects, which was required for the code materials: This problem is partially due to the point that in developed countries, most of the urban utilities are owned by private parties, so they try to keep their technical knowledge confidential, which is somehow necessary for surviving in a competitive environment. However, as mentioned in the previous section, it was necessary to start some new research subprojects to find the required solutions or

know-how in some parts of the project, of which some were accomplished before the main project deadline (which was extended by NIGC based on the IIEES request), and their results were used in the code, but some are still ongoing.
- Delay in approving the code to be put in practice: There are two reasons for this problem. One is the low knowledge of the gas system technical staff. In fact, they are presently busy with upgrading their earthquake knowledge to be able to use the prepared code. The second reason for this delay is the similar lack of knowledge as well as the low attention paid to the earthquake issues in the National Institute of Standards. Fortunately, there is some hope that this institute takes the case into consideration more seriously than before, as the government assigned a great amount of money to seismic upgrading of government facilities as well as rural houses last year, and has doubled that amount this year.

Comparison between Developed and Developing Countries

Considering the encountered problems stated in the previous section, one can realize the basic differences between developed and developing countries in a code development project. These differences and their causes can be listed as follow:
- In developed countries such projects are assigned to several teams, while in developing countries they are assigned to just a few people, because there are just few experts in each field in most of the developing countries.
- In developed countries several people form various organizations cooperate in such cases which are considered national, while in developing countries cooperation can take place very rarely, partially due to lack of expertise (in many cases just one organization has it), and partially because of the badly guided competition, which does not distinct between a national case and a private problem.
- In developed countries there is a very remarkable contribution by the corresponding industries in these cases, while in developing countries they can rarely contribute. This lack of contribution is mainly because of the low earthquake knowledge in the industry bodies, and somehow due to non-originality of most of industries, which results in a kind of ignorance with respect to the problem to which the industrial body may face in the case of a natural disaster like earthquake.
- In developed countries there are original codes in various branches of engineering, which have been developed by their own specialists, while in developing countries engineers of the same branch may employ different design codes for similar design cases. In fact, engineers use the code which they have been taught by their instructors in universities, which itself is in turn dependent on the country from which those instructors have graduated. This non-uniformity of the existing codes which are usually used by the industry bodies, makes the use of any national code problematic for the design engineers, and consequently makes it difficult for the code developers to prepare the code in such a form that all engineers can use it in the same way.
- Finally, in developed countries almost all of the specialists in any field have the same education background, as they have studied in their own countries, while in developing countries the experts in each field may have quite different education background depending on the country from which they have graduated. This discrepancy in educational background makes it difficult to find a common language

for any joint project, particularly a national code development project. Fortunately, as the Ph.D. courses are being developed more and more in Iranian universities, this shortcoming will be hopefully resolved in near future.

Conclusions

The most important difficulty in developing the aforementioned code has been the lack of sufficient technical co-operation between the earthquake experts and the gas system specialists. The know-how gained from this first experience of developing a new code can be used by other researchers and experts, who are responsible for similar tasks, to be more successful in their work. The result of this investigation can be very helpful in developing other future National Codes in developing countries, as these countries are still in the beginning, or at most in the mid-way, of applied researches in the fields of seismology and earthquake engineering. Finally, by paying attention to the difficulties mentioned in this paper, the specialists of developed countries can give better helps to their counterparts in developing countries to prepare better codes and standards.

References

1. American Society of Civil Engineers, "*Guidelines for the seismic design of oil and gas pipeline systems*", Technical Council on Lifeline Earthquake Eng., Committee on Gas and Liquid Fuel Lifelines, New York, 1984, 473 pages.
2. Charlwood, R. G.; Salt, P. E.; and Atkinson, G. M., "Seismic design of LNG storage facilities--current Canadian practice", *Proceedings of the Fourth Canadian Conference on Earthquake Engineering*, Univ. of British Columbia, Vancouver, Canada, 1983, pages 11-18.
3. Chen, Xiangqiu, "Seismic design for oil and gas pipeline in north-west of China", *Proceedings of the Third China-Japan-US Trilateral Symposium on Lifeline Earthquake Engineering*, Kunming, People's Republic of China, Aug. 1998, pages 253-259.
4. Czarnecki, R. M.; Sliter, G. E., "Development of seismic anchorage guidelines for nuclear plant equipment", *Nuclear Eng. and Design*, 107, 1-2, Apr. 1988, pages 27-41.
5. Hamada, Masanori, "Seismic code development for civil infrastructures after the 1995 Hyogoken-nanbu (Kobe) earthquake", *Proceedings of Int'l Workshop on Annual Commemoration of Chi-Chi Earthquake*, National Center for Research on Earthquake Engineering, Taipei, Taiwan, September 18-20, 2000, pages 282-290, Vol. II - Technology Aspect.
6. Havel, Radim, "Summary report on new regulatory codes and criteria and guidelines: IAEA guidelines for WWER reactor coolant system integrity assessment", *Nuclear Eng. and Design*, 196, 1, March 2000, pages 93-100.
7. High Pressure Gas and Explosive Safety Council, Sub-Committee on Design Code for High-Pressure Gas Manufacturing Facilities, Draft of "*Anti-earthquake design code for high-pressure gas manufacturing facilities,*" ERS III-5, Earthquake Research Centre, University of Tokyo, Tokyo, July 1981, 30 pages.

8. Kawashima, Kazuhiko; et al., "Guidelines for disaster information systems for important infra-structures, Technical Memorandum of PWRI 3217, Wind and Seismic Effects", *Proceedings of the 25th Joint Meeting of the U.S.-Japan Cooperative Program in Natural Resources Panel on Wind and Seismic Effects*, Public Works Research Institute, Tsukuba, Japan, 1993, pages 475-486.
9. Lu, C.-H.; et al., "Seismic analysis of a regulator in Memphis gas distribution systems", *PVP, Vol. 256-1, Seismic Engineering: Technology in a Global Society, Pressure Vessels and Piping Conference*, American Society of Mechanical Engineers, New York, 1993, pages 65-70.
10. McDonough, P. W. (ed.),"*Seismic design guide for natural gas distributors*", Technical Council on Lifeline Earthquake Engineering Monograph 9, American Society of Civil Engineers, New York, Aug. 1995, 96 pages.
11. McLaren, M. K.; Savage, W. U.; and Corollo, F. A., "Computerized databases for earthquake hazard analysis of gas and electric systems", *Special Publication 113, Proceedings of the Second Conference on Earthquake Hazards in the Eastern San Francisco Bay Area*, California Div. of Mines and Geology, 1992, pages 491-495.
12. Nakane, Hiroyuki, "Recommended practice for earthquake resistant design of gas pipelines considering liquefaction", *Proceedings of the 3rd U.S.-Japan Workshop on Earthquake Disaster Prevention for Lifeline Systems*, Tsukuba, Japan, 1989, 12 pages, Paper No. J-11.
13. Saito, K.; Nishio, N.; and Katayama, T., "Recommended practice for earthquake resistant design of medium and low pressure gas pipelines", *PVP-77, Earthquake Behavior and Safety of Oil and Gas Storage Facilities, Buried Pipelines and Equipment*, American Society of Mech. Eng., New York, 1983, pages 340-348.
14. Shinozuka, M., "Systems analysis for Memphis Light, Gas and Water", *Research Accomplishments, 1986-1994*, The National Center for Earthquake Engineering Research, State Univ. of New York at Buffalo, Sept. 1994, pages 197-206.
15. Shinozuka, Masanobu; Eguchi, Ronald T., "Seismic risk analysis of liquid fuel systems: a conceptual and procedural framework for guidelines development", *Proceedings of the NEHRP Conference and Workshop on Research on the Northridge, California Earthquake of January 17, 1994*, California Universities for Research in Earthquake Engineering (CUREe), Richmond, California, Vol. III-B, 1998, pages III-789 -- III-796.
16. Toki, K.; et al., "Recommended practice for earthquake resistant design of high pressure gas pipelines", *PVP-77, Earthquake Behavior and Safety of Oil and Gas Storage Facilities, Buried Pipelines and Equipment*, American Society of Mechanical Engineers, New York, 1983, pages 349-356.
17. Yokel, F. Y. and Simiu, E., "Application of Risk Analysis to Offshore Oil and Gas Operations", *Proceedings of an International Workshop, NBS Special Publication 695*, U.S. National Bureau of Standards, Gaithersburg, Maryland, May 1985, 209 pages.
18. Zarea, M.; et al., "Seismic design methods for oil and gas transmission pipelines: a comparative study", *Technical Council on Lifeline Earthquake Engineering Monograph No. 6, Proceedings of the Fourth U.S. Conference on Lifeline Earthquake Engineering*, American Society of Civil Engineers, New York, Aug. 1995, pages 168-175.

The American Lifelines Alliance Approach: Four Years of Progress and Future Directions

Edward M. Laatsch*, Douglas G. Honegger**, and Timothy D. Sheckler***

* Federal Emergency Management Agency, Federal Insurance Administration and Mitigation Directorate, 500 C Street S.W., Washington, D.C. 20472; PH 202-646-3885; Edward.laatsch@fema.gov
** ALA Principal Investigator, D.G. Honegger Consulting, 2690 Shetland Place, Arroyo Grande, CA 93420; PH 805-473-0856; dghconsult@aol.com
*** Federal Emergency Management Agency, Federal Insurance Administration and Mitigation Directorate, 500 C Street S.W., Washington, D.C. 20472; PH 202-646-2834; tim.sheckler @fema.gov

Abstract

The American Lifelines Alliance project (ALA) was initiated in 1998 with support from the Federal Emergency Management Agency (FEMA) with the primary goal of improving the performance of key utility and transportation lifeline systems (electric power, telecommunication, water, waste water, oil, natural gas, rail, and shipping ports) in natural hazards. ALA has not attempted to focus efforts on all lifeline systems but has instead chosen to place priority on specific topics relevant to selected lifeline systems where a need for improved hazard mitigation practices is identified. The goals and strategy adopted by ALA were introduced in a presentation at the 1999 TCLEE conference. Since 1999, ALA has successfully sought out sponsoring organizations to provide direct or in-kind support to ALA planning and project activities and expanded the ALA project team to include individuals representing a broader cross-section of lifelines stakeholders. In addition to FEMA, the number of organizations that have sponsored ALA includes the Department of the Interior, the Department of Transportation, Pacific Gas and Electric Company, Pima County Waste Management, Rohn Industries and the U.S. Geological Survey. In the four years since it's inception, ALA has made significant progress in improving the practice of engineering lifeline systems for natural hazards. Most notably, ALA projects have resulted in new, or modifications to existing, national consensus guidelines and standards from ANSI-approved standards organizations including the American Society of Civil Engineers, American Society of Mechanical Engineers, American Society of Testing and Materials, and American Water Works Association and industry standard organizations including the American Petroleum Institute and the American Railway Engineers and Maintenance-of-Way Association.

In the wake of the terrorist attack of September 11, 2001, the goals of ALA have been expanded to include consideration of man-made hazards as a component of overall hazard reduction efforts. To better address this new goal, the ALA project team has begun reaching out to man-made hazard specialists, which so far includes representation from the Structural Engineers Association of New York and Michael

Baker Jr., a consulting firm with considerable experience in addressing physical security issues for the public and private sector. This paper summarizes the accomplishments of ALA since it's inception and provides an overview of ALA's vision of future objectives and project activity.

Background

The engineering community has long worked to build safe and reliable lifeline systems -- that is, those systems necessary to provide electric power, natural gas, water and wastewater, and transportation facilities and services that are essential to the well being of the community served by these systems. Providing lifeline system function is especially important in assisting rapid recovery following natural hazards. Engineering approaches to limiting damage to lifeline systems from natural hazards have developed specifically for individual natural hazards and individual types of lifeline systems. Thus, the design of electric power transmission systems focused on loads from high wind and ice storms, while the design of natural gas transmission systems focused on landslides and fault crossings.

Using a system-based approach to assessing expected lifeline function is a relatively recent development that was largely driven by the need to prioritize efforts to improve system performance for large earthquakes capable of generating multiple hazards throughout the system. The consequences on a community that experiences simultaneous disruption of multiple lifeline systems were demonstrated during earthquakes in the US, particularly the 1906 and 1933 events in California. The severe effects of these earthquakes spurred the initial development of seismic design requirements in buildings and other structures in California. Unfortunately, lessons on the need for rapid restoration of lifeline function to aid in community response to earthquakes waned as a result of a lack of significantly damaging urban earthquakes for nearly four decades after 1933. Following the 1971 San Fernando earthquake – an event that caused catastrophic damage to virtually every type of lifeline – many new efforts were launched to better understand the causes of these failures and identify ways to mitigate future earthquake damage and disruption.

The post-1971 lifeline earthquake engineering efforts have consisted of engineering research at academic institutions and the development of improved industry practices along with local or regional regulatory requirements. These efforts have been effective in reducing lifeline earthquake risks for some lifeline systems in some areas of California and the West Coast. In particular, the importance of understanding overall system functionality, instead of just the response of isolated components, has been recognized as a key concept in developing an effective earthquake risk management program for lifeline systems.

However, identifying lifeline system risks and implementing measures to improve earthquake performance have not been uniformly carried out even in coastal California, where the seismic hazard and population density are both relatively high. In lower-hazard parts of the United States, implementation of earthquake risk management has been very and sporadic. In particular, there has been limited progress in developing and implementing consensus guidelines and standards to provide consistent improvements in the performance of new and existing utility and transportation systems in large earthquake events across the United States.

In 1998, the Federal Emergency Management Agency (FEMA) and the American Society of Civil Engineers entered into a cooperative agreement to establish the American Lifelines Alliance (ALA) to facilitate the "creation, adoption and implementation of design and retrofit guidelines and other national consensus documents that, when implemented by lifeline owners and operators, will systematically improve the performance of utility and transportation systems to acceptable levels in natural hazard events, including earthquakes." Inclusion of all natural hazards in the scope of ALA recognized the benefit of managing the risks from multiple potentially damaging natural hazards events in a balanced fashion that meets the objectives of the lifeline owner or operator for lifeline functionality and the needs of their customers who rely on the lifeline services. Decisions on where to devote resources for improving lifeline system performance should be prioritized by considering the likelihood of experiencing natural hazard events, the impact of the natural hazards on the system, and the value of improving system performance to the owners of the system and their customers. Accordingly, the need to implement system improvements that address specific natural hazards will vary depending on the likelihood and severity of natural hazards and on the operational characteristics of the system.

Following the terrorist attacks on September 11, 2001, the scope of ALA was further expanded to include man-made hazards. This change is consistent with the broader goals of FEMA and recognizes that actions that minimize the effects of natural hazards also can improve resistance of structures and systems to man-made hazards. In late 2002, FEMA brought ALA under the Multihazard Mitigation Council through a contract with the National Institute of Building Sciences.

ALA Approach to Improving Guidelines for Lifeline Systems

The approach adopted by ALA recognizes that each lifeline system is unique in its development history, and each system typically occupies a region that has geographically varying exposure to natural hazards. The evaluation of performance of the system must be able to accommodate the specific components of the system, which were designed according to different criteria at different times, and are exposed to site-specific natural hazards.

Unlike occupied buildings, damage to a lifeline system from hazards addressed by ALA does not typically present a direct threat to life safety. Most lifeline systems include networks or multiple components that provide redundancy and resilience when subjected to damage or disruption. As a consequence of this inherent functionality, a given lifeline can tolerate some level of damage while providing adequate performance as far as the users of the lifeline are concerned. The analysis of risk from any type of hazard must critically examine the hazard frequency for the design and assessment of lifeline systems and incorporate consideration of functionality as experienced by the user of the lifeline system.

The primary function of ALA is to bridge the gap between hazard mitigation practices for buildings and lifeline systems through the sponsorship of national consensus guidelines and standards specific to lifeline systems. ALA activities are focused on converting well-established practices within the utility and transportation industries into nationally applicable consensus guidelines. ALA projects take existing non-consensus guidelines, standards, and industry practices, improve upon them as appropriate, and shepherd the resulting documents through a formal consensus process. In some cases, practices are not sufficiently developed to carry out ALA's process. In these situations, the role of ALA may include funding or co-funding well-defined studies in order to improve or extend existing practices to the point where national consensus is achievable.

The mechanism used by ALA for developing national guidelines for lifeline systems is dependent upon using existing Standards Developing Organizations (SDOs) accredited by the American National Standards Institute (ANSI) as the means to achieve national consensus. Rules governing the operation of these SDOs require that the guidelines or standards they develop represent a broad range of stakeholders, including representatives from utilities, transportation organizations, and manufacturing and technical communities.

Each project involves (1) an SDO for consensus development; (2) representatives from the appropriate sector of the utility or transportation communities; and (3) representatives from the relevant manufacturing and technical communities. The interaction between ALA, SDOs, the engineering community and other relevant stakeholders is illustrated in Figure 1. To support the development of national consensus guidelines for new and existing lifelines, ALA conducts and facilitates technical projects that improve or extend industry practices to the point where national consensus is achievable. ALA then submits the results of these projects to ANSI-accredited SDOs. Using this approach, ownership of the consensus guideline remains with the SDO, along with the responsibility for maintaining and updating it. As shown below, ALA takes existing industry practice, facilitates its transformation into a guideline/ prestandard, and transfers it to an appropriate SDO. The SDO receives input from oversight/regulatory bodies, owner/operators, and all interested and affected parties, and develops a national consensus. ALA helps implement the guideline by undertaking outreach activities, technical training, and information dissemination to owners/operators.

In some cases, existing practices are not sufficiently developed to be put through a formal consensus process, or technical improvements are needed in existing consensus guidelines. In these situations, the role of ALA may include funding studies in order to improve or extend practices to the point where national consensus is achievable. ALA will continue to actively seek out partnerships with public and private organizations to assist in identifying relevant in the lifeline hazard mitigation needs and coordinating efforts to promulgate consensus guidelines to improve lifeline system performance.

Utilizing existing SDOs has a significant advantage over other approaches that ALA could have adopted. First, using an SDO to process guidelines development assures well-balanced representation of stakeholders directly affected by the guideline. Second, the SDO process creates a "living document" as the SDO is required to review the adequacy of the guidelines at intervals no greater than five years. Finally, turning over draft guidelines developed under ALA sponsorhip to SDOs for consensus development transfers responsibility for maintaining the guideline to the private sector and reduces the need for long-term financial commitments on the part of FEMA.

Identification of Topics of Interest for ALA

It is necessary to clearly identify the current status of guidance documents that are available for lifeline systems. To this end, ALA has prepared a matrix of existing standards and guidelines for each lifeline system and each hazard or threat. The complete matrix as maintained on the ALA web site (www.americanlifelinesalliance.org) identifies ANSI-approved guidelines and non-consensus guidelines, standards, and industry practices. This matrix serves as a reference for lifeline engineering practitioners and is used by ALA to identify gaps in guidance for each lifeline, and to prioritize future ALA projects. Figure 2 shows the natural hazards matrix for electric power systems (excluding power generation), with accompanying explanatory notes. Specific references to guidelines and standards are provided in the web-based matrix on the ALA web site. A corresponding matrix for human threats was recently completed by ALA and an excerpt is shown in Figure 3.

In addition to the topics identified using the guidelines matrix, ALA seeks input from individuals and organizations on potential ALA projects. Individuals and organizations are invited to participate in ALA activities by becoming Corresponding Advisors. ALA interacts with over 150 Corresponding Advisors using e-mail and postings on the ALA web site. Corresponding Advisors have been used to identify and prioritize potential projects and are used to gain feedback from the professional community on ALA activities.

ALA Progress to Date

ALA has sponsored eleven projects to date as discussed in a companion paper at TCLEE conference. Several recent ALA projects focus on the definition of what steps need to be taken to assure that necessary information is available when making decisions on what level of reliability is appropriate for a particular lifeline system. This is an especially important topic that addresses the fundamental difference between the performance requirements of a lifeline system and an isolated facility such as a building.

The consensus guidelines that result from ALA activities are being prepared to be used on a voluntary basis by owners and operators of lifelines; but they could also be adopted by lifeline regulators to provide suggested or required guidance for achieving acceptable performance in response to natural hazards or human threats. Products of ALA projects are displayed and downloadable while being refined in the SDO process.

Future Directions for ALA

The near-term efforts of ALA will continue to be focused on implementing projects to provide guidance in areas where a need is identified by the ALA or suggested by outside individuals or organizations. ALA will also be placing more effort on outreach activities to encourage system owners to implement hazard risk reduction measures. At present these efforts emphasize presentations and exhibits at engineering and lifeline conferences to inform the appropriate communities about ALA products and projects. As more guidance is developed, ALA will establish more direct interactions with lifeline system managers and operators to develop demonstration projects and other applications of ALA guidance.

ALA will continue to expand partnerships and its involvement with man-made hazards. ALA is working closely with the Lifelines Subcommittee of the Interagency Committee on Seismic Safety in Construction, which is charged with assisting Federal departments and agencies to develop and incorporate earthquake hazard reduction measures in their ongoing construction programs. ALA's efforts to develop national consensus guidance are aligned with many of the objectives of the Lifelines Subcommittee. ALA products will provide appropriately qualified seismic guidance, and the Lifelines Subcommittee can help in the preparation and adoption of such guidance by Federal agencies.

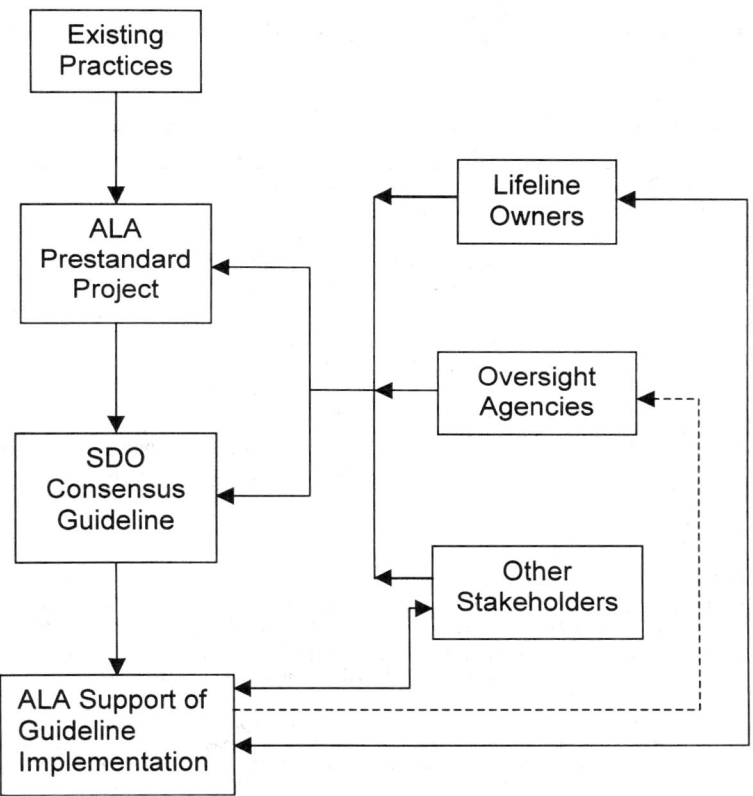

Figure 1. ALA Process for Developing Consensus Guidelines

ELECTRIC POWER SYSTEMS		NATURAL HAZARD PROVISIONS[5]		
COMPONENT	GUIDELINE or STANDARD[1]	LOADING	DESIGN	EXISTING[4]
System Reliability[3]				
Substations	IEEE-693	earthquake	earthquake	yes
	RUS 1724e300	wind/ice/eq[6]	wind/ice/eq[6]	
Transmission Towers & Poles	ASCE-10	wind/ice/eq[6]	wind/ice	yes
	ASCE Manual 74	wind/ice/eq	wind/ice	yes
	IEEE-691	none	wind/ice/eq[6]	yes
	ASCE Manual 72	wind/ice/eq[6]	wind/ice/eq[6]	yes
	ASCE Manual 91	wind/ice/eq[6]	wind/ice/eq[6]	yes
	ASCE Conc. Poles	wind/ice/eq[6]	wind/ice/eq[6]	yes
	RUS 1724e-200	wind/ice/eq[6]	wind/ice/eq[6]	yes
	PCI Prest. Conc. Poles	wind/ice/eq[6]	wind/ice/eq[6]	yes
	IEEE	wind/ice/eq[6]	wind/ice/eq[6]	yes
	NESC	wind/ice/eq	wind/ice	yes
Distribution Poles	NESC	wind/ice/eq	wind/ice	yes
	RUS 160-2	wind/ice/eq[6]	wind/ice/eq[6]	yes
Buried Conduits				

NOTES

1. Documents in **bold italics** indicate that the guidelines were not produced by a consensus process as defined for SDO's approved by the American National Standards Institute.
2. none applies if a guideline or standard does not specifically identify how loads are to be obtained; if a group of standards is referenced, the natural hazard listed may be only covered in one document.
3. System Reliability is a component of design referring to practices that are specifically developed to provide reasonable assurance that consequences of a natural hazard on system service will meet the goals established by stakeholders (owners, operators, regulators, insurers, customers, and users). Consequences are defined by multiple performance requirements but typically include impact on public safety, duration of service interruption, and costs to repair damage.
4. **EXISTING** indicates that analysis or design procedures (NOT LOADS) could be applied for existing components..
5. **LOADING** refers to whether or not specific loads for various natural hazards are defined; **DESIGN** refers to the existence of design and/or analysis procedures that account for loads arising from natural hazards.
6. This document refers to NESC and ASCE Manual 74.

Figure 2. Sample of ALA Natural Hazards Matrix for Electric Power

ELECTRIC POWER		MANMADE HAZARD PROVISIONS		
COMPONENT	GUIDE/STANDARD	LOADING	DESIGN	EXISTING
System Reliability	$ ∅	Radiological, Blast, Cyber	Biological, Blast, Cyber	
Transmission Towers	∅	Blast	Blast	
Distribution Poles	∅	Blast	Blast	
Buried Conduits	∅	Radiological	Radiological	
Substations	IEEE (1) $ ∅	Chemical Radiological	Radiological	
Elect./Mechanical Equipment	∅	Radiological, Cyber	Radiological, Cyber	

_ = Empty box indicates guidelines and standards related to the specified hazards are not available.

$ = Standards have been identified, but not reviewed in detail

∅ = Government standards exist, but are issued fro a controlled or sensitive source.

Design: Existence of design and/or analysis that account for loads arising specified hazards.

Existing: Analysis or design procedures (not loads) could be applied for existing components.

IEEE (1): Guide for Containment and Control of Oil Spills in Substations.

Figure 3. Sample of ALA Man-Made Hazards Matrix

Improving Natural Gas Safety In Earthquakes – California Recommendations

Fred Turner[1]

Douglas Honegger[2]

Abstract

This paper summarizes the efforts of the California Seismic Safety Commission and the American Society of Civil Engineers to develop comprehensive policies for improving natural gas safety in California residences. In the summer of 2002, two products were developed. The first is a report titled *Improving Natural Gas Safety in Earthquakes*, (CSSC, 2002a) The second product is new gas safety advice included in recent revisions to the Commission's *Homeowner's Guide to Earthquake Safety* (CSSC, 2002b). For more information, both are available online for free downloads at www.seismic.ca.gov.

An ad hoc task committee formed under the American Society of Civil Engineers (ASCE) committee for standard ASCE 25, *Earthquake Actuated Automatic Gas Shutoff Devices* prepared these products on natural gas safety in earthquakes. This task committee, chaired by Seismic Safety Commissioner Stan Moy, was formed in the spring of 2001 with the goal of providing information and advice to the California Seismic Safety Commission on the potential benefits and drawbacks associated with a wide range of measures to limit post-earthquake fire ignitions related to natural gas usage. The preparation of these products is in response to Initiative 8.2.2 of the *California Earthquake Loss Reduction Plan* to: "Educate local governments and the public on the application of gas safety devices such as automatic shut-off valves." (CSSC, 2002c)

The main message from these new policies is that no single risk management strategy for natural gas safety will address the needs of all individuals or jurisdictions throughout California. Those interested in more effectively managing the risk of gas systems and earthquakes are encouraged to learn about the issues and choose from the variety of options excerpted from the above reports and summarized in this paper.

---

[1] Stafff Senior Structural Engineer, California Seismic Safety Commission, 1755 Creekside Oaks Dr. #100 Sacramento, CA 95833 Fturner@quiknet.com www.seismic.ca.gov

[2] D.G. Honegger Consulting, ASCE 25 Ad Hoc Committee Facilitator, 2690 Shetland Pl., Arroyo Grande, CA 93420 dghconsult@aol.com

Introduction

The use of natural gas, like any flammable fuel, carries some risk of fire or explosion. The history of natural gas use throughout the world has shown it to be a safe fuel for consumer and industrial applications when buildings, natural gas systems, and appliances are constructed, installed, and maintained properly.

Gas utilities deliver natural gas to customers through gas distribution piping connected to a gas meter typically installed near the customer's facilities. The gas meter assembly has a manual gas service shutoff valve, in some cases a pressure regulator to reduce high pressure from the gas main to a standard delivery pressure, a gas meter to measure the volume of gas, and a service tee that allows the utility to bypass the meter set without entering the structure. The customer's gas houseline piping is attached to the service tee, which is typically considered the utility point of delivery and defines the physical boundary between utility-owned and customer-owned facilities. Typical meter installations are shown in Figures 1 and 2.

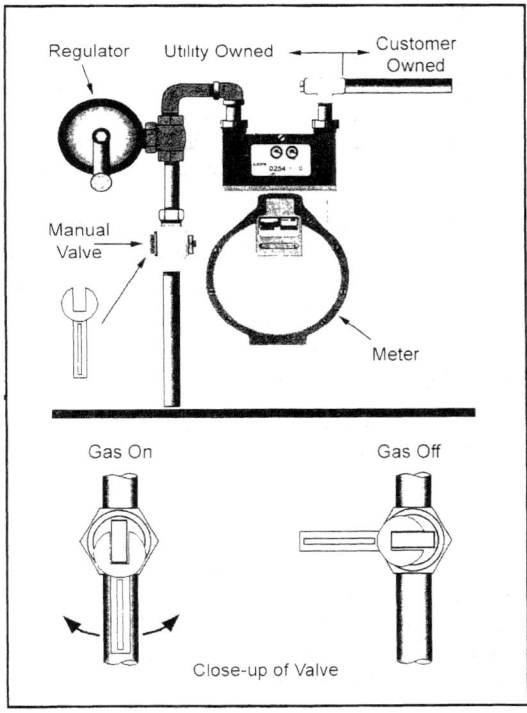

Figure 1. Typical Residential Meter with Pressure Regulator

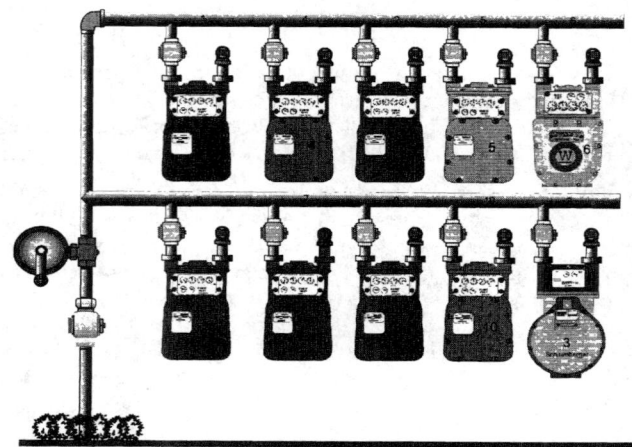

Figure 2. Multiple Meters –Typical for Multi-Unit Housing

Fires following the 1906 San Francisco earthquake are a vivid reminder to California communities of the potential consequences of post-earthquake fires. The combination of fire ignitions with conditions amenable to rapid fire growth and spread can greatly increase the level of post-earthquake fire damage. Past earthquake experience in California provides a basis for identifying characteristics of post-earthquake fire ignitions related to natural gas systems and demonstrates that natural gas is an important contributor to post-earthquake fire risk.

Several common characteristics of earthquakes and their impacts on natural gas safety are summarized below:

- Earthquake ground shaking will generally lead to substantially more instances of building damage than fire ignitions.

- Ground motions that are sufficient to damage buildings are most likely to impact utility and customer gas systems and create a potential for gas-related fire ignitions.

- The consequences of post-earthquake fire ignitions for residential gas customers are largely financial. Fire ignitions in residences only become life safety concerns when inhabitants are unable to exit buildings following earthquakes. Experience in past earthquakes indicates that egress from earthquake-damaged single-family homes is generally possible because of the limited structure height, low numbers of occupants, and multiple direct escape paths through doors and windows.

- The potential risks to life from post-earthquake fires are considerably more serious in seismically vulnerable apartment or condominium buildings since they provide a greater chance for damaging structures and trapping occupants.

There are many potentially beneficial alternatives for individuals to improve natural gas safety in future earthquakes. They include anchoring appliances, improving structural integrity and installing gas shutoff or warning devices. Each alternative has advantages and disadvantages related to the costs of implementation, level of safety improvement, and collateral benefits for non-earthquake emergencies. Because every situation is different, deciding which alternative is best suited to improve safety is best done on a case-by-case basis.

While several community-based actions to improve gas safety in earthquakes can be recommended, these actions need to be considered as just one part of a comprehensive earthquake risk management strategy. Determining which community actions are appropriate for a specific community requires a specific objective, a clear understanding of earthquake risks relative to other risks faced by the community, and potential drawbacks associated with a particular proposal to improve safety. Determining which actions are appropriate for a specific community should be made on a case-by-case basis with a clear understanding of the potential benefits associated with the costs of implementing any measures. The relative rarity of damaging earthquakes and the uncertainty in quantifying the likelihood, location, and severity of earthquake hazards require that earthquake risks be addressed in a balanced fashion considering other potential natural and man-made hazards.

Natural Gas Performance in Past Earthquakes

Experience from recent earthquakes in California is particularly useful in examining the performance of natural gas systems designed and operated according to typical practices in the United States. Table 1 summarizes fire statistics from previous earthquakes and others in the United States over the past four decades. Data from earlier U.S. earthquakes is typically lacking in detail regarding the specific causes of ignition. And it is difficult to draw meaningful conclusions from earthquakes that occur outside the U.S. because of differences in the gas systems, population density, and building construction.

The three most recent California earthquakes to strike in or near an urban region serve as examples of what might be expected in future earthquakes in the United States. Ground motions sufficient to damage buildings are most likely to impact utility and customer gas systems and create the potential for gas-related fire ignitions. Although people are advised in an emergency to shut off their gas service only when they observe or suspect gas appliance or structural damage, or can hear or smell leaking gas, many customers overreact and shut off their gas as a precaution even if they don't smell gas, which increases service restoration demands. Gas restoration efforts following strong or major earthquakes in urban areas can require massive mobilization of properly trained service personnel.

Natural gas is a significant contributor to the post-earthquake fire risk. Gas-related fire ignitions can be expected to be 20% to 50% of all post-earthquake fire ignitions. The percentage of fire ignitions in future earthquakes can be larger or smaller than in past earthquakes. While earthquakes may produce numerous leaks in a gas system, the potential for fire ignitions from natural gas will be low compared to the total number of leaks.

Earthquake	Magnitude	Earthquake Fire Ignitions	Gas-related Fire Ignitions
1964 Alaska	9.2	4-7	0
1965 Puget Sound	6.7	1	?
1971 San Fernando	6.6	109	15
1983 Coalinga	6.2	1-4	1
1984 Morgan Hill	6.2	3-6	1
1986 Palm Springs	6.2	3	0
1987 Whittier	5.9	6	3
1989 Loma Prieta	7.2	67	16
1994 Northridge	6.7	97	54
1995 Kobe	6.9	205	36

Table 1. Summary of Building Fire Ignitions for Recent Earthquakes

How Earthquakes Damage Gas Systems and Cause Losses

The most common types of earthquake damage to gas systems are described below:

- Shifting or toppling of water heaters, boilers, furnaces, dryers, and stoves is the principal cause of most gas-related, post-earthquake fire ignitions. (71% in the Northridge Earthquake)
- Large deformations or collapse of building framing can damage gas lines and meters.
- Objects or debris falling on gas systems.
- Ground displacements rupturing buried or above-ground pipelines.
- Ground shaking, particularly for gas lines weakened by corrosion or brittleness.

Such damage to gas systems can cause gas leaks, lack of fuel for heating, cooling and cooking, business interruption, loss of habitability and fires. Extensive delays in relighting gas systems may also compel non-qualified individuals to attempt to relight pilots without taking necessary safety precautions.

Figure 3. The unbraced water heater in this home fell during the 1990 Upland Earthquake. The resulting fire destroyed the home. (CSSC, 2002b, OES Photo)

As shown in Table 2, the number of days to restore gas service can be significant and varies widely. Factors influencing restoration time include the severity of damage, availability and effectiveness of personnel supplementing utility restoration efforts (from other natural gas utilities) and the sizes of the regions impacted by the earthquakes. It is also important to note that the outages in Table 2 are a relatively small fraction of total customers in the shaken regions and do not reflect the impact of mandates recently passed by local jurisdictions (including the City of Los Angeles) requiring customers to install automatic earthquake actuated gas shutoff devices on their gas systems.

Earthquake	Number of Customer Outages*	Restoration Time
Northridge	120,000	12 days
Loma Prieta	156,355	9 days
Whittier	20,600	10 days
*Does not include customers affected by the additional time needed to reconstruct gas distribution facilities or structures		

Table 2. Service Restoration Times for Three Recent Earthquakes

Life safety consequences from post-earthquake residential fires depend on the ability of individuals to evacuate buildings following earthquakes. Building layouts differ as to whether occupants must use shared or lengthy paths of emergency egress or by more direct, unshared routes. In multi-unit occupancies (R-1 occupancies), common paths of egress and limited means of escape make it more likely that persons can be trapped after earthquakes. The greater the number of occupants in a building, the greater is the likelihood they will be trapped in emergencies. Damage to exterior doors of apartment and condominium units often prevent occupants from exiting safely. In buildings of more than two or three stories, the escape paths usually include enclosed stairways whose doors can be jammed by the racking deflections of the doorframes caused by earthquakes. Frequently, elevators in buildings can be damaged and unusable.

Single-family residential units (R-3 occupancies), on the other hand, cannot, by code, be more than three stories high. Their windows are usually constructed in such a way that they can serve as secondary exits. So more and easier pathways exist for escape in R-3 occupancies than in R-1 occupancies. In addition, if R-3 structures are properly connected to their foundations, they are less likely to lose their means of escape than larger and more complex R-1 structures. (ICBO, 2001)

Options to Reduce Fires and Service Disruptions Following Earthquakes

Several options are available to improve the earthquake performance of natural gas systems and increase public safety. For communities, the Seismic Safety Commission and ASCE's Task Committee recommend that low-cost steps to improve gas safety be considered first before considering to adopt ordinances that encourage or require new devices to limit gas flow. These options include:

- Provide information to the public through government offices, mail inserts, or the Internet.
- Organize neighborhood groups and provide training to assist in simple earthquake response measures such as proper procedures for manual gas shutoff, and restraining appliances.
- Define high-risk fire areas within communities and hold workshops to publicize and manage the potential risk.
- Assess the costs and the benefits of proposed gas safety policies.
- Develop, communicate, and implement effective earthquake risk management plans.

More detailed options, recommendations and benefit-cost considerations for individuals and communities can be found in *Improving Natural Gas Safety in Earthquakes* (CSSC, 2002a).

Single-family homeowners owners should consider the advice on gas safety excerpted from the *Homeowner's Guide to Earthquake Safety* (CSSC 2002b) as shown in Figures 4 and 5 on the next pages.

Earthquake Weakness	**Natural Gas Safety**

The Problem

Natural gas piping and appliances in homes can be damaged resulting in releases of natural gas that can lead to fires if ignition sources are present. Natural gas is typically a factor in about one out of four fire ignitions following earthquakes. However, the total number of earthquake-related fires, their sources and amounts of destruction can vary greatly depending upon a number of factors. Structural weaknesses or the absence of appliance anchors and flexible pipe connections can lead to a greater possibility of gas leaks following earthquakes. Experience has shown that living in a seismically active area increases the risk of fire from all causes by a small amount. Since residential dwellings generally have several safe exit paths, the potential for life loss is limited. Therefore the primary concern for homeowners is property loss from fire damage.

How to Identify It

Examine all natural gas appliances (water heaters, dryers, stoves, ovens, furnaces) to see if they are anchored to the floor or walls and have flexible pipe connections.

What Can Be Done

Relying on manual gas shutoff valves is an effective means to stop the flow of gas if persons are present after earthquakes. If you smell gas, hear gas escaping, or suspect a broken gas pipe, appliance, vent, or flue, use a wrench to turn off the gas valve located near the gas meter. In addition, options such as earthquake actuated valves, excess flow valves, methane detectors, and hybrid systems can further reduce the risk of gas leaks and ignitions. However, once the gas has been shut off, service can only be restored by utility personnel or qualified plumbers. Demands for qualified personnel following earthquakes may lead to substantial delays in restoring service.

Homeowners should consider their specific circumstances and the suitability of these options. See Figure 5 for the benefits and drawbacks of various options. Earthquake Actuated Valves and Excess Flow Valves should be certified by the State Architect. Some installations will require building permits, so consult your local jurisdiction. Homeowners should be aware that some jurisdictions have adopted ordinances requiring gas shutoff devices at the time of sale or when significant renovations are undertaken.

Figure 4. Advice to Owners of Single-Family Homes in California from the *Homeowner's Guide to Earthquake Safety*

Table 2: Gas Shutoff Option Costs

Device[1]	Hardware Cost	Installation Cost[2]
Restrain individual gas appliance	$15-$50	$0 - $100
Manual shutoff valve & wrench	$5-$20	$0
Earthquake actuated valve	$100 - $300	$100 - over $300 [3,4,5]
Excess flow valve at meter	$20 - $100	$100 - over $300 [3,4]
Excess flow valve at appliance	$5 - $15	$0 - $100
Methane detector	$25 - $75	$0
Hybrid system	$150 - over $500 [6]	$100 - over $500 [7]

NOTES:
1. There are significant differences in the operation of the various devices listed
2. All costs are approximate and do not include permit and inspections fees that may range from $25 to cover $100 depending upon the local jurisdiction. Installations that can be performed by the building owner are assumed to have no cost.
3. Installation costs do not include a survey of the gas system that can cost over $200.
4. Higher Installation costs may occur if substantial modifications of plumbing are necessary
5. Higher installation costs may occur if substantial modifications to attach the valve to the building are necessary
6. Costs for hybrid systems depend on the number and type of components installed
7. Higher installation costs can be incurred for hybrid systems that require installation of wiring to connect multiple sensing units.

Table 3: Gas Shutoff Comparisons

Consideration	Manual Shutoff Valve and Wrench	Earthquake Actuated Valve	Excess Flow Valve	Methane Detector	Hybrid System
Basis of Operation	Utilities have installed manual shutoff valves near gas meters allowing owners with proper wrenches to shutoff gas in emergencies	Senses shaking in a building that is above a design level of shaking and automatically shuts off gas	Senses gas flows that are above a design shutoff flow rate and automatically shuts off gas	Senses the presence of natural gas in the air and triggers an alarm	A variety of modular devices that could include a main control unit, shake sensors, excess flow sensors, methane detectors, valves, and alarms.
Benefits	All gas services already have valves installed. Guidance for occupants is currently provided in many public information documents like the phone book	Actuates only in cases when building shaking may be sufficient to cause damage to the gas system. Someone does not need to be present to ensure shutoff	Actuates only in cases when excess gas flows downstream of the device. Someone does not need to be present to ensure shutoff	Alerts occupants when detectable gas concentrations are present before they reach hazardous levels, allowing time for shutoff and evacuation.	Systems are modular and can be customized for desired applications. Each module has benefits associated with specific action (e.g., motion sensing, flow sensing, methane detection)
Potential Drawbacks	Only effective if someone is present, knows the valve location, has access to the valve, and has a wrench suitable to close the valve	Can actuate even if damage and hazards do not exist. Aftershocks can cause the device to actuate after service has been restored. May actuate from shaking not related to earthquakes	Will not shut off gas if leakage is below the design shutoff flow rate, even if a slow leak exists. May not activate if the occupant changes gas systems downstream without modifying the device	Someone needs to be present to respond to the alarm. Alarm may trigger for other flammable vapors in addition to natural gas.	Each module has drawbacks associated with specific actions (e.g., motion sensing, flow sensing, methane detection)

Figure 5. Gas Shutoff Options, Benefits, Drawbacks and Costs from the *Homeowner's Guide to Earthquake Safety*

Conclusions and Recommendations

Individual natural gas customers should become familiar with their natural gas system and assess their need to implement measures to improve natural gas safety in future earthquakes.

Communities should take steps to understand their post-earthquake fire risk and implement measures to reduce this risk to an acceptable level. Decisions should be made with a clear understanding of the potential benefits associated with the costs of implementing specific measures.

Acknowledgements

The authors thank the ASCE Committee members and the Seismic Safety Commission for volunteering their time and the State of California, Pacific Gas and Electric and Southern California Gas for funding this effort.

References

American Society of Civil Engineers (1997). *Earthquake Actuated Automatic Gas Shutoff Devices*, ASCE Standard 25.

California Seismic Safety Commission (2002a) *Improving Natural Gas Safety in Earthquakes.* July 2002, SSC 02-03, www.seismic.ca.gov

California Seismic Safety Commission (2002b). *The Homeowner's Guide to Earthquake Safety*, SSC 02-04, www.seismic.ca.gov

California Seismic Safety Commission (2002c). *The California Earthquake Loss Reduction Plan, SSC 02-02, www.seismic.ca.gov*

Cote A.E. (ed.) (1996). *Fire Protection Manual*, National Fire Protection Institute, Quincy, Massachusetts.

Federal Emergency Management Agency (1997). *Multihazard Identification and Risk Assessment: A Cornerstone of the National Mitigation Strategy.*

Honegger, D.G. (1991). "Evaluation of Automatic Earthquake Shutoff Valve Performance and Recommendations for Future U.S. Standards," 3rd U.S. National Conference on Lifeline Earthquake Engineering, August 22-23.

Industrial Press (1965). *Gas Engineers Handbook*, First Edition.

Scawthorn, C.G. (1987) "Fire Following Earthquake: Estimates of the Conflagration Risk to Insured Property in Greater Los Angeles and San Francisco," All-Industry Research Advisory Council, Oak Brook, Illinois.

Williamson, R.B. and N. Groner (2000). "Ignition of Fires Following Earthquake Associated with Natural Gas and Electrical Distribution Systems," PEER.

International Conference of Building Officials (2001). *Guidelines for the Seismic Retrofit of Existing Buildings.* www.icbo.org

Seismic Design and Retrofit of Piping Systems
Overview of a Recent ALA Report

George Antaki
Aiken, South Carolina

Abstract

This paper summarizes the recent technical report "Seismic Design and Retrofit of Piping Systems", July 2002, sponsored by the American Lifelines Alliance (ALA). The ALA report describes, in sequence and in detail, the steps necessary for the seismic qualification of new or existing (retrofit) above ground piping systems (www.americanlifelinesalliance.org).

The ALA report compiles years of experience in design and retrofit of piping systems, based on analysis, testing and the study of earthquake performance.

The ALA report was prepared to help expedite a long overdue ASME B31 standard for seismic design of piping systems. The report contains, in appendix, the proposed standard, which is currently under review within the ASME B31 Mechanical Design Technical Committee.

But the ALA report also serves a second purpose: to provide in a single document a comprehensive, yet readable, guide to model, analyze and seismically qualify a piping system.

Assembling Input

The ALA report first describes the type of information that must be assembled for a new system or an existing system. This includes materials, fittings and component data, layout, operating conditions concurrent with the design earthquake, and – for existing systems – the material condition of the installation.

The level of detail and accuracy of the required input will depend on the method of seismic analysis. More detailed input will be required for a dynamic analysis compared to a static analysis, since the dynamic results will depend on mode shapes and modal frequencies, which in turn depend on stiffness and mass distribution of pipe spans, components acting as concentrated weights, and seismic restraints.

Preliminary Design

This section of the report describes how to select a preliminary seismic support and bracing configuration. The preliminary design considers equipment anchorage, mechanical joints, spacing of seismic restraints, the consideration of relative movement, and – for the retrofit of an existing system – the adequacy of existing supports. This section includes illustrations of typical piping failures in real earthquakes.

On the basis of past seismic performance and analysis experience, it is recommended to first evenly support the pipe weight, by selecting a constant span between weight supports. The rule of thumb for span length (feet) for liquid service is "size (inches) + 10", which means – for example – that a 4" pipe will be supported every 4 + 10 = 14 feet. The span is longer for gas or steam service.

Weight support is followed by expansion flexibility analysis to design for the expansion of hot lines or contraction of cold lines that will be concurrent with the postulated earthquake. In this case, there are no rules of thumb, but rather the designer must rely on experience and piping stress analysis software.

Following the weight and expansion design, the objective of the preliminary seismic design is to spot the best locations for seismic restraints. There are no simple rules to spot seismic restraints, possibly with the exception of the fire sprinkler piping rules of NFPA-13, which call for lateral sway bracing every 40 feet and longitudinal sway bracing every 80 feet. The ALA report proposes, as a preliminary step, to limit the lateral span to a length L_{max} given by

$$L_{max} = \min\{1.94 \frac{L_T}{a^{0.25}}; 0.0175 L_T \sqrt{\frac{S_Y}{a}}\}$$

L_{max} = maximum length of span between lateral restraints, ft
L_T = span between weight supports (the "size + 10" rule above), ft
a = lateral acceleration, g
S_Y = yield stress of piping system material at operating temperature, psi

For example, a 4" system made of carbon steel, operating at 70°F (S_Y = 30,000 psi), with a weight span L_T = 14 feet, and designed for a lateral seismic acceleration of a = 1(1g), will be restrained laterally every 27 feet, or – practically speaking – at every other weight support.

The other important consideration in a preliminary seismic design is the differential movements of the building or structure to which the pipe is attached. Large differential movements will tend to tear open pipe joints, particularly non-welded (mechanical) joints.

Seismic Analysis Techniques

In most cases, piping systems are seismically designed by analysis. The analysis technique is either static or dynamic. This section of the ALA report describes the advantages and shortcomings of each analysis technique: "cook book" (also referred to as the span table approach), static hand calculation of simple subsystems, computerized static analysis of a system, response spectra analysis, and time history analysis, including discussions on modal and directional combinations.

The ALA report also explains the background to the International Building Code (IBC) seismic accelerations.

The International Building Code (IBC) 2000 rules are applied to an example to illustrate the development of static coefficients for piping analysis, including in-structure amplification. This application of IBC also addresses the practical meaning of the various coefficients used in the building code equations (such as $0.4S_{DS}$, a_P, and R_P). A flow chart is presented to help the reader develop the applicable coefficients for seismic analysis for piping systems at grade or mounted in a structure.

Modeling for Analysis

This is a topic seldom documented in the past, yet of great practical interest to the designer. The topics addressed in the ALA report include:

(a) How to breakdown a large model into smaller subsystems for the purpose of seismic analysis. The practice has been to decouple relatively small branch lines from the header analysis. Small branch lines may be defined as pipes with a moment of inertia smaller than 25 times the header's.

(b) What should be the accuracy of the model (lengths, weights, angles, etc.). The ALA report provides guidance on angular accuracy in the order of $10°$, restraint locations accuracy within one pipe diameter or 1-ft (whichever is greater), and span accuracy ranging from 3" for 5 foot long spans to 2 feet for spans over 25 foot in length.

(c) How to include the flexibility of equipment (local shell flexibility as well as equipment swaying flexibility). The ALA report compares the results of a dynamic analysis in which a vessel-to-pipe nozzle is modeled as a fixed point (infinitely rigid) to the case where the vessel shell flexibility is included in the model. Including the vessel stiffness significantly reduced the pipe natural frequency (which is to be expected) but also reduced the loads imparted by the pipe on the nozzle.

(d) How to account for support gaps and stiffness, how to account for the flexibility of standard components (elbows, bends, tees, etc.) and specialty fittings (swage couplings, mechanical joints, etc.). These questions relate to how best (and most simply) to account for non-linearity, rattles and impacts.

While these questions can be solved analytically, these solutions are too complex for most practical applications. The ALA report relies on earthquake experience to note that pipe-support rattle is of no significant consequence if (1) the pipe span does not contain impact sensitive components (such as instruments), (2) an impact dent (deformation, indentation) or sharp gouge (sharp cut at point of impact between the pipe and support frame) will not cause a failure (as could be the case for high pressure oil or gas pipelines), and (3) the restraint will not rupture under the impact force.

Qualification

Seismic qualification is described in terms of seven key criteria

(1) Stresses in the piping, with either primary or fatigue limits.

(2) Loads on equipment.

(3) Loads on supports and restraints.

(4) Loads and deflections at mechanical fittings.

(5) Accelerations and loads on valves and operators.

(6) The operability of active components (pumps, valves, compressors, etc.), including a description of how to conduct seismic tests on active equipment.

(7) The interaction of non-seismically qualified components on the system being qualified. This is an important topic when we consider that many earthquake failures are due to interactions (something falling on or swinging into the piping system). Often times the cost of fixing the interactions around the qualified system may exceed the cost of qualifying the piping system. A simple example is a critical piping system in a room with block walls, where it will be costlier to brace the wall than to seismically support the system. This section of the report addresses how to perform the interaction review, and make upgrade decisions. As mentioned earlier, seismic interactions are a significant source of earthquake failures, yet they have not received sufficient attention in the national standards. This section of the ALA report has served as the basis for an update to the NEHRP provisions, introduced in the early part of 2003.

Advanced Analysis Techniques

An effort has been made to describe in simple terms what advanced seismic analytical techniques exist, beyond linear response spectra analysis, and what the acceptance criteria are in each case. This includes elastic analysis (most common, with limits on yield or ultimate stress), plastic analysis (non-linear stress-strain model, with limits on strain or ultimate stress), limit analysis collapse load (elastic-perfectly plastic model, with limit on large deflection collapse load), and plastic instability load (elastic-plastic model, with limit on instability load).

Seismic Restraints

The various types of seismic restraints are described, with reference to the applicable design codes: standard catalog components, steel frames, and concrete anchor bolts. The section includes illustrations of seismic support arrangements. Regarding anchor bolts, the ALA report presents a simple approach by which the bolt capacity is reduced by a series of penalty factors that account for embedment depth, concrete edge distance, spacing of anchor bolts, concrete cracking, and concrete strength.

Terms and Definitions

This section of the report defines commonly used terms such as peak spectral acceleration, peak ground acceleration, seismic interaction, etc. This clarification of nomenclature and acronyms is particularly useful to help clarify communications and exchange of data since often times seismic information is supplied in the form of "PGA", "ZPA", "PSA", etc.

Proposed Seismic Design Standard - ASME

In appendix to the ALA report is the first draft of a seismic standard that is currently under review and vote within the ASME B31 Mechanical Design Technical Committee. At the time of writing of this paper (February 2003) the proposed ASME seismic standard had undergone a first round of reviews and votes, and a second draft was under preparation for vote. The proposed ASME standard is a graded approach that depends on three key attributes:

(1) Required function. A piping system is seismically qualified for one of three reasons (called functions): Position retention (braced to avoid fall or excessive swing), leak tightness (can not leak out, or leak in for a vacuum system), or operability (must deliver, control, throttle or isolate flow).

(2) Level of seismic input acceleration. Systems that experience lower seismic acceleration will require less rigorous analysis.

(3) The pipe size. The smaller sizes (lighter systems) will require less analysis.

In this graded approach, teher are instances where no seismic analysis is required (required function is only position retention, the input acceleration is low, and the pipe is small bore). At the other extreme of the spectrum, there are systems that will require analysis and component testing (required function is operability, the input acceleration is large, and the pipe diameter is large with heavy components). A matrix helps define the graded approach.

A commentary accompanies the proposed ASME standard, to explain the basis and intent of the proposed rules.

Proposed Seismic Design Standard - NEHRP

In addition to ASME, the ALA report serves as the basis for a revision to section 6.4.7 regarding seismic qualification of piping systems, currently under review. The proposed seismic qualification is a graded approach, similar to the ALA report, and is summarized in the Table 1.

Table 1 Extent of Seismic Qualification (IBC nomenclature)

Seismic Design Cat.	Non-Critical $I_P = 1.0$		Critical $I_P = 1.5$	
	Pipe Size ≤ 4"	Pipe size > 4"	Pipe size ≤ 4"	Pipe size > 4"
A or B	Interactions	Interactions	Bracing Restraints Interactions	Bracing Restraints Interactions
C or D	Interactions	Interactions	Bracing Restraints Operability Interactions	Analysis Restraints Operability Interactions
E or F	Bracing Restraints Interactions	Bracing Restraints Interactions	Analysis Restraints Operability Interactions	Analysis Restraints Operability Interactions

"Seismic Design Categories A to F" are defined in the Building Code (IBC 2000) and refer to earthquake and soil conditions, in increasing order of severity.

"I_P" is the system importance factor defined in the building code (IBC 2000).

"Interactions" refers to the need to evaluate the potential adverse interactions (spatial or system interactions) between non-seismically qualified items and the system being qualified. It is self evident that such an evaluation is required in all cases, as interactions are a primary cause of failure in real earthquakes.

"Bracing" refers to the need to add restraints at fixed intervals, this is often referred to as a "cook book" approach.

"Analysis" refers to a detailed, static or dynamic, stress analysis of the piping system, and placement and design of supports accordingly. The rules of ASME B31.1 are typically used since they provide explicit equations for longitudinal stress due to occasional loads, they account for stress intensification factors, and they are standard in pipe stress analysis software. However, the B31.1 allowable stress (1.2S) has remained neglected and unchanged for decades, ignoring the tests, analyses and studies conducted through the 1980's and 1990's that justify a significantly larger allowable seismic inertia stress, in the order of 2.4S to 3S. This is recognized by

ASME B31 and ASME B31 Mechanical Design Committee is currently balloting changes to the seismic design rules.

"Operability" refers to the proof of seismic function of active equipment (pumps, compressors, valve operators) after an earthquake, by analysis, testing, or – in some cases – by comparison to earthquake performance.

"Restraints" refers to the quantitative check of demand vs. capacity of seismic restraints, typically following the rules of AISC Manual of Steel Construction, ACI 318-02 Building Code Requirements for Structural Concrete Appendix D Anchoring to Concrete, MSS-SP-58 Pipe Hangers and Supports – Materials, Design and Manufacture, MSS-SP-127 Bracing for Piping Systems Seismic – Wind – Dynamic Design, Selection, Application, IBC-2000 International Building Code, ASCE 7-02 Minimum Design Loads for Buildings and Other Structures, AISC N-690 Specification for the Design, fabrication, and Erection of Steel Safety-Related Structures for Nuclear Facilities, ASME Boiler & Pressure Vessel Code, Section III Subsection NF Supports.

Examples

Nothing helps better clarify the intent of rules as a practical example. Two examples are developed in the report. The first example is the design of a new piping system operating at 350 psi and 435°F. The second example is the seismic retrofit of a system supplying essential gases to a facility, from a cryogenic tank (liquid gas storage), through an evaporator, and to the building. The latter example (the retrofit of the existing gas supply system) includes check-list documentation of the field inspection and accompanying photographs.

Conclusion

The ALA report is a complete yet practical guide for the seismic design of new piping systems, and the seismic retrofit of existing ones. It includes examples and applications. It is the foundation for a new ASME B31 standard for seismic design, currently under review within ASME, and updated NEHRP provisions.

Analyzing the Seismic Performance of Wharves, Part 1: Structural-Engineering Approach

W.H. Roth[1], E.M. Dawson[1], M. Mehrain[2], A. Sayegh[2]

Abstract

The standard "structural-engineering approach" to wharf design relies on soil-structure-interaction (SSI) models where the complex soil continuum is replaced by 1-D springs which are fixed in space. A wharf design is considered satisfactory if the displacement capacity, as determined from a pushover analysis, is equal to, or greater than, the displacement demand. The latter is obtained either from a displacement-response spectrum, or by performing a time-history analysis. Considering a generic wharf structure, this simplified approach is contrasted with performing a SSI analysis where the soil is simulated by an elasto-plastic continuum. The results of this exercise highlight the inherent limitations of using 1-D springs where shaking-induced permanent displacements of the underwater slope/dike are completely ignored.

Introduction

The seismic performance of wharf structures is routinely analyzed with simplified procedures originally intended for structures founded on level, stable ground. The objective in preparing this paper was to critically evaluate the suitability of these techniques for pile-supported wharves, particularly when they are embedded in slopes that undergo shaking-induced permanent deformations. The down-slope drag acting on the wharf piles as a result of these deformations can be an order of magnitude larger than the transient dynamic forces generated by the structure itself. Yet, these drag forces are simply ignored in conventional structural analysis procedures where the soil continuum is replaced by 1-D springs.

The seismic analysis of wharf structures falls into a gray area between structural and geotechnical engineering. Cooperation between these two disciplines is required to produce meaningful results. This is easier said than done, however, since they use different analysis procedures and numerical modeling tools, each focused on their respective media. Curiously, while geotechnical programs often include at least some structural modeling capabilities, structural programs are rarely equipped to properly represent the soil as a continuum. As wharves are usually being designed by structural engineers, this creates an intriguing problem. Used to thinking of soil as a mere external load or boundary condition to be applied to a structure, structural engineers may have difficulties with the concept that it could actually be the most important element of an overall system. As a result, there is great reluctance to accept

---

[1] Geotechnical Engineering Group, URS Corporation, Los Angeles, CA.

[2] Structural Engineering Group, URS Corporation, Los Angeles, CA

structural results from a geotechnical program. Hence, a benchmark exercise was performed at the outset of this study to demonstrate the structural capabilities of the widely used geotechnical program, FLAC (Itasca, 2001).

Problem Setting and Analysis Tools

Generic Wharf Structure. A cross section of the generic wharf structure analyzed in this study is shown in Figure 1. The wharf deck is supported on 24-inch diameter precast concrete piles spaced 17 feet on center in both directions, transverse and longitudinal. To provide adequate support for the crane rails, the pile rows along the land- and waterside boundaries of the wharf are spaced 8.5 ft on center in the longitudinal direction. The 1.5:1 (H:V) underwater slope consists of hydraulic fill sands covered with quarry rock. The latter acts as a stabilizing "blanket" placed on a cut slope, or could also be seen as an approximation of a multi-stage rock dike constructed during hydraulic-fill placement. With the soil properties listed in Figure 1, the static factor of safety of the wharf slope (without the piles) was computed to be 1.6.

Design Earthquake. Dynamic analyses were performed with two acceleration histories spectrally matched to a magnitude 7.5 (M_w) design earthquake, which had been developed for a stiff soil site near the Port of Oakland (URS, 2002). The design spectrum and matched ground motions are shown in Figure 2.

The Geotechnical Program, FLAC. Dynamic simulations were performed with the 2-D explicit finite difference code FLAC (Itasca, 2001). Soil behavior is represented by a Mohr-Coulomb, linear elastic/perfectly plastic model. For simplicity, pore pressure generation and liquefaction were not considered in this exercise. The wharf piles and deck are modeled by elasto-plastic beam elements capable of developing plastic hinges at pre-determined yield moments. The structural nodes of the piles are connected with the soil mesh through elasto-plastic p-y springs representing the lateral load-displacement behavior of single piles (Figure 3). The parameters for these springs were derived using the procedures recommended by the American Petroleum Institute (1987). Both, the piles' section properties and the p-y springs were adjusted to account for pile spacing, so that they represent equivalent properties per unit length of wharf.

The FLAC model shown in Figure 4 was subjected to horizontal shaking via an input acceleration history applied at the base of the model through an absorbing boundary. In order to minimize lateral wave trapping and interference, the left and right sides of the mesh are also absorbing boundaries, but with real-time feedback from 1-D "free field" computations simulating level ground conditions. The elasto-plastic soil constitutive model produces hysteretic damping at large strains upon reaching the yield strength. To provide damping for small strains in the elastic range, Rayleigh damping with a critical damping ratio of 5% at a center frequency of 3 Hertz was used for all simulations.

The Structural Programs, SAP2000 and Perform-2D. Structural analyses were performed with the widely used program, SAP2000 (CSI, 1997), and also with Perform-2D (Ram International, 2002). These programs have capabilities for both static and dynamic, as well as linear and nonlinear analyses.

Pseudo SSI Analyses with 1-D Soil Springs.

The first part of the study consisted of evaluating the seismic performance of the wharf structure with the standard "structural engineering approach" utilizing the programs SAP2000 and Perform-2D to analyze the model shown in Figure 1. With this approach, a wharf design is considered satisfactory if the displacement capacity, as determined from a pushover analysis, is equal to, or greater than, the displacement demand. The latter is obtained either from a displacement-response spectrum, or by performing a time-history analysis.

In order to demonstrate the structural capabilities of the geotechnical program FLAC, the analyses above were repeated with this program, and the results compared with SAP and Perform-2D. To simulate the soil-spring approach with the FLAC model, all nodes of the soil mesh shown in Figure 4 were fixed in space.

Pushover Analyses. The results from the pushover analyses with SAP2000 and Perform-2D closely matched those obtained with FLAC. The analysis results are plotted in Figure 5 for both non-degrading (left) as well as degrading plastic hinges (right). The plastic yield moment of the piles was assumed to be 600 kip-ft. The plastic moment of the degrading hinges was assumed to be a function of hinge rotation as shown by the inset in the right-hand plot. Based on the results of the pushover analyses, and using an acceptable limit on plastic rotations of 0.024 radians, the displacement capacity of the deck was determined to be 0.3 feet.

Natural Period and Spectral Displacement. The natural period of the wharf structure was computed with SAP2000, Perform-2D, and with FLAC. For SAP2000 and Perform-2D, the period was computed through an eigenvalue analysis. For FLAC, the natural period was determined in a time domain analysis by applying a horizontal force to the deck and measuring the time between displacement peaks. Table 1 compares the natural periods computed with the three programs, along with the corresponding spectral accelerations and displacements obtained from the design spectrum shown in Figure 2.

Table 1. Mode-1 natural period, spectral acceleration and displacement

	Natural Period (sec)	Natural Frequency (1/sec)	Spectral Acceleration (g)	Spectral Displacement (feet)
FLAC	0.36	2.80	1.57	0.20
SAP	0.37	2.74	1.56	0.21
Perform-2D	0.36	2.80	1.57	0.20

Time-History Analyses. Maximum dynamic displacements were also computed by applying the two earthquake records shown in Figure 2. These analyses were carried out with Perform-2D and FLAC. The input acceleration time-histories were applied to the anchor points of the p-y springs. As shown in Table 2, the displacements computed with both programs are reasonably close to the spectral displacement at the natural frequency of the wharf structure.

Table 2. Maximum dynamic displacements from time-history analyses

Time History	Spectral Displacement at Natural Frequency (feet)	Computed Max. Displacement (feet)	
		FLAC	Perform-2D
Kocaeli/Duzce/180	0.20	0.21	0.16
Kocaeli/Yarimca/000		0.17	0.17

Seismic Performance Based on Structural Analysis. In summary, the "structural-engineering approach" utilizing a pseudo SSI analysis with 1-D soil springs leads to a displacement capacity of 0.3 feet (from pushover analysis) and a displacement demand of about 0.2 feet (from spectral and time-history analyses). Because the displacement capacity is greater than the demand, the seismic performance of the subject wharf would be deemed satisfactory.

SSI Analysis with Soil Continuum

Up to this point, our analyses have neglected the effects of shaking-induced down-slope soil movements. For further analyses, the soil mesh in the FLAC model was released to allow deformation of the soil continuum in response to earthquake shaking. Figure 6 compares the deck-displacement history computed with Perform-2D (History A) with displacement histories computed with FLAC for the following three cases: Linear-elastic soil with elasto-plastic piles (History B); elasto-plastic soil with linear-elastic piles (History C); and elasto-plastic soil with elasto-plastic piles (History D).

As seen in Figure 6, even with a linear-elastic soil slope, computed transient deck displacements are roughly twice of those obtained from the Perform-2D analysis with 1-D springs fixed in space. When the slope is allowed to undergo shaking-induced permanent deformations (elasto-plastic soil), deck displacements are even larger. They reach about 1 foot with linear-elastic piles, and are larger with elasto-plastic piles. It is noteworthy that the transient deck displacements in the latter case are larger than the permanent deck displacement at the end of shaking. This suggests that transient rotations of plastic hinges could be greater than rotations derived from the permanently deformed piles at the end of shaking.

The above observation calls into question the rationale for evaluating the seismic performance of a wharf by combining the results from a 1-D spring model with those obtained from a SSI analysis with soil continuum. Since the dynamic characteristics of the two models are drastically different, there is no justification for trying to synchronize and/or superimpose transient loading conditions of these two models. Even if the peak inertial loading in the 1-D spring model has subsided before the piles are dragged down-slope by permanent soil displacements in the SSI model, they may not represent sequential load cases. In reality, these two loading conditions may well interact with, or amplify, each other as shown in the example above.

To further elucidate the complex interaction between slope and wharf structure, simulations were also run without the wharf. Shaking-induced slope displacements with and without the wharf are shown in Figure 7. The deformed slope shapes for the two cases are plotted in Figure 8, and a summary of computed shaking-induced wharf and slope displacements is presented in Table 3. Maximum slope displacements without the wharf are approximately twice as large, indicating that the piles are supporting the slope by acting as a retaining structure.

Table 3. Computed displacements for SSI analysis with soil continuum.

Time History	Maximum Lateral Deck Displacement (ft)	Permanent Lateral Deck Displacement (ft)	Maximum Permanent Slope Displmt (ft)	Max. Permanent Slope Displmt. w/o Wharf (ft)
Kocaeli/Duzce	1.7	1.0	1.0	2.2
Kocaeli/Yarimca	1.7	1.6	1.5	3.1

In addition to producing much greater displacements than the 1-D spring model, the SSI analysis with soil continuum also imposes different deformation patterns on the piles. This results in different strength and ductility demands, as well as different locations of the plastic hinges. This huge discrepancy in behavior is demonstrated in Figure 9 which shows snapshots of the deformed models corresponding to the maximum deck displacement, both plotted at an exaggerated scale. In the Perform-2D analysis, most of the pile deformation occurs above ground, with only a single plastic hinge forming at the top of the most landside pile. In the FLAC analysis, the piles are mostly deformed within the slope, with plastic hinges forming in all of the piles.

Conclusions

Shaking-induced permanent deformations of underwater slopes/dikes have a significant effect on, and most often govern, the overall seismic performance of pile-supported wharves. Even though both structural and geotechnical modeling techniques have undergone tremendous advances over the last 30 years, most wharf designers still disregard the latter. While working with increasingly sophisticated state-of-the-art structural models, they keep relying on simple 1-D springs to simulate the complex dynamic behavior of soils.

A generic pile-supported wharf was analyzed with different modeling techniques, with the objective of evaluating the reliability of the "structural-engineering" approach to SSI analysis. The vastly different results obtained from this exercise highlight the serious deficiencies of over-simplified SSI models. Specifically, it is shown that shaking-induced down-slope deformation of the soil mass may cause significant lateral wharf displacements which are ignored when utilizing 1-D springs. These displacements result in vastly different strength and ductility demands on the piles than those obtained from the simple structural models. Enter the added complexity of shaking-induced pore-pressure generation and associated strength loss of saturated soils, and further refining the structural models while ignoring the soil continuum would appear to be barking up the wrong tree.

Acknowledgments

The writers wish to thank Philip Meymand of URS for assistance in obtaining the earthquake time histories. Many of the issues explored in this paper were inspired by discussions with Jeff Aldrich of CH2MHill and Said Salah-Mars of URS.

References

American Petroleum Institute, (1987). *Recommended practice for planning, design and construction of fixed offshore platforms*. API Recommended Practice 2A (RP2A). Seventeenth Edition.

CSI (1997). *SAP2000 - Integrated Structural Analysis & Design Software*. Computers and Structures, Inc., Berkeley, California, 1997.

Itasca Consulting Group (2001). *FLAC Ver. 4.0, Fast Lagrangian Analysis of Continua, User's Manual*. Minneapolis, Minnesota, USA.

Ram International (2002). *Perform-2D, User's Guide*. Ram International, Carlsbad California.

Roth, W.H., Fong, H., and de Rubertis, C. (1992*), Batter Piles and the Seismic Performance of Pile-Supported Wharves*. ASCE Specialty Conference, Ports'92, Seattle, Washington, USA.

URS Corporation (2002), *Development of foundation springs and ground motions, Mac Arthur Avenue On-Ramp*. Report to Caltrans Office of Structure Contract Management, URS, Roseville, California, Nov. 6.

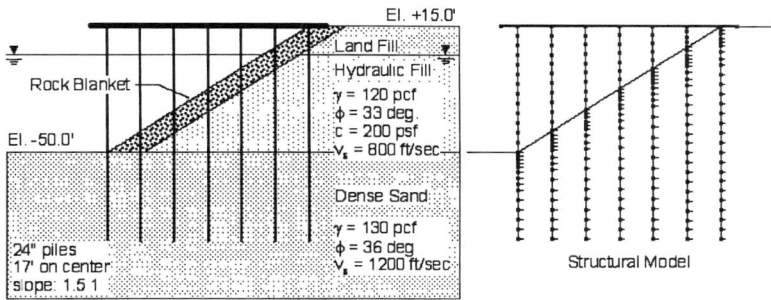

Figure 1. Wharf cross section and structural model with nodes and p-y springs.

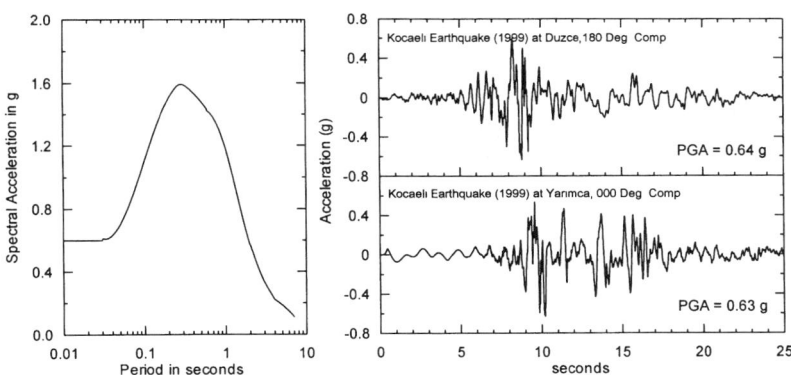

Figure 2. Design spectrum and matched time histories.

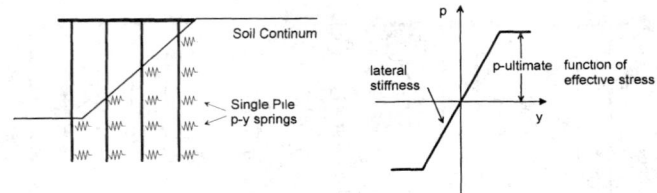

Figure 3. P-y springs connecting piles with FLAC mesh.

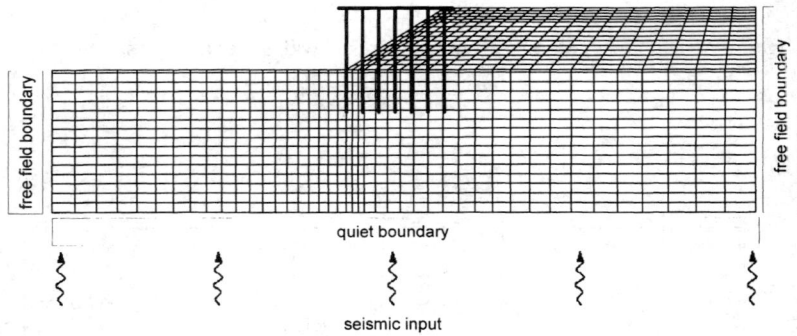

Figure 4. FLAC mesh with boundary conditions for dynamic analysis.

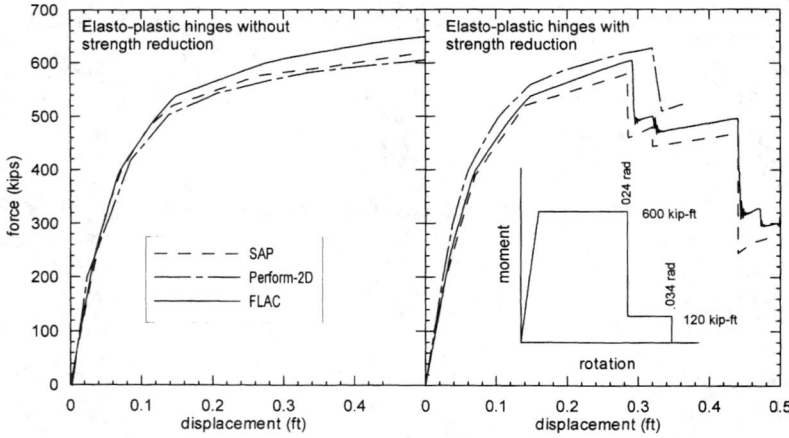

Figure 5. Pushover analyses with and without degrading hinges.

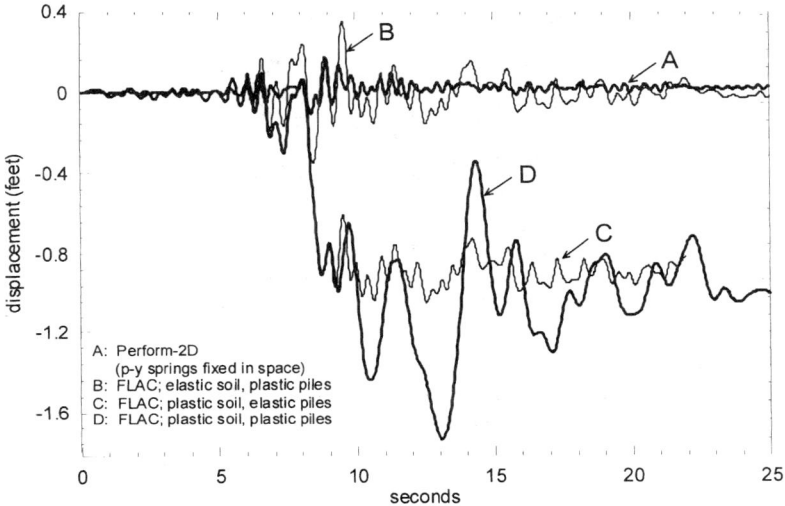

Figure 6. Deck displacement versus time for Kocaeli-Duzce-180 time history.

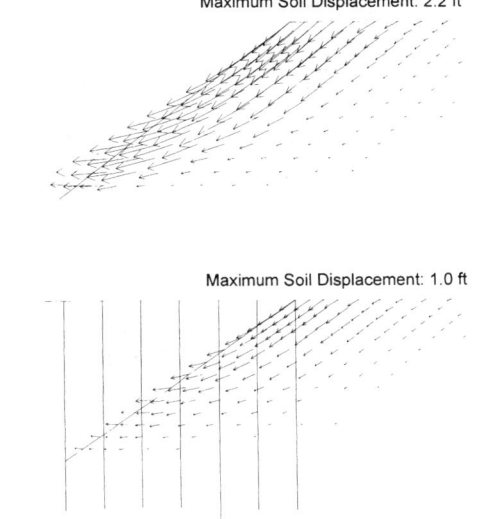

Figure 7. Shaking-induced displacement of slope, with and without wharf: Kocaeli-Duzce-180 time history.

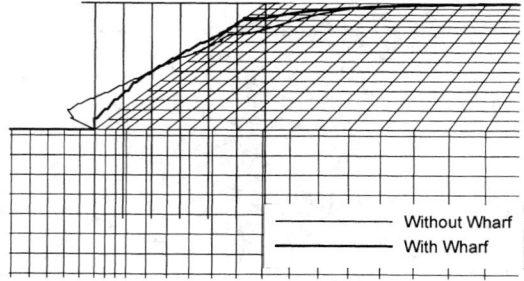

Figure 8. Outline of deformed mesh, with and without the wharf: Kocaeli-Duzce-180 time history.

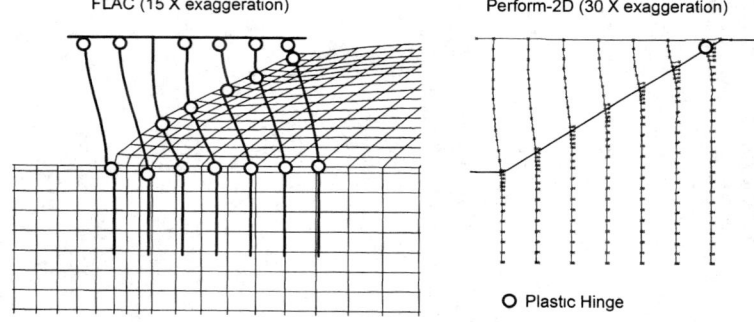

Figure 9. Deformed mesh for FLAC and Perform-2D: Kocaeli-Duzce-180 time history.

Analyzing the Seismic Performance of Wharves, Part 2: SSI Analysis with Non-Linear, Effective-Stress Soil Models

W.H. Roth[1] and E.M. Dawson[1]

Abstract

The 1989 Loma Prieta earthquake provided a unique opportunity to test the credibility of nonlinear, effective-stress based SSI analysis techniques. Three pile-supported wharves at the Port of Oakland, which suffered various degrees of structural damage, were analyzed with FLAC using a nonlinear soil model with fully coupled pore-pressure generation. Wharf piles were modeled with elasto-plastic beam elements connected to the soil grid with nonlinear springs. The wharf models were subjected to seismic motions derived from nearby seismometer recordings of the 1989 Loma Prieta earthquake. A comparison of computed and observed seismic behavior confirmed the viability of using nonlinear, effective-stress SSI analysis for predicting the seismic performance of wharf structures.

Introduction

In Part 1 of this paper (Roth, et al, 2003) it was shown that the seismic performance of pile-supported wharves is governed by the dynamic response of the foundation soils. This seemingly obvious fact is ignored by conventional structural analyses where the complex soil continuum is simply replaced by 1-D springs. Particularly, where underwater slopes and/or hydraulic-fill backlands experience shaking-induced strength loss and permanent deformations, this approach produces questionable results at best. Seismic analysis in such cases must consider coupling the structural wharf model with a soil-continuum model which is capable of plastic yielding as pore pressures are building up during shaking.

While geotechnical engineers have begun utilizing nonlinear effective-stress soil models for analyzing the seismic performance of wharf slopes/dikes and backlands, the wharf structure proper usually is not included in their models. Instead, wharf designers still perform their own structural analyses, all but ignoring the dynamic response of the soil. A precedent-setting departure from this approach was the fully integrated soil-structure interaction (SSI) analysis performed for the design of a pile-supported wharf at Pier J in the Port of Long Beach (Roth, et al, 1992). The outcome of this work challenged conventional wharf-design philosophy counting on a stable underwater slope/dike to provide lateral support for the wharf piles. Being able to account for and accommodate shaking-induced slope displacements, it was possible to avoid costly ground-improvement measures in this case.

---

[1] URS Corporation, Los Angeles, CA.

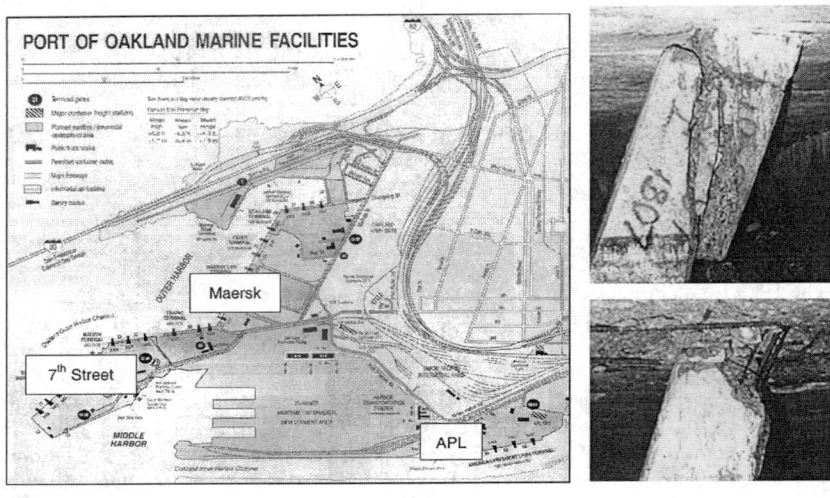

Figure 1. Port map; and pile damage at 7th Street terminal

Seismic Analysis of 3 Case Histories

The 1989 Loma Prieta earthquake provided a unique opportunity to test the applicability of nonlinear, effective-stress based soil models for SSI analyses of wharves. The three wharf structures selected for this exercise include the Seventh Street terminal at Berths 34-37, the APL terminal at Berths 60-63, and the Maersk terminal at Berths 23-25 of the Port of Oakland (Figure 1), which experienced severe to minimal structural damage, respectively (Port of Oakland, 1990). Photographs of typical pile damage suffered at the Seventh Street wharf are also shown in Figure 1.

The wharf sections and subsurface conditions shown in Figures 2 and 3 were developed based on as-built construction plans and recent geotechnical investigations (URS, 2000). Generally, the port backland areas consist of loose, hydraulically placed sand fills over soft Young Bay Mud overlying dense Merritt/Posey Sands and stiff Old Bay Mud. Sand fills placed above the water level are denser as a result of being compacted during placement.

Dynamic simulations were performed with the 2-D explicit finite difference code FLAC which has been verified extensively against closed-form solutions, physical models, and field-testing (Itasca, 2000). FLAC and its predecessor codes have been utilized in the last 20 years for analyzing a wide spectrum of static and dynamic problems in geotechnical engineering.

Soil Model. Soil behavior is represented by a Mohr-Coulomb, linear elastic/perfectly plastic model. For the submerged hydraulic-fill sands, this model was coupled with an empirical pore pressure generation scheme. Pore pressure is generated in response to shear-stress cycles using the cyclic-stress approach (Seed et al., 1976; Seed, 1979),

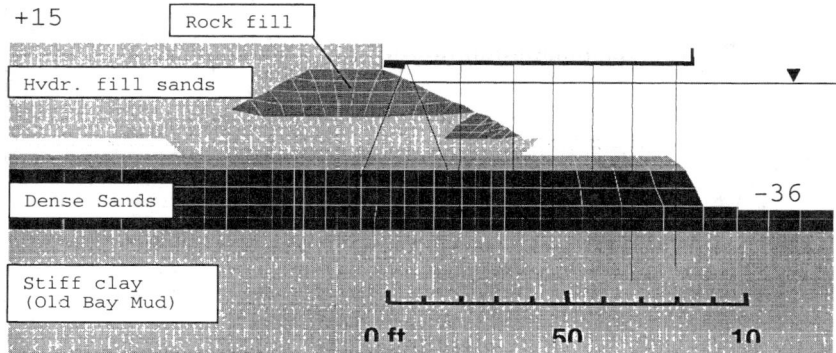

Figure 2. Typical wharf section, Seventh Street Terminal.

which was modified so that pore pressures are generated incrementally during shaking.

This practice-oriented nonlinear, effective-stress based model was first developed in 1981 for the seismic-performance evaluation of Pleasant Valley dam for the Los Angeles Department of Water & Power (Roth, et al, 1991). It has since been tested by means of centrifuge shaking tests (Roth and Inel, 1993), and full-scale earthquake case histories involving dams (Roth et al, 1993; Inel et al, 1993; Bureau et al, 1996). A detailed discussion of this model was presented most recently by Dawson et al (2001).

Soil properties used in the FLAC simulations are shown in Figure 4. The cyclic-strength curve shown on the left was estimated based on average SPT blow counts, $(N1)_{60} = 12$ for the hydraulic fill sands. This N-values was also used to estimate the post-liquefaction residual shear strength (S_{ur}) of the loose hydraulic fill after Seed and Harder (1990).

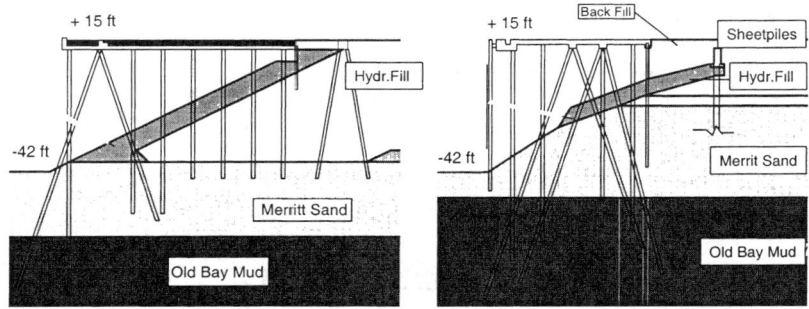

Figure 3. Typical wharf sections for APL (left) and Maersk (right) Terminals

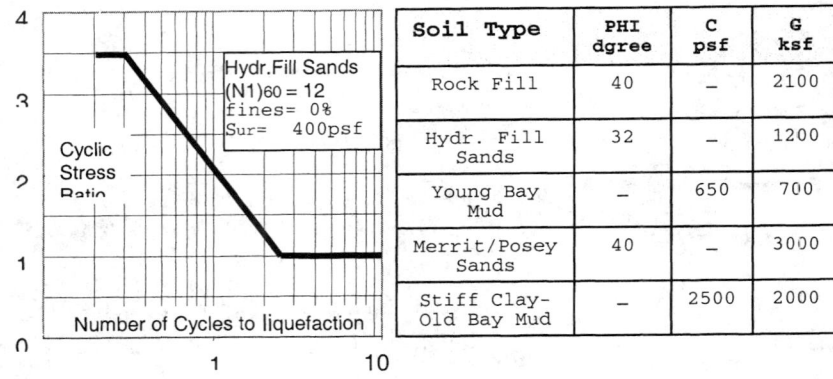

Figure 4. Soil properties.

Structural Model. The wharf piles and deck are represented by elasto-plastic beam elements capable of developing plastic hinges at pre-determined yield moments. The yield moments for the 16- and 18-inch square pre-cast concrete piles of the 7th Street wharf were estimated at 230 and 320 kip-ft, based on as-built plans and field inspections. Pile-deck connections were estimated to have a bending-moment capacity of about 140 kip-ft. These moment capacities and the piles' section properties (axial and bending stiffness) had to be adjusted to account for pile spacing, so that they represent equivalent properties per unit length of wharf.

The structural nodes of the piles are connected with the soil mesh through elasto-plastic p-y springs representing the lateral load-displacement behavior of single piles (Figure 5). The parameters for these springs were derived using procedures recommended by Reese, et al. (1974), and Matlock (1970). The springs' ultimate (yield) strength automatically decreases during shaking, as effective stresses decrease in the surrounding soil elements due to pore-pressure buildup. Thus, the soil continuum is allowed to flow past the beam elements, thereby simulating actual conditions of relatively widely spaced piles.

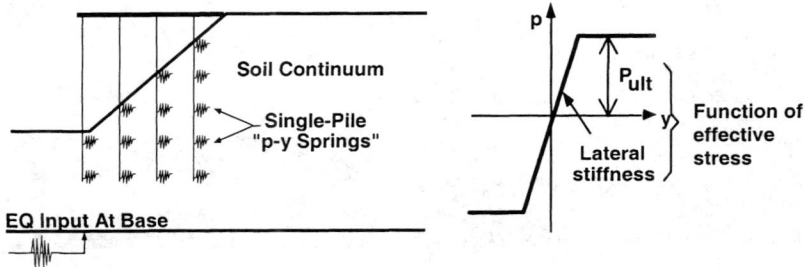

Figure 5. Nonlinear soil springs connect the piles with the soil mesh.

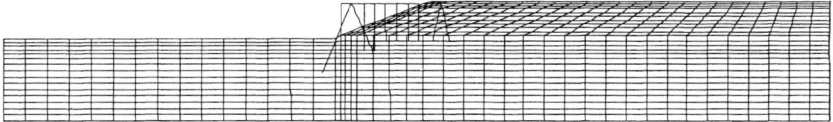

Figure 6. Typical FLAC model mesh (shown here is the APL-wharf model)

Boundary Conditions and Damping. The bottom boundary of the numerical mesh is fixed vertically and is subjected to horizontal shaking via an input acceleration history. In order to minimize lateral wave trapping and interference, the left and right sides of the mesh are absorbing boundaries with real-time feedback from 1-D "free-field" computations simulating level ground conditions.

The linear-elastic/perfectly plastic Mohr-Coulomb constitutive model produces hysteretic damping at large strains upon reaching the yield strength. To provide damping for small strains in the elastic range, Rayleigh damping with a critical damping ratio of 0.05 at a center frequency of 4 Hz was used for all simulations.

Model Setup. Figure 6 shows the FLAC model mesh for the APL terminal. This mesh, which is typical for all 3 models analyzed, consists of approximately 1,000 quadrilateral soil elements. Free water was represented by hydrostatic boundary loads acting on the channel bottom and underwater slope. First, gravity in situ stresses were computed for the existing slope, and then the wharf deck and piles were installed. After reaching equilibrium, the model was subjected to seismic shaking by applying an input motion at the bottom.

Input Motions. The horizontal acceleration histories applied to the models, were derived from seismometer records of the 1989 Loma Prieta earthquake obtained at Berth 25 in the Outer Harbor. With its epicenter located about 50 miles south of the port, this Magnitude-7 event produced about 8 seconds of strong shaking with a peak ground acceleration (PGA) of 0.29g at the site. The records shown in Figure 7 were decomposed to obtain motions perpendicular to the respective pierhead lines, and then deconvoluted with SHAKE to obtain equivalent base-input motions for the various models.

Analysis Results

Seventh Street Terminal. Displacement-vector plots depicting computed wharf-structure and dike deformations at the end of shaking are shown in Figure 8. Accurate measurements of actual wharf displacements suffered during the 1989 Loma Prieta earthquake do not exist. However, estimates based on adding up crack widths and offsets mapped in the pavement behind the wharf suggest that the wharf deck and dike moved outward about 4 to 6 inches. This compares with the computed lateral deck displacement of 3 inches and displacements of the embankment ranging from 8 inches at the toe to about 3 inches at the top.

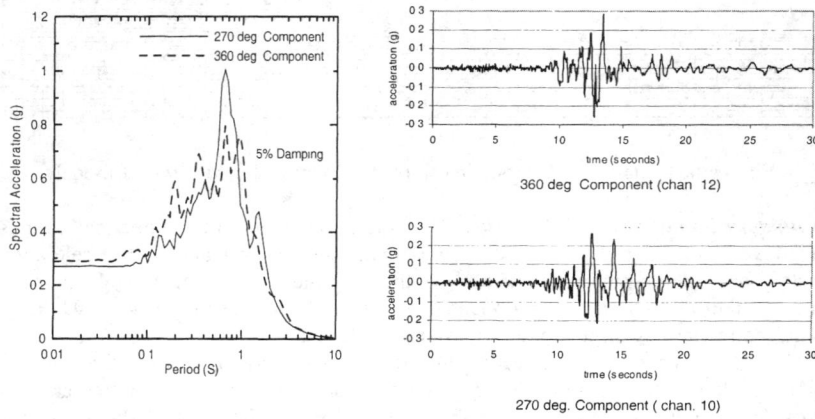

Figure 7. Outer Harbor record of 1989 Loma Prieta earthquake

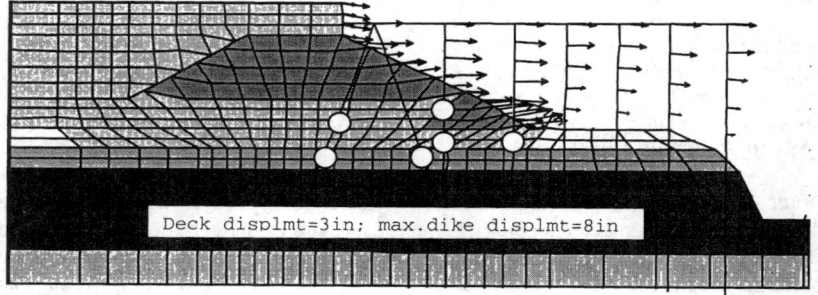

Figure 8. Shaking-induced displacements (white circles indicate plastic hinges)

Figure 9 presents a contour plot of shaking-induced pore-pressure ratios. The ratio of 1.0 within the hydraulic-sand fill beneath the landside half of the rock-fill dike indicates that 100% liquefaction has been reached in this zone. The buildup of pore pressure versus time in this particular zone is shown in the inset of this figure along with the acceleration history of the input motion.

Time histories of shaking-induced bending moments computed at the pile-deck connections are shown in Figure 10. For the batter-pile connections (Row G) the estimated moment capacity of 140 kips-ft was exceeded, which would account for the heavy damage observed at these locations as shown in Figure 1. In addition to yielding at the batter-pile connections, the analysis results also indicated the formation of several plastic hinges in the piles below grade, which are marked with the white circles in Figure 8.

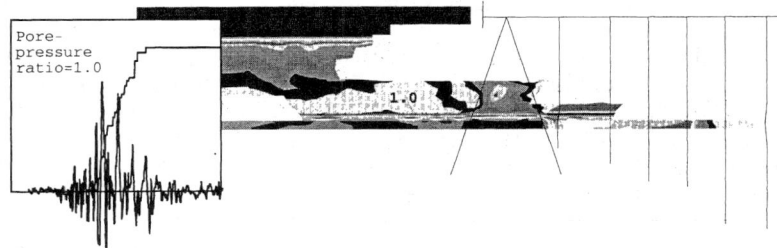

Figure 9. Contours of shaking-induced pore pressures – 7th Street Terminal
(Inset: histories of acc. input and PP build-up beneath the dike)

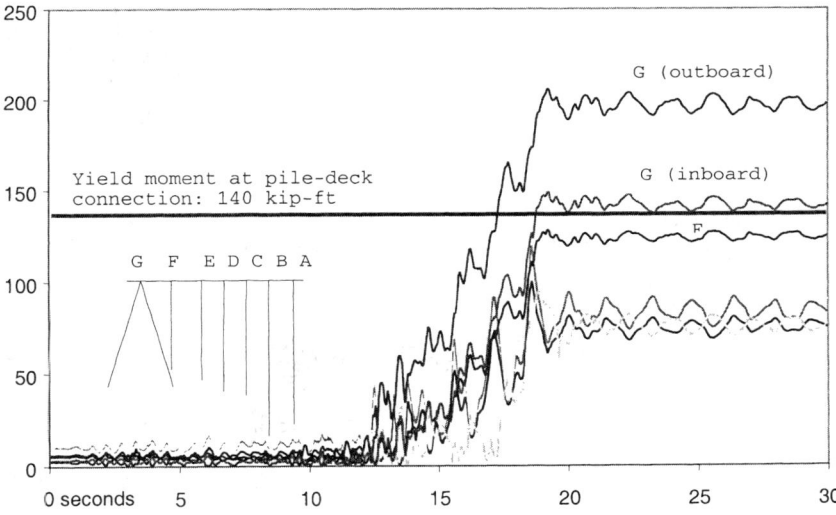

Figure 10. Time histories of bending moments at pile-deck connections

APL and Maersk Terminals. Shaking-induced deck displacements computed for the APL (Berths 60-63) and Maersk (Berths 23-25) terminals were only about 1 and 0.2 inches, respectively, without any plastic hinges developing in the piles. Again, these results agree with observations made in the field where no significant structural damage was found at these terminals.

The superior seismic performance of these two wharves has less to do with the quality of the wharf structures per se, than with the strong seismic performance of the slope and backland soils. As can be seen from the contour plot and time-history inset shown in Figure 11, for the APL wharf, the maximum shaking-induced pore-pressure ratio at the end of shaking only reached 0.4. This means that no portion of the slope or backland came even close to liquefaction.

The backland soils at the Maersk terminal were significantly more susceptible to liquefaction than at APL. However, as shown in Figure 12, the soils which suffered pore-pressure ratios as high as 0.9 and, hence, came close to liquefaction, were too shallow to have any appreciable adverse impact on the wharf structure.

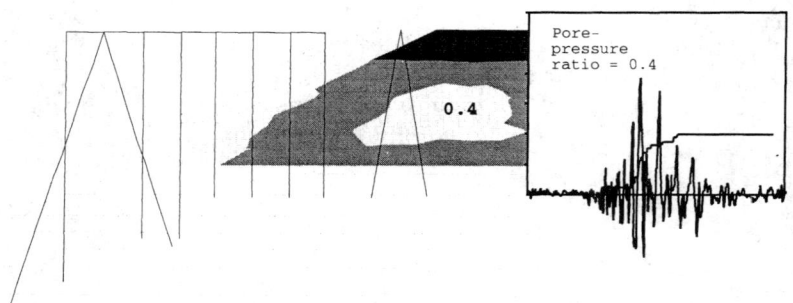

Figure 11. Contours of shaking-induced pore pressures – APL Terminal
(Inset: histories of acc. input and PP build-up in the backland behind the wharf)

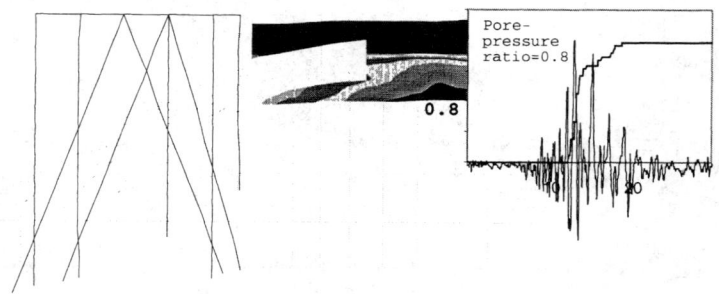

Figure 12. Contours of shaking-induced pore pressures – Maersk Terminal
(Inset: histories of acc.input and PP build-up in the backland behind the wharf)

Conclusions

The case histories discussed in this Part 2 of our paper demonstrate that it is possible with rather simple models to capture the essence of the complex dynamic interaction of wharves with their liquefaction-susceptible foundation soils. Particularly, for pile-supported wharves with relatively weak multi-lift dikes, thin rock blankets on cut slopes, and/or hydraulically filled backlands, it is the soil, rather than the structure, which governs the seismic performance. Even so, most wharf structures, to this day, are still being designed and analyzed by structural engineers without properly accounting for the soil.

Acknowledgments

The authors thank Mr. Thomas LaBasco of the Port of Oakland for the opportunity to perform the analyses reported herein as part of the overall WESP program, and for providing the photos of the damaged piles. We also acknowledge the entire WESP team for contributing to a rich exchange of information during the various workshops held for this project.

References

Bureau, G., Inel, S., Davis, C. A., Roth, W.H., (1996), "Seismic Response of Los Angeles Dam during the 1994 Northridge Earthquake." USCOLD 1996 Annual Meeting, Los Angeles, CA..

Dawson, E. M., Roth, W., Nesarajah, S., Bureau, G., Davis, C., (2001), "A practice oriented pore-pressure generation model." 2nd Int. FLAC Symp. Lyon, France.

Inel, S., Roth, W.H., and de Rubertis, C., (1993), "Nonlinear Dynamic Effective-Stress Analysis of Two Case Histories." 3rd Int. Conf. Case Histories in Geotech.l Eng., St. Louis, MO.

Inel, S., deRubertis, C., Roth, W.H., (1993), "Nonlinear Dynamic Analyses of Two Case Histories." 3rd Int. Conf. Case Histories in Geotech. Eng, St. Louis, Missouri.

Itasca Consulting Group, (2000), "FLAC, Fast Lagrangian Analysis of Continua, Version 4.0." Itasca Consulting Group, Minneapolis, Minnesota.

Matlock, H. (1970), "Correlations for design for design of laterally loaded piles in soft clay." Offshore Technology Conf., Houston, Texas. Paper No. OTC 1204.

Port Of Oakland (1990), "Loma Prieta Earthquake-After Action Report." Oakland, CA.

Reese, L.C, Cox, W.R. and Koop, F.B. (1974), "Analysis of laterally loaded piles in sand." Offshore Technology Conf., Houston, Texas. Paper No. OTC 2090.

Roth, W., Bureau, G., and Brodt, G., (1991), "Pleasant Valley Dam: An Approach to Quantifying the Effect of Foundation Liquefaction." ICOLD Conf. Vienna, Austria.

Roth, W.H., Fong, H., deRubertis, C., (1992), "Batter Piles and the Seismic Performance of Pile-Supported Wharves." ASCE Conf. Ports '92, Seattle, WA.

Roth, W.H., Inel, S., (1993), "An Engineering Approach to the Analysis of VELACS Centrifuge Tests." Int. Conf., Verification of Numerical Procedures for the Analysis of Soil Liquefaction Problems, Davis, CA.

Roth, W.H., Inel, S., Davis, C., Brodt, G., (1993), "Upper San Fernando Dam 1971 Revisited." 10th Annual Conf., Assoc.State Dam Safety Officials, Kansas City, MO.

W.H. Roth, Dawson E., Mehrain M., Sayegh A., (2003), "Analyzing the Seismic Performance of Wharves, Part 1: Structural-Engineering Approach." 6th U.S. Conference on Lifeline Engineering (TCLEE2003), August, Long Beach, CA.

Seed, H.B., Martin, P.P., and Lysmer, J. (1976), "Pore-Water Pressure Changes During Soil Liquefaction." J. Geotech Eng. ASCE, Vol. 102, No. GT4.

Seed, H.B., (1979), "Soil Liquefaction and Cyclic Mobility Evaluation for Level Ground During Earthquakes." J. Geotech Eng. ASCE, Vol. 105, No. GT2.

Seed, H.B., Harder, L., (1990), "SPT-based Analysis of Cyclic Pore Pressure Generation and Undrained Residual Strength." Seed Memorial Symp. Berkeley, CA.

URS (2000), "Geotechnical Report on Ground Motions, Phase I, prepared for Port of Oakland, CA, Berths 23-25 and Berths 60-63." WESP Project.

Seattle Alaskan Way Seawall
Emergency Earthquake Evaluation and Repair

Farhad Rowshanzamir, P.E., Senior Engineer, Reid Middleton Inc., Everett, WA
Dave Swanson, P.E., S.E., Director/Structural, Reid Middleton Inc., Everett, WA
John Buswell, Bridge Engineer, City of Seattle Department of Transportation
Daniel Mageau, P.E., Principal, GeoEngineers Inc., Seattle, WA

The History and The Present

The History

The City of Seattle began the construction of the Alaskan Way Seawall in 1934which would complete a system of seawalls spanning 2,400 m (7,900 ft) along the waterfront. The November 8, 1934, Engineering News Record featured Seattle's challenge of building a seawall. The challenge was described as: "The need for a modern type of seawall construction along this harbor front thoroughfare has been recognized as a desirable improvement for more than twenty years: a deterrent factor has been the high cost entailed by the physical problems of soft bottom, and a 16-ft tide range, with the attendant marine-borer menace to timber construction."

A design concept developed in Denmark and used in many European countries (but only once in the United States), consisting of concrete and steel vertical face tied to and integral with a pile-supported relieving platform, was adopted. The timber platform is located approximately 4 m (13 ft) below the street level of Alaskan Way. Steel H-pile or timber piles were driven into the dense soil layer below the Elliot Bay seafloor soft mud. A pre-cast concrete section was set on top of the piling and extended up to the roadway elevation. The platform was constructed of untreated timber and it was believed that, with the timber elements safely buried in soil, marine borers would not constitute a threat.

The portion exposed to the inter-tidal splash zone was constructed of pre-cast concrete sections 2.4 m wide by 6 m high and 0.5 m thick (8 ft x 20 ft x 18 in). The panel is "L" shaped with the horizontal section containing block outs that allow the timber relieving platform to key into the concrete panel. One of the key structural features of the seawall system is this connection of the cap beams, running perpendicular to the wall, and the heel of the wall. These beams are notched to receive the tops of the batter support piles. They also serve as the support for the 100 mm (4") planking that makes up the relieving platform and supports the weight of the roadway fill. Finally, the cap beams work as tie backs for the seawall system by means of a grouted key way in the horizontal section of the slab.

An important feature to the long-term service of the wall was the performance of the concrete face exposed to constant tidal action. In 1934, this was not easy to attain. Maximum concrete cover was provided between the steel and seawater and the casting procedures required one of the first uses of vibrators to consolidate the concrete and improve its density. The ready-mix concrete design required a finely ground cement and a well graded sand and gravel aggregate with a low water cement ratio and a slump of 90 mm (3-½") provided high quality, low permeability concrete.

The Present Lifeline System

Although presently the emphasis has moved from a working waterfront with primarily shipping and fishing business to a more mixed use area, the importance of the Alaskan Way Seawall is no less today than 70 years ago. The waterfront now has an Alaska cruise ship terminal; Washington State Ferry terminal; it services local sightseeing vessels, restaurants, retail shops, and a world class aquarium.

The Seawall provides support for street surface as well as the Alaskan Way Viaduct (Figure 1). Together they account for more than 120,000 vehicle trips per day, the second largest average daily traffic count within the City. The Viaduct provides access to local freight centers, Port of Seattle container ship terminals, links to Sea-Tac Airport and connections to Seattle City center. Recent engineering studies, following the 2001 Nisqually earthquake, have established that failure of the Seawall would likely result in loss of lateral support to the Viaduct foundations and may contribute to its collapse. Computer modeling of the traffic corridor indicates that in the event the viaduct became unusable, travel through downtown Seattle would double, with the very likely effect of system gridlock. The gridlock formed on the downtown streets would back up onto the Interstate-5 (I-5) ramps, and thereby reduce I-5 travel speed through the downtown Seattle area from 25 mph to 10 mph during peak periods. This, in turn, would affect traffic operation of I-405 and other parallel arteries. Region wide, traveling public would encounter an additional 15,499 hours of delay in the peak periods per day following an earthquake.

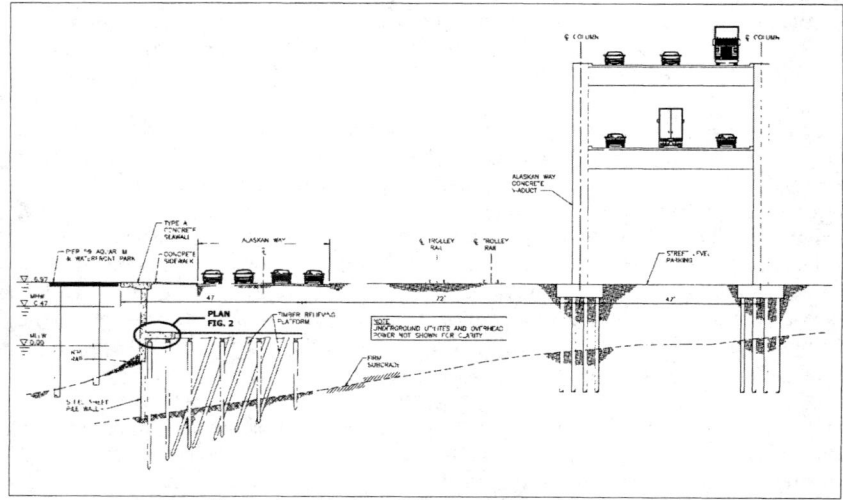

Figure 1 – Alaskan Way Seawall and Viaduct

The fill section between the relieving platform and the surface of Alaskan Way is utilized by many of the local utilities to place service infrastructure. Major utilities include: telephone, 12" high pressure gas, 4" gas, 6" steam duct, (2) 115 kv transmission lines, 13.6kv electrical distribution line, (4) fiber optic duct banks, 20" water main, 36" storm drain, 18" sanitary sewer, 36" sanitary sewer and a 48" combined sewer. The two 115kv transmission lines are major regional systems that provide primary power to the central area of the City. The 48" combined sewer is the main north/south pipe that carries effluent to the West Point Sewer Treatment plant. Disruption of these utilities for any length of time would cause significant economic and quality of life hardship to the customers they serve.

The Nisqaully Earthquake & The Seattle Seawall

The Seismic Event

At 10:54 a.m. on February 28, 2001, a magnitude 6.8 earthquake occurred with its epicenter located between Olympia and Tacoma, Washington. Named the Nisqually Earthquake for its proximity to the Nisqually River delta, the earthquake caused varying degrees of damage throughout the Puget Sound region. The hypocenter for that magnitude is estimated to have been at a depth of approximately 51 kilometers.

Post Earthquake Observations

Roadway Settlement

Immediately following this event, City of Seattle Department of Transportation (SDOT) conducted a visual inspection of Alaskan Way and discovered pavement settlement in the vicinity of Pier 59 at the north end of Waterfront Park. Topographic survey showed movements approximately 30 m (100 ft) long and up to 75 mm (3") deep and a horizontal spreading of up to 40 mm (1-½") along the southbound traffic lanes.

Underwater Survey

Underwater survey was conducted in the vicinity of pier 59 Waterfront Park in an attempt to determine the possible cause of the observed settlements and to assess possible underwater damage that may have resulted from the earthquake. The underwater investigation included the toe of approximately 150 m (500 ft) of the wall centered on the area of the roadway settlement and revealed no submarine movement of the toe material. However, lack of barnacles and other marine growth at several locations near the mud line pointed to the possibility of movements in the concrete wall and/or the riprap.

Seaside Investigation

Preliminary investigation, coupled with the Seawall's age, condition, and proximity to the Alaskan Way Viaduct structure led to a more thorough investigation. Approximately 115 m (380 ft) of the seawall was examined at minus tide. Seabed profile was reviewed in detail and it was concluded that significant waterside deviation from the expected norm was evident in the form of underwater settlements of the seabed in the immediate vicinity of the maximum road surface settlement. The riprap elevation also appeared to be lower in that area; and some additional concrete panels, which were not shown in any of the documents, appeared to have been put in place to provide cover and protection for the steel sheet pile wall. Some of those concrete panels were displaced, thus raising the possibility of increased intrusion of seawater in that region. Such suspected free flow of seawater through the wall would provide an environment for marine borers to degrade the untreated timber platform and piles as well as enable the loss of fines from behind the concrete seawall.

Roadway Side Test Pit and Boring
The roadway side at the location of maximum observed settlement was then investigated. Presence of an elaborate array of lifeline utilities presented a challenge in the location of the test pit. Added to this challenge was the need to accommodate the high daily traffic count of that corridor while minimizing the disruptive effect of construction activities to the local businesses. The test pit also needed to be positioned on top of the of saw tooth connection point between the cap beams and the concrete slab at the heel of the wall. That particular detail (Figure 2) was identified at an early stage as a possible weak link as it constituted the main shear transfer mechanism between the concrete retaining wall and the timber relieving platform.

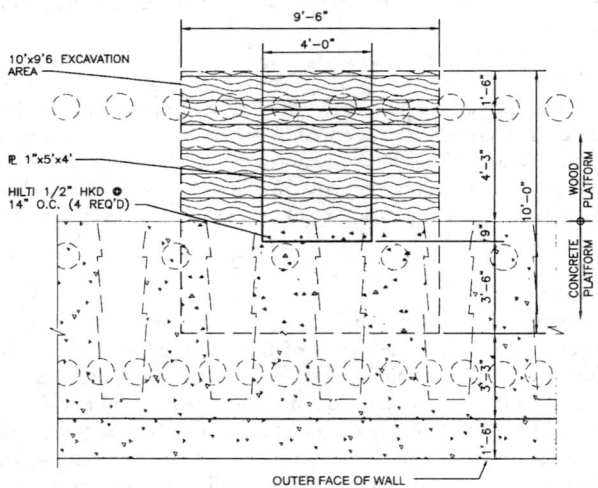

Figure 2 - Seawall and Timber Cap Beam Connection

A 3 m x 3 m (10 ft x 10 ft) excavation pit was excavated behind the westerly curb line on Alaskan Way. The sidewalk pavement was removed and fill material was machine excavated down approximately 4 m (12.5 ft) to the elevation of the top of concrete seawall footing. Hand excavation was performed below the platform and within the pile cap areas. A 1.2 m x 1.2 m (4 ft x 4 ft) hole was dug to approximately 1 m (3 ft) below the top of the concrete seawall footing. Observed conditions were:

- The fill material above the platform elevation was loose sand and gravel. In-place density test measurements were very low for these types of soils and caving of the fill material was observed to as much as 1 m – 1.2 m (3 ft – 4 ft) behind the shoring.

- No "intact" timber platform was found. Sand and gravel with brown organic in-fill material was observed between the footing joint of the concrete seawall and within the area where the platform would have been. This material appeared to be the remnants of the decayed timber-relieving platform.

- No timber piling below the seawall footing was found and a 0.3 m to 0.6 m (1 ft to 2 ft) deep void space immediately below the concrete footing surface was observed. Only a corroded drift pin was discovered.

Upon discovering that no "intact" timber platform remained, a series of test borings extending out from the test excavation pit were conducted. The objective of the test borings was to further identify fill material density surrounding the settlement area, collect material samples, and define approximate limits of decomposed or damaged relieving platform. The results from the boring samples confirmed presence of loose sand and gravel above the platform, existence of voids throughout and at the platform level, and remnants of highly decomposed, moderately decomposed, or intact wood.

Geo-technical Analysis

Liquefaction-Induced Settlement
Liquefaction-induced ground settlement that could have occurred as a result of ground shaking during the Nisqually Earthquake, assuming no pre-earthquake voids in the ground, were analyzed using the methods developed by Takimatsu and Seed (1987). As conditions favorable to liquefaction namely, loose to medium dense sand and gravel below the ground water level, were present at the site, a high potential for liquefaction exists.

Typically, the liquefaction potential of a site is evaluated by comparing the cyclic shear stress ratio induced by an earthquake with the cyclic shear stress ratio required to cause liquefaction. The latter was estimated using an empirical procedure based on the in-situ static ground stresses, the blow count data obtained during sampling in the borings, and the design earthquake magnitude (Youd and Idriss, 2001). The former utilized the characteristics of the Nisqually earthquake, which was centered approximately 58 km (36 mi) southwest of the site. Based on the PNSN (Pacific Northwest Seismograph Network) Peak Acceleration Map developed for that event, the PGA (peak ground acceleration) at the Seawall was estimated to be between 0.11g and 0.17g.

Results of the liquefaction indicate that the very loose to medium dense sand between depths of about 3 to 10 m (10 to 33 ft) liquefied and settled during the earthquake while the denser soils below a depth of 10 m (33 ft) did not liquefy and settle. The 3m (10 ft) thick cap of non-liquefiable soils and rigid concrete pavement at the ground surface can arch over looser zones of soil or void beneath it, thus reducing observable settlement at the ground surface. Arching may account for the small liquefaction-induced settlements observed outside the study area. However, even though little to no settlement was observed away from the depressed zone within the study area to date, it is likely that surface reflections will develop over time or during the next earthquake as the effect of arching diminishes. Such settlements would adversely affect travel over Alaskan Way as well as cause damage to the numerous utilities buried in the street.

Liquefaction induced settlements during the 2001 Nisqually earthquake increased the size and caused the collapse of the voids created by the deteriorated condition of the timber platform. This increase in the extent of the settlement is depicted in Figure 3 and the risk of continued settlement remains high until remedial actions are taken.

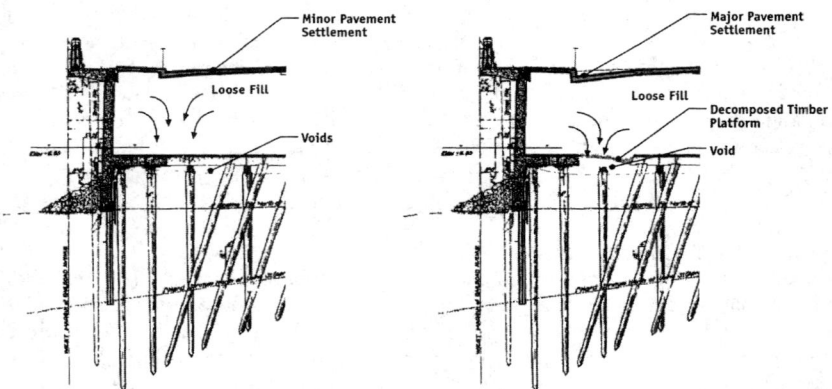

Figure 3 – Surface Settlement

Settlement Estimates Using FLAC

The effects of arching at the site were evaluated using the computer program FLAC (Fast Lagrangian Analysis of Continua). FLAC is a two-dimensional explicit finite difference program developed and licensed by Itasca Consulting Group, Inc. (Itasca, 1995). Mechanical properties used to represent the soils in the FLAC analyses for this project are summarized in the table below. The strength parameters (friction and cohesion) of the loose sand fill are estimated from SPT values and soil type. Shear and bulk moduli were estimated using published correlation with soil index properties (Bowles, 1986).

Table 1 - Soil Parameters Used in FLAC Analyses

Soil Type	Unit Density (pcf)	Friction Angle (psf)	Cohesion (psf)	Shear Modulus (psf)	Bulk Modulus (psf)
Loose Sand Fill	120	26	0	$7e^4$	$2e^5$
Timber Platform	68	40	$1e^5$	$4e^6$	$1e^7$
Concrete Wall/Ftg.	150	36	$4e^5$	$4e^7$	$8e^7$

Note: 100 pcf = 1600 kg/m3 and 21 psf = 1 kPa

Settlements calculated by FLAC for the 1.8 m (6 ft) wide void case decrease from 1.0 m (3.3 ft) just above the void near the 3.7 (12 ft) depth to less than 0.4 m (1.3 ft) at the ground surface. This 0.6 m (2 ft) reduction in settlement throughout the soil profile above the platform illustrates the effect of arching, which diminishes as the timber platform continues to decay and voids continue to form and widen at that level. Calculated ground settlements that may become evident over time (or during the next earthquake) as the void/loose soil zone widens and the overlying soils weaken.

Structural Evaluation and Analysis

Estimate of Applied Forces
The earth pressures (triangular for the Active condition and rectangular for the Seismic condition) that were used in the structural stability evaluations (where H represents wall height) are:

- Active earth pressure: 6.3H kPa [40H psf]
- Seismic earth pressure: 0.8H kPa [5H psf] for PGA=0.1g
                                              1.4H kPa [9H psf] for PGA=0.2g
                                              2.5H kPa [16H psf] for PGA=0.3g

The seismic earth pressures for a PGA range of 0.1g to 0.2g are considered to approximate the conditions experienced during the Nisqually Earthquake. The seismic earth pressure for a PGA of 0.3g is considered appropriate for an earthquake event with a return interval of about 500 years.

It is important to note that the above seismic earth pressures assume no lateral spreading. Lateral spreading could extend back as far as 30 m (100 ft) from the seawall, thus impacting the existing Alaskan Way Viaduct structure.

Estimate of Resisting Forces
The resisting forces contributing toward the overall stability of the Seawall are:

- Weight of granular fill material, the concrete wall, and cantilevered walkway slab.
- Passive resistance of soils in front of the steel sheet piles.
- Shear resistance from the timber platform at the connection point to the pre-cast concrete wall. This parameter is considered as a variable based on the degree of timber deterioration up to the maximum 100 percent decomposition as observed.
- Withdrawal resistance from the timber pile drift pins at the connection point to the pre-cast concrete wall. This parameter is a variable based on the degree of timber deterioration up to the maximum 100 percent decomposition as observed.
- Load bearing capacity of the steel sheet piling under combined axial and flexural loads. This parameter is considered as a variable based on the degree of corrosion of steel as directly proportional to the degree of decomposition of timber, but only up to a maximum of 40 percent as documented for walls at other locations.

Possible Failure Mechanisms & Their Combination
Anticipated failure mechanisms are depicted in Figure 4.

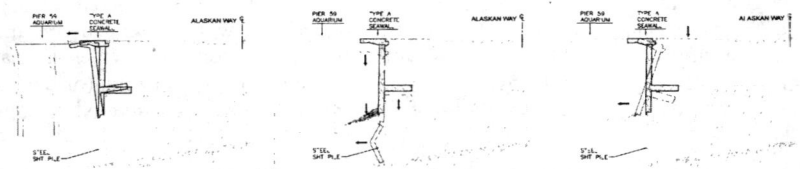

Figure 4 – Failure Modes

Stability Analysis and Assessment of System Reliability
The stability of the Seawall is maintained by the following main parameters in the order of significance:
1. The density of the fill material retained by the bottom concrete slab portion of the wall.
2. The connection to the timber relieving platform and the edge of the concrete slab portion of the wall.
3. The connection to the two rows of the timber piles below the bottom concrete slab portion of the wall.
4. The combined axial and flexural load bearing capacity of the steel sheet piling at the toe of the wall.
5. Passive pressure on the sheet pile.

Variations in the second and third parameters were quantified by assigning a percentage of timber decomposition to the general descriptions used in the geotechnical report. Table 2 shows the decomposition rates that are used:

Table 2 - Levels of Decomposition

Description	Timber Deterioration	Steel Corrosion
Highly Decomposed	75% - 100%	40%
Moderately Decomposed	25% - 75%	20%
Intact	Less than 25%	0%

A variation in the loss of support by the steel sheet pile due to corrosion or other factors was more difficult to quantify with the information available.

The governing rotational factor of safety is a function of the degree of deterioration of the supporting timber and steel elements as well as the loading imposed. The probability of the simultaneous deterioration of timber and corrosion of steel is calculated as the product of each individual degree of deterioration or corrosion (cited in Table 2). Expected overall system safety levels associated with different degrees of timber decomposition are shown in Figure 5. In that figure, the factor of safety during the Nisqually Earthquake is shown as a reference; the factor of safety of 1.00 as the minimum for maintaining stability is highlighted, and the factor of safety of 1.50 commonly used in design practice is also highlighted. It can be seen that the original design (circa 1930), without any deterioration, would have had a factor of safety slightly below the traditional value of 1.5. The reason for this lower than present day design factor of safety may be that the presently required 250 psf surface surcharge was not included in the original design.

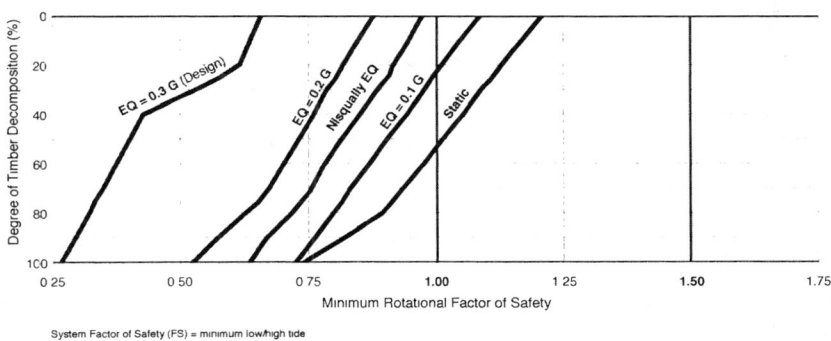

Figure 5 – System Factors of Safety

Figure 6 is a plan of the same factors of safety as a measure of reliability associated with the observed levels of timber decomposition along the wall length. In the areas where the factor of safety falls below 1.00, alternate load paths would have to be developed to maintain stability. As such, the wall may rotate and lean on the Waterfront Park deck and pier to the west. Based on our analysis, this area of the wall, with an apparent 100 percent deterioration of its connection to the relieving platform as its tieback, may have pushed against the deck of that pier to maintain stability during the earthquake.

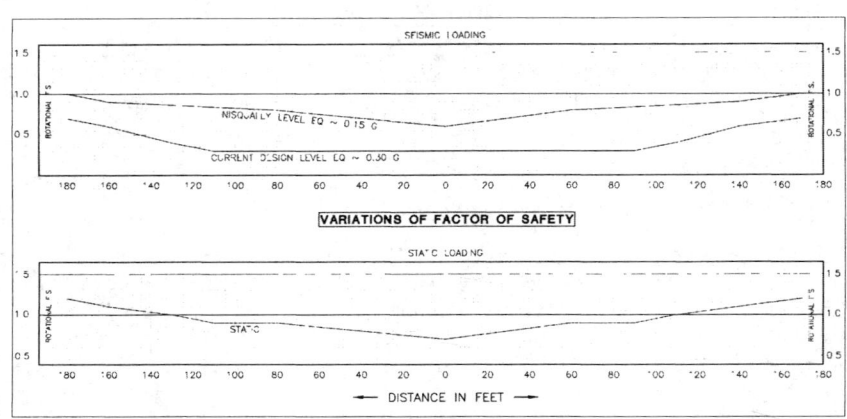

Figure 6 – Risk Assessment along the Length of Seawall

Remedial Measures
Based on the above evaluation, an area with the lowest factor of safety along the length of the wall was identified for underground chemical and cement grouting. For the purpose of grouting, a total of 124 holes are scattered through a 9 m wide x 100 m (30 ft x 325 ft) length of the pavement centered on the point of maximum deflection. The project is still ongoing at the time of this writing and additionally includes TV inspection of sewer and storm drainage, inlet adjustment, and pavement restoration. Addition of riprap is also scheduled to proceed in the fall of 2003.

References
Bowles, J.E. (1986). "Foundation Analysis and Design," *The McGraw-Hill Companies*, 5th Edition, 1175 pp.

Itasca Consulting Group. (1995). "User's Manual for FLAC," *Itasca*, Volume 3.3, Minneapolis, Minnesota.

McDonough, P. W. (2002). "The Nisqually, Washington, Earthquake of February 28, 2001 – Lifeline Performance" *Technical Council on Lifeline Earthquake Engineering*, ASCE Monograph No. 20, February 2002.

Takimatsu, K. and Seed, H.B. (1987). "Evaluation of Settlements in Sand Due to Earthquake Shaking," *Journal of Geotechnical Engineering*, ASCE Volume 113, Number 8, pp. 861-878.

Youd, T.L. and Idriss, I.M. (2001). "Liquefaction Resistance of Soils: Summary Report from the 1996 NCEER and 1998 NCEER/NSF Workshop on Evaluation of Liquefaction Resistance of Soils," Journal of Geotechnical and Geoenvironmental Engineering, ASCE Volume 127, Number 4, pp. 297-312.

CONTAINER WHARF UPGRADE AND SEISMIC STRENGTHENING GUIDELINES AT THE PORT OF LOS ANGELES

Peter Yin, SE; Stacey Jones, PE; Max Weismair, SE

ABSTRACT

The Port of Los Angeles (POLA) is one of the largest container ports in the United States. As such, POLA is proactive in implementing programs to ensure reliable, safe and efficient operations with a focus on disaster reduction and sustainability of existing facilities.

The purpose of this paper is to outline the POLA's approach to developing a wharf upgrade and seismic strengthening program. This program is driven by three important POLA factors:

a) The POLA in cooperation with the United States Army Corps of Engineers is currently deepening the main channel from −13.7 m (-45') to −16.2 m (-53'). This will require many of the existing berths designed for -13.7 m (-45') or less to be strengthened and upgraded to accommodate −16.2 m (-53'). The upgrading responds to the increasing trends to accommodate larger cranes and to provide sufficient seismic resistance. The channel deepening is required to accommodate the larger, deeper draft vessels that are anticipated to call in the future.

b) Evaluation of acceptable technical and economic/operational risks to assist the POLA in strategic decision-making as to the type of seismic retrofit necessary.

c) Establishing a seismic strengthening strategy and design criteria taking into consideration the Federal Emergency Management Agency (FEMA) requirements for reimbursement for repair or replacement costs resulting from a seismic event.

This report will serve as a blueprint for future container wharf upgrades and seismic strengthening guidelines at the POLA.

1.0 INTRODUCTION AND BACKGROUND

The POLA is comprised of 3,035 hectares (7,500 acres) of land and water and 50 km (31 miles) of waterfront in the western side of San Pedro Bay and is situated in a seismically active zone. The major earthquake faults affecting the area are the San Andreas, Newport-Inglewood Fault and the Palos Verdes Fault.[1]

Current Design Criteria

Recognizing the potential hazard risks associated with seismic events the POLA developed seismic design guidelines for new wharf construction in 1990.[2] Since 1990, modifications to the seismic design guidelines have been made to consider new data for the Palos Verdes Fault, progress in the analysis techniques of wharf structures and improvements of structural detailing based on wharf performance in earthquakes as well as testing of critical wharf components. The new updated guidelines will also be used for all new strengthening projects.

Concrete wharves shall be designed according to a two level seismic criteria described below:

<u>Operating Level Earthquake (OLE):</u>

Hazard criteria: 50% probability of exceedance in 50 years exposure. (72-year recurrence interval). Performance criteria: Forces and deformations, including permanent deformations of the embankment, shall not result in more than insignificant damage to wharf structure. Wharf shall be remaining in service during the repair. All minor damages shall be visually observable and accessible for repair. Operations shall not be disrupted.

<u>Contingency Level Earthquake (CLE):</u>

Hazard criteria: 10% probability of exceedance in 50 years. (475 years recurrence). Performance criteria: Forces and deformations (including permanent embankment deformations) shall not result in collapse of the wharf. Damages shall be economically repairable and shall be visually observable and accessible for repair. Disruption of operations may range from several months to more than a year, depending on individual berth and exact amount of damage.

<u>Consideration for Reduction in Peak Ground Acceleration for Existing Wharf</u>: Recent studies were performed for Berths 145-147 Wharf Improvement Project[3] demonstrated that most of the wharf structures that can be economically seismically retrofitted were built after the 1980's. In addition, reductions of service/exposure time [4] from 50 years to 30 years would have an insignificant effect. It would reduce peak ground acceleration (PGA) by approximately 15% in most areas of the POLA. While the cost in wharf seismic strengthening is significant, the scaling down of the PGA by 15% does not provide significant amount in construction cost savings. Further, the losses due to interruption of business far out-weigh the construction savings on a scaled down seismic design load. Therefore, the same seismic criteria should apply to both the new wharf construction as well as for seismic strengthening of existing wharves.

Other Design Considerations: Any wharf or embankment/dike upgrade shall be designed so as not weaken the seismic resistance of the existing structures. It will be the decision of the POLA as owner to fully strengthen the wharf and dike, or to maintain the structures' existing ability to resist seismic forces when modifying the dike or structures.

When wharves are damaged by an earthquake, the following Repair and Replacement Guidelines should be utilized:

- When embankment displacements cause excessive deformation on piles and significant structural deficiencies are demonstrated, the entire wharf and accompanying embankment shall be removed and replaced according to current criteria regardless of the repair costs.

- When seismic load resisting piles (generally, 1st and/or 2nd row from landside, which have a short distance between the deck and the rock dike) show significant damage, then the seismic resisting system of the wharf shall be strengthened or replaced to meet the seismic design criteria identified.

- When the wharf structure and embankment elements are significantly damaged by a seismic event and render the facility non-operational: In addition to the required Seismic strengthening described above, the facility should be reconstructed to meet current operational standards. For container wharves, 34.8 m (100') gauge crane rail, -16.2 m (–53') draft at pier head line and 74.4 tonne/m capacity (50 klf) for crane girders are recommended.

Categorization of Existing Facilities

To better characterize the POLA's container wharves, they have been divided into four distinct categories as follows: (See Figure 1)

Category 1: Container wharves designed and built after 1996. These structures were designed and built according to current seismic criteria. There may be some insignificant damages to wharf or/and dike under moderate earthquake/OLE event. Repair work shall not interrupt operations. Wharf and/or dike may have permanent deformations during a strong earthquake/CLE event, but they will not collapse and should be repairable within a reasonable period of time. Operations may be interrupted during this period. Wharves in this category are: Berths on Pier 400, Berth 144, and Berth 100.

Category 2: Container wharves built after 1990 but before 1996 were designed to criteria acceptable at the time. However, there are areas of the structure that are at higher risk to damage than Category 1. Under the OLE event, minor damage can be anticipated with minor operation disruptions, and under the CLE event may experience more damage than Category 1 structures, but still will not collapse and can be repaired. Wharves belonging to this category are: Berths 302-305.

Category 3: Container wharves built after the early 1980's supported by vertical piles and with conforming dikes. Conforming dikes have limited seismic displacements, which do not comprise the structural performance of piles. These wharves generally can be seismically strengthened from a structural and geotechnical point of view, unless exceptionally poor soil condition renders the retrofit impractical. Berths in this category are: B122-125, B126 (South 440 ft), B145-146, B174-176, B212-218, and B226-229.

Category 4: Container wharves built prior to1980's and some in the earlier1980's generally consist of battered piles and substandard dikes. Substandard dikes have seismic displacements of a magnitude for which adequate structural pile performance cannot be guaranteed. Structural conditions vary from poor to fair. If a moderate to strong earthquake occurred, these wharves could be out of operation for a long period of time due to likely significant damage, or become irreparable. Due to the short remaining life of these structures, the complexity in engineering design and high construction cost; a seismic retrofit of these wharves would be least preferable. Removal and replacement of these wharves is recommended in most cases if it is subject to major functional upgrades. Container wharves in the POLA that belong to this category are: B126-131 (except S. 440 ft on B126), B136-139, B142-143, B206-209, B219-224, B230-236.

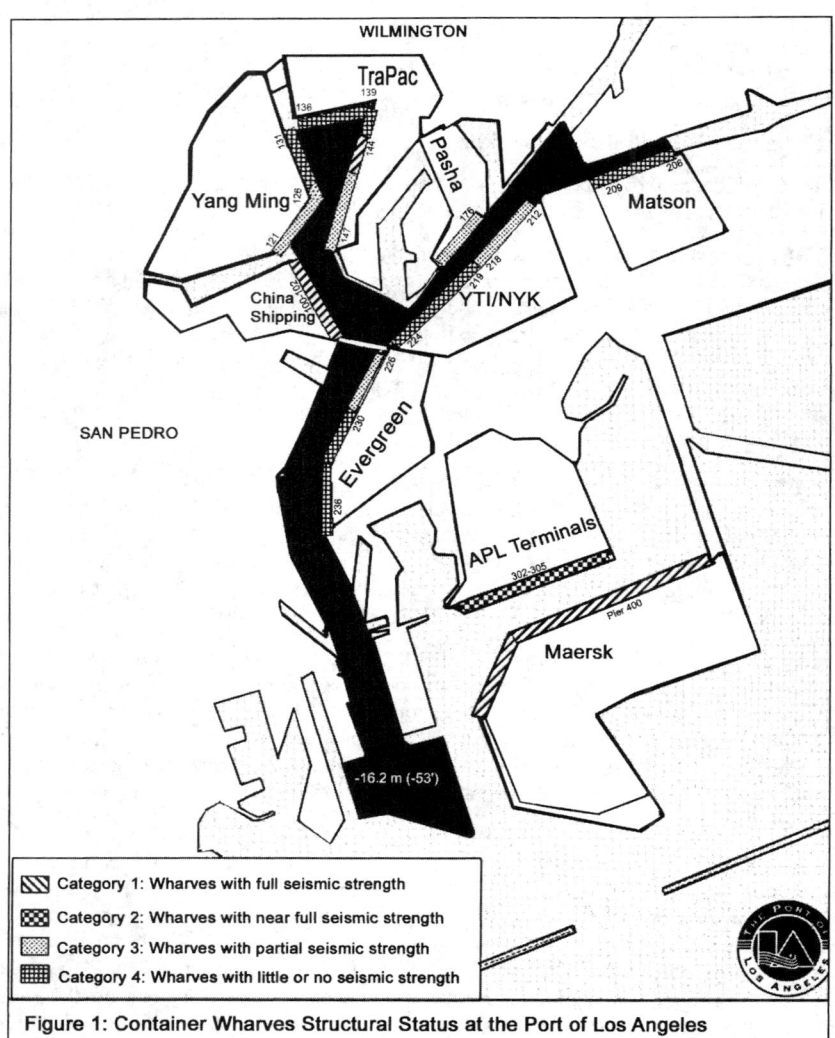

Figure 1: Container Wharves Structural Status at the Port of Los Angeles

Current Challenge

Seismic Considerations

Many of the wharves constructed prior to the 1980's, cannot be guaranteed to survive even moderate earthquakes (OLE) without major damage, let alone a major earthquake. Some similar wharves at the Port of Oakland suffered heavy damage during the 1989 Loma Prieta Earthquake, which was only a moderate earthquake event at Oakland. Some batter pile wharves in the POLA suffered significant damage during the Northridge earthquake at levels considerably below OLE seismic hazard criteria.

Because of the potential risk of damage and operational disruption due to an earthquake, a close evaluation of the existing container wharves built prior to 1990 is warranted and guidelines should be developed.

Upgrade Considerations

Due to the age of some of the existing wharves in the POLA, many of the POLA's tenants are requesting construction of new 34.8 m (100') gauge crane rail to replace existing 15.2 m (50'). gauge rail. This will also require review of the structural capacities of the existing waterside crane rail girders of each wharf. In addition, the POLA in conjunction with the Army Corps of Engineers is currently deepening the navigation channels to –16.2m (–53'). This will require deepening the berths to –16.2 m (–53') when currently the design depth ranges from –10.7m (–35') to –13.7 m (–45'). The static and dynamic stability of the embankments beneath the wharves will require analysis taking into consideration the deeper berth at –16.2 m (–53') plus 0.6 m (2') over-depth.

2.0 EVALUATION CONSIDERATIONS

In developing the Container Wharf Upgrade and Seismic Strengthening Guidelines the following were taken into consideration:

Revenue Losses - Direct

Damaged wharves would have to be taken out of service for an uncertain period of time for repairs or replacement. Depending upon the level of damage, the operation of the entire terminal could be impacted up to 2 years or more. Down time of the wharf translates to revenue losses. Using an average area of 73 hectares (180 acres), with average annual revenues at approximately $325,000 per hectare ($130,000 per acre), revenue losses could be as high as $23,000,000 or more per year at some of the most active terminals.

Past experience has indicated any major repair on a long wharf of 610 m (2,000') or more could be 18 to 24 months. For example, in the Port of Oakland, after the 1989 Loma Prietta Earthquake, the repair work at the 7th Street Terminal (480 additional piles and the replacement of 1,097 m (3,600') of crane girder reconstruction, partial dike strengthening and other repairs on the approximately 670 m (2,200') long wharf took 22 months to complete the project including design and construction. The Port of Oakland paid a large portion of the $12,000,000 in reconstruction costs prior to reimbursement from the Federal Emergency Management Agency (FEMA) and other agencies.

Permanent Business Losses - Direct

The high demand from the shipping lines and the competition among West Coast ports and other ports in the nation may force shipping lines to leave, putting the POLA at potential risk of losing permanent business if any of the container terminals are impacted for a unacceptable period of time. In Japan, the 1999 Hyogo Prefectural Government Report indicated that the Port of Kobe had only recovered 80.4% of its monthly amount of imports and exports as compared to before the 1995 earthquake. The permanent loss of business occurred even though it had recovered 75% of its cargo-handling capacity within one year of the earthquake.[5]

Economic Impacts - Indirect Losses

Indirect economic losses and impacts to the region and the nation could be as much as $100 million per day (includes industry sales, salaries, and generated taxes). As was demonstrated in the recent labor lock out on the west cost, the economic impacts were significant. It is estimated that the indirect losses in Japan due to the earthquake at the Port of Kobe in 1995 was $5.5 billion while only the first year after the quake the indirect losses already reached $6 billion.[6]

Capital Costs

This section identifies costs developed for various upgrade options including berth depth of –16.2 m (–53'), upgrades for waterside and landside crane rail girders and removal and replacement of entire existing wharf sections. In addition, selective seismic strengthening options will be examined and a comparison made of the costs of wharf upgrades verses both wharf upgrades and seismic strengthening addressed. Refer to Table 1 and Figure 2.

Cost Comparison

With average annual revenues for a 73-hectare (180-acre) terminal estimated to be $23,000,000, any wharf repairs requiring the terminal to be placed out of service for a long period must be seriously considered. The potential of some terminals this size to be severely damaged without seismic strengthening during a CLE and in some instances during an OLE earthquake will require the terminal to be out of service up

Table 1: Approximate Cost of Various Types of Wharf Upgrades			
Wharf Upgrades	Depth/Style	Cost/Meter	Cost/Foot
Channel Deepening (underwater bulkhead, inclined)	-13.7 m to -16.2 m (-45' to -53')	$ 16,404	$ 5,000
Channel Deepening (underwater bulkhead, vertical)	-13.7 m to -16.2 m (-45' to -53')	$ 11,483	$ 3,500
Channel Deepening (underwater bulkhead, inclined)	-12.2 m to -16.2 m (-40' to -53')	$ 32,809	$ 10,000
Waterside Crane Girder Upgrade (pile caps + piles)	-29.8 tonne/m to -52.1 tonne/m (20 kips/ft to 35 kips/ft)	$ 8,202	$ 2,500
New Waterside Crane Girder (removal & replacement)	74.4 tonne/m (50 kips/ft capacity)	$ 19,685	$ 6,000
New Landside Crane Girder (on existing steel pile/no seismic)	74.4 tonne/m (50 kips/ft capacity)	$ 20,013	$ 6,100
New Landside Crane Girder & Seismic (w/existing steel pile)	74.4 tonne/m (50 kips/ft capacity)	$ 30,184	$ 9,200
New Landside Crane Girder & Seismic (without steel pile)	74.4 tonne/m (50 kips/ft capacity)	$ 24,974	$ 7,600

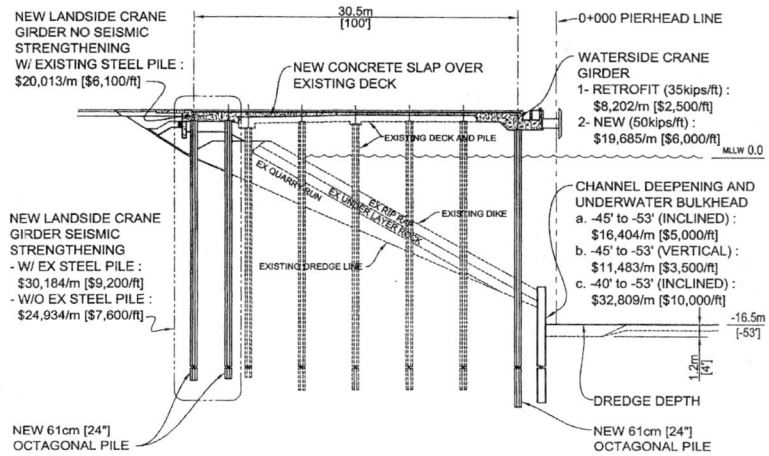

Figure 2: Typical Wharf Upgrade Cross Section with Seismic Strengthening

to 24 months for repairs. Considering the financial and economical implications, it is prudent and cost effective to have seismic strengthening and upgrades performed selectively on some wharves.

The Table 2 shows the comparison of wharf baseline upgrade, seismic strengthening, and complete removal and replacement:

	Table 2: Cost Comparison on Wharf Upgrade Options at Selected Terminals					
	(1) Baseline (-16.2m draft & 22-wide crane)	(2) Seismic Strengthening (2 berths) + Baseline	(3) Complete Removal and Replacement	(2) – (1) Seismic versus Baseline	(3) – (1) Replacement versus Baseline	(3) – (2) Replacement versus Seismic
Yang Ming[7]	$55,100,000	$65,700,000	$79,800.000	$10,600,000	$24,700,000	$14,100,000
TraPac	$22,000,000	$42,000,000	$79,000,000	$20,000,000	$57,000,000	$37,000,000
YTI	$42,780,000	$65,980,000	$99,200,000	$23,200,000	$56,420,000	$33,220,000
Evergreen	$37,700,000	$73,600,000	$100,200,000	$35,900,000	$62,500,000	$26,600,000

Redundancy - Terminal Sharing Arrangements

The POLA should consider the prospect of utilizing non-preferential berthing assignments during an earthquake emergency and initiating terminal sharing arrangements. However, in doing so, current and future berth occupancy rates need to be considered. A recent terminal capacity analysis study completed for the POLA concluded that two berths at each of the container facilities are occupied a majority of the time.[8] The study also indicated that by 2010 container terminal operations would very likely be berth-constrained. Meaning that the projected number of berthing facilities will be insufficient to handle the projected throughput or future demand. This would render berthing sharing arrangements ineffective and most likely unacceptable to the POLA's tenants. Due to this fact, and in the absence of a detailed probabilistic risk analysis, strong consideration must be given to upgrading and strengthening at least one or two berths at each container terminal facility, depending on the incremental increase in cost of any proposed upgrades vs. seismic strengthening. Although this approach may seem conservative, the financial impact can be managed to minimize undue financial risk while maintaining a high degree of assurance that our facilities will be able to continue operating with a minimum of disruption after an OLE.

3.0 RELEVANT REGULATIONS

FEMA will provide funding to state and local agencies when natural disasters occur. In March 1998, Title 44 of the Code of Federal Regulations Section 206.226 covering the restoration of damaged facilities was amended. Costs eligible for reimbursements are:

No Wharf Codes in Place by Local Authority Prior to Earthquake

- FEMA will pay for repair of the damage only if the repair cost does not exceed 50% of the replacement in kind cost as originally constructed.
- FEMA will pay for full replacement in kind cost if the repair cost is more than 50% of the replacement cost as originally constructed.

Wharf Codes in Place by Local Authority Prior to Earthquake

- FEMA will pay for repair or replacement costs to current codes adopted by local authority.

4.0 CONCLUSIONS

Cost-effectiveness: It is more cost-effective to seismically strengthen existing wharves constructed after 1980's. These wharves typically are in better physical condition. Wharves built prior to the 1980's (Category 4) cannot economically be seismically strengthened due to deficiencies in the existing design and current conditions of wharf and/or dike embankments. However, because of operational demands, these wharves may be considered for baseline upgrade improvements only to meet the needs of the tenant. These wharves can be repaired, removed, and replaced over time as part of future expansion plans as the operational throughput demands it.

When considering seismic strengthening or wharf upgrade, the following factors should be considered:

Seismic retrofit should be less expensive than replacement: When the strengthening and/or upgrade of existing structures is proposed, the age and structural condition of the existing wharves and dikes should be considered. Costs for seismic strengthening and potential functional upgrades, and future maintenance expenses should not exceed the cost of replacement of the facility. Studies on Berths 145-147 Wharf Improvements Project concluded that the average annual maintenance cost would be around 0.75% of the original construction cost or less.[9] For wharves built in 1980's that are in good structural condition, the cost of seismic strengthening and functional upgrades should not exceed 80% the cost of removal and replacement.

Shorter construction schedule: When the schedule is a factor, a shorter construction duration may be necessary. Seismic strengthening and a functional upgrade may be chosen over removal and replacement when it can deliver the project sooner but should take into consideration the requirement addressing minimum numbers of berths to remain in operation in the event of a CLE.

Life cycle: Necessary wharf upgrades are determined by the operational needs or life cycle, rather than the structural life expectation. Past experience has shown that the need for upgrades may occur every 20 to 25 years to meet the ever-changing demands of the shipping industry.

Funding reimbursement: FEMA may reimburse the cost of repairs for damaged facilities after an earthquake. This would allow a strategy whereby wharves built prior to 1980 should not be upgraded or strengthened, where feasible from an operations standpoint. However, our new guidelines should require that these wharves be upgraded and strengthened to make them eligible for FEMA reimbursement at the time of an earthquake. This strategy will have to be reviewed on a case-by-case basis for each terminal.

It should be feasible from a structural engineering point of view: Many wharves were built on batter piles, smaller piles or on a system that is extremely difficult or too costly to retrofit. The retrofit should only proceed when an engineering study demonstrates the subject structures (deck, pile, etc.) are fit for structural reconstruction. Wharves built in the 1980's have vertical piles that are better candidates for the retrofit than the batter pile wharves constructed prior to this date.

It should be feasible from a geotechnical engineering point of view: Many dikes constructed prior to 1980 do not have a full section per current standards or were subsequently dredged below the quarry run and may be unstable during seismic events. Without strengthening these dikes, they would fail during a seismic event and would severely damage the wharf structure. In these cases, a removal and replacement may be a better solution, as retrofitting on existing dike below an existing wharf is almost economically impossible.

Other Considerations

As new scientific seismic source information becomes available through studies conducted by state, federal and academic institutions, periodic reviews of POLA seismic source characterizations should be conducted and adjustments made if warranted.

Ultimately, individual organizations must decide how much of their respective resources they should expend to protect themselves, the public and their assets against natural hazard events and how they should spend that money. Recognizing that all risks from earthquake damage cannot be eliminated, the POLA's goal is to develop a strategy that minimizes the risks.

It is recommended that upgrading and seismic strengthening be considered for a two-berth minimum at each container terminal facility to meet the Contingency Level Earthquake (CLE) performance criteria contingent upon the following site-specific considerations:

- The incremental increase in cost of any proposed upgrades vs. seismic strengthening.
- Excessive construction costs.
- Operational interruptions during repair, upgrades and strengthening.
- Technical complications and condition of existing structures.
- Alternate criteria allowing at least one berth to meet the CLE performance criteria, and the second berth to meet the OLE performance criteria.

---

[1] Fugro West, Inc. (2001). *Geotechnical Framework Report Pier 400 Landfill Project Port of Los Angeles, California*

[2] Port of Los Angeles (1990) *Proceedings of the Port of Los Angeles Seismic Workshop on Seismic Engineering, Port of Los Angeles, San Pedro, CA*

[3] The Port of Los Angeles (2002). *B145-147 Wharf Improvements Upgrade Evaluation Report*

[4] International Navigation Association (2001). *Seismic Design Guidelines for Port Structures"*, Swets & Zeitlenger B.V., Lisse

[5] Pachakis, D. & Kiremidjian, A. (2002). *"Estimation of Down-time Related Revenue Losses in Seaports Following Scenario Earthquakes"*, Earthquake Spectra, EERI (DRAFT)

[6] Werner, Stuart D. (1998), *Seismic Guidelines for Ports,* Technical Council on Lifeline Earthquake Engineering Monogramph No. 12, ASCE, Reston, VA

[7] DMJM+HARRIS, Inc.(2003). *Berths 121-131 Upgrade Evaluation Report*

[8] JWD Group, Inc. (January 2003). *Port of Los Angeles Capacity Analysis*

[9] Gaythwaite, J. W. (1990). *Design of Marine Facilities for Berthing, Mooring and Repair of Vessels,* Van Nostrand Reinhold, New York, NY

THE USE OF SIMULATION IN DISASTER RESPONSE PLANNING AND RISK MANAGEMENT OF PORTS AND HARBORS

Dimitris Pachakis[1] and Anne S. Kiremidjian[2]

ABSTRACT

Seaports are exposed to various risks when subjected to large earthquakes. These include direct physical damage to port facilities, damage to stored cargo and other contents, revenue losses due to closure of wharfs or docks, and liability losses. In order for provide a rational approach to risk mitigation and transfer, the various risks are discussed and mechanisms for mitigating these risks are summarized. A model for estimating physical losses is presented and a simulation model for evaluating revenue losses from wharf closure until repaired is described. Data from a US port is used to estimate the simulation of revenue loss model. A simple example is used to demonstrate how the simulation model can be used to evaluate post-event operations alternatives and to determine which alternative is the most effective in reducing the downtime revenue losses.

Keywords: Risk management, seaports, earthquake, financial risk, simulation, disaster response planning

INTRODUCTION

Seaports are major centers of commerce, handling large volumes of cargo that is imported to and exported from the country. As such, they are critical to the economic wellbeing not only of the region that they serve, but to the entire country. They are also critical components of the transportation network of a region because they function as the sources and sinks of cargo transported by freight traffic. In order to reduce the consequences of port damage from disasters, it is necessary to evaluate the performance of these critical facilities. Risk assessment methods provide the tools for rational decisions for rehabilitation of such facilities and for developing emergency response capabilities (Frankel 1987). Furthermore, planning of recovery operations is best performed with information on the potential losses that may be incurred from such

---

[1] Research Assistant, The John A. Blume Earthquake Engrg. Center, Stanford University, Stanford, CA 94305-4020. E-mail: dpach@stanford.edu

[2] Professor, Department of Civil and Environmental Engineering, Stanford University, Stanford, CA 94305-4020. E-mail: ask@stanford.edu

disasters. The costs from disaster are divided into costs resulting from physical damage to port facilities and revenue losses due to ceased or reduced functionality of those facilities, while repairs are taking place. In this paper, the various aspects of financial exposure of the port to seismic hazard are explained and various methods of estimating these costs are discussed. Then, a methodology is presented to estimate the expected revenue losses and the uncertainty of these expected costs incurred by a multi-terminal port after an earthquake. The functionality losses are obtained by simulating the port operations that generate revenues and monitoring the differences in revenues over a long period with and without a seismic event. Through a multi-terminal example, it is demonstrated that proper post-earthquake operation strategies can minimize the total revenue loss.

TYPES OF EXPOSURE FROM SEISMIC RISK

According to (U.S. Maritime Administration 1985) report, earthquakes are classified in the natural hazards, "acts of God " category for the purposes of insurance coverage. Ports that are situated in regions of high seismicity are particularly vulnerable. Such ports are currently under careful review of their exposure and may require extensive risk mitigation measures. For this purpose it is important to assess the potential losses resulting from extreme events and consider possible mitigating actions. The types of losses that can be identified as a result of damage to a port from an earthquake include: direct property losses, net income losses, liability losses to third parties and employees, and indirect losses. It should be noted here that losses due to fire and environmental impacts such as oil spills after an earthquake although important, are beyond of the scope of this study. The different types of losses are further discussed in the following sections.

Direct property loss includes the repair or replacement costs for the damaged facilities. These facilities are the ports wharves and docks, damaged by liquefied soils, cranes that can topple or collapse from lateral spreading of their legs and buckling, office buildings and warehouses, liquid storage tanks which can sustain loss from collapse or cracking and failure of utility lines. Moreover, ports sometimes own various types of bridges. Depending on the magnitude of a seismic event the design characteristics of these facilities, the cost from repair of these structures can be excessively high imposing significant financial difficulties to the port's authority.

Net income losses are the losses that accrue due to the loss of revenue if facility damage causes interruption of the port's operations. Since most of the revenues of a port come from the transfer of cargo on and off ships, if the wharves are unusable for a period of time, the revenue loss can be significant. These losses are also described as losses due to downtime. In the net income losses one can add the extra expenses that will occur when the operations continue in an emergency mode, e.g., the rental costs for contingency equipment and temporary space.

Liability losses losses occur when port damage causes harm to another party's property or income. An example of such liability is when the power blackout caused by an earthquake results in deterioration of perishable cargo stored in refrigerated containers. Workers' compensations and tenants' losses of revenues could be classified in this category as well.

Indirect property loss arises as a result of a direct property loss. For example, when half of a building has collapsed in an earthquake and the rest has to be demolished and rebuilt because it does not comply with the codes, then the cost to reconstruct the remaining half is considered indirect property loss.

Assessing the risk from market share loss to competitors is considerably more difficult. It is generally admitted that once a ship gets diverted successfully to another port, it never comes back. Several scenarios are considered to evaluate the likelihood that a ship will be diverted under the assumption that the queue is too long and the ship will not wait an extended period of time.

RISK MANAGEMENT

In the previous section the various contributors to seismic risk to a port were identified and discussed. The core of the problem lies in that most existing facilities are designed according to older standards and are quite vulnerable not only to direct losses but also to operational losses. Facilities designed under current standards are also expected to sustain some degree of damage because design criteria are postulated primarily for life safety rather than for different performance requirements. Continued functionality after different size earthquakes, for example has not been considered until recently, as performance-based design criteria become better understood and accepted. Under certain conditions, seismic retrofit of these facilities can cost more than the expected losses. Thus, it is necessary for port management to find ways not only to minimize the losses from direct physical damage but also to plan for quick recovery. If mitigation measures are not taken prior to an event to increase the seismic resistance of port facilities, repair of these facilities after an event would require a significant amount of capital in order for recovery actions to be taken and the port can return to normal operation as soon as possible. Typically, a combination of mitigation through loss control and risk financing would provide the best approach to reducing the overall risk exposure of a port ((U.S. Maritime Administration 1985)). In the following sections these two major components of risk reduction are discussed in greater detail.

Loss Control

Loss control is a general term describing a variety of techniques available to ports in order to limit the losses that are identified from the quantitative and qualitative part of risk analysis. By examining the various consequences of the catastrophic events and the chain of events that follow, one can identify the components that can contribute the most in losses. A proper risk management program not only can eliminate or reduce the severity of unfavorable consequences but also reduce the amount of financing that will be required following and event. The cost of such a program can be easily justified by the avoidance of losses and many times can be partially undertaken by the parties responsible for loss financing, i.e., insurance companies. The mechanisms available for loss control to the ports include avoidance, prevention, reduction and separation. Risk avoidance means that the port would stop the operations whose exposure to risk is substantial and of high frequency. Also, it could mean that it would avoid taking responsibility for the consequences towards third parties. This technique is very useful for minimizing the liability exposure of the port to its tenants and workers. The downside of this option, however, is that tenants may choose other ports that provide more favorable terms such as lower risk exposure. Risk reduction is a technique that pertains to engineering the various processes so that there are redundancies in the system and further loss is avoided. For example, having a backup power system in standby mode would prevent data loss and deterioration of refrigerated cargo. Similarly, seismic upgrading of existing facilities to higher standards can reduce the risk exposure. This option would be much more appealing to a tenant than a "hold harmless" agreement for example. Other prevention measures would be plans to avoid fire after an earthquake etc. Loss reduction usually pertains to reducing the losses after the event has occurred. One example of such loss reduction is the quick settlement of any reasonable claims. Another risk reduction technique would be to change the operation mode of the port so that it can continue its operations even with reduced capacity. This can be achieved by diverting ships that were to go to damaged terminals to other terminals that are still operating. Simulation of port operation can be particularly effective in identifying the most effective post-event ship traffic management that will reduce operating losses. In this paper, a simulation model is formulated to address this particular issue. The simulation model is illustrated through an example for a port with eight terminals.

Risk Financing

The repair and reconstruction of port facilities right after a catastrophic event can be a difficult, time consuming and very costly task. In order to plan for risk financing, the port should have the ability to estimate the potential losses in future events. The extent of damage is difficult to assess and is poorly predicted by the current state of engineering. Consequently, repair costs for labor, material and repair duration are highly variable. Furthermore, the availability of materials, staff and construction crews is expected to be limited in the aftermath of an earthquake event because of the high demand for many repair projects at the same time. These factors increase significantly the uncertainty in the project costs and completion times. Uncertainty in recovery time also results in uncertainty in revenue losses. Therefore, even if its possible to obtain an estimate of the mean value of the losses through simpler analysis methods, the uncertainty on these losses, expressed in the form of variance, will be significant. Funding repairs and reconstruction in a port after a serious earthquake can be a difficult task unless these are planned for prior to the event. Usually government emergency agencies can provide some financial relief, but port needs would have to compete with other emergencies such as bridge, highway, buildings and other facility repairs. This problem can further be exacerbated by the bureaucracy involved until the funds are finally allocated. The amounts of funding that would be required to repair or replace cranes, wharves and other structures are significant and they have to be available without any delay so that the operations can resume quickly. Funding such repairs can be greatly facilitated through a combination of insurance purchase and direct reserves usually referred to as self-insurance as discussed in the next section.

Insurance

Insurance is the standard method of protection against natural hazards and other extreme events. Insurance policies would typically cover property damage up to a specified amount, liabilities and losses of income. However, there are serious limitations in the coverage that can be provided for earthquakes. Buildings, contents, equipment and vehicles can be insured readily against standard property insurance perils. Surprisingly, although damage to the port wharves after an earthquake is quite common, this type of exposure is not typically covered by insurance. To quote the (U.S. Maritime Administration 1985) report "flood damage, collapses, wave damage, or loss of piling below the water line, unreported dock damage and earthquakes are all examples of exposures that are typically not covered by insurance". The same holds true for storage tanks. Moreover, there are a number of costs associated with the repair and reconstruction that are non-insurable, such excavation costs, site preparation etc., as well as costs that occur for rebuilding to higher seismic standard for code compliance. Most frequently, the position of the insurers is that they are responsible only for the direct damage losses. Recently, some insurers would cover increases in pay out due to higher seismic code requirements than those existing at the time of the original construction of the facility. In the case of business interruption, the losses are covered only if they come from an insured peril and earthquake is typically excluded from those perils. Hence, earthquakes have to be specifically accounted for in insurance policies purchased by the port. Furthermore, insurance policies should be carefully drafted to include the various losses that can result from an earthquake.

Finding insurance coverage against earthquakes in highly seismic regions, such as the Pacific Coast of the US has become increasingly more difficult. This difficulty was the result of insurance payouts after the Northridge earthquake that were significantly higher than the premiums collected from that area creating major financial problems for many of the insurance companies((Rayland 2000)). Furthermore, competition and government regulation would not allow the premiums to rise to levels deemed necessary by the insurance companies. As a result, insurance companies have been reluctant to underwrite earthquake protection policies in California.

Appropriate determination of a premium and of a total coverage amount is a difficult issue, especially in the case of revenue losses. In order to obtain an appropriate level of revenue loss coverage, it is necessary to estimate the mean value of revenue losses and its variance at the very least. Simulation of port operations and revenues appears to be an suitable method for calculating these quantities in helping port management negotiate a reasonable insurance contract.

Self Insurance - Investment

The second financial instrument that can be used for effective recovery following an earthquake combined with commercial insurance coverage, is allocation of reserves. Such reserves are often referred to as self insurance. An efficient way to do so would be to invest the reserved funds in liquid assets that are protected from market fluctuations. For example, CAT bonds would provide such a safe investment. Again, proper estimation of not only the mean value of the losses but also of their variance through simulation, can serve as the basis for designing the appropriate investment portfolio that would help the port to hedge their risk after a disastrous event.

Revenue Loss Estimation Methodology using Simulation

The methodology presented here and described in more detail in (Pachakis and Kiremidjian 2003a) and shown on Fig. 1 is conditioned on specific scenario events with known anticipated characteristics. Also, the expected losses come from the revenues that are generated by the cargo handling on and off the ships. In order to estimate those revenue losses, there are two necessary interrelated components needed: a methodology to predict the damage state of the port facilities after a seismic event, named vulnerability model thereafter, and a methodology to relate the damage state with the monetary loss. The vulnerability model is used to connect the characteristics of the scenario earthquake to the functional state of the port system afterwards. It should be noted hereby that in case of disasters other than earthquakes, such as hurricanes or terrorist actions, the vulnerability model for earthquakes can be replaced with another appropriate model that would generate events and their probabilities.

In the case of seismic hazard, Deterministic or Probabilistic Hazard Analysis can be used for determining the ground motions at the port site. It takes as input the geographical (port location), geological (soil formations on the site) and seismological data and calculates the site ground motion intensity measure, for example, Peak Ground Acceleration or Spectral acceleration, with the use of readily available attenuation functions. Given these intensity measure, the damage state probabilities can be calculated through fragility curves. At the present, fragility functions can be found, among other sources, as part of HAZUS, a hazard and loss estimation tool developed by the (National Institute of Building Sciences 1999). Usually, five discrete states are assumed to describe the damage of structures: no damage, slight, moderate, extensive and complete. With regards to their operational status, the facilities can be considered either fully or non-operational, depending on their damage state. Then, the damage states of the port components (buildings, cranes, wharves and utilities) and their probabilities can be calculated. The determination of the functional state of the port terminals after the earthquake is critical for assessing the revenues from post-event operation of the ports. It requires a system approach, where the damage states of the port components that contribute to the cargo handling operations are combined through fault trees and event trees to produce the functional states of the terminals and their associated probabilities.

From the definition of the damage states, one can associate the repair cost as a function of the replacement cost with the damage state and generate the so-called loss curves. If the structural damage states and their probabilities can be calculated, then the total loss for a given event can be estimated.

In order to compute the revenue losses, a model is developed to simulate the port operations and track the revenues from a given instance T_0 and for an interval Δt. The initial tracking time T_0 and the interval Δt should be chosen carefully to capture all the transient effects until operations recover fully. The port can be modelled as a queueing system where the customers are the ships and the servers are the berths. The details of the ship traffic that gets processed by the port can be described by a random vector process $\mathbf{I} \equiv \{\mathbf{I}_n\}$, $n \in N$, whose components are the ship arrival times, terminal of destination, ship type, length and cargo to be loaded and unloaded. It is assumed throughout the simulation that the process $\{\mathbf{I}_n\}$ is stationary and ergodic and unaffected by the earthquake. This assumption might not be necessarily always true but the evaluation of the change of traffic due to the scenario events is beyond the scope of this study. When the port operates from time T_0 until time $T_0 + \Delta t$, the part of the input process that passes through the model (input stream), denoted as $\mathbf{I}_{\Delta t} = \{\mathbf{I}_{n_i}, n_i : t_{n_i} \in [T_0, T_0 + \Delta t), i \in N\}$ generates a port revenue $R_\mathbf{I}(t)$. In the case that an earthquake has occurred, the port with modified operating capacity would generate a revenue $R_{\mathbf{I}|\mathbf{d}}(t)$ conditional on the particular damage state. The expected loss due to downtime L_d over a period Δt can be estimated as the difference in these two revenues as follows:

$$E[L_d] = E_\mathbf{D}[E_\mathbf{I}[L_d|\mathbf{D}]] = \sum_\mathbf{d} E_{\mathbf{I}_{\Delta t}} \left[\sum_{t=T_0}^{t=T_0+\Delta t} \left(R_\mathbf{I}(t) - R_{\mathbf{I}|\mathbf{d}}(t) \right) \right] P[\mathbf{D} = \mathbf{d}] \quad (1)$$

The above equation states that the expected revenue loss due to downtime is the expectation over all possible damage states of the revenue loss conditional on a particular damage state. For a particular damage state, the expected revenue loss $E_\mathbf{I}[L_d|\mathbf{D}]$ is the expected value of the difference in cumulative revenues for a period of length Δt over all possible input streams $\mathbf{I}_{\Delta t}$. The expectation over $\mathbf{I}_{\Delta t}$ is calculated through multiple runs of the port operations simulation model with different input streams. To reduce the variance of the revenue losses $\sum_{t=T_0}^{t=T_0+\Delta t} \left(R_\mathbf{I}(t) - R_{\mathbf{I}|\mathbf{d}}(t) \right)$, the same stream $\mathbf{I}_{\Delta t}$ is run through the model, once without considering an earthquake and the second time with the effects of the earthquake on port operations. The variance of the losses due to downtime $Var[L_d]$ can be calculated from the conditional expectations and variances of the losses given the damage states:

$$\begin{aligned} Var[L_d] &= E_\mathbf{D}[Var[L_d|\mathbf{D}]] + Var_\mathbf{D}[E_\mathbf{I}[L_d|\mathbf{D}]] \\ &= \sum_\mathbf{d} Var[L_d | \mathbf{D} = \mathbf{d}]P[\mathbf{D}=\mathbf{d}] + \left(\sum_\mathbf{d} E_\mathbf{I}^2[L_d|\mathbf{D}=\mathbf{d}]P[\mathbf{D}=\mathbf{d}] - E^2[L_d] \right) \end{aligned} \quad (2)$$

From which the mean square value of the loss can be obtained:

$$\begin{aligned} E[L_d^2] &= Var[L_d] + E^2[L_d] \\ &= \sum_\mathbf{d} \{Var[L_d | \mathbf{D} = \mathbf{d}] + E_\mathbf{I}^2[L_d|\mathbf{D}=\mathbf{d}]\}P[\mathbf{D}=\mathbf{d}] \end{aligned} \quad (3)$$

The discrete-event port operations simulation program that was used, was written using the commercial software GPSS (Schriber 1991), which is specifically designed for simulation of manufacturing and queuing systems. It was calibrated and validated with traffic and operations data from an actual West Coast port (Pachakis and Kiremidjian 2003b). More details about the algorithms and the simulation set up are described in (Pachakis and Kiremidjian 2002).

APPLICATION OF SIMULATION IN LOSS CONTROL

One of the advantages of the proposed methodology is that one can built a multi-terminal port model and study the effects of different emergency operating modes. Interactions between the terminals can be observed so that a sound emergency planning strategy can emerge.

In the following scenario, the ship traffic of 1999 of a US port is used to estimate the differences in revenues if earthquake closed one of its eight terminals for six months. The ships that cannot be serviced get diverted either to another port or to each of the other terminals. Due to decreased capacity in the whole port, some ships would have to be diverted eventually and the revenues would decrease. Simulation was used to determine the expected revenue loss for each of the different diverting options. Moreover, the variance of the losses was calculated. The warm up period was 1 year and the total run time after warm up was another 2 years. The event is assumed to take place on day 290 (October of the first year after warm up) and the downtime was 182 days. The terminal opens to full operational status following this time period.

The calculated losses reflect the difference in total revenues for the two-year period. One hundred replications were necessary to achieve loss convergence and small standard error. The statistics of the simulated revenue differences are shown on Table 1. Fig. 2 shows the expected losses together with uncertainty bounds (plus and minus one standard deviation) for the different diversion options. It can be seen that even though the losses could be significant if all the traffic was diverted to another port (20.6M), it can be minimized (2.2M) if the ships are diverted to terminal 7 instead, which appears to be the least busy terminal. The differences in losses from diverting the ships to different terminals within the same port can also be significant (to the order of eleven million dollars). The variance of the revenue losses ranges in the order of 10^{12}, which shows that the interpretation of the results should be done with caution. However, the standard error of the expected losses is relatively small. The losses are also compared with the total revenues for the two-year period and they are found to range from 1 to 7 percent. This means that even if one of its eight terminals closes for six months, the port has enough operational capacity, compared to its traffic, that the losses due to diverted ships will be relatively small.

CONCLUSIONS

Ports are exposed to significant financial losses after catastrophic events for which they have to prepare. Appropriate operating strategies can minimize the revenue losses after the event and these strategies can be evaluated using a simulation model for the port operations. Moreover, by obtaining the statistical characteristics of the revenue losses for each response scenario, port management can plan the financial recovery through insurance and investment in a rational manner.

ACKNOWLEDGEMENTS

This project is partially supported by the Pacific Earthquake Engineering Research Center through Grant No. SA1831JB. The authors also wish to thank the Port Authorities for providing the data used in developing the model parameters and for extensive discussions on port operations.

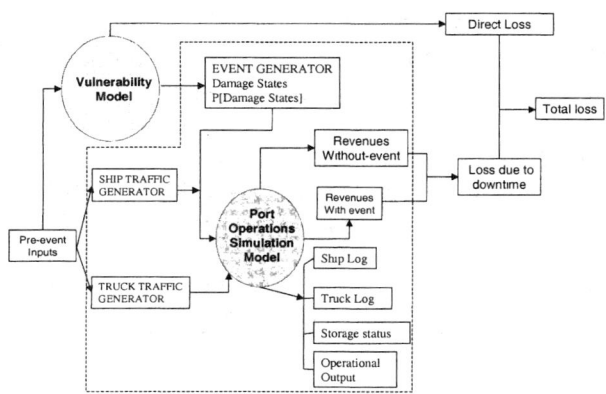

FIG. 1. Components of the loss estimation methodology

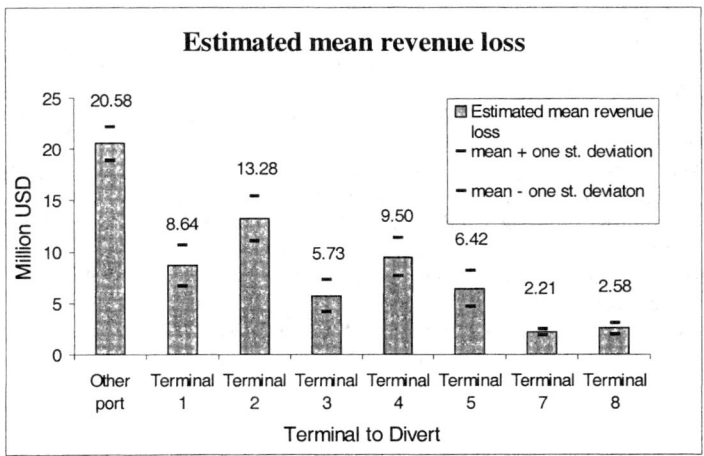

FIG. 2. Estimated mean revenue differences after closing Terminal 6 for six months

TABLE 1. Statistics of revenue losses after closing Terminal 6 for six months

Divert to:	Mean [Mill.USD]	Std error [USD]	Variance [USD]	% Difference	C.O.V.
Other port	-20.6	164,922	2.72E+12	-0.07	0.08
Terminal 1	-8.6	197,737	3.91E+12	-0.03	0.23
Terminal 2	-13.3	220,213	4.85E+12	-0.05	0.17
Terminal 3	-5.7	156,087	2.44E+12	-0.02	0.27
Terminal 4	-9.5	180,449	3.26E+12	-0.03	0.19
Terminal 5	-6.4	169,846	2.89E+12	-0.02	0.26
Terminal 7	-2.2	29,352	8.62E+10	-0.01	0.13
Terminal 8	-2.6	53,452	2.86E+11	-0.01	0.21

REFERENCES

Frankel, E. G. (1987). *Port Planning and Development*. John Wiley and Sons, New York.

National Institute of Building Sciences (1999). *HAZUS, Technical Manuals*. National Institute of Building Sciences, Washington, D.C.

Pachakis, D. and Kiremidjian, A. S. (2002). "Discrete-event simulation as a tool for risk analysis of seaports." *2002 Summer Computer Simulation Conference*, San Diego, California.

Pachakis, D. and Kiremidjian, A. S. (2003a). "Estimation of operating losses from earthquake damage to major sea ports." *The Ninth International Conference on Aplications of Statistics and Probability in Civil Engineering (ICASP9)*, Berkeley, California. accepted.

Pachakis, D. and Kiremidjian, A. S. (2003b). "Ship traffic modelling methodology for ports." *Journal of Waterway, Port, Coastal and Ocean Engineering*. in print.

Rayland, H. (2000). "A piece of the puzzle: Insurance industry perspective on mitigation." *Natural Hazards Review*, ASCE, 1(1), 43–49.

Schriber, T. J. (1991). *An Introduction to Simulation using GPSS/H*. John Wiley and Sons, New York.

U.S. Maritime Administration (1985). *Port Risk Management Guidebook*. U.S. Department of Transportation, Washington, D.C.

Economics of Seismic Retrofit of Water Transmission and Distribution Systems

John Eidinger[1]

1 Introduction

This paper examines the economic basis for seismic retrofit of water transmission and distribution systems. An estimate is made as to the size of the "marketplace" for seismic retrofit of water systems in the United States. An economic analysis is then presented for the seismic retrofit of the Hetch Hetchy water system.

2 The Marketplace for Seismic Retrofit of Water Systems

One way to gage the need for seismic retrofit of water systems is to examine the case evidence as to how much has been already spent on such endeavors. In the United States, there are more than 10,000 individual water system operators. Of these, perhaps a few dozen or so have embarked on some sort of system-wide seismic retrofit. This is not to say that the other water utilities have ignored seismic issues: in fact, the vast majority of water system operators follow codes like the UBC 97 for design and construction of new buildings. But the fact of the matter is that much of the water infrastructure currently (year 2003) in place has been designed and constructed either to no seismic standard (as is the case for 99.9%+ of all buried water pipelines and redwood tanks); out-dated seismic standards (as is the case for most pre-1973 steel and concrete tanks); arguably inadequate seismic standards for steel and concrete tanks built post-1972 in high seismic regions; lack of attention to seismic detailing for many types of non-structural items such as anchorage of motor control centers, restraint of emergency generator batteries, use of vibration isolators for diesel generators and air compressors, use of flexible suspended t-bar ceilings over operator work areas, lack of restraint of glassware and equipment in water quality laboratories, etc. About the only type of component that is consistently built to relatively good seismic standards are building structures, likely because the UBC (and similar) codes of the past many years are reasonably good and quite rigorously followed.

On a percentage basis of the value of all installed assets (including buildings, pipelines, tanks, water treatment facilities and wells), perhaps only 10% to 20% of the existing inventory has been built to modern-day concepts of earthquake-resistant and/or earthquake-reliable design. This excludes dams, the vast majority of which are well designed for earthquake loads. The reason dams are reasonably well built for earthquake loads lies in their importance; their obvious potential for large life-safety threat should they fail; and in many cases, careful regulatory oversight.

The remaining inventory of water systems (pipes, tanks, non-structural components, etc.) has little direct-life safety threat should there be failures. For this reason, up to the early 1990s, the remaining inventory has not had much attention with regards to seismic issues.

---

[1] G&E Engineering Systems Inc., 6315 Swainland Rd, Oakland, CA 94611; eidinger@earthlink.net

Since the early 1990s, a growing number of U.S. and Japanese water utilities have examined their seismic vulnerabilities. These efforts have in part been promulgated by the poor performance of a few water systems in the 1989 Loma Prieta, 1994 Northridge and 1995 Kobe earthquakes. Invariably, when utilities consider economic impacts of water outages to their customers, they come to the realization that some type of seismic mitigation is economically warranted. Table 1 lists a few such examples.

Water Utility	Population Served	Capital Cost ($)	Cost Per Person ($)
East Bay Municipal Utility District	1,200,000	$240,000,000	$200
San Diego Water Department	1,200,000	$46,000,000	$40
Los Angeles Dept of Water and Power	3,500,000		$25
Contra Costa Water District	430,000	$120,000,000	$280
Portland, Oregon	800,000		$10 est
Seattle, Washington	1,300,000	>$20,000,000	$20
St. Louis, Missouri		$20,000,000	$30
Memphis Tennessee	800,000	$20,000,000	$25
San Francisco Public Utilities Commission (Hetch Hetchy)	2,400,000	$1,300,000,000	$540
San Diego County Water Authority	2,400,000	$700,000,000	$290

Table 1. Capital Cost for Seismic / Reliability Retrofits – United States

A few notes are made for Table 1. The capital costs shown are not in constant dollars. The data used to develop these costs are based on discussions with each utility or consultants working for each utility. The costs reflect expenditures made in the mid-1990s through 2003, or planned by the year 2015. For EBMUD, the costs include upgrades for its treated water and raw water systems. For the SFPUC, the costs are only for seismic upgrades for its Hetch Hetchy transmission system. The costs sometimes reflect upgrades made for both *seismic* and *reliability* upgrades. While much of the upgrades can be identified with specific seismic-only upgrades (like anchoring a tank), it is more imprecise to state that the installation of a new pipeline or new reservoir is made only for seismic issues, as the decision might have also been influenced for non-seismic issues like drought, system build out, maintenance or other issues. However, it would be reasonable to say that without the underlying seismic threat, most of these reliability-based projects might not have been implemented.

There is a striking difference in cost per person for difference utilities. For example, in high seismic regions like the San Francisco Bay Area, the cost per person is $200 or higher per person. In contrast, in lower seismic regions like San Diego, Memphis and Seattle, the cost per person is more like $20 to $50 per person. This large difference in cost is not to say that the existing infrastructure in Memphis is better than that in Oakland, but rather that the likelihood of large earthquakes in Memphis is lower than in Oakland or San Francisco.

Table 2 provides the corresponding costs being budgeted by a variety of Japanese utilities. The data in Table 2 is taken from discussion with engineers at each utility, and is converted to US dollars at a rate of about 110 Yen = $1 US.

Water Utility / City	Population Served	Capital Cost ($)	Cost Per Person ($)
Hiroshima	1,140,000		>$300
Tokyo	11,000,000		~$300
Kobe	1,500,000	$1,360,000,000	$900
Yokohama	3,374,000		~>$300
Osaka	2,600,000	$1,000,000,000	$380
Hanshin	2,000,000	$16,000,000	$8
Hachinohe	338,000	$50,000,000	$150

Table 2. Capital Cost for Seismic / Reliability Retrofits – Japan

Table 2 shows that Japanese utilities are spending more on average than US water utilities, with $300 per person (or more) being the norm. The Hanshin water utility is a wholesaler, and the relatively low cost ($8 / person) would be added to the final cost per person for people in Kobe (Kobe City buys water from Hanshin). The typical cost for water for a Japanese resident is reasonably similar to the typical cost for a US resident.

Table 3 provides the projected US-wide "marketplace" for cost-effective seismic retrofit of water systems. The "High Risk" regions includes much of coastal California, parts of Alaska, Hawaii, coastal Oregon and Washington, etc. The "Moderate Risk" regions include areas near the New Madrid fault zone, Charleston, Salt Lake City, etc. The "low risk" regions include New York City, Boston, etc. The "very low risk" regions include most of Texas, Florida, etc.

US – Seismic Region	Population	Cost per Person ($ 2003)	Total Cost
High Risk	40,000,000	$225	$9,000,000,000
Moderate Risk	15,000,000	$30	$450,000,000
Low Risk	68,000,000	$5	$340,000,000
Very Low Risk	157,000,000	$0	$0
Total	280,000,000		$9,790,000,000

Table 3. US Capital Budget for Seismic Retrofit of Water Systems

Through 2001, about $1,500,000,000 had already been budgeted towards various seismic retrofit programs in the US. In 2002, the San Francisco water system (covering the Hetch Hetchy transmission system serving 30 different distribution systems) began a new $3.6 billion dollar program. As of mid-2003, it would appear that about 10% of the cost-effective seismic retrofits within the USA had been completed, with the bulk (90%) yet to be done.

In lesser developed regions of the world, earthquakes impact water systems at least as badly as they have done in the US and Japan. Recent earthquakes in Izmit Turkey (1999), Bhuj India (2001) and Moquegua Peru (2001) have led to widespread and long term (months) disruptions of piped potable water. Post-earthquake investigations of these earthquakes by TCLEE and others have shown that the affected water system owners had done essentially nothing to improve their systems for earthquakes, even after the compelling evidence available from the 1989 Loma Prieta, 1994 Northridge and 1995 Kobe events. Why have water system operators not done anything? Certainly not because they wished to have an earthquake cause several month water outages; but more probably because they a) were not aware of the seismic hazard; b) they knew about the seismic hazard, but had no direction (codes, standards, guidelines) to direct them what to do; or c) they knew about the seismic hazard and they knew about how to mitigate the risk, but they were not economically inclined to do anything about it; perhaps their impression was that "it was not worth the money".

3 Economics of the Seismic Upgrade of the Hetch Hetchy Water System

If one approaches the typical water utility owner, and asks them if "it is worth it to upgrade their water system for earthquakes?", one will most often get one of the following five stages of response:

1. I don't have a problem...
2. I did not know I had a problem...
3. I sense that there might be some type of problem, but I don't know how to quantify it...
4. I am pretty sure I have a problem, so I will take a shotgun approach and fix / improve as many parts of the system as my (regulators / city council / rate payers) are willing to pay for...
5. I know I have a problem, so I will study it and develop a rational and cost effective approach to address it....

In high seismic regions like Coastal California, the author has experience with various water utilities that have provided all of these five stages of response. Some of the larger water utilities serving a million or more people (like the City of San Diego, the Santa Clara Valley Water District, the East Bay Municipal Utility District) have adopted approaches consistent with response 5. Many other water utilities, serving populations from 15,000 people to millions of people have adopted any or all of responses 1, 2, 3 and or 4, with the result that some utilities are spending too little and some are spending too much. One intriguing example is currently taking shape: the seismic and reliability upgrade of the aging SFPUC Hetch Hetchy water system.

The Hetch Hetchy system is a water transmission system delivering water from Yosemite National Park (and a few other local supply sources) to about 2,400,000 people in the San Francisco Bay Area. These 2,400,000 people are served by 30 separate water distribution systems, the largest of which (770,000 people) is the City of San Francisco's own distribution system. Ownership, operation and maintenance of the Hetch Hetchy system is by the San Francisco Public Utilities Commission (SFPUC). The remaining 29 water distribution systems (the so-called "suburban customers") purchase water from the SFPUC, and pay for about 70% of the cost to operate, maintain and upgrade the Hetch Hetchy system. At times, the wishes of the 29 suburban customers do not line up exactly with the wishes of the SFPUC.

Since the late 1990s, the SFPUC has been studying seismic and other reliability aspects of the Hetch Hetchy system. In January 2000, the SFPUC completed their "SFPUC Facilities Reliability Program" (2000). This effort simulated the overall SFPUC water system reliability in the event of a major earthquake on the San Andreas, Hayward, Calaveras or Great Valley faults. The effort reportedly used the "most current understanding of effects of infrastructure from ground shaking, fault crossing and liquefaction". The analyses resulted in a recommended program of seismic improvements to increase overall SFPUC system reliability. The overall cost of this program was estimated at $3.5 Billion, of which $1.3 Billion was for seismic improvements, and the remainder for reliability improvements. These amounts include no funds to make improvements in the San Francisco City and 29 suburban distribution systems.

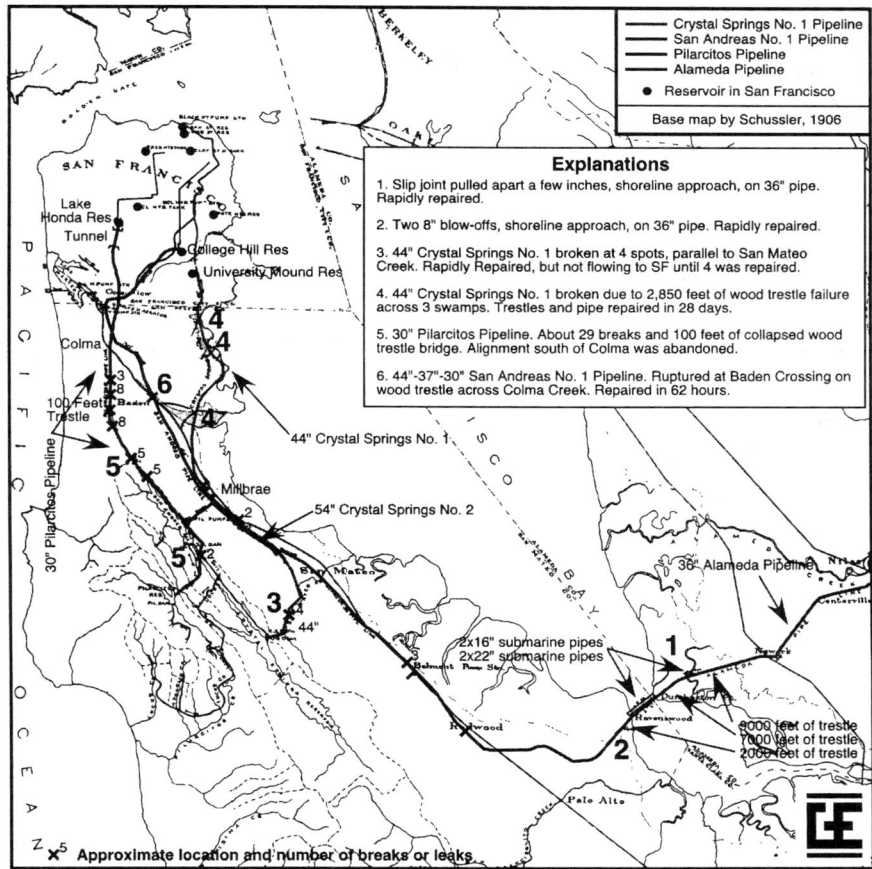

Figure 1. Damage to the SFPUC (Spring Valley Water Company) Transmission System, 1906

Figure 1 shows a map of the SFPUC transmission system as it existed in 1906 and the damage it suffered in the 1906 Great San Francisco earthquake. The modern (year 2003) SFPUC transmission system has about 3 times as many pipelines, many of which follow similar alignments as the pipelines did in 1906, except that newer pipes bypass the marshy area marked by the number "4" in Figure 1.

It remains unclear as of 2003 as to exactly how much the 29 suburban agencies are happy to pay for this program. The final cost of the Hetch Hetchy system seismic reliability upgrades will roughly triple the cost to purchase SFPUC water.

4 What About the Suburban Customers?

With large potential rate increases facing the suburban customers of the SFPUC, the level of awareness about seismic issues has risen from "about" stages 1 or 2, and most are now thinking about responses at stages 4 or 5. A series of seismic vulnerability analyses have been performed for many of the suburban customers.

Item	Amount	Note
Average Day Demand	286 MGD	81% of total system demand
Number of Pump Stations	151	
Number of Storage Tanks	192	
Miles of Distribution System Pipelines	3,713	Mostly 4" to 27" pipe
Wells	85	
Treatment Plants	6	
Emergency Generators	63	
Pipe Repairs, San Andreas M 7.9 Earthquake	2,400 to 5,000	Lower value is more likely
Pipe Repairs, Hayward M 7.1 Earthquake	1,400 to 3,600	Lower value is more likely
Seismic Improvement Program	$25 to $44 million	

Table 4. Statistics of 18 Suburban Customer Water Systems

The 18 suburban customers that have had seismic vulnerability analyses performed (Hayward, Alameda County Water District, City of Santa Clara, Mountain View, Purissima Hills, Palo Alto, Stanford University, Bear Gulch, Redwood City, San Carlos, San Mateo, Foster City, Coastside County, Mid-Peninsula, Burlingame, South San Francisco, Brisbane, Daly City) represent about 76% of the total suburban customer demand; or in conjunction with the City of San Francisco, about 81% of total Hetch Hetchy system demand. Table 4 provides some overall statistics for these 18 suburban customers.

The modern Hetch Hetchy water system has about 220 miles of large diameter (mostly 60" to 96" diameter) pipelines within the greater San Francisco Bay Area. In consideration of faulting, liquefaction, landslide and ground shaking, these pipes are expected to suffer between 16 and 23 repairs following Hayward M 7.1 and San Andreas M 7.9 earthquakes, respectively. The bulk of these repairs will likely manifest themselves as leaks at air valves or blow offs, but a few full breaks are likely at fault crossings, creek crossings or at unexpected locations. There is even a chance that a major tunnel might collapse. With available in-house repair crews, the SFPUC might be able to patch up the major breaks in 4 to 12 days, and repair all leaks within 1 to 2 months. If the unlikely but not impossible event that a major tunnel should collapse, repairs of the tunnel could last months, in the meantime the water supplies might have to be restricted to no more than about 80% of maximum winter time demands.

Given these scenarios, the following seismic improvement have been proposed:

- o $25 to $44 million of seismic improvements within the 18 suburban customer distribution systems.
- o $1.3 to $3.5 billion of seismic and reliability improvements within the Hetch Hetchy transmission system.

As of mid-2003, there remains much work to coordinate the overall transmission / distribution seismic upgrade programs. For example, should a small suburban customer invest $800,000 to construct a well, thereby providing an alternate source of water should all Hetch Hetchy water be lost for days to weeks after a major earthquake? And if that small suburban customer builds that well, should it also accept the allocated cost to improve the major water pipeline transmission system? What might be most cost effective for that one suburban customer might not be the most cost-effective for other suburban customers, or for the SFPUC as a whole, and this brings up difficult political and policy issues.

Item	EBMUD	SFPUC + 18 Suburban Customers
Miles of Transmission Pipelines	200	220
Miles of Distribution Pipelines	3,900	3,700
Tunnels	16	20
Treatment Plants	6	8
Storage Tanks	175	192
Pump Stations	125	151
Small Pipes that cross major active faults (≤ 18" diameter)	178	66
Large Pipes that cross major active faults (≥ 20" diameter)	27	11
Tunnels that cross major active faults	2	0
Pipe Repairs, Loma Prieta M 7.1	135	< 400
Pipe Repairs, San Andreas M 7.9	< 1,000	2,400 to 5,000
Pipe Repairs, Hayward M 7.1	3,300 to 5,000	1,400 to 3,600
Seismic Upgrade, Transmission System	$140 million	$1,300 million
Seismic Upgrade, Distribution System	$100 million	$25 to $44 million
Seismic Improvements, Total	$240 million	$1,325 to $1,340 mil.
Ratio, Distribution to Total	42%	2% to 4%
Population served	1,200,000	2,400,000
Cost per person	$200	$555

Table 5. EBMUD and SFPUC / Suburban Customer Cost Allocation

To provide some insight to these issues, one can examine the allocation of seismic upgrade cost made by EBMUD in their $240,000,000 seismic upgrade program. EBMUD is a utility that owns and operates both a raw water transmission as well as a large potable water distribution system. For EBMUD's case, if one sums up all costs associated with raw and treated water pipelines of 36" diameter and larger (cumulatively, the "transmission system"), EBMUD has spent about $140,000,000 on transmission upgrades. The remaining $100,000,000 was allocated to upgrades of smaller diameter pipelines (generally 12" to 30" diameter), water treatment plants, pump stations, storage tanks and emergency response. Table 5 highlights the differences in upgrade costs between EBMUD (actual) and SFPUC / Suburban customers (projected).

The age of infrastructure in the EBMUD and SFPUC transmission systems is quite similar. The original EBMUD transmission pipelines and tunnels were put into service in 1929 (Mokelumne 1, Claremont Tunnel); the original Hetch Hetchy pipelines and tunnels were put into service in 1923 to 1933 (BDPL 1 and 2, Coast Range Tunnel). EBMUD's first major transmission pipeline system upgrade was put in service in ~1948 (Mokelumne 2); similar for Hetch Hetchy (BDPL 3). EBMUD's most recent major transmission pipeline system upgrade was put in service in ~1965 (Mokelumne 3); similar for Hetch Hetchy (BDPL 4).

5 Economic Impacts to Suburban Customers

A series of seismic vulnerability analyses were performed for 18 water distribution systems that are served by the Hetch Hetchy transmission system. These 18 systems have a combined average day demand of 228 MGD, and serve a population (year 2020) of 1,419,000 people. Allowing for 20 to 30 day outages from the SFPUC transmission system (probably upper bound, more likely 4 to 12 days), and a variable amount of impacts to the local distribution systems (pipe repairs, damaged tanks, failed wells, power outages, etc.), and using the Fire Ignition and Spread models by Eidinger (1996), the following statistics (medians only) are developed:

Item	San Andreas M 7.9	Hayward M 7.1
Economic Losses, Year $2003	$1.4 to $1.6 billion	$250 to $610 million
Fire Ignitions	95	73
Fire Losses, Calm Winds	$85 to $142 million	$65 to $110 million
Fire Losses, Light Winds	$200 to $342 million	$153 to $262 million
Fire Losses, High Winds	$1.1 to $1.4 billion	$0.9 to $1.1 billion

Table 6. Impacts to 18 Distribution Systems in Scenario Earthquakes (As Is System)

Item	San Andreas M 7.9	Hayward M 7.1
Economic Losses	$93 to $333 million	$127 to $535 million
Fire Losses, Calm Winds	$14 to $57 million	$11 to $44 million
Fire Losses, Light Winds	$85 to $114 million	$66 to $88 million
Fire Losses, High Winds	$1.0 to $1.1 billion	$0.8 to $0.9 billion

Table 7. Impacts to 18 Distribution Systems in Scenario Earthquakes (Upgraded System)

The "upgraded system" evaluation is performed for the same 18 distribution systems, but this time with the assumption that seismic upgrades are in place to reliably assure that no more than a 24 hour outage of delivery of maximum winter time demand rate water from the Hetch Hetchy transmission system to each distribution system.

By comparing the difference in losses (economic and fire) from Tables 6 and 7, we can estimate the net benefit (scenario earthquake basis) of the retrofit program. Using the midpoint values, and assuming the light wind scenario, the net reduction in losses (ie, the benefit) is:

Item	San Andreas M 7.9	Hayward M 7.1
Benefit, Economic Impacts	$1,250 million	$99 million
Benefit, Fire Impacts	$172 million	$131 million
Benefit, Other Impacts	$200 million	$40 million
Total Benefit (Scenario Based)	$1,622 million	$270 million

Table 8. Net Benefits of Seismic Upgrade, Scenario Based

Allowing that there is about a 1% chance of occurrence of either of these two or similar scenario earthquakes (San Andreas M 6.8 to 7.9 event that includes the Peninsula fault segment, Hayward M 6.8 to 7.3 event that includes the southern Hayward fault segment), and allowing for other earthquakes on other faults and for smaller earthquakes, and assuming a 5.5% discount rate, and using the benefit cost model for water systems outlined in (Eidinger and Avila, 1999), the net present value of the benefits of seismic upgrades are calculated as follows (all monetary values in millions, year $2003):

- San Andreas M 6.8 – M 7.9: $1,622. Annual chance: 0.01. Annual benefit: $16.22
- Hayward M 6.8 – M 7.3: $270. Annual chance: 0.01. Annual benefit: $2.7
- Calaveras, Rodgers Creek, Great Valley, background and smaller earthquakes: Cumulative annual benefit = $9.7
- Total annual benefit over all faults, all magnitudes = $28.6
- Net present value of benefits, 5.5% discount rate, 100 year project life = $28.6 x 18.1 (NPV factor) = $518

In other words, the rate payers of the 18 distribution systems should be willing to pay, in year 2003 dollars, up to about $518,000,000 to seismically retrofit the Hetch Hetchy water system to the point where it can reliably restore water to each system within 24 hours after any earthquake, at maximum winter demand rate or higher.

6 Conclusions and Observations

The estimated size of the marketplace for cost effective seismic upgrade of water systems in the United States is about $10 Billion (year 2003 dollars). About 10% of this has been spent through mid-2003.

A comparison is made between the (almost completed) EBMUD seismic upgrade program and the (recently started) SFPUC seismic upgrade program. While there are a number of similarities between the age and quantity of infrastructure between of the two sets of water systems, the cost of the programs is quite different, as well as the ratio of cost between distribution and transmission upgrades.

By performing seismic vulnerability analyses for 18 suburban distribution systems served by the SFPUC Hetch Hetchy system, and then performing economic analyses as to the value of seismic upgrades, this paper shows that these suburban customers should be willing to pay up to about $518,000,000 to achieve a no-more than one-day outage of the Hetch Hetchy transmission system after any earthquake. Retrofits and improvements beyond this cost could be justified for non-seismic reliability issues.

7 Units and Abbreviations

All monetary values are in year 2003 U.S. dollars, except as noted.

BDPL = Bay Division Pipeline

EBMUD = East Bay Municipal Utility District

Inches (") = 25.4 millimeters

M = moment magnitude

Miles = 1.609 kilometers

MGD = Million Gallons per Day (US liquid measure). 1 MGD = 43.8 liters per second

NPV = Net Present Value

SFPUC = San Francisco Public Utilities Commission

TCLEE = Technical Council on Lifeline Earthquake Engineering

UBC = Uniform Building Code

8 References

Eidinger, J., and Avila, E., Eds., Guidelines for the Seismic Upgrade of Water Transmission Facilities, ASCE, TCLEE Monograph No. 15, January 1999.

Eidinger, J., Lifeline Considerations and Fire Potential, *in* Seismic Safety Manual, D. Eagling editor, Lawrence Livermore National Laboratory, September 1996.

SFPUC Facilities Reliability Program, Phase II – Regional System Overview, Final Report, CH2M-Hill, Olivia Chen Consultants, Montgomery Watson, EQE International, January 2000.

Seismic Reliability of Urban Pipeline Network Systems

Han Yang[1] Sun Shaoping[2]

Abstract

In this paper, the probabilistic model of seismic damage prediction of buried pipeline is presented; in which both the seismic response and the resistance of pipe structures are treated as random variables, and their statistical characteristics were discussed. Then a disjoint algorithm for reliability analysis of network is presented; in which sharp-product operation is adopted to construct a disjoint minimal path set of network. Combined with Topological Classification method, a large and complex network can be easily calculated. These presented methods have been programmed and their feasibility and effectiveness have been verified and tested in practice of prediction to seismic damage of water supply network in some cities.

Introduction

Urban pipelines include water, gas, heat transmission systems, sewer systems and so on. They are essential for sustaining the normal operation of a community. After an earthquake attacked, the damage of the pipelines and the secondary disaster are more serious in modern cities. Most of pipeline network systems in China have not been passed through normal aseismic design; many big and medium cities may be in the seismic risk background. So, it is necessary to study the seismic reliability of pipeline systems for the needs of earthquake hazard mitigation. This paper focus on water supply networks and main contents includes:

The probabilistic model of seismic damage prediction of buried pipeline is presented by taking the breakage of the pipe joints caused by seismic wave propagation as main damage mode. Both the seismic response and the resistance of --

---

[1]Senior Engineer, Dalian University of Technology, Henan Building Research Institute, Zhengzhou, P. R. China, hanyang@hnjky.com.cn
[2]Professor, Beijing Municipal Engineering Research Institute, Beijing, P. R. China, sun00570@sohu.com

pipe structures are treated as random variables, and their statistical characteristics were discussed.

An effective algorithm for reliability analysis of network is presented; in which cubic notation is used to describe the logic function of a network in a well-balanced state, and then the sharp-product operation is adopted to construct a disjoint minimal path set for the network. Combined with Topological Classification method, a large and complex network can be easily calculated.

These presented methods have been programmed and their feasibility and effectiveness have been verified and tested in practice of prediction to seismic damage of water supply network in some cities of China.

Seismic damage behavior of buried pipeline

The main damaged types of water supply pipelines under earthquake in China are: (1) Joint damage: mortar filler fall off in joint, spigot pull out or bell broken for segmented pipeline; the weld crack or the bolts of pipe-flanges loosen for steel pipeline. (2) Longitudinal and oblique crevices appear in pipe bodies of reinforced concrete, asbestos concrete and cast iron. Small diameter steel and cast-iron pipe body break when corrosion severity. (3) Tee joint, elbow, valves and the link places of pipelines with structures broken since stress concentration and motion phase difference.

In these three kind of broken types, pipe body broken generally caused by ground rift, landslip that is critical earthquake calamity in sites or owing to pipes own defects and critical corrosion, but joint broken are most common. For example, 79% joints of water supply cost iron pipes destroyed at Tang Shan earthquake in 1976 (Sun, 1994).

Most earthquake calamity indicate that at the same condition, flexible joints are much better than rigid joints for anti-seismic, because former absorb more site strain. Moreover the pipes of larger diameter destroyed less than the small ones, it indicates that the stiffness of the pipeline may restrain the deformation of the surrounding soil. Other factors as pipeline design, construction and service life, affect the anti-seismic of the pipelines to a certain degree.

We conclude the research of domestic and abroad about the characteristics of buried pipeline under the seismic wave motion as following (Wang et al., 1985; O'Rourke et al., 1999)

(1) Pipelines are mainly affected by seismic wave and under the influence of fluctuant deformation of the surrounding soil.

(2) The deformation of pipeline is less than the strain of free sites, the quantity reduced mainly decided by the ratio of stiffness between pipe and soil. When the deformation of soil causes the strain of soil nearby the surface of pipeline and the frictional force in soil exceed limit, slip will appear between pipeline and soils around or a certain soil layer interface nearby.

(3) The axial strain is major in pipelines. When pipes slide with soils each other, the places of tee joints and elbows will concentrate stress.

(4) On the effect of axial seismic wave motion, range of half of apparent wavelength is in tension and other half is in compression. For segmented pipelines, joints absorb the deformation.

Based on the past earthquake experiences and the experiment on pipe joints in China, three damage states of buried pipeline under earthquake are defined as follow:

Undamaged: No damage on pipe body, deformation at rigid joints is in elasticity and its relative displacement S is less than the fissure limit R_1. Joints may have very small fissure. There might be very little leakage at joints.

Slight damage: Deformation at rigid joints is in the stage of elasticity to plasticity, and the relative displacement S exceeds the fissure limit R_1. There is slippage between rubber ring and pipe body at flexible joints. Water pressure in pipelines may be reduced.

Serious damage: The relative displacement S of the joint exceeds the leaking limit R_2, the filler is loosened with serious leakage even out of operation.

Probabilistic prediction model of buried pipeline

Let the relative displacement of the joint be S as the earthquake effect, and the allowable displacement be R as the structural resistance of the pipeline, then the effective function Z of the variables S and R can be obtained:

$$Z = f(S, R) = R - S \tag{1}$$

S and R are both random variables, and then the failure probability is

$$P_F = P(Z < 0) \tag{2}$$

Assume that R and S follow the random variables of the two normal distributions, $N(\mu_R, \sigma_R)$ and $N(\mu_S, \sigma_S)$ respectively, then Z is a normal random variable, and follows the normal distribution $N(\mu_Z, \sigma_Z)$, in which $\mu_Z = \mu_R - \mu_S$, $\sigma_Z = (\sigma_R^2 + \sigma_S^2)^{0.5}$. $\mu_R, \sigma_R, \mu_S, \sigma_S, \mu_Z, \sigma_Z$ are the expected values and standard deviations of the random variables R, S, and Z respectively. These values were determined by experiments.

So, the probabilistic prediction model of seismic damage of the pipeline is:
1) The basic perfect state

$$P_{F1} = P(Z_1 > 0) = \phi(\frac{\mu_1}{\sigma_1}) \tag{3}$$

2) The serious damaged state

$$P_{F3} = P(Z_2 \leq 0) = \phi(-\frac{\mu_2}{\sigma_2}) \tag{4}$$

3) The moderate damaged state

$$P_{F2} = 1 - P_{F1} - P_{F3} \tag{5}$$

in which $\mu_1 = \mu_{R1} - \mu_S$, $\sigma_1 = (\sigma_{R_1}^2 + \sigma_S^2)^{0.5}$, $\mu_2 = \mu_{R2} - \mu_S$, $\sigma_2 = (\sigma_{R2}^2 + \sigma_S^2)^{0.5}$.

By means of these formulas and test statistical data, seismic risk analysis and pipeline data, the probability can be found out.

Analyses of the network reliability

In system reliability analysis, it is customary to represent the system by a probability graph, G, in which each vertex v_i and each edge e_i (directed or undirected) has a given probability. For example, the water supply plant corresponds to the source, the intersection of the pipelines to the vertex, the end-user to the terminal, the pipeline to the edge, and the probability of the edge corresponds to the probability that the pipeline can be operated during an earthquake. A probability digraph G, as an example, is given in Figure 1.

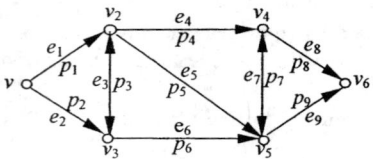

Figure 1. Probability digraph G

For a graph G containing N paths, the terminal-pair reliability $R(G)$ can be written as:

$$R(G) = P_r\{\bigcup_{i=1}^{N} P_i\} = P_r\{P_1 \cup P_2 \cup \cdots \cup P_N\} \qquad (6)$$

where $P_r(\cdot)$ is the probability, and P_i is the ith path of G. To evaluate the above-mentioned expression, the following principle of inclusion-exclusion for the probability of the union of k events $(A_1 \cup A_2 \cup \cdots \cup A_N)$ is often used.

$$P_r\{\bigcup_{i=1}^{N} P_i\} = \sum_{i=1}^{N} P_r\{P_i\} - \sum\sum_{1 \le i < j \le N} P_r\{P_i P_j\} + \sum\sum\sum_{1 \le i < j < l \le N} P_r\{P_i P_j P_l\}$$
$$- \cdots + (-1)^{N-1} P_r\{P_1 P_2 \cdots P_N\} \qquad (7)$$

The number of terms in the explicit expression (7) is 2^{N-1}. The computation using (7) is clearly not feasible for networks of even moderate complexity (e.g., if $N=20$, then the number would be nearly 10^6)(Barlow et al.,1980). So, the path or cut-set enumeration method needs to be improved.

Fundamental concept of the disjoint algorithm. The fundamental concept of the disjoint algorithm is to construct a set of some minimal paths to a disjoint events set, thus, the logic operation becomes an arithmetic operation. If the logic function S of system reliability can be expressed as a sum of some disjoint events $T_j (j = 1, 2, \cdots, I)$, i.e.

$$S = \bigcup_{i=1}^{N} P_i = \sum_{j=1}^{I} T_j \qquad (8)$$

Then, the reliability can be expressed as follows:

$$R(G) = P_r\{\sum_{j=1}^{I} T_j\} = \sum_{j=1}^{I} p_r\{T_j\} \qquad (9)$$

Based on $A \cup B = A + \overline{A}B$, the following process is used to form Eq. (8). Due to

$$\bigcup_{i=1}^{N} P_i = P_1 + \overline{P_1} \bigcup_{i=2}^{N} P_i \tag{10}$$

where $\overline{P_1}$ is the complementary set of P_1. It is evident that the two items on the right side are disjoint. Let P_1 be the first disjoint item T_1, and simplify the second item by following basic logic operation rules:

$$AB \subset A; \quad A\overline{A} = \Phi; \quad A \bigcup AB = A;$$

where $A \& B$ are arbitrary sets. After simplification, one obtains

$$\overline{P_1} \bigcup_{i=2}^{N} P_i = \bigcup_{l=1}^{n} P_l \tag{11}$$

Repeating the steps on the right side of Eq. (11), $\bigcup_{i=1} P_i$ becomes a sum of disjoint products after finite steps.

As mentioned above, the disjoint process needs different kinds of logic operations and increases the difficulty in programming. Therefore, cubic notation is used to describe the logic function of network reliability, and then the objective of sum-of-product is attained through a series of cubic operations.

Logic function expressed by cubic notation. At first, each edge of the network is expressed by a logic variable, where X_i expresses edge e_i to be reliable, $\overline{X}_i$ expresses edge e_i to be invalid $(i = 1,2,\cdots,m)$. Then, one minimal path P_i of a network is expressed by one product of some logic variables; the event of reliability of a network consisting of m edges is an m-element logic function $R = f(x'_1,\cdots,x'_i, \cdots,x'_m)$. For a logic variable x'_i, its value has only two states: X_i and $\overline{X}_i$ (1 and 0), therefore, m logic variables, their values have 2^m states, which correspond to the 2^m vertices of an m-dimension cubic. So, any m-element logic function can be expressed by the cubic set in m-dimension space(Shen S.C., 1987):

$$T(R) = \{c^1, c^2, \cdots, c^k, \cdots, c^N\} \tag{12}$$

where $c^k = c_1^k c_2^k \cdots c_i^k \cdots c_m^k (i=1,2,\cdots,m)$ is a r-cubic in m-dimension space, corresponding to a minimal path of network; $c_i^k \in \{0,1,x\}$ is an element of cubic c^i, defined as follows:

$$\begin{cases} 0, & \text{if } x'_i = \overline{X}_i; \\ c_i^k = 1, & \text{if } x'_i = X_i \\ x, & \text{if } x'_i = 1 \text{ (absence)}. \end{cases} \tag{13}$$

For example, a 4-element logic function:

$$R = X_1 X_2 \overline{X}_4 \bigcup \overline{X}_1 X_2 X_3 \bigcup X_1 X_2 X_4,$$

Its cubic set is expressed by cubic notation as follows:

$$T(R) = (11x0, 011x, 11x1)$$

Realization of sharp-product operation in computer program. Sharp-product operation is also called complement; symbol #. It is the most complex calculation and also the most useful one in a cubic operation. The sharp-product operation between two cubes a and b is based on the sharp-product table (Shen S.C.,1987) and the regulation is as follows:

$$\begin{cases} a, \text{if any } k, a_k \# b_k = \Phi; \\ a\#b = \Phi, \text{for each } k, a_k \# b_k = \varepsilon; \\ \bigcup_k \{a_1, a_2, \cdots, a_k, \cdots, a_n\}, \text{Otherwise.} \end{cases} \quad (14)$$

By the property of sharp-product:

$$A\#b = (a^1 \# b) \cup (a^2 \# b) \cup \cdots \cup (a^m \# b) \quad (15)$$

where A is the cubic set, b is cube, and $a^1, a^2, \cdots, a^m$ are cubes, the elements of A.

Taking an arbitrary cube as b and the others as A from the cubic set $T(R)$ corresponding to Eq.(8), then using Eq.(15), two disjoint cubes can be obtained: b and $A\#b$. Let b as the first disjoint term T_1 and $A\#b$ as $\overline{T}_1 \cup A$, we obtain a new cubic set $T(R_1)$ after simplifying $A\#b$. Repeat this process until $A\#b$ becomes one cube.

For convenience of computer programming and saving EMS memory, the binary code is adopted to express cubic notation. Each cube is written in one or several words in the computer.

For example, for cube $11x0$, its binary code is 10101101 and can be written in one word as integer 173. Then, these code operations turn into the basic operation of a machine instruction or language; here we use the FORTRAN90 language.

Decomposition of network by topological classification method

Lifeline networks are usually very large and complicated systems, often having hundreds of nodes and edges. The traditional graphic analysis method can't meet the need of application in practice. On the other hand, it is not precise regarding the seismic damage mechanism in different parts of the lifeline network, and there are uncertainties and inaccuracies in resistance and computational modeling, so it is practical to develop an approximate algorithm which should be simple, time-saving, and can deal with the large and complicated network, and can obtain results with enough precision.

For the above reason, the topology methodology and classification method are used to obtain the reliability of the upper bound in a large and complicated network. This method is simple and practical, and can obtain the reliability of the lower bound by dual theory (Han and Sun, 2001).

The basic idea of the classification method is as follows: according to the topology of the node set, the network is classified into ordered subsets, then the reliability of every sub-network is calculated by this method from the source to the terminal, step by step.

Example

Example 1: Consider the network shown in Figure 1. Using the DFS subroutine, we obtain 12 minimal paths from source vertex v_1 to terminal vertex v_6. They are:

$$e_1,e_3,e_6,e_7,e_8; \quad e_1,e_3,e_6,e_9; \quad e_1e_4e_7e_9; \quad e_1e_4e_8; \quad e_1e_5e_7e_8; \quad e_1e_5e_9;$$

$$e_2e_3e_4e_7e_9; \quad e_2e_3e_4e_8; \quad e_2e_3e_5e_7e_8; \quad e_2e_3e_5e_9; \quad e_2e_6e_7e_8; \quad e_2e_6e_9;$$

The logic function S of system success is established as the union of all paths:

$$S = \bigcup_{i=1} P_i = X_1X_3X_6X_7X_8 \cup X_1X_3X_6X_9 \cup X_1X_4X_7X_9 \cup X_1X_4X_8$$
$$\cup X_1X_5X_7X_8 \cup X_1X_5X_9 \cup X_2X_3X_4X_7X_9 \cup X_2X_3X_4X_8$$
$$\cup X_2X_3X_5X_7X_9 \cup X_2X_3X_5X_9 \cup X_2X_6X_7X_8 \cup X_2X_6X_9$$

Then, the logic function is expressed by a cubic set. And using binary code to express this cubic set, it is written by the computer in integers as follows:

$$T(R)' = (192427, 192446, 195566, 195579, 196331, 196350,$$
$$240622, 240635, 241387, 241406, 245675, 245694)$$

Then, using the SHARP subroutine, we get only 18 disjoint products as follows:

$$S = \sum_{i=1} T_i = X_2X_6X_9 + X_1X_4X_8\overline{X}_9 + X_1X_5\overline{X}_6X_9 + X_1\overline{X}_2X_3X_6X_9$$
$$+ X_1X_4\overline{X}_5\overline{X}_6X_7X_9 + X_2\overline{X}_4X_6X_7X_8\overline{X}_9 + \overline{X}_1X_2X_3X_5\overline{X}_6X_9$$
$$+ \overline{X}_1X_2X_3X_4X_8\overline{X}_9 + X_1\overline{X}_2\overline{X}_3X_5X_6X_9 + X_1X_4\overline{X}_5\overline{X}_6\overline{X}_7X_8X_9$$
$$+ X_1\overline{X}_2X_3\overline{X}_4X_6X_7X_8\overline{X}_9 + X_1\overline{X}_2\overline{X}_3X_4\overline{X}_5X_6X_7X_9$$
$$+ \overline{X}_1X_2\overline{X}_3X_4X_6X_7X_8\overline{X}_9 + \overline{X}_1X_2X_3X_4\overline{X}_5\overline{X}_6X_8X_9$$
$$+ X_1\overline{X}_4X_5\overline{X}_6X_7X_8\overline{X}_9 + \overline{X}_1X_2X_3\overline{X}_4X_5\overline{X}_6X_7X_8\overline{X}_9$$
$$+ 2\overline{X}_1X_2X_3X_4\overline{X}_5\overline{X}_6X_7\overline{X}_8X_9 + X_1\overline{X}_2\overline{X}_3X_4\overline{X}_5X_6\overline{X}_7X_8X_9$$

the number of terms is far below 4095 ($2^{12}-1$), and also lesser than the other similar methods. For the same problem, the number of the disjoint products was 26 by the method presented in Li *et al.*(1993); after optimizing the process, the number of disjoint terms was reduced to 23.

Since these terms exclude each other, the reliability of the network can be obtained easily by arithmetic operation. For simplicity, assuming the reliability of every edge is the same, $p_i = 0.5$, we obtain the reliability of the network:

$$R(G) = P_r\left\{\sum_{i=1}^{18} T_i\right\} = \sum_{i=1}^{18} P_r\{T_i\} = 0.398$$

Example 2: Consider a directed network with 210 vertexes and 319 edges; assuming the reliability of every edge is the same.

By the classification method, this network was decomposed into 25 sub-networks: then, the presented method was used to calculate the reliability of the network, step by step.

Table 1 shows the upper and lower bounds of the reliability of the terminal vertex. We can see that, in the high component reliability region, the deviation in system reliability from the approximation method and exact method is very small and the reduction in computer time is quite significant.

Table 1 System reliability of example 2

Component reliability	System reliability		
	Upper bound	Lower bound	Average
0.6	0.54450	0.30865	0.42658
0.7	0.81624	0.66642	0.74132
0.8	0.93750	0.87891	0.90821
0.9	0.98765	0.97546	0.98155
0.95	0.99723	0.99447	0.99585
0.99	0.99990	0.99980	0.99985

Application

Consider a water supply network in a city. The whole network is simulated as a directed graph. It has 142 vertices; among them are four source vertices, 22 sink vertices, and 194 edges. This network can be simplified as a directed graph only with one source and one terminal by constructing the dummy vertices. The direction of the edges can be simplified according to the following principles: the main pipeline near the source and the branches to the end are set as directed edges. The others can be determined according to factors such as flowing from large pipes to small ones, distance to the source, height of vertices, etc. Factors of pipeline that are uncertain can be designated as undirected edges.

By using the program YU3 as performed by the present authors (Han & Sun 2001), a 21 sub-network of this network can be established. Table 2 shows the calculated results of part of the vertices and compares it to the Monte Carlo simulation, from which we can see that its accuracy is satisfactory.

Table 2 Reliability analysis of water supply network

Node Sort	Classification method	Monte carlo simulation times		
		1000	2000	3000
5 (N_8)	0.984	0.983	0.984	0.984

9 (N_{16})	0.849	0.845	0.843	0.845
10 (N_{16})	0.876	0.862	0.853	0.850
29 (N_6)	0.978	0.983	0.980	0.979
39 (N_7)	0.975	0.971	0.976	0.976
53 (N_{17})	0.775	0.721	0.723	0.725
55 (N_{19})	0.496	0.463	0.474	0.479
79 (N_{13})	0.825	0.828	0.822	0.823
131 (N_5)	0.645	0.645	0.645	0.645
139 (N_{17})	0.545	0.543	0.541	0.543

Conclusions

The main advantage of the procedure is as follows: (1) The probabilistic prediction model of pipeline is improved, its feasibility has been verified in practical engineering. (2) Employing the sharp-product operation to obtain the disjoint sum of a logical product, the programming is simplified and the number of disjoint terms is less than the other similar methods. (3) The presented method has been programmed and combined with decomposition technology; and its feasibility and effectiveness has been verified and tested in practice.

References

Barlow, R. E., Kiureghian A and Satyanarayana A (1980). "New methodologies for analyzing pipeline and other lifeline networks relative to seismic risk." *Contributed by the Pressure Vessels & Piping Division of ASME for presentation at the Century 2 Pressure Vessels & Piping Conference*. Calif., 80-C2/PVP-65, 1-9.

Han Yang and Sun ShaoPing (2001), "Seismic reliability of lifeline systems." *Earthquake Engineering Frontiers in the New Millennium*, Spencer & Hu, ISBN 90 2651 852 8, 209-213.

Li Guiqing, Huo Da and Wang Dongwei (1993) *Analysis for Seismic Reliability of Urban Construction Network systems*, Beijing, Earthquake Publishing Company.

O'Rourke, M.J., Liu, X.(1999). Response of buried pipeline subjected to earthquake effects." *MCEER Monograph*, No.3 ISBN 0-9656682-3-1.

Shen S. C. (1987) *Computer assistant logical synthesis*, Beijing, Science Publishing Company.(in Chinese)

Sun Shaoping.(1994), "A review of buried lifeline earthquake in China." *Proceedings of Second China-Japan-US Trilateral Symposium on Lifeline Earthquake Engineering*. Xian:, 17-42.

Wang, L.R. L., Sun Shaoping, and shijie shen (1985)."Seismic damage behavior of buried lifeline systems during recent severe earthquakes in U.S.,CHINA and other countries.", *Tech. Report*, No. ODU LEE-02. 2-16.

Guidelines for Defining Natural Hazards Performance Objectives for Water Systems

W. P. Graf[1], C.E. Taylor[2], J.H. Wiggins[3], L. Lund[4], T. Volz[5]

Abstract

In July 2001, the American Lifelines Alliance (ALA) requested that the project team develop guidelines for assisting water utility decision-makers establish performance objectives for their water systems subjected to natural hazards. The goal is to provide these decision-makers with means for a defensible basis for risk management decisions with respect to these hazards. This paper outlines how the project team undertook this guideline development in accordance with the general ALA decision-making framework for establishing acceptable performance requirements for lifelines systems subjected to natural hazards. Discussed briefly in this paper are the basic steps of defining the water system to be evaluated, identifying natural hazard effects, evaluating system component performance, and evaluating system performance. These steps along with the final report were critically reviewed by five water system managers. Of special importance is that this ALA methodology is more complete than performance-based methodologies that ignore or render subjective many of the factors involved in these risk management decisions - and especially system performance considerations.

1.0 Introduction

1.1 Project Objective

The document provides guidelines for developing cost and risk information for various system performance options that may be considered for water utility systems subjected to natural hazards.

This guideline is intended to provide clear, concise guidance on what to consider in performing an evaluation of a water system before, during or after potential natural hazard events. The goal of the application of this guideline is to assist water system owners and operators in defining what approaches are necessary to characterize the anticipated performance of their systems, and to provide a defensible basis for risk management decisions. Implementation of the approaches recommended in this guideline will allow these owners and operators to define the scope of activities necessary to determine appropriate risk management actions to reduce the impact of natural hazards on water systems to acceptable levels.

1.2 Natural Hazards and Water Utility System Facilities Included

The guidelines cover the following natural hazards: earthquakes, floods, windstorm (including hurricane and tornado), and ground movements (landslide, frost heave, and settlement). By implication, liquefaction, tsunamis, and seiche are covered.

These guidelines do not cover dams or hydroelectric plants. Buildings and other water facilities are covered only insofar as they play significant roles in water utility operations. These guidelines are not designed to replace codes designed for buildings. Finally, risks from human-generated hazards (e.g., war, terror, etc.) were not specifically included in the scope of the guidelines.

---

[1] URS Corporation, 911 Wilshire Blvd., Suite 700, L.A., CA 90017; 213-996-2381
[2] Principal, Natural Hazards Mgmt, Inc., 5402 Via Del Valle, Torrance CA 90505; 310-791-0043
[3] President, JH Wiggins Co., 1650 S. Pacific Coast Highway, Suite 300, Redondo Beach CA 90277
[4] Consultant, 3245 Lowry Road, Los Angeles, CA 90027; 323-664-4432
[5] URS Corporation, 9960 Federal Drive, Colorado Springs, CO 80921; 719-531-0001

The guidelines cover water system facilities insofar as they are operationally important. Specific guidelines are designed for the following potable water facilities:
- Steel and concrete distribution reservoirs
- Transmission pipelines, tunnels, aqueducts, and canals
- Treatment plants
- Booster pumping plants
- Wells and sumps
- Pressure vessels (surge tanks, etc.)
- Inlet/outlet piping
- Distribution piping
- Service connections
- Fire hydrants
- River diversions

2.0 Layout of the Guidelines

Chapter 1 provides an introduction and overview of the document. Four steps are involved in the process: defining system performance requirements, identifying risk reduction options, defining the system to be evaluated, and identifying natural hazard events to be considered. The guidelines provide only a framework to exercise system performance requirements and evaluate natural hazard risk reduction options. Appendix A provides a full commentary on Chapter 1, and includes a detailed account of the decision procedure considered by American Lifelines Alliance.

In Chapter 2, Definition of System and Hazards to be Evaluated, guidelines are developed for defining the water system to be evaluated. To define the system, an inventory of components and data relevant to component functionality are assembled, through available documents, drawing reviews, field observations, historical operational experience, and interaction with water utility personnel. Forms, checklists, and other materials are provided to facilitate this inventory assessment procedure. Simplified procedures may be used in circumstances under which only a portion of the water system needs to be inventoried to address a specific decision pertaining to natural hazards.

Supplementing Chapter 2 is Appendix B. For the illustration of inventory procedures, Appendix B contains an idealized water utility system that contains virtually all components of potential interest, and which one can imagine being subjected to all pertinent natural hazard events.

Chapter 3, Modeling Natural Hazards, surveys the appropriate procedures for evaluating natural hazards events and their consequences at specific sites within a water utility system. Appendix C supplements this Chapter with a discussion of the phenomenology of natural hazards.

Modeling for each natural hazard begins with scenarios, defined in terms of initiating locations and severities. Purely deterministic methods use individual scenarios. Probabilistic methods will consider randomness and uncertainties in a more comprehensive selection of scenarios.

In Chapter 4, Evaluation of Component Performance, guidelines are developed for determining the appropriate procedures to be used in establishing (a) the damage state of the component (b) how this damaged component will be repaired or replaced, (c) repair or replacement costs and times, and (d) the degree of functionality of the component during repair. Alternative procedures for component performance evaluation are identified and evaluated. Examples are also provided.

Levels of component performance evaluation are provided depending on the criticality of the components to the water system decision being made. In addition, guidelines are provided for excluding specific components relative to selected natural hazards. For instance, below-ground facilities may generally be ignored if severe winds are being analyzed. Appendix D provides additional information to assist in the development of component vulnerability models.

Chapter 5, Evaluation of System Performance, surveys appropriate procedures to be used in evaluating the performance of the system relative to performance metrics established by decision-makers. The guidelines address (a) how the system state may vary with time after the occurrence of the event, (b) how damage to various links in the system affect the system's ability to provide water service to customers, (c) possible economic impacts of loss of service to customers, and (d) incremental costs of providing the means to achieve various performance objectives. Appendix E provides additional information helpful in developing systems evaluations. Various potential simplifications of systems evaluations are also discussed depending on the nature of the decision being made and the type of the water system.

Chapter 6, Example System Evaluations, provides example system evaluations and displays typical results. Models and outputs vary with the selection of system performance metrics by decision-makers. The example system evaluations are applied to an idealized system, with simplified procedures illustrated in special cases.

Additional Material--Appendices, References, and Nomenclature -- are contained at the end of this guidelines. A technical commentary, consisting of Appendices, is provided under a separate cover.

3.0 Framework for A Decision Process

3.1 Background to the American Lifelines Alliance Decision Process

Figure 1 provides a framework for defensible water utility agencies decisions regarding options for reducing risks from natural hazards. The unshaded boxes in Figure 1 are those outside the explicit scope of this project, which serves as a framework to examine potential natural hazard risk reduction options, system performance requirements, and stakeholders in the process.

3.2 Sample Natural Hazards Risk Reduction Options

Decisions for which a water system risk assessment tool may be used include individual initiatives, such as the redesign of a water distribution reservoir. Alternatively, water utilities may wish to consider a comprehensive program to address the entire range of natural hazards and practical decision alternatives and schedules to reduce their system risks. Comprehensive alternatives may involve many diverse activities designed to reduce natural hazards risks over time. For purposes of categorizing types of risk and decision alternatives, water utility decision-makers may consider:

1. *Engineering measures* - such as the design and construction of new facilities, the retrofit of existing facilities, or geotechnical remediation.

2. *Land use measures* - such as through alternative siting or reduction of exposures in building structures that may be damaged.

3. *System enhancement* - the use of multiple pathways and nodes (system redundancy) in order to assure that system performance goals are met.

4. *Emergency response* - the immediate response to emergencies / disasters.

5. *Disaster recovery and restoration* - the long-term restoration to normalcy after a large emergency or disaster, again through cooperative activities and strategic planning.

6. *Risk transfer* - the use of insurance or other liability transfers in order to limit the utility's post-disaster liabilities and assure adequate recovery funds, and

7. *Financial Reserving* - such as retaining funds for emergency response and recovery contingencies.

3.3 Pertinent System Performance Metrics

For evaluating risk and decision alternatives, water utility managers may use a variety of metrics. In general, health, safety, and welfare are the overarching goals to be evaluated. The primary system metrics for evaluating these decisions will pertain to "welfare" (broadly speaking, economic) metrics of:

- percent served (in total or by sector) within a specific number of days with raw water with adequate fire flow pressures, and/or
- percent served (in total or by sector) within a specific number of days with fully treated water.

Sample Welfare System Performance Metrics

The system performance metrics emphasized are welfare metrics - with health and safety standards and procedures presupposed as being stringent. Very typically, metrics will be of the following forms:

Metric (target): Z% of C served in W days with raw water with adequate fire flow pressures

Metric (target): X% of C served in Y days with fully treated water

In these generalized forms, "C" can stand for the entire system, or for selected stakeholders within the system. (Stakeholders are discussed in section 3.4 below.) Alternative metrics could be set in terms of number of service connections, populations served, or volume of water served (i.e., cubic feet or gallons).

In the above forms, one can use existing financial and economic data to convert such metrics into dollar terms. These would include water utility revenues lost, business interruption losses, and other higher order effects of such financial and productivity losses. One can also add probabilities to the above metrics.

From a practical standpoint, deciding in advance of a water system evaluation how reliable the water system should be is likely to be short-sighted, especially if costs are high to achieve the pre-specified level of reliability. The acceptability of the pre-specified metric may well change as one considers existing technologies to reduce risks and who pays for their incorporation into the system.

Noticeably absent from these system decision metrics are references to illness and life-safety. This is for reasons given above -- that the extremely important considerations of warding off disease, injury, and deaths are accounted for in existing standards and procedures. Absent as well are qualitative system performance factors. Full-scale decision-making will involve not merely "welfare" considerations of the financial or economic kind, but considerations pertaining to administration, social impacts, psychological impacts, political and legal concerns, and a host of other issues that are not explicitly covered in this guidelines.

In addition, the guidelines document is designed to accommodate decision metrics that are (a) scenario-based *(deterministic)* or (b) risk-based *(probabilistic)*. Scenario-based methods rely on the evaluation of a water system subjected to a small number of natural hazard scenarios. These, for instance, could include the repetition of past floods, hurricanes, severe rains, earthquakes, and so on. Or, specific scenarios may be devised to represent "large" or "maximum" events, consistent with the latest scientific and engineering knowledge of the natural hazards phenomena. A risk-based method for evaluating water systems, as described in the remainder of the guidelines, will again be based on individual scenarios, but enough of these will be modeled through random processes to provide results. Familiar versions of cost-benefit and related financial methods typically require a risk-based approach.

3.4 Basic Stakeholders in the Decision

A key factor in decision-making is consideration of who pays and who benefits from the decision. Basic stakeholders in the decision to reduce water utility system risks from natural hazards may involve:

- the water utility itself
- related water wholesalers or distributors
- municipal governments with jurisdictions affected by the water utility, or with financial interest in the utility
- other water utilities associated through mutual aid agreements
- dependent fire departments concerned to assure that fire flows are adequate
- various categories of customers (e.g., industrial, institutional, commercial, and residential) and/or specific lists of customers (e.g., health-care facilities, emergency operating and public safety facilities, special manufacturers)
- insurers, bond-holders, bond rating agencies, and lending institutions
- federal and state agencies that may provide disaster assistance
- other federal, state, and local agencies that may incur additional expenses during disruptions to the water system
- other infrastructure systems (e.g., energy, wastewater, communications) that may be affected by disruption to the water system, and
- federal, state and local agencies that regulate health effects (water quality) and/or that are involved with proactive antiterrorism programs.

In addition to various basic stakeholders, entire communities are affected by disruption to a water system. For instance, the tourist industry may be harmed, out-migration may be increased, general contractors may have additional work, and so on. Higher-order ripple effects of damages to potable water systems are beyond the scope of these guidelines, but there are expected to be many.

4.0 Multiple Levels of Analysis

4.1 Background to Analysis Steps

As elaborated in the guidelines, the basic iterative steps in a water systems risk evaluation for natural hazards consist of inventorying pertinent water system components, defining natural hazard scenario events and their natural consequences, evaluating the response of water system components to these natural hazard scenario events, and evaluating the system response to damages to the water utility components (see the shaded steps in Figure 1).

The approach to modeling risks from natural hazards has many common elements among the great variety of water utilities. Previous, small water agencies were limited in the scope of evaluations they could approach. However, aided by current advancements in software technology, even very small water systems can perform hydraulic evaluations to evaluate system-wide impacts from natural hazards events.

4.2 Characterizing an Advanced Level of Analysis

The most advanced level of analysis would consist of the following features:

- The analysis covers major natural hazards affecting the water utility system and its major facilities

- The analysis avoids obvious biases such as conservatism
- The analysis treats the system through hydraulic models
- The analysis considers various stakeholders, such as through the evaluation of results at various service zones or for various classes of customers
- The analysis treats natural hazards probabilistically, through the random selection of initiating events, some of which (e.g., earthquakes, hurricanes) may affect the entire system more or less all at once
- Special facilities are given special scientific and engineering evaluations, and
- A significant set of decision alternatives is postulated, along with their costs.

4.3 Simplifications: For Simpler Systems and for Less Advanced Analysis

The guidelines document recognizes the great differences in technical capacity and availability among water utilities. Some water utilities are very large, covering hundreds of thousands of customers. In contrast, most water utilities cover a much smaller number of customers, with tens of thousands being more normal, and even fewer customers in many cases. A paramount concern for most water utilities is to maintain low rates, which limit the development of technical capacity and availability. In addition, not all decisions require that a full-scale evaluation be undertaken of the entire system and for all natural hazards.

Given these wide ranges in technical capacity and availability, and the wide range of possible decisions for which the guidelines must serve, simplified procedures are suggested wherever appropriate. These simplifications will be divided into those that render an analysis less advanced and those that recognize special features of the system so that an advanced analysis may be performed with less effort.

Simplifications that can be used that can still lead to an advanced analysis include:

- The water system is of limited spatial extent, allowing simplification of such natural hazards as hurricane and earthquake (e.g., uniform hazards)
- One or more of the natural hazards has an insignificant potential effect on the water system (e.g., hurricanes in Montana)
- The system is primarily gravity-flow, allowing simple hydraulic evaluation
- The affected portion of the system is linear, allowing simplified systems methods to be used
- The system contains major components that are impervious to the natural hazards under evaluation (e.g., buried pipelines relative to severe winds).

The following simplifications could lead to an intermediate analysis (depending on circumstances - a scoping study could defend some of these simplifications):

- The guidelines are to be used to assist in developing a scoping study, that is, a study of what natural hazards and components should be evaluated
- The guidelines are to be used to undertake a decision that involves only a sub-system of the entire water system (e.g., a sub-system consisting of a booster pumping station and a distribution reservoir)
- The guidelines are to be used for a decision that involves only selected natural hazards (e.g., severe winds only)

- The guidelines are to be used only to assess system performance from an operational standpoint, such as through the use of pre-selected (as opposed to randomly selected) natural hazards scenarios, and
- The guidelines are to be used to evaluate a specific stakeholder interest (e.g., distribution systems served by a wholesaler, residential customers, a specific large manufacturer, emergency and critical health facilities) and so do not require that the full system be evaluated.

Note that in some of these simplified cases, the evaluation may still be very advanced, at least in some respects.

4.4 Below Intermediate Analyses

Often it is desirable simply to obtain some very initial evaluation of the water system performance subjected to natural hazards. Some of the following features characterize a less than an intermediate analysis:

- Conservatism is used in various analysis steps (as for immediate post-disaster evaluations used for response and recovery); no sensitivity evaluations are performed to evaluate the impacts of this conservatism on the risk results
- A geographically large system is evaluated using simplified or uniform initiating events (e.g., where hazard intensity is tabulated by ZIP code)
- Very coarse assumptions are used, with little examination regarding the coarseness of the assumptions and their impacts on results
- The evaluation is performed to promote actions based on demonstrating extreme natural hazard risk, rather than putting the natural hazard risk and costs of reducing it into perspective.

5.0 Preliminary Study Scope of Work: A Sample Phase I Study

A 'phase 1' study provides an overview of the seismic risks facing the utility. Further study for selected sites, systems or components identified as significant contributors to seismic risks, retrofit design or other risk-reduction steps may follow. A phase 1 project scope and sequence may be varied in proportion to hazard levels or water agency needs. Scoping study approaches may be modified for other specific needs (e.g., anti-terrorism concerns may be combined with natural hazards concerns).

Task 1 -- Data Gathering. Assembling basic documents, maps, and past studies.

Task 2 -- Asset Inventory Development. An inventory (often in database format) for agency buildings and equipment provides a vehicle for prioritizing site visits, design document review and other data gathering.

Task 3 -- Develop Operational Importance Ratings. Water agency operations managers assign operational criticality ratings to major components by judgment, considering criticality of system, facility, and component.

Task 4 -- Site Visits. The project team and agency representatives conduct walk-through surveys, discussing operational importance, design basis and master plans for the facilities.

Task 5 -- Review Natural Hazards. Geologic hazards, weather-related hazards and other natural hazards are characterized throughout the water system. Hazards without significant frequency or severity are eliminated, and the significant hazards are evaluated using simplified methods.

Task 6 -- Vulnerability Modeling. Brief reviews of selected design documents can be conducted, prioritized by hazard level and criticality. Simple financial loss models can be used to estimate damage as a fraction of replacement cost, and component operability may be crudely assessed.

Task 7 -- Risk Analysis. This may utilize individual scenarios (events), or sets of scenarios. It may be event-driven, using conventional event tree methods, or component-driven. System-wide consequences may then be evaluated formally (using a water system model) or informally (using judgment). Alternatively, for this preliminary phase, the risk analysis can be component-driven.

6.0 Decisions Under Both Risk and Uncertainty

The goal of an evaluation of a water system subjected to natural hazards events is to gather and synthesize information that assists decision-making. In such a decision under risk, there is still an element of chance, but ideally this is fully quantified through the risk evaluation process. For instance, in a deck of cards, the chance of picking a heart is one-in-four -- as long as there are no jokers in the deck. Taking a chance of picking a heart can be a decision under risk -- as long as one knows what the chances are of picking the heart. Through the synthesis of information in a water system evaluation, one can remove uncertainty and ignorance. The systems approach in Figure 1 implies that one puts together piecemeal information from the system at risk, natural hazards that may impact it, the vulnerabilities of its components to these natural hazards, and the response of the system to damages to these components. This systems approach implies that decisions based only on piecemeal information have greater uncertainty than those based on a more global approach.

In contrast, decisions under uncertainty - at the extreme - lack key information. For instance, one may be forbidden to know how many cards are in the "deck" and one may not know what proportion of the cards in the deck are hearts. In this case, one's wager on picking a heart would be a decision under uncertainty - or abject ignorance.

Ideally, the evaluation of a water system subjected to natural hazards produces a decision under risk -- not a decision under uncertainty. In a decision under risk, to repeat, all key factors bearing on the decision would be fully and adequately quantified. A systems approach to water systems moves in this direction. Ignorance of the system at risk, natural hazards that may affect it, the vulnerability of its components, and the potential response of the system are removed. Nonetheless, the state-of-the-art in this type of evaluation does not permit one to remove all uncertainties and unknowns. This is chiefly a result of the uneven quality of data and models used in such an evaluation.

Considering natural hazards events alone, and not the uncertainties in estimating water facility and system response to them, one may be guided by earlier words on nuclear power studies:

> The ANS-2.12 [American Nuclear Society] Working Group wishes to clearly state that it is difficult to precisely establish the probability of occurrence of natural and external man-made hazards. The phenomena are complex and the probability of each is a function of parameters such as geographical location, time of year and nature of the hazards (ANS, 1978, foreword)

In the guidelines, such cautionary remarks are scattered throughout. The goal of an evaluation of a water system subjected to natural hazards is to develop systematic information for a decision both under risk and uncertainty. Uncertainty and ignorance are reduced, but almost never to a point of certainty. Virtually all models used in this evaluation procedure suffer from aspects of ignorance and uncertainty. An evaluation of a water system subjected to natural hazard events thus produces bounded patterns, not estimates that can be trusted at several decimal places.

7.0 Acknowledgements

In addition to the authors, the following people contributed to the guidelines:

Danna Judish	URS Corporation, Colorado Springs, CO
Garry Lay	URS Corporation, Los Angeles, CA
Aziz Alfi	Seattle Department of Public Utilities, Seattle, WA
Chuck Call	Salt Lake Department of Public Utilities, Salt Lake City, UT
Bill Nabak	Green Bay Water Utility, Green Bay, WI
Dave Putnam	Metropolitan Water District of Southern California, Fontana, CA
Robert Gordon	Hillsborough County Public Works Department, Tampa, FL

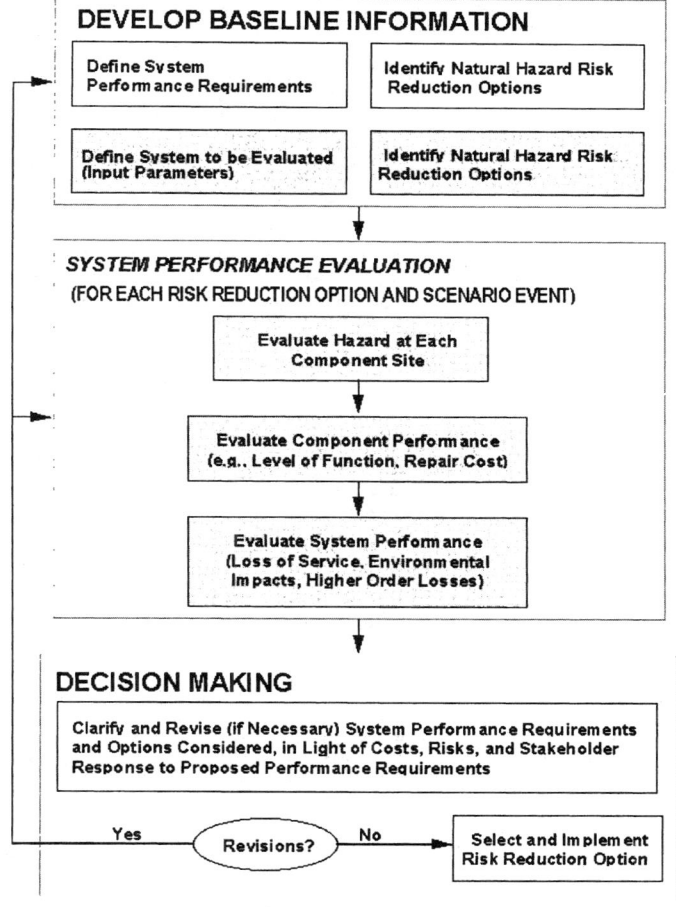

Figure 1: A Decision-Making Framework for Establishing Acceptable Performance Requirements for Water Utility Systems Subjected to Natural Hazards (shaded boxes are within scope)

Online Monitoring of Seismic Damage in a Water Delivery System

Jianwen Liang[1]

Abstract

Urban water delivery systems can be damaged by earthquakes, and the damage cannot easily be detected and located, especially immediately after the events. This paper proposes a methodology to detect and locate the damage in a water delivery system by monitoring water head online at some selected nodes in the water delivery system. For the purpose of online monitoring, emerging supervisory control and data acquisition (SCADA) technology can well be used. A neural network-based inverse analysis method is developed for detecting the location and extent of the damage based on the water head variations. It is found that the method provides a quick, effective, and practical way in which the seismic damage in a water delivery system can be detected and located.

Introduction

Urban water delivery systems can be damaged under strong earthquakes, and damage cannot easily be located, especially immediately after the events. In recent years, real-time damage estimation and diagnosis of buried pipelines has attracted much attention of researchers focusing on establishing the relationship between damage ratio (breaks per unit length of pipe) and ground motion with taking the soil condition into consideration (Nishio, 1994; Takada and Ogawa, 1994; Yamazaki et al, 1994). Due to the uncertainty and complexity of the parameters that affect the pipe damage mechanism, it is not easy to locate damage and estimate the extent of damage only with a few parameters (Liang and Sun, 2000). Eguchi et al (1991) put forward a method in which a nominal damage estimated through a few earthquake parameters is updated gradually based on post-earthquake observation data.

In recent years, supervisory control and data acquisition (SCADA) technology has been applied to water delivery systems, in which water head and/or flow rate at some selected nodes such as water reservoirs, pumps and pipes are monitored online by remote sensors and transmitted back by radio or internet. However, because the number of monitoring stations is limited, and more

---

[1] Professor, Dept of Civil Engineering, Tianjin University, Tianjin 300072, China; jwliang@tj.cnuninet.net

importantly, there are few effective methods available, it is difficult to locate seismic damage precisely and timely in practice.

This paper proposes a methodology to locate seismic damage in a water delivery system by monitoring water head online at some nodes in the water delivery system, in which, a neural network-based inverse analysis method is developed to estimate the water head variations at all non-monitoring nodes based on the water head variations at the monitoring nodes.

Methodology

For a given water delivery system, any break at pipe links or nodes will affect the water head at all other links or nodes, and different extents of the break will result in different effects, and the same breaks will bring different effects for links or nodes at different positions. Based on this principle, this paper proposes to locate damage in a water delivery system by monitoring the most damage-sensitive parameter --- water head variations before and after the damage at some nodes, then estimating the water head variations at all non-monitoring nodes.

Damage location based on water head variations actually is a nonlinear identification problem, which needs to establish the relation between water head variations at the monitoring nodes and those at all non-monitoring nodes. This relation can be established by other methods, but using neural network technique may utilize its advantage in artificial intelligence, i.e., developing database for water head variations at the monitoring nodes and all non-monitoring nodes for various representative damage and training the relation offline, then the damage may be detected and located online based on the water head variations at the monitoring nodes. Depending on the water delivery systems, this process may only need several seconds to several minutes. Secondly, using neural network technique, it only needs the relation between water head variations at the monitoring nodes and those at all non-monitoring nodes, which can utilize its advantage in nonlinear mapping to overcome the difficulty of other mathematical methods, e.g., regression analysis, in dealing with strongly nonlinear problem, so as to detect and locate damage more accurately. Finally, using neural network technique can also fully utilize the SCADA technology in urban water delivery systems available.

Artificial neural network technique. A back-propagation neural network is used in this paper, and sigmoid function is used for output function. In order to normalize the influence of input data with different cells and to prevent the saturation of the output function, in this paper the input and output data are scaled to [-1.0, 1.0] and [0.2, 0.8], respectively.

Database development by hydraulic analysis. To establish the relation between water head variations at the monitoring nodes and all non-monitoring nodes in a water delivery system, sufficient and well-distributed data for neural network training are needed. As a preliminary study, this paper produces all the data by hydraulic analysis.

For a water delivery system at any node i, the continuity condition requires that (Takakuwa, 1978)

$$\sum_j Q_{ij} + Q_{id} + Q_{il} = 0 \tag{1}$$

where Q_{ij} is the flow rate in link ij, Q_{id} is the flow rate of demand at node i, Q_{il} is the flow rate of leakage at node i. Assuming the length and diameter of link ij to be L_{ij} and D_{ij}, respectively, the Hazen-William equation between the head loss H_{ij} and the flow rate Q_{ij} in each link ij is

$$Q_{ij} = R_{ij} |H_{ij}|^{-0.46} H_{ij} \tag{2}$$

where

$$R_{ij} = 0.27853 C_{ij} D_{ij}^{2.63} L_{ij}^{-0.54} \tag{3}$$

$$H_{ij} = E_i - E_j \tag{4}$$

where E_i is the water head at node i, C_{ij} is the roughness coefficient of link ij.

The flow rate of demand Q_{id} at node i is associated with the water head at the node: when the relative water head is not less than a given design value $H_{\min}$, the flow rate of demand is the normal flow rate Q_{nor}; and when the relative water head is less than $H_{\min}$ but greater than 0, the flow rate is (Takakuwa, 1978)

$$Q_{id} = Q_{nor}(E_i - G_i)^{0.5} H_{\min}^{-0.5} \tag{5}$$

where G_i is the elevation at node i; and when the relative head is less than or equal to 0, the flow rate of demand is 0 at node i.

The flow rate of leakage at node i may be estimated by (Takakuwa, 1978)

$$Q_{il} = \begin{cases} d_i (E_i - G_i)^k & (E_i > G_i) \\ 0 & (E_i \le G_i) \end{cases} \tag{6}$$

where d_i is the leakage coefficient. In the paper k=1.15. It should be noted that the leakage along all links should be converted into leakage at nodes.

If the number of the nodes is N, the following equation is a set of nonlinear equation with N unknowns $E_1, E_2, \cdots, E_N$

$$\sum_j R_{ij} |E_i - E_j|^{-0.46} (E_i - E_j) + Q_{id} + Q_{il} = 0 \tag{7}$$

since Q_{ij}, Q_{id} and Q_{il} all are nonlinear functions of water head.

Equation (7) can be expressed in iterative form

$$\sum_j \{R_{ij} |e_{i,k} - e_{j,k}|^{-0.46} (e_{i,k} - e_{j,k}) + 0.54 R_{ij} |e_{i,k} - e_{j,k}|^{-0.46} (\Delta E_{i,k} - \Delta E_{j,k})\}$$

$$+ Q_{nor}(e_{i,k} - G_i)^{0.5} H_{\min}^{-0.5} + 0.5 Q_{nor}(e_{i,k} - G_i)^{-0.5} H_{\min}^{-0.5} \Delta E_{i,k}$$

$$+ d_i (e_{i,k} - G_i)^{1.15} + 1.15 d_i (e_{i,k} - G_i)^{0.15} \Delta E_{i,k} = 0 \tag{8}$$

where subscript k indicates the value after k-th iteration. Staring with the initial value $e_{i,0}$ for E_i ($i=1,2,\ldots,N$), a set of linear equation (8) is solved for $\Delta E_{i,0}$, thus getting the first iterative results $e_{i,1} = e_{i,0} + \Delta E_{i,0}$. The procedure is repeated until $|e_{i,k+1} - e_{i,k}|$ ($i=1,2,\ldots,N$) less than a given small value, then $E_i = e_{i,k+1}$ may be considered as the solution at node i.

When a break is occurred to a pipe in a water delivery system, water is discharged at the break. A break can be various shapes and different extents, but all can be expressed as the size of the opening area. The discharge flow rate at a break may be estimated by

$$q_d = C_d A_d \sqrt{2gH} \qquad (9)$$

where C_d is the discharge coefficient associated with the break shape, A_d is the opening area, H is the water head at the break, g is the acceleration of gravity. By adjusting the discharge coefficient, various openings can be simulated. Equation (9) can be written as

$$q_d = C_d A_d \sqrt{2g}(e_i - G_i')^{0.5} + 0.5 C_d A_d \sqrt{2g}(e_i - G_i')^{-0.5} \Delta E_i \qquad (10)$$

where G_i' is the elevation of the opening.

Example

Figure 1 shows a water delivery system with one reservoir, 30 nodes and 50 links. The length and diameter of pipe links are listed in Table 1. It is assumed that the roughness coefficients for all links are 140, the demand flow rates are uniformly distributed with 0.05m³/sec at all nodes, and the leakage coefficients at all nodes are 2.0×10^{-5}. The elevation of all links is 48.0m, and the water head at resource is 100.0m. For the purpose of demonstrating the efficiency of the proposed methodology, this example considers only one location of seismic damage, and multiple-damage case will be a straightforward extension.

At first, develop database for training of neural network by hydraulic analysis of the water delivery system with one break. Calculate the water head at all nodes (including three monitoring stations) without any damage and with one break in sequence at the middle of each link. The damage extent is described by the ratio (A_d / A_0) of discharge area at the break to the cross area of the pipe, and the ratios are chosen as 0.01, 0.02, 0.1, 0.2 and 0.5, respectively, and the discharge coefficient uses 0.64, then total 250 sets of data for the water head variations can be obtained for 50 links and 5 damage states for each link. The database is not shown here for the limited space. In order to make the 5 damage states well distributed, take the logarithm of the ratios, and then normalize them. Use the 250 sets of data to train the neural network: 3 neural cells at input layer for the water head variations at 3 monitoring stations; 27 cells at output layer for the water head variations at non-monitoring nodes. Figure 2 shows that the neural network tends to converge after 10000 cycles training, and the RMS error is 0.00177 after 100000 cycles training.

Next, test the neural network to see if it can give us the results that we expect. Input the data used for training and Table 2 shows the water head variations at non-monitoring nodes, e.g., node A, B, C and D for break (A_d / A_0 =0.1) at link 9, 20, 24, 25, 26, 27, 28, 31 and 42 (in profile 1-1 and 2-2), respectively. The results in Table 2 are the normalized water head variations, and the values in brackets are the relative error with the targets. From the results it is known that the relative errors are very small, with the minimum 0.05% and the maximum 2.83%, and most are below 1%, which shows the neural network training is successful. This may indicate that there indeed exists an inherent relation between water head variations at different nodes in a given water delivery system, which is the key to the methodology, and 3 monitoring stations are sufficient for the purpose of damage location in the example. As the back-propagation neural network technique is based on the decrease of RMS error of all training data, the relative error of water head variation for a specific node varies around the final training error 0.00177.

Input data that have never been used for training to test the neural network. Table 3 and Table 4 show the water head variations at non-monitoring nodes, e.g., node A, B, C and D for breaks A_d / A_0 =0.05 and A_d / A_0 =0.15 at link 9, 20, 24, 25, 26, 27, 28, 31 and 42 (in profile 1-1 and 2-2), respectively. Same as before the results in Table 3 and Table 4 are the normalized water head variations, and the values in brackets are the relative error with the targets. From these results it is known that the relative errors are larger than those in Table 2, and the maximums are 9.03% and 9.41% for Table 3 and Table 4, respectively, but most are below 5%. This is due to that for non-training data the neural network gives the interpolations, and their precisions depend on the RMS error and the distance away from the training data, and the errors tend to the largest at the middle of two training data. Table 3 and Table 4 further show that there indeed exists the inherent relation between water head variations at different nodes in a given water delivery system, and further proves the methodology: the water head variations at 27 nodes (non-monitoring nodes in the paper) can well be predicted from the water head variations at 3 nodes (monitoring stations in the paper).

At last, discuss the detection of location and extent of the damage. Figure 3 shows the water head variation contour lines at all non-monitoring nodes for the break A_d / A_0 =0.05 at link 28, which are corresponding to the maximal relative error 9.03% in Table 3. In Figure 3, the upper figure (a) is the target water head variation contour lines, and lower figure (b) is the diagnosis water head variation contour lines. From the figures, the diagnosis water head variations are in good agreement with the target water head variations, and it is easy to locate the damage: at the position ⊕ with maximal water head variations, i.e., at link 28, and it is easy to know the damage extent: the maximal value of water head variation 0.02< A_d / A_0 <0.1. Figure 4 shows the water head variation contour lines at all non-monitoring nodes for the break A_d / A_0 =0.15 at link 28, and again the diagnosis water head variations are in good agreement with the target water head variations, and it is easy to locate the damage: at the position ⊕ with maximal water head variations, i.e., at link 28, and it is easy to know the damage extent: the maximal value of water head variation 0.1< A_d / A_0 <0.2. As for Figure 5 and Figure 6, it is easy to locate the damage (link 42) in the same way, and to know the damage extents (0.02< A_d / A_0 <0.1 and 0.1< A_d / A_0 <0.2, respectively).

Conclusions

A methodology is proposed to detect and locate seismic damage in a water delivery system by monitoring water head variations online at some nodes in the water delivery system, in which, a neural network-based inverse analysis method is developed to estimate the water head variations at all non-monitoring nodes based on the water head variations at the monitoring nodes. It is found that the methodology provides a quick, effective, and practical way in which damage in a water delivery system can be located. It should be pointed out that the method may also be used for diagnosis of water reservoirs and pumps, etc., and for location of a fire fighting.

Acknowledgements

This work was supported by the National Natural Science Foundation of China under grant 59878032.

References

Eguchi, R. T., Chrostowski, J. D., Till, C. W., et al. (1991). A rapid post-earthquake damage detection method for underground lifelines. Proceedings of 3rd U.S. Conference on Lifeline Earthquake Engineering, ASCE, 714~724.

Liang, J., and Sun, S. (2000). Site effects on seismic behavior of pipelines: A review. J. of Pressure Vessel Technology, ASME, 122, 469~475.

Nishio, N. (1994). Damage ratio prediction for buried pipelines based on the deformability of pipelines and the nonuniformity of ground, J. of Pressure Vessel Technology, ASME, 116, 459-466.

Takada, S., and Ogawa, Y. (1994). Seismic monitoring and real time damage estimation for lifelines. Proceedings of 4th U.S. Conference on Lifeline Earthquake Engineering, ASCE, 224-231.

Takakuwa, T. (1978). *Pipeline network analysis and design*, Morikita Publishing Company, Tokyo. (in Japanese)

Yamazaki, F., Katayama, T., and Yoshikawa, Y. (1994). On-line damage assessment of city gas networks based on dense earthquake monitoring. Proceedings of 5th U.S. National Conference on Earthquake Engineering, EERI, 829-837.

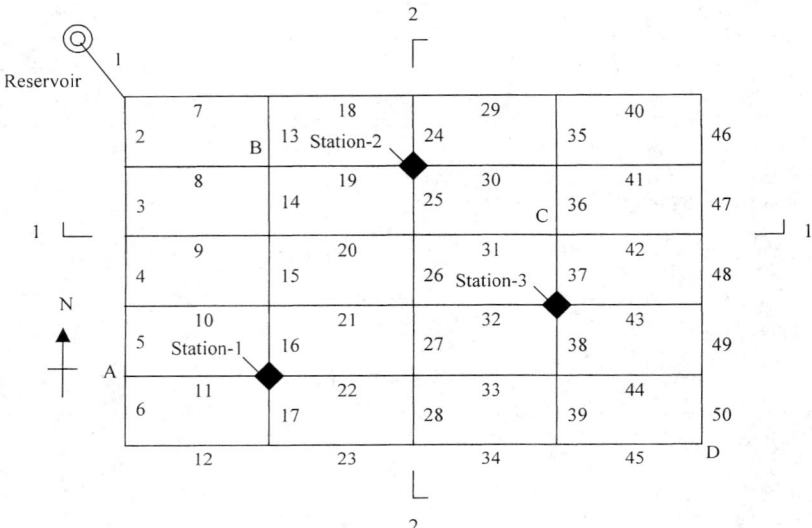

Figure 1. The example water delivery system with one reservoir, 30 nodes and 50 pipe links, where water head at three nodes are monitored.

Table 1. The diameter and length of pipe links in example water delivery system.

Link No.	D (m)	L (m)	Link No.	D (m)	L (m)	Link No.	D (m)	L (m)	Link No.	D (m)	L (m)
1	0.80	50	14	0.50	1000	27	0.40	1000	40	0.50	2000
2	0.60	1000	15	0.50	1000	28	0.35	1000	41	0.40	2000
3	0.60	1000	16	0.40	1000	29	0.50	2000	42	0.40	2000
4	0.50	1000	17	0.40	1000	30	0.50	2000	43	0.35	2000
5	0.50	1000	18	0.60	2000	31	0.40	2000	44	0.35	2000
6	0.40	1000	19	0.50	2000	32	0.40	2000	45	0.30	2000
7	0.60	2000	20	0.50	2000	33	0.35	2000	46	0.40	1000
8	0.60	2000	21	0.40	2000	34	0.35	2000	47	0.40	1000
9	0.50	2000	22	0.40	2000	35	0.50	1000	48	0.35	1000
10	0.50	2000	23	0.35	2000	36	0.40	1000	49	0.35	1000
11	0.40	2000	24	0.50	1000	37	0.40	1000	50	0.30	1000
12	0.40	2000	25	0.50	1000	38	0.35	1000			
13	0.60	1000	26	0.40	1000	39	0.35	1000			

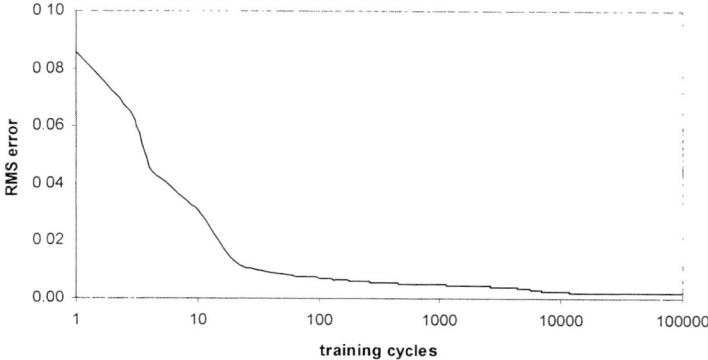

Figure 2. The training curve of artificial neural network where the RMS error tends to converge after 10000 cycles training, and 0.00177 after 100000 cycles training.

Table 2. The diagnosis water head variations at non-monitoring nodes A, B, C and D for break $A_d / A_0 = 0.1$ at link 9, 20, 24, 25, 26, 27, 28, 31 and 42.

Break at link No.	Water head variation at node A	Water head variation at node B	Water head variation at node C	Water head variation at node D
9	0.5813 (0.41%)	0.4763 (0.13%)	0.5333 (0.39%)	0.5565 (0.80%)
20	0.5753 (0.79%)	0.4832 (0.10%)	0.5805 (0.19%)	0.5857 (0.10%)
24	0.5227 (0.48%)	0.4874 (1.29%)	0.5712 (0.21%)	0.5499 (0.48%)
25	0.5510 (0.56%)	0.4975 (2.83%)	0.5986 (2.18%)	0.5797 (0.05%)
26	0.4682 (0.26%)	0.3758 (0.66%)	0.4813 (2.36%)	0.4904 (0.22%)
27	0.5125 (0.18%)	0.3731 (0.11%)	0.4696 (0.06%)	0.5482 (0.59%)
28	0.4612 (0.81%)	0.3297 (0.57%)	0.4049 (0.15%)	0.5086 (1.88%)
31	0.4412 (1.12%)	0.3755 (1.00%)	0 5141 (2.24%)	0.4796 (0.25%)
42	0.4506 (0.74%)	0.3779 (0.56%)	0.5553 (1.24%)	0.4982 (0.50%)

Table 3. The diagnosis water head variations at non-monitoring nodes A, B, C and D for break A_d / A_0 =0.05 at link 9, 20, 24, 25, 26, 27, 28, 31 and 42.

Break at link No.	Water head variation at node A	Water head variation at node B	Water head variation at node C	Water head variation at node D
9	0.5625 (3.71%)	0.4792 (6.73%)	0.4970 (1.23%)	0.5231 (0.65%)
20	0.5440 (0.13%)	0.4697 (3.16%)	0.5460 (0.22%)	0.5480 (0.24%)
24	0.5057 (3.31%)	0.4944 (8.90%)	0.5416 (0.95%)	0.5215 (1.44%)
25	0.5231 (0.58%)	0.4967 (8.85%)	0.5608 (0.32%)	0.5422 (0.17%)
26	0.4346 (2.05%)	0.3802 (5.26%)	0.4538 (2.12%)	0.4578 (1.29%)
27	0.4927 (2.26%)	0.3608 (1.32%)	0.4443 (0.05%)	0.5056 (1.27%)
28	0.4399 (1.59%)	0.3306 (3.60%)	0.3882 (0.54%)	0.5132 (9.03%)
31	0.4160 (1.61%)	0.3564 (1.57%)	0.4785 (3.24%)	0.4459 (1.35%)
42	0.4186 (1.23%)	0.3487 (2.84%)	0.5044 (4.36%)	0.4576 (3.09%)

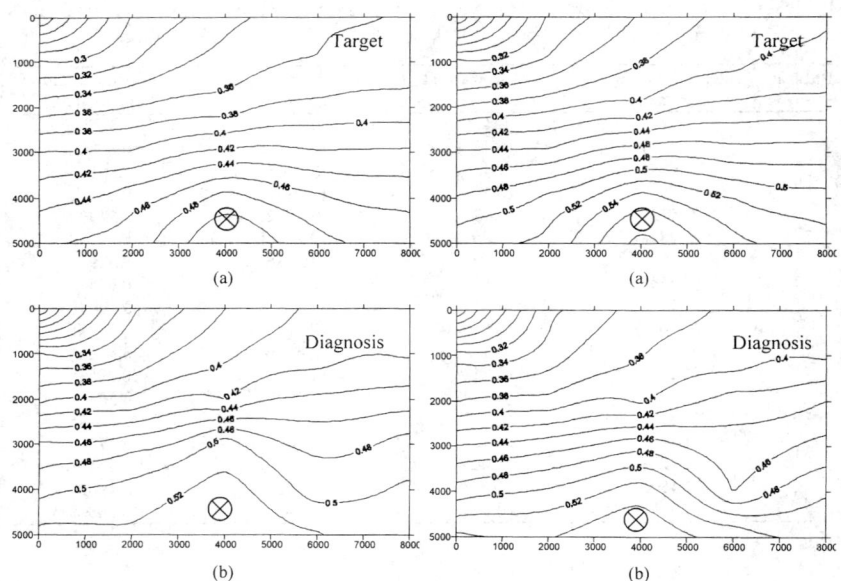

Figure 3. Location of damage at link 28 for A_d / A_0 =0.05.

Figure 4. Location of damage at link 28 for A_d / A_0 =0.15.

DISASTER RESPONSE FOR LIFELINE SYSTEMS 473

Table 4. The diagnosis water head variations at non-monitoring nodes A, B, C and D for the break A_d / A_0 =0.15 at link 9, 20, 24, 25, 26, 27, 28, 31 and 42.

Break at link No.	Water head variation at node A	Water head variation at node B	Water head variation at node C	Water head variation at node D
9	0.6139 (0.23%)	0.4952 (1.43%)	0.5670 (0.11%)	0.5924 (0.50%)
20	0.6200 (0.60%)	0.5278 (3.49%)	0.6095 (1.41%)	0.6249 (0.27%)
24	0.5459 (0.89%)	0.4701 (7.55%)	0.6031 (0.84%)	0.5751 (0.91%)
25	0.5932 (0.88%)	0.5593 (9.41%)	0.6426 (1.12%)	0.6169 (0.21%)
26	0.5102 (3.07%)	0.4107 (3.87%)	0.5122 (3.23%)	0.5208 (0.33%)
27	0.4962 (8.33%)	0.3554 (8.68%)	0.4809 (2.97%)	0.5369 (7.08%)
28	0.4889 (1.45%)	0.3489 (1.39%)	0.4294 (1.06%)	0.5142 (2.52%)
31	0.4840 (3.09%)	0.4188 (5.62%)	0.5535 (0.65%)	0.5222 (3.45%)
42	0.4781 (1.57%)	0.4036 (2.80%)	0.6025 (0.89%)	0.5445 (2.87%)

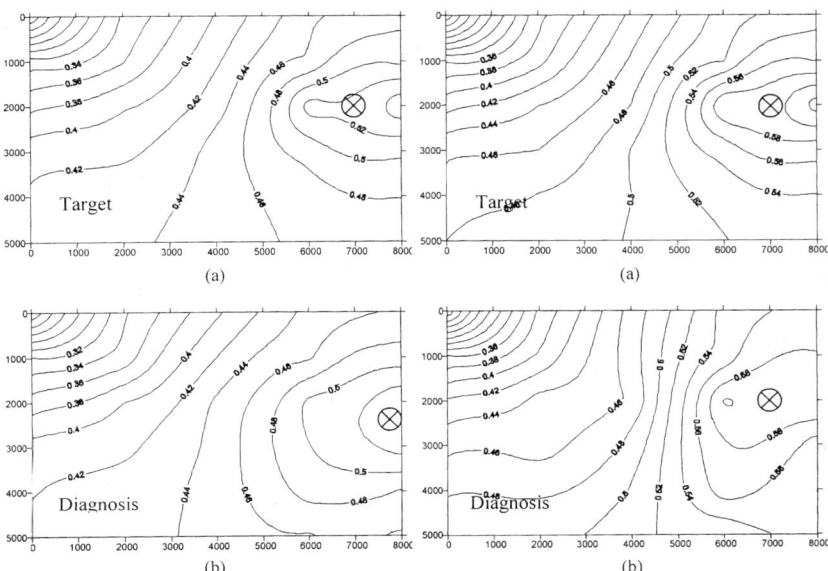

Figure 5. Location of damage at link 42 for A_d / A_0 =0.05.

Figure 6. Location of damage at link 42 for A_d / A_0 =0.15.

Evaluating Mitigation of Urban Infrastructure Systems: Application to the Los Angeles Department of Water and Power

Stephanie E. Chang[1] and Hope A. Seligson[2]

Abstract

This paper describes the application of a life cycle cost methodology for evaluating seismic mitigations for urban infrastructure systems. In contrast to more traditional benefit-cost analysis, the life cycle cost framework readily and transparently accomodates costs that may change over time. For infrastructure systems, this is advantageous for addressing such issues as infrastructure deterioration and urban growth. Moreover, the framework developed here emphasizes costs that would be imposed on the community, as well as the lifeline agency, in the event of infrastructure failure in a disaster. The framework therefore allows a comprehensive assessment of mitigation benefits.

The methodology is applied to the case of the Los Angeles Department of Water and Power (LADWP) electric power system. The case study examines mitigation of high voltage transformers using base isolation. The impact of the mitigation on network performance was assessed by collaborating researchers at the University of Southern California, using power flow modeling software. Direct losses to the utility agency were estimated as transformer repair costs, as well as revenue losses resulting from customer power outage. Costs to the community were evaluated in terms of direct economic losses from business interruption. The probabilistic analysis evaluates multiple Monte Carlo simulations of 47 scenario earthquakes. Expected annual losses vary over time as a result of projections in urban and economic growth over a 50-year period. Results indicate that while the seismic retrofit is not cost-effective from the standpoint of utility agency costs alone, it is highly cost-effective if societal impacts are also considered. Direct economic losses to the community are found to outweigh utility agency costs by over 50 times.

---

[1] Research assistant professor, Department of Geography, Box 353550, University of Washington, Seattle, WA 98195-3550; phone 206-616-9018; sec@u.washington.edu.
[2] Principal engineer, ABS Consulting, 300 Commerce Drive, Suite 200, Irvine, CA, 92602; phone 714-734-4242; hseligson@absconsulting.com.

Introduction

For urban infrastructure systems, evaluating the benefits and costs of hazard mitigation measures is complicated by several important considerations. First, electric power, water, and other infrastructure networks are spatially distributed across a wide area; hence, evaluations must account for system functionality as well as the spatial correlation of hazard (such as earthquake ground motion) across the urban space. Second, these systems have long functional lives. Analyses must therefore consider how the system itself, as well as the potential impacts of its failure, might change over a period of many decades. Third, infrastructure systems are vital to all sectors of the urban community. Mitigation assessments should therefore include the benefits to both the lifeline infrastructure provider and society as a whole.

This paper applies a methodology for evaluating infrastructure mitigations to the case of seismic retrofit of the electric power system in Los Angeles. The methodology adopts a life cycle cost approach. Details of the methodology, along with an application to the Portland, Oregon, water delivery system, can be found in Chang (forthcoming). In this approach, the options of "mitigation" and "no mitigation" are compared in terms of their total life cycle costs. Total life cycle costs include repair costs, utility revenue losses, and societal losses in future earthquake disasters. These costs are evaluated for a period of N years, with expected annual costs discounted to their present value. For the mitigation case, life cycle costs also include the cost of the retrofit itself. As a variant of benefit-cost analysis, the life cycle cost framework readily and transparently accomodates costs that may change over time. For infrastructure systems, this is advantageous for addressing such issues as infrastructure deterioration and urban growth.

LADWP Electric Power System

The LADWP power division provides electricity within the City of Los Angeles and to some neighboring areas. LADWP's 1.4 million customers (households and businesses) use more than 22 million megawatt (mw) hours of electricity each year, with business and industry consuming about 70% of the total (www.ladwp.com/aboutdwp/history/allabout/allabout.htm).

Within LADWP's service area, there are 21 receiving stations (substations) that receive power at moderate or high voltage (generally at 138 kV or 230 kV, with a few at 500 kV), and step it down in voltage for further local distribution. For the purposes of this analysis, one service zone, as mapped by LADWP, is associated with each substation. It should be noted that the City of Glendale, while outside the City of Los Angeles, currently imports the majority of its power and is assumed here to be dependent on LADWP for service continuity. The analysis of earthquake impacts in this study focuses on potential damage at the 21 receiving stations.

Probabilistic Earthquake Scenarios

In evaluating the costs associated with potential future earthquakes, the performance of the power system in 47 deterministic earthquake scenarios was evaluated. These scenarios were developed by applying a loss estimation software tool, EPEDAT, which was used to generate regional ground motion patterns for a given earthquake epicenter, magnitude, and depth. The 47 events included 13 maximum credible earthquakes on various faults in the Los Angeles region and 34 smaller events of magnitude 6.0 or higher. These 47 events are associated with "equivalent probabilities" so that collectively, they represent the full range of the local seismic hazard curve. For details, see Chang et al. (2000).

Mitigation, Damage and Outage Modeling

Historically, electric power system outage in earthquakes has been caused predominantly by damage to substations and their equipment (primarily damage to porcelain equipment at older 500 and 220 - 230 kV substations), while power-generating facilities have proved to be less vulnerable. In California, substation damage occurred in the 1986 North Palm Springs earthquake (M 5.9), the 1987 Whittier Narrows earthquake (M 5.9), the 1989 Loma Prieta earthquake (M 7.1), and the 1994 Northridge earthquake (M 6.7).

This study considers base isolation of transformers and their bushings in substations as a mitigation measure. The modeling of mitigation, damage, and power outage was conducted by M. Shinozuka and colleagues at the University of Southern California (USC) (now at the University of California at Irvine). In addition to empirical damage data from the 1994 Northridge earthquake, they used data from shake-table tests of a hybrid base-isolation system (including both sliding bearings and rubber bearings) to develop transformer fragility curves. Three sets of fragility curves were developed; as-is, transformers with performance enhanced by 50%, and transformers with performance enhanced by 100%. Based on the test results, the researchers determined with 92% confidence that base-isolation could improve performance by 50%, and with 80% confidence that performance could be improved by 100% (Shinozuka and Dong, in press).

The collaborating researchers at USC further assessed electric power system performance in the 47 scenario earthquakes, using multiple Monte Carlo simulations and the results of electric power flow software IP-FLOW (see Shinozuka and Dong., in press, for more detail). Twenty Monte Carlo simulations were performed for each condition (as-is, 50% enhancement, and 100% enhancement) for each of the 13 maximum credible earthquake (MCE) scenarios, while 10 simulations were performed for each of the smaller earthquakes. The results provided by USC for use in the current analysis included an estimate of the total number of transformers system-wide that failed in each simulation, and an assessment of power outage (i.e., outage or no outage) in each service zone in each simulation.

Repair Costs

Electric power transformers have a number of different failure modes (Anagnos, 1999), including bushing gasket leakage, bushing porcelain breakage, radiator leakage, radiator breakage, anchorage failure, and foundation failure. In general, bushing failure is the most common failure mode. Radiator damage is considered to be a more rare occurrence, and for transformers within the LADWP system, anchorage failures are not expected, as most transformers are considered adequately anchored. Similarly, foundation failure is not expected (Merz, 2002).

Repair data from the 1994 Northridge earthquake indicate that LADWP had to replace ninety (90) 230 kV bushings at a total cost of $970,000, approximately $11,000 for each bushing (LADWP, 1995). It should be noted that three phase transformers require six bushings. For the purpose of estimating approximate repair cost for failed transformers in the scenario earthquakes, we have assumed that on average, a failed transformer will have 3 failed bushings, with a repair cost of $35,000 (the estimated cost to replace with a composite bushing).

Direct damage in each of the 47 earthquake scenarios, under each mitigation condition (as-is, 50% enhancement and 100% enhancement), and for each simulation was estimated as the number of "failed" transformers multiplied by the typical repair cost of $35,000. The earthquakes which result in the highest average transformer repair costs are the Elysian Park MCE (M 7.1; $1.1 million) and the Newport-Inglewood North MCE (M 7.0; $1.0 million).

Revenue Loss

In addition to the cost to repair damage, costs to the utility company in a seismic event would include potential revenue losses. Results from the USC system analysis included identification of the service areas that would be without power in each earthquake simulation. Evaluating the associated revenue loss required data on customers, power usage, rates, and the duration of outage.

The analysis included an assessment of how urban growth over the 50-year analysis period would change the number of customers affected and the utility's revenue loss over time. Growth forecasts were based on current and projected energy consumption data for LADWP by customer type from the California Energy Commission (CEC, 2002). It was assumed that typical electricity usage per customer (e.g., kWh/year) would remain constant over time, while the total number of customers would grow over time. Rate forecasts were also available from the CEC.

Residential customer data were derived from the 2000 population census by zipcode. A geographic overlay was performed to associate zipcodes (by centroid) with LADWP service zones. Population projections for the City and County of Los Angeles have been developed by the Southern California Association of Governments (SCAG) as part of their Regional Transportation Plan (RTP) Socioeconomic Forecast (SCAG, 2001a). Projections for cities and counties within the SCAG region are available through 2025. For this analysis, they were extrapolated through the year 2050. Between 2000 and 2050, the population in the LADWP service area is projected to increase by 53 percent, from 4.0 million to 6.1 million people.

Business customer data were obtained from SCAG's information on employment by detailed economic sector by zipcode. Two- and 4-digit Standard Industrial Classification (SIC) code data were grouped into ten economic sectors comprising the economy. SCAG has developed total wage and salary employment projections for the City and County of Los Angeles through 2025 (SCAG, 2001a), as well as the distribution of projected employment by sector for 2025 (SCAG, 2001b).

These data were used to project employment growth through 2050. Estimates for the 10 sectors were further aggregated into "commercial" and "industrial" sectors for association with electric power rates for these categories. Between 2001 and 2050, commercial employment in the LADWP service area is projected to increase by 96 percent from 2.3 million to 4.5 million, while industrial employment is anticipated to decline by 8 percent from 305,000 to 279,000. Table 1 shows the service area shares of population in 2000 and commercial and industrial employment in 2001.

Table 1. LADWP Service Area Shares of Population and Employment

LADWP Service Area	Population	Commercial Employment	Industrial Employment
Rinaldi	3.1%	1.9%	0.5%
Northridge	3.0%	4.3%	8.7%
Tarzana	4.5%	3.6%	1.0%
Olympic	6.0%	12.6%	3.1%
Scattergood/Airport	1.3%	3.1%	1.3%
Valley	5.8%	2.6%	5.2%
Toluca	7.1%	4.8%	3.8%
Hollywood	5.2%	7.5%	7.9%
Fairfax	10.2%	8.2%	3.3%
Glendale	8.0%	8.7%	8.3%
Atwater	9.3%	4.4%	4.7%
St.John	3.2%	6.3%	6.0%
River	2.0%	6.9%	3.7%
Velasco	1.6%	3.9%	12.2%
Century	15.9%	7.9%	13.6%
Halldale	1.0%	1.1%	3.4%
Wilmington	2.0%	1.0%	2.0%
Harbor	2.0%	1.1%	1.8%
Canoga	3.4%	4.7%	5.1%
Van Nuys	5.4%	5.2%	4.5%
TOTAL	100%	100%	100%
Total Number	3,981,000	2,298,000	305,000

The duration of any outage in all simulation cases was assumed to be one day. Recent California earthquakes have demonstrated that while full restoration may require days to weeks, service is generally reinstated quite rapidly, usually within one day. In the Loma Prieta earthquake, power to most of San Francisco was restored within 7 hours, and most of the remaining customers had service within 2 days. In many cases, power restoration was accomplished with emergency repairs, bypassing damaged components and operating at a reduced level of circuit protection (Eguchi and Seligson, 1994). In the Northridge earthquake, power to 93 percent of the LADWP service area was restored within 24 hours, and power to virtually all of the service area restored within 72 hours. All of Southern California Edison's (SCE) service territory was restored within 20 hours (EERI, 1995).

Revenue losses due to customer outage were estimated under current and future conditions. Service zones suffering outage were identified in the results of the power flow analysis conducted by the collaborating researchers at USC. The number of customers suffering outage in each service zone was determined from the ratio of population or employees to customers, using the zipcode based population and employment data. Revenue losses were estimated for all customer types (residential, commercial and industrial) for each benchmark year (2001, 2025, 2050) and transformer condition (as-is, 50% enhancement and 100% enhancement), averaged over the 10 or 20 Monte Carlo simulations. Results for each year were linearly interpolated between the three benchmark years. For 2001 conditions, the largest revenue losses without mitigation are suffered in the Malibu Coast MCE (M 7.3; $4.7 million). Note that, depending on the spatial pattern of power outage relative to the customer base, earthquake scenarios that caused the most damage were not necessarily associated with the greatest revenue loss.

Direct Economic Loss

Societal impacts for the earthquake scenarios were evaluated in the form of direct economic loss, or direct business interruption loss. This loss is estimated for each earthquake simulation and mitigation condition as follows:

$$L_t = \sum_s \sum_j l_j \cdot d_s \cdot e_{s,j,t} \qquad (1)$$

where L_t is direct economic loss ($) for analysis year t, l_j is a loss factor for industry j ($0 \leq l \leq 1$), d_s is a disruption indicator for service area s ($d=1$ in case of power outage, $d=0$ in case of no outage), and $e_{s,j}$ is daily industry j economic activity in area s ($). The disruption indicators d_s were obtained directly from the USC modeling results.

The loss factors l_j reflect the dependency of each industry on electric power. The loss factors were empirically developed on the basis of survey data collected following the Northridge earthquake. Specifically, a large survey of over 1100 businesses was conducted by K. Tierney and colleagues at the Disaster Research Center of the University of Delaware (see Webb et al., 2000). Data from this survey that were used in the current study included information on whether a business lost electric power, for how long, the level of disruptiveness associated with this outage, and whether or not the business closed temporarily in the disaster. Data on other sources of disruption (e.g., building damage, loss of water, etc.) were also used to estimate the net effect of electric power outage. For details on the methodology, see Chang et al. (2002). Table 2 shows the loss factors, which range from a low of 0.39 for mining and construction to a high of 0.60 for manufacturing. These factors pertain to a one-day power outage.

Data on industry economic activity by service area and analysis year, e_{sjt}, were estimated from several sources. Information on California gross state product (GSP) and employment for the years 1990 and 2000 were obtained from the Bureau of Economic Analysis. The annual growth in productivity, or output per job, was calculated for each of the ten major industries. This annual growth was assumed to hold through 2050. Annual productivity growth rates ranged from close to no increase (construction, government, services) to as much as 6.3 percent (manufacturing) in real terms, net of inflation. Industry employment forecasts for the LADWP service area, described above, were multipied by the productivity estimates to obtain a rough forecast of gross regional product (GRP) by industry, service area, and benchmark year. Data for each year were linearly interpolated between benchmark years. The highest direct economic loss in the 2001 benchmark year for the unmitigated case was associated with the Malibu Coast MCE event ($207 million).

Table 2. Loss Factors for Electric Power Outage

Industry	Loss Factor
Agriculture, Forestry, and Fisheries	43%
Mining	39%
Construction	39%
Manufacturing	60%
Transportation, Communication, and Utilities	39%
Wholesale Trade	51%
Retail Trade	51%
Finance, Insurance, and Real Estate	56%
Services	53%
Government	53%

Total Life Cycle Costs

Results from the analysis described above were brought together in the assessment of total life cycle costs for the unmitigated and mitigated cases. The mitigated case was represented by a weighted average of the results from the "as-is," "50% performance enhancement," and "100% performance enhancement" analyses. The respective weights are 0.08, 0.12, and 0.80. These weights were based on the confidence levels associated with each performance level in the USC base isolation testing.

Total life cycle costs C are the sum of the discounted stream of future costs:

$$C = C_s + \sum_t (C_{r,t} + C_{v,t} + C_{e,t}) \cdot (1+y)^{-t} \tag{2}$$

where C_s is the cost of the seismic mitigation measure (assumed to take place in the initial year of analysis), $C_{r,t}$ is expected annual earthquake repair costs in year t, $C_{v,t}$ is expected annual earthquake revenue loss to the utility in year t, C_e is expected annual direct economic loss to the community in year t, and y is a discount rate (assumed to be a real rate of 3 percent). The total timeframe of analysis is 50 years. All costs are evaluated in constant dollars, net of inflation. $C_{r,t}$, $C_{v,t}$, and $C_{e,t}$ are each probabilistically aggregated over the 10 or 20 simulations of the 47 earthquake scenarios, as described above.

The mitigation case consists of base isolation of 109 transformers. Although actual data on the mitigation cost was not available, based on consultation with USC researchers and LADWP, a rough estimate of $60,000 per transformer was used for this analysis. Seismic mitigation cost C_s is therefore $6.54 million.

Table 3 summarizes the four cost components for the mitigated and unmitigated cases. Mitigation does lead to considerable reductions in expected repair costs and revenue loss in future earthquakes. Total discounted costs to the utility agency are reduced from $1.67 million to $0.21 million. However, these savings do not outweigh the cost of mitigation. If societal impacts are disregarded, total life cycle costs with mitigation ($6.75 million) are four times as large as costs in the do-nothing case ($1.67 million), and mitigation does not appear advisable.

Table 3. Life Cycle Cost Results
(2000$ millions)

	Unmitigated Case	Mitigated Case
Mitigation Cost	$ 0.00	$ 6.54
Repair Cost (discounted)	$ 0.39	$ 0.07
Revenue Loss (discounted)	$ 1.28	$ 0.14
Direct Economic Loss (discounted)	$ 95.52	$ 10.45
Total Life Cycle Costs	$ 97.19	$ 17.19
Total Life Cycle Costs without Direct Economic Loss	$ 1.67	$ 6.75

However, if direct economic losses are considered, mitigation appears very cost-effective. Direct economic losses are on the order of 50 times as large as the utility's repair and revenue losses combined. If they are included in the analysis, total life cycle costs without mitigation ($97.19 million) are 5.7 times as large as costs with mitigation. Mitigation could reduce discounted, expected direct economic loss by some $85 million. Base isolation of the transformers appears highly cost-effective.

Table 4 presents the results if it is assumed that population and economic activity remain constant over the 50-year period (i.e., no urban growth or productivity increase). The basic finding – that mitigation is cost-effective if societal impacts are considered, but not if they are excluded – is unchanged. In this case, direct economic losses are on the order of 30 times as large as repair and revenue losses combined.

Table 4. Life Cycle Cost Results without Urban Growth
(2000$ millions)

	Unmitigated Case	Mitigated Case
Mitigation Cost	$ 0.00	$ 6.54
Repair Cost (discounted)	$ 0.39	$ 0.07
Revenue Loss (discounted)	$ 0.89	$ 0.09
Direct Economic Loss (discounted)	$ 39.38	$ 4.29
Total Life Cycle Costs	$ 40.66	$ 11.00
Total Life Cycle Costs without Direct Economic Loss	$ 1.28	$ 6.70

Conclusions

This paper demonstrates how life cycle cost analysis can be applied to evaluate seismic mitigations for an urban electric power system. It showed that mitigations that do not appear cost-effective from the viewpoint of the utility agency can be very cost-effective if societal impacts are also considered. This general finding is similar to that of a previous case study of the Portland, Oregon, water delivery system (Chang, forthcoming). In that case, direct economic losses were found to outweigh utility agency costs by 100 times. Seismic upgrading of tanks and pump stations was found to be cost-effective if these losses were considered.

Several differences between mitigation analyses of electric power and water systems are, however, worth noting. First, the collateral benefits of seismic mitigations may be more significant for water than for power systems. In the case of water, some mitigations such as pipe replacement also entail substantial savings in future maintenance costs. This is because infrastructure deterioration appears to be more significant for vulnerable components such as pipes than for transformers. Second, catastrophic failure or interruption in one part of a power network can bring the entire network down within seconds, as was demonstrated by the immediate blackout in the Northridge earthquake. With water systems, catastrophic failure of a system component may cause localized loss of water (e.g., tank failure, subsequent drainage and loss of supply), but instantaneous system-wide outage is unlikely. Third, initial power restoration can often be accomplished rapidly by rerouting power around damaged components and operating the network at reduced levels of circuit protection. As a result, customer outage and revenue losses are minimized. No similar expedient restoration methods exist for water utilities. Outages – and the associated consequences for customer loss-of-service, utility revenues, and societal impacts – are therefore likely to be much greater for water than for power systems. That is, in an earthquake, water disruptions may affect fewer customers but for longer periods of time than power outages.

The LADWP case study has indicated several data gaps and areas for further research. Data on the actual distribution of customers by type within each service zone were unavailable. Also, projected growth of customers has been approximated by population and employment growth. Better projections of electric power customers by zone would enhance the results of the analysis. Improved data are also needed on post-disaster restoration times for lifeline outages, costs of mitigation and repair, and the social and economic impacts of outages on the community.

Acknowledgments

The authors are grateful to M. Shinozuka and X. Dong at the University of California at Irvine (formerly at the University of Southern California) for sharing findings from their LADWP systems analysis. We also thank R. Tognazzini at LADWP for sharing background and data, K. Tierney at the University of Delaware for providing data from her Northridge business survey, and A. Rose at the Pennsylvania State University for assistance with economic data. This material is based upon work supported by the National Science Foundation under grant no. 9802151.

References

Anagnos, T. (1999) "Development of an Electrical Substation Equipment Performance Database for Evaluation of Equipment Fragilities", paper prepared for Pacific Gas & Electric and the Pacific Earthquake Engineering Center (PEER).

California Energy Commission (CEC). (2002) "California Energy Demand 2002," www.energy.ca.gov/energyoutlook/documents/2001-10-04_demand_forecast.pdf

Chang, S.E. "Evaluating Disaster Mitigations: A Methodology for Urban Infrastructure Systems," *Natural Hazards Review*, forthcoming.

Chang, S.E., Shinozuka, M., and Moore, J.E. II. (2000) "Probabilistic Earthquake Scenarios: Extending Risk Analysis Methodologies to Spatially Distributed Systems," *Earthquake Spectra*, Vol.16, No.3, pp.557-572.

Chang, S.E., Svekla, W.D., and Shinozuka, M. (2002) "Linking Infrastructure and Urban Economy: Simulation of Water Disruption Impacts in Earthquakes," *Environment and Planning B*, Vol.29, No.2, pp.281-301.

Earthquake Engineering Research Institute (EERI). (1995) "Northridge Earthquake Reconnaissance Report, Volume 1", *Earthquake Spectra*, Supplement C to Volume 11, Earthquake Engineering Research Institute, Oakland, California.

Eguchi, R.T. and Seligson, H.A. (1994) "Lifeline Perspective," *Practical Lessons from the Loma Prieta Earthquake*, Washington, D.C.: National Research Council.

Los Angeles Department of Water and Power (LADWP). (1995) "Intermediate Term Plan for Seismically Hardening the Los Angeles Transmission Level Power Facilities," LADWP, June.

Merz, K. (2002), personal communication.

Shinozuka, M., and Dong, X. "The Seismic Performance Analysis of Electric Power Systems," Multidisciplinary Center for Earthquake Engineering Research (MCEER) Technical Report, forthcoming.

Southern California Association of Governments (SCAG). (2001a) "2001 RTP Growth Forecast: City Projections Report," Southern California Association of Governments, www.scag.ca.gov/growthforecast/

SCAG (2001b), "2001 RTP Social Economic Forecast Report", Southern California Association of Governments, www.scag.ca.gov/growthforecast/

Webb, G.R., Tierney, K.J., and Dahlhamer, J.M. (2000) "Businesses and Disasters: Empirical Patterns and Unanswered Questions," *Natural Hazards Review*, Vol.1, No.2, pp.83-90.

URAMP (Utilities Regional Assessment of Mitigation Priorities)– A Benefit-Cost Analysis Tool for Water, Wastewater and Drainage Utilities: Software Development

Charles K. Huyck[1]
Ronald T. Eguchi[2]
Reid M. Watkins[3]
Hope A. Seligson[4]
Stephen Bucknam[5]
Edward Bortugno[6]

Abstract

This paper discusses the development of Utilities Regional Assessment of Mitigation Priorities (URAMP), a GIS-based software package for estimating earthquake-induced damage to water, wastewater, and drainage systems. The package was created for the California Governor's Office of Emergency Services (OES) by ImageCat Inc. and ABS Consulting, under the guidance of a steering committee composed of engineers, local utility representatives, and OES personnel. This paper is submitted with a companion paper that focuses on methodology development (Seligson et al., 2003).

Introduction

Utilities Regional Assessment of Mitigation Priorities (URAMP) is a GIS-based software package that estimates earthquake-induced damage to water, wastewater,

---

[1] Senior Vice President, ImageCat, Inc., 400 Oceangate, Suite 1050, Long Beach, CA 90802; ckh@imagecatinc.com
[2] President and CEO, ImageCat, Inc., rte@imagecatinc.com
[3] Programmer Analyst, ImageCat, Inc., rmw@imagecatinc.com
[4] Principal Engineer, ABS Consulting, hseligson@absconsulting.com
[5] President, Bucknam Associates, csbjr@prodigy.net
[6] Senior Geologist, Governor's Office of Emergency Services, edward_bortugno@oes.ca.gov

and drainage systems as well as the costs and benefits associated with specific mitigation activities. ImageCat Inc. and ABS Consulting developed the package for the California Governor's Office of Emergency Services (OES) in consultation with a steering committee composed of engineers, local utility representatives, and OES personnel. Working with the steering committee to conduct a needs assessment, we were able to develop an application that was both powerful and easy to use. The software allows the user to quantify the effectiveness of various mitigation strategies, identify vulnerable system components, and prioritize capital improvement programs.

This paper will highlight some of the issues that emerged during the software development process. We will begin by discussing the user needs assessment and the technical solutions that met these needs. An import wizard that allows users to interactively assign water network attributes to of vulnerability is a key element in URAMP's usability. We will discuss the import wizard in detail. We will then present how the software works, focusing on the interaction of the modules.

Assessing the User Requirements

We faced several technical challenges when we began the development of URAMP. The integration of a hydraulic network analysis into a GIS application development environment was a primary concern. Although the budget for the project would not accommodate creation of new hydraulic analysis model, we realized that hydraulic modeling output is crucial to estimating losses associated with fire-following. Using a hydraulic model to calculate the pressures throughout a water network allows an assessment of water service availability after a major earthquake including an evaluation of whether adequate pressures will exist to suppress ensuing fires. In addition, by providing information about pressures at nodes throughout a water network which can be used to calculate the percentage of an area that will receive at least some level of minimum service, the model allows us to estimate losses from business interruption . These measurements directly impact the level of loss to be estimated for each earthquake scenario.

In addition to the challenge of integrating hydraulic modeling capability, we realized that the process of inputting data would need to be simple. Because water utilities are complex systems, complex data was required to estimate the benefits of mitigation. Since the input data required to assess mitigation are held by local utilities, it was not possible to deliver the software with a default data set that could be used anywhere it the state. The user must be able to import local data. Because hydraulic network models do not normally contain the attributes required to allow assessment of system vulnerability, links to attributes such as pipeline vulnerability, the seismic design of tanks, and other important features must be established.. The lack of database utilities available within hydraulic network modeling tools complicates this process. We saw that the additional attributes would have to either be assigned in a GIS program, or in a customized interface.

To assess user requirements, the project team began by distributing a survey to URAMP steering committee members. The survey included questions about the extent and format of digital data available in the utility, whether or not the utility had a GIS capability, whether or not the digital network files were in a real world projection, and other questions concerning GIS expertise. The responses showed that few cities had an active GIS capability. In several cases, users had GIS software, but analysts or data were not available. We realized that URAMP would have to accommodate users with no GIS or database skills and that our users would not have the resources to link a hydraulic network model to the additional attributes necessary to assess vulnerability.

Although we saw that we could not expect users to handle complex GIS tasks, it was evident through the survey as well as discussions with the steering committee, that the water utility engineers often knew the additional information necessary to assess the vulnerability of water networks. In other words, engineers could be expected to remember the necessary parameters -- for example, tank size, replacement cost, and the construction parameters of tanks -- without referring any documentation.

Technological solutions

Based on the survey and several meetings with the steering committee, we found acceptable technological solutions to meet the challenge of developing a program that would be both useful and usable. We concluded that URAMP would be a GIS-based application, but that no GIS expertise would be required of the user. Developed in Visual Basic with a MapObjects GIS component, the user would only need to pan, zoom, select, and use the information tool. An import wizard simplifies the process of entering information. The program requires no additional GIS data, although the user may update the liquefaction data.

To incorporate hydraulic analysis capabilities, we integrated "EPANET", a program developed by the Environmental Protection Agency, into URAMP. The EPANET model has several advantages.. Because the EPANET hydraulic model is public domain software, no licensing fee is required. Additionally, using EPANET allows URAMP to integrate hydraulic modeling capability that appears seamless to the user. Most of EPANET's programmatic functions can be used outside of the EPANET program by calling subroutines in a "dynamic link library" or DLL. A DLL allows the integration of a particular software functionality into another software program without requiring the user to navigate through two software interfaces. In this way, the user benefits from EPANET's analytical capabilities without ever having to see it. Many commercially available water network modeling programs utilize EPANET in this fashion.

While integrating EPANET into URAMP enabled the use of more sophisticated fire following algorithms, it, at the same time, simplified the import process for the user. Through several postings to a list serve (EPANET-L), we determined that most commercially used hydraulic modeling applications, including H2OMAP, H2ONET, MIKENet, and WATERCAD, could export to a text file format that could be read by EPANET. Through use of EPANET files as the input, we assured the largest possible group of users.

During software development, we placed a high priority on simplifying the data input process. Working with the steering committee on a monthly basis throughout the project, we were able to develop an "Import Wizard" that allows users to define the attributes of their hydraulic networks through a customized graphical user interface (GUI), rather than through standard GIS and database programs.

The import wizard has several dialogs in which the user defines the parameters of the water network. An initial dialog determines the UTM projection zone and approximate age of the water system, information which is used to populate default vulnerability classes on the subsequent import dialogs. The age category options are pre-1960, 1960-1974, 1975-1990, and post-1990. The user then progresses through the wizard and enters the extra attributes needed for each component type in turn.

Figure 1 is a sample dialog in the import wizard for pipelines illustrating how the user adds attributes to the baseline system. Pipelines are the first water system component in the import wizard. The user has several standard GIS tools available to select multiple or individual pipelines. With these tools, the user establishes the vulnerability of the pipelines, assigning them each to one of several categories (Very High, High, Moderate, or Low vulnerability). The user is provided with examples of material types corresponding to each of the vulnerability categories. The color of each pipeline on screen reflects the user's progress in updating the pipeline information. The user will not be able to continue through the wizard without updating the classifications of all pipes.

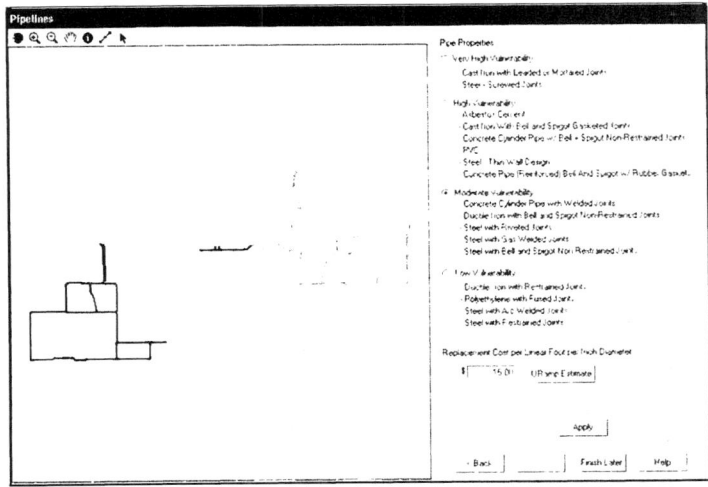

Figure 1. Import Wizard Dialog for Pipelines

Figure 2 is an image of the next import wizard dialog for water tanks. The user selects a specific tank on screen, and then specifies the type of tank. Steel tank at grade, concrete reservoir at grade, elevated tank, hydropneumatic tank, and buried concrete reservoir are the tank types available. When the tank type is selected a new frame appears on the right with the specific attributes that the user enters for the type of tank selected. With the exception of pipes, categories generally follow a three level classification system. Level 0 corresponds to no seismic design, Level 1 have some seismic design, such as flexible joints and anchored equipment, and Level 2 generally has more significant seismic design retrofitting, such as a seismically designed superstructure. The classifications for each facility type correspond to vulnerability functions. The software presents the user with the same options on similar dialogs to assess mitigation of existing pipelines and facilities. Additional import dialogs for water pumps, wells, and treatment facilities follow. For each component, the user navigates a simple point-and-click interface to denote water-network attributes.

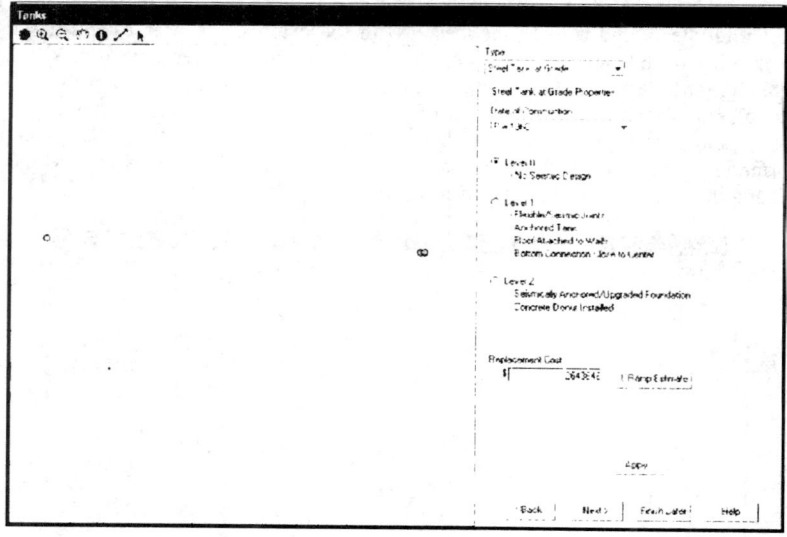

Figure 2. Import Wizard Dialog for Water Tanks

URAMP Model flow

In the following section we will walk the reader through the various components of the model to illustrate how the user can assess the cost effectiveness of a mitigation option. Figure 3 represents a flow chart of the model components in URAMP. The first step is to import a georeferenced hydraulic model file in INP format from EPANET or other hydraulic modeling program. The file must be georeferenced (use-real world, geographic coordinates) to overlay with hazard and census data. Users who do not have a georeferenced file must scale the coordinates of the water network outside of URAMP, for example, using a Microsoft Excel spreadsheet. Once the user has imported the EPANET input file into the URAMP model, the process of defining the baseline system through the import dialog begins. The baseline model reflects the current state of the system before mitigation measures are assigned.

After the additional vulnerability and cost attributes have been assigned to the hydraulic network, the URAMP software converts the network to an ESRI shapefile which is duplicated internally. The second file tracks the mitigation options. When assessing the benefits of mitigation, URAMP runs both files through the analysis in tandem, and the difference in estimated earthquake performance is compared with the cost of mitigation to assess cost effectiveness. The user can specify mitigation of all pipelines or facilities, of specific pipelines or facilities, or of additional pipelines or facilities added. Through a standard map interface and point-and-click tools, the user can create a modified network model representing the mitigated network. For each mitigation activity, the user has the choice of allowing URAMP to estimate the cost of mitigation, or entering an estimate manually. The recommended initial step for assessing mitigation is to run a significant earthquake scenario to reveal the vulnerabilities of the system, and then begin drafting mitigation options. The current version of URAMP does not recommend mitigation options. The user has the ability to save project workspaces representing various mitigation options. Each project workspace can be retrieved and modified or compared to other mitigation strategies.

Once the mitigation parameters have been defined, the user selects the mode of the seismic hazard calculations. The user can select whether to run a Maximum Credible Earthquake (MCE), which represents the largest scenario earthquake from a single fault source, or a probabilistic analysis, which evaluates many MCEs. An underlying database stores ground shaking information for each MCE on a census tract basis. This allows the analysis to run quickly. Details on the seismic hazard model development can be found in the following paper contained in these conference proceedings "Quantitative Method for Developing Hazard-Consistent Earthquake Scenarios" by Kenneth W. Campbell and Hope A. Seligson (TCLEE, 2003).

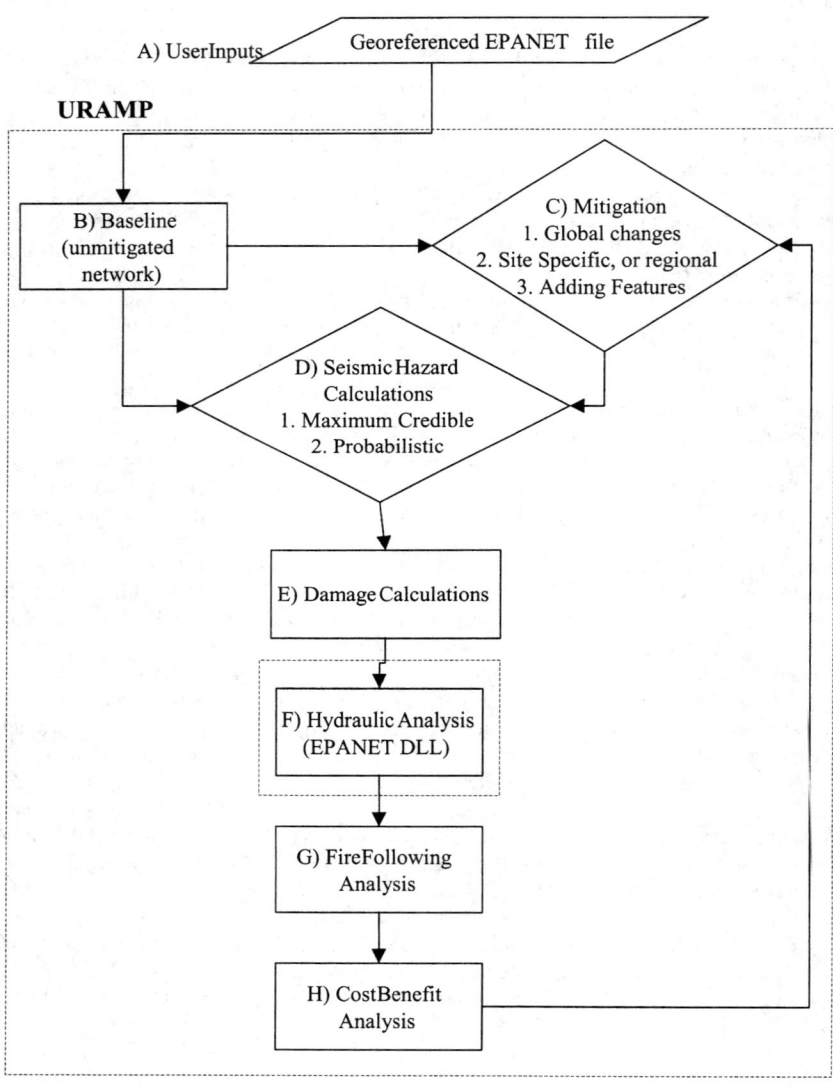

Figure 2. Flow Chart of the Model Components in URAMP

When choosing the seismic hazard option, the user can also enter several economic and demographic parameters for the analysis such as average water price, average water demand, alternative living expenses, number of available water repair workers, and number of households in service zone. The economic module uses these parameters to calculate the indirect losses. Additionally, the fire following earthquake loss module uses the demographic parameters to calculate density of building stock for assessing vulnerability to fire following events. The next three modules, damage calculations, hydraulic analysis, and fire following analysis are transparent to the user (modules E, F, and G in Figure 3).

URAMP calculates the direct damage to the various network components and the expected damage is used to change parameters in the hydraulic analysis. To model the effects of damaged pipes within the network, a weighted average demand is placed on the nodes at either end of the damaged pipeline. To represent a pipeline failing, we used a demand of 3,000 gallons per minute. The project team assumed that 20% of the repairs would be due to broken pipes, having a demand of 5,000 gallons per minute, and 80% of the pipes repairs would be leaks, which would have an average rate of 2,500 gallons per minute. URAMP modifies most of the components in the hydraulic model to reflect direct damage. When the program estimates damage at a water treatment plant, the pipes connected to the plants are closed in the underlying EPANET file. When a tank is estimated to be significantly damaged, the initial level of the tank is set to the minimum level, and the connecting pipes are closed.

The mitigated and baseline EPANET hydraulic models are then modeled at the critical time period of 25 minutes after the event. After this amount of time, the pressure should be representative of the water system stabilized within the first 6 hours, which is the time frame important for fire fighting. The percentage of junctions with water pressure greater than 20 PSI is used to determine a Water Service Reliability factor (WSR) for the fire following module. This threshold is an accepted minimum pressure within the fire fighting industry.

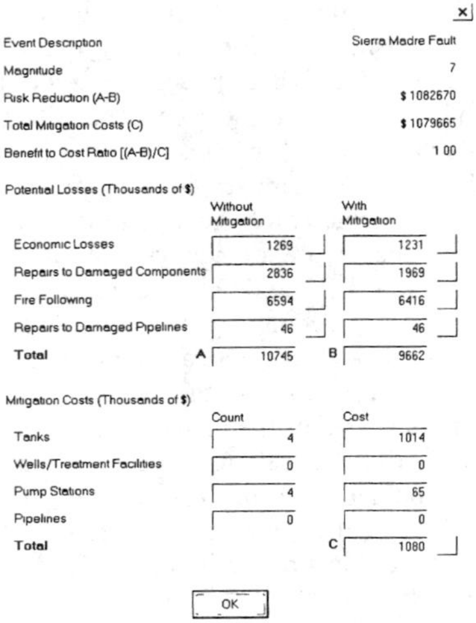

Event Description		Sierra Madre Fault
Magnitude		7
Risk Reduction (A-B)		$ 1082670
Total Mitigation Costs (C)		$ 1079665
Benefit to Cost Ratio [(A-B)/C]		1.00

Potential Losses (Thousands of $)

	Without Mitigation	With Mitigation
Economic Losses	1269	1231
Repairs to Damaged Components	2836	1969
Fire Following	6594	6416
Repairs to Damaged Pipelines	46	46
Total	A 10745	B 9662

Mitigation Costs (Thousands of $)

	Count	Cost
Tanks	4	1014
Wells/Treatment Facilities	0	0
Pump Stations	4	65
Pipelines	0	0
Total		C 1080

Figure 4. URAMP Results

Using URAMP Results

Figure 4 shows the final results menu presented to the user after running a scenario earthquake. The top portion of the dialog presents the final results of the cost benefit analysis, including the net risk reduction, the total mitigation cost, and the resulting benefit cost ratio. The "Potential Losses" section of the dialog provides details of the; the difference in performance of the mitigated and unmitigated network. The losses are divided into four categories: economic losses, repairs to damaged components, fire following losses, and repairs to damaged pipelines.

Following each category of loss is an ellipse. The user can click on the ellipse to obtain specific information on the damage at each pipeline, component, or census tract (in the case of economic and fire following losses). These specific dialogs give the user the information they need to assess the effects of mitigation at each point in the water system. The mitigation costs portion of the results presents a summary of the costs by facility type. The user can click on the ellipse to see a full description and cost of each mitigation option defined for the active results.

The information in the extended mitigation and loss dialogs can be used to prioritize various mitigation options for the network, and the analysis can be run until a combination of mitigation options yields a cost-effective solution. Mitigation options are defined and assessed in iterations. At any point in the process, the user can revert to a previously saved project workspace.

Summary

The URAMP programming effort was a tremendous success. Often, loss estimation programs are hampered by data processing requirements. URAMP, like many programs, required some very sophisticated capabilities and data, such as hydraulic modeling, probabilistic ground shaking, and the linking of hydraulic networks to additional attributes. Several problems arose very early in the development process, including a user base that was currently not GIS capable. By working directly with the end users, assess their needs through frequent meetings that included demonstrations, we overcame many software development obstacles. The result is an easy to navigate interface in which a user establishes the baseline system in categorical terms that are related to the vulnerability functions analyzing the components. Replicating these categories in the mitigation dialog, gives users a simple interface that can be used to create a digital mitigated water system. Additionally, the seamless integration of a hydraulic modeling capability that is invisible to the user enabled a more accurate assessment of fire following losses. By focusing on the end user, we developed a loss estimation model that can be implemented without compromising capability. During every step in the development process, we presented preliminary dialogs to the steering committee to assure that the solutions could be implemented and that no lingering concerns remained.

URAMP (Utilities Regional Assessment of Mitigation Priorities) – A Benefit-Cost Analysis Tool for Water, Wastewater and Drainage Utilities: Methodology Development

Hope A. Seligson, ABS Consulting[1]
Donald B. Ballantyne, ABS Consulting[2]
Charles K. Huyck, ImageCat Inc.[3]
Ronald T. Eguchi, ImageCat Inc.[4]
Stephen Bucknam, Bucknam & Associates[5]
Edward Bortugno, California Governor's Office of Emergency Services[6]

Abstract

This paper describes the methodology underlying the URAMP software (Utilities Regional Assessment of Mitigation Priorities), developed for the California Governor's Office of Emergency Services (OES) by ABS Consulting and ImageCat Inc. The software development project was conducted under the guidance of a steering committee made up of engineers, local utility representatives, and OES personnel. URAMP is a GIS-based software package that estimates earthquake-induced damage and the benefits of mitigation for water, wastewater, and drainage utilities on a regional basis. The software allows the user to assess various mitigation strategies, identify vulnerable system components, and prioritize capital improvement programs. While this paper focuses on the underlying methodology, a companion paper focuses on the software development (Huyck et al., 2003).

Introduction

The URAMP software builds on earlier work performed for OES - the development of the RAMP (Regional Assessment of Mitigation Priorities) software. RAMP was developed following the 1994 Northridge earthquake to streamline cost-effectiveness determinations for non-structural mitigation of schools and hospitals. Infrastructure

---

[1] Principal Engineer, ABS Consulting, 300 Commerce Drive, Suite 800, Irvine, California 92602; phone (714) 734-4242; hseligson@absconsulting.com
[2] Vice President, Lifeline Services, ABS Consulting, Seattle, WA; dballantyne@asbconsulting.com
[3] Senior Vice President, ImageCat, Inc., Long Beach, CA; ckh@imagecatinc.com
[4] President and CEO, ImageCat, Inc., Long Beach, CA; rte@imagecatinc.com
[5] President, Bucknam Associates, Laguna Niguel; CA, csbjr@prodigy.net
[6] Senior Geologist, Governor's Office of Emergency Services; edward_bortugno@oes.ca.gov

systems, such as water, sewage and drainage were considered excellent candidates for a regional "RAMP-type" approach, because many of the elements are similar in design and construction, and extensive failure could result in high direct and indirect costs. Furthermore, extensive damage to these systems could hinder long-term recovery efforts.

The URAMP software is GIS-based, with a user-friendly interface that is described in detail in the companion paper on software development (Huyck, et al., 2003). The URAMP software is unique in that it incorporates the functionality of EPANET, a widely used hydraulic modeling software package distributed by the U.S. Environmental Protection Agency (EPA), to assess the impact of earthquake damage on water network performance. A simplified version of URAMP allows for the analysis of wastewater and drainage systems, which do not require the same sort of network analysis. URAMP allows the user to assess the benefits of individual mitigation strategies, and to determine which combination of mitigation measures best enhances water network performance in the aftermath of an earthquake.

The URAMP analysis modules include state-of-the-art seismic hazard models to facilitate regional network analysis (described in another related paper in these proceedings, Campbell & Seligson, 2003) and consider both deterministic (e.g., scenario earthquakes) and probabilistic seismic risk assessments. The damage estimation modules utilize engineering-based damage and mitigation models, based, whenever possible, on published methodologies. In addition to direct damage to utility facilities, URAMP estimates potential damage and economic losses due to fire-following earthquake, and other economic impacts on the provider utility and community, such as lost revenue, business impacts of fire, and cost of sewage clean-up. Many of the economic models are based on earlier "Seismic Reliability Assessment Studies" performed by the U.S. Army Corps of Engineers (see, for example, USACE, 1996). URAMP represents an integration of the original RAMP approach with the economic loss methods developed in USACE studies, updated with fire-following models and the latest seismic hazard information.

The URAMP Methodology

The technical approach for calculating losses and assessing the impacts of mitigation within URAMP is outlined in Figure 1. The approach consists of five basic steps that are implemented under the "without project" (baseline case) and "with project" (mitigated) conditions.

Seismic Hazard Assessment. The first step in the process is the seismic hazard determination. Seismic hazards may be analyzed using either probabilistic or deterministic methods. Probabilistic methods consider potential earthquakes from all sources, incorporating each earthquake's probability of occurrence to derive site ground motions with a particular probability of exceedance. Deterministic seismic hazards are determined as the likely ground motions from a particular earthquake. To

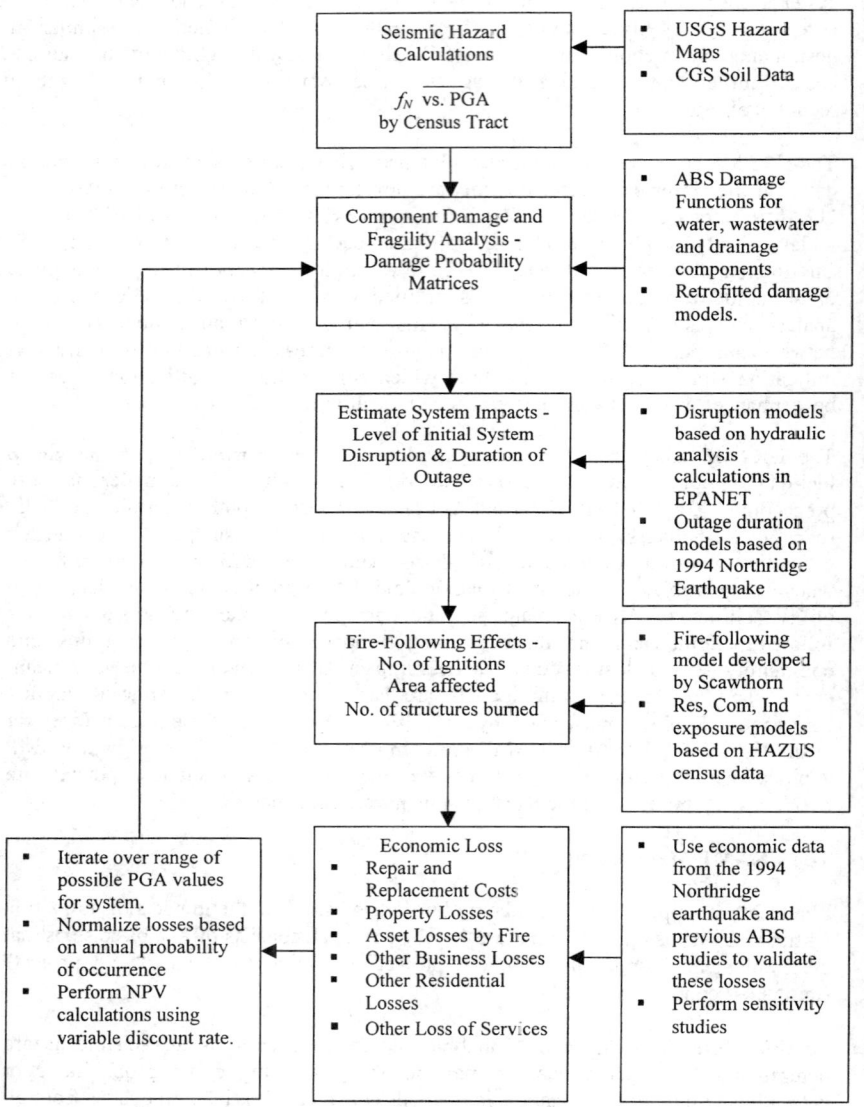

Figure 1: URAMP Approach for Estimating Potential Losses and the Benefits of Mitigation

meet a variety of user needs, URAMP includes all data required to assess losses using both probabilistic and deterministic hazards anywhere within the State of California. The gridded seismic hazard data developed by the U.S. Geological Survey (USGS) served as the starting point for the probabilistic seismic hazard models. In addition, the digital statewide geologic map developed for California by the California Geological Survey (CGS), has been utilized to assess local site conditions (for additional information on the seismic hazard models, see Campbell & Seligson, 2003).

Component Modeling within URAMP. The second step in the URAMP methodology is the modeling of lifeline system components, including both the assessment of potential damage and the modeling of repair, replacement and mitigation costs.

Facility Classification Scheme. URAMP includes damage, mitigation and cost models for five basic types of water, wastewater and drainage facilities: water supply facilities, water storage facilities, pumping stations, treatment facilities, and pipelines. Within the four groups of non-pipeline facilities, there are a total of eleven different classes of facilities (pipelines are classified by material type and joint type, as related to earthquake vulnerability), as follows:
- Water storage facilities (5 classes): buried concrete reservoirs, concrete tanks at grade, steel tanks at grade, elevated steel tanks, and hydro-pneumatic tanks.
- Pump stations (1 class): water booster pumps, sewage/drainage lift stations
- Water supply facilities (2): wells and groundwater collection facilities/tunnels.
- Treatment facilities (3): chemical feed treatment plants, conventional filtration plants, and pressure treatment plants

Common earthquake damage mechanisms and associated mitigation strategies were identified for each facility class. A sample, for steel tanks at grade, is provided in Table 1.

Table 1: Damage Mechanisms and Mitigation Options for Steel Tanks at Grade

Damage Mechanism	Required Data	Mitigation Strategy
Outside pipe connection failure	Anchored, pipe flexibility	Add flexible joint
Bottom pipe connection failure	Tank anchorage (none, inadequate, adequate), height, diameter, seismic design, date of construction, connection distance from edge (required for all connections).	Anchor tank, move bottom connection closer to center.
Tank uplift, wall buckling	Tank anchorage (none, inadequate, adequate), height, diameter, seismic design, and date of construction.	Anchor/upgrade foundation, install concrete donut inside tank
Foundation settlement	Liquefaction susceptibility, liquefaction considered in the design.	Soil stabilization
Roof sliding/collapse	Roof attachment	Attach roof to walls

In addition, facility data required to model damage for the various damage mechanisms were identified. These parameters provide the fundamental basis for the four design parameters utilized in URAMP:

- Design level:
  - Tanks – Level 2 (e.g., tank is anchored, flexible connections installed), Level 1 (e.g., tank is unanchored, but has flex connections), and Level 0 (e.g., tank is unanchored, no flex connections).
  - Other Facilities – Level 1 (equipment is anchored, and flexible connections have been installed), or Level 0 (no anchorage, no flexible connections).
- Seismic design (for facilities with superstructures, i.e., pumps, wells and treatment facilities) – whether the facility design considered seismic loads, generally determined by a facility's age (yes/no)
- Liquefaction Design – whether the risk of liquefaction was explicitly considered in a facility's design (yes/no).
- Emergency generator (for facilities requiring electricity, e.g., pumps, treatment plants) – present or absent (yes/no). (It should be noted that emergency generators do not improve a facility's physical performance in terms of damage reduction, but do improve a facility's functionality and continued operation when not significantly damaged.)

Accordingly, various combinations of design parameters may be associated with each facility class, resulting in multiple facility models for each facility class and yielding a total of 108 different facility models within URAMP. The URAMP construct of various design levels readily facilitates the modeling of mitigation strategies that are considered upgrades (i.e., anchoring equipment or an unanchored tank, installing flexible connections etc.), rather than replacements. For a given facility with known design parameters, candidate upgrade mitigations are those that "improve" one or more of the design parameters, i.e., improving a treatment facility from design level 0 to design level 1 by anchoring equipment and installing flexible connections, or instituting liquefaction countermeasures. These types of improvement mitigations are available within URAMP, in addition to replacement, removal, or installation of new components.

Damage Functions. In general, damage functions estimate percent damage and functionality relative to input ground motion. For URAMP, peak ground acceleration (PGA) is used as the input ground motion, except for pipelines, which used peak ground velocity (PGV).

As noted above, upgrade mitigation strategies have been modeled as an improvement in one or more design parameters. Therefore, one set of damage functions (108 functions) is utilized to represent the full range of facility conditions (e.g., both unmitigated or "without project", and mitigated or "with project"). In essence, damage under the "with project" conditions is represented by stand-alone damage functions (e.g., a damage function for an anchored tank with flexible connections, where the connections were installed as a mitigation measure), rather than as a function representing effectiveness of mitigation (percent reduction in damage).

Numerous damage functions for water, wastewater and drainage facilities are available from published literature (e.g., ATC, 1985; ASCE, 1991; NIBS/FEMA, 1999 and 2002; ALA, 2001) as well as from internal project team project files and reports. Two published methods provide the basis for estimating damage to non-pipeline facilities within URAMP; the ATC-13 damage state and damage probability matrix (DPM) approach (ATC, 1985), and the library of fragility functions utilized by HAZUS (NIBS/FEMA, 1999 and 2002). The fragility functions used within HAZUS, relating likelihood of a facility being within a given damage state (as defined by a range of percent damage) for a given input ground motion, were used to refine many of the damage functions within URAMP. The lognormal HAZUS fragility functions were converted to mean damage relationships for comparison to the preliminary URAMP mean damage estimates generated from DPMs developed by the Project Team. The final URAMP vulnerability models consist of damage probability matrices associated with each of the 108 URAMP facility classes.

The final URAMP pipeline damage assessment models are based on the approach developed by the American Lifelines Alliance (ALA, 2001). The ALA methodology classifies pipelines into vulnerability categories according to material and joint type, and estimates the number of repairs caused by ground shaking (in terms of peak ground velocity) and ground failure (in terms of the amount of permanent ground deformation).

Pipeline repairs can take the form of both leaks and breaks. The breakdown of repairs into leak repairs and break repairs can be estimated from the total predicted number of repairs with a commonly-used (e.g., NIBS/FEMA, 1999), simple "rule-of-thumb":
- Repairs caused by ground motion will be approximately 80% leaks and 20% breaks
- Repairs caused by ground failure will be approximately 20% leaks and 80% breaks.

In general, pipeline leaks diminish flow, but are not likely to be isolated to maintain water supplies for fire-fighting. Breaks are assumed to require isolation to prevent extensive water loss, and service is assumed to be restored once the breaks are repaired.

Functionality Models. Functionality models for URAMP (models that indicate the level of damage at which a facility is no longer operational) were developed in terms of the underlying damage states. For each facility type, a physical description of expected damage in each damage state is provided, as well as indication of the damage state for which the facility is no longer functional. An example is provided for steel tanks at grade in Table 2. In general, facilities (including steel tanks at grade) in the "moderate" damage state are expected to lose function.

Table 2: Damage State Descriptions for Steel Tanks at Grade

Damage State	Damage Description
Slight	Minor anchorage distress
Moderate	Connecting piping failure
Extensive	Shell buckling, severe roof damage
Complete	Tank bursts, not repairable

Replacement Cost Models. Component costs for repair, replacement and mitigation have been developed from published literature (e.g., NIBS/FEMA, 1999, ASCE, 1991), recent earthquake experience, internal project files, and project team expert opinion. Facility replacement cost models have been developed on a "per unit" basis, to facilitate automatic application to facilities of a variety of sizes. For example, replacement costs for steel tanks at grade is taken as $1.00 per gallon, based on typical costs for a 2 million gallon tank. In general, the user has the ability to override the URAMP default replacement cost when better information is available.

Mitigation Cost Models. Mitigation cost models for facility upgrades have been developed in terms of percent of total replacement cost for that facility (e.g., the cost to mitigate a steel tank at grade from Level 0 to Level 1 by installing flexible connections is 4% of the total replacement cost). As with replacement costs, the user can enter actual mitigation cost estimates in place of the default URAMP cost estimates. (Mitigation costs for full facility replacement are assumed to be the same as the replacement costs)

System Impact Assessment. The third phase in the analysis entails the estimation of system impacts. To facilitate the network analysis of damaged and undamaged networks, URAMP incorporates the functionality of EPANET, a widely used hydraulic modeling software package distributed by the U.S. Environmental Protection Agency (EPA). The application of EPANET within URAMP required development of methods for representing physical damage to the water system caused by the earthquake within the hydraulic model (see Huyck, et al., 2003 for more information).

Fire-Following Earthquake. The fourth element is the estimation of effects and losses associated with post-earthquake fire, and reduction in fire-fighting capabilities related to damage and outage in the water supply system. Fire-following earthquake (FFE) losses have been significant in several notable historic earthquakes, when fire-related losses greatly exceeded the cost of direct damage to buildings, including the 1906 San Francisco earthquake and the 1923 Kanto earthquake in Japan. A number of FFE loss-estimation methodologies exist, developed primarily for the insurance industry (Scawthorn, 1987). FFE losses are also included within FEMA's nationally applicable earthquake loss estimation methodology and HAZUS software (NIBS/FEMA, 1999).

Data required for implementation of a FFE loss assessment includes data on the local building inventory (exposed building area, construction material, and occupancy), fire-fighting resources (number and location of fire engines and crews), meteorological conditions (wind speed and direction), and estimated earthquake ground motions. Estimation of resulting fire damage (in the form of number of ignitions and estimates of the burned area and burned inventory), considers fire ignition, fire spread, and fire suppression. Determining the precise location of fire ignitions is a difficult task, with the resulting uncertainty often addressed either by random placement of postulated ignitions, or by multiple simulations.

The FFE loss estimation methodology consists of simplified models relating mean burnt area as a function of peak ground acceleration, building density (i.e., total building area per square mile for low, medium or high density residential, commercial or industrial construction), and water supply availability (determined from the results of the EPANET analysis). Required aggregate exposure databases for California have been developed on a census tract basis, and are provided within the URAMP software.

Economic Loss. The final step in the URAMP approach is the calculation of the various economic losses associated with earthquake damage to water, wastewater and drainage utility systems. In addition to losses associated with fire following earthquake (FFE) and direct damage to water, wastewater and drainage facilities, URAMP estimates economic losses associated with utility service interruption, business losses associated with both utility disruption and fire-following earthquake (FFE), and clean-up costs associated with sewage overflow. The economic loss estimation methodologies adopted within URAMP are based on models developed by the USACE for the water system evaluation studies performed in southern California (e.g., USACE, 1996).

Determine Cost-Effectiveness of Mitigation. In general, the determination of cost-effectiveness involves comparing the costs associated with mitigation (direct construction or installation costs, increased O&M costs, etc.) with the benefits or avoided losses. Depending on the model construct (annualized losses vs. deterministic or scenario-based losses), this comparison may take the form of a typical benefit/cost ratio, or an avoided cost comparison. The recent USACE studies in southern California have utilized an economic analysis based on avoided cost, which estimates potential damage and losses avoided through mitigation, assuming that a single seismic event has occurred. In this case, the computation is simple; avoided damage from the scenario earthquake is compared to the total cost of mitigation. Both the avoided loss and the benefit/cost ratio approaches have been incorporated in URAMP, allowing the user to determine avoided losses in individual scenario events, or determine the B/C ratio from an annualized risk assessment.

<u>Annualized Risk Assessment.</u> The net present value (NPV) of future benefits (i.e., the reduction in damage in future earthquake events due to mitigation or the projected difference between damage to the unmitigated facility and the mitigated facility) is

estimated according to FEMA's standard methodology (FEMA, 1992). In addition to direct repair cost savings, reductions in associated losses, such as utility revenue loss or fire-following earthquake loss attributable to seismic rehabilitation are also included. A time horizon of 50-years is assumed, reflecting the effective life of many engineered structures, and a discount rate of 7% is employed (although URAMP software does allow input of other time horizons and discount rates by the user). Because the seismic hazard model is time-independent, annual benefits are constant over time, simplifying the benefit/cost calculation. The final net present value equation is:

$$NPV = -INV + B\left[\frac{1-(1+i)^{-T}}{i}\right] \quad (1)$$

where *INV* is the investment, or the cost of the mitigation measure, *B* is the expected annual benefit (avoided loss) attributed to the rehabilitation, *T* is the time horizon, and *i* is the discount rate.

Conclusions

The URAMP Software and Methodology was developed under the guidance of a Steering Committee of potential users, and was successfully pilot tested for a small water system in Southern California. For that system, avoided FFE losses represented a significant portion of the benefits of mitigation. This is a likely scenario for many water distribution systems subjected to earthquakes. Future research should include additional pilot testing to determine the impact of various modeling assumptions to allow the developers to further enhance the product, making it more useful to the utility community.

In addition to its primary use for mitigation benefit/cost analysis, URAMP may be used to assess facility performance (to assess the earthquake performance of existing facilities, identify components critical to system performance, and identify candidate facilities for mitigation or replacement as part of the Utility's Capital Improvement Program) and to facilitate pre-disaster planning (stock-piling of critical components, arranging for required personnel and resources, and for earthquake exercises). These uses could also be demonstrated with additional real-world pilot studies.

References

ALA (2001), "Seismic Fragility Formulations for Water Systems: Part I – Guideline", prepared by the American Lifelines Alliance, a public-private partnership between the Federal Emergency Management Agency (FEMA) and the American Society of Civil Engineers (ASCE).

ASCE (1991), "Seismic Loss Estimates for a Hypothetical Water System – a Demonstration Project", Monograph No. 2, American Society of Civil Engineers,

Technical Council on Lifeline Earthquake Engineering, Water and Sewage and Seismic Risk Committees, New York, New York.

ATC (1985), "Earthquake Damage Evaluation Data for California", Applied Technology Council Report ATC-13, Redwood City, California.

Campbell, K.W. and H.A. Seligson (2003), "Quantitative Method for Developing Hazard-Consistent Earthquake Scenarios", Proceedings of the Sixth U.S. Conference on Lifeline Earthquake Engineering, American Society of Civil Engineers, Technical Council on Lifeline Earthquake Engineering (ASCE TCLEE), August 10 – 13, 2003, Long Beach, California.

FEMA (1992), A Benefit-Cost Model for the Seismic Rehabilitation of Buildings, FEMA-227, Federal Emergency Management Agency.

Huyck. C.K., R.T. Eguchi, R.M. Watkins, H.A. Seligson, S. Bucknam, and E. Bortugno (2003), "URAMP (Utilities Regional Assessment of Mitigation Priorities)– A Benefit-Cost Analysis Tool for Water, Wastewater and Drainage Utilities: Software Development", Proceedings of the Sixth U.S. Conference on Lifeline Earthquake Engineering, American Society of Civil Engineers, Technical Council on Lifeline Earthquake Engineering (ASCE TCLEE), August 10 – 13, 2003, Long Beach, California.

NIBS/FEMA (1999), HAZUS®99 Earthquake Loss Estimation Methodology, Service Release 1 (SR1) Technical Manual, Developed by the Federal Emergency Management Agency through agreements with the National Institute of Building Sciences, Washington, D.C.

NIBS/FEMA (2002), HAZUS®99 Earthquake Loss Estimation Methodology, Service Release 2 (SR2) Technical Manual, Developed by the Federal Emergency Management Agency through agreements with the National Institute of Building Sciences, Washington, D.C.

Scawthorn, C., 1987 Fire Following Earthquake - Estimates of the Conflagration Risk to Insured Property in Greater Los Angeles and San Francisco. Prepared for the All-Industry Research Advisory Council, Oak Park IL.

USACE (1996), Final Special Study Report: Southeast Los Angeles County Water Conservation and Supply Study, prepared for the U.S. Army Corps of Engineers, Los Angeles District by CH2M Hill.

Standard Guidelines to Assess the Seismic Fragility of Water Transmission Systems

Ronald T. Eguchi[1]
Douglas G. Honegger[2]

Abstract

In an effort to reduce the vulnerability of our nation's water supply systems in future earthquakes, ASTM, with the support of the Federal Emergency Management Agency and the American Lifelines Alliance, has established Subcommittee 36.30 to create standard guidelines to assess the seismic fragility of water transmission systems.

Introduction

Recent earthquakes have underscored the need to address the vulnerability of our nation's lifelines, that is, those systems necessary to support post-earthquake response and recovery activities. For example, the latest estimates of lifeline damage and loss as a result of the 1994 Northridge Earthquake are in excess of $2 billion. While this amount may appear low relative to other types of losses, namely damage to buildings, it only reflects those costs associated with the repair of damaged systems. Other costs which may more accurately reflect the impact of damaged or inoperable systems, such as business losses due to lifeline disruption, may be several factors higher than these repair costs. Also, it must be recognized that the Northridge earthquake was a moderate-sized event and that the Los Angeles area is capable of generating earthquakes of much larger magnitude. Therefore, the relatively good performance of lifelines compared to buildings in the Northridge earthquake should not promote complacency in current or existing design measures for lifelines systems.

---

[1]President and CEO, ImageCat, Inc., 400 Oceangate, Suite 1050, Long Beach, CA 90802; rte@imagecatinc.com
[2]President, D.G. Honegger Consulting, 2690 Shetland Place, Arroyo Grande, CA 93420-5351; dghconsult@aol.com

The engineering community has long recognized the importance of lifeline systems during disasters. After the 1971 San Fernando earthquake – an event that caused catastrophic damage to virtually every type of lifeline – many efforts were launched to better understand the cause of these failures and ways to mitigate future earthquake damage. Lifeline damage in the San Fernando earthquake was the main catalyst in prompting the American Society of Civil Engineers to establish the Technical Council on Lifeline Earthquake Engineering (TCLEE) in 1974. The goal of TCLEE is to advance the state-of-the-art and practice of lifeline earthquake engineering in the U.S. Since its formation, TCLEE has sponsored six national conferences (including this one), published numerous monographs and reports, and participated in over a dozen reconnaissance investigations of earthquakes worldwide. TCLEE continues to serve as an important forum for discussing all issues that affect the seismic design, construction, operation and maintenance of lifeline systems. Table 1 shows other important milestones or events in the history of lifeline earthquake engineering in the United States.

The American Lifelines Alliance

Despite the efforts of TCLEE and other organizations with similar goals and objectives, we have yet to develop and implement consensus standards and guidelines that will dramatically improve the performance of existing utility and transportation systems in large earthquake events. This need for uniform guidance has been repeatedly supported in numerous workshops held over the last 10 years and was explicitly identified in the 1990 reauthorization of the National Earthquake Hazards Reduction Program. In 1998, the Federal Emergency Management Agency and the American Society of Civil Engineers (ASCE) entered into a cooperative agreement to establish the American Lifelines Alliance (ALA), a public-private partnership project whose goal is to reduce risks to lifeline systems from natural hazards. ALA's charter is to facilitate the "creation, adoption and implementation of design and retrofit guidelines and other national consensus documents that, when implemented by lifeline owners and operators, will systematically improve the performance of utility and transportation systems to acceptable levels in natural hazard events, including earthquakes." The general approach adopted by ALA is to fund projects that lead to the generation of draft guidelines and standards that allow further development into national consensus standards by established standard developing organizations accredited by the American National Standards Institute.

One of the first guideline products sponsored by ALA addressed uniform methods for estimating the seismic damage to components of water conveyance systems. While seismic measures have generally been considered by most of the larger west coast water utilities, they have yet to be implemented on a regular basis in other parts of the country. Furthermore, what measures are being implemented are usually done so on an uneven or inconsistent basis. One reason for inconsistent implementation of seismic mitigation measures is the lack of a consistent set of standards or guidelines that quantify the level of seismic vulnerability of water system components. The goal of ALA is to facilitate a set of consensus-based guidelines to allow water utilities and others to gauge the relative exposure to seismic damage and the need to implement retrofit strategies for various regions of the United States.

An ALA project completed in 2001 was undertaken by a team of practicing engineers, water utility personnel, and academics under contract to G&E Engineering Systems, Inc. The product of this ALA project was *Seismic Fragility Formulations for Water Systems*, a report that has been available online on the ALA website, www.americanlifelinesalliance.org. This report covers six different water system components: aqueducts, distribution pipelines, storage tanks, tunnels, canals, and valves and SCADA system components. In addition, the report provides special guidance on how to evaluate or assess seismic vulnerabilities based on different hazard effects, i.e., fault offsets, liquefaction and lateral spread, landslides, and strong ground shaking.

Key Role for ASTM

To help facilitate translation of the ALA report into a national consensus standard, ALA sought out an appropriate standard developing organization to further refine the concepts in the ALA report and was pleased when ASTM contacted ALA with an offer of support. In late 2001, an official ASTM activity entitled "Seismic Fragility of Water Transmission Systems," was begun to define standardized methods and procedures for assessing the seismic vulnerability of a broad range of water system components. This activity – formally referred to as F36.30 – falls under ASTM Committee F36 on Technology and Underground Utilities. In total, 15 experts in the water and earthquake areas comprise this subcommittee. The Chair of this subcommittee is Ronald T. Eguchi, of ImageCat, Inc. in Long Beach, California. Also playing a major role in this initial standards development activity is Douglas G. Honegger, a consultant in Arroyo Grande, California, who is currently overseeing the technical activities of ALA.

Sample Fragility Model

Figure 1 shows an example of the type of approach the ALA report uses to quantify the seismic vulnerability of a major water system component. In this case, the component is an anchored, steel water tank. The schematic, often referred to as a *Fault Tree*, identifies several tank damage states and the failure mechanisms that can lead to these states. For example, Damage State 4 describes a condition where the tank is damaged, requires major repairs and is functionally out of service. The figure shows those failure mechanisms that are likely to lead to Damage State 4, i.e., damage to inlet and outlet pipes, tank wall uplifts leading to leakage, "elephant foot" buckling leading to tank leakage, and hoop overstress to the tank wall. Experiencing any of these failures will result in the tank being shut down for repairs. The benefit of using a fault tree approach is that it allows the engineer to break down the assessment of vulnerability into a series of steps or calculations that can be easily combined at the end to arrive at a failure rate or damage state probability.

Figure 2 shows a set of fragility curves for an anchored, steel water tank, as presented by the ALA report. The ordinate is measured in terms of probability; the abscissa indicates severity of ground shaking in terms of peak ground acceleration, or PGA. Four damage states are specified in the figure: minor, moderate, extensive and complete. By having such curves, an engineer is able to estimate the probability or likelihood of surviving given different levels of ground motion. This information can be used in selecting the appropriate design specifications of a new tank or deciding whether an existing tank should be upgraded or not. When combined with economic information or data, these probabilities can also be translated into expected repair costs over the life of the tank.

F36.30 Subcommittee

Over the next year, the F36.30 subcommittee will be reviewing the ALA report, *Seismic Fragility Formulations for Water Systems*, as well as other relevant documents and preparing a set of standard guidelines for assessing the seismic fragility of water transmission systems. Some of the key questions and issues that will be addressed by the F36.30 subcommittee, particularly as they relate to the ALA report, are:

- Were the reviews that were conducted in the development of this report comprehensive enough, i.e., are there reports, data or studies missing?
- Are the models in the report technically defensible?
- Are there technical areas where improvements are needed given the existing data?
- Are data needed in order to use or implement these models clearly defined?
- Are the models easily understood by 1) water agency personnel, 2) consultants, and 3) regulatory officials?
- Are the models transferable to 1) different users, and 2) different regions of the country?
- Are the models flexible enough to incorporate new data or datasets?

When completed, these guidelines will allow engineers to use a common set of tools and databases to determine what the seismic design level should be for new and existing water facilities

Summary

So what are the benefits in establishing standard guidelines for assessing the seismic fragility of large water supply systems? First, these guidelines will provide a uniform and consistent means to quantify the expected seismic performance of major water transmission system components across the nation. We know from past experience that failure of key components can shut down an entire water system. If we are to be successful in ensuring the continuity of service after a major earthquake event, we must identify critical components within the system, assess their likelihood of failure in a series of events, and either design or retrofit these facilities to perform to some acceptable level of performance. Second, knowing the expected performance for all critical water system components is essential to performing a systems analysis to assess the level of service that can be provided by the water system following an earthquake. Third, understanding the improvements in component performance that can be gained by implementing mitigation measures, is important to properly prioritize investments in water system upgrades that will help to assure that adequate water supplies are available to fight post-earthquake fires and minimize health risks. Finally, establishing a set of standard guidelines to evaluate the seismic fragility of water systems will provide a common framework for a comparison of the relative level of effort undertaken to assess potential water system damage by utilities across the country. Only when we have such standards in place can we understand where resources should be focused to assure the reliability of our nation's water supply systems in major earthquakes.

Acknowledgements

The authors would like to thank Drew Azzara of ASTM for his help in organizing and supporting the activities of this subcommittee. In addition, we would like to acknowledge the financial support of the American Lifelines Alliance and the Federal Emergency Management Agency. We would also like to acknowledge the efforts of the members of Subcommittee 36.30, all of whom are participating on a voluntary basis.

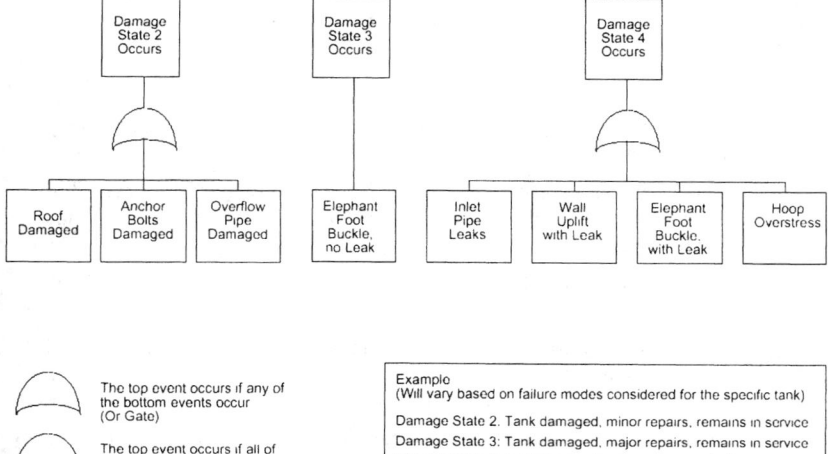

Figure 1. Example Fault Trees for Evaluation of an Anchored Steel Tank
(Source: G&E Engineering Systems, Inc., 2001)

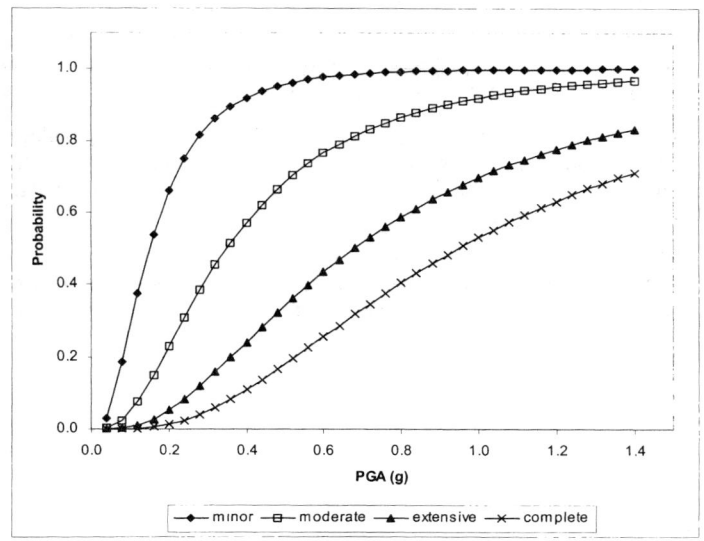

Figure 2. Fragility Model for an Anchored, Steel Water Tank
(Source: G&E Engineering Systems, Inc., 2001)

TABLE 1. Brief Chronology of Lifeline Earthquake Engineering in the U.S

Year	Event/Milestone
1971	San Fernando Earthquake (M6.4) causes catastrophic damage to every lifeline system
1974	The Technical Council on Lifeline Earthquake Engineering formed to address issues regarding the state-of-the-art and practice of lifeline earthquake engineering in the U.S.
1977	National Earthquake Hazards Reduction Program established by Congress
1989	Loma Prieta Earthquake (M7.1) reaffirms the need to assess and improve seismic design and construction standards for all lifeline systems
1990	Public Law 101-614 (Reauthorization of the National Earthquake Hazards Reduction Program) requires FEMA and NIST to submit to Congress a plan for developing and adopting seismic design and construction standards for all lifelines.
1991	NIST conducts workshop on developing and adopting seismic design and construction standards for lifelines (NISTIR 5907)
1994	Northridge Earthquake causes damage to some lifeline systems; significant improvement in performance from previous earthquakes
1996	FEMA Plan for developing and adopting seismic design guidelines and standards for lifelines published (FEMA 271)
1997	Interagency Committee on Seismic Safety and Construction conducts lifeline policymakers workshop to develop recommendations for developing national guidelines for lifelines (ICSSC TR-19/NISTIR 6085)
1998	The American Lifelines Alliance project is initiated under a cooperative agreement between FEMA and ASCE with the goal of achieving the major objectives identified in the 1996 FEMA Plan.

TABLE 2. Subcommittee F36.30 Members

Committee Member	Organization
Ballantyne, Donald B.	ABS Consulting
Davis, Craig	City of Los Angeles
Edwards, Curtis L.	Pountney & Associates
Eguchi, Ronald T., Chair	ImageCat, Inc.
Eidinger, John	G & E Engineering Systems
Elliott, William	Elliott Consultants, LLC
Ford, Duane B.	D.B. Ford & Associates
Goodson, Mary	CH2M Hill
Honegger, Douglas G.	D.G. Honegger Consulting
Lew, Desmond H.	City of Los Angeles
Niles, Harry O.	Ductile Iron Pipe Research Association
Nisar, Ahmed	MMI Engineering, Inc.
O'Rourke, Michael	Rensselaer Polytechnic Institute
Power, Maurice S.	Geomatrix Consultants
Taylor, Craig E.	Natural Hazards Management, Inc.

Comparison of Mitigation Alternatives for Water Distribution Pipelines Installed in Liquefiable Soils

Donald Ballantyne[1]
William Heubach[2]

Abstract

Water pipeline distribution system mitigation measures are being evaluated: pipeline replacement, automated pipeline control systems, and planned manual valve actuation response. Seattle Public Utility's distribution system is being used as model for this evaluation. The most vulnerable part of the distribution system is located in the highly liquefiable Duwamish River Valley.

A decade ago, it was estimated that it would cost in excess of one billion dollars to replace all of the cast iron pipe founded in liquefiable soils in Seattle's system. This cost was too expensive particularly when considering the other financial demands confronting the utility such as water treatment and reservoir upgrades.

Seattle has joined forces with other water utilities in seismically active parts of the world to fund research under the American Water Works Association Research Foundation to evaluate alternate, less expensive, means of mitigating pipeline damage to improve the post-earthquake functionality of the system.

The concern is that many pipelines will break, draining water from storage reservoirs, and also affect areas without significant pipeline damage. Following the Kobe earthquake, as well as many other earthquakes, this scenario has resulted in loss of water for fire suppression. Damage in the 2001 Nisqually Earthquake, as well as similar events in 1965 and 1949 was limited to less than 40 failures. However, a Cascadia subduction or Seattle Fault event will be much more damaging, and Seattle wants to be ready.

The general strategy is to quickly isolate the damaged sections of the distribution network, or to replace pipe in the network so it will not fail. For each of these options, scenarios are being developed estimating pipeline performance and the resulting network hydraulic performance. The estimated costs and expected losses are being developed for a range of mitigation options.

---

[1] Donald Ballantyne, General Manager and VP Lifeline Services, ABS Consulting, 1411 4th Ave., #500, Seattle, Washington; 206-226-7496;
dballantyne@absconsulting.com
2 William Heubach, Senior Civil Engineer, Seattle Public Utilities, Key Tower, 700 Fifth Avenue, # 4900, Seattle, Washington 98104-5004; (206) 386-1389;
Bill.Heubach@seattle.gov

Introduction

This paper presents a project to develop an estimate of the damage from a future earthquakes and develop a cost effective mitigation strategy to minimize service disruption.. The project focuses on performance of the water distribution pipeline system in Seattle, Washington.

One of the most significant issues following an earthquake is fire. Fires can be ignited by electrical shorts and fueled by natural gas leaks. It is a critical time for water to be available for fire suppression. Water systems have performed poorly in past earthquakes. In the 1989 Loma Prieta, 1994 Northridge and 1995 Kobe earthquakes, pipeline failures have been the primary reason for loss of water system pressure.

Regionally there were approximately 40 pipelines failures resulting from the 2001 magnitude 6.8 Nisqually Earthquake, about half of which of which occurred in the Seattle water system. In this "moderate" earthquake, loss of service was limited to about 40 customers. Similar damage occurred in Seattle region earthquakes in 1949 and 1965. Hazard and pipeline damage data from these three events was collected.

In larger events such as the 1995 Kobe Earthquake, and the 1994 Northridge Earthquake where there were in excess of 1,000 failures each, the systems drained and could not provide adequate water for fire suppression. In the 1994 Northridge Earthquake, approximately two-thirds of the San Fernando Valley water system drained. In 1995 Kobe Earthquake, the entire system serving the urban area drained within six hours. Post-earthquake performance and pipeline damage was gathered obtained for these events to augment information from the Seattle earthquakes.

Methods have been developed and applied estimating expected damage from selected earthquakes, such as to the Seattle water system (Ballantyne et al, 1991). These methods are dependent on an accurate mapping representation of geologic hazards. In the Seattle case, it appears that there may be subtle features that may have resulted in pipeline damage that are not evident in geologic hazard mapping. Even if we have an understanding of the expected level of pipeline damage, mitigation can be expensive. One mitigation approach is to replace all cast iron pipe buried in soils susceptible to liquefaction, with restrained joint ductile iron. However, such an approach is prohibitively expensive.

Isolation of heavily damaged pipeline, alternate system control strategies, and development of emergency plans for pipeline repair offer a promising ways to mitigate damage at more reasonable costs than pipeline replacement. There are many issues that still need to be resolved such as isolation and control strategy development, including inadvertently isolating an area from needed fire-suppression water, and selection of appropriate hardware.

Seismicity and Ground Motions

The Seattle region is vulnerable to earthquake from three different source zones (see Figure 1): A Benioff Intraplate Zone, a Benioff Interplate Zone and a shallow fault zone. The three significant historic earthquakes, 1949, magnitude 7.1 Olympia; 1965, magnitude 6.5 Seattle-Tacoma; 2001, magnitude 6.8 Nisqually, were deep (50 to 60 kilometers) Benioff Intraplate Zone events. These events occur directly below the Puget Sound region, can recur as often as every 25 to 35 years and may produce firm ground accelerations as high as 0.3g in epicentral areas.

A Benioff Interplate Zone extends from the Northern California coast to the Southern British Columbia coast. These Cascadia interplate subduction earthquakes may exceed Magnitude 9.0. However, because the epicenter would be located further away from the Puget Sound than intraplate events, peak firm soil ground accelerations are not expected to exceed 0.3g in the Puget Sound region. However, strong ground motion from these events may last more than two minutes and will likely trigger extensive liquefaction. The last Cascadia Subduction event occurred in 1700, and has an average recurrence of 500 to 600 years.

There are also a series of active shallow faults in the Puget Sound region. The Seattle Fault runs east-west through Seattle and passes within a kilometer of the southern edge of downtown Seattle. A Seattle Fault event, potentially of Magnitude 6.5 or 7.0, is thought to have occurred about 1,100 years ago with a recurrence interval as short as 700 years. A "moderate" Seattle Fault event is expected to produce PGA's on the order of two-thirds times gravity on firm soils.

Project Approach

The following steps were used to develop the mitigation strategies:

1. Map peak ground velocity and permanent ground acceleration during the 1949, 1965 and 2001 Puget Sound earthquakes.

2. Document and map the pipeline damage that occurred during these earthquakes.

3. Compare the peak ground velocity and permanent ground displacement maps with the pipeline damage maps to verify pipeline vulnerability assumptions.

4. Modify pipeline vulnerability models, as necessary to reflect the findings in Step 3.

5. Prepare GIS-Based maps that predict peak ground velocity and permanent ground displacement for design-level ground motions (in this case, 10% probability of exceedance in 50 years).

6. Estimate the number and location of pipe breaks from the design-level earthquake.

7. Use hydraulic models to evaluate the system effect of the pipeline damage.

8. Identify mitigation measures to reduce damage effects on system operation.

Peak Ground Velocity and Permanent Ground Displacement Hazards

Ground motion mapping from the 1949 and 1965 events is limited. We used Modified Mercalli Index (MMI) data to estimate ground motions for pipeline damage in those events. In the 2001 Nisqually Earthquake, better PGA mapping was available due to the much higher density of ground motion instruments. However, Peak Ground Velocity (PGV) mapping was still too course to be of much value. PGV is the preferred ground motion intensity parameter used to evaluate pipeline damage due to wave propagation.

Liquefaction and associated lateral spreading and landslides are both considered to be forms of permanent ground deformation (PGD). PGD results in significantly higher pipeline unit failure rates than wave propagation, so it becomes an important parameter when evaluating pipeline damage.

Detailed liquefaction potential maps were developed by Shannon & Wilson, under subcontract to the U.S. Geological Survey for the City of Seattle, which have subsequently been published by the USGS (Grant, et al., 1998). The maps are based on an extensive subsurface boring database and more quantified liquefaction analysis of the data in the database. The hazard mapping and studies identify a large portion of south Seattle (the Duwamish River Valley) as underlain by extensive deposits of sandy Holocene alluvium and hydraulically placed fill that have a high susceptibility to liquefaction (see Figure 1).

Reports of ground-deformation related damages to pipelines are known for the 1949 and 1965 events (Chleborad and Schuster, 1998). Reconnaissance work by Shannon & Wilson personnel and others have identified areas of liquefaction and ground deformations as a result of the recent 2001 Nisqually Earthquake. Liquefaction and lateral spreading occurred primarily in hydraulic fill and Holocene alluvium in the Seattle area during all three events. The majority of reported instances of liquefaction and permanent ground deformations occurred in the filled tideflats in and around the port areas in Seattle. During the 2001 Nisqually earthquake, liquefaction and lateral spreading was again observed in the hydraulic fills in the port and filled tideflats area in Seattle and farther south in the Duwamish Valley, particularly in the vicinity of Boeing Field. This information was used to correlate the cause of pipeline damage for these three events.

The City of Seattle and the US Geological survey collaborated to map areas susceptible to landslides. Historically, these landslides are activated during extended periods of rainfall. However, they did not appear to be a significant cause of pipeline failures in the three historic Seattle earthquakes. It should be noted that 2001 was a particularly dry year that may have limited the number of landslides that occurred in the Nisqually Earthquake.

Historic Pipeline Damage

Pipeline damage for the three earthquakes was gathered for analysis (refer to Figure 1). As expected, many of the pipeline failures occurred in liquefaction zones along the Duwamish River. However, there were clusters of failures in other areas where soils are not liquefiable. Refer to locations A, B, and C on Figure 1. It is known that the topography and geology of the Seattle Basin amplify ground motions in some of the hilly areas with firm soils adjacent to liquefiable lowlands. iAt this time it is unclear whether these clusters of failures occurred due to an area of one type of particularly vulnerable pipe, ground motion amplifications, a result of unidentified/unmapped geotechnically vulnerable deposits or a combination of these factors..

Pipeline Seismic Vulnerability Model Development

Although mechanical pipeline and soil models may be practical for site-specific studies, it is not practical to use theoretical analysis techniques to assess the seismic vulnerability of large pipeline networks. Researchers have developed relationships between pipeline damage and various seismic and geotechnical parameters using empirical data. The most commonly used are relationships between shaking intensity (PGA or PGV) and failures per unit length, and permanent ground deformation (PGD) and failures per unit length. Eguchi developed damage relationships for shaking in the early 1980's. Relationships for PGD were developed for application to the San Francisco water system after the Loma Prieta earthquake (Harding and Lawson, 1991). These relationships are sometimes used in GIS-based computer models that are used to estimate the post-earthquake damage state of water system following an earthquake.

The accuracy of these models are all limited by relatively small data sets, accuracy of the data, data interpretation by the researchers, and the inherent variability of pipeline condition and construction quality. Perhaps the best empirical data and pipe damage databases, are from the Northridge Earthquake and the Kobe Earthquake (T. O'Rourke, et al, 1996; Shirozu, et al, 1996; Takada, et al, 1996). This data, and the damage relationships developed from this data were used to augment the information form the three Seattle events.

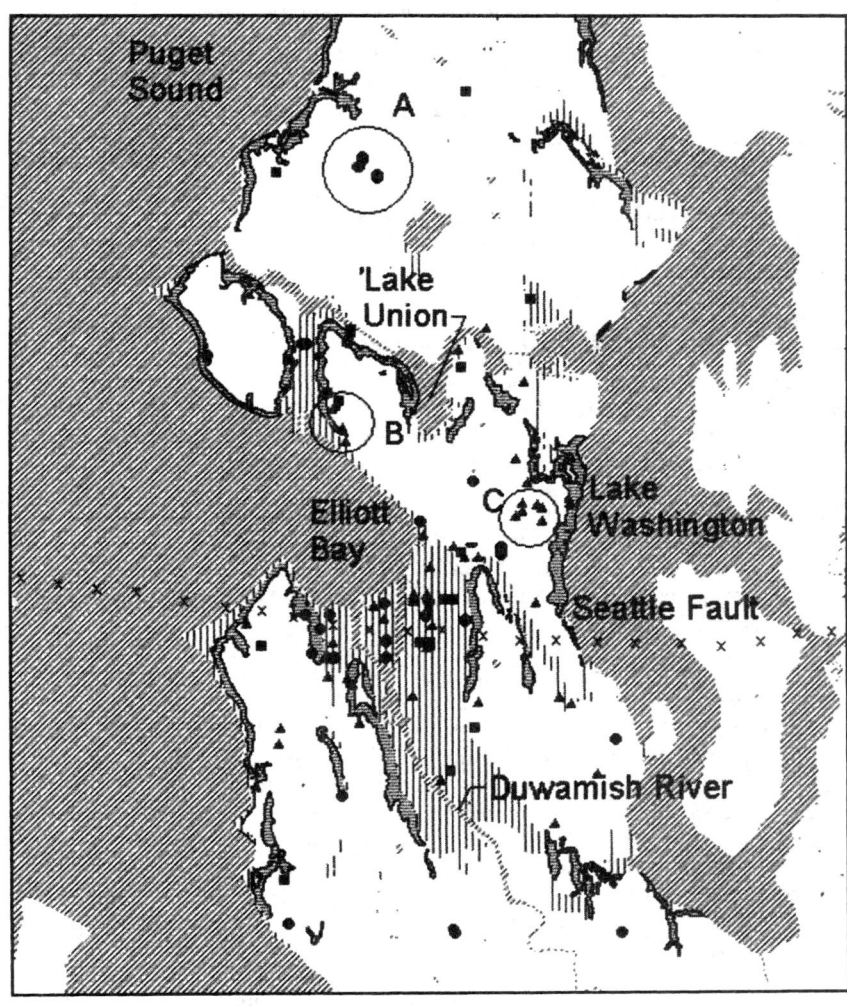

Figure 1. 1949 (triangles), 1965 (circles) and 2001 (squares) earthquake pipeline damage overlaid on liquefaction (vertical hatching) and landslide (horizontal hatching) susceptible areas. Pipeline failure clusters A, B and C are outside PGD-susceptible areas.

Peak Ground Velocity and Permanent Ground Displacement Maps

Zipper-Zeman Associates prepared ground hazard maps for 10% probability of exceedance in 50 year ground motions. These maps were developed from supplementing USGS geologic hazard maps with soil boring data and from USGS ground shaking intensity maps. Peak ground velocity was computed by propagating the bedrock ground motions up through the soil column. Liquefaction-induced ground deformation was computed using the models developed by Bardet (Regional Modeling of Liquefaction-Induced Ground Deformation, Jean-Pierre Bardet, et al, Earthquake Spectra, February, 2002).

Estimate Operational State in Hours Following Event

Immediate post earthquake functionality of the water system is important to provide water for fire suppression. The operational state was modeled using a hydraulic network analytical tool, EPANET. The model results were supplemented with performance of water pipeline networks in previous earthquakes.

The most relevant information was provided by representatives of the Los Angeles Department of Water and Power and the Kobe Water Department for the 1994 Northridge and 1995 Kobe earthquakes, respectively. They provided inside into the failures that limited system functionality, and the potential for use of the strategies considered in this project.

For example, in the Northridge Earthquake about 2/3 of the San Fernando Valley (West Valley) was without water in the hours following the earthquake. This was not due to distribution pipe failures, but to transmission line failures primarily in and just downstream of the Van Norman Complex. Every pumped transmission line leaving the Complex had damage. The was compounded because the Los Angeles Aqueduct also went down. The trunk lines leaked but stayed in service. The result was problems with the pumped zones on the southern edge of the Valley, the opposite side from the source/treatment plant. Isolation of "damaged" sections of pipe would not have helped. Further, leaking water is critical for locating leaks. If the damaged lines are isolated (shut down), it would be nearly impossible to locate the leaks (interview with Marty Adams, LADWP).

One strategy that worked in Kobe was to place seismic-actuated shutoff valves on storage tanks that operated in pairs. The strategy was to allow water for fire-fighting also save some water for post-earthquake use in the event of extensive pipeline failures. Although unvalved reservoirs did drain quickly after the earthquake, water in approximately 30 valved reservoirs was saved and used as a drinking water supply after the earthquake.

Mitigation Strategy

The relative effectiveness of pipeline upgrade, isolation and control strategies, and emergency response to reduce the effect of pipeline failures on water system functionality is being examined. We are developing plans for: 1) pipeline replacement, 2) control valve implementation, and 3) emergency response. Specific hardware and strategies are being identified for typical water system configurations, and ranges of costs for the three alternatives provided.

The pipeline replacement program would propose replacement for all vulnerable pipe (e.g. – all cast iron pipe in liquefiable areas or at least increasing the replacement priority), and as an alternative, replacement of all backbone pipe in the same areas. Additional appurtenances may also be required if only the backbone was upgraded. Pipe replacement would minimize the recovery time.

Another potential option may be to use more seismic-resistant pipe for new and replaced lines. This option could range from always using already available restrained joint ductile iron pipe to exploring the feasibility of importing the Japanese S-Joint which has performed extraordinarily well in earthquakes. The S-Joint permits both rotation and axial movement beyond the normal construction tolerances (see Figure 2).

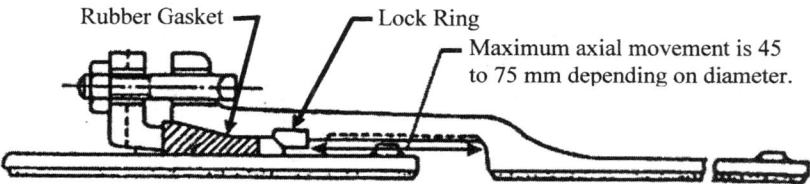

Figure 2. Japanese S-II Joint (100-450 mm Diameter) permits both rotation and axial movement.

Control valve implementation could include of a system to isolate vulnerable areas of the distribution system to keep the system from draining. The control valve system may be significantly less expensive to implement and would provide for water for fire suppression, but would still require a significant recovery period. The emergency response plan would require no capital improvements (stockpiling of spare parts and flexible hose that could be used to temporarily span broken mains is an option), and rely on manual operation of valves following the earthquake. Such a system may be too slow to effectively maintain system operation. The control valve system alternative would address immediate post-earthquake system operation, but would not help in reducing the overall system recovery time.

The envisioned control valve system could make use of pressure reducing valves (PRVs) that feed water into the pressure zone. PRVs could be operated by installing a solenoid valve on the control loop. The decision to close could be based on an earthquake ground motion threshold (e.g. – 20% g), excess flow, rate of change of flow, etc. Each control valve could close automatically or could require interaction by an operator.

Summary

This paper discusses an ongoing Seattle Public Utilities project that is evaluating alternative methods for mitigation of earthquake damage to the pipeline distribution system. Empirical data from three historic earthquakes are used as input for the analysis. Mitigation alternatives include wholesale replacement of the pipe and/or incorporating seismic factors in pipeline replacement prioritization, automated system control to allow quick isolation of damaged areas in the system, use of more earthquake-resistant pipe joints such as the Japanese S-Joint, or application of an emergency plan where response staff would evaluate and isolate damaged system components as appropriate. The project is scheduled for completion by the end of July, 2003

Acknowledgements

The authors thank the American Water Works Association Research Foundation for their interest and financial support of the project. We thank James Doane, David Lee, and Charles Pickel, members of the AWWARF Project Advisory Committee, for their direction. Further, the authors would like to express appreciation for the participating utilities that also provided financial support for the project: Seattle Public Utilities, Washington; City of Everett, Washington; Greater Vancouver Water District, British Columbia, Canada; Los Angeles Department of Water and Power, California; San Francisco Public Utilities, City of St. Louis, Missouri; Tacoma Public Utilities, Washington; and the Thames Water Company, Great Britain.

References

Ballantyne, D.B.; Taylor, C.; 1990; *Earthquake Loss Estimation Modeling of the Seattle Water System*, USGS Grant Award 14-08-0001-G1526, Kennedy/Jenks/Chilton Report No. 886005.00, Federal Way, Washington.

Chleborad, A.F., and Schuster, R.L., 1998, Ground failure associated with the Puget Sound region earthquakes of April 13, 1949, and April 29, 1965, chap. earthquake hazards *of* Rogers, A.M., Walsh, T.J., Kockelman, W.J., and Priest, G.R., Assessing Earthquake Hazards and Reducing Risks in the Pacific Northwest: U.S. Geological Survey Professional Paper 1560, v. 2, p. 373-439.

Grant, W.P., Perkins, W.J., and Youd, T.L., 1998, Evaluation of liquefaction potential in Seattle, Washington, chap. earthquake hazards *of* Rogers, A.M., Walsh, T.J.,

Kockelman, W.J., and Priest, G.R., Assessing Earthquake Hazards and Reducing Risks in the Pacific Northwest: U.S. Geological Survey Professional Paper 1560, v. 2, p. 373-439.

O'Rourke, T.D, S. Toprak, and Y. Sano, "Los Angeles Water Pipeline System Response to the 1994 Northridge Earthquake", *Proceedings from the Sixth Japan-U.S. Workshop on Earthquake Resistant Design of Lifeline Facilities and Countermeasures Against Soil Liquefaction*, edited by Hamada and T. O'Rourke, National Center for Earthquake Engineering Research, Buffalo, N.Y., 1996

Shirozu, T., S. Yune, R. Isoyama, and T. Iwamoto, "Report on the Damage to Water Distribution Pipes Caused by the 1995 Hyogo-ken-Nanbu (Kobe) Earthquake", *Proceedings from the Sixth Japan-U.S. Workshop on Earthquake Resistant Design of Lifeline Facilities and Countermeasures Against Soil Liquefaction*, edited by Hamada and T. O'Rourke, National Center for Earthquake Engineering Research, Buffalo, N.Y., 1996

Takada, S., J. Ueno, and S. Goto, "Damage Features of Buried Water Pipelines Related to Active Fault Geography During the 1995 Kobe Earthquake", *Proceedings from the Sixth Japan-U.S. Workshop on Earthquake Resistant Design of Lifeline Facilities and Countermeasures Against Soil Liquefaction,* edited by Hamada and T. O'Rourke, National Center for Earthquake Engineering Research, Buffalo, N.Y., 1996.

PERFORMANCE OF THE TRANS-ALASKA PIPELINE IN THE NOVEMBER 3, 2002 DENALI FAULT EARTHQUAKE

William J. Hall[1], Douglas J. Nyman[2], Elden R. Johnson[3], and J. David Norton[4]

Abstract

The magnitude 7.9 earthquake that occurred in south-central Alaska on November 3, 2002 ruptured a 336-km long segment of the Denali Fault. The epicenter was located about 88 km west of the Trans-Alaska Pipeline, and the rupture propagated to the east across the pipeline right-of-way. The above-ground segments of the pipeline were subjected to violent near-fault ground shaking approaching or exceeding design criteria, and liquefaction was observed at a number of locations along the pipeline, including a remote gate valve location. The performance of the pipeline was in line with original project design requirements, and there was no oil leakage. The paper presents a high-level overview of the seismic design of the Trans-Alaska Pipeline, performance of the pipeline system during the magnitude 7.9 event, and a brief commentary on post-event emergency response.

Introduction

The Atlantic Richfield Company discovered oil at Prudhoe Bay, on the Alaskan North Slope, in 1968. Construction of the Trans-Alaska Pipeline System (TAPS) was proposed in early 1969, but controversies over Alaska native land claims and environmental issues delayed pipeline construction until 1974 (Roscow, 1977). On June 20, 1977, after three years of construction by 70,000 men and women and an investment of eight billion dollars, oil began flowing through the pipeline. Since then, this pipeline system has safely transported over 14 billion barrels of oil from Alaska's North Slope to the Port of Valdez. Currently the pipeline transports approximately 17 percent of the crude oil produced in the United States. In 2003, the pipeline owners received state and federal right-of-way approval for another 30 years of operation. The pipeline is operated by Alyeska Pipeline Service Co (Alyeska) for its owners.

Over its 1,287-km (800-mile) route from Prudhoe Bay to a marine tanker terminal at the port of Valdez, the 1,219-mm (48-inch) diameter Trans-Alaska Pipeline passes through areas with high potential for significant seismic activity. To safeguard the fragile arctic environment, major seismic design requirements were imposed on the TAPS, namely that the entire pipeline system should be capable of withstanding all reasonably anticipated effects of earthquakes without impairing the structural integrity of the oil pipeline or the associated pressure containing system components.

---

[1] Professor Emeritus of Civil Engineering, University of Illinois, Urbana, IL, H.M. ASCE.
[2] Consulting Engineer, D.J. Nyman & Associates, Houston, TX, F. ASCE.
[3] Engineering Advisor, Alyeska Pipeline Service Co., Fairbanks, AK, M. ASCE.
[4] Principal, J.D. Norton and Associates, Anchorage, AK, M. ASCE.

With the exception of nuclear power plants, the attention given to the seismic design of TAPS rivals that for any other critical facility in the United States.

For many aspects of TAPS design, just as in the case of nuclear power plants, no seismic criteria, standards or codes existed at the time of design. Alyeska's adoption of seismic criteria for pipeline design was the first major action of this kind in the pipeline industry. Even so, during that time period, TAPS design and construction was benefited immensely by the research and development under way for seismic design of nuclear power plant facilities. Consequently, the design of TAPS was state-of-the-art for its time and has remained remarkably consistent with current practice in earthquake engineering.

The attention given to seismic design paid off on November 3, 2002, with the occurrence of the magnitude 7.9 Denali Fault earthquake. Ground motions approached the seismic design criteria for the section of the Trans-Alaska Pipeline passing through the Alaska Mountain Range in the vicinity of Pump Station 10 and the Denali fault. The surface rupture passed across the Trans-Alaska pipeline right-of-way, producing approximately 5.5 m of right-lateral offset and 0.6 m of up-to-the-north vertical displacement. There was no damage to the pipeline or release of crude oil; but there was incidental damage to the above-ground pipeline support hardware where violent pipe shaking and ground motion apparently took place. Limited displacement of the below-ground pipeline occurred in liquefaction areas, but no damage to the pipe occurred as verified by an in-line inspection device ("smart pig").

Seismic Hazard along the TAPS Route

The pipeline route begins in a region of low seismicity on the North Slope of Alaska and terminates at Valdez, which is one of the most seismically active regions of the world owing to its location along the Pacific Rim. In 1964 one of the largest earthquakes ever recorded in North America, with a moment magnitude of 9.2, occurred in Prince William Sound, Alaska. The epicenter of this earthquake was about 65 km west of the yet-to-be-built Valdez Marine Terminal (VMT). The city of Valdez and its harbor facilities, at that time at the eastern end of the Valdez Arm, were severely damaged by wave run-up caused by a massive submarine landslide. The city of Valdez was subsequently relocated to a site on the north shore of the Valdez Arm that offers natural protection against a similar occurrence. Prior to the November 3, 2002 Denali Fault earthquake, the largest earthquakes near the pipeline corridor were of modest magnitude and at significant distances from the TAPS route. There had been no reports of earthquake damage to TAPS facilities prior to the Denali Fault event.

The seismic design criteria for the Trans-Alaska Pipeline System (TAPS) were defined as stipulated design earthquakes for five seismic zones along the 1,287-km route as shown in Figure 1. The delineation of seismic zones was based on studies by the U.S. Geological Survey (Page, et al., 1972) and independent assessments performed by consultants to the pipeline project. A Design Contingency Earthquake (DCE), with an estimated return period of 300 years or more, was established for each of the five seismic zones according to Richter magnitude. The areas along the pipeline route of principal interest are the magnitude 8.5 Richter zone at the southern end of the pipeline (Valdez) and the magnitude 8.0 Richter zone in the Alaska Range in the vicinity of the Denali Fault and Pump Station 10 (about 5 km north of the pipeline fault crossing).

The seismic design ground motion criteria employed at the time of original design are shown in Table 1 as a function of earthquake magnitude (Newmark and Hall, 1974). By way of example, for the magnitude 8.5 seismic region, the criteria called for 0.60 g Zero Period Acceleration (ZPA) for free field ground motion and 0.33 g ZPA for structural design. Both values were selected on the basis of the concept of "effective acceleration" in conjunction with discussions with the USGS. Based on the ZPA values of Table 1, design spectra were constructed for each magnitude and for different soil conditions.

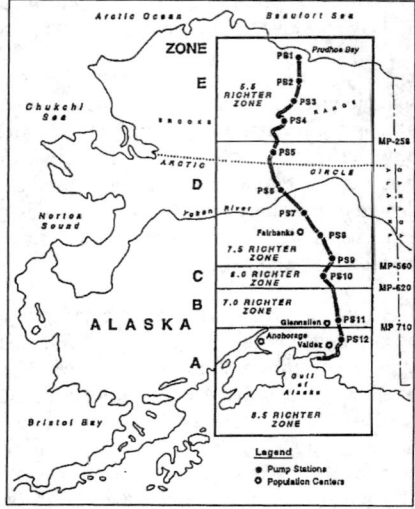

Figure 1. Seismic Zones along the Trans-Alaska Pipeline

Table 1. Design Ground Motions for TAPS Seismic Zones

Richter Magnitude	Acceleration (g)		Velocity (in/sec)	
	Free Field	Structures	Free Field	Structures
8.5	0.60	0.33	29	16
8.0	0.60	0.33	29	16
7.5	0.45	0.22	22	11
7.0	0.30	0.15	14	7
5.5	0.12	0.10	6	5

A structural geologic and engineering geologic study and field survey to identify and locate possible active surface faults that may affect the trans-Alaska pipeline system was carried out by Woodward-Lundgren Associates in 1973-74 (Cluff et al., 2003). Three potentially active fault zones that cross the TAPS route were identified in the fault study: Denali, McGinnis Glacier, and Donnelly Dome faults. The McGinnis Glacier and Denali fault crossings are located within 5 km to the south of Pump Station 10. The Donnelly Dome fault crossing is located on the pipeline alignment where it passes to the east side of Donnelly Dome, about 24 km south of Pump Station 9. The design displacements were taken as approximately two-thirds of the maximum credible displacement, as shown in Table 2.

Table 2. Design Displacements at Fault Crossings

Fault	Maximum Credible, m		Design Value, m	
	Right Shift	Dip Shift	Right Shift	Dip Shift
Denali	9.1	2.1	6.1	1.5
McGinnis Glacier	4.0	3.0	2.4	1.8
Donnelly Dome	1.5	4.6	0.9	10.0
Note: Displacements were originally specified in units of integer feet.				

Seismic Design and Observations on Earthquake Performance

During design, it was recognized that the major seismic hazards that could potentially affect TAPS were fault movement, liquefaction, landslides, ground motion (shaking), seismic wave propagation and tsunami. Of these, the first three hazards, faulting, liquefaction, and landslides, were the most serious concerns for the buried portion of the pipeline (about 612 km). Ground shaking was a major concern for the above-ground pipeline (about 675 km)[5], buildings, structures, vessels, and similar components. Seismic wave propagation was also considered, but resulting pipe strains are normally not high enough to cause damage if the pipeline is constructed of ductile steel with good quality girth welds (which was the case for TAPS). At the Valdez Marine Terminal, in addition to ground shaking, slope and rock instability and the potential effects of tsunami were important considerations.

The philosophy underlying the original design of TAPS was that for a high-level earthquake with a recurrence interval of about 300 to 500 years, referred to as the "Design Contingency Earthquake" (DCE), considerable inelastic behavior and limited damage would be permitted, but that there would no structural collapse, loss of function of essential facilities, or release of crude oil or hazardous substances. The amount of permissible damage varies according to the type of structure or component and its function. The functionality of essential control, communications and emergency systems was specified to be maintained without interruption during and after a DCE. For a lower level earthquake having ground motions one-half the DCE, referred to as the "Design Operating Earthquake" (DOE), the pipeline and facilities were to be capable of withstanding the prescribed seismic motions without damage, significant deformation, or interruption in operation.

Fortunately, the timing of the development of the seismic criteria for the TAPS project and design-related analyses was coincidental with the early, formative stages of the ATC-3-6 effort (Newmark, 1975; Applied Technology Council ATC-3-6, 1978). As such, not only the seismic loading criteria employed in the TAPS design and analysis approaches were "modern," but equally important, the design and construction employed high quality materials and construction practices that were at the forefront of practice at that time, and almost identical to that which would be employed today.

---

[5] The pipeline was constructed above ground in areas of unstable and/or ice-rich permafrost.

The November 3, 2002 Denali Fault Earthquake, magnitude 7.9, was the largest strike-slip earthquake in North America since the 1906 San Francisco earthquake. Surface rupture on the Denali, Totschunda, and Susitna Glacier faults extended 336 km (USGS, 2003). The rupture passed beneath the Trans-Alaska pipeline producing approximately 5.5 m of right lateral offset and 0.6 m of up-to-the-north vertical displacement, as confirmed by geodetic and GPS surveys. The fault displacement was distributed over a zone approximately 200 m in width, with approximately 50 to 70 percent of this displacement occurring as an abrupt offset. The largest maximum measured offset along the fault was 8.8 m. Peak recorded free-field acceleration at Pump Station 10, about 5 km north of the Denali fault, was 0.34 g, with a significant velocity component of about 114 cm/sec (45 in/sec). Strong ground motions lasted about 90 seconds.

Above-Ground Pipeline. To accommodate the effects of thermal expansion and contraction, the above-ground portion of the pipeline is configured in a trapezoidal or "zigzag" configuration on sliding pipe supports spaced at 18-m (60-ft) intervals as shown in Figure 2. The 18-m support spacing was selected to provide adequate support for the pipeline in the event of a loss-of-support condition at any two adjacent supports. A typical above-ground support is shown in Figure 3. The pipe supports are "H-type" supports

Figure 2. Trapezoidal or "zigzag" above-ground pipeline configuration.

consisting of a cross-beam supported by two 457-mm (18-inch) diameter pipe piles (actually drilled piles) referred to as vertical support members, or VSMs. At each support a sliding shoe assembly is clamped to the 1,219-mm pipeline, and is allowed to slide on the support cross-beam to accommodate thermal expansion or seismic movement. The bearing surface between the pipe and the beam is a low-friction Teflon interface. Thermal transmission protection from the pipeline to the VSMs is accomplished through use of Micarta bearings in the assembly.

The above-ground pipeline is anchored against movement at intervals of 200 to 600 m by anchor assemblies supported by four VSMs (see Figure 4). The dynamic response

Figure 3. Sliding shoe support, used at 18-m intervals.

of the pipeline is limited at the anchors by a pretensioned longitudinal slipping mechanism and crushable sacrificial honeycomb energy absorber and by specially designed bumpers placed at selected H-type supports (see Figure 5) to limit lateral movement of the pipeline within the configuration.

The dynamic response of the above-ground pipeline to earthquake motions is highly non-linear and the design requires a non-linear transient dynamic analysis to simulate the interaction at the soil-VSM interface, support stiffness, frictional sliding of the pipe shoes, bumpers that stop pipe movement, non-linear reaction characteristics at the anchors, and the spatial variation of ground motion inputs due to seismic wave propagation. The insulation modules, provided to retain heat, also provide some degree of damping to the system.

Figure 4. Anchor support for above-ground pipeline. Longitudinal capacity of 667 kN (150 kips). Frictional slipping occurs at 467 kN (105 kips).

The magnitude 7.9 Denali Fault earthquake produced ground motions near the fault rupture that generally approach design criteria in the low to mid-frequency range of the design spectrum and exceed the TAPS seismic criteria at frequencies shorter than about 1.0 Hz as can be observed in Figure 6. Considering that the natural frequency of the above-ground pipeline system generally lies in a range from about 0.5 Hz to 2.0 Hz, this level of ground shaking provided a full-scale test of the pipeline design for ground motions essentially equivalent to those considered in the original design analyses.

Most importantly, and consistent with the original design premise, pressure integrity of the pipeline system was maintained. There was no damage to pipeline, i.e., no wrinkling, buckling, or excessive curvature (strain) conditions resulting from earthquake ground shaking. There was

Figure 5. Bumper restraint to control lateral seismic movement.

incidental damage to the above-ground support system within about one km either side of the Denali Fault crossing. This consisted of structural damage to eight above-ground supports and separation[6] of eight support shoes from the pipe. The worst of this damage is illustrated in Figure 7. The frictional slipping mechanisms at nine anchors "tripped" to absorb excess energy from seismic motion. The segment of TAPS that experienced support damage is generally considered to be the zone over which near-fault violent motion exceeded design criteria.

All things considered, the above-ground pipeline performed as required by the design. However, the damage to the supports was more extensive than expected, mainly due to the effects of near-fault violent motion. Studies are in-progress to evaluate this behavior in detail and to determine if any retrofits are advisable.

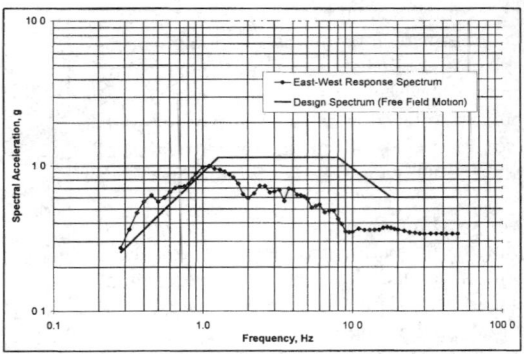

Figure 6. Spectra for ground motion recorded at Pump Station 10, 5 km north of Denali Fault, N39°E. Note that the free-field design spectrum has been exceeded below 1.0 Hz.

Figure 7. Loss of two adjacent supports on above-ground pipeline near Denali Fault. The possible loss of two adjacent supports was a fundamental design assumption.

Buried Pipeline. Nearly one-half of the 1,287-km Trans-Alaska Pipeline System (TAPS) is buried in a standard pipeline trench with a minimum depth of cover of 0.9 m. The buried pipeline (Figure 8) must be able to deform axially and in bending in order to accommodate ground strain and curvature associated with seismic wave propagation and permanent ground deformation related to liquefaction, slope movements, surface fault effects, and normal settlement. In the buried mode it was

---

[6] Separation of a pipeline support shoe means that the support dropped away from the pipe when the shoe slid longitudinally off the cross-beam. This is an intentional design feature that precludes pounding of the support against the cross-beam.

assumed that the pipeline would interact axially and in bending so as to be fully compliant with the ground deformation, i.e., the pipeline will deform axially and in bending to have the same curvature and longitudinal strain as the ground.

The duration of strong shaking was approximately 90 seconds, and this undoubtedly contributed to widespread liquefaction of subsurface deposits along the pipeline, as evident from numerous sand blows.

Figure 8. Buried TAPS pipeline section being lowered into trench during original construction in 1976.

In some areas in proximity to the Denali Fault crossing, moderate lateral spread movements were observed. Landsliding did not occur along or across the pipeline right-of-way, although there a number of landslides and rockfalls that occurred along the highways in the area.

The long period nature of the ground motion, a relatively high ground velocity of approximately 114 cm/sec (45 in/sec)[7], and soil liquefaction along the TAPS route provided significant opportunity for developing large bending and axial strain in the buried pipeline. The pipeline was excavated at a remote gate valve location where surface ground movement was apparent (see Figure 9), but there was no evidence of damage in the form of wrinkling or buckling of the pipe wall or significant displacement of the pipe. Approximately one month after the earthquake, an instrumented pig ("smart pig") was run through the pipeline. The pig data indicated that the pipeline had moved in some of the liquefaction areas, but the resulting pipe curvatures were well within acceptable limits. In fact, curvatures were reduced in some cases, apparently due to some redistribution of the bedding and padding material within the pipe trench (Johnson et al., 2003). Liquefaction also occurred around the VSM foundation of the RGV equipment shelter (see Figure 10), resulting in minor structural damage.

Figure 9. Pipeline and remote gate valve excavated at liquefaction zone. There was no evidence of damage or movements were well within tolerable limits.

---

[7] The maximum computed velocity is affected by high-pass filtering at 0.1 Hz. Based on preliminary results from in-progress studies by the USGS, it is believed that the actual peak velocity could have been about 50 percent higher than the calculated value of 114 cm/sec, i.e., about 170 cm/sec.

Fault Crossings. Fault crossings are handled by various modes, namely elevated on pipe supports, laid at the ground surface on "sleeper" supports, or buried (Newmark and Hall, 1975; ASCE, 1984). At the Denali Fault crossing, the pipeline is supported in a special above-ground configuration. Support beams were placed on grade on gravel berms for approximately 600 m, with the ability to slide to accommodate thermal movement and surface fault rupture to prevent excessive pipeline strain. The crossing was designed to accommodate a fault displacement offset of approximately 6 m right-lateral strike-slip and 1.5 m vertical slip, without exceeding permissible strain limits.

Figure 10. Evidence of liquefaction at remote gate valve equipment shelter. Some structural damage occurred.

The magnitude 7.9 earthquake that occurred in south central Alaska on November 3, 2002 ruptured a 350-km long segment of the Denali Fault. The epicenter was located about 88 km west of the pipeline, and the rupture propagated to the east. The fault displacement at the pipeline crossing was about 5.5 m, and reached a maximum of nearly 9 m about 120 km southeast of the pipeline.

The pipeline withstood the Denali Fault rupture without damage. Since the fault trends generally northwest and crosses the pipeline at an angle of approximately 60°, right-lateral movement will place the pipeline in compression. The zigzag configuration in the fault zone compresses like a coiled spring. This can be observed by comparing the two photographs in Figure 11. The photograph on the left in Figure 11 was taken prior to fault movement, and the photo to the right was taken after the occurrence of the 5.5-m fault displacement. The net compressive displacement applied to the pipeline in the fault zone (due to the 55° crossing angle) was approximately 3.5 m. Not only did the pipeline slide laterally toward the outside of the bends to accommodate the compressive displacement, but the initially straight segments of the pipeline took on a bowed shape.

Following the earthquake, it was necessary to re-center eleven support shoes in the fault crossing that had reached the end of their travel tolerance (see Figure 12) to restore fault displacement capacity to an additional 2 m of strike slip, which enveloped geologists' estimates of further short term fault displacements that could reasonably be expected. An evaluation of permanent design retrofit alternatives is in

(a) before (b) after

Figure 11. TAPS crossing of Denali Fault before and after fault slip, looking south. Note movement and bowed segment after fault displacement, which acts to compress the pipeline crossing segment.

progress, and a final determination of a suitable retrofit will be made after the completion of a post-event field geologic investigation of the Denali Fault.

Piping and Equipment. The DOI Stipulations and U.S. DOT requirements through Title 49 of the Code of Federal Regulations, Part 195, and thereby ANSI Standard B31.4, provide regulations for the design, construction and operation of crude oil pipelines. However, the regulations at that time were essentially "quiet" as to specific requirements for seismic design. As a result Alyeska undertook major effort to adopt applicable American Society of Mechanical Engineers loading and code provisions, with seismic and faulting input added in much the same way as was occurring in the nuclear power industry, the offshore pipeline industry, and the military establishment. Software such as DRAIN-2D was then under development and was employed, to the extent possible, as part of the design and analysis process to validate the behavior of large diameter piping on sliding supports. Equipment qualification was carried out in accordance with recognized procedures of the period

Figure 12. Pipe shoe at Denali Fault crossing perched at edge of grade beam support following fault displacement.

(Hall et al., 1975; Anderson and Nyman, 1979). The same general philosophy was undertaken for tankage and the other many ancillary portions of the system.

Although Pump Station 10 was near the Denali Fault, it had been taken out of service over six years prior to the event. The "mothballed" process and control facilities withstood the earthquake ground shaking with no significant damage, but design seismic loading was not attained during the earthquake, because the piping, vessels and tanks were empty of fluid, and there were no snow loads on the building roofs. Certain critical communications and control systems (including electronic components, cabinets, battery racks, inverters, cable trays, etc.) were in operation at Pump Station 10 during the earthquake and performed as intended without damage or malfunction.

Post-Earthquake Response. The Denali Earthquake in November 2002 provided a "live" test of the TAPS' earthquake emergency response plan. An Oil Spill Contingency Plan coupled with a general emergency management system known as ICS (Incident Command System) provides an emergency management structure for responding to pipeline system emergencies, road and highway closures, commercial power outage, loss of communications, and general disruption including oil spills. During the earthquake, particular emphasis was afforded to the reconnaissance of fault displacements, liquefaction and landslide areas, including the evaluation of the pipeline for excessive strain conditions and/or wrinkling.

An earthquake monitoring system (EMS) serves as a cornerstone for planning and guiding field reconnaissance. The EMS has been included as part of the pipeline control system since the start-up in 1977 (Nyman et al., 1981; Nyman et al., 1999). It was developed pursuant to stipulated Federal and State governmental requirements, and consists of eleven remote digital strong motion accelerograph stations distributed over the 1,287-km pipeline route. Six of the eleven stations triggered into an alarm status when the Denali earthquake occurred. The EMS automatically processed ground motion data immediately after the event, which helped system operators evaluate the severity of the earthquake ground shaking and, in turn, to make a general assessment of the potential for damage to the pipeline and supporting facilities. A key function of the EMS was that it delineated inspection requirements (via detailed checklists) for the affected portion of the route. The use of a comprehensive and focused field inspection effort based on estimates of ground motion shaking severity permitted the pipeline operation to resume after a 66-hour shutdown (Nyman et al., 2003).

Conclusion

This is the first case, known to the authors, where a large-diameter crude oil pipeline system, specifically designed for major earthquake hazards, has been subjected to design level earthquake shaking and fault displacement. The magnitude 7.9 Denali Fault earthquake that occurred on November 3, 2002 approached design criteria for the section of the pipeline passing through the Alaska Mountain Range in the vicinity of Pump Station 10 and the Denali Fault crossing. There was no damage to the pipeline or release of crude oil, although there was incidental damage to the above-ground support hardware and some movement of the below-ground pipeline in liquefaction areas. The pipeline system withstood the earthquake in a manner consistent with the original design premise.

Acknowledgment

This summary paper is based on the authors' personal knowledge and experiences as part of their affiliation with the Alyeska Pipeline Service Company, and in no way reflects on policies or practices of Alyeska with respect to the items presented. Acknowledgment is gratefully made to Alyeska for permission to publish this summary paper.

References

1. Anderson, T.J., and D.J. Nyman (1979). "Lifeline Engineering Approach to Seismic Qualification," *Journal of the Technical Councils of ASCE*, Proc. ASCE, 105:TC1, April, pp. 149-161.

2. Applied Technology Council ATC-3-06 (1978). Tentative Provisions for the Development of Seismic Regulations for Buildings, National Bureau of Standards Special Publication 510.

3. ASCE (American Society of Civil Engineers) (1984). *Guidelines for the Seismic Design of Oil and Gas Pipeline Systems,* Technical Council on Lifeline Earthquake Engineering, Committee on Gas and Liquid Fuel Lifelines, D. J. Nyman, Principal Investigator, New York, 473 p.

4. Cluff, L.S., R.A. Page, D.B. Slemmons, and C.B. Crouse (2003). "Seismic Hazard Exposure for Trans-Alaska Pipeline," *Proceedings of the 6th U.S. Conference on Lifeline Earthquake Engineering (Advancing Mitigation Technologies and Disaster Response),* Long Beach, CA, TCLEE, ASCE.

5. Hall, W.J., V.J. McDonald, D.J. Nyman, and N.M. Newmark (1975). "Observations on the Process of Equipment Qualification," *Proceedings U.S. National Conference on Earthquake Engineering*, Univ. of Michigan, Ann Arbor, MI, pp. 495-501.

6. Johnson, E.R., M.C. Metz, and D.A. Hackney (2003). "Assessment of the Below-Ground Trans-Alaska Pipeline Following the Magnitude 7.9 Denali Fault Earthquake," *Proceedings of the 6th U.S. Conference on Lifeline Earthquake Engineering (Advancing Mitigation Technologies and Disaster Response),* Long Beach, CA, TCLEE, ASCE.

7. Newmark, N.M. (1975). "Seismic Design Criteria for Structures and Facilities -- Trans Alaska Pipeline System," *Proceedings of U.S. National Conference on Earthquake Engineering*, Ann Arbor, MI, June 1975, pp. 94-103.

8. Newmark, N.M., and W.J. Hall (1974). "Seismic Design Spectra for the Trans-Alaska Pipeline," 5th World Conference on Earthquake Engineering, IAEE, Rome, Italy, v. 1, pp. 554-557.

9. Newmark, N.M. and W.J. Hall (1975). "Pipeline Design to Resist Large Fault Displacement," *Proceedings U.S. National Conference on Earthquake Engineering*, University of Michigan, Ann Arbor, MI, pp. 416-425.
10. Nyman, D.J., E.R. Johnson, C.H. Roach (2003). "Trans-Alaska Pipeline Emergency Response and Recovery Following the November 3, 2002 7.9 Denali Fault Earthquake," *Proceedings of the 6th U.S. Conference on Lifeline Earthquake Engineering (Advancing Mitigation Technologies and Disaster Response)*, Long Beach, CA, TCLEE, ASCE.
11. Nyman, D.J., V.J. McDonald and G.G. Simmons (1981). "Earthquake Monitoring System Trans-Alaska Pipeline," *Lifeline Earthquake Engineering – the Current State of Knowledge 1981*, Oakland, CA Conference, TCLEE, ASCE, pp. 139-151.
12. Nyman, D.J., E.L. Nelson, C.H. Roach (1999). "Earthquake Monitoring System Trans-Alaska Pipeline," *Proceedings of the 5th U.S. Conference on Lifeline Earthquake Engineering (Optimizing Post-Earthquake Lifeline System Reliability)*, Seattle, WA, TCLEE, Monograph No. 16, pp. 897-906, ASCE.
13. Page, R.A., D.M. Boore, W.B. Joyner, and H.W. Coulter (1972). "Ground Motion Values for Use in the Seismic Design of the Trans-Alaska Pipeline System," U.S. Geological Survey Circular 672, 23 pp.
14. Roscow, J.P. (1977). *800 Miles to Valdez – The Building of the Alaskan Pipeline*, Prentice-Hall, Inc.
15. USGS (U.S. Geological Survey) (2003). "Rupture in South-Central Alaska – the Denali Fault Earthquake of 2002," USGS Fact Sheet 014-03, Menlo Park, CA, http://geopubs.wr.usgs.gov/fact-sheet/fs014-03/.

SEISMIC HAZARD EXPOSURE FOR THE TRANS-ALASKA PIPELINE

Lloyd S. Cluff,[1] Robert A. Page,[2] D. Burton Slemmons,[3] and C. B. Crouse[4]

Abstract

The discovery of oil on Alaska's North Slope and the construction of a pipeline to transport that oil across Alaska coincided with the National Environmental Policy Act of 1969 and a destructive Southern California earthquake in 1971 to cause stringent stipulations, state-of-the-art investigations, and innovative design for the pipeline. The magnitude 7.9 earthquake on the Denali fault in November 2002 was remarkably consistent with the design earthquake and fault displacement postulated for the Denali crossing of the Trans-Alaska Pipeline route. The pipeline maintained its integrity, and disaster was averted. Recent probabilistic studies to update previous hazard exposure conclusions suggest continuing pipeline integrity.

Introduction

When oil was discovered on the North Slope of Alaska in the late 1960s, the Alyeska Pipeline Service Company, a newly formed consortium of oil companies, envisioned a major pipeline to transport the crude oil from Prudhoe Bay to the ice-free Port of Valdez. The proposed Trans-Alaska Pipeline System (TAPS) would be a 1,219-mm- (48-inch-) diameter pipe that would travel a distance of 1,287 km (800 mi). The pipeline would traverse spectacular wilderness country, cross three mountain ranges, 350 rivers, and numerous faults, some of which are active.

The proposed pipeline route would cross mostly federal lands, requiring a right-of-way permit from the U.S. Department of the Interior. Alyeska requested the permit in June of 1969. The newly enacted National Environmental Policy Act of 1969 required an environmental impact analysis to address the unavoidable and threatened impacts that would result from construction and operation of the TAPS. Thus, public safety and environmental preservation were important issues from the start, and needed to be addressed before the Secretary of the Interior would issue a permit. A Technical Advisory Board was installed by the Secretary to review pipeline design criteria submitted by Alyeska.

U.S. Geological Survey scientists in Menlo Park, California who had expertise in permafrost and Alaskan geology, and independent consultants were selected to study the perceived hazards, including disturbing the thermal equilibrium in permafrost and earthquakes. A temporary seismograph network was installed by the USGS in 1970, centered on the pipeline's proposed intersection with the Denali fault, and USGS

---

[1]Director, Geosciences Department, Pacific Gas & Electric Company, San Francisco, California
[2]Geophysicist Emeritus, U.S. Geological Survey, Menlo Park, California
[3]Professor Emeritus, Mackay School of Mines, University of Nevada at Reno, Nevada
[4]Principal Engineer, URS Corporation, Seattle, Washington, M. ASCE

scientists resurveyed the first-order geodetic triangulation arc established in the Delta River canyon in 1941-1942. Only four of the thirty-three microearthquakes recorded by the network were along the trend of the fault valley (Page, 1971), and the geodetic survey showed no evidence of slip on the Denali fault since the 1942 survey (Page, 1972). Although the Denali fault had geomorphic evidence indicating it was as dangerous as the San Andreas, the last major displacement apparently predated 1800, strongly suggesting significant strain accumulation since the last earthquake.

The Final Environmental Impact Statement was issued by the Department of the Interior in 1972, and included Stipulations governing aspects of the design and operation of the pipeline that were to be attached to the right-of-way permit (U.S. Department of the Interior, 1972). Design earthquakes, considered the maximum likely to occur in 100 to 300 years or more, were established for each of five seismic zones along the pipeline route, and defined in terms of Richter magnitude (the accepted magnitude scale at the time) (Table 1).

Table 1
Stipulated Earthquake Magnitudes along the Trans-Alaska Pipeline Route
(U.S. Department of the Interior, 1972)

Zone	Richter Magnitude
Valdez to Willow Lake	8.5
Willow Lake to Paxson	7.0
Paxson to Donnelly Dome*	8.0
Donnelly Dome to 67 degrees north	7.5
67 degrees north to Prudhoe Bay	5.5

*The Denali earthquake zone.

During the pipeline permitting process, a M_W 6.7 earthquake on February 9, 1971 struck the San Fernando Valley near Los Angeles, California. The fault that released the earthquake, later named the San Fernando fault, had not been recognized previously. Detailed studies along the mountain front, where the fault had ruptured for 16 km, were conducted following the earthquake (Oakeshott, 1975). They revealed abundant geomorphic and stratigraphic evidence that the fault was indeed active, and could have been recognized as active before the earthquake if experienced geologists had looked in the right places. This surface fault rupture eventually led to the creation of the Alquist-Priolo Earthquake Fault Zoning Act of 1972, requiring special studies to accurately locate faults in California, assess their activity, and regulate building within the Alquist-Priolo Zones. The San Fernando earthquake also produced an unusually high peak ground acceleration of 1.25 g, recorded on an abutment of the Pacoima dam on the up-faulted block in the mountains directly north of and near to the causative fault. This acceleration was significantly higher than the 0.5 g many earthquake engineers thought at that time to be the maximum design value; however, up to that time, no strong ground motions had been recorded within 40 km of a magnitude 7 earthquake or within more than 100 km of a magnitude 8 event (Page and others, 1972), and estimates had to be extrapolated from known data. The surface-faulting and strong-ground-motion implications were very controversial, and several large engineering projects, including the TAPS, were put on hold until this and other environmental issues could be accommodated or resolved.

Earthquake Ground Motions

Robert Page and others at the USGS in Menlo Park recognized that high accelerations needed special consideration for important critical engineering projects, including the proposed TAPS. In 1972, the USGS published Circular 672, *Ground Motion Values for Use in the Seismic Design of the Trans-Alaska Pipeline System* (Page and others, 1972). Circular 672 characterized the earthquake ground motions for the stipulated design earthquakes in terms of peak values of near-fault horizontal acceleration, velocity, and displacement, as well as the duration of shaking. They estimated that Pump Station 10, located 4.8 km north of the Denali fault crossing, could experience free-field accelerations as large as 1.2 g in a magnitude 8 earthquake on the Denali fault. Nathan M. Newmark and William J. Hall used the concept of "effective acceleration" (Newmark and Hall, 1975; Hall and others, 2003) to argue for lower accelerations. The agreed-upon acceleration for the magnitude 8.0 earthquake was 0.60 g for the free-field ground motion, considered to affect slope stability, liquefaction, and the below-ground pipe. This value is similar to the 0.7 g estimated by Bolt (1972) for Alyeska. Newmark and Hall (1974) recommended a lower acceleration, 0.33 g, for structural design (considered to affect structures and the above-ground pipe). The same ground motions were recommended for the more distant magnitude 8.5 earthquake in the southernmost seismic zone. The rationale for these ground-motion values was the collective experience of the earthquake engineers.

Surface Fault Displacement

Circular 672 (Page and others, 1972) also pointed out that the proposed pipeline would unavoidably cross active faults as it traversed Alaska. The significance of faults to the design of the pipeline also was called out in the Environmental Impact Statement:

> *The proposed pipeline route intersects several recognized major faults in the active seismic regions; but, except for the Denali fault, which shows evidence of large relatively recent (Holocene) offset, the risk of significant movements on the other faults is essentially unknown.* (U.S. Department of the Interior, 1972, vol. 2, p. xxii-xxiii)

Acknowledging the lack of information at that time, the Stipulations included the requirement that studies be conducted to identify and delineate "all recognizable or reasonably inferred faults or fault zones" along the route. The objective was to assure that the "risk of oil leakage resulting from fault movement and ground deformation has been adequately assessed and provided for in the design of the pipeline."

A fault evaluation project was developed and undertaken as part of the effort by Alyeska to identify all the factors that had to be considered in developing designs that would ensure the structural integrity of the pipeline at the active fault crossings. The TAPS Fault Evaluation Project was led by Lloyd S. Cluff and David B. "Burt" Slemmons, co-principal investigators. Cluff was Chief Engineering Geologist at Woodward-Lundgren and Associates and also a visiting Associate Professor at the University of Nevada at Reno. Slemmons was Professor of Geology and Geophysics at the Mackay School of Mines, University of Nevada at Reno. They were assisted by George E. Brogan and Marjorie K. Korringa, and fifteen other specially selected Woodward-Lundgren earthquake geologists and geophysicists (Figure 1).

Figure 1. Trans-Alaska Pipeline Fault Study Team (1972-1974). Standing, left to right: Burt Slemmons, Tom McCarthy, Sue McCarthy, Richard Hardyman, Tom Welsh, Marjorie Korringa (deceased), Phil Watson (deceased), Lloyd Cluff, Kerry Sieh, David Schwartz, Cheri Carver, Gary Carver, George Brogan. Seated, left to right: Norma Bigger, Linda Hadley, Dan Collins, Marc Seeley.

The fault study was to satisfy Alyeska's compliance with the Stipulations appended to the Final Environmental Impact Statement issued by the U.S. Department of the Interior; specifically, Stipulation 3.4.2 and its subsections, which dealt with fault displacements (U.S. Department of the Interior, 1972). The fault evaluation project consisted of three phases specifically focused on identifying, delineating, and characterizing active faults crossed by the pipeline route, presenting design values for these faults, and making general recommendations for monitoring at active-fault crossings. The results of the studies are reported in *Summary Report, Basis for Pipeline Design for Active Fault Crossings for the Trans-Alaska Pipeline System* (Cluff and others, 1974).

Active Fault Study. Identification and characterization of active faults was a relatively new field that required first-hand knowledge and experience of how active fault features may be expressed in the topography and their geomorphic expression as they traversed the countryside. Most of the field team members were former students who had been enrolled in Slemmons' geology and Cluff's engineering geology courses at the University of Nevada, at Reno, between 1967 and 1973. An important part of the course work focused on earthquake hazards, including the identification and characterization of active faults. Experience was gained during field trips to observe faults in California and Nevada. The students were exposed to various methods of evaluating fault activity, including interpretation of aerial photographs and special low-sun-angle aerial photographs (Slemmons, 1969; Cluff and Slemmons, 1972) and interpretation of geomorphology. They also learned to interpret historical seismicity and earthquake intensity reports.

Although the members of the TAPS active fault team understood the state-of-the-art principles, methods, and techniques required to evaluate fault activity (Sherard and others, 1974), it was also considered important, at the beginning of the study, for the team to have first-hand experience observing features associated with active faults in Alaska. Therefore, one of the initial work efforts was to observe features representative of active faults in the Alaskan environment to "calibrate the eyeballs" of the fault study team. During the winter months, while preparing for the 1973 field season, aerial photographs were obtained of areas in Alaska traversed by known active faults, and the team studied these photos to gain relevant experience. Faults that were studied included the Fairweather fault, the location of the 1958 magnitude 7.8 earthquake and associated surface fault rupture; the Patton Bay and Hanning Bay faults, which experienced surface rupture on Montague island during the 1964 magnitude 9.2 earthquake in the Prince William Sound region; the Ragged Mountain fault; the Castle Mountain fault; and the Long Glacier fault. Not only were these faults observed and interpreted on aerial photographs, they were studied during reconnaissance flights and ground studies at the beginning of the 1973 field effort.

The field season began in May 1973, and lasted about five months. The "calibration" training prompted insightful evaluations of approximately 8,000 lineaments and faults. Aerial reconnaissance was conducted of the entire pipeline route, especially along the Fairweather, Castle Mountain, Donnelly Dome, McGinnis Glacier, Denali, Totschunda, Stevens Creek, Kobuk, Clearwater Lake, Tintina, and Kaltag fault zones, then considered among the most important potentially active faults. Interpretation of aerial photographs, radar images, and satellite images continued at various scales along the pipeline route. Low-sun-angle aerial photographs were taken along features of interest. Detailed field evaluations using helicopters and fixed-wing aircraft were conducted of all identified lineaments and faults, and hundreds of fault displacements were measured. Along the Denali fault, evaluations at 84 locations along a 277-km length of the fault from near McKinley Park on the west to near the Denali/Totschunda fault juncture on the east included measurement of the width of the most active zone, and the amount of horizontal and vertical surface displacements.

Some geologic materials were dated, and limited trenching and geophysical surveys were conducted; however, these studies were constrained by time and logistics.

From the fall of 1973 until the summary report was completed in January 1974 (Cluff and others, 1974), the team worked on the final collation, evaluation, and interpretation of all the field data. The pipeline route was found to cross three active faults: the Donnelly Dome fault, the McGinnis Glacier fault, and the Denali fault. The locations of these faults were mapped, and their behavioral and other characteristics important to pipeline designers, such as the attitude of the faults and their angle of intersection with the pipeline, were described. Finally, the fault displacement parameters to be used in pipeline design at the active fault crossings were estimated, and the areas potentially influenced by fault displacement were zoned to guide the pipeline designers in designing for fault displacement. Because the Denali fault was considered the most active, and is the fault that released the November 3, 2002 earthquake, we focus, in this paper, on the Denali fault crossing.

Design Fault Displacement. The fault study concluded the 2,150-km-long Denali fault had the potential of releasing an earthquake as large as magnitude 8.0, agreeing with and verifying the Stipulations. The maximum surface displacement along the entire fault zone was estimated to be 9.1 m (30 ft) horizontal and 2.1 m (8 ft) vertical.

The principal investigators' experience during the mid to late 1960s included investigations of large strike-slip faults, such as the San Andreas and related faults in California, the Bocono fault in northwestern Venezuela, the Alpine and related faults in New Zealand, and the North Anatolian fault in Turkey. These investigations and the available literature showed that major strike-slip surface faulting events worldwide could range from 200 to 400 km in length, and the maximum surface displacements could be as large as 9 m. However, the maximum displacement was observed to occur at only a few locations along the total rupture length. The fault study team reasoned it would be unlikely for the maximum Denali fault displacement to occur at the pipeline crossing. An average displacement amount, usually about one-third of the maximum, was considered a more reasonable estimate for random locations along a total rupture segment; however, the team thought 3 m was not sufficiently conservative for the pipeline crossing. The recommended horizontal displacement was double the hypothetical average displacement; a design displacement of 6.1 m (20 ft) horizontal and 1.5 m (5 ft) vertical, up on the north side, were selected for the Denali fault crossing.

Design Displacement Zone. Selecting the location and width of the zone of future fault rupture at the pipeline crossing may sound like a simple task; however, more than 100 m of alluvial deposits overlie the bedrock at the Denali fault crossing. These alluvial deposits are a product of very active post-glacial erosion and deposition, thus evidence of surface faulting at this location has been severely modified or destroyed.

The design displacement zone at the Denali fault crossing had to be located based on projections from confident fault locations to the west and the east. Of the 58 measured data points, the closest to the west was 2.4 km distant. In the glacial valleys to the east of the pipeline, the Denali fault is covered for about 58 km by glacial ice; the closest data point was 64 km distant. Because differential erosion occurs more easily along the sheared rock of a fault zone, the team assumed the glacial valley was fault-controlled and that it delineated the fault zone. Subtle geomorphic features suggestive of faulting in the post-glacial outwash deposits immediately east of the pipeline crossing also helped "connect the dots."

A width of 76.2 m (250 ft) was assigned to the active-fault zone at the pipeline. The zone was extended 152.4 m (500 ft) to the north and to the south to provide a margin of safety. The northern margin was then extended another 198.1 m (650 ft) to include an escarpment that was a natural limit to the zone. The recommended width of the design displacement zone at the Denali fault crossing was 579.1 m (1,900 ft).

Independent Review. John C. Crowell, then Professor of Geology, University of California at Santa Barbara, provided Alyeska an independent review of the fault evaluation report. Alyeska's objective was to ensure high technical quality and compliance with the Stipulations with respect to fault displacements at pipeline fault crossings. Bruce A. Bolt, then Professor of Seismology and Director of Seismographic Stations, University of California at Berkeley, reviewed the report for consistency of the design values for surface fault displacements with the design values for earthquake ground motions.

Innovative Design. The TAPS design team, Nathan M. Newmark, William J. Hall, and James Maple, assisted by Douglas J. Nyman, developed an innovative design to accommodate the expected surface fault displacements at the above-ground sections of the pipeline (Hall and others, 2003). The design consists of a series of 12-m-long Teflon-coated steel beams, which support the pipeline on Teflon-coated shoes. The above-ground pipeline is articulated in a zigzag fashion. This combination of design characteristics allows the pipeline to move freely, horizontally and vertically, in response to fault displacements, without disrupting the integrity of the pipeline.

November 3, 2002 Earthquake

The November 3, 2002, M_W 7.9 earthquake was released due to rupture on three tectonically related faults: the Susitna Glacier thrust fault, the Denali strike-slip fault, and the Totschunda strike-slip fault. The fault rupture began on the Susitna Glacier fault about 72 km east of Cantwell, propagated northeastward along the Susitna Glacier fault for 51 km, joined the Denali fault and continued rupturing southeastward for 216 km, where it branched onto the Totschunda fault, and propagated southeastward for 74 km. (D. P. Schwartz and T. Dawson, personal communication, 2003). The total rupture length along all three faults was about 350 km. Investigations by the U.S. Geological Survey document that vertical fault displacements along the Susitna Glacier thrust fault ranged from 0.5 to 6.2 m. Along the western part of the Denali fault, about 80 km west of the pipeline fault crossing, right-slip displacements ranged from 0.3 to 6.2 m, increasing as the fault ruptured southeastward to maximum right-slip displacement of 8.8 m about 120 km southeast of the pipeline crossing. Right-slip displacements along the Totschunda fault ranged from 1.0 to 2.1 m (D. P. Schwartz and T. Dawson, personal communication, 2003).

At the pipeline crossing, the Denali fault experienced 5.5 m of right-slip displacement, and about 0.8 m of vertical slip, up on the north. The zone of ground disturbance was 200 m wide, and occurred near the southern edge of the design displacement zone, overlapping part of the 76.2-m-width assigned to the active-fault zone at the pipeline.

Comparison of Results

There was minor damage to the pipeline; however, the pipeline performed as designed and not a drop of oil was spilled. This excellent performance was due to accurate and conservative ground motion estimates, accurate and conservative fault displacement parameters, and innovative structural design—all developed 30 years ago. Tests of quantitative estimates of strong ground motions and fault displacements are rare. It is therefore of both historical and methodological interest to compare some of these estimates with the recorded ground motions and fault displacements near the pipeline (Table 2).

Table 2
Comparison of Denali Fault Parameters

Denali Fault Parameters	Estimated		Design	3 November 2002	
Earthquake magnitude	8.0		8.0	7.9	
Horizontal Acceleration at Pump Station 10	Page (1972) 1.2 g	Bolt (1972) 0.7 g	0.6 g	0.34 g	
Horizontal Velocity at Pump Station 10	145 cm/s	-	74 cm/s	114 cm/s	
Maximum right slip	9.1 m		6.1 m	Denali rupture 8.8 m	At pipeline 5.5 m
Maximum vertical slip	2.1 m		1.5 m	Denali rupture 2.0 m	At pipeline 0.8 m
Displacement zone width	579.1 m		610 m	Rupture within zone	
Fault rupture width	76.2 m		Included	200 m	

The peak horizontal ground acceleration recorded at Pump Station 10 was less than the design value and much less than the USGS estimate, whereas the recorded peak velocity exceeded the design value and was midway between the design value and the USGS estimate. Both seismic source modeling and the observed distribution of shaking-induced ground failure strongly suggest that near-fault shaking was more intense well to the east of the pipeline crossing than that measured at Pump Station 10.

Recent Seismic Hazard Analyses

Alyeska and its consultants conducted a reassessment of the seismic design criteria for the Trans-Alaska Pipeline in the mid 1990s. Following the reassessment, the USGS (Wesson and others, 1999a; 1999b; Frankel and others, 2000) published ground-motion maps for the State of Alaska that have been incorporated into the 2000 National Earthquake Hazard Reduction Program (NEHRP) seismic provisions, and into the 2000 International Building Code (IBC). To update Alyeska's seismic criteria and design procedures to be consistent with the design provisions of the 2000 IBC, a site-specific seismic hazard analysis was conducted. C. B. Crouse directed the analysis of the TAPS route to update the earlier seismic hazard work. The new study provided the ground-motion parameters (S_s and S_1), defined in the IBC. The study also provided IBC Site Class B response spectra for discrete locations along the pipeline route, which included the twelve pump stations (several of which are no longer in operation), selected milepost locations near the Denali fault, and the Valdez Marine Terminal.

A key aspect of the study was the modeling of the Aleutian Interplate Megathrust zone, the source of the magnitude 9.2 Prince William Sound earthquake of 1964. The important issues were the likelihood of a similar size event during the remaining lifetime of the pipeline, and the recurrence model for the intracycle seismicity between the giant megathrust events.

The results of geologic studies for the Copper River delta area (Plafker and others, 1992; Dames & Moore, 1991a) indicate the recurrence time of great megathrust earthquakes in the Prince William Sound region is approximately 720 ±200 years. Analyses of these recurrence data by Donovan, Tang, and Wen and Tang (included in Dames & Moore, 1991b) indicate the likelihood of the repeat of an event similar to the 1964 earthquake is very small (probability significantly less than 10^{-4}/year for the next 100 years). Consequently, this type of event was not included in the hazard analysis for the computation of ground motions.

In contrast, in its seismic hazard analysis, the USGS (Wesson and others, 1999a; 1999b) did include the magnitude 9.2 event, and assigned it an average annual probability of occurrence of 1/700, where the denominator is the USGS estimate of the average recurrence time for this event, in years. Not surprisingly, for the 2-percent-exceedance-in-50-years probability level, the USGS response spectral accelerations at the Valdez Marine Terminal, which is approximately 20 km above the megathrust, were higher than the values computed in the reassessment by roughly 30 to 60 percent.

The Gutenberg-Richter magnitude/frequency relation was selected by Crouse to model the recurrence of the intracycle seismicity associated with the Aleutian

Interplate Megathrust. The maximum intracycle magnitude associated with this model is debatable. Dames & Moore (1991a) argue that magnitude 7.5 to 8.0 is a reasonable range, based on a time-dependent model of strain accumulation on the megathrust since the 1964 event, and comparisons with other strongly coupled subduction zones. On the other hand, in his review of the Dames & Moore (1991a) report, Page (1992) cites the example of intracycle earthquakes in the central Aleutians, southwest of Prince William Sound, where the 1986 magnitude 8.0 earthquake occurred within the rupture zone of the 1957 magnitude 8.6 earthquake. Page (1992) estimates that, by the year 2025, the likely maximum magnitude for the intracycle event for the Aleutian Interplate Megathrust in the Prince William Sound region would be magnitude 8.2. Despite the uncertainty, the reassessment demonstrated that the ground motions computed for the Valdez Marine Terminal were not very sensitive to the maximum intracycle magnitude selected, whether it be 7.75 or 8.25.

An additional update of seismic criteria may be performed in the near future, depending on results from the study of the November 2002 earthquake. Pipeline integrity, public safety, and environmental preservation continue to be important issues in the operation of the Trans-Alaska Pipeline System.

Acknowledgments

This paper is based on the authors' personal knowledge and experience as part of their affiliation or interaction with the Alyeska Pipeline Service Company. Acknowledgment is made to the Alyeska Pipeline Service Company for permission to publish this paper.

References

Bolt, B. A., 1972, Ground accelerations and durations of strong ground motion to be expected from earthquakes affecting the Trans-Alaska Pipeline System, in Basis for the Earthquake Design of the Trans-Alaska Pipeline System: Alyeska Pipeline Service Company, Appendix A-3.1051a, Appendix I, 14 p.

Cluff, L. S., and Slemmons, D. B., 1972, Wasatch fault zone—features defined by low-sun-angle photography: Utah Geological Association, Publication 1, p. G1-G9.

Cluff, L. S., Slemmons, D. B., Brogan, G. E., and Korringa, M. K., 1974, Summary report, basis for pipeline design for active fault crossings for the Trans-Alaska Pipeline System, prepared for Alyeska Pipeline Service Company by Woodward-Lundgren & Associates, 1,257 p.

Dames & Moore, 1991a, Geologic and seismic studies, Trans-Alaska Gas System, Anderson Bay LNG Terminal, Port Valdez, Alaska: Prepared for Yukon-Pacific Corporation, D&M Job No. 14930-018-128, 2 vol.

Dames & Moore, 1991b, Seismic hazard studies for the Anderson Bay Terminal of the Trans-Alaska Gas System: Prepared by N. Donovan for Yukon Pacific Corporation.

Frankel, A. D., and others, 2000, USGS national seismic hazard maps: Earthquake Spectra, vol. 16, p. 1-19.

Hall, W. J., Nyman, D. J., Johnson, E. R., and Norton, J. D., 2003, Performance of the Trans-Alaska Pipeline in the November 3, 2002 Denali fault earthquake: TCLEE 2003, Sixth U.S. Conference and Workshop on Lifeline Earthquake Engineering Proceedings, Technical Council on Lifeline Earthquake Engineering, American Society of Civil Engineers, 13 p.

Newmark, N. M., and Hall, W. J., 1974, Seismic design spectra for Trans-Alaska Pipeline: Fifth World Conference on Earthquake Engineering Proceedings, vol. 1, p. 554-557

Newmark, N. M., and Hall, W. J., 1975, Pipeline design to resist large fault displacement: U.S. National Conference on Earthquake Engineering Proceedings, University of Michigan, Ann Arbor, p. 416-425.

Oakeshott, G. B., ed., 1975, San Fernando, California earthquake of 9 February 1971: California Division of Mines and Geology Bulletin 196, 463 p.

Page, R. A., 1971, Microearthquakes on the Denali fault near the Richardson Highway, Alaska (abs): American Geophysical Union Transactions, v. 52, p. 278.

Page, R. A., 1972, Crustal deformation on the Denali fault, Alaska, 1942-1970: Journal of Geophysical Research, vol. 77, no. 8, p. 1528-1533.

Page, R. A., 1992, Review comments on the seismic design criteria for the Anderson Bay Terminal of the Trans-Alaska Gas System: Appendix A of FERC (1995).

Page, R. A., Boore, D. M., Joyner, W. B., and Coulter, H. W., 1972, Ground motion values for use in the seismic design of the Trans-Alaska Pipeline System: U.S. Geological Survey Circular 672, 23 p.

Plafker, G., Lajoie, K. R., and Rubin, M. 1992, Determining the recurrence intervals of great subduction zone earthquakes in southern Alaska by radiocarbon dating, in Radiocarbon After Four Decades: Taylor, R. E., Long, A., and Kra, R., ed., p. 27, University of Arizona Press, Tucson.

Sherard, J. L., Cluff, L. S., and Allen, C. R., 1974, Potentially active faults in dam foundations: Geotechnique, vol. 24, no. 3, 367-428.

Slemmons, D. B., 1969, New methods of studying regional seismicity and surface faulting (abs): EOS, American Geophysical Union Transactions, vol. 50, p. 397-398.

U.S. Department of the Interior, 1972, Final environmental impact statement, proposed Trans-Alaska Pipeline: Washington, D.C., U.S. Government Printing Office, 6 vol.

Wesson, R. L., Frankel, A. D., Mueller, C. S., and Harmsen, S. C., 1999a, Probabilistic seismic hazard maps of Alaska: U.S. Geological Survey Open-file Report 99-36.

Wesson, R. L., Frankel, A. D., Mueller, C. S., and Harmsen, S. C., 1999b, Seismic hazard maps for Alaska and the Aleutian Islands: U.S. Geological Survey Miscellaneous Investigation Series, I-2679.

EFFECT OF THE DENALI FAULT RUPTURE ON THE TRANS-ALASKA PIPELINE

Steve P. Sorensen [1] and Keith J. Meyer [2]

Abstract

The Trans-Alaska Pipeline winds its way south from the Alaskan North Slope, transecting the state for a distance of 1,287 km (800 miles), to the Marine Terminal at Valdez. At peak throughput, the 1,219-mm (48-inch) diameter pipeline transported 2.1 million barrels of warm North Slope crude oil per day, and currently delivers about 1.0 million barrels per day. Alyeska Pipeline Service Company has operated the pipeline for its owner companies since startup in 1977.

On November 3, 2002 the Denali Fault, which intersects the pipeline route near Milepost 589 in central Alaska, ruptured over a distance of 336 km, producing the largest earthquake from a continental strike slip fault in North America since the 1906 San Francisco earthquake. This paper describes the design of the special above-ground pipeline segment in the Denali Fault zone and its response to violent, near fault shaking and displacement during the November 3rd, magnitude 7.9 Denali Fault event.

Introduction

Prior to construction of the Trans-Alaska Pipeline System (TAPS), a geologic study and field survey was conducted to identify active surface faults crossing the proposed route of the pipeline. Three potentially active fault zones were identified: Denali, McGinnis Glacier, and Donnelly Dome. In addition, over half of the route traverses areas of thaw unstable permafrost necessitating an above-ground mode on piling supports. A special design utilizing long grade beam supports was utilized within the Denali Fault zone.

The magnitude 7.9 Denali Fault earthquake produced ground motions that slightly exceeded the TAPS seismic criteria at periods longer than about 1.0 second and generally approached design criteria at shorter periods. The duration of shaking was approximately 100 seconds, with violent shaking in the near fault region of the pipeline. In near proximity to the Denali Fault crossing, large fault movements occurred.

The long period nature of the ground motion, produced a maximum ground velocity of approximately 114 cm/sec (45 in/sec)[3] coupled with violent near fault movement in

---

[1] Engineering Coordinator, Alyeska Pipeline Service Company, Fairbanks, AK, M. ASCE.
[2] Vice President, Michael Baker Jr., Inc., Anchorage, AK, M. ASCE.
[3] The maximum computed velocity is affected by high-pass filtering at 0.1 Hz. Based on preliminary results from in-progress studies by the USGS, it is believed that the actual peak velocity could have been about 50 percent higher than the calculated value of 114 cm/sec, i.e., about 170 cm/sec.

the above-ground segments. The event tested the ability of the fault crossing design to withstand fault rupture displacements approaching design level with respect to pipe movement on the at-grade sliding supports and pipe stress. There was no evidence of damage to the pipeline in the form of wrinkling or buckling of the pipe, but there was some support damage as described later in this paper. Realignment of a limited number of supports in the fault zone was required following the event to restore the fault movement response capacity of the pipeline to an additional 1.7 m of strike slip. This allowance accounts for early estimates of further fault displacements that may reasonably be expected if additional faulting occurs in the short term. The standard above-ground pipeline segments immediately north and south of the fault crossing zone were also inspected and support damage was repaired as described in Sorensen et al. (2003). Subsequent smart pig runs verified this observation.

This paper provides an overview of the field reconnaissance conducted in proximity to the Denali Fault, observations of movement and damage resulting from the violent lurching motion induced by the earthquake, and a discussion of the methodology used during early adjustments to the special grade beam design for the Denali Fault.

TAPS Design Background

Seventy five percent of the TAPS route was originally underlain by permafrost, necessitating a unique above-ground design that accommodates potentially unstable, ice-rich permafrost conditions. The below-ground pipeline is a conventional buried design used in thawed soils and in permafrost soils that are defined as thaw stable. A deep burial mode (below unstable surficial soils) and a refrigerated insulated burial mode (thick non-thaw stable soils) are used in some areas. Over half the pipeline (676 km) is constructed above ground and the remainder (611 km) is buried.

Both above and below-ground designs are affected by the extreme seismic activity of Alaska. (The magnitude 9.2 Prince William Sound subduction zone earthquake of 1964 is the second largest earthquake ever recorded.) TAPS traverses a wide range of geotechnical conditions that directly affect design and operation of the pipeline and related facilities. The goal of the original geotechnical design was to provide a stable foundation for pipeline elements, both statically and dynamically, so that the system can operate for the long term in an effective and safe manner without undesirable consequences to the public or environment. Alyeska Pipeline Service Company (Alyeska) conducted extensive seismological and engineering studies during the design of the pipeline to develop seismic structural design criteria, characterize active faults crossing the pipeline, and mitigate the potential affects of geohazards such as soil liquefaction and landslides.

Faulting that results in surface rupture is an important consideration for pipelines because pipelines crossing fault zones must deform or move longitudinally in response to axial compression or tension forces and laterally in response to bending and shear forces to accommodate ground surface offsets. Three potentially active faults were identified that crossed the pipeline in central Alaska. Alyeska carried out fault studies to characterize fault length, expected rupture slip, fault zone width, and slip recurrence interval (Cluff et al., 2003). Estimated ground displacements associated with these faults are shown in Table 1.

Table 1. Design Displacements at Active Fault Crossings.

Fault	Milepost	Max Credible Slip (m)		Design Slip Value (m)	
		Strike Slip	Dip Slip	Strike Slip	Dip Slip
Donnelly Dome	556	9.1	2.1	6.1	1.5
McGinnis Glacier	587	4.0	3.0	2.4	1.8
Denali	589	1.5	4.6	0.9	10.0
Note: Displacements were originally specified in units of integer feet.					

During the November 3, 2002 magnitude 7.9 event, the Denali Fault ruptured over a distance of 336 km. The epicenter occurred near the newly discovered Susitna Glacier thrust fault, approximately 90 km to the west of the pipeline. The fault intersects TAPS at Milepost 589 in central Alaska. Average slip on the fault is estimated to be 5.5 m near the pipeline crossing with maximum slip of almost 9 m occurring 120 km to the east of the pipeline crossing.

Special Above-Ground Configuration at Denali Fault Crossing

The Denali Fault extends east to west more than 650 km through the Alaska Range. It is a right-lateral strike-slip fault with a normal-slip component. The fault plane is nearly vertical, and the up-block is to the north. The pipeline crosses the Denali Fault zone between Lower Miller Creek and Miller Creek near MP 589. The width of the fault zone used in design was 579 m, which implies that a surface rupture was possible anywhere within this zone. The limits of the fault zone were established between pipeline as-built Stations 31082+00 and 31101+00 (feet). The fault strike at the pipeline crossing is N55°W. Since the pipeline bearing is N6°32'E at the fault crossing, the fault crossing intersection angle is 61°32' counterclockwise with respect to the pipe survey centerline as shown in Figure 1.

The area is predominately a level outwash plain composed of thawed fluvial gravels and glacial till material overlaying bedrock at an undetermined depth. However, the northern 61 m of the fault zone is a relatively steep hill composed of silty gravelly glacial till and granite bedrock.

During the course of pipeline design, it was determined that the warm crude oil pipeline would be above-ground in the Denali Fault area because of the presence of thaw unstable permafrost soil. Due to the magnitude of the design fault displacements, conventional above-ground construction was judged impractical at the Denali Fault crossing. Instead, at-grade beam construction on which the pipe is free to slide was selected to provide support for the pipeline across the fault zone because it can accommodate relatively large displacements without significant pipe deformation. This low-to-the-ground construction mode has the added benefit of limiting damage to the pipeline if, for any reason, the pipe unexpectedly slips off the

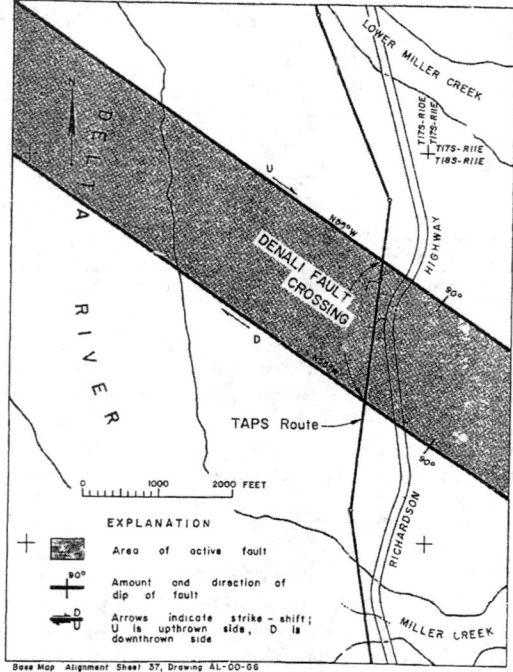

Figure 1. Pipeline crossing of Denali Fault.

support beam during a seismic event. The pipeline was designed to accommodate a right lateral strike slip of 6 m and a vertical slip of 1.5 m with the north block up. The dip plane was assumed to be vertical. The three-dimensional displacement components relative to the pipeline orientation (i.e., parallel to the pipeline bearing of N 6°32′ E) are 2.9 m longitudinal 5.4 m transverse, and 1.5 m vertical.

At the Denali Fault crossing, the pipeline is supported on 33 steel box beams and concrete beams set on-grade, at approximately 18-m intervals over a distance of 579 m. The beams are approximately 12 m long. The location of some of these beams is shown as Grade Beam Numbers 5 through 12 (Figure 2) in the immediate vicinity of the fault. These beams were sized and arranged to accommodate a fault slip. Anchors in the typical above-ground configuration bound the special Denali Fault above-ground segment north and south of the fault crossing area.

The pipeline shoes were lengthened in the fault zone to accommodate the large longitudinal and lateral pipeline movements anticipated by the largest projected fault displacement and magnitude as shown in Figure 3. Close attention was paid to the sliding surface of the crossbeam and the material lining the bottom of the pipe shoe. The pipe shoe base consisted of a 0.63-cm thick steel plate to which was bonded a

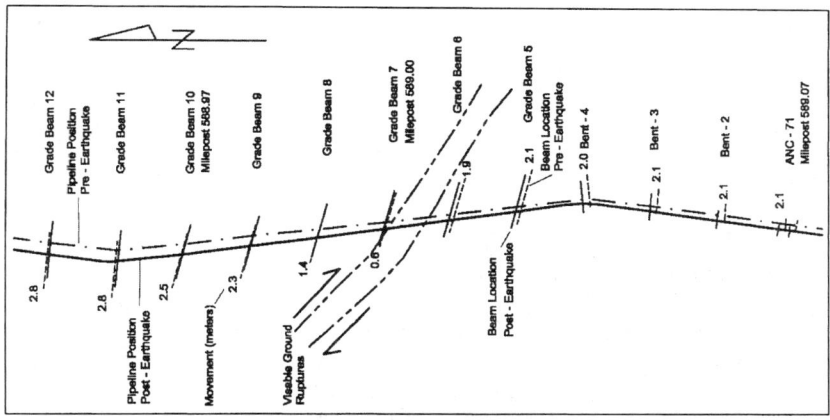

Figure 2. Denali Fault pipeline crossing schematic

Teflon pad. The crossbeam surface was sand blasted and painted with zinc rich epoxy paint. The coefficients of friction values of 0.10 static and 0.05 dynamic were specified by design and confirmed through testing.

Performance of Special Fault Crossing Configuration

As expected, significant pipeline and pipeline support movements occurred during the November 3, 2002 fault rupture. The trace of the rupture was clearly visible between grade beam supports 6 and 7 near the southern end of the special configuration shown

Figure 3. Typical special configuration bent on a concrete grade beam: Note offset shoe position by design to accommodate pipe movement in this case to the left.

schematically in Figure 2. A range of potential locations of fault surface rupture was analyzed during design. From these design studies, the most severe effect on the pipeline was determined to be when surface rupture was postulated at grade beam 7, which is only one bent away from the observed fault trace. Therefore, this event actually represented the near maximum expected movement for the design, as was indeed the observed result. The relative locations between pre-event and post-event pipeline and beam positions are shown in Figure 4.

(a) before   (b) after

Figure 4. TAPS crossing of Denali Fault before and after fault slip, looking south. Note movement and bowed segment after fault displacement, which acts to compress the pipeline crossing segment.

Anchors are designed to control longitudinal movement of the pipeline through frictional resistance. Anchors begin slipping at an instantaneous force of 475 kN (105 kips). The anchors north and south of the fault crossing slipped longitudinally and experienced further movement through slippage between the anchor frame and support brackets on the VSMs. Some deformation of the anchor frames was observed via yielding of the slotted bolted connection on the frame itself, but not at the support bracket. During repair operations on the first anchor south of the Denali Fault, the pipe clamps connecting the anchor assembly to the pipeline were released. There was no observed movement of the pipe through the anchor clamps demonstrating that there was no residual compressive stress in the pipeline.

The pipeline support shoe at the first VSM bent (immediately south of the last grade beam configuration) was severely damaged (see Figure 5). This was probably due to the vertical faulting and concurrent violent dynamic shaking both vertically and horizontally. The shoe was replaced during the field repairs after the event, and the pipe was thoroughly inspected. Based on documented inspection reports, the pipe itself had no observable damage in this region.

Figure 5. Post-event shoe damage (left) and pre-startup repair (right) on the seismic bent immediately south of the Denali Fault. Note the temporary shoring beneath the pipe used during repairs (right side of picture), the bumper beam on the pipe clamp (center), and the impact absorber on the VSM (far left side of picture).

At special configuration locations, ground movement at the grade beams was evident, although no damage to the grade beams, pipe support shoes or the pipe itself was observed (see Figure 6). Slip marks caused by the pipeline shoe sliding across the crossbeams and grade beams during faulting and associated shaking were observed throughout the fault crossing. Some of the grade beams in the near fault area exhibited evidence of minor ground plowing, probably through inertial forces in the beam and/or as a result of sliding friction resistance between the shoe and beam. Pipe shoe movement was often more than the estimated ground fault displacement, and occurred well back from the rupture trace line due to compression and pipe flexure. The final "set" of the pipeline results from friction resistance between the shoe and the cross beam, producing the illusion of residual compression. The observed movement was expected and predicted by the design analysis.

Figure 6. Post-earthquake shoe location on grade beam prior to re-centering.

Post Earthquake Assessment and Repairs

Surveillance of the pipeline in the first hours after the event clearly established that surface rupture of the Denali Fault had occurred near the south end of the fault crossing zone and that sliding pipe shoes had experienced large movements. Within hours of the event, a response coordinator was mobilized in the field to initiate damage assessments and mobilize repair crews. The fieldwork was aided by unseasonably warm (near freezing) temperatures, but hampered somewhat by short daylight hours. By early the next day, engineering assessment crews were fully engaged, and repair crews and equipment were on site.

Initial field estimates of the ground faulting at the fault scarp indicated displacements of approximately 2.3 m strike slip and 0.8 m vertical. GPS and geodetic surveys initiated within one week after the earthquake determined the fault displacements to be larger: 5.5 m horizontal and 1.5 m vertical distributed over a zone of approximately 200 m.

Several of the shoes in both the typical VSM-supported configuration and the special Denali Fault grade beam configuration were reported to be near the edge of the crossbeams. An analysis was performed to determine which pipe shoe locations needed to be adjusted to ensure the shoes would remain on the beams should additional fault displacement occur.

The Alyeska survey group collected post-earthquake longitudinal shoe position data on November 13th. Because of the onslaught of winter, a decision was made during the early repair period to try to accommodate any further short-term fault movement by leaving the support beams in place and relocating the shoes on the pipe where necessary. This would allow additional longitudinal and lateral movement capacity. By consensus among geologists it was decided that this was a major event, and that it had released most if not all of the locked-in strain energy. The geologists expected no more than a 1.7 m displacement potential in the short term. This was believed to be an upper bound limit.

The allowable extents of future longitudinal shoe movements were estimated based on the least remaining amount of lateral shoe movement available on the grade beams. This shoe position was then compared to the early estimates of a maximum near term lateral fault shift, 1.7 m. The additional expected longitudinal pipeline movement was found by multiplying the predicted full design movements at each bent by the ratio of the anticipated fault shift (1.7 m) and the design fault shift (6.25 m).

The final location of the shoes were then determined by adding the expected longitudinal movements to the post-earthquake locations for each shoe and verifying whether or not the shoe could still rest on the beam. A safety buffer of 150 mm was used; meaning that at least 150 mm of shoe had to remain overhanging the crossbeam at the shortest shoe location, or the shoe location on the pipe would require adjustment.

A total of ten shoes were estimated to have less than 150 mm of remaining overhang after calculating the projected additional fault displacement. One shoe had less than 150 mm of overhang in its post-event position. All eleven of the target shoes were re-centered on the pipeline.

Conclusion

The design of the Denali Fault crossing for TAPS was the first specially designed crossing of an active fault by a crude oil pipeline. The innovative above-ground crossing design was developed over 30 years ago when "lifeline earthquake engineering" was in its infancy. The 2002 Denali Fault earthquake provided a full-scale test of this crossing concept, and the pipeline and support system performed as expected, without damage to the pipeline or leakage of oil. It is worthwhile to note that had the pipeline had been buried in a special fault crossing trench (loose backfill and sloped sides), it would have required much heavier wall pipe and local buckling likely would have occurred, hence requiring pipe repair and more extended downtime.

Additional studies are in progress to characterize the potential for future displacement on the Denali Fault during the remaining life of the pipeline. The pipe and supports in the fault crossing zone will be further realigned if required.

Acknowledgment

This summary paper is based on the authors' personal knowledge and experiences as part of their affiliation with Alyeska, and in no way reflects on policies or practices of Alyeska with respect to the items presented. All figures presented are courtesy of Alyeska. The authors would like to thank to Alyeska for permission to publish this summary paper.

References

1. Cluff, L.S., R.A. Page, D.B. Slemmons, and C.B. Crouse (2003). "Seismic Hazard Exposure for Trans-Alaska Pipeline," *Proceedings of the 6th U.S. Conference on Lifeline Earthquake Engineering (Advancing Mitigation Technologies and Disaster Response),* Long Beach, CA, TCLEE, ASCE.

2. Sorensen S.P., K.J. Meyer, P.A. Carson, and W.J. Hall (2003). "Response of the Above-Ground Trans-Alaska Pipeline to the Magnitude 7.9 Denali Fault Earthquake," *Proceedings of the 6th U.S. Conference on Lifeline Earthquake Engineering (Advanced Mitigation Technologies and Disaster Response)*, Long Beach, CA, TCLEE, ASCE.

RESPONSE OF THE ABOVE-GROUND TRANS-ALASKA PIPELINE TO THE MAGNITUDE 7.9 DENALI FAULT EARTHQUAKE

Steve P. Sorensen[1], Keith J. Meyer[2], Paul A. Carson[3], and William J. Hall[4]

Abstract

The Magnitude 7.9 Denali Fault earthquake produced ground motions that exceed the Trans-Alaska Pipeline System seismic criteria at periods longer than about 1.0 second and approach design criteria at shorter periods. Consequently, this ground motion provided a test of the pipeline design for ground motions essentially equivalent to those considered in the original seismic criteria and design analyses. This paper compares the actual (measured) ground motions to those used in design, discusses the response of the above-ground pipeline to the ground motion inputs, and describes the repairs necessary to restore the pipeline support hardware and pipeline alignment to a satisfactory operating condition.

Introduction

Over one-half of the 1,287-km (800-mile) Trans-Alaska Pipeline System (TAPS) is elevated above ground on support bents spaced at 19 m (60 feet) with a frictional bearing surface between the pipe and the bent. Bents are arranged in zigzag configurations that terminate at specially designed slipping anchors. Since the pipeline passes through active seismic zones, the above-ground pipeline and support system has been designed to withstand ground shaking effects.

The dynamic response of the above-ground pipeline to earthquake ground motions is highly nonlinear and requires a nonlinear transient analysis of some form to simulate the soil-pile interaction known as Vertical Support Member (VSM) interaction, the pipe movement with the anchors, bumpers, and frictional supports, and the spatial variation of ground motions due to seismic wave propagation. The dynamic response of the pipeline is limited by the geometry of the zigzag configuration, by pretensioned slip at anchors, by crushable sacrificial honeycomb energy absorbers at the anchors, and by specially designed bumpers placed at selected VSMs to limit lateral movement of the pipeline. The flexibility and the tipping and tilting capability of the pipe supports are considered, as well as the resilience and flexibility of the pipeline and its nonlinear frictional interaction between the shoe base and the crossbeam bearing surface. The insulation modules on the pipe serve to provide additional damping for the pipeline configuration. The above-ground sections of the pipeline

---

[1] Engineering Coordinator, Alyeska Pipeline Service Company, Fairbanks, AK, M. ASCE.
[2] Vice President, Michael Baker Jr., Inc., Anchorage, AK, M. ASCE.
[3] Structural Lead, Michael Baker Jr. Inc., Anchorage, AK.
[4] Professor Emeritus, Department of Civil Engineering, University of Illinois, Urbana, IL, H.M. ASCE.

were designed to provide adequate support without damage to the pipe in the event two adjacent intermediate supports were lost.

Pipeline Description

TAPS is a 1,287-mm (48-inch) diameter crude oil pipeline that has transported over 14 billion barrels of oil from the oil fields in and around Prudhoe Bay, Alaska to the Valdez Marine Terminal, where the oil is loaded on ocean-going tankers for delivery to refineries. The pipeline was constructed in the mid-1970's with the first oil reaching Valdez, Alaska in July 1977.

The pipeline crosses three major mountain ranges, several major rivers (including the Yukon River), and three fault zones. The subsurface conditions range from continuous permafrost for the first 451 kilometers (280 miles) to discontinuous permafrost for the next 741 kilometers (460 miles) across the interior of Alaska to sporadic permafrost through the Chugach Mountain Range as the pipeline traverses the last 96 kilometers (60 miles) to its terminus. The pipeline was generally buried wherever thawed or thaw-stable soils were encountered. Where this criterion could not be met, the pipeline was constructed on specially designed above-ground supports.

This above-ground mode is built in a zigzag configuration to allow for expansion or contraction of the pipe due to temperature changes by converting changes in pipeline length into lateral (sideways) movements. To permit lateral motion, the pipe is mounted on a structure that can slide on crossbeams installed between VSMs. This design also allows for motions caused by an earthquake.

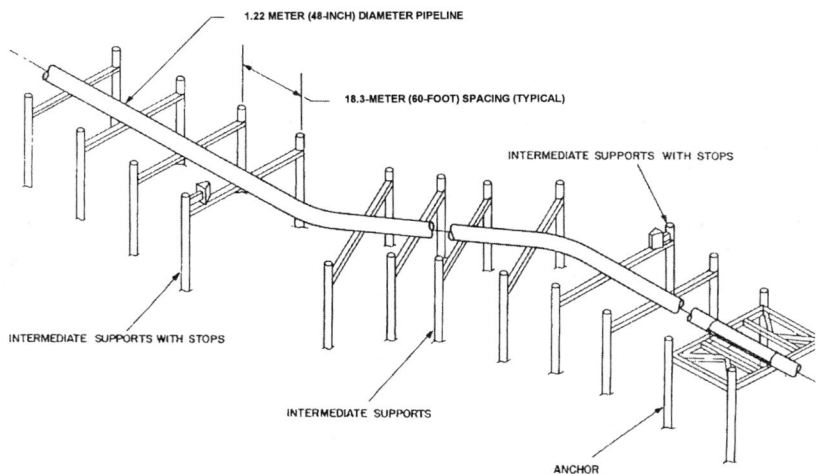

Figure 1. Schematic of standard Zee Configuration.

The above-ground system is arranged into a series of connected segments that are 244 to 549 m long (800 to 1,800 feet). The geometry of each segment is defined by one of five types of zigzag "configurations". The standard zee configuration (Figure 1) is the most common. At each end of a typical configuration, an anchor support assembly restrains the pipe. The anchor supports consist of four VSMs, a structural steel platform and a friction slide plate assembly. The friction slide assembly is design to resist an initial differential force in the longitudinal direction (along the pipe) of 467 kN (105 kips) before slipping occurs.

The pipe rests on intermediate supports (Figure 2) at approximately 18.3-meter (60-foot) spacing between the anchors. Intermediate supports typically consist of two VSMs and a structural steel crossbeam. In a relatively few cases, where the height of the pipe above the ground was high, a third VSM was installed and braced back to the standard two-bent configuration in order to increase resistance to lateral loads.

At the Denali Fault crossing itself, 33 supports in a 579-meter (1,900 feet) section consisting of either steel box beams or concrete beams set on-grade support the pipeline. These beams were sized and arranged to accommodate a fault slip of 6.1 meters (20 ft) of horizontal and 1.5 meters (5 ft) of vertical fault displacement. The response of the pipeline in this segment is described in Sorensen and Meyer (2003).

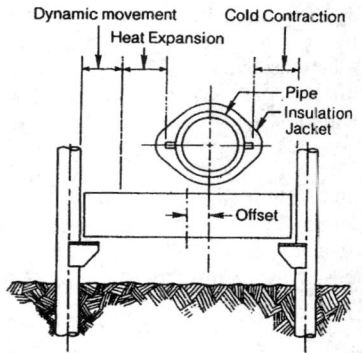

Figure 2. Cross section of intermediate support VSM.

Original Design Analysis

In the original design, all above-ground configurations for TAPS were analyzed under static conditions (gravity, design pressure and design temperature differential, Figure 3). In addition, dynamic analyses were conducted to determine the movement of the pipe on the supports during a seismic event. The results of these analyses were used to size the support hardware and determine crossbeam length. Crossbeams of varying length were installed to accommodate the predicted lateral movements of the pipeline within each configuration. Prior to startup and in accordance with design specifications, the cross beams were positioned off-center with respect to the pipe

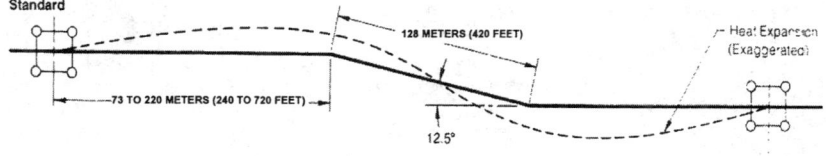

Figure 3. Movement of pipeline in a standard Zee Configuration.

centerline to allow for thermal expansion during startup as the pipe warmed to the design operating temperature. The side to which the pipe was expected to move once it was warm is called the "hot" side, while the opposite side was designated the "cold" side.

Seismic bumpers were designed for selected locations to control lateral seismic movement of the pipeline within the selected bents where contact was possible (Figure 4). At these same bumper bents, bolts were added between the support bracket on the VSMs and the crossbeams to further strengthen the beam-VSM connections and provide a mechanism for sharing lateral impact loads among both VSMs in the bent.

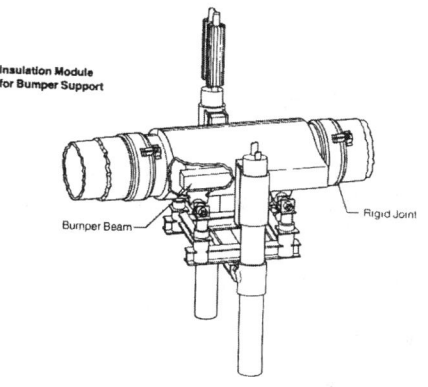

Figure 4. Intermediate support with bumper beam.

Seismic Analysis

The pipeline alignment is divided into five distinct seismic zones. The northern part of the line is designated as a relatively low, 5.5 Richter Magnitude (RM) zone, while the two highest (8.0 and 8.5 RM zones) are located near the fault crossing zones close to the Alaska Mountain Range and at the extreme southern end of the line, respectively.

Design response spectra were used to define seismic loading on the above-ground pipeline. There are two basic spectra for the TAPS project, developed during the original design, which correspond to the "Operational" and "Contingency" loading. Operational loading defines the excitation level which the project structure must withstand without damage, significant deformation, or interruption of operation. Contingency loading is the excitation level which the structure must withstand without structural collapse, loss of function of essential facilities, or release of hazardous substances that could damage the environment. At the contingency level, it is recognized that some damage is permissible, and that a safe shutdown of the structure may be required for inspection and repair. Often, the operating level is used to define structure excitation under combined loads such that members remain within allowable stress limits and well below the minimum specified yield value. Contingency allowable stresses are typically set at, or slightly beyond the minimum specified yield value.

There are several specific techniques to approach the analysis of structures subjected to dynamic loading, but they can all be loosely categorized into three main approaches: Equivalent Static techniques, Response Spectrum/Modal techniques, and Time History techniques. The Equivalent Static Procedure and Response Spectrum Procedure, which are the easiest to apply and also the most commonly used procedures, have implicit assumptions which make them only somewhat applicable to the TAPS above-ground pipeline. First, these techniques assume the structure has a linear response, or at least linear about a defined equilibrium position. The presence of friction and nonlinear stops and gaps violates this assumption. Also, these techniques typically assume that the excitation at the base of the structure is uniformly applied in time, i.e., all supports are excited in phase by the same time loading function. However, ground excitation is best characterized by a traveling wave, which imparts motion to the surface as it travels at a seismic wave velocity. This wave velocity (typically 1000 m/sec or 3283 ft/sec) imparts a lag to excitation for structures that have a considerable length relative to the wave velocity. For these reasons, time history techniques are most applicable for the analysis of TAPS above-ground configurations.

The time history technique describes the seismic motion as a function of excitation versus time to the structure. An analysis of structural response during each discrete time interval is performed, followed by an evaluation of the new state of the structure. This structural state then forms the initial basis for the next discrete excitation interval and the process is repeated for all time steps. Generally, this technique has the advantage of incorporating nonlinear events directly into the analysis using commercially available programs. However, it is unusual to find a commercially available analysis program that will readily analyze the lag in excitation for long-axis structures, such as the above-ground pipeline configurations of TAPS.

The analytical technique requires input of a time history of the ground motion and subsequent integration of the response. The design basis spectra for the TAPS project formed the basis for the above-ground seismic loading. Earthquake time history motions were artificially generated using random generation routines governed by statistical procedures consistent with the overall frequency-response characteristics of the design basis spectra. In order to ensure that the response is bounded by the analysis, it is standard practice to generate an ensemble of time histories, each having some differences in amplitude-frequency characteristics. Five such earthquake time histories were generated for the TAPS design. Each record is an acceleration time history with accelerations recorded at a constant time interval of 0.03 seconds. There are a total of 1,664 accelerations for each record, so each record simulates an earthquake history with duration of 49.89 seconds. It should be noted that an alternate approach with some merit involves selecting a suite of actual earthquake records. This approach may be considered for comparative use in the future.

The analytical technique utilizes one of two approaches. In the first approach all five earthquake records for each direction of motion are run, while running another one of the five time histories in the orthogonal direction. The second approach identifies a combination, that by experience using the first approach, produces the dominate response envelope for the configuration under consideration. At every time step, the input excitation is applied to the base of each pipe support. The excitation input value for each support at any one time step is generally different from the value at the same time step for another support, because the phase lag of the wave as it traverses the structural axes causes the excitation record to "start" at different times.

In addition, it is recognized that the angle between the seismic wave front and the structure axes, or "attack angle" should be specified in order to produce the maximum response. During the design phase it was determined that an attack angle along the longitudinal axes of the typical above-ground configuration was the most conservative case.

Analysis Tools

During the original design, the TAPS project recognized that time history analysis was applicable to the analysis and design of the TAPS pipeline. Lacking an applicable computer analytical tool at the time, a unique program titled DrainPIPE was developed specifically for the TAPS project by Graham Powell, then Professor at the University of California, Berkeley. This program incorporates friction, gaps, and stops at pipeline supports, as well as traveling wave excitation. The DrainPIPE program was originally written in FORTRAN and ran on a CDC 6600 mainframe computer. The DrainPIPE program was recently converted to Visual Basic with Excel serving as the IO interface.

The same synthetic earthquake records were used for all RM zones and for both the Operational and Contingency load cases. The records were scaled to reflect the different excitation levels in each RM zone, while still preserving the frequency content of the records. The Operational load case was taken as 50% of the corresponding contingency scale factors for acceleration.

Viscous damping values for typical structural analyses use estimates of 2% to 5% for analysis. However, DrainPIPE explicitly models energy dissipation, using friction elements and stop elements with slip after yield. Therefore, only a nominal value of 0.5% damping is input to DrainPIPE to control the numerical integration procedure.

Structural pipe support elements are modeled to incorporate not only initial elastic resistance, but also yield points and resistance after yield. Successive yield points are input to simulate successive failures of the components, for example insulation crushing followed by bending failure of the VSM. The pipe element, however, is linear-elastic so the results must be examined to ensure that the analytical results conform to this assumption. All analysis performed to date indicate that the need for inelastic pipe elements has not been a concern because calculated pipe stresses have remained within elastic limits. The focus of the analysis thus becomes pipe movement prediction, and/or the reaction of the pipe support system.

November 3, 2002 Earthquake

All of the synthetic time histories used for TAPS utilized the design basis spectra as "target" envelopes. On the other hand, the 2002 Denali Fault earthquake showed a significant deviation from the design basis spectra in the longer period ranges of 1.0 second and above (or frequency range less than 1.0 Hz). A comparison of the structural design spectrum and the November 3rd earthquake spectrum is illustrated in Figure 5. Although the nonlinear frictional effects of the support hardware dominate pipe response, it is nevertheless instructive to note that "linear" models of typical above-ground configurations are in this longer period range.

An extensive damage survey and evaluation of the pipeline and its supports was started by APSC the morning following the earthquake. Because post-event

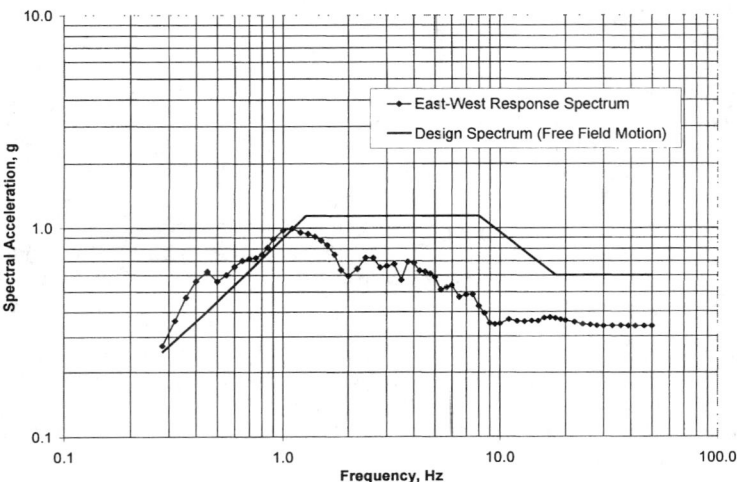

Figure 5. TAPS Design Spectrum and 2002 Denali Fault earthquake spectrum.

surveillance conducted within hours of the earthquake indicated substantial violent pipe movement near the fault zone, the pipe configurations within and adjacent to the Denali Fault were deemed the highest priority items for this assessment.

There was no damage to the mainline pipe itself. The Denali Fault-crossing segment moved on its supports per design. Damage to the VSM pipe support system hardware and crossbeams was limited to within approximately 2 kilometers of the Denali Fault zone. The predominant damage occurred during pipe impact with some VSMs.

Figure 6. Support damage near the Denali Fault – loss of support and tilted VSM.

Indentation or localized crushing of insulation modules was clearly caused by contact with VSMs. The most extensive damage to the support system was found in the above-ground configuration just to the south of the Denali Fault-crossing zone, although the region to the North was also affected.

At several locations the 25 mm (1-inch) connection bolts holding the pipe support crossbeam to the support bracket on the VSM were sheared. In five cases this allowed the crossbeams to dislodge from the support brackets and fall to the ground (Figure 6). In most cases, although crossbeam connection bolts and seismic bolts either yielded plastically or failed at the support boss (Figure 7), the beams remained supported on the VSM support brackets. The failure of the connection bolts were most often the only observed damage to the hardware. At some locations, the VSMs were tilted from vertical to an estimated 7 to 10 degrees due to impact with the pipe (Figure 6). At these locations the VSMs were notably flattened at the contact points but no damage was observed on the oil pipeline wall itself (Figure 8). It should be noted that the original TAPS project criteria included a provision for the complete loss of two consecutive supports without damage to the pipeline. Analysis conducted just prior to the earthquake, using a theoretical pre-event loss-of-support model coupled with seismic dynamic loading, verified this design parameter. The Denali earthquake event tested this analysis and validated the results.

Figure 7. Deformation of seismic bolts tying the beam to the support bracket (left). Seismic bolt connection failure (right).

Immediately following the damage assessment in the near fault region, and concurrent with ongoing assessments for all other areas of the pipeline, repair efforts were initiated to restore pipe support near the fault zone. Timber cribbing was placed beneath the pipe at locations where the crossbeam connections had been compromised. The VSMs were brought back to vertical with jacks and heavy equipment where required, crossbeam support connections were restored, and the temporary shoring, jacks and air bags removed. At most locations the original crossbeam could be re-installed after inspection. At two locations the support hardware was completely replaced. At the station locations that exhibited the greatest insulation damage, the insulation module was completely removed for a detailed inspection of the pipe itself. No damage to the pipe was found at these locations; hence, further inspection of the pipe was postponed to after restarting the pipeline.

Figure 8. VSM damage south of the Denali Fault (left) with no damage to the pipeline or pipe clamp (right)

By the time restart of the pipeline was initiated, approximately 66 hours after the earthquake, the above-ground configuration supports at all critical locations were fully restored with permanent structural hardware.

Evaluation Status

Pipe movement and associated hardware damage in the near fault zone area exceeded the analytical estimates based on the design criteria. This is possibly due to both the November 3rd spectra exceeding the design spectral values in the longer period range. The damage assessment will provide boundary values for the structural response forces for evaluation of amplification effects. This work is ongoing.

Conclusion

The results of the 2002 Denali Fault earthquake confirmed that the TAPS above-ground configurations are able to withstand expected seismic forces sufficiently to maintain integrity of the pipeline. The above-ground pipe was inspected at severe contact locations with no observable evidence of pipe damage. On the other hand, damage to the above-ground support hardware exceeded expectations in the near-fault zone. The observed contact of the insulation modules with VSMs on both sides of the pipe, especially at support locations without seismic bumpers, failure of some of the crossbeam connection and seismic bolts, and tilting of some VSMs through impact with the pipe demonstrated that near fault ground movement was violent. Seismically induced lateral displacement of the pipe on the crossbeams, as well as lateral load reactions at the VSMs were obviously greater than expected in some cases. However, it is noteworthy that while damage to supports and support connections occurred, there was no indication that the pipeline itself experienced stresses beyond elastic limits. Documented post event inspections demonstrated that the pipeline remained undamaged and fully serviceable, even with loss of two consecutive supports. This was predicted during design and later by a dynamic modeling study. Further clarification of the near-field response of the pipeline and supports is the focus of ongoing engineering efforts by APSC.

Acknowledgment

This summary paper is based on the authors' personal knowledge and experiences as part of their affiliation with the Alyeska Pipeline Service Company (APSC), and in no way reflects on policies or practices of APSC with respect to the items presented. All figures presented are courtesy of APSC. The authors would like to thank to APSC for permission to publish this summary paper.

References

1. Sorensen, S. P., and K. J. Meyer (2003). "Effect of the Denali Fault Rupture on the Trans-Alaska Pipeline," *Proceedings of the 6th U.S. Conference on Lifeline Earthquake Engineering (Advanced Mitigation Technologies and Disaster Response)*, Long Beach, CA, TCLEE, ASCE.

ASSESSMENT OF THE BELOW-GROUND TRANS-ALASKA PIPELINE FOLLOWING THE MAGNITUDE 7.9 DENALI FAULT EARTHQUAKE

Elden R. Johnson[1], Michael C. Metz[2] and David A. Hackney[3]

Abstract

Nearly half of the 800-mile (1,287-km) Trans-Alaska Pipeline System (TAPS) is buried in a standard pipeline trench with a minimum depth of cover of 0.9 m (3 feet). The magnitude 7.9 Denali Fault earthquake produced peak ground motions of 0.34 g at a location 5 km from the fault. The duration of shaking was approximately 90 seconds, contributing to liquefaction of subsurface deposits along the pipeline as evident from numerous sand boils. In some areas in proximity to the Denali Fault crossing, moderate lateral spread movements occurred. Landsliding did not occur along or across the pipeline right-of-way itself, although there were a number of landslides and rock falls that occurred proximate to the Denali Fault zone several tens of kilometers west of the pipeline.

The near-source violent motions coupled with soil liquefaction provided the opportunity for developing significant bending and axial strain in the buried pipeline. The pipeline was excavated at a location where evidence suggested lateral spreading or subsidence due to liquefaction, but no damage to the pipe was observed. Approximately one month after the earthquake, the pipeline was inspected using an instrumented in-line monitoring device ("smart pig") capable of detecting pipeline curvature and deformation. No evidence of pipe deformation, strain increases, or curvature changes in excess of acceptable limits were observed.

This paper provides an overview of the field reconnaissance of the below-ground pipeline segments in proximity to the Denali Fault, observations of liquefaction and ground failure, and a discussion of the use of an instrumented pig to validate pipeline structural integrity.

Introduction

The Trans Alaska Pipeline System (TAPS) zigzags off the North Slope in its unique above-ground configuration, transecting the State of Alaska 1,287 km (800 miles), south to its marine terminal in Valdez (Figure 1). At peak throughput, the 1,219-mm (48-inch) diameter pipeline transported 2.1 million barrels[4] per day (mbpd) of warm North Slope crude oil. Due to declining North Slope production, it currently delivers about 1.0 mbpd, or about 16 percent of America's domestic supply. The pipeline has

---

[1] Engineering Advisor, Alyeska Pipeline Service Company, Fairbanks, AK, M. ASCE.
[2] Consulting Geologist, Anchorage, AK.
[3] Engineering Coordinator, Alyeska Pipeline Service Company, Fairbanks, AK, M. ASCE.
[4] One barrel equals 42 gallons.

been operated by Alyeska Pipeline Service Company (Alyeska) for its owners since startup in 1977.

On November 3, 2002 the Denali Fault, which intersects the pipeline route at Milepost 588 in central Alaska, ruptured over a distance of 336 km, producing the largest earthquake from a continental strike-slip fault in the United States since the 1906 San Francisco earthquake. This paper describes the design and performance of buried portions of TAPS near the fault during the magnitude 7.9 Denali Fault event. The above-ground pipeline design and its performance during the Denali Earthquake are described in a paper by Sorensen, et al. (2003).

TAPS Design

TAPS traverses a wide range of geotechnical conditions that directly affect design and operation of the pipeline and related facilities. The goal of the original geotechnical design was to provide a stable foundation for pipeline elements under both static and dynamic conditions. The pipeline design was affected by the extreme seismicity of Alaska, acknowledging that the magnitude 9.2 Prince William Sound subduction zone earthquake of 1964 remains the second largest earthquake ever recorded worldwide. Prior to construction, Alyeska conducted extensive seismological and engineering studies (Cluff et al 2003) to characterize active subduction zone and continental faults crossing the pipeline, develop ground motion design criteria, and mitigate the potential affects of geohazards such as soil liquefaction and landslides.

Three potentially active faults that crossed the pipeline in central Alaska were identified and fault studies were carried out to characterize fault length, expected rupture slip, fault zone width, and slip recurrence interval. Large rupture displacements up to 6m/1.5m (strike/dip) were anticipated on the Denali Fault, with 4m/3m on the McGinnis Glacier, and 2m/5m on the Donnelly Dome Fault.

Approximately 75 percent of the 1,287 km TAPS route was originally underlain by permafrost, necessitating a unique above-ground design in many places to accommodate potentially unstable, ice-rich permafrost conditions. Over one-half the pipeline (676 km) is constructed above ground, with the remainder (611 km) buried. The below-ground pipeline is, for the most part, a conventional buried design used in thawed soils and in permafrost that is defined as thaw stable (less than 6 percent fine grained soil). However, TAPS unique terrain and seismicity conditions required two additional options – a deep burial mode for use below unstable surficial soils, and a refrigerated insulated burial mode for use in thick non-thaw stable soils. As expected, the warm buried pipeline has thawed most permafrost along the right-of-way (where permafrost was initially present within 35-50 m below pipe), except for a relatively short (6-km) portion of the buried pipeline where a specially insulated and refrigerated system keeps the foundation frozen.

During the development of the design concept for TAPS, it was presumed that a buried construction mode would be incapable of withstanding the large ground ruptures that might occur at the three identified active fault crossings. Hence, all known active faults were crossed above-ground (Sorensen et al., 2003; Hall et al., 2003). As an additional design safeguard, the standard trench design was validated for a generic condition of 0.6 m (2 ft) of differential movement due to ground rupture.

Seismic design ground motions for the Trans-Alaska Pipeline System are based on stipulated design earthquakes for five seismic zones along the route as delineated by the U.S. Geological Survey (USGS) (Cluff et al., 2003). A maximum credible earthquake, with an estimated return period of 300 years or more, was specified for each of the five seismic zones and defined in terms of Richter magnitude (used at the time of pipeline design) as depicted in Figure 1. Alyeska established design ground motions for earthquakes in each seismic zone as delineated in Table 1.

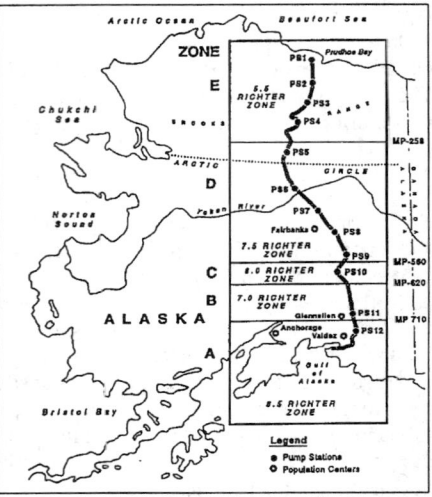

Figure 1. Seismic zones along the Trans-Alaska Pipeline

Table 1. Free Field Design Ground Motions

Seismic Zone according to Pipeline Milepost	Acceleration (g)	Velocity (cm/sec)
0 – 258	.12	15
258 – 560	.45	56
560 – 620	.60	74
620 – 710	.30	36
710 – 800	.60	74

November 3, 2002 Magnitude 7.9 Event

The Denali Fault ruptured over a distance of nearly 350 km. The epicenter occurred near the newly discovered Susitna Glacier thrust fault, approximately 90 km to the west of the pipeline. The fault intersects the TAPS route at Milepost 588 in central Alaska. Slip on the fault was measured at 5.5 m near the pipeline crossing with maximum slip of almost 9 m occurring 120 km to the east of the pipeline crossing.

Ground Motion during the Nov 3 event was recorded by Digital Strong Motion Accelerograph (DSMA) units maintained by Alyeska at Pump Stations 7, 8, 9, 10, 11, and 12 as part of its Earthquake Monitoring System (EMS). Peak ground acceleration measured at these stations is given in Table 2. The highest ground motions were recorded at Pump Station 10, 5 km north of the Denali Fault. Maximum measured peak ground acceleration was 0.34 g, and a maximum ground velocity of 114 cm/sec (45 in/sec) was computed from the acceleration record. Considering that the acceleration signal was high-pass filtered at 0.1 Hz, the maximum ground velocity actually may have been about 50 percent higher than the value computed from the accelerograms[5].

At Pump Station 11, 155 km south of the Denali Fault pipeline crossing, measured peak ground acceleration was about 0.1 g whereas at Pump Station 9, 64 km to the north, the measured peak ground acceleration was about 0.08 g. Measured horizontal and vertical ground accelerations were less than design ground accelerations at all measurement stations, but the peak ground velocity was three to five times higher than would normally be associated with this level of acceleration due to the near-source violent ground motions associated with fault rupture.

Table 2. Peak Ground Accelerations vs. Design Accelerations

Pump Station	Pipeline Milepost	Measured Peak Accel.		Design Accel.	
		Horiz	Vert	Horiz	Vert
PS 07	414	0.018	0.010	0.450	0.300
PS 08	489	0.046	0.024	0.450	0.300
PS 09	549	0.074	0.053	0.450	0.300
PS 10	586	0.337	0.238	0.600	0.400
PS 11	686	0.087	0.033	0.300	0.200
PS 12	735	0.039	0.024	0.600	0.400

Below-ground Pipeline Design and Performance

The below-ground pipeline design and operation considers the affects of a number of geohazards including thaw induced subsidence, landslides, soil liquefaction, fault movement, and axial strain associated with seismic wave propagation. Subsidence caused by thawing permafrost was the key factor influencing the TAPS below-ground design, and while not directly related to seismic aspects of design per se, it precipitated the development and use of in-line inspection technology that proved very effective in post-earthquake damage assessment of the buried pipeline.

Since 1992 Alyeska has used an inertial navigation inline inspection device ("smart pig") to obtain measurements that can be used to detect pipe deformation and other conditions that approach operational tolerance limits (Figure 2). The pig uses three-axis gyroscopes and accelerometers to determine the position of the pig in three-

---

[5] The maximum computed velocity is affected by high-pass filtering at 0.1 Hz. Based on preliminary results from in-progress studies conducted by the USGS, it is believed that the actual peak velocity could have been about 50 percent higher than the calculated value of 114 cm/sec, i.e., about 170 cm/sec.

dimensional space, and thus the pipe at 5-cm intervals along the pipeline. This position data is post-processed to calculate bending radius and bending strain. The pig also uses 64 radius-measuring fingers to determine circumferential deformation.

Slope Stability

The federal and state Right of Way Agreements stipulated that unstable slopes shall be avoided, or where not possible, the design shall mitigate "harmful effects". Alyeska slope stability evaluations considered all types of soil mass movement including liquefaction. Near the Denali Fault, the pipeline design avoids potentially unstable slopes.

Figure 2. Inertial navigation tool used to determine pipeline curvature and deformation

Field surveillance was initiated immediately following the November 3rd earthquake. No evidence of slope instability affecting the pipeline was observed. Shallow soil detachments on ridge crests and rockfall and raveling on talus slopes was noted in the vicinity of the pipeline corridor, but these were too distant from the pipeline to constitute a threat. Significant (massive) landslides were noted by USGS observers where the fault zone parallels the Black Rapids Glacier, but these areas were several tens of kilometers to the west of the pipeline route.

Soil Liquefaction

Seismic liquefaction describes the behavior of certain cohesionless soils which, under saturated conditions, may lose a large portion of their shear strength as a result of earthquake shaking and may acquire characteristics of a viscous liquid mass with flow capabilities. The potential for soil liquefaction during a seismic event was one of the most important considerations in the design of the pipeline. Approximately 160 km (100 miles) of the below-ground pipeline is located in areas that were judged to be potentially liquefiable.

A pipeline buried in level ground subjected to liquefaction will tend to "float" or "sink" depending on buoyancy considerations. Lost ground associated with sand boils can also contribute to subsidence. When liquefaction occurs in a soil mass not confined by adjoining stable soil strata, lateral sliding or motion in the unconfined direction may occur.

For potential liquefaction areas that could not be avoided in construction, design mitigation measures were employed to assure long-term integrity for the pipeline. For below-ground sections of the pipeline, the most common mitigation is burial below the potentially liquefiable soils. Other buried pipeline mitigations include grading of slopes to less than 2 percent, removal and replacement of potentially liquefiable soils and, special design with insulated pipe and heat pipes (thermosiphons). On flat ground (less than 2 percent grade) no protective measures were required, but buoyancy of buried pipe sections was considered in the design.

Field inspection after the November 3 earthquake showed liquefaction events in high groundwater areas of floodplains, especially in the Delta River and Phelan Creek areas up to 30 km south of the Denali Fault. All of the observed liquefaction events are in relatively flat areas (less than two percent grade) with shallow groundwater conditions. Four locations were observed with a length totaling about 2 km (Table 3).

Table 3. Areas of Observed Soil Liquefaction

Location (Milepost)	Length, m	Slope
590.7 – 591.1	640	<2 percent
592.7 – 592.9	320	<2 percent
599.4 – 599.6	320	<2 percent
606.4 – 606.8	640	<2 percent

Observed liquefaction events resulted in sand boils, surface cracking with liquefied soil flow, and surface cracking resulting from lateral spreading of surface fill. Figures 3 and 4 show typical liquefaction sites in the Delta River floodplain south of the Denali Fault.

Ground cracking occurred parallel to the sides of the original pipeline ditch where select clean backfill surrounds the pipe. No ground cracking was observed over the pipe. Liquefied soils are thought to have used these parallel cracks as an avenue for escape, causing sand boils at the ground surface (Figure 5).

Figure 3. Surface evidence of liquefaction near Remote Gate Valve 91. south of Denali Fault.

Figure 4. Surface evidence of liquefaction in the Delta River floodplain.

Fault Movement

As mentioned earlier, the Denali Fault was crossed in an above-ground mode. Thus, the Denali Fault rupture did not affect the buried pipeline in a direct sense, i.e., due to abrupt surface offset and localized ground deformation in close proximity (100 m) to the rupture. The response of the above-ground fault crossing is discussed by Sorensen and Meyer (2003).

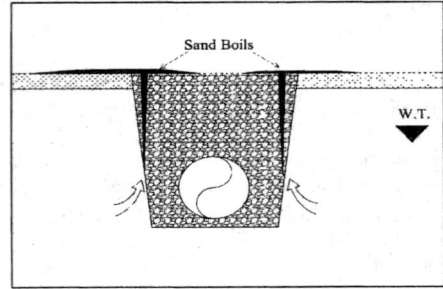

Figure 5. Diagram of typical liquefaction event

Surface Wave Propagation

A pipeline buried in soil that is subject to the passage of seismic waves (compression, shear and surface waves) will incur longitudinal and bending strains as it conforms to the associated ground strains. In most cases, these strains are relatively small, and welded pipelines in good condition, which is the case with TAPS, typically do not incur damage.

A relatively simple method to estimate axial ground strain from wave propagation is based on an approach developed by Newmark (1968) can be used to obtain an upper-bound estimate of ground strain due to a propagating seismic wave with a constant shape (sine wave). The maximum axial ground strain, ε_g, is given by:

$$\varepsilon_g = \frac{V_{max}}{\alpha_\varepsilon c}$$

where V_{max} is the maximum horizontal ground velocity in the direction of wave propagation, c is the apparent propagation speed of seismic wave, i.e., the component of wave speed parallel to the pipeline under consideration, and α_ε is the ground strain coefficient corresponding to the most critical angle of incidence and type of seismic wave and ranges from 1.0 for compression and Raleigh waves to 2.0 for shear waves (ASCE, 1984). Estimates of wave travel speeds have not been made for the Denali Fault area, at least not by the authors; however, an apparent propagation velocity of 900 m/sec can generally be taken as a lower bound estimate (ASCE, 1984) to provide a maximum estimate of ground strain from body waves.

Assuming a maximum ground velocity of 170 cm/sec, a propagation speed of 900 m/sec, and a ground strain coefficient of 1.0, the maximum ground strain, , is 0.187 percent. If there is no slippage of the pipeline relative to the surrounding soil, then the maximum axial strain in the pipeline can be taken equal to maximum ground strain, an upper-bound, conservative assumption. However, if it is assumed that 100 percent of the ground strain is transferred to the buried pipeline, this would relate to a pipe stress of 387 MPa (56 ksi), which is about 85 percent of the specified minimum yield stress for Grade X65 steel pipe, well within elastic limits for buried pipe.

Ground curvature resulting from wave propagation will also induce bending in a buried pipeline, but the resulting pipe strains are generally so small compared to the direct axial strain effect that they can be neglected.

Curvature/Deformation Pig Results

The buried pipeline, located outside the fault zone, is typically less vulnerable to seismic hazards than the above-ground design. Hidden damage to the buried pipeline however, could not be visually inspected and, thus, required use of in-line inspection (ILI) methods (Figure 2).

Immediately after the occurrence of the Earthquake, planning began for the running of a smart pig. Before the pig could be run, it was necessary to run five cleaning pigs through the line to assure it was clear of wax accumulation which could affect the instrumented pig readings. Two of these cleaning pigs had a greater hard diameter thereby assuring clear passage for the smart pig. Following the smart pig run, the measurement data acquired between Pump Stations 9 and 11 was processed and reviewed to assess pipeline integrity. This covered a pipeline distance of approximately 220 km (137 miles).

If the pipe were to change position as a result of the earthquake, the change would first be apparent in the strain or curvature data. The software allows data from one year to be subtracted another and thus differences in values other than zero indicate a change. Example output from the smart pig run is shown in Figure 6. Changes were noted in the below-ground pipe in the area of Remote Gate Valve 91 (RGV91) and other areas with observed surface disturbance. Interestingly, in each case the change

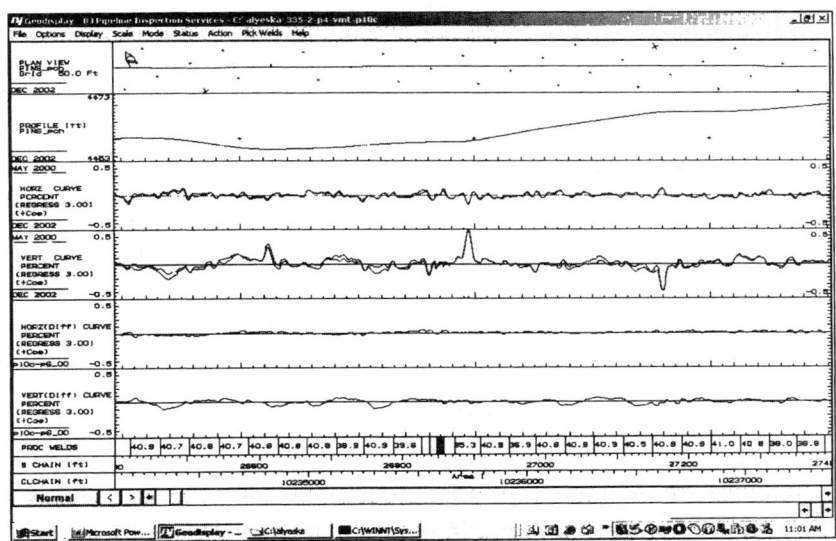

Figure 6. Smart pig data

was to a lower strain level. For example, one area north of RGV91 experienced a curvature change from 60 to 41 percent of critical bending curvature. Another area south of RGV91 experienced a curvature change from 56 to 34 percent of critical bending curvature. Both these areas demonstrated some indication of liquefaction, which apparently allowed the pipe to relax in the ditch during the shaking. Between Pump Station 9 and Pump Station 11 there are 15 dents and 18 other target areas of known bending curvature. Each of these locations were checked and found not to have changed as a result of the earthquake.

Summary of Observations and Findings

The buried pipeline met all performance expectations associated with the magnitude 7.9 earthquake of Nov 3, 2002. This section summarizes some of the key observations and findings.

1. Ground motions of 0.34 g maximum acceleration, velocities perhaps as much as 170 cm/sec, and a ground shaking duration of 90 seconds caused liquefaction to occur in loose, saturated cohesionless soils. A design strategy of crossing potentially liquefiable areas either on essentially flat terrain (less than 2 percent slope), or buried below liquefaction depth successfully avoided harmful effects of subsidence or spreading.

2. Avoidance of unstable slopes in the vicinity of the Denali Fault proved to be an effective slope stability design strategy. Actual earthquake ground motions (0.34g), however, were less than ground motions used to assess potential geohazards (0.60g).

3. As has been observed worldwide for buried steel pipelines with good quality welds, seismic wave propagation did not damage or permanently deform the below-ground pipeline. Estimated maximum pipe strain was less than the specified minimum yield strength of the pipeline.

4. Post-event inspection of the pipeline using the instrumented curvature/deformation monitoring pig was an effective and conclusive technique for verifying pipeline integrity. The pig data provided a continuous record of pipe displacement and curvature through the area affected by the earthquake.

5. The usual design approach for pipeline fault crossings is to construct the pipeline in shallow, sloped-wall trench with loose backfill while permitting large strains and permanent deformation to occur, provided pipe rupture is prevented. In other words, the risk of damage to the pipe requiring repair is generally acceptable, so long as leakage is prevented. Had this concept been used on TAPS, an extensive pipe repair operation would have been required. However, since the pipeline crosses the Denali Fault above grade on sliding shoe supports, the pipeline was not subjected to the large bending strains and local buckling that would have resulted in a shallow buried pipeline. While

fault rupture is a rare event, it certainly happened in the case of TAPS, and as planned, the above-ground crossing mode prevented damage that would have resulted in a more significant interruption in throughput.

Acknowledgment

Alyeska Pipeline Service Company is gratefully acknowledged for its role in supporting field reconnaissance of the TAPS route following the Denali Fault earthquake and for permission to present this paper. Any opinions expressed herein are solely those of the authors.

References

1. ASCE (American Society of Civil Engineers) (1984). *Guidelines for the Seismic Design of Oil and Gas Pipeline Systems,* Technical Council on Lifeline Earthquake Engineering, Committee on Gas and Liquid Fuel Lifelines, D. J. Nyman, Principal Investigator, New York, 473 p.

2. Cluff, L.S., R.A. Page, D.B. Slemmons, and C.B. Crouse (2003). "Seismic Hazard Exposure for Trans-Alaska Pipeline," *Proceedings of the 6th U.S. Conference on Lifeline Earthquake Engineering (Advancing Mitigation Technologies and Disaster Response),* Long Beach, CA, TCLEE, ASCE.

3. Hall, W.J., D.J. Nyman, E.R. Johnson, and J.D. Norton (2003). "Performance of the Trans-Alaska Pipeline in the November 3, 2002 Denali Fault Earthquake," *Proceedings of the 6th U.S. Conference on Lifeline Earthquake Engineering (Advancing Mitigation Technologies and Disaster Response),* Long Beach, CA, TCLEE, ASCE.

4. Newmark, N.M. (1968). "Problems in Wave Propagation in Soil and Rock," *Proceedings, International Symposium on Wave Propagation and Dynamic Properties of Earth Materials,* University of New Mexico, pp. 7-26.

5. Sorensen, S.P., and K.J. Meyer (2003). "Effect of the Denali Fault Rupture on the Trans-Alaska Pipeline," *Proceedings of the 6th U.S. Conference on Lifeline Earthquake Engineering (Advancing Mitigation Technologies and Disaster Response),* Long Beach, CA, TCLEE, ASCE.

6. Sorensen, S.P., K.J. Meyer, P.A. Carson, and W.J. Hall (2003). "Response of the Above-Ground Trans-Alaska Pipeline to the Magnitude 7.9 Denali Fault Earthquake," *Proceedings of the 6th U.S. Conference on Lifeline Earthquake Engineering (Advancing Mitigation Technologies and Disaster Response),* Long Beach, CA, TCLEE, ASCE.

TRANS-ALASKA PIPELINE EMERGENCY RESPONSE AND RECOVERY FOLLOWING THE NOVEMBER 3, 2002 DENALI FAULT EARTHQUAKE

Douglas J. Nyman[1], Elden R. Johnson[2], and Christopher H. Roach[3]

Abstract

The Denali Fault earthquake on November 3, 2002 provided a "live" test of the earthquake emergency response plan and earthquake monitoring system for the Trans-Alaska Pipeline, which transports crude oil from Alaska's North Slope to a marine tanker terminal at Valdez, Alaska. The use of a comprehensive and focused field inspection effort based on estimates of ground motion shaking severity allowed pipeline operation to resume after a 66-hour shutdown. This paper provides a summary of the post-earthquake emergency response by Alyeska Pipeline Service Company (Alyeska), operator of TAPS, including the use of ground motion data and reports generated by the Earthquake Monitoring System. The response provides an example of how an earthquake monitoring system, coupled with a focused field response effort, can be used to limit the disruption associated with a large earthquake.

Introduction

The Trans-Alaska Pipeline System (TAPS) was designed to withstand the effects of earthquake ground shaking and permanent ground deformation set forth in the seismic criteria for the project. Criteria for the 1,219-mm (48-inch) diameter crude oil pipeline were defined as stipulated design earthquakes for five seismic zones along the 1,287-km (800-mile) route as shown in Figure 1. A Design Contingency Earthquake (DCE), with an estimated return period of 300 years or more, was established for each of the five seismic zones according to Richter magnitude. Included in the characterization of the TAPS seismic hazard were three potentially active faults that crossed the pipeline in central Alaska, one of which – the Denali Fault – was the source for the November 3, 2002 earthquake. The TAPS seismic criteria are described in a companion paper in these Proceedings (Cluff et al., 2003).

The magnitude 7.9 earthquake that occurred in south-central Alaska on November 3, 2002 ruptured a 336-km long segment of the Denali Fault (USGS, 2003). The epicenter was located about 88 km west of the Trans-Alaska Pipeline, and the rupture propagated to the east across the pipeline right-of-way, reaching a maximum of 8.8 m about 120 km east of the pipeline. At the pipeline crossing of the Denali Fault, the fault rupture displacement was approximately 5.5 m of right lateral offset and 0.6 m of up-to-the-north vertical displacement, which is approximately equal to the design fault displacement (Hall et al., 2003). The fault displacement was distributed over a

---

[1] Consulting Engineer, D.J. Nyman & Associates, Houston, TX, F. ASCE.
[2] Engineering Advisor, Alyeska Pipeline Service Co., Fairbanks, AK, M. ASCE.
[3] Consulting Engineer, Anchorage, AK, M. ASCE.

zone approximately 200 m in width, with approximately 50 to 70 percent of this displacement occurring as an abrupt offset.

The ground motions and surface fault rupture produced by the Denali Fault earthquake approached design criteria for the section of the pipeline passing through the Alaska Mountain Range in the vicinity of Pump Station 10 and the Denali Fault crossing. Liquefaction was observed at a number of areas along the pipeline, including a remote gate valve location. There was no damage to the pipeline or release of crude oil, although there was incidental damage to the aboveground support hardware and some movement of the belowground pipeline in liquefaction areas. The pipeline system withstood the earthquake in a manner consistent with the original design premise, which permitted limited damage to the pipeline, support system, and facilities provided that there is no release of crude oil or hazardous substances or threat to safety.

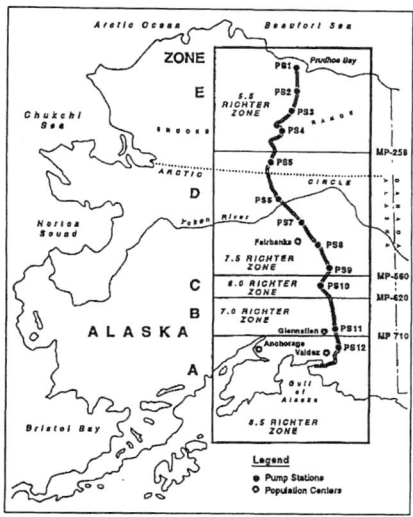

Figure 1. Seismic zones along the Trans-Alaska Pipeline

The pipeline was shutdown for assessment and repair for a period of only 66 hrs. This rapid recovery to such a large seismic event is attributed to Alyeska's earthquake preparedness, response and recovery plan. The plan guided the mitigation of the effects of earthquake damage and helped limit the economic impact on pipeline stakeholders. The elements of this plan and its implementation following the Denali Fault earthquake are presented in the sections that follow.

Earthquake Preparedness

Recognizing that it is not possible to achieve total mitigation of the seismic damage risk through design, TAPS utilizes a number of management and information systems. These include surveillance and maintenance programs, earthquake preparedness and response plans, an oil spill response plan, an earthquake monitoring system, and an incident command system. Each of these components is essential to earthquake preparedness and post-earthquake response as explained below.

Surveillance and Maintenance. The satisfactory performance of the pipeline during the Denali Fault earthquake is attributed largely to its design for seismic ground motions and surface fault displacement that are roughly equivalent to those that actually occurred during the event. Operation and maintenance also played a key part, namely in that pipeline structural integrity and functional reliability had been well-maintained over its 25-year life.

Alyeska conducts routine aerial surveillance at least once every two weeks as required by regulation. Detailed ground surveillance is conducted at least four times per year over the entire length of the pipeline. Additionally, periodic monitoring of the above-ground pipeline support system is conducted to measure and observe performance details such as support system movement, slope stability, and thermal conditions. A maintenance data base and repair inventory is used to track surveillance and monitoring observations. Engineering assessment and necessary maintenance is performed annually, usually over the short summer season.

Earthquake Preparedness Plan. Following the design and construction of the Pipeline System, Alyeska developed an Earthquake Preparedness Program (EPP) tailored for the crude oil pipeline operating environment. The objectives of the EPP were to assure that TAPS facilities will function as required during and after earthquakes to minimize capital loss and business interruption, prevent environmental damage, and provide for life and safety protection. The EPP encompasses all aspects of seismic design as it relates to modification, maintenance, and operation of facilities in the pipeline system, as well as response to a major earthquake. The four elements of the EPP have been developed over the 25-year operating history of TAPS and include lessons learned from construction, operation, and maintenance of pipeline system facilities:

1. Design control to assure that new work will be designed and installed per applicable seismic criteria and performance objectives.

2. Configuration management to manage change (non-engineered maintenance and modification) of existing facilities and preserve the requisite level of seismic integrity.

3. Seismic housekeeping to maintain the workplace in a manner that precludes the introduction of interaction hazards between non-engineered items or furnishings with essential equipment and minimizes impediments to post-earthquake response.

4. Earthquake response plan to provide for organizational earthquake preparedness, rapid post-earthquake damage assessment, repair as needed, and resumption of normal oil movements at the earliest opportunity (assuming a shutdown is required).

Prior to the occurrence of the November 3 earthquake, actions had been taken to train engineering personnel on post earthquake damage assessment; to develop forms and placards specific to Alyeska facilities and equipment; and to accommodate the timing, resource requirements, and distribution of a potential earthquake response. Forms were developed for the building evaluation, equipment and non-structural items. Recognizing that building occupancy and personnel safety would be an early response objective, weather-proof adhesive placards were developed to delineate "inspected" (unrestricted entry and occupancy permitted), "restricted use," and "unsafe" (do not enter or occupy). The forms and placards were customized for the types of facilities and equipment typically found on the pipeline system and were based on earlier forms and placards developed by the Municipality of Anchorage and ATC-20 (Applied Technology Council, 1989).

Oil Spill Response Plan. An Oil Spill Contingency Plan (OSCP) is used by Alyeska to provide operations and support personnel with information to assist in the response to oil spills should they occur. While not specifically related to earthquakes, the OSCP provides basic preplanning and information resources needed for response to a broad range of emergencies that could result in a leak, including earthquakes. The OSCP information is used as a starting point, supplemented by more specific action plans directly related to seismic events. The general sections of the OSCP relevant to earthquakes include:

1. Specific information related to equipment requirements, response action plan and checklists, strategies and tactics, communications, etc.

2. Supplemental information such as storage tank information, type and amounts of oil, command system, operating conditions, logistical support, equipment inventory and training and exercises.

3. Site-specific spill plans for river drainages that include potential spill volumes, reconnaissance actions, pre-identified containment sites, pre-staged equipment, priority control actions, access points, anticipated oil migration routes, and sensitive area locations.

Earthquake Monitoring System. An earthquake monitoring system (EMS) has been included as part of the pipeline control system since the start-up of the crude oil pipeline operation in 1977 (Nyman et al., 1981, 1999). The EMS consists of eleven PC-computer-based digital strong motion accelerograph stations (Figure 2) located along the pipeline alignment (Figure 1) at Pump Station 1, Pump Stations 4 through 12, and the Valdez Marine Terminal. The eleven PC stations are networked via a TCP/IP-based network. The purpose of the EMS is to detect strong earthquake activity along the pipeline, send alarms to the Pipeline Control System, post-process ground motion data to evaluate the severity of earthquake ground shaking (i.e., peak ground motions and response spectra), and assess the potential for damage to the pipeline and supporting facilities. The most important objectives of this assessment are to determine whether the pipeline should be shut down in response to earthquake activity and to delineate inspection requirements for the affected portion of the route.

Using the post-processed data from each active station, an evaluation program known as "DrQuake" estimates event severity and possible damage along the pipeline route. This evaluation utilizes an interpolation scheme based on estimated earthquake location and acceleration attenuation to calculate seismic event parameters (peak accelerations and velocities, and spectral response coefficients) at intermediate locations along the pipeline from those computed at the individual recording stations Nyman et al. (1981). Next, a report describing earthquake severity along the pipeline is generated to assist the pipeline controller in decision-making and to guide post-earthquake inspection efforts. Specific information of interest in the earthquake report includes the assessment of design margin, i.e., the comparison of computed earthquake response parameters to design limits. If inspection of the shaken area is required, a detailed checklist of physical inspection attributes is generated to guide field response teams in their inspection efforts. An example of an inspection checklist is presented later in this paper.

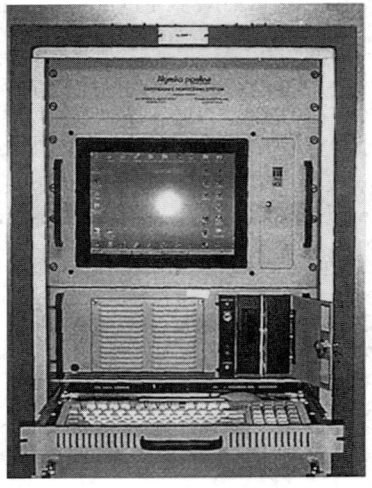

Figure 2. DSMA cabinet

Incident Command System. The Incident Command System (ICS) is an emergency management structure originally developed for fighting forest fires. It was adapted by Alyeska for pipeline emergencies, including oil spill response. Oil spills are similar to wild fires in that they involve the mobilization of labor and equipment in response to a catastrophic event spread by the affects of weather, and they harm the natural and manmade environment.

ICS provides management structure and discipline related to emergency events. The emergency organization is divided into Command, Plans, Operations, Logistics and Finance Sections, each having prescribed functions, roles and procedures. The Command Section sets objectives, monitors performance and controls the response. In multi-jurisdictional events, Command is often referred to as Unified Command and includes representatives of affected agencies working in cooperation to provide clear, non-conflicting direction to the response. The Plans section provides the daily work plan to Operations in response to command objectives. Plans administer a disciplined planning function, defining the situation, keeping track of resources, and producing a daily Incident Action Plan. The Operations Section carries out the action plan with labor and equipment in the field. Logistics orders and tracks requested resources, and manages communications, transportation, and support functions. Finance tracks costs, and administers claims.

Response to the Magnitude 7.9 Denali Fault Earthquake

The Magnitude 7.9 Denali Fault earthquake on November 3, 2002 provided an unscheduled test of the TAPS' seismic emergency response plan and EMS. The surface rupture displacement and the level of earthquake ground shaking in the region of the fault crossing approached design limits. In the sense of earthquake severity, the Denali Fault earthquake represented a major test of the pipeline design and emergency response plan. The EMS generated seismic alarms at six locations, Pump Stations 7 through 12[4], over a pipeline route distance of 517 km (Pipeline Milepost 414 to 735). All six stations recorded acceleration time histories, and the EMS generated an evaluation report for the earthquake, i.e., the "DrQuake" report. A summary of peak ground motions measured by the EMS are provided in Table 1. Ground motions were considerably larger at Pump Station 10, which is about 5 km north of the Denali Fault. A "quick-look" graphic produced by DrQuake is shown in Figure 3. This graphic provides a rough indication of the severity of ground motions along the pipeline compared to design limits.

Table 1. Ground Accelerations, Measured vs. Design

Pump Station	Pipeline Milepost	Measured Peak Accel.		Design Accel.	
		Horiz	Vert	Horiz	Vert
PS 07	414	0.018	0.010	0.450	0.300
PS 08	489	0.046	0.024	0.450	0.300
PS 09	549	0.074	0.053	0.450	0.300
PS 10	586	0.337	0.238	0.600	0.400
PS 11	686	0.087	0.033	0.300	0.200
PS 12	735	0.039	0.024	0.600	0.400

Referring to Figure 3, the distribution of ground motions was predicted reasonably well between Pump Stations 7 and 9 and Pump Stations 11 and 12. However, between Pump Stations 10 and 11, the ground motions were over-predicted. This occurred because the DrQuake computational algorithm attempts to find a central location of the energy release and then interpolates to estimate ground motions between the measurement stations. Since the source of energy release was a fault crossing the pipeline, and since the Denali Fault "unzipped" from west to east-southeast, DrQuake was unable to make an accurate prediction of ground motions in close proximity to the fault rupture. Adding to the uncertainty was that peak ground motions at Pump Station 11, about 155 km south of the Denali Fault crossing, were higher than the ground motions at Pump Station 9, which is only 64 km north of the fault. Consequently, DrQuake interpreted the energy release to be well south of Pump Station 10, and hence, the gross overestimate of ground motions between Pump Stations 10 and 11.

Despite this shortcoming, the EMS provided a quick indication that severe ground motion had occurred in this region of the pipeline. Once magnitude and epicenter information were available from the Alaska Earthquake Information Center (AEIC), the DrQuake report was quickly screened to provide a more accurate indication of what had happened and to identify physical pipeline system attributes (potential liquefaction areas, slopes, aboveground pipeline segments, remote gate valve

---

[4] Pump Stations 8 and 10 were taken out of service in 1996. Limited communications and control system hardware were still in operation at these sites, but the process facilities were no longer functional. Pump Station 11 does not contain pumping equipment, only communications and control systems.

locations, etc.) requiring field inspection. This screening process reduced the list of over 400 attributes initially flagged for inspection to approximately 150 attributes between Pump Stations 9 and 11.

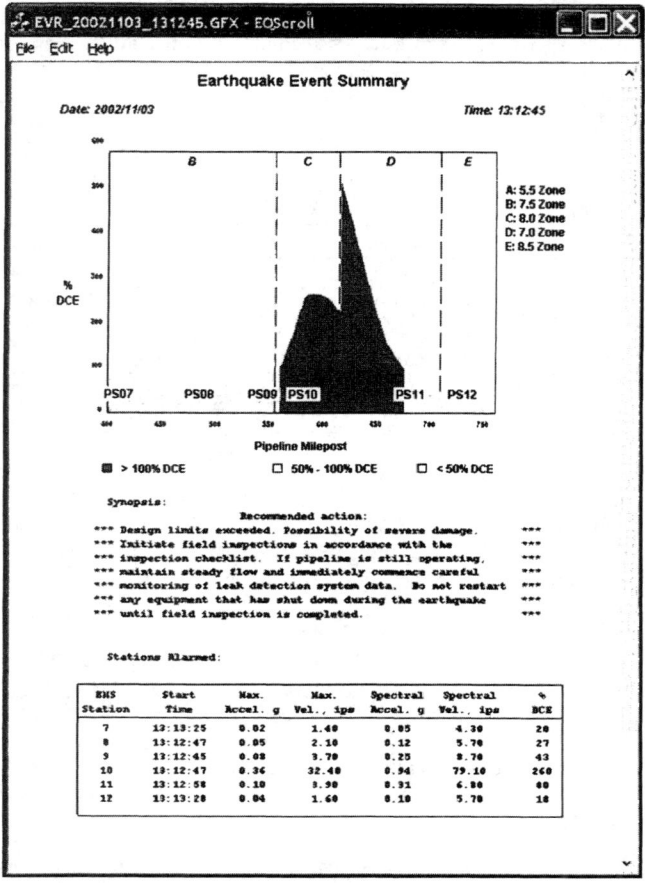

Figure 3. DrQUAKE visual summary of magnitude 7.9 Denali Fault earthquake

Alyeska Management declared an emergency response immediately after the earthquake, and ICS was initiated. As mentioned earlier in this paper, a limited number of above-ground supports were damaged, resulting in loss of support at several locations. At the Denali Fault crossing, some of the support shoes had experienced their full range of allowable movement due to the 5.5-m strike-slip fault displacement. Through all this, the pipeline leak detection system indicated normal conditions, i.e., no detectable leaks. Nevertheless, about 45 minutes after the earthquake, Alyeska elected to shut down the pipeline as a precautionary measure to assess numerous damage reports and to evaluate pipeline structural integrity.

Unified Command was established between Alyeska and its regulatory agencies. A command center was set up at pipeline headquarters in Fairbanks with a field operations center at Pump Station 9 near Delta Junction, about 64 km north of the pipeline crossing of the Denali Fault. All field resources necessary for damage assessment and repair were assigned to the field office. Response objectives were established early to assess pipeline damage, prioritize repairs, and repair as necessary and to return the pipeline to operation in the shortest time possible. At then-current crude oil prices, the shutdown of the pipeline represented an economic impact of approximately $1,000,000 per hour.

The DrQuake inspection checklist served an important role in the post-earthquake field inspection. The checklist contained a comprehensive list of identifiable physical attributes along the pipeline that could have been affected by seismic hazards such as ground shaking, surface fault rupture, liquefaction-induced ground movement, or slope instability. The physical attributes were clearly identified in the checklist according to type, location, and general instructions for field identification and evaluation of potential damage. An excerpt of the DrQuake inspection checklist in the vicinity of the Denali Fault is provided in Figure 4 for the Denali Fault earthquake. The checklist attributes were sorted according to technical discipline, prioritized according to importance, and assigned to the various damage assessment teams for field follow-up inspection. The principal objectives of damage assessment were to determine where repairs were needed, provide a repair inventory list, and to document inspection progress and repair status.

Damage assessment teams were formed by pairing engineering resources with local maintenance coordinators responsible for right-of-way maintenance. In this manner, the engineering expertise needed for conducting a thorough damage assessment was combined with maintenance personnel who were familiar with the pipeline, right-of-way, facilities and with their general condition prior to the earthquake. Assessment teams were on-site at the field ICS office at Pump Station 9 by the morning of November 4. Objectives for damage assessment teams included evaluating facilities for damage that would degrade the safety or operability of the pipeline system and assessing the readiness of the system for restart. Additionally, engineering resources within the damage assessment teams served in a support role as necessary for implementing repairs.

Information packages were assembled for each damage assessment team to provide the basic data and documentation materials necessary for conducting and reporting Rapid Evaluations. This material included the EMS Post-Earthquake Data Evaluation Report; and EMS Inspection Attribute Checklists (Figure 4). Additional information included damage assessment forms and placards; a preliminary engineering analysis memorandum delineating likely areas of potential damage based on recorded ground motions, a Pipeline Atlas, excerpts from maintenance and repair manuals, and information resources provide by the Oil Spill Response Plan.

```
******************************
*      Milepost 589.0        *
******************************
     Denali Fault Crossing
          Check aboveground segments for
          - anchors slipped with respect to the frames and energy
            absorbers crushed
          - anchor frames slipped with respect to the VSMs
          - crushed insulation resulting from impact between pipe and
            bumper stop
          - broken bumper beam and bumper stops
          - crushed insulation resulting from impact between pipe and
            non-bumpered VSM
          - damaged shoes - shoe legs buckled, shoe jacks cocked, teflon
            on slide plates scratched
          - leaning VSM
          - cocked or sagging beams
          - broken or deformed stud bolts or bolts at beam to bracket
            connection
          - damage to valves and valve support structures
          - soil disturbance or damaged pipe near ground point at
            transitions
          - visual indications of buckling in the pipeline

******************************
*      Milepost 589.0        *
******************************
     Denali Fault Crossing
          Check fault zone for
          - large displacements of shoes on beams which might indicate
            surface faulting
          - ground surface indication of fault movement
          - damage to concrete and steel sleepers
          - damage to the special fault crossing shoes
          - abrupt, uncharacteristic lateral shift between adjacent
            sets of VSMs
          - use standard aboveground checklist below to inspect other
            items in the fault crossing

******************************
*      Milepost 591.0        *
******************************
     Remote Gate Valve - RGV91 - B/G
          Check valve for
          - apparent damage or leaks
          - damage to motor operator
          - damage to electrical conduit, especially at terminations
          Check RGV Building for        [!!follow safe entry procedures!!]
          - damage to batteries, battery racks, and rack anchorage
          - damage to comm. and control circuits and cabinet/rack anchorage
          - all cabinet front panels/doors securely latched
          - structural damage to building - damage at foundation connections,
            cracks of exterior panels or joints, integrity of appurtenances,
            antennas, ORMAT converters, exchangers, and supply piping, etc.
```

Figure 4. Excerpts from DrQUAKE Inspection Checklist, 2002 Denali Fault Earthquake

Access for the Rapid Evaluations conducted over the first few days of the response was provided by both vehicle and helicopter. Weather in the area was unseasonably warm and dry, which allowed the evaluations and repairs to proceed much more efficiently than if typical snow cover and cold temperatures were present. If significant snow cover had been present ground access would have been limited to snowmobile and tracked vehicles.

The EMS damage check list provided a complete and measurable inspection inventory used by the field damage assessment teams. Damage assessment teams documented findings on Rapid Evaluation forms and forwarded findings through the Incident Command System for disposition. Once assessments were completed, and repairs identified, they were turned over to maintenance crews for repair. Repairs were made and reported back to the ICS command center. Progress was measured by the percentage of assessments and designated repairs completed.

Repairs were completed and the pipeline returned to operation after a total downtime of 66 hrs. Following pipeline restart, a limited number of facilities were selected for detailed evaluations. Selected sites included mainline valves, vehicle bridges, communication equipment, and operating facilities within about 10 km of the Denali Fault crossing. Detailed evaluations revealed a few minor items that had not been identified in the initial rapid evaluations, but generally it was determined that the rapid evaluations were conducted in a thorough manner and had identified all significant issues.

Concluding Remarks

Alyeska was praised by a member of the U.S. Senate, the Governor of Alaska, its Owner Companies, regulatory agencies, professional organizations, and the press for professional and proficient handling of the earthquake response. The U.S. Department of Interior had scheduled renewal of the pipeline right of way grant for an additional 30 years in early January of 2003. Alyeska's exemplary response to Mother Nature's Final Exam of November 3, 2003 allowed this important milestone to continue on schedule.

After years of standing by "just-in-case," the Earthquake Monitoring System fulfilled a vital role of instantaneously alarming for a significant seismic event, recording ground motions, and generating a detailed assessment and inspection checklist. Developed nearly 25 years ago and updated about five years ago, this system is a rudimentary version of the Shake Map system recently implemented by Trinet (www.trinet.org/shake.html) in Southern California and the Advanced National Seismic System (www.anss.org). The recent experience with the Denali Fault earthquake illustrates the value of using seismic data on a near-real time basis to warn of potential earthquake damage and guide response efforts.

Alyeska Pipeline Service Company is gratefully acknowledged for its role in supporting field reconnaissance of the TAPS route following the Denali Fault earthquake and for permission to present this paper. Any opinions expressed herein are solely those of the authors.

References

1. Applied Technology Council (1989). Procedures for Post earthquake Safety Evaluation of Buildings, ATC-20, Redwood City, CA.
2. Cluff, L.S., R.A. Page, D.B. Slemmons, and C.B. Crouse (2003). "Seismic Hazard Exposure for Trans-Alaska Pipeline," *Proceedings of the 6th U.S. Conference on Lifeline Earthquake Engineering (Advancing Mitigation Technologies and Disaster Response),* Long Beach, CA, TCLEE, ASCE.
3. Hall, W.J., D.J. Nyman, E.R. Johnson, and J.D. Norton (2003). "Performance of the Trans-Alaska Pipeline in the November 3, 2002 Denali Fault Earthquake," *Proceedings of the 6th U.S. Conference on Lifeline Earthquake Engineering (Advancing Mitigation Technologies and Disaster Response),* Long Beach, CA, TCLEE, ASCE.
4. Nyman, D.J., V.J. McDonald and G.G. Simmons (1981). "Earthquake Monitoring System Trans-Alaska Pipeline," *Lifeline Earthquake Engineering – the Current State of Knowledge 1981,* Oakland, CA Conference, TCLEE, ASCE, pp. 139-151.
5. Nyman, D.J., E.L. Nelson, C.H. Roach (1999). "Earthquake Monitoring System Trans-Alaska Pipeline," *Proceedings of the 5th U.S. Conference on Lifeline Earthquake Engineering (Optimizing Post-Earthquake Lifeline System Reliability),* Seattle, WA, TCLEE, Monograph No. 16, pp. 897-906, ASCE.
6. USGS (U.S. Geological Survey) (2003). "Rupture in South-Central Alaska – the Denali Fault Earthquake of 2002," USGS Fact Sheet 014-03, Menlo Park, CA, http://geopubs.wr.usgs.gov/fact-sheet/fs014-03/.

SERA II

Dennis K. Ostrom[1]

Abstract - Southern California Edison conducted a System Earthquake Risk Assessment (SERA) on its bulk power delivery system in 1990. The approach is a statistical modeling approach on a component and installation detail level and uses Monte Carlo simulations to produce failure sets on a component level. Since the SERA conducted in 1990, Edison has made improvements to its system, the SERA approach has been improved, and more has been learned about the performance of utility systems under earthquake loading. Recognizing this, Edison is conducting another SERA on its bulk power system. This paper reports on this second SERA and focuses on the data gathering (including conductor slack data and format) and insights gained during the second risk assessment.

Introduction - Southern California Edison's (Edison's) efforts to gain knowledge of its system's earthquake risk utilizes the System Earthquake Risk Assessment (SERA) approach that was developed in-house. Edison conducted a SERA on its bulk power delivery system in 1990 (SERA I), (Burhenn, 1992). Since SERA I, Edison has made changes to its system, the SERA approach has been improved, and more has been learned about the performance of utility systems and their components under earthquake loading. Recognizing this, Edison is conducting another SERA (SERA II) on its bulk power system. This paper reports on SERA II and focuses on the system-modeling/data gathering (recording and formatting), review of the latest simulations on the Edison System conducted during SERA II, and how the results of the SERA are being used to evaluate the performance of the system.

In the context of this paper, risk refers to Edison's bulk power transmission system's risk of function failure after an earthquake. Based on information from its SERA capability, Edison believes that it has been and is still able to make rational choices and priorities for earthquake risk mitigation programs that improve functional reliability while working with a finite budget. The function of Edison's bulk power transmission system is to transmit high quality bulk power to Edison's sub-transmission and distribution systems and to and from interties with generating plants and other utilities. Earthquake risk to life safety of Edison employees and customers is a great concern to Edison, but is not the focus of this paper.

SERA - The Edison SERA approach develops realistic damage profiles that are statistically consistent with past earthquake experience of the Edison and other similar systems after they have been subjected to an earthquake. The approach incorporates statistical modeling on a component level and uses Monte Carlo

---

[1] Dennis K. Ostrom, ASCE Member, PhD, Lifelines Consultant, 16430 Sultus St., Canyon Country, CA 91387, Phone - 661 251 6113, FAX 661 251 2842.

simulations to produce damage profiles on a component level. From the many failure profiles obtained, patterns of failure are identified and their impacts on system function are assessed. From this, priorities for mitigation strategies and programs can be developed. As such, the performance of the utility system to a future earthquake is not predicted but is forecasted similar to a meteorologist forecasting the weather.

All perceived vulnerabilities cannot be mitigated efficiently before an earthquake; some vulnerabilities can be handled acceptably and more efficiently after an earthquake. This latter type of vulnerability can also be identified and mitigated effectively with knowledge gained from a SERA.

The SERA process is surprisingly simple to conceptualize. It consists of taking available data of the seismic hazard and of the system and modeling both. Of course, the more accurate and complete the models are the more reliable will be the conclusions. The hazard model includes faults in terms of their location, attenuation relationships and seismicity. The model of the system includes the site geographic locations, site hazards, motion amplification characteristics, component inventory, connectivity, conductor slack, earthquake performance functions and installation details. The seismological community has provided various seismicity and ground motion models and Edison personnel working with other major California electric utilities have developed the information to model an electrical system

Using the approach shown in Figure 1, SERA can simulate the hazard and provide information that can lead to the development of an itemized list of components that are forecast as either damaged (along with the damage state) or undamaged. The Edison System disposition (transmission lines connectivity, switchyard functionality, transformer function and return to service levels times) after a simulated damaging earthquake can then be developed. Edison then has a picture of what it will face in terms of realistic post-earthquake situations that are consistent with up-to-date knowledge on the regional and site specific earthquake hazards and actual equipment vulnerabilities, connectivity and installation details.

SERA I - Edison conducted the SERA I on its bulk power delivery system in 1990[Ref. 1]. The SERA I demonstrated that the primary vulnerability of the Edison Bulk Power System to earthquakes was a subset of the total population of live tank circuit breakers that Edison had in service at that time. Since then, Edison has replaced these vulnerable Live-tank circuit breakers with earthquake resistant Dead-tank circuit breakers.

Since the SERA I, Edison has made improvements (installed better equipment with better installation details and improved the installation details of a large number of existing equipment) to its system beyond those originally initiated based on SERA I. In the mean time, the SERA approach has been improved (better data gathering and modeling), and more has been learned about the performance of utility components under earthquake loading (there are more experience data and recorded experience data has been better analyzed).

SERA II - Recognizing these improvements, Edison is in the process of conducting SERA II on its bulk power system. During 1997, emphasis was placed on accurately documenting component fragilities, updating its component installation data files format and data gathering procedures so that data can be gathered that will more accurately reflect the existing Edison system. This included reprogramming the SERA software to consider conductor slack and accepting the newly formatted component data taken in the field. During 1998, emphasis was placed on visits to all substations to validate and enhance company inventory data sets and gauge and document actual installation details. During 1999, emphasis was placed on introducing this information into the SERA process. The SERA calculations that forecast component performance were completed during 1999. In 2000, the Seismic Review Team (an internal panel made up of individuals who have the combined expertise identified as necessary to conduct the assessment) began studying the failures (failure patterns) forecast by the SERA II. Emphasis was placed on understanding the results of the study as well as evaluating whether other major subsystems (protection, communication, transmission lines, and low voltage switchgear) of the transmission system should be included in SERA II; return-to-service times were also developed.

The same system that was modeled in SERA I, i.e. the 220kV and 500kV bulk power transmission systems, was modeled in SERA II. In SERA II the model is considerably more comprehensive. Conductor slack, component installation ratings and more equipment types were considered in SERA II. The performance algorithms used in SERA II had the benefit of being based on more earthquake experience than those used in SERA I. Component connectivity and system connectivity were also modeled in SERA II.

The data gathering process for system modeling was far more comprehensive in SERA II than in SERA I. SERA II data was gathered in the field. It became apparent that data gathering could consume considerable time and so a system of data entry that was more intuitive and easier to up-date was developed. This approach has the added benefit of being suitable for entering data via a Personal Data Assistant (PDA). This could speed up data collection and keep data handling to a minimum.

Data Gathering and Recording – The most important step in any effort to model a utility system is the gathering of accurate component and site data. Some data can be obtained in company inventory data sets (power transformer and circuit breaker types and positions), but the majority of the data needs to be collected in the field. The data contained in the company inventory data sets about transformers and circuit breakers were found to contain errors because components had been changed out since the data file was last updated. Modeling details such as equipment anchorage, component interaction, transformer radiator type and installation and other equipment must be determined in the field along with actual system and component connectivity.

SERA I component data for each substation was basically a summary table of component types and a matrix that identified their location order. This was very cumbersome and difficult to update or to validate when in the field. SERA II utilized the new inventory recording scheme that matches the actual physical installation. The substation single line diagram location and identification scheme is center to the data documentation. One beneficial byproduct of this approach was that it was intuitive to Edison personnel that were brought in to interpret the results. The recording of a hypothetical "double breaker" and a "breaker and a half" position shown in Figure 2 is given below:

#,POSITION,220,5;
P,652N,BS1999,DS31333,CB01333,DS11;
L,Line 1-Sub A,DS11333,DS31333,CC1333;
P,452S,DS31333,CB01333,DS11999,BS1;
#,POSITION,220,6;
P,662N,BS1999,DS31333,CB01333,DS11;
L,Line 2-Sub B,DS11333,DS31333,CC1333,WT1SXX;
P,562S,DS31333,CB01333,DS11;
L,Line 3-Bank 1,DS11333,DS31333,CC1333;
P,462S,DS31333,CB01333,DS11999,BS1;

The data of a typical (large) bulk power switching station can be developed within a day or two on site.

Conductor Slack – High voltage substation components are typically connected to other high voltage substation components via a conductor (flexible or rigid). In the higher voltages, above 220kV, component motion at the conductor attachment points can range from a fraction of an inch to a few feet. Evidence from several recent earthquakes suggests that loads arising from component interaction are one of the contributors to component failure. To determine the risk of failure due to component interaction through the conductor, the amount of conductor slack that is available to take up this relative motion between any two connected components must be determined and compared with the possible relative deflections of these same components. This information must be entered into a model that will forecast whether the connected equipment is damaged by the interaction

Conductor slack (excess conductor between components) can be determined several ways: direct measurement (too time consuming and impractical while system is energized), three-dimensional surveying using laser distance finders (too time consuming and varying degrees of accuracy) and estimated (least accurate but efficient).

The first two methods were ultimately dismissed because they were either impractical or too costly (time consuming). The third approach, estimation, at first seems difficult if not impossible. After reviewing many substations, slack estimations fall under three obvious categories, way too little slack, way too much slack and all the rest. The first and third categories comprise about 50% of the cases and some means of measurement must be made to work for the other 50%

Casual estimations of slack usually result in amounts of slack that are greater than actual measurements. The author in field situations on several occasions has verified this. The author has also run tests using a lightweight chain of known length between points of varying distances and relative elevations. The chain was strung from the points and the slack was estimated and compared against the measured value (length of chain suspended between two points minus distance between the two points). After many attempts and repetitions, the author believes that one can become fairly accurate (+- one or two inches over several feet) at making these estimates.

The fragility or likelihood of failure of components due to the components interacting (insufficient slack) is not known and is presumed to be different for different equipment, equipment combinations and conductor type. It is only assumed that when two components are vibrating and there is no more slack available that impact loading will occur and this could lead to one or both components failing. It is also assumed that this is more likely with components that have substantially different frequencies than when they are similar. Experience data demonstrates that all cases of insufficient slack do not result in component failure. SERA II provides for a risk of failure when the ratio of required slack to available slack equals one and increases as the ratio increases.

Fragilities – Equipment fragilities are necessary for determining the performance of each individual component during each scenario. Equipment fragilities can be directly determined by detailed analysis. Ideally, these component models should be verified by test. Each component model could be evaluated in real time during each scenario. Another way is to evaluate component performance (preferably for motion shaking in all three directions) to a statistically significant number of earthquakes of varying intensities and frequency content. The statistics of failure could be determined to a very high level of confidence provided that proper failure/motion parameters are developed. Unfortunately, the former approach is too expensive and the data and experience doesn't exist for the latter. There are multitudes of component types in existence that have never experienced any significant shaking or for which there is not any reported damage during an earthquake. Yet, equipment fragilities are needed for substation components in order to obtain system performance forecasts that are useable.

The author has used a "best guess" approach for developing component fragilities. SERA I utilized Edison staff experience and estimates for component fragility. Basically, these values were obtained informally, by consensus and judgment. In this process, Edison staff developed first-order estimates of such models solely from experience. The models included estimates of peak ground acceleration levels at which 0%, 16%, 50%, 84% (rough bell curve with a lower end cut off) of the population of like components would fail due to a particular failure mode. For some components, several failure modes were proposed. Later (in 1993), experts from

SCE, LADWP and PG&E held a one-day meeting at Edison to review and improve the SCE component fragility estimates. The value of these fragilities was that vulnerable components could be readily identified and such knowledge could be used in any form of risk assessment. One additional benefit of the meeting at Edison was that fragility estimates were developed for more component types.

Analytical Process – The analytical approach amounts to listing all the equipment and determining, via, the Monte Carlo approach whether they fail during several scenarios of specified earthquakes. Each component's fragility is evaluated using Monte Carlo techniques against a motion intensity that is consistent with its substation's site conditions and location with respect to the fault and the Magnitude of the earthquake. Each equipment function in the system can be determined by virtue of its type and location in the substation and the substation's connectivity with the balance of the system. Assuming that when a component is damaged (defined by the equipment failure modes) the flow of electricity through the node where that equipment is positioned is interrupted either during or immediately after the earthquake, the functional state of the system can be determined for each scenario.

System Stability – The key findings of a SERA is information about whether the system will be able to continue to deliver power and to what portion of the system will service be interrupted and how long will it take to return the system to some level of service.

Electrical System "Load Flow" programs have been developed to determine whether the defined power source, the defined load and the existing functional system are a stable operable combination. In the case of forecasting the recovery from a disaster of a large utility with many sources of power and load, the problem soon becomes one of what are the defined load(s) and the defined power source(s) for which to focus on at a target service level.

Edison took the approach of evaluating each substation's basic function: power transformation between voltages, switching and connectivity and determining whether basic substation (normal or emergency) performance goals could be developed. For example, emergency levels of transformation (one goal) service could likely be met if a prescribed number of A (220kV) and/or the same or different number of AA (500kV) transformer banks are operable at the substation. Generic emergency level of service configurations of substation switchyards could not be determined, instead, representative switchyard damage configurations for each substation were developed based on inspection of the whole system performance for several simulations. System operation personnel viewed these and established what mitigation efforts were required at each substation, on the component level, to bring the system to pre-established service level (e.g. emergency service level) goals.

Return to Service - Edison staff, in SERA I and II, evaluated return to service times. The purpose of the SERAs was to forecast whether return to service goals could be met. Edison has developed component downtime estimates for various failure modes, operation states and emergency conditions. These were developed in a similar manner as were the performance algorithms, that is, informally and heavily based on judgement. The downtime modeling is in terms of "crew-hours" and estimates were made for all the failure modes of all components. With these downtime model estimates, damage forecasts as well as returns to service estimates were developed for several earthquake scenarios.

Ref (1) identified four service levels that a utility eventually maintains after a significant earthquake. Edison still concludes that "emergency level" and "full capacity with diminished redundancy and reliability" are the service levels that this type of analysis should target.

Early analyses were very detailed in that specific service crews were "followed" throughout the analysis and the total service crew resource was applied to the workload optimally. This was a substantial amount of effort and the results were close to the simple calculation of total work resource requirement divided by total resource available. The results are in the units of hours. The assumption was made that each day is 12 hour until the first milestone, emergency level of service, is reached and 10 hours until the second milestone, full capacity with diminished redundancy and reliability, is reached.

SERA II Results - The Edison system has indeed improved in the last decade and the SERA II demonstrates this. While considerable damage still occurs from a M8.0 San Andreas earthquake (located to cause the most damage) to the Edison system, the damage is not to expensive (dollars and time to repair) circuit breakers, but to disconnect switches. More damage occurs (than in SERA I) to transformers, disconnect switches and protection equipment, due to the fact that these components are more vulnerable that previously though (Northridge Data and careful study of existing data). With a better understanding of component fragility data, more components modeled and several components more vulnerable that previously thought, the Edison System has been found to perform significantly better than during the SERA I evaluation.

Note on SERA use. A lot of attention is paid toward accuracy of results in developing and evaluating this SERA process. The rational is more accurate models can provide more accurate results. Utility personnel, while looking for accuracy, should understand that a SERA, as described in this paper, provides way of integrating the fragmented data and intuitions about a utility system with complex seismicity/attenuation and local effect models. Having done this, they may be better able to obtain a picture of the implications to a electric utility system of some future earthquake. With or without a SERA, as described in this paper, if the risk to a utility is perceived as possibly too high, mitigation programs in response to the utility's earthquake hazard must be made. In other words, a more realistic way to view a SERA is as a rational way to assess the system performance in terms of the best information at hand rather than as a way to provide an accurate prediction of the utility's system performance to a future earthquake. This way of viewing a SERA also leads to a more appropriate appreciation for the results coming from such a study.

Conclusion - The SERA I and II evaluations provide Edison with an insight into the function of its bulk power system during a severe earthquake. This capability is unique in the electric industry. The improvements in SERA II over SERA I are substantial, but more remains to be done. Better component fragilities would increase confidence in results. Better ways of determining and representing system performance/stability are needed to measure this key parameter.

Acknowledgements - The author would like to acknowledge the important contributors who along with the author really developed, provided information, insight and collaboration, and broke standard industry barriers in communication to cause, in what was really a grand-team effort, SERA II to be able to form and be utilized. These individuals are: Ed Matsuda of Bay Area Rapid Transit, Woody Savage of United States Geological Survey (both formerly with Pacific Gas and Electric), Ron Tognazzini of Los Angeles Department of Water and Power, and Jim Kennedy (now deceased) of Southern California Edison.

Reference –

Burhenn, T. A., Hawkins, H. G., Ostrom, D. K., Richau, E. M., (1992) "The Earthquake Vulnerability of a Utility System", Proceedings of the American Power Conference, V-54-1, P.224-227.

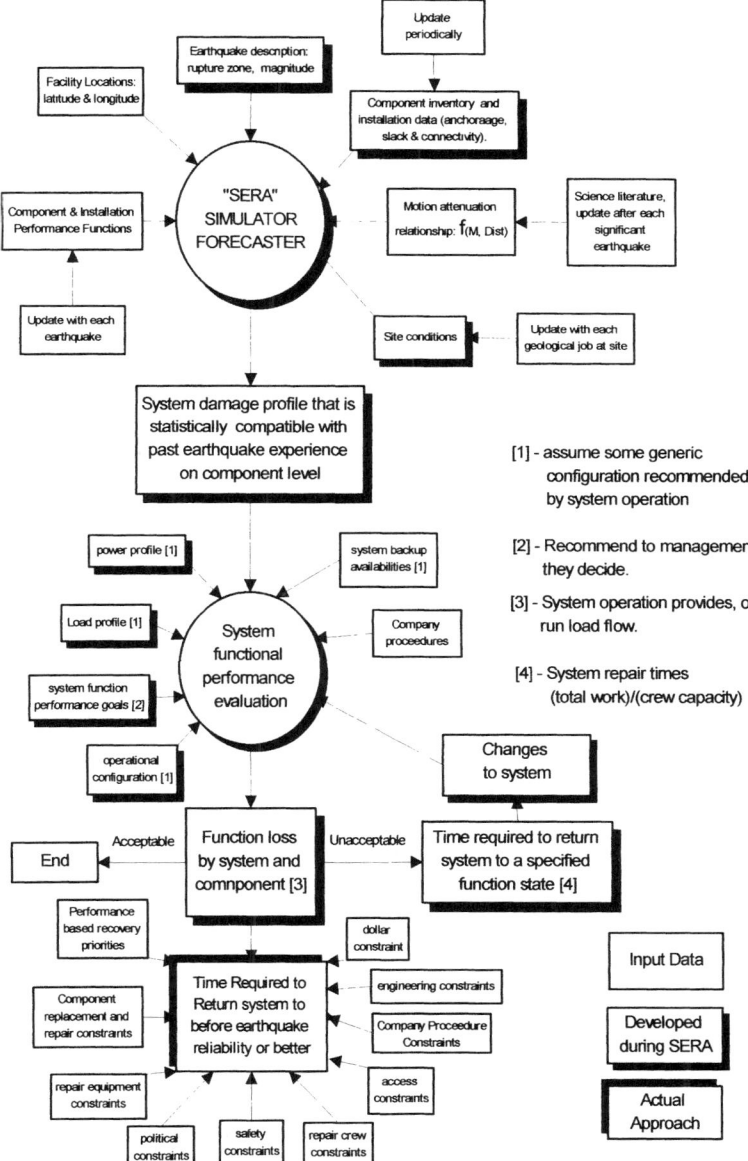

Figure 1 – SERA Approach

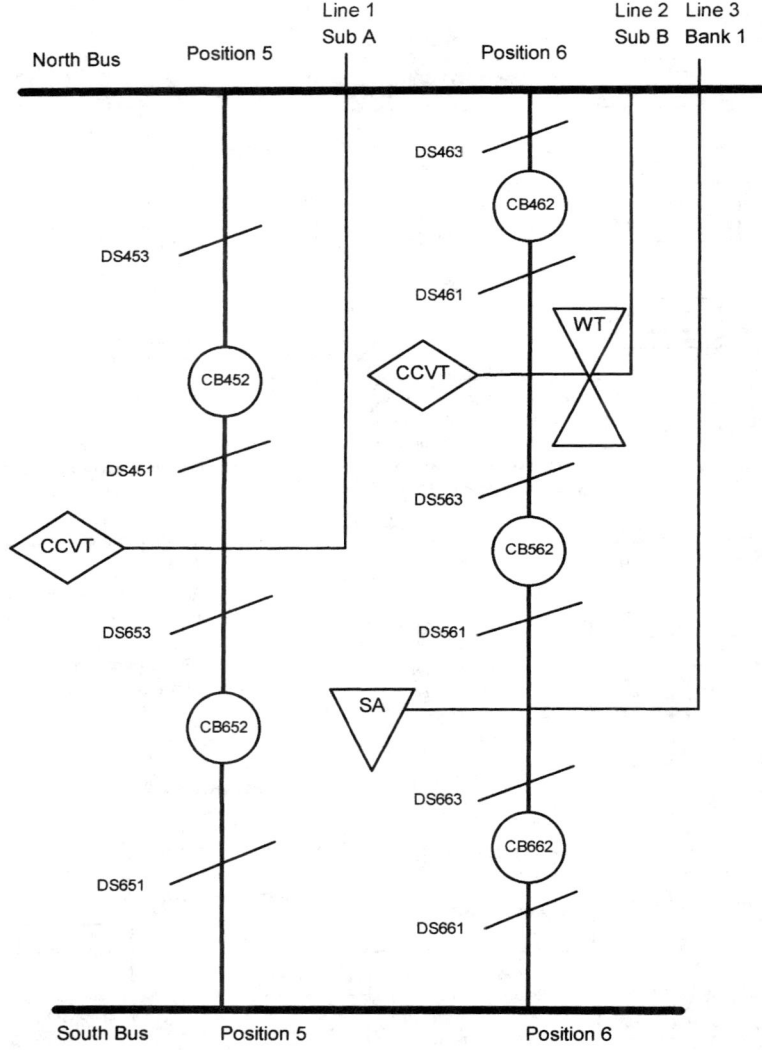

Figure 2 – Hypothetical Position 5 (double breaker) and Position 6 (breaker and a half).

Seismic Displacement at Interconnection Points of Substation Equipment

Jean-Bernard Dastous[1] and Andre Filiatrault[2]

Abstract

During an earthquake high-voltage substation equipment may experience substantial displacement at interconnection points, resulting in relative displacement between interconnected components. Sufficient slack must therefore be provided in the conductor between the moving components to prevent the impact loading that can occur as a result of a shortage of slack. The first step in estimating the required conductor slack is to determine the individual displacement that each interconnected equipment may experience. Once this is known with sufficient accuracy, methods to estimate the required slack such as prescribed in the IEEE 693 standard may be used.

This paper presents a simple and useful method to estimate the expected displacement at the interconnection point of substation equipment under seismic excitation based on knowledge of the fundamental frequency of an equipment, its damping and the use of a response spectrum. A survey and analysis of qualification reports on standalone equipment revealed that for many types of equipment, the displacement is composed almost entirely of the contribution of the first vibration mode. The generalized single-degree-of-freedom method is therefore recommended to evaluate the displacement using the first modal-participation factor and the spectral displacement from the required response spectrum. Mean and bound values of this factor were determined and are presented for most classes of equipment used. Shake table tests were performed to investigate the effect of the flexible connection that is present a the top of equipment on this factor. It was found that, when sufficient slack is provided, its main effect, on average, is to decrease the standalone modal-participation factor. The method presented is thus of practical use, specially when no data is available on the expected displacement of a given type of equipment.

Introduction

During an earthquake, high-voltage substation equipment may experience substantial displacement at interconnection points, resulting in relative displacement between interconnected components. Sufficient slack must therefore be provided in the conductor between the moving components to prevent the impact loading that can

---

[1] Research Scientist, Institut de recherche d'Hydro-Québec (IREQ), 1800 Lionel-Boulet, Varennes, Qc, Canada, J3X 1S1; phone: 450-652-8341; Dastous.jean-bernard@ireq.ca.

[2] Professor, Department of Structural Engineering, University of California in San Diego, La Jolla, CA 92093-0085; phone: 858-822-2161; afiliatrault@ucsd.edu.

occur as a result of a shortage of slack. The first step in estimating the required conductor slack is to determine the individual displacement that each interconnected equipment may experience. Once this is known with sufficient accuracy, methods to estimate the required slack such as described in current standards (IEEE 1998, IEEE 2002) may then be applied.

There is still little data readily available to estimate displacement at the interconnection points, which is understandable since it is only recently that issues related to conductor slackness have been formally considered in seismic standards. The objective of the study reported in this paper was to fill the gap in displacement data by undertaking a review of available qualification reports. In the course of this review, a simple method to estimate the displacement was identified. This method is based on knowledge of the fundamental frequency, the related damping and uses a response spectrum technique. This paper presents the method and demonstrates that it can be used to predict bound and mean values for the expected displacement.

Methods to compute displacement

Single-degree-of-freedom methods. With the use of a response spectrum, the Single-Degree-Of-Freedom (SDOF) method can first be used (Chopra, 1981). However, its application is limited to equipment with most of the mass essentially lumped at its top. A more accurate approximation is to assume that the system behaves as a generalized SDOF system and that it oscillates principally in its first cantilever flexural vibration mode $\psi(y)$ (denoted in what follows as the first mode), as shown in Figure 1. The maximum displacement, x_{max}, for such a system using the response spectrum method is given by (Clough and Penzien, 1975):

$$x_{max} = \frac{L}{m^*} \cdot S_d(f, \zeta) = \frac{L}{m^* \omega_n^2} \cdot S_a(f, \zeta) \qquad (1)$$

where L/m^* is the modal-participation factor of the first mode with L the effective mass producing the external inertia loading and m^* the generalized mass, f is the first mode undamped frequency (also denoted as the fundamental frequency), ω_n is the associated circular frequency, and ζ the equivalent viscous damping of the first mode. The modal-participation factor L/m^* characterizes the difference of the generalized method with the SDOF method for which this factor is implicitly equal to unity in (1).

Multi-degree-of-freedom method. An even better approximation is to model the equipment as a Multi-Degree-Of-Freedom (MDOF) system using the finite-element method. In the case of a linear model (displacements and deformations assumed small and material properties assumed constant), the displacement can be obtained by summing up the contribution from a sufficient number of modes, by using an appropriate combination method such as the SRSS (square root of sum of squares) (Clough and Penzien 1975) or the CQC (complete quadratic combination) method (Der Kiureghian, 1980).

Recommended method for substation equipment. In a recent study, it has been demonstrated for candle-like and slender frame types of substation equipment on support, that the displacement is essentially composed of the first-mode contribution only, even when its frequency does not lie in the maximum amplification range of the spectrum used (Dastous et al. 2003). For such types (which encompass the majority of equipment) the recommended displacement estimation method is therefore the generalized SDOF method, whereas not much more accuracy would be gained using the MDOF method. For equipment for which only the first mode frequency and damping are known or estimated, the generalized SDOF method can be used as long as the value used for the first modal participation factor is representative. Theoretical values for equipment on support have been evaluated to be mostly in the range of 1 to 1.6, with extreme values up to approximately 2 (Dastous et al. 2003). Values close to 1 apply to equipment with important mass at their top, such as live-tank circuit breakers. Values close to 1.6 apply to equipment on support with mass and rigidity more evenly distributed, such as measurement transformers.

Survey of actual values of modal-participation factors from qualification reports

Extraction of first modal-participation factor from reports. Since the contribution of the first mode is predominant, it is justified to extract and approximate the first modal-participation factor from reports using (1) by:

$$\frac{L}{m^*} \cong \frac{x_{max}}{S_d(f,\zeta)} \tag{2}$$

In (2), the tip displacement x_{max} is the value actually measured during a test or evaluated analytically. The value of S_d can be obtained with the response spectrum used in the analysis or the spectrum that corresponds to the shake table motion feedback during a seismic test, using the first mode frequency and damping of the equipment.

Database of substation equipment used. An electronic database of substation equipment data was established by the authors and other contributors by surveying actual qualification reports (PEER 2002). Over 300 reports were surveyed. Of these, approximately 200 gave explicit information about the displacement at the attachment point. In over 90% of the reports, the qualification had been done by analysis, using the finite-element method and a design response spectrum. In the others, the qualification was done by means of shake table tests. It should be noted that, before the publication of the IEEE-693-1997 standard (IEEE 1998), there was no formal standard way to prepare a qualification report and the type of information obtained in each report surveyed therefore varied widely. While some reports were very detailed, others were too sketchy to be of value and were rejected.

Data retained. After examination of the modal-participation factors obtained by means of analysis and testing, it was decided to retain only those obtained by analysis based on the finite-element method for the following reasons:

First, a theoretical and representative value of the modal-participation factor can be obtained if the stiffness and mass distribution are carefully modeled. Since the finite-element method provides an especially good estimate of the first mode and its frequency, even with a very simplified model, it can therefore be concluded that the factors obtained by analysis are indeed representative.

Second, the survey of qualification reports based on test results revealed many uncertainties. One uncertainty is related to the measurement of damping, a determining quantity in establishing S_d in (2). Even a small variation in damping (e.g. from 3 to 5% of critical) causes a large variation in S_d and, hence, in the modal-participation factor. From our experience in measuring actual values on equipment, this quantity varies with the amplitude of testing and, also, possibly, with the method used to evaluate it. Since, in most reports, details were not given as to how damping was measured, considerable uncertainty was felt about the values obtained.

Another uncertainty is related to the actual attainment of the targeted design response spectrum in the tests. The observation that in certain frequency ranges there is often a large discrepancy between the test response spectrum and the required response spectrum also adds uncertainty to the determination of the spectral displacement.

Type of equipment considered. The type of equipment selected was high-voltage apparatus mostly rated over 120 kV and usually interconnected with flexible connections. Equipment items that can be assimilated as generalized SDOF systems (e.g. candle-like or frame types) were considered. Dissimilar equipment items such as dead-tank circuit breakers or equipment with dampers (due to complex modes) were not considered.

Survey of modal-participation factors. A total of 181 modal-participation factors were obtained from the survey. Of these, 161 were obtained with equipment on a support and 20 for non-supported equipment. The main results of the survey for each class of equipment are presented in Table 1. It is observed that the modal-participation factor for all equipment is in the range 0.65 to 1.98 (with most values over 1) with a mean of 1.34. On average, it is observed that equipment on supports has a lower modal-participation factor, although this observation is based on a limited sample of equipment items without support. Comparison with the theoretical values as discussed previously thus shows good agreement.

Figure 2 presents the distribution of the modal-participation factor with fundamental frequency, for the data contained in Table 1. In this figure, the fundamental frequency of the equipment surveyed varies from 0.4 to 17 Hz with an average of 3.9 Hz. It is readily observed that the modal-participation factor is not correlated with frequency. Therefore, using a single average value and building confidence intervals from the data gathered would theoretically apply to all classes of equipment, independently of the frequency. A similar study of the variation of modal-participation factor with equipment voltage led to the same conclusion.

Bound value for the expected displacement using the IEEE-693 Spectrum

The data presented in Table 1 can be used to obtain reference and bound values for the modal-participation factor for different types of equipment. It is also possible to use the mean value and the estimate of the standard deviation from the data collected to estimate bound values that would cover all classes of equipment. Based on this, a statistical analysis was performed (Dastous et al., 2003). It was found that the statistical distribution of the modal-participation factor is almost *normal* (Gaussian). Using this distribution and its parameters, tolerance limits for the expected displacement were constructed. The upper-bound tolerance limit for 95%, $x_{max}|_{95\%}$, at a 95% confidence level was found to be:

$$x_{max}|_{95\%} = 1.62 \cdot S_d(f,\varsigma) \tag{3}$$

By analogy with (1), (3) indicates that a value of 1.62 for L/m^* should cover 95% of all cases when used to compute the expected displacement, according to the database compiled in this study. It is noteworthy that this value is close to the upper bound value determined theoretically as discussed. Therefore, when only f and ς are known or estimated, (3) can be used with confidence to obtain a bound value on the expected displacement (for candle-like and frame types of structures).

Using (3), Table 2 presents the values of the expected displacement for 2% damping and for a range of equipment frequencies from 0.5 to 10 Hz, which should cover 95% of the equipment surveyed, using the IEEE-693-1997 response spectrum for the high qualification level (1g). Also presented in this table is the bound value of the expected displacement using the entire database where 2 (1.98 rounded) was the maximum value observed.

Influence of flexible connections on modal-participation factors

In this section, the influence of flexible connections between pairs of equipment is evaluated on the basis of the results of well controlled shake table tests. The influence of flexible connections is characterized by a calibration factor to be applied to the modal-participation factor discussed above for stand-alone equipment. With this approach, the interaction effect can be taken into account approximately without recourse to a full dynamic analysis of the interconnected equipment.

Scope of shake table tests. Shake table tests were performed on five pairs of generic substation equipment connected by three different flexible connections made of aluminum cables (Filiatrault and Stearns, 2002). Simulated horizontal ground motions were applied in the longitudinal direction of the flexible connections by the uniaxial earthquake simulation facility at the University of California, San Diego. The variables considered in the tests were the dynamic characteristics of the generic equipment, the types and slackness of flexible connections, the simulated ground motions, and the intensities of the simulated ground motions.

Description and properties of generic equipment. Five different pairs of generic substation equipment were considered for these tests. Each pair of generic equipment was designed to be representative of the dynamic properties of actual interconnected substation electrical equipment based on the data contained in the database discussed earlier. Each equipment specimen comprised of 4.3 m tall cantilevered tubular steel columns with a lumped mass at the top (Filiatrault and Stearns, 2002).

Description of flexible connection specimens. Three 4.6-m steel-cored aluminum cable assemblies were used to connect the various pairs of generic equipment during the shake table tests. Three different levels of slackness of connection were considered for interconnecting the generic equipment. The slackness s is defined as:

$$s = (l_c - l_{ch})/l_{ch} \qquad (4)$$

where l_c is the cable length and l_{ch} the chord length, which for a pair of equipment of similar height is the horizontal distance between the ends of the cable.

Slackness values of 2, 5 and 10% were considered in the shake table tests. At 5 and 10% slackness values, the individual fundamental frequencies of connected equipment were not affected significantly by the connections (the available slack not being all used up during the tests). At 2% slackness, significant nonlinear interaction occurred between the equipment, even at low amplitude, which drastically changed their dynamic characteristics. Therefore, the data for 2% slackness was not included.

Earthquake Ground Motions. Two recorded components of near-field earthquake ground motions were used for the seismic tests on the shake table: Tabas (1978 Iran earthquake) and Newhall (1994 Northridge, California, earthquake). These two records are representative of earthquakes known to have a high potential for damaging structures and equipment.

The Tabas record was modified using a non-stationary response-spectrum matching technique (Abrahamson, 1997) to match the IEEE 693 standard target response spectrum for testing. The record was further high-pass filtered using a cut-off frequency of 1.5 Hz so as not to exceed the displacement limit of 150 mm of the shake table. The two records were scaled at different amplitudes during the tests (Filiatrault and Kremmidas, 2000).

Computation of correction factors for first-modal participation factor of interconnected equipment. For each shake table test conducted at 5% and 10% slackness, (2) was used to estimate the first modal participation factor for each item of interconnected equipment. A correction factor CF was then computed by normalizing each modal-participation factor of interconnected equipment by the corresponding standalone first modal-participation factor. The latter was obtained from the shake table tests on the standalone equipment. Note that all spectral values in (2) were taken at 2% damping, which was representative of the measured damping ratios. The resulting correction factors are plotted in Figure 3 for the two levels of connection slackness considered in the study (5% and 10%).

The correction factors were found slightly higher for the low-frequency equipment than for the high-frequency equipment. Most of the computed CF values are less than unity, indicating that the first modal-participation factor of the equipment is reduced by the presence of the connection. This is caused by the added mass of the connection applied at the top of the equipment, which effect was shown to decrease the modal-participation factor (Dastous et al., 2003). Also shown in Figure 3 are the mean values of CF for the low-frequency equipment (0.90) and the high-frequency equipment (0.75). Until more data becomes available, these values could be used to correct the first modal-participation factor obtained by (2) for equipment analyzed as standalone in order to account approximately for the interaction effect of flexible connections. Note that this approach can be used only if sufficient slackness is included so that no significant nonlinear interaction occurs between connected equipment.

Conclusions

The standalone displacement at the interconnection point of substation equipment during an earthquake was found to be almost composed entirely of the contribution of its first vibration mode only. This applies to candle-like and slender frame types of structures. The displacement of such structures, can therefore be obtained by considering them as generalized single-degree-of-freedom systems. For a given earthquake record or response spectrum, the displacement can be obtained simply by using the corresponding first modal-participation factor and multiplying it by the spectral displacement related to its first mode frequency and damping.

Practical values were obtained from a survey of qualification reports done by analysis. Average values were presented for the different types of equipment surveyed. For all equipment surveyed, an average value of 1.34 was found, while a value of 1.62 was estimated as a bound covering 95% of the values surveyed here.

The influence of flexible connection on the standalone modal-participation factor was investigated through shake table tests on a limited number of pairs of interconnected generic equipment. When sufficient slack was provided so that no nonlinear interaction effects occurred, it was observed that, in general, the stand-alone modal-participation factor was lower in the presence of the flexible conductor. It was also found that the equipment's first mode frequency in such conditions was not modified substantially. An average correction factor of 0.90 was obtained for the stand-alone modal-participation factor of the low-frequency equipment of the pair tested, while an average value of 0.75 was obtained for the higher-frequency equipment.

The proposed generalized single-degree-of-freedom method applied to standalone equipment can therefore provide a good estimate of the displacement for interconnected equipment. For more accuracy, the average correction factors obtained here can be used.

Acknowledgments

The authors gratefully acknowledge Mr. Rulon Fronk from Fronk Consulting, Mr. Eric Fujisaki from PG&E, Mr. William Gundy and Ms. Kathe Matheson from Gundy and Associates, and Mr. Peter Nguyen and Mr. Robert Stewart from B.C. Hydro, for their contribution in gathering a significant portion of the data used in this study.

References

Abrahamson, N. (1997). Private Communication, Pacific Gas and Electric Company, San Francisco.

Clough, R.V. and Penzien, J. (1975). *Dynamics of Structures*, McGraw-Hill, New York.

Chopra, A.K. (1981). *Dynamics of Structures*, Earthquake Engineering Research Institute, Berkeley.

Dastous, J.B., Filiatrault, A. and Pierre, J.R. (2003). "Estimation of Displacement at Interconnection Points of Substation Equipment Subjected to Earthquakes", accepted for publication in *IEEE Transactions on Power Delivery*.

Der Kiureghian,A. (1980). "Structural Response to Stationary Excitation," *J. Eng. Mec. Div.*, ASCE, 106, pp. 1195-1213.

Filiatrault, A. and Kremmidas, S. (2000). "Seismic Interaction Between Components of Electrical Substation Equipment Interconnected by Rigid Bus Conductors," *J. Structural Eng.*, ASCE, 126, pp. 1140-1149.

Filiatrault, A. and Stearns, C. (2002), *Electrical Substation Equipment Interaction – Experimental Flexible Conductor Studies*, Report No. SSRP-2002/09, Department of Structural Engineering University of California, San Diego.

IEEE Recommended Practice for Seismic Design of Substations (1998). IEEE Standard 693-1997.

IEEE Recommended Practice for the Design of Flexible Buswork Located in Seismically Active Areas (2002). IEEE Standard 1527/Draft 10, 2002 (in preparation).

PEER, (2002). http://seismic.ucsd.edu/peer/substation.html

DISASTER RESPONSE FOR LIFELINE SYSTEMS 605

Table 1: Results of survey on first modal participation factors

Equipment type	No. of items	No. with support	No. without support	L/m all values			L/m With support only			L/m Without support only		
				min	max	mean	min	max	mean	min	max	mean
Current transformers	19	14	5	1.18	1.84	1.43	1.18	1.84	1.44	1.34	1.51	1.40
Circuit breakers/live tank	38	38	0	0.65	1.82	1.25	0.65	1.82	1.25			
Circuit switcher	1	1	0	-	-	1.44	-	-	1.44			
Disconnect switches: hinge end open	8	8	0	0.88	1.82	1.21	0.88	1.82	1.21			
Disconnect switches: jaw end open	8	8	0	0.71	1.82	1.15	0.71	1.82	1.15			
Disconnect switches: hinge end closed	9	9	0	1.16	1.91	1.45	1.16	1.91	1.45			
Disconnect switches: jaw end closed	9	9	0	1.09	1.91	1.42	1.09	1.91	1.42			
Ground disconnect switches (column type)	1	1	0	-	-	1.44	-	-	1.44			
Lightning/surge arresters	24	17	7	1.08	1.59	1.43	1.08	1.59	1.40	1.49	1.56	1.51
Reactors	3	3	0	1.28	1.29	1.29	1.28	1.29	1.29			
Shunt capacitors on rack	49	49	0	0.80	1.98	1.32	0.80	1.98	1.32			
Voltage transformers	12	4	8	1.3	1.63	1.45	1.44	1.60	1.52	1.30	1.63	1.41
All values	**181**	**161**	**20**	**0.65**	**1.98**	**1.34**	**0.65**	**1.98**	**1.33**	**1.30**	**1.63**	**1.44**

Table 2: Bound values for the expected displacement for 2% damping using the IEEE Response spectrum at 1 g

f (Hz)	x_{max} (cm) L/m =1.62	x_{max} (cm) L/m =2
0.5	243	300
1	122	150
2	32	40
3	15	18
5	5.2	6.4
8	1.9	2.4
10	0.97	1.2

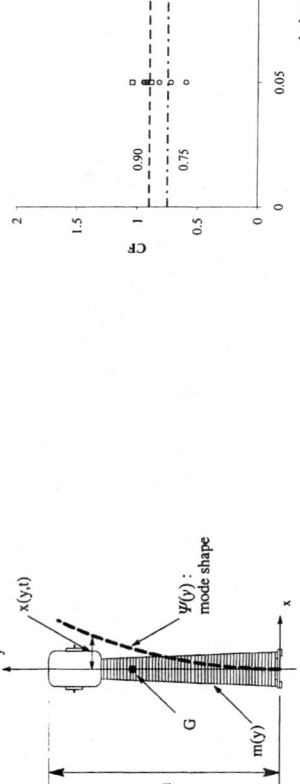

Figure 1. Generalized single degree of freedom.

Figure 3. Correction factors for the first modal-participation factor.

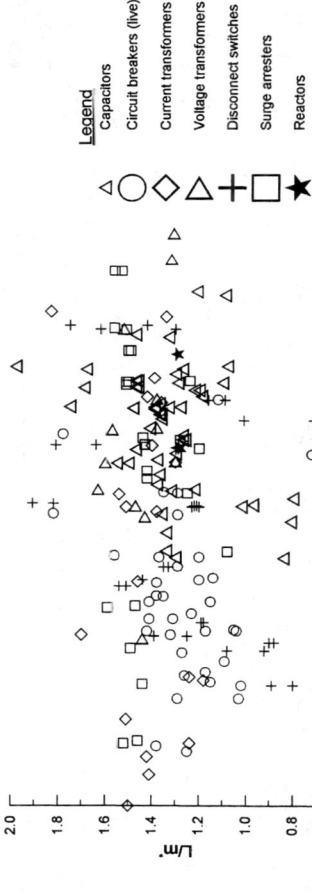

Figure 2. First modal participation vs. frequency for all equipment surveyed

Issues and Guidance for IEEE 693 Equipment Qualification Tests

Anshel J. Schiff[1] and Leon Kempner, Jr.[2]

Abstract

This paper addresses issues associated with seismic qualification testing. Recommendations are made to reduce over testing and help to provide a valid equipment qualification.

Introduction

Observations made during equipment qualification tests to meet IEEE Standard 693 (1997), which will be referred to as IEEE 693, and experimental research conducted at commercial, government, and university testing facilities have identified potential difficulties in procedures, equipment characteristics, and analysis techniques that can lead to over-testing or inadequate equipment evaluation.

Some of these observations were obtained during a research project associated with the Electric Power Research Institute (EPRI) Seismic Qualification Consortium (10 utilities and the California Energy Commission) and from research and qualification tests conducted at several facilities by others involving several manufacturers over a period of several years. Issues addressed are test procedures, use of instrumentation, shake-table characteristics and limitations, and methods of data analysis. For some of the issues, methods are recommended to mitigate the noted concerns.

Shake-Table Rotation

IEEE 693 typically requires equipment with an operating voltage of 161 kV and above to be qualified by shake-table testing. IEEE 693 recommends that equipment be qualified using its structure, which are usually from 2.5 m (8') to 4.3 m (14') tall. The height of 500 kV equipment above the support structures may range from 4.6 m (15') to 12.2 m (40') and they may weigh as much as 2250 kg (4963 pounds). The fundamental frequencies of equipment installed on shake tables have ranged from 1.25 Hz (disconnect switch) to 20 Hz (transformer bushing). The large weight and high center of gravity of this equipment can induce very large overturning moments on the shake tables during testing.

The shake tables generally used for qualification testing have hydraulic actuators. These systems exhibit, at some frequencies, oil-column resonance making it difficult to control the table motion. Some tables that are capable of generating multiple-degrees-of-freedom excitation have mechanical couplings that can also introduce uncontrolled motions. These effects cause differences between the target and test

---

[1] PMI, 27750 Edgerton Rd, Los Altos Hills, CA 94022, schiff@ce.stanford.edu
[2] BPA, P.O. Box 61409 (TNFC-TPP3), Vancouver, WA 98666-1409, lkempnerjr@bpa.gov

response spectra. The test response spectra are generated using measurements from accelerometers that are usually mounted near the center of the table. Test data (Gilani 1998) in Figure 1 shows how the table response can affect the test results. Figure 1a shows the response spectrum of the electronic input before it is filtered to reduce the low frequency content of the record. Figure 1b shows the response spectrum of the modified signal to accommodate displacement and velocity limitations of the table. Figure 1c shows the test response spectrum. All plots show the IEEE 693 target High Required Response Spectrum. While the input signals closely match the target spectrum, the test response spectrum has large deviations from the target spectrum. Generally, the test response spectrum more closely follows the target spectrum.

The rotational motions, pitch, yaw, and twist, of the table are usually not measured when equipment is qualified. This would require three additional accelerometers, vertical accelerometers in the $x = 0$, and $y = 0$ planes and a horizontal accelerometer with a measuring axes offset from the center of the table. These measurements were made in the PEER research project (Gilani 1998) and the resulting pitch, yaw, and twist plus the associated power spectral densities are shown in Figures 2 and 3, respectively. In the report it was estimated from the test result that the added acceleration at the top of the bushing was about 1 g. The peak acceleration observed at the top of the bushing during the test was about 3.5 g (Gilani 1998). Two characteristics of the spectral distribution of the rotational motion should be noted. The pitch and yaw spectra, that is rotations about the horizontal axes tend to be high frequency and may be related to interaction of the shake-table system with the natural frequency of the item being tested. The frequency content of the twist spectrum, that is rotations about a vertical axis, is broadband and extends to low frequencies. In the case of the tall, slender bushing that was being tested, which is centered on the table, these motions would probably have little effect on the response. Had the test equipment been a 500kV disconnect switch that spans the table, the effect on the equipment is not clear.

Table rotations can contribute to over testing equipment. The height of the support structure coupled with the pitch and yaw introduces an added horizontal acceleration at the top of the support structure that is not part of the intended input excitation. It may be possible to partially compensate for the rotational effects. While one could estimate the increase in the peak acceleration and then lower the amplitude of the input accordingly, such a procedure would be inappropriate. As Figure 3 shows, rotation effects are distributed unevenly over a broad frequency range. Reducing the level of the entire spectrum would have an unknown effect and may result in under testing. For records that are derived from broadband noise, there is a technical basis for reducing the power spectral density of the input signal by a scaled factor of the power spectrum of the rotational input. It should be noted that the measured test response spectrum would no longer satisfy the Required Response Spectrum (RRS). Also, because the response of equipment subjected to large qualification inputs may respond nonlinearly, an exact correction may not be possible. One approach would be to test the equipment at half the ultimate test level to determine the extent of the rotational motion of the table, and use this data to correct the input motions. After the test at the full level, the rotational spectra should be evaluated to check the quality of the correction. It is not clear if this approach can be used for records that are not based on broadband random inputs.

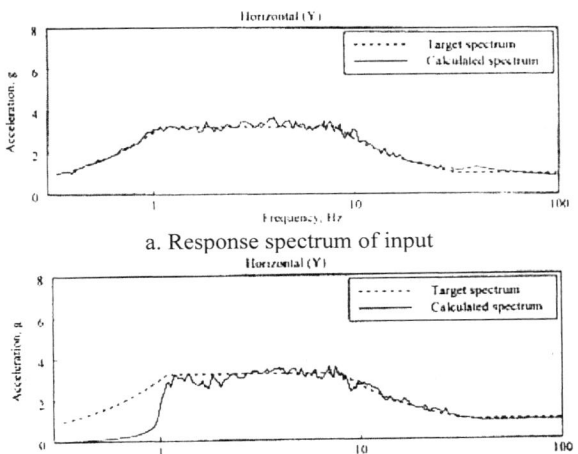

a. Response spectrum of input

b. Response spectrum of input modified to be compatible with table limitations

c. Test response spectrum

Figure 1 - Response spectra of the input and test signals.

Two current research projects (PEER and EPRI) are attempting, in part, to determine if some equipment can be qualified without support structures. Clearly, motions associated with table rotations will complicate the evaluation process. Even if the translations at the base of the equipment due to table rotations can be corrected, this procedure will not correct for the angular accelerations introduced by table rotations. The significance of these motions on the equipment response and performance is not clear. Several earthquake simulators are designed to generate rotational as well as rectilinear motions. These could theoretically be used to correct for table rotation. However, these features are seldom used so that the accuracy and precision of this type of correction is not known. Also, since the rotations may be associated with oil-column resonance, it is not clear if the table will respond appropriately to corrective control signals.

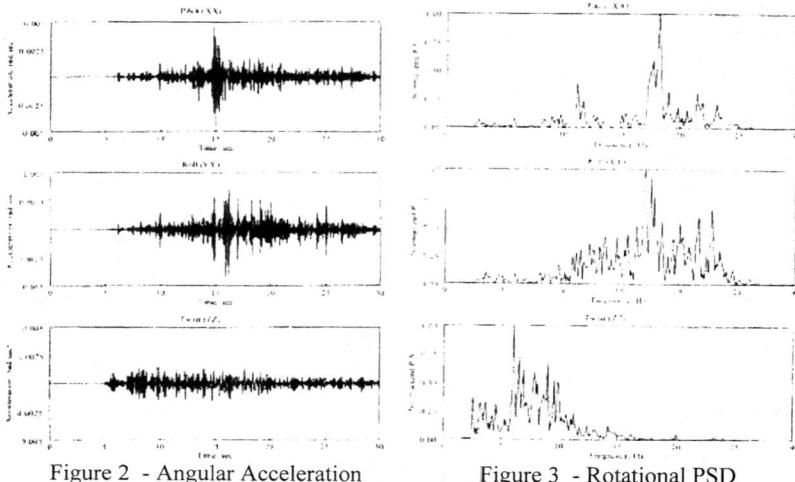

Figure 2 - Angular Acceleration   Figure 3 - Rotational PSD

- It is recommended that as a minimum, testing facilities quantify their table's rotation effects by measuring the rotational spectra while supporting similar equipment. The contribution of table rotation at the base of equipment adds horizontal motion at the top of the supports, which should be evaluated.

- If there is significant table rotational inputs that add to the structure's test level, it is recommendation that a test run be conducted at 1/2 of the target test level and this data be used to correct the input to reduce over testing as suggested above. The rotational effects should be reevaluated using data from the target level tests.

Qualification to the High Performance Level

IEEE 693 provides criteria to set testing levels at the Moderate- or High-Required Response Spectra. It allows manufacturers to test to a High Performance Level (HPL), that is, twice the High-Required Response Spectra. This provides higher confidence that the equipment will perform well in earthquakes with motions up to the HPL. This is because unanticipated and nonlinear problems will be identified and the functionality can only be assured to the level that it is tested. For these reasons, the authors encourage testing to the HPL even though this increases the risk of damage.

A comparison of the calculated response spectra of the simulated earthquake electronic inputs to the RRS shows that well matched input typically fluctuate 2 % to 5% about the target levels. The test response spectra typically fluctuate 10% to 15% in the region of 1 to 8 Hz. At higher frequencies, above 14 to 18 Hz, the test response spectrum often exceeds the RRS by 50% or more. Since IEEE 693 requires that the test response spectrum envelop the RRS, the test response spectrum can be expected to exceed the RRS by 20% to 30% in parts of the well controlled portion of the spectrum. Testing to the HPL, an option allowed in IEEE 693, is expected to expose the equipment to near the ultimate stress levels. Indeed, at the HPL permanent deformations are allowed in parts that do not affect the equipment function. It is the authors' view that requiring the test response spectrum to envelop twice the RRS if testing is done to the HPL, may result in over testing by 30%. This is not a procedure that encourages testing to the HPL.

- It is recommended that when testing to the HPL the response spectrum need only equal twice the RRS at the equipment fundamental frequency and natural frequencies below 12 Hz. At other frequencies in the range from 1 to 12 Hz the test response spectrum should not be less than 80% of twice the RRS.

Post-Test Frequency Search

Shake-table testing of substation equipment with input motions that conforms to the High Seismic Performance Level or even to the High Required Response Spectrum can place high demands on the equipment. In the case of porcelain members, failure is usually obvious and catastrophic. There have been tests where a cast support housing, gusset or weld has been damaged. Some of this damage can be difficult to detect, particularly when it is at an unanticipated location. For this reason, after the target test has been performed a post-test frequency search should be conducted. Shifts in an equipment natural frequency could be indicative of the presence of undetected damage and the equipment should be inspected for damage. A difference in the natural frequencies found in the pre- and post-frequency searches would not in itself invalidate the qualification. Large shifts in frequencies, say larger than 20% should be explained.

- It is recommended that a post-test frequency search be conducted. It can provide an indication of the presence of unexpected damage and the need for careful examination.

Low Frequency Equipment

Many equipment items have very low fundamental frequencies. For example instrument transformers that use composites and 500 kV disconnect switches have had frequencies as low as 1.25 Hz. At these low frequencies it can be difficult to excite the equipment to meet the RRS, because of inherent limits on shake-table amplitudes and velocities associated with low frequencies. Filtering of the signal supplied to the table controller is used to reduce the low frequency content. The upper value of the frequency at which these reductions are needed is a function of the excitation level and the specific capabilities of the shake table; inputs must be reduced in the range from 1 to 2 Hz. This may require sine-beat testing at the natural frequencies where the test response spectrum is below the target response spectrum. It is suggested that the sine-beat be added to the test signal so that multiple frequency affects are addressed. This can usually be accomplished by raising the lower limit of the high-pass filter above what is requires to meet displacement and velocity limits so that the sine-beat can be added to the signal.

It should be noted that extra care is needed when estimating the natural frequency and damping at low frequencies, as closely spaced data points are needed in the frequency plots. In some cases, manually shaking the item being tested can provide good estimates of frequency and damping at low excitation levels.

• It is recommended that when equipment frequencies approach the lower cutoff frequency used to modify shake-table inputs to meet excitation limits of the shake table, the response spectrum should be checked to assure that the intent of the IEEE 693 is being met. A Sine-beat(s) can be added to the signal so testing can be performed to satisfy the target response spectrum.

• It is recommended that data analysis methods at low frequencies have sufficient spectral resolution to characterize the results that they are to portray and allow accurate frequency and damping estimates.

Estimating Support Structure Anchor Loads

IEEE 693 recommends that load bolts be used to measure anchor loads during shake-table tests. On columnar support structures two strain gages near the base of the support column have also been used. It is important that they be installed so that they accurately characterize the loads. The gages should be installed above stiffening gussets often used at the base of columns.

Evaluation of Shake-Table Excitations

Test facility use time histories that meet the requirements of IEEE 693, that is, the input to the shake table has a response spectrum that exceeds the RRS and the duration of strong shaking is at least 20 seconds. As noted above, test response spectrum can be significantly above the RRS at frequencies above 12 Hz. While this has resulted in over testing the equipment, it does not appear to have prevented equipment from being qualified. Most commercial testing laboratories derive time histories by starting with stationary, band-limited "white" noise, which is then filtering and the amplitude reduced at the start and end to adjust the length of the record.

There has been a research effort to select a time history to be incorporated into IEEE 693. This effort has focused on starting with a real earthquake strong-motion recording from a near-field, large-magnitude earthquake. Two arguments have been put forward for taking this approach. First, records derived from random noise do not look like real earthquake records. Second, near-field records often contain a large velocity spike. A case can be made from observations of damage to buildings that this velocity pulse can be very damaging to low-frequency structures. While the latter may be true for buildings and for substation equipment, the limitations of shake-table velocity require that this velocity pulse be eliminated from the input before it is applied to the equipment being tested.

A detailed evaluation of time histories to be used for testing is beyond the scope of this paper. However, it should be noted that the evaluation of a record to be used in IEEE 693 has not taken into account the nonlinear character of the response of substation equipment when it is tested at the levels required for qualification. Since most substation equipment that is qualified by testing, such as instrumentation transformers, surge arresters and transformer bushings is characterized by a dominant single-mode response in any given direction, a linear response would only require that the excitation meet the required response spectra. There would be no need for a 20-second duration. There has been no suggestion that the duration requirement is to address low-cycle fatigue concerns.

While any record that has a response spectrum that conforms to the RRS will generate the same peak response in a linear system. The frequency content of test records derived from real earthquake record may change over the course of the record. Records derived from stationary noise will tend to have broader and consistent frequency content over the duration of the record. For any given test, it is not clear which would be more severe for a nonlinear system, but this is an effect that should be evaluated.

- It is recommended that any time history record that is incorporated into IEEE 693 be evaluated by comparing the nonlinear response of the equipment using the time history record and that obtained by a noise generated record. The compatibility of the record to commercial test facilities should also be evaluated.

Observations of Peak Response

IEEE 693's acceptance criterion for some equipment requires that the peak response not exceed certain specified values. For example, this is required for the stresses on porcelain and the displacement of composite members. This requires the vector value of the response variable to be measured. Calculating the square root of the sum of the squares (SRSS) of the peak response in each direction is often done. This can be quite conservative and the instantaneous peak value should be considered as an alternative. That is, the SRSS at each data sample should be calculated and the peak of this value over the duration of the test should be used. Using the peak response of one component of the response, such as a strain gage, would be inappropriate,

Serviceability and Seismic Qualification of Equipment

In the course of seismic qualification testing a question arose about the serviceability of equipment being degraded as a result of the seismic qualification, and by implication as a result of an earthquake. Such an example might be the detectable degradation of a seal, where structurally and electrically the tested unit could pass the seismic qualification. There are parts of IEEE 693 that imply that serviceability is an issue, although this is not explicitly stated in IEEE 693.

- It is recommended that IEEE 693 explicitly state that damage that compromises the long-term serviceability of the equipment would be grounds for not qualifying the equipment.

Initial Pull Test of Composite Members

IEEE 693 establishes a procedure for qualifying composites. Prior to shake-table tests a unit is subjected to a cantilever pull test to a load of 1/2 the Specified Mechanical Load (SML). The manufacturer of the composite element specifies the SML. It corresponds to a bending moment at the base of the member at which irreversible visible damage may be evident. From experience, some manufacturers indicate that the 1/2 SML is below the threshold of damage of the composite member.

The displacement at the top of the composite member is measured during the pull test at the 1/2 SML and this value establishes the acceptance criteria that can not be exceeded during shake table tests. In addition, the residual deflection, that is the deflection that remains after the 1/2 SML is removed, must be less that 5% of the 1/2 SML deflection. This section discusses several issues related to the pull test.

For some equipment, such as an instrumentation transformer, manufacturers install composite member on top of a box that contains system components. For the pull test, the box is anchored to a rigid base support. The possibility of additional flexibility associated with the anchorage, connection between the box and the rigid base, in the gasket between the box body and its lid, and deformation of the lid could affect the pull test results. Based on these parameters, there could be an increase in the measured deflection at the top of the composite member introduced by these possible flexibilities. These flexibilities would also contribute to added deflections during the shake table test, so that their effect would not significantly change the results of the test. However, the acceptable 1/2 SML deflection would be exaggerated and this would increase the allowable residual deflection.

- It is recommended that rotation at the base of composite members be measured during the initial pull test. The use of a simple laser pointer attached to the base of the composite members can be used to measure and correct for the rotations.

Use of the SML Deflection as an Acceptance Criterion

During shake-table tests to the High RRS, the deflection at the top of the composite member relative to its base must be less than the measured 1/2 SML deflection. During a shake-table test the relative deflection is typically measured in orthogonal planes that pass through the longitudinal axis of the composite member. The instantaneous peak displacement, that is the peak over the duration of the test of the square root of the sum of the squares (SRSS) of the two deflections, or the SRSS of the peak values in each direction can be used. The latter is more conservative.

There are two issues associated with this procedure, one adding conservatism and the other being unconservative. The deflection at the top of the composite member is a segregate for the moment at the base of the composite member. The deformation from the concentrated load at the top during the pull test is different from that caused by inertial loads during the shake-table test. When the deflection during the shake-table test equals that of the pull test, the moment during the shake-table test will exceed that observed in the pull test, so the criteria is unconservative. An evaluation of deformation shapes indicates that the size of the overload is relatively small.

The test that determines the SML requires that the load be applied for 1 minute. The failure mechanism of a composite is highly time-dependent, so that a load substantially below the SML will cause failure if applied for a long duration. During shake-table tests, the time that the moment is applied is related to the equipment's natural frequency, but even for very low frequency equipment, the duration that the large moments are applied is a small fraction of a second. In this respect the acceptance criteria will be overly conservative.

Evaluating the Acceptance Criterion

There are several potential issues associated with the measurement of the deflections during shake-table tests. One of two methods is typically used to measure the relative deflection in a composite member. One method is to use displacement-measuring devices, such as string-type potentiometric transducers, at the base and top of the composite member. Alternatively accelerometers could be used and their output integrated twice to obtain the deflections. In both cases, the relative deflections are obtained by subtracting the bottom deflection from that at the top. One issue that is common to both measuring methods is that any rotation at the base of the composite member, which is not measured by these methods, will contribute to a displacement at the top that will be inappropriately interpreted as a moment at the base. The use of accelerometers to get the displacements can be a issue, as any offset in the output voltage, when integrated twice will give large deflections, so that special signal conditioning will be needed to eliminate the erroneous large deflection. Also, on accelerometers there is an issue with the rotation of the axis of sensitivity. If the horizontal axis of sensitivity rotates slightly, say that it shifts so that it points down by one degree. The effect will reduce the amplitude of the horizontal component by the cosine, which will be small. But the accelerometer will now be sensitive to vertical input motions that will be interpreted as horizontal accelerations. A 1-degree rotation will reduce the horizontal signal by 0.02% but add 1.7 % of the vertical accelerations.

Finally, any difference in the phase shift in the signal conditioning will introduce an error that will be difficult to correct for. An alternative approach is discussed below.

Use of Strain Gages

As part of a research project, strain gages were installed on the lower flange of the composite member, as illustrated in Figure 4. These gages were calibrated during the pull test to relate moment at the base of the composite member to observed strain. The strain gage dynamic output during the shake-table test can be used as an alternative acceptance criterion to assure that the moment did not exceed that due to the 1/2 SML. An additional advantage for using the strain gage configuration, shown in Figure 4, is that strain measurements can be used to determine if any debonding within the flange has occurred during testing. The lower gage would indicate the extent of debonding, if it occurs. There can be several indicators of partial debonding.

- It is recommended that, in addition to measuring deflections during seismic qualification testing, strain gages be installed low on the flange. The use of strain gages on the flange should be evaluated as an alternative to dynamic deflection measurements as an acceptance criterion.

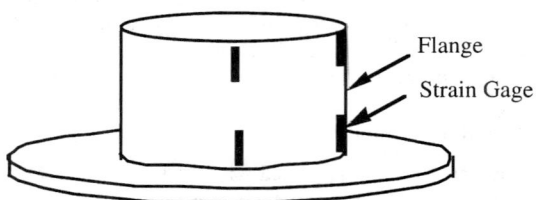

Figure 4 -Composite flange showing strain gages

- It is recommended that a pull test be performed after the shake-table tests as part of the acceptance criteria. The residual deflections should be determined with pull load lines removed or unloaded to remove any extraneous load or deflection.

Conclusions

A review of qualification tests and research studies conducted over several years has identified several potential difficulties in conducting qualification tests. These have been discussed and in some cases solutions or mitigation actions are suggested. In addition, recommendations are made to the IEEE 693 committee for their consideration.

References

IEEE Standard 693 (1997), Recommended Practice for Seismic Design of Substations, Institute of Electrical and Electronic Engineers, Inc.

Gilani A. S., Chavez J. W., Fenves G. L., and Whittaker A. S. (1998), Table 4-4, "Seismic Evaluation of 196 kV Porcelain Transformer Bushings," PEER 98/02.

Interaction between Electrical Substation Equipment Connected by Rigid Bus Slider

Junho Song[1], Armen Der Kiureghian[2], and Jerome L. Sackman[3]

Abstract

Electrical substation equipment items connected by a rigid bus may experience strong dynamic interaction under seismic excitations. In practice, a flexible connector is used to reduce the amplification of the equipment response due to the interaction. In order to investigate the effectiveness of one such connector, the bus slider, its hysteresis loop is first described as a bilinear model using differential equations. The model, fitted with hysteretic loops from quasi-static tests, provides good prediction of the dynamic responses of connected equipment items as compared to shake-table tests conducted by other investigators. Furthermore, the model is convenient for nonlinear random vibration analysis by use of the equivalent linearization method. This analysis method is used to compare the performance of the bus slider with that of other rigid bus connectors in terms of reducing the adverse interaction effect.

Introduction

Electrical substations are critical nodes within the power distribution system. In case of an earthquake, the functionality of an electrical substation depends on the functionality of its constituent equipment items, such as transformers, disconnect switches,

---

[1] Doctoral Student, Dept. of Civil and Environmental. Engrg., Univ. of California, Berkeley, CA94720; phone 510-642-0749; jhsong@ce.berkeley.edu
[2] Taisei Prof. of Engrg, Dept. of Civil and Environmental. Engrg., Univ. of California, Berkeley, CA94720; phone 510-642-2469; adk@ce.berkeley.edu
[3] Prof. Emeritus, Dept. of Civil and Environmental Engrg., Univ. of California, Berkeley, CA94720; phone 510-642-2121; sackman@ce.berkeley.edu

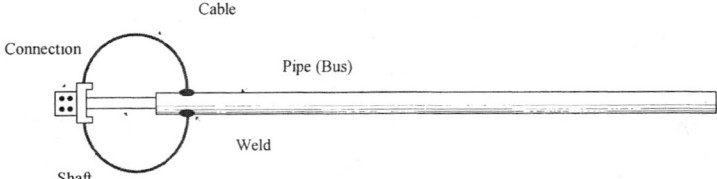

Figure 1. Rigid bus slider

circuit breakers and surge arresters. During the Loma Prieta (1989), Northridge (1995) and Kobe (1996) earthquakes, significant damages to electrical substation equipment were observed, causing long interruptions in the delivery of electrical power to affected areas. It is believed that an important factor contributing to this damage was dynamic interaction between connected equipment items with dissimilar characteristics. It is noted that, in practice, equipment items are usually qualified by shake-table testing in their stand-alone configuration. Unfortunately, such qualification does not account for the possibility of amplification in the equipment response due to interaction with other connected equipment.

To investigate the effect of interaction between the equipment items, we determine the ratio of the root-mean-square responses of each equipment item in its connected and stand-alone configurations. For the i-th equipment item, the response ratio is defined as $R_i = \text{rms}|u_i(t)|/\text{rms}|u_{io}(t)|$, $i = 1, 2$, where $\text{rms}|u_i(t)|$ is the root-mean-square displacement of the equipment in the connected system and $\text{rms}|u_{io}(t)|$ is the root-mean-square displacement of the equipment in its stand-alone (not connected) configuration. Clearly, a value greater than unity for the response ratio indicates amplification of the equipment response due to the interaction effect. Previous research with a linear connecting element (Der Kiureghian et al., 2001) has shown that the response ratio is usually greater than unity for the higher-frequency equipment item.

A rigid conductor bus typically consists of an aluminum pipe with rigid connection pads at both ends. To allow thermal expansion and reduce the interaction effect, a bus slider is inserted at one end of the rigid bus. Figure 1 illustrates a typical rigid bus with a bus slider. The bus slider shown (Type 221A, 30-4462 of Pacific Gas & Electric Company) is made of a 3.05m long SPS aluminum pipe having an outside diameter of 11.4 cm and a thickness of 1.2 cm. A shaft slides against the inside surface of the pipe producing a friction force. Two looped aluminum cables are welded to the pipe and to a terminal pad attached to the end of the shaft, providing elastic resisting forces.

In this paper, we investigate the effectiveness of the bus slider in terms of reducing the adverse interaction effect between two connected equipment items. First, the hysteresis loop of the bus slider is described as a bilinear model using nonlinear differential equations. The model is fitted with hysteretic loops from quasi-static tests conducted by Filiatrault *et al.* (1999). The fitted model is used to predict the dynamic responses of connected equipment items by nonlinear time-history analysis. The predicted responses are compared to those from shake-table tests by Filiatrault *et al.* (1999). In order to investigate the interaction effect for an ensemble of earthquake ground motions, we use the bilinear hysteresis model in conjunction with nonlinear random vibration analysis using the equivalent linearization method. The performance of the bus slider is compared to that of other rigid bus connectors commonly used in practice.

Analytical Model of Bus Slider

Filiatrault *et al.*, (1999) performed a quasi-static test to investigate the hysteretic behavior of the bus slider assembly in Figure 1. The specimen was subjected to cyclic displacements in the axial direction of the pipe within the range ± 8.89 cm. The dashed line in Figure 2 shows the hysteresis loop of the bus slider as obtained by Filiatrault *et al.* (1999). It is observed that the hysteresis loop has a nearly perfect bilinear shape. This is predictable since the slider bus can be considered as a Coulomb-

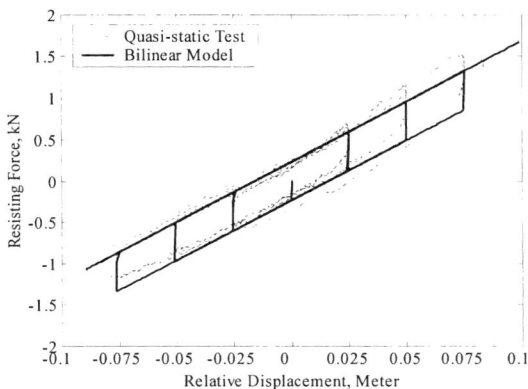

Figure 2. Hysteresis loops of bus slider

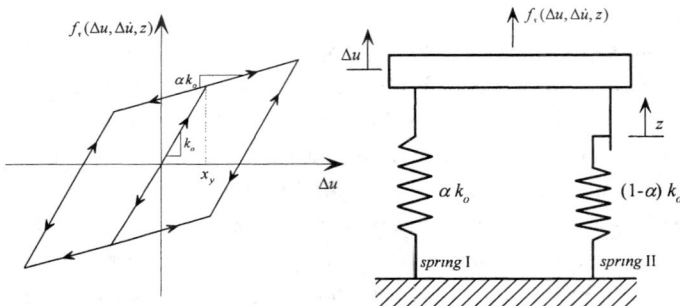

Figure 3. Bilinear hysteresis loop and its mechanical model

friction element coupled with elastic springs. The slight stiffening observed in tension is due to the geometric nonlinearity of the connecting cables.

Several analytical models exist for describing a bi-linear hysteresis behavior characterized by an initial stiffness k_o, yield displacement x_y, and post- to pre-yield stiffness ratio α, as shown in Figure 3. One possible method of analysis of such a system is to use nonlinear differential equations to describe the hysteresis loop. Consider the parallel assembly of a linear spring of stiffness αk_o (*spring* I) and a Coulomb friction slider in series with a second linear spring of stiffness $(1-\alpha)k_o$ (*spring* II), as shown in Figure 3. With Δu and z denoting the total displacement of the assembly and the relative displacement of *spring* II, respectively, the resisting force f_s is given by (Asano and Iwan, 1984)

$$f_s(\Delta u, \Delta \dot{u}, z) = \alpha k_o \Delta u + (1-\alpha)k_o g(\Delta \dot{u}, z) \tag{1a}$$

$$\begin{aligned} g(\Delta \dot{u}, z) = & z[1 - U(z - x_y)U(\Delta \dot{u}) - U(-z - x_y)U(-\Delta \dot{u})] \\ & + x_y[U(z - x_y)U(\Delta \dot{u}) - U(-z - x_y)U(-\Delta \dot{u})] \end{aligned} \tag{1b}$$

where $U(\cdot)$ is the unit step function. The relative displacement z should equal the total displacement x before sliding occurs, i.e., while $-x_y < \Delta u < x_y$, and it should equal x_y or $-x_y$ when sliding occurs in either direction. These conditions are satisfied by an auxiliary differential equation (Kaul and Penzien, 1974):

$$\dot{z} = \Delta \dot{u}[U(z + x_y) - U(z - x_y) + U(z - x_y)U(-\Delta \dot{u}) + U(-z - x_y)U(\Delta \dot{u})] \tag{2}$$

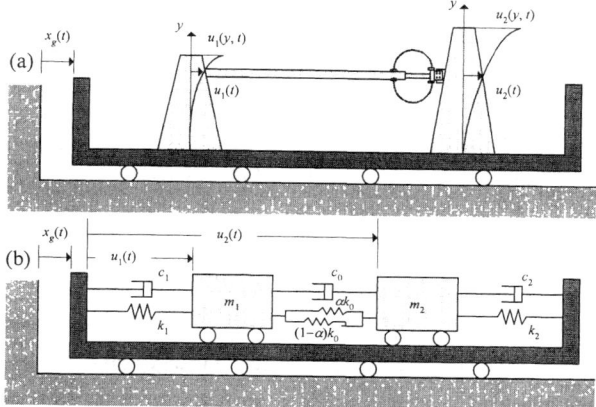

Figure 4. (a) Electrical equipment connected by rigid bus slider, (b) idealized system with SDOF equipment and bilinear connecting element

The above model is fitted to the parameter values measured by Filiatrault et al. (1999), i.e., $x_y = 0.2$ mm, $k_o = 1,160$ kN/m, and $\alpha = 0.0125$. Figure 2 compares the theoretical hysteresis loop obtained by this model (solid line) under the same quasi-static loading as the test. The numerical result was obtained by an adaptive Runge-Kutta-Fehlberg method (Fehlberg, 1969) with a relative tolerance of 10^{-6}. It is observed in Figure 2 that the theoretical model provides a reasonably accurate representation of the hysteresis behavior of the bus slider. This modeling approach allows us to avoid complicated and algorithmic models in static or time-history analysis of the connected equipment. Furthermore, the analytical model makes it possible to perform nonlinear random vibration analysis by use of the equivalent linearization method (ELM).

Dynamic Analysis of Equipment Items Connected by Rigid Bus Slider

Consider two equipment items connected by a rigid bus slider and subjected to the base acceleration $\ddot{x}_g(t)$, as shown in Figure 4a. In the figure, $u_i(y,t)$, $i=1,2$, denotes the displacement of the i-th equipment item at the coordinate y and time t, while $u_i(t)$ denotes its displacement at the attachment point. We idealize each equipment item as a single-degree-of-freedom oscillator by employing an appropriate displacement "shape" function (Der Kiureghian et al., 1999). Figure 4b shows the simplified analytical model in conjunction with the bilinear model of the bus slider introduced in the previous section. Let m_i, c_i, k_i and l_i respectively denote the effective mass, damping coefficient, stiffness and influence factor of the i-th equipment item. The equations of motion of the system are

$$\begin{bmatrix} m_1 & 0 \\ 0 & m_2 \end{bmatrix} \begin{Bmatrix} \ddot{u}_1 \\ \ddot{u}_2 \end{Bmatrix} + \begin{bmatrix} c_1 + c_o & -c_o \\ -c_o & c_2 + c_o \end{bmatrix} \begin{Bmatrix} \dot{u}_1 \\ \dot{u}_2 \end{Bmatrix} + \begin{Bmatrix} k_1 u_1 - f_s(\Delta u, \Delta \dot{u}, z) \\ k_2 u_2 + f_s(\Delta u, \Delta \dot{u}, z) \end{Bmatrix} = -\begin{Bmatrix} l_1 \\ l_2 \end{Bmatrix} \ddot{x}_g \quad (3)$$

where $\Delta u = u_2 - u_1$, c_o is the viscous damping coefficient of the bus slider, and f_s is the resisting force of the bus slider, as given by (1) and (2). For a numerical solution of the above equations, in this study we first reduce the second-order differential equations to first order and then use the Runge-Kutta algorithm.

The analytical bilinear model of the bus slider is examined by comparing predicted dynamic responses of the equipment items with measured responses in the shake-table tests by Filiatrault et al. (1999). Test RB-78 is selected for this purpose (see Filiatrault et al. 1999). The equipment specimens used by Filiatrault et al. (1999) were typical steel tubular columns with steel weights attached at their tops. The natural frequencies and viscous damping ratios of the individual equipment items were measured during the test. The effective mass, m_i, and influence factor, l_i, for each equipment item are computed employing the shape function $\psi(y) = 1 - \cos(\pi y / 2L)$ and the actual mass distribution of each specimen. In the test, the lower frequency

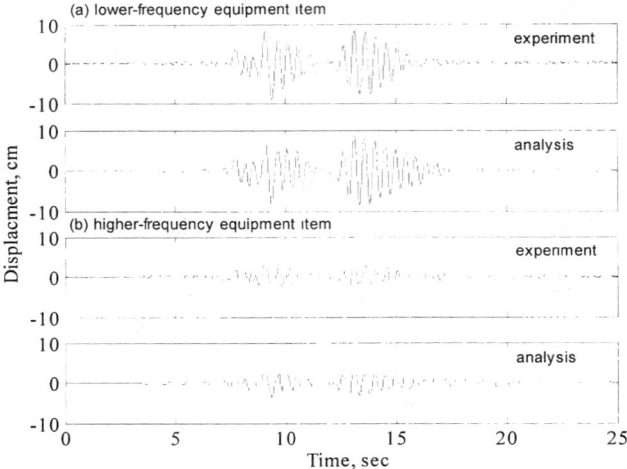

Figure 5. Displacement time histories for the 1994 Newhall-Northridge record

equipment item had measured frequency $\omega_1 = 12.5$ rad/sec and damping ratio $\xi_1 = c_1 / (2\omega_1 m_1) = 0.0042$. The effective mass and influence factor are computed as $m_1 = 360$ kg and $l_1 = 372$ kg. The corresponding values for the higher-frequency item are $\omega_2 = 25.8$ rad/sec, $\xi_2 = c_2 / (2\omega_2 m_2) = 0.0041$, $m_2 = 67.1$ kg and $l_2 = 82.1$ kg. The equipment items were connected by a rigid bus slider assembly, which was identical to the one used for the quasi-static test. Therefore, for the parameters of the bilinear model, i.e., k_o, x_y and α, the values measured in the quasi-static test are used in the dynamic analysis. No viscous damping is assumed for the bus slider, i.e., $c_o = 0$. This system was subjected to a modified version of the Newhall record of the Northridge (1994) earthquake ground acceleration. The dynamic analysis employed the input motion recorded on the shake table during the test. Figure 5 compares the measured and predicted displacement time histories of the two equipment items. The close agreement between the analytical and test results clearly indicates that the adopted bilinear model accurately characterizes the hysteresis behavior of the bus slider. Although not shown here, similar results were obtained for the tests RB-15, RB-18, RB-47, RB-49, RB-79 and RB-112, as reported in Filiatrault et al. (1999).

Nonlinear Random Vibration Analysis of Connected Equipment Items

The equivalent linearization method is employed to determine the response of the system defined by (1)-(3) to a zero mean, stationary Gaussian acceleration process. Using the method by Atalik and Utku (1976), the nonlinear differential equations (1b) and (2) are replaced by the linear equations

$$g(\Delta \dot{u}, z) = a_1 \Delta \dot{u} + a_2 z \tag{4a}$$
$$\dot{z} = b_1 \Delta \dot{u} + b_2 z \tag{4b}$$

where the coefficients a_1, a_2, and b_2 are determined by minimizing the mean-square errors between the original nonlinear equations (1b) and (2) and the linearized equations (4a) and (4b), subject to the response having the Gaussian distribution. The analytical expressions for the coefficients were derived by Asano and Iwan (1984). Since these coefficients are nonlinear functions of the statistical moments given by the random vibration analysis of the linearized equations, an iterative solution method is employed (Wen, 1980).

As an example, consider a bus-slider-connected equipment system having the parameters $m_1 / m_2 = 2.0$, $\omega_1 = \sqrt{k_1 / m_1} = 2\pi$ rad/sec, $\omega_2 = \sqrt{k_2 / m_2} = 10\pi$ rad/sec,

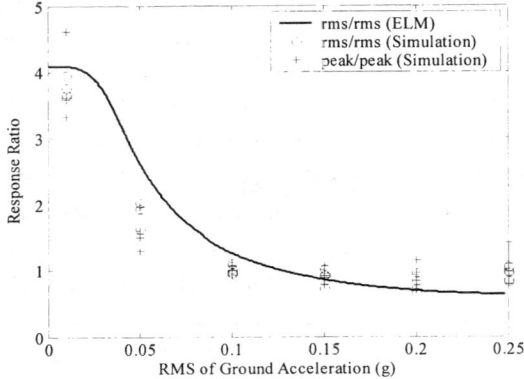

Figure 6. Response ratio of higher-frequency equipment by ELM and simulations

$\kappa = k_o / (k_1 + k_2) = 0.5$, $\xi_i = c_i / (2\omega_i m_i) = 0.02$ and $l_i / m_i = 1.0$ for $i = 1, 2$, and $c_o = 0$. The bus slider is as described above. For the ground acceleration, we consider stationary, filtered white-noise process defined by the well-known Kanami-Tajmi power spectral density (Clough and Penzien, 1993). The present analysis uses $\omega_g = 5\pi$ rad/s and $\xi_g = 0.6$ as the frequency and damping ratio of the filter. The amplitude of the process is varied to examine the variation in the nonlinearity of the system and the interaction effect with increasing intensity of the ground motion.

Figure 6 shows plots of the response ratio of the higher-frequency equipment item, R_2, versus the rms value of the ground acceleration. The response ratio is evaluated by two different approaches: 1) nonlinear random vibration analysis by use of the ELM, 2) nonlinear time-history analyses by use of five simulated ground motions based on the specified power spectral density. In the latter case, temporal averaging over a long response interval is used to estimate the rms values. The results based on the ELM clearly show how the softening and energy dissipation of the bus slider reduces the adverse interaction effect on the higher frequency equipment item and how this reduction is a function of the intensity of the ground motion. When the intensity is low, the response ratio is much greater than unity indicating strong amplification of the response due to the interaction. As the intensity grows and the shaft starts to slide, the interaction effect is quickly reduced. The ELM results show fairly good agreement with the time history simulations. This stochastic approach allows us to evaluate the effect of interaction for a class of ground motions, rather than arbitrarily selected time histories.

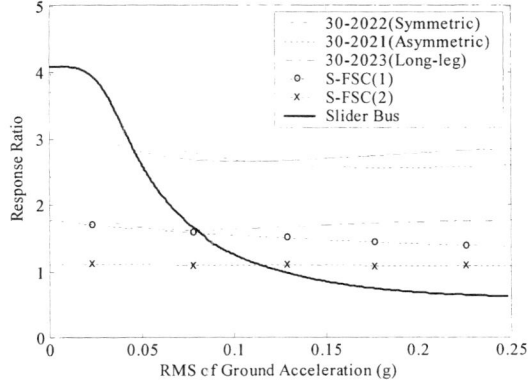

Figure 7. Response ratios by ELM for various flexible connectors

To evaluate the performance of the bus slider, we estimate the response ratios of the higher-frequency equipment item in the same configuration but connected by other types of rigid bus conductors. These include three existing PG&E flexible strap connectors (FSC), PG&E 30-2021, 2022 and 2023, and two new FSC designs, denoted S-FSC's (Song et al., 2002), which are all modeled using Bouc-Wen-type hysteresis models (Wen 1980). Figure 7 compares the response ratios estimated by ELM. It can be seen that, for high intensity ground motions, the bus slider provides an advantage in significantly reducing the adverse interaction effect due to its flexibility and energy dissipation.

Acknowledgment

This study was supported primarily by the Pacific Gas & Electric Company and the California Energy Commission through a grant to the Pacific Earthquake Engineering Research Center. Partial support was also provided by the Earthquake Engineering Research Center Program of the National Science Foundation under Award No. EEC-9701568.

References

Asano, K. and Iwan, W. D. (1984). An alternative approach to the random response of bilinear hysteretic systems, *Earthquake Engineering & Structural Dynamics*, 12, 229-

236.

Atalik, T. S. and Utku, S. (1976). Stochastic linearization of multi-degree-of-freedom non-linear systems, *Earthquake Engineering & Structural Dynamics*, 4, 411-420.

Clough, R. and Penzien, J. (1993). *Dynamics of Structures*. Prentice Hall, Englewood Cliffs, NJ.

Der Kiureghian, A., Sackman, J. L. and Hong, K.-J. (1999). Interaction in interconnected electrical substation equipment subjected to earthquake ground motions, *Report PEER 1999/01*, Pacific Earthquake Engineering Center, University of California, Berkeley, CA.

Der Kiureghian, A., Sackman, J. L. and Hong, K.-J. (2001). Seismic interaction in linearly connected electrical substation equipment, *Earthquake Engineering & Structural Dynamics*, 30(1), 327-347.

Fehlberg, E. (1969). Klassische Runge-Kutta formeln fünfter und siebenter ordnung mit schrittweitenkontrolle. *Computing*, 4, 93-106.

Filiatrault, A., Kremmidas, S., Elgamal, A. and Seible, F. (1999). Substation equipment interaction – rigid and flexible conductor studies, *Report No. SSRP-99/09*, Division of Structural Engineering, University of California, San Diego, CA.

Kaul, M. K. and Penzien, J. (1974). Stochastic analysis of yielding offshore towers, *Journal of Engineering Mechanics Division*, ASCE, 100, EM5, 1025-1038.

Song, J., Der Kiureghian, A. and Sackman, J. L. (2002), Reducing the Effect of Interaction between Electrical Substation Equipment Connected by Rigid Bus, *Proc. of 7NCEE*, July 21-25, 2002 7NCEE; 7th U.S. National Conference on Earthquake Engineering Boston, Massachusetts, USA.

Wen, Y. K. (1980). Equivalent linearization for hysteretic systems under random excitation, *Transaction of the ASME*, ASME, 47, 150-154.

Seismic Evaluation of Hybrid Bus Connections

L. Kempner Jr., PE[1], W. H. Mueller III, PE[2], N. J. Hutson, PMP[3]

Abstract
Shake table testing was performed at Portland State University (PSU), Seismic Testing and Applied Research (STAR) Laboratory, for the Bonneville Power Administration (BPA) to evaluate the seismic performance of 230 kV and 500 kV hybrid bus connections. The scope of the testing was to determine the significance of the interaction forces applied to the electrical equipment connected to the hybrid bus connections during a major earthquake. These tests provided the BPA with important data about existing equipment connections in the field and prototype connections that have yet to be installed. The hybrid bus connections provide electrical continuity and mechanical de-coupling between high-voltage equipment.

Introduction
High-voltage (115 kV to 500 kV and above) electrical equipment is connected to provide for the flow of energy in and out of the substation. The types of connections typically used within a high-voltage substation are rigid bus conductors, flexible conductors [IEEE 2002], and hybrid bus connections consisting of both rigid bus and flexible conductor components. The rigid bus and hybrid conductor connections are typically used to span large distances between high-voltage equipment. These distances range from 6.1 m to 8.2 m (20 – 27 ft.) between the equipment connection terminals. The large spans are necessary to accommodate the passing of maintenance and transportation vehicles with adequate safety clearances. Figure 1 shows a 500 kV hybrid bus installation.

Figure 1 – 500 kV Hybrid Bus Equipment Connection

---

[1] Structural Engineer, Bonneville Power Administration, P.O. Box 61409, Vancouver, WA 98662
[2] Professor of Civil and Environmental Engineering, Portland State University, P.O. Box 751, Portland, OR 97207
[3] Electrical Engineer, Bonneville Power Administration, P.O. Box 61409, Vancouver, WA 98662

In high seismic regions, rigid bus connections can make the equipment vulnerable to damage by large interaction forces transferred between the equipment through the rigid bus. The seismic interaction forces may be minimized with terminal expansion devices, such as sliders, installed along the axis of the rigid bus.

BPA has been developing hybrid bus connections to span long distances between equipment and to minimize the equipment interaction forces. Hybrid bus connections are fabricated using extruded aluminum alloy, 6061-T6, tubing with diameters of 9.2 cm and 15.3 cm (3 in. and 5 in.), and one or both end connections of multiple stranded conductors. Figure 2 shows one design for the end connection of multiple stranded conductors.

Figure 2 – Three Hood, Multiple Stranded Conductor End Connection

Shake Table Test Criteria

The required response spectrum used for all the shake table tests was based upon IEEE Std 693 [1997], High Level qualification at 2% damping. The test configuration was designed to simulate both flexible and rigid power circuit breaker bushings and disconnect switch end connection boundary conditions.

Boundary Conditions

The hybrid bus test specimen end boundary conditions attempted to simulate the fundamental frequency and displacement parameters for a power circuit breaker bushing (B) and disconnect switch (D). The end connection boundary conditions were designed using estimated displacements and fundamental frequencies for typical equipment response based on IEEE Std. 693, section 6.9.2. Table 1 shows the displacements and frequencies for two bushings (B1 and B2) and four disconnect switch models (D1, D2, D3, and D4) used during the shake table tests. The combination of B2 and D4 represent stiff end boundary conditions, whereas the combination of B1 and D1 represent the more flexible end boundary conditions. The boundary conditions were fabricated using steel rods and tubes with end attachment plates, Figure 3.

Table 1 – Boundary Conditions

Boundary Condition		Modeled Condition	Displacement Range (cm)	Frequency (Hertz)		Stiffness (N/cm)	
				OP	IP	OP	IP
Bushing:	B1	Flexible	+/- 20	5.5	5.5	167	167
	B2	Rigid	----------	58	58	2189	2189
Disconnect:	D1	Flexible	+/- 40	3.8	3.8	130	130
	D2	Semi-Flexible	+/- 20	5.3	8.3	247	735
	D3	Semi-Rigid	+/- 10	7.4	18.9	503	2010
	D4	Rigid	----------	84	84	8756	8756

Notes: (1 in. = 2.54 cm, 1 lbs. = 4.45 N), OP = Out-of-Plane, IP = In-Plane

(a) (b)
Figure 3 – Typical Configurations (a) for D2 and D3, (b) for B1, B2, D1, and D4

Flexible End Connector

BPA's high-voltage equipment connections are designed to provide flexible interconnections with the target displacements shown in Table 2. These displacement, (slack) ranges are based on industry recommendations and engineering judgment. Static load-displacement tests are performed on each jumper to confirm that the jumpers are within load and displacement limits. Load limits for the hybrid bus seismic connections are discussed in the results section of the paper.

Table 2 – Equipment Connection Target Displacement Ranges

Voltage	Displacement Range
115 kV	23 – 31 cm
230 kV	31 – 41 cm
500 kV	41 - 46 cm

Note: (1in. = 2.54 cm)

Hybrid bus connections for 230 kV, 2000 amps. (A), and 500 kV, 3000 and 4000 A, have been tested, [Albi 1999, Gardner 2002, Hamel 2002, Nuno 2000, Starkel 1998, Van Dyke 2003]. These hybrid bus connections typically consist of a single rigid bus component. A

500 kV, 4000 A, hybrid connection with twin rigid bus was also tested. Table 3 lists the general parameters of the different hybrid connections tested. Table 4 lists the parameters of the flexible end connectors and Figure 5 shows the four different connector configurations. The Hood conductor, AAC/TW, used for the end connectors is manufactured with trapezoidal wires (TW). Round-rod wires are used to manufacture the Arbutus conductor, AAC, Figure 6. The round-rod conductor is more flexible than the TW conductor.

Table 3 - Hybrid Bus Connection Parameters

Voltage	Amp.	Rigid Bus Parameters			Bus Orientation	Flexible Conductor
		Diameter	Schedule	Length		
230 kV	2000	8.9 cm	40	6.5 m	Inclined	2 Hood, ea. End
230 kV	2000	8.9 cm	80	6.5 m	Inclined [1]	2 Hood, ea. End
500 kV	3000	14.13 cm	40	7.6 m	Horizontal	3 Hood, DS End
500 kV	4000	14.13 cm	80	8.1 m	Horizontal	4 Hood, ea. End
500 kV	4000	2-8.9 cm	40	8.0 m	Horizontal	6 Arbutus ea. End

Notes: (1 in. = 2.54 cm, 1 ft. = 0.305 m), DS = Disconnect Switch
1. Tested hybrid bus for horizontal angle of 20°, 25° and 30°.

Table 4 - Flexible End Connector Parameters

Voltage	Amperage	Flexible Connector Parameters			Conductor Orientation	Total Weight
		Conductor	Diameter	Length[3]		
230 kV	2000	2 Hood	3.3 cm	74.2 cm	Inclined [1]	445 N
230 kV	2000	2 Hood	3.3 cm	66.0 cm	Inclined [2]	498 N
500 kV	3000	3 Hood	3.3 cm	91.4 cm	Vertical	1050 N
500 kV	4000	4 Hood	3.3 cm	51.0 cm	Vertical	1095 N
500 kV	4000	6 Arbutus	2.5 cm	42.0 cm	Vertical	1041 N

Note: (1 in. = 2.54 cm, 1 lbs. = 4.45 N)
1. Tested three different configurations of in-plane end boundary conditions.
2. Bushing flexible end connector inclined out-of-plane at 24 degrees from vertical.
3. Conductor free length

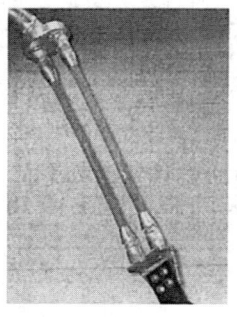  

(a)　　　　　　　　　　(b)　　　　　　　　　　(c)
Figure 5 – Connectors (a) 2 Hood, (b) 6 Arbutus, (c) 4 Hood and 3 Hood

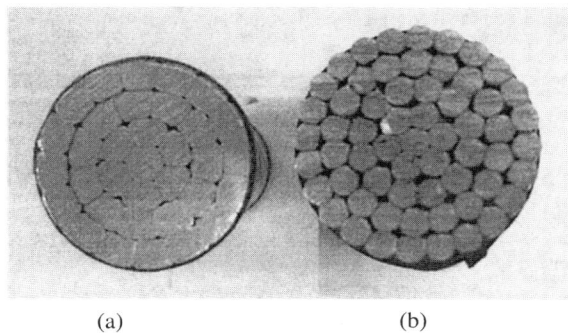

(a) (b)
Figure 6 – (a) Trapezoidal and (b) Round-Rod Wire Conductor, not actual size

Test Configurations
The Portland State University, Seismic Testing and Applied Research Laboratory, has a one-dimensional MTS system Control Corporation shake table. The table is driven by two 3.79 L/sec (60 gpm), Vickers hydraulic pumps, with an operating pressure of 20.7 MPa (3000 psi). The table weighs approximately 60 kN (13,500 lbs.) and has a specimen platform 3 m by 3 m (9.84 ft. x 9.84 ft.). The table hydraulic actuator provides up to 222 kN (50,000 lbs.) force. The hydraulic actuator has a static stroke of 349.2 mm (13.8 inches), and a dynamic stroke of 304.8 mm (12 inches). The servo-value operates at frequencies between 0.0 – 33 Hz. The shake table is mounted on four hydraulic bearings. The basic limitations of the shake table are ± 15.24 cm (± 6 inches) stroke, 0.9 m/s (2.95 ft./sec.) velocity, 3g's bare table acceleration, and 1g acceleration with an 89 kN (20,000 lbs.) specimen and frequencies up to 33 Hz.

The data acquisition system consists of two Pentium processor computers with National Instruments Corporation AT-MIO 16x multi-function I/O boards. LabVIEW® for Windows software is used for data transmission and acquisition.

The hybrid bus is mounted to the table as shown in Figure 7. Wide flange beam extensions were necessary to accommodate the required length of the test specimen. The test specimen was mounted diagonally, at 45 degrees, to the shake table drive direction. This arrangement was used to accommodate the length of the rigid bus and to allow testing with 0.5g simultaneously in two orthogonal directions, both Out-of-Plane (OP) and In-Plane (IP). The arrangement shown in Figure 7 is the 500 kV, 3000 A, hybrid bus connection. The other tested hybrid bus connections were obtained by rising the disconnect switch and power circuit breaker bases by installing rigid supports to obtain the heights required.

Test instrumentation consisted of load cells, strain gauges, and accelerometers. Vertical forces at the interface between the hybrid bus connector and the disconnect switch boundary condition and bushing boundary conditions, were measured utilizing a load cell assembly. Three load cells were placed between plates mounted at the top surface of the

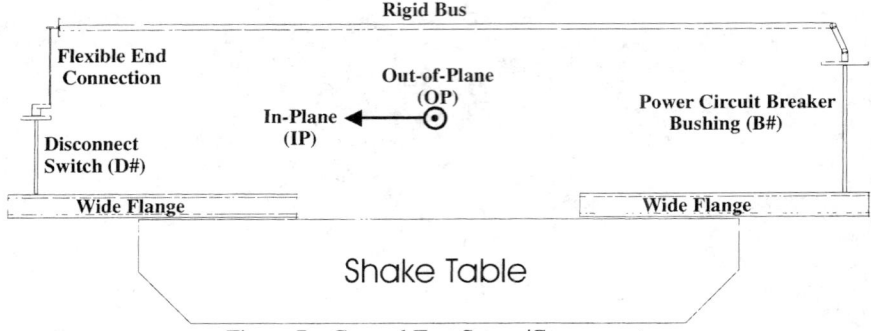

Figure 7 – General Test Set-up/Components

disconnect switch and power circuit breaker bushing mounting plates. The three cells were located 12.7 cm (5 in.) from the center of the circular plates set apart at 120° angles. The load cells were used to measure the vertical forces and bending moments. The power circuit breaker support tube was instrumented with strain gages installed 10.2 cm (4 in.) above the bottom of the bushing boundary pedestal base plate to measure the base moment, which is used to determine the shear forces, cantilever load, at the top surface of the power circuit breaker bushing mounting plate. Accelerometers were mounted on the rigid bus ends and at the center of the rigid bus to measure the out-of-plane and in-plane accelerations. Using the instrumentation and the concept of dynamic equilibrium the simulated disconnect switch and power circuit breaker bushing terminal loads were determined.

Time History Testing
The required response spectrum (RRS) used for these tests is based upon IEEE's Std 693 High Qualification Level for 2% damping. A time history was generated that provided a test response spectrum (TRS) that enveloped the RRS. Due to the size of the hybrid bus seismic connection and the testing lab, it was desirable to test in both the longitudinal and transverse directions simultaneously. This was accomplished by orienting the jumper at a 45° angle to the drive direction of the shake table. In order to achieve the full 100% IEEE signal in both orthogonal directions, IP and OP, of the rigid bus, it was necessary to overdrive the shake table signal to a level of 141%. The ZPA for the resulting shake table drive direction for all tests ranged between 0.7 – 0.9g. At a 45° angle to the table, the hybrid bus connection would feel approximately 71% of the table motion, 0.5g, in both IP and OP directions.

Test Results
The equipment connection interaction forces were evaluated at the connecting interface between the hybrid bus and the disconnect switch, and bushing. The maximum test specimen forces were obtained using dynamic equilibrium. The maximum allowable

horizontal shear forces for the disconnect switch and power circuit breaker bushing terminal loads were evaluated. Table 5 shows the recommended equipment terminal maximum operational load limits. These loads are specified in ANSI C37.04-1999 and ANSI C37.32-1996. Table 6 shows the equipment connection damage state force limits determined for BPA's applications. These values were established after discussions with BPA's manufacturers for power circuit breakers and disconnect switches.

Table 5 –Equipment Connection Maximum Operational Force Limits

Voltage	Disconnect (kN)			Bushing		
	IP	OP	Vertical	IP	OP	Vertical
230 kV	1.02	0.33	3.04	1.25	1.00	1.25
500 kV	2.00	0.67	3.33	1.75	1.25	1.25

Notes: (1 lbs. = 4.45 N), OP = Out-of-Plane, IP = In-Plane

The bushing load limits are based on tests conducted by the manufacturer of BPA's power circuit breakers. It was determined that for a six hole terminal connection pad, the load limit in the strong axis is 5.79 kN (1300 lbs) applied 51 cm (20 inches) between the rigid bus center-line and the base of the bushing terminal. For the same moment arm of 51 cm (20 inches) distance in the weak axis the load limit is 2.00 kN (450 lbs.). These load limits have to be adjusted for different moment arm distances. The force limits for 500 kV shown in Table 6 were corrected for the flexible connector length. The load limits on a four hole terminal pad are different, lower levels, than that for a six-hole pad.

The force limits shown in Table 6 for the disconnect switch are based on the cantilever strength of the disconnect switch insulators. BPA uses 1550 BIL insulators with minimum cantilever strength of 13.80 kN (3100 lbs.) for 500 kV disconnect switches, and 900 BIL insulators with cantilever strength of 6.45 kN (1450 lbs.) for 230 kV disconnect switches. The in-plane load limits are based on 40% of the rated cantilever strength. The out-of-plane load limits are based on 27% of the rated cantilever strength. This percentage was based on the 40% manufacturer strength limit and the terminal load distribution to the disconnect switch insulators. Based on discussions with a disconnect switch manufacturer, the cantilever strength was used for determining the load limits. The insulators are considered to be the weakest disconnect switch component.

Table 6 – Equipment Terminal Connection Maximum Damage State Force Limits

Voltage (kV)/ Amperage	Disconnect (kN)		Bushing (kN)		
	IP	OP	Flex Connector-Length	IP	OP
230/2000	2.58	1.74	91 – 100 cm	TBD	TBD
500/3000	5.52	3.72	50 cm Terminal Extension	5.79	2.00
500/4000 [1]	5.52	3.72	81.3 cm	3.62	1.25
500/4000 [2]	5.52	3.72	66.7 cm	4.41	1.53

Notes: (1 lbs. = 4.45 N, 1 in. 2.54 cm), OP = Out-of-Plane, IP = In-Plane,
TBD =To Be Determined
1. Four Hood Flexible Connectors
2. Six Arbutus Flexible Connectors

The hybrid bus connections tested had fundamental frequencies in the range of 1 – 3 Hz in the out-of-plane direction and 2 – 4 Hz in the in-plane direction. The different boundary conditions, D4B2 (Rigid – Rigid), D3B1 (Flex – Flex), D3B2 (Flex – Rigid), and D4B1 (Rigid – Flex) produced different equipment terminal loads, both magnitude and distribution between the power circuit breaker bushing and disconnect switch. In general, the Flex – Flex boundary condition produced the largest terminal loads, 25 –50 % for the bushing and 200 – 300 % for the disconnect switch. The flexible disconnect switch boundary condition allows the hybrid bus connector to whip out-of-plane, which resulted in minimum relative action of the hybrid conductors, thereby providing minimum damping action to reduce the terminal loads, Table 8.

A study of Table 7 leads to several conclusions. The 500 kV, 4000 A, hybrid bus connection has flexible connectors, four Hood conductors positioned in a square configuration, at both ends. The 500 kV, 3000 A, hybrid bus connection has a flexible connector only at the disconnect switch end. The flexible connector consisted of three Hood conductors positioned in a line normal to the rigid bus. Thus the 3000 A hybrid connection is stiff when loaded normal to the axis of the rigid bus. Comparing the loads for boundary condition D4B2 shows that the loads tend to be lower at the equipment that is connected with the flexible connector when compared to an end without a flexible connector. The loads also tend to increase when the stiffness of the end of the hybrid bus connection increases. This increase in stiffness may be due to the configuration of the cables in the flexible connector or the fact that there is no flexible connector used. This follows the logic that says that loads flows to stiffness.

Table 7 – 500 kV Hybrid Bus Shear Forces, D4B2

Voltage (kV) /Amperage	Connector	Disconnect Switch				Bushing			
		IP (kN)	SF	OP (kN)	SF	IP (kN)	SF	OP (kN)	SF
500/3000	3 Hood	0.11	50.0	0.61	6.1	2.91	2.0	1.76	1.1
500/4000	4 Hood	0.20	27.6	0.61	6.1	1.37	2.6	2.81	0.5
500/4000	6 Arbutus	0.47	11.7	1.31	2.8	1.23	3.6	1.46	1.1

Notes: (1 lbs. = 4.45 N), OP = Out-of-Plane, IP = In-Plane, SF = Safety Factor

The 500 kV, 4000 A, four Hood conductor connector configuration exceeded the allowable load limit of 1.25 kN (281 lbs.) for the bushing out-of-plane load of 2.81 kN (632 lbs.). An alternate design using six Arbutus conductors was tested and acceptable load limits were obtained. The tested out-of-plane load was 1.46 kN (328 lbs.) versus the 1.53 kN (344 lbs.) allowable damage state load. The load reduction between these two hybrid bus connections is contributed to shortening the length of the flexible conductor connector, reducing the weight of the hybrid bus connection, and slightly higher damping with the round-rod wire. The damping and fundamental frequencies for the 500 kV, 4000 A, hybrid bus connections are shown in Table 8.

The 230 kV, 2000 A, hybrid bus connection tests compared the affect of different two Hood conductor flexible connector configurations. One of the boundary conditions used the straight conductor connector configuration, Figure 5, discussed in previous sections of this paper. The second flexible connector consisted of flexible conductor extending along the rigid bus axis, which formed a curved conductor connector. The test configuration consisted of the disconnect switch connection tab being 3.5 m (11.4 ft) above the bushing terminal connection tab. The three flexible connector boundary configurations tested were bushing straight – disconnect switch curved conductor boundary conditions (Jumper 1), bushing curved – disconnect switch straight conductor boundary conditions (Jumper 2), and bushing straight – disconnect switch straight conductor boundary conditions (Jumper 3). Table 9 shows that the curved conductor boundary condition did affect the loads seen by the equipment.

Table 8 – Damping and Frequencies for the 500 kV, 4000 A, Hybrid Connection

Boundary Condition		Four Hood Conductors		Six Arbutus Conductors	
Out-of-Plane		Frequency	Damping	Frequency	Damping
D4B2	Rigid -Rigid	1.43 Hz	2.3 %	2.28 Hz	2.6 %
D3B1	Flex-Flex	0.71 Hz	1.1 %	0.98 Hz	0.5 %
D4B1	Rigid - Flex	1.08 Hz	1.8 %	1.76 Hz	1.3 %
D3B2	Flex - Rigid	1.11 Hz	3.3 %	1.83 Hz	2.1 %
In-Plane					
D4B2	Rigid -Rigid	1.85 Hz	3.0 %	3.21 Hz	4.7 %
D3B1	Flex-Flex	1.65 Hz	3.1 %	2.15 Hz	1.5 %
D4B1	Rigid - Flex	1.69 Hz	3.3 %	2.22 Hz	3.0 %
D3B2	Flex - Rigid	1.83 Hz	1.7 %	3.18 Hz	3.9 %

It was recognized that for 230 kV three phase power circuit breakers the hybrid bus connector for the outer phases are sometimes positioned out-of-plane, with reference to the vertical axis, due to the slope of the bushing. This situation raised the question as to the influence of vertical seismic motion. With the limitation of the one-dimensional shake table, a test series was performed in an attempted to include vertical motion by positioning the test specimen, rigid bus, on a slope with reference to the motion of the shake table. The rigid bus horizontal angles tested were 20, 25 and 30 degrees. The results of these tests suggest that vertical motion does not produce significant changes in the forces seen by the equipment connection tabs..

Table 9 – 230 kV Hybrid Bus Shear Forces, D4B2

Voltage (kV)/	Test Level	DC (kN)		Bushing (kN)	
Amperage	Configuration	IP	OP	IP	OP
230/2000 Jumper No. 3	20 Degrees	0.37	0.70	0.11	0.29
	25 Degrees	0.48	1.01	0.10	0.25
	30 Degrees	0.48	0.96	0.12	0.20
230/2000	Jumper No. 1	1.00	1.11	0.05	0.08
	Jumper No. 2	0.84	0.92	0.22	0.21
	Jumper No. 3	0.85	1.42	0.17	0.16

Notes: (1 lbs. = 4.45 N), OP = Out-of-Plane, IP = In-Plane

Conclusion

Shake table tests of hybrid bus connections were tested and shown to provide adequate seismic slack with acceptable equipment connection tab loads. The most difficult task in the development of high-voltage equipment connections is the determination of the allowable equipment connection tab loads. Equipment connection operational tab loads are very restrictive as the load levels have significantly low magnitudes. An effort to evaluate the equipment connection interaction forces based on the damage states caused by equipment connection tab loads was attempted. Based on damage state loads, the BPA hybrid bus connections will provide electrical continuity and mechanical de-coupling during a major earthquake event. Additional effort will be devoted to defining acceptable equipment connection tab loads. These load magnitudes are important for evaluating all potential high-voltage substation equipment connection seismic designs.

References

Albi F.M., Mueller W.H, and Kempner L. Jr., 1999, "Seismic Evaluation: 500 kV, 3000 Amperage, Seismic Connection with Rigid Bus Conductor," Research Report, Portland State University, OR.

Gardner I., Mueller W.H, and Kempner L. Jr., 2002, "Seismic Performance Evaluation of 500kV, 4000 Amperage, Rigid Bus Conductor," Draft Research Report, Portland State University, OR.

Hamel C., Mueller W.H, and Kempner L. Jr., 2002, "Seismic Performance Evaluation of 230kV, 2000 Amperage, Rigid Bus Conductor," Research Report, Portland State University, OR.

IEEE 693, 1997, Recommended Practice for Seismic design of Substations, Institute of Electrical and Electronic Engineers, Inc., 445 Hoes Lane, P.O. Box 1331, Piscataway, NJ.

IEEE C37.04, 1999, Standard Rating Structure for AC High-Voltage Circuit Breakers, Institute of Electrical and Electronic Engineers, Inc., 445 Hoes Lane, P.O. Box 1331, Piscataway, NJ.

IEEE C37.32, 1996, High-Voltage Air Switches, Bus Supports, and Switch Accessories – Schedules of Preferred Ratings, Manufacturing Specifications, and Application Guide, Institute of Electrical and Electronic Engineers, Inc., 445 Hoes Lane, P.O. Box 1331, Piscataway, NJ.

IEEE P1527, 2002, Draft 10, "Recommended Practice for the Design of Flexible Buswork Located in Seismically Active Areas, Institute of Electrical and Electronic Engineers, Inc., 445 Hoes Lane, P.O. Box 1331, Piscataway, NJ.

Nuno J.C., Mueller W.H, and Kempner L. Jr., 2000, "Seismic Evaluation: 230 kV, 2000 Amperage, Seismic Connection with Rigid Bus Conductor," Research Report, Portland State University, OR.

Starkel D.L., Mueller W.H, and Kempner L. Jr., 1998, "Seismic Evaluation: 500 kV, 3000 Amperage, Seismic Connection with Rigid Bus Conductor," Research Report, Portland State University, OR.

Van Dyke, R.W., Mueller W.H, and Kempner L. Jr., 2003, "Seismic Evaluation: 500 kV, 4000 Amperage, Six Arbutus Connector, Seismic Connection with Rigid Bus Conductor," Draft Research Report, Portland State University, OR.

Seismic Design of Secondary Systems

T.S. Aziz[1], D. Sc., P. Eng., M. ASCE

Abstract

Historically, building codes evolved to deal with the seismic design requirements of building structures with very little attention given to seismic design of equipment, parts and portions housed inside. While ductility is routinely used for building structures, it is rarely considered for equipment housed in lifeline facilities.

Studies dealing with the dynamic interaction between inelastic primary and secondary systems are somewhat limited.

The objective of the current investigation is to study the seismic response of Primary-Secondary systems in critical facilities when the frequencies of free vibration of the elastic uncoupled systems coincide.

Presented in this paper is a study of the inelastic seismic behavior of tuned Primary-Secondary (P-S) systems. The Primary-Secondary system is represented by a coupled non-linear (elastic-plastic or bilinear) two degree of freedom non-linear system. The response of both the coupled and the uncoupled systems are obtained by solving the governing differential equations numerically. The important parameters considered are the tuned frequency of the system, the mass ratio, damping and the yield levels for the structure and the equipment that reflects ductility incorporated in the design.

Based on the numerical results obtained, improvements for code design provisions of equipment seismic design for lifeline facilities are proposed.

Introduction

Natural hazards such as earthquakes can cause considerable loss of life and property damage. Improved seismic design practices leading to structures and equipment better able to resist seismic loads are effective means of reducing these losses. Civil structures typically contain a large number of equipment and components in the form of architectural, mechanical, and electrical systems among others. These systems can be treated for seismic analysis purposes as a dynamic system attached to the main structural system and receiving its seismic input motion from it. Unfortunately, these equipment systems which may affect life safety and whose failure may be costly, have received very little attention in the seismic design process. The damage for the non-structural systems in recent earthquakes was in the billions of dollars and by far exceeded the damage to the buildings themselves.

Review of international codes indicates that the differences in equipment seismic design requirements between codes are significant.

---

[1] Professor (Adjunct), Civil Engineering Department, McMaster University, Hamilton, Ontario; Principal Civil Engineer, AECL, Mississauga, Ontario, CANADA

Detailed seismic design provisions coverage for equipment and non-structural components in buildings was first proposed in the 1978 Applied Technical Council (ATC-03) Report. These have been adopted with some minor changes in the 1991, 1994 and finally the 1997 NEHRP Recommended Provisions. These with minor revisions now form the basis of the ASCE 7-98 Provisions by the American Society of Civil Engineers. The current provisions in Canada are represented by the 1995 NBCC. Critical examination of these current provisions reveals the following:

1) The dynamic characteristics of the supporting structure as well as those of the non-structural component or equipment are not considered.
2) The effect of the inelastic behaviour and ductility of the supporting structure or the supported component, is not accounted for
3) The expected performance of the non-structural component or equipment; in particular in the inelastic range is not stipulated.
4) The current code provisions are largely empirical.
5) There is no predictive model upon which the current code provisions are based. This shortcoming limits the future evolution of these code provisions.

Objectives

The primary objective of the current research program is to develop an improved code type approach for the seismic design of equipment and components of buildings and important civil structures. While a code type approach exists for buildings (e.g. NBCC 1995, UBC, New Zealand Building Code and the Japanese Building Codes), it is believed that the current approach may impose a large design penalty as a result of the inherent simplification process. There is also a possibility that the current approach may lead to unsafe designs.

The specific objective of the paper is to compare the various recent codes concerning seismic design of non-structural components with analytical calculations obtained from the non-linear coupled time history analysis that accounts for interaction effects.

Previous Work

Traditionally, architectural, mechanical and electrical systems and components of buildings have been designed with little, if any, regard to seismic forces. Equipment supports have been generally designed for gravity loads only, and attachments to the structure itself were often deliberately designed to be flexible to allow for vibration isolation or thermal expansion. The majority of the work on seismic design of equipment has been done in the context of seismic qualification of nuclear plants.

For dynamic analysis purposes, a component can be modeled as a dynamic system in isolation of the supporting building structure and therefore receiving its motion from it. Alternatively, the dynamic model of the equipment or component can be included with the building model and the combined Primary-Secondary (P-S) system model solved in either the time-domain or the frequency-domain. Due to many practical difficulties in carrying out a coupled dynamic analysis, an uncoupled approach is customarily adopted. Equipment and a structure can be treated as two separate uncoupled systems under the conditions developed by Aziz and Duff (1978). These conditions, which are known as decoupling criteria, have been implemented already in the Canadian CSA N289.3 practice for Nuclear Power Plants.

Observed Non-structural Damage

Numerous instances of damage to non-structural components during past earthquakes have been reported. The most widely reported type of non-structural damage in earthquakes is the failure of suspended ceiling systems even under moderate shaking. A more serious type of damage is the collapse of light fixtures and air diffusers incorporated into the ceiling system. This can pose a significant safety hazard because of falling debris. It can hinder evacuation and rescue efforts, and can render a building unusable following the earthquake. A classic example of a falling plaster ceiling is the collapse of the ceiling in the Geary Theatre during the 1989 Loma Preita Earthquake.

Another example is the performance of the Olive View Hospital during the 1994 earthquake. The building structure performed extremely well from the structural stand point of view, even though the ground accelerations recorded near and on the structure were unusually high. The hospital had to be abandoned, however, following the earthquake, in part because the fire sprinkler system and chilled water lines had been damaged, resulting in water leaks.

Examples of damage due to inadequacy of non-structural components to sustain seismic lateral force include failure of anchors to hold equipment in place, such as water tanks and boilers. This results in equipment sliding off supports, spilling of contents, and disruption of service. Other non-structural damage of this type includes failure of ceilings at points of connection, and cracks in partition walls.

National Building Code Of Canada (NBCC) – 1995

According to the Canadian NBCC 1995, parts of building as described in Tables 4.1.9.1D and 4.1.9.1E of the Code and their anchorage shall be designed to accommodate the code stipulated deflections for the building. The building component is to be designed for a lateral force V_p as given by the following equation:

$$V_p = v \cdot I \cdot S_p \cdot W_p \tag{1}$$

v = zonal velocity ratio

I = seismic importance factor of the structure.

S_p = horizontal force factor for part or portion of a building and its anchorage.

W_p = the weight of a part or portion of a structure.

The values of S_p for architectural components shall confirm to Table 4.1.9.1D. This gives a wide range for S_p values (ranging from 0.7 to 15).

The values of S_p for mechanical/electrical components shall be equal to

$$S_p = C_p \cdot A_r \cdot A_x \qquad (2)$$

Where:

C_p = seismic coefficient for components for mechanical/electrical components as given in Table 4.1.9.1E of the code (range 0.7 to 1.5).

A_r = 1.0 for components that are both rigid and rigidly connected and for non-brittle pipes and ducts
= 1.5 for components located on the ground that are flexible or flexibly connected except for non-brittle pipes and ducts
= 3.0 for all other cases

A_x = $1.0 + (h_x / h_n)$.

This gives a coefficient ranging between 1 and 2; depending on equipment height.

h_x = the height above the base to level x

h_n = the nth uppermost level of the building

American Society of Civil Engineers ASCE 7-98 Practice

The seismic force (F_p) on a part shall be determined in accordance with the equation:

$$F_p = 0.4 \, a_p \, S_{DS} \, W_p \, [\, 1 + 2 z/h \,] \, I_p / R_p \quad \text{(see Note 1)} \qquad (3)$$

Note 1: F_p should not be less than $0.3 \, S_{DS} \, W_p \, I_p$ or greater than $1.6 \, S_{DS} \, W_p \, I_p$

Where:

F_p = seismic design force centred at the component's centre of gravity and distributed relative to component's mass distribution

S_{DS} = design, 5 % damped, spectral acceleration at short periods (.2 s)

a_p = component amplification factor. (range from 1.00 to 2.50)

I_p = component importance factor. (range from 1.00 to 1.50)

W_p = component operating weight

R_p = component response amplification factor. (range from 1.50 to 5.00)

z = height of structure at point of attachment of component h   =   Average roof height of structure relative to the grade elevation

Table 1 compares the seismic coefficients of the American ASCE 7-98 and the 1995 Canadian NBCC provisions for selected building components.

Non-linear Time-history Analysis of Equipment-Structure Systems

The equipment-structure system is represented by a coupled elasto-plastic (or bilinear) two-degree-of-freedom spring mass model. If the system can be decoupled, the two-degree-of-freedom system can be divided into two uncoupled single degree of freedom systems. The response of both the coupled and the decoupled systems were obtained by solving the governing differential equations numerically using the Wilson-θ method. The important parameters studied are the tuned frequency of the system, the mass ratio, damping and yield levels for the structure and for the equipment. In the analysis, one or both of the system components may behave inelastic ally. The characteristics of the inelastic behaviour of the system will depend on the degree of ductility incorporated into the design.

The input ground motion used in the equipment-structure systems analysis is taken as actual strong motion earthquake records normalized to a spectral acceleration of 1.0g at the period of tuned equipment-structure system. Various strong motion records are used in order to account for the variations in the actual earthquake characteristics. The ground motions used in the current study are given in Table 2.

The response of the coupled equipment-structure system is obtained by numerical integration of the coupled equations of motion. The peak displacement and acceleration response data of the coupled and uncoupled systems when subjected to the three different normalized earthquakes are averaged for each set of parameters considered.

The damped, tuned two-degree-of-freedom system is defined by the period T, mass ratio 'μ' which is the ratio of the secondary system mass to the primary system mass, the damping ratio 'β', and the yield levels R_p and R_s for the primary and secondary systems respectively. For simplicity and to maintain accuracy, damping in the coupled system is taken as the average damping of the components of the system. The periods of the primary and secondary systems are set to be equal which represents the case of a tuned system. The yield level forces in the primary and secondary springs are determined by multiplying the maximum elastic spring force by the yield factors R_p and R_s for the primary and secondary systems respectively. The maximum elastic spring force is obtained from the uncoupled elastic analysis for each specific case. The range of values of the parameters under study are chosen to represent practical cases. The range of fundamental periods encountered in structural and equipment design is found to be from 0.1 to 10.0 seconds. Five representative mass ratios are selected for numerical calculations with the values 0.1, 1.0, 2.0, 5.0, 10 %. Damping ratios of 3 and 5 % are selected to correspond to steel and concrete structures respectively. The values of the yield level factors R_p and R_s representing the primary and secondary systems respectively are taken as 1.0, 0.75, 0.50 and 0.25.

Table 3 gives secondary system response accelerations for different parameters. The table can be translated directly to forces on equipment "V_p" by multiplying by the component weight "W_p". This can be compared to the building codes design levels.

As an illustration for the effect of the different parameters, the response of the secondary system is plotted in Figure 1 for the case of an inelastic secondary system on an elastic primary system for different secondary system yield levels (R_s values of 1.00, 0.75, 0.50 and 0.25). Response acceleration can range between approximately 1.3 to 8 g depending on the dynamic characteristics of the systems.

Figure 2 demonstrates the effect of the mass ratio for the elastic systems. Figure 3 demonstrates the effect of the mass ratio for the inelastic systems. Three tuned periods of 0.2, 1.0 and 10 seconds are displayed which covers a wide range of interest. It is clear that the mass ratio is a significant parameter; in particular for an elastic system.

Comparison Example

An example is chosen here to demonstrate the typical differences between the two building codes and the results of non-linear time-history coupled analysis. The Spectral accelerations for the time-history analysis are all normalized to 1.0g. This would imply an S_{DS} = 0.7g for 5 % damping as required by ASCE 7-98 . It also implies a v value per the NBCC 1995 of 1.0.

Seismic design forces were calculated for each code using an importance factor of unity and a height factor of unity. The ratio between the forces for the two codes varies approximately between 0.7 to 13. Considering the variation of the design force level with height being 1 to 2 in Canada and 1 to 3 in USA, another 50 % variation due to height effect between the two codes is expected. The force levels used in USA. can be very different by comparison to the force levels used in Canada. Both the two codes show large differences from the theoretical values calculated from non-linear dynamic analysis. Figure 4 compares the secondary system accelerations for the elastic systems. Figure 5 compares the secondary system accelerations for the inelastic systems with force yield level (ductility) of 0.5 for the primary and secondary systems. It can be observed from the Figures that the present codes fail to recognize the effect of the key parameters (e.g. the mass ratio, damping, period and ductility) due largely to the empirical nature of these codes.

Conclusions

The most important parameters that have a significant effect on equipment response are the mass ratio, yield levels of the components, damping and the tuned frequency.

Numerical design charts such as those given in Table 3 are essential as a first step for codifying equipment seismic design. These charts are most useful to the designer to assess the effect of the different parameters as well as the merits of undertaking a time-history non-linear coupled seismic analysis.

Future building codes for equipment seismic design should account for the different physical parameters as those identified in this study. At present, lack of such quantification makes equipment seismic design largely empirical.

DISASTER RESPONSE FOR LIFELINE SYSTEMS 643

Table 1. Seismic Design Coefficients

Parts or Portions of Buildings	NBCC-1995	ASCE 7-98	
	S_p or C_p	a_p	R_p
Architectural Components			
Exterior Wall Panels	1.5	1.0	2.5
Suspended Ceilings	2.0	1.0	2.5
Masonry veneer connections	5.0	1.0	2.5
Mechanical Components			
HVAC Ductwork -Vibration Isolation	1.0	2.5	2.5
HVAC Ductwork -Non-vibration Isolation	1.0	1.0	2.5
Piping System - High Deformability	1.0	1.0	3.5
Piping System - Limited Deformability	1.0	1.0	2.5
Piping System - Low Deformability	1.0	1.0	1.25
Electrical Components			
Bus ducts, conduits, cable trays	1.0	2.5	5.0
Electrical Equipment	1.0	1.0	2.5
Lighting Fixtures	2.0	1.0	1.25

Table 2. Ground Motion Input Data

Earthquake Location and Component	Magnitude	Total Duration (s)	Maximum Displacement (cm)	Maximum Acceleration (cm/s^2)
El-Centro (1940) S90W	6.3	53.48	19.8	210.1
Parkfield (1966) Temblor, N65W	5.6	30.42	4.7	264.3
San Fernando (1971)- Pacomia Dam, S74W	6.6	41.74	10.8	1054.9

Table 3. Secondary System Acceleration (g)
(Average of three Normalized Earthquakes - Damping = 3 %)

Primary System Yield Level	Mass Ratio	Secondary System Yield Level										
		1			0.75			0.5			0.25		
		T = 0.2	T = 1	T = 10	T = 0.2	T = 1	T = 10	T = 0.2	T = 1	T = 10	T = 0.2	T = 1	T = 10
1	10	2.036	2.742	2.205	2.036	2.742	2.205	2.036	2.742	2.192	1.973	2.19	1.501
	5	2.668	3.322	3.145	2.668	3.322	3.145	2.668	3.322	2.876	2.1	2.317	1.611
	2	4.26	4.355	4.45	4.26	4.355	4.383	4.013	3.894	3.18	2.132	2.34	1.625
	1	5.468	4.963	5.19	5.472	4.963	4.684	4.187	4.287	3.191	2.144	2.347	1.629
	0.1	7.406	8.08	5.996	6.234	6.799	4.733	4.212	4.582	3.201	2.152	2.354	1.632
0.75	10	2.017	2.562	2.188	2.017	2.562	2.188	2.017	2.562	2.187	1.956	2.09	1.502
	5	2.513	3.047	2.925	2.513	3.047	2.925	2.513	3.047	2.824	2.091	2.295	1.605
	2	4.258	3.578	3.985	4.258	3.578	3.985	4.011	3.576	3.16	2.132	2.324	1.617
	1	5.265	4.283	4.479	5.265	4.283	4.479	4.182	4.134	3.171	2.14	2.332	1.618
	0.1	6.293	6.777	4.915	6.074	6.659	4.915	4.195	4.56	3.177	2.142	2.335	1.619
0.5	10	1.569	1.997	1.609	1.569	1.997	1.609	1.569	1.997	1.609	1.569	1.925	1.426
	5	2.25	2.302	2.24	2.25	2.302	2.24	2.25	2.302	2.24	2.044	2.08	1.59
	2	3.652	3.01	2.834	3.652	3.01	2.834	3.59	3.01	2.834	2.119	2.299	1.598
	1	4.072	3.562	3.098	4.072	3.562	3.098	3.934	3.562	3.04	2.123	2.307	1.618
	0.1	4.706	4.763	3.315	4.706	4.763	3.315	4.139	4.482	3.103	2.125	2.309	1.602
0.25	10	1.117	1.159	0.828	1.117	1.159	0.828	1.117	1.159	0.828	1.117	1.159	0.828
	5	1.725	1.374	1.107	1.725	1.374	1.107	1.725	1.374	1.107	1.725	1.374	1.107
	2	2.197	1.866	1.362	2.197	1.866	1.362	2.197	1.866	1.362	2.042	1.866	1.362
	1	2.309	2.136	1.472	2.309	2.136	1.472	2.309	2.136	1.472	2.076	2.112	1.458
	0.1	2.606	2.448	1.565	2.606	2.448	1.565	2.606	2.448	1.565	2.085	2.266	1.515

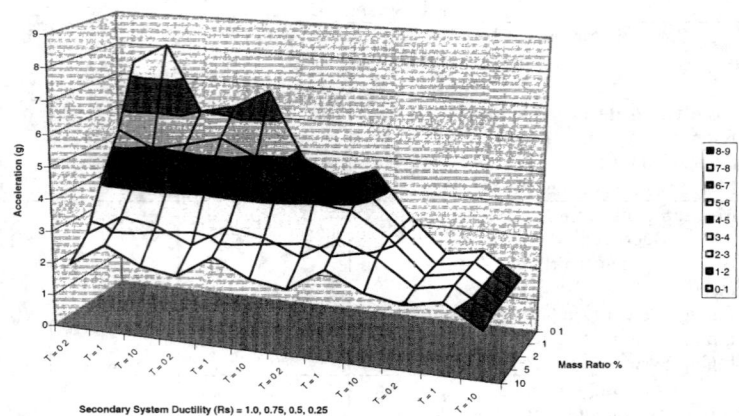

Figure 1. Acceleration of Secondary System (g) - Elastic Primary and Inelastic Secondary Systems

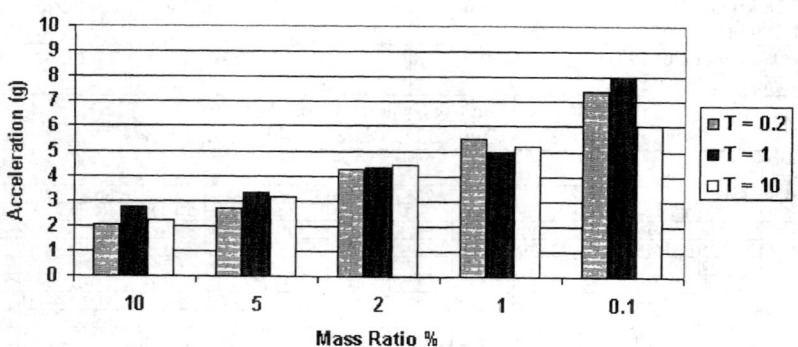

Figure 2. Acceleration of Secondary System (g) – Tuned Elastic Systems Period (T) of 0.2, 1.0, 10 Seconds

DISASTER RESPONSE FOR LIFELINE SYSTEMS 645

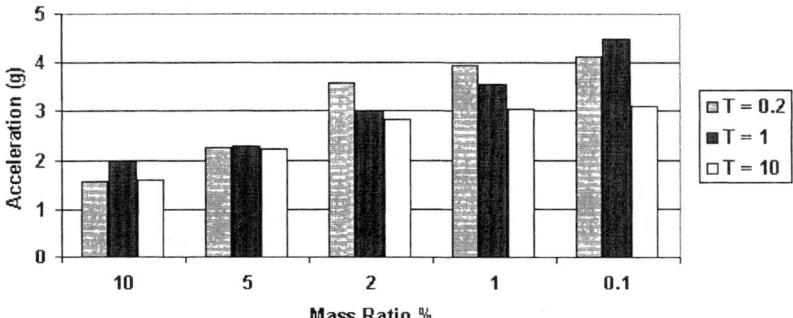

Figure 3. Acceleration of Secondary System (g) – Tuned Inelastic Systems (Rp=Rs=0.5) - Period (T) of 0.2, 1.0, 10 Seconds

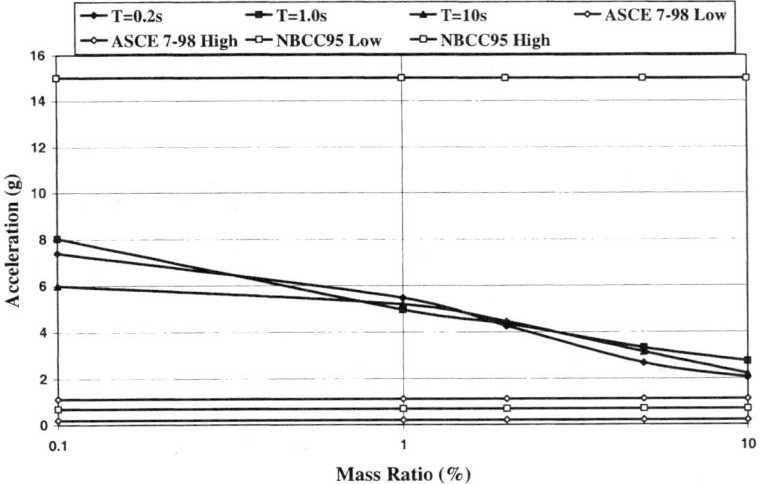

Figure 4. Elastic Primary and Secondary Systems

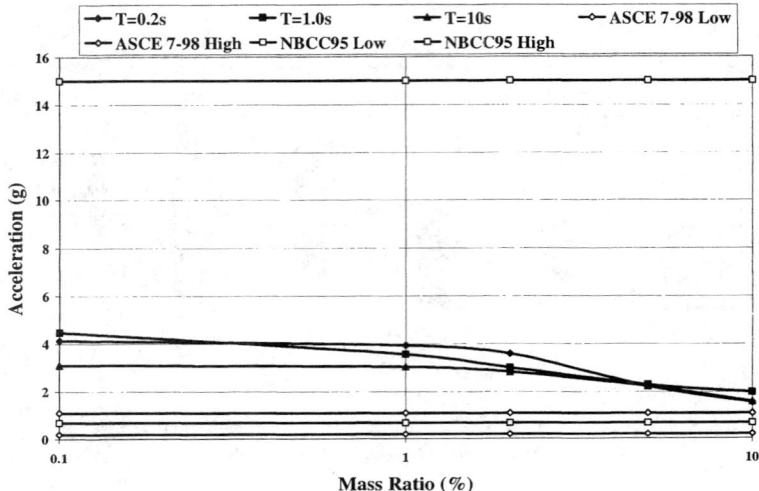

Figure 5. Inelastic Primary and Secondary Systems $R_p = R_s = 0.50$

References

American Society of Civil Engineers (ASCE), 1998, Minimum Design Loads for Buildings and Other Structures, (ASCE 7-98), New York, N.Y.

Aziz, T.S. and Duff, C.G.(1978) Decoupling criteria for seismic analysis of nuclear power plant systems, ASME paper 78-PVP-27, ASME/CSME Pressure Vessels and Piping Conference with Nuclear Materials Division, Montreal, Canada, June 25-29.

Building Seismic Safety Council (BSSC), 1998, 1997 Edition NEHRP Recommended Provisions for Seismic Regulations for New Buildings and Other Structures.

Canadian Standard Association (CSA), CAN3-N289.3 - (R-1998) Design Procedures for Seismic Qualification of CANDU Nuclear Power Plants.

National Research Council of Canada, National Building Code of Canada (NBCC), NRCC No. 32379, Ottawa, 1995.

Seismology Committee, SEAOC (1999), 7th Edition, Recommended Lateral Force Requirements and Commentary, San Francisco, CA.

Seismic Response of High Voltage Transformers

Howard Matt[1] and André Filiatrault[2]

Abstract

The IEEE 693-1997 standard is a guideline for the seismic qualification of electrical substation equipment. High voltage bushings, typically attached to the top of voltage transformers, are one electrical component highly susceptible to damage during strong shaking. For seismic qualification of bushings, it is assumed that the dynamic amplification that occurs from the ground to the base of the bushing is a constant factor of 2.0. Despite the use of this standard, recent seismic events have resulted in a large number of high-voltage bushing failures, thereby raising doubts on the accuracy of the assumed amplification factor of 2.0. The main objectives of this analytical study are to identify the critical parameters of supporting transformer structures that affect the seismic response of bushings and to quantify and compare the dynamic response of voltage transformers with the predictions of the IEEE-693 1997 standard. Three-dimensional finite element models were developed for five different voltage transformers that varied in voltage rating and manufacturer. Time history analyses were then performed on each of the models in order to determine the dynamic response characteristics as well as the amplification that occurs at the base of the bushing. The results of the analyses show that the transformer structures were much more flexible than currently assumed. In addition, the local flexibility of the transformer top upon which bushings are attached significantly reduces the natural frequency of the bushings. The dynamic amplifications occur predominately at two frequencies: the natural frequency of the transformer frame and the natural frequency of the bushing. The highest amplifications occurred when these two natural frequencies were close to each other. For all transformers, the lower frequency transverse direction consistently resulted in larger amplifications than the higher frequency longitudinal direction. Only one transformer had an amplification value greater than 2.0 at the bushing frequency.

Introduction

Electrical Substations are critical components of an electrical network that supplies power for industrial, business and residential use. Several large structural systems exist within these substations, each contributing to the overall functionality of the substation. Like any structural system located in an earthquake prone region, seismic vulnerability must be addressed. After major seismic events such as the 1989 Loma Prieta and 1994 Northridge earthquakes, it was observed that bushings attached to voltage transformers have a high susceptibility to damage under strong shaking.

Graduate Student Researcher[1] and Professor[2], Department of Structural Engineering, University of California, San Diego, 9500 Gilman Drive, Mail Code 0085, La Jolla, CA 92093.

Voltage transformers are essential pieces of equipment in any electrical substation. The function of these transformers is to step up or step down the voltage within the transmission lines. Bushings are cantilever like components that protrude vertically, or at a slight angle from the top of the transformer and are used to connect the electrical coils within the transformer to the external power lines. The only structural integrity of bushings is provided by the external insulating material, which typically is made up of porcelain or some other lighter composite material.

Due to the size and complexity of the transformer-bushing system, repair to these items can prove to be a timely and very expensive task. In addition, power transmission may be halted during repair time resulting in localized blackouts. In recognition of these potential problems, several related research tasks funded by Pacific Gas and Electric (PG&E), Pacific Earthquake Engineering Research Center (PEER) and the California Energy Commission (CEC) have been recently initiated with the purpose of providing knowledge of the seismic response of various substation equipment, as well as methods to ultimately reduce their seismic vulnerability. Along with this, there has been certain seismic design criteria adopted when qualifying substation equipment for field use. These criteria are included in the IEEE-693-1997 Standard (IEEE, 1997). The IEEE 693 document was written to provide a recommended practice for the seismic design of substations.

Seismic Standards Included in the IEEE-693-1997

As a general rule, the higher the voltage rating of the bushing is, the heavier and taller it is and therefore the more susceptible it is to damage during seismic loading. Because of this, the majority of research interests and seismic concerns are focused on bushings rated at voltages exceeding 161kV. The IEEE 693 standard states that bushings with voltage ratings exceeding 161kV must be qualified by time history shake table tests. Since placing a full-scale transformer-bushing system upon a shake table is highly unpractical for these bushing qualification tests, the bushings are placed upon a rigid frame as a replacement of the transformer body itself. Although the transformer body is assumed to be fairly rigid, it is acknowledged that the supporting structure of the bushing, consisting of the turret and transformer frame, amplifies the ground acceleration. Knowledge of exactly how much amplification, and what key parameters of the supporting structure affect this amplification is limited. Despite this, the IEEE 693 assumes that the motion at the base of the bushing is equal to the ground motion multiplied by a factor of 2. Therefore, during bushing qualification through shake table testing, the rigid frame is subjected to ground motions that match a specified response spectrum scaled such that it accounts for this amplification.

Objectives of Analytical Study

Previous shake table tests performed in related PEER projects (PEER, 2002) as well as the large number of bushing failures that occurred during recent seismic events have raised doubts about the adequacy of the bushing qualification tests and more specifically whether this amplification factor of two is indeed valid. The analytical study presented herein is a portion of an ongoing larger study aimed at gaining a better understanding of the bushing-transformer system response, and to eventually improve the qualification requirements for transformer bushings that are currently defined. This analytical portion has been completed in order to identify the critical parameters of supporting transformer structures that affect the seismic response of bushings and to quantify and compare the dynamic response of voltage transformers with the predictions made in the IEEE-693 standard.

Scope of Analytical Study

For each of the five transformers used in this study, a three-dimensional finite element (FE) model was developed using the structural analysis program SAP 2000 (Computers and Structures Inc., 2002). Time-history analyses were then performed on these finite element models using 20 different strong ground motion time histories scaled to the 2% damped high performance level response spectrum shown in Figure 1. This response spectrum is defined in the IEEE-697 standard and is used to qualify bushings located in regions of high seismicity. The 20 earthquake records were separately run in both the longitudinal and transverse directions; after which, the dynamic amplification that occurs between the base of the bushing and the input ground motion was found and compared with the IEEE assumed amplification value of two. This amplification was quantified by taking the ratio of the mean 2%-damped response spectrum computed at the base of the high voltage bushing to the corresponding mean response spectrum of the ground acceleration considered. This spectral ratio is defined as a "spectral amplification" which explicitly gives the dynamic amplification as a function of frequency:

$$Spectral\ Amplification = \frac{Mean\ Response\ Spectrum\ at\ Base\ of\ Bushing}{Mean\ Response\ Spectrum\ of\ Ground\ Motion} \quad (1)$$

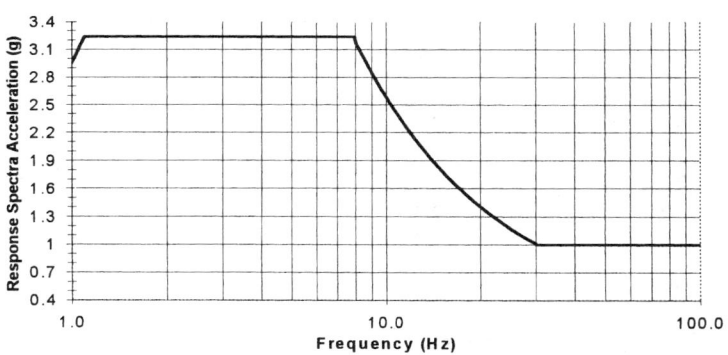

Figure 1. High performance level response spectrum as defined in the IEEE-693

Descriptions of Transformers used in FE Analysis

There are many different types of voltage transformers that vary greatly in weight, size and geometry. Therefore, the dynamic response of transformers, even of the same voltage rating, can widely differ. In an attempt to capture these variations, five different voltage transformers were modeled by finite elements in order to gain a better understanding of the supporting structure's seismic response for various transformer sizes and manufactures. The five transformer models were: a Westinghouse 525 kV, a Siemens 500 kV, a Brown Boveri 500 kV, a Siemens 230 kV, and a Waukesha 69 kV transformer. The Westinghouse 525 kV transformer weighs 2060 kN, and has dimensions 2.69 m x 3.02 m x 6.96 m. The Siemens 500 kV transformer weighs 2995 kN, and has dimensions 3.3 m x 7.92 m x 5.13 m. The Brown Boveri 500 kV transformer weighs 1340 kips, has dimensions 1.73 m x 3.48 m x 4.83 m. The Siemens 230 kV transformer weighs 2125 kN, and has dimensions 3.05 m x 7.37 m x 4.39 m. Te Waukesha 69 kV transformer weighs 665 kN, and has dimensions 1.83 m x 4.06 m x 1.12 m. The finite element models and actual photographs the Westinghouse 525 kV, Brown Boveri 500 kV, and Siemens 230 kV transformers are shown in Figures 2 to 4. Note that all transformers shown below have been stripped of their radiators,

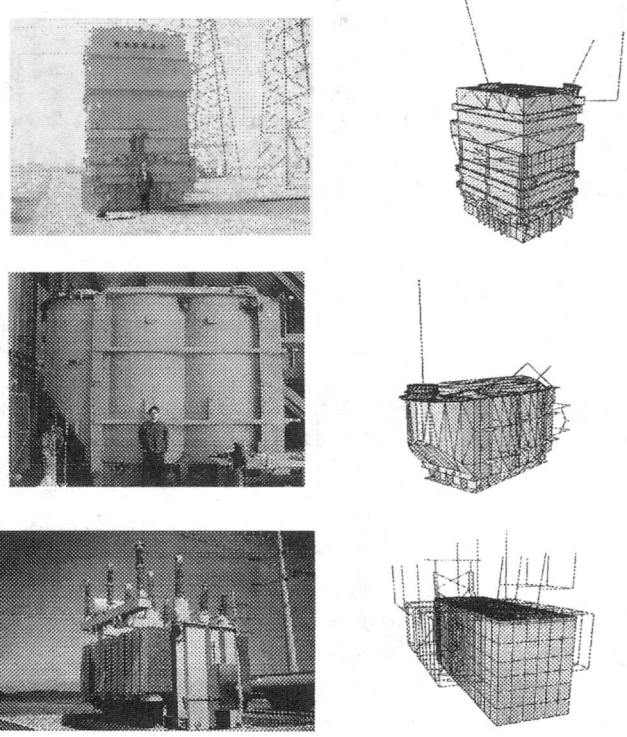

Although most voltage transformers vary greatly in weight, size, and geometry, they all contain a certain number of key components: the transformer frame or tank, the core and coil contained within the tank, radiators attached to the outside, bushings mounted on the top of the tank, oil contained within the tank, and often an oil conservator tank also attached to the top of the tank. The majority of a transformer's mass is comprised of the core, copper coils and oil. The majority of the lateral stiffness of a transformer tank is provided by the tank walls made of steel and having a typical thickness of 12 mm, and stiffeners such as channels, I-beams, or plates attached to the tank walls.

Description of FE Models

Before building the finite element models within the SAP program, certain modeling assumptions had to be made. The first of which was how to model the oil. Oil contained within transformers is generally filled up to the top of the tank. For this condition, oil-sloshing effects can be ignored and were not accounted for in modeling. To account for the mass of the oil within the tank, additional mass was symmetrically added to the vertical perimeter of the steel tank walls leading to an appropriate center of gravity of the oil.

Due to the nature of its design, the core and coil can safely be assumed as rigid. However, one significant issue is if, and how, it may be braced to the interior walls of the transformer. If it is not braced, then the core and coil provides no stiffness to the tank. The tank and core will act as two separate and independent structures; therefore the mass of the core should not be included in the dynamic analysis. On the other hand, if the core and coil is rigidly braced to the core and coil, then the whole transformer will be essentially rigid and the mass of the core should be included in the dynamic analysis. Although some cores and coils are lightly braced to the transformer body, most are not. Even the ones that are braced may only be braced at mid-height or braced by shimming a piece of wood between the core and transformer shell. Therefore it was decided to not include any stiffness from the core and coil as well as neglect this mass in the dynamic modeling.

Another assumption made was that the radiators and oil conservator tank are rigidly attached to the transformer frame. This allowed for simplification of the model and eliminated some of the non-critical modes of vibration. To ensure the validity of this assumption, comparison of the transverse and longitudinal frequencies of the transformers were done when allowing for flexibility of these components. It was concluded that by making them rigid had no significant effect on the longitudinal and transverse modes of both the bushing and transformer body. The appropriate masses of each radiator and oil conservator tank were added at their respective center of gravities.

All bushings were modeled as single beam elements with the appropriate height, thickness and stiffness. Bushing models were based upon information from available bushing qualification reports and structural drawings included in the transformer manufacturer's reports. The final assumption was made in the support conditions of the transformers. Pin supports were used at bolt locations, fully fixed conditions were used at weld locations, and roller supports were added under the tank base to prevent out of plane bending of the bottom shell.

Transformer frames were modeled as shell elements with the appropriate thickness. The shell elements allowed for bending in and out of plane. Beam elements as well as appropriate shell elements were used for the stiffeners attached to the tank sides. The geometry, thickness and location of all walls, plates and beams were obtained by manufacturer's structural drawings, thorough surveying, as well as previous static models (Gundy, 2002).

Earthquake Ground Motions

The 20 ground motions chosen for this study are representative of feasible events that could occur within the California region (Krawinkler et al., 2001). These strong ground motions were recorded from various recent events with varying fault mechanisms. All ground motions are such that the location of measurement was far enough from fault rupture to be free of any near-fault pulse conditions. Of the 20 ground motions, three are from the Superstition Hills 1987 earthquake, seven were recorded during the Northridge 1994 earthquake, six are from the Loma Prieta earthquake 1989, two are from the Landers 1992 earthquake, and the last two are from the Cape Mendocino 1992 earthquake.

Modal Analysis Results

Due to the complexity of the finite element models, the number of modes to be considered during analysis had to be greatly reduced. For each model, roughly the first 20 modes were considered such that at least 90% of the total modal mass participation was accounted for. The first few modes generally corresponded with extraneous elements such as the oil conservator, surge arrestors and bushings. The modes that contributed the largest percentage of total modal mass participation were that of transformer frame in the transverse and longitudinal direction.

For each model, the transverse (narrow) direction had a lower natural frequency than in the longitudinal direction. These frequencies, summarized in Table 1 below, are 9.8, 13.6, 17.0, 11.2 and 14.9 Hz in the transverse direction and 13.0, 22.4, 27.2, 20.6 and 29.2 Hz in the longitudinal direction for the Westinghouse, Siemens 500 kV, Brown Boveri, Siemens 230 kV and Waukesha transformers, respectively.

The high voltage bushing natural frequencies also shown in Table 1 are 2.8, 2.3, 2.6, 8.1 and 9.0 Hz in the transverse direction and 3.3, 6.2, 4.2, 9.7 and 11.6 Hz in the longitudinal direction for the Westinghouse, Siemens 500 kV, Brown Boveri, Siemens 230 kV and Waukesha transformers respectively. These frequencies are representative of the modes when the bushing is attached to the transformer supporting structure. These frequencies will be naturally larger when the bushings are analyzed as having a rigid support.

Table 1. Bushing and transformer tank natural frequencies

Transformer	Object	Transverse Frequency (Hz)	Longitudinal Frequency (Hz)
Westinghouse 525 kV	Tank	9.8	13.0
	Bushing	2.8	3.3
Siemens 500 kV	Tank	13.6	22.4
	Bushing	2.3	6.2
Brown Boveri 500 kV	Tank	17.0	27.2
	Bushing	2.6	4.2
Siemens 230 kV	Tank	11.2	20.6
	Bushing	8.1	9.7
Waukesha 69 kV	Tank	14.9	29.2
	Bushing	9.0	11.6

Spectral Amplification Results

The spectral amplification results for each of the transformers had three common trends. First, for a given transformer in a given direction, there were two peaks in the spectral amplification. These two peaks corresponded with the natural frequency of the transformer tank and the natural frequency of the bushing in the given direction. In addition, the magnitude of the amplification at the transformer frequency was consistently higher than the magnitude at the frequency of the bushing. Also, the amplification that occurred in the transverse direction for each transformer was larger than that of the longitudinal direction. These trends can be observed in the spectral amplification results shown in Figures 5 and 6. These figures show the mean spectral amplification for the Siemens 500 kV transformer in the transverse and longitudinal direction respectively. It can be seen in both directions that the amplification at the transformer frequency is larger than the IEEE assumed value of 2, yet slightly smaller than 2 at the bushing frequency.

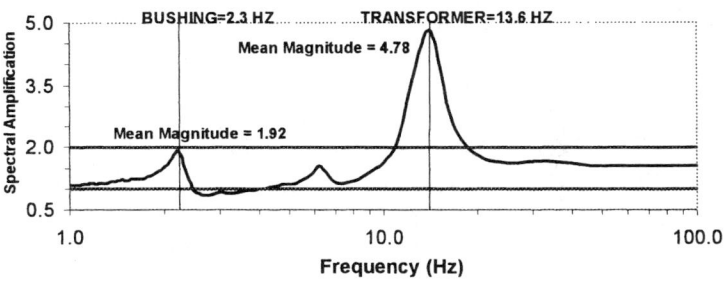

Figure 5. Mean spectral amplification results for Siemens 500 kV transformer, transverse direction.

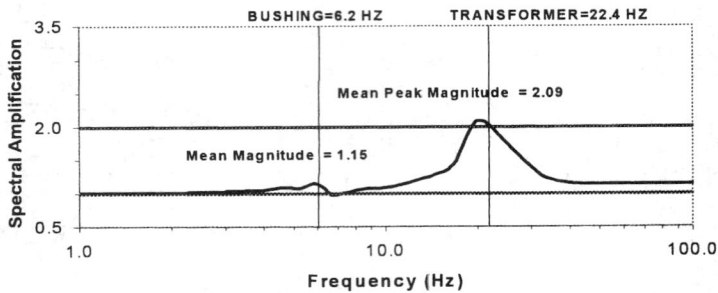

Figure 6. Mean spectral amplification results for Siemens 500 kV transformer, longitudinal direction.

Although the amplification at various frequencies is of interest, ultimately the main concern for this study is the amplification that occurs at the frequency of the bushing, since this will govern the behavior of the bushing under seismic excitation. The mean spectral amplification at the bushing frequency in the transverse direction was found to be 1.2, 1.9, 1.06, 7.6, and 1.3 for the Westinghouse, Siemens 500 kV, Brown Boveri, Siemens 230 kV, and the Waukesha transformer respectively. The mean spectral amplification at the bushing frequency in the longitudinal direction is 1.1, 1.15, 1.04, 3.6, and 1.1 for the Westinghouse, Siemens 500 kV, Brown Boveri, Siemens 230 kV, and the Waukesha transformer respectively. Table 2. below summarizes these results and also shows the spectral amplification that occurred at the transformer tank frequencies.

Table 2. Spectral amplification results at transformer tank and bushing frequency

Transformer	Object	Transverse Direction		Longitudinal Direction	
		Frequency (Hz)	Spectral Amplification	Frequency (Hz)	Spectral Amplification
Westinghouse 525 kV	Tank	9.8	11.2	13.0	7.0
	Bushing	2.8	1.2	3.3	1.1
Siemens 500 kV	Tank	13.6	4.8	22.4	2.1
	Bushing	2.3	1.9	6.2	1.2
Brown Boveri 500 kV	Tank	17.0	6.2	27.2	1.6
	Bushing	2.6	1.0	4.2	1.0
Siemens 230 kV	Tank	11.2	3.0	20.6	2.4
	Bushing	6.1	7.6	9.7	3.6
Waukesha 69 kV	Tank	14.9	3.0	29.2	1.4
	Bushing	9.0	1.3	11.6	1.1

Analyses and Conclusions

The presumption about transformer frames is that they are essentially rigid. However the results of the analyses show that transformer structures, provided the core is unbraced, are much more flexible than currently assumed. The majority of the flexibility seems to occur on the transformer top plate upon which bushings are attached. This local flexibility significantly reduces the natural frequency of the bushings, thereby changing the response of the bushing during seismic loading. Current bushing qualification tests performed according to the IEEE-693 standard do not take into account this reduced natural frequency during shake table testing. As a result, the bushing response from qualification testing is not necessarily representative of the response that will occur in the field. An example of this may be when considering a 230 kV bushing. Its fixed condition natural frequency is roughly 20 Hz, but reduced to about 8 Hz when attached to the top of the transformer. The IEEE pseudo-acceleration response spectrum for the bushing attached to the transformer is more than two times that of the bushing when rigidly attached. Therefore, this reduced bushing natural frequency should be accounted for during seismic qualification tests.

Regarding spectral amplification, the largest amplifications occur at two predominant frequencies: the natural frequency of the transformer frame and the natural frequency of the bushing. In addition, the transverse, lower frequency, direction consistently resulted in larger amplifications than in the longitudinal, higher frequency, direction. Although the amplifications were generally significantly larger than the assumed amplification of two at the transformer frequencies, only the Siemens 230 kV transformer had mean amplifications larger than two at the bushing frequencies. This result can be justified when comparing the bushing and transformer frequencies. In the case of the Siemens 230 kV transformer, these two natural frequencies were much closer to each other than for all other transformers. The fact that the highest amplifications occurred when these two natural frequencies were relatively close to each other seems rather intuitive; however, nothing in the IEEE-693 standard accounts for such a situation. In general, the assumed amplification value of two appears to be conservative for 500 kV transformers or larger. Despite this, seismic qualification tests should still consider cases such as this Siemens 230 kV transformer where the bushing and tank frequencies are close enough to significantly increase the amplification that occurs between the ground and bushing base.

Acknowledgements

The research project described herein was funded by the Pacific Earthquake Engineering Research (PEER) Center Lifeline Directed Studies Program. The authors greatly appreciated the input and coordination provided by Dr. Michael Riemer from PEER and Mr. Eric Fujisaki from the Pacific Gas and Electric Company (PG&E) during the development of this research project. The supports of Mr. David Chambers from the California Energy Commission and of Mr. Craig Riker from the San Diego Gas and Electric Company (SDG&E) are also gratefully acknowledged. The authors graciously acknowledge Mr. William Gundy for the donation of three original transformer models as well as his professional input. Finally, the authors extend their sincere appreciation to Mr. Chris Stearns, Graduate Student Researcher at UC-San Diego, for his assistance in surveying the transformer specimens.

References

Computers and Structures Inc. (2002). SAP2000 Integrated Software for Structural Analysis & Design. Berkeley, CA.

Gundy, W. (2002). Private Communication, W.E. Gundy & Associates, Inc., Hailey, ID.

Institute of Electrical and Electronics Engineers. (1997). *Recommended practices for Seismic Design of Substations*, IEEE-693 Standard, IEEE Standards Dept., Piscataway, NJ.

Krawinkler, H., Parisi, F., Ibarra, L. Ayoub, A. Medina, R. (2001). Development of a Testing Protocol for Woodframe Structures, Curee Publication No. W-02, Department of Civil and Environmental Engineering, Stanford University, Stanford, CA, 46, 22-24.

Pacific Earthquake Engineering Research (PEER) Center. (2002). PEER Lifelines Program, Web Site: http://peer.berkeley.edu/lifelines/index.html.

Interpretation and Application of Hilbert-Huang Transformation for Seismic Performance Analyses

J. Jerry Shen[1], W. Phillip Yen[2], and John O'Fallon[3]

Abstract

In consideration of the increasing significance of nonlinearities in seismic analysis and design, Federal Highway Administration (FHWA) had investigated numerical tools that can handle general behavior of dynamic systems beyond elastic and stationary limits. The typical tools currently used to study the ground motion and ductile structural response are mostly Fourier-spectra-based, which can do well for stationary oscillations. They may misinterpret the seismic input and response due to the time-variation of frequency characteristics in non-stationary processes. The Hilbert-Huang Transformation (HHT) is a combination of the Empirical Mode Decomposition (EMD) and the Hilbert Transformation (HT). It has been one of the modern tools that enable engineers to analyze non-stationary oscillation systems. The Hilbert spectra resulted from HHT can provide significantly more detailed description on time-varying frequency composition. The utilization of HHT in seismic engineering has a relatively short history. The physical interpretation and the most effective application in seismic engineering are deemed in need of clarification. This paper presents the physical interpretation from the highway structure dynamics point of view and demonstrates the possible applications on seismic design and analysis.

Introduction

As the requirements on structural performance and cost-efficiency become stricter along time, there is an increasing demand on better understanding on structural behavior under various types of seismic loading. Earthquake ground motions are produced by a series of non-stationary processes such as the ground rupture within finite time and space as well as reflection and refraction. The ductile structures produce nonlinear response to the seismic input. The popular Fourier approach for studies of vibration phenomena decomposes nonlinear phenomena into linear combinations. It can provide valuable information on the source and response to a certain extend. The Hilbert-Huang Transformation (HHT) method based on the Empirical Mode Decomposition (EMD) is a new tool available for studies of oscillations of nonlinear systems. Due to some of the highly acclaimed features of the new method, it has been applied to a wide range of dynamic problems in geophysics, biomedical engineering, oceanic science (Huang et al,

---

[1] LENDIS/FHWA, 6300 Georgetown Pike, McLean, VA 22101, Tel:202-493-3098; E-mail: Jerry.Shen@fhwa.dot.gov (through LENDIS Corporation)
[2] Office of Infrastructure, R&D Federal Highway Administration, 6300 Georgetown Pike, McLean, VA 22101, Tel:202-493-3056; E-mail: Wen-huei.Yen@fhwa.dot.gov
[3] Office of Infrastructure, R&D Federal Highway Administration, 6300 Georgetown Pike, McLean, VA 22101, Tel:202-493-3051; E-mail: John.O'Fallon@fhwa.dot.gov

1999), and mechanical engineering (Feldman and Seibold, 1999). Attempt has been made to apply Hilbert spectrum method to seismic engineering problems (Loh et al, 2001).

The primary objective of this study is to review and explore applications of HHT in bridge seismic analysis and design. Issues in existence and other mathematic derivation and proof are omitted. Physical interpretation and potential applications are discussed and demonstrated.

Hilbert transform

In the complex coordinates in which horizontal coordinate (x) represents real value and vertical coordinate (y) represents imaginary value, a complex function $\Psi(z)$, or $\Psi(x,y)$ (z=x+iy, i is square root of -1), is analytic or holomorphic if the derivative

$$\Psi'(z) = \lim_{h \to 0} \frac{\Psi(z+h) - \Psi(z)}{h}$$

exists in the complex domain. It has been shown (Hahn, 1996) that the Hilbert Transformation can find the imaginary component that can form an analytic complex function when combined with the original real curve. That is

$$\Psi(t) \text{ is analytic} \Leftrightarrow \Psi(t_r) = \psi(t_r,0) + iH(\psi(t_r,0)) \quad (4)$$

in which time $t = t_r + it_i$ and H(.) represents Hilbert transform of the enclosed function. Practically, imaginary part of time is meaningless. It is therefore set to be zero in the above equation. The amplitude $|\Psi(t_r)|$ and phase angle $\angle\Psi(t_r)$ of this analytical signal are defined and differentiable everywhere on the real axis. The instantaneous amplitude $A(t_r)$ (envelope of a vibration) and instantaneous frequency $\omega(t_r)$ can be obtained.

$$A(t_r) = |\Psi(t_r)|$$

$$\omega(t_r) = \frac{d\angle\Psi(t_r)}{dt_r}$$

Physical links to mechanical systems

In an engineering oscillation problem of a single-degree-of-freedom system, the displacement response x is governed by the equation of motion:

$$m\ddot{x} + c\dot{x} + kx = f(t)$$

The general (homogeneous) solution of an underdamped system is

$$x_g(t) = Ce^{\lambda_1 t} + C^* e^{\lambda_2 t}$$

in which C and C* are complex conjugates obtained from initial conditions and

$$\lambda_{1,2} = \frac{-c \pm i\sqrt{4mk - c^2}}{2m}$$

The particular (nonhomogeneous) solution for a periodical f(t) with period T can be reached by expanding the Fourier Series of f(t):

$$f(t) = \sum_{n=-\infty}^{\infty} D_n e^{i\frac{2n\pi}{T}t}$$

in which D_{-n} is the complex conjugate of D_n when f(t) is real-valued. The particular solution is

$$x_p(t) = \sum_{n=-\infty}^{\infty} \frac{D_n}{-\left(\frac{2n\pi}{T}\right)^2 m + k + ic\frac{2n\pi}{T}} e^{i\frac{2n\pi}{T}t}$$

Using Cauchy-Riemann equations, it can be shown that the above general and particular solutions are analytic everywhere in the complex domain. Consequently, being the sum of these terms, the complete solution x (t) is analytic in the complex domain.

Because the forced vibration and free vibration are both real-valued, each term in the solutions has a conjugate term that cancels the imaginary part. Due to the lack of imaginary terms, it is impossible to tell whether one term has a positive frequency (n>0) or negative frequency (n<0) by looking at the real part. If we look at the dynamic response data without knowing anything about the imaginary part, there is no way to know if we are looking at the real component of a complex signal (Fig. 1) or the sum of two complex conjugate signals (Fig. 2). For example, an oscillatory component cos(ωt) can be represented by either real($e^{i\omega t}$) or ($e^{i\omega t}+e^{-i\omega t}$)/2.

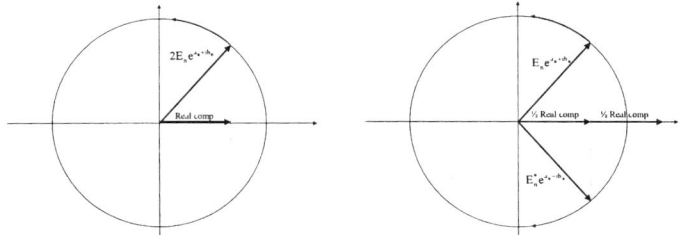

Fig. 1 Single complex signal representation   Fig. 2 Conjugate signal representation

Since all complex functions that construct the real response signal are analytic, Hilbert transform can reconstruct the imaginary part of the functions from the real response signal. That is

$$H(x(t)) = \sum (\text{imag}(E_n e^{a_n + ib_n}) + (-\text{imag}(E_n^* e^{a_n - ib_n}))) = \sum \text{imag}(2E_n e^{a_n + ib_n})$$

For a linear system, the amplitude of the analytic function with certain frequency is twice of corresponding Fourier coefficient E_n and E_{-n} (or E_n^*).

$$X_n(t) = x_n(t) + iH(x_n(t)) = 2E_n e^{i\frac{2n\pi}{T}t}$$

The Fourier decomposition of the signal is unique and contains the assumption of linearity and stationarity. Each frequency component remains same amplitude through the entire time domain. For signal produced by nonlinear process, the analytic signal can be decomposed in other ways that are more realistic in an engineering sense.

Empirical Mode Decomposition

The Hilbert transform vanishes when the subject function is a constant. Therefore, an oscillatory signal with an offset results in a circle or spiral centered at a nonzero point on real axis. This can produce a few different situations. If the vibration has different amplitude and period on positive and negative sides, the offset (nonzero-mean) properly represents the asymmetry of the signal. For a frequency variation within 100% (e.g. $\omega^+ = 2\omega^-$, see Fig. 3), the error of the instantaneous frequency (Fig. 4) found through the Hilbert transform is within 10% (Feldman and Seibold, 1999). This signal has a zero mean within each half cycle of oscillation, for which Hilbert transform provides accurate frequency estimation.

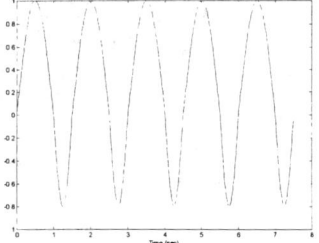
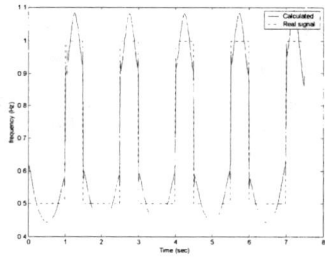

Fig. 3 Asymmetric signal   Fig. 4 Frequency by Hilbert transform

Another situation is that the signal is symmetric but offset by a constant value (Fig. 5) or riding on very low frequency oscillation. Directly applying Hilbert transform can result in serious fluctuation of calculated frequency (Fig. 6). Negative frequency may occur in some cases. An adequate decomposition is needed to separate the offset or low frequency oscillation from the interested signal. EMD is developed to separate the riding oscillations from the lower frequency or DC signal. Since the decomposition is empirical, the result is not necessarily unique and depends on the algorithm used to decompose the signal. The EMD used in HHT produces Intrinsic Mode Functions (IMF) that are adequate for application of Hilbert transformation. This algorithm also successfully separates low-frequency oscillations from high frequency riding signals. Detailed procedure can be found in Huang et al (1998).

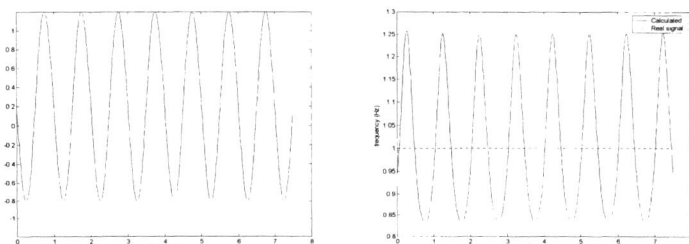

Fig. 5 Offset harmonic signal Fig. 6 Unwanted frequency fluctuation

Applications on ground motion

Properties of earthquake ground motions can normally be revealed through time history envelope and Fourier spectrum (response spectrum is another useful representation of the ground motion but it involves structural properties and is not mathematically operable). Since the time span of an earthquake is limited, the Fourier spectrum is in fact a Fourier series that decomposes the ground motion signal into harmonic components. Each of these harmonic components has constant amplitude and frequency through the time span of the earthquake. For a ground motion record with nearly random motions, this decomposition provides a convenient representation of frequency distribution.

Latest experiences in large earthquakes worldwide (Chang et al, 2000; Ghasemi et al, 2000; Loh et al, 2002) demonstrated that many earthquake ground motions contain special features such as long period velocity pulse. These features appear for only a short time but can produce significant structural damage. For example, the synthetic ground motion shown in Fig. 7 consists of random signal and a long period pulse in the middle section of the record. Fig. 8 shows another synthesized record with the same peak ground acceleration (PGA) but without the pulse. Figs. 9 and 10 are the Fourier amplitude spectra of the two records. The Fourier amplitude spectra of the two records are nearly identical. The difference of the time history is solely due to the difference in phase angle of each frequency component. The pulse in the first record is caused by coincide phase of multiple frequency components at certain stage in the time history. Figs. 11 and 12 show the response acceleration of a structure with period of 1 sec and damping of 10%. The record with the pulse results in 78% higher acceleration response. Figs. 13 and 14 show the Hilbert spectra of the two records. In Fig. 13, a high amplitude single frequency component (1 Hz) is observed at the same moment where the pulse occurs. In Fig. 14, this frequency component is distributed through the most part of the time span. This example demonstrates the capability of HHT in identifying the occurrence and properties of the destructive near-fault ground motions. When specific structural parameters are concerned, a more quantitative approach based on input energy of each EMD can be used to identify the near-fault pulse motion (Loh et al, 2001).

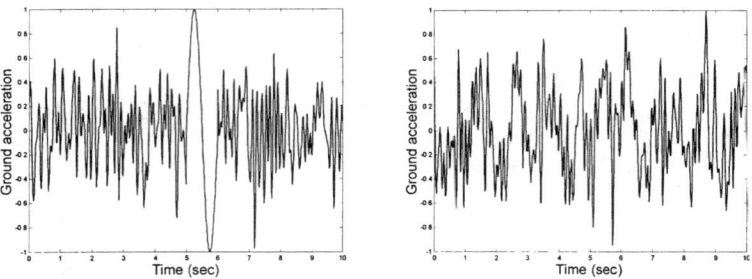

Fig. 7 Ground acceleration A with long period pulse   Fig. 8 Random acceleration B

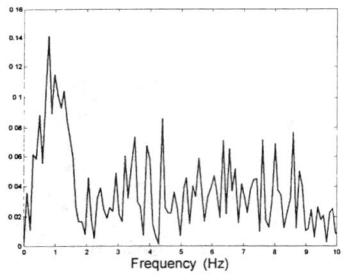
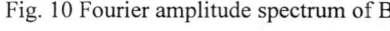

Fig. 9 Fourier amplitude spectrum of A   Fig. 10 Fourier amplitude spectrum of B

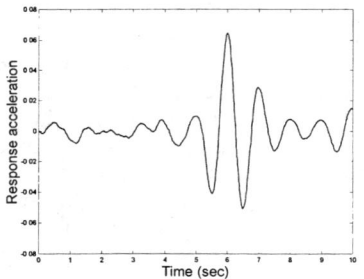
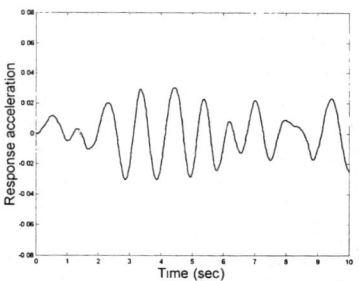

Fig. 11 Response acceleration from A   Fig. 12 Response acceleration from B

These methods are currently used for demonstration of the capability of Hilbert spectrum approach on analyzing special ground motions. Result does not yet directly assist estimation of seismic loading in bridge design. Some research on near-fault effect showed that structural response can be amplified by certain features of ground motions, which are not completely attributed in current spectrum approach for bridge design (Somerville, 2002). New or improved design procedures based on the new view through Hilbert spectrum should be developed in the future.

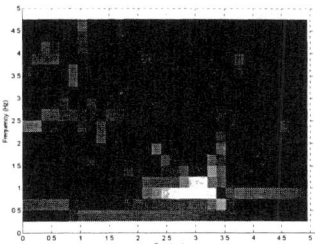 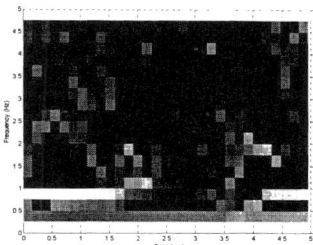

Fig. 13 Hilbert amplitude spectrum of A   Fig. 14 Hilbert amplitude spectrum of B

Application on structural response

The Hilbert spectrum can be used to identify variation of structural response along time. These time-dependent parameter changes are normally indications of structural damage. Hilbert spectrum provides a quick observation of system parameter change and, consequently, indication of damage level (Feldman and Seibold, 1999). Change of modal damping (or loss factor) is very commonly used to indicate the level of damage. Stiffness change, and consequently frequency change, can also indicate system degradation. An example is shown in Fig. 15. A bilinear single-degree-of-freedom system is subjected to a random signal with an envelope increasing along time (Fig. 16). The system has an initial frequency of 2 Hz (fundamental period 0.5 sec). Fig. 17 shows the Hilbert spectrum when the input is scaled down to avoid yielding of the system. The largest response component appears at the resonant frequency. The amplitude increases along time, proportionally with input. Fig. 18 shows the Hilbert spectrum of the response from full-scale input. When the input gradually increases, the frequency components become increasingly scattered as a result of increasing level of nonlinearity.

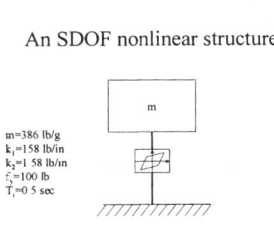 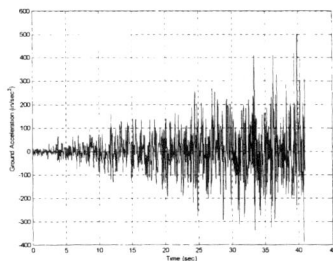

Fig. 15 a bilinear system   Fig. 16 Random load with increasing envelope

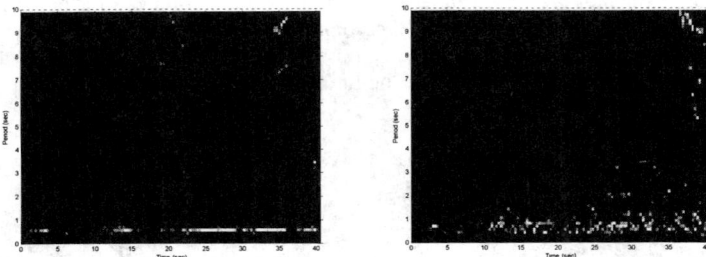

Fig. 17 Response from scaled input   Fig. 18 Response from full-scale input

System identification

The above sections discussed the application of HHT on input and output of nonlinear structure systems under seismic loading. Combination of these techniques can provide assistance to identification of parameters of nonlinear structural systems. Studies showed that characteristics and special events occurrence (e.g. impact at abutment) of isolated bridge under moderate earthquake loads can be identified using Hilbert amplitude spectrum and Hilbert damping spectrum (Loh et al, 2001). Free vibration tests, usually conducted by an impact or suddenly released force, can be analyzed with Hilbert spectrum as well. Fig. 19 shows a free vibration of a nonlinear system with an exponential decay and amplitude-dependent frequency:

$$f(t) = \frac{1}{2\sqrt{a(t)}}$$

in which f(t) is the damped frequency associated with the amplitude envelope a(t). The frequency increases from 0.5 Hz to 2.4 Hz within 10 sec and damping ratio drops from 10% to 2%. The Hilbert amplitude spectrum (Fig. 20) and Hilbert damping spectrum (Fig. 21) agree with the real parameters.

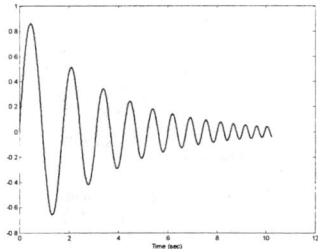 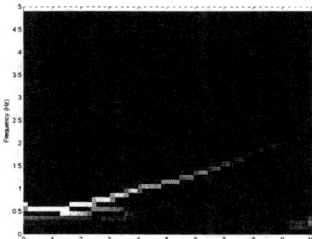

Fig. 19 Free vibration of a nonlinear structure   Fig. 20 Hilbert amplitude spectrum

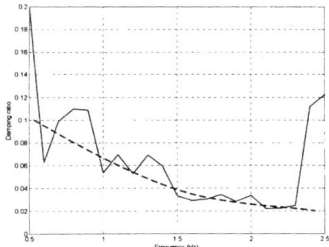

Fig. 21 Damping vs. frequency

Summary

(1) Studies conducted in Federal Highway Administration (FHWA) and the review of previous researches revealed great potential of the Hilbert spectrum produced through EMD in seismic research and seismic design of bridges.
(2) Three types of applications have been identified: energy concentration in ground motion, structural damage assessment, and structural characteristics identification.
(3) Due to the incompatibility to the current seismic design methodology, none of the applications of HHT can promptly assist seismic analysis and design at this moment. However, the large response from the near-fault ground motion manifests the necessity of more sophisticated criteria to handle transitory features in the seismic load. New seismic analysis and design methodology that can utilize Hilbert spectrum needs to be developed before substantial contribution can be made by HHT.

Reference

Chang, K. C., Shen, J., Tsai, M. H., and Lee, G. C. (2000). "Performance of A Seismically Isolated Bridge under Near-Fault Earthquake Ground Motions," *Proceedings of the Taiwan-US-Japan Workshop on Seismic Mitigation of Highway Bridges*, Taipei, Taiwan, 280-290.

Feldman, M. and Seibold, S. (1999). "Damage Diagnosis of Rotors: Application of Hilbert Transform and Multihypothesis Testing", *Journal of Vibration and Control*, 5: 421-442.

Ghasemi, H., Cooper, J. D., Imbsen, R., Piskin, H., Inal, F., and Tiras, A. (2000). *The November 1999 Duzce Earthquake: Post-Earthquake Investigation of the Structures on the TEM*. Publication No. FHWA-RD-00-146, Federal Highway Administration, U. S. Department of Transportation, Washington, D.C.

Huang, N. E., Shen, Z, and Long, S. (1999). "A New View of Nonlinear Water Waves: The Hilbert Spectrum", *Annu. Rev. Fluid Mech.* 31: 417-457.

Huang, N. E., Shen, Z., Long, S., Wu, M. C., Shih, H. H., Zheng, Q., Yen, N. C., Tung, C. C., and Liu, H. H. (1998). "The Empirical Mode Decomposition and the Hilbert Spectrum for Nonlinear and Non-stationary Time Series Analysis," *Proc. Royal Society Lond. A*, 454, 903-995.

Loh, C. H., Liao, W. I., Chai, J. F. (2002) "Effect of Near-Fault Earthquake on Bridges: Lessons Learned from Chi-Chi Earthquake." *Proc. Third National Seismic Conference & Workshop on Bridges and Highways*, FHWA Western Resource Center and MCEER, Portland, Oregon, 211-222.

Loh, C. H., Wu, T. C., and Huang, N. E. (2001). "Application of the Empirical Mode Decomposition-Hilbert Spectrum Method to Identify Near-Motion Characteristics and Structural Responses," Bulletin of the Seismological Society of America, 91, 5, 1339-1357.

Somerville, P. (2002). Characterizing near fault ground motion for the design and evaluation of bridges. *Proc., The 3rd National Seismic Conference and Workshop on Bridges and Highways*, April 28-May 1, 2002, Portland, OR, 137-148.

An Experimental Study on the Seismic Response of Electrical Substation Equipment Interconnected by Flexible Conductors

André Filiatrault[1] and Christopher Stearns[2]

Abstract

This paper investigates experimentally the dynamic interaction between components of electrical substation equipment interconnected by flexible (cable) conductors. Shake table tests were conducted on five different pairs of generic substation equipment specimens interconnected by three different flexible conductors with three different levels of slackness. No damage to any of the three flexible conductors was observed during all the seismic tests conducted. Two different types of dynamic responses were observed during the seismic tests. The first type involves low interaction between the interconnected equipment due to a large slack and/or low intensity ground motions. For this case, very different frequency contents between the relative displacement, absolute acceleration, and force response of each equipment item were observed. Also, the large horizontal movement of the more flexible equipment was transferred almost entirely into a vertical motion of the conductor with little transmission to the more rigid equipment. The second type of dynamic response involves high interaction between the interconnected equipment due to a small slack and/or high intensity ground motions. For this case, large vertical acceleration pulses were observed at mid-span of the conductor.

Introduction

Electric power distribution and transmission systems are particularly vulnerable to earthquake loading. In North America, most of these systems were constructed in the 1950s and 1960s, and incorporate several pieces of equipment, such as porcelain bushings or poorly anchored transformers that are particularly vulnerable to earthquake damage. Furthermore, electrical bus conductors (rigid and flexible) are used to interconnect substation equipment components, thereby complicating their structural dynamic response. During recent earthquakes in California, it is believed that significant structural dynamic interaction and equipment damage due to forces transferred through the conductors occurred. The main objective of this paper is to shed some light on this issue by characterizing, through shake table testing, the seismic response substation equipment interconnected by flexible conductors.

---

[1]Professor and [2]Graduate Student Researcher, Department of Structural Engineering, University of California, San Diego, 9500 Gilman Drive, Mail Code 0085, La Jolla, CA 92093.

Scope of Experimental Study

Shake table tests of five pairs of generic substation equipment interconnected by three different flexible conductors with three different levels slackness were performed to evaluate the influence of different conductor assemblies on the structural dynamic response of interconnected substation equipment components. Simulated horizontal ground motions were applied in the longitudinal direction of the bus assemblies by the uniaxial earthquake simulation facility at UC-San Diego (Filiatrault et al., 2000).

Description of Generic Substation Equipment Components

Five different pairs of generic substation equipment were considered for the shake table tests. Each pair of generic equipment was designed to be representative of the range of dynamic properties of actual interconnected substation electrical equipment. Table 1 presents the target dynamic characteristics of the five pairs of generic equipment.

Table 1. Target Dynamic Characteristics of Pairs of Generic Equipment.

Pair	Equipment A			Equipment B		
	Equipment No.	Seismic Weight (kN)	Natural Frequency (Hz)	Equipment No.	Seismic Weight (N)	Natural Frequency (Hz)
1	1	4.45	1.5	3	1.11	5
2	1	4.45	1.5	4	1.56	7.5
3	2	0.89	1.5	3	1.11	5.0
4	2	0.89	1.5	4	1.56	7.5
5	1	4.45	1.5	5	1.56	12.0

For simplicity, steel cantilevered tubular columns, of appropriate stiffness and strength, were anchored to the shake table to represent the equipment components (Filiatrault and Stearns, 2002).

Description of Flexible Conductor Specimens

Three different flexible conductor assemblies (2300 MCM, MCM 1113C and Lupine) were tested with the five pairs of interconnected equipment. The properties of the flexible conductor specimens are given in Table 2. The MCM 1113C conductor assembly incorporated a pair of bundle conductors, while the other two assemblies were tested as single conductors.

Three different slackness values were considered for interconnecting the generic equipment with the flexible conductors. The slackness s is defined as:

$$s = \frac{l_c - l_{ch}}{l_{ch}} \quad (1)$$

where l_c are the conductor length and l_{ch} the chord length, which for a pair of equipment of similar height is the horizontal distance between the ends of the conductor.

Table 2. Characteristics of Flexible Conductor Specimens.

Designation	Construction	Conductor Diameter (mm)	Strand Diameter (mm)	Lay Angle (degree)	Number of Strands
2300 MCM	Steel core + 4 aluminum strand layers	44	4.9	10	61
MCM 1113C	Steel core + 4 aluminum strand layers	32	3.6	10	61
Lupine	All aluminum	46	4.2	10	91

Slackness values of 2, 5 and 10% were considered in the shake table tests, as shown in Fig. 1.

Slackness = 10%　　　　Slackness = 5%　　　　Slackness = 2%

Figure 1. Equipment pair 4 connected by 1113C conductor.

Shake Table Test Program

Three different types of shake table tests were conducted on the pairs of generic equipment models interconnected by flexible conductor assemblies: 1) Frequency Evaluation Tests, 2) Damping Evaluation Tests, and 3) Seismic Tests.

The purpose of the frequency evaluation tests was to identify the natural frequencies and mode shapes of the various pairs of interconnected generic equipment. For this purpose, a low-amplitude 0-40 Hz, clipped-band, and flat white noise excited each configuration. The purpose of the damping evaluation tests was to estimate the first equivalent modal viscous damping of each equipment configuration. In these tests, each pair of generic equipment was excited by a low-amplitude base sinusoidal input at its previously identified fundamental frequency. When a steady-state response was obtained, the input was suddenly stopped and the absolute accelerations at the top of the equipment were recorded during the subsequent free vibration. The first modal damping ratio of the structural configuration was then established by the logarithmic decrement method (Clough and Penzien, 1993). In the seismic tests, the earthquake ground motions excited the pairs of interconnected equipment.

Earthquake Ground Motions

Two recorded components of near-field earthquake ground motions were used for the seismic tests on the shake table: Tabas (1978 Iran earthquake) and Newhall (1994 Northridge, California, earthquake). These two records are representative of earthquakes known to have a high potential for damaging structures and equipment.

The Tabas record was modified using a non-stationary response-spectrum matching technique (Abrahamson, 1997) to match the IEEE 693 standard (Institute of Electric and Electronics Engineers, 1997) target response spectrum for testing. The record was further high-pass filtered using a cut-off frequency of 1.5 Hz so as not to exceed the displacement limit of 150 mm of the shake table.

Preliminary nonlinear dynamic time-history analyses were performed to estimate the response of the interconnected equipment. Based on the results of these preliminary analyses, different intensities, expressed as a percentage of full span (100% span corresponds to the full-scale amplitude), were retained for each ground motion record (Filiatrault and Stearns, 2002).

Results of Frequency Evaluation Tests

Table 3 summarizes the results of the frequency evaluation tests on the stand-alone (unconnected) generic equipment specimens. The fundamental frequencies of all equipment items agree reasonably well with the target frequencies shown in Table 1.

Table 3. Measured Fundamental Frequencies of Generic Equipment Specimens.

Equipment	Fundamental Frequency (± 0.04 Hz)
1	1.60
2	1.56
3	5.00
4	7.38
5	11.33

For all tests at 5 and 10% slackness, the fundamental frequencies of the equipment items are not affected significantly by the presence of the conductor assemblies (Filiatrault and Stearns, 2002). The coupled fundamental frequency of each equipment item was reduced slightly when a conductor assembly was introduced. This reduction in fundamental frequency can be attributed to the added mass of the conductors on both interconnected equipment. This reduction in fundamental frequency was more pronounced for the more rigid interconnected equipment item. This reduction in fundamental frequency was also more pronounced for the equipment connected by the Lupine conductor, which represents the heaviest of the three conductor assemblies used in the tests.

The change in fundamental frequencies was not as consistent for the tests at 2% slackness. For these tests, the behavior of the interconnected equipment was highly nonlinear and depended heavily on the excitation amplitude of the random white noise input signal. For low amplitudes, the behavior was uncoupled and similar to the 5 and 10% slackness tests. When the amplitudes increased, the slackness in the conductor was taken up and the conductor became very tight and transmitted axial vibrations between the interconnected equipment.

Results of Damping Evaluation Tests

For each damping evaluation test, the logarithmic decrement method was applied to a succession of pairs of adjacent response cycles in order to obtain the variation of equivalent damping ratio with displacement amplitude. For this purpose, the displacement amplitude was defined as the mean amplitude of two adjacent response cycles.

The presence of the conductor assemblies increased significantly the damping ratios (typically from below 0.5% of critical to above 1% of critical) of both interconnected equipment for all tests but one (Filiatrault and Stearns, 2002). The higher damping values were obtained for 2% slackness in the conductors. Note, however, that for this slackness value, the vibrational response of both interconnected equipment is nonlinear and a function of the amplitude of the response. Therefore, these apparent damping ratios at 2% slackness may not be representative of the actual free vibration responses of interconnected equipment items.

Results of Seismic Tests

No damage to any of the three flexible conductors was observed during all the seismic tests conducted. Two different types of dynamic response were observed during the seismic tests. The first type involves low interaction between the interconnected equipment due to a large slack (5 and 10%) and/or low intensity ground motions. This low interaction response could be observed by the completely different frequency contents between the relative displacement and absolute acceleration response of each equipment item. Also, for this low interaction response, the absolute vertical acceleration at mid-span of the connector was in phase with the absolute horizontal acceleration at the top of the most flexible equipment. This result indicates that the large horizontal movement of the most flexible equipment was transferred almost entirely into a vertical motion at mid-span of the conductor with little transmission to the most rigid equipment.

The second type of dynamic response involves high interaction between the interconnected equipment due to a small slack (2%) and/or high intensity ground motions. This high interaction phenomenon could be observed by the large vertical acceleration pulses (above 10 g) observed at mid-span of the conductor. These pulses induced peak horizontal forces simultaneously at both ends of the conductor.

Another interesting effect noted during the seismic tests is the transmission of horizontal forces at both ends of the conductor as a function of its slackness. Figure 2 illustrates this observation by showing the net horizontal force time-histories measured by load cells at the top of each equipment items for a series of tests involving equipment Pair 2 connected by a pair of MCM 1113C conductors and excited by the Tabas record at 25% of its amplitude. The results are presented for the uncoupled equipment and for the coupled equipment at 10 and 2% slackness, respectively. When the two equipment items are uncoupled, the force at the top of each equipment item is obviously independent, involving very different frequency contents with peak values occurring at different times. When the conductor is introduced with a slackness of 10%, the force time-histories are still very different because of the lack of strong interaction, but some similar modulations begin to appear with peak values occurring at similar times. When the slackness is reduced to 2%, the two force time-histories have similar frequency contents and modulation. The peaks occur at similar times. Note also that much larger forces occur in tension than in compression since the conductor acts as a tension-only longitudinal spring, since it is much tighter.

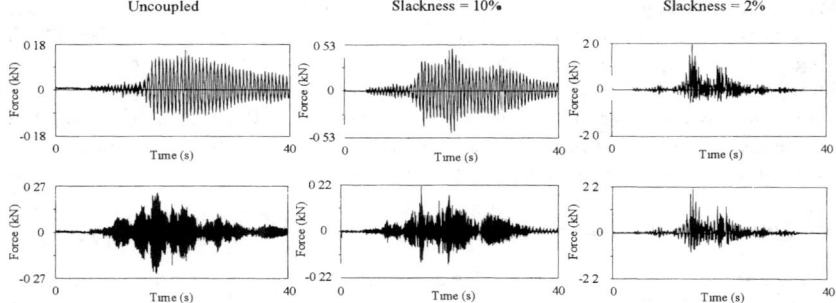

Figure 2. Net horizontal force time-histories at top of equipment items, equipment pair 2, pair of MCM 1113C conductors, Tabas record, 25% span (Note: different vertical scales).

The effect of the various flexible conductors on the dynamic response of the generic equipment specimens can be evaluated by defining a Displacement Amplification Factor (DAF) and an Acceleration Amplification Factor (AAF) as (Der Kiereghian et al., 1999; Filiatrault et al., 1999; Filiatrault and Kremmidas, 2000):

$$DAF = \frac{\text{Maximum Relative Displacement of Interconnected Equipment}}{\text{Maximum Relative Displacement of Stand Alone Equipment}} \qquad (2)$$

$$\text{AAF} = \frac{\text{Maximum Absolute Acceleration of Interconnected Equipment}}{\text{Maximum Absolute Acceleration of Stand Alone Equipment}} \quad (3)$$

The DAF and AAF values computed at the top of equipment pair 5 during the seismic tests are presented in Figures 3 to 8. The results are presented for each ground motion, intensity level, conductor type, and slackness.

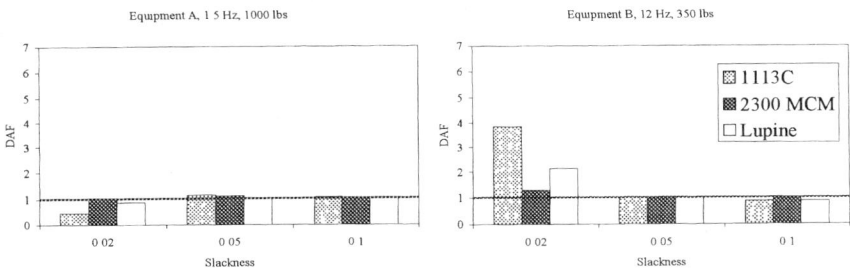

Figure 3. Displacement Amplification Factor (DAF), equipment pair 5, Newhall ground motion, 30% span.

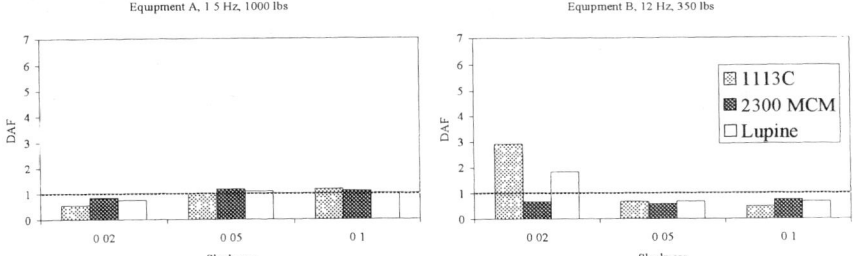

Figure 4. Displacement Amplification Factor (DAF), equipment pair 5, Tabas ground motion, 25% Span.

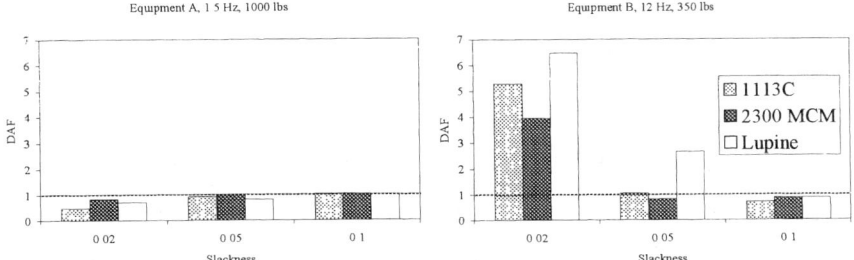

Figure 5. Displacement Amplification Factor (DAF), equipment pair 5, Tabas ground motion, 50% Span.

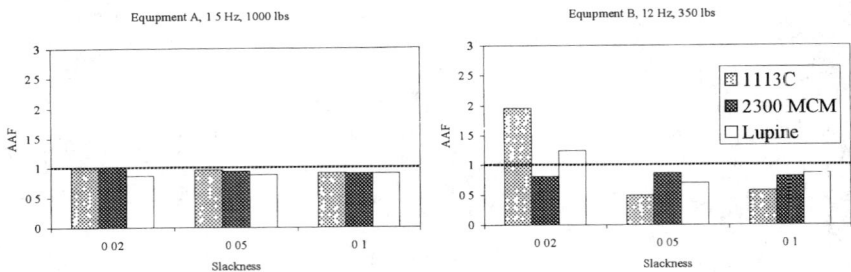

Figure 6. Acceleration Amplification Factor (AAF), equipment pair 5, Newhall Ground Motion, 30% Span.

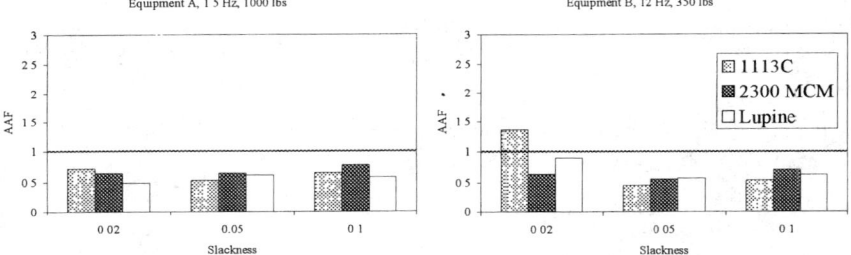

Figure 7. Acceleration Amplification Factor (AAF), equipment pair 5, Tabas ground motion, 25% Span.

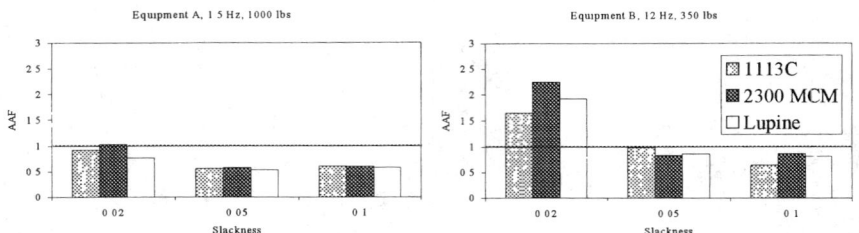

Figure 8. Acceleration Amplification Factor (AAF), equipment pair 5, Tabas ground motion, 50% Span.

Although the presence of the flexible conductors can amplify or reduce the dynamic response of equipment components depending on their dynamic characteristics, slackness of the conductor and the frequency content and intensity of the earthquake ground motion input, the results presented in Figs. 5 to 10 show several clear trends:

- For all slackness values, the dynamic response of the flexible equipment A is not affected appreciably by any of the three flexible conductors tested. The DAF and AAF values are less than equal than unity for almost all cases.
- At 5 and 10% slackness, the dynamic response of the rigid equipment B is generally reduced by the presence of the conductor assemblies. The reduction in AAF is more important than the reduction in DAF.
- At 2% slackness, the dynamic response of the rigid equipment B can be significantly increased once the conductor assemblies become tight and act as tension-only springs. This amplification of the dynamic response increases with the ground motion intensity. The maximum DAF and AAF values measured for the rigid equipment B was over 6 and 2, respectively, for the Equipment Pair 5 under the Tabas ground motion at 50% span.

At 5 and 10% slackness, the forces generated at the top of the interconnected equipment were small. The maximum measured force was 1.43 kN at the top of the stiff equipment B of equipment pair 5 interconnected by the Lupine conductor with 5% slackness and excited by the Tabas ground motion at 50% span. As noted previously, in most cases, the force at the top of the flexible equipment A was larger than the force at the top of the stiff equipment B.

At 2% slackness, the forces generated at the top of the interconnected equipment were an order of magnitude higher that that measured at 5 and 10% slackness. The maximum measured force was 13.1 kN at the top of the stiff equipment B of equipment pair 5 interconnected by the 2300 MCM conductor and excited by the Tabas ground motion at 50% span. For this slackness, the forces at the top of both interconnected equipment were similar.

Conclusions

The shake table testing reported in this paper has provided an opportunity to evaluate the structural dynamic interaction between components of substation equipment interconnected by flexible conductors. In particular, from the results of the shake table tests, it has been observed that in some cases the dynamic response of interconnected substation equipment can be amplified over the response of individual equipment. This conclusion is important since substation equipment items are currently qualified on an individual basis (that is unconnected).

Acknowledgments

The research project described in this report was funded by the PEER Lifeline Directed Studies Program. We greatly appreciated the input and coordination provided by Dr. Michael Riemer from PEER and Mr. Eric Fujisaki from the Pacific Gas and Electric Company (PG&E) during the development of this research project. The supports of Mr. David Chambers from the California Energy Commission and of Mr. Craig Riker from the San Diego Gas and Electric Company (SDG&E) are also gratefully acknowledged. Finally, the authors extend their sincere appreciation to Mr. Howard Matt, Graduate Student Researcher at UC-San Diego, for his assistance in conducting the shake table tests.

References

Abrahamson, N. 1997. *Private Communication*, Pacific Gas and Electric Company, San Francisco, CA.

Clough, R.W., and Penzien, J. (1993). *Dynamics of Structures*, Second Edition, McGraw-Hill, New York.

Der Kiureghian, A., Sackman, J.L., and Hong, K.J. (1999). *Interaction in Interconnected Electrical Substation Equipment Subjected to Earthquake Ground Motions*, Report PEER 1999/01, Pacific Earthquake Engineering Research Center, University of California, Berkeley, Berkeley, CA.

Filiatrault, A., Kremmidas, S., Elgamal, A. and Seible, F. (1999). *Substation Equipment Interaction – Rigid and Flexible Conductor Studies*, Structural Systems Research Project Report No. SSRP-99/09, Department of Structural Engineering, University of California, San Diego, La Jolla, CA, 218 p.

Filiatrault, A. and Kremmidas, S. (2000). "Seismic Interaction Between Components of Electrical Substation Equipment Interconnected by Rigid Bus Conductors", *ASCE Journal of Structural Engineering*, 126(10), 1140-1149.

Filiatrault, A., Kremmidas, S., Seible, F., Clark, A.J., Nowak, R., and Thoen, B.K. (2000). "Upgrade of First Generation Uniaxial Seismic Simulation System with Second Generation Real-Time Three-Variable Digital Control System", *12th World Conference on Earthquake Engineering*, Auckland, New Zealand, Paper # 1674, on CD-ROM, New Zealand Society for Earthquake Engineering, Upper Hutt, New Zealand.

Filiatrault, A., and Stearns (2002). *Electrical Substation Equipment Interaction – Experimental Flexible Conductor Studies*, Structural Systems Research Project Report No. SSRP-02/09, Department of Structural Engineering, University of California, San Diego, La Jolla, CA, 105 p.

Institute of Electrical and Electronics Engineers. 1997. *Recommended practices for Seismic Design of Substations*, IEEE-693 Standard, IEEE Standards Dept., Piscataway, NJ.

Seismic risk management system for electric power facilities

Yoshiharu Shumuta[1]

Abstract

This paper presents a support system for seismic countermeasures of power facilities based on a risk management concept. When determining the priority for the renewal of substation equipment, the risk due to an earthquake is not usually taken into consideration. However, it is emphasized that earthquake risk should also be considered in order to maintain the appropriate performance level for the entire electric power system. The system performance level against an earthquake is compared with that against multi-hazards in order to discuss an acceptable system performance level. The proposed system also includes a model to evaluate business interruption losses to the power industry due to the decrease in power consumption and the reduction of the electric energy supply capacity caused by the earthquake. The proposed system enables us to discuss not only the risk control issues but also the risk financing issues.

Introduction

In Japan, even if the seismic design codes for power facilities are revised after a large earthquake, it is not legally necessary to upgrade existing facilities which are not in accord with the present design code. As a result, many old facilities still remain in existing power systems due to budget constraints. For example, in the heavily damaged area due to the 1995 Hyogo-ken Nanbu earthquake, more than 50% of the total number of substation equipment was constructed before the present seismic design code, which was formally applied to all power industries in Japan. The earthquake damage was concentrated on such old equipment. Some developed countries also have the same problem associated with the renewal of old equipment (Shumuta, 2001).

On the other hand, during a seismic event, the demand side, which receives electric power, such as residential homes and industrial factories also suffers their own seismic damage at the same time. Therefore, not only the seismic risk on the supply side but also that on the demand side should be considered in order to mitigate the management risk of the electric power industries (Shumuta, 2002).

This paper presents an outline of a seismic risk management system for

---

[1] Research Engineer, Geotechnical & Earthquake Engineering Department, Central Research Institute of Electric Power Industry, 1646 Abiko, Abiko-shi, Chiba, Japan, 270-1194; phone +81 4 7182-1181; shumuta@criepi.denken.or.jp

electric power facilities. The proposed system supports the discussions associated with the reasonable seismic performance level of an electric power system and the prioritization for renewal equipment considering the economic impact in the seismic damaged area. The proposed system enables us to discuss not only the risk control issues but also the risk financing issues. Particular attention is paid to the following:
  (i) Identifying exposures and hazards to which the power system potentially might be subject
  (ii) Identifying component and system vulnerabilities
  (iii) Ranking components and substations critical for the functioning of the entire system
  (iv) Estimating potential losses and their associated uncertainties
  (v) Analyzing the benefit cost for upgrading and risk transfer.

Section 2 presents the general concept of the proposed risk management system. Section 3 provides the discussion associated with a reasonable system performance level and with the cost-effectiveness for the seismic upgrading of power facilities based on the proposed system. Section 4 contains the conclusion and identifies areas for further research.

Framework of seismic risk management system

In order to rationalize the seismic countermeasures including the renewal of existing electric power facilities, the data associated with the risk assessment of electric power facilities need to be efficiently collected. **Fig. 1** shows a conceptual figure of the proposed seismic risk management system for electric power facilities. It is an integrated system that links the database, system performance analysis against an earthquake scenario and other multi-hazards, risk control analysis and risk financing analysis programs. The proposed system provides a framework to construct the database for the risk assessment. It also enables us to qualitatively understand the cost-effectiveness of seismic upgrading, power outage cost and business interruption loss due to a seismic event.

Fig. 2 shows the flowchart of the proposed. It consists of five steps.

STEP1 Selection of multi-risks. Electric power facilities are exposed to not only seismic risk but also multi-risks over their life cycles. When determining the seismic upgrade priority of power facilities, facility planners have to consider the balance between the seismic risk and other risk factors such as a typhoon or facility deterioration. The characteristics of multi-risks depend on their local area conditions. For example, a power company distributing power in Tokai area, Japan, is exposed to a high seismic risk. On the other hand, Kyusyu, Japan, has a higher typhoon risk but lower seismic risk. Therefore, in **STEP1,** multi-risks to be considered in the upgrade planning are selected according to the condition of the target area though the proposed system mainly focuses on the seismic upgrading of existing facilities.

STEP2 Data Collection. The proposed system provides the data formats associated

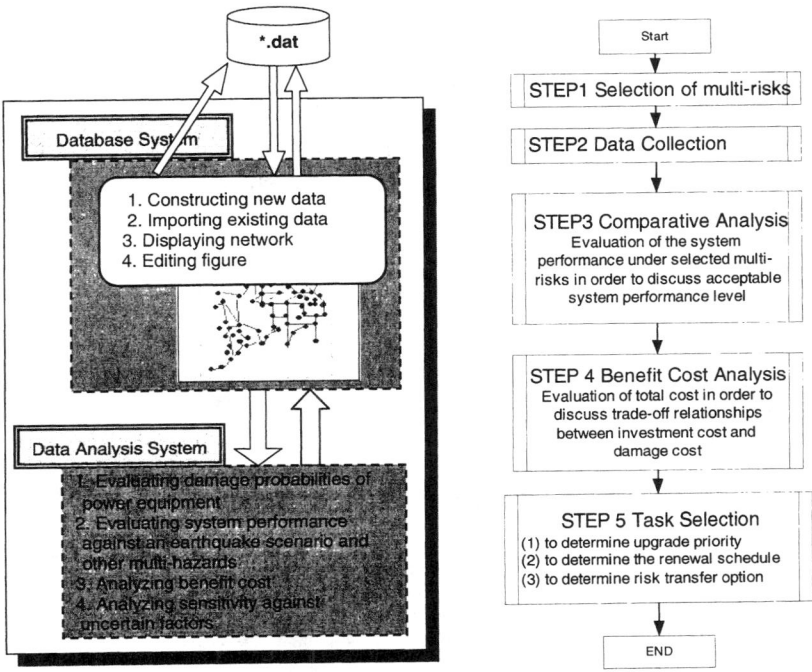

Fig. 1 Conceptual figure of seismic risk management system

Fig. 2 Flow-Chart of seismic risk management system

with the risk assessment and the benefit cost analysis for the seismic countermeasures including upgrade planning. The data consist of some files related to the risk assessment and the benefit costs analysis. **Fig. 3** shows an interface of the database system, which is a part of the proposed system. Its functions are as follows:

(i) The database system enables us to edit and display the transmission network and the substation skeleton. The transmission network is displayed by nodes as substations and links as transmission lines. The substation skeleton is also represented by a node for equipment and links for electric current lines. In order to facilitate the data editing, it is easy to customize the shape of the nodes and links for end users.

(ii) From a transmission network window in **Fig. 3**, we can search the substation skeleton data. For example, if we click on a node in a transmission network window, the specified substation skeleton is displayed in another window.

(iii) The attribute data associated with transmission network and substation skeleton are also displayed and linked with their figures.

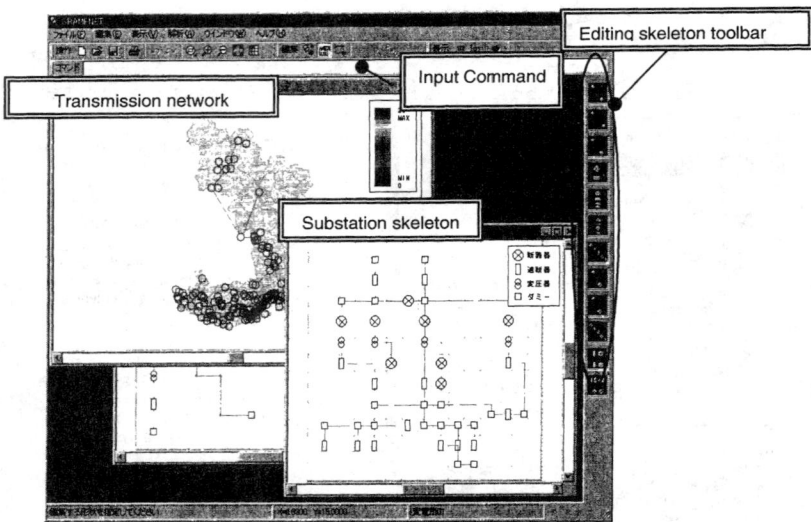
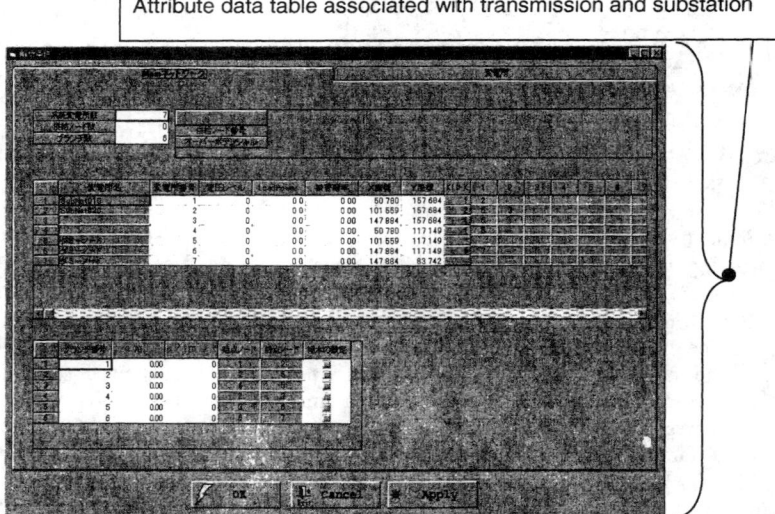

Fig. 3 Interface of the proposed database system

STEP3 Comparative Analysis. The system performance level of an entire power system against multi-hazards is evaluated. The system performance level against multi-risks is defined as

$$SPL^{NT}(k) = \left(1 - \frac{\sum_{k=1}^{nf}\sum_{t=1}^{NT} Po^t(k) \cdot f(P_1^t(k),\cdots P_j^t(k),\cdots P_n^t(k))}{\sum_{t=1}^{NT} DP^t} \right) \times 100 \quad (1)$$

Where NT = total time period; nf = total number of hazards; $SPL^{NT}(k)$ = system performance level against hazard k in NT, $Po^t(k)$ = annual occurrence probability of hazard k at time period t; $P_j^t(k)$ = damage probability of equipment j against hazard k at t. DP^t=total demand power without multi-hazards at t; $f(P_1^t(k),...P_j^t(k),...P_n^t(k))$ = power failure function of target power system at t; The power failure function presented in Eq.(1) can be evaluated only on the condition that $P_j^t(k)=1$ or 0, which simulates the condition of damaged or undamaged power equipment, respectively. Thus, in order to evaluate $f(P_1^t(k),...P_j^t(k),...P_n^t(k))$, it is linearly interpolated by a Monte Carlo simulation proposed by Shumuta (1998).

In **STEP3**, it is discussed whether the system performance level against multi-risks is acceptable or not. In order to determine its acceptable performance level, the final judgment of stakeholders is usually needed. However, the proposed system supports the judgment by comparing the system performance level against the seismic risk with that against other risks.

STEP4 Benefit cost analysis. Fig. 4 shows a benefit cost analysis concept, which presents the trade-off relationship between investment cost and damage cost. The ordinate indicates cost such as present values and the abscissa indicates the performance level of an entire power system against multi-risks. The increment of the investment cost reduces the damage cost and improves the performance level. The proposed model enables us to find an optimum condition to minimize the total cost.

Fig. 5 shows the cost structures to be considered in the benefit cost analysis. The investment cost consists of the prior countermeasure cost (Ex. seismic upgrade cost), the post countermeasure cost (Ex. post disaster mitigation training) and risk transfer cost (Ex. earthquake insurance fee). The damage cost, on the other hand, can be divided into two major categories associated with the damage to the supply side and that to the demand side. The damage cost associated with the supply side consists of the repair and revenue loss costs. The damage cost pertaining to the demand side indicates the power interruption cost. In this paper, the total cost is further divided into the two types shown in **Fig. 5**. The total cost associated with the power industry is defined as the sum of the investment cost and the damage cost pertaining to the supply side. Similarly, that associated with the power market consists of the investment cost and all the damage cost's contents related to the supply and demand sides. The above total costs can be reduced by the feedback cost due to the risk transfer options. The benefit cost analysis tries to find a reasonable condition to minimize the total cost considering the benefit cost of the power industry and power market.

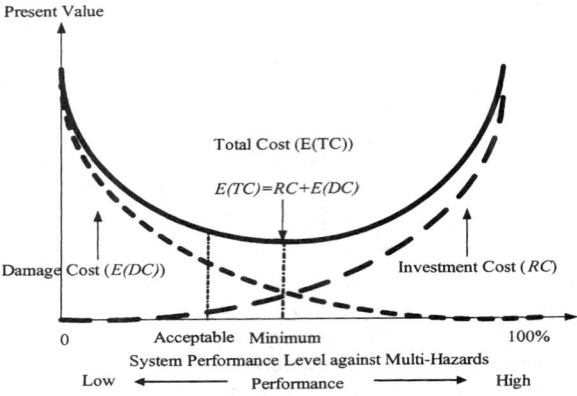

Fig. 4 Benefit cost analysis concept

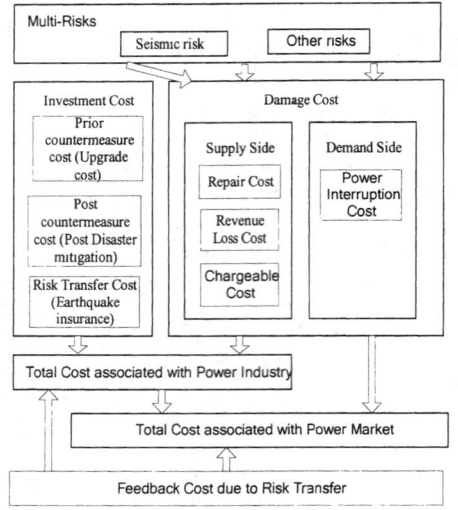

Fig. 5 Target cost to be considered in benefit cost analysis

The earthquake reduces the power sales income, which the power industry should have obtained under normal conditions. In this paper, the power sales income loss allocated as a revenue loss cost in **Fig. 5** is called the Business Interruption Loss (BIL). During an earthquake, the BIL is caused by both damage to the supply and demand sides. In general, the supply and demand sides are simultaneously damaged during an earthquake disaster. To evaluate the BIL, the damage to both sides should be simultaneously considered. **Fig. 6** shows the BIL in local region i caused by the simultaneous seismic damage to both sides. The BIL is determined by the seismic damage conditions on both sides. X_i and X'_i indicate the supply power capacities in local region i before and after the earthquake, respectively. x_{0i} and x_{1i} indicate the electric energy demand in local region i before and after the earthquake, respectively. In **Fig. 6(a)**, if X'_i is smaller than X_i, which indicates that the supply power capacity becomes smaller than the electric energy demand, the BIL ($A'ADD'$) is mainly caused by the damage to the supply side. On the other hand, as shown in **Fig. 6(b)**, if X'_i is larger than X_i, which indicates that the supply power capacity becomes greater than the electric energy demand, the BIL ($A'ADD'$) is mainly caused by the damage to the demand side. The business interruption loss in local region i (BIL_i) is defined as

$$\begin{aligned} BIL_i &= (p - MC) \times (x_{0i} - \min\{x_{1i}, X'_i\}) \\ &= \max\{\underbrace{(p - MC) \times (x_{0i} - x_{1i})}_{\text{BIL due to the damage to demand side}}, \underbrace{(p - MC) \times (x_{0i} - X'_i)}_{\text{BIL due to the damage to supply side}}\} \end{aligned} \quad (2)$$

(a) BIL mainly caused by seismic damage to supply side

(b) BIL mainly caused by seismic damage to demand side

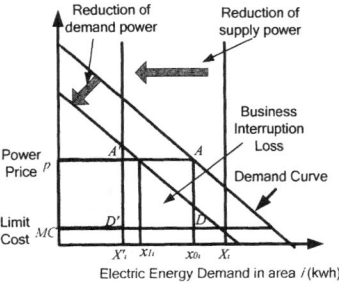
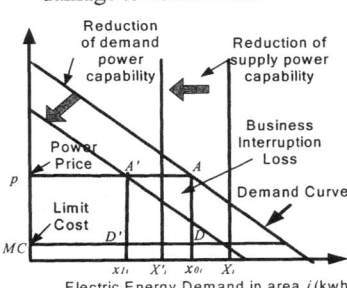

Fig. 6 Business Interruption Losses (BILs) due to damage to supply and demand sides

STEP5 Task Selection. The following decision making is supported based on the results of **STEP3** and **STEP4**.

(i) to determine the upgrade priorities
(ii) to determine the renewal schedule
(iii) to determine the risk transfer option

Numerical examples

Discussion of a reasonable system performance level. **Fig. 7** shows the comparison of the annual System Performance Levels (SPLs) defined by Eq. (1) with the change in the elapsed time from the latest earthquake event in an actual area, Japan. SPL is evaluated under the following three hazard conditions; 1) the Earthquake (M7.0), 2) environmental conditions associated with the deterioration, and 3) the salt stain caused by a typhoon. Note that it is assumed that power facilities progressively deteriorate with increasing age and operating times based on Hoshiya (1989). The abscissa shows the elapsed time from the occurrence of the latest earthquake. The ordinate indicates the annual system performance. In **Fig. 7**, it was assumed that the occurrence probability of the earthquake was in accord with the Brownian Passage Time (BPT) distribution. Therefore, the SPL for the earthquake becomes lower with an increase in the elapsed time. On the other hand, the SPLs for the other two hazards always become constant regardless of the elapsed time. This is because it is assumed that three multi-hazards independently occur and there are no relationships among the elapsed time and the other two hazards. In the target area, it has been 406 years since the last earthquake occurred (at the time of 2002).

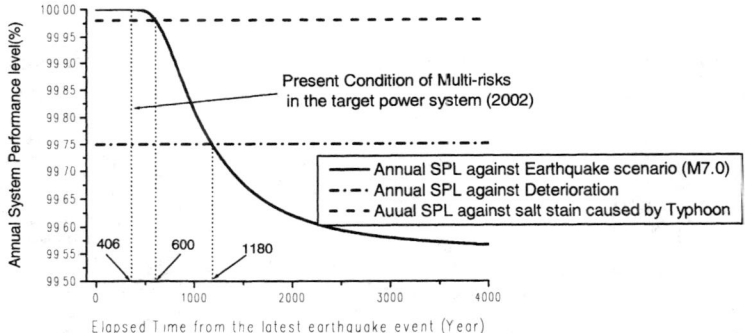

Fig. 7 Comparison of the system performance levels (SPLs) against the multi-risks with change in the elapsed time from the latest earthquake event

Fig. 7 suggests that at the present condition of the target power system, the seismic risk is the lowest level in the three risks. However, after 194 years, at which the elapsed time is 600 years, the seismic risk is higher than the salt stain risk. Furthermore, after 774 years, at which the elapsed time reaches 1180 years, the seismic risk becomes the highest of the three risks. To determine the renewal priority of the existing power facilities, the existing renewal planning usually neglects the seismic risk. This result supports this renewal policy if the acceptable performance level is lower than the SPL under the salt stain risk. However, if the elapsed time exceeds about 1180 years, the earthquake causes the highest functional loss of the three hazards. In this case, the seismic risk should be considered for the renewal planning to maintain an acceptable performance level of the target power system.

Discussion of trade-off relationships between investment cost and damage cost.
Fig. 8 and **Fig. 9** show tradeoff relationships between the investment cost and the damage cost for the renewal planning of 125 actual substations against the three hazards over 25 years. The total analytical time frame, NT, is assumed to be 25 years. **Fig. 8** focuses on the cost components associated with the power industry in **Fig. 5**. On the other hand, **Fig. 9** considers the cost components associated with the power market. As mentioned before, the differences between **Fig. 8** and **Fig. 9** is whether the power interruption cost is considered or not as one of the cost components of the damage cost. **Fig. 8**, which does not include the power interruption cost, suggests that in order to minimize the total cost, no renewal cost should be invested in the 25 years.

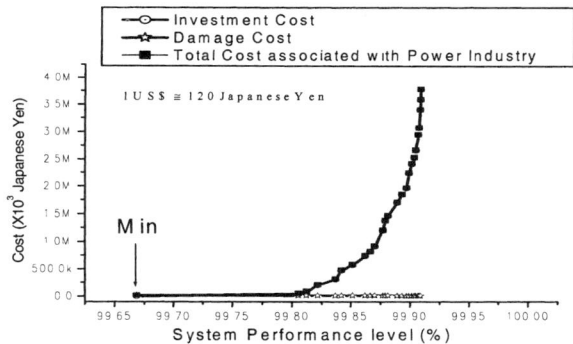

Fig. 8 Tradeoff relationship between renewal cost and damage cost associated with Power Industry over 25 years

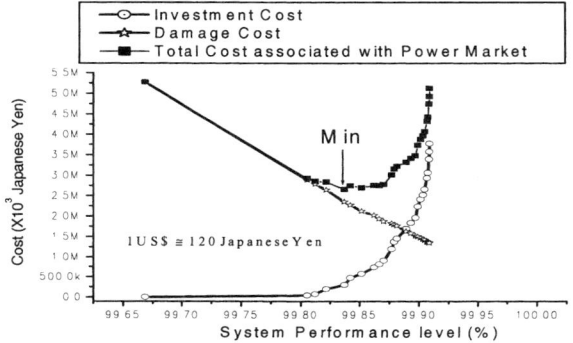

Fig. 9 Tradeoff relationship between renewal cost and damage cost associated with Power Market over 25 years

From an economic point of view for the power industry, the present system performance level is too reliable to upgrade any power facilities for 25 years. In contrast, **Fig. 9**, which indicates the power interruption cost, shows that 300 million Japanese yen should be invested as the renewal cost to minimize the total cost associated with the power market.

Concluding remarks

This paper presented the seismic risk management system for electric power facilities. The proposed system enables us to discuss not only the risk mitigation strategies including renewal planning of old power equipment but also the issue associated with business interruption losses (BIL). In order to discuss the reasonable system performance level of the entire power system, the proposed system was provided from the standpoint of the comparison of the system performance during multi-hazards. In numerical examples, it was demonstrated that the seismic risk might be higher than the other scheduled risks according to the local condition though the seismic risk has not considered as an important risk factor in actual renewal planning. As an example of the benefit cost analysis, the target power system has no need to upgrade all substations to minimize the total cost associated with the power industry. On the other hand, to minimize the total cost associated with the power market, it should upgrade them that were appropriately selected by the proposed model.

As a future subject, the unit power interruption cost should be improved. In Japan, there are few studies related to the unit power interruption cost, especially during and after an earthquake event.

References

Hoshiya, M. et al., (1989) Study on life cycle estimation of substation equipment, Technical Report, Association for the Development of Earthquake Prediction (in Japanese).

Shumuta,Y., (1998) Study on seismic retrofit planning method of substation facilities on the basis of the seismic risk assessment of electric power system, CRIEPI report U33 (in Japanese).

Shumuta,Y., and Tohma, J,. (2001) Long-term infrastructure renewal planning with a focus on power lifeline –Application to Japan-, Proceedings of the 4th Multi-lateral Workshop on Development of Earthquake and Tsunami Disaster Mitigation Technologies and Their Integration for the Asia-Pacific Region (EqTAP), EDM.

Shumuta, Y., Tatano, H., Yamano, N., and Kajitani, Y., (2002) Measuring business interruption losses in electric power industry caused by an earthquake, The 49th annual North American meetings of the regional science association international, November 14-16,2002, San Juan, Puerto Rico.

A Comparison of Seismic (Dynamic) and Static Load Cases for Electric Transmission Structures

Michael J. Riley, EIT[1], Leon Kempner Jr., PE, PhD[1]
Wendelin H. Mueller III, PE, PhD[2]

Abstract

This paper reports on research in progress. A preliminary investigation into this topic occurred in 1999, [Gobo 1999]. A previous publication reported on two towers and one seismic event, [Riley 2002]. This paper expands the study to three towers and 4 seismic events.

Introduction

The electric transmission grid is, by its nature, a distributed system. Thus, it is reasonable to assume that a transmission structure will be in the vicinity of strong ground motion when a seismic event occurs. Because of this, there has been some interest in adding a seismic load case to the traditional load cases already required for the design of electric transmission structures. Structural engineers, familiar with the design of electric transmission structures, contend that this is not necessary, because traditional load cases control the design and produce a safe and reliable structure. It is argued, from a practical point of view, that transmission structures must be the most common structure to be subjected to seismic loads and yet there has been no reported earthquakes that have caused transmission structures to collapse, because of member overstress due to ground motion. Earthquake related transmission line problems have been associated with foundation failures as a result of large permanent ground displacement or ground failures. However, transmission structures have failed due to traditional load cases, e.g. extreme wind, extreme ice, and extreme unbalanced conductor tension.

The purpose of this study is to determine under what circumstances, if any, a seismic (dynamic) load case controls the design of transmission structures, relative to traditional load cases. Lattice steel towers and steel poles are the transmission structures studied. The failure mode selected for the lattice steel tower is the tension and compression capacity of the primary leg members. The tower primary leg member, which is the main support for this structure, was chosen because the concern is the collapse of the tower. The failure mode selected for the steel pole is the base moment capacity of the pole, where again the collapse of the tower is the concern.

---

[1] Structural Engineer, Bonneville Power Administration, PO Box 61409, Vancouver, WA 98662
[2] Professor of Civil Engineering, Portland State University, PO Box 751, Portland, OR 97207

Towers Studied

The electric transmission structures used in this study are a 500 kV double circuit Suspension River Crossing Tower (RCT), a 500 kV double circuit Standard Suspension Transmission Tower (STT) and a 500 kV double circuit Suspension Transmission Pole (TP). The RCT and STT towers are lattice steel structures and the TP is a steel tubular tower. Figure 1 shows a sketch of the transmission structures modeled in this study. These transmission structure types were chosen because they are tall structures that will result in low natural frequencies and they support conductor systems with significant mass. The low natural frequency parameters of the selected transmission line components could interact and produce a worse case scenario for design.

Suspension River Crossing Tower (RCT)

The RCT tower is 143.3 m (470 ft.) tall supporting a triple bundle, 4.45 cm (1.752 in) diameter, Special AACSR conductor system. The tower weight is 2,491 kN (560 kips) and the conductor weighs 37.5 N/m (2.57 lbs./ft.). The tower is used by the Bonneville Power Administration (BPA) in a river crossing application. However, the analysis was done on a tower/conductor configuration that is not a specific river crossing, but was modeled as a worse case that could possibly be found crossing large rivers along the west coast.

A RCT is a unique structure, within a distributed system, that is located in a very specific location. This configuration was included in this study because it was reported in literature [Li 1991] that, for the design of river crossing towers with long spans, earthquake loading could be significant. The lowest natural frequencies of the tower/conductor system in the transverse and longitudinal directions are 0.87 Hz and 0.94 Hz respectively. The lowest natural frequency of the conductor/tower system is 0.08 Hz. The tower fundamental frequency without the conductor is 1.4 Hz.

Standard Suspension Transmission Tower (STT)

The STT tower is 70.1 m (230 ft) tall supporting a triple bundle, 4.06 cm (1.6 in.) diameter, ACSR Chukar conductor system. The tower weight is 218 kN (49 kips) and the conductor weighs 30.1 N/m (2.06 lbs/ft.). The tower and conductor system is used by BPA in the Pacific Northwest electric transmission grid. The configuration was chosen to determine the effects of a typical tower and conductor system compared to the RCT configuration. The lowest natural frequencies of the tower/conductor system in the transverse and longitudinal directions are 0.75 Hz and 2.9 Hz respectively. The lowest natural frequency of the conductor/tower system is 0.18 Hz. The tower fundamental frequency without the conductor is 1.3 Hz.

Standard Suspension Transmission Pole (TP)

The TP tower is 53.3 m (175 ft.) tall supporting a triple bundle, 4.06 cm (1.6 in.) diameter, ACSR Chukar conductor system. The pole weight is 302 kN (68 kips) and the conductor weighs 30.1 N/m (2.06 lbs/ft.). The pole and conductor system is used by BPA in the Pacific Northwest electric transmission grid. The configuration was chosen to expand the scope of the study to include steel poles. The lowest natural frequencies of the pole/conductor system in the transverse and longitudinal directions are 0.16 Hz and 0.18 Hz respectively. The lowest natural frequency of the conductor/pole system is 0.18 Hz. The tower fundamental frequency without the conductor is 0.58 Hz.

Computer Model

The computer program used was SAP2000 Non-linear Version 7.4, [SAP2000 2000]. This program is a standard structural application tool for BPA.

Tower Models

The analytical models consisted of two transmission structures and three equal spans of conductor. The towers had pinned base supports and the poles had fixed base supports. The conductors were hung from the transmission structures with insulators in a "V" string configuration. Pinned supports were modeled as the boundary conditions at the ends of the conductor system. This boundary condition represents the attachment to the deadend towers. Figure 2 gives an overview of the transmission structure/conductor model and the specific dimensions for the conductor spans. This configuration was chosen as a possible worse case configuration that could be expected in the field. It should be noted that the symmetry of the model could result in higher loads in the transmission structures, because the loads can be amplified by vibration coupling of the conductor spans.

The towers, poles and conductor system were modeled using three dimensional beam elements with lumped mass at each of the end joints. The insulators were included in the model as beam elements with end releases. The conductor spans were divided into 35 beam elements per conductor span for the RCT and 20 beam elements for both the STT and TP. A P-Delta force equal to the initial tension was used to pretension the conductor elements.

Seismic

The earthquake time histories selected for this study were obtained from a collection of time histories used by the Oregon Department of Transportation and developed by Geomatrix, [ODOT 1995]. Earthquake time histories for four seismic zone Site Groups (SG) were developed that represent the contributions from different seismic sources: Cascadia interface subduction zone, Cascadia interslab, and crustal events. Site Group 1 represents the interface source. Site Group 2 has a significant contribution from crustal sources, but with a long period contribution from the interface source, source distance greater than 50 km (31 miles.). Site Group 3 has a large contribution from crustal sources with a significant contribution from the interface source, particularly at long periods. Site Group 4 contains mostly contributions from crustal sources. Time histories for all four Site Groups were developed for 500, 1000, and 2500-year return periods. The time histories are scaled to provide equal-hazard spectra, equal probability of being exceeded, over the spectra period of interest (0.3333 to 33.33 Hertz). These time histories were obtained from actual earthquake records modified to produce equal-hazard spectra.

One 2500-year return period time history from each of the four Site Groups was selected for this study. Table 1 describes the selected records. A time history trace and response spectrum for each record is shown in Figures 3 through 6. The response spectra are plotted with respect to IEEE 693 High Seismic Qualification Level response spectrum, [IEEE 693, 1997].

Table 1. Time History Parameters (2.54 cm = 1 in.)

Record ID	Earthquake	Yr 19	SG #	M	Time Step	No. of Pts	Peak Accel. (g)	Peak Displ. (cm)	Time (sec.)
sf279254	San Fernando	71	4	6.6	0.02	849	+0.79 -1.02	+11 -7	17.0
lpg1000	Loma Prieta	89	3	7	0.02	1074	+1.46 -1.26	+14 -21	21.5
m039180	Michoacan	85	2	8	0.01	4955	+0.89 -0.99	+36 -26	49.5
ccuF070	Valparaiso	89	1	8	0.005	13158	+0.79 -0.73	+14 -14	65.8

The ground motion was applied in the transverse direction to the transmission line. This direction was chosen, because it was considered to be the worse case, in that the "V" string insulator is stiffer in this direction, allowing it to transfer energy from the motion of the conductors to the transmission structure.

The results of the analysis are based on the 2,500-year return period event that was reduced by 2/3 to change the return period of the earthquake to an equivalent 500-year return period, [NEHRP 1997].

Dynamic Analysis

The dynamic analysis used the Ritz technique and 5% damping. The Ritz technique was chosen because it searches for and uses the critical "modal vectors" thus avoiding the requirement that the user select which "vectors" to include in the analysis, [Wilson 2000]. The Ritz vectors used were the accelerations in the transverse, longitudinal, and vertical directions to the transmission line. Again, this choice minimized the number of decisions by the analyst, in that no specific load cases were needed to base the choice of the Ritz Vectors. The results of the analysis were accepted when the participating mass was above 90% in each of the principal directions. This level of participating mass is acceptable for obtaining reasonable results.

The purpose of this study was to determine if there would be any increased reliability of an electric transmission grid by adding a seismic load case to the traditional load cases already required for the design of transmission structures. The potential collapse of the transmission structure was chosen as the determining criterion. Therefore, for the lattice towers, the axial loads in the main leg and for the pole, the base moments are the values used in comparing load cases.

Observations and Recommendations

The work presented herein was an effort to determine if there was a need to include a seismic load case in the design of electric transmission structures. As described above, two conductor-tower systems and one conductor-pole system were included in the analysis. Table 2 shows the load demand, based on traditional load cases, for the towers and pole. Table 3 gives the seismic load demand, due to each seismic event, for the towers and pole. Negative values represent compression and positive values represent tension. The seismic loads are given for 2,500-year return period and a 500-year return period. A 2/3 factor was used to convert from one return period to the other. Table 4 gives the comparison between traditional load case demand and seismic load case demand. A demand ratio (DR) is given in the table. The DR is the ratio of traditional load case demand to seismic load case demand. For a DR greater than one, the traditional load case controls the design.

Table 2. Traditional Load Case Demand (4.45 kN = 1 kip, 0.305 m = 1 ft.)

Transmission Structure	Traditional Load Case Demand	Controlling Load Case
RCT	-6145 kN / +4646 kN	Extreme Wind on Bare Wires
STT	-1130 kN / +837 kN	Extreme Wind on Bare Wires
TP	10571 kN-m	Extreme Wind on Bare Wires

Using results reported in Table 4, traditional load cases control the design of 500 kV double circuit standard suspension transmission towers, whereas a seismic load case controls the design of 500 kV double circuit suspension river crossing towers and suspension transmission poles for the transmission line parameters studied.

Table 5 shows the transition g-level, for the 500-year return period, when the seismic demand controls each of the tower/conductor systems for the four time histories used. The g-levels were obtained by linearly scaling the results shown in Table 3.

Table 3. Seismic Load Case Demand (4.45 kN = 1 kip, 0.305 m = 1 ft.)

Transmission Structure	Seismic Load Case Demand 2,500-year Return Period	Seismic Load Case Demand 500-year Return Period	Seismic Load Case
RCT	-3289 kN / +2114 kN	-2194 kN / +1411 kN	sf279254
	-5976 kN / +4762 kN	-3983 kN / +3173 kN	lpg1000
	-9483 kN / +8353 kN	-6323 kN / +5567 kN	m039180
	-4187 kN / +2977 kN	-2790 kN / +1985 kN	ccuF070
STT	-476 kN / +271 kN	-316 kN / +182 kN	sf279254
	-788 kN / +583 kN	-525 kN / +387 kN	lpg1000
	-1295 kN / +1090 kN	-863 kN / +725 kN	m039180
	-574 kN / +369 kN	-383 kN / +245 kN	ccuF070
TP	26,232 kN-m	17,488 kN-m	sf279254
	50,769 kN-m	33,845 kN-m	lpg1000
	85,627 kN-m	57,085 kN-m	m039180
	34,464 kN-m	22,977 kN-m	ccuF070

Table 4. Traditional Load Case Demand vs. Maximum Seismic Load Case Demand (4.45 kN = 1 kip, 0.305 m = 1 ft.)

Structure	Traditional Load Case Demand	Seismic Load Case Demand 2,500-year Return Period Demand	DR	Seismic Load Case Demand 500-year Return Period Demand	DR
RCT	-6145 kN / +4646 kN	-9483 kN / +8853 kN	0.57	-6323 kN / +5567 kN	0.83
STT	-1130 kN / +837 kN	-1295 kN / +1090 kN	0.77	-863 kN / +725 kN	1.15
TP	10,571 kN-m	85,627 kN-m	0.12	57,085 kN-m	0.19

Table 5. Seismic Demand 500-year Return Period g-Level Control Limit

Transmission Structure	sf279254 Control	lpg1000 Control	m039180 Control	ccuf0707 Control
RCT	1.9 g	1.5 g	0.6 g	1.2 g
STT	2.5 g	2.0 g	0.9 g	1.6 g
TP	0.4 g	0.3 g	0.1 g	0.3 g

The preliminary result of this study suggests that, for 500 kV double circuit suspension river crossing towers (RCT) and standard suspension transmission towers (STT), traditional load demand provides reasonable capacity to resist the seismic load demand. The 500 kV suspension transmission pole (TP) is the more seismically sensitive structure of the three towers studied. The controlling seismic load demand g-levels are significantly lower relative to the other two towers. The TP tower/conductor system used in the study represents the application limits of a single pole 500 kV tower. Additional TP configurations will be studied to determine the sensitivity of this tower type to the seismic demand of an earthquake.

The authors recognize that this paper reports on a limited number and type of transmission structure/conductor systems and only a select few of the members in the towers. This work is not comprehensive, therefore is considered preliminary. Work is continuing in such areas as: studying the load demand in all members, incorporating a cable element and non-linear effects, applying the ground motion in multiple directions and using real seismic events vs. modified seismic events.

Although the worldwide performance of electric transmission towers has been very good, this issue needs further study. Historically, damage has been limited to foundation failures caused by landslides, ground fracture, slope failures, and liquefaction. The final goal of this research is to clarify the significance of seismic affects on electric transmission structures and develop a simplified analysis method that can be used to evaluate seismic load effects in comparison to traditional load cases, e.g. extreme wind, extreme ice, and extreme unbalanced conductor tensions.

This research paper represents the continued effort [Gobo 1999, Riley 2002] by BPA to provide engineering information to advance the knowledge of electric transmission line tower seismic performance.

References

Gobo T.G., Kempner L. Jr., and Mueller W.H. III, 1999, "Performance of Electric Transmission Structures During Earthquakes", Proceeding: Structural Engineering in the 21st Century, American Society of Civil Engineers, Reston, VA.

Riley M.J., Kempner L. Jr., and Mueller W.H. III, Gobo T.G., 2002, "A Comparison of Seismic (Dynamic) and Static Load Cases for Lattice Electric Transmission Towers", Proceeding: Electrical Transmission in a New Age, American Society of Civil Engineers, Structural Engineering Institute, Omaha, NB.

ODOT, 1995, "Seismic Design Mapping State of Oregon, Project No. 2442", Oregon Department of Transportation, Salem, OR 97310.

NEHRP, 2001, "Recommended Provisions for Seismic Regulations for New Buildings and Other Structure", 2000 edition, FEMA 368.

Li H, Wang S, Lu M., and Wang Q., 1991, "Aseismic Calculation for Transmission Towers, Lifeline Earthquake Engineering," Monograph No. 4, American Society of Civil Engineers, Reston, VA.

SAP2000, 2000, NonLinear Version 7.40, Computers & Structures Inc., Berkeley, CA 94704.

IEEE Std 693, "Recommended Practice for Seismic design of Substations", Institute of Electrical and Electronics Engineers, Inc., 1997.

Wilson, et.el., 2000, "Three Dimensional Static and Dynamic Analysis of Structures", Computers & Structures Inc., Berkeley, CA 94704.

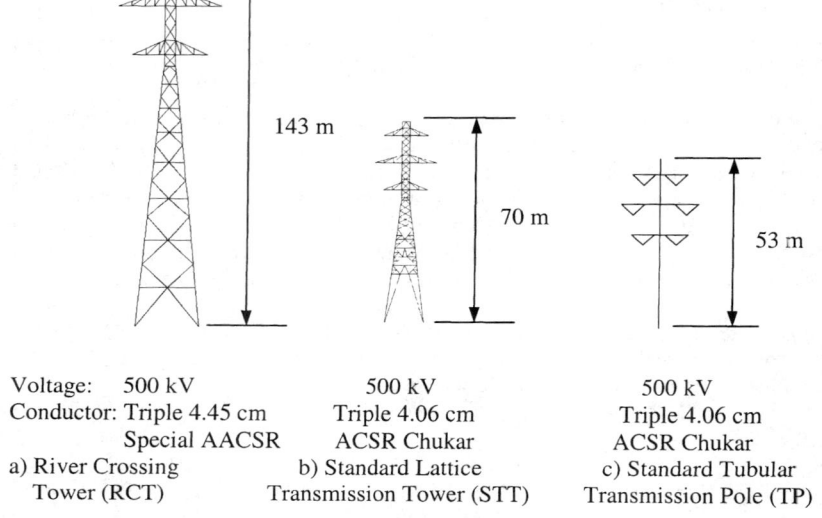

Voltage: 500 kV	500 kV	500 kV
Conductor: Triple 4.45 cm	Triple 4.06 cm	Triple 4.06 cm
Special AACSR	ACSR Chukar	ACSR Chukar
a) River Crossing	b) Standard Lattice	c) Standard Tubular
Tower (RCT)	Transmission Tower (STT)	Transmission Pole (TP)

Figure 1. Tower Configurations (0.305 m = 1 ft.)

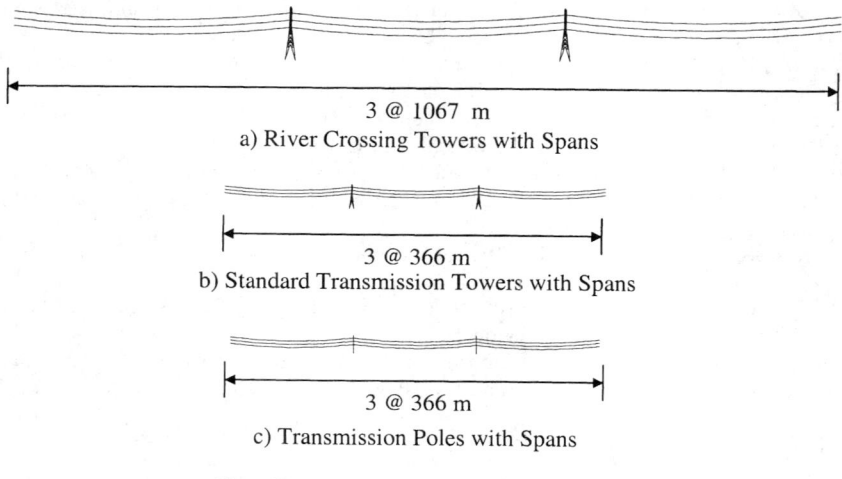

3 @ 1067 m
a) River Crossing Towers with Spans

3 @ 366 m
b) Standard Transmission Towers with Spans

3 @ 366 m
c) Transmission Poles with Spans

Note: Free ends of Conductor are Pinned.

Figure 2. Modeled Transmission Line Spans (0.305 m = 1 ft.)

DISASTER RESPONSE FOR LIFELINE SYSTEMS 695

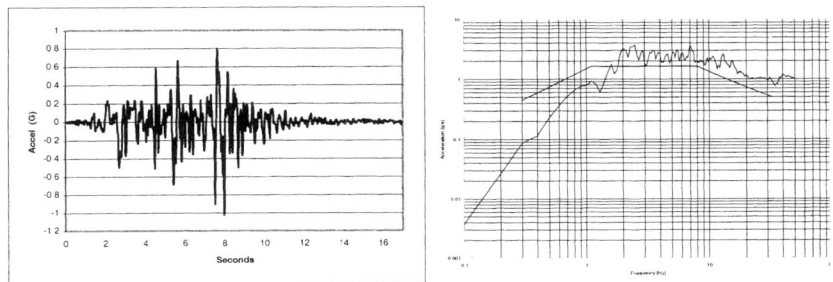

(a) Acceleration Time History   (b) Acceleration Response Spectrum

Figure 3 - sf272954 Site Group 4

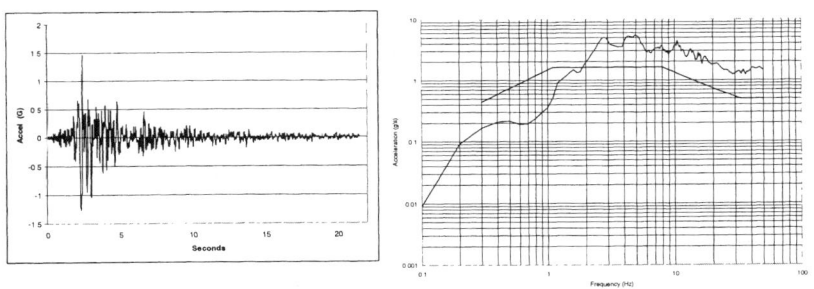

(a) Acceleration Time History   (b) Acceleration Response Spectrum

Note: The previous paper [Riley 2002] reported a peak acceleration of approximately 3g, which was a clerical error

Figure 4 - lpg1000 Site Group 3

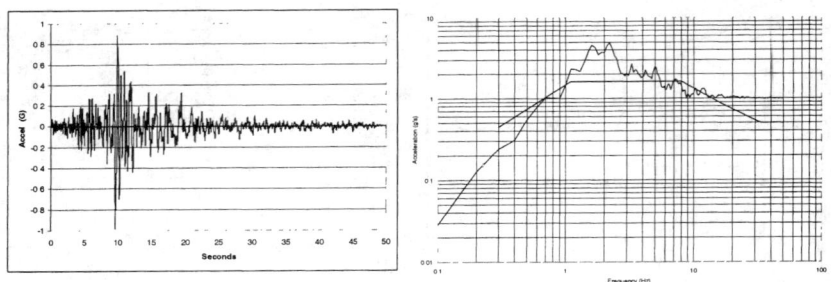

(a) Acceleration Time History (b) Acceleration Response Spectrum

Figure 5 – m039180 Site Group 2

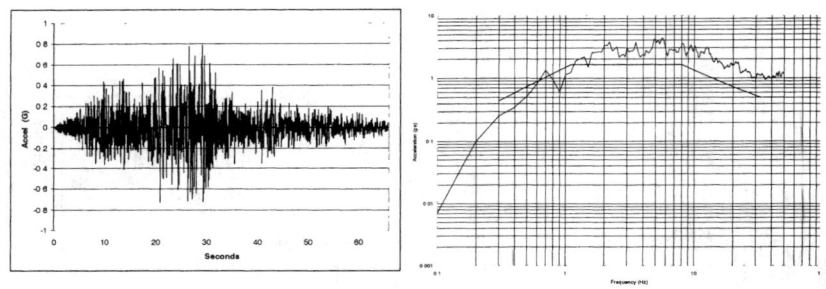

(a) Acceleration Time History (b) Acceleration Response Spectrum

Figure 6 – ccuf070 Site Group 1

Simplified Seismic Calculation Method for Coupled System of Transmission Lines and Their Supporting Tower

Hong-Nan Li [1] Wen-Long Shi [2] Su-Yan Wang [3]

Abstract

In this paper, according to the simplified model of aseismic calculation of the transmission tower presented by the authors, a computer program compiled is used to calculate the seismic responses of high voltage transmission tower under the action of earthquake. The results show that the effects of the lines on the high voltage transmission tower increase as the span between two towers becomes large. And the limit span of lines that should be considered to be important in the seismic response of high voltage transmission tower is determined based on the analyses of earthquake responses of lots of transmission towers. At the end of this paper, the simplified method of aseismic calculation is presented for the transmission tower-lines coupled system.

Introduction

The high voltage system consisting of transmission lines and their supporting towers is an important lifeline engineering one. The previous researches about high voltage systems mainly focused on the static loading, impulsive loading caused by the break of conductor and equivalent static wind load (ASCE, 1982, 1991; DL/T 5092, 1999; Li, 1990). However, the studies on the response of system by dynamic load are very little. The calculation method of considering the effect of transmission lines for the system is not included in the *Design Regulation for 110~500KV Aerial Cables* (DL/T 5092-1999). The effect of transmission lines can be neglected for the dynamic analysis of short span transmission tower for the mass of lines is very small

---

[1] Prof., Ph.D., Dean, School of Civil and Hydraulic Engineering Dalian University of Technology, Dalian 116024, P. R. China; Phone: 86-411-4708512-208; hnli@dlut.edu.cn
[2] Ph. D. Student, Dept. of Civil Engineering, Tongji University, Shanghai 200092, China; swlsxf@sina.com
[3] Prof., School of Civil and Hydraulic Engineering Dalian University of Technology, Dalian 116024, P. R. China

compared with the one of tower. But the mass of conductor is very considerable for long-span transmission lines system. The authors (1990, 1997, 2003a, 2003b) has presented the reasonable aseismic calculation method and analyzed the seismic response of the system, which indicated the important effect of conductor on the dynamical characteristic of transmission tower. But there is no definite conclusion about the limitation of span for which the effect of conductor can be neglected.

In this paper, the effect of the conductors with various spans on the aseismic performance of the transmission tower system is investigated and the limitation for considering the effect of conductor is determined. The simplified aseicmic calculation method is also presented for the transmission tower-lines coupled system in this paper.

Model of Seismic Calculation and Equation of Motion

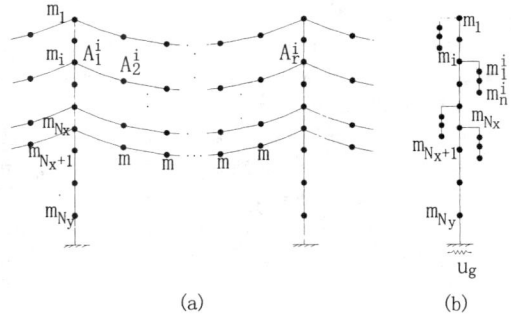

Figure 1. Diagram of transmission tower-lines coupled system

The vibration model of one-span transmission tower system in lateral and longitudinal direction can be simplified as shown in Figure 1. The vibration of transmission lines and tower system is elasticity-gravity coupled under seismic. Considering the interaction of transmission lines and tower, the equations of motion for the system in the lateral and longitudinal directions can be derived as:

$$[M]_{(x)}\{\ddot{x}\}_{(x)} + [K]_{(x)}\{x\}_{(x)} = -[M]_{(x)}\{E\}\ddot{x}_g(t) \tag{1}$$

$$[M]_{(y)}\{\ddot{x}\}_{(y)} + [K]_{(y)}\{x\}_{(y)} = -[M]_{(y)}\{E\}\ddot{y}_g(t) \tag{2}$$

where, $\{x\}_{(x)}$ and $\{x\}_{(y)}$ are the displacement vectors of the system in lateral and longitudinal direction, respectively; $\{E\}$ denotes the unite vector; $\ddot{x}_g$ and $\ddot{y}_g$ express the ground acceleration in the later and longitudinal directions, respectively;

$[M]_{(x)}$ and $[M]_{(y)}$ are the mass matrix of system in lateral and longitudinal, respectively; $[K]_{(x)}$ and $[K]_{(y)}$ mean stiffness matrix of the system in the lateral and longitudinal directions, respectively. The detailed expressions of the above matrixes are shown in References (2003a, 2003b).

State-Space Equation of System

The program for calculating the response of the transmission tower system is compiled in MATLAB. The time history analysis method is used in the program and the damping is assumed to be Rayleigh form. To express the equation (1) and (2) with same form, the subscripts which denote the direction are removed and the ground motion acceleration is expressed as $\ddot{u}_g$. After the damping is introduced, the Equation (1) and (2) can be written with an uniform equation as follow:

$$[M]\{\ddot{x}\} + [C]\{\dot{x}\} + [K]\{x\} = -[M]\{E\}\ddot{u}_g \tag{3}$$

where, $[C]$ is the damping matrix of the system. The equation (3) can be rewritten as:

$$\{\ddot{x}\} = -[M]^{-1}[C]\{\dot{x}\} - [M]^{-1}[K]\{x\} - [M]^{-1}[M]\{E\}\ddot{u}_g \tag{4}$$

Let $\{z\} = \begin{Bmatrix} x \\ \dot{x} \end{Bmatrix}$, $\{\dot{z}\} = \begin{Bmatrix} \dot{x} \\ \ddot{x} \end{Bmatrix}$, and

$$\{\dot{z}\} = \begin{Bmatrix} \dot{x} \\ \ddot{x} \end{Bmatrix} = [A_1]\{z\} + [B_1]\{v\} = [A_1]\begin{Bmatrix} x \\ \dot{x} \end{Bmatrix} + [B_1]\{v\} \tag{5}$$

The following equation can be deduced from Equation (4) and $\{\dot{x}\} = \{\dot{x}\}$:

$$\begin{Bmatrix} \dot{x} \\ \ddot{x} \end{Bmatrix} = \begin{bmatrix} 0 & E \\ -[M]^{-1}[K] & -[M]^{-1}[C] \end{bmatrix} \begin{Bmatrix} x \\ \dot{x} \end{Bmatrix} + \begin{Bmatrix} 0 \\ -\{E\} \end{Bmatrix} \ddot{u}_g \tag{6}$$

So, $[A_1] = \begin{bmatrix} 0 & E \\ -[M]^{-1}[K] & -[M]^{-1}[C] \end{bmatrix}$; $[B_1] = \begin{Bmatrix} 0 \\ -\{E\} \end{Bmatrix}$; $\{v\} = \{\ddot{u}_g\}$

The state equation can be expressed as:

$$\begin{cases} \{\dot{z}\} = [A_1]\{z\} + [B_1]\ddot{u}_g \\ \{z\} = [A_2]\{z\} + [B_2]\{v\} \end{cases} \tag{7}$$

where, $[A_2]=[E]$, $[B_2]=0$.

Calculation and Analysis

Five transmission towers with various shapes are considered in this paper to analyze the effect of conductors on the dynamical response of the system, as shown in Figure 2. All of the structural systems of the transmission towers in Figure 2 are steel truss with square cross section. Each tower is simplified as seven lumped masses and the conductors of each span are simplified as four lumped masses. The parameters of the transmission towers are shown in Table 1.

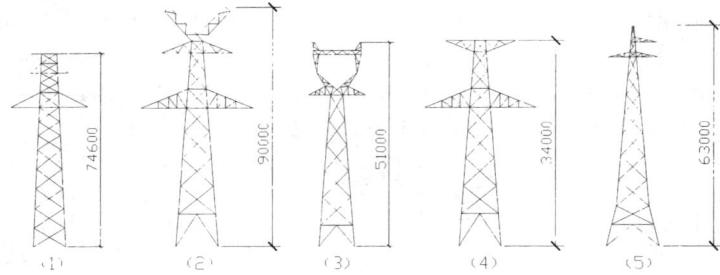

Figure 2. Instances of transmission tower

Table 1. Parameters of Transmission Towers

	Height (m)	Mass (Kg)	Type of Cable	Type of Lightning Line	Mass of Cable (Kg/km)
Tower 1	74.6	33427	LGJ-500/45	GJ-70	1688
Tower 2	90	40430	LGJ-500/45	GJ-70	1688
Tower 3	51	15067	LGJ-400/50	GJ-70	1511
Tower 4	34	11390	LGJ-400/50	GJ-70	1511
Tower 5	63	16434	LGJ-400/50	GJ-70	1511

Considering the hard site, medium site and soft site, three different seismic records for each type of site are selected to calculate the response of system. The peak values of seismic records are adjusted to be 0.2g. The information about the selected seismic records is shown in Table 2.

Two types of models are considered for the calculation. The first model is to neglect the effect of conductors and the corresponding shear and moment are V1 and M1, respectively. The second model is to consider the effect of conductors and the corresponding shear and moment are V2 and M2, respectively. To investigate the of effect of conductors with various spans on the seismic response of the transmission tower system, the maximum values of shear ratio (V2/V1) and moment ratio (M2/M1) are calculated with the increasing of span between two towers. The response of the system for each tower is marked by the average results of three seismic records. The maximum values of shear ratio and moment ratio with the spans are shown for tower 3 in lateral and longitudinal direction in Figure 3 and Figure 4. The horizontal coordinate means the span of the tower from 100m to 1000m with the interval 100m. The vertical coordinate denotes the maximum values of shear ratio and moment ration for seven lumped masses of each tower.

Table 2 Earthquake acceleration records

Site Type	Earthquake Name	Magnitude	Time	Record Place	Acceleration Peak (gal)
Soft site	San Fernando	6.6	1971.2.9	Navy Laboratory	25.91
	San Fernando	6.6	1971.2.9	University Avenue	56.36
	Tangshan Aftershock	7.1	1976.11.16	Tianjin hospital	104.18
Medium site	Imperial Valley	6.7	1940.5.18	El-Centro	341.7
	Kern County	7.7	1952.7	Taft Lincoln school	152.7
	Imperial Valley	5.6	1951.1.23	El Centro	30.35
Solid site	Landers	7.5	1992.6.28	Baker Fire	105.58
	Landers	7.5	1992.6.28	Fort Irwin	119.85
	Tangshan Aftershock	6.3	1976.11.16	Qian An	118.91

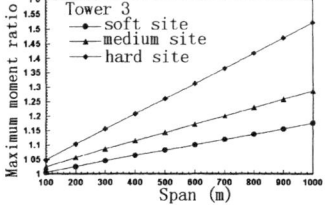 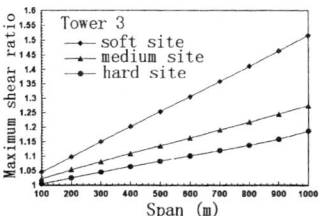

Figure 3. Diagram of average lateral seismic response for Tower 3

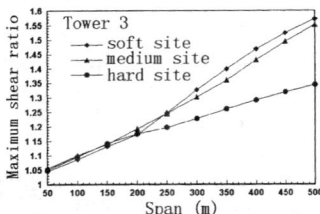 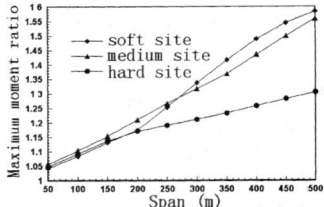

Figure 4. Diagram of average longitudinal seismic response for Tower 3

Table 3 Limit span of lateral tower considering conductor

Site Type		Soft	Medium	Hard
Limit span l_0 (m)		300	200	150
Mean Error	Shear Ratio	4.93%	3.57%	4.34%
	Moment Ratio	4.77%	3.53%	3.65%
Root Mean Square	Shear Ratio	0.001034	0.000837	0.000554
	Moment Ratio	0.000995	0.001443	0.001185

Table 4 Limit span of longitudinal tower considering conductor

Site Type		Soft	Medium	Hard
Limit span l_0 (m)		300	200	150
Mean Error	Shear Ratio	4.7%	5.8%	4.8%
	Moment Ratio	4.5%	5.9%	4.6%
Root Mean Square	Shear Ratio	0.001136	0.004930	0.000932
	Moment Ratio	0.001294	0.005428	0.000868

It can be seen from the Figure 3 and Figure 4 that the effect of conductors on the seismic internal force increases with the increment of the span. The influence of conductor on the seismic internal force of transmission tower system is within 5% when the span of conductors reaches the value in the Table 3. It can be concluded that the effect of the conductor on the dynamical performance of transmission tower system should be considered when the span of conductors is greater than the values given in the Table3 and Table 4. The detailed aseismic calculation is shown in next section. The limit spans given in the second line of Table 3 and Table 4 are marked by l_0. The mean errors and root mean squares of calculation results for the five towers with limit spans are also given in Table 3 and Table 4. It can be seen from the tables that the root mean squares of calculation results are very small, which indicate that the data is low in dispersion.

Suggestion on Simplified Method of Seismic Calculation

According to *Aseismic Design Regulation for Electric Power Equipment* (GB 50260-96)(1996), the modal analysis response spectra method is applicable for calculating the response of large span tower and self-support tower with the height more than 50m under horizontal seismic; The mass of conductors and lightning lines can be neglected in calculating the dynamical performance of pole tower. The seismic response of the system can be calculated through the following equation for model analysis response spectra method:

$$F_{ji} = \zeta \alpha_j \gamma_j X_{ji} G_i \quad (i = 1,2,\cdots n; j = 1,2,\cdots m) \tag{8}$$

$$\gamma_j = \frac{\sum_{i=1}^{n} X_{ji} G_i}{\sum_{i=1}^{n} X_{ji}^2 G_i} \tag{9}$$

where, F_{ji} is the standard value of horizontal seismic for the ith lumped mass of the jth model; ζ means the structural coefficient; α_j denotes the influence coefficient of horizontal seismic for the jth model; γ_j is the participant coefficient of the jth mode; X_{ji} expresses horizontal relative displacement for the ith lumped mass of the jth model in X direction; G_i denotes the representative value of gravity for the jth model including overall dead load, gravity of fixed equipment and other additional gravities on the lumped mass.

From the analysis on the response of the transmission tower with various types and spans in the previous section, it is indicated that the tower system designed according to *Aseismic Design Regulations for Electric Power Equipment* (GB 50260-96) is unsafe when the spans of conductor exceed the limit spans. Hence, the simplified aseismic method is to add additional mass Δm to G_i to consider the effect of conductor. The value of Δm can be determined by the following equation:

$$\Delta m = f(l_x) \times l_x \times q \tag{10}$$

where, Δm is the additional mass to consider the effect of conductor; l_x is the span of conductor; q is the mass of conductor per one thousand meters; $f(l_x)$ is additional mass coefficient, which is determined by the following equation for lateral direction:

$$f(l_x) = \begin{cases} 0.17 + \dfrac{3l_x}{200l_0} & soft \quad site \\ 0.21 + \dfrac{l_x}{100l_0} & medium \quad site \\ 0.35 + \dfrac{l_x}{20l_0} & hard \quad site \end{cases} \quad \text{(For } f(l_x)>0.7, \; f(l_x)=0.7\text{)} \quad (11)$$

$f(l_x)$ is determined by the Equation (12) for longitudinal direction:

$$f(l_x) = 0.5 + \dfrac{l_x}{200l_0} \quad \text{(For } f(l_x)>1.0, \; f(l_x)=1.0\text{)} \quad (12)$$

The responses of transmission tower systems with various spans are calculated by the suggested simplified method and the results are compared with the one calculated by the integral model, which shows that they fit well. The comparisons of results for tower 3 in the lateral and longitudinal direction are shown in the Figure 5 and Figure 6. It can be seen from the figures that the results by the simplified method and integral model are very close for soft and hard site, with maximum error approximate 3%. The errors are a little great for medium site, but the maximum error is only approximate 6%. This is mainly because of the resonance for the period of tower 3 is near to the characteristic period of the site. The error is acceptable in engineering.

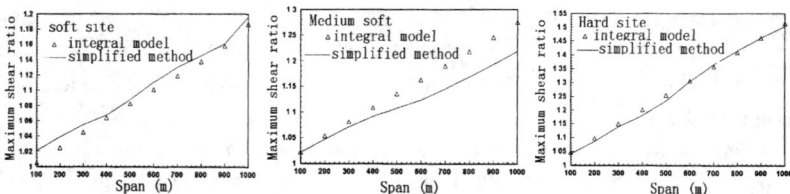

Figure 5. Comparisons of integral model and simplified method for lateral direction

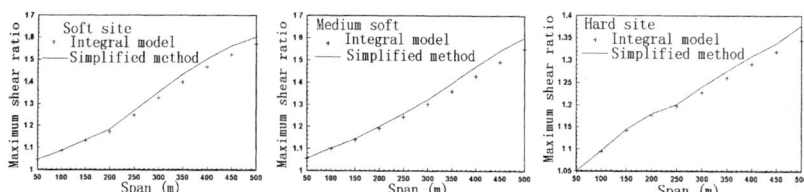

Figure 6. Comparisons of integral model and simplified method
for longitudinal direction

Conclusions

From the study here, the conclusions can be drawn as follows:
(1) The effect of conductors on the dynamical response of transmission tower increase with the increment of spans;
(2) For different sites, the effect of conductor should be considered when the span of conductor exceeds the limit span shown in the Table 3 and Table 4;
(3) The suggested simplified aseismic calculation method can be applied to the design of engineering project.

The results can also be used for compiling the design regulation.

References

American Society of Civil Engineers Committee on Electrical Transmission Structures (1982). "Loadings for Electrical Transmission Structures." *Journal of Structural Division*, ASCE, 108(5).

American Society of Civil Engineers (1991). "Guideline for Electrical Transmission Line Structural Loading." *Manuals and Reports on Engineering Practice*, ASCE, No.74, New York.

East China Electric Power Design Institute of State Power Corporation of China (1999). *Design Regulations for 110~500KV Aerial Cables* (DL/T 5092-1999), China Electric Power Press.

Li, H. N., Lu, M. and Wang, Q. X. (1990). "Simplified Aseismic Calculations for High Voltage System Consisting of Long-span Transmission Lines and Their Supporting Towers." *Earthquake Engineering and Engineering Vibration*, 10(3), 73-87

Li, H. N., and Wang, Q. X. (1997) "Dynamical Characteristics of Large-span

Transmission Tower." *China Civil Engineering Journal*, 30(5), 28-36

Li, H. N., Shi, W. L. and Jia, L. G. (2003a, In press). "Lateral Simplified Aseismic Calculations for High Voltage System Considering the Effect of Conductor." *Journal of Vibration Engineering*.

Li, H. N., Shi, W. L. and Jia, L. G. (2003b, In press). "Limit for the Influence of Conductor on the Longitudinal Vibration of High Voltage System and Simplified Aseismic Calculations." *Journal of Vibration Engineering*.

Minister of Electric Power Industry of the People's Republic of China (1996). *Aseismic Design Regulation for Electric Power Equipment* (GB 50260-96). China Plan Press.

Development of a Probabilistic Assessment Model for Post-Earthquake Residual Capacity of Utility Lifeline Systems

Nobuoto Nojima[1] and Masata Sugito[2]

Abstract

A probabilistic assessment model for post-earthquake residual capacity of utility lifeline systems is proposed. On the basis of the damage statistics obtained in the 1995 Hyogoken-Nanbu earthquake, Japan, two empirical models have been derived for simple estimation of outage probability and duration of disruption as functions of JMA seismic intensity. Residual capacity model is then derived using the two elements of outage and duration estimation in a probabilistic manner. Simple estimation of post-earthquake residual capacity can be made only with the information of seismic intensity.

Introduction

Estimation of post-earthquake residual capacity of utility lifeline systems is one of important issues in seismic risk assessment and emergency response planning. ATC-25 (ATC, 1991) is a compilation of organized studies integrating practical methods associated with vulnerability of lifeline facilities, physical damage, interruption of service, and eventual economic impact. As an important part of estimates of indirect economic losses, ATC-25 provides a consistent method to estimate interruption of lifelines as a result of direct damage in a straight-forward manner.

In Japan, less attention had been paid to restoration estimates on a practice level. However, intensive damage to lifeline facilities in the 1995 Hyogoken-Nanbu earthquake gave rise to strong research needs of practical estimation of lifeline

---

[1] Dr. of Eng., Associate Professor, Dept. of Civil Engineering, Gifu University, Yanagido 1-1, Gifu 501-1193, JAPAN, Phone/FAX: +81-58-293-2416, E-mail: nojima@cive.gifu-u.ac.jp

[2] Dr. of Eng., Professor, River Basin Research Center, Gifu University, Yanagido 1-1, Gifu 501-1193, JAPAN, Phone/FAX: +81-58-293-2420, E-mail: sugito@cive.gifu-u.ac.jp

outage time from overall perspective especially for the purpose of seismic risk management. Besides, the earthquake provided great amount of data that facilitates revision of existing methods for lifeline functional restoration.

Straight-forward procedures for restoration estimates of utility lifelines in Japan require a variety of information on the network facility inventory, fragility relationships, anticipated seismic intensity maps, network configuration, physical-functional damage relationships, resources available for recovery works, and restoration strategy. For practical application, such a conventional method involves several problems: data availability, high uncertainty in fragility relations, the difference between planned and actual restoration strategy, etc.

As an alternative way of estimation, this study presents a probabilistic method to outline the post-earthquake functional performance of utility lifelines in a simple manner. It is widely accepted that the dominant factor accounting for the physical damage is seismic intensity. Besides, ordinary restoration processes suggest that repair work generally proceeds from areas of slight damage to those of heavy damage. Understanding that the chance of outage and its duration of lifelines strongly depend on the seismic intensity, the proposed estimation model is mainly based on the seismic intensity distribution at the area concerned.

The assessment method proposed in this study follows five major steps:
(1) Estimate the probability of lifeline outage based on the seismic intensity.
(2) Estimate the duration of lifeline disruption under the condition that the outage occurred.
(3) Estimate the residual capacity for given seismic intensity using the two elements above.
(4) Estimate the overall residual capacity considering seismic intensity distribution.
(5) Improve the estimate according to the physical vulnerability of network facility.

In the present paper, focus is placed on the procedure of model development regarding the former three steps. The latter two steps will be described in the future study.

GIS database of 1995 Hyogoken-Nanbu earthquake for model development

Spatial distribution of seismic intensity.
Yamaguchi and Yamazaki (1999) evaluated a high-density seismic intensity distribution in damaged area of 1995 Hyogoken-Nanbu earthquake mainly based on damage data of low-rise building structures. JMA seismic intensity was estimated at the level of cho-cho-moku which is a minimum unit of Japanese census tract. Takada et al. (1996) investigated the intensity distribution in the southern Hyogo Prefecture using the questionnaire method proposed by Ohta et al. (1979). Tsurugi et al. (1999) also conducted the questionnaire survey to evaluate the seismic intensity in Osaka Prefecture. In this study, detailed JMA intensity map covering both of the Hanshin and Osaka area was developed by integrating the above-mentioned three studies. The total number of cho-cho-moku counts was 9937. Figure 1 shows the distribution map of JMA intensity.

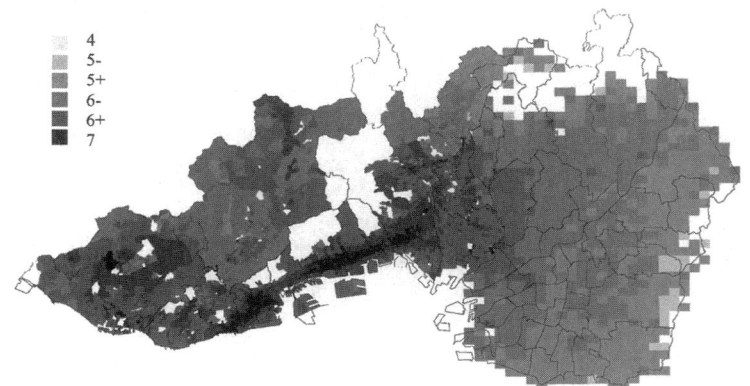

Figure 1 JMA seismic intensity distribution map of the 1995 Hyogoken-Nanbu Earthquake, Japan.

Spatial distribution of lifeline outage and duration of disruption.

Kameda et al. (1998) developed a GIS database of damage and restoration process of water and gas supply systems in the 1995 Hyogoken-Nanbu earthquake. The database contains number of days required for full restoration in the hardest-hit cities (Kobe, Nishinomiya and Ashiya) on the cho-cho-moku basis. For this study, additional data from the surrounding areas including five cities in Hyogo Prefecture (Amagasaki, Takarazuka, Itami, Kawanishi, and Akashi) and 19 cities in the northern Osaka Prefecture was compiled based on materials provided by related city offices and lifeline companies. Figures 2-4 show the extent of outage and time required for full restoration for electric power supply, water delivery, and city gas supply systems.

Spatial distribution of JMA intensity (Figure 1) was combined with the functional damage data (Figures 2-4) to establish their one-to-one correspondence. Since both datasets were compiled on the same cho-cho-moku unit, overlay and recompilation was easily done.

Two-step model for functional performance estimation

On the basis of the GIS database, a two-step model are developed for evaluation of post-earthquake serviceability of utility lifelines in terms of an estimate of seismic intensity at the site concerned. The first model is a logistic model for probabilistic assessment of occurrence of lifeline disruption. The second model is a statistical prediction model for the evaluation of outage time under the condition that outage has occurred.

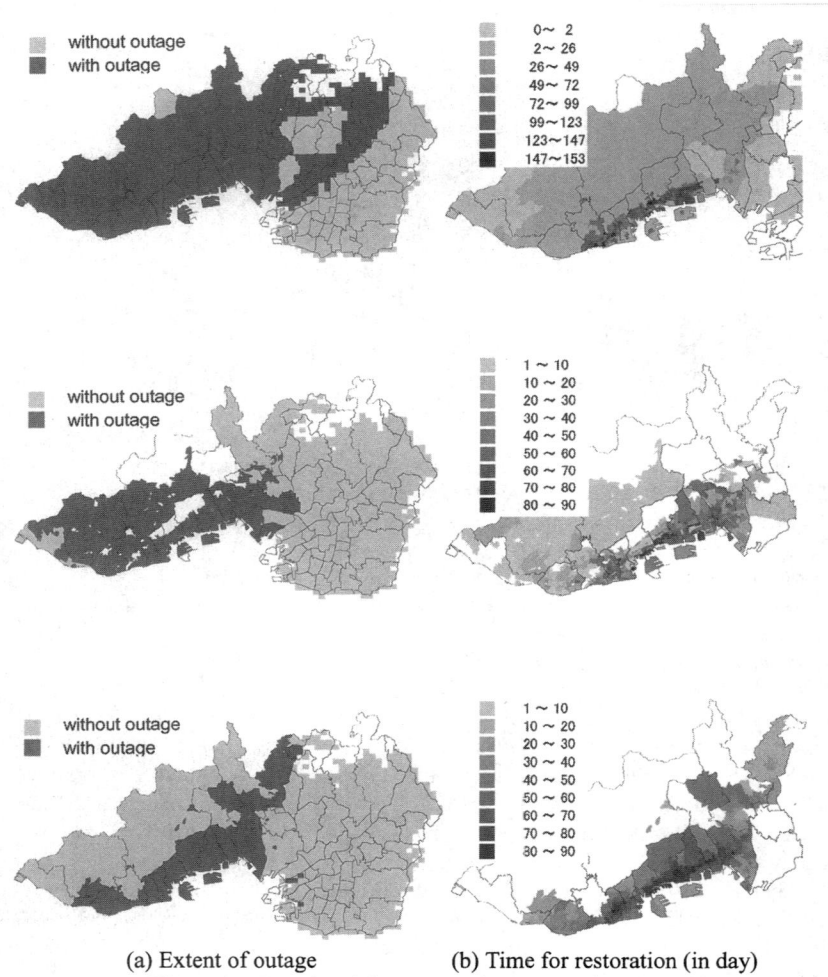

(a) Extent of outage   (b) Time for restoration (in day)
Figure 4   Functional damage to city gas supply system.

Estimation of the probability of lifeline outage based on seismic intensity.
Figure 5 (a)-(c) show the conditional histograms of JMA intensity in the light of the occurrence of lifeline outage. The number of cho-cho-moku counts are : electric power supply 9935 (with outage : 4450 + without outage : 5485), water delivery 9817 (with outage : 6392 + without outage : 3425), gas supply 9937 (with outage : 6985 + without outage : 2952).

Using a logistic model, the probability of lifeline outage p can be modeled in terms of the JMA intensity I as the following equation.

$$p(I) = \frac{\exp[b_0 + b_1 \cdot I]}{1 + \exp[b_0 + b_1 \cdot I]} \tag{1}$$

Two parameters b_0 and b_1 were identified using the maximum likelihood method (See Table 1).

Figure 6 shows the derived logistic models for probabilistic evaluation of utility lifeline outage, which serve as functional fragility relations. For reference, fragility curves for low-rise building damage (Yamaguchi and Yamazaki, 1999) are also drawn. It is observed that the utility lifeline functions are more susceptible to seismic intensity compared to the physical damage to buildings. In particular, the electric power supply function is the most vulnerable to strong motion. The average and standard deviation of JMA intensity are 5.26 and 0.48 for electric power supply system, 5.71 and 0.38 for water delivery system, 5.86 and 0.42 for gas supply system (See Table 1).

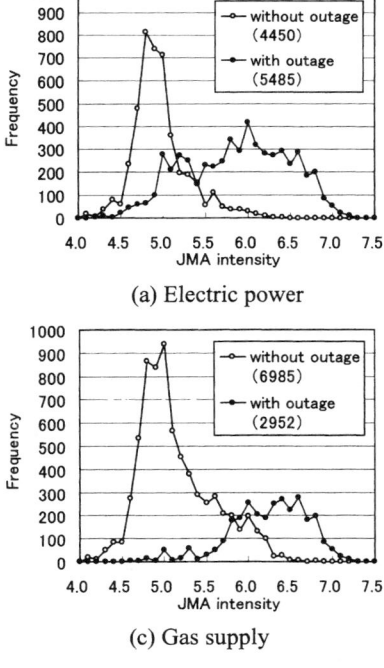

Figure 5 Histograms of JMA intensity with/without lifeline outage.

Table 1 Parameters for logistic models.

	Electric power	Water delivery	Gas supply
b_0	-19.72	-26.98	-25.08
b_1	3.75	4.72	4.28
Average	5.26	5.71	5.86
Std.Dev.	0.48	0.38	0.42

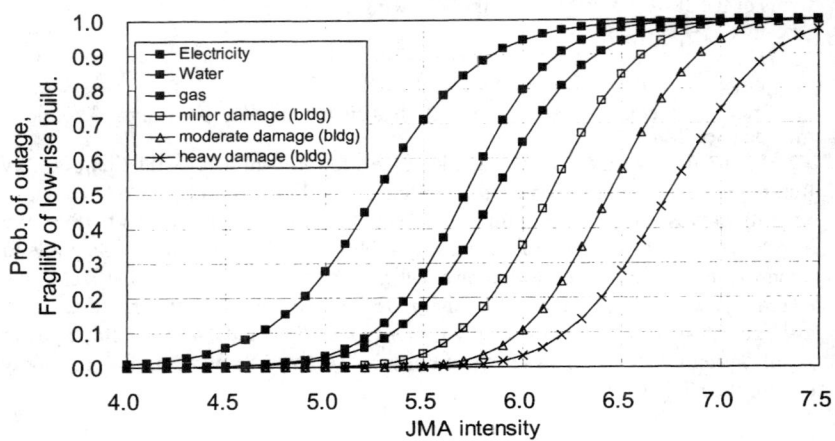

Figure 6 Logistic models for probabilistic evaluation of utility lifeline outage and fragility curves for low-rise building.

Estimation of the probability of lifeline outage time based on seismic intensity.

Figure 7 (a)-(c) show the scattergrams representing the relationship between the JMA intensity and the duration of outage. Applying the smoothing method of moving window, it was found that both the moving average and moving standard deviation can be appropriately modeled by quadratic curves in terms of the JMA intensity I.

$$\mu(I) = a_0 + a_1 I + a_2 I^2 \qquad (2)$$

$$\sigma(I) = c_0 + c_1 I + c_2 I^2 \qquad (3)$$

Six parameters are determined using the least square method. On this basis, probabilistic models using gamma distribution were derived to evaluate the duration of lifeline disruption for given value of the JMA intensity I.

$$f(t \mid I) = \frac{t^{\alpha(I)-1} \exp\left(-\dfrac{t}{\beta(I)}\right)}{\beta(I)^{\alpha(I)} \Gamma(\alpha(I))}, \text{ where } \alpha(I) = \left(\frac{\mu(I)}{\sigma(I)}\right)^2, \ \beta(I) = \frac{\sigma^2(I)}{\mu(I)} \qquad (4)$$

Figure 8 (a)-(c) show the evaluation model representing the probability of lifeline restoration within a certain time period elapsed after the earthquake. It can be observed that the general tendency in Figure 7 is reflected in Figure 8. The duration of outage dramatically increases in the range of JMA intensity between 5.5 and 6.5.

DISASTER RESPONSE FOR LIFELINE SYSTEMS 713

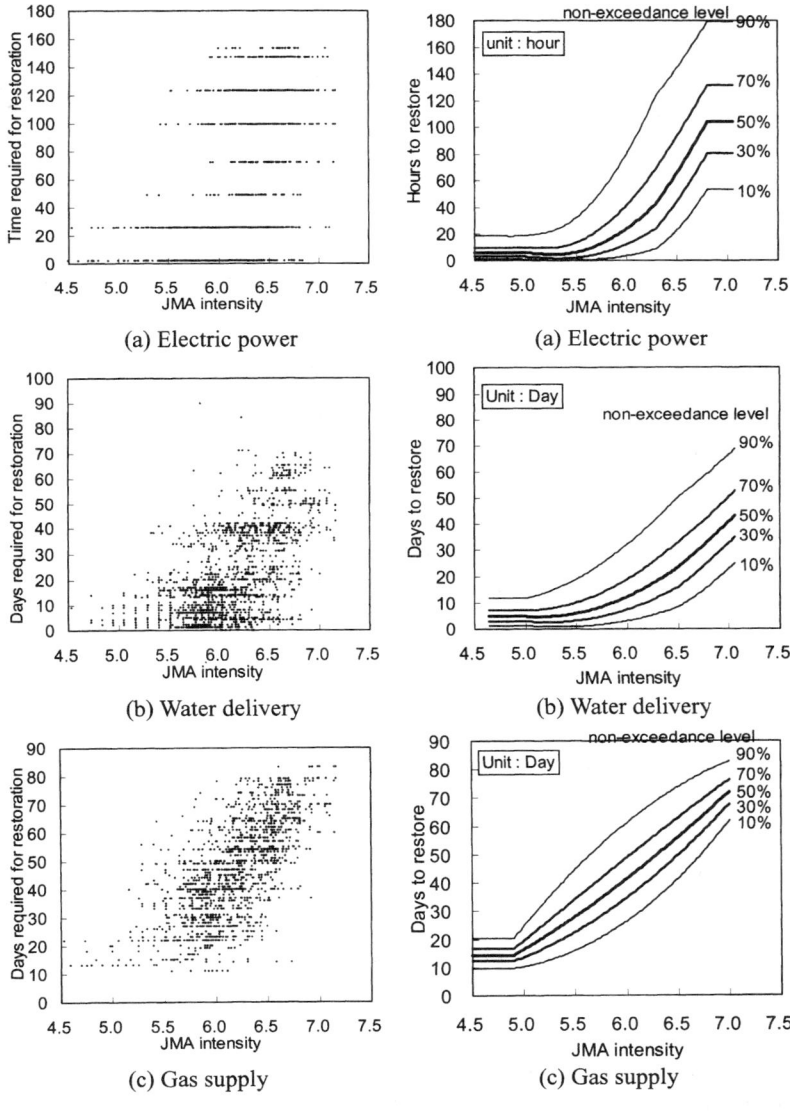

Figure 7 Relationship between the JMA seismic intensity and the duration of outage.

Figure 8 Probability of lifeline restoration within a certain time period.

Residual capacity estimation using the two-step model

The two-step empirical model described above are combined to construct the residual capacity estimation model for given seismic intensity, providing a probable pathway of functional restoration of utility lifelines for individual customer. By integrating Figures 6 and 8, residual capacity curve at an arbitrary site that goes through a certain level of the JMA intensity I can be derived. Using Eqs(1) and (4), residual capacity $D(t \mid I)$ can be expressed as :

$$D(t \mid I) = 1 - p(I) + p(I) \int_0^t f(\tau \mid I) d\tau \qquad (5)$$

Figure 9 shows the residual capacity curves for three utility lifeline systems. Using this model, one can conveniently perform pre- and post-event evaluations of serviceability of electric power, water, and gas supply systems only on the basis of JMA intensity. Although the estimates are considered to be highly uncertain, the proposed model is capable of providing decision makers with rapid damage estimation which is necessary to conduct emergency response. For citizens, estimation of availability of lifeline functions for scenario earthquakes is useful for preparedness purpose such as emergency stockpile.

Concluding remarks

This paper presented a probabilistic assessment model for post-earthquake residual capacity of utility lifelines. On the basis of damage statistics obtained in the 1995 Hyogoken-Nanbu earthquake, Japan, a two-step empirical model was developed to evaluate the probability of lifeline outage and duration of disruption in terms of JMA intensity. By combining the two-step model, residual capacity estimation model was derived.

The proposed model provides vital information of serviceability of utility lifelines at an arbitrary site. In order to aggregate the effect of lifeline disruptions to the overall customers, population exposure to seismic intensity are to be considered for scenario earthquake. The overall residual capacity is consequently derived for the entire disaster-stricken area.

A consideration of network vulnerability is an essential way to improve the present model. Since the estimation model is based on the GIS database of lifeline damage due to 1995 Hyogoken-Nanbu earthquake, model parameter reflect the network vulnerability in the Hanshin region. In a practical application of the residual capacity model, it is important to take account of seismic vulnerability of the network system under consideration. Post-earthquake functional performance may be better than the estimate suggested by the model because of intensive replacement of weak pipes during the past several years. Or it may be worse because of poor effort of seismic improvement. For this purpose, comprehensive indices representing the network vulnerability are now being developed.

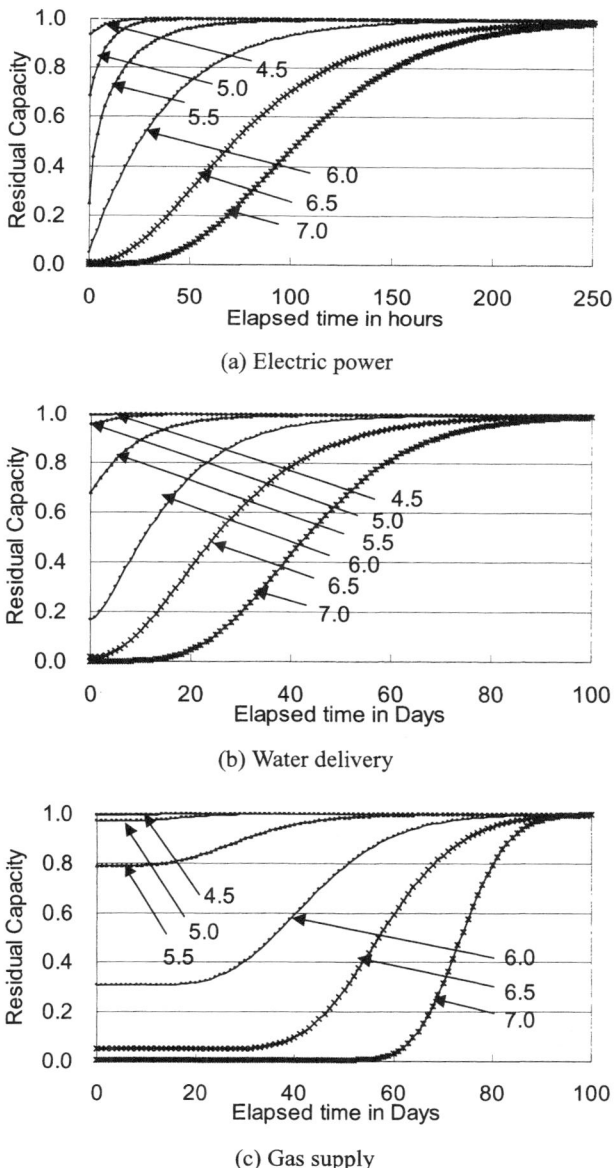

Figure 9 Post-earthquake residual capacity curves for various JMA intensity.

The future study following the present one will focus on the model improvement together with numerical examples for the scenario of anticipated off-shore huge events along the Nankai trough, i.e., Tokai, Tonankai and Nankai earthquakes. Since intensive damage to utility lifelines are expected in several prefectures surrounding the source regions, the damage estimation outlined by the proposed model provides useful information for pre- and post-earthquake disaster response planning in a broad area.

Acknowledgements

The authors wish to thank Dr. Yutaka Ishikawa and Mr. Toshihiko Okumura of Izumi Research Institute, Shimizu Corporation for their kind support to this study. The assistance of Mr. Yasuo Suzuki, graduate student of Gifu University, is also acknowledged with gratitude.

References

Applied Technology Council. (1991) "Seismic Vulnerability and Impact of Disruption of Lifelines in the Conterminous United States," ATC-25, Redwood City, California.

Kameda, H., Iwai, S., Usui, T., Nojima, N., Tsuboi, K., Kotoh, T., Ogawa, Y., Matsushita, M., Fujita, Y. and Hashigami, S. (1998) "GIS Analysis of Lifeline Recovery Process and Daily-Life Inconvenience in the Great Hanshin-Awaji Earthquake Disaster," IMDR Report No.6, DPRI, Kyoto University (in Japanese).

Ohta, Y., Goto, N. and Ohashi, H. (1979), "A Questionnaire Survey for Estimating Seismic Intensities," Bulletin of the Faculty of Engineering, Hokkaido University, No.92, pp.117-128 (in Japanese).

Takada, S., Kashima, T., et al. (1996) "Questionnaire Study related to Hyogoken-Nambu Earthquake," Faculty of Engineering, Kobe University (in Japanese).

Tsurugi, M., Sawada, S., Irikura, K. and Toki, K. (2000) "Distribution of Seismic Intensity in Osaka Prefecture during the 1995 Hyogo-ken Nanbu Earthquake Based on Questionnaire Survey," *Journal of Structural Mechanics and Earthquake Engineering,* Japan Society for Civil Engineers, No.612/I-46, pp.165-179 (in Japanese).

Yamaguchi, N. and Yamazaki, F. (1999) "Estimation of Strong Ground Motion in the 1995 Hyogoken-Nanbu Earthquake Based on Building Damage Data," *Journal of Structural Mechanics and Earthquake Engineering,* Japan Society for Civil Engineers, No.612/I-46: 325-336 (in Japanese).

Numerical Simulation of the Behaviour of Buried Jointed Pipelines under Extremely Large Fault Displacements

Radan Ivanov[1] and Shiro Takada[2]

Abstract

A comprehensive numerical method for limit state simulation of buried jointed pipelines is presented. The pipe body is modelled by line elements and plastic deformation within it considered by plastic hinges. The basic mechanical properties of all common pipe joints e.g. stiffening in compression, detachment in tension, fracture in bending can also be considered. Soil-pipe interaction is modelled by elasto-plastic direct and drag springs. The method is capable of tracing the behaviour of pipelines until and after failure of the pipe body and detachment at joints, thus allowing the physical damage to a pipe network to be better estimated. The method is three-dimensional, thus allowing the interaction of pipeline segments running in different directions to be evaluated as well. The solution algorithm is based on the direct integration of uncoupled equations of motion, which ensures that the behaviour of a pipeline can be investigated for any magnitude of ground displacements, revealing the whole spectrum of events occurring in the pipe-soil system with the increase of earthquake induced ground displacements: elastic deformations, plastic deformations in the soil, plastic deformations in the pipe body, rupture and detachment at joints, failure and rupture within the pipe body. The method is implemented into a computer program and several trial runs are presented.

Introduction

A number of particularly destructive earthquakes hit the world in recent years, e.g. the 1995 Kobe earthquake, the 1999 Chi-Chi and Kocaeli earthquakes. Along with the usual heavy damage to building structures, these earthquakes exposed the seismic vulnerability of utility pipelines. The reasons for this can be put into two categories; first, the fact that the earthquakes struck urban areas with well developed underground infrastructure and second, the still insufficient understanding of the behaviour of underground pipelines and the process of their failure, resulting in inadequate design codes or lack of such codes in the first place. Field surveys of damages pipelines have repeatedly confirmed that a large proportion of the failures are due to joint failures that lead to discontinuities in the pipeline body. Such joint detachments are often accompanied by large deflections inducing material and geometrical nonlinearity effects in the pipe behaviour. The damage is particularly severe when pipes cross fault dislocations.

---

[1] Assistant Professor, Dept. of Civil Engineering, Kobe University, Kobe, Japan;
phone: ++81-78-803-6047; radan@kobe-u.ac.jp

[2] Professor, Dept. of Civil Engineering, Kobe University, Kobe, Japan;
phone: ++81-78-803-6037; takada@kobe-u.ac.jp

In order to reliably evaluate the overall pipeline behaviour, a comprehensive numerical method for limit state simulation of buried pipelines was developed to facilitate the drafting of new codes for design of such pipelines.

The analysis of the behavior of buried pipelines has been done by many methods in the past.

Many closed form solutions to the pipe–soil interaction problem based on beam on elastic foundation have been proposed (Kennedy, 1977; Wang, 1995). Such methods are excellent for grasping the nature of the problem but cannot be applied when large deflections or material nonlinearities are present.

The Finite Element Method (FEM) has been routinely applied for pipe analysis; numerous examples of beam, shell or combined models can be found in the literature (Takada, 2001). Transfer matrix methods have also been developed (Takada, 1987). While the FEM analyses are successful in revealing most of the features of pipeline behaviour, rupture and detachment at joints cannot be dealt with; for the shell version the computation effort is too great to allow large models consisting of many pipe segments to be analysed. The main purpose of the proposed method is to extend the domain of pipeline analysis through the failure and post-failure stages, while taking advantage of the efficacy of simple elements in the modelling.

Modelling and solution method

Modelling of the pipe body. The model consists of lumped masses (elements) connected by sets of springs as shown in Figure 1. Each spring set consists of an axial, bending and torsional components derived from beam theory. Figure 2 shows the forces and displacements at the ends of a 3-D beam segment with Young's modulus E, second moment of area I and length l which are linked by these springs. A plane projection is shown for simplicity.

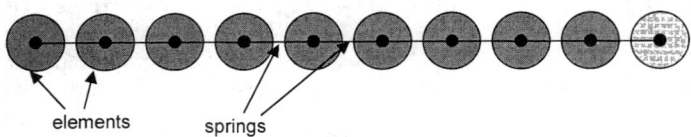

Figure 1. Model of a pipe

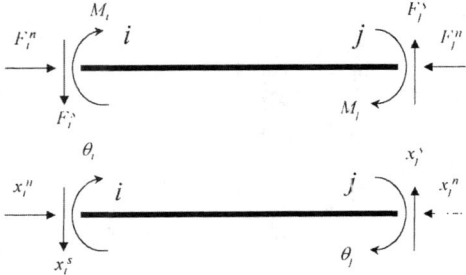

Figure 2. Forces and displacements at the ends of a beam segment

Solution procedure. The solution is based on the double integration of the Newton equations of motion for each element. The motion of an element is considered uncoupled from the motion of the rest during a single time step, thus eliminating the need to assemble a stiffness matrix. Coupling, which of course exists, is accounted for by updating the spring forces at every time step. An element then moves due to the out-of-balance force appearing on summation of the forces in all springs originating from it. For the system to reach equilibrium two types of damping are applied. First, the relative motion between elements is damped by a coefficient of damping calculated from the critical damping ratio of the material of the structure (steel, plastic, etc.). This is termed local damping. In addition a small amount of viscous damping works on the absolute velocities of the objects. This is referred to as global damping and represents the resistance of the medium in which the structure stands (soil). The method is inherently dynamic, geometrically non-linear, and can accommodate easily arbitrary amounts of rigid body motion. The equations for translational and rotational motion of a single element are,

$$\ddot{x}_i + \alpha \dot{x}_i = F_i / m_i + g \tag{1}$$
$$\dot{\omega}_i + \alpha \omega_i = M_i / I_i \tag{2}$$

where x_i is the position vector, ω_i the rotational velocity of element i, m_i and I_i are its mass and mass moment of inertia respectively, and g the gravity acceleration. Integration in time is designated by dots. F_i and M_i are the total force and moment acting on lumped mass i and are formed by contributions from all springs originating from the lumped mass, including soil springs. New and old values are designated by superscripts plus and minus. A centred finite difference is used to integrate the equations of motion. The expressions for translational and rotational velocities are,

$$\dot{x}_i = \frac{1}{2}\left[\dot{x}_i^- + \dot{x}_i^+\right] \tag{3}$$
$$\omega_i = \frac{1}{2}\left[\omega_i^- + \omega_i^+\right] \tag{4}$$

and the expression for translational and rotational accelerations,

$$\ddot{x}_i = \frac{1}{\Delta t}\left[\dot{x}_i^+ - \dot{x}_i^-\right] \tag{5}$$
$$\dot{\omega}_i = \frac{1}{\Delta t}\left[\omega_i^+ - \omega_i^-\right] \tag{6}$$

Inserting these expressions in the equation of motion Eq. 1 and 2 and solving for new values of velocities results in,

$$\dot{x}_i^+ = [D_1 \dot{x}_i^- + (F_i / m_i + g)\Delta t]D_2 \tag{7}$$
$$\omega_i^+ = [D_1 \omega_i^- + (M_i / I_i)\Delta t]D_2 \tag{8}$$

where $D_1 = 1-(\alpha\Delta t/2)$, $D_2 = 1/[1+(\alpha\Delta t/2)]$, and Δt is the time step.

Having obtained the new velocities we can move forward to update the spring forces. In what follows, the interaction of a pair of elements with an spring between them is considered. First we calculate the incremental displacements of the two elements Δx_i and Δx_j.

$$\Delta x_i = v_i^+ \Delta t, \qquad \Delta x_j = v_j^+ \Delta t \qquad (9)$$

where v_i^+ and v_j^+ are the velocities of elements i and j respectively. Next, we calculate the relative displacement between the two elements Δx_{ij} and its normal component Δx_{ij}^n

$$\Delta x_{ij} = \Delta x_j - \Delta x_i \qquad (10)$$
$$\Delta x_{ij}^n = (\Delta x_{ij} n_{ij}) n_{ij} \qquad (11)$$

where n_{ij} is the unit normal vector pointing from element i to element j. Then we compute the increment of spring force,

$$\Delta F = k_0 \Delta x_{ij} \qquad (12)$$

The direction of the force in the spring has meanwhile changed, so we first update it to mach the latest normal between the two element centres,

$$F^- = F^- n_{ij} \operatorname{sgn}(F n_{ij}) \qquad (13)$$

where F^- is the force in the spring from the previous time-step, n_{ij} is the normal vector between elements i and j and sgn( ) designates the sign function (yielding + 1 or -1). Finally we add the increment to yield the provisional updated value of the force in the spring.

$$F^+ = F^- + \Delta F \qquad (14)$$

where F^+ is the force in the spring at the end of the current time step. The force so calculated is sufficient to carry out elastic analysis. A strain hardening hysteretic algorithm for the axial springs is also implemented. It keeps track of the elastic-plastic state and the accumulated plastic deformation ensuring that the value of the force is at all times between lines BE and DC, see Figure 3, and also breaks the spring if the ultimate plastic deformation is exceeded. In addition two other versions of the strain hardening algorithm with excluded tension or compression are used to simulate members that do not carry tension or compression respectively. The displacement limit up to which a spring is inactive is specified at the program input.

After having been modified, the force in the spring is added to the resultants of forces acting on the elements that the spring connects,

$$F_i = F_i + F_{\text{mod}} \tag{15}$$
$$F_j = F_j - F_{\text{mod}} \tag{16}$$

where F_{mod} the modified value of F^+ as yielded by the hysteretic algorithm.

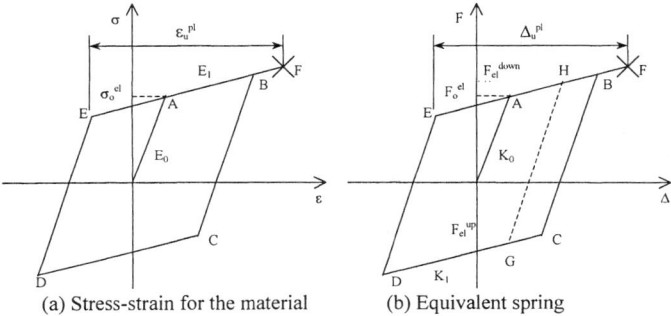

(a) Stress-strain for the material (b) Equivalent spring

Figure 3. Constitutive behaviour; axial springs

The shear forces and bending moments resulting from the bending components of the springs are computed from the following equations,

$$F_i^s = \frac{12EI}{l^3}(x_i^s - x_j^s) - \frac{6EI}{l^2}(\theta_i - \theta_j) \tag{17}$$

$$M_i = \frac{6EI}{l^2}(x_i^s - x_j^s) - \frac{4EI}{l}(\theta_i - \frac{\theta_j}{2}) \tag{18}$$

where x and θ are the translational and rotational displacement increments of a lumped mass and subscript s stands for shear direction. The equations are solved every time step and provide contributions to the total driving force and driving moment acting on the elements the spring set connects. These are added to the force and moment which enter Eqs. 1 and 2.

Modelling of soil-pipe interaction. The soil surrounding the pipeline is modelled by pairs of springs having axial stiffness only, with one end attached to a lumped mass and the other end fixed. One spring is perpendicular to the pipe representing the direct contact between pipe and soil, and the other tangential to it representing the drag between pipe and soil. Input displacements are specified at the fixed ends to simulate fault displacements. The direct soil springs are active only in compression, whereas the drag springs are active in both directions of relative displacement between pipe and soil. The direction of all soil springs is modified at each time step to preserve the angle they made with the pipe in the initial undeformed configuration.

Modelling of joints. A joint is introduces to the pipe model in the following way. Two elements instead of one are specified at nodes where joints are present, e.g. the lighter elements in Figure1 have been added since there are joints at these locations and the continuity of the pipe is broken there. The forces and moments acting on the elements due to the joint are computed according to the mechanical behaviour of the joint, and added to the total driving force and moment acting on the joint elements. Force contributions are computed according to the axial behaviour of the joint, and moment contributions according to it's bending behaviour. The relative rotation of the joint elements needed for computing the moment contributions is obtained directly as the difference of the absolute rotation angles of the two elements.

For obtaining the relative expansion or contraction an assumption needs to be made for the axial direction of joint displacement. This has been assumed to be the axis of one of the adjoining pipe segments, which is called the *reference beam*. In has been further assumed that a relative displacement between the joint nodes in the lateral direction does not occur. The mechanical behaviour of a GM-II joint obtained by experiment is shown in Figure 4 (Takada, 1983). It is a good example of the complexity of behaviour of a mechanical joint. The features important to the incorporation of such a constitutive behaviour into a computer program are as follows:

    1. The behaviour in the direction of pipe axis (tension/compression) as well as in bending is characterised by several distinct stages, each stage corresponding to a particular mechanical event taking place inside the joint. The program algorithm is so designed as to allow the treatment of a sequence of as many linear stages as necessary to fully describe the joint behaviour in a particular. This approach provides generality of treatment at no added computational cost. Notably curved stages can always be interpolated in a piecewise linear manner.

    2. Since the mechanical events in the tension and compression regimes are entirely different the load displacement curves for tension and compression are also different. To this end the program algorithm allows for separate treatment of tension and compression.

    3. Experimental data is normally available for continuous loading only, so unloading curves as well as points of transition between tension and compression or clockwise bending and counter-clockwise bending will not be known so reliable input data cannot generally be available for these items. The situation is further complicated by the unavailability of data for the interaction between the axial and bending behaviour of joints. In the program algorithm the axial and bending behaviour are considered independent, and gradients for unloading are specified for each stage. The transition between loading direction is set at the zero point, and reloading does not occur before the distance/angle between the joint elements reaches the amount of previously accumulated plastic flow displacement/angle.

    4. One or more stages of the constitutive behaviour paths may have large gradients of loading or unloading. These stages correspond to situations where the

two pipe segments meeting at the joint come in direct contact or in contact via a stiff joint part, i.e. become interlocked. In fact, since an experiment on a joint alone is practically impossible, such stages may be thought to correspond to deformations in the adjoining pipe segments used in the experiment. Given the time-stepping nature of the algorithm, that would necessitate a decrease of the time step needed for stable computation. This is computationally inefficient, so a special handling technique was developed for these stages. The displacement or rotation of one of the joint nodes is constrained to this of the other node until the maximum or minimum force or moment for the stage is reached. The values for comparison to the ultimate values of the stage are taken from the *reference beam* of the joint. Such stages are termed *locked*.

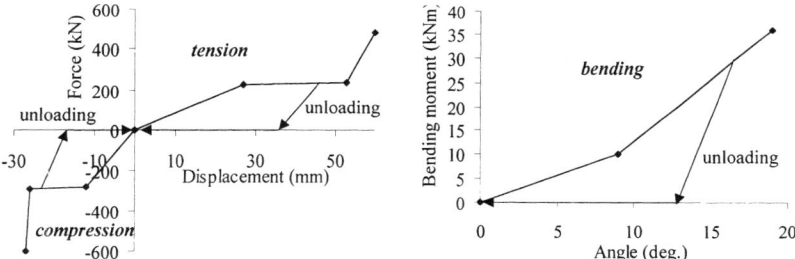

Figure 4. Mechanical behaviour of GM-II joint (150mm)

Plastic hinges in the pipe body. Apart from the joints, failure may occur due to excessive deformations within the pipe body. The handling of plastic deformations and failure due to the axial forces in the pipe has been explained in *Solution procedure* above. The formation of plastic hinges in the pipe body is handled by introducing joints within a pipe segment. For these joints, only one stage is specified for the constitutive behaviour in the axial direction (tension and compression), and is set to *locked*. The bending behaviour is made-up of three stages; the first is locked, the second is the transition between the yield bending moment M_y and the ultimate bending moment M_u of the section, and the last has a decaying gradient that represents the gradual failure of the cross section at the location of plastic hinge formation. Interaction between bending moment and axial force is presently not considered.

Case studies

Analysis setting. A pipeline crossing a fault was analysed to test the integrity of the developed program. The geometry of the model, as well as the properties of the pipe and the soil springs (JGA, 1982) are shown in Figure 5. A ductile iron pipe with $E=1.57 \times 10^8 \text{kN/m}^2$ and yield stress $\sigma_y=1.92 \times 10^5 \text{kN/m}^2$ is considered. The model consists of five 5m pipe segments interconnected by four joints of type GM-II, with constitutive behaviour as shown in Figure 4. Plastic hinges within the pipe body are not considered, but plastification due to axial forces is. The fault displacement of 2m is applied upwards to the right-hand side of the model over time of 2sec. The fault intersects the pipe 1m to the right of joint 2, and the crossing angle α is varied between 30° and 120°. The analysis cases are shown in Table 1. The bending moment diagrams along the pipe length are shown in Figure 6 and the axial forces in Figure 7. The final deformed shapes for all analyses are shown in Figure 8.

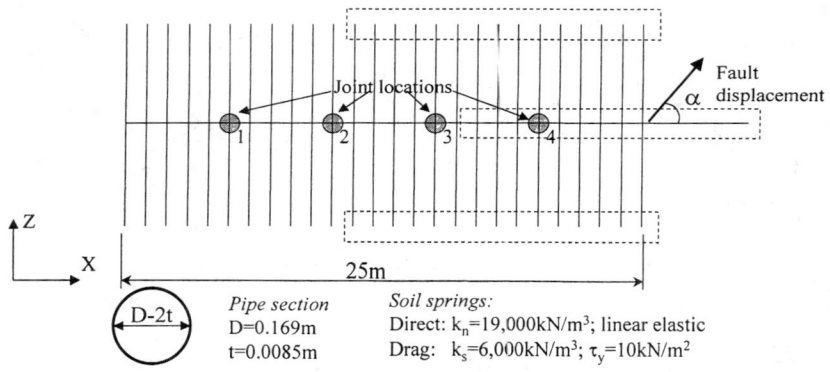

Figure 5. Analysis model

Table 1. Analysis cases

ID	Fault crossing angle (deg.)	Joint failure		
		Joint number	Failure mode	Fault slip (m)
A1	30	2	Tension	0.292
		3	Tension	0.293
		4	Tension	0.294
A2	60	3	Tension	0.467
		4	Tension	0.468
		2	Bending	1.08
A3	90	2	Bending	0.97
A4	120	2	Compression	0.12

Discussion. The joint failure modes shown in Table 1 agree well with the direction of fault displacement, i.e. failure in tension occurs under predominantly tensile loading, A1 and A2; failure in bending under predominantly bending loading, A3; and failure in compression under predominantly compressive loading.

The joints are more vulnerable to axial failure than to bending one. This is especially so for compression because of the small ultimate joint displacement in this regime.

The values in Figure 6 confirm that the axial forces in the pipe are relatively uniform along the pipe length before failure occurs, and much smaller than the axial yield force for the pipe N_u=866kN . This indicates that damage to the pipe body is not likely under axial deformations, but rather joint failure will occur since the ultimate force of the joint in both tension and compression is smaller that the yield force of the pipe.

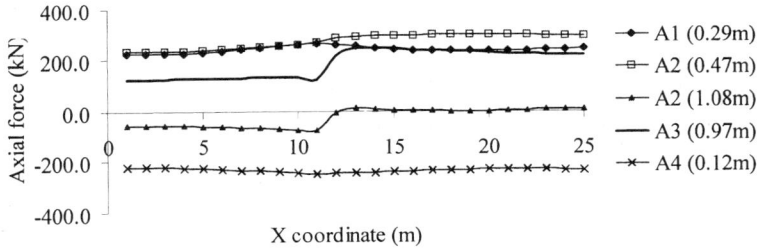

Figure 6. Axial forces in the pipe immediately before joint failure

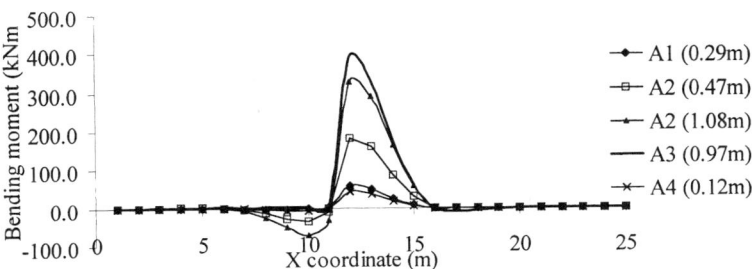

Figure 7. Bending moments in the pipe immediately before joint failure

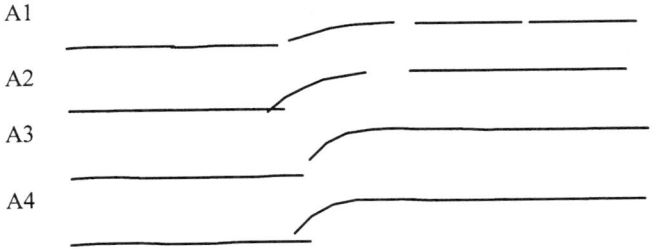

Figure 8. Final deformed shapes of the pipe

In bending, the situation is entirely different. As shown in Figure 7, the pipe segment crossed by the fault develops extremely large bending moments, which exceed the ultimate plastic moment of the section, M_u=38kNm by as much as ten times. This indicates that this segment is likely to sustain damage even if no joints have been damaged.

Conclusions

A comprehensive numerical method for limit state simulation of buried jointed pipelines is presented.

The method is capable of tracing the behaviour of pipelines until and after failure of the pipe body and detachment at joints, thus allowing the physical damage to a pipe network to be better estimated.

In order to minimize the computation time the pipe body is modelled by line elements. The basic mechanical properties of all common pipe joints e.g. stiffening in compression, detachment in tension, fracture in bending can be considered.

The solution algorithm is based on the direct integration of the uncoupled equations of motion of the elements. The implementation of such solution algorithm ensures that the behaviour of a pipeline can be investigated for any magnitude of ground displacements, revealing the whole sequence of events occurring in the pipe-soil system with the increase of earthquake induced ground displacements: elastic deformations, plastic deformations in the soil, plastic deformations in the pipe body, rupture and detachment at joints, failure and rupture within the pipe body.

References

Kennedy, R. P., Chow, A. W., and William, R. A. (1977) *Fault movement effects on buried oil pipeline*, Transportation Engineering J., ASCE, Vol. 103, No. TES, 617-633.

Wang, L. L. R., and Wang, L. J., (1995) *Parametric study of buried pipelines due to large fault movement*, ASCE, TCLEE, No.6, 152-159.

Takada, S., Hassani, N., and Fukuda, K. (2001) *A new proposal for simplified design on buried steel pipes crossing active faults*, Journal of Structural Mechanics and Earthquake Engineering, JSCE, Vol. 668, No. 54, 187-194.

Takada, S., and Tanabe, K. (1987) *Three dimensional seismic response analysis of buried continuous or jointed pipelines*, Journal of Pressure Vessel Technology, ASME, Vol. 109, No. 54, 80-87.

Takada, S., Tsubakimoto, T., and Hori, K. (1983) *Earthquake behaviour and safety of oil and gas storage facilities, buried pipelines and equipment – PVP-77*, ASME, 357-364.

Japan Gas Association (1982) *Recommended standards for earthquake resistant design of gas pipelines*

Southern Loop Pipeline – Seismic Installation in Today's Urban Environment

Tom Shastid, Javier Prospero[1]
John Eidinger[2]

1 Introduction

The Southern Loop Pipeline is a key element of the 10-year Seismic Improvement Program (SIP) of the East Bay Municipal Utility District (EBMUD). The newly constructed 11-mile long Southern Loop Pipeline connects the Northern California cities of San Ramon and Castro Valley and is designed for flow in both directions so that it can provide an emergency water supply following major seismic events on either the Hayward or the Calaveras Faults or other kinds of emergency events that could disrupt the normal flow of water to these cities. The portion of the pipeline within San Ramon crosses the Calaveras Fault. This paper discusses how the EBMUD Southern Loop Pipeline (Figure 1) addressed the challenges in constructing a large diameter pipeline across a major fault with an anticipated magnitude 7± earthquake. The fault was crossed using a design that incorporates thick-walled pipe using non-standard pipeline steel (both minimum and maximum yield strengths had to be considered), specialty pipeline coating and custom backfill material into a system that accommodates the predicted ground movement.

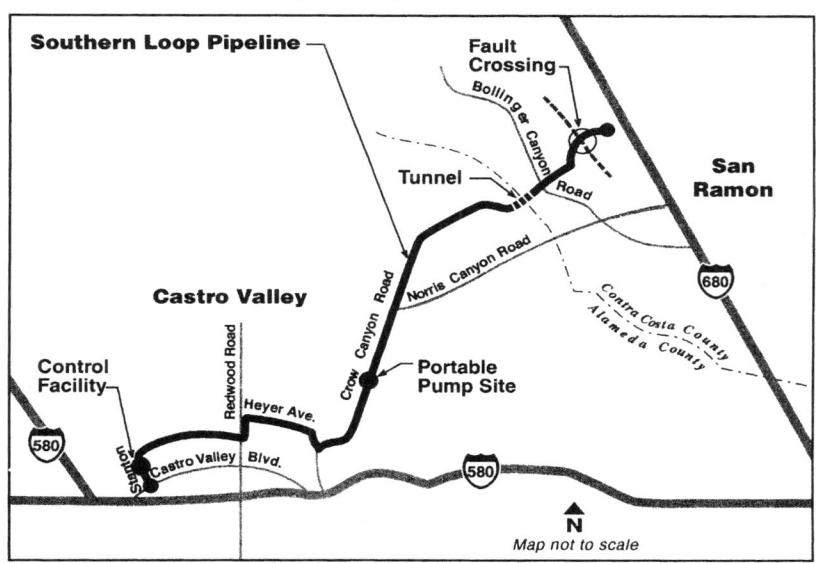

Figure 1. Southern Loop Pipeline, Showing Location of Calaveras Fault in San Ramon Area

---

[1] EBMUD, 375 Eleventh Street, Oakland, CA 94607; tshastid@ebmud.com, jprosper@ebmud.com
[2] G&E Engineering Systems Inc., 6315 Swainland Rd, Oakland, CA 94611; eidinger@earthlink.net

2 Fault Offset Design Criteria

The pipeline alignment crosses the active trace of the Calaveras fault. As the primary function of this pipeline is to deliver water in the event of a major earthquake (or other event) that might damage the other pipelines that bring water to the service areas, it was decided that the Southern Loop Pipeline should be designed to have a high reliability of remaining functional immediately after the earthquake (in other words, the pipe should not break).

As the first step in the design process, it was necessary to establish a suitable amount of fault offset. The amount of fault offset is an uncertain parameter, even if one can characterize the magnitude of the earthquake. Further, the magnitude of future large earthquakes is also uncertain. We combined these two sources of uncertainty to establish a probability of offset exceedence, given the occurrence of a characteristic earthquake. Figure 2 shows the result. Moment magnitude 6.9 to 7.1 earthquakes dominate the risk, although there is some chance (under 5%) that the magnitude could be has high as 7.4 to 7.6.

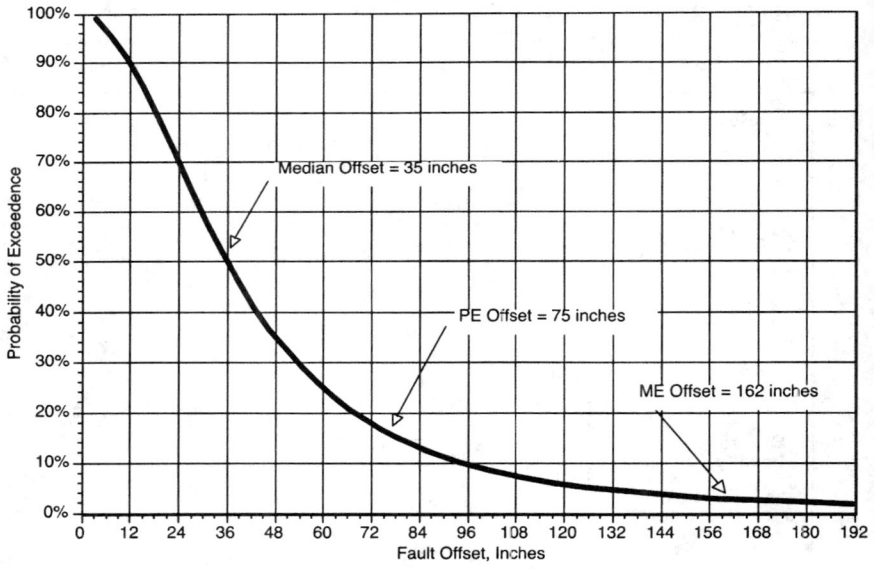

Figure 2. Range of Possible Offsets of the Calaveras Fault

Based on these findings, we decided to design the pipeline for reliable operation with a right lateral fault offset of 75 inches plus simultaneous vertical offset of 7.5 inches (16% chance of exceedence). We also recognized that fault offset could exceed this amount (about a 2% chance of offset of 162 inches or more), so the design concept included components to address these less likely but still possible occurrences.

3 Design Concept

As the Calaveras Fault trace within San Ramon is well established it was projected that the bulk of the 75" of design fault offset would occur within a relatively short fault zone, approximately 10 feet wide. The location of this 10 foot wide fault zone was somewhat uncertain, and this was reflected in the design. Rather than trying to accommodate this offset within the fault zone, the design concept adopted was to allow the strain to be distributed over an extended length of the pipeline. This would be achieved by providing a slick polyurethane coating for the pipeline within and near to the fault zone combined with a sand backfill. The combination would allow the pipeline to slip within its bed during the predicted fault offset, distributing the strain over a greater distance. Outside of the fault zone the pipeline would be anchored using a controlled-density fill to prevent the strain from extending too far into the remainder of the pipeline. Within this anchor region, the pipe has exterior steel rings welded on to provide 'teeth' to grab onto the controlled density fill. The length of the 'slip zone' is around 600 feet and each of the anchor zones are approximately 300 feet long. Thus the 75" of design fault offset will be absorbed within a total design length of around 1,200 feet.

As noted above, fault offset prediction is not an exact science and there is a probability that the actual fault offset will exceed the design offset. The design concept addressed this in two ways. First, the pipe in the anchorage zones was designed to yield if the offset exceeds the design assumptions, allowing the strain to be distributed over a greater length. An additional element of design redundancy was included in the form of manual shut-off valves and bypass manifolds as part of the design. These components (see Figure 3) allow for bridging of the flow across the fault in the event the actual strains exceed the design assumptions and the pipeline leaks or breaks.

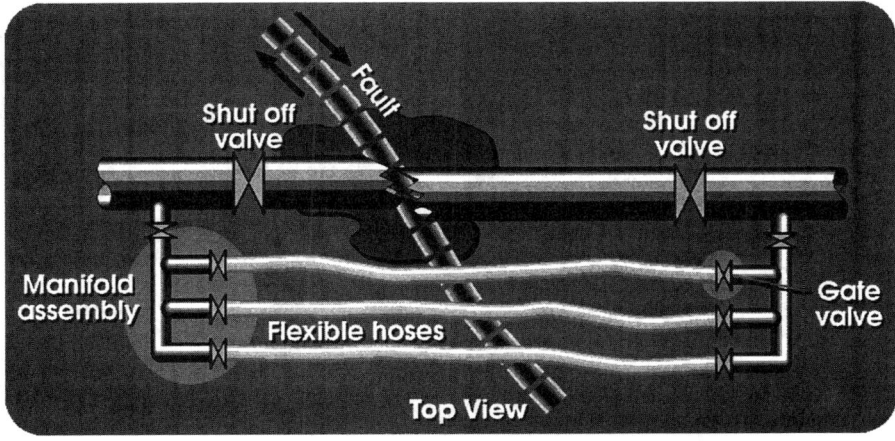

Figure 3. Bypass Design Concept

4 Alignment Selection Across the Fault Zone

During the preliminary design phase of this pipeline, several possible alignments were considered for the pipeline. From an earthquake / fault crossing point of view, the optimal alignment was to install the pipe in a straight alignment for several hundred feet either side of the fault, oriented such that right lateral offset would produce net tension everywhere in the pipeline. In order to use this type of alignment, it would have been necessary to purchase a right-of-way from private property owners through undeveloped land. From a construction and ownership point of view, EBMUD desired to keep the pipeline within the right-of-way of an existing street.

After much discussion it was decided to keep the pipeline within the public right-of-way. This necessitated a pipeline design that would include a few minor to medium bends near the primary fault crossing zone, in order to accommodate the right-of-way and the presence of other utilities already in the street (Figures 4a and 4b). As will be further discussed, this caused somewhat higher strains in the final design than would otherwise have been possible, and contributed to the decision to include a set of manual shutoff valves and bypass outlets on either side of the fault.

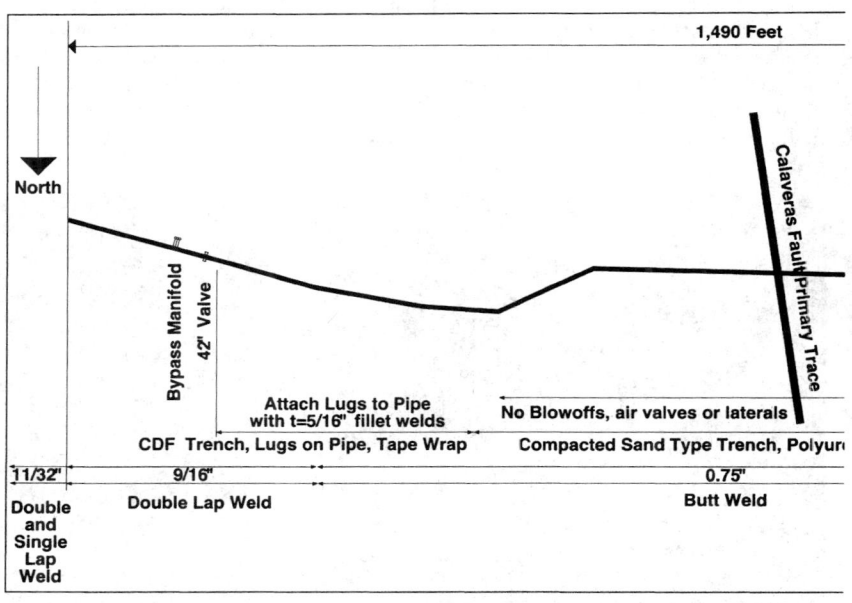

Figure 4a. Pipe Alignment With Main Design Features (East of Fault)

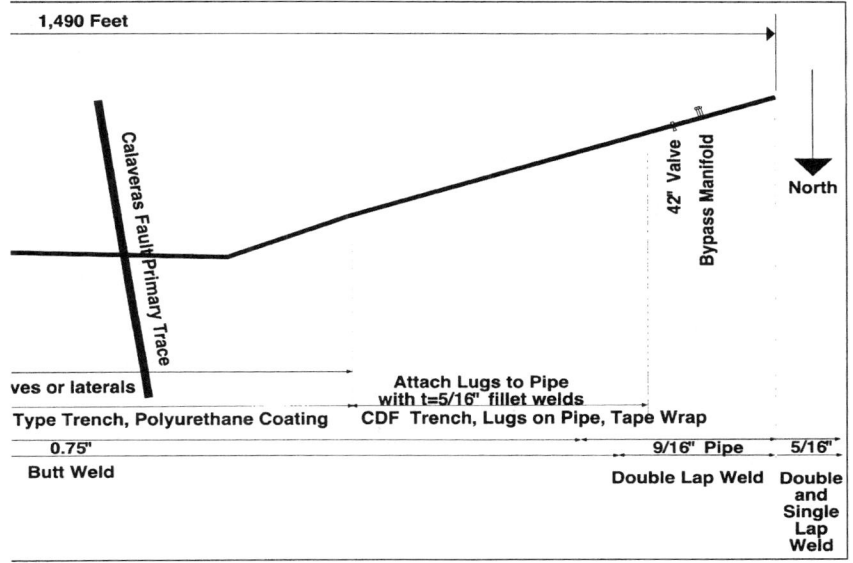

Figure 4b. Pipe Alignment With Main Design Features (West of Fault)

5 Pipeline Design

A series of nonlinear structural analyses of the pipeline in its final alignment were undertaken. Of most importance was the prediction of high tensile and compressive strains. After iteration on the design wall thickness, it was decided to use a 42" diameter butt welded steel pipeline, with wall thickness of 0.75" in the immediate fault crossing zone, tapering off to 3/16" wall thickness at some distance from the primary trace of the fault. The wall thickness was adjusted along the length of the alignment to minimize the cost of installation, while maintaining pipe strains at all locations within tolerable levels. Figure 5 shows typical results from the nonlinear analyses. Highlighted in Figure 5 are the strains at points A, B, C, D and E.

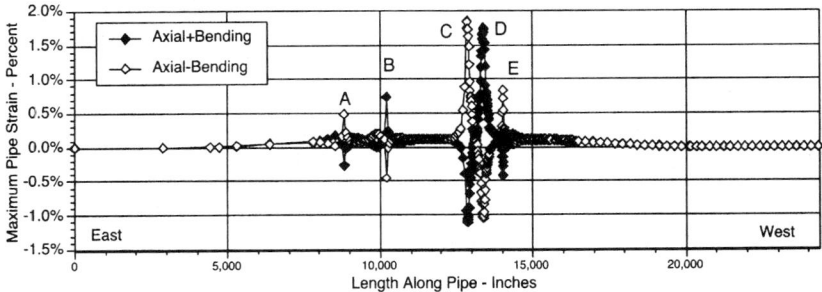

Figure 5. Pipe Strains Due to Fault Offset

Points A, B and E correspond to bends in the pipeline that could not be avoided due to interferences from existing utilities within the right of way. Points C and D represent the highest strain points in the pipeline immediately either side of the fault. As can be seen in Figure 5, the slight acute orientation of the pipeline from a 90 degree crossing results in a slight net tensile strain over about 15,000 inches of alignment (1,250 feet). The strains are greatly magnified at the three bend points, reaching about +0.8% / -0.5%. The strains at the fault crossing itself are highest, at about +1.9% / -1.1%.

When considering the level of strain, it is important to reflect that the tensile strain is usually easiest to resist, with a well installed butt welded pipe with ductile steel able to accommodate about 5% tensile strain with a small chance of failure. In compression, a 42" diameter x 0.75" thick pipe wall will begin to wrinkle somewhere about 0.6% to 0.8% compression, and reach a full wrinkled state somewhere about 1.5% to 2.0% compression. The analytical prediction of -1.1% strain suggests that there would be modest factor of safety against large wrinkling, given a 75 inch fault offset. Even with major wrinkling, the potential for a pipe tear is modest, and even with a pipe tear, the leak rate would likely be small enough to allow the pipe to be kept in service. Still, the urban requirement to keep the pipe within the right of way results in a much higher compressive strain than could have been achieved with a better alignment. (Note: all strains mentioned here exclude additional strains due to localized wrinkling).

6 Material Specification

During the course of the design, it became apparent that a ductile steel with yield stress of about 36 to 40 ksi and ultimate of about 60 to 78 ksi or so would result in the lowest cost installation. As yield stress goes up, the force generated in the pipeline due to fault offset goes up, and hence a longer anchorage length is needed of thicker wall pipe. Accordingly, the pipeline steel purchase specification listed both minimum *and* maximum yield and ultimate stresses. Several iterations were required with the pipeline vendor in order to assure that the actual steel delivered was not stronger (higher yield stress) than what the design would allow for (see also Section 7). It was found that steel vendors often produce steels with actual yield strengths on the order of 54 ksi to 62 ksi, even if the minimum specified is just 36 ksi to 40 ksi; this would not have been the usual case 30 years ago, but apparently the modern decision to re-use scrap steel as part of the make-up of new steel has resulted in limiting steel manufacturers' economic capability of producing steel within a relatively tight range for its actual yield strength.

In addition, the controlled density fill placed around the pipe within the anchor zone required a minimum compressive strength in order to function effectively as an anchor for the pipe elongation and a maximum compressive stress in order to ensure that the anchor failed before the pipe was overstressed . Thus the controlled density fill was specified with a minimum compressive strength of 70 psi and a maximum compressive strength of 150 psi.

7 Steel Procurement

The above described design process produced a design that addressed the inherent uncertainty of fault offset in an innovative and cost-effective manner. However, construction of the fault crossing identified several real world constraints that need to be addressed when implementing a design solution of this type. Procurement of the steel to be used in pipe fabrication was the first of the challenges encountered. Working with the selected pipeline vendor, it became apparent that the vendor had little experience with specification of both minimum *and* maximum yield and ultimate stresses for pipeline steel. Their internal quality control processes were based upon assuring that the minimum strength was achieved but they had no data on what the maximum yield and ultimate strengths were (perhaps the engineer should have watched over the unusual steel specification and procurement process a little closer). A representative set of steel samples were sent to a testing laboratory and it was determined the supplied steel material did not comply with the maximum stress allowed by the specifications. After several iterations a specific heat of steel was identified that met both the minimum and maximum strength criteria. Unfortunately, when several cylinders had to be rejected due to weld defects, it was determined that the specified heat was from a mill that had just gone out of business (thanks in part to globalization) and no additional steel was available from that vendor. After several weeks of querying steel vendors throughout the United States a supplier was located in Alabama that could produce a steel that was within 5% of the maximum specified. After several reiterations of the design calculations it was determined that this new steel could be used in the areas outside of the primary fault zone. Ultimately, in order to save on construction installation cost and schedule, individual spool pieces had to be carefully marked and installed, to assure that the spool pieces of pipeline nearest the fault did not have excessively high yield stress. While careful factory and on-site inspection ensured that the material installed matched the design criteria, the experience re-emphasizes that designers need to be address the potential procurement issues, when non-industry standard materials are specified.

8 Controlled Density Fill Quality Control

The specification of the Controlled Density Fill with both a maximum and minimum strength also presented several challenges. While Controlled Density Fill, or CDF, has seen widespread acceptance over the last decade as a backfill material, it is not typically utilized as a structural element and therefore does not have the extensive quality control standards that have been established for Portland cement concrete. When asked for quality control history of CDF compressive strengths, the local cement vendors had no data but indicated strength variations between 50 psi and 500 psi are not uncommon. In addition, the standard cylinder breaks used to monitor concrete performance cannot be utilized for CDF. Due to its low strength, there is a high potential for cylinder fractures during transport and bands of aggregate can produce anomalous breaks. Also, the strength vs. cure time relationship established for concrete are not applicable to low-strength CDF and predicting final strengths from early cylinder breaks was not possible. Eventually, a specialized testing protocol was established, a series of test batches was run and strict quality control instituted at the batch plant to ensure that this atypical strength criteria was met. If future designs anticipate utilizing CDF as a structural fill then a similar testing and quality control program should be included in the specifications. Photo 1 shows the installation of the CDF around the pipeline. Several of the exterior steel rings that provide the "teeth" for the CDF to grab onto can also be seen in the foreground.

Photo 1. Placement of CDF Around the Pipeline

9 Utility Realignment

The other challenge that arose during the construction of the pipeline was the conflict with an unanticipated utility. Every effort was made to locate known utilities during the design phase. However, the designers were really not thinking much about *new* utilities still being built in the area. During the actual construction sequence, as the pipeline construction got closer and closer to the fault crossing zone, it was determined that only four weeks earlier a new sewer had been installed that conflicted with the proposed pipeline alignment at the fault crossing zone (a few choice words were muttered by the field team upon discovery of this). The problem required quick resolution, as the pipeline installation crew would reach the fault zone within a matter of weeks. Since the installation crew and equipment were costing on the order of $16,000 per day, the cost of delays would mount quickly. The initial approach was to include changes in pipeline angles within the design length so as to go around the sewer. However, even minor additional bends would increase the stresses and strains in the pipeline to unacceptable levels. Similarly, lowering the pipeline by the 12" to 18" required to avoid the sewer would compromise the pipeline performance as the amount of overburden affected the pipeline's ability to 'slip' within its sand bed.

Ultimately a fast track design was developed in cooperation with the sewer agency in order to realign the recently-installed sewer line.

Towards the end of the pipeline installation within the fault zone a buried concrete vault was discovered directly in the pipeline alignment, see Photo 2. As discussed in the previous paragraph, modifying the pipeline alignment to the degree necessary to avoid the vault would have a major impact on pipeline performance. Relocating the vault, if at all possible, would have had major cost and schedule impacts. Surprisingly, the owners of the vault quickly determined that its removal would have minimal impacts on their operation and the vault was demolished within one day. Future designs with similar tight dimensional constraints within an urbanized area should anticipate the potential for similar conflicts.

Photo 2. Large Concrete Vault Discovered in Pipeline Alignment

10 Acknowledgements
The authors would like to acknowledge the contributions and efforts of both Kennedy/Jenks Consultants, the Lead design firm on the project, as well as Ranger Pipelines Inc., the construction General Contractor. Construction of a major pipeline is never a solo effort, and the teamwork exhibited by the personnel involved greatly contributed to a successful installation of a state-of-the-art design for crossing a critical water lifeline across a major earthquake fault.

Pipeline Seismic Mitigation Using Trenchless Technology

Le Val Lund, P.E., M. ASCE-TCLEE [1]

Abstract

Water, wastewater and natural gas utilities are usually limited in funds for seismic mitigation of their pipelines. A number of methods are discussed which can be used for lifeline pipeline rehabilitation and enhancing the seismic performance of pipelines used for water, wastewater and natural gas, as well as improving the flow capacity and in the case of a water pipeline water quality. These methods use the trenchless technology, which minimizes disruption to traffic, business and residents; impact on the environment; reduces costs and the time of construction. All the methods use the annular space in an existing pipe for the pipeline rehabilitation avoiding trench excavation.

Introduction

Lifelines. A lifeline pipeline can convey water, wastewater, storm drainage, natural gas and liquid fuels and a lifeline conduit can contain power and communication cables. Lifelines are those systems and facilities necessary for the functioning of an industrialized society and also necessary for emergency response and recovery after a natural disaster, such as an earthquake. The seismic hazards that may impact pipelines and conduits are faulting, shaking, liquefaction, lateral spreading, and landslides. Also they may be damaged by another lifeline system, which is damaged and is co-located with the pipeline or conduit.

Seismic Mitigation. The pipeline seismic mitigation is based on the facts that using trenchless technology an almost complete overhaul is made of the pipeline by the following activities:

1. All the old fittings (bends, tees, crosses, valves and other obstructions) are removed from the pipeline.
2. After the pipeline is cleaned, a thorough physical and closed circuit television inspection is made of integrity of the pipeline at the launch pits and locations where the fittings have been removed.

---

[1]Civil Engineer, 3245 Lowry Road, Los Angeles, CA 90027, telephone 323-664-4432 lundasan@earthlink.net

3. Laboratory testing of the materials may be made to determine there physical characteristics.
4. All the fittings are replaced with new units and in some cases this involves replacing the fire hydrants, service connections and meters.
5. The insertion of a liner of generally flexible materials provides for some seismic movement of the pipeline.
6. In some cases the liner provides for the seismic strengthening of the pipeline and make it available for higher internal pressure.

Temporary Repairs. Temporary repair of pipelines and conduits are necessary to restore utility service to medical facilities, police and fire stations, and emergency operating centers. Also necessary is water for public fire protection, natural gas for cooking and space heating and liquid fuel for operation of emergency equipment and lifeline facilities. Temporary repairs are made to prevent floods from broken water lines, fires and explosions from broken gas and fuel lines, and sewers and storm drains to prevent pollution and the spread of disease. Restoration of power and communications facilities is necessary for emergency response and recovery. Eventually pipelines and conduits are repaired to provide service to homes, business and industry.

Permanent Replacement. Permanent pipeline replacement is made to increase seismic resistance, increase capacity, improve water quality and minimize leaks of gas, fuel oil, water and wastewater. These leaks result in unaccounted for gas, fuel oil and water and result in loss of revenues. This report is concentrating on water and wastewater systems; however, the methods can be applied to the other lifeline systems.

Financial Impact. Difficulties in implementing seismic mitigation or rehabilitation of pipelines and conduits are normally financial, which requires management and sometimes public and political support. However, in some cases the lack of good information on the condition of pipes, and the operational restraints necessary to maintain service may impact the decision process. Financing the program may come from normal rates charged for the quantity of product used, emergency surcharges, taxes approved by the governing body, general obligation or revenue bonds, insurance, government grants or loans. Underground permanent replacement, if repair is not feasible, and seismic mitigation if desired, can be done by two methods, trench excavation or trenchless technology.

Trench Excavation

Replacement of pressure pipelines using trench excavation method, using welded steel pipe with welded joints and ductile iron pipe with rubber gasket joints has shown good performance in past earthquakes. Various types of mechanical couplings, which provide flexibility during a seismic event, can join both of these types of pipe. Trench excavation in an urban area with busy roadways causes disruption of traffic and disturbs the residents and businesses and may have an impact on the environment. The type of joints used in this pipe is important for good seismic performance. Joint types will be discussed later in this report.

Trenchless Technology

Trenchless technology rehabilitation of pipe has increased in popularity in urban areas, because of its lessened impact on traffic, residents, businesses and environment. The pipe or conduit rehabilitation has some value towards seismic mitigation; however, it cannot be specifically quantified. Also important is the fact generally new pipe occupies the existing space among other substructures in a somewhat crowded area with other utilities. Also trenchless technology can reduce the cost and time of construction. All of the methods require a drained and cleaned pipe and temporary supply of water to implement. The rehabilitation of existing water mains by cleaning and lining insitu permanently improves the hydraulic capacity, seals joints and pinhole leaks and improves water quality by minimizing taste and odor problems associated with unlined pipe. Included in this report is some samples of trenchless technology; however, there are others and some others under development.

Temporary Bypass. Temporary bypass or sideling systems maintain customer water service prior to the cleaning and lining operation used in trenchless technology. The quick connect pipeline is usually laid in the gutter along the curb line on each side of the street. Connections to the active water system are made at the fire hydrants or in some cases by a wet tap to the pressurized water main. Normal installation requires 50 and 100-millimeter (mm) (2 and 4-inch) diameter lines. Temporary hose connections are made to the existing meters for residential, commercial and in some cases for fire service. Wastewater system can be bypassed by pumping from an upstream manhole to a downstream manhole. Temporary supply of water, although rarely done, can be provided by tank trailers, tank trucks, bottled water or portable hose.

Pipe Cleaning. After the bends, valves and other obstructions are removed, the pipe is cleaned by using one of the following methods:

1. Hydraulic-A steel frame in the form of a piston with protruding metal scrapper blades is propelled through the pipeline by water pressure.
2. Mechanical-The cleaning scrappers are pulled through the pipe by a winch. The scrapper is pulled back and forth until the pipe wall is cleaned for lining. In both cases water is used to flush debris out of the pipe.

Closed Circuit Television Inspection. Closed circuit television inspection is done to observe the condition of the interior of the pipe and identify any unknown obstructions or connections. After the pipe has been cleaned a self-propelled television camera photographs the interior of the pipe. The camera has lights to illuminate the interior of the pipe. The image is transmitted to a monitor in van located outside the launching pit, where an operator observes the condition of the pipe and makes videotape for future reference.

Rehabilitation methods

Cement Mortar Lining. After inspection and the pipeline is cleaned, a cement-mortar is premixed above ground and pumped to the lining machine through high-pressure rubber hoses inside the pipe. The cement mortar lining machine consists of a rapidly spinning dispensing head that centrifugally applies a uniform coating of cement, sand and water to the pipe wall as the lining machine is slowly winched through the pipe. As the cement-mortar is applied, a flexible conical troweling device follows behind to produce a smooth hydraulically efficient surface. In large diameter pipes the smoothing of the cement-mortar is done with rotating trowels instead conical troweling device. Cement mortar lining is specified by American Water Works Association (AWWA) Standard AWWA C-602. The mortar is 1:1 mixture of Portland cement and silica sand with water added for proper placement. Cement-mortar lining is generally not considered a seismic improvement; however, the fact the fact that old valves, fittings and service connections are upgraded during rehabilitation, it provides some seismic improvement. In the 1994 Northridge earthquake, field personnel observed old cast iron and steel pipe which had been cement mortar lined in place appeared to have a lower frequency of leaks; however, this has not been scientifically documented.

Pipe Insertion. Pipe insertion method is done in existing larger diameter pipes, where retaining the diameter for flow capacity is not critical. The existing pipe requires inspection, cleaning and the removal of valves and sharp bends. Prefabricated welded steel pipe is installed within the annular space of the old pipe and joints are welded. The annular space between the old pipe and the new pipe is filled with cement grout.

Slip Lining. Slip lining method using high density polyethylene (HDPE) pipe is used on existing pipe in both the small and large sizes mainly for gas, water and wastewater lines. The HDPE pipe is laid out along a roadway, butt joints are fused together under high temperature and pressure The HDPE pipe is pulled through the existing host pipe. Connections are made to existing pipe by a specially fabricated bolted flanged connection. The HDPE pipe can either be smaller than the existing pipe or through a special process (pipe bursting or splitting to be discussed later) maintain the same inside or increase the inside diameter the host pipe.

The British gas industry has been the pioneer since the 1970's, in using slip-lining technologies to rehabilitate gas mains. Later wastewater agencies started using the slip lining method for sewer mains. Gas mains rehabilitated with HDPE pipe performed well in the 1994 Northridge earthquake. The use of HDPE pipe for water service has lagged due to the approval process for the material to be used in potable water systems.

Expanded Molecular Reoriented PVC. The Expanded Molecular Reoriented Poly Vinyl Chloride (PVC) method is a patented system for water pipelines, which guarantees a working water pressure of 1000 kilopascal (kPa) (150 pounds per square inch (psi)). The manufacturer "DURALINER" ® of this pipe claims that the reoriented pipe will test for 5 to 6 times this pressure and the resulting pipe is a "structural stand alone pipe". The method lines the host pipe with a standard AWWA C-900 class PVC pipe.

After the existing pipe has been cleaned and inspected, a slightly smaller diameter standard new PVC pipe is pulled through the host pipe and each end is sealed. Joints are fused together with heat and pressure similar to the joints in the slip lining method. A boiler makes steam and a compressor injects the high-pressure steam into the PVC pipe, gradually expanding the PVC, so that fits tight against the interior of host pipe. The integrity of the host pipe is not important for maintaining pressure, since the PVC pipe is structurally sound.

Insituform Technologies, Inc. Methods. There are several methods available for gravity and pressure pipe systems, which are trade names of the Insituform Technologies, Inc.

The Paltem System ® uses continuous woven polyester hose with an elastomer coating and can be used in pipelines up to 1000-mm (40-in) in diameter and through bends up to 90 degrees. The uncoated side of the liner is covered with an epoxy resin. Compressed air or water pressure is then used to invert (turn inside out) and propel the liner through the pipe from the access pit. This is called the inversion process. Heat or ambient temperature is used to cure the epoxy resin and adhere the liner to the inside of the pipe. The ends of the liner are then cut off and the end seals are installed before the line is placed back into service.

The Pressure Pipe Liner ® is like the Paltem System for structurally sound pipe system. The system uses a reinforced felt tube and modified resin system. It can be installed in diameters 200- to 1200-mm (8- to 48-in) in diameter and uses the same inversion process as the Paltem.

The Thermopipe System ® is used for rehabilitating small distribution mains, 100- to 200- mm (4- to 8-in), and has a long-term hydraulic pressure rating of 1000 kPa (150 psi). The system is polyethylene tube reinforced woven polyester fiber and is factory folded in a "C" shape. It is then wound onto a reel, which enables it to be transported to a job site. The liner is winched into the pipe and is re-rounded using steam and air pressure, so that it closely fits the existing inside diameter of the pipe. End seals are installed before the line is placed back into service. It can be used to rehabilitate pipes of all common materials and can be used through slight bends.

Pipe Bursting. The pneumatic pipe bursting process simultaneously breaks the old pipe pushing it into the surrounding soil while the new pipe is pulled into place. This method is mainly used for cast iron, concrete, and vitrified clay pipe. Again this is done after a television inspection and removal of gate valves and sharp bends and the cleaning of the pipe. An expander on the pipe-bursting tool increases the diameter of the hole. The pneumatic tool (Grundocrack ®) is guided through the old pipe by a cable, and an internal reciprocating piston breaks the pipe and supplies the force for most of the forward motion.

The HDPE pipe is laid out along a roadway; joints are fused together under high temperature and pressure as in the slip lining process. The HDPE pipe is attached and immediately follows the expander tool into the existing host pipe. A large capacity portable compressor is used to drive the pneumatic tool. Sometimes an extension in front of the expander tool, called a schnoze, is used to help guide the expander tool in the host pipe. A bentonite solution is sometimes used to reduce friction for larger diameter HDPE and to maintain the annular space created as the tool travels through the host pipe.

Connections are made to existing pipe by a specially fabricated bolted flanged connection. The new pipe is a high-density polyethylene (HDPE) material, which can be up to 50-mm (2-in) thick depending on system pressure. Even with the same diameter this new pipe reduces the hydraulic friction and automatically increases the flow capacity. Also this process provides an opportunity to increase the pipe diameter from 0 to 25% to further increase flow capacity of the pipe.

While most pipe bursting jobs use the pneumatic type of bursting tools there are other types. There is static bursting which pulls a splitting head into the host pipe and fractures the host pipe by the constant pulling by the chain or cable attached to the splitting head. The other method used is the hydraulic actuated bursting head to fracture the pipe. A cable attached to the front of the bursting head pulls the devise through the pipe. Again the new HDPE is immediately pulled through the expanded pipe.

Pipe Splitting. The pipe splitting system is somewhat similar to the pipe bursting process, which simultaneously breaks the old pipe pushing it into the surrounding soil while the new pipe is pulled into place. This method is mainly used for steel pipe. Again this is done after a television inspection and removal of gate valves and sharp bends and the cleaning of the pipe. A launch pit is required at each end of the pipeline to be lined. The Grundoburst ® System consists of a hydraulic power unit, hydraulic flow control and Grundoburst ® hydraulic bursting unit.

Quicklot ® bursting rods are pushed through the host pipe with the Grundoburst ®, until they reach the launch pit at the other end of the pipeline. A flexible guide rod out front helps the rods navigate the existing line. Once at the launching pit, the guide rod is removed and bladed cutting wheels, a bursting head, an expander and new HDPE pipe are attached. The entire configuration is pulled back through the host pipe. The bladed cutting wheels split the existing pipe. The bursting head and expander displace the fragmented host pipe into the surrounding soil while the new HDPE is pulled through the host pipe. An expander on the pipe bursting tool increases the diameter of the hole. The especially design bladed rollers split the host pipe instead of ripping or tearing it. Various bladed rollers are available to split a wide range of pipe diameters.

The HDPE pipe is laid out along a roadway; joints are fused together under high temperature and pressure as in the slip lining process. The HDPE pipe is attached and immediately follows the expander tool into the existing host pipe. Connections are made to existing pipe by a specially fabricated bolted flanged connection. The new pipe is a high-density polyethylene (HDPE) material, which can be up to 50-mm (2-in) thick depending on system pressure. Even with the same diameter this new pipe reduces the hydraulic friction and automatically increases the flow capacity. Also this process provides an opportunity to increase the pipe diameter from 0 to 25% to further increase flow capacity of the pipe.

Joint System

In the trenchless technology method to improve the seismic performance of the joint system requires the excavation of each joint and may prove to be as much inconvenience to the residents, business and the environment. The costs and time of construction would increase.

The joint system is critical in the seismic performance of segmented piping systems. Butt-welded joints have shown better performance in past earthquakes than bell and spigot welded joints. One of the better joint systems is one that permits both transverse and longitudinal movement of the joints for water and wastewater pipelines. (Examples are in USA FlexTend ® and in Japan SuperFlex ®) This type of joint is available in sizes up to 1500-mm (60-in) in diameter, and is about 30% more expensive than conventional pipe joints. It performed very well in the 1995 Kobe earthquake. This joint is normally used for inlet/outlet connection to tanks and reservoirs, and at large service connections, but it may be used at other critical locations, such as, fault crossings or areas subject to liquefaction.

The butt-welded joint system avoids the eccentricity of the conventional bell and spigot joints. Butt-welded joints on gas lines performed well in the 1994 Northridge earthquake. Studies at the Cornell University, New York are being performed to improve the performance of the bell and spigot joint, by changing the configuration of the bell and spigot or by applying a reinforcing plastic wrap around the joint. This study is funded by the Multidisciplinary Center for Earthquake Engineering Research, Los Angeles Department of Water and Power and others. The ductile iron seismic joint (S1 and S2) with rubber gaskets used in Japan performed very well in the 1995 Kobe earthquake.

Conclusions

Pipeline rehabilitation using trenchless technology is used by utilities to minimize the impact of construction on residents, business and the environment at a reduce cost and time of construction to provide improved flow capacity, water quality and enhance the seismic resistance of pipelines.

Implementation of seismic mitigation program should be done after completion of a seismic vulnerability analysis of the entire utility system. The mitigation program also requires an updated and practiced emergency response and recovery plan. Pipeline utilities in areas of seismic activity should maintain an inventory list and continuously update emergency resources of temporary bypass piping, pipe, fittings, repair clamps, equipment and pipeline construction industry specialized trained personnel.

Related References

American Society of Civil Engineers, Pipelines in the Constructed Environment, Proceedings of the 1998 Pipeline Division Conference, August 23-28, 1998, San Diego, California, Edited by Joseph P. Castronovo and James A. Clark, Reston, VA, 1998

Insituform Technologies, Inc., Technical Bulletins, Chesterfield, MO, 1998

Insituform Technologies, Inc., Grundocrack ® Pneumatic Pipe Bursting System, Aurora IL 1996

J. Fletcher Creamer and Son, Inc., Pipe Cleaning and Lining Services, Hackensack, NJ, 2000

Insituform Technologies, Inc., Grundoburst ® Static Pipe Bursting System, Aurora IL 2001

Personal contact with Richard Nieman and Mark Smith, Duraliner, New Orleans, LA; Welfredo Paz and Kathy Harada, Los Angeles Department of Water and Power; George Mallakis, J. Fletcher Creamer and Son, Inc., Sylmar, CA; Collins K. Orton, TT Technologies, Inc., Aurora, IL; 2002.

LL 01-27-03 6USCLEE

CONSIDERATIONS FOR THE DESIGN OF BURIED NATURAL GAS AND LIQUID HYDROCARBON PIPELINE FAULT CROSSINGS

Douglas J. Nyman[1], Douglas G. Honegger[2], and Paul C. Thenhaus[3]

Abstract

This paper presents a summary of the approach currently used by the authors on high-pressure natural gas and liquid hydrocarbon pipeline projects for the analysis and design of buried fault crossings. The paper describes the investigation of fault crossing hazards, presents a strain-based analysis and design methodology, and identifies key areas of uncertainty that affect determination of fault crossing design margins. Recommendations for provisional strain criteria are included along with a discussion of several materials and welding issues of current interest to practitioners.

Introduction

For the evaluation and design of a buried fault crossing by a high-pressure natural gas or liquid hydrocarbon pipeline, it is first necessary to delineate its location, orientation, slip characteristics, and zone of disturbance for the fault and to estimate the amount and type of potential displacement that may occur. Equally important, it is necessary to define acceptable performance for the pipeline. Since fault rupture is a rare event, most pipeline companies are willing to accept the risk of damage provided pressure integrity is maintained, i.e., no leaks. Next, compressive and tensile strain criteria commensurate with the desired level of pipe performance, materials and welding must be established. Compressive strain criteria are primarily a function of the depth-to-thickness ratio and, thus, can vary with selection of a pipe cross-section. The design of the fault crossing then involves the appropriate selection of crossing angle, trench configuration and backfill, burial depth, and pipe wall thickness. Various permutations of crossing parameters can be investigated via non-linear finite element analysis until a suitable crossing configuration is determined.

Fault Characterization

The level of geologic investigation required for fault crossings depends on the requirements of project-specific design and risk criteria. Risk criteria are developed from a number of economic, environmental, and social factors that are external to the geological investigation, but yet they determine the level of refinement required of the geological investigations. The effort and cost of field investigations increases sharply with increasing requirements for precise information regarding age of latest fault displacement and the nature, timing, and amount of expected future displacement. At some point in every investigation, there are practical limitations on

---

[1] Consulting Engineer, D.J. Nyman & Associates, Houston, TX, F. ASCE.
[2] D.G. Honegger Consulting, Arroyo Grande, CA, M. ASCE.
[3] Director of Seismic Hazard Services, ABS Consulting, Inc., Lakewood, CO

the amount and precision of the information required for adequate engineering of a pipeline-fault crossing.

Common to all geological investigations are the collection and review of available geological literature and maps for the project region. The quality of this general information is highly variable and depends on the remoteness of the region, the economic significance of the region relative to resource development, and geological uniqueness of the region (which may have attracted prior research studies). Typically, the general fault information gained from the literature search forms the preliminary inventory of fault sites relative to the pipeline route. However, the precision of this fault information is unknown, and inferred ages of displacement from standard geological maps are too broad to confidently identify specific faults that are active in the contemporary tectonic stress regime.

Stereo aerial photography is perhaps the most cost-effective and useful tool for identification and prioritization of faults according to their surface expression, which, in turn, is an indication of their activity. With good quality images at a scale of 1:24,000 or larger, an experienced interpreter can identify vertical offsets of surface-breaking faults on the order of 0.3 m or less. Particular attention is paid to projections of fault traces through Quaternary-age deposits (1.6 million years before present), such as alluvial fans at the foot of mountain fronts and alluvium deposited in stream valleys. The youngest surface sediment in these geological settings is likely deposited within Holocene time (the last 10,000 years). If offset of these young materials is observed in the aerial photography, the fault can tentatively be defined as an active fault[4]. Just as important, if young geological deposits are undisturbed along the length of an identified fault, the fault is possibly older and inactive, or active at such a low rate as to be irrelevant to project-specific design criteria.

The next step in the fault investigation is to "ground-truth" the aerial photo interpretations. It involves field reconnaissance visits to the identified fault locations to: (1) confirm that lineaments identified in the aerial photos are actually faults; (2) establish as precise a location as possible at the pipeline crossing; (3) establish the widths of ground deformation zones either side of the fault plane location; (4) examine exposures of the fault plane, either naturally occurring or man-made; (5) establish the displacement sense in three-dimensional space; (6) examine sites of surface offset through the youngest geological materials to constrain the most likely age of latest displacement and the likely average annual rate of slip; and (7) establish error bounds on interpretations of location, displacement sense, and rate of fault slip.

The geological investigation scope described above, when performed by an experienced professional, is usually sufficient to identify faults with recurrence intervals of at least 10,000 years. However, refinement of earthquake magnitudes and recurrence intervals may require more detailed geological field investigations. These

---

[4] Generally, a fault is considered active if it can be demonstrated to have displaced the surface of the ground during the Holocene Epoch, i.e., within the past 10,000 years. This, of course, assumes that faults with evidence of youngest geological movement will be the ones that move again, which has not always been the case.

additional investigations typically consist of excavating 2 to 3-m deep trenches across the surface fault trace at strategic locations to permit the geologic mapping of soil horizons observed in the walls of the trench and age-dating of carbonaceous deposits to determine fault displacement episodes that have occurred in the past. The geological data obtained from such specialized investigations are generally the best constraints available on the timing and magnitude of past surface displacements along a fault, but at a dramatic increase in cost. Shallow, high-resolution geophysical techniques in conjunction with reconnaissance geological investigations are sometimes useful either to refine a particularly critical fault location or to provide insight to ambiguous geomorphic features that cannot be definitively proven to be a fault.

Fault Displacements

Ideally, characteristic fault displacements would be established through detailed field studies and knowledge of historical seismicity. In most cases, such information is absent or impractical to obtain; hence, empirical relationships are typically used to determine mean average and mean maximum displacement. Two regressions on fault rupture length are available from Wells and Coppersmith (1994):

$$\log(MD) = -1.38 + 1.02 \cdot \log(SRL) \qquad [\log(\sigma_{MD}) = 0.41]$$

$$\log(AD) = -1.43 + 0.88 \cdot \log(SRL) \qquad [\log(\sigma_{AD}) = 0.31]$$

where MD is the mean maximum net surface fault displacement in meters, AD is the mean average net surface fault displacement in meters, SRL is the surface rupture length[5] in km, σ_{MD} is the standard deviation of the maximum displacement regression, and σ_{AD} is the standard deviation of the average displacement regression. Wells and Coppersmith also provide regressions of displacement as a function of moment magnitude, M, which could be used instead of the above regression equations based on surface rupture length.

The fault displacement values in the Wells and Coppersmith relationships represent resultant fault displacements in a vertical plane[6]. These resultant displacements should be transformed into three-dimensional displacement components for use in estimating fault displacements for normal or reverse faults. This transformation should be computed through appropriate consideration of the geometry resulting from the normalized Wells and Coppersmith displacement, the angle of intersection between the regional stress azimuth and the fault strike, and the fault dip angle.

---

[5] The surface rupture length is the maximum length of the fault segment crossing the pipeline that, by reasonable and qualified judgment, can be expected to rupture during an earthquake.
[6] The Wells and Coppersmith regression does not compute resultant fault displacements in three-dimensional space and, as such, is not suitable for direct use in determining a design displacement. Their database does not include the vertical and horizontal components of displacement in the vertical plane that intersects the strike, nor does it include the transverse component of reverse or normal slip components of displacement perpendicular to the fault strike.

Mathematical frameworks have been suggested for defining fault displacements in a probabilistic manner (e.g., Youngs et al., 2003). However, implementation requires an extensive effort to characterize fault displacement for various levels of earthquake events and the distribution of ground displacement with location along the fault. Normally, this is feasible only for highly critical projects with operational lives much greater than the expected recurrence intervals of surface faulting (e.g., nuclear waste repository).

Design Fault Displacements

There are a number of factors that influence the determination of design fault displacements. Preferably, the annual probability of exceedance of a design fault displacement should be consistent with the acceptable risk levels established by the pipeline owner for pipeline damage and consequences. In practice, design fault displacements are most often based on a characteristic earthquake occurring on the fault in question, because reliable information on fault rupture recurrence intervals is often not available.

The distribution of fault displacement varies considerably along the length of surface rupture, particularly for reverse faults. The maximum surface displacement typically occurs over a relatively small portion of the total fault length and far exceeds the median or mean displacement. This is a particularly important point for pipelines since they typically cross faults only at a single discrete location. For example, fault displacement measurements along the rupture for the 1983 Borah Peak earthquake indicated only a 3% to 6% chance of experiencing a displacement within 90% of the maximum observed fault displacement if a location on the fault were selected at random (McCalpin, 1996). Given the irregular distribution of fault displacement and the uncertainty regarding the spatial distribution of fault displacement, basing pipeline crossing design on the mean maximum fault displacement would seem to be overly extreme. Simply stated, this implies that fault-crossing design should be based on what is likely to occur if a fault ruptures the ground surface, and not merely by what is possible. As a general guideline, we recommend that design displacements for natural gas and liquid hydrocarbon pipelines be based on mean maximum and mean average displacements as given in Table 1.

Table 1. Recommended Guideline for Establishing Design Displacements

Crossing Type and Location	Design Displacement
Gas pipelines and flammable or explosive liquid hydrocarbon pipelines in Location Class 4 areas (ASME B31.8)	MD
Gas pipelines and flammable or explosive liquid hydrocarbon pipelines in Location Class 3 areas (ASME B31.8)	$2/3 \times MD$
Liquid hydrocarbon pipelines located in environmentally sensitive areas	$2/3 \times MD$
Other gas or liquid hydrocarbon pipelines not included above	AD

Prior to use, the above guidelines should be carefully reviewed and evaluated by the pipeline owner or its representatives for consistency with project objectives and regulatory requirements. A reduction in these displacements may be appropriate for cases where the return period for the design event is 5 to 10 times shorter than the estimated recurrence for the characteristic earthquake on the fault. However, caution should be exercised when historical movement on the fault has not been well-defined.

Dual-Level Design Earthquake Criteria

Absent from the previous discussion of design fault displacement is the concept of a dual-level earthquake definition. The use of dual-level earthquake design criteria is an outgrowth of practices developed for the nuclear power industry and other projects (e.g., offshore structures) where seismic damage has the potential for severe safety or environmental consequences. The lower level event is often viewed as a threshold for continued operation of the pipeline within accepted safety margins, and little, if any, permanent deformation of the pipeline is permitted. For the higher level event, the objective is to prevent rupture, although there may be significant pipe deformation requiring repair (e.g., wrinkling or buckling) and possible interruption of throughput.

A dual-level design is generally not recommended for pipeline fault crossings for two reasons. First, the dual-level earthquake approach for fault crossing design can unnecessarily increase the conservatism in the definition of the upper-level fault displacement. Unlike structures and equipment responding to inertial loads, the upper-level fault displacement capacity of a buried pipeline is reduced because of the strains produced by lower-level earthquake fault displacements. If the intent is to maintain pipeline operation following the lower-level fault displacement, one could argue that this displacement should be added to the suggested design fault displacements in Table 1. Second, as previously stated, there are no practical methods for defining fault displacement probabilistically. This results in either an arbitrary definition of the lower-level fault displacement (e.g., one-third or one-half of the design fault displacement in Table 1) or estimating the lower-level fault displacement by the product of the fault slip rate and a particular recurrence interval (e.g., 200-year recurrence interval at 2 mm/year).

Buried vs. Above-Ground Crossings

The usual design approach for pipeline fault crossings is to construct the pipeline in a shallow, sloped-wall trench with loose backfill while permitting large strains and permanent deformation to occur, provided pipe rupture is prevented. In other words, the risk of damage requiring repair is generally acceptable provided that the integrity of the pressure boundary is maintained. Buried crossings are normally preferred, because it avoids technical issues associated with a long run of unrestrained pipe, and it limits exposure to third-party damage, particularly in politically unstable areas of the world.

The design of the Denali Fault crossing for the Trans-Alaska Pipeline System (TAPS) was the first specially designed crossing of an active fault by a crude oil pipeline, but this crossing was made above ground with pipeline support cross-beams constructed essentially at grade level to accommodate large movement of the pipeline on sliding pipe shoes (Hall, et al., 2003). An above-ground crossing was selected for TAPS for a number of reasons: (1) it would have been difficult at the time to obtain thicker wall pipe for a diameter of 1,217-mm (48-inches); (2) a stress-based design approach was used instead of a strain-based design, and (3) the exposure of the pipeline above ground at the fault crossing did not present additional third-party risk or technical issues, since over one-half of the pipeline was to be built above ground. During the 2002 magnitude 7.9 Denali Fault earthquake, the pipeline and support system performed as expected, without damage to the pipe or leakage of oil (Hall et al., 2003). However, it is worthwhile to note that had the pipeline had been buried, local buckling likely would have occurred (Sorensen and Meyer, 2003). Damage of this type would have required pipe repair and probably an extended downtime. Therefore, drawing from the TAPS experience, pipeline owners should recognize the consequences of accepting the risk of fault displacement-related pipeline damage when establishing pipe performance requirements and selecting between buried and aboveground fault crossing concepts.

Analysis Methodology

The analysis of a buried pipeline subjected to surface fault rupture requires a non-linear finite element model that can account for inelastic pipeline behavior, the nonlinear behavior of the surrounding soil mass, and large displacement effects. The segment of the pipeline used in the analysis model should extend away from the fault at least to points of virtual anchorage of the pipeline in the soil. If pipe bends or other discontinuities exist near the fault zone, they could cause a virtual anchor point that should be included in the analysis. Pipe elements are typically made shorter in regions of critical interest near the fault where high bending will occur, and longer in segments of the pipeline that are more remote from the region of high bending.

Inelastic pipe behavior is simulated by specifying a nonlinear stress-strain curve for the pipeline steel. It is important to note that pipe elements account for nonlinear stress-strain behavior, but do not include the effect of compressive buckling or wrinkling of the pipe wall. Consequently, it is necessary to utilize test-based compressive strain criteria that normalize axial strains over a particular length of pipeline encompassing the wrinkled or buckled sections (usually over one to two pipe diameters). It is recommended that pipe element lengths be limited to one diameter in regions of high strain to maintain correlation between the finite element analysis results and the compressive strain limits based on test data.

Soil-pipeline interaction can be modeled by discrete nonlinear springs oriented in the axial, horizontal, and vertical directions. The methodology for calculating soil springs is well-established (ASCE, 1984; ALA, 2001). In some cases, such as pure strike-slip faulting, the model may be simplified by a two-dimensional representation, but three-dimensional models are generally preferred, especially for crossings involving three significant components of fault displacement or crossings that induce axial compression into the pipeline. The fault displacement is applied to the model as displacements of the base of the soil springs on one side of the fault as shown in Figure 1. The definition of soil (spring) restraint properties must be consistent with field conditions. In particular, for displacement of the pipeline in a transverse horizontal direction, the soil failure wedge must be enveloped by the limits of the excavated pipe trench that is backfilled with the selected material. If these trench excavation and backfill requirements are not satisfied, soil parameters applicable to in situ soil conditions must be considered in the development of soil restraints for the pipeline. Similarly, for vertical displacement, the upward breakout will occur within the designated backfill.

Strain Criteria for Buried Pipelines at Fault Crossings

A pipeline responding to fault displacement experiences soil loads generated by ground movement relative to the

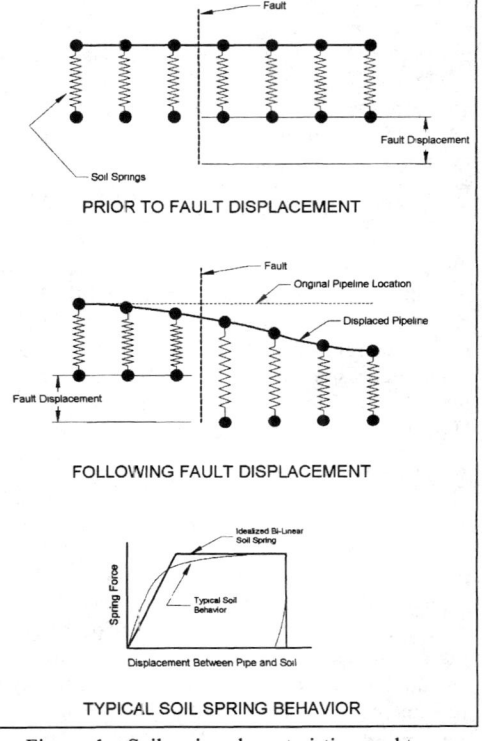

Figure 1. Soil spring characteristics used to represent soil restraint.

pipeline. The pipeline sees no further soil load once it has deformed sufficiently to match the ground movement. This type of loading is commonly referred to as being displacement-controlled. Another type of displacement-dependent load condition may arise when axial compression effects due to fault displacement or thermal expansion cause upheaval buckling whereby the buried pipeline displaces vertically upward out of the soil. In some cases, the elastic component of compressive strain in the pipeline will provide additional energy to increase the post-buckled pipeline deformation and may require a more detailed investigation of local deformation effects. Nevertheless, soil restraint limits the maximum amount of displacement the pipeline can experience. This loading condition is referred to as displacement-limited. For pipelines experiencing displacement-controlled or displacement-limited loading, it is appropriate to base the design on strain limits as opposed to stress limits.

Strain limits for determining the fault displacement capacity of the pipelines are based on allowing yielding and distortion of the pipe wall while maintaining pressure boundary integrity. In other words, failure in a strain-based design is taken to mean loss of pressure boundary integrity, but not necessarily deformations requiring repair. The recommended tensile and compressive strain limits discussed below are intended to be used for the design of pipeline fault crossings and other earthquake-triggered ground movements where the recurrence interval of the earthquake corresponds to the acceptable recurrence interval for loss of pressure integrity. These strain limits may not be appropriate for other, ground deformation hazards associated with seasonal or more frequent events (e.g., frost-heave, thaw settlement, mine subsidence, expansive soil).

Tensile Strain Limits. Small homogenous test specimens of typical pipe material in a tensile test fail at a total tensile strain on the order of 20% to 25%. However the strain between first yield and maximum load (onset of necking and plastic instability) is only 5% to 10% and may be even smaller on specimens from large weldments, especially if the weld material strength is less than the pipe material or the welds contain surface-breaking flaws. However, there is extensive successful experience with highly strained welded line pipe installed from reel barges for subsea pipelines in diameters up to 457 mm (18 inches). Nominal bending strains (tensile and compressive) in the coiled pipe on the reel are on the order of 2% to 4%, and only infrequent failures have been reported. For new pipelines constructed of moderate strength pipe steel (X65 and below) using welding and inspection specifications similar to those used in offshore applications, tensile strain capacities of 4% prior to loss of pressure integrity is generally achievable.

For projects where pipeline failure may have unusually severe consequences, it may be necessary to demonstrate the ability of pipe welds to achieve the desired level of tensile strain. The preferred approach for confirming tensile strain capacity would be to perform a statistically significant number of wide-plate tensile tests on samples randomly selected from girth welds that have been fabricated to specifications (i.e., specifications for pipe material, welding, and inspection) identical to those to be used on the project.

Compressive Strain Limits. A considerable number of full-scale combined axial compression and bending tests and supplementary finite element analyses have been conducted by various universities and test organizations. The maximum strain attained in these tests has been mined from publicly available papers and reports and plotted against the diameter-to-thickness ratio, D/t (see Figure 2). Direct comparison of this variety of tests is complicated by differences in load conditions, measurement techniques, and test objectives. In particular, it appears that no controlled tests have been performed that imposed axial compression and bending deformations to the point of loss of pressure integrity. In addition, there appears never to have been a test program that examined the variation of behavior for identical pipeline load conditions, welding properties, pipe sizes, and limit states. For these reasons, it is impossible to provide an estimate of the reliability associated with loss of pressure integrity for a specific level of strain or the margin of safety against loss of pressure integrity. In other words, it is not possible to quantify, in a statistical fashion, a specified minimum strain capacity such as is done for specifying pipe material yield strength. Therefore, the definition of appropriate relationships for seismic assessment of pipeline response is based largely on judgment.

In a report of tests at the University of Alberta (Mohareb et al., 1994), a compressive bending strain relation corresponding to 15% ovalization was observed to be a conservative lower bound of most of the publicly available test results and is recommended as an appropriate pressure integrity compressive strain limit for fault crossing evaluations. This lower-bound strain criterion is given as:

$$\varepsilon_c = 1.76 \frac{t}{D} \leq .04$$

where t is pipe wall thickness, and D is pipe outside diameter. The not-to-exceed limit of 0.04 strain (4% strain) has been added by the authors. In the absence of project-specific pipe bending tests, this strain criterion is recommended by the authors as a reasonable lower bound pursuant to satisfying important requirements for materials and welding as discussed later in this paper. This strain criterion is shown as function of D/t as a solid line plot in Figure 2.

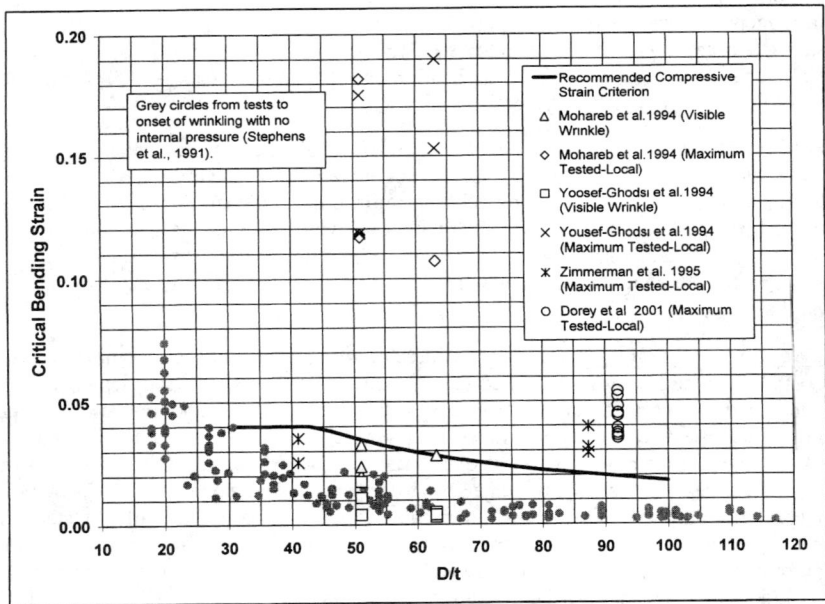

Figure 2. Comparison of recommended compressive strain limit with test data from various sources.

Pipelines strained in axial compression or bending to the above limit will likely exhibit a well-formed outward buckle in the pipe wall. If pipeline response is displacement-limited, as might occur if the pipe experiences upheaval buckling under axial compression loading, the potential for localization of strain in a buckle should be evaluated. Sensitivity analyses examining the impact of alternate soil-spring definitions or fault displacement patterns can often help to assess the potential for strain localization. A second option is to perform a shell analysis of a short length of pipe with applied loads and displacements determined from fault crossing analysis using pipe elements. A third option is to couple the fault crossing analysis with pipe elements to a shell model in the region of high strain.

Welding and Weld Inspection. Strain acceptance limits are based on two important assumptions: (1) the pipeline is subjected to thorough field inspection including 100% radiographic or essentially equivalent ultrasonic weld inspection through the fault zones, and (2) the specification of weld materials and procedures that assures sufficient weld strength and toughness to develop gross section yielding of the pipe wall (i.e., failure of the pipe occurs before failure in the weld or the weld heat affected zone). The specification of a welding procedure must also account for the expected range of actual yield strengths of the pipe steel.

In addition to strength, fracture toughness, and weldability, the most important pipe property for strain-based design is the yield to ultimate strength ratio (Y/U) of the pipe, heat-affected zone (HAZ), and weld deposit. This ratio together with the shape of the stress-strain curve determines the amount of plastic strain that can be tolerated without failure of the material in, and adjacent to, the girth weld. A ratio on the order of 75% is preferred and is common on low yield strength pipe. (The low Y/U ratio should also apply to the longitudinal seam weld of the fabricated pipe.) Higher strength pipe (grade X70 and higher) that is normally specified to achieve minimum pipe weight and price is likely to have a Y/U ratio of 85% to 90% or perhaps higher. Selection of a lower strength pipe such as X52 or X60, with a necessarily thicker wall, is a reasonable alternative to assure a low Y/U ratio.

A welding process should be chosen to minimize both the number and size of imperfections. The presence of small imperfections or flaws in the weld deposit, particularly those having a planar geometry and located near the root or cap of the weld, severely degrade the ability of the material to distribute plastic strain uniformly. Specifying a weld metal having higher yield strength than the pipe metal minimizes the opportunity for concentration of strain within the weld material.

The experienced-based workmanship requirements (flaw acceptance criteria) contained in stress-based welding codes may not be adequate for strain-based designs. Specifying an allowable flaw size presupposes an inspection system capable of detecting and sizing any flaws remaining in the weld. It is well established that flaw height and the distance to the pipe surface are the critical parameters. Conventional radiographic techniques are not suitable for either of these measurements, and the use of a mechanized ultrasonic procedure such as that commonly used for offshore pipe laying operations is recommended. An accuracy of about one mm is achievable for flaw height measurement in pipe with wall thickness less than about 25 mm. The net effect of a special welding procedure and extensive weld inspection may lead to a high weld rejection rate, thereby reducing productivity and increasing cost per weld.

However, this should be economically practicable considering the relatively limited length of pipeline requiring a strain-based design.

Tensile Strain Criteria and Fracture Mechanics. In recent years, there have been efforts to validate tensile strain limits with fracture mechanics methods and analyses. Most fracture mechanics methods rely on semi-empirical relationships developed for use in determining pipeline fitness for continued operation. Tests on specimens containing welds with machined (intentionally introduced) flaws are highly variable, and the adequacy of the assumptions used in a particular fracture mechanics approach are typically judged by the ability of the fracture mechanics results to match a lower-bound strain capacity from available laboratory tests. This is appropriate given that the primary use of fracture mechanics is to provide a conservative estimate of strain capacity for operational fitness assessments, in which case thousands of welds must be demonstrated to have a very low probability of individual weld failure (e.g., 10^{-7} to 10^{-6}) under normal operating stresses to assure continued safe operation.

More liberal strain limits than what would be considered for an operational fitness assessment are appropriate for fault crossing design given that only a limited number of pipe welds will be exposed to high strain. Consideration also needs to be given to the probabilities associated with (1) a individual weld being exposed to a high strain level, (2) a flaw existing in a highly strained weld, (3) the size and type of the flaw, (4) the location of the flaw relative to the location of high strain around the circumference of the pipe, and (5) the variation of weld and pipe material strengths that could lead to the weld being weaker than the pipe. In the opinion of the authors, the probability of an individual weld failure might be taken as 10 percent or higher for fault crossing applications.

Conclusions

The methodology for the analysis of high-pressure natural gas and liquid hydrocarbon pipelines at buried fault crossings is well established. It involves the use of non-linear finite element analysis procedures and strain-based design criteria for displacement-controlled or displacement-limited load conditions. It sometimes requires consideration of strain localization associated with upheaval buckling due to compression resulting from fault movement or thermal expansion. There are a number of uncertainties that must be considered in the design of pipeline fault crossings. Included among these are:

1. Characterization of fault displacement and recurrence is often based on limited field investigation.
2. Tensile and compressive strain limits are based on pipe bending test programs that focused on the initiation of wrinkling/buckling rather than strain capacity at rupture.
3. Strain limits, especially tensile strain, are affected by materials, welding procedures, and weld inspection, i.e., the pipeline girth welds must "overmatch" the tensile strength of the pipe to assure ductile behavior.

The design of fault crossings relies heavily on the combined experience and judgment of the engineer, geologist, and pipeline owner to overcome the uncertain nature of the parameters and characteristics listed above. It is important that these uncertainties be managed in a balanced manner without compounding a series of conservative judgments. It is also important that the pipeline owner understand and be able to tolerate the risk of pipe damage and service interruption if surface fault rupture were to occur. Further research is needed to:

1. Quantify strain conditions associated with the loss of pressure integrity under various combinations of axial compression and bending.
2. Determine actual variation in tensile strain capacities at girth welds for the high quality construction practices typically specified for fault crossings.
3. Develop fracture mechanics methodology that provides probabilistic strain capacities as opposed to conservative lower-bound strain capacities.

References

1. ALA (American Lifelines Alliance) (2001). Guidelines for the Design of Buried Steel Pipe, G.A. Antaki and J.D. Hart, Co-Chairmen, July, www.americanlifelinesalliance.org, 76 p.
2. ASCE (American Society of Civil Engineers) (1984). *Guidelines for the Seismic Design of Oil and Gas Pipeline Systems,* Technical Council on Lifeline Earthquake Engineering, Committee on Gas and Liquid Fuel Lifelines, D. J. Nyman, Principal Investigator, New York, 473 p.
3. Dorey, A.B, J.J.R. Cheng, and D.W. Murray (2001). "Critical Buckling Strains for Energy Pipelines," University of Alberta Department of Civil and Environmental Engineering, Structural Engineering Report No. 237, April.
4. Hall, W.J., D.J. Nyman, E.R. Johnson, and J.D. Norton (2003). "Performance of the Trans-Alaska Pipeline in the November 3, 2002 Denali Fault Earthquake," *Proceedings of the 6th U.S. Conference on Lifeline Earthquake Engineering (Advancing Mitigation Technologies and Disaster Response),* Long Beach, CA, TCLEE, ASCE.
5. McCalpin, J.P., ed. (1996). *Paleoseismology,* Vol. 62 in the International Geophysics Series, Academic Press, 588 p.
6. Mohareb, M.E., A.E. Elwi, G.L. Kulak, and D.W. Murray (1994). Deformational Behavior of Line Pipe, Structural Engineering Report 202, Department of Civil Engineering, University of Alberta, Canada.
7. Sorensen, S.P., and K.J. Meyer (2003). "Effect of the Denali Fault Rupture on the Trans-Alaska Pipeline," *Proceedings of the 6th U.S. Conference on Lifeline Earthquake Engineering (Advancing Mitigation Technologies and Disaster Response),* Long Beach, CA, TCLEE, ASCE.
8. Stephens, D.R., R.J. Olson, and M.J. Rosenfeld (1991). "Pipeline Monitoring – Limit State Criteria," Battelle NG-18 Report No. 188, September.

9. Wells, D.L. and K.J. Coppersmith (1994). New Empirical Relationships among Magnitude, Rupture Length, Rupture Width, Rupture Area, and Surface Displacement, Bulletin of the Seismological Society of America, Vol. 84, No. 4, pp. 974-1002.
10. Yoosef-Ghodsi, N., G. L. Kulak, and D. W. Murray (1994). Behavior of Girth-Welded Line Pipe, Structural Engineering Report 203, Department of Civil Engineering, University of Alberta, Canada.
11. Youngs, R.R., W.J. Arabasz, R.E. Anderson, A.R. Ramelli, J.P. Ake, D.B. Slemmons, J.P. McCalpin, D.I. Doser, C.J. Fridrich, F.H. Swan III, A.M. Rogers, J.C. Yount, L.W. Anderson, K.D. Smith, R.L. Bruhn, P.L.K. Knuepfer, R.B. Smith, C.M. dePolo, D.W. O'Leary, K.J. Coppersmith, S.K. Pezzopane, D.P. Schwartz, J.W. Whitney, S.S. Olig, and G.R. Toro (2003). A Methodology for Probabilistic Fault Displacement Hazard Analysis (PFDHA), *Earthquake Spectra*, Vol. 19, No. 1, February, Earthquake Engineering Research Institute, Oakland, CA.
12. Zimmerman, T.J.E., M.J. Stephens, D.D. De Geer, and Q. Chen (1995). "Compressive Strain Limits for Buried Pipelines," 1995 Offshore Mechanics and Arctic Engineering Conference, American Society of Mechanical Engineers, Vol. V, pp. 365-378.

Centrifuge Modeling of Buried Pipelines

Michael O'Rourke[1], Vikram Gadicherla[2] and Tarek Abdoun[3]

Abstract

Fault crossing, lateral spreads and other types of permanent ground deformation (PGD) are arguably the most severe seismic hazards for continuous buried pipelines. Current analysis and design procedures, to a great extent, are based upon Finite Element (FE) modeling. There are, unfortunately, relatively few full- scale case histories which could be used to benchmark or confirm the applicability of FE assumptions. That is, full scale field verification of the predicted behavior of buried pipes is, at best, sparse.

In this paper, a new centrifuge based method for determining the response of continuous buried pipeline to PGD is presented. Laboratory equipment, experimental procedures, similitude relations as well as sample results are presented. Specifically, physical characteristics of the Rensselaer centrifuge are described, as well as those for our current lifeline experiment split-box. The split-box contains the model pipeline and surrounding soil and is manufactured such that half can be offset in flight, simulating PGD. Governing similitude relations which allow one to determine the physical characteristics (diameter, wall thickness, material modulus of elasticity) of the model pipeline are presented.

Finally the recorded strains induced in a prototype 0.64 m (25 in.) diameter, 0.013 m (0.5 in) wall thickness, steel pipe by 0.80 m (2.6 feet) and 2.0 m (6.6 feet) of full scale fault offset for a prototype steel pipe 0.95 m (37 in.) diameter, 1.9 cm (.75 in.) wall thickness are presented and compared to corresponding FE results.

----

[1]Professor, Civil and Environmental Engineering, Rensselaer Polytechnic Institute, Troy, NY 12180-3590
[2]Graduate Student, Civil and Environmental Engineering, Rensselaer Polytechnic Institute, Troy, NY 12180-3590
[3]Research Assistant Professor, Civil and Environmental Engineering, Rensselaer Polytechnic Institute, Troy, NY 12180-3590

Introduction

Buried pipelines are commonly used to transport oil, water, sewage and natural gas. These pipelines are sometimes referred to as lifelines as they are essential for the support of life and maintenance of property. Pipelines in seismic zones are prone to surface faulting and wave propagation hazards. Though the surface faulting hazards are limited to small areas within the pipeline network, the potential for damage is high since surface faulting imposes large deformation on the pipelines. Hence fault rupture is one of the most severe seismic hazards for the buried pipes.

The analysis of a buried pipe subject to fault movement is complex and currently no closed form solutions are available. Of late, finite element models have been used to predict the behavior of buried pipelines subjected to fault movements. A different approach to the problem employs centrifuge modeling of buried pipelines, and is discussed as follows.

Centrifuge Modeling

For problems in which soil-structure interaction forces are dominate, small-scale models by themselves cannot replicate similar stresses at comparable points on the model and the prototype. This is because soil loading and stiffness at a particular depth are related to the weight of the soil above. For example, the vertical stress at a depth of 10 m for a soil having a density of 10 KN/m^3, would be 100 KPa. However, in a tenth scale model, the stress at the corresponding location (i.e. 1 m depth in the model) in the same soil is only 10 KPa. As a result, the similitude of soil –structure interaction forces is not maintained. This difficulty can be overcome by effectively increasing the weight of the soil by a factor of 10. In a centrifuge, this is accomplished by spinning at a speed corresponding to a centrifugal acceleration of 10 times the earth's gravity (10 g). This is the underlying concept behind centrifuge modeling of soil and sub-structure systems. A partial list of scaling laws in terms of a centrifugal acceleration Ng is presented in Table I.

This concept has been used extensively to study the effect of explosions on soil, the response of soils subjected to earthquake effects and a number of other geotechnical problems. The purpose of experiments described herein, is to study the response of a continuous buried pipeline subjected to fault movements.

When modeling the response of the buried pipe to fault movement, care must be taken to maintain similitude. In terms of soil, soil similitude is maintained by using the same type of soil in the model and the prototype. This ensures that both the soil density and friction angle in model and prototype match. For buried pipe, the longitudinal forces at the soil-pipe interface are influenced by the roughness of pipe surface. Hence, this quantity should be similar in both model and prototype. Finally, since the soil forces acting on the pipe are linearly proportional to pipe diameter, the pipe diameter needs to scale as N.

In relation to the buried pipe, the similitude of axial force effects is maintained if the axial rigidity, EA, (the product of elasticity and cross sectional area) scales by a factor of $1/N^2$, when the model is subjected to a centrifugal acceleration of Ng. Note, that stress has same dimensions or units as modulus of elasticity (psi, ksi, etc.) and hence from table I, area (having units of length squared) scales as N^2. Hence, axial rigidity, which is the product of modulus of elasticity and area, scales as

$$\frac{E_p A_p}{E_m A_m} = N^2 \qquad (1)$$

where E is the modulus of elasticity, A is the cross-sectional while m and p refer to the model and prototype respectively. When the wall thickness to diameter ratio, t/d, is small, the area may be approximated as $A = \pi d t$. Hence, the scaling relation to maintain similitude for EA becomes

$$\frac{E_p t_p d_p}{E_m t_m d_m} = N^2 \qquad (2)$$

As noted above, the ratio of diameters d_m/d_p scales with N to ensure similitude of soil forces on the pipe interface, hence equation (2) becomes,

$$\frac{E_p t_p}{E_m t_m} = N \qquad (3)$$

which is the scaling relation for EA.

In order to maintain the similitude of bending moment in the buried pipe the flexural rigidity, EI needs to be scaled as

$$\frac{E_p I_p}{E_m I_m} = N^4 \qquad (4)$$

where I is the moment of inertia of the pipe. When the t/d ratio is small, the moment of inertia may be approximated as $I = \pi d^3 t/8$. Hence, EI similitude requires

$$\frac{E_p t_p d_p^3}{E_m t_m d_m^3} = N^4 \qquad (5)$$

However as noted before, d_m/d_p scales as N, hence the similitude relation for EI is identical to that for EA as given in equation (3).

Experimental Equipment

The Rensselaer Geotechnical centrifuge facility is located at the basement level of the Jonsson Engineering Center. . The centrifuge itself is located in a belowground closed circular chamber. Operation of centrifuge and data acquisition is performed in the control room. The centrifuge is an Acutronic model 665-1.

The in-flight radius of the centrifuge is 3m. The centrifuge is capable of carrying a maximum payload of 1 metric ton at 100 g (i.e. 100 g ton). The maximum speed of the centrifuge corresponds to an acceleration of 200 g. The soil sits on the testing platform which is located at the end of the centrifuge arm. The maximum model or payload dimensions are 100cm x 80 cm x 80 cm. The transfer of data is enabled by wireless DAQ system and the control signals is enabled by 64 electric slip rings; 50 are for analog signals, 12 for power and two for video signals. The hydraulic rotary joints have a total of six passages, two of which are hydraulic oil passages rated at 3000 psi and the remaining four are air/water passages rated at 300 psi.

Split box. As noted above the split box allows simulation of fault offset or the relative horizontal displacement at the margin of a lateral spread. The split box has inside dimensions of 1 m x 0.354 m x 0.203 m (39.4" x 14" x 8"). It consists of two halves, one is fixed and the other can move horizontally on rollers to simulate an offset. The moving portions of the container are supported and guided using roller bearings to provide precise movement with minimal friction. The sliding interface between fixed and movable portions of the container utilizes low friction Teflon seals protected by steel shields.

A hydraulic cylinder is used to displace the movable half of the split box. The driving shear force is provided by a 3000 psi hydraulic actuator system, which includes a flow-metering valve, a solenoid valve for remote operation, and hoses for connecting to the centrifuge's quick connects. The load cell, located between the actuator and the movable portion of the split box, measures the force applied by the actuators. The maximum relative displacement of the movable section is 8cm, simulating, 4 m offset at 50 g. The flow-metering valve controls the rate of movement of the split box. The motion of the actuator is controlled by a servo valve and a feedback control system, while an LVDT measures the offset.

Anchor points. The connection between the pipe model and the split box end walls consists of a steel rod that sits in one of the metal plates attached to the split box wall. This assembly allows the pipe to rotate freely about the axis. The assembly also has space for 2 LVDTs, which can measure the rotation of the pipe model at the support. This assembly is covered by a plastic enclosure and rubber sheeting to prevent the entry of sand near the LVDTs.

Instrumentation and data acquisition system. Strain gauges were installed on the pipe model to measure the axial and bending strains at various points. The strain gages were model CEA-032UW-120 from the Measurements Group Inc. Ten strain gauges were set in a quarter bridge configuration and the remaining four strain gauges in a half bridge configuration.

The data acquisition system used is capable of recording 128 channels of data at a 10 kHz sampling rate per channel. A Pentium-4 PC generates the digital input signal, which causes the split box to move by the desired offset. The servo controller on the centrifuge arm receives the signal through the slip rings. A hydraulic pump is used to pressurize the actuator in the split box before sending the signal.

Centrifuge Tests

The two pipelines were tested in the Rensselaer centrifuge. The diameter and wall thickness of the steel prototype lines are listed in Table II. The diameter and d/t ratios are common for larger gas and liquid fuel pipe in the U.S.

Commercially available small diameter pipe typically do not have such large d/t ratios. For that reason, aluminum was chosen as the model pipe material. That is, since, E_m/E_p for the aluminum model and steel prototype is about 0.3, the scale factor for wall thickness was 0.3*50" or 15', in order that equations (3) and (5) are satisfied. The resulting diameter and wall thickness for the models are also shown in Table II.

Figures 1 through 4 show the axial and bending strains along the pipe measured by strain gages on Pipeline #1 model. For the given fault displacement, the pipe is in the elastic range. The strains in Figure 1 are for an offset of 0.4 cm while the offsets for figures 2 through 4 are 0.8, 1.2 and 1.6 cm respectively. The largest of these correspond to a prototype offset of 0.8 m at 50 g.

Figures 5 through 8 show the bending and axial strains for the second model pipe subjected to fault displacements of 1 cm, 2 cm, 3 cm and 4 cm respectively. The pipe is in the elastic range for the offsets up to 2 cm, which corresponds to prototype fault displacement of 1 m. For the maximum fault offset of 4cm, corresponding to a prototype fault displacement of 2 m at 50 g, the pipe strains are well beyond the yield strain.

As one would expect, the bending and axial strains are increasing functions of the offsets. For these tests with a 90° interaction angle between the pipe axis and the fault trace, the bending strains dominate. Also note that the fault location is a point of counter flexure (i.e. zero bending moment) due to the asymmetric nature of the offset.

Finite Element Model

The stress strain diagram for the aluminum model pipe material is shown in Figure 9 and was based upon laboratory testing at Rensselaer. Note that the yield strain is about 0.004(i.e. 4000 microstrain), and that the post yield curve looks parabolic.

A finite element idealization of the centrifuge model was constructed. The pipe was modeled with beam elements while the soil was modeled as elasto-plastic springs. The burial depth for the pipeline models was 2.4 cm (little under 1 inch) which corresponds to 1.2 m (~4 feet) in prototype scale. The corresponding maximum soil spring resistances, with units of force per unit length of pipe, and yield displacements were based upon the ASCE Guidelines [1]. Since the burial depth was constant across the model, and the offset had no vertical component, there was no need for vertical soil springs. That is, FE model was two dimensional in nature. The pipe model was taken to be pinned at the split box wall (i.e. end of FE model located

40 cm (20 m in prototype scale) each side of fault location). The offset was simulated by displacing the base of all the soil springs located on one side of the fault as well as the pin end on that side of the fault.

The results of the FE simulations are also shown in Figures 1 through 8. For pipeline #1 (see figures 1 through 4) the axial and bending strains from the FE simulation match well with the measured strains. The match is remarkably good for the offsets of 1.2 m and 1.6 m (0.6 m and 0.8 m in prototype scale) while somewhat less remarkable for lower offsets. For pipeline #2 (see figures 5 through 8), the match between measured strains in the centrifuge experiments and corresponding values from the FE simulation is quite good for offsets of 2 cm or less (1m in prototype scale). As the offset increases beyond 2 cm, the correspondence between the measured and simulated strains deteriorates somewhat. In this range, where the pipe material is beyond yield, the bending strains compare reasonably well, but the measured axial strains are somewhat lower than those from the FE simulation It is thought that this mismatch may be due to compliance or inward movement at the anchor point. That is, in the FE model the anchor points do not move while in the centrifuge model, they may move slightly, resulting in less measured axial stress.

Conclusions

The paper describes the first known attempt to use centrifuge modeling to determine the seismic response of buried pipeline. The experiments were successful in the sense that the experimental equipment functioned well and the recorded strains were generally in good agreement with those predicted by Finite Element models.

The agreement between pipe strain values measured in the centrifuge and those predicted by FE modeling was quite good for small offsets where the pipe remained elastic. At larger offsets, the agreement between the measured and simulated bending values remained good. However, the measured axial strains were less than those predicted by the FE method. It is thought that the inward movement of the split box wall at the pipe anchor points may account for these differences.

Finally it must be noted that FE simulation of large offsets (i.e. pipe in the inelastic range) requires knowledge of the actual stress-strain behavior of the centrifuge pipe material.

Acknowledgements

The research work described herein was sponsored by the National Science Foundation through Award No. CMS-0085256. The original NSF program manager was Vijaya Gopu, who was succeeded by Peter Chang. The construction of the split box was sponsored by the National Science Foundation through George E. Brown, Jr. Network for Earthquake Engineering Simulation program (NEES). The support is gratefully acknowledged. However, all statements, results and conclusions are the authors' and do not necessarily reflect the views of NSF.

References

[1] American Society of Civil Engineers (ASCE), 1984, Guidelines for the Seismic design of Oil and Gas Pipeline Systems, Committee on Gas and Liquid Fuel Lifelines, ASCE.

Table I Select similitude relations for centrifuge modeling at centrifugal acceleration Ng

Parameter	Model Units	Prototype units
Length	$1/N$	1
Strain	1	1
Axial Rigidity	$1/N^2$	1
Flexural Rigidity	$1/N^4$	1

Table II Pipeline Properties

Property	Pipeline 1		Pipeline 2	
	Prototype	Model	Prototype	Model
Diameter	0.64 m (25")	1.27 cm (.5")	0.95 m (37")	1.9 cm (.75")
Wall thickness	1.1 cm (0.43")	0.71 mm (.028")	1.85 cm (0.73")	1.27 mm (.05")

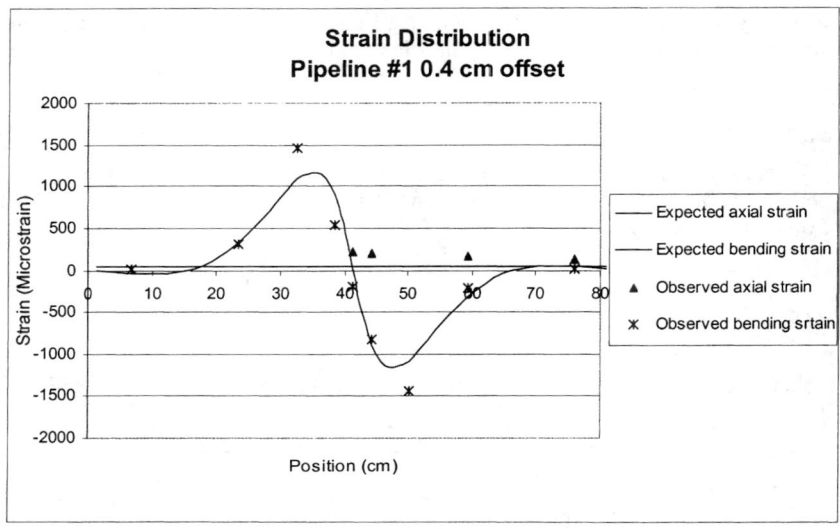

Figure 1. Measured and simulated strains for Pipeline #1 model subject to a 0.4 cm offset

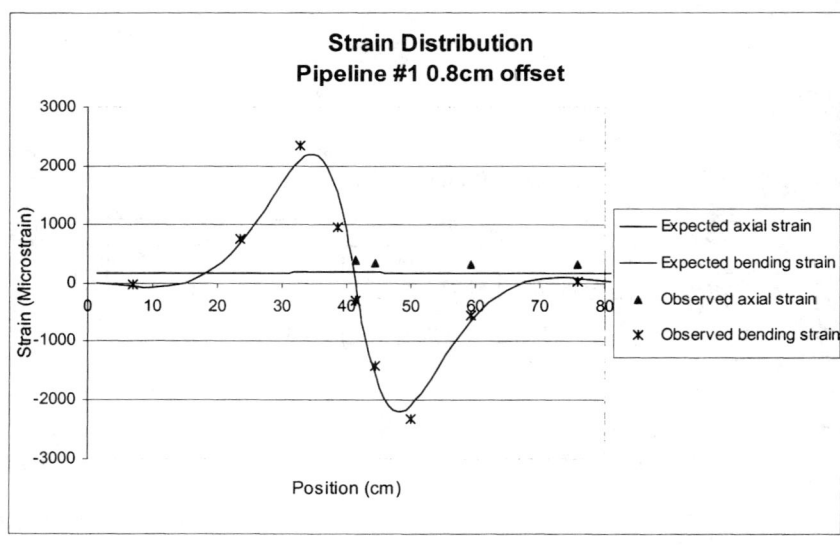

Figure 2. Measured and simulated strains for Pipeline #1 model subject to a 0.8 cm offset

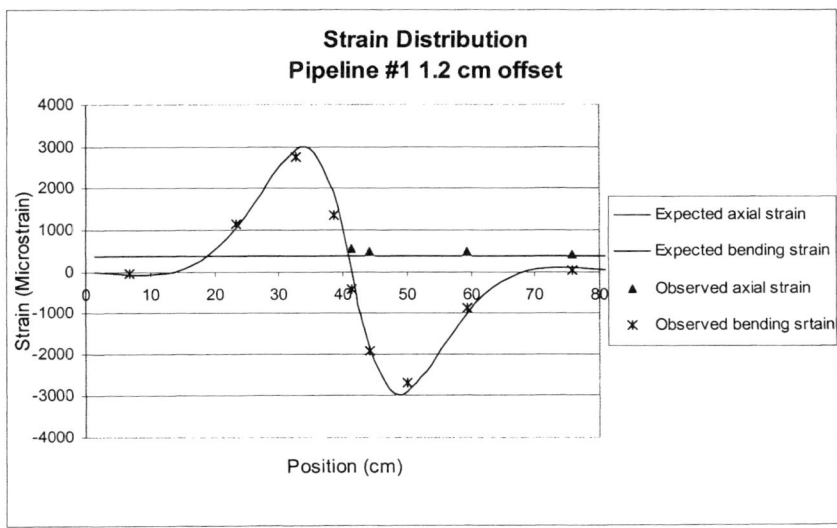

Figure 3. Measured and simulated strains for Pipeline #1 model subject to a 1.2 cm offset

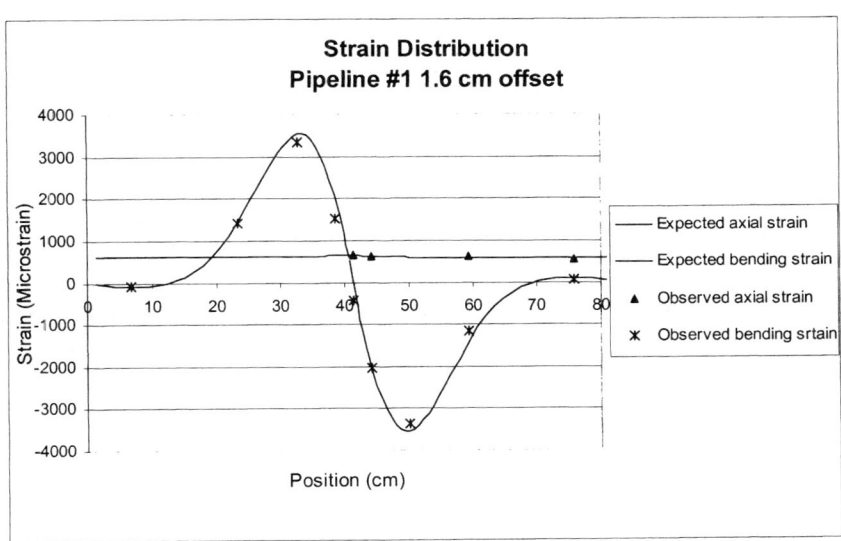

Figure 4. Measured and simulated strains for Pipeline #1 model subject to a 1.6 cm offset

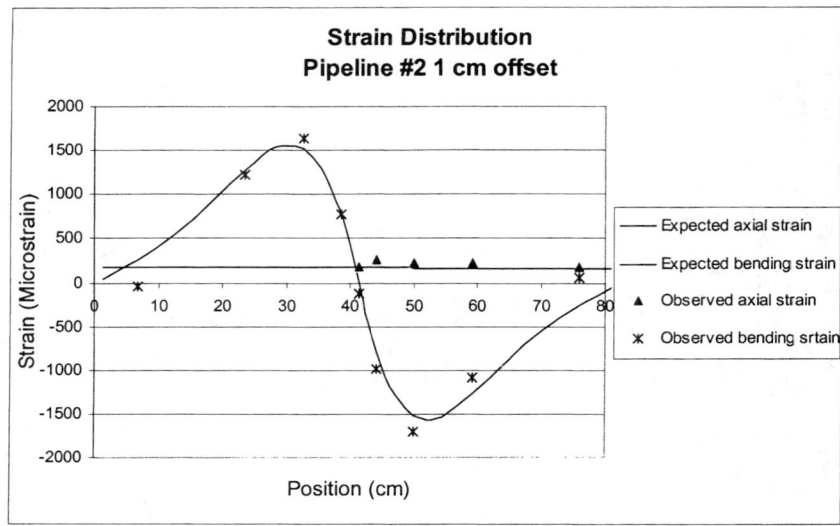

Figure 5. Measured and simulated strains for Pipeline #2 model subject to a 1 cm offset

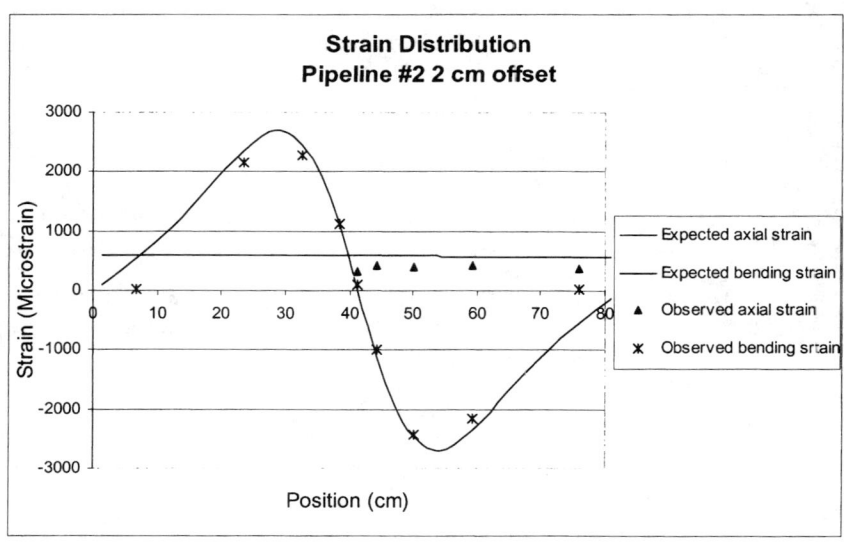

Figure 6. Measured and simulated strains for Pipeline #2 model subject to a 2 cm offset

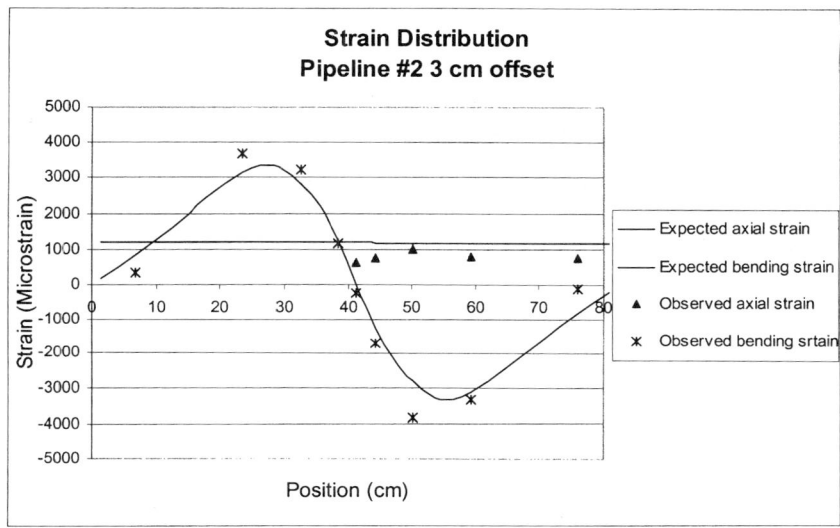

Figure 7. Measured and simulated strains for Pipeline #2 model subject to a 3 cm offset

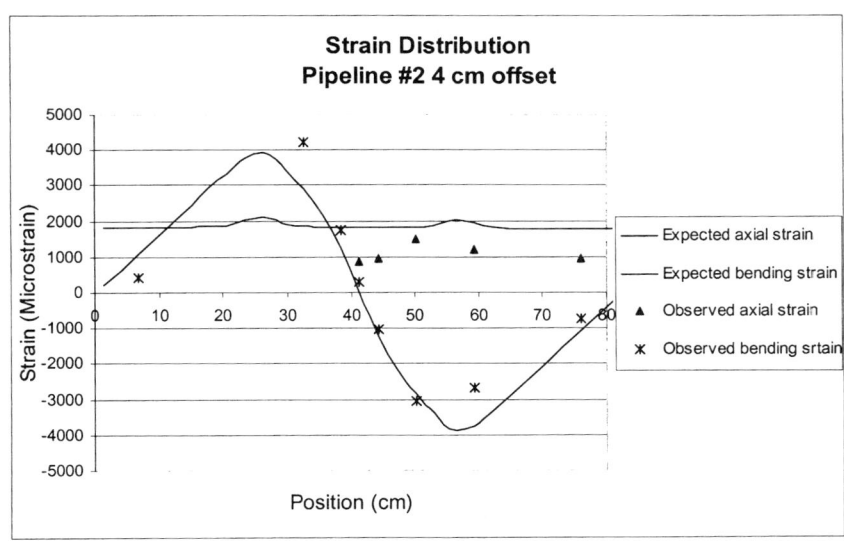

Figure 8. Measured and simulated strains for Pipeline #2 model subject to a 4 cm offset

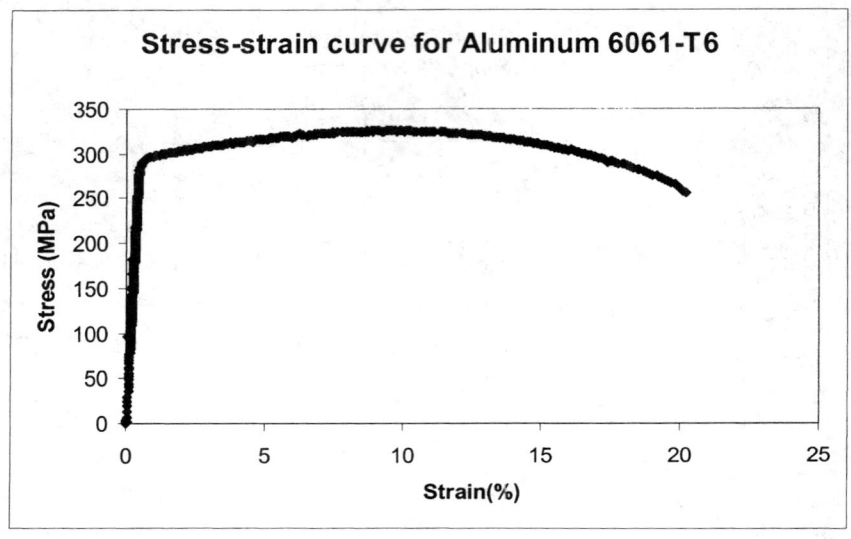

Figure 9. Stress-strain curve for Aluminum 6061-T6 obtained from tests at Rensselaer.

Modeling of Phase Spectra for Simulation of Near-fault Design Earthquake Motions

T. Sato[1], Y. Murono[2] and M. Murakami[3]

Abstract

Modelling a phase of earthquake ground motion is essential to synthesize design earthquake ground motion. We use the concept of group delay time which is defined by the derivative of a phase spectrum with respect to circular frequency. We find that it is strongly affected by the rupture directivity in the near source regions and derive regression equations for mean and standard deviation of group delay time. Simulated sample phase spectra were used to simulate earthquake motions assuming that Fourier amplitude spectra are given at observation stations. Their time histories are compared with the observed ones. To demonstrate the efficiency of propose model of the group delay time we simulate response spectrum compatible design earthquake motions.

Introduction

The classical procedure for simulating artificial earthquake motion is the superposition of harmonic waves of different amplitudes and phase angles. To generate a non-stationary time history, an envelope function is multiplied to a stationary time history simulated using random phase criteria [Shinozuka and Jan, 1972]. We have pointed out that a phase characteristic of earthquake motion strongly controls the non-stationary nature of earthquake motions and developed the regression equations of mean and standard deviation of group delay time of earthquake motions were derived as functions of earthquake magnitude and epicentral distance [Sato, Murono *et al*,1999]. These equations however cannot use for estimating phase spectra at near source regions because the database used dose not contain earthquake motions observed near source regions and an assumption of the single source mechanism is adopted. We therefore derive here regression

---

[1] Proffesor, Disaster Prevention Research Institute, Kyoto University, JAPAN
phone 81-774-38-4065; sato@catfish.dpri.kyoto-u.ac.jp
[2] Railway Technical Research Institute, JAPAN
phone 81-425-73-7262; murono@rtri.or.jp
[3] Railway Technical Research Institute, JAPAN
phone 81-425-73-7262; muramasa@rtri.or.jp

equations for mean and standard deviation of group delay time that can take the rupture directivity effect into account using records observed in near fault regions.

Analytical Method

Definition of Group Delay Time

The group delay time is defined by the derivative of the Fourier phase spectrum $\phi(\omega)$ with respect to circular frequency ω [Papoulis,1962];

$$t_{gr}(\omega) = \frac{d\phi(\alpha)}{d\alpha} \tag{1}$$

The mean of the group delay time within a certain frequency band with the central frequency ω expresses the arrival time of a wave component with frequency α. The distribution width of the group delay time is related to the duration of the time history of the wave component. Because of these characteristics of group delay time, its modelling is much easier than direct modelling of the phase spectrum.

Average Group Delay Time and Its Standard Deviation

The concept of wavelet transformation is used to show a frequency dependency of group delay time. Although there are several ways to define an analyzing wavelet $\varphi(t)$, we used the method of Meyer [1992] to compose $\varphi(t)$. The Fourier transformation of $\varphi(t)$ has a compact support for each scale factor j (named the j-th compact support) defined by

$$\{2^j/3T_d \le f \le 2^{j+2}/3T_d\} \tag{2}$$

in which T_d is the duration of earthquake motion.

For all oberved earthquake motions with a sampling interval of 0.01(sec) we added zero data until the total number of sampling data for each earthquake motion, N, became 131071 (=2^{17}). Using the wavelet transformation of each earthquake motion $x(t)$ we decomposed each time history of earthquake motion to a component time history of each scale factor j (j=1-17). We called this the j-th component time history, $x^{(j)}(t)$. The group delay time of this time history, $t_{gr}^{(j)}(\omega)$, was calculated. Its mean, $\mu_{tgr}^{(j)}$, and standard deviation, $\sigma_{tgr}^{(j)}$, on the j-th compact support were then obtained from Eqs.(3) and (4):

$$\mu_{tgr}^{(j)} = \frac{1}{E^{(j)}} \int_{\omega_l^{(j)}}^{\omega_u^{(j)}} A^{(j)}(\omega) \cdot t_{gr}(\omega) d\omega \tag{3}$$

$$\sigma_{tgr}^{(j)} = \sqrt{\frac{1}{E^{(j)}} \int_{\omega_l^{(j)}}^{\omega_u^{(j)}} A^{(j)^2}(\omega) \cdot [t_{gr}(\omega) - \mu_{tgr}^{(j)}]^2 d\omega} \tag{4}$$

Table 1: Earthquake records used for analyses

Earthquake	M Magnitude	Number of records
Hyogoken Nambu EQ (1995)	7.3	18
Kagoshima-Ken Hokuseibu EQ (1997)	6.3	18
Taiwan Chi-Chi EQ (1999)	7.7	146
Tottori-ken Seibu EQ (2001)	7.3	28

in which $\omega_l^{(j)}$ and $\omega_u^{(j)}$ are the lower and upper limits of compact support of the j-th component, respectively, $A^{(j)}$ is Fourier amplitude of the j-th component, $E^{(j)}$ is the power of the j-th component, and $t_{gr}^{(j)}(\omega)$ is the group delay time of the j-th component time history at a circular frequency of ω as defined by Eq.(5);

$$t_{gr}^{(j)}(\omega_i) = \frac{d\phi}{d\alpha} = -\frac{\phi(\omega_i) - \phi(\omega_{i+1})}{\Delta\alpha} \qquad (5)$$

The calculated Fourier phase spectrum $\phi(\omega)$ defined only principal values within the range from $[-\pi,\pi]$, therefore we must unlap it to obtain the group delay time based on Eq.(5). We applied the method of Sawada *et al* [1998] to unlap $\phi(\omega)$.

Data Used for Analyses

The data used for regression analyses are listed in table1. We selected 210 horizontal earthquake motions observed during 4 earthquakes. The shortest distances from the surface fault line to the observation stations are within 30km. Using observed NS and EW components of earthquake motions we calculate the fault normal and fault parallel components.

Numerical Examples

Probabilistic distribution characteristic of group delay time
The record observed during Taiwan Chi-Chi earthquake (1999) at TCU052 observation station, which is located north end of the surface fault, is used to calculate the distribution of group delay time on a compact support defined for $j=11$ as shown in Figure 1. The distribution of group delay time is expressed by a student distribution with freedom of $\phi = 3$ (t-distribution) of which density distribution function $f_X(x)$ is defined by

$$f_X(x) = \frac{\Gamma[(\phi+1)/2]}{\sqrt{\pi\phi}\,\Gamma(\phi/2)}\left(1 + \frac{x^2}{\phi}\right) \qquad (6)$$

$$-\infty < x < \infty$$

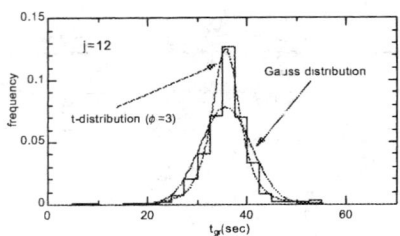

Figure 1: Probabilistic distribution characteristic of the group delay time $t_{gr}^{(j)}(\omega)$

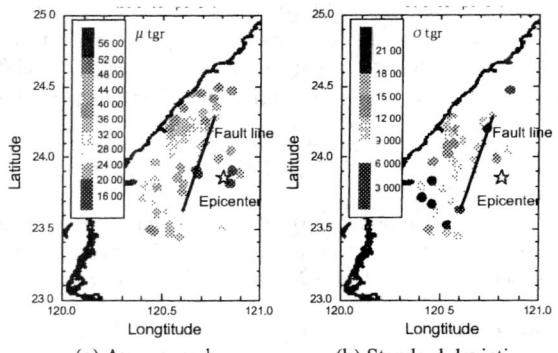

(a) Average value   (b) Standard deviation

Figure 2: Spatial distribution of mean and standard deviation of group delay time for $j=13$

in which ϕ is a positive integer and called the freedom. We compare both of normal and t-distributions with the distribution of calculated group delay time in Figure 1. From this figure we conclude that the t-distribution expresses well the probabilistic characteristic of group delay time.

Spatial distribution of average and standard deviation of group delay time

The mean group delay time, $\mu_{tgr}^{(j)}$, and its standard deviation, $\sigma_{tgr}^{(j)}$, are calculated for all the observed earthquake motions listed in Table 1. Spatial distributions of $\mu_{tgr}^{(j)}$ and $\sigma_{tgr}^{(j)}$ ($j=10$ and 13) defined for Taiwan Chi-Chi earthquake are shown in Figure 2 as examples. The mean group delay time, $\mu_{tgr}^{(j)}$, is distributed over a concentric circle with the epicenter and $\mu_{tgr}^{(j)}$ becomes larger as the distance from the epicenter becomes longer. The average group delay time strongly depends on the epicenter distance. On the other hand, the spatial distribution of the standard deviation of

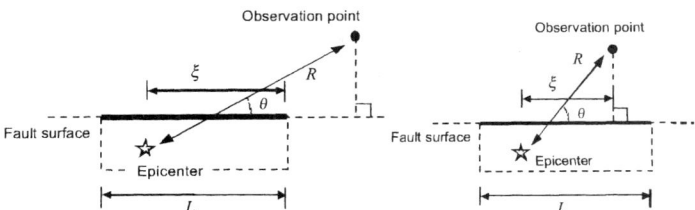

Figure 3: Concept of parameters used in regression equations

group delay time, $\sigma_{tgr}^{(j)}$, is different from that of $\mu_{tgr}^{(j)}$. The standard deviation, $\sigma_{tgr}^{(j)}$, at the forward rupture direction site is smaller, especially at the both ends of the fault. The $\sigma_{tgr}^{(j)}$ is strongly affected by the rupture directivity of the fault. This means that the duration time of earthquake motion becomes shorter when the observation point is located in the direction of fault rupture propagation and it becomes longer when the observation point is located in opposite direction.

Modeling of Probabilistic Characteristics of Group Delay Time

The mean group delay time, $\mu_{tgr}^{(j)}$, and its standard deviation, $\sigma_{tgr}^{(j)}$, were calculated for all the observed earthquake motions summarized in Table 1. Because the trigger time of each earthquake motion is recorded and the rupture stating time is given we shifted the origin time of each earthquake motion to the rupture staring time. The concerned period range in standard aseismic design standards is 0.1-5 sec. The regression analyses of $\mu_{tgr}^{(j)}$ and $\sigma_{tgr}^{(j)}$ therefore were conducted for j=7-15.

At first, a regression analysis was conducted using only two parameters, the earthquake magnitude M and the epicentral distance R [Sato, Murono et al, 1999]. But correlation coefficients for $\sigma_{tgr}^{(j)}$ were very low (less than 0.2). This result suggests that the standard deviation $\sigma_{tgr}^{(j)}$ (i.e. corresponding to the duration of earthquake motion) is affected by the fault rupture directivity and the assumption of single source mechanism cannot be adopted for modeling the group delay time of near source region. We, therefore, introduce the directivity model into the regression equations of mean and standard deviation of group delay time, $\mu_{tgr}^{(j)}$ and $\sigma_{tgr}^{(j)}$.

$$\mu_{tgr}^{(j)} = \alpha_1^{(j)} \times 10^{\beta_1^{(j)} M} \times R^{\gamma_1^{(j)}} \times \eta_{dir\,\mu}^{(j)} \tag{7a}$$

$$\sigma_{tgr}^{(j)} = \alpha_2^{(j)} \times 10^{\beta_2^{(j)} M} \times R^{\beta_2^{(j)}} \times \eta_{dir\,\sigma}^{(j)} \tag{7b}$$

The parameter $\eta_{dir}^{(j)}$ is the coefficient to express the effect of fault extent and rapture propagation and given by

Table 2: Regression coefficients

j	Average value $\mu_{tgr}^{(j)}$				Correlation coefficient	Standard deviation $\sigma_{tgr}^{(j)}$				Correlation coefficient
	α	β	γ	κ		α	β	γ	κ	
7	7.136	0.005	0.428	-0.136	0.628	2.369	0.085	-0.027	0.067	0.114
8	5.724	0.022	0.437	-0.226	0.732	0.204	0.187	0.216	-0.411	0.429
9	12.387	-0.016	0.399	-0.165	0.741	0.725	0.064	0.501	-0.560	0.532
10	19.548	-0.042	0.380	-0.135	0.781	0.453	0.049	0.706	-0.723	0.651
11	18.157	-0.037	0.349	-0.075	0.789	0.081	0.157	0.608	-0.695	0.757
12	13.651	-0.014	0.304	-0.037	0.837	0.011	0.279	0.544	-0.609	0.790
13	12.724	-0.005	0.258	0.043	0.826	0.002	0.403	0 380	-0.398	0.840
14	3.457	0.067	0.261	0.019	0.856	0.001	0.438	0.439	-0.417	0.840
15	15.027	-0.019	0.260	0.061	0.700	0.003	0.399	0.403	-0.398	0.803

$$\log \eta_{dir\,\mu}^{(j)} = \kappa_1^{(j)} \times (\xi/L) \times \cos^2 \theta \qquad (8a)$$
$$\log \eta_{dir\,\sigma}^{(j)} = \kappa_2^{(j)} \times (\xi/L) \times \cos^2 \theta \qquad (8b)$$

where θ is the angle between the fault strike and epicentral azimuth, ξ is the distance from the projection point of epicenter on the surface fault to the point where the shortest distance calculated and L is the fault length. The definition of these parameters is illustrated in Figure 3. Parameters, $\alpha^{(j)}$, $\beta^{(j)}$, $\gamma^{(j)}$, $\kappa^{(j)}$ are the coefficients of regression equations obtained by regression analyses for the j-th component time history of earthquake motion. The results of regression analyses are shown in Table2. The correlation coefficients for $\mu_{tgr}^{(j)}$ are 0.63-0.84 and the correlation coefficients for $\sigma_{tgr}^{(j)}$ are 0.43-0.84, except for $j=7$.

The simulated mean value of group delay time, $\mu_{tgr}^{(j)}$, and standard deviation, $\sigma_{tgr}^{(j)}$, by our newly developed model are compared with those of observed records (Figure 4). As for the $\mu_{tgr}^{(j)}$ simulated values show very good agreement with observed values. As for the $\sigma_{tgr}^{(j)}$, there are some dispersion between simulated values and observed values.

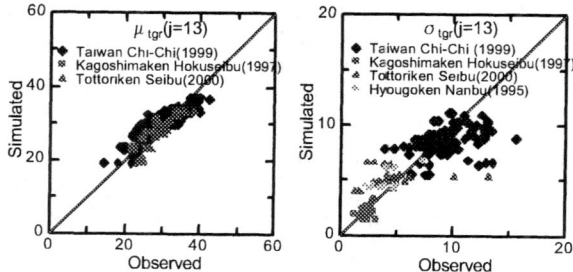

Figure 4 Simulated $\mu_{tgr}^{(j)}$ and $\sigma_{tgr}^{(j)}$ are compared with observed values

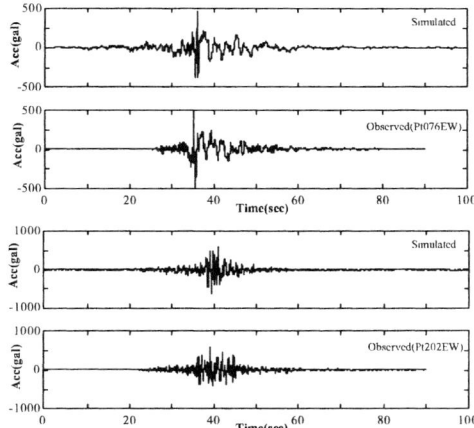

Figure 5 Re-simulated time histories compared with observed records

Simulation of Earthquake Motion Using Proposed Group Delay Time Model

To check the applicability of obtained regression relations to simulate group delay times we resimulate earthquake motions observed during Taiwan Chi-Chi EQ (1999) and compared with observed records. We use the observed Fourier amplitude spectrum at each observation point to resimulate an earthquake motion because the purpose of this paper is to model phase spectra. The group delay time, $t_{gr}^{(j)}(\omega)$, for the j-th component time history on the compact support $\{2^j/3T_d \le f \le 2^{j+2}/3T_d\}$ is generated by assuming the t-distribution with mean, $\mu_{tgr}^{(j)}$, and standard deviation, $\sigma_{tgr}^{(j)}$. Integration gives the phase spectrum, $\phi^{(j)}(\omega)$, for the j-th component time history, $x^{(j)}(t)$. For this integration the phase at the frequency, $2^j/3T_d$, which is the initial value in the j-th compact support, is chosen to be equal to the end phase of $(j-1)$-th compact support. The Fourier amplitude spectrum on the j-th compact support, $A^{(j)}(\omega)$, is assumed to coincide with the Fourier amplitude of the observed earthquake motion, $A(\omega)$, in the frequency range $\{2^{j-1}/T_d \le f \le 2^j/T_d\}$ and to be zero outside of this region.

Using the calculated $\phi^{(j)}(\omega)$ and modeled $A^{(j)}(\omega)$ for $j=7\sim15$, we resimulated each component time history and obtained the earthquake motion by summing up all the j-th component time histories. One of the results is shown in Figure 5. The several characteristics of earthquake motions such as the arrival time and the duration time are expressed well for different earthquake motions. In particular, the arrival time of earthquake motion shows good agreement with that of the observed motion.

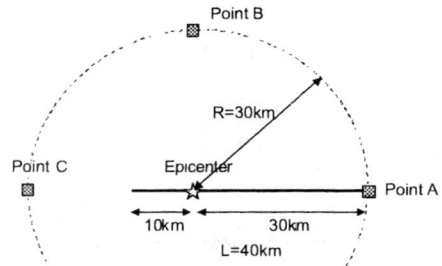

Figure 6 Condition for simulating artificial motions

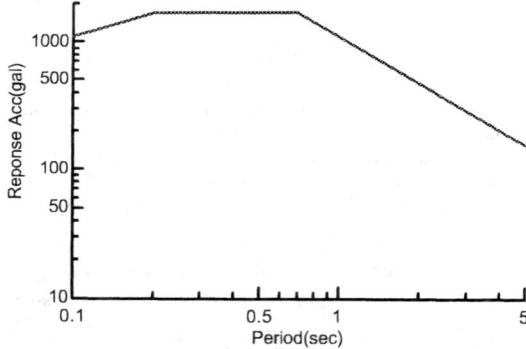

Figure 7 Design response acceleration spectrum for near fault region used in the seismic design code of Japanese facilities

Examples of Artificial Ground Motion Compatible with Seismic Design Spectrum

In seismic design specifications for structures, generally, design earthquake motions are often defined as the acceleration or velocity response spectra. For an example, the design acceleration response spectrum is shown in Figure 7 which is used in the seismic design code for Japanese railway facilities. This is the design spectrum to be used in a near fault region where sever earthquake ground motions are expected. Artificial earthquake motions, therefore, are simulated which are compatible with the design acceleration spectrum by using the proposed model of the group delay time.

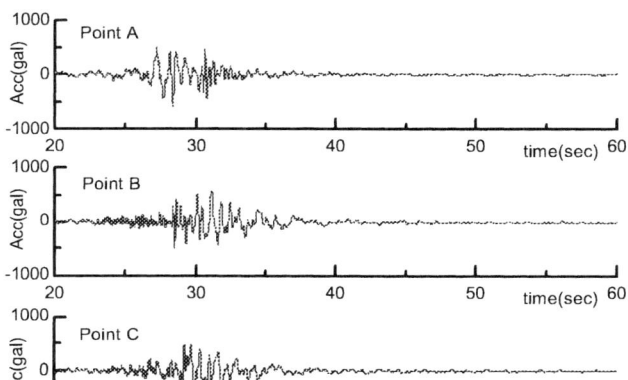

Figure 8 Simulated earthquake motions compatible with design response spectrum by using proposed model

An seismic fault of an earthquake magnitude 7 with the 40km fault length and with the 90 degree dip angle is assumed here, as shown Figure 6. The sign ☆ denotes a epicenter of the earthquake. And three locations with epicentral distances of 30km are assumed.

Figure 8 shows results of simulated motions. Even though the epicentral distances are same, the waveforms of motions are quite different due to the directivity effects. The wave form at point A shows pulse-like shape and the duration becomes the shortest because it locates in the rupture propagate direction. The duration of simulated motion at point B is the longest and doesn't show the pulse-like shape. The proposed phase model is effective to simulate design ground motion compatible with design spectrum.

Concluding Remarks

A method to model phase spectra near source region was developed using the concept of the group delay time. The group delay time at the near source region was strongly affected by a rupture directivity of the fault. Regression equations for mean of group delay time and its standard deviation on compact supports of mother wavelet were derived using the data-set of observed earthquake motions. As well as the epicenter distance and earthquake magnitude, we selected two regression parameters that are the angle between the fault strike and epicentral azimuth and the distance from the projection point of epicenter on the surface fault to the point where the shortest distance calculated, in order to take the rupture directivity effects into account. The effectiveness of this regression model was evaluated by re-simulating earthquake motions of near source region by using this model.

References

Katukura, K. and Izumi, M.(1983) "A fundamental study on the phase properties of seismic waves" *Journal of Structural and Construction Engineering, Transactions of AIJ.* 327, 20-27 (in Japanese).

Meyer, Y.(1992) *Wavelets and Operators*. Cambridge University Press.

Papoulis, A.(1962) *The Fourier integral and its application*. McGraw-Hill.

Sato, T., Murono, Y., and Nishimura (1999). "A Modeling of Phase Spectrum to Simulate Design Earthquake Motion" *Elliott, W.M. and McDonough, P, Editors. Optimizing Post-Earthquake Lifeline System Reliability. Proceedings of the 5th US Conference on Lifeline Earthquake Engineering*, Seattle, ASCE, 804-813.

Sawada, S., Morikawa, H., Toki, K. and Yokoyama, K.(1998), "Identification of path and local site effects on phase spectrum of seismic motion" *Pcoceedings of 10th Japan Earthquake Engineering Symposium*. 915-921 (in Japanese).

Shinozuak, M and Jan C.-M.(1972). "Digital simulation of random processes and its applications" *Journal of Sound and Vibration*, 25(1), 111-128.

The Role of Urban Planning and Design in Lifeline-Related Seismic Risk Mitigation

Mahmood Hosseini[1] and Leila Niazi Shemirani[2]

Abstract

This paper discusses how urban planners and designers can play a great role in seismic risk mitigation by considering the risk creation/mitigation capabilities of lifeline systems in large cites. At first, some of the previous researches performed by urban planners and designers and also lifeline earthquake engineering specialists with regard to seismic risk mitigation are reviewed. Then, the roles of urban planning and urban design are explained separately in the two cases of existing cities, particularly large ones or metropolises, as well as the new cities. These roles are discussed with respect to the urban development and land-use, lifeline earthquake engineering, and sustainable development issues. Following the recommendations given in this paper can help significantly the seismic risk mitigation in urban areas[*].

Introduction

Many large and populated cities in the world, including the capital cities of several countries are located in highly seismic regions, and many of these cities have suffered extensively during the past earthquakes. On the other hand, urban planners and designers, who are the main responsible people for locating the urban components, as cities are renewed or developed, and also locating new cities, particularly satellite towns around big cities, can play a very great role in decreasing the seismic risk in urban areas. This can be done through either decreasing the hazard level by performing a proper site selection for urban components or decreasing the vulnerability of these components by choosing proper orientation and configuration based on appropriate design criteria. Although many researchers have pointed out the

---

[1] Associate Professor and Head of Lifeline Engineering Department, International Institute of Earthquake Engineering and Seismology (IIEES), P.O.Box 19395-3913 Tehran, Iran; Phone: (98) 21 283 1116-9 and 283 3634, Fax: (98) 21 229 9479, Email: hosseini@dena.iiees.ac.ir
[2] Graduate Student, Department of Civil and Environmental Engineering, Iran Uiversity of Science and Technology, Tehran, Iran.
[*] The main idea of this paper was developed in a new course entitled "Earthquake Considerations in Architectural and Urban Design", proposed and taught by the first author in Cornell University, USA, in Spring 2002 semester.

major role of Urban Planning And Design (UPAD) in Seismic Risk Mitigation (SRM), still more attention should be paid to the role of planners and designers in planning design of city components by a risk mitigating approach. In this approach, regarding the two critical characteristics of lifeline systems with regard to earthquake, namely risk creation and risk mitigation aspects of their reaction to a major earthquake in an urban area, particular attention should be paid to these very important urban components. In fact, as presented in the following sections of the paper, there are several publications with regard to the role of UPAD as well as the role of lifeline earthquake engineering in SRM separately, but studies in which attention has been paid to both UPAD and lifeline earthquake engineering at the same context is very rare.

Therefore, in this paper after a review on some of the researches performed separately by UPAD experts and lifeline earthquake engineering specialists, it is tried to show the role which urban planners and designers can play in SRM by considering the lifeline systems components and their seismic performance in their plans and designs. The role of urban planners and designers are dealt with in two different scales. One is the country or state scale, which can be done by using the macrozonation maps, and is mainly the responsibility of urban planners. The other is the city scale, which can be done by using the microzonation maps, and is mainly the responsibility of urban designers. A very important problem, which can be considered as a significant factor in the extent of earthquake damages and losses in urban areas, is the "adverse interaction of main urban components", in which the lifeline systems are of great concern.

Researches by UPAD Experts with Regard to SRM

Since early 80s urban planners and designers started to pay attention to their great role in seismic risk mitigation in urban areas. One of the first works in this regard has been done following the 1976 Tangshan earthquake in China, which has used the lessons learned form architectural damage and other urban disasters caused by that earthquake as well as other earthquakes around that time in China (Ye, 1981). Ye has examined the earthquake performance of prevailing building systems in China, vulnerable building types, and conventional building code requirements. The influence of building configuration, plan, and elevation on earthquake performance has been then considered. Next, interaction between the structural and non-structural elements and damage to non-structural elements has been treated and aseismic measures specified. Finally,

The application of physical development planning measures at the national, regional, and local levels have been also studied (Ciborowski, 1982). In that study basic definitions of hazard, vulnerability, sensitivity, and risk have been given, and a check list of optimum physical planning measures, the goals to be achieved, and the level at which planning is envisaged has been included. Measures have been suggested to control vulnerability, in particular, the use of low-density development and surrounding open space in densely populated urban areas.

A land use planner's handbook for earthquake risk reduction strategy has been also published (Heikkala et al., 1984). In that handbook a decision-making framework for site selection strategies has been described. The framework involves a series of analytical stages in which the following factors are assessed: the seismic hazard in a community, the physical development context and the range of available planning tools, the feasibility and cost of each of several applicable planning tools, and the overall effectiveness of any particular tool.

Physical planning methodology has been studied also after 1979 Montenegro earthquake in Yugoslavia (Pavichevich, 1986). It has been described that studies which define a physical planning methodology in seismic zones include identifying concepts and interpreting methods for definition and analysis of seismic hazard, seismic vulnerability, and seismic risk. The paper has discussed the many studies carried out within this framework after the 1979 Montenegro, Yugoslavia, earthquake, and by classifying 64,000 buildings immediately after that earthquake according to type and building materials a study of vulnerability analysis for defining an acceptable level of seismic risk has been preformed.

A few years later another study has been done on land use planning with regard to seismic risk reduction (Mader, 1991). That paper has discussed land use planning measures that can be taken to reduce the risks associated with the major earthquake hazards of surface faulting, ground shaking, landslides, and liquefaction. Briefly mentioned are the landslides and liquefaction that occurred as a result of the 1989 Loma Prieta earthquake. The key tools used by local governments in dealing with earthquake hazards have been described. These include seismic safety elements for city and county general plans to identify seismic hazards and to set forth policies for minimizing risk, zoning ordinances, subdivision ordinances, the need for city or county staff geologists, the Uniform Building Code (UBC), and the California Environmental Quality Act.

An investigation has been also done on the relationship between risk assessment and land use planning based on some experiences in Italy (Menoni and Pergalani, 1995), in which a framework to assess urban and regional vulnerability has been shown. The framework takes into account both induced hazards that may occur as consequences of earthquakes, and physical and systemic vulnerabilities. The authors have focused on the town of Toscolano Maderno in the province of Lombardy, Italy. A more general study has been also done in this regard which has emphasized on the complexity of land use planning in seismic areas (Mihailov and Talaganov, 1995). They have stated that land use planning becomes a more complex task when applied in seismically active areas, and that the seismic conditions of the area must be determined and included in the development plan.

Recently, a methodology for assessment seismic risk in urban planning has been applied to Regione Lombardia, Italy (Belloni, 1998). The aim of that study has been to set up a methodology for providing a risk matrix to be used for drafting town plannings in the 41 seismic municipalities of the Regione Lombardia. The study has been divided into three phases: 1) determination of probabilistic distributions of occurrences and intensities; 2) definition of geomorphological and geological situations and estimation of related coefficients of amplification using a finite element method; and 3) evaluation of vulnerability of buildings. This type of methodology has been applied to the town planning of Vallio Terme, Province of Brescia.

Researches in Lifeline Earthquake Engineering with Regard to SRM

Although the preliminary works on "lifeline earthquake engineering" goes back to early 70s, studies on lifelines with regard to urban planning and risk mitigation started more than a decade later. In early 80s a state-of-the-art has been published on lifeline hazard mitigation analysis (Taylor et al., 1982). In that report it has claimed that because so many low-cost measures await implementation, lack of complete network analyses may not hinder many significant amelioration efforts. By emphasis on connectivity modeling, they have discussed systematic approaches incorporating flow analyses and operational strategies. They have stated that component vulnerability models are greatly needed if much guesswork is to be removed.

Another preliminary study, performed on lifeline systems from the urban planning point of view, has been on post-earthquake reconstruction and seismic protection of historic town centers (Lagorio, 1985). A few years later lifeline hazard mitigation through strategic industrial planning has been also studied (Phipps, 1989). Phipps has stated that the vitality of a community in the aftermath of an earthquake will be largely dependent on the recovery of industry, so through the development and implementation of comprehensive hazard mitigation and disaster preparedness programs, industry can greatly enhance the physical well being of its employees after an earthquake and improve its chances for expedient business recovery as well. He has cited specific examples of industry's earthquake preparation related to power and communications in Northern California, and has presented general recommendations for strategic industrial planning related to lifeline hazard mitigation.

Following the 1985 Mexico Earthquake a study has been done on lifelines and disaster response/mitigation (Krimgold and Gelman, 1989). The set of research projects undertaken by Krimgold and Gelman has fallen under two general headings: disaster response management and impact assessment. Under the first heading they have studied transportation lessons learned, earthquake injuries, search and rescue, and organizational and public response. Under the second heading nonstructural damage (glass), damage assessment of buried pipelines, and seismic damage recognition systems have been studied.

Following the 1989 Loma Prieta Earthquake, interaction among damaged lifelines was paid attention by some researchers (Isenberg, et al., 1990). They have stated that due to disruption of essential utility services, or lifelines, in Loma Prieta earthquake, communities such as Watsonville, California, suffered economic losses, and their emergency response staffs were severely challenged to provide essential public safety and health services. Although that study has been on the small town of Watsonville in California with 6 square miles area, 29,000 population, and 1.4 billion dollars total assessed valuation of property, the idea could have been helpful for considering the case from the urban planning point of view.

Later on, lifeline interaction has been taken into consideration more seriously in post-earthquake urban system reconstruction (Zhang, 1992). In that study the restoration of each lifeline system has been modeled as a Markov process. The degree of effect that a certain lifeline has on the restoration of some other lifeline has been considered as a function of its critical state and its current expected state. A computer simulation has been carried out to show the feasibility of the approach.

Results from the simulation have shown significant effects of the interaction on the speed and cost of the urban system reconstruction.

A case study has been also performed on lifeline interaction effects on the earthquake emergency response of fire department in Tehran metropolis (Hosseini and Mirza-Hessabi, 1999). The main problems considered in that study include bridge failure, slope instability in high embankments or deep trenches, overturning of high retaining walls and electricity towers, breakage in water mains and wastewater pipes, gas pipeline explosion, collapse of high-rise buildings, and, finally, the public reaction to the event, which can result in traffic congestion and have a very adverse effect on the emergency response activities. The results have shown that the water system has the most adverse interaction effect on the function of the transportation system. It has been also concluded that in emergency planning special attention should be paid to rushing of people into streets as a crucial consequence of an earthquake in large cities such as Tehran.

A study has been done also on partnerships for regional lifeline earthquake mitigation (Avila and Welch, 1999). They have stated that performance relationships between facilities are important to consider when planning for their respective sizes, locations, and other characteristics. They have also expressed that analyzing how a facility can also support or interact with an adjoining lifeline agency can significantly increase the operational and emergency reliability it provides both agencies. The Contra Costa Water District has initiated and executed this type of approach on existing and future facilities that lie next to other lifelines, including the East Bay Municipal Utility District, and several other cities in the East San Francisco Bay Area. This strategy has also encouraged pre-emergency planning and inter-agency communication and significantly improved the mutual aid capabilities of participating agencies. This partnership approach now benefits over 400,000 residents in the San Francisco Bay Area.

Recently, a study has been also performed on modeling earthquake impact on urban lifeline systems to be used in loss estimation for future earthquakes and risk mitigation (Chang et al., 2000-2001). They have summarized the development and application of a so-called advanced, integrated earthquake loss estimation methodology for urban lifeline systems. The methodology, which evaluates direct and indirect economic losses from lifeline failures, provides a means for assessing both expected losses from future earthquakes and potential loss reduction from mitigation alternatives. That effort has built on and coordinated contributions from lifeline earthquake engineering, geography, sociology, and economics by researchers of the Multidisciplinary Center for Earthquake Engineering Research (MCEER). The methodology combines Monte Carlo simulations, GIS, business resiliency questionnaire surveys, and economic computable general equilibrium modeling. They have applied their proposed method to the water delivery system of Memphis, Tennessee, the major city in the New Madrid earthquake zone.

Relations between Lifeline Systems, UPAD, and SRM

Lifeline systems are related to urban planning and particularly urban design in a few ways as follow:

- The site selection for the key components of many lifeline systems like dams, power plants, refineries, long-span bridges, water and specially wastewater treatment plants, and so on is a matter of UPAD.
- The transportation system planning in the country or state scale and the urban transportation network planning inside and in the vicinity of cities, especially metropolises, in the city scale are both major parts of urban development issues.
- Making decision on the type(s) of energy resource(s), and consequently the energy transmission system is, in both country and city scales, an UPAD issue.

By consideration of the mentioned relations between lifelines and UPAD issues now the relation between UPAD and lifeline-related SRM is quite clear. However, as it is seen in the review of previous works, although SRM issue has been a matter of interest and concern for many urban planners and designers on the one hand, and lifeline systems and their components have been paid attention from various aspects by several researchers with regard to SRM issues on the other, still the lifeline-related aspects of SRM has not been taken into consideration by urban planners and designer. In the following section it is tried to discuss this subject thoroughly.

The SRM Roles of Urban Planning and Urban Design with Regard to Lifelines

Considering the relation between lifeline systems and UPAD, mentioned in the previous section, urban planners and urban designers can have greater roles in SRM. These roles for the case of existing cities are somehow different from the case of new cities or under-development urban areas. Therefore, it is tried to describe these roles separately hereinafter.

The SRM Role of Urban Planners in the case of New Cites - Urban planners who are the main decision makers on the location of urban components, including those of lifeline systems, can play their SRM role in the following ways:
- Site selection for lifeline systems key components like dams, power plants and so on not only based on the general urban and regional planning criteria, but also by paying attention to the seismic hazard level considered in the region based on the regional seismic hazard macrozonation maps; this means that the lifelines key components will be less likely to get damaged in future earthquakes, and consequently their adverse effects on the whole country situation, particularly form the sustainable development point of view will decrease.
- Considering more access ways, particularly ways with lower seismic hazard level, for cities specially the major cities, which have been planned, by some rationalization, to be located in highly seismic areas; this will make it possible to have more certain rescue and relief activities for serving the help-needing people in those area, regarding the vital role of transportation system in the successfulness of rescue and relief teams activities in the case of a major seismic event in an urban area.
- Considering the less seismically vulnerable energy supply system for under-planning urban areas; this decision making can be performed effectively by a close cooperation of urban planners and lifeline engineering specialists.

- Considering more appropriate large open areas around densely populated districts of the city by predicting enough urban utilities access for temporary settlement after the seismic event; this will decrease the probable life loss in the case of a major earthquake, will hinder somehow the social impact of the event.
- Considering the compatibility between site selection for future satellite town around a big city and the lifeline route to those town by the proper use of the seismic hazard maps; this is a very important point with regard to site selection for future development that even when the selected site for a town is seismically safe the routes of lifelines to that town can be improper, namely the utilities have to pass hazardous areas to reach that town. In such a case an economic study may become necessary to make a final decision on the site selection for the satellite town.

The SRM Role of Urban Planners in the case of Existing Cites – In the case of existing the urban planners' responsibilities can be divided into two main categories. One is with regard to those parts of the existing cities, which are going to be renewed. In this case their role is basically similar to their role in the case of new cities as the old parts of the cities are renewed. Just, some modifications may be necessary because of the interaction between the old and new parts of the urban area. The other category of urban planners' responsibility in the case of existing cities is related to the changes or additions which they can offer to make the existing cities safer against the future earthquakes. These changes or additions can be as follow:

- Considering an Emergency Management Center in the safest location of the city (adding it to the city, if not existing, or changing its location if not located in a safe place), and placing all the crisis management authorities offices of the city lifeline systems around it, possibly as a complex, or in such a way that they have access to each other quickly; this will make the communications or gathering of all the crisis management authorities much easier and safer in the time of earthquake. Obviously, it has been assumed that all of the city authorities and decision makers, including lifeline system authorities, have been trained for emergency management actions, and all lifeline system organizations have their corresponding crisis management office.
- Modifying the access ways form densely populate districts to large open areas of the city; this will facilitate the evacuation of the stricken people if it becomes necessary after a major earthquake.
- Adding some urban utilities access in the large open areas of the city to be used during the temporary settlement after the probable earthquake; a good example of this additions is considering some elevated water tanks in open areas of the city, so that they can be used by the stricken people without any need to pumping (regarding that electric power shortage is very likely after major earthquakes).
- Modifying the access ways to hospitals, particularly the larger ones to expedite the rescue and relief teams activities; this modification is also desired in the access way of fire departments to densely populated areas.
- Changing the energy transmission system if it is realized highly vulnerable; for example if the electric power towers are located on instable slopes they should be relocated or changed into underground transmission system.

The SRM Role of Urban Designers in the case of New Cites - urban designers, who are responsible persons and decision makers for configuration and orientation of urban components, can play the SRM role in the following ways:
• Considering the compatibility between site selection for the city components and the lifeline route to that component by the proper use of the seismic hazard microzonatoin maps; this is a very important point with regard to site selection for future urban components that even when the selected site for a component is seismically safe the routes of lifelines to that component can be improper, namely the utilities have to pass hazardous areas to reach that component. In such a case an economic study may become necessary to make a final decision on the site selection.
• Considering the proper configuration and orientation for the city lifeline components, such as bridges and tunnels, and also proper route for the passage of extended lifeline components such as power transmission lines based on the detailed seismic hazard maps
• Designing the major city component, including the lifeline key components in such a way that the adverse interaction between different lifeline systems is kept at the possible minimum level; this needs a close cooperation of urban designers and lifeline engineering specialists.
• Designing the urban transportation network, as the most important lifeline for rescue and relief activities, in such a way that it is not likely to get damaged by other interactive lifeline systems and other city components as described by Hosseini and Mirza Hessabi (1999).
• Designing the large public complexes and buildings such as sport complexes, malls and so on in such a way that they can give the sufficient service, up to their capacity, to the probable stricken people; in other words these places should have a flexible architecture giving them the possibility of presenting a dual performance, one as their routine daily services, and the other as their emergency action services. This will facilitate the temporary settlement of the stricken people and will decrease the adverse mental consequences of the event in the society.
• Designing a crisis management complex, comprising all of the offices mentioned before, in the safest place of the city, with highly assured inter-access and communication possibilities, and considering the required security issues.

The SRM Role of Urban Designers in the case of Existing Cites - As mentioned in the case of urban planners' SRM role in existing cities, urban designers' responsibilities with regard to existing cities can be also divided into two main categories; one with regard to those parts of the existing cities, which are going to be renewed, and one related to the changes or additions which they can offer to make the existing cities safer against the future earthquakes. In the latter category the following modification can be proposed:
• Redesigning of Emergency Management Complex based on the aforementioned points
• Upgrading the safety of the access ways from densely populate districts to large open areas of the city
• Designing some safe urban utilities access to be added to the existing system in the large open areas of the city to be used during the temporary settlement

- Upgrading the safety of the access ways to hospitals, and the access way of fire departments to densely populated areas
- Redesigning the major city component, including the lifeline key components for decreasing the adverse interaction between different lifeline systems
- Upgrading the seismic safety, or altering the performance of unsafe important buildings which are related to lifeline systems

Conclusions

Based on the descriptions on the roles of urban planners and designers in lifeline-related seismic risk mitigation, it is recommended that these specialists improve more and more their knowledge on the earthquake issues, particularly lifeline earthquake engineering, and apply and implement the proposed ideas and know-how in their professional activities. In fact, without implementing these provisions the sustainable development can not be achieved. Some of the proposed items is the roles of urban planners and designers need further joint research by these experts and lifeline earthquake engineering specialists, which include:

- Identifying the less seismically vulnerable energy supply system for under-planning urban areas
- Design of the lifeline key components in such a way that the adverse interaction between different lifeline systems is kept at the possible minimum level
- Flexible architecture of public complexes and buildings which gives them the possibility of presenting a dual normal/emergency performance

References

1. Avila, Ernesto A.; Welch, Stephen J., "Partnerships for regional lifeline earthquake mitigation", *Proceedings of the 5th U.S. Conference on Lifeline Earthquake Engineering*, American Society of Civil Engineers, Reston, Virginia, Aug. 1999, pages 917-921.
2. Belloni, A.; Padovani, N.; Pergalilni, F.; and Petrini, V., "An application of a methodology for assessment seismic risk in urban planning (Regione Lombardia, Italy)", *Proceedings of the 11th European Conference on Earthquake Engineering*, A. A. Balkema, Rotterdam, 1998.
3. Chang, S. E.; et al., "Modeling earthquake impact on urban lifeline systems: advances and integration in loss estimation", *Proceedings of the China-U.S. Millennium Symposium on Earthquake Engineering*, Beijing, 8-11 November 2000, A. A. Balkema, Lisse, The Netherlands, 2001, pages 195-201
4. Ciborowski, A., "Physical development planning and urban design in earthquake-prone areas", *Engineering Structures*, 4, 3, July 1982, pages 153-160.
5. Heikkala, S.; et al., "A land use planner's handbook to developing an earthquake risk reduction strategy", *Proceedings of the Eighth World Conference on Earthquake Engineering*, Prentice-Hall, Inc., Englewood Cliffs, New Jersey, 1984, pages 729-735, Vol. VII.
6. Hosseini, Mahmood and Mirza-Hessabi, Ali, "Lifelines interaction effects on the earthquake emergency response of fire department in Tehran metropolis",

Proceedings of the 5th U.S. Conference on Lifeline Earthquake Engineering, American Society of Civil Eng., Reston, Virginia, Aug. 1999, pages 731-740.
7. Isenberg, J. and Phipps, M. T.; Scawthorn, C., "Watsonville regional study: interaction among damaged lifelines", Putting the Pieces Together: the Loma Prieta Earthquake One Year Later, *Proceedings of the Bay Area Regional Earthquake Preparedness Project*, Oakland, California, 1990, 10 pages.
8. Krimgold, F.; Gelman, O., "Working group conclusions on lifelines and disaster response/mitigation", *EERI 89-02, Lessons Learned from the 1985 Mexico Earthquake*, Earthquake Engineering Research Inst., El Cerrito, California, Dec. 1989, pages 202-204.
9. Lagorio, Henry J., "*Importance of lifelines in post-earthquake reconstruction and seismic protection of historic town centers: urban planning aspects*", CEDR-WP04-85, Center for Environmental Design Research, University of California, Berkeley, Calif., 1985, 12 leaves.
10. Mader, G. G., "*Land use planning to reduce seismic risk*", Future Earthquakes in the San Francisco Bay Area: a forum for corporate, commercial and governmental decision-makers, Seismological Society of America, El Cerrito, California, 1991, 7 pages.
11. Menoni, S.; Pergalani, F., "Exploring the relationship between risk assessment and land use planning: some recent experiences in Italy", *Proceedings of the Fifth International Conference on Seismic Zonation*, October 17-18-19, 1995, Nice, France, Ouest Editions, Nantes, France, Vol. I, 1995, pages 826-834.
12. Mihailov, V. and Talaganov, K., "Seismic risk control in urban and land use planning", *Proceedings of the Fifth International Conference on Seismic Zonation*, October 17-18-19, 1995, Nice, France, Ouest Editions, Nantes, France, Vol. I, 1995, pages 835-840.
13. Pavichevich, B. S., "The issues of mitigation and seismic risk control in physical and urban planning after Montenegro 1979 earthquake", *Proceedings of the 8th European Conference on Earthquake Engineering*, Lab. Nacional de Engenharia Civil, Lisbon, 1986, 2.4/7-14, Vol. 1.
14. Phipps, Maryann T., "Lifeline hazard mitigation through strategic industrial planning", *Proceedings of the 3rd U.S.-Japan Workshop on Earthquake Disaster Prevention for Lifeline Systems*, the Workshop, Tsukuba, Japan, 1989, 16 pages, Paper No. U-6.
15. Taylor, C. E.; Eguchi, R. T.; Wiggins, J. H., "Lifeline earthquake engineering: state-of-the-art of hazard mitigation analysis", *Proceedings of the Third International Earthquake Microzonation Conference*, Univ. of Washington, Seattle, 1982, pages 1599-1627, Vol. III.
16. Ye, Y., "Architectural design and urban planning for seismic region", *Proceedings of the P.R.C.-U.S.A. Joint Workshop on Earthquake Disaster Mitigation through Architecture*, Urban Planning and Engineering, Office of Earthquake Resistance, State Capital Construction Commission, P.R.C., Beijing, 1981, pages 7-21.
17. Zhang, R. H., "Lifeline interaction and post-earthquake urban system reconstruction", *Proceedings of the Tenth World Conference on Earthquake Engineering*, A. A. Balkema, Rotterdam, Vol. 9, 1992, pages 5475-5480.

Earthquake and Terrorism Risk Assessment: Similarities and Differences

Stephanie A. King[1], Hamid R. Adib[2], John Drobny[3], James Buchanan[4]

Abstract

This paper describes a risk assessment method that was developed to help determine how to optimally allocate resources for mitigating the adverse effects of terrorist acts on critical transportation facilities and the occupants of those facilities. The method provides a rational and systematic approach that considers, for each facility, the combination of hazard occurrence likelihood, consequences given the occurrence, and socio-economic importance of the facility. Relying on a classical risk formulation, the method has several aspects that are derived from well-established natural hazard (i.e., seismic) risk assessment techniques. The key similarities and differences between terrorism risk and seismic risk are discussed in the paper. The method was recently utilized in a mitigation project benefit-cost analysis by an owner of critical and highly visible public transportation infrastructure in the eastern United States. A brief summary of the procedure and results is included.

Introduction

The terrorist attacks of September 11, 2001 and subsequent potential threats have created a need for the development of techniques for rationally and systematically assessing the risk of terrorism to the nation's infrastructure. The application of these techniques facilitates critical decision making for risk management personnel to help answer questions such as: What is the risk to my facilities? Which facility should I address first? How can I reduce or transfer the risk most cost effectively? The latter question becomes more important as significant amounts of federal and state mitigation funds are being directed toward protecting critical facilities and their occupants.

Probabilistic risk assessment methods for natural and accidental hazards are fairly well developed; however, the application of these methods to man-made hazards, i.e., acts of terrorism, is relatively new. Following the classic approach of characterizing risk as the multiplication of the probability of occurrence by the

---

[1] Director of Risk Analysis, Weidlinger Associates, Inc., Los Altos, CA; 650-949-3010; king@hart.wai.com; Member ASCE.
[2] Principal, Weidlinger Associates, Inc., New York, NY; 212-367-3000; adib@wai.com; Fellow ASCE.
[3] Assistant Director - Infrastructure Management Division, Port Authority of New York & New Jersey, New York, NY; 212-435-4887; jdrobny@panynj.gov.
[4] Manager - Contract Services, Port Authority of New York & New Jersey, New York, NY; 212-435-4921; jbuchanan@panynj.gov.

consequences given the occurrence, terrorism risk can be quantified and used as a means for prioritizing resources to reduce the adverse effects of potential threats on critical infrastructure. Although the formulation of the probabilistic risk assessment approach is similar for most types of hazard, there are significant differences between natural hazards, e.g., earthquakes, and man-made hazards, e.g., terrorist attack, that necessitate refinements to the classic approach.

This paper discusses a terrorism risk assessment method that was developed to provide a rational and systematic approach for making decisions regarding how best to spend mitigation funds. The scope and limitations of the method are based on the needs of the primary end-user of the method: an owner of critical and highly visible public transportation infrastructure in the eastern U.S. The development of the method was designed to make use of the valuable knowledge and experience of the facility owner managers and engineers, and to produce one final ranked priority list of mitigation projects (physical security) for all facilities evaluated.

The paper begins with an overview of the terrorism risk assessment approach. The following sections discuss each of the main contributors to the risk characterization, i.e., occurrence, vulnerability, and importance. The differences and similarities between earthquake and terrorism hazard are highlighted throughout this discussion. The remainder of the paper provides a case study illustration of a cost-benefit analysis utilizing the terrorism risk assessment method for a portfolio of eight facilities owned and operated by the sponsoring agency. For obvious reasons, specific details regarding the name, location, operation, and vulnerability of the facilities, as well as the threats considered in the analysis, are not included.

Overview of Risk Assessment Approach

In developing the basis for the terrorism risk assessment approach, several key sources of information were consulted, including:

- Prior work in seismic risk assessment for retrofit prioritization, providing methods for overall characterization of risk (Maroney, 1990; Sheng and Gilbert, 1991; Kim, 1993; Babaei and Hawkins, 1993; Hart Consultant Group et al., 1994; King and Kiremidjian, 1994; Basoz and Kiremidjian, 1995; Audigier et al., 2000)
- U.S. Department of Defense procedures for addressing physical threats in facility planning, providing methods for occurrence and vulnerability modeling (U.S. Department of Defense, 1994)
- AASHTO Guidelines for highway vulnerability assessment, providing methods for occurrence and importance modeling (AASHTO, 2002)
- U.S. Department of Justice state preparedness support program, providing methods for importance modeling (U.S. Department of Justice, 2000)

The risk to a facility due to terrorism hazard (Risk Score, RS) is written as:

$$RS = IF \times \Sigma [OF_i \times VF_i] \qquad (1)$$

where IF is the Importance Factor, a measure of the socio-economic impact of the facility's operation; OF_i is the Occurrence Factor, a measure of the relative probability or likelihood of threat i occurring; VF_i is the Vulnerability Factor, a measure of the consequences to the facility and the occupants given the occurrence of threat; and Σ denotes the summation over all considered threats to the facility.

The format of Eq. 1 follows the classical method of characterizing risk as the multiplication of likelihood of occurrence (OF) by the consequences given the occurrence (VF); and for eventual ranking among alternatives, the product is multiplied by a measure of the relative importance of each facility (IF). This format has been used extensively in assessing risk due to seismic hazard. The key differences in applying this format to terrorism risk are in the modeling of hazard occurrence (OF_i) and facility vulnerability (VF_i). Each of the factors in Eq. 1 is discussed in more detail below.

Occurrence Factor

Although there are efforts underway to develop probabilistic models for occurrence of terrorism acts (e.g., Woo, 2003), practical quantitative models for use in risk assessment are not yet publicly available. For seismic hazard, probabilistic occurrence models are very mature. In addition, there is general consensus among the scientific community as to the location of active sources and the size of events those sources are capable of generating, information that is needed for defining scenarios for deterministic occurrence modeling. For terrorism hazard, no such scientific basis exists – occurrence in terms of size, location, and frequency of event is primarily a function of an aggressor's intent, capabilities, and belief that the desired outcome will be achieved. Another key difference between seismic and terrorism hazard occurrence modeling is that mitigation efforts (such as increased security) can reduce the probability of occurrence of an act of terrorism, an important issue that needs to be considered in the formulation of the risk assessment approach.

In Eq. 1, the Occurrence Factor (OF_i) for a given threat i is represented as a number between 0 and 1, computed as a weighted sum of utility values as follows:

$$OF_i = \Sigma[w_j \times v_j(x_j)] \qquad (2)$$

where x_j is the value of attribute j; w_j is the weighting factor on attribute j; $v_j(x_j)$ is the function or table that maps x_j to a utility value (between 0 and 1); and Σ denotes the summation over all considered attributes for the factor. Note that the $\Sigma w_j = 1$.

The attributes that are considered important to threat occurrence likelihood are:

- Level of access for attack at the facility
- Level of security against attack at the facility
- Visibility or attractiveness of the facility as a target for attack
- Level of publicity if the attack on the facility were to occur
- Number of times the facility has been threatened or attacked in the past

For each facility and each threat considered in the analysis, a group of experts comprised of the facility owner personnel and Weidlinger Associates engineers assigned (via round-table discussion) a value for each of the five attributes listed above. Table 1 shows the attribute utility value table for the level of access for attack at the facility. In most cases, the tables used in this step were derived from information in AASHTO (2002) and U.S. Department of Defense (1994). Note that

Table 1 illustrates how a mitigation scheme to restrict vehicle access reduces the likelihood of occurrence for a given threat.

The weighting factors used for combining the attribute utility values (w_j in Eq. 2) were developed using the pair-wise comparison procedure of the Analytic Hierarchy Process (Satay, 1980), whereby each member of the decision making group assigns a numerical value to the relative influence of one attribute over another. The scores are averaged and used to compute the weighting factors, which are then reviewed by the group as a whole and revised until all members of the group are satisfied with the results.

It should be emphasized that the likelihood of occurrence, characterized by Eq. 2, does not represent an actual probability; it is a measure of subjective expectation that threat i will occur relative to the other threats considered in the analysis. For this reason, it is important that the group of experts assigning the values of the attributes (x_j in Eq. 2) remain consistent throughout the risk assessment process.

Table 1. Attribute Utility Value Table for Level of Access for Attack at the Facility

Access proximity (x_j)	Rating Factor ($v_j(x_j)$)
Asset with no vehicle traffic and no parking within 50 feet.	0.2
Asset with no unauthorized vehicle traffic and no parking within 50 feet.	0.4
Asset with vehicle traffic but no vehicle parking within 50 feet.	0.6
Asset with vehicle traffic but no vehicle unauthorized parking within 50 feet.	0.8
Asset with open access for vehicle traffic and vehicle parking within 50 feet.	1

Vulnerability Factor

In assessing terrorism risk, the vulnerability of a facility, i.e., the consequences given the occurrence of an act of terrorism, follows directly from the most general case of earthquake vulnerability modeling. The consequences of an occurrence (either terrorist attack or seismic event) are the damage (structural and non-structural), loss of facility use or downtime, and casualties (fatalities and injuries). For regional seismic risk analysis, other social and economic consequences are often included; however, for site-specific evaluation, these three are most often used.

In seismic risk assessment, when consequences must be combined into one characterization of vulnerability for benefit-cost purposes, utility value theory is often used. The same approach is used to compute the Vulnerability Factor (VF_i) in Eq. 1 for a given threat i as a number between 0 and 1 as follows:

$$VF_i = \Sigma[w_j \times v_j(x_j)] \qquad (3)$$

where x_j is the value of attribute j; w_j is the weighting factor on attribute j; $v_j(x_j)$ is the function or table that maps x_j to a utility value (between 0 and 1); and Σ denotes the summation over all considered attributes for the factor. Note that the $\Sigma w_j = 1$.

The attributes that are used to characterize the facility vulnerability given the occurrence of threat i are:
- Expected damage to the facility
- Expected downtime or closure of the facility
- Expected number of casualties at, on, or in the facility

For each facility and each threat considered in the analysis, a group of Weidlinger Associates experts in protective design and blast evaluation of critical facilities assigned (via round-table discussion) a value for each of the three attributes listed above. In some cases, the consensus-based values were replaced with engineering analysis of key components of facilities for the identified threats. The values were reviewed by the facility owner managers and engineers and refined if necessary based on specific facility knowledge. Table 2 shows the attribute utility value table for the expected downtime or closure of the facility. This table was derived from information in TM 5-853-1 (U.S. Department of Defense, 1994). Utility value tables for the expected damage and expected casualties were derived from information contained in the standardized earthquake loss estimation methods, ATC-13 (ATC, 1985) and HAZUS (FEMA, 1999). The weighting factors used for combining the attribute utility values (w_j in Eq. 3) were developed using the same process as described above for the Occurrence Factor.

Similar to the likelihood of occurrence, the facility vulnerability, characterized by Eq. 3, does not represent a physical quantity; it is an aggregate measure of the adverse consequences for a specific facility given that threat i occurs, relative to other facilities and threats considered in the analysis. A key difference between earthquake vulnerability modeling and terrorism vulnerability modeling is worth noting here. The effects of earthquake shaking on facilities and their occupants have been the subject of much study over the past several decades; there are nationally accepted standardized models for estimating damage, downtime, and casualties for various classes of structures for various levels of ground shaking (e.g., ATC, 1985; FEMA, 1999). Similar models for evaluating vulnerability to blast effects are not yet available to the general public.

Table 2. Attribute Utility Value Table for Expected Downtime or Closure of Facility

Expected Downtime (x_j)	Rating Factor ($v_j(x_j)$)
Very Low (~ 0 to 5 days)	0.1
Low (~ 6 to 30 days)	0.3
Moderate (~ 31 to 90 days)	0.5
High (~ 91 to 180 days)	0.7
Very High (more than ~180 days)	0.9
Indefinite	1

Importance Factor

For assessing risk and performing benefit-cost analysis of mitigation options for a single facility, social and economic importance of the facility need not be considered.

However, when a diverse portfolio of facilities is to be evaluated and a mitigation prioritization developed, the importance of the facility in terms of value to the owner, the region, and the society at large must be included. The characterization of facility importance for terrorism risk assessment follows directly from earthquake risk assessment in terms of the attributes considered and the use of utility theory. In fact, the Importance Factor is hazard-independent, as it is a function only of the facility's economic and social status. Once the factor is computed for each facility in a given portfolio it can be used in risk assessment methods addressing almost any type of natural or man-made hazard.

In Eq. 1, the Importance Factor (IF) for a given facility is represented as a number between 0 and 1, computed as a weighted sum of utility values as follows:

$$IF = \Sigma[w_j \times v_j(x_j)] \qquad (4)$$

where x_j is the value of attribute j; w_j is the weighting factor on attribute j; $v_j(x_j)$ is the function or table that maps x_j to a utility value (between 0 and 1); and Σ denotes the summation over all considered attributes for the factor. Note that the $\Sigma w_j = 1$.

The attributes that comprise the facility importance are:
- Historical and symbolic importance
- Replacement value
- Importance as an emergency evacuation route
- Importance to the regional economy
- Importance to the regional transportation network
- Annual revenue value
- Criticality of the utilities attached to the facility
- Military importance
- Exposed population on or in the facility

For each facility considered in the analysis, a group of experts comprised of the owner's facility managers and engineering staff assigned (via round-table discussion) a value for each of the nine attributes listed above. In all cases, the attributes are characterized by subjective values ranging from very low or insignificant to very high or highly critical. Although the subjective values can be replaced with actual data for each facility, the level of effort associated with collecting and calibrating these data can be enormous. In most cases, the tables used in this step were derived from information in AASHTO (2002) and U.S. Department of Justice (2000).

The weighting factors used for combining the attribute utility values (w_j in Eq. 4) were developed using the same process as described above for the Occurrence and Vulnerability Factors. Also, similar to the Occurrence and Vulnerability Factors, the facility importance, characterized by Eq. 4, does not represent a physical quantity; it is a measure of subjective social and economic importance of a given facility relative to the other facilities considered in the analysis. Note, however, that while the Occurrence and Vulnerability Factors change as a result of a mitigation action, the Importance Factors remain constant throughout the analysis.

Application of Risk Assessment Method for Benefit-Cost Analysis

As discussed in the Introduction section, the terrorism risk assessment method outlined in this paper was utilized for a security mitigation benefit-cost analysis of a portfolio of eight critical and highly visible public transportation infrastructure facilities owned and operated by an agency in the eastern United States. The analysis included the following basic steps that were performed for each facility, following the pair-wise comparison computation of the weights needed for Eqs. 2, 3, and 4:

1. Compute the Importance Factor (Eq. 4)
2. Identify vulnerable components or target areas (the term "asset" was used in the project to describe the combination of a component critical to the operation of a facility located in an area or region of the facility that could possibly be attacked)
3. Identify (and define parameters of) credible threats to each asset
4. Compute the Occurrence and Vulnerability Factors (Eqs. 2 and 3) for each threat
5. Compute the baseline or pre-mitigation Risk Score for the facility (Eq. 1)
6. Identify the mitigation projects for the facility, their costs (in terms of capital expenditure, operation and maintenance, and disruption), and the threats they are designed to address
7. For each mitigation project, re-compute the Occurrence and Vulnerability Factors (Eqs. 2 and 3) given the presence of the mitigation action, and then re-compute the Risk Score (Eq. 1)
8. For each mitigation project, compute the ratio of benefit, i.e., the reduction in Risk Score (as compared to the baseline Risk Score computed in Step 5) to cost

The result of the analysis described by the steps above is a list of mitigation projects ranked in order of priority based on the ratio of benefit (reduction in facility Risk Score) to cost (estimated net present dollar value of the mitigation project). In total, for the eight facilities evaluated, 97 threats were considered and 31 mitigation projects proposed. Most of the consensus-based decision work was facilitated through approximately four two-day group meetings of all project participants and several smaller meetings with specific groups of experts.

One of the key advantages of the risk assessment formulation given in Eq. 1 is that each component of the total risk is clearly identified, there are no pre-determined weighting factors or black-box type of calculations hidden from the user. This advantage serves several purposes: it allows for validation of intermediate results as the analysis progresses; it provides a means for tracking the influence or sensitivity of subjectively-assigned attribute values; and, most importantly, it allows the users or decision-makers to play an active role in the entire process, ensuring feelings of control and understanding of the process, as well as ownership of the final results.

Because of the sensitive nature of terrorism risk assessment, specific details about the facilities, threats, and mitigation projects cannot be discussed in this paper. Figure 1 shows an example summary of the 31 mitigation projects in terms of benefit (reduction in Risk Score of the facility) versus project cost, with numbers indicating overall rank. Although benefit-cost analysis provides a systematic and rational method for prioritizing the mitigation projects, other issues such as optimal construction scheduling, financing, disruption, and public acceptance of architectural alterations also need to be considered in the final decisions regarding which projects will be completed first. It should be emphasized that the results shown in Figure 1 are only an example using preliminary data and are not intended to represent the final project selection.

Summary

This paper describes the development and utilization of a terrorism risk assessment method that is based on the classical probabilistic risk analysis approach of combining likelihood of occurrence with expected consequences given the occurrence. Several aspects of the method are taken from the well developed field of natural hazard risk assessment, in particular site-specific seismic risk assessment. The method outlined in this paper is intended to provide the user or decision-maker with a clear understanding of, and in many cases a role in developing, the various factors that characterize the risk to a given facility. In addition, the risk assessment approach is modular in design. This means that the existing models for computing the Occurrence and Vulnerability Factors, which are based primarily on an expert opinion utility value approach, can be replaced with more analytically derived models as they are developed and accepted by the scientific community.

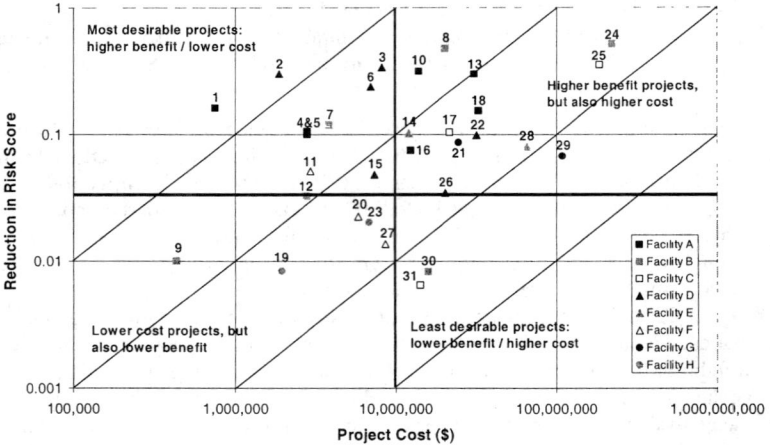

Figure 1. Summary of benefit-cost comparison of mitigation projects.

Acknowledgements

Several engineers and facility managers employed by the owner of the critical infrastructure described in this paper played key roles in accomplishing the risk assessment work discussed in this paper. These include Albert Terriego, Michael DeGidio, Gerald DelTufo, Kenneth Sagrestano, Robert Mckee, Roger Prince, Salvador Gonzales, Harendra Patel, Sergio Martinez, and Mohammed Mohib. Several Weidlinger Associates engineers contributed to the technical aspects of the project through analysis and/or expert opinion. These include Abdol Hagh, Peter Quigley, Ray Daddazio, Ivan Sandler, Bob Smilowitz, Mrinal Bose, Sary Malak, Colleen Blackwell, Eric Hoyt, Jake Wolff, Sam Summerville, Sante Camo, and Herb Rothman. Medhat O'Kelly of Parsons Brinkerhoff provided the cost estimating information for the mitigation projects as well as valuable input on other aspects of the risk assessment process.

References

AASHTO, 2002, *A Guide to Highway Vulnerability Assessment for Critical Asset Identification and Protection*, Prepared by SAIC, Washington, DC.

ATC, 1985, *Earthquake Damage Evaluation Data for California*, ATC-13 Report, Applied Technology Council, Redwood City, CA.

Audigier, M.A., Kiremidjian, A.S., Chiu, S.S., and King, S.A., 2000, "Risk Analysis of Port Facilities," *Proceedings of the 12th World Conference on Earthquake Engineering*, paper no. 2311.

Babaei, K. and Hawkins, N., 1993, *Bridge Seismic Retrofit Planning Program, Report WA-RD 217.1*, Washington State Department of Transportation, Olympia, WA.

Basoz, N. and Kiremidjian, A.S., 1995, *Prioritization of Bridges for Seismic Retrofitting*, Report No. 114, John Blume Earthquake Engineering Center, Department of Civil Engineering, Stanford University, Stanford, CA.

FEMA, 1999, *HAZUS: National Standardized Earthquake Loss Estimation Methodology*, prepared by National Institute of Building Sciences, Washington, DC.

Hart Consultant Group et al., 1994, *Seismic Risk Decision Analysis for Kaiser Permanente Pasadena*, Final Project Report, Santa Monica, CA.

Kim, S.H., 1993, *A GIS-Based Risk Analysis Approach for Bridges Against Natural Hazards*, Ph.D. Dissertation, Department of Civil Engineering, State University of New York, Buffalo, NY.

King, S.A. and Kiremidjian, A.S., 1994, *Regional Seismic Hazard and Risk Analysis Through Geographic Information Systems*, Report No. 111, John Blume Earthquake Engineering Center, Department of Civil Engineering, Stanford University, Stanford, CA.

Maroney, B., 1990, *CALTRANS Seismic Risk Algorithm for Bridge Structures*, Division of Structures, California Department of Transportation, Sacramento, CA.

Saaty, T.L., 1980, *The Analytic Hierarchy Process*, McGraw-Hill, New York, NY.

Sheng, L.H. and Gilbert, A., 1991, "California Department of Transportation Seismic Retrofit Program: The Prioritization and Screening Process," *Proceedings of the Third U.S. National Conference on Lifeline Earthquake Engineering*, pp. 1110-1119.

U.S. Department of Defense, 1994, *ARMY TM 5-853/AFMAN 32-1071, Security Engineering Project Development,* Volume 1, Chapter 3, Washington, DC.

U.S. Department of Justice, 2000, *Fiscal Year 1999 State Domestic Preparedness Support Program,* Washington, DC.

Woo, G., 2003, "Quantifying Insurance Terrorism Risk," to be published in *Alternative Risk Strategies,* M. Lane, ed., Risk Publications, Ltd. (also available at: http://www.rms.com/NewsPress/Quantifying_Insurance_Terrorism_Risk.pdf)

PREVENTION AND REPAIR MEASURES FOR INFRASTRUCTURE NATURAL DISASTER RISK MANAGEMENT

Jacob Greenstein-Inter-American Development Bank-jacobg@iadb.org
1300 New York Avenue N.W. Washington, DC 20577

ABSTRACT

Reliable risk assessment is crucial to building emergency response and risk management capabilities before the disaster strikes (preparedness). National risk management of possible catastrophic events includes the development of susceptibility risk-maps, strategy planning of preparedness activities, design guidelines of affordable cost effective and cost reliable prevention works. Local risk management assessment includes mapping of populated areas (affected people and poverty classification), determination of the type, severity and probability of possible disaster impacts (land slide, debris flows, flooding, structure failure), and using appropriate engineering tools of preparedness works (soils, geology, drainage, topography, earthquake shaking characteristics, meteorological and rainfall records). Key preparedness activities include research, training, development of quantitative monitoring and alert indicators of possible catastrophic events, and cost effective engineering solutions. Under the current budget constraints of many of the Inter-American Development Bank (IDB) member countries, a rational priority rating is necessary to allocate scares resources among competitive needs. Urgent or emergency job order contracting procedures are useful to combine many urgent works into one project administered by one project team. These projects may cover all type of works, reconstruction, repair and rehabilitation of different facilities (roads, bridges, slopes, seawalls, docks, water supply) under a single contract. The projects are competitively bid with fixed costs and with performance indicators. A new research in Israel has indicated that world wide risk management has failed and partial privatization is recommended. This paper emphasizes engineering planning of affordable preventive and remedial measures, vulnerability reduction and lessons learned from Hurricane, tropical storm and El Niño impacted areas.

INTRODUCTION

World wide trends of natural disasters such as Hurricanes, tropical storms, floods and earthquakes show increase in frequency and severity that have caused irreversible impacts[1-11]. Inadequate surface and subsurface drainage capacities is one of the most important factors of the 45,000 fatalities and more than $20 billion in damages in the Latin America and the Caribbean (LAC) region during the 1990s, or about forty significant disasters per annum. Key issues are on time availability of adequate institutional and technical capabilities, sustainable funding, on time affordable and achievable designs, clear definition of the role and responsibilities of national and international agencies. Finance ministries (FM) are important players in disaster management systems in terms of economic planning and shaping the private and the public sectors financial decisions. FM can decide whether to invest effectively in disaster preparedness or in disaster prevention rather than in costly works during or after the occurrence of the emergency. Strategic plans of preparedness emphasize culture of prevention, reduction of the vulnerability of the poor, provision of reliable risk information for decision making, fostering leadership and partnership of the private sector, the public sector and affected communities.

To support the culture of preparedness and prevention, this paper presents engineering-planning tools, socio economic considerations, issues and challenges of preparedness, risk management aspects of mountainous landslides, and vulnerability reduction considerations of catastrophic events.

MOST COMMON NATURAL DISASTERS

With an average of 40 disasters a year, the Latin America and the Caribbean (LAC) region ranks second only to Asia in terms of the frequency of disaster occurrence. For some of the LAC region countries, certain natural disasters had a devastating effect on their populations, welfare, and development prospects. As an example, the maximum of 285 km/h winds and rainfall intensity of over 600 mm developed during the 1998 Hurricane Mitch, from October/23 to November/3, affected 1.92 million people of whom 9214 died in Costa Rica, El Salvador, Guatemala, Honduras and Nicaragua. The total estimated damage was $ 6.0 billions, and the balance of payment effect was $1.6 billions. Natural disasters have greater devastating effect on developing countries than on developed countries, and it was no coincidence that 95% of the death due to natural disasters in 1998 were in developing countries. During the 10-year period of the 1990 decade, disasters had killed more than 45,000 people in the LAC region, affected another 40 million, and caused over $20 billion in direct damages [6-9]. About seventy percent of the 1970-1999-period of natural disasters in the LAC region were meteorological, such as rainfall, hurricanes and strong windstorms. The remainder natural disasters are related to geological phenomena such as earthquakes and volcanic eruptions [1,6,7]. The most costly catastrophic events in the LAC region were the 1985 Mexico City earthquake and the 1988 El Niño-related flooding in Argentina, Peru, and Ecuador. Each one of these two events caused an estimated US$6 billion in damages (in 1999 dollars). However, a number of hurricanes have also caused sever flood related damages. Hurricane Mitch caused an estimated $5 billion in damages in five Central American Countries, and hurricane George caused $2 billion in damages in the Dominican Republic. More recent flooding and related land-slide damages from tropical storms were recorded in Belize (October 2000) and in Jamaica (May/ June 2002), with estimated damages of $260 and US$20 millions. In conclusion, while the frequency of excessive rainfall and flooding related catastrophic events appear to be on the rise, the severity of the economic damages from these events is also increasing[1,6,7]. Therefore, the development and implementation of affordable and cost effective preparedness tools is essential[5,16].

SOCIO ECONOMIC CONSIDERATIONS

While the disaster events are natural in origin, the extent and severity of the damages are mostly a consequence of human activity and inactivity[6-8]. Poor planning of affordable prevention works has aggravated the socio-economic impacts of these catastrophic events. These impacts are measured in terms of damages in economic and social infrastructure, a destruction of economic assets, environmental deterioration, negative income distribution and acceleration of inflationary processes. The result is a negative long-term macroeconomic impact that is reflected on a downtrend in per capita income.

The experience of the LAC region confirms that there is a high correlation between the gross domestic product (GDP) growth and the frequency, magnitude and severity of natural disasters [1,6-12]. One of the most important impacts is the worsening of the national living standards. This effect usually affects a country's entire population and in some cases affects neighboring countries in terms of migration, vector transmission, deterioration of watersheds, reduced demand for imports, and interrupted communications.

In the aftermath of a large disaster, countries often face declining exports and rising imports, a deceleration of economic growth, and a reduction of per capita income. A decline in tax revenues can prolong fiscal instability and increase the country's level indebtedness. Therefore the IDB has aimed to help countries move from the emergency to the reestablishment of their development trajectory as efficiently as possible[8-11]. First, the IDB financing is aimed to benefit the poor. It includes smaller projects to repair and reconstruct water and sanitation infrastructure, stabilize unsafe slopes and flood control in low-income communities. The IDB has also financed programs that protect public expenditures and improve living conditions and economic opportunities for the poor. These programs have traditionally been replaced by reconstruction needs when disaster strikes. For example, a program in the Dominican Republic after Hurricane George made up the fiscal shortfalls to safeguard programs aimed at children's welfare. There, the IDB financing has also helped countries to mange their adverse macroeconomic impacts, including financing to shortfall in recurrent public expenditures for vital social programs, for balance of payment support, and for restructuring debt. Therefore, to reduce the long-term socio-economic damages of catastrophic events it is essential to consider the following strategies: (1) classifying financial resources earmarked for preventing and mitigating the impact of natural disasters as high-yield and long term investment in economic, social and political terms; (2) Developing catastrophic risk classification maps and affordable national and regional natural disaster preparedness plans that progressively reduce the degree of vulnerability and therefore improve the prospects for future development.

The reasons for the high vulnerability are varied and complex. In many of the LAC region countries, the poor, and among these, women, children, and ethnic minorities, are the most fragile and vulnerable population groups. The poor live in greater risk areas, use environmentally damaging farming activities or work marginal land, and have limited access to information, basic services and pre and post disaster protection. In many ways, poverty exacerbates the vicious circle of disasters. Therefore, effective vulnerability reduction must follow the strengthening of the national and the regional macroeconomic capacity, the introduction of active policies that reduce social distortions, poverty alleviation, and improving the coordination of regional policies and of international aid.

ISSUES AND CHALLENGES OF PREPAREDNESS

Preparedness involves building an emergency response and management capability before a disaster occurs. Key disaster preparedness activities include training programs, citizens information and education programs, reliable hazard detection and warning systems and proper engineering tools for risk assessment and prevention designs. Key issues and challenges of preparedness activities are:

- **Inter-institutional and inter-disciplinary coordination,** to break the cycle of destruction and reconstruction. The long term objective is more institutional preparedness, more investments in preventive works and less repair or remediation works[10-13]. The challenge is to consider prevention as an high-return investment and to achieve a significant reduction of the vulnerability in terms of controlling urban expansion, land use planning, productive activities, and misuse and degradation of natural resources.
- **Private-Public sector Collaboration,** among financial institutions, insurance agencies, governmental and inter-governmental entities. In this aspect, Finance ministries are important players in terms of economic planning and shaping private and public financial decisions. The leadership of finance ministries helps to ensure funding, facilitate the incorporation of disaster management into development policy, and provides incentives for financing preparedness activities.
- **Development of susceptibility risk maps,** for preparedness planning and prediction of lifeline disruption in future earthquake or heavy flooding. These maps define 5 to 6 levels of risk-susceptibility, considering flooding, hydro-geological and earthquake-shaking characteristics. As an example, along the Ecuadorian mountainous area of the Cuenca-Molleturo-Empalme (C-M-E) highway the slope failure risk varied with extremely stable slopes with a Factor of Safety (FS) of over 2.0 to unstable slopes with FS of less than 1.2 (minimum recommended FS value of low risk conditions)[18,19]. Slope susceptibility risk data is integrated into digital-layer mapping of topographical (1:100,000 scale), land use; hydro-geological; rainfall, flooding, and earthquake characteristics (location, peak acceleration, intensity). To achieve high reliability, this mapping procedure uses data of each 10-meter-square- cells of the high-risk areas [20].
- **Cost effectiveness and Urgent Job orders contracting procedures,** of many urgent works into one contract administrated by one multi-skill project team. These rehabilitation and operation activities are competitively bid with indefinite quantity, indefinite delivery and fixed unit prices or fixed costs with performance indicators. The Urgent Job Order (UJO) contracting is different from traditional contracts in terms of being in place before the completion of the designs, using more affordable and more achievable quality indicators [16,17]. Urgent job orders may cover all types of works, construction, repair, maintenance and rehabilitation of different facilities (roads, bridges, slope stability, seawalls, docks, and water supply) under one single contract. Advantages of using UJO contracts are: competitive procedures of fixed unit prices, performance specifications, transparent pay equations and quality control procedures. In addition, contractors have incentives to produce good quality products in order to receive more works contracts, and there are more participation opportunities for small businesses that can not compete for larger projects.
- **Ideals of Preparedness Capabilities are:** (1) Agency with adequate institutional capabilities for administration of planning, programming, construction, maintenance, quality control (QC), reporting and monitoring procedures of cost effective prevention activities. Most of these activities are contracted out. (2) Agency, contractors and consultants with effective evaluation, prediction and engineering methodology, procedures and preciseness tools, to ensure cost effectiveness and cost reliability of preparedness works. A good reference is Japan(JICA-Japan: 2001). The

number of fatalities from typhoons in Japan had gone down from approximately 45,000 to almost zero during the last 50 years.
- **Partial Privatization of Disaster Risk Management,** to support transparency, reliable reporting and cost effectiveness[24]. This 2-year research recommends to transfer to the local private sector more responsibilities in terms of emergency evacuation; supply of power, water and food, and ambulance and medical trauma services. An Israeli survey shows that a third of the population is willing to pay directly for these services.

RISK MANAGEMENT CONSIDERATIONS OF MOUNTAINOUS LANDSLIDES

Thunderstorms, excessive rainfall and quick raise of ground water often trigger avalanche or debris-flow that caused the closure of major highways in the Andeans mountains area in Ecuador and on Interstate 70 in Colorado mountainous area[14,19,21]. In both cases the mountainous instability occurred at altitudes between 2200 and 3500 meters above mean sea level. Interstate 70 was closed for 25 hours and there were no fatalities. In Ecuador, the Cuenca-Molleturo-Empalme (C-M-E) highway was closed for several days to several weeks. The rainfall and hydro-geological characteristics of both cases indicate that they are susceptible to debris-flow hazards. Typically, debris flows are initiated in tributary drainage basins with accumulation and acceleration occurring at the basin mouth. In the Ecuadorian case, the debris flow covered extensive lower areas, clogged rivers and creeks and caused heavy agricultural damages. Historic observation of debris flows along the C-M-E highway and Interstate 70 have shown that the mean recurrence interval in years (y) is approximately: $y = 19,400/EXP^{(4.67x)}$, where x is the Melton's number[22,23], defined as: $x = H/(A)^{0.5}$, where H is the basin height above the fan and A is the basin area above the fan. Basins that are small and steep have higher Melton values than basins that are large and have low to moderate H values (moderate relief). In the C-M-E case, the Melton numbers along fifteen locations, during the 1998-9 El Nino rainy season, varied between 2.0 and 2.6 indicating that mean recurrent interval of an avalanche of debris flow varies between one to eighteen months. Eventually about fifteen major debris occurred during this period and blocked the highway for several days to a couple of weeks. The probability (P), defined as the percentage chance of one or more major debris flows occurring on an individual location during a specific time (t) and a mean recurrence interval (y), is calculated using the Poisson probability model, $P = 1 - e^{-(t/y)}$. The probability in terms of the percentage chance of occurrence of one or more debris flows is shown in table 1. Considering these probability indicators together with other affordability concerns during the maintenance and remediation works of the C-M-E highway, drove the selection of a mean recurrent interval of 50 years with an occurrence probability of less than 40% during a life expectancy of 25 years. This slope stability design assumes that the new hazard monitoring procedures along the road will further reduce accidents and major highway closure.

The most common methodology used to determine the factor of safety (FS) of the stability of mountainous slopes, considers stress-strain and limit equilibrium relationships. However, viscous-plastic engineering laws, critical velocities and critical accelerations of the rocks, soils and debris materials are the newer procedures to evaluate the probabilities and predict catastrophic landslides[15,18,19-22]. The definition of critical values of velocities (v_{crit}) and of accelerations (a_{crit}) of catastrophic landslides are as follows: v_{crit} = 5 cm/day, and a_{crit} = 5 mm/day^2. Therefore, when the v_{crit} or when the a_{crit} reaches these values, an uncontrolled or catastrophic landslide could be imminent. Therefore, in populated mountainous area, when early signs of possible landslide appear, monitoring of slope movement is critical. Possible landslide early signs include, development of tensile cracking, and Soil/ Rock-Mass Movement (SRMM) of 2 mm/day to 5 mm/day. In these cases, table 2 recommends a slope evaluation of at least once a month. This table presents a useful guideline of monitoring and alert procedures of possible catastrophic landslides. For example, when the SRMM is between 5mm/day and 15mm/day, a daily monitoring and first alert is recommended. Evacuation of the affected population should be considered when SRMA>50 mm/day or when SRMA > 5 mm/day^2. Also, when the Soil/ Rock-Mass Acceleration (SRMA) is over 2-3 mm/day^2, the monitoring, and the alert procedures should be more conservative and an hourly monitoring should be considered. The most practical procedures to monitor the SRMM and the SRMA values is using a satellite based Very Small Aperture Terminal (VSAT) with centrally managed hub. The VSAT system monitors and reports movements and changes of inclinometers, piezometers, and other defined benchmarks. The VSAT uses remote site dish antenna with a diameter of less than 1.2m and the centrally management unit can monitor unlimited location of unstable slopes. The VSAT technology is independent from terrestrial infrastructure and provides consistence and continuous ground and slope movements. The Chinese bureau of seismology has implemented a 150-site VSAT network for earthquake monitoring. The VSAT network provides also Intranet information exchange and voice applications, essential for risk management of catastrophic events.

Mean Recurrent Interval, (y= years)	Table 1: Probability (percentage chance of one or more debris flows on a individual slope location during specified time)				
	1 yr.	10 yr.	25 yr.	50 yr.	100 yr.
10	9.52	63.21	91.79	99.33	100.00
50	1.98	18.13	39.35	63.21	86.47
100	1.00	9.52	22.12	39.35	63.21
200	0.5	4.88	11.75	22.12	39.35

Table 2: Monitoring and alert indicators of possible landslides (SMRA<2 mm/day^2)			
Alert level	SRMM (mm/day)	Status	Monitoring
0	2<	Normal	1/ month
0	2-5	Normal	1/ 10 days
1	5-15	1st alert	1/ day
2	15-25	2nd alert	2/ day
3	25-40	1st warning	(2- 5) / day
4	40-50	2nd warning	1/ hour
5	>50	Consider evacuation	Continuously

In critical conditions, the relationship linking the critical time (Tc) before failure occurrence, to the SRMM is: LOG(Tc) = (A-LOG (SRMM))/ B, [14,15]. Where, A is a site specific technical parameter and B is an engineering correlation factor that varies between 0.5 and 1. In the specific C-M-E highway in Ecuador, during the 1998-9 El-Niño heavy rains, when SRMM was about 25 to 45 mm/day, the estimated Tc values were approximately 50 to 200 hours. In this specific geological area of cohesive-friction soils, the volume of the landslide (V in cubic meters-M^3) is usually a function of the friction factor (FF) of the local soils. The approximate relation between FF and V is:

V $(M^3)*10^5$	1	2.5	5.0	10.0	25.0	50.0
FF	0.79	0.77	0.75	0.74	0.72	0.71

FINANCIAL PREPAREDNESS: RISK-TRANSFER CONSIDERATIONS

Effective risk-transfer mechanism may support the national or the regional financial preparedness to catastrophic events. Risk exposure associated with natural catastrophes is generally characterized by low frequency and high impact[2]. This reference provides in-depth presentation of how losses due to catastrophes are insured and who absorbs the costs of compensating the insured assets. The IDB is undertaking further studies in the areas of financial planning and risk transfer instruments that will help to understand the mechanism of these instruments that provide financial protection for the private and public sectors in the LAC region[3,4,8-11]. The principal objectives of catastrophic risk transfer instruments are[2-5,17]:

(1) To support affordable insurance premiums for damages associated with natural disasters. This is a realistic objective that requires collaboration among governments, academia, and insurance companies in the development and administration of a reliable database, emphasizing code and safety compliance, probability, magnitude and severity of expected catastrophic events. This database will address local, national and regional risk assessments and will be available to practitioners and researchers;

(2) To cover catastrophic risk exposures in individual investment projects. A project level approach to manage catastrophic risks through mitigation or specific insurance that will reduce specific project exposures;

(3) To facilitate country risk management plans and establish cover for higher catastrophe risk layers. Using reliable countywide risk management plans to mitigate risks and attract insurance mechanisms of risks transfer cover. The lower level risk layers could be covered by tax-funded calamity funds as the main source for disaster relief and rehabilitation. Cover for higher risk layers could be obtained through the international financial markets in terms of cat-bonds, risk-swaps, and contingent capital.

(4) To establish national insurance pools that require mandatory insurance policies. Governments must enforce stringent initiatives of risk management, such as enforcing building cods and effective property registration. Once these policies are implemented, local insurance companies can support local market involvement of the lower risk layers. Insurance pool could cover part of the higher risk layers in

international financial markets, through reinsurance contracts, risk-linked securities, contingent and surplus notes.
(5) To combine risk exposure across several countries. This international insurance pooling collaboration can provide a natural first line risk diversification that engages local primary insurance companies in the development of regional insurance companies. In addition, it may provide scale economies to risk financing arrangements in the international financial markets.

SUMMARY: VULNERABILITY REDUCTION AND ENGINEERING RISK MANAGEMENT

Vulnerability measures the probability of a catastrophic events on people and on infrastructure due to insufficient prevention, mitigation and warning procedures. The development pattern of most countries with high rates of poverty, socio economic exclusion and environmental damage, is the leading factor of high vulnerability. This vulnerability has revealed itself during natural disaster events, when not enough has been invested in prevention works and alert procedures. Catastrophic failures of mountainous slopes, roads, bridges, and earth dams are usually associated with uncontrolled surface and underground water levels. Once, large masses of saturated soils and rocks generate an accelerated speed of 40mm/day to 55mm/day a catastrophic event could be imminent and evacuation of affected communities should be considered. Risk reduction of such events requires cost effective and affordable planning, engineering, monitoring and reporting procedures. The planning activity is associated with the ex-ante development and use of susceptibility flooding and slope-failure risk-maps. Theses risk maps are essential to:(1) reduce the probability occurrence of a possible catastrophic event; (2) reduce its extension or its magnitude; and (3) reduce its severity. In engineering terms, improvement of the factor of safety to at least 1.2 for an affordable future life expectancy, which usually varies in the LAC region from 5 years to 50 years.

To reduce vulnerability, the IDB has financed projects aimed at building countries' risk management capacity[3-5, 8-11]. For example, between 1990 and 1999, the IDB financed in Central America disaster mitigation, prevention and preparedness activities accounted for $350 million. For the most part, this disaster-related lending was not part of emergency or reconstruction lending, but rather for programs that reduce vulnerability in the absence of an event. To further reduce vulnerability the current projects financed by the IDB take advantages of the best practice, under affordable financial conditions, to resolve critical problems of risk reduction. The current priorities have changed as follows: (1) From reconstruction projects to ex-ante risk reduction and management, including institutional development (legal and institutional frameworks, technical and operational capacities), codes and enforcement, and land-use planning; (2) From centralized administration of emergency management to decentralized or de-concentrated administration, including empowering constituents, and building knowledgeable consumers and a reliable database for risk monitoring and reduction; (3) From a focus on large national disasters that overwhelm a countries' capacity to handle smaller, cumulative events; and (4) From public sector financing of losses to the mobilization of new private resources for risk reduction. To further reduce the vulnerability of catastrophic events, the IDB introduced in 2001 a sector facility of disaster prevention[10]. This financial instrument assists countries to take integrated approach to reducing and managing their risk to natural hazards before a disastrous event, through the following components: (1) Risk identification and forecasting, to understand and quantify vulnerability and disaster risks; (2) Mitigation, to address the structural sources of vulnerability; (3) Preparedness, to enhance a country readiness to cope quickly and effectively with an emergency; (4) Risk transfer measures to spread financial risks over time and among different actors, including insurance and capital market schemes; and (5) Transition to new policies, legal and regulatory frameworks, and institutions to build effective national systems for risk reduction. The IADB has implemented a streamlined financial procedure that can quickly provide up to $5 million per operation to put high-impact measures of vulnerability reduction.

REFERENCES
(1) Charveriat, C. 2000. "Natural Disaster Risk in Latin America and the Caribbean" Washington, D.C: Inter-American development Bank.
(2) Andersen J.A. 2002. "Innovative Financial Instruments for Natural Disasters Risk Management" Washington, D.C: Inter-American development Bank.
(3) IADB-,BL-0018, 2000. "Emergency Reconstruction Facility Following Hurricane Keith" Washington, D.C: Inter-American development Bank.
(4) IADB-,PR-2690,-2002. "Jamaica Emergency Reconstruction Facility Following the torrential rains in Jamaica".
(5) Kari K., Justin T. 2002. "Planning and Financial Protection to Survive Disasters" Washington, D.C: Inter-American development Bank.
(6) ECLAC, 1999 a. América Latina y el Caribe: el impacto de los desastres naturales en el desarrollo, 1972-1999. LC/MEX/L.402.MÉXICO.
(7) ECLAC, 1999 b. Centroamérica: Evaluación de los daños causados por el Huracán Mitch en 1998. LC/MEX/L.375.MÉXICO.
(8) Iglesias, E. V. 2001. "Text for the International Aid & trade Review".
(9) IDB. 2000. "The Challenge of Natural disasters in Latin America and the Caribbean: IDB Action Plan". Sustainable Development Department. Washington, D.C.
(10) IDB. 2001. " Sectorial Facility for Disaster Prevention". Document GN-2085-5. Washington, D.C.
(11) IDB. 2002. " Information and Indicators Program for Disaster Risk Management", Document AT-1259 Sustainable Development Department. Washington, D.C.
(12) ECLAC, IDB. 2000. "How to reduce the vulnerability in the face of natural disasters". Mexico.
(13) Scott Solberg, 2002 "Institutional analysis of natural disaster management in infrastructure in Ecuador.,
(14) Riemer W. & Locher T", 1988. "Mechanics of deep seated mass movements in metamorphic rocks of the Ecuadorian Andes, Fifth International conference on landslides,
(15) Oboni F., 1988. "General report: Analysis methods and forecasting of behavior, Fifth International conference on landslides,
(16) Greenstein J. 1997 "Issues Related To The Performance & Management of Pavement Systems"3rd International Symposium on Pavement Evaluation and Overlay Design Belém-Pará Brazil.

(17) Greenstein J. 2001 "Nuevas Técnicas Para Administración De Proyectos De Inversión En Infraestructura De Transporte Y Mantenimiento De Pavimentos" Expovial Argentina.

(18) Hunt E. R. "Geotechnical Engineering Techniques and Practices" McGraw-Hill Book Company, 1986.

(19) Hunt E. R. 2000 "Slope Failures, Carretera Cuenca-Molleturo-Elpalme-Ecuador" MOP, Quito Ecuador.

(20) USGS 2002, Documentation for the 2002 Update of the National Seismic Hazard Maps.

(21) Coe J.A. Gogt J.W. Hencerton A.J. 2002, "Debris Flows along the Interstate 70 Corridor-Colorado" USGS.

(22) Croveli, R.A., 2000, "Probability models for estimation of number and costs of landslides: U.S. Geological Survey Report 00-249, 23p.

(23) Melton, M.A., 1965, "The Geomorphic and Paleoclimatic Significance of Alluvial Deposits on southern Arizona" Journal of Geology, v.73, p. 1-38.

(24) Kirshenboim A., "Privatization of Disaster Risk Management Responsibilities", Technion Magazine, Israel, Winter 2003.

Regional Assessment of Earthquake Hazard in Japan
(Part 1 Overall Framework)

Takayuki Shimazu [1] Haruki Shimazu [2] and Naoki Shimazu [3]

Abstract

Japan is administratively divided into 47 prefectures. Since the 1995 Hyogoken Nanbu earthquake (the Kobe earthquake), each prefecture has conducted the task on the new assessment of earthquake hazard on its entire region. The overall framework of this assessment basically consists of the following, regardless of prefectures. 1. Geology and Seismicity   2. Prediction of Physical Phenomina due to Earthquakes a. ground motions b. soil liquefaction c. landslides d. tsunami 3. Prediction of Damage Potential   a. buildings b. lifeline c. transportation d. fires and spreads   4. Impact on Population   a. deaths   b. injuries c. suffers.   5. Countermeasures   a. emergency response and recovery   b. mitigation of earthquake risk   This paper (Part 1) presents the above content.

Introduction

The Kobe earthquake occurred at 5.46 A.M. local time on January 17, 1995.   It was asseigned a magnitude 7.2 , on the Richter scale by Japan Meteorological Agency (JMA).   This earthquake was devastating as its epicenter was located close to a densely populated area; in fact the fault ruptured through the central part of Kobe. The earthquake produced severe building damage and over 6000 people were killed due to building collapse , resulting from severe shaking and fire.   As a result, important questions have been raised about earthquakes

---

[1] Professor Emeritus of Hiroshima University, 2-15-22 Ushitaasahi, Higashiku Hiroshima 732-0067, Japan, Phone(Fax) 81-82-228-2741, shimazu@hicat.ne.jp

[2] Civil Engineer (Ph.D), Hiroshima City Office, Hiroshima, Japan
  hshimazu@do.enjoy.ne.jp

[3] Urban Planner (M.D.), LaCanada Flintridge City Office, California, U.S.A.
  naoki7la@earthlink.net

preparedness,disaster response and design & upgrading earthquake resistant structures among administrative, academic and industrial sides. Particularly, each prefectural government started the task on the new assessment of earthquake hazard on its region,in financial support of the national government.The assessment mostly covers all the aspects from the geology & seismicity of prefectural region up to the prefectural countermeasures to be taken (1). It seems. that these content is "crucial " to all the countries in earthquake- prone region from the earthquake hazard mitigation point of view. However, little literature on these synthetic treatment of disaster reduction has been reported. It should be added that the total amount of removed debris after the event reached about half of all the wastes treated in Japan that year. The mitigation of earthquake hazard is also quite important from the environmental point of view.

Geology and Seismicity
Japan is situated at the intersection of the North American, Pacific, Eurasian and the Philippine plates, as shown in Fig. 1 of the page before the last page. The North American,Pacific and Philippine plates are being subducted beneath Japan and a number of large damaging earthquakes have occurred along these tectonic boundaries. The 1944 and 1946 Nankaido M = 8 events in southern Japan, occurred as a result of subduction of the Philippine plate. Other major interplate events include the 1923 Great Kanto earthquake M = 7.9, which devastated the city of Tokyo , and several large and damaging earthquakes along faults near Hokkaido ,the northen island of Japan, since January 1993. In contrast, the Kobe earthquake was an intraplate event that occurred at relatively shallow depth within the Eurasian plate.The Eurasian plate ishighly fractured and a quite number of Quatemary faults has been identified.The magnitude of of the fault fracture type is much less than that of sea plate type, but fault fracture types have brought the worst disaster to near by cities although periodical intervals of occurrence of this type is considerably much longer than those of sea plate type with the intervals of one hundred years or more. *In the regional assessment of earthquake hazard of Japan*, each prefecture has adapted a couple of intraplate types of fault fracture, existing in the region of a prefecture and adjoining prefectures, in addition to one or more interplate of a sea plate for the prediction of earthquake sources.
Mountains account for 70 percent of total area of Japanese Islands. Accordingly populations concentrate on alluvial plains. These areas are usually in many soil

layers, which is assumed to cause the amplification of earthquake waves. The total numbers of types of soil layers dealt with in the assessment of earthquake hazard for a prefecture reach a couple of hundreds, reflecting a great number of combination of soil layer adapted for each unit mesh of grounds (500m*500m).

Prediction of Physical Phenomina due to Earthquakes
a. ground motions

For earthquakes of interplate type, occurring at deep fault plane, Kobayashi-Midorikawa model (1) is often used as source wave function at hypocenter. In this model, pulses are assumed to be generated subsequently from each subelement of a fault plane when the plane slips and to form the waves from the source A fault plane is usually assumed to be a rectangle with moment magnitude, having a strike, a dip and a rake angle. On the other hand, " Shake model (Seeds et. al -1) is often used above the upper level from hard rock with 3000 m/sec. of shear velocity ,considering equivalent stiffness & equivalent viscous damping for each of considerably many layers up to ground surface. For earthquake due to shallow hypocenter, transmission functions methods are often used.

The JMA (Japan Meterological Agency) instrumental intensity scale was revised in October of 1996. To calculate JMA intensity, Fourier transform is applied to each of three component acceleration time history.   Then a band-pass filter, consisting of three component acceleration time history. The JMA seismic intensity I is obtained as a real number, using a referenence acceleration value of a_0

$$I = 2.0 \log a_0 + 0.94$$

JMA intensity scale(S) is also provided using this I value as follows ;   S = 4 for 3. 5 $\leqq$ I < 4.5, S = 5 lower for 4.5 $\leqq$ I < 5.0, S = 5 upper for 5.0 $\leqq$ I < 5.5 , S = 6 lower for 5.5 $\leqq$ I < 6.0, S = 6 upper for 6.0 $\leqq$ I < 6.5,   S = 7 for 6.5 $\leqq$ I For instance, the ground motions measured at the Sylmar County Hos. During the 1994 Northridge earthquake, having 826.8 cm/s2 of PGA and 128.9 cm /sec. of PGV ,larger of two components, are determined to be a_0 = 478.6 cm/sec2 I = 6.3 and JMA intensity scale = 6 upper.

b. soil liquefaction
Extensive liquefaction of natural and artificial fill deposits occurred along much of the north side of Osaka Bay.   On the Kobe mainland ,evidence of liquefaction

extended along the entire length of the waterfront, east and west of Kobe, for a distance of about 20 km, while the liquefaction failures of relatively modem fills on Rokko and Port Islands were the most notable with ground settlements on the order of as much as 0.5 m. Overall, liquefaction was a principal factor in the extensive damage experienced by the poor facilities in the affected region Compaction was generally only applied to materials placed above water level. As a result, liquefaction occurred within the underwater segments of these fills, causing settlement in their interior region and lateral spreading along the margins. On the other hand, the ground settlements caused surprisingly little damage to high- and low-rise buildings, bridges,tanks and other structures supported on deep foundations, although lifeline system suffered serious damage at the joints with structures. ***For the prediction of soil liquefaction in regional assessment of earthquake hazard***,the design code of highway bridges is used. After the Kobe earthquake, the design code of highway bridges (2) was revised from the previous one based on analytical results on damages by the 1964 Niigata earthquake. The current code takes account of not only alluvial but dilluvial areas for soil liquefaction. In this code FL values (dynamic shear strength ratio / shear stress ratio at earthquake) are first calculated for 1 m unit depth and PL values are calculated using the following equation for soil with FL value of less than 1, integrated with depth z from ground level up to underground in 20 m depth

$$PL = \int FL \, W(z) \, dz$$

in which $W(z)$ is the weighting function along the depth ($W(z) = 10-5z$). In the prediction of earthquake hazard, measures for soil liquefaction are taken for a mesh containing the area ratio of 70 % with Pl values more than 5.

c. landslides

Japan is a mountainous country .Therefore,the geological susceptibility to landslides is very high throughout the country. The 1995 Kobe earthquake caused significant landslide to Nishinomia area . ***In the prediction of landslides in reassessment of earthquake hazard***, the following factors are considered ; 1 slope angle   2. slope height   3. depth of surface soil   4.slope shape   5. transverse shape of slope   6. crack of rock   7 spring. Using these factors, the index $Y = \Sigma ai \cdot Xi / (\Sigma ai)$ is calculated with ai = relative importance coefficient and Xi=rank level. Finally the hazard level of landslides for each mesh is evaluated, .

combining the above Y values ,with the seismic intensity scale predicted for each mesh.

d. Tsunami

Unusual water wave motions were not remarkably observed, following the Kobe Earthquake.However,tsunami run-up have historically been observed up to 2m along the entire length of the water front of inland sea and up to 10 m along much of pacific shoreline,in the event of major interplate earthquakes. *In the regional assessment of earthquake hazard*, tsunami run-ups are usually considered to be relatively slight in an intraplate event, even by calculation confirmation.

Prediction of Damage Potential

a. Buildings

A large number of traditional wooden structures collapsed , during the Kobe earthquake causing loss of property and death. Row houses, residences and stores performed poorly, causing most the six thousand earthquake casualties On the other hand, engineered buildings were most heavily damaged in the narrow zone adjacent to the JR railway Line running through Kobe and along the edge of Osaka Bay. Some damage statistics in Chuo ward of Kobe showed that many pre-1971 reinforced and steel reinforced concrete buildings collapsed or were severely damaged but no such damage was observed in buildings designed based on post-1971code, particularly on 1981 Building Standard Law. A similar trend is evident for steel buildings. *In the regional assessment of earthquake hazard* (Mochizuki-Emoto, Okada-NakanoMds-5) classfication has been made on the wooden or non-wooden structures and also made on the terms of codes of practice enforced at the time of construction, regarding each mesh (500m*500m), based on the books of fixed property tax kept by each city. Wooden house structures are assumed to have strength simply dependent on their natural period which are determined , based on the term of construction as well as numbers of story. Then, this strength is compared with response value , corresponding to the natural period in spectral curve given for a mesh. If the response value is three times greater than the strength, the mesh is judged to be severely damaged area, while if the response value is six times greater than the strength, the mesh is judged to be collapse area.. For the liquefaction, damage ratio of a mesh with PL value greater than 15 is judged to be 10%

for collapse and 20% for severely damaged.

For non-wooden structures (reinforced concrete, and steel reinforced concrete and steel structures), a little different method is adapted. First, the distribution of the numbers of buildings in a mesh is made, in the relation with their ultimate strength for each number of story, and for each term of construction as well as the structural type. Then the response value is determined from the spectral curve for the mesh, using the natural period corresponding to the the number of story. Finally damage ratio is calculated as the ratio of the number of buildings with ultimate strength less than $k_1 \alpha$ and $k_2 \alpha$ to the total number of buildings of each class, in which k_1, k_2 and α are collapsed level severely damaged level, and elastic response level of buildings, respectively. For the liquefaction, damage ratio of a mesh is determined from the type of foundation and the number of story in a mesh, judged to be a mesh with 30 % area of liquefaction, having PL values more than 15. On the other hand, there is another prediction method using directly the relations between the damage ratio of buildings and peak ground velocity of a mesh. This was derived from the analyses on the structural damage suffered during the Kobe event for each term of construction and structural type, regardless of numbers of story

b. lifelines

Extensive damage to lifeline systems occurred throughout the epicentral area of the Kobe earthquake. Liquefaction was a major factor, involved in the failures of lifeline systems due to geotechnical causes, such as the damage to underground utilities, underground transit and port facilities. There was pervasive disruption of underground utilities caused by ground deformations. Most common failures occurred, due to differential movements between foundation elements and the surrounding soil at points of entry and other structures. Damage to bridges was widespread. Full or partial collapse occurred in highway bridges, railway bridges and train stations. *In the regional assessment of earthquake hazard* (1) (2) (4) each method based on these various damages suffered during the Kobe earthquake is used for water system, wastewater system gas & fuel, electric power, telecommunications and also transportation & ralated structures, highways, railways, harbors and airports as explained in Part 2.

Fires and spreads

Fires have often destroyed much of cities from old times. Fire was still a major problem in Tokyo after the 1923 great Kanto earthquake which burned through the traditional structures of that city. During the Kobe earthquake more than 7000 houses burned in fires, with 5000 ones in the Nagata ward of Kobe city. *In the regional assessment of earthquake hazard*, a revised method (Pasco- 1) from the one established based on the 1978 Miyagiken-oki earthquake damage which occurred in northen Japan is often used,in expression of the following

$$Y = (\alpha N y) * 0.211$$

in which Y = numbers of fires, N = numbers of households, α = compensation coeffcient due to hour at which the earthquake occurs., y = fire- break out ratio expressed with collapse ratio of buildings and the season at the event   In this. expression, the multiplier of 0.211 was introduced after the Kobe earthquake based on the fact that the fire damage of the event was much less than those of the previous earthquakes.This is mainly attributed to the widespread of tremor sensor built in fot heaters in the affected area.   On the other hand, fire spread ratios are calculated by the equation established based on the past earthquakes which is functions of fire spread speed, wind speed and non-inflammable zone, such as parks in a mesh.

Impact Population

By the Kobe earthquake, over 6000 people were killed and more than 40000 were injured with over 800,000 people left homeless.

a. deaths

In the assessment of earthquake hazard, the following Ota et.al. model(1) is often used for estimating the numbers of deaths D by an earthquake.

$$D = 1.45 \, N^{0.94} \cdot \alpha \, \beta \, \gamma$$

in which   N = numbers of collapsed houses, α = fire scale coefficient (0.12 ~ 1.00), β = day (0.73) or night (1.00) of the event, γ = year of the event (0.22 ~ 1.00,   0.22 after 1955)

b. Injuries

The following Shiono-kosaka model (1) is often used for predicting the numbers ratio of injuries (R %)..

$$R = 0.039 H^{0.676}$$

with H = damage ratio (%) of houses

Countermeasures.

In advance of countermeasures, the comparisons of the predicted results are shown in Table among the three prefectures. In this table, the values of Hyogo prefecture are those of the 1995 Kobe earthquake. The Shizuoka prefecture is mostly anticipated to undergo an attack of large earthquakes in the very near future, while Hiroshima prefecture is also anticipated to have an earthquake, the type and size of which will be similar to the 1905 Geiyo earthquake that an interplate fault fracture caused. Table 2 shows the scenario of emergency response, and recovery immediately after earthquakes to be taken usually by a prefecture office. Table 3 shows the preventive measures against earthquakes of a prefecture., which aims at mitigation of earthquake risk with a long view Hiroshima city, the largest city with the population of more than one million, located in the south-west part of Hiroshima prefecture is one of the largest cities of Japan, designated by ordinance. This city has an independent countermeasures based on its own assessment of earthquake hazard. The countermeasures also consists of two elements of preventive measures and emergency & recovery measures. However, the constitution of the preventive measures of this city is more detailed with an emphasis on urban planning.

Conclusions, Acknowledgements and References

These are presented in the last part of Part 2.

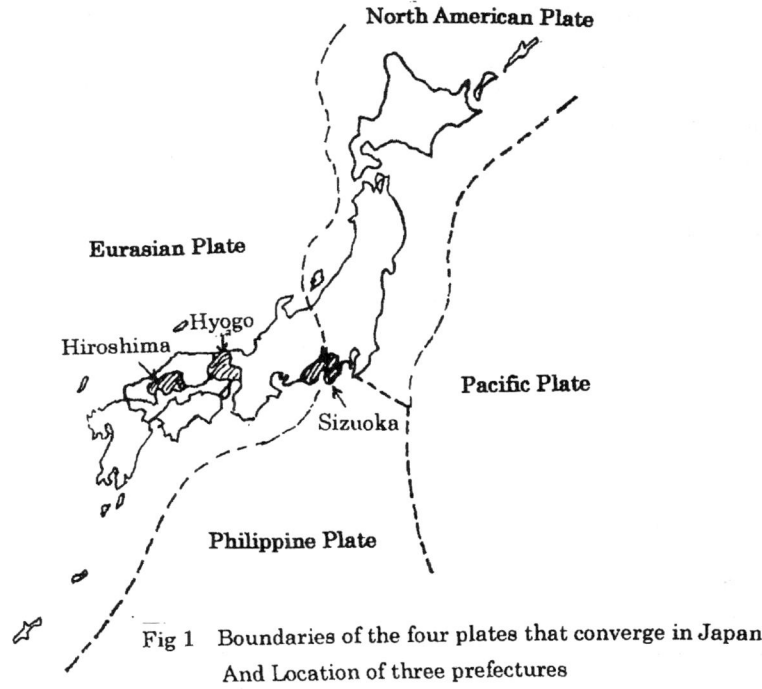

Fig 1   Boundaries of the four plates that converge in Japan And Location of three prefectures

Table 1. Predicted earthquake damage on Hiroshima and Shizuoka Prefectures with the 1995 Hyogoken-Nanbu Earthquake damage

Prefecture	Hiroshima	Hyogo (actual)	Shizuoka
Earthquake	Geiyo	Hyogoken-Nanbu	Tokai
Magnitude	7.25	7.2	8.0
Max. Accel. (gal)	500	818	Over 600
Max. I. S (JMA)	6 upper	7	6 upper~7
`Area (km^2)	8474	8385	7779
Population	2,882,000	5,431,000	3,868,000
Buildings	1,438,000	2,402,000	1,425,000
Damaged Bs.	82,000	257,890	294,404
Deaths	650	6425	2576
Serious Injuries	2800	8763	9300
Sufferers	215,000	Over 800,000	93,920

Table 2 Scenario of emergency response and recovery after an earthquake

	0.5 hour	2 hours	3 days	0.5 month	3 months	1 year
Operation systems	1. ──────→					
Formation		2. ─────────────→				
Information Manage.	1. ───────────────────→					
Life rescue	1. ────────────	2. ──────────→				
Fire-fighting Ope.	1. ───────>					
Evacuation	1. ──────────→					
Shelters, toilets		2. ──────────────────→				
Rescue of injuries	1. ──────────→					
Medical treatment	2. ──→					
Transportation	1. ──────→ 2. ──────────→					
Harbors, railways &	1. ──────────→					
Airports		2. ─────────────────────────────────→				
Urgent supply	1. ──────────────────→					
Debris removal		2. ────────────────→				
Water recovery			2 ────────────────→			
Gas recovery			2 ───────────────→			
Wastewater re.			2. ──────────→			
Electric powers re.			2. ─────>-			
Telecommunication re.			2 ──────────→			

1. emergency response   2. urgent recovery

Table 3 Preventive measures for earthquakes

1. preservation of territorial integrity of prefecture for disaster
2. promotion of earthquake-proof structures and facilities
3. promotion of investigation and research for disaster prevention
4. enhancement of people awareness for disaster prevention
5. reservation of materials and implements in an emergency
6. establishment of funds for disaster prevention
7. environmental preparation for the weak for disasters

Regional assessment of Earthquake Hazard
(Part 2   Lifeline and Transportation)

Takayuki Shimazu[1] Naoki Shimazu[3] and Haruki Shimazu[2]

Abstract
In this paper, focus is placed on the detailed aspects on lifelines and transportation & related structures, following the part 1 (overall framework).The subitems dealt with are as follows. 1. Lifeline   a. water system   b. wastewater system   c. gas system d. electric power system e. telecommunication system 2. transportation & related structures   a. highways and bridges   b.railways c. harbors d. airports.   In each subitem, prediction methods are explained on damages due to earthquakes and recovery after earthquakes. The calculated results using these methods are also presented on Hiroshima prefecture having the land area of about 9000 km2 & the population of about three millions Finally the relations with adjoining prefectures are discussed.

Introduction
Lifeline are civil infrastructure systems that are essential to the robust functionning of a modern society, they include water and wastewater, gas and liquid fuel, electric power, telecommunications   and transportation & related structures.   The Kobe earthquake caused severe lifeline damage and disruption, some of which has not been observed in other earthquake. In this paper assessment method on each subitem are presented based on each evaluation method modified from the previous one after the Kobe earthquake.

---

[1] Professor Emeritus of Hiroshima University, 2-15-22 Ushitaasahi,Higashiku Hiroshima Japan 732-0067Phone(Fax) 81-82- 228-2741shimazu@hicat.ne.jp

[2] Civil Engineer (Ph.D) Hiroshima City Office, Hiroshima Japan
 hshimazu@do.enjoy.ne jp

[3] Urban Planner (M.D), LaCanada Flintridge City Office, California U.S.A.
 naoki7la@earthlink.net

a water system

In the nine cities and the five towns of Hyogo prefectures, including Kobe city, about 85 % of the customers were without water service, immediately after the Kobe earthquake. In the Kobe city, the primary facilities were immediately recovered but it took 90 days to resume full supply for the whole city, due to the difficulty of finding the numerous places of water leakage of underground pipes as well as damage to access roads.

In the regional assessment of earthquake hazard, the prediction method is based on the previous one established by Kubo-Katayama (3) from the damages suffered by the 1971 San Fernando earthquake. In this method the relations between damage ratios (numbers of damaged places of pipelines /km) of cast iron pipes and peak ground acceleration is adapted. However, damage ratios was three times higher than those predicted by this method in the affected area of liquefaction during the Kobe earthquake. Thus the following expression is used.

$$Rfm = Cg\ Cp\ Cd\ Rf$$

in which Rf = standard damage ratio, expressed with $1.7 A^{6 \cdot 1} \cdot 10^{-16}$ (A =PGA) Cg = liquefaction correction coefficient (0.5 ~ 3.4), Cp = pipe type coefficient (0.2 ~ 2.0) and Cd = pipe diameter coefficient (0.2 ~ 2.0). On the other hand the relationship between damage ratio of pipelines and service loss ratio often used (Kawakami model –1) is as follows.

$$\tan\{\pi(Y-0.5)\} = 4.8 \log(X/0.18)$$

in which Y = service loss ratio and X = damage ratio(numbers of damage /km). Days required for recovery is estimated through the following process. During the Kobe earthquake, the damage of filtration plants and reservoirs was relatively slight and they were repaired in short terms, while numbers of damage of pipelines were enormous (principal trunk; 1757 and sublines ; 62300 in Kobe city) and it took long time to recover as mentioned above. Based on the past experiences, the following is used for estimating days for recovery ; 3 ~ 6 person day / place for principal pipelines and 1.5 person day / place for others. For emergency immediate after earthquake, water supply is planned with 3 L for the first days and 20 L from the fourth day for a person.

a. wastewater system

During the Kobe earthquake, wastewater facilities, such as wastewater treatment plants sustained miner damage, while pipeline and manholes suffered serious damages. Most pipeline damage is due to ground acceleration and deformation. *In the regional assessment of earthquake hazard*, the same type of expression, as explained in the item of water system is used as follows.

$$Rfm = Cg\ Cp\ Cd\ Rf$$

in which Rf is the same as suggested in a . water system ,.Cg = liquefaction correction coeficient (0.5 ~ 7.0), Cp = pipe type coefficient (0.1 ~2.0), Cd =pipe diameter coefficient (00.5 ~ 1.2).   Days required for recovery is estimated through the following process.   After the Kobe earthquake, wastewater treatment plants were recovered in a month and pumping stations were recovered in half a month , while the numbers of damage of pipelines were enormous (numbers of sewage damage 16086) and it took four and half months to recover even temporarily, due to the difficulty of finding the damaged places. The next expressions are used for predicting the months M of starting recovery work & days D required for recovery.

$$M = 10.24 \cdot \log Nh - 27.25 \quad (Nh > 1000)$$
$$= 0.0034 \cdot Nh \quad (Nh \leqq 1000)$$
$$D = 0.0084 \cdot Nh$$

in which Nh = numbers of damage of wastewater pipes.

c. gas and fuel systems

During the Kobe earthquake, the gas transmission system suffered serious damage, particularly in the area of liquefaction. Immediately after the earthquake, local distribution companies shut off their entire systems, because strong-motion instrument recordings at gas storage sites exceeded preset level. Service to approximately 850000 customers was interrupted and it required 90 days to get complete recovery.

In the prediction of earthquake hazard, the relationship between damage ratio of pipelines and peak ground velocity is used.   This was established after the Kobe earthquake, including the several past earthquake experiences. The dam-. age ratio Rfm (numbers of damage / km) is expressed with,

$$Rfm = Cg\ Cp\ Cl\ Rf$$

in which Cg = soil coefficient ($1.0 \sim 2.5$), Cp = pipe type coefficient ($0 \sim 1.0$) Cl = liquefaction correction coeffcient ($1.0 \sim 5.0$) and Rf = standard damage ratio (numbers of damage / km), being expressed with peak ground velocity SI as follows ; 0 (SI $\leq$ 20 kine), 0.025 SI – 0.5 (20 $\leq$ SI $\leq$ 90 kine) and 1.75 (90 $\leq$ SI). Gas distribution companies shut off their entire systems, when SI = PGV is more than 60 kine or supply continuation become difficult due to overflow at gas plants or substations, dividing their supply area into several blocks with the aim of minimalization of the shut-off area and days for recovery

d . electric powers system

During the Kobe earthquake, there was damage to generating stations, substations and transmission systems. Particularly, numbers of damage of transmission system were enormous and it required a plenty of work to recover. Immediately after the earthquake, service to approximately 2600000 customers was interrupted and six days later, emergency supplies were available for all households. *In the regional assessment of earthquake hazard*, the damage ratios of poles P in power transmission system due to the Kobe earthquake is used This was summarized by Source & Energy Agency (1996) ,.as follows ; P = 0% (0%) for 5 JMA instrumental intensity scale, P = 0.5% (0.65%) for 6 lower ~ 6 upper of JMA and P = 6.7% (8.71 %) for 7 of JMA. in which values in parenthesis means the ones in liquefied area.(PL > 15) The damage of poles in the area of fire spreads is also included in the above. For the calculation of service loss ratio, the following equation (3) is used.

$$Y = 19.5 \cdot X^{0.35}$$

in which Y = service loss ratio (%) and X = damage ratio of poles (%). Days required for recovery is based on the following estimation.; 3.6 person day / a pole and 4.6 person day/ a span of cables.

E. telecommunications

During the Kobe earthquake, damage ratios of aerial cables, poles and underground cables were 1.7 %, 1.5 % and 0.23 % in the affected area respectively.

Service loss ratios of call attempts were 13.4 % (193000 / 14430000). Call attempts were 50 times than usual on the crucial first day of the event and 20 times than usual on the second day. On the other hand,base stations for wirelesss performed well. The assessment of earthquake hazard on telecommunications system was made based on the above facts.

Transportations & related structures

a. highways and bridges

During the kobe earthquake, extensive damage to or collapse of several highways and bridges caused widespread disruption after the earthquake. *In the regional assessment of earthquake hazard*, seismic safety is first examned on bridges, cut grounds and mounted grounds. After that, networks are evaluated.(2). Damage ratio of bridge are calculated based on the relationship between rank of seismic safety level of bridges and JMA instrumental intensity scale determined for a mesh. These values are 0.00 (5 lower)~0.05 (7) for upper safety rank. 0.00 (5 lower) ~ 0.16 (7) for middle safety rank and 0.00 (5 lower) ~ 0.63 (7) for lower safety rank. On the other hand ,seismic safety rank of bridges is determined from each safety level of upper parts (girders) ,lower parts (piers) and foundations. Damage ratios of mounted grounds are calculated only for lower rank as follows; 0.00 (5 lower) ~0.44 (7) based on the damage ratios by 1978 Miyagikenoki earthquake.Damage ratios ratios of cut grounds or landslides are calculated as follows; 0.00 (5 lower) ~ 0.02 (7) for higher safety rank, 0.00 (5 lower) ~ 0.12 (7) for middle safe rank and 0.00 (5 lower)~ 0.54 (7) for lower safety rank., based on damage ratios by 1986 Izuoshima earthquake. Using these values of damage ratios, the probability that a road suffer no damage is expressed as follows.

$$Pf = \amalg \; Pb \cdot \amalg \; Pm \cdot \amalg \; Pc$$

in which P b, m,c means each summation of probability of no damage for bridges, mounted grounds and cut grounds or slides for a highway. $Pf < 0.3$. means high probability of damage, $0.3 \leqq Pf < 0.7$ means middle probability and $0.7 \leqq Pf$ means low probability.

b. railways

During the Kobe earthquake, bridges,tunnels and mounted grounds suffered damage. Service loss of long term was unavoidable due to damage of electric facilities and rail beds, in addition to these damage. Railways are quite important transportations particularly in Japan, in respect to being not only transtation with high speed, high safety, and mass transport, but also emergency port, together with highways. *In the regional assessment of earthquake hazard* (4), damage probability of railways is calculated as summation of both the damage probability of bridges and tunnels. The relationship between damage probability of railway bridges and JMA intensity scale is expressed as follows; 0.00(4), 0.05 (5 upper), 0.08 (6 upper), 0.16 '(7) for large-scale damage and 0.00 (4), 0.09 (5 upper), 0.28 (6 upper), 0.34 (7) for small-scale damage This expression is based on the damage ratios of the 1978 Miyagikenoki earthquake on railway bridges   On the other hand, the relationship between damage probability of tunnels and JMA intensity scale is expressed with 0.00 (4), 0.03 (5 upper), 0.05 (6 upper) and 0.07 (7), which is based on the damage by the 1978 Miyagikenoki earthquake on tunnels.   By adding these values, the evaluation on service loss probability of railway is obtained as follows.

$$B = \Sigma PBi + \Sigma PTi$$

in which PBi = damage probability of bridges, PT_i = damage of probability of tunnels for a railway.

a. Harbors

During the Kobe earthquake, a large numbers of harbors suffered serious damage. Liquefaction was a principal factor in the extensive damage experienced by the port facilities in the affected region.   *In the regional assessment of earthquake hazard* (3) , the following items are considered for the stability of port facilities, particularly, for quays,

1. seismic force levels Ke is set up as follows.

$$Ke = \alpha/g \quad (\alpha \leq 200 \text{ gal})$$
$$Ke = 1/3\,(\alpha/g) \quad (\alpha > 200 \text{ gal})$$

in which α = peak ground acceleration at sites, g = gravity acceleration
2—1. stability of quay   Fsd (Ke) is caluculated , using the above Ke values as follows.

$$Fsd(Ke) = \beta \cdot Fso$$

in which β = reduction coefficient ; β = 0.67(Ke = 0.1), 0.5(Ke = 0.2),0.38 (Ke = 0.3) and Fso = static safety factor, provided for each type of facilities
2—2. stability of quay, considering the hydrauric pressure due to over saturation of soils

$$Fsd(\Delta u) = \gamma \cdot Fso$$

in which v = reduction coeffcient; 0.25 (FL $\leqq$ 1.0), 1.0 (1.5 < FL) for sand soils and 0.4 (FL $\leqq$ 1.0) 1.0 (1.5 < FL) for clay soils.

a. airports

In the earthquake hazard assessment, seismic safety of the building structure and runways as airport facilities is evaluated in relations of the acceleration level and liquefaction predicted at the sites

Calculated results on Hiroshima Prefecture

Fig 1 shows the map of Hiroshima prefecture, indicating the cited four cities of thirteen ones in total and a town located in Seto inland sea, selected from more than a half hundred towns in Hiroshima prefecture This picture also shows the trunk of highways, JR (Japan Railway) and main horbors & airports   Furthermore, .four faults, assumed as earthquake sources in the regional assessment of earthquake hazard are shown without the fifth fault of Nankai trough earthquake , assumed along the boundary between Eurasian and Philippine plates , although this fifth fault turned out not to affect Hiroshima prefecture.   Table 1 shows the predicted results on transportations & related structures.   The 2~4 shows the predicted results on water system, wastewater system and electric power system for the two earthquakes. One is Geiyo earthquake which is of interplate type,   assumed to generate under Aki-Iyonada zone of the Seto inland sea ,due to the subduction of the Pilippine plate.   The fault line & the magnitude were assumed to be the same as the 1905 Geiyo earthquake.   Chuokozosen earthquake is of intraplate type the fault line of which actually exists in the northern part of Shikoku island

Cooperativre relationship with adjoining prefectures

Earthquakes have dramatic effects on widespread regions. Cooperative agreement was made among all the five prefectures in Chugoku district in Jul.1995 a half year after the Kobe earthquake and among all the nine prefectures in Chugoku district and in Shikoku district across Seto inland sea, in Dec. 1995. Main items of cooperative agreement consist of the following cooperative service in an emergency ; necessities of life, materials and facilities, vehicles & helicopters and staffs and so on.   Cooperation is also arranged among the all the administrative organs,including the Self-Defence Force, various public & incorporate bodies and self-imposed private bodies existing in a prefecture.

Conclusions

Regional assessment of earthquake hazard is crucial to the seismic safety of various aspects of a regional society due to high probability of earthquake occurrence in Japan. Both the Part 1and Part 2 were presented,using the prediction method established through the analyses of the greatest damage that an earthquake ever caused to one of the most modernized countries in the world. The authors will be honoured if the basic philosophy underlying this method is well understood , and put to the most practical use , in earthquake-prone countries.

Acknowledgements

The information provided by many prefectural reports on earthquake hazard assessment is acknowledged,with special thanks to the head committee on disaster prevention of Hiroshima prefecture.

References

(1) Aomori (1997), Fukuoka (1997), Hiroshima (1997), Hyogo (2001) Kyoto (1998), Miyagi (1997), Osaka (1997), Shizuoka (1996, 2002) and Tokushima (1997) prefectures ." Report of investigation on the assessment of earthquake hazard in prefectural region."
(2)   Japan highway Institute (1986). "   Seismic measures of highways"
(3) Kanagawa Prefecture (!986, 1993). " Report of investigation on the assessment of earthquake hazard in the region. "
(4) Saitama Prefecture (1983)." Report of investigation on earthquake hazard assessment and the seismic plan in its region "
(5) Tokyo Metropolis(1993)." Report of investigation on the assessment of earthquake hazard in the region".

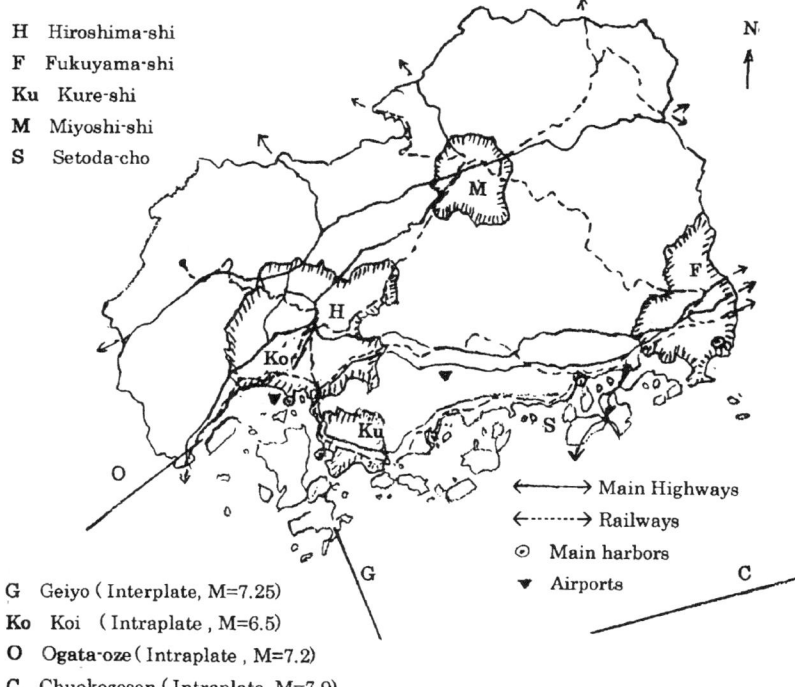

Fig. 1 Hiroshima Prefecture with the locations of four cities & one town and of assumed four faults

Table 1 Predicted damage of highways, railways, harbors and airports

	Outline of predicted damage
Highways	Little damage to principal highways and damage to coastal local highways and mountain highways
Railways	Damage to one or two places on Japan Railway Sanyo-sen having many tunnels and mounted grounds
Harbors	Damage to almost all harbors
Airports	Damage to the runways of local airport in Hiroshima city and little damage to principal airport

Table 2 Water system damage predicted

	Total length of Distribution(km)	Numbers of Damage	Households of Service loss	Days for Recovery
Hiroshimashi	3873.85	829(1272)	156572(283199)	15 (26)
Fukuyamashi	176.15	3 (327)	8709 (119745)	1 (21)
Kureshi	176.83	331(444)	80153 (80753)	22 (41)
Miyoshishi	35.18	1(8)	0(6693)	1 (7)
Setodacho	56.83	1(402)	2416 (3348)	19 (73)
Prefecture	6690.48	2344(7421)	405158(753579)	--------

Geiyo earthquake(Chuokozosen earthquake)

Table 3 Wastewater system damage predicted

	Total length of Distribution(km)	Numbers of Damage	Days for Recovery	Months for Starting
Hiroshimashi	2213.87	1448 (2541)	13 (22)	5(7.5)
Fukuyamashi	112.33	1(102)	1(1)	0.5(0.5)
Kureshi	57.45	116 (203)	1(2)	0.5(1.0)
Miyoshishi	2.35	0(0)	1(1)	0.5(0.5)
Setodachou	5.55	1(140)	1(2)	o.5(0.5)
Prefecture	2596.22	1966(3753)	----------	----------

Geiyo earthquake (Chuokozosen earthquake)

Table 4 Service loss of electric power predicted

	Numbers of Poles	Numbers of Damage	Households supplied	Households of Service loss
Hiroshimashi	77177	230(342)	436772	55754(64059)
Fukuyamashi	49124	0(326)	130905	0(22114)
Kureshi	16002	121(292)	85634	15142(20611)
Miyoshishi	12106	0(0)	14270	0(0)
Setodachou	2419	0(13)	3618	0(576)
Prefecture	437107	426(1287)	1077611	85813(158427)

Geiyo earthquake (Chuokozosen earthquake)

Quantitative Method for Developing Hazard-Consistent Earthquake Scenarios

Kenneth W. Campbell, M.ASCE[1] and Hope A. Seligson, A.M.ASCE[2]

Abstract

This paper presents a quantitative method for developing hazard-consistent earthquake scenarios that can be used to probabilistically evaluate the seismic performance of a geographically distributed lifeline system. These earthquake scenarios are defined to be consistent with the average probabilistic hazard for a given range of ground-motion values defining a specified damage state of the system and its components. The result is a different set of scenarios for each damage state. Probabilistically defined earthquake scenarios rather than point estimates of probabilistically defined ground shaking, such as portrayed on the 1996 and 2002 USGS seismic hazard maps, are used in order to account for the correlation of ground shaking between geographically diverse system components during a damaging earthquake. The procedure is applied to two hypothetical lifeline systems in southern California, one geographically concentrated and the other geographically diverse, and is shown to give a reasonable set of hazard-consistent earthquake scenarios under both conditions. The number of scenarios defined by this method ranged from just a few to as many as 12 depending on the particular damage state and size of the system.

Introduction

The probabilistic evaluation of ground motion for a single-site facility requires only a point estimate of seismic hazard. Such hazard estimates can be derived from a probabilistic seismic hazard analysis (PSHA) (e.g., Thenhaus and Campbell, 2002) or from seismic hazard maps derived from such analyses (Frankel et al., 2000, 2002). In contrast, a geographically distributed lifeline system requires the evaluation of ground motion from a single earthquake over the entire system in order to quantify the earthquake's impact on the system's performance. This precludes the use of point estimates of hazard as this hazard will vary geographically and will usually not represent the ground motion from a single event at different points along the system. This poses a challenge when attempting to evaluate the performance of a lifeline system probabilistically, e.g., for benefit-cost, economic, or other risk analyses.

Because of the difficulty in probabilistically evaluating the performance of a lifeline system, such a system is usually evaluated deterministically using earthquake scenarios that are perceived to be the main contributors to the probabilistically defined hazard in the region. However, this perception is often far from reality and will depend on the level and type of ground motion of interest. Chang et al. (2000) proposed a methodology for defining regional earthquake scenarios probabilistically

---

[1]Director, ABS Consulting and EQECAT Inc., Beaverton, OR 97006; kcampbell@absconsulting.com
[2]Principal Engineer, ABS Consulting Inc., Irvine, CA 92602; hseligson@absconsulting.com

based on system performance degradation, but the development of these scenarios were somewhat subjective and not necessarily reproducible by another analyst or for another region. The method for developing hazard-consistent earthquake scenarios proposed in this paper is absolutely quantitative, meaning that it can be computerized, does not depend on the person doing the analysis, and will produce consistent results from system to system and region to region.

Mathematical Formulation

There are two major assumptions that are necessary to develop the hazard-consistent earthquake scenarios using the methodology proposed in this paper. The first assumption is that the majority of the seismic risk to the system comes from the major faults in the region. Distributed sources such as area sources and smoothed seismicity could be incorporated, but that would require a modification of the methodology proposed here. This assumption is needed in order to restrict the scenarios to major earthquakes, defined has having moment magnitudes (M_W) of 6.5 or greater, which are considered to be the Maximum Credible Earthquakes (MCE) on these known faults. System-wide damaging events, which are of most concern to system performance, are expected to be restricted to earthquakes of these magnitudes. The methodology requires that the magnitude, fault geometry, style of faulting, and recurrence frequency be available for each MCE

The second assumption is that the relative contribution of each of the faults to the total system risk can be defined in terms of two parameters: (1) the probability of observing the range of ground-motion values of interest averaged over the grid of points that define the system, and (2) the recurrence frequency of the MCE. A test case showed that this assumption is reasonable for a system that spans an area as large as Los Angeles County. Nonetheless, it might be necessary to break larger systems into smaller parts for analysis in order to limit the number of selected scenarios to a reasonable number.

The methodology requires that the following information be available for each grid point defining the system: (1) a seismic hazard curve for the ground-motion parameter of interest for a reference site condition, (2) the median value of ground motion and its standard deviation for the reference site condition from each MCE in the region, (3) the predominant or average site conditions, and (4) the distance from the rupture plane of the MCE.

Grid-Point Data. Given a seismic hazard curve on a reference site condition, the hazard (frequency) that a specified ground-motion parameter will have a value between Y_{k1} and Y_{k2} is calculated as the difference in the exceedance frequencies of these values. The resulting incremental hazard is given by

$$H_{jk} = H_{jk1} - H_{jk2} \tag{1}$$

where k represents the specified range of ground-motion values (called a "bin", e.g., 0.145–0.284g where g = 981 cm/sec^2) and j represents the particular grid point.

Attenuation relationships are used to calculate the average value of the natural logarithm of ground motion ($\ln y_{ij}$) and its logarithmic standard deviation (σ_{ij}) at the

jth grid point for the ith MCE given its moment magnitude (M_i), style of faulting (F_i), and distance to the grid point (R_{ij}). Assuming a Gaussian distribution for $\ln y_{ij}$ (i.e., a lognormal distribution for y_{ij}), the total probability that the ground-motion parameter for the ith MCE will fall in the kth ground-motion bin is given by

$$P_{ijk} = RF_i \left[LN_{jk2}(\ln y_{ij}, \sigma_{ij}) - LN_{jk1}(\ln y_{ij}, \sigma_{ij}) \right] \quad (2)$$

where RF_i is the recurrence frequency of the ith MCE and

$$LN_{jk1}(y_{ij}, \sigma_{ij}) = \int_0^{y_{k1}} \frac{1}{\sigma_{ij}\sqrt{2\pi}} \exp\left(-\frac{1}{2\sigma_{ij}^2}(\ln u - \ln y_{ij})^2\right) du \quad (3)$$

The integral on the right-hand side of (3) is the Gaussian cumulative distribution function with mean $\ln y_{ij}$ and standard deviation σ_{ij}.

System Data. For a given system or subsystem, the hazard-compatible earthquake scenarios that contribute the most to the probabilistic performance of that system are defined in terms of the values of the grid-point data H_{jk}, P_{ijk} and $\ln y_{ij}$ averaged over the grid points that define the system. These data can be calculated ahead of time and stored in a look-up table. Weights are used to place more or less emphasis on some grid points to reflect the importance of their system components and facilities to the overall performance of the system.

The weighted average system-wide probability of experiencing a ground-motion value in the kth bin from the ith MCE is given by

$$P_{ik} = \sum_{j=1}^{n} w_j P_{ijk} \quad (4)$$

where n is the number of grid points and w_j is the weight assigned to the jth grid point. Note that the weights must sum to 1. Similarly, the corresponding weighted average system-wide hazard for the kth ground-motion bin is given by

$$H_k = \sum_{j=1}^{n} w_j H_{jk} \quad (5)$$

To select only those earthquake scenarios that contribute significantly to the weighted average system-wide hazard for a particular ground-motion bin, each MCE is ranked in order of decreasing probability (P_{ik}) and only those events that contribute at least 5% to the running cumulative total of these ranked probabilities for a given ground-motion bin are retained. Mathematically, an MCE is selected if its meets the criterion

$$\left(P_{ik} \bigg/ \sum_{i=1}^{m} P_{ik} \right) \geq 0.05 \quad (6)$$

where i now represents the event's rank and m is the total number of events that eventually meet the specified criterion, such that when $i = m+1$, the ratio on the left-hand side of (6) will be less than 0.05. Those events that do not meet this criterion are considered to contribute only negligibly (in this case only 5%) to the total weighted average system-wide hazard and are neglected. The value on the right-hand side of (6) could be set to a lower or higher number, in which case a smaller or larger number of scenarios would be selected. If this value is too large, some critical scenarios might be neglected. If it is too small, too many scenarios might be selected. A few test cases suggested that the 5% criterion was a reasonable tradeoff between accuracy and number of scenarios.

The weighted average system-wide hazard associated with each selected scenario and ground-motion bin is given by

$$H_{ik} = H_k \left(P_{ik} \bigg/ \sum_{i=1}^{m} P_{ik} \right) \tag{7}$$

where, again, m is the total number of selected scenarios. The summation in the denominator is needed to ensure that the selected scenarios account for the total weighted average system-wide hazard (H_k) for the particular ground-motion bin.

The evaluation of system performance for each of the selected hazard-compatible earthquake scenarios requires an estimate of ground motion at each grid point in the system, collectively referred to as the scenario's footprint. The footprint is defined as the set of ground-motion values ($\ln y_{ij}$) that have been adjusted such that their average value is equal to the average of the values that define the particular ground-motion bin.

If the weighted average system-wide value of $\ln y_{ij}$ for the ith MCE is given by

$$\ln \bar{y}_i = \sum_{j=1}^{n} w_j \ln y_{ij} \tag{8}$$

then the corresponding adjustment factor for each scenario and ground-motion bin is calculated from

$$\varepsilon_{ik} = \tfrac{1}{2}(\ln Y_{k1} + \ln Y_{k2}) - \ln \bar{y}_i \tag{9}$$

The resulting ground-motion footprint, adjusted for local site conditions, is given by

$$y_{ijk} = S_{ijk} y_{ij}^{\varepsilon} \tag{10}$$

where S_{ijk} is a site factor that adjusts the ground motion on the reference site condition to the local site conditions at the grid point and

$$y_{ij}^{\varepsilon} = \exp(\ln y_{ij} + \varepsilon_{ik}) \tag{11}$$

If the site factor incorporates nonlinear soil behavior, it will necessarily be dependent on the value of the ground motion on the reference site condition (y_{ij}^o).

Example for a Hypothetical Lifeline System

To demonstrate the methodology for a concentrated lifeline system, a small cluster of grid points were selected in the southeast corner of the Los Angeles Basin, near the city of Huntington Beach. Peak ground acceleration (PGA) was chosen as the ground-motion parameter. The grid points were defined as census tract centroids (there are 40 of them). The MCE events were taken from the inventory of faults compiled by Petersen et al. (1996). These faults were used in the development of the 1996 USGS seismic hazard maps (Frankel et al., 2000). The 1996 seismic hazard curves for each census tract for NEHRP site class BC were taken from a database compiled by Frankel and Leyendecker (2001).

For consistency, PGA and its logarithmic standard deviation were estimated from the same three attenuation relations that were used in the 1996 USGS study (Boore et al., 1997; Campbell, 1997; Sadigh et al., 1997). The geometric mean of the PGA and the arithmetic mean of the standard deviation given by these three relations were used to define the ground-motion distribution at each census tract centroid and the resulting ground-motion footprint. The BC site class used as the reference site condition in the 1996 USGS study represents a site profile with an average shear-wave velocity in the top 30 meters (100 feet) of 760 m/sec (2500 ft/s). To represent the ground motion for this site condition, the USGS evaluated the Sadigh et al. (1997) relation for rock, the Boore et al. (1997) relation for a 30-meter shear-wave velocity of 760 m/sec, and the Campbell (1997) relation for soft-rock. Although it is the opinion of the writers that the Campbell and Sadigh et al. relations as they were evaluated in the 1996 USGS study are not consistent with the assumed BC site class, they were evaluated the same way in this study to be consistent with the definition of the 1996 hazard estimates.

The adjustment of PGA for local site conditions was done using a modified version of the site coefficient (F_a) defined for use in the U.S. seismic building codes (Dobry et al., 2000). This coefficient is amplitude dependent to account for the known nonlinear behavior of short-period ground motion. It is defined in terms of five site classes, designated A through E, each representing a specified range of 30-meter shear-wave velocities. A sixth site class (F) is used to identify certain soft soils that require special studies. The only soils in southern California that fall into this latter category are those considered to have high or very high liquefaction potential. Liquefaction is not considered in the development of the hazard-consistent earthquake scenarios in this example, however, it can be taken into account by use of special failure probability distributions.

The geographic distribution of site classes B through E (A and F are not mapped in California) was taken from the GIS site conditions map of Wills et al. (2000). In addition, this map defines site classes that are intermediate to those defined above, namely BC, CD and DE. Site coefficients for these intermediate site classes were assumed to be the average of those for the bounding site classes. A description of the mapped site classes is given in Table 1. The corresponding site coefficients are listed in Table 2. The site coefficients in Table 2 were modified from those given by Dobry et al. (2000) to reflect their use with PGA. A site coefficient of 0.9 was placed in the lower right-hand cell as originally recommended by Borcherdt (1994) and adopted by the Building Seismic Safety Council (2001). Also, in no case was the value of PGA on site class E allowed to fall below 0.45g for values of PGA on site class BC exceeding 0.5g (Dobry et al., 2000). Lower values might be expected only if the site were to liquefy. The reference site condition in the original table was listed as B, but the value of the ground motion for site class BC was actually used to represent this site condition in the building codes as a matter of conservatism (E.V. Leyendecker, personal comm., 2002). For purposes of developing the hazard-consistent earthquake scenarios in this study, this conservatism was removed by re-normalizing the site coefficients to 1 for site class BC.

Since the census tracts in the urbanized area of Los Angeles, as in most such areas, are small, the hazard and ground-motion values were defined at their centroids and the local site conditions were defined as the predominant site class within their boundaries. On the other hand, the USGS hazard curves are given for a series of grid points spaced at intervals of 0.05 degrees of latitude and longitude. Since these grid points do not necessarily correspond to the census tract centroids, the hazard curve was taken from the grid point that was located closest to the centroid. Because of the dense grid of hazard curves, in no case was the closest census tract centroid farther than 0.025 degrees (approximately 2 km) from one of these grid points.

The hazard-consistent earthquake scenarios for each of the specified PGA bins are listed in Table 3. The location of the hypothetical system and the faults that are responsible for the scenarios are shown in Figure 1. By definition, the sum of the weighted average system-wide hazard for each bin equals the total weighted average system-wide frequency from the 1996 USGS seismic hazard curves. The PGA values defining the bins are listed in the note at the bottom of the table. As the values of PGA increase (increasing bin numbers) there are fewer and fewer scenarios that contribute at least 5% to the total hazard. Also, the smaller PGA values (lower bin numbers) are associated with regional scenarios that generally have larger magnitudes and higher recurrence frequencies than the larger PGA values. In this particular example, there are 12 regional scenarios that contribute to the hazard in the smallest PGA bin (0.0738–0.145g) and there are 3 local scenarios that contribute to the hazard in the largest PGA bin (0.778–2.13g). The total hazard also decreases with increasing PGA values (increasing bin numbers), although this decrease will not always be monotonic, since it also depends on the specific definition of the PGA bins.

The method was also applied to three sets of census tracts scattered throughout the populated area of Los Angeles County to see how it might work for a more distributed lifeline system. In this case, there were 8 selected regional scenarios for the smallest PGA bin and 8 selected local scenarios for the largest PGA bin. Unlike the more concentrated system, a larger number of local scenarios was needed to represent the large, infrequent ground motions from local faults in the vicinity of each of the regionally diverse sets of census tracts.

References

Boore, D.M., Joyner, W.B., and Fumal, T.E. (1997). "Equations for estimating horizontal response spectra and peak acceleration from western North American earthquakes: a summary of recent work." *Seism. Res. Lett.*, 68, 128–153.

Borcherdt, R.D. (1994). "Estimates of site-dependent response spectra for design (methodology and justification)." *Earthquake Spectra*, 10, 617–653.

Building Seismic Safety Council (2001). "The 2000 NEHRP recommended provisions for new buildings and other structures." FEMA-368, Federal Emergency Management Agency, Washington, D.C.

Campbell, K.W. (1997). "Empirical near-source attenuation relationships for horizontal and vertical components of peak ground acceleration, peak ground velocity, and pseudo-absolute acceleration response spectra." *Seism. Res. Lett.*, 68, 154–179.

Chang, S.E., Masanobu, S., and Moore II, J.E. (2000). "Probabilistic earthquake scenarios: extending risk analysis methodologies to spatially distributed systems." *Earthquake Spectra*, 16, 557–572.

Dobry, R., Borcherdt, R.D., Crouse, C.B., Idriss, I.M., Joyner, W.B., Martin, G.R., Power, M.S., Rinne, E.E. and Seed, R.B. (2000). "New site coefficients and site classification system used in recent building seismic code provisions." *Earthquake Spectra*, 16, 41–67.

Frankel, A.D., and Leyendecker, E.V. (2001). "Seismic hazard curves and uniform hazard response spectra for the United States." *U.S. Geol. Surv. CD-ROM Ver. 3.10.*

Frankel, A.D., Mueller, C.C., Barnhard, T.P., Leyendecker, E.V., Wesson, R.L., Harmsen, S.C., Klein, F.W., Perkins, D.M., Dickman, N.C., Hanson, S.L., and Hopper, M.G. (2000). "USGS national seismic hazard maps." *Earthquake Spectra*, 16, 1–19.

Frankel, A.D., Petersen, M.D., Mueller, C.S., Haller, K.M., Wheeler, R.L., Leyendecker, E.V., Wesson, R.L., Harmsen, S.C., Cramer, C.H., Perkins, D.M. and Rukstales, K.S. (2002). "Documentation for the 2002 update of the national seismic hazard maps." *U.S. Geol. Surv. Open-File Rept. 02-420.*

Petersen, M.D., Bryant, W.A., Cramer, C.H., Cao, T., Reichle, M.S., Frankel, A.D., Lienkaemper, J.J., McCrory, P.A., and Schwartz, D.P. (1996). "Probabilistic seismic hazard assessment for the State of California." *U.S. Geol. Surv. Open-File Rept. 96-706.*

Sadigh, K., Chang, C.Y., Egan, J.A., Makdisi, F., and Youngs, R.R. (1997). "Attenuation relationships for shallow crustal earthquakes based on California strong motion data." *Seism. Res. Lett.*, 68, 180–189.

Thenhaus, P.C., and Campbell, K.W. (2002). "Seismic hazard analysis." *Earthquake engineering handbook*, Chapter 8, CRC Press, Boca Raton, Florida, 1–50.

Wills, C.J., Petersen, M., Bryant, W.A., Reichle, M., Saucedo, G.J., Tan, S., Taylor, G., and Treiman, J. (2000). "A site-conditions map for California based on geology and shear-wave velocity." *Bull. Seism. Soc. Am.*, 90, S187–S208.

Table 1. Description of Mapped NEHRP Site Classes for California

Site Class	V_{30} (m/sec)	Geologic Description
B	1130	Plutonic and metamorphic rocks, most volcanic rocks, coarse sedimentary rocks of Cretaceous age and older.
BC	760	Franciscan Complex rocks except "melange" and serpentine, crystalline rocks of the Transverse Ranges which tend to be more sheared, Cretaceous siltstones, or mudstone.
C	560	Franciscan melange and serpentine, sedimentary rocks of Oligocene to Cretaceous age, or coarse-grained sedimentary rocks of younger age.
CD	360	Sedimentary rocks of Miocene and younger age, unless formation is notably coarse grained, Plio-Pleistocene alluvial units, older (Pleistocene) alluvium, some areas of coarse younger alluvium.
D	270	Younger (Holocene) alluvium.
DE	180	Fill over bay mud in the San Francisco Bay Area, fine-grained alluvial and estuarine deposits elsewhere along the coast.
E	150	Bay mud and similar intertidal mud

Note: V_{30} is the nominal value of the average shear-wave velocity in the top 30 meters (100 ft) of a soil profile defined as the mid-point of the range of velocities defining the given NEHRP site class or the value for the bounding NEHRP site classes as defined by Dobry et al. (2000). The value of V_{30} for site class E is that recommended by Borcherdt (1994). Boundary site class designations and geologic descriptions are from Wills et al. (2000).

Table 2. Site Coefficients for Peak Ground Acceleration (PGA)

Site Class	Site Coefficient for Specified Value of PGA on NEHRP BC				
	≤0.1g	0.2g	0.3g	0.4g	≥0.5g
B	0.91	0.91	0.95	1.00	1.00
BC	1.00	1.00	1.00	1.00	1.00
C	1.09	1.09	1.05	1.00	1.00
CD	1.27	1.18	1.10	1.05	1.00
D	1.45	1.27	1.14	1.10	1.00
DE	1.86	1.41	1.14	1.00	0.95
E	2.27	1.55	1.14	0.90	0.90

Note: Modified from Dobry et al. (2000). See text for explanation.

Table 3. Hazard-Consistent Earthquake Scenarios for Hypothetical Lifeline System in Southeast Los Angeles Basin

Fault and MCE Name	M	RF	F	PGA (g)	Hazard for Specified PGA Bin					
					Bin 1	Bin 2	Bin 3	Bin 4	Bin 5	Bin 6
San Andreas (1857 rupture)	7.8	0.00485	S	0.071	0.005323	0.000838	—	—	—	—
Elsinore (Glen Ivy)	6.8	0.00294	S	0.087	0.003845	0.001035	—	—	—	—
Sierra Madre	7.0	0.00260	R	0.081	0.003206	0.000730	—	—	—	—
San Andreas (southern)	7.4	0.00454	S	0.054	0.002894	—	—	—	—	—
San Jacinto (S. J. Valley)	6.9	0.01205	S	0.038	0.002888	—	—	—	—	—
Elsinore (Temecula)	6.8	0.00417	S	0.049	0.002251	—	—	—	—	—
Coronado Bank	7.4	0.00153	S	0.100	0.002194	0.000774	—	—	—	—
Whittier	6.8	0.00156	S	0.117	0.002134	0.001144	—	—	—	—
San Jacinto (S. B. Valley)	6.7	0.01000	S	0.035	0.001859	—	—	—	—	—
Cucamonga	7.0	0.00154	R	0.075	0.001760	—	—	—	—	—
Anacapa-Dume	7.3	0.00189	R	0.064	0.001744	—	—	—	—	—
Chino-Central Ave.	6.7	0.00113	S	0.085	0.001447	—	—	—	—	—
Newport-Inglewood	6.9	0.00099	S	0.517	—	—	0.000340	0.000392	0.000308	0.000237
Compton Thrust	6.8	0.00148	R	0.455	—	—	0.000307	0.000285	0.000189	0.000125
Newport-Inglewood (off.)	6.9	0.00154	S	0.250	—	0.001792	0.000630	0.000263	0.000080	0.000021
Palos Verdes	7.1	0.00154	S	0.221	—	0.002091	0.000548	0.000161	0.000031	—
Elysian Park Thrust	6.7	0.00182	R	0.187	—	0.001175	0.000222	0.000057	—	—
TOTAL	—	—	—	—	0.031545	0.009579	0.002047	0.001158	0.000608	0.000383

Note: For each MCE, M is moment magnitude, RF is recurrence frequency, F is faulting style (S = strike slip, R = reverse), PGA is average system-wide peak ground acceleration (g) on reference site condition BC, and hazard is weighted average system-wide occurrence frequency for PGA values that fall within the following ranges: 0.0738–0.145g (Bin 1), 0.145–0.284g (Bin 2), 0.284–0.397g (Bin 3), 0.397–0.556g (Bin 4), 0.556–0.778g (Bin 5), and 0.778–2.13g (Bin 6).

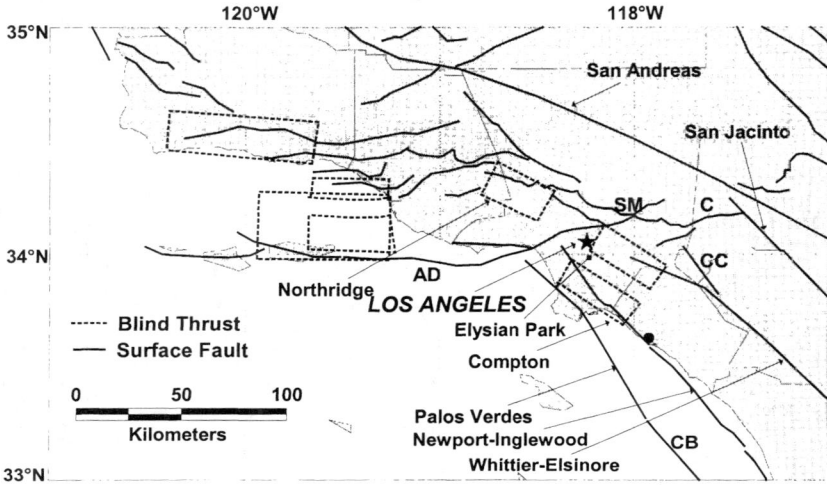

Figure 1. Map of southern California showing the location of the faults used to define the MCE events in this study. The approximate location of the 40 census tracts that define the hypothetical lifeline system in the southeast Los Angeles Basin is marked by the solid circle at the intersection of the onshore and offshore segments of the Newport-Inglewood fault. AD, Anacapa-Dume fault; C, Cucamonga fault; CB, Coronado Bank fault; CC, Chino-Central Ave. fault; SM, Sierra Madre fault. County boundaries are shown for reference.

Seismic Hazard Analysis and Developing the Uniform Hazard Spectra for an Under-Construction Railroad Bridge

Fariborz Yaghoobi Vayeghan[1], Maryam Firoozi Nezamabadi[2] & Mahmood Hosseini[3]

Abstract

A site investigation as well as a Seismic Hazard Analysis (SHA) have been carried out for a railroad bridge, which crosses Sephidrood River in Qazvin-Rasht-Anzali route in the northern part of Iran and is presently under-construction. The aim has been founding out the level of seismic hazard for the bridge and developing the Uniform Hazard Spectra (UHS) for verifying the performed seismic design of the bridge. At first, it was tried to recognize all the active faults around the bridge. Secondly, by using the appropriate attenuation laws, the PGA values on the site were estimated. These values obtained for the site vary between 0.169g and 0.296g depending on the applied attenuation laws. Finally, the UHS, which are more reliable for design purposes, were constructed for the 2% and 10% probability of exceedence in 50 years ground motions.

Introduction

As Iran is located in the high seismic area, reduction of seismic risk in different parts of the country by controlling the behavior of structures, particularly the key structures of lifeline systems is necessary. Seismic design of railroad system as an important transportation lifeline in the one hand, and the key role of the main bridges of the system in its function on the other hand, require a reliable seismic hazard analysis for the system and its components. The best way for performing a reliable seismic hazard analysis is using both deterministic and probabilistic methods. This paper reports an actual case of applying these methodologies for an under-construction river railroad bridge. At first, it was tried to recognize all the seismic sources (faults) in a radius of 110 km around the bridge, and to evaluate their seismic potential based on the seismic activities in recent century. Secondly, by using the appropriate attenuation relationships, the PGA values on the site were estimated by considering the focal

---

[1] Ph.D. Candidate, IIEES, and Member of Road and Transportation Research and Training Center (RTRTC), Tehran, Iran; Email: fyaghoobi@yahoo.com
[2] Ph.D. Candidate and Instructor, Tehran South Branch of the Islamic Azad University (IAU), Tehran, Iran; Email: m_firoozi_n@yahoo.com
[3] Associate Professor and Head of Lifeline Engineering Department, International Institute of Earthquake Engineering and Seismology (IIEES), P. O. Box 19395-3913, Tehran, Iran; Email: hosseini@dena.iiees.ac.ir

depths of recorded earthquakes, horizontal site-to-source distance and the local soil conditions. Then the PGA values were calculated by using deterministic method and hazard curves for the site were prepared by using probabilistic method. Finally, the UHS were constructed for the 2% and 10% probability of exceedence in 50 years ground motions based on spectral acceleration curves.

Site Location and Seismic Sources Parameters

The studied site in this paper corresponds to an under-construction railroad bridges which crosses Sephidrood River in Qazvin-Rasht-Anzali route in the northern part of Iran (49.5^0 T and 36.9^0 L). This bridge is located in a distance of about 6 km from Lahijan fault. Some important faults around the site in an area with radius of about 110 km are Astara and Soltanieh. By using Iran Earthquake Catalogue all of the ground motions with magnitude of more than 4.0, which were related to nearest linear faults (i.e. Lahijan, Astara and Soltanieh), or area fault, including small faults in north-east of the site were considered for hazard analysis. Earthquake data are shown in Table 1, and faults and site locations are shown in Figures 1 and 2 respectively.

Table 1. Earthquake data for the bridge site

Fault	Date	Longitude	Latitude	Depth	m_b	M_s
Lahijan	1983/7/22	49.18	36.948	41	5.6	5.0
	1990/6/20	49.35	36.99	19	6.2	7.4
	1991/11/28	49.58	36.86	50	5.6	5.0
	1998/9/28	48.80	36.80	33	4.9	---
Astara	1917/6/2	48.50	38.00	---	5.5	---
	1959/5/31	48.99	37.70	---	5.0	---
	1965/10/29	48.70	37.90	33	4.6	---
	1968/6/4	49.00	37.50	50	4.5	---
	1970/7/11	49.00	37.60	65	5.1	---
	1972/1/18	48.70	37.50	---	4.9	---
	1984/3/18	48.868	37.678	33	4.4	---
	1985/11/2	49.68	37.521	33	4.5	---
Soltanieh	1927/10/31	49.00	36.50	---	4.5	---
	1984/6/17	49.00	36.50	---	5.2	---
	1951/6/5	48.50	36.00	---	5.2	---
	1962/9/6	49.50	36.00	---	4.0	---
	1986/9/10	48.968	36.129	33	4.2	---
Area	1952/7/18	50.10	37.50	---	4.7	---
	1956/4/12	50.20	37.30	---	5.5	---
	1980/7/22	50.201	37.19	62	5.4	---
	1989/2/15	50.304	37.29	53	4.6	---
	1994/12/3	49.35	37.64	33	4.8	---
	1998/6/29	49.09	37.20	33	4.7	---

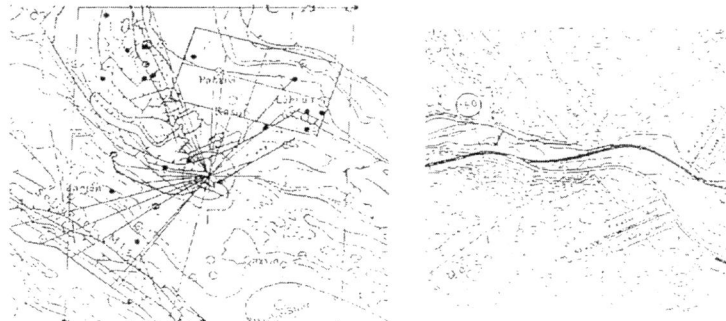

Figure 1. Faults location around the site   **Figure 2.** The bridge site location

Attenuation Relationships

The general form of attenuation expression used in most investigation can be characterized by the expression:

$$y = b_1 \cdot f_1(M) \cdot f_2(R) \cdot f_3(M,R) \cdot f_4(P_i) \varepsilon \tag{1}$$

Where y is the strong motion parameter to be predicted, b_1 is a constant and

$$f_1(M) = e^{b_2 M} \tag{1a}$$

$$f_2(R) = e^{b_4 R}[R+b_5]^{-b_3} \quad \text{or} \quad f_2(R) = e^{b_4 R}\left[\sqrt{R^2+b_5^2}\right]^{-b_3} \tag{1b}$$

$$f_3(M,R) = [R + b_6 e^{b_7 M}]^{-b_3} \tag{1c}$$

$$f_4(P_i) = \sum e^{b_i P_i} \tag{1d}$$

In expression (1a) to (1d) b_6 is a constant and M, R, b_2, b_3, b_4, b_5, b_7, P_i, and ε are respectively magnitude, site-to-source distance, magnitude attenuation rate, geometrical attenuation rate, the coefficient of elastic attenuation, the coefficient that limits the value of y at zero distance, negative coefficient that reduces the amount of magnitude scaling at short distances, site effect, random variable that is usually assumed to be log-normally distributed [1]. Although an attenuation relationship that include all of the above factors are theoretically possible, two factors that are often represented in attenuation expressions are geometric spreading and magnitude.

In this study the following attenuation relationships have been used for the bridge site.

1) Boore, Joyner and Fumal [2]

$$\begin{aligned} LogPGA = b_1 &+ b_2(M-6) + b_3(M-6)^2 + b_4\sqrt{R^2+h^2} + b_5\left(Log\sqrt{R^2+h^2}\right) \\ &+ b_6 G_B + b_7 G_c + \sigma \end{aligned} \tag{2}$$

where R, M and σ are site-to-source distance, moment magnitude and standard error respectively; b_{1-7} and h are constants, and G_b and G_c are site classification coefficients, and their values are as follow:
$b_1 = -0.038$, $b_2 = 0.216$, $b_3 = 0$, $b_4 = 0$, $b_5 = -0.777$, $b_6 = 0.158$
$b_7 = 0.254$, $h = 5.48$, $G_B = 0$, $G_C = 0$, $\sigma_{Log(y)} = 0.205$

2) *Nuttli and Herrmann [3]*

$$PGA = 3.79 \times 10^{-3} \times e^{1.15 M_b} \left[\sqrt{R^2 + (0.018 e^{1.05 M_b})^2} \right]^{-0.83} \times e^{-0.00159 R} \quad (3)$$

where M_b is the body waves magnitude.

3) *Battis [4]*
$$PGA = 0.0239 e^{1.24 m_b} (R+25)^{-1.24} \quad (4)$$

4) *Donovan and Bornstein [5]*
$$PGA = \frac{[2154000(R)^{-2.1} e^{(0.046+0.445 LogR)M} (R+25)^{-(2.515-0.486 LogR)}]}{R} \quad (5)$$

5) *Crouse [6]*

$$LnPGA = 6.36 + 1.76 M - 2.73 \ln(R + 1.58 e^{0.608 M}) + 0.00916 H + 0.773 \quad (6)$$

where H is the focal depth.

6) *Campbell and Bozorgnia [7]*
$$LnPGA = -3.512 + 0.904 M - 1.328 \ln\left[\sqrt{R^2 + (0.149 e^{(0.647 M)})^2}\right]$$
$$+ [1.125 - 0.112 \ln R - 0.0957 M] F + [0.44 - 0.171(\ln R) Sar + 0.405 - 0.222(\ln R) Shr + 0.05] \quad (7)$$

where Sar equals zero and Shr and F equal one.

Calculating PGA by (Deterministic Seismic Hazard Analysis (DSHA) Method

For using this method, PGA values were obtained from designated attenuation relationships. The used site-to-source distance, R and the maximum moment magnitude of occurred earthquakes, M_{max} are presented in Table 2, and PGA values in Table 3. The maximum calculated PGA value is 0.738g, which is obtained by using Battis attenuation relationship.

Table 2. The values of R and M_{max} for all sources

Seismic Sources	Lahijan	Astara	Soltanieh	Area
R (km)	6	33	95	40
M_{max}	7.4	5.5	5.2	5.5

Table 3. The PGA values obtained from various attenuation relationships

Attenuation Relationship	Seismic Source			
	Lahijan	Astara	Soltanieh	Area
Boore, Joyner and Fumal	0.361	0.047	0.018	0.040
Nuttli and Herrmann	0.529	0.109	0.029	0.092
Battis	0.738	0.142	0.040	0.124
Donovan and Bornstein	0.604	0.055	0.011	0.044
Crouse	0.579	0.121	0.023	0.105
Campbell and Bozorgnia	0.610	0.035	*----	0.025

* This Relationship is used only for distances at least 60 km far from a source

Hazard Estimation by (Probabilistic Seismic Hazard Analysis (PSHA) Method

This method considers all earthquakes with possible magnitude, on all significant sources, at all possible distances from the site, considering the likelihood of each combination. Therefore, using PSHA allows a desired facility to be designed for ground motion with a specified probability of exceedence [8].

Steps Involved in a PSHA: In the first step, all seismic sources that can produce damaging ground motion at the site were identified. Then each line source was divided into 4 or 5 segments. Distances of center of each segment to the site are given in Table 4.

Table 4. Distances between the centers of line source segments to the site

Seismic Sources	No. of Segments	Site-to-Segment Distance (Km)				
		1	2	3	4	5
Lahijan	5	10	27	36	50	70
Astara	4	45	77	122	164	--
Soltanieh	5	95	101	109	127	147
Area	4	53	68	75	89	--

The second step was the establishment of earthquake recurrence relationships, magnitude distribution and average occurrence rates which were obtained from equations (8) to (10).

$$\ln N = \alpha - \beta M \quad \text{or} \quad N(m) = e^{(\alpha - \beta M)} \tag{8}$$

$$\vartheta = N(m_o) - N(M_{max}) \tag{9}$$

$$f(M) = C\beta e^{-\beta(M - m_o)} \tag{10}$$

where α and β are Gutenburg-Richter coefficients, N is the number of earthquakes of magnitude greater than or equal to m_0 (the lower magnitude limit, was supposed 4.0), M is the magnitude, and C is as follows:

$$C = \frac{1}{1 - e^{-\beta(M_{max} - m_o)}} \tag{11}$$

Figures 3 and 4 indicate Gutenburg-Richter relationship and $f(M)$ vs. magnitude respectively.

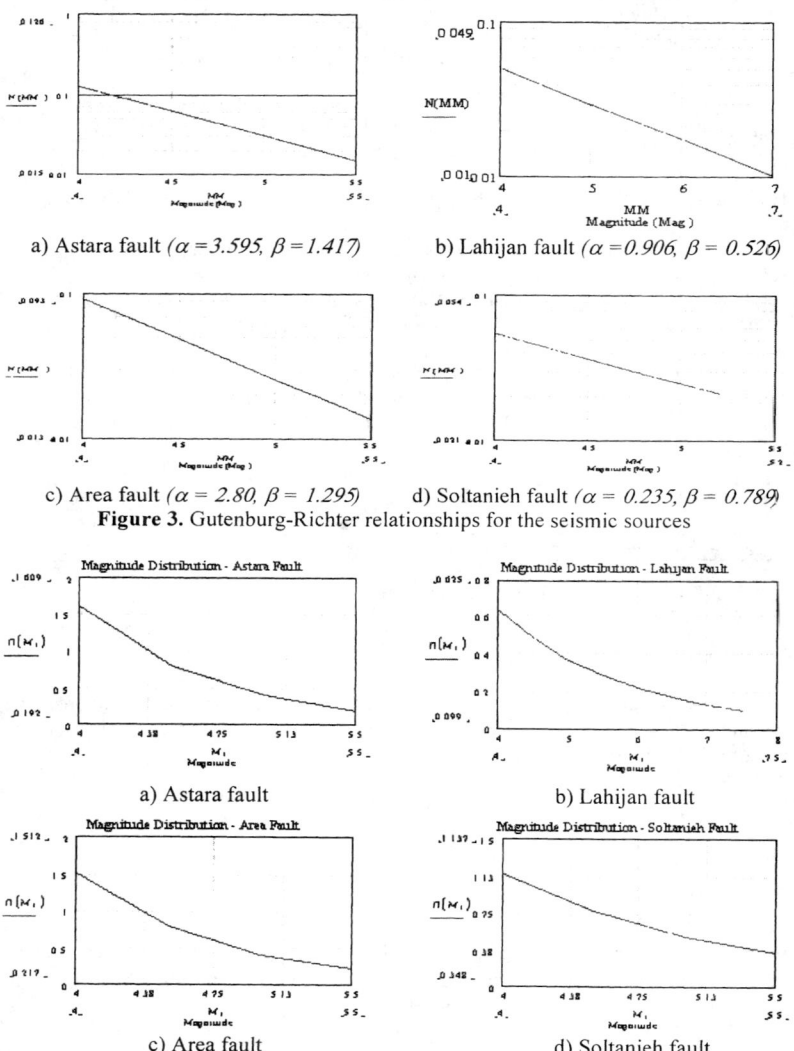

a) Astara fault $(\alpha = 3.595, \beta = 1.417)$
b) Lahijan fault $(\alpha = 0.906, \beta = 0.526)$
c) Area fault $(\alpha = 2.80, \beta = 1.295)$
d) Soltanieh fault $(\alpha = 0.235, \beta = 0.789)$

Figure 3. Gutenburg-Richter relationships for the seismic sources

a) Astara fault
b) Lahijan fault
c) Area fault
d) Soltanieh fault

Figure 4. Variations of $f(M)$ vs. M for the seismic sources

In the third step, the PGA values were calculated from equations (2) and (5), for various amount of R, given in Table 4, and M, between m_0 and M_{max} with a value of 0.5 for Δm. For example, Figure 5 indicates the PGA values obtained from equation (2) for Lahijan fault.

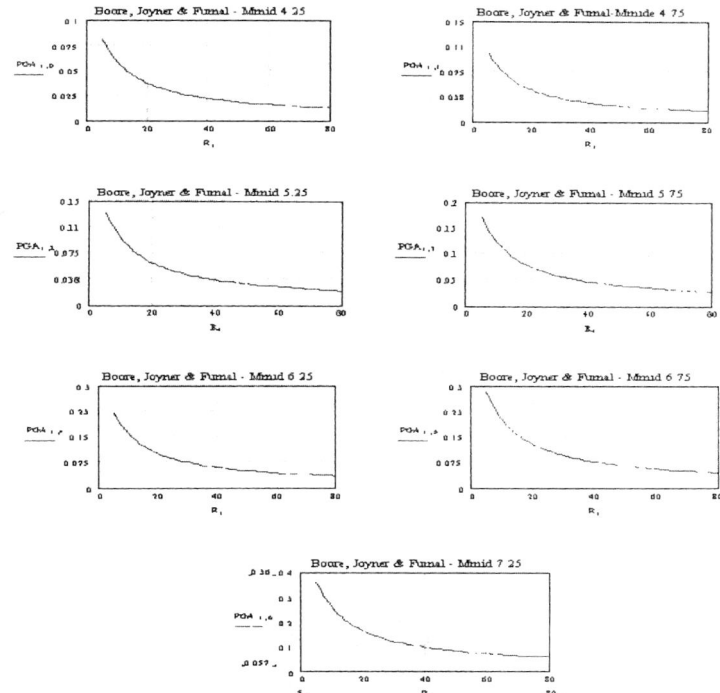

Figure 5. The PGA values obtained from equation (2) for Lahijan fault

Given the occurrence rate of an earthquake, υ, the probability that the site PGA will exceed an acceleration value acc of interest were determined for every combination of discretized magnitude and distance for each source by using equation (12).

$$P(PGA > acc | EQ : M, R) = 1 - \bar{\phi}\left(\frac{\ln(acc) - \lambda}{\zeta}\right) \quad (12)$$

where acc, varies from 0.5g to 0.65g with Δacc equal to 0.05g and

$$\lambda = E[\ln(PGA)] = mean \text{ value of } \ln(PGA) \quad (13a)$$

$$\zeta = \sigma_{\ln(PGA)} \quad (13b)$$

In the forth step, by using equation (14), the probability of exceedence for each fault was obtained.

$$P(PGA > acc | EQ) = \sum_R \sum_M P(PGA > acc | EQ : M, R) f(M).\Delta m. f(R).\Delta R \qquad (14)$$

where the values of $f(R)$, ΔR are 0.2 for Lahijan and Soltanieh faults, and 0.25 for Astara and the area faults. The annual probability of exceedence for each fault was calculated by equation (15).

$$P(PGA > acc) = 1 - \exp[-\vartheta t. P(PGA > acc | EQ)] \qquad (15)$$

where t, equals 1.0 and υ, the average occurrence rate of earthquake for Lahijan, Astara, Soltanieh and area fault is 0.041, 0.111, 0.037 and 0.079 respectively. Figure 6 indicate the annual probability of exceedence obtained by Eq. (2). Similar curves were obtained by Eq. (5), which can not be shown here because of lack of space.

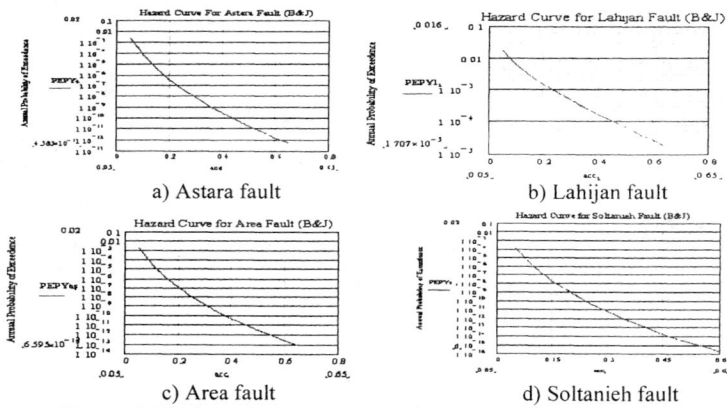

a) Astara fault   b) Lahijan fault

c) Area fault   d) Soltanieh fault

Figure 6. The hazard curve obtained for the seismic sources by Eq. (2)

Finally, as the fifth step, the results from the line and area faults were combined by Eq. (16), [8], which are shown here in Figure 7.

$$P(PGA > acc) = 1 - \Pi[1 - P(PGA > acc)] \qquad (16)$$

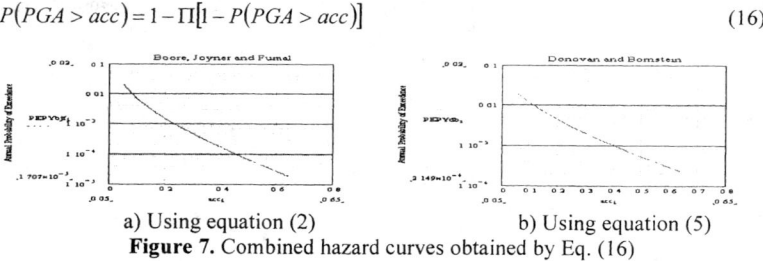

a) Using equation (2)   b) Using equation (5)

Figure 7. Combined hazard curves obtained by Eq. (16)

Table 5 presents the PGA values for 10% probability of exceedence in 50 years using linear interpolation.

Table 5. PGA values for 10% probability of exceedence in 50 years

Attenuation Relationship	PGA
Boore, Joyner and Fumal	0.169
Donovan and Bornstein	0.296

Developing the Uniform Hazard Spectra (UHS)

By definition the response at each discrete frequency of a UHS has an equal probability of being exceeded. The steps involved in computing a UHS are the same as those for the probabilistic hazard curve described above, except that the steps are repeated several times using different coefficients corresponding to each discrete frequency. The Boore, Joyner and Fumal spectral attenuation expression was used to compute the S_{pv}. Each curve in Figure 8 shows the S_a (spectral acceleration) values for the period range of 0.1s to 2.0s. Figure 9 indicates the UHS curve for the 10% probability of exceedence in 50 years (Life Safety Level).

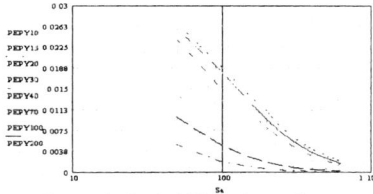

Figure 8. Probabilistic hazard curves vs. S_a for various periods

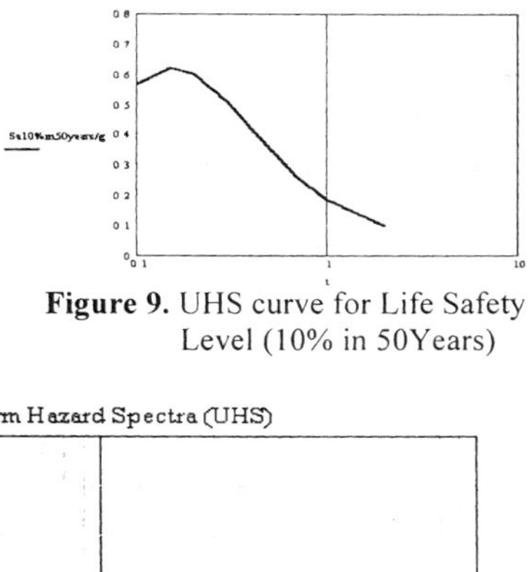

Figure 9. UHS curve for Life Safety Level (10% in 50Years)

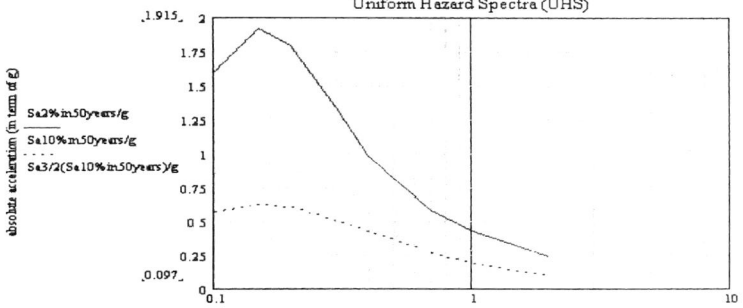

Figure 10. UHS curves for 2%, 10% and 1.5 times the 10% probability of exeedence in 50 years ground motion

In Figure 10, UHS curves were drawn for 2% in 50 years (Collapse Prevention Level), 10% in 50 years and 1.5 times the 10% in 50 years ground motions. Comparison of the 1.5 times the 10% in 50 years and 2% in 50 years spectra for the site indicates that if the bridge is designed for a 10% in 50 years ground motion, it would be much less likely to survive the 2% in 50 years ground motion.

Conclusions

In this study the PGA values obtained for the site from DSHA method was 0.738g and from PSHA method varied between 0.169g and 0.296g for 10% exeedence in 50 years ground motion depending on the applied attenuation relationship. In general, the results of DSHA method are over estimated, because it uses not only the minimum site-to-source distance, but also the maximum magnitude of ground motions. The results of PSHA method are more reliable, because this procedure uses seismicity parameter and several site-to-source distances. Comparison of the 1.5 times the 10% in 50 years and 2% in 50 years UHS spectra for the site indicates that if the bridge is designed for a 10% in 50 years ground motion, it would be much less likely to survive the 2% in 50 years ground motion.

Acknowledgments

The authors wishes to acknowledge Prof. M. Ghafory-Ashtiany, the president of International Institute of Earthquake Engineering and Seismology (IIEES), Dr. M. Zare, assistant Prof. Of Engineering Seismology Dept., IIEES and Mr. M. Monajemi, from ministry of Road and Transportation of Iran for their valuable supports.

References

[1] Campbell, K.W., "Strong Motion Attenuation Relations: A Ten Year Perspective", *Journal of Earthquake Spectra*, vol.1, No.4, August 1985, pp. 759-804.
[2] Boore, D.M., Joyner, W.B., and Fumal, T.E., *"Estimation of Response Spectra and Peak Acceleration from Western North America Earthquakes"*, an interim Report, USGS Open-File Report 93-509, 1993.
[3] Nuttli O.W. and Herrmann, R.B., "Ground Motion of Mississippi Valley Earthquakes", *J. of Tech. Topics in Civil Engineering*, ASCE, 110, 1984, pp. 54-69.
[4] Battis, J., "Regional Modification of Acceleration Functions", *Bulletin of Seism. Soc. Am.*, 71, 1981, pp. 2629-2653.
[5] Donovan N.C., and Bornstein, A.E., "Uncertainties in Seismic Risk Procedures", *J. of the Geotechnical Eng. Div.*, ASCE, Vol. 104, No. GT7, 1978, pp. 869-887.
[6] Crouse, C.B., "Ground-Motion Attenuation Equations for Earthquakes on the Cascadia Subduction Zone", *Earthquake Spectra*, Vol. 7, No.2, 1991, pp. 201-236.
[7] Campbell, K.W., and Bozorgnia, Y., "Near-Source Attenuation of Peak Horizontal Acceleration from Worldwide Accelerograms Recorded from 1957 to 1993", *Proc. of 5th U.S. National Conf. on Earthquake Eng.*, Chicago, U.S.A., 1994.
[8] Green, R.A., and Hall, W.J., *"An Overview of Selected Seismic Hazard Analysis Methodologies"*, Civil Engineering Studies, Structural Research Series No: 592, University of Illinois at Urbana-Champaign, Urbana, U.S.A., 1994.

Current Developments and Future Directions for Seismic Risk Analysis of Highway Systems

Stuart D. Werner[1]

Abstract

A methodology for seismic risk analysis of highway systems has been developed, and is now being programmed into a public-domain software package. As part of this work, our development team has been meeting with potential users of this software to discuss how they envision using it in the future and how the methodology may be further improved to facilitate its future application to highway systems nationwide.

Introduction

A highway research program that is being sponsored by the Federal Highway Administration (FHWA) and implemented by the Multi-disciplinary Center for Earthquake Engineering Research (MCEER) has included a task to develop tools for: (a) estimating how earthquake damage to a highway system can affect post-earthquake traffic flows; and (b) enabling users to consider these effects during their evaluation and selection of strategies for seismic risk mitigation and post-earthquake emergency response. This task has led to a new methodology for deterministic and probabilistic SRA of highway systems that has been applied to highway systems in Shelby County, TN and Los Angeles CA. The methodology is now being programmed into a public-domain software package named REDARS (Risks due to Earthquake DAmage to Roadway Systems) that will be beta tested during 2004 and publicly released in 2005. This paper summarizes this methodology, and also discusses future SRA applications and research needs that have been identified thus far during an ongoing series of workshops with potential future users of REDARS.

Methodology

The SRA methodology (Fig. 1) consists of input data and analysis setup (Step 1), analysis of the performance of the highway system for each scenario earthquake established in Step 1 (Steps 2 and 3), and aggregation of the results from each of these applications (Step 4). Key features of this methodology are summarized below.

Modules. The methodology is organized into a series of modules that contain the input data, models, and analysis procedures necessary to characterize the highway system, the seismic hazards to which the system will be subjected, the fragility of the various components that comprise the system, and the economic losses due to earthquake-induced damage and traffic disruption (Fig. 2). This modular organization will facilitate the future inclusion of new improvements to hazard, component and

---

[1] Principal, Seismic Systems & Engineering Consultants, 8601 Skyline Boulevard, Oakland CA 94611; phone 510-531-7489; sdwerner@ix.netcom.com

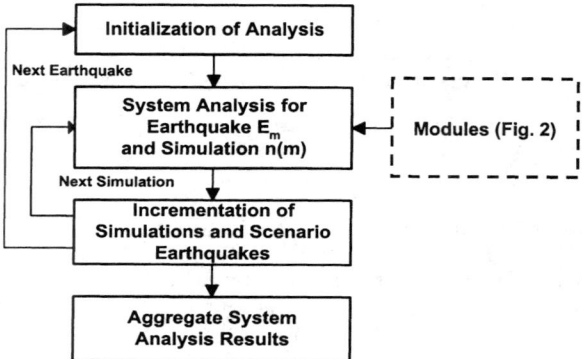

Figure 1. Methodology for Seismic Risk Analysis of Highway Systems

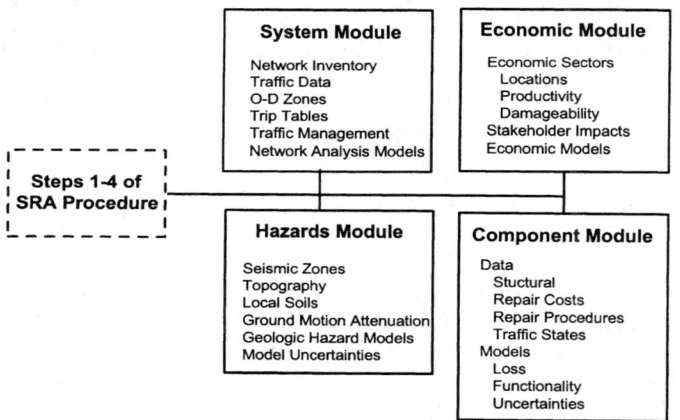

Figure 2. Modules included in Seismic Risk Analysis Methodology

network models as they are developed from future research (Werner et al., 2000). Figure 3 shows how these models are used to carry out a highway system analysis for a single simulation. (Note that a simulation is defined as one complete set of system SRA results for one particular set of uncertain input and model parameters.)

Walkthrough Process. The methodology incorporates a walkthrough process, which enables uncertainties to be treated by carrying out the SRA for a specified walkthrough time duration, exposure time, and time increment. For each successive time increment, this process consists of: (a) random selection of the number of earthquakes that will occur, along with the magnitude and location of each earthquake event; (b) random selection of the uncertain parameters included in the hazards and component fragility models; and (c) estimation of the resulting seismic hazards, component damage states, system-wide traffic flow disruptions, and corresponding economic losses. As described in Taylor et al. (2001), the method can account for all quantifiable model and input parameter uncertainties, and facilitates the calculation of nominal confidence levels and limits of the risk results that are produced.

Decision Guidance. A new post-processor has been developed to enable results from the SRA methodology to be readily included into an "acceptable risk" framework that will guide seismic risk reduction decision making. In this, transportation department decision makers would first establish risk reduction alternatives and seismic performance requirements to be considered for the highway system, and would then use SRA methodology to assess and compare relative costs and risks associated with each alternative. From this, a preferred alternative would be selected. Stakeholder interaction in evaluating system performance goals relative to this overall decision-making process should be an important element of this process.

Probabilistic SRA Application

Shelby County, Tennessee Highway System. The first probabilistic application of the SRA methodology was to the highway system in Shelby County, Tennessee (Fig. 4). Its purpose was to show how the methodology can be used to analyze an actual highway system. This SRA application, which is described in detail elsewhere (e.g., Werner et al., 2000), is summarized below.

Input Data. The following input data is needed to carry out the SRA: (a) a highway-roadway network model, that includes the network's geometry, and each roadway's lanes traffic carrying capacities, free-flow speeds, and component locations (Fig. 5a); (b) for each component (bridge, roadway segment, tunnel, etc.), a fragility model that characterizes its configuration and structural attributes; (c) soil conditions throughout the network, as needed to estimate site-specific ground motions (Fig. 5b) and permanent ground displacement hazards at each component site (Fig. 5c); (d) trip demands on the network, in the form of origin-destination (O-D) zones (Fig. 5d) and trip tables that define the number of pre-earthquake trips from each zone to all other zones in the region; and (e) economic parameters needed to estimate economic losses

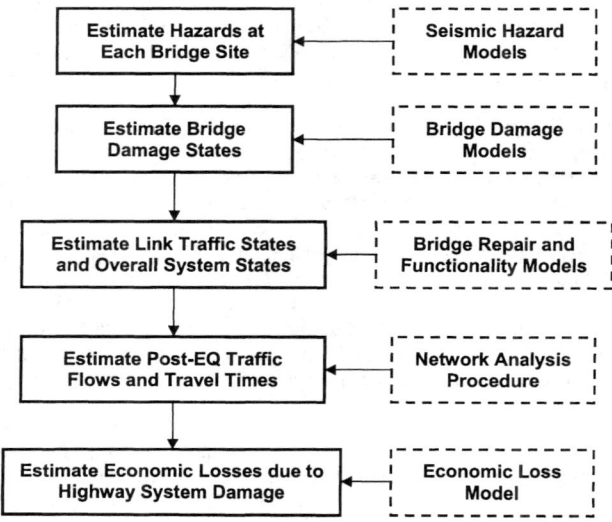

Figure 3. Analysis Procedure for Each Scenario Earthquake and Simulation

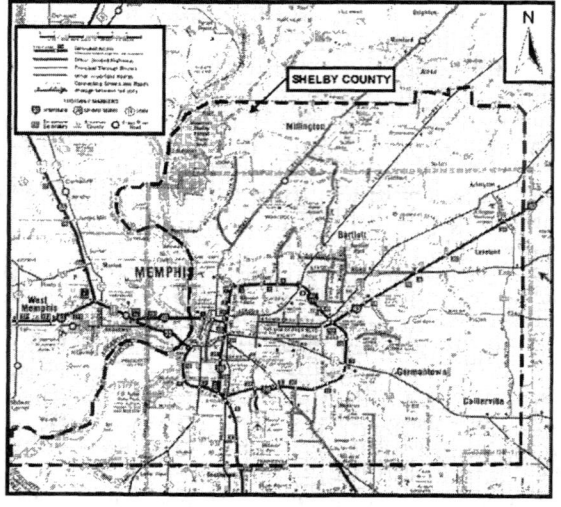

Figure 4. Shelby County, Tennessee Highway-Roadway System

due to travel time delays that result from earthquake damage to the highway network. The above O-D zones and trip tables (Item d above) are projections for the year 2020 that were provided by the Shelby County Office of Planning and Development.

Analysis. This SRA was conducted by using a walk-through with a duration of 50,000 years. Earthquakes occurring during each year of the walkthrough were estimated by applying the Frankel et al. (1996) models for the Central United States. This generated 2,321 earthquakes with moment magnitudes ranging from 5.0 to 8.0. The network model included 7,807 links, 15,614 nodes, and 384 bridges. The system analysis for each simulation was carried out according to the procedure illustrated in Figure 3. The analysis incorporated ground shaking and liquefaction hazard models by Hwang and Huo (1997) and Youd (1998), bridge fragility models by Jernigan and Hwang (1997) and by Werner et al. (2000), an approach fill settlement model by Youd (1999), and network analysis procedures described in Moore et al. (1997).

Results. The results of the analysis included: (a) probabilistic estimates of economic losses for various exposure times (Fig. 6) and also for access and egress times to/from selected key locations in Shelby County; (b) for selected simulations, tabular displays of economic losses and access or egress times to/from selected locations (Table 1), and graphical displays of minimum-time travel paths between selected locations and traffic volumes along selected roadways (Werner et al., 2000). In addition, GIS displays of system-wide ground shaking and liquefaction hazards, bridge damage states, and post-earthquake system states were generated for selected simulations.

Feedback from Potential Future Users

Objective. During the initial stages of the REDARS software development, we are soliciting feedback from potential future users nationwide regarding possible SRA applications, output needs, and research recommendations. This feedback will guide our development of the software to enable it to better meet user needs.

Caltrans Mini-Workshops. This process was initiated by mini-workshops with maintenance, earthquake engineering, transportation-engineering/planning, and emergency response personnel from the California Department of Transportation (Caltrans) District 7 Los Angeles CA), District 4 office (Oakland CA), and Headquarters office (Sacramento CA). Feedback from these discussions is summarized below. It is noted that, since REDARS is to be applicable to highway-roadway systems throughout the United States, additional mini-workshops with other state transportation agencies nationwide are being planned to enable us to obtain similar SRA-related feedback from these agencies.

Possible Future Applications. The following possible future applications of SRA and REDARS were identified during the various Caltrans mini-workshops:

- *Critical Lifeline Routes.* The goal of Caltrans' recent statewide seismic retrofit program was to enable their bridges to provide adequate life-safety protection

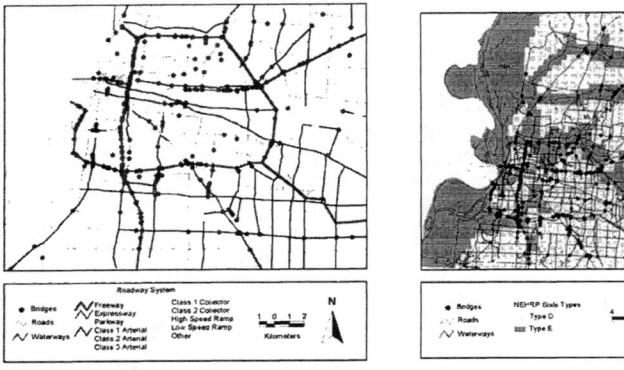
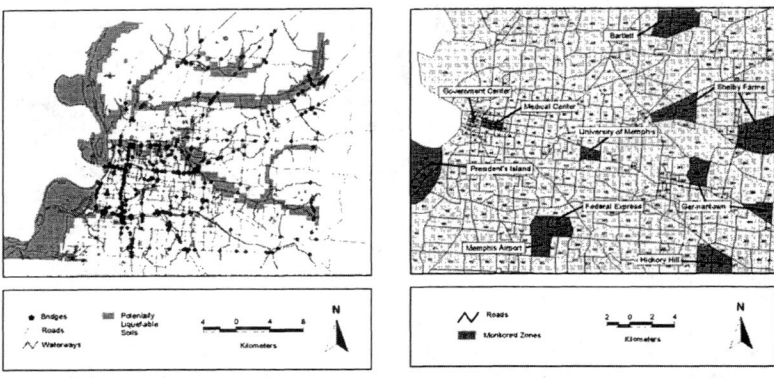

a) Highway System and Bridges b) NEHRP Soil Types (Ground Motions)

c) Liquefiable Soils (Geologic Screening) d) Origin-Destination Zones

Figure 5. Input Data for SRA of Shelby County, Tennessee Highway System

during a major earthquake. However, this does not preclude a bridge from suffering damage that results in a loss of traffic-carrying functionality over an extended time. Such damage will be unacceptable for critical lifeline routes -- whose ability to carry traffic after an earthquake will be essential to emergency response and recovery. SRA can enable Caltrans to: (a) assess whether these lifeline routes will, in their current state, meet acceptable post-earthquake functionality requirements; and (b) evaluate various options for mitigating inadequate seismic performance along these routes (e.g., identifying suitable seismic upgrade strategies for bridges along these routes, identifying alternative routes with higher levels of seismic performance, etc.)

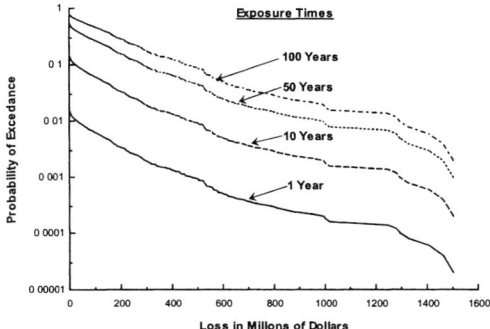

Figure 6. Economic Losses due to Travel Time Increases caused by Earthquake Damage to Shelby County TN Highway System

Table 1. Increase in Access Time to Selected Locations in Shelby County TN due to Earthquake Damage to Highway System

Origin-Destination Zone (see Fig. 3)	Post-Earthquake Access Time		
	7 Days after EQ	60 Days after EQ	150 Days after EQ
9 (Government Center in downtown Memphis)	43.8%	5.8%	2.0%
28 (Major Hospital Center, just east of downtown Memphis)	44.6%	6.7%	2.0%
205 (Memphis Airport and Federal Express transportation center, south of beltway)	53.7%	4.0%	1.6%
73 (University of Memphis campus in central Memphis)	21.6%	4.3%	1.5%
310 (Germantown, residential area east of beltway)	2.9%	0.9%	0.4%
160 (President's Island, Port of Memphis at Mississippi River)	34.9%	6.1%	1.6%
246 (Hickory Hill, commercial area southeast of beltway)	3.9%	1.9%	1.1%
335 (Shelby Farms residential area northeast of beltway)	28.4%	4.8%	1.6%
412 (Bartlett, residential area north of beltway)	13.2%	3.0%	1.3%

- *Pre-Earthquake Assessment of Available Repair Resources.* Caltrans' ability to quickly shore and/or repair damaged bridges after a major earthquake will have an important effect on economic losses and emergency response. SRA can enable Caltrans to: (a) assess their current repair resources, the corresponding time required to repair the highway damage and restore the system's traffic carrying capacity, and the economic losses due to travel time delays while repairs are proceeding; (b) assess various strategies to increase repair resources and reduce repair times, the costs of these strategies, and the degree to which they will reduce post-earthquake repair times and economic losses due to travel time delays; and (c) from these assessments, select a preferred strategy for expanding repair resources and using them to reduce post-earthquake repair times.

- *Post-Earthquake Assessment of Alternative Traffic Management Strategies.* After an earthquake, Caltrans may wish to deploy various traffic management strategies to reduce congestion near damaged areas. Such strategies could include requirements for certain roads to carry one-way instead of two-way traffic, restricting parking along such roads, and identifying (or constructing) appropriate detour routes. SRA can enable Caltrans to estimate the relative effectiveness of such alternative strategies in real time after an actual earthquake.

- *Post-Earthquake Coordination of Emergency Response Activities between Agencies.* Post-earthquake emergency response plans typically include meetings of various agencies (e.g., Caltrans, government, utilities, police and fire departments, local and urban mass transit agencies, etc.) in order to coordinate emergency response activities. The ability of SRA to provide real-time estimates of traffic congestion locations due to highway system damage after an earthquake will provide valuable input to these emergency response activities.

- *Pre-Earthquake Identification of Vulnerable Sections of Highway Systems.* In the past, some Caltrans offices have been able to identify vulnerable sections of the highway system where there will be a high likelihood of major traffic congestion after an earthquake (e.g., due to a lack of redundancy of the roadways in this section.) The ability of SRA to clearly demonstrate the existence of such problem areas and the potential benefits of alternative system enhancement strategies (such as construction of parallel roadways alongside non-redundant roadway sections) would facilitate the procurement of funding to implement such strategies.

Decision Variables. During the mini-workshops, Caltrans personnel identified the following decision variables that could be helpful for guiding future seismic risk mitigation and post-earthquake emergency response decisions: (a) component damage states; (b) post-earthquake system states (which will vary with time after the earthquake to reflect the estimated rate at which the highway system can be repaired and traffic carrying capacity restored;) (c) probabilistic estimates of economic losses; (d) system-wide travel time increases; (e) estimates of earthquake effects on access and egress times to/from locations within the region that are essential to post-earthquake emergency response or economic recovery; (f) minimum-time travel paths and associated travel times and distances between selected locations in the region; and (g) traffic volumes along selected links in the highway-roadway network. Currently, all of these variables can be provided as output from REDARS.

Research Recommendations. Caltrans personnel provided the following research recommendations to further improve the current SRA methodology:

- *Rapid Post-Earthquake Analysis Updates.* After an earthquake, REDARS could facilitate emergency response by providing initial real-time estimates of highway system damage and congested traffic flows, and then updating its traffic flow estimates as actual damage information is received from the field. Research to enable REDARS to develop such real-time updates was recommended.

- *Improved Fragility Models for Bridges.* Current bridge fragility models are primarily for new bridges subject to ground motion hazards. Further development of bridge fragility models should focus on: (a) consideration of retrofitted bridges; (b) consideration of permanent ground displacement as well as ground shaking hazards; and (c) development of a family of bridge repair cost and functionality models (as a function of damage state) that reflect regional differences in bridge construction practices, repair resources, and post-earthquake repair experience.

- *Fragility Models for Non-Bridge Components.* Although some fragility models for tunnels and pavements have been developed (FEMA, 1999), there is a need to review these models and upgrade them if appropriate.

- *Network Analysis Procedures.* Caltrans recommended further validation of current network analysis methods to demonstrate their ability to estimate traffic flows in real time after an actual earthquake. They also recommended research to develop improved estimates of post-earthquake trip demands.

Closing Comments

SRA Methodology. The SRA methodology summarized in this paper will enable users to carry out either deterministic vulnerability analysis (for a specified earthquake and fixed input-parameter values) or probabilistic risk-based analysis that accounts for model and input-parameter uncertainties. It is the basis for the REDARS software package now under development. This development is benefiting from interaction with potential future users nationwide, from which feedback on user applications, needs, and recommendations will guide our development of REDARS into a practical, technically sound, and useful SRA software package. The feedback from Caltrans that is reported in this paper is an example of the type of interaction we are soliciting from other potential users as well.

Input Data. A key issue in the future applicability of REDARS will be the availability of suitable input data. Because of the many components and sites within a highway system, development of such input data can be labor-intensive. For this reason, our SRA research work will include a task to develop an input "wizard" that will facilitate input-data compilation by enabling users to directly access available electronic databases of highway network attributes, soil conditions, bridge attributes, and O-D zones (ImageCat, 2002). However, the planning of research to improve existing models for hazards and for bridges and other components that will be built into REDARS must consider that these available databases may not contain all of the attributes needed to apply these improved models (e.g., the FHWA National Bridge Inventory database is intended for bridge maintenance applications, and does not contain many of the structural attributes ordinarily needed for seismic analysis.) Therefore, future development of improved hazard and component-fragility models for REDARS (which is a clearly a desirable objective) may require parallel efforts to develop electronic databases that contain the input data needed for these models.

Acknowledgements

The author acknowledges FHWA and MCEER for their support of the SRA methodology and software development summarized in this paper, and the Pacific Earthquake Engineering Research Center for their support of the mini-workshops with Caltrans. In addition, the author acknowledges the significant contributions to this work by the other members of the SRA-REDARS development team, who are Craig Taylor of Natural Hazards Management Inc., Jim Moore of the University of Southern California, Ron Eguchi, Charles Huyck, Sungbin Cho, and Shubharoop Ghosh of ImageCat Inc., and Jean-Paul Lavoie and Chip Eitzel of Geodesy.

References

Federal Emergency Management Agency (FEMA) (1999). *Earthquake Loss Estimation Methodology, HAZUS99 Technical Manual,* Washington D.C.

Frankel, A., Mueller, C., Barnhard, T., Perkins, D., Leyendecker, E.V., Dickman, N, Hanson, S., and Hopper, M. (1996)., *National Seismic Hazard Maps, Documentation June 1996,* Denver: United States Geological Survey, Open-File Report 96-532, June.

Hwang, H.H.M and Huo, J.-R. (1997). "Attenuation Relations of Ground Motion for Rock and Soil Sites in Eastern United States", *Soil Dynamics and Earthquake Engineering,* Volume 16, pp 363-372.

ImageCat Inc. (2002). *Year 4 Research Plan for Task B1-3 of FHWA-MCEER Highway Project,* Long Beach CA, November.

Jernigan, J.B. and Hwang, H.H.M. (1997). *Inventory and Fragility Analysis of Memphis Bridges,* Center for Earthquake Research and Information, University of Memphis, Memphis TN, Sept. 15.

Moore, J.E. II, Kim, G., Xu, R., Cho, S., Hu, H-H, and Xu, R., *Evaluating System ATMIS Technologies via Rapid Estimation of Network Flows: Final Report,* California PATH Report UCB-ITS-PRR-97-54, December 1997.

Taylor, C.E., Werner, S.D., and Jakubowski, S. (2001). "Walkthrough Method for Catastrophe Decision Making", *Natural Hazards Review, American Society of Civil Engineers,* Volume 2, Number 4, November, pp 193-202.

Werner, S.D., Taylor, C.E., Moore, J.E. II, Walton, J.S., and Cho, S. (2000). *A Risk-Based Methodology for Assessing the Seismic Performance of Highway Systems,* Report No. MCEER-00-0014, Multidisciplinary Center for Earthquake Engineering Research, Buffalo NY, December 31.

Youd, T. L. (1998). *Screening Guide for Rapid Assessment of Liquefaction Hazard at Highway Bridge Sites,* Technical Report MCEER-98-0005, Multidisciplinary Center for Earthquake Engineering Research, Buffalo NY, June 16.

Earthquake occurrence modeling for evaluating seismic risks to roadway systems

by
David Perkins[1] and Craig Taylor[2]

Abstract

In an earlier study on risk from earthquake damages to roadway systems (REDARS), a brute-force Monte Carlo time series method ("walkthrough") was used to simulate annual losses for 50,000 years for a prototype Memphis roadway system. The walkthrough method made it possible to distinguish, in a transparent way, the annual rate of damaging events and the conditional loss given a damaging event. Initial efforts to calculate annual losses and confidence limits for the mean annual loss used an application of the binomial distribution.

With the computer times required for roadway systems loss evaluations, there was a need to find ways to reduce the number of simulations required for a probabilistic evaluation in order to achieve the same mean and confidence limits. Accordingly, we report here the results of the application of a variety of techniques, including bootstrap sampling, the use of antithetic variates, the use of Latin Squares (or permutation) sampling, the use of control functions, a compound Poisson approach, and importance sampling.

We found that extremely large reductions in the number of simulations needed could be achieved for the mean and confidence limits of the conditional loss distribution (the loss distribution given some loss in a specific year). However, for the unconditional, annual-loss distribution, the reduction of the number of simulations achieved through post-sampling techniques was only a multiplicative reduction factor of slightly above 3.2. Thus, the large number of years in which no losses are expected reduces the power of variance reduction techniques for low annual probability but high consequence networks. As a result, further efforts are underway to explore whether or not greater successes lie, paradoxically, for various pre-sampling techniques (e.g., geographic importance sampling), to estimate annual and conditional means, their confidence limits and the means and confidence limits of various fractile estimates, such as the 500-yr return period loss.

[1] United States Geological Survey, Golden CO; perkins@usgs.gov
[2] Natural Hazards Management Inc., Torrance CA and University of Southern California; cetaylor@earthlink.net

Background and Introduction

In earlier studies on risk from earthquake damages to roadway systems (REDARS) Werner et al. (2000) and Taylor et al. (2001), losses for a prototype Memphis roadway system were determined by various deterministic and probabilistic methods.

In these studies, *scenarios* are defined in terms of specific fault rupture centers and earthquake magnitudes. *Simulations* are defined as random evaluations of network performance for some specific scenario, that is, the simulations may take into account various uncertainties (e.g., attenuation uncertainty, bridge structure vulnerability models, and system vulnerability models). In effect, because of these uncertainties, there may be many diverse simulations for a specific scenario, and one may capture the results of these many simulations in a loss distribution conditional upon the specific scenario.

Of special interest, though, is how scenarios are selected. Scenarios that are user-specified yield a *deterministic* analysis—even if a loss distribution conditional upon the specific scenario is developed. In contrast, scenarios that are randomly chosen yield a *probabilistic* analysis. The use of random processes in scenario selection is a powerful means to derive a total loss distribution, not merely a loss distribution conditional upon a specific scenario. See Figure 1 for a prototype total loss distribution for the Memphis roadway network.

The original and current versions of baseline earthquake scenarios for REDARS have been constructed principally on the basis of models used in the United States Geological Survey (USGS) national probabilistic ground motion hazard maps. Although users may specify alternative earthquake scenarios, construction of the USGS-based baseline earthquake scenarios benefits from over thirty years of continuous work that involves periodic public updates after many reviews and revisions. Randomly selected baseline earthquake scenarios are developed through magnitude and occurrence probabilities from USGS source data. These magnitudes, locations and annual probabilities of occurrence are utilized in a "walkthrough" input table in which year of occurrence (starting from year zero) and rupture location and earthquake magnitude are specified through random selection. For instance, 50,000 annual trials were selected in a demonstration project for the Memphis roadway system. The user may then choose to subdivide these 50,000-year trials into, say, 1000 50-year exposure periods. Figure 2 shows example walkthrough plots for a few 1000-yr exposure periods.

These random processes permit the calculation of confidence levels and limits for the resulting network loss distribution. In previous research, it was assumed that the total number of years simulated constituted Bernoulli trials. As a result, the binomial distribution was used to estimate confidence limits for the mean annual loss. In particular, for the mean loss (the AAL, or average annualized loss), one can use the following formula to derive the nominal confidence limit interval (Hogg and Klugman, 1984; Law and Kelton, 1991):

$$AAL \pm t_o \sigma/(n-1)^{1/2} \tag{1}$$

in which

n = the number of Bernoulli trials ($n - 1$ = the number of degrees of freedom);

t_o = the value of Student's t distribution corresponding to any designated nominal confidence level and value of n; and
σ = the adjusted standard deviation of the loss distribution.

From expression (1), one can derive the following 95th centile confidence limits for the AAL in the distribution in Figure 1:

($2.17 ± $0.274) x 10^6

The binomial distribution, however, cannot be used to obtain mean and confidence limits for various annual centile estimates (e.g., the 500-yr return period loss), which depend on the shape of the upper fractiles of the annual loss distribution rather than mean and standard deviation. We return to this issue, below.

Recent developments have led to the need to improve these raw Monte Carlo estimates. In particular, the network vulnerability evaluation in REDARS (as opposed to the hazards assessment and component vulnerability evaluations procedures) has proven to be very time-consuming for the computation of many scenarios. Accordingly, there has been the need to determine whether or not statistical techniques can somehow reduce the number of simulations needed for REDARS. This reduction in scenarios needed is here posed as the question as to whether or not fewer simulations can be used to achieve the same nominal confidence limits derived in the previous research.

First Explorations: Post-Sampling Variance-Reduction Techniques

The focus of these first explorations has been on developing effective variance reduction techniques in the estimate of confidence limits for the AAL, or mean loss. Early in these explorations—if only to reduce the computational time of all the hypotheses tested—the actuarial literature was followed in treating the total loss distribution as being a compound Poisson process. (see Daykin et al., 1994) The *Poisson parameter (λ)* designated the probability of some network loss. The *conditional loss distribution* hereafter refers to the distribution of network losses given that some loss occurs.

In order to utilize a smaller number of simulations to achieve a comparable level of confidence, the smaller samples were subjected to bootstrap sampling. (Bootstrap sampling consists of sampling with replacement from the smaller set of simulations.) A large number of techniques were tested singly and in combination. These include:

- Antithetic variates—here, selecting the uniform random variate, $(1 - U)$, in the trial immediately following the selection of the uniform random variate U. This assures that each fractile sample is balanced by a complementary fractile sample.
- Latin squares or permutation sampling—here, assuring over many trials that a specific data item is selected (more or less) in accordance with its probability of occurrence in the original dataset;

- Fractile-based sampling—here, sampling a distribution at equally based intervals (e.g., the 1st centile, 3rd centile,..., 97th centile, and 99th centile);
- Control functions—parametric distributions used as a first-order approximation of the actual distribution, and the residuals of the two are bootstrapped to obtain a correction to the approximation. Importance sampling uses such functions in a weighting method.
- Stratification techniques—use of mutually exclusive groups to characterize the network loss data.

After many efforts, a post-sampling post-processor was developed for REDARS that employs a combination of bootstrap sampling, Latin squares or permutation sampling, fractile-based sampling, and control functions (with an emphasis on residuals, although importance sampling was also found to be very effective—but with little marginal gain for the total distribution). For the annual loss distribution, these results are illustrated in Table 1. In particular, Table 1 shows how the efficacy of variance reduction techniques depends heavily on lambda (λ). For the Memphis roadway system, with a probability of some network loss calculated as 0.01568, these variance reduction techniques reduced the number of simulations needed by a multiplicative factor of slightly above 3.2. This means, for instance, that the user of REDARS will need only 16,000 years of trials (240 scenarios for the distribution in Figure 1) to achieve the same nominal confidence limits as achieved through 50,000 years of trials in previous efforts.

Table 1 further suggests that as lambda approaches unity, variance reduction techniques become extremely efficacious (the more extreme the reduction—the more the need for careful scrutiny). As well, these investigations have shown that variance reduction techniques are vastly more effective for the conditional loss distribution. This result is explained in detail in a working paper in progress. The conclusion derived is that variance reduction techniques appear to be designed principally for distributions which have few or no zeroes. But, earthquake losses in regions of moderate seismicity but high catastrophic loss potential have many years in which no losses occur.

We have also examined the determination of the 500-yr return period loss. The so-called 500-yr loss, l_{500}, is defined as the loss that has one chance in 500 of being exceeded annually. The earthquake loss, l, that has a certain probability of not being exceeded in time, T, can be found from the equation

$$F_{\max,T}(l) = e^{-\phi T(1-F(l))}$$

where ϕT is the expected number of losses in time, T, and $F(l)$ is the cumulative distribution function for losses given that a loss has occurred. Thus, the expression in the exponent is the expected number of exceedances of loss, l, in time T. The expression e^{-x} is the Poisson probability of getting none, when x are expected. Hence the equation gives the probability that loss, l, is not exceeded in time, T.

The value of l corresponding to some probability in time T can be found from the appropriate fractile of the probability $F(l)$. For a single sample of 50,000 years, the 500-year loss should be the value that has been exceeded 100 times, and is the

$(50,000-100)/50,000 = 0.998$

fractile of the sample loss distribution. That value from the REDARS simulation is 305×10^6. (This calculation corresponds to the analogous calculation in probabilistic seismic hazard analysis for mean ground motion having 1/500 of being exceeded annually.)

An interesting question is to what extent do we understand this to be a *mean* for simulations in which the number of annual samples is much less than 50,000. Also what does it mean for such a sample to have a *median* 500-yr value, and 5 and 95 percentiles? We have just begun exploring these issues.

One interesting case is to take 500 annual samples. The number of times l_{500} is expected to be exceeded in 500 years is 1. This means that in a series of simulations of losses in 500 years, there is a probability of $e^{-1} = 0.37$ that this level of loss will not be exceeded.

For the Memphis model, 7.68 loss events are expected in 500 years. Figure 3 shows two realizations of samples of losses in nominal 500-yr histories. In one case we assumed exactly 8 losses are experienced in each sample. (This is the expected number in 520 years.) In the other case we assumed that the number of losses experienced had a Poisson distribution with expected value 7.68. In both cases, sets are derived from successive events in the REDARS simulation. From each set, the *largest* loss is extracted. In both cases, the figures show that the value of the largest observed loss in the sample that has a 37 percent chance of not being exceeded is 316×10^6, a number only slightly larger than the value 305×10^6 determined from the 50,000-year series. It is interesting to note that the distributions in the two cases differ very little above the 20 percentile.

It should be clear that fractile estimates for the 500-yr return period loss will have to depend on the size of the sample used for estimation and the exceedance probability of the fractile as well as the shape of the distribution function itself in its upper fractiles. We expect that pre-sampling will improve the resolution of the latter, and other bootstrapping methods will permit more accurate determinations of the former.

References

Ang, Alfredo H.-S. and Wilson H. Tang, 1975, *Probability Concepts in Engineering Planning and Design, Volume I, Basic Principles*, New York: John Wiley & Sons.

Davison, A.C. and D. V. Hinkley, 1998, *Bootstrap Methods and their Application*, Cambridge, U.K.: Cambridge University Press.

Daykin, C. D., T. Pentikainen, and M. Pesonen, 1994, *Practical Risk Theory for Actuaries*, London: Chapman & Hall.

Efron, Bradley and Robert J. Tibshirani, 1993, *An Introduction to the Bootstrap*, New York: Chapman & Hall.

Hastings, N. A. J. And J. B. Peacock, 1974, *Statistical Distributions*, London: Butterworth & Co (Publishers) Ltd.

Hogg, R. V. And S. A. Klugman, 1984, *Loss Distributions*, New York: John Wiley & Sons.

Law, A. M. And W. D. Kelton, 1991, *Simulation Modeling and Analysis*, New York: McGraw-Hill.

Lemaire, J., C. Taylor, and C. Tillman, 1993, "Models for Earthquake Insurance and Reinsurance Evaluations," in *Proceedings of the Second International Symposium on Uncertainty Modeling and Analysis*, Los Alamitos, CA: IEEE Computer Society Press, April.

Liu, Jun S., 2001, *Monte Carlo Strategies in Scientific Computing*, New York: Springer-Verlag.

Panjer, Harry H. And Gordon E. Willmot, 1992, *Insurance Risk Models*, Schaumburg, Il: Society of Actuaries.

Taylor, Craig E., Stuart D. Werner, and Steve Jakubowski, 2001, "Walkthrough Method for Catastrophe Decision Making," *Natural Hazards Review*, November, pp. 193-202.

Taylor, C. E., J. Lemaire, and C. Tillman, 1994, "A New Earthquake Insurance and Reinsurance Index: Uncertainties and Future Developments," *Uncertainty Modeling and Analysis: Theory and Applications*, B. M. Ayyub and M. M. Gupta, eds., Elsevier Science BV., pp. 497-514.

Werner, Stuart D., Craig E. Taylor, James E. Moore III, Jon S. Walton, and Sungbin Cho, 2000, *A Risk-Based Methodology for Assessing the Seismic Performance of Highway Systems*, Buffalo, NY: Multidisciplinary Center for Earthquake Engineering Research, Technical Report MCEER-00-0014, FHWA Contract Number DTFH61-92-C-00016.

Table 1

Multiplicative Factor for Reduced Simulations (Reduction In Variance for the Estimate of the Unconditional Mean Loss) as a Function of Lambda

Lambda (n/N)	Multiplicative Factor for Reduced Simulations
0.000154	2.93
0.000768	3.17
0.001536	3.20
0.003072	3.22
0.006144	3.23
0.009216	3.24
0.012288	3.25
0.01568 (test case)	3.25
0.018432	3.26
0.019968	3.27
0.024576	3.28
0.029184	3.29
0.04608	3.33
0.06144	3.37
0.07680	3.41
0.10752	3.49
0.13824	3.58
0.15360	3.63
0.30720	4.20
0.46080	5.12
0.61440	6.76
0.76800	10.58
0.84480	15.32
0.92160	29.35
0.96768	69.77

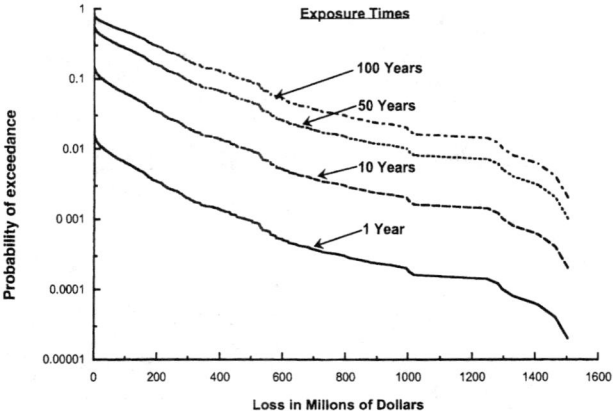

Figure 1. Economic Losses due to Increases in Commute Times caused by Earthquake Damage to Highway-Roadway System in Shelby County, Tennessee

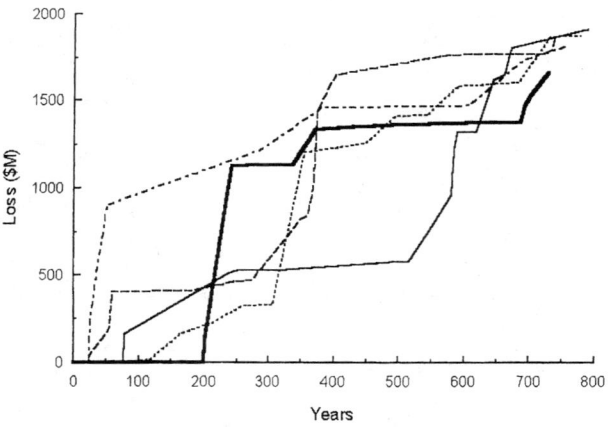

Figure 2. Random Walk Plots showing Variations of Cumulative Economic Losses over Different 1000-Year Time Segments

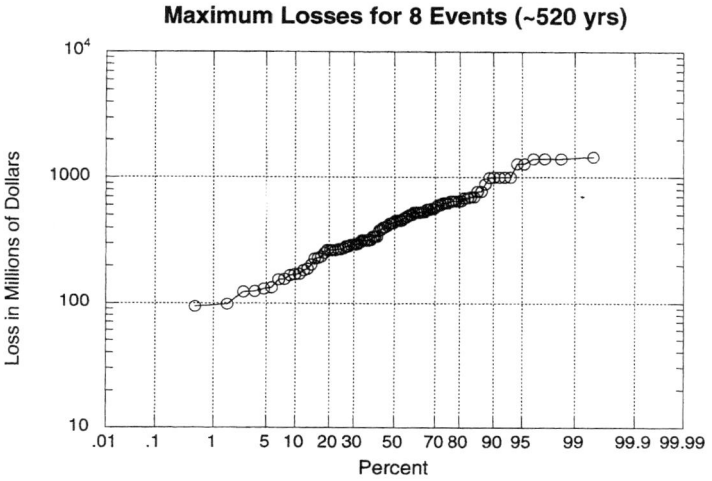

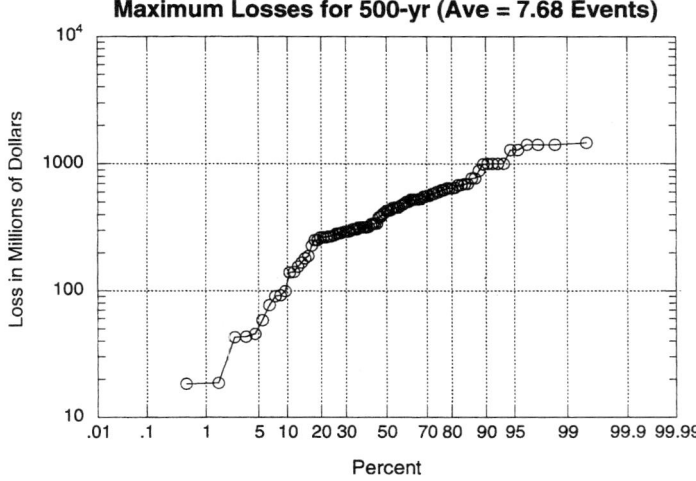

Figure 3. (Upper) Distribution of Largest Losses for Exactly 8 Random Loss Events. (Lower) Distribution of Largest Losses for Exactly 500 years. (Horizontal Axis is Normal Probability Scale)

Modeling Transportation Network Flows as a Simultaneous Function of Travel Demand, Earthquake Damage, and Network Level Service.

Sungbin Cho[1], Yue Yue Fan[2], and James E. Moore, II, A.M. ASCE[3]

Abstract

This paper summarizes the results of a transportation analysis incorporating a novel approach to modeling variable transportation demand. This approach links the demand for destinations to the level of service available on the transportation network. The paper includes applications to a set of predicted, post-earthquake transportation network configurations for the city of Memphis, Tennessee, and to network damage resulting from a large scenario earthquake in the San Francisco Bay Area.

Introduction

In addition to replacement and repair cost of damage to transportation structures, large earthquakes produce an increase in time delay resulting from these network components' loss of function. Earthquake losses due to travel time increases may be evaluated by examining the difference between baseline conditions and network performance following a scenario earthquake.

Theoretical Background of Variable Demand Model.

Competitive Flows in Transportation Networks. The user equilibrium network flow model is one of the most useful transportation analysis models. In a standard, fixed demand model, total demand for travel between an origin and destination (OD) pair does not vary with travel cost between the origin and destination. Route selection is a function of route delay, but the propensity to travel is not.

More realistically, trip rates are influenced by the level of service available on the transportation network. Following a large earthquake, the congestion level would increase because of reductions in network capacity. The increased congestion level would then induce a reduction in travel demand. Figure 1 shows this relationship. Before the earthquake, the transportation system supplies service according to the function S_1, and flows d_1 use the system at cost of p_1. After the earthquake damages the network, the level of service drops to S_2, and demand for travel decreases in

---

[1] PhD Candidate, Urban Planning, School of Policy, Planning, and Development, University of Southern California; and ImageCat, Inc., 400 Oceangate, Suite. 1050, Long Beach, CA 90802; phone 562-628-1675; sungbinc@rcf.usc.edu, sc@imagecatinc.com.
[2] Research Associate, Department of Civil and Environmental Engineering, School of Engineering, University of Southern California, KAP 210 MC-2531, Los Angeles, CA, 90089-2531; phone 213-740-0575; yueyuefa@usc.edu.
[3] Professor of Civil Engineering; Public Policy and Management; and Industrial and Systems Engineering, University of Southern California, KAP 210 MC-2531, Los Angeles, CA, 90089-2531; phone 213-740-0595; jmoore@usc.edu.

response to this drop in level of service. Flows d_2, occur at (delay) cost of p_2'. This effect can be accounted by an appropriate function relating trip rates to travel times.

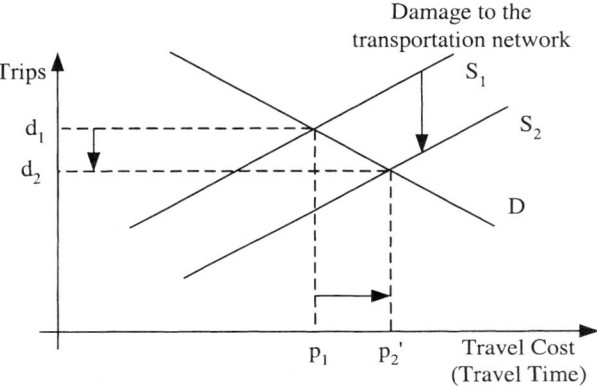

Figure 1. Transportation demand, supply, and user equilibrium delay costs in a network with reduced capacity and variable demand for travel: Movement along the travel demand curve.

User equilibrium flows subject to variable demand for travel occur when the travel times on all used paths between any pair of origin-destination zones are equal, and are also equal to or less than the travel times on any unused paths connecting this pair. In addition, the trip requirements for all of the OD pairs in the network should conform to the travel demand function. The problem addressed in this paper is that of finding a set of link volumes, link travel times, and OD requirements that simultaneously satisfy the conditions for user equilibrium on the transportation network. The mathematical form of the static, path flow version of this problem is (Beckmann, McGuire, and Winston, 1956),

$$\max \ z(\mathbf{x},\mathbf{q}) = \sum_a \int_0^{x_a} t_a(w) \ dw - \sum_{rs} \int_0^{q_{rs}} D_{rs}^{-1}(w) \ dw \tag{1}$$

subject to

$$\sum_k f_k^{rs} = q_{rs} \qquad \forall \, r,s \tag{2}$$

$$f_k^{rs} \geq 0 \qquad \forall \, k, r, s \tag{3}$$

$$q_{rs} \geq 0 \qquad \forall \, r,s \tag{4}$$

$$q_{rs} = D_{rs}(u_{rs}) \quad \forall r,s \tag{5}$$

$$x_a = \sum_{rs}\sum_{k} f_k^{rs} \cdot \delta_{a,k}^{rs} \quad \forall a \tag{6}$$

where

- t_a = link performance function of link a,
- D = travel demand function,
- D^{-1} = inverse of the travel demand function,
- f_k^{rs} = flow on path k connecting OD pair r-s,
- q_{rs} = trip rate between OD pair r-s,
- u_{rs} = travel time between OD pair r-s,
- x_a = flow on link a, and
- $\delta_{a,k}^{rs}$ = 1 if link a is on path k between OD pair r-s, 0 otherwise.

The first term on right-hand side of Equation (1) ensures that link volumes and travel times meet user equilibrium conditions. The second term adjusts trip rates between OD pairs so that the travel demand loaded on to network is an appropriate function of travel time.

Solution Procedure. We developed a numerical algorithm based on the secant method to solve this highly endogenous nonlinear programming problem. The algorithm is summarized as follows.

Step 0: Initialization.

Find an initial feasible flow pattern $\{x_a^n\}$, $\{q_{rs}^n\}$. Set iteration counter $n:=1$.

Step 1: Update link travel times as a function of the feasible set of link flows update and the equilibrium travel times as a function of trip making requirements.

Set $t_a^n = t_a(x_a^n) \,\forall a$; compute $D_{rs}^{-1}(q_{rs}^n)\,\forall r,s$. \tag{7}

Step 2: Find auxiliary link volumes and trip rates.

Compute the shortest path, m, between each O-D pair r-s based on link travel time $\{t_a^n\}$, $c_m^{rs^n} = \min_{\forall k}\{c_k^{rs^n}(t_a^n)\}$. \tag{8}

Find auxiliary trip rates.

If $c_m^{rs^n} < D_{rs}^{-1}(q_{rs}^n)$, set $g_m^{rs^n} = \overline{q_{rs}}$ where m is shortest path, and $\overline{q_{rs}}$ is the upper bound on the trip rate.

If $c_m^{rs^n} > D_{rs}^{-1}(q_{rs}^n)$, set $g_k^{rs^n} = 0 \,\forall k$. \tag{9}

If $\left|c_m^{rs^n} - D_{rs}^{-1}(q_{rs}^n)\right| < \varepsilon$, set $g_m^{rs^n} = g_m^{rs^{n-1}}$. (10)

Compute auxiliary link volumes $y_a^n = \sum_{rs}\sum_{k} g_k^{rs^n} \cdot \delta_{a,k}^{rs} \quad \forall a$. (11)

Compute auxiliary trip rates $v_{rs}^n = \sum_{k} g_k^{rs^n} \quad \forall r,s$. (12)

Step 3: Find the best direction of search.

Solve following system for α.

$$\min z(\alpha) \sum_{a} \int_0^{x_a^n + \alpha(y_a^n - x_a^n)} t_a(w)dw - \sum_{rs} \int_0^{q_{rs}^n + \alpha(v_{rs}^n - q_{rs}^n)} D_{rs}^{-1}(w)dw \quad (13)$$

subject to $\quad 0 \leq \alpha \leq 1$

Step 4: Update the feasible set of link flows.

$$x_a^{n+1} = x_a^n + \alpha_n(y_a^n - x_a^n) \quad (14)$$

$$q_{rs}^{n+1} = q_{rs}^n + \alpha_n(v_{rs}^n - q_{rs}^n) \quad (15)$$

Step 5: Test for convergence.

If the following inequality holds for sufficiently small κ, stop. Otherwise, set iteration counter $n:=n+1$ and go to step 1.

$$\sum_{rs} \frac{\left|D_{rs}^{-1}(q_{rs}^n) - u_{rs}^n\right|}{u_{rs}^n} + \sum_{rs} \frac{\left|u_{rs}^n - u_{rs}^{n-1}\right|}{u_{rs}^n} \leq \kappa \quad (16)$$

This algorithm is an extension of the standard Frank-Wolfe algorithm used to solve the fixed-demand version of the user equilibrium problem. The key extension is the auxiliary trip rates found in Step 2 (Sheffi 1985).

Calibrating the Travel Demand Function. The travel demand function might be expected to be strictly decreasing with respect to the travel time between zone-pairs. In reality, however, the distribution of trip rates with respect to interzonal travel time is peaked at a positive value. For modeling purposes this nonmonotonic relationship must be estimated with a best fitting monotone form.

We specified and applied the following demand function. The function is a version of the gravity model (Wilson, 1970, Putnam, 1983), which explains changes in the way spatial activities interact,

$$q_{rs} = \frac{O_r \cdot D_s \cdot A_r \cdot B_s}{1 + \exp(\alpha + \beta \cdot u_{rs})} \quad (17)$$

where

q_{rs} = trip rate between OD pair r-s,
u_{rs} = travel time between OD pair r-s,
O_r = trip production from origin zone r,
D_s = trip attraction to destination zone s,
A_r = origin zone r specific coefficient to be estimated,
B_s = destination zone s specific coefficient to be estimated, and
α, β = model parameters to be estimated.

The model coefficients, and parameters are estimated iteratively. The user equilibrium model estimates a matrix of zone-to-zone travel times $[u_{rs}]$, based on a matrix $[q_{rs}]$ describing OD requirements. An econometric model estimates the coefficients α, β from OD requirements and zone-to-zone travel times. Given travel times and estimated parameters, the gravity model is applied to estimate zone specific coefficients (A_r, B_s). Once initial estimates have been obtained for all unknowns, an updated set of OD requirements $[q_{rs}]$ is estimated. These steps are repeated until the estimated values α, β stablize. This procedure converges quickly.

A Medium-Scale Example: Applying the Variable Demand Model to the Memphis, Tennessee Network

Mississippi River detours. The representation of Memphis network was specified to introduce the option of a wide detour for traffic flowing into the region from West of Mississippi River. See Figure 2. If the Interstate Route 40 bridges bridges across the Mississippi river are eliminated due to damage, travel demand detours and enters the region at several locations in the south of Memphis.

Three examples. A walk-through, Seismic Risk Analysis (SRA) of Memphis has been conducted by Werner, et al., (2000, 2003) for the Multidisciplinary Center for Earthquake Engineering Research (MCEER) and the Federal Highway Administration. The analysis applied the Frankel, et al., (1996) model for the Central United States to generate 2,321 earthquakes with moment magnitudes ranging from M 5.0 to M 8.0. From these, we selected three representative cases of network damage. These are cases CC40962, JJ9295, and JJ40962. For each case, network damage is assumed to be repaired incrementally; and in each of these three cases network configurations are reported for three points in time, seven days, 60 days, and 150 days following the earthquake.

These inputs were use to model network flows under assumptions of both fixed and variable travel demand. As shown in Table 1, the loss and restoration of network capacity has a substantial impact on system-wide travel times. In this exercise, the increases in travel time ranges from 39% to 145% depending on the realized level of damage and the amount of capacity that has been restored following the earthquake.

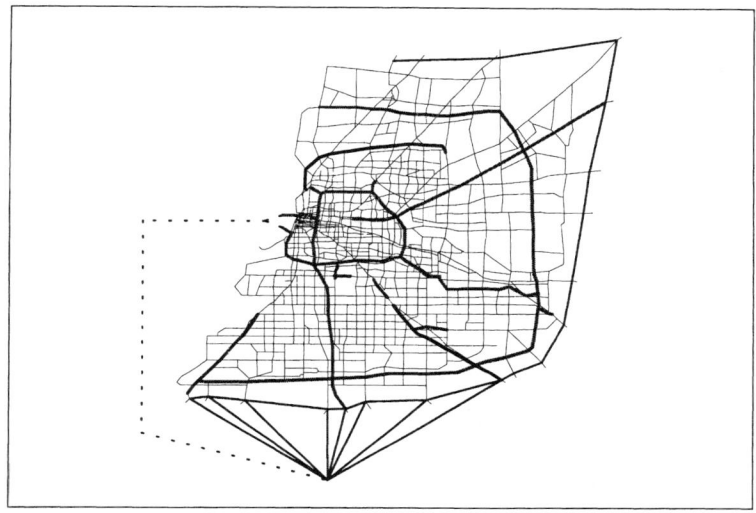

Figure 2. Memphis transportation network: Mississippi River detour scheme.

Table 1. Travel time estimates for the Memphis network: Fixed travel demand.

Senario		Total System Delay (PCU[a] • Hours)	Average Travel Time (minutes)	System Travel Demand (PCU)
Baseline		15,134,460.43	18.748	4,978,307
CC40962	7 days	42,414,441 (180.3%)	45.981 (+145.3%)	4,978,307
	60 days	29,862,091 (+97.3%)	32.210 (+71.8%)	4,978,307
	150 days	25,269,027 (+67.0%)	26.900 (+43.5%)	4,978,307
JJ9295	7 days	26,604,318 (+75.8%)	28.421 (+51.6%)	4,978,307
	60 days	25,553,932 (+68.8%)	27.150 (+44.8%)	4,978,307
	150 days	24,583,169 (+62.4%)	26.042 (+38.9%)	4,978,307
JJ40962	7 days	31,828,428 (+110.3%)	34.367 (+83.3%)	4,978,307
	60 days	28,200,775 (+86.3%)	30.305 (+61.6%)	4,978,307
	150 days	25,179,945 (+66.4%)	26.804 (+43.0%)	4,978,307

Note: a. Passenger Car Units. Freight and passenger flows are measured in terms of PCUs.

Table 2, in contrast, is based on an assumption of variable demand, and accounts for the way changes in network capacity induces changes in both travel time and trip rates. These effects both influence total system delay, but in opposing directions. Average travel times increase as network capacity is reduced. At the same time, the model predicts the travel demand (trip rates) will be reduced. These net changes in system delay combine reductions in demand and increases in link travel times, and may be either positive or negative. In this exercise, total system delay is reduced, because travel is reduced.

Table 2. Travel time estimates for the Memphis network: Variable travel demand.

Senario		System Cost (PCU^a • Hours)	Average Travel time (minutes)	System Travel Demand (PCU)
Baseline		15,134,460	18.748	4,978,307
CC40962	7 days	13,146,626	21.555 (+15.0%)	3,900,357 (-21.7%)
	60 days	14,154,427	19.911 (+6.2%)	4,347,444 (-12.7%)
	150 days	14,814,855	19.146 (+2.1%)	4,761,775 (-4.3%)
JJ9092	7 days	14,559,738	19.499 (+4.0%)	4,533,609 (-8.9%)
	60 days	14,717,086	19.185 (+2.3%)	4,711,870 (-5.3%)
	150 days^b	14,718,516	18.810 (+0.3%)	4,804,661 (-3.5%)
JJ40962	7 days	13,146,626	20.145 (+7.5%)	4,302,625 (-13.6%)
	60 days	14,154,427	19.767 (+5.4%)	4,355,090 (-12.5%)
	150 days	14,814,855	19.089 (+1.8%)	4,774,186 (-4.1%)

Note: a. Passenger Car Units. Freight and passenger flows are measured in terms of PCUs.
b. Results required 100 iterations of the solution algorithm. All other examples required 25.

The variable demand model is intended to adjust OD requirements in response to changes in the level of service available on the network. These adjustments do not necessarily mean that the longest trips are eliminated first. The largest share of the trips eliminated by reductions in network level of service are trips ranging from 10 to 30 minutes in length. The relative trip rate reductions tend to be greatest for those zone pairs showing the greatest relative increase in travel times. Interzonal travel times range from 85% of baseline times to 200%. Increasing travel time by 20% results in travel demand that is about 80% of the baseline value.

Trips Forgone. Time has value. In most network flow models, cost and delay are treated as interchangeable quantities. This standard perspective is deficient in the context of this application. While an earthquake may well result in lower total system delay because few trips are taking place, the trips that have been forgone also have value, and their absence is not a cost accounted for by network delay costs.

Characterizing total transportation system cost in a variable demand context requires accounting for network congestion costs and the opportunity costs associated with trips forgone. Total system delay is the sum across all links of the total travel time accumulating on each of link. In addition to network delay, demand eliminated from the system due to increased congestion implies additional costs. This opportunity cost associated with these trips forgone is more difficult to calculate. The value of these trips is at least as great as the baseline delay associated with making them, otherwise these baseline trips would not occur. Consequently, the baseline delays associated with trips forgone after the earthquake provide a useful lower bound on value of trips forgone.

The lower bounds on the total cost associated trips forgone are of the same order of magnitude as total congestion costs. When the opportunity cost of trips forgone is accounted for and combined with delay costs, the total daily transportation cost accumulating as a result of the earthquake ranges from $311 million to $358 million across the 4 cases. These costs are based on a unit value of time estimate that incorporates several value of assumptions for travelers and freight.

A Large-Scale Example: Applying the Variable Demand Model to the San Franciso Bay Area Network

The variable demand results associated with the Memphis application are encouraging, but Memphis is a relatively small city. Extending standard network equilibrium flow models to account variable demand in a much larger network presents a further challenge. The Metropolitan Transportation Commission (MTC) highway network model for the San Francisco Bay Area consists of 1,120 traffic analysis zones and 26,904 network links. The MTC also provides a 1998 matrix of OD requirements for this network model. The California Department of Transportation (Caltrans) District 4 provides a bridge inventory, which we combined with the MTC network as part of the Pacific Earthquake Engineering Research Center's (PEER) Highway Demonstration Project (Kiremidjian, et al. 2003).

The PEER Highway Demonstration Project examines four scenario events, including a moment magnitude M 7.5 earthquakes on the Hayward fault that is used here. The vulnerability of bridges to ground shaking, liquefaction and landslides is evaluated considering each hazard separately and in combination. Moderately and severely damaged bridges are assumed closed immediately after the earthquake. These results make it possible to calibrate and test the variable demand model on the Bay Area network, and to compare these results to network flow and delay estimates premised on fixed demand.

The cumulative link counts in Figure 3 summarize and compare the results for the fixed and variable demand models across all of the links in the network. Results providing volume/capacity (v/c) ratios above 1.0 indicate that network has been swamped by infeasibly large, unrealistic travel demands. In the fixed demand results, post-earthquake v/c ratios routinely exceed 1.0, in the most extreme examples by an order of magnitude. The network capacity losses associated with the earthquake are so extreme that any conventional network flow model is left with no substantive predictive power: The conditions being modeled are too extreme for the model to represent meaningfully. In contrast, the results for the variable demand formulation are much more realistic with respect to both traveler behavior and the technical performance of the network. Predicted volume/capacity ratios slightly exceed 1.0 in a few cases. Average ratios within link categories never do. As a result of the very substantial loss in network capacity associated with this event, congestion and total network delay costs increase in the variable demand case despite substantial reductions in travel demand. See Figure 4 for relative changes in trip origins across Bay Area traffic analysis zones.

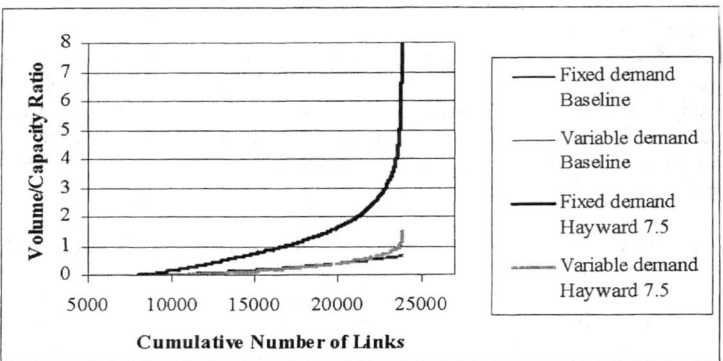

Figure 3: Comparison of v/c ratios predicted by the fixed demand and variable demand models for transportation flows associated with the Bay Area baseline and Hayward M 7.5 earthquake scenario.

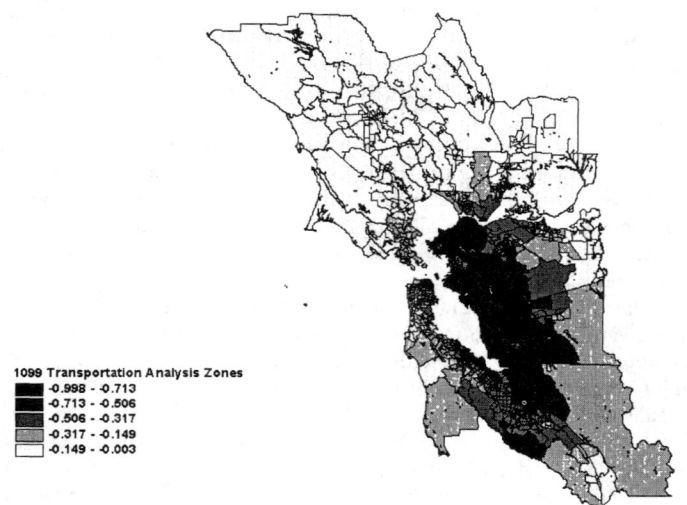

Figure 4: Relative differences across travel analysis zones in trip productions predicted by the variable demand model for the Hayward M 7.0 earthquake scenario.

Conclusions

Ideally, post-earthquake transportation flows would be model based on estimated damage to both the transportation network and the urban activity system (Cho, et al., 2000). This is computationally feasible, but challenging. The alternative approach presented here is a convergent, computationally efficient approximation that appears to offer much in terms of economic meaning, and relevant results.

Transportation is an induced demand driven by demand for other goods and services. This research extends and integrates standard network modeling procedures to predict changes in travel demand in addition to route choice. The possibility of substantial, sudden losses of transportation network supply makes it important to look beyond network delays and account for the economic value of trips forgone due to decreases in network level of service. These costs can be very significant. Improvements are certainly possible with respect to estimating these opportunity costs, and should be investigated.

References

Beckmann, M., McGuire, C. B., and Winsten, C. B. (1956) *Studies in the economics of transportation*, Yale University Press, New Haven.

Cho, S.B., Gordon, P., Moore II, J.E., Richrardson, H.W., Shinozuka, M., and Chang, S.E. (2001) "Integrating transportation network and regional economic models to estimate the costs of a large urban earthquake," *J. of Regional Sci.*, 41(1), 39-65.

Frankel, A., Mueller, C., Barnhard, T., Perkins, D., Lyendecker, E.V., Dickman, N., Hanson, S., and Hopper, M. (1996) *National seismic hazard maps, documentation June 1996*, United States Geological Survey, Open-File Report 96-532, Denver.

Kiremidjian, A., Moore II, J. E., Basoz, N., Williams, M., Fan, Y. Y., Yazlali, O., Roblee, C., and Wang, Z. (2003) "The PEER Highway Demonstration Project," paper presented at the ASCE Technical Council on Lifeline Earthquake Engineering's Sixth U.S. Conference, August 10-13, 2003, Long Beach.

Putnam, S. H. (1983) *Integrated urban models: Policy analysis of transportation and land use*, Pion, Ltd., London.

Sheffi, Y. (1985) *Urban transportation networks-equilibrium analysis with mathematical programming methods*, Prentice-Hall, Englewood Cliffs.

Werner, S. D. (2003) "Current developments and future directions for seismic risk analysis of highway systems," paper presented at the ASCE Technical Council on Lifeline Earthquake Engineering's (TCLEE) Sixth U.S. Conference, August 10-13, 2003, Long Beach.

Werner, S.D., Taylor, C.E., Moore, J.E. II, Walton J.S., and Cho, S.B. (2000) *A risk-based methodology for assessing the seismic performance of highway systems*, Report No. MCEER-00-0014, Multidisciplinary Center for Earthquake Engineering Research, Buffalo.

Wilson, A.G. (1970) *Entropy in urban and regional modeling*, Pion Ltd., London.

A Validation Study of the REDARS Earthquake Loss Estimation Software Program

Sungbin Cho[1], Charles K. Huyck[2], Shubharoop Ghosh[3] and Ronald T. Eguchi[4]

Abstract

This paper presents interim results on a study to validate an earthquake loss estimation methodology for highway systems. Over the past five years, the Multidisciplinary Center for Earthquake Engineering Research has funded the development of a methodology and software program to quantify the effects of earthquakes on highway bridges and systems. This program – entitled REDARS for *Risks from Earthquake Damage to Roadway Systems* – estimates damage to highway bridges, changes in traffic patterns, and economic losses associated with the repair of highway elements and traffic congestion. The current paper describes steps and analyses employed to validate the REDARS methodology. Data and information from the recent 1994 Northridge Earthquake are used in this validation process.

Key Words: Earthquake loss estimation, highway systems, bridges, traffic congestion, economic losses, REDARS

Introduction

For the last six years, the Multidisciplinary Center for Earthquake Engineering Research (MCEER) has been sponsoring the development of an earthquake loss estimation program for transportation systems. This program, entitled REDARS, analyzes the risk and loss to transportation networks as a result of large and moderate earthquakes. Losses are described as a combination of repair costs and costs associated with delayed or disrupted trips. The lead researcher for this development has been S.D. Werner of Seismic Systems and Engineering Consultants. As part of the overall validation process, a special project team established by MCEER has been 1) reviewing the technical details of the loss estimation methodology, 2) testing the user-friendliness of the software program, and 3) validating the methodology using data from the 1994 Northridge

---

[1]Senior Transportation Planner, ImageCat, Inc., 400 Oceangate, Suite 1050, Long Beach, CA 90802, sc@imagecatinc.com
[2]Senior Vice President, ImageCat, Inc., ckh@imagecatinc.com
[3]Transportation Planner, ImageCat, Inc., sg@imagecatinc.com
[4]President and CEO, ImageCat, Inc., rte@imagecatinc.com

Earthquake. This validation study will be completed in 2003, after which modifications to improve both the usability of the REDARS program and the reliability of the estimates will be made.

Some preliminary results indicate that the reliability of the traffic-modeling module is surprisingly good. When compared against actual Northridge Earthquake data, the model provides good agreement with regional transportation flows after the earthquake. During the Northridge Earthquake, significant damage was observed at over a dozen sites in southern California. Three bridge sites (I5/14 interchange; the Santa Monica Freeway, and I118), in particular, were of interest to the validation team because they suffered complete collapse of highway spans and disrupted local traffic for several weeks. In this study, we compared the performance of traffic systems around these three sites and identified several limitations in the current methodology. These include 1) unconstrained flow capacities on arterial streets, and 2) unchanging origin-destination requirements. In a final report to MCEER, the project team will identify possible solutions to overcome some of these limitations.

Methodology for Validating Redars

The following flowcharts illustrate our approach for validating the REDARS program. We are performing our validation work at two levels: component and system. The purpose of the component evaluations is to identify the extent to which each module deviates from actual earthquake experience. The figure below shows the procedure for comparing REDARS output (at the component level) to actual earthquake data.

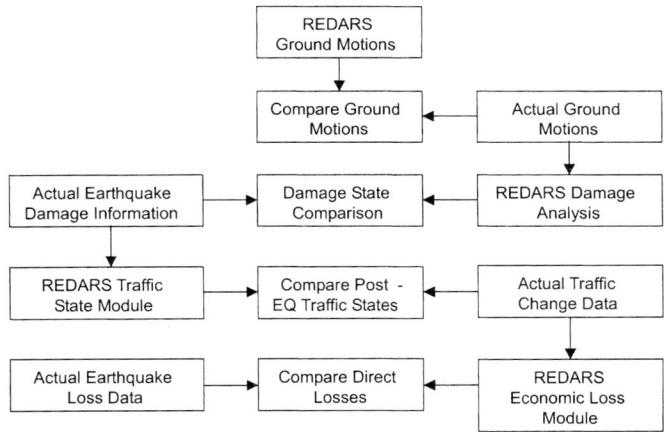

Figure 1. Approach for Assessing the Accuracy of Individual Modules

As our initial work plan stated, we are using data from the 1994 Northridge earthquake to perform these validations. If it is determined that improvements are necessary to better match actual earthquake experience, the data from these component evaluations will help decide where to focus these efforts. Currently, we are validating the following modules: damage state calculations; traffic state analysis; and direct economic loss. The economic loss evaluation is focusing on two areas: repair costs of bridges and increase travel costs due to delays.

As part of the validation effort, we are also performing a number of additional simulations in order to test the sensitivity of the final results to changes in the individual modules. By conducting these sensitivity studies, we will be able to recommend to the methodology developer (Werner) which modules should be reviewed for possible improvements. Figure 2 outlines some basic steps that we are following in developing this sensitivity matrix.

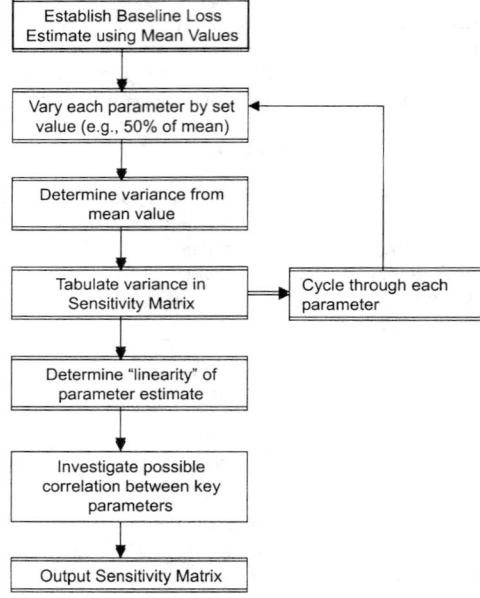

Figure 2. Approach for Developing Parameter Sensitivity Matrix

Validation Of Traffic Congestion Model

REDARS was initially developed to analyze earthquake damage to large roadway transportation networks. The transportation model that is utilized in REDARS is based on a User Equilibrium (UE) methodology. Currently, the different modes of system disruption include:

- Capacity reduction
- Reduced or disrupted demand
- Changes in route choice
- Lack of route information

It is critical that the performance of the UE model be assessed relative to actual transportation data. In this study, we are using traffic data before and after the 1994 Northridge earthquake as a "ground truth" dataset. The following discussion highlights some of the major findings from this study in the area of traffic modeling.

Figure 3 shows the locations of severely damaged bridge sites in the Northridge earthquake (shown in circles), as well as key points within the highway transportation system where traffic flows were monitored (identified by letters A through G). In the cases where traffic was monitored, both before-earthquake and after-earthquake traffic counts were available.

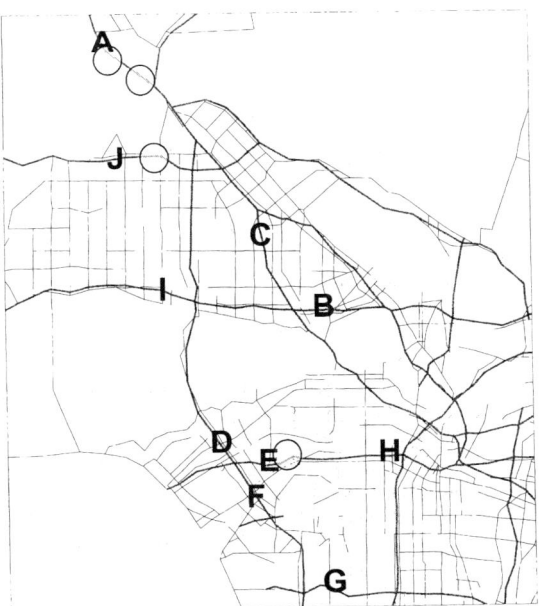

Figure 3. Locations of Damaged Bridges after 1994 Northridge Earthquake (shown in circles) and Traffic Monitoring Points Before and After the Earthquake (shown in letters)

Figure 4 shows a comparison between estimated versus observed traffic volumes for the key traffic points in Figure 3. In the case of this comparison, the ratio of after-to-before traffic volumes is being used to measure the efficacy of the traffic model. The 45 degree line shown in the figure represents the "perfect" match line, that is, if the points all fell on this line, there would be a one-to-one correspondence of the before and after traffic data. However, the figure shows that this is not the case, and in fact, there is a tendency for the relationship between the estimated and observed data to be quite off for low ratios of after/before traffic volumes. In essence, this comparison is telling us that the traffic volumes after the earthquake – as calculated by the REDARS model – are substantially higher than what was actually observed after the earthquake. As we shall see, one possible explanation for this is that the traffic demand model after the event may not be representing the actual situation properly.

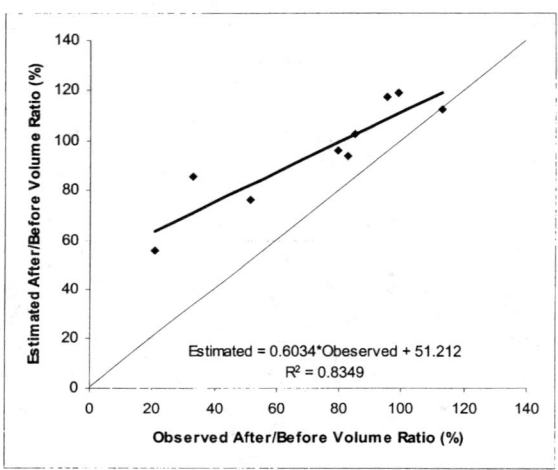

Figure 4. Comparison of Estimated (REDARS) versus Observed Traffic Volume Ratios (After/Before) For Key Transportation Points in Network

Whereas the example above focused on regional traffic flows, Figure 5 concentrates on one particular highway segment, i.e., the Santa Monica Freeway, I10. The reader may recall that it was on this segment that several highway spans completely collapsed after the earthquake. This particular freeway is also considered the busiest freeway in the U.S.

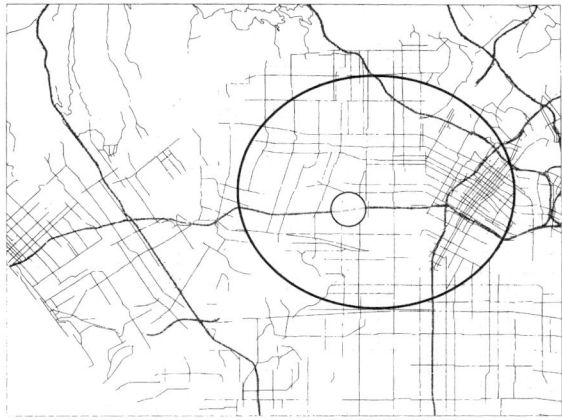

Figure 5. Detailed Transportation Network for Central Los Angeles Area

Figure 6 shows a comparison between after/before traffic volumes from REDARS and the actual Caltrans database. The observed curve shows that only minor changes in traffic volume occurred after the earthquake; whereas, REDARS estimates large changes based on current models. Again, more traffic volume is being estimated by REDARS in the after-earthquake case for local roadway network with high redundancy.

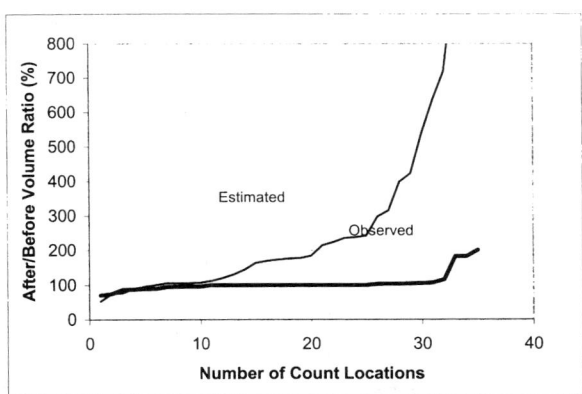

Figure 6. Comparison of Traffic Volumes from REDARS and Caltrans Database

There is also a significant difference in the detour routes chosen by REDARS and what actually occurred after the Northridge Earthquake. Figure 7 shows a diagram that maps out the estimated main detour route and the observed detour route around the damaged Santa Monica freeway structure.

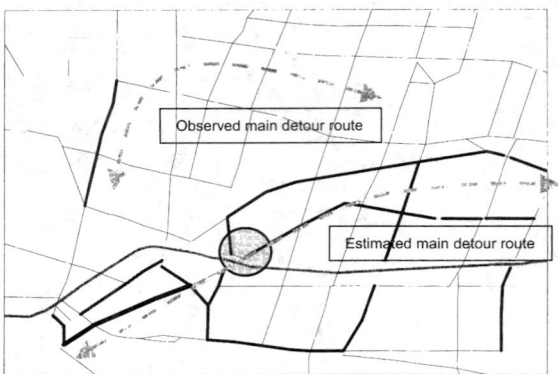

Figure 7. Comparison of Estimated vs. Observed Detour Routes Around Damaged Santa Monica Freeway.

As Figure 7 points out, there were some significant differences between the model predictions and the actual data from Northridge. The actual data showed that traffic volumes along Robertson and Pico had noticeable increases of traffic volume after the earthquake. In reality, drivers seemed to take longer detours in order to avoid the damaged area. The model, on the other hand, allocates demand to immediate detours – such as Washington Blvd. and Venice Blvd. This is because these routes – based on the models used by REDARS – have a "comparative advantage" against other routes in terms of travel time. Since the actual data used in this comparison was for the day after the earthquake, it is assumed that drivers did not have enough information regarding route conditions to make the best selection of alternative routes. The UE model, however, makes the assumption that drivers have perfect information about traffic conditions, thereby creating marginal changes in travel time when a driver changes his/her route.

Future Work

In the next stages of this project, we will be examining the efficacy of the bridge fragility models used in REDARS in order to assess whether modifications in this part of the program are necessary. We will also be investigating the differences between estimated and observed economic losses. In this study, economic losses are generated for 1) repair of damaged structures, 2) disruptions caused by traffic congestions, and 3) extended outages or disruptions to the transportation system.

Conclusions

The following conclusions are made with respect to this validation study:

1. In general, the estimated traffic volume demonstrates a linear relationship with counted traffic data during pre- and post-earthquake conditions.

2. The model is very sensitive to post-earthquake conditions. In some cases, link volumes are increased by more than 10 times relative to pre-earthquake volumes, which does not match the real situation.

3. The traffic model is sensitive to changes in link attributes, like free flow speed and capacity.

4. In general, the performance of the network, represented by fewer links, is more effective at capturing the real distribution of traffic flows after an earthquake.

5. Since these findings are for one controlled situation (Northridge Earthquake), caution must be applied before making generalizations for other areas and events.

Acknowledgments

The authors would like to thank Stuart Werner of Seismic Systems and Engineering Consultants for his cooperation, leadership and guidance in this validation study. Without his vision of REDARS, this study would not have been possible. The authors would also like to thank Ian Buckle, Ian Friedland, Nesrin Basoz, George Lee and Jerry O'Connor for providing overall program guidance throughout this study.

References

California Department of Transportation (Caltrans), 1995, Northridge Earthquake Recovery Report, Final Comprehensive Transportation Analysis, District 7 (Los Angeles and Ventura Counties), August

Werner, S.D., Taylor, C.E., Moore III, J.E., Walton, J.S. and S. Cho (2000). *A Risk-Based Methodology for Assessing the Seismic Performance of Highway Systems,* Technical Report MCEER-00-0014, December.

Walton, J.S., Cho, S., Werner, S.D, Moore III, J.E., and C. Taylor (2000). REDARS 1.0 - A Computer Program For Evaluation of Risks from Earthquake Damage to Roadway Systems, MCEER, December.

Application of Seismic Risk Assessment Procedures to the Performance-Based Design of Highway Systems

Ian G. Buckle[1]

Abstract

In recent years, seismic design methods for highway bridges have been extensively reviewed in the United States and a move towards performance-based, multi-level seismic design of bridges has begun. In a parallel exercise, a risk-based methodology has been developed for assessing the performance of highway systems taking into account the seismic fragility of bridges and their interconnectivity. These efforts have opened the door to performance-based seismic design of highway systems, in which system-level performance criteria, such as maximum acceptable restoration times, are targeted for highway systems immediately following earthquakes of varying size. This paper first reviews progress made with the applying performance-based design to individual bridges and then suggests the same philosophy may be applied to complete highway systems. Performance criteria for these systems are discussed and verification of these objectives is illustrated using REDARS, a seismic risk assessment tool recently developed for highway systems.

Background

The seismic performance of highway systems in recent earthquakes has been less than satisfactory. In the last decade, systems have been closed in California, Japan, Turkey and Taiwan due to earthquake damage, and although life-safety was maintained with few exceptions, indirect economic losses have been unacceptably high.

Just as with many other lifeline and infrastructure systems, highways are rarely designed for seismic loads and there are no known codes or specifications for the seismic design of highway systems. Instead most of the progress that has been made towards reducing the vulnerability of these systems has been directed towards the performance of bridges, essential components of most highway systems. But despite the widespread use of seismic bridge codes and specifications, many of the highway closures in recent earthquakes have been due to bridge damage and collapse.

Historically, the United States and many other countries have used a single-level earthquake to seismically design bridges and other structures. Usually called the *design* earthquake, both the *Standard Specifications for Highway Bridges* and the *LRFD Bridge Design Specifications* (AASHTO 1992 and 1998) define this event as one that has a 15% probability of exceedance in a 75-year exposure period, i.e. a return period of about 500 years. In addition, a set of performance criteria are given which are required to be satisfied for this and larger earthquakes, should they occur in the life of the bridge. These criteria are summarized as follows:
- Hazard to life to be minimized
- Bridges may suffer damage but have a low probability of collapse
- Function of essential bridges to be maintained, and

---

[1] Professor, Department of Civil Engineering, MS 258, University of Nevada, Reno, 89557 (775) 784-1519; igbuckle@unr.edu

- Ground motions used in design should have a low probability of being exceeded in the normal lifetime of the bridge.

Characterized by a lack of specificity, these criteria were nevertheless a significant advance over earlier editions of the AASHTO Specifications. Furthermore, in the last five years a significant effort has been made by AASHTO to develop rigorous, performance-based, multi-level, seismic design provisions for bridges. Funded through the National Cooperative Highway Research Program, this effort is known as NCHRP Project 12-49. Recommendations made in the Final Report of this Project (NCHRP, 2002) currently await adoption by the AASHTO Bridge Subcommittee.

By contrast, little has been achieved in the assessment of performance of a highway system, which may be viewed as a collection bridges interconnected by a network of roads. Nor have other components of highway systems (retaining walls, slopes, tunnels, culverts and the like) been systematically studied and their contribution to system vulnerability determined. Applications of seismic risk assessment (SRA) procedures to water supply systems and other utilities have been developed, but until very recently their application to highway systems has not been attempted. One of the most comprehensive efforts in this regard, has been the development of REDARS, a risk-based methodology and analysis tool for assessing the seismic performance of highway systems (Werner et al, 2000).

Application of REDARS to the performance-based design of complete highway systems now seems feasible and this paper develops a framework for such an application. First the paper summarizes how performance-based design is applied to bridges and then translates this experience to highway systems.

The Case for Performance-Based Design of Bridges

The assumption is made in single-level design of a bridge that if performance at the design event is satisfactory, it will be satisfactory at all other levels, both smaller and larger. Such an assumption is generally not true, as seen in recent earthquakes in California and elsewhere. It would be true for smaller events if elastic performance was required at the design event, and it may also be true for larger events, if the design event was sufficiently large and a generous degree of conservatism used in the design. But under the design event, inelastic performance (damage) is explicitly intended (in most bridges), and provided life safety is preserved, the consequential restrictions on access are considered to be tolerable.

However, these restrictions become unacceptable, if they were to occur on a more frequent basis such as during a smaller earthquake. Since this is a nonlinear problem, assurances regarding performance during smaller earthquakes cannot be obtained simply by scaling performance at the design event and thus explicit design (or at least a design check) should be made at this level, to gain this assurance.

Similarly, performance during a larger event cannot be estimated by scaling upwards and relying on reserve strength. Without explicit quantification, this approach is unreliable because it is based on engineering judgment and an experience database that is thin and largely unverified, especially in the central and eastern United States (CEUS).

The argument is thus made, that to avoid adverse performance, such as seen in recent earthquakes, explicit consideration of bridge performance during at least two levels of earthquake (and perhaps more) should be undertaken. Furthermore, the expected level of performance during these earthquakes should be stated with a greater level of specificity than has been the case in the past, and assurances given that these performance levels will be met. This argument leads to the consideration of performance-based engineering for the seismic design and retrofit of bridges.

Performance based engineering (PBE) has been defined as consisting of the selection of design criteria, structural systems (layout, proportioning and detailing), and the assurance and control of construction quality and long-term maintenance, such that at specified levels of ground motions, and with defined levels of reliability, *the structure will not be damaged beyond certain limiting states or other usefulness limits.* (SEAOC 1995). This definition has been paraphrased from that developed for buildings in the SEAOC Vision 2000 Project where PBE was explored and its potential for improving the seismic performance of new buildings was clearly demonstrated.

Application of the design phase of PBE requires two fundamental issues to be addressed. These are:
- Selecting the ground motions (hazard levels) and corresponding performance objectives, and
- Verifying that these performance objectives will be met using rational analytical procedures.

The first of these bullets requires that ground motions be known with a large degree of confidence and that realistic and meaningful objectives can be defined. The second bullet requires the availability of sophisticated methods of analysis that can be implemented with ease and reliability.

Hazard Levels and Performance Objectives For Bridges

Factors to be considered when selecting hazard levels and setting performance objectives include:
- The number of earthquake levels to be used. Ideally it should be many, but in practice two or three levels should sufficient to assure that the desired range of performance is achieved. These might be *small, moderate and large* if three events are favored, or *small and large* if only two events are considered. In the latter case, they might also be referred to as *frequent and rare* events.
- The number of different kinds of bridges to be considered. It is unreasonable to expect that all bridges should have the same performance criteria for the same earthquake. More important bridges for example, might be expected to perform to a higher level than less important bridges. Temporary bridges and those under construction might also have specific criteria. Setting aside these special cases, two or three categories should be again be sufficient, and these might be based solely on *importance,* although it might be preferable to use *expected performance level* as the differentiating parameter.
- The specification of performance requirements. It is not a simple matter to measure performance and be able to specify it. One measure might be the number of days a bridge is closed for repair following an earthquake, or has restricted access (lane reduction or weight reduction or both). Another measure might be the extent of damage as given by residual displacements or offsets, crack widths, extent of spalled concrete and exposed rebar, number of misaligned or unseated bearings, settlement of approach fills, distress to expansion joints and vehicle barriers, and the like. Neither measure is particularly satisfactory, and in practice both are used to complement each other. In this case both a performance level (PL) and a damage level (DL) is used to set the performance riteria.

If dual events are considered (rather than three levels) and two bridge types identified, the above performance criteria may be formatted in a 2 x 2 matrix with the rows assigned to the earthquake level and the columns to bridge type. Elements within the matrix are the required performance and damage levels. Table 1 shows such a performance criteria matrix.

Four performance levels and four damage levels are shown in Table 1 corresponding to two earthquake levels and two bridge types. If more hazard levels and/or more bridge types are to be considered, the number of performance and damage levels (PL, DL) would, in principle, increase. But in practice duplication among the PLs and DLs is common and the number of separate and distinct levels may not even be as many as shown in Table 1.

Table 1. Performance Criteria Matrix For Highway Bridges

EARTHQUAKE	BRIDGE TYPE 1 (e.g. Standard Bridges)	BRIDGE TYPE 2 (e.g. Important Bridges)
Frequent Earthquake	PL1 DL1	PL2 DL2
Rare Earthquake	PL3 DL3	PL4 DL4

where PL1 through PL4 is Performance Level 1 through 4
and DL1 through DL4 is Damage Level 1 through 4.

Verification of Performance Objectives For Bridges

Traditional methods of seismic design are force-based using either modified forces from elastic models (including R-factor methods) or nonlinear capacity and demand analyses (including pushover methods). Some of these methods are listed below in order of increasing rigor and complexity (MCEER 2001):

- *Capacity spectrum method*, in which demand and capacity evaluation are combined in a single procedure. Method is restricted to very regular structures, which can be modeled as single degree-of-freedom systems; is the basis of the AASHTO guide specification for isolated bridges.
- *Elastic response spectrum methods*, in which demands are calculated from response spectrum analysis using elastic spectra and single- or multi-mode techniques depending on the complexity of the structure. R-factors are used to obtain design forces based on assumed capacity of the structure for inelastic action. Design displacements are set equal to elastic displacements.
- *Nonlinear static displacement capacity verification methods* (pushover analysis), in which the displacement capacities of individual bridge substructures are determined from lateral load-displacement analyses taking into account the nonlinear behavior of their components.
- *Nonlinear dynamic analysis methods*, in which force and displacement demands are found from step-by-step time-history analyses using ground motion records and taking into account the nonlinear behavior of various bridge components.

Although the development of capacity-spectrum and capacity-verification methods have greatly improved the analyst's ability to directly address various damage states, and by implication, various performance objectives, they are essentially force-based and appear to be less powerful than the newer *displacement-based methods* which use nonlinear displacement spectra rather than acceleration spectra to characterize the earthquake loads.

In these latter methods, displacements and deformations are calculated directly and forces follow from the displacements. Since many damage states and performance objectives are in fact displacement states, a displacement-based method allows these damage states to be targeted directly.

The Case for Performance-Based Design of Highway Systems

The methodology described above has opened the door to implementing performance-based seismic design for highway systems. As for highway bridges, the goal of such a design approach is to satisfy certain specified performance criteria following earthquakes of different sizes, but in this case, the objectives are set for a highway network or subset thereof. Following the approach for bridges, successful application will require two issues to be addressed:
- Establishment of realistic and meaningful performance objectives at various hazard levels, and
- Verification of these objectives using rational procedures.

Hazard Levels and Performance Objectives For Highway Systems

Performance objectives for highway systems might simply be related to changes in total system travel times (Table 2) for emergency traffic should a small, medium or large earthquake occur in the region. More stringent criteria might be imposed for small and more frequent earthquakes, than for the large and rare events. Alternatively, performance might be measured by system restoration time, which is the time required to restore a system back to full capacity (or some fraction thereof) following an earthquake. For small earthquakes this might less than a day, but for larger events restoration times might be measured in months. Table 2 presents a possible set of criteria based on maximum acceptable restoration times, using two sets of times corresponding to 80 and 100% restoration respectively.

Verification of Performance Objectives For Highway Systems

Until very recently, methods of analysis for assessing highway performance were not available and verification of performance objectives was simply not possible by analytical means. However, the recent development of seismic risk assessment (SRA) procedures for highway systems shows great promise as a potential verification tool. This procedure is outlined in the following steps and a case study is given in a subsequent section.

SRA Overview. An outline of a seismic risk assessment (SRA) procedure for a highway system is shown in Figure 1, where it is seen to involve four main steps. These are: (1) initialization of the SRA; (2) development of system SRA results for scenario earthquake and simulation specified under Step 1; (3) repeat of Step 2 for suite of simulations and scenario earthquakes; and (4) aggregation of results for all earthquakes and simulations.

Table 2. Performance Criteria Matrix for Highway Systems Based on Restoration Times

EARTHQUAKE	HIGHWAY SYSTEM TYPE 1 Standard Operating Requirements	HIGHWAY SYSTEM TYPE 2 Essential Operating Requirements
Frequent Earthquake (FEE)	$T_{80} \leq 2$ days $T_{100} \leq 7$ days	$T_{80} < 1$ day $T_{100} < 1$ day
Rare Earthquake (SEE)	$T_{80} \leq 30$ days $T_{100} \leq 90$ days	$T_{80} \leq 7$ days $T_{100} \leq 30$ days

NOTES:
1. A 'highway system' may be a subset of a larger highway network, subdivided according to operational requirements.
2. Two classes of operating requirements are defined: standard and essential. 'Essential' requirements are more rigorous than 'standard' requirements.
3. System performance is measured by time required to restore network to given percentage of traffic capacity before earthquake.
4. T_{80} and T_{100} are times required to restore system to 80% and 100% of capacity before earthquake, respectively.

This procedure has several desirable features. First, it may be carried out within a geographical information systems (GIS) framework, which enhances data management, improves the efficiency of the analysis, and enables the immediate display of analysis results. Second, if the GIS database is modular, the addition of improved data, procedures, and models as they are developed from future research and development efforts is facilitated. Third, the procedure enables the effects of uncertainties in the earthquake characterization, hazard models, and vulnerability models to be considered, and has the capability of developing aggregate SRA results that could be either deterministic or probabilistic, depending on user needs.

Four modules comprise the SRA procedure as shown in Figure 1. These are:
- *system module*: network inventory, traffic data, origin-destination zones, trip tables, traffic management, network analysis models
- *hazards module*: seismic zones, topography, local soils, ground motion attenuation, geologic hazard models, model uncertainties
- *component module*: structural data, repair costs, repair procedures, traffic states, loss models, fragility models, model uncertainties
- *economic module*: economic sectors (locations, productivity, damageability), stakeholder impacts, economic models

Werner et al (2000) have developed a computer program, called REDARS, for the implementation of this methodology and demonstrated its application in several case studies. On of these is summarized below.

SRA Case Study. The City of Memphis is located in the southwestern corner of Tennessee, just east of the Mississippi River and just north of the Tennessee-Mississippi border. The potential seismic risks to the Memphis area are well recognized because of its proximity to the New Madrid seismic zone. The system

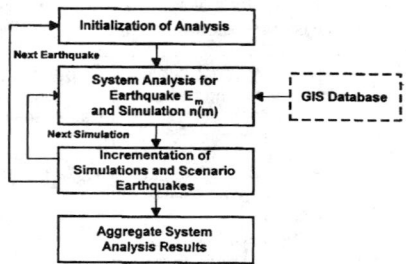

(a) Outline of four-step procedure

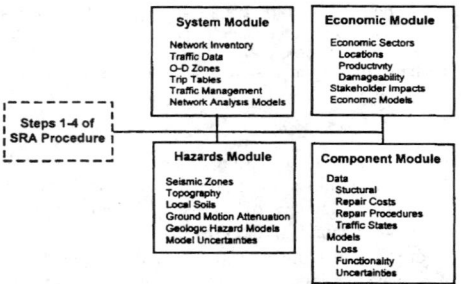

(b) Modules comprising GIS database

Figure 1. Seismic Risk Analysis Procedure for Highway Systems
(Werner et al 2000)

evaluated in this case study includes the beltway of interstate highways that surrounds the city, the two crossing of the Mississippi River (at Interstate Highways 40 and 55), major roadways within the beltway, and highways just outside of the beltway that extend to important transportation, residential and commercial centers to the south, east, and north. The system contains a total of 286 bridges.

Earthquake Event. For the purpose of this example, a scenario earthquake is chosen with a moment magnitude = 5.5, and epicentral distance between 35 and 50 km from the closest and furthest points of the Memphis highway system

Assumptions. This example makes several simplifying assumptions, which include:
- traffic flow and volume data, roadway capacities and O-D zones as provided by the Memphis and Shelby County Office of Planning and Development (OPD); traffic flow data were based on OPD's 1988 traffic forecasting model
- ATC-25 loss models for conventional highway bridges that differentiate between simple spans and bridges with continuous girders but do not consider the influence of other structural attributes (ATC 1991)

- simplified functionality models for estimating closure impacts and restoration times for simple span and continuous girder bridges; post-earthquake traffic management was not considered, and
- MINUTP traffic forecasting models for O-D times pre- and post- earthquake.

Analysis. The analysis consisted of two parts. First, the peak ground accelerations for the scenario earthquake were used with the bridge fragility models to estimate the state of the system at times of three days and six months (in terms of the number of available lanes along each roadway in the system). Then, the effects of any reductions in the available lanes (due to earthquake damage) on traffic flows throughout the system were estimated by using the MINUTP transportation forecasting software, together with a regional traffic capacity and flow data base developed at the Memphis and Shelby County OPD. From this data, travel times and distances throughout the system after each earthquake may be compared to pre-earthquake travel times and distances (in which all travel times and distances are average values for a 24 hour period). Overall travel time and distance for the entire system may then be compared, for pre- and post-earthquake conditions.

Results. Detailed results of this study are given by Werner et al (2000). Since many of the bridges in the city are located on the freeways, the primary impact of the earthquake is a reduction in the capacity of the high volume routes. In this short paper only the overall travel times and distances are presented but these are sufficient to illustrate the effect of network redundancy and the calculation of system capacity.

System Travel Times. Table 3 shows that overall system travel times three days after the earthquake are nearly 34 percent larger than the pre-earthquake values. Six months after the earthquake bridge repairs have reduced the overall travel time, but it is still nearly 20 percent larger than the pre-earthquake value.

Table 3. Total System Travel Times and Distances and System Capacity

PARAMETER	Pre-earthquake value	Value @ T=3 days	Change from pre-earthquake value	Value @ T=6 months	Change from pre-earthquake value
Total vehicle time traveled in 24-hr period (hours)	3.73×10^5	4.99×10^5	+33.8%	4.46×10^5	+19.6%
Total travel distance in 24-hr period (miles)	15.5×10^6	15.6×10^6	small	15.6×10^6	small
System speed (mph)	41.6	31.3	-24.7%	35.0	-15.9%
System capacity as fraction of pre-earthquake value[1]		75%		84%	

NOTE: 1. Capacity ratios are based on changes in system speed

System Travel Distances. Table 3 also shows that overall system travel distances are not sensitive to the bridge damage due to the earthquake, despite the fact that the total number of trips estimated by MINUTP was nearly the same for the pre- and post-earthquake conditions. This lack of change in travel distance is due to the availability of parallel surface routes with few, if any, damaged bridges. These routes are however heavily congested leading to a marked increase in travel time and a 25% decrease in system speed (at 3 days). If system speed is used to indicate system capacity, this drop in speed also indicates the system is at 75% of its pre-earthquake capacity at 3 days and 84% of its pre-earthquake capacity at 6 months.

Conclusions

Performance-based, multi-level seismic design of bridges has gained wide support in the U.S. and other seismically active countries. A logical next step is to apply performance-based design (PBD) to complete highway systems. The goal of such an approach is to satisfy specified performance levels for highway systems immediately following earthquakes of different size. Such criteria might be minimum delay times for emergency traffic for a small, medium or large earthquakes. More stringent criteria might be imposed for smaller and more frequent earthquakes, than for larger and rare events. Alternatively, performance might be measured by system restoration time, which is the time required following an earthquake to restore the system back to full capacity (or some fraction thereof). For small earthquakes this might less than a day; for large events restoration times might be measured in months.

But a critical step in the move towards PBD for highway systems is the ability to verify that expected performance objectives can be achieved, and/or identify retrofit strategies necessary to achieve these objectives. A risk-based methodology, such as that implemented by REDARS, is an attractive solution to this problem.

As with the performance-based design of bridges, consequential issues arise when considering application to highway systems. For example, the uncertainty in the ground motion needs to be reduced and the relationship between component damage states (e.g. bridge column crack widths) and overall system performance (e.g. travel times to emergency care facilities) needs to be better understood. Nevertheless the above tools show great promise and deserve further study.

Acknowledgements

The development of a performance-based seismic design methodology for bridges in the United States has been the work of many people over the last decade. Principally funded by the California Department of Transportation (Caltrans), the American Association of State Highway and Transportation Officials (AASHTO), and the Federal Highway Administration (FHWA), the author is grateful for this support and for the opportunity to contribute to this effort.

The development of a seismic risk assessment procedure for highway systems has been the achievement of Stuart Werner at Seismic Systems and Engineering Consultants, and has also been funded by the Federal Highway Administration. The assistance of the City of Memphis and the State of Tennessee with the analysis of Shelby County is gratefully acknowledged.

References

AASHTO (1992), *Standard specifications for highway bridges*, 15th Edition, American Association State Highway and Transportation Officials, Washington DC

AASHTO (1994), *LRFD bridge design specifications*, First Edition, American Association of State Highway and Transportation Officials, Washington DC

ATC (1991), *Seismic vulnerability and impact of disruption of lifelines in the contiguous United States*, Report ATC-25, Applied Technology Council, CA

MCEER (2001), *Recommended LRFD guidelines for the seismic design of highway bridges*, Part I: Specifications, Part II: Commentary, MCEER Report, Buffalo, NY

NCHRP (2000), *Comprehensive Specification for the Seismic Design of Bridges*, National Cooperative Highway Research Program, Report 472, Washington D.C

SEAOC (1995), *Performance based seismic engineering of buildings*, Structural Engineers Association of California, Vol. 1, Sacramento, CA

Werner, S.D., Taylor, C.E., Moore III, J.E., Walton, J.S., and Cho, S. (2000), *A risk-based methodology for assessing the seismic performance of highway systems*, Technical Report MCEER-00-0014, Buffalo NY

The PEER Highway Demonstration Project

Anne Kiremidjian[1,1], James Moore[2], Yue Yue Fan[3], Ozgur Yazlali[4], Nesrin Basoz[5], and Meredith Williams[6]

ABSTRACT

When evaluating the earthquake risk to transportation system it is important to take into account the integrated effect of ground motion, liquefaction and landslides on the network system. In this paper, the contribution of each of the site effects to the loss from damage to bridges is estimated using the San Francisco Bay area as a test bed. Four scenario earthquakes are considered for the analysis. Damage and loss to bridges from ground shaking and ground displacements (vertical and horizontal) from liquefaction and landslides are estimated. It is found that liquefaction damage is the largest contributor to the direct repair cost.

INTRODUCTION

Transportation systems are spatially distributed systems whereby components of the system are exposed to different ground effects due to the same earthquake event. The ground effects that various components of the system are subjected include ground shaking, vertical displacements due to settlement and horizontal displacements due to lateral spreading and sliding. The ground displacements occur because severe ground shaking causes liquefaction and landslides under the appropriate environmental conditions. Bridges are key components of transportation systems and are particularly susceptible to liquefaction and landslides as they are located over streams and rivers with piers situated over sandy saturated deposits; or they may be over canyons with high slopes that may result in

---

[1] Professor, Civil & Envir. Engng., Stanford University, Stanford, CA, 94305 ask@stanford.edu
[2] Professor, Civil & Envir. Engng.; Public Policy & Managmt.; Industrial & Systems Engng., University of Southern California, Los Angeles, CA 90089, jmoore@usc.edu
[3] Doctoral Student, Dept. Civil & Envir. Engng., University Southern California, Los Angeles, CA 90089
[4] Doctoral Student, Dept. Managmnt. Science & Engng. Stanford University, Stanford, CA 94305
[5] Director, CATRisk Technology Research, St. Paul Fire & Marine Insurance, Concord, CA 94520
[6] GIS Consultant, Earth Science Library, Stanford University, Stanford, California, mjwilliams@stanford.edu

slope instability. Thus, it is important to integrate the effect of each site effect in the overall earthquake risk of a transportation system.

Consideration of the spatial dependence of individual components is an important factoring the evaluation of the network system connectivity and traffic flow through the system. Risk assessment methods require that not only the component performance is assessed, but the overall system performance is evaluated. Most recently, Werner et al. (2000) and Basoz and Kiremidjian (1996) considered the problem of transportation network systems subjected to earthquake events. In both of these publications, the risk to the transportation system is computed from the direct damage to major components such as bridges and the connectivity between a predefined origin-destination (O-D) set. Basoz and Kiremidjian (1996) also consider the time delay and use the information primarily for retrofit prioritization strategies. The current software HAZUS (1999) for regional loss estimation developed by the National Institute for Building Standards (NIBS) for the Federal Emergency Management Agency (FEMA) considers only the direct loss to bridges in the highway transportation network. The connectivity and traffic delay problems resulting from damage to components of the system are not presently included in that software. Chang et al. (2000) propose a simple risk measure for transportation systems to represent the effectiveness of retrofit strategies by considering the difference in costs associated with travel times before and after retrofitting.

In a current study by the authors, a framework for risk assessment of a transportation system is postulated that considers the direct cost of damage and costs due to time delays in the damage system. The site hazards include ground shaking, liquefaction and landslides. In this paper, the effect of each ground hazard on the direct damage to bridges is evaluated. The effect of these hazards on the transportation network is also being investigated, but it is expected that the primary conclusions based on the component analysis will hold for the network analysis as well.

MODEL FORMULATION

The risk to transportation network systems is defined as the expected cost of damage and loss of functionality of the system when subjected to a severe earthquake, denoted by $E[Loss]$. For a given earthquake event Q_i, the expected loss from the system can be estimated as:

$$E[Loss \mid Q_i] = \int_0^1 l(D \mid Q_i) f_D(d \mid Q_i) dd + \int_0^1 l(t \mid D, Q_i) f_D(d \mid Q_i) dd \quad (1)$$

where

$l(D \mid Q_i)$ = cost of repair of individual components of the system at damage D due to an event Qi, where the damage is $0 \leq D \leq 1.0$,

$f_D(d \mid Q_i)$ = probability density of damage D due to an event Qi,

$l(t \mid D, Q_i)$ = costs associated with time delays due to detours of route closures per event Qi.

The annualized risk of loss for the transportation system from all possible events Q_i

that may affect the system, occurring with rates v_i, is:

$$E[Loss] = \sum_{allevents} E[Loss|Q_i]v_i$$

$$= \sum_{allevents} v_i \left\{ \int_0^l l(D|Q_i)f_D(d|Q_i)dd + \int_0^l l(t|D,Q_i)f_D(d|Q_i)dd \right\} \tag{2}$$

The direct loss functions $l(D|Q_i)$ in equations 1 and 2 include losses due to damage from ground shaking and ground deformations such as those due to liquefaction, landslides and differential fault displacements. For a given event Q_i, losses due to time delays arise from delays in commuter and freight traffic. The time delays result from closure of particular routes because of excessive damage to key components such as bridges, or due to reduced flow capacity (either from imposed lower speed limit or closure of number of available traffic lanes) due to minor or moderate damage. Figure 1 summarizes the major components of the overall risk assessment methodology.

The focus of this paper is on the computation of direct damage to bridges and losses resulting from this damage due to earthquake ground shaking, landslides and liquefaction. Thus, only the fist integral in equations 1 and 2 is considered. Expanding this integral to take into account ground shaking, liquefaction and landslides, the equations become:

$$E[Loss|Q_i] = I_A \int_D \int_A l(D|A,Q_i)f_D(d|A,Q_i)f_A(a|Q_i)dadd$$

$$+ I_L \int_D \int_{S_H} l(D|S_H,Q_i)f_D(d|S_H,Q_i)f_{S_H}(s_H|Q_i)ds_H dd \tag{3}$$

$$+ I_L \int_D \int_{S_i} l(D|S_V,Q_i)f_D(d|S_V,Q_i)f_{S_i}(s_V|Q_i)ds_V dd$$

where,

$$I_A = \begin{cases} 1 & \text{if there is no liquefaction or landslides at a site} \\ 0 & \text{if there is liquefaction or landslide at a site} \end{cases} \tag{4}$$

$$I_L = \begin{cases} 1 & \text{if there is liquefaction or landslides at a site} \\ 0 & \text{if there is no liquefaction or landslide at a site} \end{cases} \tag{5}$$

A = ground shaking severity and can represent either peak ground acceleration or response spectral acceleration, or another appropriate parameter;
S_H = horizontal ground displacement due to either liquefaction or landslides
S_V = vertical ground displacement due to either liquefaction or landslides.

It is assumed in this formulation that either liquefaction or landslides occur at a site but not both. Similarly, if there is either liquefaction or landslide, they govern the damage and preempt any damage due to ground shaking alone.

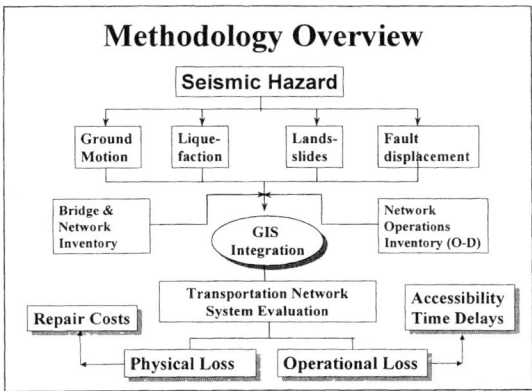

Figure 1. Risk assessment methodology for highway network systems

The total risk has to take into account all possible events Q_i, $i=1,2,...N$ that can occur in the region of the transportation network and is given by the sum of the losses from all events weighted with the likelihood of occurrence of each event. The assessment of time delays requires extensive network analysis, which may prove to be unwieldy and computationally expensive if performed for all possible events. Thus, for the purposes of illustrating the methodology, the analysis is performed for four scenario earthquakes. They include magnitudes 7.5 and 8.0 events on the San Andreas Fault and 7.0 and 7.5 events on the Hayward Fault. In this paper, only the results for the magnitude 7.0 event on the Hayward fault are included.

In order to evaluate the contribution of each hazard, it is necessary that an appropriate system be in place with the various risk analysis components integrated within the system. Geographic information systems (GIS) provide the tools for information storage, overlay, integration and display that are particularly suitable for application to the problem of transportation network risk assessment. ARC/INFO^{TM} GIS is used to develop the different components of the hazard and loss estimation.

The bridge inventory for the San Francisco Bay region was obtained from the California Department of Transportation (CalTrans). There are 2,640 bridges in five counties in the study area. Information in the database that is particularly important for risk analysis includes bridge location, bridge superstructure and substructure type, number of bridge spans, type of connections (simple or continuous), skew angle and design date. The information, however, is not complete for all bridges, and it had to be inferred. Furthermore, the inventory is for pre-retrofitted bridges. *Thus, all results shown in this paper are for pre-retrofitted bridges.*

Peak ground accelerations and spectral accelerations are estimated for the four scenario earthquakes using the Boore at al. (1997) attenuation function. The geologic map for the Bay Area was obtained from the California Geological Survey and the ground motions were amplified according to the local soil at the site of the bridges. Basoz and Mander's (1999) fragility functions are used to estimate the damage to the bridges for the different scenario events resulting from ground shaking. The fragility functions define the probability of being or exceeding one of five damage states for a given ground motion level. The five damage states are: 1) no damage, 2) minor, 3) moderate, 4) major and 5) complete. Figure 2 shows the distribution of peak ground acceleration for the Hayward 7.0 earthquake and the resulting damage state for each bridge in the database. From the figure it can be observed that the ground shaking varies from 0 g to 0.7 g with the largest shaking near the Hayward fault. As expected, bridges near the fault are also found to have the highest damage.

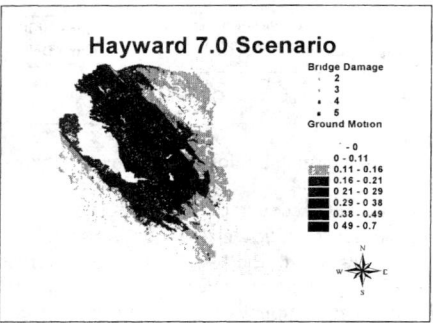

Figure 2. Distribution of bridge damage from ground shaking resulting from a magnitude 7.0 earthquake on the Hayward fault in the San Francisco Bay area

The liquefaction analysis follows the formulation presented in HAZUS (1999). The liquefaction susceptibility map for the region is shown in Figure 3 with the highest liquefaction potential along the bay. There are six liquefaction susceptibility categories included in the analysis. The transportation network is overlaid on the liquefaction susceptibility map identifying the sections of the network that are most likely to be subjected to liquefaction failure. Using the liquefaction susceptibility information, the magnitude of the event, and the peak ground acceleration at the site of a bridge, the horizontal displacement from lateral spreading is estimated. Similarly, the vertical displacement from settlement due to liquefaction is evaluated using the same parameters. The maximum of the two displacements is used to determine the damage state to a bridge resulting from liquefaction.

The distribution of bridge damage from liquefaction resulting from a magnitude 7.0 scenario event on the Hayward fault is shown on Figure 4. As can be seen from this figure, there appear to be significantly more bridges in damage state 4 and 5 due to liquefaction than there are from ground shaking alone. This result is expected in general, but to a great extent is most likely a function of the ground deformation assessment method. A review of the ground motion displacements predicted by the liquefaction analyses revealed that indeed some of these displacements may not be very realistic or at least difficult to substantiate with actual observations. An additional investigation on this subject is deemed necessary to obtain more reliable results.

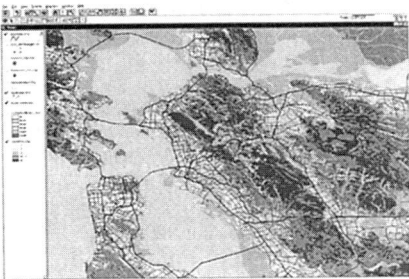

Figure 3. Liquefaction potential and the transportation network system

Analysis for landslides also follows the HAZUS (1999) formulation. The landslide susceptibility map was obtained from the California Geological Survey which identifies eleven severity categories. This information is combined with the predicted ground motion data and the magnitudes of the event to estimate the amount of ground deformation. Damage to bridges is evaluated based on the predicted ground displacements. Figure 5 shows the distribution of bridge damage resulting from landslides. The number of damaged bridges is significantly smaller than that due to liquefaction. This result is expected since the landslide potential is high only in the hilly regions of the Bay Area that have recent geologic deposits. Many or these regions fall outside of the study area.

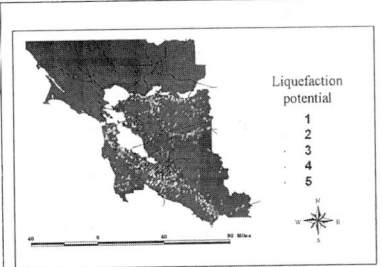

Figure 4. Distribution of bridge damage due to liquefaction in the San Francisco Bay Area from a 7.0 magnitude event on the Hayward fault

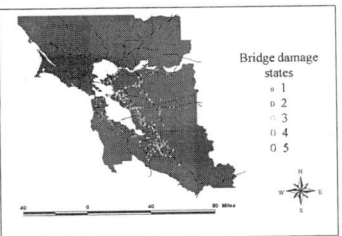

Figure 5. Distribution of pre-retrofitted bridge damage due to landslides in the San Francisco Bay Area from a 7.0 magnitude event on the Hayward fault

Loss estimates presented in this paper are limited to repair costs due to damage to bridges. Losses due to time delays in traffic are currently being investigated. Repair cost depends on the size of the bridge and the expected damage state of the bridges. The expected damage state for each bridge is evaluated by computing the probability that a bridge will be in each of the five damage states and then computing the expected value of damage for that bridge. These are the damage states shown in Figures 2, 4 and 5. Repair cost for a given bridge is given by:

Repair Cost = Repair Cost Ratio * Area * Cost (6)

where the *Repair Cost Ratio (RCR)* is a function of the damage state of the bridge. The *RCR* values are given in Basoz and Mander (1999). Since these values are difficult to obtain, a best estimate, high and low values are provided. Repair cost estimates are provided with all three values for the *RCR*, however, only the best estimate is reported here for brevity.

The area of the bride is computed using the following simple formula:

Area = bridge length * bridge (deck) width (7)

where information on the bridge length and width is obtained from the CalTrans bridge database. The repair cost for different types of bridges was provided by Jack T. Young (personal communication, CalTrans, Jan 2000). The average repair costs vary from $117.5 per square foot to $165 per square foot of bridge deck depending on the bridge type.

Table 1 provides the repair cost estimates for all the bridges in the study area for the four scenarios. Repair costs are obtained for damage due to ground shaking, ground shaking and liquefaction, ground shaking and landslides, and the total due to ground shaking, liquefaction and landslides. From this table it can be observed that the losses due to liquefaction dominate. This corresponds to the high damage distribution observed with liquefaction occurrence. The losses from liquefaction, however, are significantly higher primarily because if liquefaction occurs the bridge is considered to be in damage state 4 or 5 resulting in very large repair costs. Landslides do not appear to have a major contribution to the overall repair cost which is consistent with the estimated damage states for this hazard.

Table 1 Summary of losses from ground shaking, liquefaction and landslides to bridges in the San Francisco Bay Area from the four scenario events *(times 1,000)*

	Ground Shaking Only	Ground Shaking + Liquefaction	Ground Shaking + Landslides	Ground Shaking + Liquefaction + Landslides
Hayward 7.0	$ 494,046	$ 1,392,593	$ 571,497	$ 1,416,405
Hayward 7.5	$ 594,894	$ 1,855,247	$811,580	$ 1,861,046
San Andreas 7.5	$ 517,164	$ 1,686,116	$ 677,670	$1,704,257
San Andreas 8.0	$ 799,343	$ 2,188,848	$ 1,060,300	$2,233,668

TRANSPORTATION NETWORK ANALYSIS

Information on the highway transportation network for District 4 in California, which corresponds to the San Francisco Bay Area, was obtained from the Metropolitan Transportation Commission (MTC). The MTC Bay area highway network model consists of 1,120 zones and 26,522 links. These links are defined by 15,582 nodes with geographic coordinates. Each node corresponds to a traffic analysis zone.

A significant effort was devoted to importing the highway network information within the $ARC/INFO^{TM}$ GIS. The bridge data were then linked to the highway network and corrected to match bridge locations with network locations. Baseline analysis was conducted on the transportation network pre-earthquake scenario. The post-earthquake scenario for a magnitude 7.0 event on the Hayward fault was modeled in EMME/2, a transportation systems network analysis software. Based on this analysis closed links within the system were identified, shown in Figure 6. Table 2 summarizes the vehicle hours by link congestion status. The baseline calculations correspond to the pre-event conditions and demands. The post-event analysis results are listed under the Hayward (HA1) row. Fixed travel demand is assumed in these analyses.

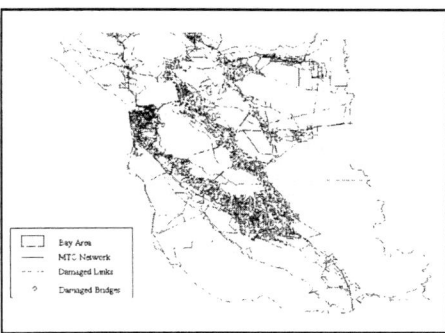

Figure 6. Closed highway links for pre-retrofit bridge damage in the San Francisco Bay Area for a scenario earthquake of moment magnitude 7.0 on the Hayward Fault.

Table 2. Summary of Vehicle Hours by Link Congestion Status, Fixed Travel Demand

Total Vehicle Hours		Frwy to Frwy Ramps	Freeways	Expressways	Collectors	On/Off Ramps
BASELINE	V/C<1	223,448	6,633,475	1,539,282	2,500,326	681,289
	V/C>1	37,519	8,235,451	236,547	1,261,654	464,950
	TOTAL	260,967	14,868,927	1,775,829	3,761,980	1,146,239
HW1	V/C<1	349	13,415	1,248,891	189,264	1,619
	V/C>1	1,936	8,322,682	1,375,029,766	17,563,861,312	86,816
	TOTAL	2,285	8,336,098	1,376,278,658	17,564,050,576	88,435

Table 2 (cont'd). Summary of Vehicle Hours by Link Congestion Status, Fixed Travel Demand

Total Vehicle Hours		Centroid Connectors	Major Roads	Metered Ramps	Golden Gate Bridge	Grand Total
BASELINE	V/C<1	850,444	7,580,948	32,903	47,866	20,089,981
	V/C>1	8,320	2,195,583	41,591	0	12,481,616
	TOTAL	858,764	9,776,532	74,494	47,866	32,571,596
HW1	V/C<1	6,744	7,433,922	52	0	8,894,256
	V/C>1	0	84,905,952,296	0	0	103,853,254,808
	TOTAL	6,744	84,913,386,218	52	0	103,862,149,065

A method was developed in this project to treat variable travel demand. The results from the variable demand model are summarized in Table 3 for the four scenario earthquakes. Again the base line analysis corresponds to the pre-event conditions and the subsequent columns summarize the analysis for the vehicle hours after damaged bridges are closed following the scenario event.

Table 3. Summary of Total Vehicle Hours by Link Type, Variable Travel Demand

TYPE	BASELINE	HW1	HW2	SA1	SA2
Frwy to Frwy Ramps	3,510	273	747	34	50
Freeways	133,228	5,948	6,308	1,375	1,397
Expressways	22,176	234,966	9,910,183	2,629	25,979
Collectors	28,650	974,053,825	293,379,322	2,195,236	175,811
On/Off Ramps	12,387	38,367	2,051	873	865
Centroid Connectors	10,540	4,674	4,975	3,826	4,076
Major Roads	99,142	55,629,479	27,334,268	122,997	95,273
Metered Ramps	1,195	42	22	10	13
Golden Gate Bridge	562	0	0	0	0
Grand Total	311,390	1,029,967,575	330,637,877	2,326,980	303,464

CONCLUSIONS

A method is presented for evaluating the direct losses from damage to bridges in a highway transportation network. This method is used to investigate the contribution of ground shaking, liquefaction and landslide hazard to the total repair costs. For this purpose, the repair costs for four scenario events are evaluated in the San Francisco Bay area. Damage distributions for each hazard are reported only for the magnitude 7.0 event on the Hayward fault. From the example analyses, it is observed that damage to bridges is the greatest due to liquefaction. Thus, the repair costs are also the highest from liquefaction. In comparison, landslides appear to have a very small contribution to both the damage estimates and the repair cost estimates. In general, the contribution of liquefaction hazard to the repair cost is region dependent, however, in this analysis it is attributed to the method used for estimating the ground deformations. A more robust model for liquefaction displacement assessment and associated fragility functions is needed in order to obtain reliable damage and loss estimates.

The transportation network was evaluated for changes in vehicle travel times under two assumptions – constant post-event demand and variable post-event demand. The total vehicle hours increase in post-earthquake networks relative to the baseline network, but not as dramatically in the variable-demand case as for the fixed-demand model. The variable demand model assigns fewer trips to the network. This results in fewer total vehicle hours of travel. However, less total time delay does not indicate lower costs. The trips being eliminated because of high travel costs have value. These absences impose an opportunity cost. Therefore, the total losses should count both the total observed delay and the value of the trips forgone. As in the case of the fixed-demand model, some freeway links are isolated by network damage, even though they are otherwise fully functional. The Golden Gate Bridge is a consistent example across all earthquake scenarios.

ACKNOWLEDGMENT

This research was supported by the PEER Center project SA2401JB. The help of Caltrans personnel in providing the bridge and transportation network databases is gratefully acknowledged. We also express our gratitude to TeleAtlas Corporation for providing the street based network.

REFERENCES

Basoz, N. and Kiremidjian, A. (1996). "Risk Assessment of Highway Transportation Systems", *The John A. Blume Earthquake Engineering Center Report No. 118*, Department of Civil and Environmental Engineering, Stanford University, Stanford, CA 94305.

Basoz, N., and Mander, J. (1999). "Enhancement of the Highway Transportation Module in HAZUS", in *Report to National Institute of Building Sciences*. Washington, D.C.: [GET PUB].

Boore, D., Joyner, W., and Fumal, T. (1997). "Equations for Estimating Horizontal Response Spectra and Peak Acceleration from North American Earthquakes: A Summary of Recent Work", *Seism.Res.Let.*, 687(1), 128-154.

Chang, s., Shinozuka, M., and Moore, J. (2000). "Probabilistic Earthquake Scenarios: Extending Risk Analysis Methodologies to Spatially Distributed Systems", *Earthquake Spectra* 16(3), pp.557-572.

HAZUS (1999). *"Earthquake Loss Estimation"*, Technical Manual, National Institute of Building Sciences, Washington, DC.

Kiremidjian, A.S., Moore J., Basoz, N., Burnell K., Fan Y. and Hortacsu A. (2002). "Earthquake Risk Assessment for Transportation Systems: Analysis of Pre-Retrofitted System", Proceedings for the 7th *National Conference on Earthquake Engineering*, EERI, Boston, July 21-25.

Werner, SD., Taylor, C.E., Moore, J.E., Walton, J.S. and Choet, S. (2000). "Risk-Based Methodology for Assessing the Seismic Performance of Highway Systems." MCEER-00-0014, SUNY Buffalo, Buffalo, NY. December.

Fragility Curves for Concrete Bridges Retrofitted by Column Jacketing and Restrainers

Sang-Hoon Kim[1] and Masanobu Shinozuka[2]

Abstract

The Northridge earthquake inflicted various levels of damage upon a large number of Caltrans' bridges not retrofitted. In this respect, this study represents results of fragility curve development for two (2) sample bridges typical in southern California, strengthened for seismic retrofit by means of steel jacketing of bridge columns and restrainers at expansion joints. Monte Carlo simulation is performed to study nonlinear dynamic responses of the bridges before and after retrofit. Fragility curves in this study are represented by lognormal distribution functions with two parameters and developed as a function of PGA. The sixty (60) ground acceleration time histories for the Los Angeles area developed for the Federal Emergency Management Agency (FEMA) SAC (SEAOC-ATC-CUREe) steel project are used for the dynamic analysis of the bridges. The improvement in the fragility with retrofit is quantified by comparing fragility curves of the bridge before and after retrofit. In this first attempt to formulate the problem of fragility enhancement, the quantification is made by comparing the median values of the fragility curves before and after the retrofit. Under the hypothesis that this quantification also applies to empirical fragility curves developed on the basis of Northridge earthquake damage, the enhanced version of the empirical curves is developed for the ensuing analysis to determine the enhancement of transportation network performance due to the retrofit.

Introduction

Several recent destructive earthquakes, particularly the 1989 Loma Prieta and 1994 Northridge earthquakes in California, and the 1995 Hanshin-Awaji (Kobe) earthquake in Japan, caused significant damage to a large number of highway structures that were seismically deficient (Basoz and Kiremidjian 1998, Buckle 1994). The investigation of these negative consequences gave rise to serious discussions about seismic design philosophy and extensive research activity on the retrofit of existing bridges as well as the seismic design of new bridges. In this respect, this study presents an approach for the seismic assessment of older bridges retrofitted by steel jacketing of the columns having substandard seismic characteristics and by restrainers at expansion joints to prevent bridge decks from unseating. The main objective of the study is focused to evaluate the effects of column retrofit with steel jacketing on the ductility capacity of bridge columns.

The Caltrans' seismic retrofit program was underway prior to the 1994 Northridge earthquake and was accelerated after the Northridge event. This resulted in implementation of steel and composite jacketing of the columns, and of installing and

---

[1] Research Associate, Dept. of Civil and Environmental Engineering, University of California, Irvine, CA 92697; sanghk@uci.edu
[2] Professor, Dept. of Civil and Environmental Engineeering, University of Califorma, Irvine, CA 92697; shino@uci.edu

upgrading of the restraining devices at expansion joints for many bridges for which the seismic retrofit was deemed necessary. Therefore, it is most timely to assess the engineering significance and benefit from such retrofit. This study first develops moment-curvature curves of bridge columns and then performs nonlinear dynamic time history analyses producing fragility curves for two (2) sample bridges before and after retrofitting their columns with steel jacketing. The effect of retrofit is demonstrated by means of the ratio of the median value of the fragility curve for retrofitted column to that of the column before retrofit. This ratio is referred to as fragility enhancement. The fragility enhancement is found to be more significant for more severe state of damage. It is then assumed that the same fragility enhancement is applicable to the empirical fragility curves developed by Shinozuka *et al.* (2000a) based on the Northridge damage data. The fragility curves for one of two (2) sample bridges are also developed before and after retrofitting its expansion joints with restrainers.

This physical improvement of the seismic vulnerability due to steel jacketing becomes evident in terms of enhanced fragility curves shifting those associated with the bridges before retrofit to the right when plotted as functions of PGA (Peak Ground Acceleration). Thus, this study makes it possible to evaluate the improvement of the highway network performance resulting from such retrofit by providing basic information for fragility enhancement.

Retrofit of Concrete Column

Concrete columns of earlier design often lack flexural strength, flexural ductility and shear strength. One of the main causes for these structural inadequacies is lap splices in critical regions and/or premature termination of longitudinal reinforcement. A number of column retrofit techniques, such as steel jacketing, wire pre-stressing and composite material jacketing, have been developed and tested. Although advanced composite materials and other methods have been recently studied, the steel jacketing has been widely applied to bridge retrofit as the most common retrofit technique.

Chai *et al.* (1991) observed that confinement of the concrete columns can be improved if transverse reinforcement layers are placed relatively close together along the longitudinal axis by restraining the lateral expansion of the concrete. It makes it possible for the compression zone to sustain higher compression stresses and much higher compression strains before failure occurs. Obviously, however, this is for original design and construction, but not applicable to existing bridges, to enhance the performance of columns by adding transverse reinforcement layers. In this respect, this study focuses on the steel jacketing technique for retrofitting existing bridge columns to improve their seismic performance.

Bridge Analysis

Description of bridges. Two (2) sample bridges used for example analysis are shown in Fig. 1. Bridge 1 has an overall length of 226 m with five spans, consisting of three frames separated by two expansion joints. The columns have varying lengths with longer ones in the center span and shorter ones near the abutments. The superstructure consists of a reinforced concrete box girder to the left of the left expansion joint and to the right of the right expansion joint, and a prestressed box girder in the central span. The deck has a 6-cell box girder section 20 m wide and 2.6 m deep, and the column section is oblong.

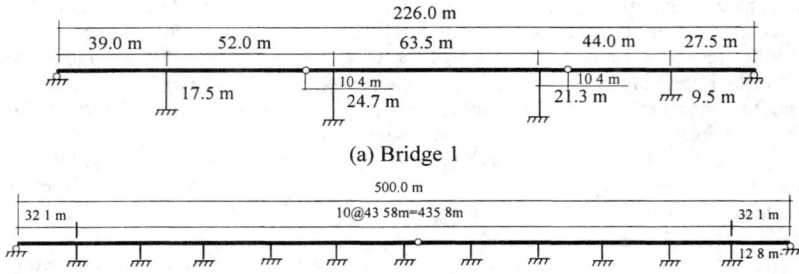

(a) Bridge 1

(b) Bridge 2

Figure 1. Elevation of sample bridges

The bridges are modeled to exhibit the nonlinear behavior of the columns. A column is modeled as an elastic zone with a pair of plastic zones at each end of the column. Each plastic zone is then modeled to consist of a nonlinear rotational spring and a rigid element depicted in Fig. 2. The plastic hinge formed in the bridge column is assumed to have bilinear hysterestic characteristics. Furthermore, pounding effect at the expansion joint of the sample bridges is reflected in the structural response analysis, so that the fragility information of the structure becomes more realistic. In this respect, the expansion joint is constrained in the relative vertical movement, while freely allowing horizontal opening movement and rotation. The closure at the joint, however, is restricted by a gap element when the relative motion of adjacent decks exhausts the initial gap width of 2.54 cm leading to deck pounding. A hoop element sustaining tension only is used for the bridge retrofitted by restrainers at expansion joints and the opening is restricted by the element when the relative motion exhausts the initial slack of 1.27 cm. Springs are also attached to the bases of the columns to account for soil effects, while two abutments are modeled as roller supports. To reflect the cracked state of a concrete bridge column for the seismic response analysis, an effective moment of inertia is employed, making the period of the bridge longer.

The column ductility program developed by Kushiyama (2002) is used to model the moment-curvature relationship of plastic hinges for columns. The critical parameter used to describe the nonlinear structural response in this study is the ductility demand. The ductility demand is defined as θ/θ_y, where θ is the rotation of a bridge column in its plastic hinge and θ_y is the corresponding rotation at the yield point.

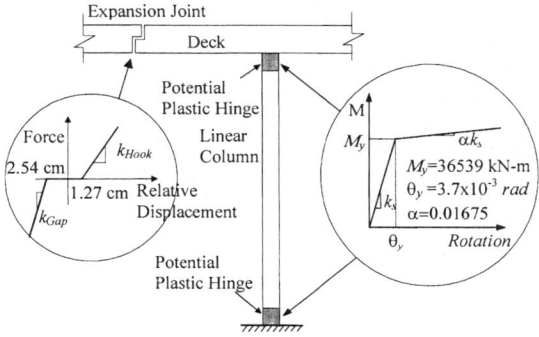

Figure 2. Nonlinearities in bridge model

Damage states. A set of five (5) different damage states recommended by Dutta and Mander (1999) are introduced in Table 1 which displays the description of these five damage states and the corresponding drift limits for a typical column. For each limit state, the drift limit can be transformed to peak ductility demand of the columns for the purpose of this study. Table 2 lists the values of these ductility demands for two (2) sample bridges.

Table 1. Description of damaged States

Damage state	Description	Drift limits
Almost no	First yield	0.005
Slight	Cracking, spalling	0.007
Moderate	Loss of anchorage	0.015
Extensive	Incipient column collapse	0.025
Complete	Column collapse	0.050

Table 2. Peak Ductility demand of columns of sample bridges

Damage state	Bridge 1		Bridge 2	
	before retrofit	after retrofit	before retrofit	after retrofit
Almost no	1.0	1.0	1.0	1.0
Slight	1.3	2.1	1.7	2.5
Moderate	2.2	6.4	4.3	8.3
Extensive	3.5	11.7	7.5	15.7
Complete	6.5	25.2	15.7	34.0

Response analysis of bridges. The *SAP2000/Nonlinear* finite element computer code (Computer and Structures, 2002) is utilized for the extensive two-dimensional response analysis of the bridge under sixty (60) Los Angeles earthquake time histories (http://quiver.eerc.berkeley.edu:8080/studies/system/motions) to develop the fragility curves before and after column retrofit with steel jackets.

Nonlinear response characteristics associated with the bridges are based on moment-curvature curve analysis taking axial loads as well as confinement effects into account. The moment-curvature relationship used in this study for the nonlinear spring is bilinear without any stiffness degradation. Its parameters are established according to the equations in Priestley *et al*. (1996).

Section of the column, stress-strain relationship, distribution of axial force, P-M interaction diagram, moment-curvature curve and moment-rotation curve for a column before and after retrofit are plotted in Fig. 3. The result shows that the moment-curvature curve after retrofit gives a much better performance than that before retrofit by 4 times based on curvature at the ultimate compressive strain and by 1.6 times at the ultimate moment.

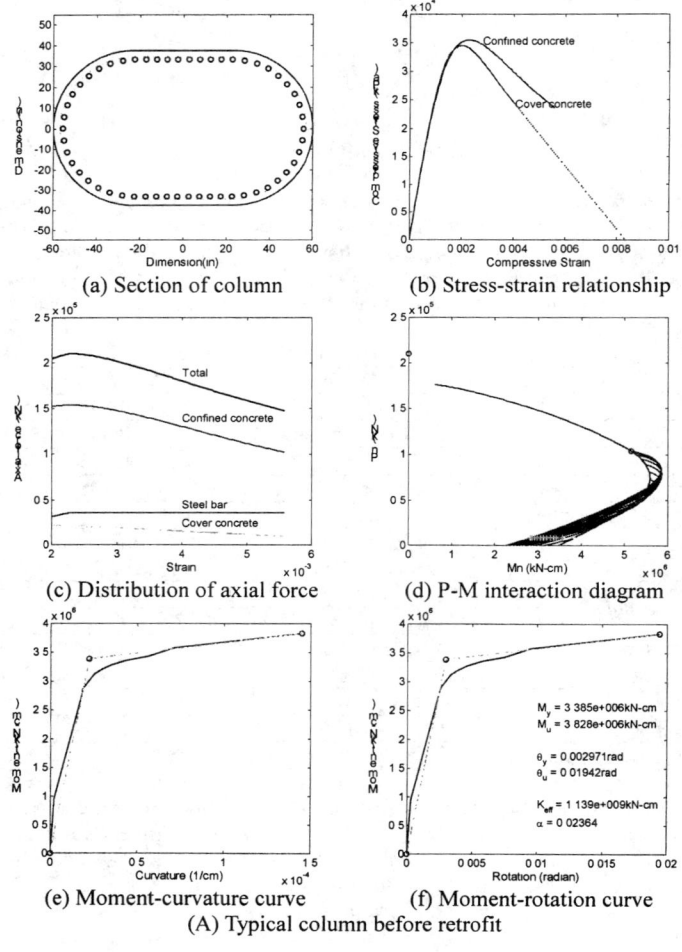

(a) Section of column (b) Stress-strain relationship
(c) Distribution of axial force (d) P-M interaction diagram
(e) Moment-curvature curve (f) Moment-rotation curve
(A) Typical column before retrofit

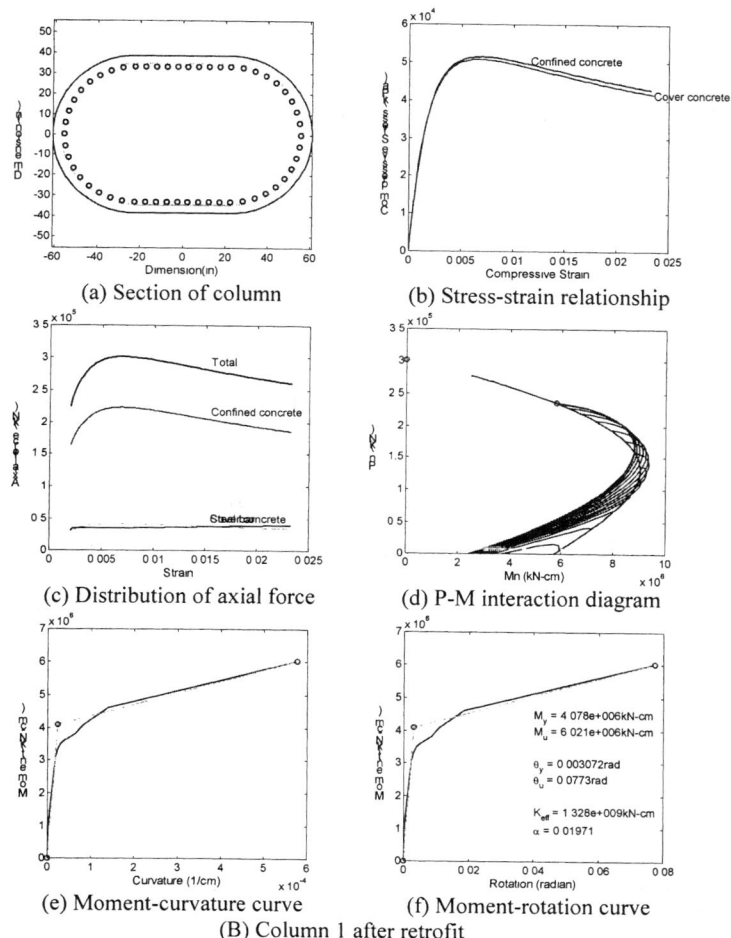

(B) Column 1 after retrofit
Figure 3. Moment-curvature analysis of Bridge 1

Fragility Analysis of Bridges

It is assumed that the fragility curves can be expressed in the form of two-parameter lognormal distribution functions, and the estimation of the two parameters (median and log-standard deviation) is performed with the aid of the maximum likelihood method. A common log-standard deviation, which forces the fragility curves not to intersect, can also be estimated. The likelihood formulation for the purpose of this method was described by Shinozuka et al. (2000b).

Fragility curves. The fragility curves for Bridge 1 associated with those damage states are plotted in Fig. 4 for the cases before retrofit and after retrofit as a function of peak ground acceleration. It is noted here that the log-standard deviation for the pair of fragility curves in each of Fig. 4 is obtained by considering both two cases (before and after retrofit) together. This is for the reason that the bridge with jacketed columns is expected to be less vulnerable to ground motion than the bridge with the columns not jacketed and therefore we expect that the pair of these fragility curves should not theoretically intersect.

Table 3. Number of Damaged Bridge 1 sample size=60

Damage States	before Retrofit	After Retrofit
Almost No	58	57
Slight	56	44
Moderate	47	28
Extensive	41	17
Complete	28	4

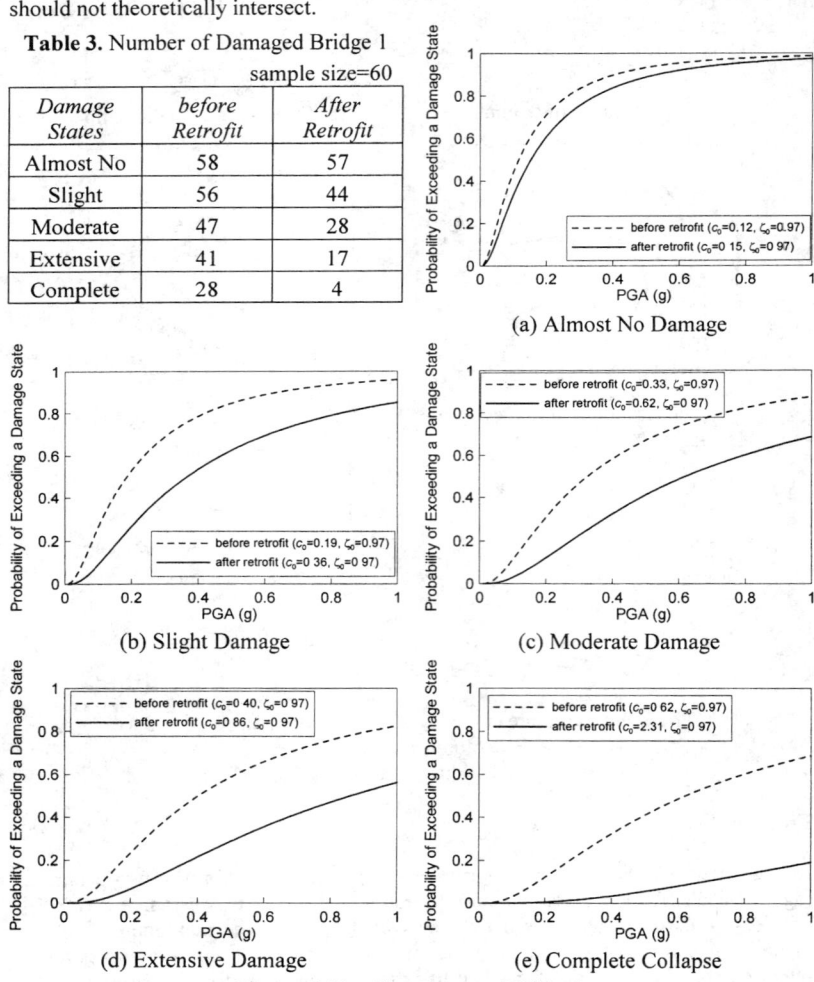

Figure 4. Fragility Curves of Bridge1

Effect of steel jacketing and enhancement after retrofit. The damage state of a bridge in this study is defined in terms of the maximum value of the peak ductility demands sustained by all the column ends. In this context, comparison between the two curves in each of Fig. 4 indicates that the bridge is less susceptible to damage from the ground motion after retrofit than before. The simulated fragility curves in this study demonstrate that, for all levels of damage states, the median fragility values after retrofit are larger than the corresponding values before retrofit. This implies the following: if the number of bridges suffering from a certain state of damage is counted, on average, the damage is smaller when the bridge is subjected to these sixty (60) earthquakes after retrofit than before retrofit. The number is listed in Table 3 for before and after retrofit to Bridge 1. The result in Table 3 is consistent with the observation that the fragility enhancement is found to be more significant for more severe state of damage. This is not unexpected because the ductility demands for more severe states of damage increase after retrofit by much larger multiples than those that occurred before retrofit. It shows that the effect of column retrofit on the seismic performance is excellent in explaining that the bridges are up to 3.7 times less fragile for Bridge 1 (complete damage) after retrofit compared to the case before retrofit in terms of the median values.

Considering these two (2) sample bridges with oblong columns and corresponding sets of fragility curves before and after retrofit, the average fragility enhancement over these two (2) sample bridges at each state of damage is computed and plotted as a function of the state of damage. An analytical function is interpolated and the "enhancement curve" is plotted through curve fitting as shown in Fig. 5. This curve shows 20%, 34%, 58%, 99% and 170% improvement for each damage state described on the x axis in Fig. 5.

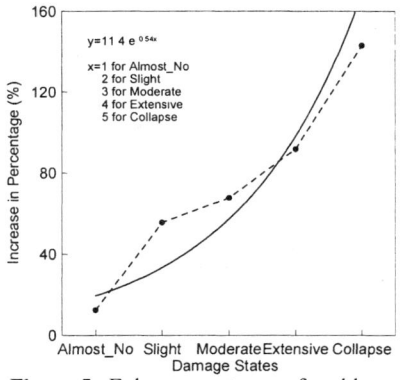

Figure 5. Enhancement curve for oblong columns retrofitted by steel jacketing

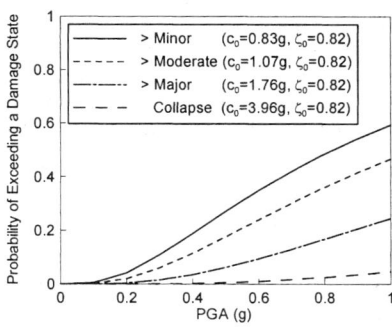

Figure 6. Empirical fragility curves of Caltarns' bridges (Shinozuka *et al.* 2002b)

It is assumed that the fragility enhancement obtained from this function also applies to the development of the fragility curves after the retrofit from the empirical fragility curves (Fig. 6) for the Expressway bridges in Los Angeles and Orange

County, California subjected to the Northridge earthquake. Under the assumption that Dutta and Mander's damage states (1999) are interchangeable with the Caltrans definitions so that "slight=minor", "moderate=moderate", "extensive=major" and "complete=collapse", two enhanced empirical fragility curves after retrofit for minor, and moderate damage are plotted in Figs. 7 and 8, respectively, to be used in ensuing expressway network performance analysis in future.

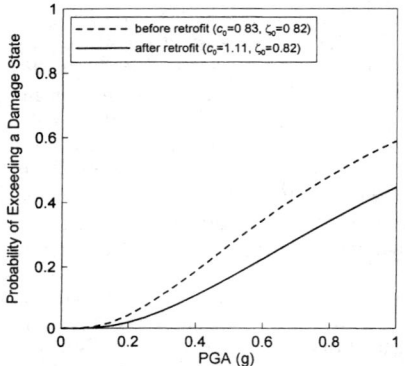 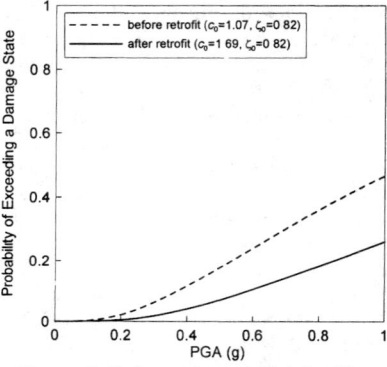

Figure 7. Enhanced empirical fragility curves for Minor damage after retrofit

Figure 8. Enhanced empirical fragility curves for Moderate damage after retrofit

Fragility Curves for Retrofit with Restrainers

The fragility curves for Bridge 1 are also developed to demonstrate the effect of retrofit at expansion joints by extending seat width in Fig. 9 and installing restrainers (modeled as the hoop element introduced in Fig. 2) designed for anchor force capacity in Fig 10. These two figures show excellent improvement for both retrofit methods at expansion joints. However, this observation might not always apply, depending on the details of specific bridge characteristic.

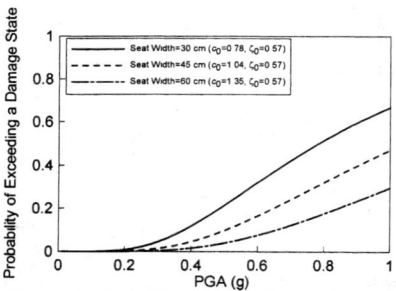 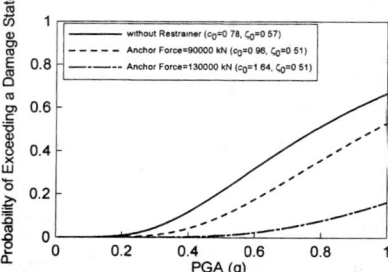

Figure 9. Fragility curves for expansion joint retrofit by extension of seat width

Figure 10. Fragility curves for expansion joint retrofit by restrainers

Conclusions

The computed analytical fragility curves corresponding to these damage states appear to make intuitive sense relative to the bridge's design, retrofit and performance in past seismic events. The following conclusions are drawn from the results of this study.
1) The simulated fragility curves after column retrofit with steel jacketing show excellent improvement (less fragile) compared to those before retrofit by as much as 3.7 times based on median PGA values simulated.
2) An "enhancement curve" is proposed and applied to develop fragility curves after retrofit on the basis of empirical fragility curves. This is done for the reason that empirical fragility curves are expected to be significantly more reliable than those derived otherwise.
3) The fragility curves developed for retrofit with restrainers provide useful information to the seismic design practice by quantifying the improvement due to retrofit.

Acknowledgement

This study was supported by the MCEER/FHWA under Contract No. DTFH 61-98-C-00094 and Caltrans under Contract No. 59A0304.

References

Chai, Y.H., Priestley, M.J.N. and Seible, F. (1991). "Seismic Retrofit of Circular Bridge Columns for Enhanced Flexural Performance," *ACI Structural Journal*, 8(5): 572-584.

Computer and Structures, Inc. (2002) *SAP2000/Nonlinear Users Manual Version 8*, Berkeley, CA, USA.

Dutta, A. and Mander, J.B. (1999). "Seismic fragility analysis of highway bridges," Proceedings of the Center-to-Center Project Workshop on Earthquake Engineering in Transportation Systems, Tokyo.

Kushiyama, S. (2002) "Calculation Moment-Rotation Relationship of Reinforced Concrete Member with/without Steel Jacket," Unpublished Report at University of Southern California, CA, USA.

Priestley, M.J.N., Seible, F. and Calvi, G.M. (1996) *Seismic Design and Retrofit of Bridges*, John Wiley & Sons, Inc.: 270-273.

Shinozuka, M., Feng, M.Q., et al. (2000a). "Statistical analysis of fragility curves," *J. Engrg. Mech., ASCE*, 126(12): 1224-1231.

Shinozuka, M., Feng, M.Q., et al. (2000b). "Nonlinear static procedure for fragility curve development," *J. Engrg. Mech., ASCE*, 126(12): 1287-1295.

Post-Earthquake Road Unblocked Reliability Estimation based on an Analysis of Randomicity of Traffic Demands and Road Capacities

Yanyan Chen[1] and Ronald T. Eguchi[2]

Abstract

In this paper, a revised four-stage traffic plan method is proposed to estimate the post-earthquake traffic flow assigned to a road network. We also suggest a post-earthquake road capacity estimation method that considers the seismic vulnerability of structures and the potential building collapse area that might occupy a road after a major earthquake. Through an analysis of the "randomicity" of traffic demands (traffic flow) and road capacities after an earthquake, and depending on reliability theory, the "unblocked" reliability of road segments and the reliability of the entire transportation network can be estimated using techniques presented in this paper. Through the component importance analysis, the optimum improvement and thus, optimum recovery plan for a transportation system that focuses on alleviating earthquake indirect losses is suggested. At the end of this paper, an example is given to illustrate the application of the unblocked reliability index in post-earthquake traffic condition estimation and in developing earthquake response strategies for urban transportation systems.

Key words: Unblocked Reliability, Earthquake, Road Network, Importance Analysis

Introduction

Transportation systems include various kinds of intersections and road segments, which can be modeled as a network with interconnected nodes and links. It is important to develop anti-earthquake plans for the system because these systems

---

[1]Professor, Transportation Research Center, Beijing University of Technology, Beijing P.R. China, 100022; cdyan@bjpu.edu.cn
[2]President and CEO, ImageCat, Inc., 400 Oceangate, Suite 1050, Long Beach, CA 90802; rte@imagecatinc.com

can be easily destroyed in an earthquake. Earthquake damage can cause not only direct losses (in the form of repairs to damaged structures), but could also cause extraordinary indirect losses because of driver delays from traffic congestion. Most rescue and recovery activities rely on the safe and efficient transportation of material and personnel. One key step in developing an anti-earthquake plan is the estimation of traffic conditions after the earthquake; this leads to optimal decision-making during the recovery period. This last topic is the major focus of this paper.

In previous research discussions, the transportation system is regarded as reliable when it is connected between origin-destination pairs. And, system reliability is decided on by the connectivity reliability of each road. In fact, when an earthquake occurs, some origin-destination pairs (i.e., links), which are structurally reliable, may be disrupted because traffic demands greatly exceed road traffic capacities. Both traffic demands and road traffic capacities are random variables, and these random characteristics becomes more obvious during an earthquake because of the unusual increase in traffic, the undirected flow of traffic, the randomicity of road damage, the uncertain process of road repairing and the interruption of pedestrians, etc. The randomicity of traffic demands and road capacity results in uncertain post-disaster traffic blockages. In this paper, depending on the randomicity analysis of post-disaster traffic demands and road capacity, and utilizing reliability theory, the unblocked reliability of road components and of the entire transportation network can be estimated. This unblocked road reliability can be taken as an estimation index of the traffic conditions after an earthquake.

Considering the fact that different effects will be observed in the system depending upon which components are improved, this paper focuses on an optimal improvement and recovery plan for transportation systems by using component importance analysis.

The Estimation and Randomicity Analysis of Traffic Demands and Road Capacities

Traffic Demand Estimation and Randomicity Analysis

Travel demands are generally embodied by traffic flow (written as V) assigned to the road network. Usually, in traffic plan theory, the traffic flows assigned to road segments and intersections are estimated using a four-stage method, i. e. traffic generation, traffic distribution, mode choice and traffic assignment. When an earthquake occurs, there will be abnormal traffic demands. Extraordinary traffic convergence will occur in places such as damaged areas, fire stations, hospitals, emergency management centers, etc. People may be more inclined to go out to see the effects of the earthquake if their relatives or friends are safe and accounted for. In the meantime, because of damage to the roadway system, people will pay more attention to road safety in their route selection. All of these reasons can result in the traffic flow patterns changing sharply as compared usual road conditions.

In this paper, considering travel behavior changes that are influenced by the earthquake, the conditional four-stage method is adjusted so that the assignment of traffic flows can meet the traffic conditions after the earthquake. A detailed set of procedures for making these adjustments is suggested below:

1) In the traffic generation stage, traffic generation in zones with different travel characteristics (utility character) should be revised according to changes in travel time and frequency as a result of the earthquake. This information can usually be obtained from historic data or experience. For simplification, the zones can be divided into 20-50 groups by their utility character. Extraordinary traffic convergence areas can be identified and the traffic generation change rate can be estimated.

2) In the traffic distribution stage, the distribution of O-D pairs should be revised according to the activity relationships between different zones after the earthquake.

3) In the mode choice stage, the safety of different traffic modes, as well as the vulnerability of infrastructure systems such as roads or underground subways should be considered.

4) In the traffic assignment stage, road damage probabilities, as well as travel costs, should be taken into consideration in deciding the road deterrence function and route choice.

In fact, the traffic flow after an earthquake has a strong randomicity character, which results from uncertainties in decisions to travel or not, destination choices, different travel modes, and travel route selections.

In this paper, traffic demands (flow) are considered as random variables with a normal distribution. The traffic flow assigned to road segments and intersections after an earthquake are estimated using the revised four-stage method. Traffic can be regarded as the median of a random flow distribution, i.e., this parameter is referred to in this paper as μ_V.

Road Capacity Estimation and Randomicity Analysis

Capacity (written as C) reflects the ability of the road to handle traffic flows. When an earthquake occurs, the road capacity is reduced because the transportation infrastructure is damaged and because collapsed buildings may fall onto the roadway system. The road capacity after an earthquake also has a strong randomicity character. There are a myriad of reasons why the capacity of a system and its components (intersections and road segments) would be subject to variations. These include the uncertainty associated with structure damage, the different types of vehicles, disturbances caused by non-vehicles and/or pedestrians, disruption of neighboring road segments and intersections, changes in driving behavior, and the uncertain process of road repairs. The random nature of these factors becomes more important when the earthquake occurs. In this paper, the actual road capacity after the earthquake is modeled as a normal random variable affected by the continuous disruption of random events.

Usually, the degree of road damage can be divided into five states: intact, slightly damage, moderate damage, serious damage and complete damage. Accordingly, the effect of road damage on road capacity can also be classified into five levels: no influence, slight influence, moderate influence, serious influence and thorough influence. Each level relates to a different degree of decrease in capacity.

Under certain damage conditions, the actual road capacity can be regarded as a normal random variable affected by the continuous disruption of different random events. The capacity associated with different damage state is computed using traditional certainty methods (Chenggang, 1994); capacity can be regarded as the median of random capacity, which is written as μ_C in this paper.

The variance of C and V can be estimated based on the certainty degree of C and V.

Unblocked Probability of Road Unit

Usually, there are six degrees of road service ranging from no traffic delay (level A) to completely blocked (level F); road service can be estimated through a saturation (μ_V / μ_C) analysis. For simplification purposes, the traffic condition after an earthquake can be divided into two states, i.e., unblocked and blocked. When a vehicle can drive above an expected service level and/or design velocity, we can say that the road is unblocked. Level C is defined as stable flow, and the driver accepts some delay. Level D is defined as partially closed because of unstable flow, and delay times are longer but the driver accepts these delays. The probability of unblocked flows at peak traffic hours during a prescribed period is called "road unblocked reliability."

Generally, the factors that influence road reliability involve two comprehensive values, i.e., capacity C and demand flow V (Chen, 2000).

Let

$$Z = g(C - V) = C - V \tag{1}$$

Z is also a random variable with the following three states:

$Z>0$ road is unblocked
$Z<0$ road is blocked
$Z=0$ road is in the margin state

Here capacity is calculated based on design velocity and expected service level. According to the value of the Z, which is also a random variable, the road condition can be estimated. The blocked probability P_F can be calculated using the following formula:

$$P_F = P\{Z < 0\} = \int_{-\infty}^{0} f_Z(Z)dz \qquad (2)$$

Since the event $\{Z<0\}$ is opposite to the event $\{Z \geq 0\}$, there is a relationship between reliability (unblocked) degree φ and blocked probability P_F, that is:

$$\varphi = 1 - P_F = 1 - P\{Z < 0\} = 1 - \int_{-\infty}^{0} f_Z(Z)dz \qquad (3)$$

In Figure 1, the shaded area is defined as the blocked probability. The left curve is a probability density function for traffic demand; the right curve is for road capacity.

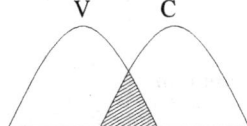

Figure 1. Reliability with random traffic demand and capacity

Depending on the distribution of C and V, the reliability degree φ and blocked probability P_F can be estimated.

Because the unblocked probability has a direct relationship with the saturation degree μ_V / μ_C, for simplicity, the unblocked reliability can also be estimated by the ratio of μ_V / μ_C. That is:

$$\varphi = \begin{cases} 1 & \mu_V / \mu_C < 0.1 \\ 1 - \mu_V / \mu_C & 0.1 \leq \mu_V / \mu_C < 0.9 \\ 0.1 & \mu_V / \mu_C \geq 0.9 \end{cases} \qquad (4)$$

Considering the possibility of road damage and subsequent changes in traffic supply/demand, the road blocked probability for different earthquake intensities (written as $P_f(I_j)$) can be estimated using the following formula:

$$P_F(I_j) = \sum_{i=1}^{5} P\{\overline{\Omega} \mid D_i\} P\{D_i \mid I_j\} \tag{5}$$

where, $\overline{\Omega}$ is the road blocked event; D_i ($i=1,2,3,4,5$) is the road damage state ranging from intact to thoroughly damaged; I_j is earthquake intensity. $P\{D_i \mid I_j\}$ is the probability of damage state D_i occurring given intensity I_j is experienced, which is calculated using a statistical method or a half theory, half empirical method (Chenggang Z., 1994). $P\{\overline{\Omega} \mid D_i\}$ is the road blocked probability when damage state D_i occurs, which is estimated using Equations 3 or 4.

Traffic Reliability (Connectivity) of O-D Pairs

The function of a reliable road network is to provide effective transportation service to the users from origination to destination. For one O-D (origination to destination) pair, when a vehicle can travel at a design speed on any route not too far from its normal, non-earthquake route, then this condition can be referred to as unblocked. And the unblocked reliability of the kth O-D pair is the probability of the unblocked event between the O-D pairs. Suppose that the blocked conditions among different routes are statistically independent, then the unblocked reliability of the kth O-D (written as ψ_k) can be estimated using the following formula (Yanyan Chen, 1998):

$$\psi_k = P\{\bigcup_{i=1}^{m} A_i\} \tag{6}$$

Here, A_i represents the ith reasonable route as well as the unblocked event of these routes, m is the number of the reasonable routes between the kth O-D pair.

Depending on graphic theory, the route set $\{A_i\}_m$ between any O-D pair can be built, and the network reliability can be computed through disjointing disposal to the route set. Then the unblocked reliability of the system can be estimated as follows (Guiqing Li, et al. 1994):

$$\psi_k = P\{\bigcup_{i=1}^{m} A_i\} = P\{\sum_{j=1}^{m'} A'_{dis,j}\} = \sum_{j=1}^{m'} P\{A'_{dis,j}\} \tag{7}$$

Here, $A'_{dis,i}$ is the ith disjoint route, m' is the number of disjoint routes. Usually,

$m' \geq m$.

$P\{A'_{dis,j}\}$ can be estimated by computing the product of the state (blocked or unblocked) probability of the components on the *j*th disjoint route (Guiqing Li, et al. 1994). In the computation, the failure dependence between some components is considered.

Suppose there are K important O-D pairs in the system, then the system unblocked reliability ψ_s can be estimated using the following equation:

$$\psi_S = \sum_{k=1}^{K} \xi_k \psi_k \tag{8}$$

In this equation, ξ_k is an important parameter in the *k*th O-D pairs, $\sum_{k=1}^{K} \xi_k = 1$.

Importance Analysis of Component

Because the ranks of components in the network are different, the contributions to the system unblocked reliability from improvements to different road segments and intersections will also be different. Through the component importance analysis, optimum system improvement and optimal recovering strategies can be suggested.

We use the degree of unblocked importance of component I_i to reflect the impact on the overall system unblocked reliability to changes in component unblocked reliability. This is estimated using the following formula:

$$I_i = \frac{\partial \psi_S}{\partial \varphi_i} = \sum_{k=1}^{K} \xi_k \frac{\partial \psi_k}{\partial \varphi_i} = \begin{cases} 0 & i \notin A_{dis,j} \\ \dfrac{A_{dis,j}}{\varphi_i} & i \in A_{dis,j} \\ -\dfrac{A_{dis,j}}{1-\varphi_i} & \bar{i} \in A_{dis,j} \end{cases} \tag{9}$$

where, the i and $\bar{i}$ represent the components with unblocked and blocked conditions separately, and φ_i represents the *i*th component unblocked reliability.

But, some components have higher unblocked reliabilities, as well as higher unblocked sensitivities. For example, it may cost more to improve the unblocked reliability of some components. In these cases, it is better to incorporate a revised importance parameter I'_i to reflect this influence, i.e., use the component unblocked reliability to improve the overall system unblocked reliability.

$$I'_t = (1 - \varphi_t)I_t \qquad (10)$$

In the development of a mitigation strategy for urban transportation system, it is advisable to improve - in order of priority - components with lower unblocked reliabilities and higher unblocked importance levels.

Example

In this section, we present an example to demonstrate the feasibility of the proposed method.

Figure 2 contains an urban transportation network graph. Here [i] represents the arc component of road segment, (i) represents the intersection vertex component. According to the component failure dependence analysis, components [1][2](1) belong to group 1, written as <1>; and components [9][13] belong to group 2, written as <2>. In this example, we will estimate the system unblocked reliability when an earthquake of 7 degree intensity occurs.

During the post-earthquake emergency period - through building the distribution curve of random traffic/supply of the road component by the method proposed before - the unblocked probability of road components is computed using Equation 3, which is shown in Table 1. The unblocked reliability of O-D pairs can be computed using Equation.7, which is shown in Table 2. Suppose the importance of all O-D pairs is the same in this example. Then, system reliability can be computed as 0.8823 based on Equation 9. The unblocked importance degree of component I_t and revised unblocked importance parameter I'_t can be computed using Equation 7, which is shown in Table 2.

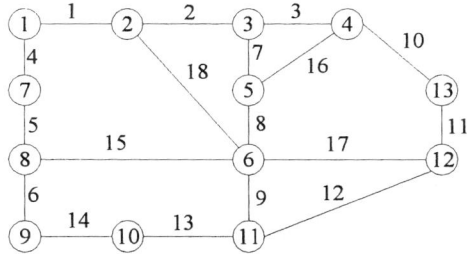

Figure 2. An Urban Transportation Network Graph

Table 1 The Unblocked Reliability of Components

No.i	φ_i	No.i	φ_i	No.i	φ_i
(1)	0 87	(4)	0.78	(5)	0.87
(7)	0.90	(13)	0.83	[3]	0.76
[4]	0.75	[5]	0.79	[6]	0.78
[7]	0.85	[8]	0 75	[10]	0 80
[11]	0.86	[12]	0.67	[14]	0.80
[15]	0.76	[16]	0.78	[17]	0.77
[18]	0.78	<1>	0.78	<2>	0.76

Table 2 The Importance Degree and Revised Importance Degree of Components

No.i	I_i	$I_i{'}$	No.i	I_i	$I_i{'}$	No.i	I_i	$I_i{'}$
(1)	0.0113	0 0015	(4)	0.0329	0.0072	(5)	0.0521	0.0068
(7)	0.0109	0.0011	(13)	0.0301	0.0051	[3]	0.0329	0.0079
[4]	0.0049	0.0012	[5]	0.0124	0.0026	[6]	0.0971	0.0214
[7]	0.0533	0.0080	[8]	0.0605	0.0151	[10]	0.0313	0.0063
[11]	0.0291	0.0041	[12]	0.0882	0.0291	[14]	0.1045	0.0209
[15]	0.0706	0.0169	[16]	0.0296	0.0065	[17]	0.1646	0.0379
[18]	0.0000	0.0000	<1>	0.1830	0.0403	<2>	0.1332	0 0320

Depending on the revised unblocked importance parameter I'_i, it is recommended that to improve the reliability of the system that components with higher revised unblocked importance levels be selected.

Conclusions

An effective way to control or reduce indirect losses after an earthquake is to improve the ability of transportation systems to function effectively after the event. In this paper, the unblocked reliability is taken as an index that reflects the road systems ability to survive an earthquake. The traffic demands (flow) after an earthquake is estimated by using a revised four-stage method that depends on modeling travel behavior after an earthquake. The post-event road capacity is estimated by performing a vulnerability analysis and by considering the building collapse area that occupies the road. Depending on an analysis of the randomicity of post-earthquake traffic demands (traffic flow) and road capacities, and using reliability theory, the unblocked reliability of the road and transportation network can be estimated. This information can be translated into indirect losses, and measures of anti-earthquake strength, and post-earthquake traffic control. Through a component importance analysis, optimum improvements to recovery strategies can be produced using this methodology.

Acknowledgment

The primary author would like to thank the sponsoring organizations for this research: National and Beijing National Science Foundations.

References

Chen, Anthony (2000), Hai Yang, and Hong K. Lo, "Wilson H. Tang. A Capacity Related Reliability For Transportation Network: an assessment methodology and numerical results". *J. Profl. Issues in Transportation Research Part B, 301(1)*.

Chen, Yanyan (1997). "The Aseismic Reliability Analysis of Urban Transportation System." *J. Profl., Issues in Nature Disaster, Supplement*, 68-74.

Li, Guiqing et al. (1994). *The Anti-earthquake Reliability Analysis of Civil Building Network System*, Press of Earthquake, Beijing, China.

Zhao, Chenggang (1994). *Anti-earthquake of Lifeline Engineering*, Press of Earthquake, Beijing, China.

A Gis-Based Emergency Response System for Transportation Networks

Nesrin Basoz[1], Meredith Williams[2] and Anne Kiremidjian[3]

Abstract:

A GIS-based emergency response network analysis program, T-RoutER, is developed for post earthquake traffic routing. For this purpose a detailed highway network is linked with the Caltrans bridge database. Critical paths are identified between origins and destinations based on Durkstra's shortest path algorithm. The program is able to identify critical paths for multiple-origins and multiple-destinations that are of particular importance following catastrophic earthquakes. T-RoutER is can be linked to any hazard analysis program for pre-event planning. For post-event emergency response purposes a ground motion maps such as Shake Maps from Tri-Net can be imported for identification of damaged bridges and post event routing.

Introduction

Traffic routing is one of the many activities of emergency response planning and management, used both for purposes of pre-event preparedness and mitigation and post-event response and recovery (Miller and Shaw, 2001). Finding the best routes for emergency response vehicles in emergency situations, such as after bridge closures due to damage from an earthquake, is challenging.

Highway network analysis for daily traffic routing, in general, has been intensively researched. Integration of highway network analysis in emergency response planning and management, however, is a more recent research topic. The use of GIS for seismic risk assessment has been widely accepted within the last decade. However, the use of GIS for highway network analysis for post-earthquake emergency response planning and management has been mostly limited to the display of road and bridge closures and alternative routes. Until now, network analysis itself has largely been performed outside of GIS, mainly due to the lack of compatibility between

---

[1] Consulting Professor, Department of Civil and Environmental Engineering, Stanford University Stanford, California, 94305
[2] GIS Consultant, Earth Science Library, Stanford University, Stanford, California, , 94305 mjwilliams@stanford.edu
[3] Professor, Department of Civil and Environmental Engineering, Stanford University Stanford, California, 94305 ask@stanford.edu

infrastructure inventory and detailed highway network data necessary for GIS-based network analysis. The few GIS applications that are available for post-earthquake traffic routing have been developed for demonstration purposes or are in a prototype stage. Most of these applications, if not all, need to translate a highway network database to a non-GIS platform, in which the graph theory-based operations are performed. Usually the non-GIS platforms require a cumbersome definition of connectivity between network components, such as nodes and arcs (Hobeika, 1987, Werner et al., 2000). The applications have not integrated the automatic creation of the network database. In general, building a highway network suitable for routing applications is very time consuming, and many of the commercially available products, used for daily traffic routing, are created through the labor of an intense and continuous workforce collecting and manipulating data. After a major earthquake, it is almost impossible to create a detailed highway network database fast enough to facilitate emergency response management. Therefore, it is essential to incorporate a value-added highway network database to any practical emergency response application.

This research centered on the development of T-RoutER, a GIS-based traffic routing application for emergency response planning and management used to determine the best routes for traffic before and after bridge closures due to earthquakes. The application utilizes a commercially available highway network database and can be used after an earthquake (or after any other disaster, such as hurricane, flood and fire) to assist in determining alternative detours on a near real-time basis.

The following sections present the use of GIS for network analysis, the highway network data necessary for network analysis, and the T-RoutER application developed in this research project. The components of T-RoutER, automation and customization in GIS and the issues related to linking detailed highway network data with Caltrans bridge data are also discussed.

The detail of road network coverage (database) maintained by different states varies from all public roads to no network database. As a state government agency, the California Department of Transportation (Caltrans) maintains GIS-compatible data only for the state's major highways. The surface streets are considered to be under the jurisdiction of local government agencies. Therefore, the limited information that Caltrans does maintain for surface streets isn't integrated with their state/major highway databases. Without integration of arterial streets, traffic routing will be limited to major highways. Several of the states, including California, are interested in developing road/surface street layers. In the absence of such a database, commercially available street level database can fulfil the need.

The following sections present the use of GIS for network analysis, and an application for network analysis to be used in emergency response, T-RoutER, developed in this research project. Finally, a discussion is presented on how to link detailed highway network data with Caltrans bridge data.

Defining the Highway Network System in GIS

Traffic models demand large amounts of data, including the configuration of the highway network, zone-data and trip matrices. GIS is a natural tool for handling most of these data as it can ease the work process by spatially integrating otherwise disparate data. A GIS can incorporate the highway network's descriptive information for nodes and arcs with the information on the topological relationship among the nodes and arcs, based on graph theory (see Figure 1.) A node describes the location at which two or more lines connect and the endpoints of each line. An arc is a set of ordered co-ordinates that represent the shape of linear geographic features such as: contours, county boundaries, streams or roads. An arc is synonymous with a line (AGI, 1999). Nodes and arcs can carry information about their position within the topology of a network. The topology defines the spatial relationship (i.e., the connectivity and adjacency) between features, such as roads and travel destination points. For example, the topology of a line includes its from- and to-nodes, and its left and right polygons. Arc-node topology supports the definition of linear features and analysis functions such as network tracing.

Figure 1. A generic description of highway network in GIS
http://www.ordsvy.gov.uk/gis-files/stage2/page18.htm

A GIS based routing program uses a logical highway network as an index for referencing the actual feature geometry of the roads. A logical network contains information about the topology (connectivity of a network) rather than information about the geometry of the features. A logical network is useful in GIS because many spatial modeling operations don't require co-ordinates; topological information is sufficient. For example, to find an optimal path between two points requires a list of the lines or arcs that connect to each other and the cost to traverse each line in each direction. Geographical co-ordinates are only needed for drawing the path after it is calculated. Storage of the logical network allows for fast retrieval of features in the GIS (AGI, 1999).

Analyzing the Highway Network in GIS

Network analysis engines in GIS are available to determine the shortest path, fastest path or closest facilities to a selected point. These network analysis engines use the well-known Dijkstra's shortest path algorithm (Ahuja et al., 1993), where, in general, the roadways, bridges and tunnels are modeled as arcs, and intersections and origin-destination points are defined as nodes. The travel times or the travel distances are used as the impedance factors[4] for the arcs. A similar algorithm is employed in this software.

T-RoutER - Traffic Routing for Emergency Response

T-RoutER is a GIS-based traffic routing application for emergency response planning and management, used to determine the best routes for traffic before and after bridge closures due to earthquakes. The application can be used after an earthquake (traffic incidents or any other disaster, such as hurricane, flood and fire) to assist in determining alternative detours on a near real-time basis. The status of the damaged highway network after a disaster can be either estimated based on predictive risk models (such as the one demonstrated in this project) or from data collected by field specialists, including emergency personnel and Caltrans bridge inspectors.

Various predictive models have been developed for damage assessment to transportation systems. The model developed by Kiremidjian et al. (2002) is used in this application. For a highway network, the seismic hazard has to be evaluated at the multitude of the system component locations. This process is greatly facilitated with the use of GIS. Overlaying the highway network data onto the hazard information creates a link between the hazard and the network component within the same platform (Kiremidjian et al., 2002). This information ca be generated within minutes following an earthquake event and can serve for initial estimates of identifying the area that has been affected by the earthquake. Information on damaged bridges can be entered into the system as it becomes available and should be used to augment the predicted damage database.

Components of T-Router – Data and Software

Roadway Network Database: T-RoutER uses a commercially available GIS-based highway network database developed by TeleAtlas, formerly known as ETAK (TeleAtlas, 2001). The main reason for using a commercially available highway network database is to include street level network data in order to achieve realistic traffic routing after bridge or roadway closures. When routing emergency vehicles, a detailed and accurate street database including information on the topology of the network and the network flow is necessary to find the shortest and fastest paths. After an earthquake, major highways may not be passable due to bridge and/or roadway damage, and arterial streets become critical to maintaining network flow. Modeling a network with only the interstate and state highways for post-event routing applications does not provide a realistic routing since more often than not arterial streets are used in re-routing traffic after a major earthquake.

Other commercial street databases that were available for this project required that this logical network be constructed and then reconstructed after a change is made

---

[4] Impedance is the amount of resistance (or cost) required to traverse a line from its origin to its destination. Resistance may be a measure of travel distance, time or speed of travel times the length. Impedance is used in network routing and allocation. An optimum path in a network is the path of least resistance (or lowest impedance) (ESRI, 1999).

to any arc or node, such as a bridge closure due to earthquake-induced failure. For large networks, such as the nine counties of the San Francisco Bay Area, this process takes longer than would be acceptable for near-real time applications. However, TeleAtlas street data includes a built-in logical network, which speeds up the process of post-event highway network configuration within GIS. TeleAtlas also provides a robust selection of descriptive attributes that facilitate realistic routing solutions (see Table 1)

Each bridge and road segment is represented as an arc in this database. In order to determine which arc corresponds to a specific bridge from the Caltrans database, a tedious matching algorithm is used.

Table 1 TeleAtlas Database - Attributes Used for Routing

Attribute Name	Full Name
ID	Transportation Element Identification: • Road Element Identification • Ferry Connection Element Identification • Address Area Boundary Element Identification
Name	Official Street Name
RouteNum	Blank: Not applicable (Multiple values are separated by "/.")
Meters	Feature Length (in meters)
FRC	Functional Road Class: -1: Not applicable (FEATTYP 4165) 0: Main Road: Motorways 1: Roads not belonging to 'Main Road' major importance 2: Other Major Roads 3: Secondary Roads 4: Local Connecting Roads 5: Local Roads of High Importance 6: Local Roads 7: Local Roads of Minor Importance 8: Others
FOW	Form of Way: -1: Not applicable 1: Part of Motorway 2: Part of Multi Carriageway which is not a Motorway 3: Part of a Single Carriageway – Default 4: Part of a Roundabout 6: Part of an ETA: Parking Place 8: Part of an ETA: Unstructured Traffic Square 10: Part of a Slip Road 11: Part of a Service Road 12: Entrance / Exit to / from a Car Park 14: Part of a Pedestrian Zone 15: Part of a Walkway 17: Special Traffic Figures 20: Road for Authorities
SLIPRD	0: No Slip Road – Default 1: Parallel Road 2: Slip Road of a Grade Separate Crossing 3: Slip Road of a Crossing At Grade

FREEWAY	0: No Part of Freeway – Default
	1: Part of Freeway
ONEWAY	Direction of Traffic Flow:
	Blank: Open in Both Directions – Default
	TF: Open in Negative Direction
	FT: Open in positive Direction
	N: Closed in Both Direction
F_LEVEL	Begin Level: 0: default (Range: -9 to +9)
T_ELEV	End Level: 0: default (Range: -9 to +9)
KPH	Speed Limit
MINUTES	Travel Time
RTEDIR	Route Direction (Multiple values are separated by "/"):
	Blank: Not applicable
	Direction + | + Route Directional Text where Direction is:
	FT: Route Directional in the Positive Direction
	TF: Route Directional in the Negative Direction

Caltrans Bridge Inventory: Bridges damaged in an earthquake can cause disruptions in the traffic flow, hence are critical components of the highway network system. A bridge database with appropriate attributes, such as the bridge identification features, latitude and longitude, structural characteristics, and traffic flow is among the key components for post-earthquake traffic routing. The most up-to-date Caltrans bridge database is used in this research project.

GIS Platform: T-RoutER is developed within ArcView 3.2 [5]. ArcView 3.2 has a number of built-in tools that allow visualization, query and analysis of geographic data and has its own integrated object-oriented programming (OOP) language, Avenue. ArcView does not have an integrated capability for routing analysis. However, the Network Analyst (NA) extension for ArcView includes advanced tools that can be accessed through Avenue scripts, allowing delivery of sophisticated network analysis applications. The NA extension can find the most direct route between two locations and generate detailed directions across the route. T-RoutER builds on the capabilities of the NA to expand the single-origin single-destination computations to multiple-origin multiple-destination analysis. The ArcView environment was chosen because it is a product of the predominant GIS software manufacturer, ESRI, used by most highway transportation management organizations for other data and operations management.

A Routing Algorithm: The routing module uses the "find closest facility" procedure included in the ArcView Network Analyst, which is based on the Dijkstra's shortest path algorithm (ESRI, 1998). The main benefit of using the procedure included in the ArcView Network Analyst is that updating of the logical network after changing an arc is achieved very quickly, making it possible to use the application for near-real-time situations.

Application in GIS - Automation and Customization

The ArcView application window is customized to accommodate the needs of the

---
[5] ArcView 3.2 is a product of ESRI, Inc.

project to make it easy to use:
- Application buttons are added to load themes, such as bridge inventory, origin-destination points, ground motion levels experienced from a selected event and damaged bridges (determined either by field inspection or by seismic bridge vulnerability models).
- Pull-down menu for performing pre-event and post-event network analysis. In the post-event network analysis, links that represent damaged bridges are automatically selected and closed to traffic. The logical network is then modified to register the closed links and the network analysis is performed using the modified network.

Connecting the Bridge Database and the Highway Network

Both the bridge structural information from the Caltrans database and routing information from TeleAtlas database are essential for T-RoutER to predict the performance of each bridge during an earthquake, simulate closure of bridges, and create alternate driving routes. Caltrans uses a linear referencing system to identify features along their highways, including bridges. All of the small and large bridges, ramps and connections, as point features, are referenced by their postmiles. In the TeleAtlas highway network, only links representing major bridges, such as the San Francisco-Oakland Bay Bridge, are designated as bridges. In literature, one can find many methodologies that integrate point databases and polyline network systems using dynamic segmentation in GIS, but this requires a linear referencing system such as Caltrans' postmile method. None of the detailed commercial street databases, including TeleAtlas, use the postmile referencing system. In order to provide access to both databases, bridge point features from one database must be linked to bridge line features from the other. Therefore, an algorithm needed to be developed in order to determine the arcs in the TeleAtlas database that represent the bridges from the Caltrans database. Combining features from these two databases is not a simple, automatic process.

The algorithm that was developed works by comparing attributes such as: name, route number, county and type from each database. If a match is made, the bridge ID from the Caltrans database is assigned to the corresponding TeleAtlas link ID. The Caltrans bridge ID is then used to translate the expected damage state to the network functionality level. The expected damage state for a bridge can either be estimated from a vulnerability assessment or based on damage data collected from the field. For emergency response purposes, bridges with moderate damage or worse are closed to traffic. That is, the links corresponding to these bridges are made impassable, and the logical network topology is updated to represent the damaged highway network.

The success of the matching algorithm is highly dependent upon the naming conventions used in the two databases. TeleAtlas uses names and abbreviations such as Ave for Avenue and Blvd for Boulevard. In addition, Caltrans uses a standardized naming convention for bridges within its database; for overcrossings, underpasses and overheads, the abbreviations OC, UP and OH are used. The most challenging items to determine are separations and ramps since there are no attributes (e.g., route number or ramp name) to match between the two databases. Furthermore, the TeleAtlas database does not distinguish bridges over creeks, so these cannot be matched automatically.

With over 4000 bridges in the San Francisco Bay Area, the amount of time required for quality assurance is extensive. Yet, the process needs to be performed only once for each bridge and the benefits of having the integrated information outweighs the potential costs. The automation methods developed in this research can be used as a major stepping-stone towards accomplishing this goal.

Emergency Response Scenario

In this case study, T-RoutER is used to determine alternate routes between multiple origins and destinations before and after a scenario event of moment magnitude 7.0 on the Hayward fault. Alameda County was selected as the study area because the Hayward fault runs beneath several critical highway junctures. Hence, both building damage and bridge damage are expected to be high in this region following a magnitude-7.0 earthquake on the Hayward fault. Figure 2 shows the results of a building damage assessment calculated within HAZUS 99 (NIBS, 1999).

Input: The following data sets were used as input for the case study:
Highway Network: Tele Atlas MultiNet Shapefile 4.0 road network data with built-in topology is used as the highway network. The highway network shapefiles for the ten counties of the San Francisco Bay Area were merged: Alameda, Contra Costa, Marin, Mendocino, Napa, San Francisco, San Mateo, Santa Clara, Solano, and Sonoma. The 10 county street-level network database consists of over 360,000 links.

Bridge Inventory: Bridges in Alameda County are extracted from the Caltrans bridge inventory for the San Francisco Bay area. A total of 506 state bridges and 243 local bridges are located in Alameda County, and only the state bridges are considered in this case study.

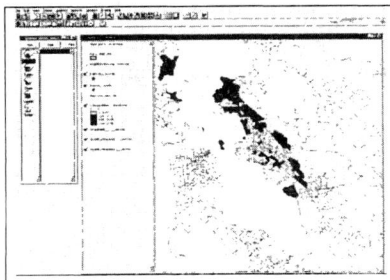

Figure 2: Damage Ratio for the Default Building Inventory in Alameda County

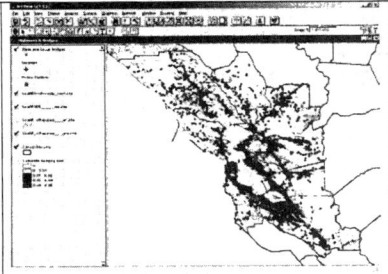

Figure 3: Highway Network Overlaid with the State and Local Bridge Inventory

Origin-destination Points: For demonstration purposes, all the hospitals and critical facilities (police stations and city buildings) in Alameda County are selected as potential origin and destination points. Figure 3 shows the bridge inventory overlaid upon the highway network. Also shown in this figure are the sets of origin-destination points, i.e., the hospitals in Alameda County.

Damaged Bridge Database: Damage calculations are based on scenario ground motion and bridge structural characteristics. The bridge inventory is then grouped by functionality level, and bridges with expected damage state 3 or higher (i.e., moderate damage, major damage and collapse) are assumed to be closed to traffic.

Pre-event Analyses First, the bridge inventory for the area is loaded to T-RoutER. The pre-event routing calculates the fastest (or the shortest) path among all pairs. T-RoutER determines all the shortest paths among all origin-destination combinations in less than a minute. The analyses produce paths for every combination of origin and destination. Each unique origin - destination route appears as a new layer in the table of contents. For example, for the six hospital origin-destination set, the routing analyses produce 30 layers. Travel times are computed based on the free volume speed and distance of a link. The free volume speed of each link is based on its functional class.

Post-Event Routing The routes among the selected origin-destinations are re-computed by selecting the *Post- Event Routing* option under the Routing pull down menu. The highway network database, the origin destination sets and the cost unit to be used in the analysis are then selected via dialog boxes. This process could be fully automated for pre-defined origins and destinations. Bridges that sustain moderate damage or worse are assumed to be inaccessible. The network topology is updated based on the damaged network.

Figure 6 shows pre- and post-earthquake routes for two-origin destination pairs. The dashed lines depict the pre-event routes and the solid lines depict the post-event routes. Blue solid and dashed lines show the routes from San Leandro Hospital (SLH) to Alta Bates Medical Center (ABMC). The red solid and dashed lines show the routes from San Leandro Hospital to Summit Medical Center (SMC).

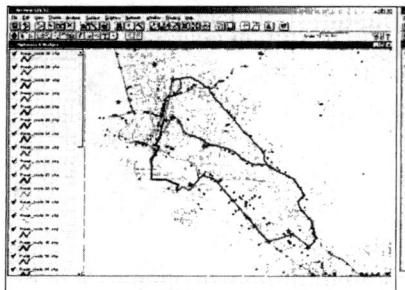	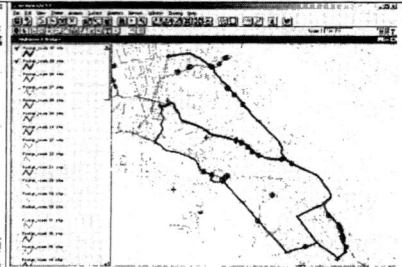
Figure 4. Pre-event routes among selected origins and destinations.	Figure 5. Pre- and Post-event Routes between Two Selected Origin – Destination Pairs (fastest path)

Table 2 lists the pre-event and post-event travel times for the selected origin-destination pairs. The effect of bridge closure is observed in the increased travel times. The origin destination pairs with increased travel time are highlighted in yellow.

Table 2. Pre-event and Post-event Travel Times for Different Origin-Destination Pairs

Origin Label	Destination Label	Pre_Event Travel Time (min)	Post_Event Travel Time (min)
SAN LEANDRO HOSPITAL	ALTA BATES MEDICAL CENTER	17	20
	SUMMIT MEDICAL CENTER	15	18
	KAISER PERMANENTE OAKLAND	14	18
	ALAMEDA HOSPITAL	13	14
	ALAMEDA COUNTY MED CENTER-HIGH	12	15

ACKNOWLEDGEMENTS

This research was partially supported by the Pacific Earthquake Engineering Center Grant No. SA3487 and the UPS Foundation Grant to Stanford University. Their support is gratefully acknowledged. The authors also express their gratitude to TeleAtlas, Inc. and ESRI, Inc. for providing data and help in the development of the software.

REFERENCES

Association for Geographic Information (AGI), GIS Dictionary (1999): http://www.geo.ed.ac.uk/agidexe/term?890

Basoz, N., and Mander, J. 1999. Enhancement of the Highway Transportation Module in HAZUS, in Report to National Institute of Building Sciences. Washington, D.C.: [GET PUB].

Boore, D., Joyner, W., and Fumal, T. 1997. "Equations for Estimating Horizontal Response Spectra and Peak Acceleration from North American Earthquakes: A Summary of Recent Work", *Seism.Res.Let.*, 687(1), 128-154

ESRI. (1998). "ArcView Network Analyst, An ESRI White Paper", http://www.esri.com/library/whitepapers/pdfs/ana0498.pdf

ESRI. (1999). Glossary. http://www.esri.com/library/glossary/i_l.html

GIS-T. (2002). "GIS-T 2002 State Summary",

Kiremidjian, A.S., Moore J., Basoz, N., Burnell K., Fan Y. and Hortacsu A. (2002). "Earthquake Risk Assessment for Transportation Systems: Analysis of Pre-Retrofitted System", Proceedings for the 7th *National Conference on Earthquake Engineering*, EERI, Boston, July 21-25.

Transportation Research Board (TRB), (2002). TRB 81st Annual Meeting, Washington, D.C., January 13-1

Large-Displacement Soil-Structure Interaction Facility for Lifeline Systems

S.L. Jones[1], T.D. O'Rourke[2], H.E. Stewart[3], and S.L. Billington[4]

Abstract

Lifeline systems for the delivery of electric power, gas and liquid fuels, telecommunications, transportation, wastewater facilities, and water supply represent some of the most critical features within our built environment. A two-year project has been funded at Cornell University and Rensselaer Polytechnic Institute (RPI) as part of the George E. Brown, Jr. Network for Earthquake Engineering Simulation (NEES) of the National Science Foundation. This project will develop advanced simulation and experimental evaluation of key lifeline components under earthquake conditions. This paper describes the experimental facilities planned at Cornell and RPI. The problems of soil-structure interaction and above-ground structural response that can be addressed through physical simulation with the facility are discussed.

Introduction

Lifeline systems are essential for civil infrastructure because they deliver the resources and services needed to sustain a modern community. Lifelines are often grouped into six principal systems: electric power, gas and liquid fuels, telecommunications, transportation, wastewater facilities, and water supply. When an earthquake strikes, life and property are threatened in the short term when functional water supply, transportation systems, electric power, and telecommunications either fail or lose their capabilities during emergency operations. In the long term, earthquake recovery is prolonged, especially when significant construction is required to rehabilitate damaged facilities.

---

[1] Research Associate, School of Civil and Environmental Engineering, Hollister Hall, Cornell University, Ithaca, NY 14853-3501; phone 607-255-3697; slj25@cornell.edu
[2] Thomas R. Briggs Professor of Engineering, School of Civil and Environmental Engineering, Hollister Hall, Cornell University, Ithaca, NY 14853-3501; phone 607-255-6470; tdo1@cornell.edu
[3] Associate Professor, Director of the Civil Infrastructure Laboratory, School of Civil and Environmental Engineering, Hollister Hall, Cornell University, Ithaca, NY 14853-3501; phone 607-255-4734; hes1@cornell.edu
[4] Clare Boothe Luce Assistant Professor, Department of Civil and Environmental Engineering, Stanford University, Stanford, CA 94305-4020; phone 650-723-4125; billington@stanford.edu

There is a compelling need in the George E. Brown Jr. Network for Earthquake Engineering Simulation (NEES) for experimental and testing facilities to evaluate lifeline earthquake behavior. Not only are experimental facilities required for investigating the aboveground response of structures, such as viaducts and bridges, but equipment is needed to investigate the soil-structure interaction of underground lifeline components. In congested urban and suburban environments, large portions of lifeline systems are buried or constructed underground. Understanding how ground deformation affects buried lifelines, therefore, is a critical aspect of earthquake engineering, which needs to be addressed in NEES by advanced laboratory experiments and computational modeling.

Potential Research Projects

The NEES facility at Cornell University and RPI has been designed to address several classes of research projects not covered by the other equipment sites in NEES. Three project classes in particular have been prominent in planning the facility; they are discussed under the subheadings that follow.

Soil-Structure Interaction Under Permanent Ground Deformation. It has long been recognized that the most serious damage to underground lifelines during an earthquake is caused by permanent ground deformation (PGD) [e.g., O'Rourke, 1998]. It is not possible to model with accuracy the soil displacement patterns at all potentially vulnerable locations. Nevertheless, it is possible to set upper bound deformation effects on buried lifelines by simplifying spatially distributed PGD as movement concentrated along planes of soil failure. Figure 1 illustrates the concept, which provides the basis for laboratory simulation of the most severe PGD effects associated with surface faulting, liquefaction-induced lateral spread, and landslides.

The laboratory equipment will have the capability of imposing abrupt soil displacements on buried lifelines consistent with the type of PGD effects at fault crossings and the margins of lateral spreads and landslides. As shown in Figure 1, relative displacement is generated along a moveable interface between two test basins, or boxes, containing soil and the buried lifeline. The lifeline is buried in soil that is placed and compacted according to field construction practice. The scale of the experimental boxes is chosen by computational modeling and previous test experience such that the soil-lifeline interaction is unaffected by the boundaries of the test facility. As shown in Figure 1, it will be possible to evaluate many different conditions with the experimental facility, including different types of pipelines, electric conduits, and telecommunication cables. Straight pipe sections and pipelines with elbows and tees can also be investigated. Steel, plastic, ductile iron, reinforced concrete piping, as well as a variety of specialized joints, coatings, and retrofitting techniques are viable candidates for testing. Soil-lifeline interaction for different soils, unit weights, and moisture contents, depths of burial, and pipeline diameters can be investigated.

Even with the advanced capabilities of the proposed facility, labor and preparation demands place practical limitations on the number of full-scale tests that can be performed. Large-scale tests are needed to benchmark performance for a

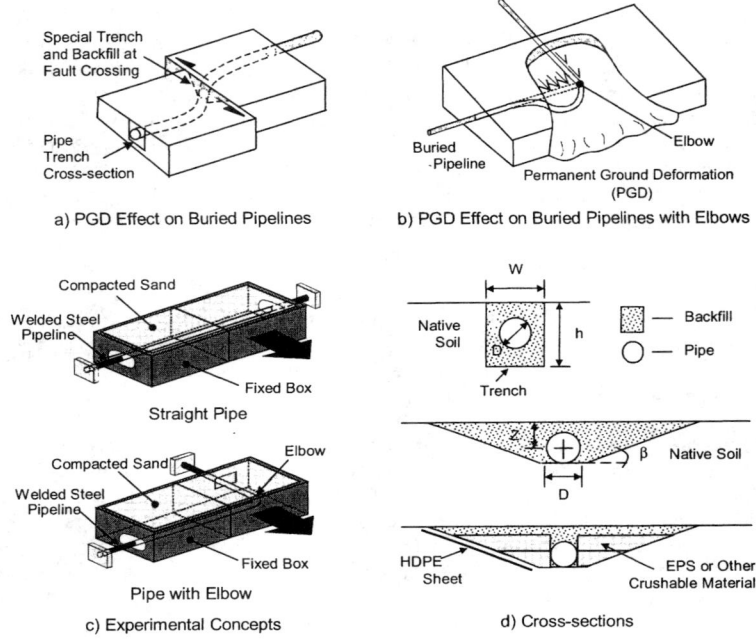

Figure 1. PGD Effects on Buried Lifelines

limited number of cases. To expand experimental capabilities for additional sensitivity studies without overloading the human resources needed for experimental preparation, this facility includes special centrifuge testing equipment that will be installed and operated at RPI. In addition to extending the number of variables that can be explored by experiment, the RPI centrifuge equipment will also be valuable for planning and designing large-scale experiments in the Winter Laboratory.

Soil-Structure Interface Interactions. Soil-structure interface problems involve locations where abrupt transitions from structure to soil create localized stresses and deformations. As illustrated in Figure 2, examples include bridge abutments where a number of different cables and conduits may transition from soil through the abutment and/or other structural elements. Additional examples include basement and vault penetrations of cable and conduits. At these locations, transient motion of the structure and adjacent soil can be significantly out of phase. Furthermore, settlement can occur in the adjacent soil, thereby imposing permanent ground deformation at the same time transient movements take place. Penetrations of structural walls and abutments have been identified as one of the most important issues for the earthquake resistant design of lifelines (e.g., Committee on Gas and Liquid Fuel Lifelines, 1984).

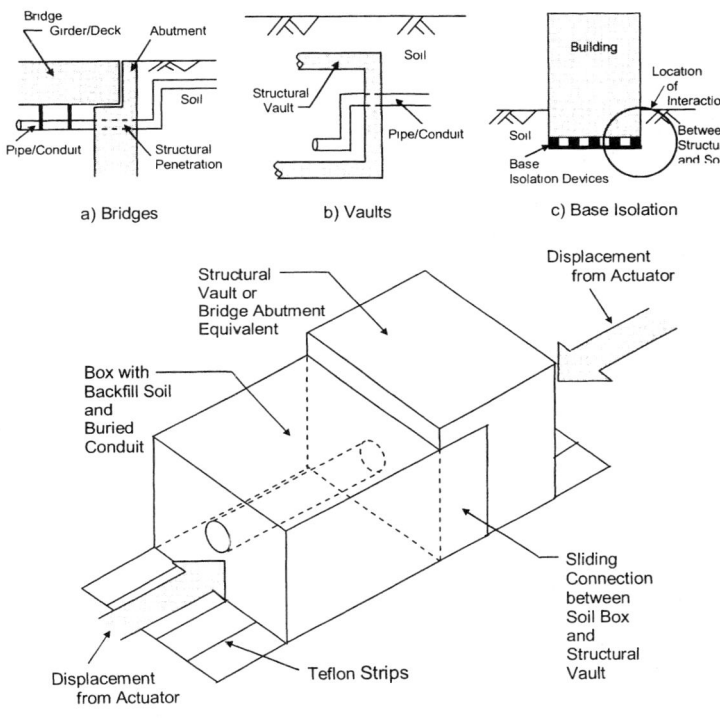

Figure 2. Soil-Structure Interface Interactions

This experimental facility will have the ability to simulate complex interactions at soil-structure interfaces. The experimental concept is shown in Figure 2d. An actuator can apply lateral displacements to a structural vault or bridge abutment element at the same time another actuator applies displacements to a test box with backfill soil and a buried conduit that penetrates the structural element. A special sliding connection can be fabricated to allow relative movement between the test box and structural element. Teflon strips will allow for low-friction sliding of the experimental members.

Highly Ductile Structural Response. The predominant emphasis in earthquake-resistant design practice is on ductility. In reinforced concrete, ductility traditionally has been achieved with monolithic construction and proper reinforcement detailing, such as adequate concrete confinement. With steel structures, moment-resisting frames with adequate connection details are a common approach to ductile design. In highway structures, innovative restrainers and isolation devices also are becoming more common. The limits of ductility are being pushed, and further improvements in ductility will continue into the future.

Highly ductile systems must be well characterized to facilitate performance-based design. Such characterization requires careful experimentation coupled with the development of appropriate simulation approaches. Many new bridge design and retrofit options that use advanced materials and new combinations of traditional materials are exhibiting large ductilities. An example of a new ductile system is that of precast segmental concrete bridge piers with unbonded vertical post-tensioning and localized use of highly ductile fiber-reinforced concrete. Unbonded post-tensioned bridge pier systems have been receiving particular attention in seismic research. The use of advanced ductile materials in combination with unbonded post-tensioning is new and has recently been explored at Cornell University.

One-fifth scale experiments on partial columns with unbonded post-tensioning and localized use of the ductile composites demonstrated the feasibility of this system (Yoon et al., 2002). As shown in Figures 3a and 3b, it was found that the highly ductile concrete located in hinge regions maintains its integrity whereas traditional concrete did not. Beyond drifts of 8%, the unbonded post-tensioning did not yield and there were minimal residual displacements. Experiments recently performed on near full-scale columns further demonstrated the feasibility of the proposed system (Rouse and Billington, 2003). The response was less ductile in the large-scale designs tested (4-5% drift). Alternate designs could ensure higher ductilities and should be tested in full pier configurations.

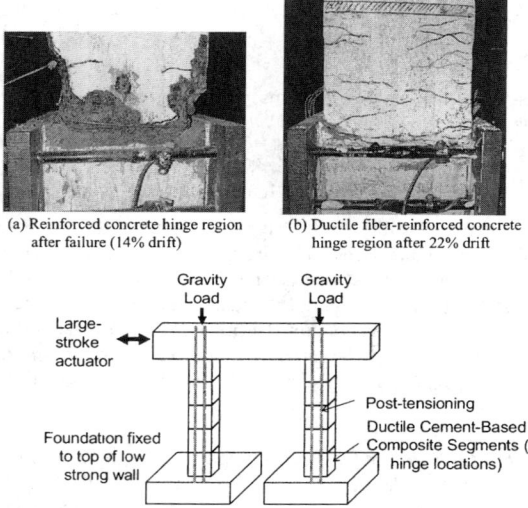

(a) Reinforced concrete hinge region after failure (14% drift)

(b) Ductile fiber-reinforced concrete hinge region after 22% drift

(c) Example bridge support system to be tested using proposed NEES equipment

Figure 3. New highly ductile materials and example of a highway structure needing experimental verification under large-displacement cycles.

Figure 3c illustrates the experimental concept of full bridge pier tests. Because the post-tensioning is unbonded, there is no localized yielding of the post-tensioned steel. Strain induced in the bridge columns is spread along the entire length of the post-tensioning. Therefore full bridge piers using the full length of strand are necessary for accurate investigation of such systems.

Experimental Facilities

The NEES equipment will be housed in the Winter Laboratory at Cornell University. Figure 4 provides an expanded view of the existing Winter Lab highbay within which elements of the NEES equipment system are shown. The strong walls are modular and can be assembled for a maximum 17-m length for the low wall and 7.2-m height for the high wall. There will be two ±0.91-m actuators and one ±0.63-m actuator. Soil will be stored in special bins recessed into the walls to conserve space. Room is available for supplemental soil storage in the highbay should a future experiment require additional volumes of soil. A portable conveyor system provides rapid movement and placement of soil. Nominal soil test boxes are shown. The dimensions of the boxes need to be chosen according to the purpose and type of experiment. Room is available for boxes as long as 20m. The remainder of this section provides a summary of the NEES equipment and its performance specifications (Tables 1-5).

Large-Displacement Actuators and Servo-Hydraulics. Servo-hydraulic actuators and ancillary hydraulic equipment are necessary to support large-displacement physical testing for lifeline systems. Recent testing at Cornell in collaboration with Tokyo Gas has involved the largest laboratory tests ever performed of pipeline response to permanent ground deformation to improve design and siting procedures for steel pipelines with elbows. The motions imposed on the test system were on the order of a meter so that full soil-structure interaction could be mobilized. Multiple actuators are proposed with one-way strokes on the order of 2m to provide unique testing equipment that can be used on a very wide range of buried and above-ground lifeline systems. These actuators and supporting hydraulic equipment will provide state-of-the-art systems not available at other experimental locations.

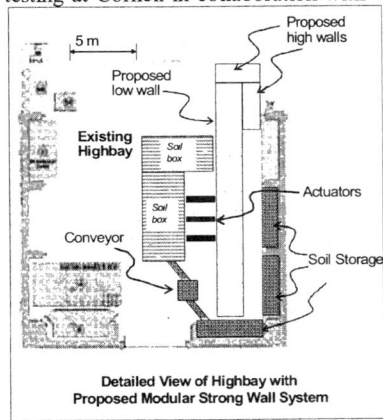

Figure 4. Expanded view of Winter Lab Highbay with NEES Equipment

Table 1. Servo-Hydraulics Performance Specifications

Large-Displacement Actuators and Servo-Hydraulics	Performance Specification
Linear Hydraulic Actuators	Two actuators with load capacities of 295 kN tension, 498 kN compression, strokes of +/- 0.91 meters. One actuator with load capacity of 445 kN tension, 649 kN compression, stroke of +/- 0.63 m.
Hydraulic Power Supplies	Servovalves, manifolds, and pump with flow rates and capacities for large actuator movements and simultaneous use of multiple actuators.
Electronic Controls	Independent control of either load or displacement on multiple actuators in simultaneous use.
Hydraulic Wedge Grips	Apply up to 220 kN tension to gripped material while ensuring a true alignment of axial force; grips should not slip in the direction of loading.

Data Acquisition and Sensors. Upgraded high-speed data acquisition systems will be assembled using a variety of components. Two Pentium 4 computers will be interfaced with high-speed multiplexers, signal conditioners, and data converter boards. The data acquisition systems will be interfaced with the servo-hydraulic system controls and connected to the Internet. The main sensors consist of an advanced fiber-optic signal conditioning unit and large-stroke displacement transducers. The fiber optic instrumentation consists of a high-resolution, high-precision system. This is a high-speed sensor conditioner that can adapt to slow or fast testing (sampling rates up to 1000Hz). All data acquisition systems will be capable of multi-channel measurements of temperature, pressure, force, displacement, or strain using a common sensor-conditioning unit with interchangeable sensors. Magneto-strictive displacement measuring devices with 2-meter ranges also will be used. These devices are a necessary measuring tool for large-displacement SFSI testing.

Modular Reaction Walls. Experiments on lifelines can be performed in numerous ways using a segmentally precast, post-tensioned concrete strong wall/floor assembly. The baseline assembly would be made up of a long, low segmental box girder along the existing lab floor with modular high walls perpendicular to each other and forming a corner on one end (see Figure 5). The low box segments would

Table 2. Data Acquisition Performance Specifications

Data Acquisition Systems	Performance Specification
Computers	High-speed, large storage capacity, Internet connectivity
A/D boards. Multiplexers	16-bit resolution, expandable for 128 to 256 data channels
Signal Conditioning	Stable power supply; low noise; independent variable gain; capable of using a wide variety of transducers
Sensors	Large displacement (up to 2 meter), precision and accuracy, compatibility with signal conditioning and other control systems, fiber-optic system capable of measuring strains up to 5000 to 10000 microstrain, laser extensometers for large displacement measurements.

Table 3. Modular Reaction Wall Performance Specifications

Modular Reaction Walls	Performance Specification
Low strong wall/box	Must resist lateral loads of 675 kN locally and 1350 kN overall anywhere along the height; must resist local vertical loads of 900 kN; must be match-cast, precast so as to be easily post-tensioned to form a long, low wall and be stackable for storage; each segment must weigh less than 89 kN to use existing overhead crane; must be hollow to allow for access from within; must be able to post-tension to both high walls.
High strong walls	Must resist lateral loads of 900 kN at a height of 5m from a fixed base; must resist vertical tensile/compressive loads of 1800 kN; must be able to post-tension to the low strong wall/box in two directions; must be able to post-tension to perpendicular high strong wall to facilitate lateral loading in two directions.
Floor anchor system	Must have a minimum shear capacity of 1350 kN to resist lateral forces from soil box experiments.

form a maximum length of 17 m off of which the soil box experiments on buried lifelines will react. Simple extension of the low strong wall to include two narrow high walls at one end broadens the possibilities of shared use of the proposed NEES site. The top surface of the low wall will be used for a variety of above-ground lifeline testing including highway component and system testing as well as structural pipe testing prior to the soil-structure interaction tests. On this surface, vertical loads can be applied to bridge girders, substructure components and bridge connections. In the raised wall portion of the assembly these components and systems can be tested with lateral loads in two directions. Vertical loads can be supported off of the low wall acting as a strong floor or off of the high walls through an attached load frame. Experiments on the top surface of the low wall can take place without interfering with

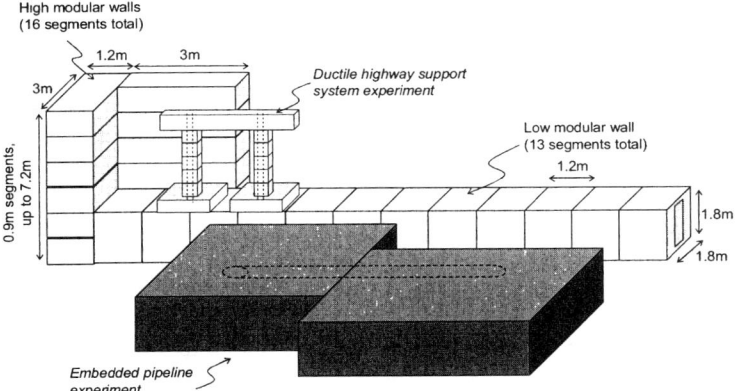

Figure 5. Schematic of modular strong wall system for large-displacement lifeline experiments

the floor space where the soil box experiments would be set up. In addition, when the floor space is not being used for soil box experiments, various structural configurations could be tested under lateral loads laying flat. A limited version of this arrangement was recently used in the Winter Lab for the research on unbonded post-tensioned concrete columns. Finally, the low wall could be built in two parts with portions of the high walls stacked on the inside of the openings to form abutments. These two abutments could then be used as reaction walls to conduct soil-structure interaction experiments in an axial configuration.

Soil Storage and Conveyance. A soil storage system capable of holding and handling large quantities of soil for full-scale and near full-scale soil-structure interaction experiments on pipelines and bridge systems will be constructed in the crane bay area of the Winter Lab. The crane bay has 5.5 m high, 0.3-m-thick concrete walls spanning the 4.5 m horizontal distances between heavy, laced, concrete jacketed columns that support the roof. The columns are jacketed for their lower 5.5 m and unjacketed for the remaining 6.7 m. Steel beams with an exposed flange were cast into the concrete columns. The flanges are used to connect other structural members to the columns. The columns are approximately 1.2 m deep and there is approximately 4 m between the inner edges of any two adjacent columns. This volume is reduced in the lower portions of the units because of the tapered sections. Reinforced steel plating will be placed between the inner steel flanges of adjacent columns to create the basic storage unit. A conveyor belt assembly with a cleated belt trough slider bed belt will be used to charge the soil bins. The front of the soil storage containment bins will have sliding steel discharge panels. Discharged soil will be moved with an existing small Bobcat loader, a trip-release concrete bucket and overhead crane, or the conveyor belt.

Centrifuge Containers. The containers at the RPI centrifuge will use two hydraulic cylinders to produce localized shear strains along one or two vertical interfaces in a soil model while being spun at centrifugal accelerations of up to 75g. Load cells directly connected between each actuator and the movable portions of the container measure the shearing force applied by the actuators. The maximum achievable displacement is 8 cm (6m prototype units). Motion of each actuator is precisely controlled using a servo-valve and feedback control system. Using a function generator or computer equipped with a DAC interface board, a variety of input strain distributions and time histories can be created. The containers will be manufactured from high-strength aluminum alloy. The moving portions of the container are

Table 4. Soil Storage Performance Specifications

Soil-Storage	Performance Specification
Soil Bins	On-site storage of on the order of 40 to 45 cubic meters of soil used in large-scale movable split soil boxes. Inside storage for moisture control and to avoid freezing. Minimize internal use of floor space in crane bay.
Conveyor System	Mechanical movement of large quantities of soil to and from storage bins. Portability. Flexible configurations.

Table 5. Centrifuge Containers Performance Specifications

Centrifuge Containers	Performance Specification
Split Boxes	Overall dimensions: 108 cm L x 69 cm W x 36 cm H;
	Inside container dimensions: Model dimensions of 100 cm L x 36 cm W x 20 cm H; Prototype dimensions at 75g of 75m x 27m x 15m
	Empty weight of 900 N
	Displacement of movable sections = 0 to 8 cm
	Operating hydraulic pressure = 8.3 MPa
	Maximum Actuator force = 8.9 kN

supported and guided using roller bearings to provide precise movement with minimal friction. The sliding interface between the fixed and movable portions of the container utilizes low-friction Teflon seals protected by steel shields. When used with a suitable Teflon sheet liner, this design effectively excludes soil from the interface, maximizing the service life of the seals. One container will have three sections having two actuators and a two-channel displacement control system. In this concept, one section will be fixed, and either one or both of the other sections can be moved. If two sections are moved, they can be moved either together or independently. In this way a wide variety of strain configurations can be modeled.

Acknowledgements

The authors gratefully acknowledge the contributions of Professors M. J. O'Rourke and T. Abdoun of RPI.

This work was supported in part by the George E. Brown, Jr. Network for Earthquake Engineering Simulation (NEES) Program of the National Science Foundation under Award Number CMS-0217366.

References

O'Rourke, T.D. (1998), "An Overview of Geotechnical and Lifeline Earthquake Engineering," *Geotechnical Special Publication No. 75*, ASCE, Reston, VA.

Committee on Gas and Liquid Fuel Lifelines (1984), *Guidelines for the Design of Oil and Gas Pipeline Systems*, ASCE, Technical Council on Lifeline Earthquake Engineering, Reston, VA.

Rouse, JM and Billington, SL (2003), "Behavior of Bridge Piers with Ductile Fiber-Reinforced Hinge Regions and Vertical, Unbonded Post-Tensioning," Proc. *fib* Symposium on Concrete Structures in Seismic Regions, Greece, in press.

Yoon, JK, Billington, SL, Rouse, JM, (2002), "Precast Segmental Bridge Piers with Unbonded Post-tensioning and Ductile, Fiber Reinforced Concrete for Seismic Applications," *Proceedings 7th NCEE*, Boston, July.

Seismically Induced Lateral Earth Pressures on a Cantilever Retaining Wall

Russell A. Green[1], C. Guney Olgun[2], Robert M. Ebeling[3], and Wanda I. Cameron[4]

Abstract
A series of non-linear dynamic response analyses of a cantilever retaining wall were performed to assess the appropriateness of the Mononobe-Okabe method for determining the seismically induced lateral earth pressures on the stem of the wall. For the wall analyzed, it was found that at very low levels of acceleration the induced pressures were in general agreement with those predicted by the Mononobe-Okabe method. However, as the accelerations increased to those expected in regions of moderate seismicity, the induced pressures are larger than those predicted by the Mononobe-Okabe method. This deviation is attributed to the flexibility of the retaining wall system and to the observation that the driving soil wedge does not respond monolithically, but rather as several wedges.

Introduction
Earth retaining structures constitute an integral part of the lifelines across the US, extensively being used for port facilities, cuts along highways, bridge abutments, etc. In current design standards and guidelines (e.g., ASCE 4-98), the Mononobe-Okabe method (Mononobe and Matsuo, 1929; Okabe, 1924) is commonly specified for determining seismic earth pressures for which the retaining structure must resist. Inherent in the Mononobe-Okabe method are the assumptions that the earth retaining structure and the driving soil wedge act as rigid bodies, which have been shown to be reasonable assumptions for large gravity type retaining structures (e.g., Seed and Whitman, 1970).

The focus of this paper is to assess the validity of the Mononobe-Okabe method for determining seismically induced lateral earth pressures on a more flexible cantilever retaining wall, particularly the stem portion of the wall. In this vein, a series of non-linear dynamic response analyses of a cantilever retaining wall were performed using the commercially available computer program FLAC (Itasca Consulting Group, Inc.). The analyses consisted of an incremental construction of the wall and placement of the backfill, followed by dynamic response analyses.

---

[1] Associate Member, ASCE; Assistant Professor, Department of Civil and Environmental Engineering, University of Michigan, 2372 GG Brown Building, Ann Arbor, MI 48109-2125; rugreen@engin.umich.edu
[2] Doctoral Candidate, Department of Civil and Environmental Engineering, Virginia Polytechnic Institute and State University, Blacksburg, VA.
[3] Senior Research Engineer, Information Technology Laboratory, US Army Engineer Research and Development Center, Vicksburg, MS.
[4] Graduate Student Research Assistant, Department of Civil and Environmental Engineering, University of Michigan, Ann Arbor, MI

In the following, the earth pressures determined from the FLAC analyses are compared with those predicted using the Mononobe-Okabe method. However, prior to the comparison, a brief review of the Mononobe-Okabe method is presented.

Mononobe-Okabe Earth Pressures

The Mononobe-Okabe method for determining seismically induced active and passive lateral earth pressures is based on limit equilibrium and is an extension of the Coulomb theory for static stress conditions. The method entails three fundamental assumptions (e.g., Seed and Whitman, 1970):

1. Wall movement is sufficient to ensure either active or passive conditions, as the case may be.

2. The driving soil wedge inducing the lateral earth pressures is formed by a planar failure surface starting at the heel of the wall and extending to the free surface of the backfill. Along this failure plane the maximum shear strength of the backfill is mobilized.

3. The driving soil wedge and the retaining structure act as rigid bodies and therefore experience uniform accelerations throughout the respective bodies.

As demonstrated by Dr. Ignacio Arango (Seed and Whitman, 1970), the dynamic earth pressures may be determined from analogous static conditions. Accordingly, the Mononobe-Okabe expressions for dynamic earth pressures can be derived from the Coulomb's expressions for static earth pressures. The analogous static conditions are achieved by rotating the wall-backfill system by an angle ψ, such that the vector sum of the horizontal and vertical inertial coefficients (k_h and k_v, respectively) is oriented vertically, where $\tan(\psi) = k_h / (1-k_v)$. This procedure is illustrated in Figures 1 and 2 for active and passive stress conditions, respectively. In regards to the mathematical expressions, the Mononobe-Okabe expressions can be derived from the Coulomb's expressions by replacing the static values for the total unit weight of the soil (γ_t), height of the wall (H), inclination of the backfill (β), and inclination of the wall face from the vertical (θ), with the corresponding dynamic values (i.e., γ_{td}, H_d, β_d, and θ_d). This substitution is demonstrated in the following set of equations.

Active case:
Static conditions (Coulomb's expression)

$$P_A = \frac{1}{2} \cdot \gamma_t \cdot H^2 \cdot K_A$$

$$K_A = \frac{\cos^2(\phi - \theta)}{\cos^2(\theta) \cdot \cos(\theta + \delta) \cdot \left[1 + \sqrt{\frac{\sin(\phi + \delta) \cdot \sin(\phi - \beta)}{\cos(\delta + \theta) \cdot \cos(\beta - \theta)}}\right]^2}$$

Dynamic conditions (Mononobe-Okabe expression)

$$P_{AE} = \frac{1}{2} \cdot \gamma_{td} \cdot H_d^2 \cdot K_A(\beta_d, \theta_d)$$

$$= \frac{1}{2} \cdot \gamma_t \cdot H^2 \cdot (1 - k_v) \cdot K_{AE}$$

$$K_{AE} = \frac{\cos^2(\phi - \theta - \psi)}{\cos(\psi) \cdot \cos^2(\theta) \cdot \cos(\delta + \theta + \psi) \cdot \left[1 + \sqrt{\frac{\sin(\phi + \delta) \cdot \sin(\phi - \beta - \psi)}{\cos(\delta + \theta + \psi) \cdot \cos(\beta - \theta)}}\right]^2}$$

Passive case:
Static conditions (Coulomb's expression)

$$P_P = \frac{1}{2} \cdot \gamma_t \cdot H^2 \cdot K_P$$

$$K_P = \frac{\cos^2(\phi + \theta)}{\cos^2(\theta) \cdot \cos(\delta - \theta) \cdot \left[1 - \sqrt{\frac{\sin(\phi + \delta) \cdot \sin(\phi + \beta)}{\cos(\delta - \theta) \cdot \cos(\beta - \theta)}}\right]^2}$$

Dynamic conditions (Mononobe-Okabe expression)

$$P_{PE} = \frac{1}{2} \cdot \gamma_{td} \cdot H_d^2 \cdot K_P(\beta_d, \theta_d)$$

$$= \frac{1}{2} \cdot \gamma_t \cdot H^2 \cdot (1 - k_v) \cdot K_{PE}$$

$$K_{PE} = \frac{\cos^2(\phi + \theta - \psi)}{\cos(\psi) \cdot \cos^2(\theta) \cdot \cos(\delta - \theta + \psi) \cdot \left[1 - \sqrt{\frac{\sin(\phi + \delta) \cdot \sin(\phi + \beta - \psi)}{\cos(\delta - \theta + \psi) \cdot \cos(\beta - \theta)}}\right]^2}$$

A plot of the Mononobe-Okabe active and passive earth pressure coefficients (K_{AE} and K_{PE}, respectively) as functions of the horizontal inertial coefficient (k_h) are shown in Figure 3a for $\beta = \delta = \theta = 0°$ and $\phi' = 35°$. As may be observed from this figure, when $k_h = 0$ (i.e., static conditions), the values of K_{AE} and K_{PE} are equivalent to Coulomb's active and passive coefficients (K_A and K_P, respectively). However, as k_h increases in value, K_{AE} becomes greater than K_A, and K_{PE} becomes less than K_P. This trend can be understood by referring back to Figures 1 and 2. In the active case

(Figure 1), as k_h increases, the analogous static condition is achieved by tilting the wall forward, thus increasing the inclination of the backfill and increasing the pressure induced on the wall. Correspondingly, in the passive case (Figure 2), as k_h increases, the analogous static condition is achieved by tilting the wall backward, thus decreasing the inclination of the backfill and decreasing the pressure induced on the wall.

The limiting pressures for both the active and passive cases occur when for the analogous static conditions, the inclination of the backfill equals the angle of internal friction (i.e., $\beta_d = \phi'$). At this point, the failure wedges become infinite in size, or synonymously, the angle of the failure planes equal the static inclinations of the backfill (i.e., $\alpha_{AE} = \beta$ and $\alpha_{PE} = \beta$, where both α_{AE} and α_{PE} decrease as k_h increases for the respective stress conditions). For the active case, no sized wall could restrain the backfill from movement, while in the passive case, the pressure induced on a wall restrained from movement becomes zero, as the backfill yields under its own inertial load.

FLAC Computed Earth Pressures on the Cantilever Wall

A series of non-linear dynamic response analyses were performed on the cantilever wall shown in Figure 4 using the finite difference program FLAC (Itasca Consulting Group, Inc.). The geometry and structural detailing of the wall analyzed were determined following the US Army Corps of Engineers static design procedures (Headquarters, US Army Corps of Engineers, 1989, 1992). Both the foundation and backfill soils were modeled as being elasto-plastic, and interface elements were used between the wall and the soil to allow relative movements and permanent displacements in the wall-soil system to occur. The wall and backfill were numerically constructed in 2 ft lifts, allowing for equilibrium of the stresses to occur between lift placements. Additional details of the wall design and FLAC modeling are presented in Green and Ebeling (2002).

Upon the completion of the numerical construction of the wall and placement of the backfill, the lateral pressures imposed on the stem of the wall were in good agreement with the active earth pressures determined using the Coulomb expressions. A series of dynamic response analyses were then performed using the same acceleration time history scaled to different peak ground accelerations (pga). For all the analyses, the computed stresses on the stem of the wall were in very good agreement with those predicted by the Mononobe-Okabe expressions at the early part of the time history where the accelerations were very low. However, at the larger levels of shaking, the Mononobe-Okabe expressions failed to predict the induced stresses on the stem of the wall. A comparison of the results is shown in Figure 3b. The data shown in this figure are the lateral earth pressure coefficients (K_{FLAC}) back-calculated from the FLAC results using the following expression:

$$K_{FLAC} = \frac{2 \cdot P_{FLAC}}{\gamma_t \cdot H^2 \cdot (1-k_v)}$$

where P_{FLAC} is the resultant of the FLAC computed stresses imposed on the stem of the wall. K_{FLAC} values were computed at times corresponding to the peaks in the time history of the horizontal inertial coefficient (k_h) acting away from the backfill (i.e., active-type conditions), wherein spurious high frequency spikes were filtered from the k_h time history.

The reason for the deviation of the FLAC computed stresses and those computed by the Mononobe-Okabe expressions can be understood from examining Figure 5. Shown in this figure is the deformed mesh from one of the FLAC analyses, wherein the deformations are magnified by a factor of 3. At large values of k_h directed away from the backfill, the induced inertial forces on the structural wedge cause it to simultaneously bend, rotate, and potentially slide away from the backfill, at which time a small wedge of soil or graben moves vertically downward. (The structural wedge consists of the cantilever wall and the backfill contained within; see Figure 4.) As the direction of k_h reverses (i.e., changes direction from away to towards the backfill), the graben prevents the structural wedge from returning to its undeformed shape, in effect locking in the elastic stresses resulting from the bending and rotation of the structural wedge.

This process is illustrated by the dashed arrows and corresponding data points in Figure 3a, wherein the initial stresses imposed on the stem of the wall correspond to active conditions. As k_h increases in the direction away from the backfill, the stresses on the stem increase according to the Mononobe-Okabe expressions for active conditions. However, upon reversal of the direction of k_h, the stresses imposed on the stem do not decrease as predicted by Mononobe-Okabe expression, but rather remain relatively constant. This stepwise increase in the locked-in stresses continues until the residual stresses imposed on the stem correspond to at-rest (or K_o) conditions, while the dynamically induced inertial stresses are superimposed on the locked-in residual stresses. The increase in residual stresses is clearly shown in Figure 6, wherein plots are shown of both the time history of k_h and of the resultant of the lateral stresses (P_{FLAC}) imposed on the stem of the retaining wall.

The locked-in residual stresses on the wall are not released by the slippage of the wall away from the backfill. This is because the "driving soil wedge" is not monolithic, but rather, in this case, consists of a graben and five driving soil wedges, with the later tending to move downward and away from the backfill as the wall slides outward. As a result, the graben "rides along" with the driving soil wedges maintaining its role of locking in the residual stresses.

Summary and Conclusions
For the cantilever retaining wall numerically modeled and analyzed, the stresses induced on the stem of the wall did not correspond with those predicted by the Mononobe-Okabe method. The reason for this deviation is attributed to the relative flexibility of the structural wedge and to the non-monolithic motion of the driving soil wedge, both of which violate assumptions inherent in the Mononobe-Okabe method. The dynamic response of the wall-backfill system was such that there was an

incremental increase from active to at-rest stress conditions in the residual stresses imposed on the stem of the retaining wall. The conclusions drawn from this study may not apply to retaining walls systems of differing geometry and/or material properties. Further research is required in order to draw more general conclusions regarding the appropriateness of the Mononobe-Okabe method to evaluate the dynamic pressures induced on cantilever retaining walls.

Acknowledgements

A portion of this study was funded by the Headquarters, US Army Corps of Engineers (HQUSACE) Civil Works Earthquake Engineering Research Program (EQEN). Permission was granted by the Chief of the US Army Corps of Engineers to publish this information. The first author benefited from several enlightening discussions with Professor Radoslaw Michalowski, University of Michigan, regarding the derivation of the Mononobe-Okabe expressions.

References

ASCE 4-86: Seismic Analysis of Safety-Related Nuclear Structures and Commentary on Standard for Seismic Analysis of Safety Related Nuclear Structures, American Society of Civil Engineers.

Green, R.A. and R.M. Ebeling (2002). Seismic Analysis of Cantilever Retaining Walls, Phase I, ERDC/ITL TR-02-3, Information Technology Laboratory, US Army Corps of Engineers, Engineer Research and Development Center, Vicksburg, MS. http://libweb.wes.army.mil/uhtbin/hyperion/ITL-TR-02-3.pdf

Headquarters, US Army Corps of Engineers (1989). Retaining and Flood Walls, EM 1110-2-2502, Washington, DC.

Headquarters, US Army Corps of Engineers (1992). Strength Design for Reinforced-Concrete Hydraulic Structures, EM 1110-2-2104, Washington, DC.

Itasca Consulting Group, Inc., FLAC (Fast Lagrangian Analysis of Continua), Minneapolis, MN.

Mononobe, N. and H. Matsuo (1929). On the Determination of Earth Pressure During Earthquake, Proceedings: World Engineering Congress, Tokyo, Vol IX, Part 1, 177-185.

Okabe, S. (1924). General Theory on Earth Pressure and Seismic Stability of Retaining Wall and Dam, Journal Japan Society of Civil Engineering, 10(6), 1277-1323, plus figures.

Seed, H.B. and R.V. Whitman (1970). Design of Earth Retaining Structures for Dynamic Loads, Lateral Stresses in the Ground and Design of Earth-Retaining Structures, ASCE, 103-147.

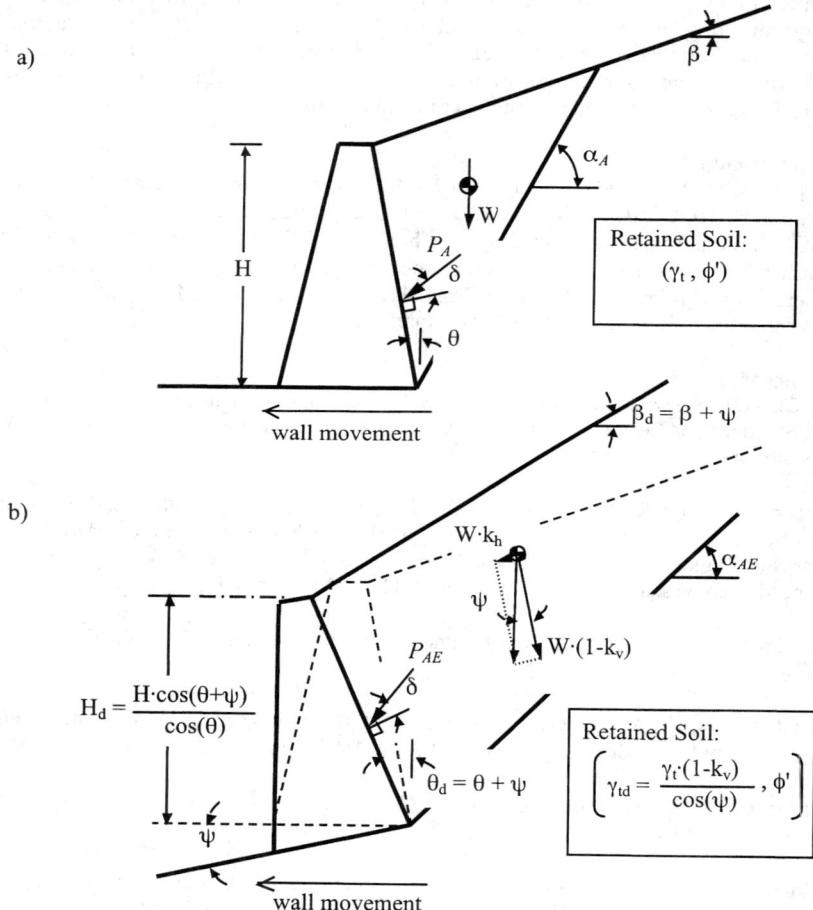

Figure 1. Active earth pressures for a) static conditions and b) analogous static conditions for the dynamic case.

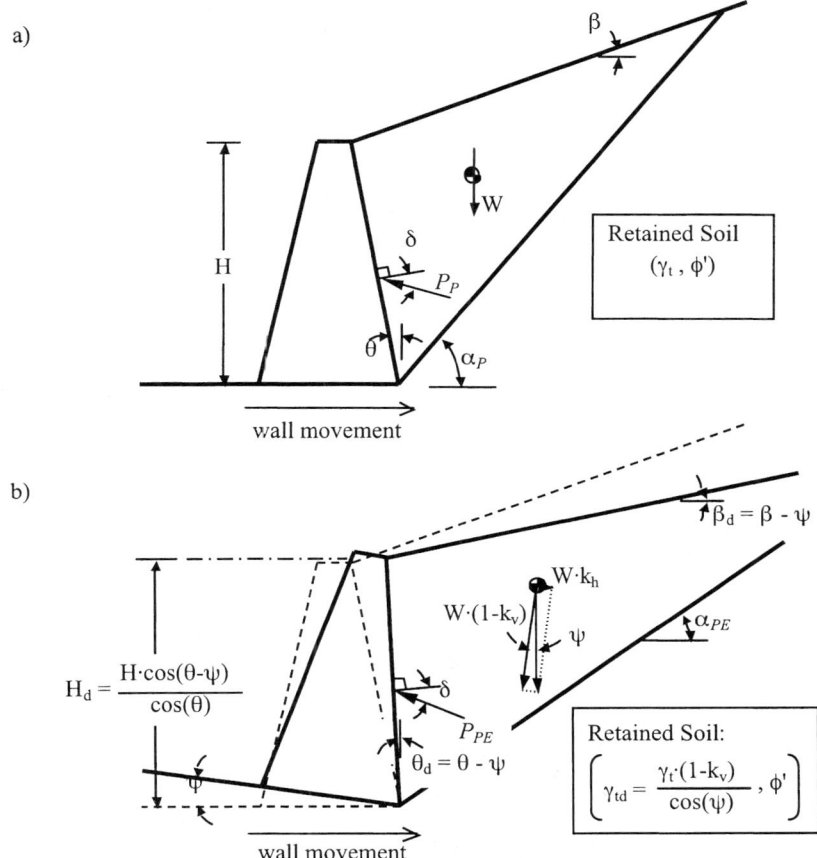

Figure 2. Passive earth pressures for a) static conditions and b) analogous static conditions for the dynamic case.

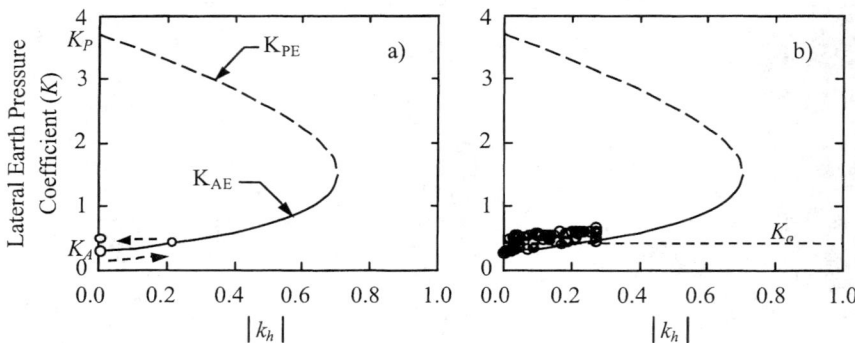

Figure 3. a) Mononobe-Okabe lateral earth pressure coefficients for $\beta = \theta = \delta = 0°$ and $\phi' = 35°$. b) Comparison of active lateral earth pressures (K_{FLAC}) back-calculated from FLAC results with values computed using the Mononobe-Okabe expressions.

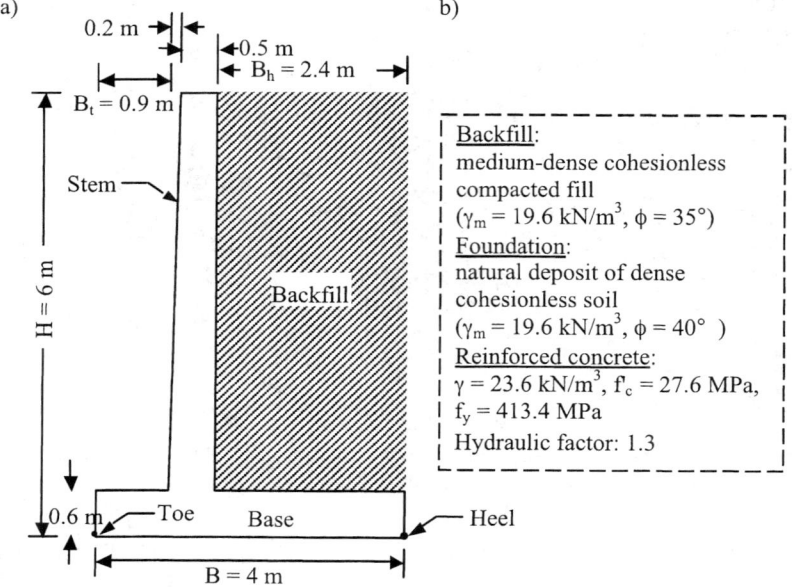

Figure 4. Cantilever retaining wall analyzed in FLAC a) geometry and b) material properties.

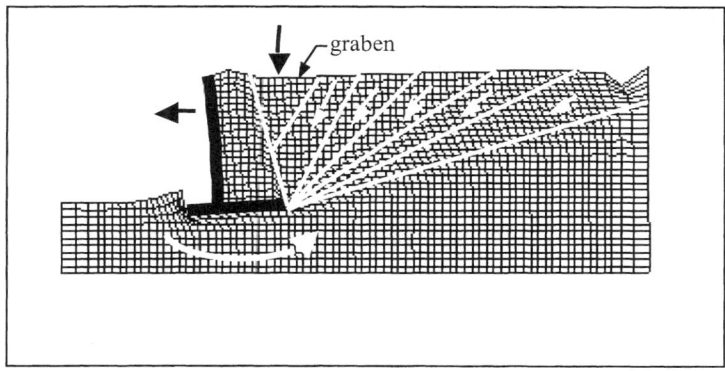

Figure 5. Annotated deformed mesh from one of the FLAC analyses; deformations magnified by a factor of 3. (Note: Toe of wall not initially embedded.)

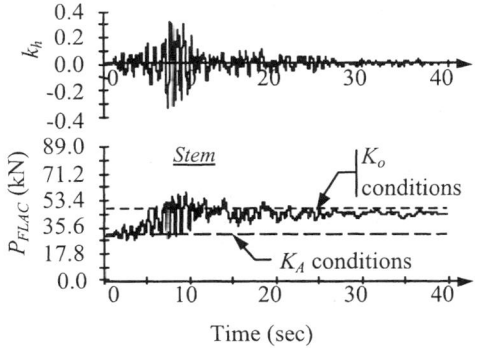

Figure 6. Time history of the horizontal inertial coefficient (k_h) at approximately the center of the structural wedge, and the time history of the resultant of the imposed stresses (P_{FLAC}) on the stem of the cantilever retaining wall.

A Simplified Two Dimensional Soil Model for the New Madrid Seismic Zone

Wei Zheng[1] and Ronaldo Luna[2]

Abstract

The New Madrid Seismic Zone (NMSZ) is one of the most seismically susceptible areas in United States. Deep alluvial soil deposit of the Mississippi Embayment overlies the NMSZ, which remains a major uncertainty for site response analysis. Based on the experiment data, a simple constitutive soil model is developed to take into account the influence of the effective confining pressure on the shear modulus degradation and the viscous damping development in soil, and also the shear strain and the plasticity index. This model is implemented into a two-dimensional finite element code in the time domain. The Rayleigh viscous damping scheme is applied in each finite element according to its shear strain level where the two significant modes of vibration are selected for damping matrix calculation. The new model is calibrated through the recorded motion at Treasure Island during Loma Prieta Earthquake (1989). Results show that this model could provide a reliable outcome with simple input parameters. The soil model is also implemented in a site response analysis in a Missouri bridge site near the NMSZ. The results are compared with those obtained from a SHAKE analysis. Based on these comparisons, the importance of the influence of the confining pressure on the seismic site response is evident.

Introduction

The New Madrid Seismic Zone (NMSZ) has experienced some of the largest magnitude (estimated 8.0 - 8.3) earthquake events in North American recorded history (1811-1812). Experts agree that similar or greater magnitude earthquakes will strike this region again. Earthquakes in the NMSZ are expected to have anomalously high frequency, long duration of strong motion and long recurrence intervals. Deep alluvial soils (up to 1000m) are widespread within the Mississippi embayment. These may amplify the seismic wave transmitted from rock to ground surface in a unique way which may lead to extensive damage. Understanding the effect of high confining pressure on the propagation of seismic waves in the NMSZ is important for the seismic behavior of the stratigraphy and geologic structure in the region.

However, the study of data on earthquake hazards in the NMSZ has been collected only in the past 20 years and no strong motion has been recorded since. Numerical models can be used to develop an understanding of wave propagation characteristics

---

[1] Graduate Student, Department of Civil, Architectural and Environmental Engineering, University of Missouri-Rolla, Rolla, 65409; phone: (573) 341-6232; wzheng@umr.edu.
[2] Associate Professor, Department of Civil, Architectural and Environmental Engineering, University of Missouri-Rolla, Rolla, 65409; phone: (573) 341-4484; rluna@umr.edu.

of the NMSZ. This paper describes a two-dimensional finite element soil model in an equivalent linear approach. The model takes into account the influence of very high confining pressure encountered in the NMSZ to the shear modulus degradation curve and the viscous damping curve. The wave propagation equations are solved in discrete time increments in the time domain. The model is verified through the case study to analyze the recorded motion at Treasure Island during Loma Prieta Earthquake (1989). The results from this model are compared with recorded motion and also that from SHAKE (Schnabel et al., 1972).

Nonlinear Site Response Analysis

The mechanical behavior of soil stress-strain relationship is quite nonlinear under seismic loading condition. Even at very small shear strain level (10^{-4}), the soils show shear modulus reduction. At the same time, the material damping is developed, which is increased with the cyclic shear strain. Numerous researchers (e.g., Hardin and Drnevich, 1972; Seed and Idriss, 1970) have performed the characterization of shear modulus degradation and damping curves for many soil types and provide an invaluable database for the analyses. Based on experimental data from 16 publications encompassing normally and overconsolidated clays (OCR=1-15), as well as sands, Vucetic and Dobry (1991) developed the charts for the relationship between G/G_{max} vs. γ_c and λ vs. γ_c (Figure 1). It is shown that the plasticity index (PI) is the main factor to control these relationships. However, Ishibashi (1992) pointed out that the method of Vucetic and Dobry did not include one of the significant parameters, the effective mean normal stress, particularly for soils of low plasticity (Figure 2). The effective mean normal stress (or effective overburden) will be a significant factor to influence the wave propagation in the NMSZ due to the very deep soil deposit.

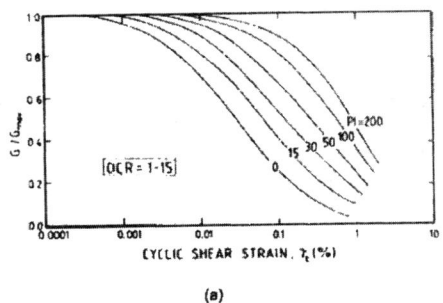
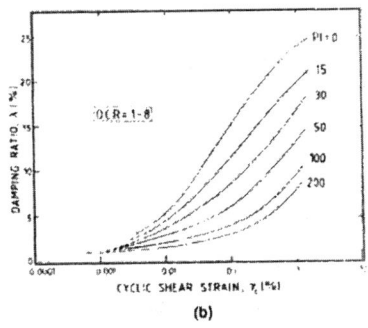

Figure 1. Relations Between G/G_{max} vs. γ_c and λ vs. γ_c Curves and Soil Plasticity for Normally and Overconsolidated Soils (Vucetic and Dobry, 1991)

Once the dynamic soil properties for the site response analysis are defined, the problem of the site response analysis is how to determine the soil response to the prescribed dynamic loading. Thus, the primary objective is to accurately model the soil non-linearity and yet accomplish this with a computationally efficient formulation. Methods used to describe nonlinear soil response for seismic site response analyses fall into one of three categories: equivalent linear models, cyclic nonlinear models and advanced constitutive models. Of these, equivalent linear models are the simplest and most commonly used but have limited ability to represent many aspects of soil behavior. On the other hand, advanced constitutive models can represent many details of dynamic soil behavior, but their complexity and difficulty of calibration currently renders them impractical for many common geotechnical earthquake engineering problems. The cyclic nonlinear models are intermediate in complexity and practicability.

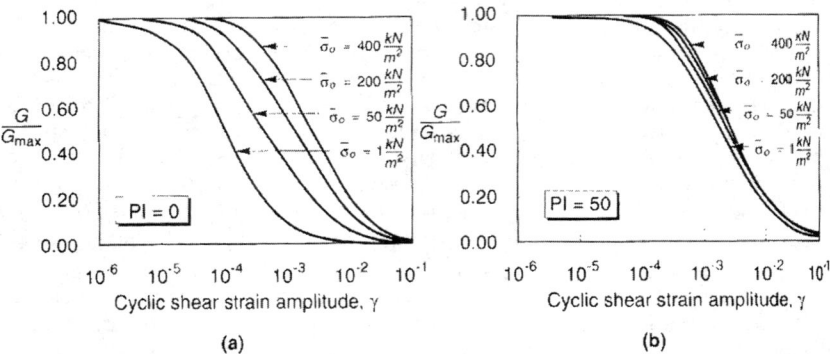

Figure 2 Influence of Mean Effective Confining Pressure on Modulus Reduction Curves for (a) nonplastic (PI = 0) soil, and (b) plastic (PI = 50) soil (Ishibashi, 1992).

Two-dimensional Wave Propagation Model

Generally, real seismic waves generated from an earthquake are propagated in a three-dimensional continuous medium. However, modeling the nonlinear soil behavior as well as the three-dimensional wave propagation is extremely difficult. In most situations the main response in the soil deposit can be adequately approximated with one or two-dimensional vertical propagation of shear waves.

This paper proposes a two-dimensional equivalent linear soil model in the time domain. The analysis of dynamic site response requires solving the global dynamic equation of motion given by the following equation in matrix form:

$$[M]\{\ddot{u}\} + [C]\{\dot{u}\} + [K]\{u\} = P(t) \tag{1}$$

where $[M]$, $[C]$ and $[K]$ are the global mass, damping and stiffness matrices for assemblage of elements, respectively; $\{\ddot{u}\}$, $\{\dot{u}\}$ and $\{u\}$ are the relative nodal acceleration, velocity and displacement vectors, respectively. $P(t)$ is the load vector, which for base excitation can be written as:

$$P(t) = -[M]\{I\}\ddot{u}_g(t) \tag{2}$$

where, $\{I\}$ is the identity vector and $\ddot{u}_g(t)$ is the input base acceleration time history. The $[M]$, $[C]$ and $[K]$ matrices are assembled using an incremental approach and are updated at every time step. The direct integration method – Newmark method (Newmark, 1959) is used to solve equation (1). The dynamic soil properties – the shear modulus and the damping ratio are obtained by using the published unified formulas (Ishibashi and Zhang, 1993). The formulas take into account the effect of the effective confining pressure, the plasticity index of the soil and the shear strain level to the shear modulus degradation curve and the damping ratio curve, which can be expressed in the following form:

$$\frac{G}{G_{max}} = K(\gamma, PI)(\sigma'_m)^{m(\gamma, PI)} \tag{3}$$

where G is the shear modulus at the shear strain γ; G_{max} is the initial shear modulus; PI is the plasticity index of the soil; σ'_m is the effective confining pressure. Based on the plasticity index, K and m are two functions used to control the shape of the shear modulus degradation curve, which can be written as:

$$K(\gamma, PI) = 0.5\left\{1 + \tanh\left[\ln\left(\frac{0.000101 + n(PI)}{\gamma}\right)^{0.492}\right]\right\} \tag{4}$$

$$m(\gamma, PI) = 0.272\left\{1 - \tanh\left[\ln\left(\frac{0.000556}{\gamma}\right)^{0.4}\right]\right\}\exp(-0.0145PI^{1.3}) \tag{5}$$

where n is coefficient to consider the influence of the plasticity index to the degradation curve. It can be determined by following equations:

$$n(PI) = \begin{cases} 0.0 & \text{for } PI = 0 \\ 3.37 \times 10^{-6} PI^{1.404} & \text{for } 0 < PI \le 15 \\ 7.0 \times 10^{-7} PI^{1.976} & \text{for } 15 < PI \le 70 \\ 2.7 \times 10^{-5} PI^{1.115} & \text{for } PI > 70 \end{cases} \tag{6}$$

Damping behavior of soil in seismic loading is also influenced by the effective confining pressure, particularly for soils of low plasticity. Ishibashi and Zhang (1993) also developed a unified formula for the damping ratio of plastic and non-plastic soils. Based on the modulus reduction factor G/G_{max} calculated above, the damping ratio λ is given by

$$\lambda = \frac{0.333(1 + \exp(-0.0145PI^{1.3}))}{2}\left\{0.586\left(\frac{G}{G_{max}}\right)^2 - 1.547\left(\frac{G}{G_{max}}\right) + 1\right\} \tag{7}$$

At very small strain level (10^{-5}), the damping ratio λ is found to be typically in the order of 0.5-1.5% for clayey soils and less than 0.5% for sands (Pestaña and Nadim,

2000). It will give a higher value for the damping ratio λ by using equation (7). In this model, the constant 1% damping ratio for clay soil and 0.5% damping ratio for sands are used when the shear strain is less than 10^{-5}. The constitutive laws presented above are implemented in a two-dimensional finite element code. The soil properties are assumed as homogenous in two dimensions. The element shear modulus and the damping ratio are determined by the shear strain of the element at each time step. An iterative process in each increment is performed until the shear modulus and the damping ratio are compatible with the shear strain level.

Rayleigh damping scheme is used to determine the element damping matrix as shown in equation (8). The global damping matrix $[C]$ is obtained through the assemblage of the element damping matrices.

$$[c]_{el} = \alpha [m]_{el} + \beta [k]_{el} \tag{8}$$

where $[m]_{el}$, $[c]_{el}$ and $[k]_{el}$ are the mass, damping and stiffness matrices for element el, and α, β are the mass and stiffness proportional Rayleigh damping coefficient, respectively. The coefficients α, β are computed based on the predominant frequency of the earthquake motion as well as the fundamental frequency of the system (Hudson et al., 1994). They are given by:

$$\alpha = 2\lambda \omega_1 \omega_2 /(\omega_1 + \omega_2) \text{ and } \beta = 2\lambda /(\omega_1 + \omega_2) \tag{9}$$

where, ω_1 is the fundamental frequency of the soil column, which is estimated by calculating a weighted average shear velocity of the soil column over the four times the thickness of the soil layer. Based on ω_1, ω_2 is determined as shown in Figure 3.

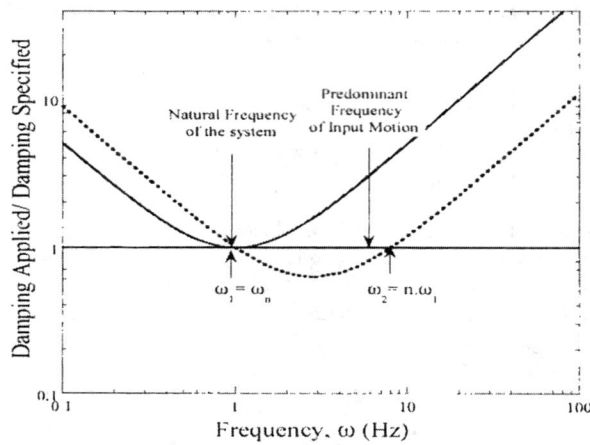

Figure 3 Variation of Damping with Frequency for Rayleigh Damping Scheme (n is the smallest odd integer) (Lok, 1998)

Validation of the New Soil Model

A case study was performed to analyze the Treasure Island soil profile for the 1989 Loma Prieta earthquake along the San Andreas faults in the Santa Cruz Area (Ms=7.1). The earthquake records were obtained at ground surface on fill material underlain by San Francisco Bay sediments and at the rock outcrop of adjacent Yerba Buena Island, which are located about 80 km northwest of the epicenter. The peak acceleration ranged from 0.067 g at the rock outcrop to 0.16 g at the soil surface.

The soil profile at the Treasure Island site consists of about 13m sandy fill, which is underlain by about 16 m thick of Young Bay Mud. Underlying the Young Bay Mud are alternating layers of dense sand and Old Bay Mud to a depth of about 89 m. Weathered shale extends from this depth to about 98 m, where the more competent sandstone is encountered (Figure 4). The Treasure Island Geotechnical Array acceleration instrumentation is located at the surface and at depths of 7, 16, 31, 44 and 104 m as denoted by the circular symbol (●) shown on Figure 4.

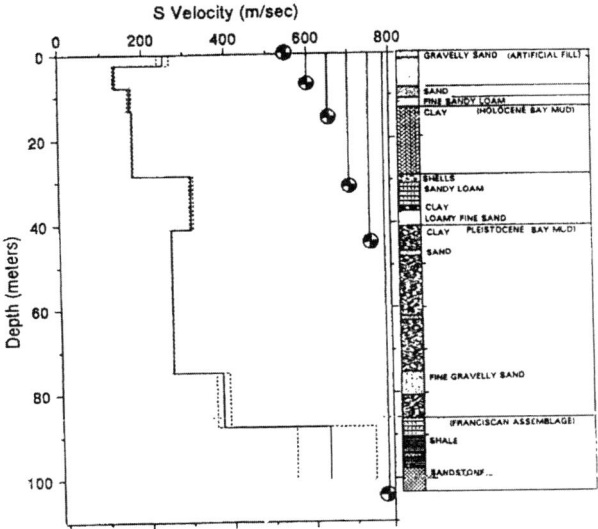

Figure 4 Soil Profile and Shear Wave Velocity Measured at Treasure Island by USGS (Gibbs et al, 1992)

The recorded motion at Yerba Buena Island is used as the input motion at the base of the soil column. The soil surface ground motion is calculated by the new ground response model using estimated parameters based on available soils. The calculated surface motion is compared with the recorded motion obtained at Treasure Island surface and also compared with the surface motion calculated using SHAKE (Idriss, 1993). The comparisons are shown as plots in 5% damping response spectra analysis (Figure 5).

The comparisons in Figure 5 show that the new soil model provides a reasonable prediction of the acceleration spectrum. It catches the predominant period very well, and gives a better prediction in predominant frequency area than that of SHAKE. Due to the influence of the effective confining stress, which gives soil stiffer modulus, less damping is involved in calculation and the new model gives a higher value than that from SHAKE. However, The new soil model also provides a higher value than the recorded motion in high frequency domain, which is attributed to not enough damping for small strain under this low level of shaking. How to accurately represent the viscous damping in the low strain should be studied.

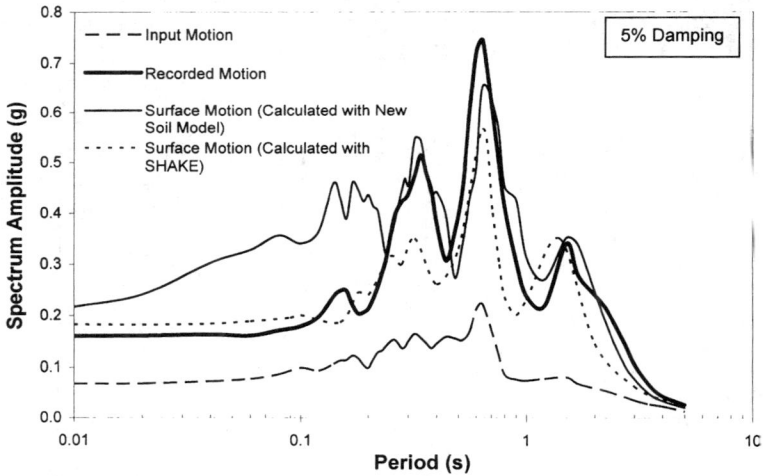

Figure 5 Comparisons of Recorded Motions at Treasure Island with Computed Response Spectra (5% damping)

Application in the New Madrid Seismic Zone

The study site is Wahite Ditch Bridge Site, which is located in southeastern Missouri near the NMSZ associated with the New Madrid rift complex. Based on generalized geologic cross-sections of the Mississippi Embayment, the soil deposits at this site could extend to depths of about 500 meters. Given the depth of these soil deposits and the fact that they cannot be reached with conventional geotechnical tools requires composing a profile based on available information. The composite soil profile and the shear wave velocity profile is referred as that one used in Memphis, which is based on a combination of surface information and a few deep wells as complied by Romero et al. (2001). The near surface soil properties are developed from the boring log at the Wahite Ditch Bridge Site (Figure 6), which consisted of approximately 6m (20 ft) of high plasticity clay and underlying approximately 53 (170 ft) of medium sand, containing numerous thin gravel lenses.

Since no strong ground motion records are available in the NMSZ, Herrmann (1999) developed a series of synthetic ground motions for use in the site response analysis for the Wahite Ditch Bridge Site. One of the synthetic ground motions, which has the magnitude 7.0 and a distance 65 km from the epicenter resulted in a maximum acceleration at rock base 0.166g, and was chosen as the input motion at the rock base. The site response analysis is performed in both the new ground response model and SHAKE analysis. The surface motions are recorded and compared in the 5% damping response spectra in Figure 7. The results from SHAKE analysis significantly underestimate the ground response when compared to the new soil model, especially for the higher frequency ranges. The comparison shows the importance of the influence of the confining pressure on the seismic site response analysis.

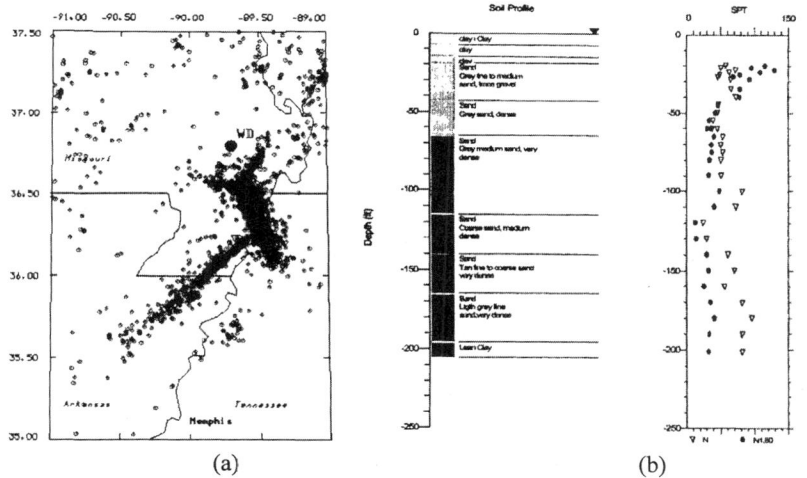

Figure 6 (a) Seismicity in the 1974 - 1995 Time Period in the Vicinity of the Wahite Ditch Site (b) Soil Profile at the Wahite Ditch Bridge Site (Anderson, et al., 2001)

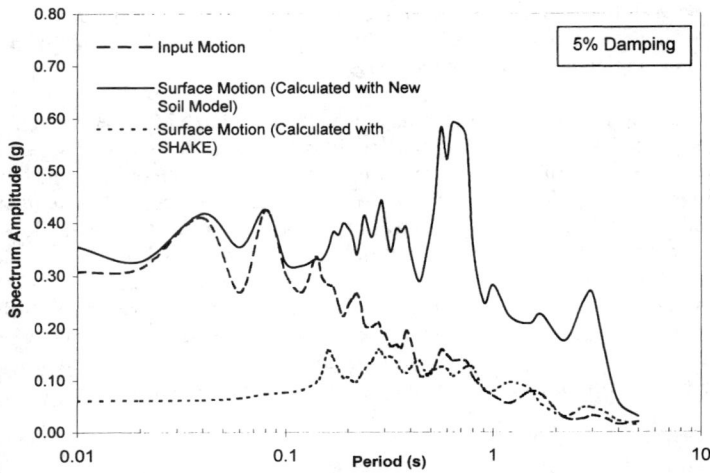

Figure 7 Comparison of the Computed Response Spectra at the Wahite Ditch Bridge Site

Conclusion

A new two-dimensional soil model is presented in this paper. The constitutive laws are implemented in a finite element code as two-dimensional plane strain elements in the time domain, which can also be incorporated into soil-structure interaction analysis. The case study showed that the new model has the capability to provide reasonable results when compared to field observations of a well documented soil profile in California, Treasure Island. The particular advantage of this soil model is the use of simple soil properties as input: the initial shear modulus and the plasticity index. The new soil model is used on a composite deep soil site response study near the NMSZ. The response spectra for the deep soil site resulted in an increased amplitude for periods around 0.5 sec and greater, which correspond to those for bridge structures. The results from this study show the importance of the influence of the confining pressure on the seismic site response analysis. For the influence of the confining pressure on wave propagation in the deep soil deposit, the ground motion data from the NMSZ are needed to improve the model calibration in future work.

Acknowledgments

Financial support for this research was provided by the Federal Highway Administration (Cooperative Agreement DTFH61-02-X-00009). The writers would also like to acknowledge the contributions of Dr. S. Prakash, University of Missouri-Rolla, Dr. R. Herrmann, St. Louis University, and Dr. B. Jeremic, University of California, Davis.

References

Anderson, N., H. Baker, G. Chen, T. Hertell, D. Hoffman, R. Luna, Y. Munaf, E. Neuner, S. Prakash, P. Santi, R. Stephenson. (2001). "Earthquake Hazard Assessment along Designated Emergency Vehicle Acess Routes." MoDOT Report RDT98-043, Missouri Department of Transportation.

Gibbs, J. F., Fumal, T.E., Boore, D.M., and Joyner, W. B. (1992) "Seismic Velocities and Geologic Logs from Borehole Measurements at Seven Strong-Motion Stations that Recorded the Loma Prieta Earthquake," USGS, Open-File Report 92-287

Hardin, B. O., and Drnevish, V. P. (1972). "Shear Modulus and Damping in Soils: Measurement and Parameter Effects," *J. Soil Mech. Found. Div.*, ASCE, 98(6), June, 603-624.

Herrmann, (1999), "Site Specific Ground Motions for St. Francis and Wahite Sites", St. Louis University, http://www.eas.slu.edu/People/RBHerrmann/MODOT.001014/

Hudson, M., Idriss, I. M. and Beikae, M. (1994) "QUAD4M: A Computer Program to Evaluate the Seismic Response of Soil Structures using Finite Element Procedures and Incorporating a Compliant Base," Center for Geotechnical Modeling, Dep. of Civil & Env. Eng., University of California, Davis.

Idriss, I. M. (1993) "Assessment of Site Response Analysis Procedures," Report of Center for Geotechnical Modeling, Dep. of Civil & Env. Eng., University of California, Davis.

Ishihara, I. (1992). Discussion to "Effect of Soil Plasticity on Cyclic Response," by Vucetic, M., and Dobry, R., *Journal of Geotechnical Engineering*, ASCE, 118(5), 830-832.

Ishihara, I. and Zhang, X. (1993). "Unified Dynamic Shear Moduli and Damping Ratios of Sand and Clay," *Soils and Foundations*, Vol. 33, No. 1, 182-191.

Lok, T.M. (1998). "Effect of Soil Nonlinearity to the Behavior of Seismic Soil-Pile-Structure Interaction," Ph. D. dissertation, University of California, Berkeley.

Newmark, N. M. (1959). "A Method of Computation for Structural Dynamics," *Journal of Geotechnical Engineering Division*, ASCE, 85, 67-94.

Pestaña, J. M. and Nadim, F. (2000), "Nonlinear Site Response Analysis of Submerged Slopes," Report UCB/GT/2000-04. University of California, Berkeley, CA.

Romero, S., Hebeler, G., Rix, G. J. (2001) Recommended reference profile for Memphis, Tennessee. http://mae.ce.uiuc.edu/Research/GT-1/RRP.htm. Engineering Geology 62, 137-158.

Schabel, P. B., Lysmer, J. L. and Seed, H. B. (1972). "SHAKE: A computer program for earthquake response analysis of horizontally layered sites." Report EERC-72/12. Earthquake Engineering Research Center, Berkeley, CA.

Seed, H. B., and Idriss, I. M. (1970), "Soil Moduli and Damping Factors for Dynamic Response Analysis," Report UCB/EERC-70/10. Earthquake Engineering Research Center, Berkeley, CA.

Vucetic, M., and Dobry, R. (1991). "Effect of Soil Plasticity on Cyclic Response," *Journal of Geotechnical Engineering*, ASCE, 117(1), 89-107.

Pipe-Soil Interaction during Transverse Permanent Ground Deformation

Moon Kyum KIM [1], Yunmook LIM[2], TaeWook KIM[3], SungHee Chang[4]

[1] *Department of Civil Engineering, Yonsei University*
134, Shinchon-dong, Seodaemun-gu, Seoul, 120-749, KOREA: applymkk@yonsei.ac.kr
[2] *Department of Civil Engineering, Yonsei University*
134, Shinchon-dong, Seodaemun-gu, Seoul, 120-749, KOREA: yunmook@yonsei.ac.kr
[3] *Track & Geotechnical Engineering Research Team, Korean Railroad Research Institute*
374-1, Woulam-dong, Uiwang-city, Kyoungi-do, 437-050, KOREA: karisma2k@krri.re.kr
[4] *Department of Civil Engineering, Yonsei University*
134, Shinchon-dong, Seodaemun-gu, Seoul, 120-749, KOREA: splendor@yonsei.ac.kr

SUMMARY: In this study, the applicability of currently used interaction force and previously proposed analytical relationship for the response analysis of buried pipeline subjected to transverse permanent ground deformation (PGD) due to liquefaction is evaluated. Based on meaningful contemplation, the improvement of interaction force and proposition of analytical relationship is made. Improved interaction force includes various patterns of PGD or spatial distributions of interaction force caused by the decrease of soil stiffness, and proposed relationship based on improved formula is applicable without regard to the width of PGD. Through the comparison of numerical results by use of commercial FEM program, the rational applicability of proposed relationship is objectively confirmed.

INTRODUCTION

The buried pipeline structure such as water, electricity, gas and oil pipeline, most of which are located at underground, is a cornerstone of the complex modern society. On natural hazards such as an earthquake, damage of an element or a part can result in the shut down of entire system and induce the outburst of magnificent economic and social losses with the elapse of time. Especially, indirect losses such as economic and social ones are likely to be magnified by the increase of complexities of society. Therefore the structural and functional safety of buried pipeline structures are urgently needed. (Hamada, 1992)

Seismic hazard of buried pipeline structures can be classified as being either permanent ground deformation (PGD) hazard or wave propagation hazard. The hazard by PGD due to liquefaction, fault movement and landslide is restricted to relatively small area. On the other hand, the hazard by seismic wave is distributed over large area. It should be generally noted that the PGD hazard or damage typically occurs in isolated areas of ground failure with higher damage rates while wave propagation damage occurs over much large areas, but with lower damage rates. So it can be said that the prediction of PGD considering local soil conditions and the establishment of an appropriate analysis method are needed for the seismic safety of buried pipeline structures. However, there are few established seismic design standards or codes due to the phenomenological perplexities of PGD. Nevertheless the importance of the research on this theme can never be exaggerated considering the national, societal and functional importance of buried pipeline structures.

With this viewpoint, in this study, the applicability of currently used pipeline-soil interaction force and previously proposed analytical relationship for the response analysis of buried pipeline subjected to transverse PGD due to liquefaction is evaluated. Based on meaningful findings, the improvement of interaction force and proposition of analytical relationship is made. Improved interaction force includes various patterns of PGD or spatial distributions of interaction force caused by the decrease of soil stiffness, and proposed relationship based on improved formula is applicable without regard to the width of PGD. Through the comparison of numerical results by use of commercial FEM program, the rational applicability of proposed relationship is confirmed.

PGD & BEHAVIOR OF PIPELINE

PGD is usually defined as large-scale ground displacement. The causes of PGD are classified as follows: soil liquefaction, sliding and fault movements. Among these three causes, sliding and fault movements have shown a low occurrence frequency. To the contrary, several cases in history have shown that soil liquefaction happens much more frequently and induces large PGD, which generates severe damage to pipeline structures.

Pipeline structures that are usually located at shallow depths have possibilities of localized or complete failure caused by PGD. Figure 1 shows schematic failure modes of pipeline structures due to PGD. For the case of Figure 1(a), perpendicular crossing or arrangement of a pipeline what we call a transverse PGD, imposes bending stresses on the pipeline. When the axis of a pipeline is parallel to the direction of ground deformation, such as in Figure 1(b), tension, compression or local bucking failure of pipeline is possible to occur. It is expected that arbitrary PGD disperse ground deformation effect. On the other hand longitudinal and transverse PGD concentrate forces on pipeline-soil interfaces causing extreme loading conditions. (Nordberg, 1991)

As important things in the response analysis of buried pipeline subjected to PGD due to liquefaction, the pattern, magnitude and width of PGD should be considered. In spite of numerous researches that have been preceded so far, there are few established design guidelines or values due to the phenomenological perplexities of PGD. Especially in Korea, due to the limitation of instrumental earthquake damage histories, it is difficult to determine the analysis and design value of PGD for the present. Therefore, in this study, various patterns of PGD such as in figure 2 are assumed to occur while the magnitude and width of PGD are assumed to change.

THE DECREASE OF SOIL STIFFNESS DURING PGD DUE TO LIQUEFACTION

Another characteristic feature of PGD due to liquefaction is the decrease of soil stiffness during liquefaction process. At the initial stage of an earthquake, soil stiffness has the original value before liquefaction. But, at the final stage, the obvious decrease of stiffness due to the generation of excess pore water pressure (PWP) is observed. That is to say, soil stiffness decreases with the increase of excess PWP from the initiation of an earthquake to the outbreak of PGD. Figure 3 is the experimental result by Ishihara (1992). As shown in Figure 3, the soil stiffness decreases definitely as the value of excess PWP ratio approaches 1.0. Besides, many researchers have tried to evaluate soil stiffness in liquefied region as the ratio of stiffness in non-liquefied region. For examples, Takada, Yoshida, Uematsu, Matsumoto, Yasuda and Tanabe concluded that the soil stiffness in liquefied region ranges from 1/1000 to 1/3000 of that in non-liquefied region. (M.O'Rourke, 1999)

Consequently, it can be said that the soil liquefaction, PGD and the generation of excess PWP have a close relationship with the decrease of soil stiffness, vice versa. Therefore, in this study, the decrease of soil stiffness due to liquefaction and increase of excess PWP is transformed into the spatial distributions of pipeline-soil interaction force. Note that the spatial distributions of pipeline-soil interaction force can be interpreted into the spatial distributions of ground displacement, which is the realistic pattern of PGD. By use of above concept, the currently used formula for interaction force is improved and the range of application is extended to the liquefied region.

PIPELINE-SOIL INTERACTION FORCE

Buried pipeline is structurally damaged by relative displacement caused by interfacial interaction between pipeline and soil. Pipeline-soil interaction force can be subdivided into two dominant cases for the sake of convenient analyses; one in the non-liquefied region and the other in the liquefied region. For the case of pipeline that is buried within non-liquefied sandy soil, a formal interaction force was proposed by TCLEE through experiments performed in many years. (M.O'Rourke, 1999) Proposed interaction force has different form according to the relative direction of ground movement to buried pipeline, that is to say longitudinal and transverse movement.

Longitudinal $\quad t_u = \mu \cdot (\gamma H) \cdot \dfrac{(1+K_o)}{2} \cdot (\pi D)$ (1)

Transverse $\quad p_u = (\gamma H) \cdot N_{qh} \cdot (D)$ (2)

Here, μ is the interfacial friction coefficient, γ means the unit weight of soil, H the burial depth, K_o the lateral coefficient of earth pressure, D the diameter of pipe, Nqh the horizontal bearing capacity factor which can be determined by experimental charts. Looking into eqn.(1) and (2), it can be known that the proposed interaction force is estimated by standards of burial depth and is the simple average value of vertical and lateral earth pressure acting on the cross sectional surface of pipe as shown in figure 4.

On the other hand, for the case of liquefied region, reasonable quantification or theoretical agreement of interaction force does not likely to be accomplished as yet.

Nevertheless the importance and need of the research on this theme are obvious. Therefore, in this study, improved formula is proposed through the modification of currently used formula using the concept of soil stiffness decrease in liquefied region.

PROPOSITION OF IMPROVED INTERACTION FORCE

Takada, Yoshida, Uematsu, Matsumoto, Yasuda and Tanabe concluded that the soil stiffness in liquefied region -- especially shows the lowest value in the central part -- ranges from 1/1000 to 1/3000 of that in non-liquefied region. This means that the confining effect of soil decreases to about 1/1000-1/3000 of the original value. Moreover they insisted that the soil stiffness in both edges, which is the boundary between the liquefied and non-liquefied region, can be magnified about the twice of original value. This "Pseudo Stiffness Effect" seems to be due to the decrease of the confining effect in the central part and the accompanying concentration in both edges.

Based on the above research result, the improved formula for interaction force is proposed in this study. At first, the change or the spatial distributions of pipeline-soil interaction force is proposed as the shape of sinusoidal function as shown in figure 5. Then, the interaction force in liquefied region is represented by the ratio of the interaction force in non-liquefied region. It has the value of 1/3000 of original in the central part and 2.0 in both edges. Defining interaction force in non-liquefied region as "$IF_{non-liquefied}$", pipeline-soil interaction force in liquefied region can be derived as follows.

$$IF_{liquefied} = 2IF_{non-liquefied} \cdot (1 - \frac{2}{\pi} + \frac{2}{3000\pi}) \quad (3)$$

Considering the similarities between the spatial distributions of pipeline-soil interaction force and the realistic patterns of PGD, various spatial distributions of interaction force or patterns of PGD can be assumed as shown in figure 5. Respective formulas are as follows. Here, p indicates a factor that determines the shape of smooth trapezoidal PGD.

Triangle $\quad IF_{liquefied} = \frac{IF_{non-liquefied}}{2} \cdot (2 + \frac{1}{3000}) \quad (4)$

Trapezoidal $\quad IF_{liquefied} = 2IF_{non-liquefied} - \frac{1}{2} \cdot (2 - \frac{1}{3000})(1 + p) \cdot IF_{non-liquefied} \quad (5)$

SEMI-ANALYTICAL RELTIONSHIP

M.O'Rourke developed a simple analytical model for the response of continuous buried pipeline to spatially distributed transverse PGD (M.O'Rourke, 1999). He considered two types of responses as shown in figure 6 and made several assumptions as follows. At first, for a wide width of PGD zone, the pipeline is assumed to be relatively flexible and its lateral displacement is closely match that of the soil. On the while, for a narrow width, the pipeline is assumed to be relatively stiff and the pipe lateral displacement is substantially less than that of the soil. Consequently, following analytical relationship is proposed. Here, δ is the magnitude of transverse PGD, D is the diameter of pipeline, W is the width of PGD, Pu is the interfacial interaction force for transverse movement, E is the elastic modulus of pipeline and t is the thickness of pipeline.

Wide zone $\quad \varepsilon = \varepsilon_b + \varepsilon_a = \pm \dfrac{\pi^2 \delta D}{W^2} + \left(\dfrac{\pi}{2}\right)^2 \left(\dfrac{\delta}{W}\right)^2$ (6)

Narrow zone $\quad \varepsilon = \varepsilon_b = \pm \dfrac{p_u W}{3\pi E t D^2}$ (7)

However, because he did not provide the appropriate critical value of PGD width that acts as a standard to application of respective equations, above equations are said to have somewhat limited applicability. Moreover, they do not reflect various patterns of ground deformation caused by the decrease of soil stiffness. For the clarification of these findings, comparative analyses between FEM and eqn.(6) are performed using various patterns of PGD as shown in figure 2. A steel pipe, generally used in Korea, is analyzed. The magnitude of PGD is set to 1.0m and the width is assumed to range 200~600m. Properties and dimensions are tabulated in Table.1.

Figure 7 shows the results of comparative analyses. For the case of smooth trapezoidal having a=0.15L, it can be said that the overall analysis result using FEM matches well with the result of M.O'Rourke. However, except for the case of a=0.15L, there are somewhat large differences between analysis results. Considering that the result of FEM represents various responses corresponding to patterns of PGD while the result of analytical relationship does only reflect the specific pattern of PGD, it can be said that the limited applicability of analytical relationship is reconfirmed.

PROPOSITION OF IMPROVED ANALYTICAL RELATIONSHIP

With the object of overcoming above-mentioned limitations of currently used analytical relationship, improved analytical relationship is proposed. The proposed one is based on improved pipeline-soil interaction force and thus it is able to reflect various patterns of PGD. Above all, it is applicable without regard to the width of PGD.

The whole system behavior of buried pipeline subjected to PGD due to liquefaction is separately considered as the behavior of cable and beam. In estimating maximum strain of pipeline, combined effect of respective behavior is included and various patterns of PGD are reflected by use of interfacial interaction shape factor β.

At first, for narrow width of PGD, the behavior of cable is considered to be dominant because response of pipeline increases with the increase of PGD width. A Middle portion of pipeline-W/2-is assumed to be subjected to axial tension and boundary part-W/4 from both ends-to inward tensile strain.

$$\varepsilon_{cable} = \varepsilon_{axial} + \varepsilon_{inward} = \dfrac{IF}{\delta} \cdot \dfrac{W^2}{16EA} + \mu \cdot \delta \cdot \sqrt{\dfrac{W}{10}} \quad (8)$$

Here, IF is interfacial interaction force in non-liquefied region such as eqn.(2) and μ is the coefficient determined by considering relative relationship between elastic modulus and inward tensile displacement.

On the other hand, for wide width of PGD, the behavior of beam is considered to be dominant because response of pipeline decreases with the increase of PGD width.

Uniform interaction force is assumed to be loaded on the beam with both ends fixed. The derived formula is based on the theory of elasticity.

$$\varepsilon_{beam} = \frac{M_{max}}{EI} \cdot C = \frac{8\delta D}{W^2} \qquad (9)$$

Summation of axial strain of cable and bending strain of beam is thought to be inappropriate. Moreover, selection of formula is considered to be inadequate because response can change according to the width of PGD and underestimation of maximum response is possible. Therefore, in this study, total response of pipeline is estimated as follows. Eqn.(10) represents the behavior of cable for narrow width of PGD while the behavior of beam for wide width of PGD.

$$\varepsilon_{total} = \frac{\beta}{1/\varepsilon_{cable} + 1/\varepsilon_{beam}} \qquad (10)$$

VERIFICATION & NUMERICAL ANALYSIS

For the objective verification of above-mentioned analytical relationship, comparative analyses are performed by use of commercial FEM program. Whole system is numerically modeled based on the beam on elastic foundation theory as shown in figure 8. The surrounding soil region is simply subdivided into liquefied and non-liquefied region. A continuous buried pipeline is assumed to pass through both regions simultaneously and extend infinitely. The length of buried pipeline is set to 5L to compensate for boundary effect of both ends and PGD is assumed to be loaded on the middle part. Soil stiffness is represented as series of axial and transverse springs estimated using following equations. (ASCE, 1984)

$$K_A = 2 \cdot \frac{t_u}{x_u}, K_L = 2.7 \cdot \frac{P_u}{y_u} \qquad (11)$$

A steel pipe, generally used in Korea, is analyzed. The magnitude of PGD is set to 0.1m, 0.2m, 0.3m and the width is assumed to range 10~150m. Properties and dimensions are tabulated in Table 1.

Figure 9 shows the response of PGD magnitude 0.1m. With the increase of PGD width, response of pipeline increases until W=30m and decreases thereafter. This phenomena might be caused by the dominant influence of axial strain until W=30m. Thus, for this case, W=30m is the critical width of PGD that acts as a standard for division between the behavior of cable and beam. Figure 10 and figure 11 shows the response of PGD magnitude 0.2m and 0.3m respectively. It can be known that the maximum strain of pipeline increases nearly proportional to the increase of PGD magnitude. Considering the difference between results of proposed analytical relationship and FEM is not more than 2%, it can be said that proposed relationship is rationally applicable.

Figure 12 presents the results of comparative analyses corresponding to various patterns of PGD. There are distinct changes of responses caused by various patterns of PGD. Especially it should be noticed that the response to sine pattern PGD is close to the lower bound of maximum strain. Even though sine pattern PGD is considered as a general mathematical representation of real phenomena, evaluation of pipeline response using it has the possibility of underestimation of maximum response. Therefore, for the structural safety of buried pipeline prone to PGD due to liquefaction environment, it can be concluded that the careful consideration of possible PGD pattern and the estimation of design strain based on the upper bound of maximum response are needed.

CONCLUSIONS

In this study, based on the evaluation of previous research results, the improvement of interaction force and proposition of analytical relationship for the response analysis of buried pipeline subjected to transverse PGD due to liquefaction is made. A number of comparative analyses are performed and following conclusions are drawn.

1. Currently used ASCE formula for pipeline-soil interaction force does not reflect the decreasing effect of soil stiffness in liquefied region appropriately. Therefore, by use of experimental results, formula is improved to reflect various spatial distributions of interaction force or various patterns of PGD.

2. Previously proposed analytical relationship of buried pipeline does not provide the appropriate critical value of PGD width that acts as a standard for application and does only reflect the specific pattern of PGD. With the object of overcoming these limitations, improvement of analytical relationship is carried out. Through the comparison of numerical results by use of commercial FEM, the rational applicability of proposed relationship is confirmed.

3. Considering that there are obvious differences of responses due to various PGD patterns, it can be concluded that the careful consideration of possible PGD pattern and the estimation of design strain based on the upper bound of maximum response are needed.

ACKNOWLEDGEMENT

This research was made possible through a grant from the Korea Earthquake Engineering Research Center (KEERC).

REFERENCES

Bartlett, S.F. and Youd, T.L.(1995), Empirical Prediction of Liquefaction-Induced Lateral Spread, *Journal of Geotechnical Engineering, ASCE*, Vol.121, No.4, pp.316-327.

Committee on Gas and Liquid Fuel Lifelines(1984), *Guidelines for the Seismic Design of Oil and Gas Pipeline Systems, ASCE*, New York.

Hamada, M. and O'Rourke, T.D.(1992), *Case Studies of Liquefaction and Lifeline Performance During Past Earthquakes*, Technical Report, NCEER-92-0001, Vol.1.

Ishihara, K., Taguchi, Y. and Kato, S.(1992), "Experimental Study on Behavior of the Boundary between Liquefied and Non-liquefied Ground", *Proceedings from the Third Japan-U.S. Workshop on Liquefaction, Large Ground Deformation and Their Effects on Lifeline*, Technical Report, NCEER-92-0032, pp.639-653.

O'Rourke, M.J. and Nordberg, G.(1991), "Analysis Procedures for Buried Pipelines Subject to Longitudinal and Transverse Permanent Ground Deformation", *Proceedings from the Third Japan-U.S. Workshop on Earthquake Resistant Design of Lifeline*

Facilities and Countermeasures for Soil Liquefaction, Technical Report, NCEER-91-0001, pp.439-453.

O'Rourke, M.J. and Liu, X.(1999), *Response of Buried Pipelines subject to Earthquake Effects*, MCEER Monograph, No.3, pp.77-87.

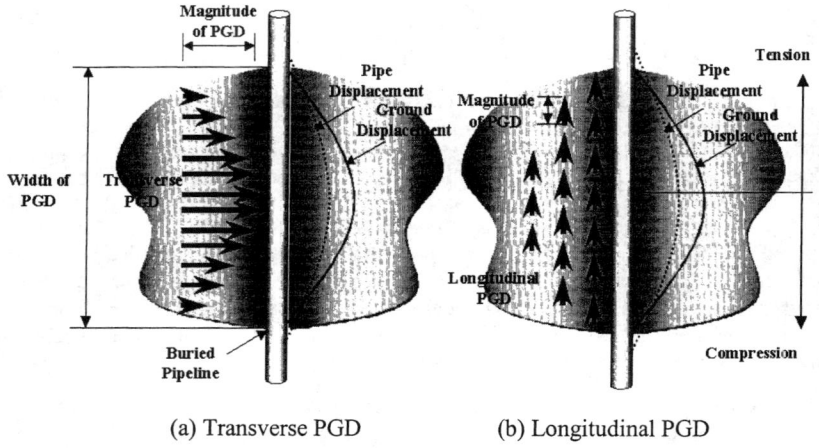

(a) Transverse PGD   (b) Longitudinal PGD

Figure 1. Schematic Failure Modes of Buried Pipeline

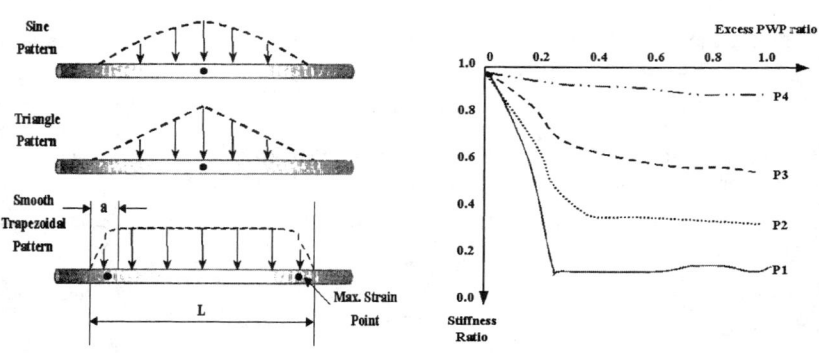

Figure 2. Various Patterns of PGD   Figure 3. The Decrease of Soil Stiffness

DISASTER RESPONSE FOR LIFELINE SYSTEMS 975

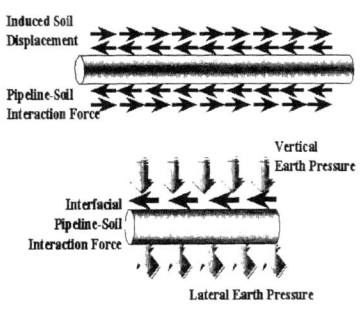

Figure 4. Interaction Force   Figure 5. Proposed Interaction Force Concept

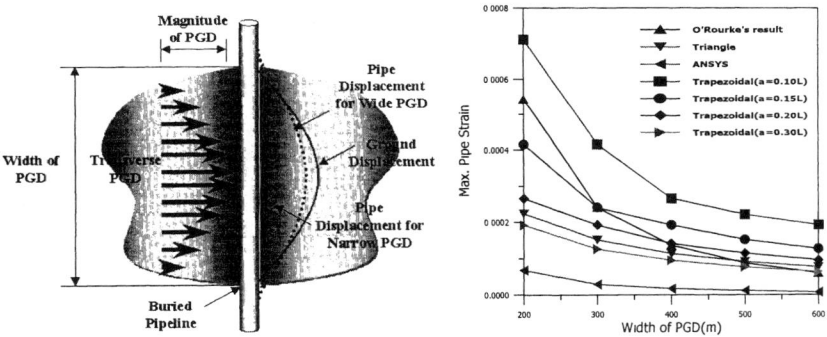

Figure 6. Concept of Analytical Relationship   Figure 7. Comparative Analysis

Table 1. Dimensions and Properties

Division	Contents		
Steel Pipeline	Elastic Modulus	210	GPa
	Yield Strength	300	MPa
	Outer Diameter	0.61	m
	Thickness	9.50	mm
Surrounding Soil	Weight Density	18.70	kN/m^3
	Burial Depth	1.20	m
	Axial Soil Stiffness	12600	kN/m^2
	Transverse Soil Stiffness	4500	kN/m^2

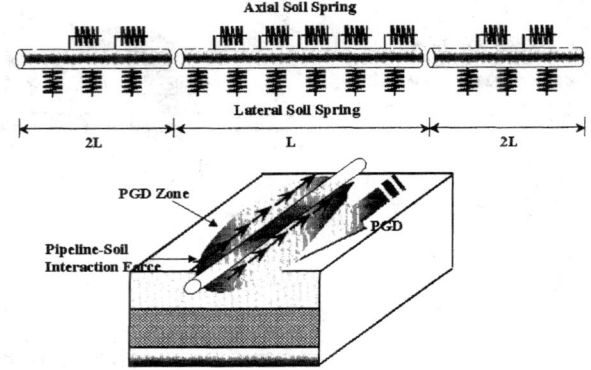

Figure 8. Numerical Modeling

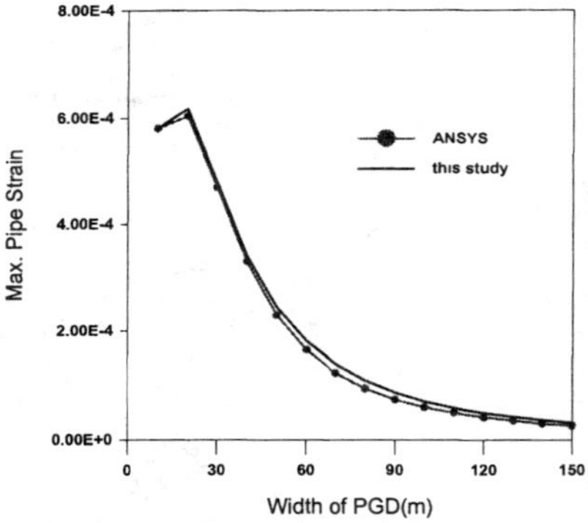

Figure 9. Response to PGD 0.1m

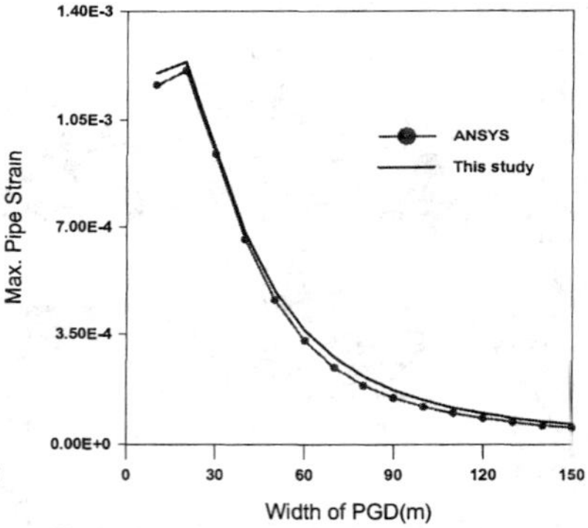

Figure 10. Response to PGD 0.2m

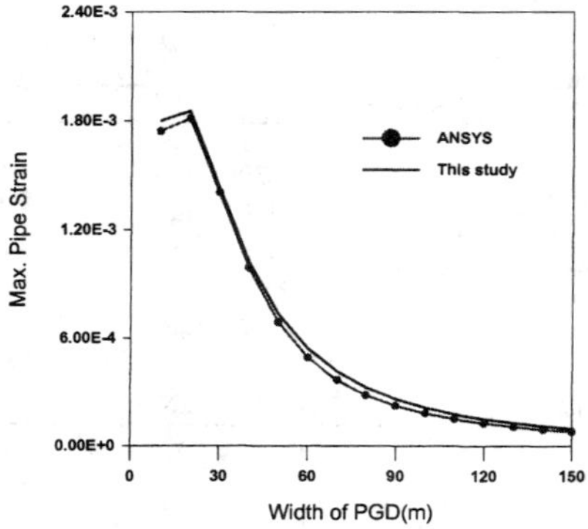

Figure 11. Response to PGD 0.3m

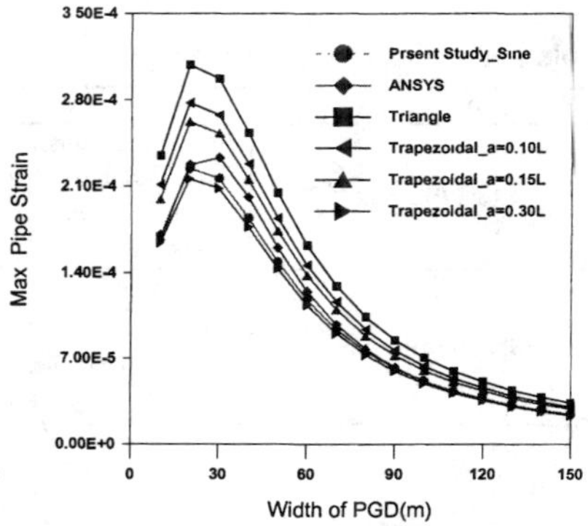

Figure 12. Response to PGD Patterns

Modeling of Unbounded Domain in Seismic Soil-Pile-Structure Interaction

Dongmei Chu[1], Kevin Z. Truman[2]

Abstract

The effects of Soil-Pile-Structure Interaction (SPSI) on the seismic behavior of bridges, dams and critical buildings are of paramount important in their seismic design. SPSI typically elongates the structural periods and degrades soil stiffness. Using a three-dimensional finite element model of a structure supported on pile foundations coupled with a nonlinear soil model provides an accurate method for assessing the seismic effects on bridges (piers), dams and critical buildings (hospitals/critical facilities). Nonlinear analysis using the Drucker-Prager soil model is performed in the time domain. The results obtained using a coarse mesh as well as a fine mesh are compared in order to verify the appropriate mesh size. To minimize the reflection of propagating waves back into the model, this paper also examines an infinite element "quiet" boundary condition and a viscous damper boundary condition, respectively. Both harmonic and El Centro seismic excitations are considered. The proposed model is validated against existing results of elastic analysis. This paper studies the effects of boundary conditions on the seismic analyses and hence the performance of the piles and superstructures. This study shows the consideration of radiation condition of the soil in seismic SPSI is significant and that appropriate boundary conditions are necessary to obtain accurate results.

Introduction

Catastrophic economic and social losses due to earthquakes demand increased research to modify design codes and building regulations. The prediction of structural performance of the structure-foundation systems is important in the seismic assessment and design of existing and new structures.

For a flexible structure supported on a stiff foundation, it is reasonable to assume that the input motion of the structure due to a design earthquake is the same as the motion of the free field. For a stiff and massive structure supported on a soft foundation, the input motion of the structure due to a design earthquake differs substantially from that of the free field due to dynamic soil-pile-structure interaction (SPSI).

The consideration of SPSI effects is required to evaluate seismic "demand" and deformation "capacity" of this soil-pile-structure system. The performance of structure-foundation systems in the earthquakes of Northridge (1994) and Kobe (1995) provided sufficient reason to believe that the SPSI effects should be investigated with greater rigor and precision. Moreover, SPSI can be included in practical engineering structures easily as computer technology advances and computer resources grow.

Graduate Research Assistant, Dept. of Civil Engineering, Washington University, St. Louis, MO 63130
Email: dc5@cec.wustl.edu
Professor and Chairman, Dept. of Civil Engineering, Washington University, St. Louis, MO 63130
Email: ktrum@seas.wustl.edu

State of the Art

The analysis of the seismic SPSI is a complex problem with highly coupled components SPSI elongates the period of a structure and tends to decrease its peak seismic respons (Guin & Banerjee, 1998; Kramer, 1996). Substantial research (Kaynia & Kausel, 198? Gazetas, 1984; Mamoon & Banerjee, 1990; Makris & Badoni, 1995) has been focused o studying and simulating the kinematic dynamic behavior of piles and pile groups subjected to seismic excitation since the early 1980s.

Matlock et al. (1978) first implemented an uncoupled approach to perform dynamic analysis of offshore structures. Subsequently, numerous researchers have studied the soil-pile-structure dynamic response using a similar approach (Wang, Kutter, Chacko, Wilson, Boulanger & Abghari, 1998; Prakash & Powell, 1993). The analysis consists of site response analyses to calculate the dynamic response of the free-field soil and dynamic p-y analyses to calculate the dynamic response of the structural models. The site response calculations are a greater source of uncertainty than the dynamic p-y calculations in the prediction of superstructure response, according to Boulanger, Curras, Kutter, Wilson and Abghari (1999).

In the coupled nonlinear Winkler foundation method, a linear elastic beam-column represents the pile and column and nonlinear p-y springs and dashpots represent the surrounding "near-field" soil (Lok, Pestana & Seed, 2000). The dynamic response of the soil-pile-structure system and the site response of "far-field" soil are calculated simultaneously.

Finite element methods (FEM), including Substructure Method and Direct Method, are employed to study the seismic SPSI performance. In the Substructure Method, the finite element (FE) region includes the structure, the pile foundation and the geometrically irregular and materially non-homogenous soil. The infinite soil is modeled as a regular layered homogeneous semi-infinite domain and is considered by a rigorous interaction force-displacement relationship. Integration of this interaction force-displacement relationship of the unbounded domain into the equations of motion of the structure constructs the dynamic analysis of soil-pile-structure systems. In the Direct Method, the FE region contains the structure, the pile foundation and soil medium up to the artificial boundary. The soil is represented by a semi-infinite half-space with approximate and artificial boundary conditions. Moreover, such boundary conditions are required to simulate the wave propagation properly so that no wave reflection exists due to the outwardly propagating waves.

The Substructure Method is generally formulated in the frequency domain, so this approach does not include the non-linear seismic response of soil-pile-structure systems and can not be used in a non-linear SPSI analysis. The Direct Method considers the non-linearity of near-site soil domain and involves a large number of degree of freedom due to a much larger FE region than that in the Substructure Method. Therefore, this approach requires much computational effort, but is popular in non-linear SPSI analysis due to its direct time solution.

Methodology

This study will focus on the nonlinear seismic response of a soil-pile-structure system, so nonlinear analysis in the time domain will be performed. A three-dimensional finite element method will be used to simulate the structure, the pile foundation and soil media enclosed inside the boundary conditions. The displacement based finite element method is employed in this study to provide the basic algorithm for the nonlinear finite element procedures.

Wave Propagation: Boundary Conditions. The energy dissipated from this soil-pile-structure system by outwardly propagating waves is recognized as *radiation damping* and necessitates the consideration of appropriate boundary conditions. From the viewpoint of wave propagation, a new part of the unbounded soil media, which is at rest and adjacent to the wave front, is excited by the propagating waves during an infinitesimal time increment. From the viewpoint of "the conservation of momentum", this new part of soil absorbs energy and is required to be incorporated into the FE zone and the dynamic analysis.

A frequency independent viscous dashpot boundary presented by Lysmer and Kuhlemyer (1969) will be incorporated on the boundary nodes to account for the radiation damping in all directions. Dashpots with coefficients ρc_p and ρc_s per unit area (ρ - soil density, c_p – compression wave or p-wave velocity, c_s -- shear wave or s-wave velocity) are used to model the energy dissipation of the unbounded medium in the longitudinal and tangential directions, respectively. Moreover, an infinite element will be introduced as another type of boundary conditions to represent the unbounded soil domain.

Radiation damping plays an important role in seismic soil-pile-structure interaction analysis. Boundary conditions are required to represent the energy dissipated by the propagating waves in the soil field properly. This study considers two boundary conditions, frequency independent dashpot and infinite element, respectively and compares the dynamic responses of the soil-pile-structure system.

Soil Constitutive Relationship: Drucker-Prager Model. Material non-linearity of the soil has significant effects on the seismic response of a soil-pile-structure system (Bentley & Naggar, 2000; Maheshwari, Truman & Gould, 2002). Bentley (2000) investigated the effects of soil plasticity on the kinematic response of single piles using Drucker-Prager perfectly plastic soil model. Drucker-Prager plastic soil model will be employed in this study. The computational model developed in this study can be applied easily to nonlinear seismic analysis of soil-pile-structure systems with different configurations and different boundary conditions. In addition, this model can be used extensively for dynamic analysis for massive and stiff structures such as nuclear facilities, lock and dams and bridges.

Analytical Model

The analytical model of a soil-pile-structure model is illustrated in Figure 1. The soil-pile-structure system with dimensions of 40m*30m*18m is modeled as eight-node hexahedral three-dimensional finite elements for the soil media with embedded 1m*1m single pile and lumped masses for the three-story structure.

Finite Element Meshes. Soil is discretized as different size meshes for near-site soil and far-site soil. As its distance to the pile is further, the soil mesh size becomes larger. Soil is considered as a plastic material with the work hardening behavior. The constitutive relationship of soil is governed by a Drucker-Prager plasticity model.

The length of the pile is the same as the depth of the soil media. The concrete pile is modeled with three-dimensional eight node hexahedral finite elements and is assumed to be linear behavior. The bottom of the soil and the single pile are fixed with both harmonic and seismic excitation in one direction.

The three-story superstructure is modeled with three lumped masses connected with the pile foundation with a rigid pile cap. Simulating kinematic effects and inertia effects of SPSI together, the seismic analysis is carried out in the time domain. The frequency independent dashpot boundary condition with the coefficients derived from the law of conservation of momentum is illustrated in Figure 2 to represent the soil-pile system without the superstructure.

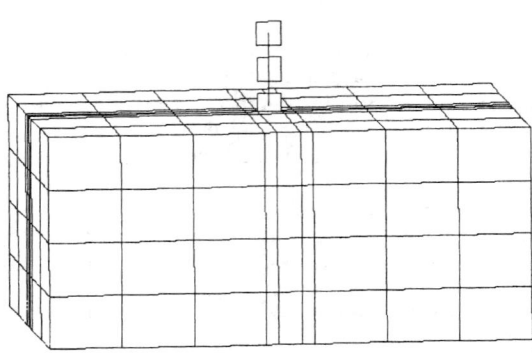

Figure 1.   The model of Soil-Pile-Structure Systems

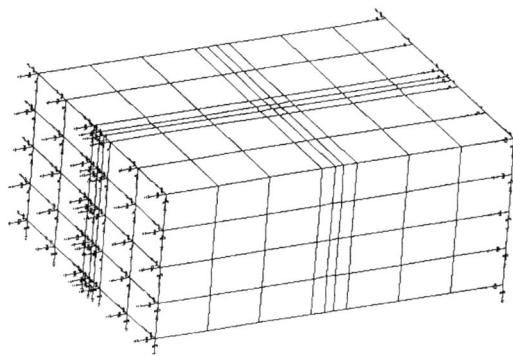

Figure 2.   Frequency Independent Dashpot Boundary Condition

Finite and Infinite Element Mesh. On the other hand, the infinite element boundary condition is illustrated in Figure 3 to represent the soil-pile system with the simulation of the unbounded soil domain as infinite elements. The infinite elements are assumed to have linear behavior, which avoids the reflection of dilatational and shear wave energy back into the finite element mesh. It is optimized for the infinite elements to transmit energy out of the FE zone (without trapping or reflecting the waves) if the interface between the finite and infinite elements is selected as close as possible to being orthogonal to the direction of wave propagation.

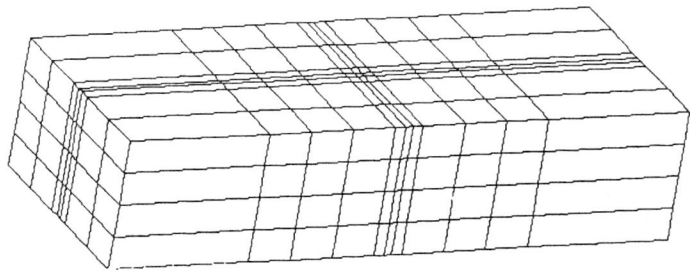

Figure 3.   Infinite Element Boundary Condition

Validation and Parametric Study

Validation. Pile head response due to harmonic excitation is validated against the existing results for linear analyses. To verify the kinematic interaction between a single pile and surrounded soil, a three dimensional finite element mesh is constructed and the radiation damping is represented by the frequency independent dashpot boundary condition. This soil-pile system is excited by a harmonic bedrock motion, which has the magnitude of $1m/s^2$ and frequency of 2 Hz. The acceleration of pile head is steady state with the amplitude of $1.6m/s^2$, which is shown in Figure 4. It can be verified with the existing time domain analysis results (Maheshwari., Truman & Gould, 2002), which suggests the amplitude of pile head response is about 1.4 times of the bedrock motion.

Mesh Size. Mesh sizes play an important role in the accuracy of finite element analyses, so two mesh sizes, coarse mesh and fine mesh, are compared to verify the appropriateness of mesh size used in this study. The coarse mesh is as shown in Figure 2, while the fine mesh divides soil elements into one fourth of the size as in the coarse mesh. Finite element meshes with dashpot boundary condition are excited by harmonic bedrock motion of $1m/s^2$ magnitude and 1Hz frequency. The pile head responses of linear analyses for two meshes are shown as in Figure 5. It can be seen the difference is insignificant. Therefore, the coarse mesh size used in this study provides fairly accurate results.

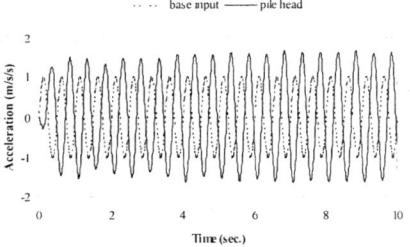

Figure 4. Elastic Pile Head Response for f = 2Hz   Figure 5. Effects of Mesh Sizes

Rayleigh Damping. In structural dynamic analysis, it is very common to use two to five percent Rayleigh damping to consider structural damping such as joint friction and heat diffusion. Rayleigh damping matrix is proportional to the mass and stiffness matrices. Considering finite element meshes with dashpot and infinite element boundary conditions with harmonic bedrock motion (amplitude=$1m/s^2$, frequency=1Hz), this study performs the pile head response analyses with Rayleigh damping of 0%, 2% and 5%, respectively.

The pile head acceleration amplitudes corresponding to different damping ratios are shown in Table 1 for a FE model with the dashpot boundary condition. The pile head acceleration response due to 5% damping is compared with the response without damping in Figure 6. The pile head response of the FE model with the infinite element boundary condition shows the same results as the model with the dashpot boundary condition. It can be seen the effects of Rayleigh damping are unimportant because the energy dissipation systems (dashpots and infinite elements) defined in the FE analytical model are predominant.

Table 1.	Pile Head Responses for f = 1Hz		
Rayleigh Damping	0%	2%	5%
Pile Head Acceleration (m/s^2)	1.304	1.275	1.226
Effects of Damping (%)	-	2.22	5.98

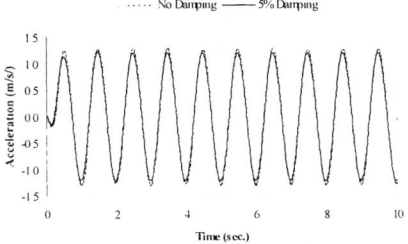

Figure 6. Effects of Rayleigh Damping

Harmonic Excitation. To study the effects of harmonic excitation frequencies on pile head responses, three pairs of simulations are performed for bedrock motions with 1m/s^2 magnitude and frequencies of 1Hz, 2Hz and 4Hz, respectively. Each pair of simulation includes elastic analysis and plastic analysis considering the Drucker-Prager soil model. For models with either dashpot or infinite element boundary conditions, the elastic pile head response is identical to the plastic response due to bedrock motion of 1Hz and 1m/s^2, as shown in Figure 7. They are steady state with the amplitude of 1.25m/s^2. This also applies to 2Hz bedrock motion, which is shown in Figure 8, except the pile head response amplitude is 1.6m/s^2.

However, plastic pile head response differs from elastic response significantly, as shown in Figure 9, for the finite element model with dashpot boundary conditions under the harmonic excitation of 4Hz and 1m/s^2. It is observed that the peak pile head plastic response in the first two seconds is 0.6m/s^2 compared with 2.6m/s^2 in elastic response. This can be explained as the results of energy absorption by the soil plastic behavior. Soil plasticity distributes the pile head acceleration history during the time period of 2~6 seconds. For the rest of the response history, the elastic response and plastic response are similar because the dashpot energy dissipation dominates as the wave propagates away from the finite element zone of soil and pile.

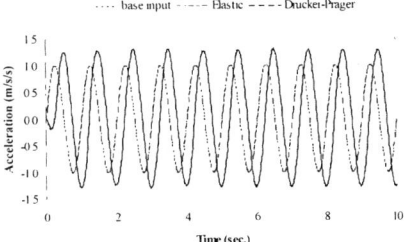 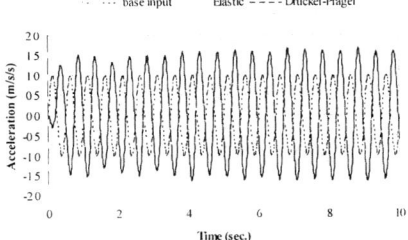

Figure 7. Elastic & Plastic Responses for f = 1Hz   Figure 8. Elastic & Plastic Responses for f = 2Hz

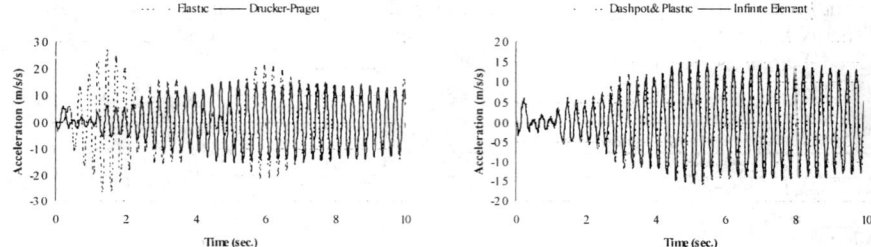

Figure 9. Elastic & Plastic Responses for f = 4Hz   **Figure 10. Comparison of Boundary Conditions**

On the other hand, it is observed that the plastic pile head response is identical with the elastic response if the infinite element boundary is combined into the finite element model with harmonic bedrock motion of 4Hz and 1m/s^2. The difference between the pile head response with infinite element boundary and the plastic response for dashpot boundary conditions is insignificant, as shown in Figure 10. The identity of elastic and plastic responses can be explained by the stiffer soil-pile system than the model with the dashpot boundary condition due to the consideration of infinite elements and the assumption of elastic behavior in infinite elements.

El Centro Seismic Excitation. El Centro 1940 N-S component, which is recorded during the Imperial Valley, California earthquake of May 18, 1940 with the peak ground acceleration 0.319g, is considered as the one-dimensional bedrock motion. Time intervals in the acceleration record are 0.02 second and the duration time is 50 seconds in the time domain analysis. The elastic and plastic responses of the pile head are investigated for a finite element model with a dashpot boundary condition. Pile head acceleration history and its relative displacement to bedrock are shown as in Figures 11 and 12, respectively. Soil plasticity decreases peak acceleration from 0.43g in elastic analysis to 0.41g and increases peak displacement from 0.074m using an elastic analysis to 0.155m. It can also be observed from Figure 12 that permanent plastic pile head displacements exist in plastic analysis by the end of the excitation comparing with no displacement in the elastic analysis with the consideration of the baseline correction. The results of a model with infinite element boundary conditions are the same as those of dashpot boundary.

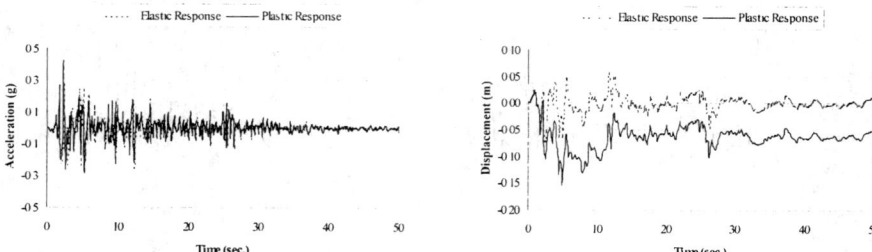

Figure 11. Pile Head Acceleration for El Centro   **Figure 12. Pile Head Displacement for El Centro**

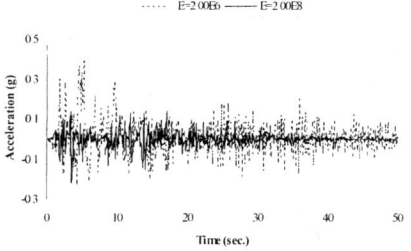 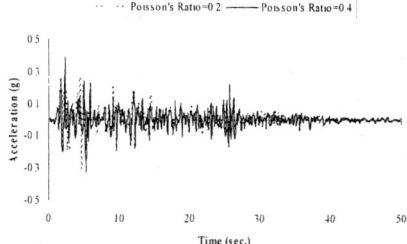

Figure 13. Effects of E on Pile Head ResponseFigure 14. Effects of ν on Pile Head Response

The effects of Young's modulus (E) and Poisson's ratio (ν) on pile head plastic response are investigated, which are shown in Figures 13 and 14, respectively. Figure 13 shows that soft soil (i.e., small E) amplifies bedrock motion during the entire period of the El Centro earthquake and stiff soil (i.e., large E) generates smaller pile head response than bedrock motion. This further states the importance of site responses. The effects of ν on pile head responses are not significant, as shown in Figure 14.

Conclusions

This paper studies the effects of soil plasticity and boundary conditions on the dynamic analyses of soil-pile-structure interaction system. Both elastic and plastic analyses are performed in time domain using a three-dimensional finite element model coupled with Drucker-Prager soil model. The pile head responses due to harmonic and El Centro excitation are investigated. Material plastic behavior plays an important role in energy dissipation mechanisms for both relatively high frequency harmonic excitation and seismic excitation.

To simulate radiation damping in an unbounded domain, frequency independent dashpot and infinite element boundary conditions are formulated into a finite element model, respectively. Results presented in this study illustrate the predominant effects of radiation damping on the dynamic performance of soil-pile-structure systems. The methodology and the analytical model proposed in this paper provide an accurate method for the seismic performance of soil-pile-structure interaction systems with the appropriate representation of soil plasticity and radiation damping. Moreover, this study has significant meanings in seismic design of new structures and retrofit strategies of existing structures.

References

Bentley, K.J. & Naggar, M.H. El (2000). Numerical analysis of kinematic response of single piles. *Canadian Geotechnical Journal*, 37, 1368-1382.

Boulanger, R.W., Curras, C.J., Kutter, B.L., Wilson, D.W. & Abghari, A. (1999). Seismic soil-pile-structure interaction experiments and analysis. *Journal of Geotechnical and Geoenvironmental Engineering*, 125 (9), 750-759.

Gazetas, G. & Dobry, R. (1984a). Simple radiation damping model for piles and footings. *Journal of Engineering Mechanics*, 110 (6), 937-956.

Gazetas, G. & Dobry, R. (1984b). Horizontal response of piles in layered soils. *Journal of Geotechnical Engineering*, 110 (1), 20-40.

Guin, J. & Banerjee, P.K. (1998). Coupled soil-pile-structure interaction analysis under seismic excitation. *Journal of Structural Engineering*, 124 (4), 434-444.

Kramer, S.I. (1996). *Geotechnical earthquake engineering*. Englewood Cliffs, NJ: Prentice-Hall Inc.

Lok, T.M. (1998). *Effect of soil nonlinearity to the behavior of seismic soil-pile-structure interaction*. Ph.D. dissertation, University of California, Berkeley.

Lok, T.M., Pestana, J.M. & Seed, R.B. (2000). Numerical modeling of seismic soil-pile-superstructure interaction. *12th World Conference on Earthquake Engineering*, Auckland, New Zealand

Lysmer, J. & Kuhlemeyer, R.L. (1969). Finite element model for infinite media. *Journal of Engineering Mechanics Division, ASCE*, 95, EM4, 859-877.

Maheshwari, B.K., Truman, K.Z. & Gould, P.L. (2002). Nonlinear kinematic response of single piles. *Proc. 7th National Conference on Earthquake Engineering*, Boston, MA

Mamoon, S.M. (1990). *Dynamic and seismic behavior of deep foundation*. Ph.D. dissertation, State University of New York, Buffalo.

Markris, N. & Badoni, D. (1995). Seismic response of pile groups under oblique-shear and Rayleigh waves. *Earthquake Engineering and Structural Dynamics*, 24, 517-532.

Novak, M. (1991). Piles under dynamic loads: State of the art. *Proc. 2nd Int. Conf. On Recent Advances in Geotech. Earthq. Engng. & Soil Dynamics*, III. Univeristy of Missouri-Rolla, 250-273.

Penzien, J. (1970). Soil-pile foundation interaction. *Earthquake engineering*. Englewood Cliffs, NJ: Prentice-Hall Inc.

Wang, S., Kutter, B.L., Chacko, M.J., Wilson, D.W., Boulanger, R.W. & Abghari, A. (1998). Nonlinear seismic soil-pile-structure interaction. *Earthquake Spectra*, 14 (2).

Wolf, J.P. (1985). *Dynamic soil-structure interaction*. Englewood Cliffs, NJ: Prentice-Hall, Inc.

Ground Improvement Effectiveness for Liquefaction Mitigation at an Existing Highway Bridge

by Harry G. Cooke,[1] Associate Member, ASCE, and
James K. Mitchell,[2] Honorary Member, ASCE

Abstract

Prior field case history, physical modeling, and numerical modeling studies have shown the effectiveness of using ground improvement for reducing liquefaction-induced ground movements due to earthquakes and their effects on structures. However, in some of these studies even the reduced permanent displacements from using improved ground were larger than those typically tolerable for most existing highway bridges, and in other cases the treatment areas were larger than feasible or desirable at an existing bridge. In addition, few of the cases involved bridges. To investigate whether improved ground zones of limited size can potentially be used to reduce liquefaction-induced bridge movements to tolerable levels, a numerical modeling study was conducted consisting of two-dimensional, dynamic, effective stress analyses using FLAC and a non-linear, elasto-plastic soil model. Limited calibration and verification analyses of six centrifuge tests and a field case history showed the modeling method predicted liquefaction-induced ground and structure displacements within a factor of approximately two. Subsequent parametric study analyses of a highway bridge test case indicated densified or cemented zones extending completely through liquefiable soils could reduce the pier and abutment movements to tolerable levels. For the pier the effective treated zone was centered under the pier itself. For the stub abutment the improved zone typically extended from a location under the approach embankment behind the abutment to some distance beyond the toe of the embankment slope. The optimum width of treatment, beyond which further increases in width did not result in substantial reductions in structure movements, was dependent on the strength and stiffness of the treated zone.

Introduction

Recent experience in the United States and abroad has shown that ground treatment can be used to improve liquefiable soils and mitigate the damaging effects of earthquake-induced liquefaction on structures and facilities. Existing highway bridges are particularly challenging for use of this approach due to the constraints posed by the bridge and site conditions in performing the mitigation work.

An example of an existing bridge where ground improvement might be used for reduction of liquefaction risk is shown in Figure 1. The bridge consists of stub abutments at the ends of approach embankments and substructure piers supported on

---

[1] Assoc. Prof., Construction Tech. & Management Dept., Ferris State Univ., 915 Campus Dr., Swan 312, Big Rapids, MI 49307; (231)591-3555; cookeh@ferris.edu
[2] Univ. Distinguished Prof., Emeritus, Dept. of Civil & Environmental Engineering, Virginia Tech, Blacksburg, VA, 24061; jkm@vt.edu

spread footings over liquefiable sands. If liquefaction of the sands occurs, there is loss of support for the pier footings. At the abutments lateral spreading and settlement may occur, with both abutments moving in towards the center of the bridge. Similar problems may exist for bridges supported on deep foundations that partially or fully penetrate liquefiable soils. Effective ground improvement measures must limit the abutment and pier movements due to earthquake-induced liquefaction to tolerable levels to prevent damage to the bridge superstructure.

The research in this study was a preliminary investigation into whether ground improvement can be effectively used to mitigate the effects of liquefaction on an existing highway bridge. A key issue evaluated is the type, size, and location of the ground treatment that should be used. Numerical analyses of the bridge shown in Figure 1 were performed after incorporating different ground improvement schemes in the underlying liquefiable ground. Particular emphasis was placed on evaluating the expected permanent displacements of the ground and supported structure to see if they were within tolerable limits. The shallow foundation case shown was selected because it generally involves more severe movements than would develop for deep foundations, and it also provides some insight into the behavior of bridges on partially-penetrating deep foundations.

A brief summary of relevant observations made in prior studies by others regarding the use of ground improvement for liquefaction mitigation is presented first. Then an overview of the numerical model used in the analyses for this research is provided, followed by results and conclusions from a parametric study performed for the bridge pier and stub abutment.

Prior Studies

Prior studies on the use of ground improvement for liquefaction mitigation performed with centrifuge and shaking table tests and numerical modeling have yielded important results including:

- a limiting width of treatment around a structure or underneath an embankment exists such that further width increases do not substantially reduce the movements of the ground and supported structure or embankment (Hatanaka et al., 1987; Riemer et al., 1996);
- a limiting depth of treatment under a footing on level ground exists such that further treatment depth increases do not substantially reduce the predicted settlement (Liu and Dobry, 1997);
- a treated zone is more effective under an embankment when located beneath the slope rather than beyond the toe of the slope (Yanagihara et al., 1991; Riemer et al, 1996); and
- the greater the stiffness of the treated zone used to contain or restrict movement of an embankment/slope or similar structure and the underlying liquefiable soils, the smaller the movement of the embankment or structure (Yasuda et al, 1991; Adalier et al., 1998).

In most of these studies the movements observed for the more effective treatments tested were larger than those generally acceptable for a highway bridge, such as the 10 centimeter movement limits adopted for this study. There is also concern that in 1-g shaking table tests, such as those performed by Hatanaka et al. (1987), Yanagihara et al. (1991), and Yasuda et al. (1991), the low effective stress levels can result in more dilative behavior of sand than seen in the field.

Mitchell et al. (1995) and Hausler and Sitar (2001) present case history information regarding the performance of improved ground in reducing the effects of earthquake-induced liquefaction on structures. However, there is some variability in the reported effectiveness of different treatment configurations presented in these case histories, as well as a lack of complete details regarding the treatment in some cases and potential site-specific factors that may have influenced performance. In addition, only one of these case histories involves a bridge. In this case, the 1989 Loma Prieta Earthquake in California produced no observed pier movements at the bridge where chemical grouting of liquefiable soils had been previously performed (Mitchell et al., 1995).

Based on the limitations of the modeling studies and case histories cited above, a numerical modeling approach was developed and used in a parametric study of the effectiveness of ground improvement for mitigating the damaging effects of liquefaction on an existing highway bridge. Particular emphasis was placed on keeping permanent movements of the bridge within tolerable limits, which can differ from the movement limits for other types of structures. Maximum foundation movement limits of 10 cm vertically and 10 cm horizontally were adopted for this study based in part on observations and comments by Youd (1998) and Youd et al. (1998) on earthquake-induced liquefaction ground displacement criteria for maintaining serviceability of more modern bridges supported on shallow foundations.

Numerical Modeling

The numerical modeling method selected for this study consisted of two-dimensional, plane-strain, dynamic, effective stress analyses using the computer program FLAC Version 3.4 (Itasca Consulting Group, 1999). The program's routine for generating pore water pressure during shaking, which is based on a partially-coupled solution, was modified to include a formulation by Byrne (1991). In addition, the linear elastic-perfectly plastic stress-strain behavior, which is used in conjunction with a Mohr-Coulomb failure envelope for the soil, was modified to incorporate non-linear, elasto-plastic behavior using the hyperbolic shear stress-strain formulation for simple shear by Pyke (1979). The Seed and Idriss (1970) equation for maximum shear modulus was used to compute the initial modulus for each element. The drained bulk modulus was computed from the maximum shear modulus and an assumed small strain Poisson's ratio of 0.05, and used in conjunction with the bulk modulus of water. Pore pressure migration and dissipation during and after shaking, and resultant changes in the strength and stress-strain properties of the soil, were also included.

The ability of FLAC and the soil model to adequately predict the behavior and performance of liquefiable ground, with and without structures and ground improvement present, was evaluated by simulating six dynamic centrifuge tests (Liu, 1992; Zeng, 1993; Adalier, 1996) after calibrating the model using laboratory test data on the centrifuge test sand. A field case history involving lateral spreading at the Wildlife Site, during the 1987 Superstition Hills Earthquake in California, was also simulated. It was found that the permanent deformations/displacements of a structure, embankment, or improved zone predicted using the numerical modeling method ranged from about 0.5 to 2.5 times the measured values. In addition, the effectiveness of different treatment types relative to each other, in reducing permanent deformations/displacements, could be predicted. The predicted pore water pressures and accelerations at specific elements or nodes varied more from the measured values than the predicted deformations. More details concerning the model, including its strengths and limitations, can be found in Cooke (2000).

Parametric Study Analyses

The numerical model was used to separately evaluate the potential effectiveness of different ground improvement schemes in mitigating liquefaction effects for one of the bridge piers and stub abutments shown in Figure 1. More details concerning the bridge can be found in Cooke (2000). The focus was on the type of treatment as well as the width, depth, and location of the treated zone within the 11-m-thick liquefiable deposits. The ground improvement effectiveness was assessed in terms of predicted permanent foundation movements. Since the model yielded predicted displacements approximately one-half of the actual values for the worst cases of underpredicted movements in the verification simulations, a goal of the parametric study was to find improved ground designs that limit the predicted vertical and horizontal foundation movements to 5 cm (i.e. − one-half of the 10-cm tolerable limit) or less.

Two-dimensional finite difference grids were constructed to numerically model a bridge pier and stub abutment independently of each other, as shown in Figures 2 and 3, respectively. The influence of the bridge superstructure, which consists of 21-meter-long simply supported spans, was included by using a beam element and point load at the top of the pier and a point load on the stub abutment. The horizontal acceleration record having a peak of 0.23 g (i.e. − 228 cm/s^2) shown in Figure 4, which was scaled down from the major component of horizontal shaking recorded at 16-m depth on Port Island during the 1995 Kobe Earthquake in Japan (Geo-Research Institute, 1995), was input at the nodes along the base and sides of the grids. Pore water pressures at the top nodes of the liquefiable sands (i.e. − top of the medium dense sand layer shown in Figure 1) were fixed to 0.

The performances of the bridge pier and stub abutment were evaluated for ground improvement methods qualitatively judged to have moderate to high applicability for use at existing bridges supported on shallow foundations. These methods included (1) densification of liquefiable sands by compaction grouting, (2) cementation by chemical grouting, (3) reinforcement and containment by jet grouting at the pier and cementation by jet grouting at the stub abutment, and (4) in-situ stress increase using a buttress fill at the stub abutment. Properties assigned to both the improved and unimproved soils are presented in Table 1. The non-linear,

Table 1. Properties of Improved and Unimproved Soils

Soil Type		D_r (%)	k (cm/sec)	ϕ (°)	c (kPa)	$\sigma_{tensile}$ (kPa)	K_{2max}[a]	ε_v Constants[b]	
								C_1	C_2
Natural Sand:	*Loose*	39	5 x 10^{-3}	33	0	0	39	0.8	0.5
	Medium Dense	53	5 x 10^{-3}	34	0	0	48	0.37	1.08
Embankment & Buttress		81	5 x 10^{-2}	38	0	0	64	0	0
Densified Sand:	$D_r = 75\%$	75	3 x 10^{-3}	36.5	0	0	59	0.156	2.56
	$D_r = 85\%$	85	2.5 x 10^{-3}	37.5	0	0	66	0.114	3.5
Chemical-Grouted:	*Pre-Failure*	-	5 x 10^{-5}	33	81.4	30	78	0	0
	Post-Failure	-	5 x 10^{-5}	33	0	0	39	0	0
Jet-Grouted:	*Pre-Failure*	-	5 x 10^{-6}	0	2250	1500	_[c]	0	0
	Post-Failure	-	5 x 10^{-6}	0	0	0	_[c]	0	0

Note: k is permeability, ϕ is friction angle, c is cohesion, and $\sigma_{tensile}$ is tensile strength.
[a]K_{2max}, used to calculate maximum shear modulus, G_{max}, based on formula by Seed and Idriss (1970): $G_{max} = 1000 K_{2max} (\sigma_m')^{0.5}$ where σ_m' and G_{max} in pounds per square foot (psf).
[b]C_1 and C_2 are volumetric strain (ε_v) constants computed from formulae by Byrne (1991): $C_1 = 7600 (D_r)^{-2.5}$ and $C_2 = 0.4 / C_1$, where D_r is relative density in percent.
[c]Shear modulus of 2 x 10^5 kilopascals (kPa) used. Treated as linear elastic until failure.

elasto-plastic soil model previously described was used to model the behavior of both the treated and untreated soils, with the exception of the jet-grouted material, which was modeled using a linear elastic-perfectly plastic stress-strain model along with the Mohr-Coulomb failure envelope. Pore pressure generation in the chemically-grouted or jet-grouted sands was not modeled, since it should not be as significant as in densified or untreated sand. Other modeling details can be found in Cooke (2000).

Two-dimensional, plane strain analyses of the bridge pier and stub abutment were made. No attempt was made to adjust the results for three-dimensional effects or the interaction of all bridge components with each other.

Results For Pier

The effects of the treated zone type and width on the predicted displacements of the pier at the end of shaking are shown in Figure 5 for densified and chemically-grouted blocks that extend completely through the liquefiable soils. The width of treatment is expressed in terms of the width-to-depth ratio (W/D), defined in Figure 5 as the width of treatment beyond the footing edge, W, divided by the depth of treatment below the footing bottom, D (i.e. – 10 meters). For each treatment type the width-to-depth ratio is varied, with it being as small as 0.07 (i.e. – treated zone extends 0.7 m beyond the footing edge) for chemical grouting and as large as 2.0 (i.e. – treated zone extends 20 m beyond the footing edge) for densification to 75 percent relative density.

As seen from Figure 5, treatment significantly reduces the magnitude of vertical (i.e. – Y direction) and horizontal (i.e. – X direction) displacements during shaking compared to no ground improvement. This reduction can be attributed in part to the higher shear stiffness and reduced tendency for volumetric strain of the improved zone, which in the case of the densification improvement results in lower compressibility and less pore water pressure development in the densified zone. As expected, increasing the width of treatment beyond the footing edge for densification or chemical grouting results in less footing settlement, which is consistent with results observed by Hatanaka et al. (1987) in shaking table tests with densified zones. The incremental decrease in pier settlement that occurs for incremental increases in the treatment width gets progressively smaller, with the settlement approaching a limiting value obtained when all of the liquefiable soil is treated. The width of treatment needed to reduce the settlement to 5 cm is less for the zone having a higher stiffness and strength, with the width-to-depth ratios being approximately 0.3, 0.5, and 1.0 for chemical grouting, densification to 85 percent relative density, and densification to 75 percent relative density, respectively. The time histories of settlement for the pier footing indicate that while the settlement increases over time during shaking, most of the movement occurs during strong shaking (i.e. − over the first 13 seconds).

No clear trend was apparent between the treated zone width and the final horizontal displacement of the pier at the end of shaking for either the densified or chemically-grouted cases. Only the accumulated back-and-forth horizontal displacement of the pier base (i.e. − the distance the pier base translates relative to the bottom of the grid during shaking calculated by summing the distances traversed between reversals in the time record of relative horizontal displacement) showed a clear trend with the densified or chemically-grouted zone width, as seen in Figure 5. The accumulated horizontal displacement of the pier decreases with increases in the treated zone width-to-depth ratio, as well as stiffness and strength. These trends are similar to those observed for the pier settlements. This measure of horizontal displacement can only provide a relative comparison of performance. It cannot be compared to conventional horizontal displacement criteria.

The impact of the treated zone depth on the bridge pier displacements was studied by varying the depth of treatment below the footing while fixing the type and width of treatment. For a zone densified to 75 percent relative density and extending 10 m beyond the footing edge, a depth of treatment of at least 9 m below the footing bottom (i.e. − 90 percent of the liquefiable soil thickness below the footing) was needed to reduce the pier settlement to about 5 cm. This treatment depth is larger than the 50 percent of the liquefiable soil thickness suggested by centrifuge test results for a shallow footing case by Liu and Dobry (1997). The trend of decreasing pier settlements with increasing treatment depth obtained from the FLAC analyses was similar to that seen for a shallowly-embedded rigid block supported by treated zones of varying thickness in liquefiable soil, in the centrifuge tests reported by Hausler et al. (2002). Although treating 90 percent of the liquefiable soil thickness for the case evaluated with FLAC might appear desirable in terms of limiting the pier settlements, while at the same time providing some isolation of the pier from high ground accelerations, any slight inclination of the ground or the presence of a nearby approach embankment could result in unacceptable movements of the pier due to lateral spreading. Therefore, it is unlikely a partial treatment depth would be used for a pier in practice.

Most of the analyses were for a time period of 26 s, which is approximately 13 s after the end of strong shaking. A few analyses performed for extended times of 120 s to 240 s indicate that pore pressure migration occurred into the densified zone after strong shaking stopped, causing an increase in pore water pressures in the zone and increasing the number of elements with fully-mobilized strength. Additional analyses indicate that using drains along the boundaries of the densified zone can reduce the pore pressure migration effect.

Analyses of jet-grouted walls placed on each side of the pier footing for reinforcement and containment of the liquefiable sands under the footing produced some inconsistent results. Further analysis of this scheme is required.

Results For Stub Abutment

The effects of densification by compaction grouting and cementation by chemical or jet grouting on the predicted stub abutment movements at the end of earthquake shaking are presented in Figure 3. In all cases the treated zone extended through the full 11-m depth of liquefiable soils beneath the embankment. The final vertical (i.e. $-$ Y direction) and horizontal (i.e. $-$ X direction) movements are for the point on the abutment where the superstructure load is applied. Movements at this point should be most directly related to the bridge performance. Horizontal movements in the underlying embankment and liquefiable soils were observed to be up to 30 to 50 percent larger than the abutment movement. For all cases the vertical and horizontal abutment movements increased with time during shaking, with most of the movement occurring during strong shaking (i.e. $-$ within the first 13 s).

As seen from Figure 3, constructing a treated zone in the vicinity of the approach embankment slope reduced the horizontal and vertical abutment displacements of +82 cm and $-$32 cm, respectively, predicted to occur with no ground improvement. The amount of reduction in the downslope movement of the abutment was dependent on the treated zone size and location, as well as the type of treatment. Treating directly under the approach embankment slope and stub abutment only, which was an approach found to be effective in other studies for limiting ground displacements of embankments (Riemer et al., 1996; Yanagihara et al., 1991), was not adequate to reduce the abutment displacements to tolerable levels for densification and chemical grouting treatments (Cases A and E in Figure 3). Extending the densified and chemically-grouted zones outward beyond the toe up to 8 m helps to further reduce the movements (Cases B and F, with F actually extending out 11 m). Further extension beyond the toe did not have a substantial effect (Cases C and G). To reduce the displacements to more acceptable levels required that the densified and chemically-grouted zones be extended underneath the embankment behind the crest (i.e. - back of the stub abutment) 19.4 m and 6 m, respectively, giving total treatment widths of 39.4 m and 26 m (Cases D and H), and the relative density of the densified zone be increased from 75 to 85 percent.

For jet-grouting, a treated block under the embankment slope and stub abutment having a 12-m width resulted in abutment horizontal and vertical displacements close to the tolerable level of 5 cm (Case I). Extending the zone 3 m behind the embankment crest reduced the movements to +1.3 cm and +3.0 cm (Case J). Using only a 5.5-m wide block near the embankment toe produced unacceptable movements (Case K) due to overturning of the intact jet-grouted block.

For both the densification and grouted zone cases, pore water pressures in the liquefiable soils under the approach embankment were similar to those generated for the case of no ground improvement. Therefore the key factor in reducing abutment movements was the increased lateral restraint for the softened soils under the embankment provided by the improved soils, with a smaller zone width required for treated material having a higher strength and stiffness. Increasing the width of the treated zone under the abutment and embankment slope in either direction helps reduce the shear deformation that occurs in the treated zone and surrounding soils. Extending the zone out beyond the embankment toe helps to reduce the potential for large downslope movements of the abutment associated with a localized slope stability failure at the abutment and slope. Expanding the treated zone behind the stub abutment reduces the potential of displacements associated with the weakened, untreated soils under the embankment being pushed laterally and upwards towards the embankment slope and toe. In the case of a high-strength grouted block that remains intact throughout shaking, adequate block width must also be provided to limit movements associated with overturning effects.

Most of the analyses performed for the stub abutment with densification and chemically-grouted zones were run for a duration of 26 s. Additional analyses having durations of 240 s and 205 s were performed for the 39.4-m-wide densified zone having a relative density of 85 percent and a 26-m-wide chemically-grouted zone, respectively. There was no increase in the movement of the abutment beyond that observed at 13 s in either case, but there was an increase in the number of elements with fully-mobilized strength. This increase almost produced a nearly continuous failure surface for the densification improvement case. This was due in part to pore pressure migration into the zone near the embankment toe after the shaking stopped. These results indicate that placement of drains at the perimeter of the densified zone near the toe might be useful to reduce the pore pressure migration effects.

Analyses of a partially-penetrating improved zone at the stub abutment indicated unacceptable performance. This is consistent with the results of numerical analyses by Yasuda et al. (1991). The analyses also indicated use of an 8-m-wide gravel berm placed at the embankment toe and extending upward along the slope to a height of 3.6 m was ineffective in reducing the abutment movements significantly.

Conclusions

The results of numerical analyses in this study indicate that improved ground zones of limited size at a bridge pier or stub abutment supported on shallow foundations can be potentially effective for reducing final horizontal and/or vertical footing displacements caused by earthquake-induced liquefaction to approximately 10 cm or less. For the bridge pier in this study the successful methods for limiting settlements were a densified or grouted block created under and around the pier for the full-depth of the liquefiable soil layer. For the stub abutment on an approach embankment successful methods included a densified or grouted block that extended under the embankment slope and abutment, as well as a limited distance beyond the toe and behind the crest, with the block fully penetrating the liquefiable soil. A

limited size buttress fill placed against the embankment slope was not successful in preventing excessive deformations.
Other observations and conclusions from this study are:

- The required width of the improved zone is dependent in part on the strength and stiffness of the treated soil, with a smaller width generally being acceptable for stronger and stiffer material.
- There appears to be a limiting width of treatment beyond which further reductions in the pier or abutment displacements are not substantial.
- Post-earthquake pore pressure migration into narrow densified zones can potentially lead to additional displacements.
- If improved zones for adjacent abutments and piers are close together or overlap, due to their required widths, they should be combined together and evaluated as one zone.
- The numerical modeling method used in this study has the potential for predicting the performance of bridges on improved ground zones, however, further development, testing, and refinement of the model is desirable.

Despite the progress that has been made in evaluating and designing improved ground zones for liquefaction mitigation, as well as the overall good performance of such zones and supported structures during actual earthquakes, further investigation is still needed in this area. For bridges topics for investigation include bridges supported on deep foundations, performance of bridge piers and abutments for different soil profiles and earthquake motions, and performance of combined improved zones supporting more than one bridge element. In addition, further work is needed on the impact (if any) of foundation width on the lateral extent of treatment, three-dimensional behavior and design of improved zones, effects of forces exerted by laterally spreading soils, pore pressure development and migration in chemically-grouted and jet-grouted zones, and continued development and improvement of numerical modeling methods for analyzing performance.

Acknowledgements

Funding for this research was provided primarily by the Federal Highway Administration through the Multidisciplinary Center for Earthquake Engineering Research in Buffalo, NY, under FHWA Contract DTFH61-92-C-00106. Support for the first author also came from a nine-month fellowship from the Federal Emergency Management Agency through the Earthquake Engineering Research Institute in Oakland, CA.

References

Adalier, K. (1996). "Mitigation of Earthquake Induced Liquefaction Hazards," Ph.D. Dissertation, Rensselaer Polytechnic Institute, Troy, New York.

Adalier, K., Elgamal, A.-W., and Martin, G. R. (1998). "Foundation Liquefaction Countermeasures for Earth Embankments," *J. of Geotech and Geoenv. Engrg.*, ASCE, Vol. 124, No. 6, June, 500-517.

Byrne, P. M. (1991). "A Cyclic Shear-Volume Coupling and Pore Pressure Model for Sand," *Proc., Second International Conference on Recent Advances in Geotechnical Earthquake Engineering and Soil Dynamics*, Vol. 1, Prakash, S., ed., University of Missouri, Rolla, Missouri, 47-55.

Cooke, H. G. (2000). "Ground Improvement for Liquefaction Mitigation at Existing Highway Bridges," Ph.D. Dissertation, Dept. of Civil & Environmental Engineering, Virginia Polytechnic Institute and State University, Blacksburg, VA, 372 pp.

Geo-Research Institute (1995). Ground motion records from vertical array in northwest corner of Port Island during 1995 Kobe Earthquake.

Hatanaka, M., Suzuki, Y., Miyaki, M. and Tsukuni, S. (1987). "Some Factors Affecting the Settlement of Structures Due to Sand Liquefaction in Shaking Table Tests," *Soils and Foundations*, JSSMFE, Vol. 27, No. 1, 94-101.

Hausler, E. and Sitar, N. (2001). "Performance of Soil Improvement Techniques in Earthquakes," *Proc., Fourth International Conference on Recent Advances in Geotechnical Earthquake Engineering and Soil Dynamics*, Prakash, S., editor, University of Missouri, Rolla, MO, 6 pp.

Hausler, E., Sitar, N., Matsuo, O., and Okamura, M. (2002). "Influence of Ground Improvement on Settlement and Liquefaction, A Comparison of Dynamic Centrifuge Tests at Two Centrifuge Centers," *Proc., International Conference on Physical Modeling in Geotechnics 2002*, Phillips, Guo & Popescu, eds., Swets & Zeitlinger Lisse, 557-562.

Itasca Consulting Group, Inc. (1999). "FLAC - Fast Lagrangian Analysis of Continua", *User's Manual - Version 3.4*, Itasca Consulting Group, Minneapolis, MN.

Liu, L. (1992). "Centrifuge Earthquake Modelling of Liquefaction and Its Effect on Shallow Foundations," Ph.D. Dissertation, Rensselaer Polytechnic Inst., Troy, NY, 303 p.

Liu, L. and Dobry, R. (1997). "Seismic Response of Shallow Foundation on Liquefiable Sand," *J. Geotech. and GeoEnv. Eng.*, ASCE, Vol. 123, No. 6, 557-567.

Mitchell, J. K., Baxter, C. D. P., and Munson, T. C. (1995). "Performance of Improved Ground During Earthquakes," *Soil Improvement for Liquefaction Hazard Mitigation*, Geotech. Special Pub. No. 49, ASCE, 1 – 36

Pyke, R. (1979). "Nonlinear Soil Models for Irregular Cyclic Loadings," *J. of Geotech. Engrg. Div.*, ASCE, 105(GT6), 715-726.

Riemer, M. F., Lok, T. M. and Mitchell, J. K. (1996). "Evaluating Effectiveness of Liquefaction Remediation Measures for Bridges," *Proc., 6th Japan-U.S. Workshop on Earthquake Resistant Design of Lifeline Facilities and Countermeasures for Soil Liquefaction,* Technical Report NCEER-96-0006, NCEER, Buffalo, NY, 441-455.

Seed, H. B. and Idriss, I. M. (1970). "Soil Moduli and Damping Factors for Dynamic Response Analyses," *Report EERC 70-10*, Earthquake Engineering Research Center, University of California, Berkeley, CA.

Yanagihara, S., Takeuchi, M., and Ishihara, K. (1991). "Dynamic Behavior of Embankment on Locally Compacted Sand Deposits," *Proc., Fifth International Conference on Soil Dynamics and Earthquake Engineering V*, Computational Mechanics Publication, Boston, and Elsevier Applied Science, New York, 365-376.

Yasuda, S., Nagase, H., Kiku, H., and Uchida, Y. (1991). "Countermeasures Against Permanent Ground Displacement due to Liquefaction," *Soil Dynamics and Earthquake Engineering V*, Elsevier Science Publishing Inc., New York, 341-350.

Youd, T. L. (1998). "Screening Guide for Rapid Assessment of Liquefaction Hazard at Highway Bridge Sites," *MCEER Technical Report MCEER-98-0005*, Multidisciplinary Center for Earthquake Engineering Research, Buffalo, NY, 58 pp.

Youd, T. L., Willey, P. S., Gilstrap, S. G., and Peterson, C. R. (1998). "Liquefaction Hazard Evaluation of Interstate, Federal, and State Highway Bridge Sites in Utah," *Report to Utah Department of Transportation*, Brigham Young Univ., Provo, Utah.

Zeng, X. (1993). "Experimental Results of Model No. 11," *Proc., Verification of Numerical Procedures for the Analysis of Soil Liquefaction Problems, Vol. 1*, Arulanandan, K. and Scott, R. F., eds., Balkema, Brookfield, Vermont, 895-908.

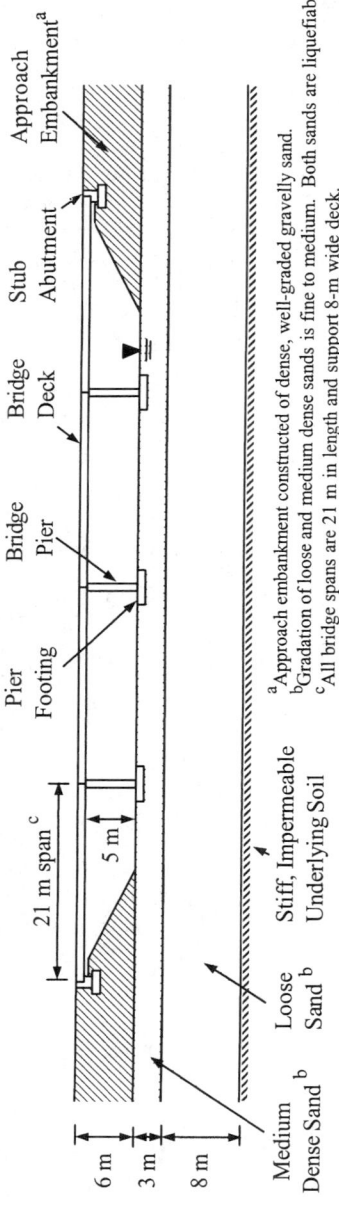

Figure 1. Elevation View of Bridge Used to Study Effectiveness of Ground Improvement in Mitigating Liquefaction-Induced Deformations

[a] Approach embankment constructed of dense, well-graded gravelly sand.
[b] Gradation of loose and medium dense sands is fine to medium. Both sands are liquefiable.
[c] All bridge spans are 21 m in length and support 8-m wide deck.

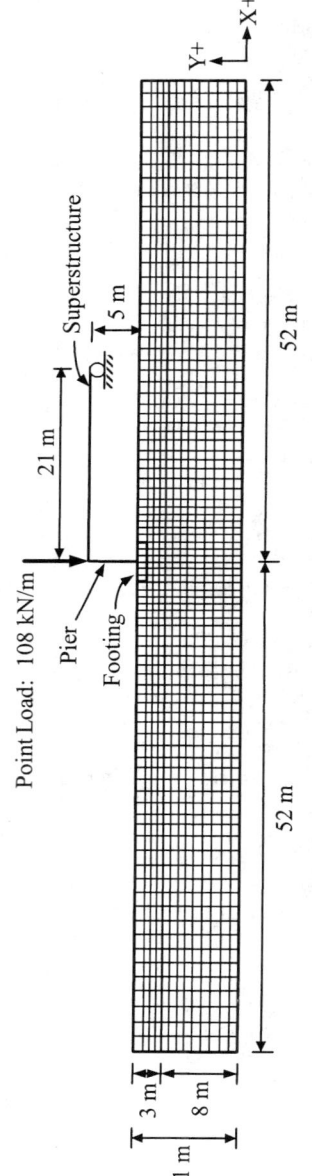

Figure 2. Grid Used for FLAC Analysis of Pier in Parametric Study

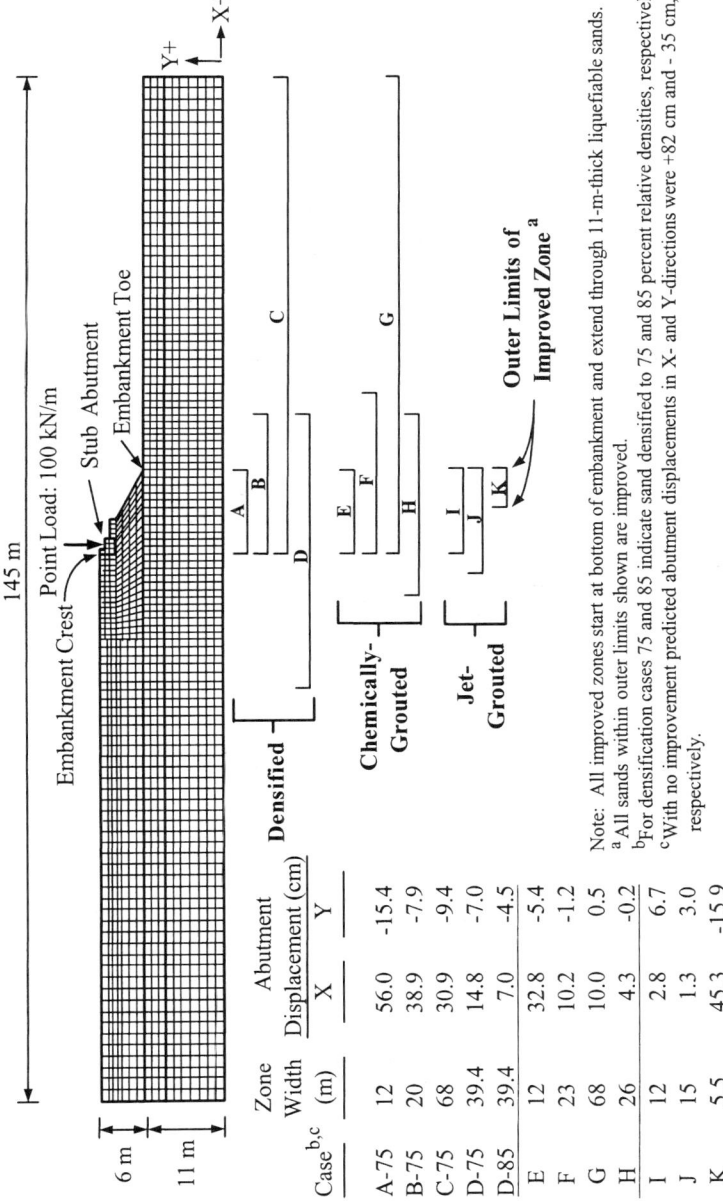

Figure 3. Stub Abutment Grid and Predicted Abutment Displacements at End of Shaking for Different Improvement Schemes

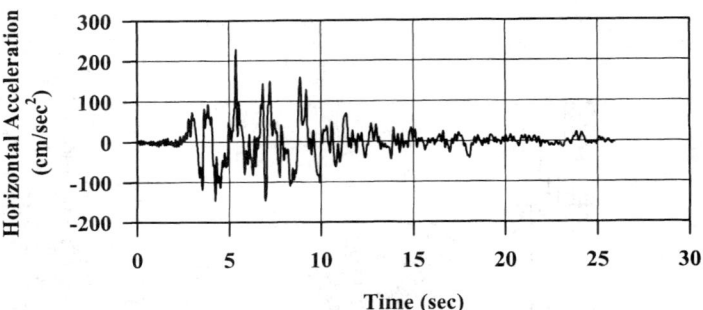

Figure 4. Input Horizontal Acceleration Record

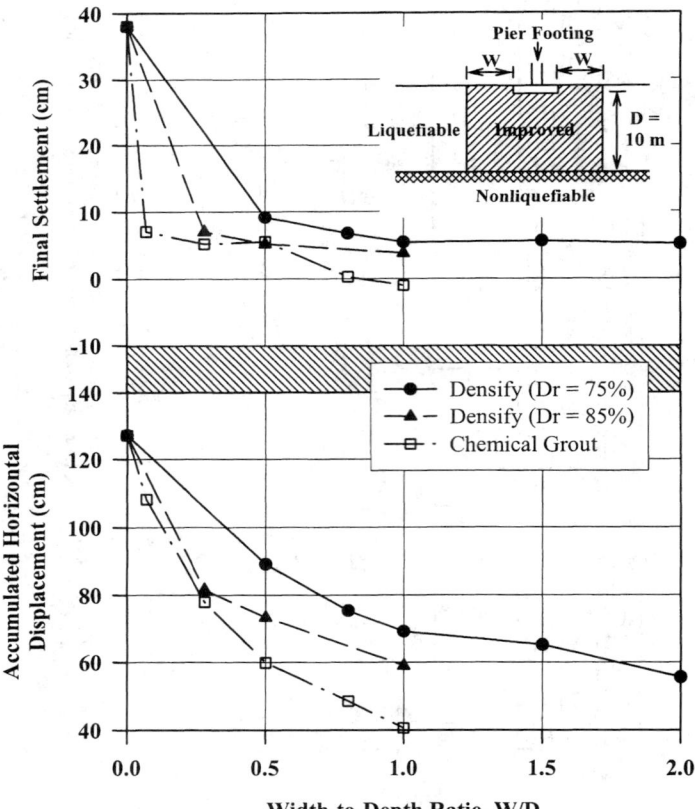

Figure 5. Effect of Treatment Width and Type on Predicted Pier Base Displacements for Fully-Penetrating Improved Zone (D = 10 m)

LATERAL SEISMIC PRESSURES FOR DESIGN OF RIGID UNDERGROUND LIFELINE STRUCTURES

Craig A. Davis[1]

Abstract

An analytical solution and simplified design procedure is presented for evaluating the lateral seismic stress distribution and resultant force on a rigid buried structure embedded within an infinitely deep soil layer subjected to vertically propagating seismic shear waves. The solution improves upon previous models, which assume the rigid retaining structure is bonded to a rigid foundation and the top of structure is coincident with the ground surface, by incorporating more realistic foundation conditions and allowing for the determination of lateral stresses on structures having the top buried a depth d below the ground surface. The soil in the structure vicinity is modeled as a continuum and incorporates soil stiffness variation with confining pressure. The proposed model provides lower stresses on rigid walls than the previous models, mainly because of different boundary conditions for site response, but provides comparable results with the previous solutions when similar soil and boundary conditions are employed. The model presented herein improves the seismic design and assessment of lifeline systems.

Introduction

Many lifeline systems utilize rigid underground structures, such as box culverts, utility vaults, retaining walls, basement walls, rectangular access tunnels, etc., that are restrained against lateral movement and significantly stiffer than the surrounding ground. Even though these structures are commonly built in highly seismic areas, and can be damaged from strong shaking (e.g., Wood, 1973; Iida et al., 1996; Davis, 2000), they typically are not adequately designed to resist lateral earthquake forces, in part because: (1) lack of appropriate design methodologies, and (2) a misunderstanding among design professionals of the forces an earthquake may impose upon an underground structure. An initial investigation was performed to assess the design needs for rigid buried structures that included a literature review, evaluation of existing methodologies and model assumptions, review of typical site and geometric conditions, and an informal survey of geotechnical and structural design professionals to asses the methodologies used in lifeline earthquake engineering practice. This initial investigation revealed the need for a better understanding of the physical problem of stresses on underground rigid structures, development of a new analytical solution to incorporate realistic conditions, and a simplified design procedure that is easily incorporated into common practices.

Some limitations with existing solutions are: (1) The Wood (1973), Scott (1973), and other similar rigid wall models assume a rigid base site and top of structure at ground surface, resulting in very large design stresses, and (2) the Hashash et al. (2001) and other similar models for large flexible subway structures require advanced evaluations and are inappropriate for design of lifeline structures of focus herein. To overcome current design limitations, an analytical solution and simplified design procedure is presented for evaluating the lateral seismic stress distribution and resultant force on a rigid buried structure embedded within an infinite half-space, below the ground surface, subjected to vertically propagating seismic shear waves.

---

[1] Waterworks Engineer, Los Angeles Department of Water and Power, Los Angeles.

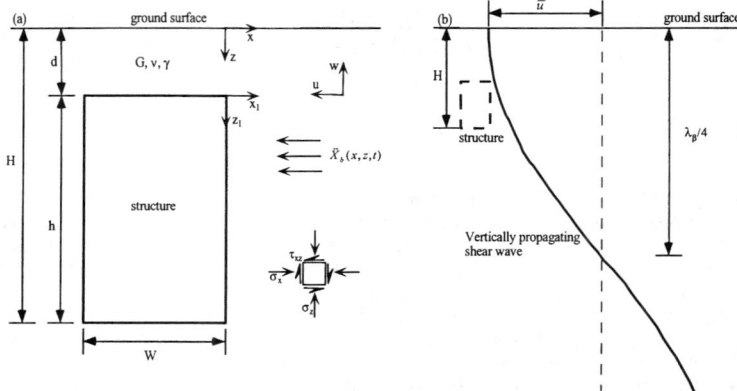

Figure 1. Underground rigid box structure model: a) structure geometry and coordinate systems, and b) wave propagation model at ground surface.

Model

Figure 1a shows the model for evaluating lateral stresses on underground rigid box structures embedded within an infinite half-space. The half-space represents a deep unsaturated soil deposit having a unit weight γ, shear modulus G, and Poisson ratio ν. The structure has a width W, vertical rigid walls of height h, and is buried a depth d with its base at depth H below the stress-free ground surface. The structure is assumed to have an infinite length L satisfying plain strain conditions. The model is defined by the x-z and x_1-z_1 coordinate systems originating at the ground surface and top of wall, respectively, which are related by $x = x_1$ and $z = z_1 + d$.

The structure is subjected to vertically incident shear waves. The horizontal and vertical motion in the soil medium is represented by displacements u and w, respectively, having positive motions as defined in Fig. 1a. Figure 1a also shows the stress sign convention where compressive stresses are defined as positive.

G is determined at any depth z from:

$$G = G_0 f_G(z), \quad f_G(z) = (z/H)^m \tag{1}$$

where G_0 is the free-field modulus at depth H, and $f_G(z)$ is a function representing the variation in shear moduli with depth with $0 \leq m \leq 1$; $m = \frac{1}{2}$ is applicable for most soils (Hardin and Drnevich, 1972; Bardet, 1997); $m = 0$ represents a material having a shear modulus independent of confining pressure; $m = 1$ after large shear strains.

The box structure sidewalls are fixed against rotation at the base and restrained against lateral movements. Actual wall movements are limited to flexural deflections, which are less than that required to develop an active pressure state, and are therefore neglected in this study by assuming perfectly rigid vertical walls. As a result, the solution is derived independent of the structures' stiffness properties.

Lateral Earthquake Stresses and Resultant Forces

The lateral seismic pressures σ_E applied to underground structure retaining walls are determined from the superposition of: (1) lateral soil stress increases in the wall vicinity $\Delta\sigma_h$ that result from densification and dilation of soil particles during shearing (e.g., Youd and Craven, 1975), and (2) transient stresses σ_x resulting from the passage of seismic waves (e.g., Scott, 1973; Veletsos and Younan, 1994a, 1994b):

$$\sigma_E = \Delta\sigma_h + \sigma_x = K_E \gamma z + k_s(u_f - u_H), \; \sigma_x = k_s(u_f - u_H) \qquad d \leq z \leq H \qquad (2)$$

where K_E is the lateral earth pressure coefficient resulting from seismic shaking, u_f is the horizontal soil free-field deformation during seismic shaking, u_H is the horizontal soil free-field deformation at depth H, and k_s is a continuous soil stiffness parameter. Equation 2 defines only the incremental components resulting from earthquake shaking; the pre-earthquake stress components must be superimposed to obtain the total stresses applied to the structure.

Youd and Craven (1975) and Sherif et al. (1982) measured permanent static lateral stress increases from applied shearing stresses. At present, no empirical methods for estimating K_E exist. Richards et al. (1999) present an analytical model for estimating K_E and $\Delta\sigma_h$ using the Mohr-Coulomb failure criteria. Unfortunately, under large a_h, such as that occurring in near-field earthquake shaking (e.g., Bardet and Davis, 1996), Richards et al. (1999) predicts very large $\Delta\sigma_h$ that is not substantiated by actual measurements or field observations. In the absence of adequate analysis methods and insufficient testing data useful for developing empirical relationships, $\Delta\sigma_h$ can be estimated from limited published test results (e.g., Youd and Craven, 1975; Sherif et al, 1982) assuming the linear relationship of Eq. 2. Under many soil-wall systems, values may fall within the range $0 \leq K_E \leq 0.5$, with lower K_E for denser, fine grained, and cohesive soils. K_E may exceed this range in loose cohesionless soil but confirmation with controlled measurements is needed for use in design. The remainder of this paper focuses only on σ_x, leaving σ_h improvements for future work.

Harmonic motion and stress

Equation 2 shows that the lateral wall stresses are a function of the free-field displacement u_f, which is dependent upon the subsurface site conditions and depth of burial. For a given site, u_f can be evaluated using common site response analysis methods such as Shake (Schnabel et al., 1972) or EERA (Bardet et al, 2000). Previous lateral stress solutions by Wood (1973), Scott (1973), Veletsos and Younan, (1994a; 1994b), and Richards et al. (1999) used a rigid rock base subsurface model that can provide unreasonably large u_f and conservative σ_x impractical for design of common underground structures embedded in soil overlying an elastic rock base. As a result, modifications are made herein to represent more realistic conditions.

The horizontal displacement solution over the region $0 \leq z \leq H$ is described in Appendix I and given by:

$$u = \overline{u}\left(1 - e^{-k_\beta x_1/\psi_e}\right)\left(\cos k_\beta(z_1 + d) - \cos k_\beta H\right), \; \psi_e^2 = \frac{2-\nu}{1-\nu}, H \leq \lambda_\beta/4 \qquad (3)$$

where $\overline{u} = -a_h/\omega^2$ is the horizontal free-field ground surface displacement, a_h the horizontal acceleration, $k_\beta = 2\pi/\lambda_\beta = \omega/\beta$ is the shear wave number, λ_β is the shear wave length, $\omega = 2\pi f$ is the angular frequency, f is the wave frequency, β is the average free-field shear wave velocity of the soil medium over H determined from:

$$\beta = \sqrt{gG/\gamma}, g \text{ is the acceleration of gravity} \qquad (4)$$

The harmonic nature of the motion described by $e^{-i\omega t}$ in Appendix I is implied in Eq. 3. The $\cos k_\beta H$ term in Eq. 3 represents the horizontal displacement at depth H, which causes a uniform ground translation over the region $0 \leq z \leq H$. The uniform translation does not result in any stress increase against the wall and is therefore subtracted out of the formulation.

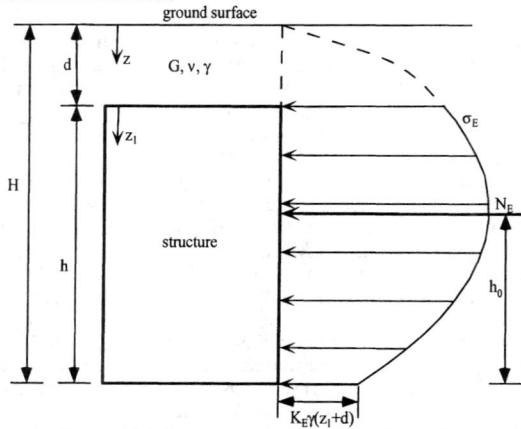

Figure 2. Lateral stress distribution σ_E, resultant force N_E, and resultant force location h_0 against rigid underground structure.

The free field displacement u_f for $0 \leq z \leq H$ is:

$$u_f = \frac{-a_h}{\omega^2}\left(\cos 2\pi \frac{z_1+d}{\lambda_\beta} + \cos 2\pi \frac{H}{\lambda_\beta}\right) \quad (5)$$

The solution for k_s is described in Appendix I and is given by the following at $x_1 = 0$:

$$k_s = \frac{2Gk_\beta}{\sqrt{(1-v)(2-v)}} = Gk_\beta \psi_\sigma, \quad \psi_\sigma = \frac{2}{\sqrt{(1-v)(2-v)}} \quad (6)$$

Substituting Eqs. 1 and 4 to 6 into Eq. 2 gives:

$$\sigma_x = \frac{a_h}{g}\frac{\gamma\psi_\sigma\beta}{2\pi f}\left(\frac{z_1+d}{H}\right)^m [\cos(k_\beta(z_1+d)) - \cos(k_\beta H)] \quad (7)$$

where σ_x is taken as a compressive stress against the structure oriented as shown in Fig. 2. Equation 7 shows that the lateral stresses applied to the structure are dependent upon the input motion frequency.

The resultant earthquake force N_E, moment about the base M_E, and resultant force location h_0 on a buried structure, as shown in Fig. 2, is determined in the frequency domain from:

$$N_E = \int_0^h \sigma_E dz_1, \quad M_E = \int_0^h (h-z_1)\sigma_E dz_1 = N_E h - \int_0^h z_1\sigma_E dz_1, \text{ and } h_0 = \frac{M_E}{N_E} \quad \text{(8a, b, and c)}$$

Unfortunately, Eqs. 8a to 8c cannot easily be solved for $m \neq 0$ or $m \neq 1$. As a result, Eqs. 8a and 8b are solved numerically using Simpson's 1/3 Rule giving:

$$N_E = K_E \gamma h \left(d + \frac{h}{2}\right) + \frac{a_h \gamma}{g} \frac{\beta \psi_\sigma}{2\pi f_a} \frac{\Delta y}{3} \left\{ \left(\frac{d}{H}\right)^m (\cos k_a d - \cos k_a H) \right.$$

$$+ 4 \sum_{r=1}^{n/2} \left(\frac{(2r-1)\Delta y + d}{H}\right)^m [\cos k_a ((2r-1)\Delta y + d) - \cos k_a H] \quad (9a)$$

$$+ 2 \sum_{s=1}^{n/2-1} \left(\frac{2s\Delta y + d}{H}\right)^m [\cos k_a (2s\Delta y + d) - \cos k_a H]$$

$$M_E = K_E \gamma h^2 \left(\frac{d}{2} + \frac{h}{6}\right) + \frac{a_h \gamma}{g} \frac{\beta \psi_\sigma}{2\pi f_a} \frac{\Delta y}{3} \left\{ h \left(\frac{d}{H}\right)^m (\cos k_a d - \cos k_a H) \right.$$

$$+ 4 \sum_{r=1}^{n/2} [h - (2r-1)\Delta y] \left(\frac{(2r-1)\Delta y + d}{H}\right)^m [\cos k_a ((2r-1)\Delta y + d) - \cos k_a H] \quad (9b)$$

$$+ 2 \sum_{s=1}^{n/2-1} (h - 2s\Delta y) \left(\frac{2s\Delta y + d}{H}\right)^m [\cos k_a (2s\Delta y + d) - \cos k_a H]$$

where $\Delta y = h/n$ and n is an even integer representing the number of increments over which the solution is to be numerically integrated. The value of n is selected large enough to minimize the numerical integration error to an acceptable level. The error decreases with increasing d/H and using $n = 10$ provides an *error* $\leq 1\%$ in most cases; moreover $n = 100$ gives very accurate results in all cases with minimal computing time using a standard spreadsheet or programming language formulation.

Application to Design

Estimating seismic transient lateral pressures from Eq. 1 requires a dynamic analysis utilizing a time history of acceleration transformed into the frequency domain to calculate the lateral pressure response, and then inversing the frequency response to determine the time history of wall pressures, similar to that presented in Veletsos and Younan (1994a). This type of analysis is far too complicated for typical structure designs and therefore a more simplified analysis method is proposed herein.

For design purposes, only the maximum stresses need be evaluated. The maximum wall pressures result from a wave having the highest acceleration with the shortest wavelength (i.e., a free-field wave with maximum differential displacements). Peak accelerations in the earthquake near-field normally occur from waves within the frequency range $1\ Hz < f < 6.67 Hz$ (Bardet and Davis, 1996). Therefore, for design purposes the maximum stresses can be conservatively assumed to result from a wave having a peak ground acceleration of a_h and $f = 6.67\ Hz$. From this, σ_x for any depth z_1 can be determined from Eqs. 2 and 7, N_E from Eq. 9a, and location h_0 from Eqs. 8c and 9b for a buried rigid structure of given geometric dimensions h and d, within a soil with known γ, β, and ν, subjected to an earthquake peak acceleration a_h.

Figure 3 shows the seismic stress distribution σ_E and the total stress distribution $\sigma_{tot} = \sigma_0 + \sigma_E$ for a rigid soil-wall system having the properties shown in the figure caption. The active σ_A, passive σ_P, at-rest σ_0, and traffic surcharge σ_{tr} pressures are also shown for comparison. σ_{tr} is taken as a 0.61m soil surcharge load recommended by AASHTO (1989). The soil friction angle ϕ is not needed to estimate the lateral earthquake stresses on rigid walls, but is necessary for determining

σ_0, σ_A, and σ_P as described in Das (1990). As indicated in Fig. 3, $\sigma_A < \sigma_{tot} < \sigma_P$ and the total area under the σ_E curve (i.e., resultant force) is comparable to that of σ_{tr}. The rigid base σ_E is shown for comparison and described in the next section. Figure 3 is presented for $d/H = 0$, however, for the analysis method presented the same stress distribution would result for any d/H ratio; the only difference is that the stress above the top of wall would not be included in calculating N_E.

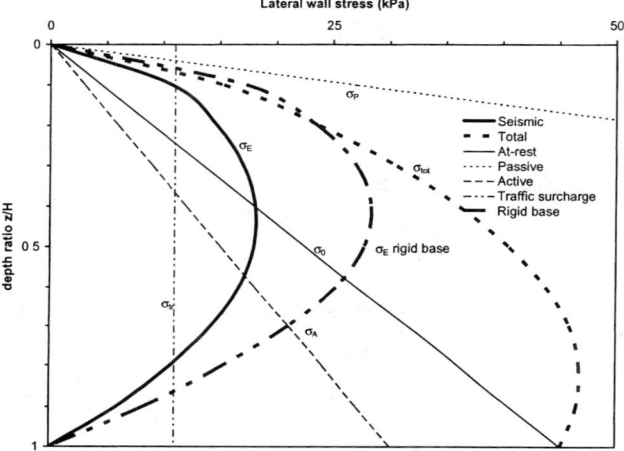

Figure 3. Lateral wall stresses σ_E and σ_{tot} compared to σ_A, σ_P, and σ_{tr} using $H=5m$, $d=0$, $m=1/2$, $\gamma=18.1$ kN/m^3, $\beta=250m/s$, $v=0.35$, $f=6.67$ Hz, $a_h=0.5g$, $K_E=0$, and $\phi=30°$.

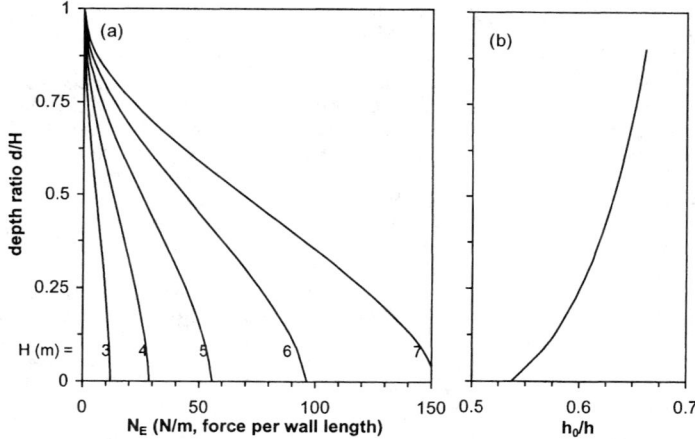

Figure 4. a) Resultant force for various H and d/H using same input parameters presented for Fig. 3, and b) resultant force location variation with depth ratio d/H.

Fig. 4 shows N_E and h_0 for a range H and d/H. These plots are useful for determining forces on structures having differing heights and burial depths for a project. Figure 4a shows the variation of N_E with depth changes with d/H and H. h/h_0 changes with d/H, as seen in Fig. 4b, but has a minor variation with H. As a result, Fig. 4b is an average of several closely spaced h/h_0 curves determined for different H.

Comparison with Existing Solutions

There are no known previously existing solutions for pressures against a rigid wall within a semi-infinite medium. The only solutions available for comparing with the present model are for a rigid base. The present solution is not constrained by a rigid base, but can approximate a rigid base condition by requiring $\lambda_\beta = 4H$ and $\omega = \pi\beta/2H$, which gives:

$$\sigma_x = \frac{2\psi_\sigma \gamma H^2}{\pi h} \frac{a_h}{g} f_G(z) \cos\frac{\pi}{2H}(z_1 + d) \tag{10}$$

Figure 5a and 5b show comparisons of Eq. 10, for a constant soil stiffness with depth (i.e., $m = 0$), with the solutions of Wood (1973) and Veletsos and Younan (1994b), respectively. The present model compares reasonably well with the previous models near the ground surface, but has lower pressures with depth, which result directly from the difference in site response models. The Wood (1973) and Veletsos and Younan (1994b) solutions, which compare very well with each other, are strictly for a rigid base model that does not allow the seismic wave energy to leave the soil stratum, whereas the present solution is for a semi-infinite medium that allows the seismic waves to radiate away from the structure. As a result, the previous solutions excite several vibration modes resulting in a greater stress while the present solution is presented only for the first (fundamental) vibration mode. As shown in Fig. 5b, the Veletsos and Younan (1994b) first mode is only 1.065 times greater than and compares very well with the present model; thus the present solution converges very closely to Wood (1973) and Veletsos and Younan (1994b) when forced into the same boundary conditions.

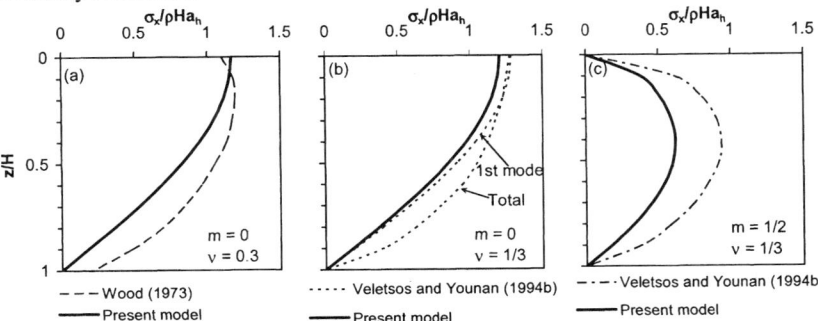

Figure 5. Comparison of present model (Eq. 10) with previous solutions of Wood (1973) and Veletsos and Younan (1994b).

Figure 5c compares Eq. 10 ($m = \frac{1}{2}$) with Veletsos and Younan (1994b) [$f_G(z) = 2z/H - (z/H)^2 \cong (z/H)^{1/2}$] for a soil stiffness varying with depth and shows that the present solution provides a lower wall pressure for inhomogeneous soils than that for a rigid base. In addition, Fig. 3 compares the present model solution from Eq. 7 with the rigid base approximation of Eq. 10 and shows that the recommended solution of Eq. 7 is significantly less than the rigid base approximations of previous models.

Discussion

The pressure distribution of Equation 7 was determined assuming the rigid wall is placed within an infinitely deep uniform material. However, this same pressure distribution results on walls within the top layer of a layered media where the depth of the top layer is greater than or equal to H (Davis, 2000; Schnabel et al., 1972). Thus, for horizontal layered sites the proposed model can be used for a_h determined from a more specific site evaluation (e.g., Shake, EERA).

The solution presented herein is for an underground structure with rigid side walls embedded within a uniform undamped soil for the purpose of developing a realistic but simple model that can be easily applied to common structures. The model can be extended to incorporate deformable walls, soil damping, horizontal layering, soil modulus reduction with shear strain, and vibration modes greater than fundamental for the evaluation of more complicated structures and subsurface conditions subjected to harmonic and transient input motions.

Conclusion

An analytical model and simple procedure is presented herein to analyze the lateral stress distribution useful for design and assessment of vertical rigid underground walls buried below the ground surface. The solution builds upon and is an improvement to the previous models of Wood (1973), Scott (1973), Veltsos and Younan (1994a, 1994b), and Richards et al. (1999), which provide very large loads that can lead to impractical design requirements inconsistent with past underground structure earthquake performances. Reasonable simplifying assumptions allow the lateral wall stresses can be estimated from easily obtained geometric and soil properties. Compared with previous models, the proposed model provides lower stresses on rigid walls, mainly because of differing site response assumptions, but converges very closely to the previous solutions when the same boundary conditions are employed. The present model is useful for understanding the physical process of how earthquakes apply lateral stresses to rigid buried structures and the results are easily expanded to incorporate wall movements and layered soils. In addition, the proposed procedure can be directly applied to normal design procedures, without the need for more complicated and time consuming evaluations, and improves the seismic design and assessment of lifeline systems.

Acknowledgements

Contributions and support from the Los Angeles Department of Water and Power are gratefully acknowledged.

Appendix I: Solution for Transient Response

The equation of motion and stress-strain relations are (Timoshenko and Goodier, 1934):

$$\frac{\partial \sigma_x}{\partial x} + \frac{\partial \tau_{xz}}{\partial z} = \rho \frac{\partial^2 u}{\partial t^2} + \ddot{X}_b(x,z,t), \quad \frac{\partial \sigma_z}{\partial z} + \frac{\partial \tau_{xz}}{\partial x} = \rho \frac{\partial^2 w}{\partial t^2} \qquad \text{(A1a, A1b)}$$

$$\sigma_x = \lambda\left(\frac{\partial u}{\partial x}+\frac{\partial w}{\partial z}\right)+2G\frac{\partial u}{\partial x}, \ \sigma_z = \lambda\left(\frac{\partial u}{\partial x}+\frac{\partial w}{\partial z}\right)+2G\frac{\partial w}{\partial z}, \ \tau_{xz} = G\left(\frac{\partial u}{\partial z}+\frac{\partial w}{\partial x}\right) \quad \text{(A2a–c)}$$

where $\ddot{X}_b(x,z,t)$ is a body force and $\lambda = 2G\dfrac{v}{1-2v}$ is the Lamé constant.

Eq. A1 must satisfy the boundary conditions: $\tau_{xz} = \sigma_z = 0$ at $z = 0$ and $u = 0$ at $x_1 = 0$ and $0 \le z_1 \le h$. Equations A2 assume the stresses in the structure vicinity are from a combination of coupled dilational and shear motion. The incident and reflected free-field motions act as a forcing function against the structure represented by $\ddot{X}_b(x,z,t)$.

Substituting Eq. A2a into Eq. A1, and accounting for the fact that dilational motion only results from wave reflections off of the structure for $0 \le z_1 \le h$, gives the equation of motion in terms of displacement:

$$G\left(\frac{\partial^2 u}{\partial z^2}-\frac{\partial^2 u}{\partial x^2}\right)-(\lambda+G)\left(\frac{\partial^2 u}{\partial x^2}+\frac{\partial^2 w}{\partial x \partial z}\right) = \rho\frac{\partial^2 u}{\partial t^2}+\ddot{X}_b(x,z,t) \quad \text{(A3)}$$

The negative component motions in Eq. A3 result from reflections off the vertical wall moving in the opposite direction as the incident wave motions. In the absence of a structure, those terms having negative components in Eq. A3 vanish and Eq. A3 reduces to the free-field equation of motion. Equation A3 assumes that inertial stresses from the structure and waves transmitted into the structure are negligible compared to stresses developed by the input and reflected waves, but structural inertial forces may be added to Eq. 7. Assuming that all motion remains horizontal (i.e., dilational waves propagate horizontally and shear waves propagate vertically) then the condition $\sigma_z = 0$ may be assumed (Veletsos and Younan, 1994a) with reasonable accuracy for estimating horizontal stresses and from Eq. A2b:

$$\frac{\partial w}{\partial z} = -\frac{v}{1-v}\frac{\partial u}{\partial x} \quad \text{(A4)}$$

Derivation of Eq. A4 with respect to x and substituting into Eq. A3 gives:

$$\beta^2\left(\frac{\partial^2 u}{\partial z^2}-\frac{\partial^2 u}{\partial x^2}\psi_e^2\right) = \rho\frac{\partial^2 u}{\partial t^2}+\ddot{X}_b(x,z,t) \quad \text{(A5)}$$

where ψ_e is given by Eq. 3. Using separation of variables and applying the boundary conditions to solve Eq. A5 gives:

$$u = \bar{u}\left(1-e^{-k_\beta x/\psi_e}\right)\left[\cos k_\beta z - \cos k_\beta H\right]e^{-i\omega t} \quad \text{(A6)}$$

the forcing function and the stiffness parameter are given by:

$$\ddot{X}_b(x,z,t) = \rho\bar{u}\omega^2 e^{-k_\beta x/\psi_e}\cos(k_\beta z)e^{-i\omega t}, \ k_s = \frac{2G}{1-v}\frac{k_\beta}{\psi_e}e^{-k_\beta x_1/\psi_e} = \psi_\sigma G k_\beta e^{-k_\beta x_1/\psi_e}$$

Equation A6 indicates that $u = 0$ at $x = 0$ for all z. Therefore horizontal displacement components need to be added into the solution for $z < d$ and $z > H$, but is beyond the scope of this work. The solution is approximate, not satisfying all the mathematical conditions due to the $\sigma_z = 0$ assumption, but is very reasonable for estimating σ_z. Equation A5 is completely satisfied, however, similar to Veletsos and Younan (1994a) Eq. A1b is not. Finally, the soil-wall contact $(x_1 = 0)$ is found to be neither fully bonded ($w \ne 0$) nor perfectly smooth ($\tau_{xz} \ne 0$).

References

AASHTO (1989), American Association of State Highway and Transportation Officials (1989) "Standard Specifications for Highway Bridges," 14th Ed.

Bardet, J.P. (1997) "*Experimental Soil Mechanics*," Prentice-Hall, N.J.

Bardet, J. P., and C. A Davis (1996) "Engineering Observations on Ground Motion at the Van Norman Complex after the 1994 Northridge Earthquake," Bull. Seism. Society of America, Vol. 86, No. 1B, pp. S333-S349.

Bardet, J. P., K. Ichii, and C. h. Lin (2000) "EERA – A Computer Program for Equivalent-Linear Earthquake Site Response Analysis of Layered Soil Deposits," *University of Southern California Department of Civil Engineering*.

Das, B. M. (1990) "Principals of Foundation Engineering," 2nd Ed. PWS-KENT Publishing Company, Boston.

Davis, C. A. (2000) "Study of Near-Source Earthquake Effects on Flexible Buried Pipes," Dissertation presented to the University of Southern California in partial fulfillment for the degree of Doctor of Philosophy, May.

Hardin, B. O., and V. P. Drnevich (1972) "Shear Modulus and Damping in Soils: Measurements and Parameter Effects," *Journal of the Soil Mechanics and Foundations Division*, ASCE, Vol. 98, No. SM6, pp. 603-624.

Hashash, Y. M. A., J. J. hook, B. Schmidt, and J. I. C. Yao (2001) "seismic Design and Analysis of underground Structures," *Tunneling and Underground Space Technology*, 16, pp. 247-293.

Iida, H, T. Hiroto, N. Yoshida, and M. Iwafuji (1996) "Damage to daikai Subway Station" Soils and Foundations, Japanese Geotechnical Society, pp. 283-300.

Richards, R. Jr., C. Huang, and K. L. Fishman (1999) "Seismic Earth Pressure on Retaining Structures," *Journal of Geotechnical and Geoenvironmental Engineering*, ASCE, Vol. 125, No. 9, pp. 771-778.

Schnabel, P. B., J. Lysmer, and H. B. Seed (1972) "SHAKE: A Computer Program for Earthquake Response Analysis of Horizontally Layered Sites," *Report No. UCB/EERC-72/12*, Earthquake Engineering Research Center, University of California, Berkeley, December, 102p.

Scott, R. F. (1973) "Earthquake-Induced Pressures on Retaining Walls," *Proc., 5th World Conf. On Earthquake Engr.*, Int. Assn. of Eq. Engrg., Tokyo, 2, 1611-1620.

Sherif, M. A., I. Ishibashi, and C. D. Lee (1982) "Earth Pressures Against Rigid Retaining Walls," *J. Geo. Engr. Div.*, ASCE, Vol. 108, No. GT5, pp. 679-695.

Timoshenko, S.P., and J.N. Goodier (1934) "Theory of Elasticity," McGraw-Hill, N.Y.

Veletsos, A. S., and A. H. Younan (1994a) "Dynamic Soil Pressures on Rigid Vertical Walls," *Earthquake Engineering and Structural Dynamics*, Vol. 23, pp. 275-301.

Veletsos, A. S., and A. H. Younan (1994b) "Dynamic Modeling and Response of Soil-Wall Systems," *J. Geo. Engr.*, ASCE, Vol. 120, No. 12, pp. 2155-2179.

Wood, J. H. (1973) "earthquake-Induced Soil Pressures on Structures," Report EERL 73-05, Earthquake Engr. Research Laboratory, California Institute of Technology.

Youd, T. L., and T. N. Craven (1975) "Lateral Stress in Sands during Cyclic Loading," *J. Geo. Engr. Div.*, ASCE, Vol. 101, No. GT2, pp. 217-221.

Characterizing the Effects of Pile Foundations for Evaluation of Performance Based Seismic Design of Critical Lifeline Structures

W. D. Liam Finn[1], N. Fujita[2], and T. Thavaraj[3]

Abstract

The effects of pile foundations on the seismic response of lifeline structures are represented in structural analysis by discrete single valued springs. The spring stiffnesses are usually determined by approximate methods which typically neglect one or more of the important factors that that affect seismic response such as inertial interaction, kinematic interaction, seismic pore water pressures, soil nonlinearity, cross stiffness coupling and dynamic pile to pile interaction. A nonlinear 3-D analysis is used to show how these factors affect pile response and to demonstrate some of the consequences of using various approximate methods.

Introduction

Performance based design of critical lifeline structures such as bridges is design for controlled levels of damage. In order to deliver the expected performance at competitive cost, it is essential to be able to assess reliably the performance of a proposed design, while taking into account all significant factors affecting structural vulnerability. The evaluation may be done using a nonlinear dynamic response analysis. The value of such an analysis depends on how well the structural model represents the real structure. A major weakness in some models is the inadequate representation of the effects of the foundations on the structure, especially of pile foundations. Usually pile foundations are replaced by discrete, single valued springs to model rotational and linear stiffnesses and any coupling between these springs is often ignored. The spring stiffnesses are frequently estimated using approximate, simplified methods of unknown reliability. This is a natural consequence of the complexity of a full 3-D nonlinear dynamic analysis of pile foundations. Even for the elastic case, only a limited number of 3-D parametric studies have been published. These have focused mainly on providing dynamic interaction factors between piles in small groups or frequency dependent stiffnesses and damping for single piles.

    A complete picture of the effects of the foundation on the structure during strong earthquake shaking requires taking simultaneously into account many factors: soil nonlinearity, seismically induced pore water pressures, kinematic interaction

---

[1]Anabuki Professor, and [2]Research Assistant, Kagawa University, 2217 Hayashi-cho, Takamatsu, 761-0396 Japan, phone 81-(0)87-864-2170, finn@eng.kagawa-u.ac.jp
[3]Klohn Crippen Consultants, 10200 Shellbridge Way, Richmond, BC V6X, 2W7, Canada

between piles and soil, inertial interaction of the superstructure with soil and piles and interaction between the piles themselves. All of these factors can be taken into account by a nonlinear, effective stress, dynamic continuum analysis that can handle piles, soil and superstructure all together. Such analyses provide time histories of direct and coupled stiffnesses and demonstrate the relative importance of kinematic and inertial interactions, the effects of pore water pressures and soil nonlinearity. One prime benefit of such analyses, in addition to their use in the context of a specific design, is that results of parametric studies provides the data base for evaluating the effectiveness of the various approximate methods in use.

The purpose of the proposed paper is to present results from one such continuum analysis and present some comparisons with analyses that neglect one or more of the following: inertial interaction, kinematic interaction, seismic pore water pressures, soil nonlinearity, and stiffness coupling. It is hoped that the paper will demonstrate what needs to be done to do an effective job of characterizing pile foundations for the evaluation of performance based design of critical lifeline structures.

Methods of Analysis

The pile foundation-structure system vibrates during earthquake shaking as a coupled system. Logically it should be analyzed as a coupled system. However this type of analysis is generally not feasible in engineering practice many of the popular structural analysis programs do not include the pile foundation directly into a computational model. Therefore the pile head stiffnesses are typically calculated by analyzing the pile foundation without any mass contribution from the superstructure.. The analysis is done usually for a single pile and the group stiffnesses are evaluated using pile interaction factors, often static factors, or a group reduction factor.

The most common approach to the analysis of pile foundations is to use Winkler springs to simulate soil-pile interaction. The springs may be elastic or nonlinear. Some organizations, such as the American Petroleum Institute (API 1995), gives specific guidance for the development of nonlinear load-deflection (p-y) curves as a function of soil properties to represent nonlinear springs. The API (p-y) curves, which are the most widely used in engineering practice, are based on data from static and slow cyclic loading tests in the field.
Murchison and O'Neill (1984) suggest that the reliability of the Winkler (p-y) model may not be high even for static analysis. Finn and Thavaraj (2001) have shown that a dynamic analysis version of the Winkler model using cyclic p-y curves may prove quite unreliable for seismic response analysis during strong shaking on the basis of centrifuge model data.

Seismic analysis of a pile foundation for design evaluation is often conducted by applying the base shears and moments from a fixed base analysis of the structure to the pile head and using a static (p-y) analysis to estimate moments, shears and displacements in the piles. This static analysis neglects many important factors that affect seismic response of the structure-soil-foundation pile system. Inertial interaction between structure and foundation are neglected. This interaction increases the nonlinear behavior of the soil and reduces pile head stiffnesses. These effects increase the period of the system and change the spectral response and hence the base shears and moments. The kinematic moments are also neglected. These moments arise from the pressures generated against the pile to ensure that the seismic displacements of soil and pile are compatible at points of contact along the pile. These moments, which can be captured by a full dynamic analysis, can be very significant. Finally the effects of high pore water pressures and liquefaction on the base moments and shears are also of neglected.

An alternative to the Winkler type computational model is to use a finite element continuum analysis based on the actual soil properties. Dynamic nonlinear finite element analysis in the time domain using the full 3-dimensional wave equations is not feasible for engineering practice at present because of the time needed for the computations. However, by relaxing some of the boundary conditions associated with a full 3-D analysis, Finn and Wu (1994) found it possible to get reliable solutions for nonlinear response of pile foundations with greatly reduced computational effort. The results are accurate for excitation due to horizontally polarized shear waves propagating vertically. Wu and Finn (1997a, b) give a full description of this method and of numerous validation studies. The method is incorporated in the computer program PILE-3D. An effective stress version of this program, PILE-3D-EFF, has been developed by Finn and Thavaraj (1999) and validated by Finn et al (1999) and Finn and Thavaraj (2001) in cooperation the geotechnical group at the University of California at Davis. Seismic response analysis is usually conducted assuming that the input motions are horizontally polarized shear waves propagating vertically. The PILE-3D model retains only those parameters that have been shown to be important in such analysis. These parameters are the shear stresses on vertical and horizontal planes and the normal stresses in the direction of shaking. The soil is modeled by 3-D finite elements as shown in Figure 1. The pile is modeled using beam or volume elements.

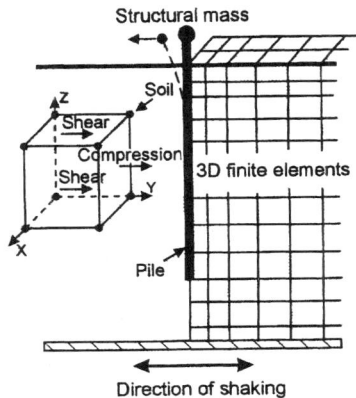

Figure 1 Soil-pile model for analysis   Figure 2 Pile foundation with structure

The pile is assumed to remain elastic, though cracked section moduli are used for concrete piles, when displacements exceed specified threshold values. This assumption is in keeping with the philosophy that the structural elements of the foundation should not yield. The constitutive soil model is equivalent linear with modulus and damping being maintained compatible with shear strain level for the duration of analysis. A yield condition is incorporated consistent with the shear strength of the soil and no tension is allowed to develop between the soil and the pile.

Pile Cap Stiffnesses

The pile cap stiffnesses of the pile foundation shown in Figure 2 will be determined to illustrate the ideas discussed above. This is one of the foundations that support the piers of the bridge used as an example in the guide to the design of bridges issued under the auspices of the American Association of State Highway and Transportation Officials (AASHTO, 1983). Lack of space precludes showing the bridge or considering the damping components of the pile head impedances. Each pier is supported on a group of sixteen (4x4) concrete piles of diameter $d = 0.36$ m and length $L = 7.2$ m, at a spacing $s = 0.9$ m. The Young's modulus and mass density are $E = 22,000$ MPa and $\rho = 2.6$ Mg/m^3 respectively. The foundation soil has a unit weight $\gamma = 18$ kN/m^3 and a Poisson's ratio $\mu = 0.35$. The low strain shear modulus, G, varies as the square root of the depth. It has a value of zero at the surface and of $G = 213$ MPa at a depth of 10 m. The variations in modulus and damping ratio with dynamic shear strains follow the recommendations of Seed and Idriss (1970) for sand. Analyses were conducted for various ratios of column stiffness, K_c, to the pile foundation stiffness, K_f. Results for $K_c/K_f = 50\%$ approximately are given below. The input motions are the first 20 seconds of the N-S component of the ground motions recorded at CSMIP Station No.89320 in Rio Dell, California during the 1992 Cape Mendocino Earthquake.

A PILE 3-D analysis is conducted first. This analysis stores time histories of modulus and damping. Then an associated program PILIMP calculates the time histories of dynamic pile head impedances using the time histories of modulus and damping. The dynamic impedances are calculated at the appropriate frequency by applying a harmonic force to the pile head at the desired frequency and calculating the generalized forces for unit displacements.

Time histories of lateral and cross coupling stiffnesses are shown in Figure 3 and for rotational stiffnesses in Figure 4. These stiffnesses were calculated for the predominant frequency of the input motions, $f = 2.2$Hz. It is clearly not an easy matter to select a single value of each stiffness to characterize the discrete single valued springs used in structural analysis to represent the effects of the foundation. In the absence of a complete analysis such as that discussed, probably a good approach to including the effects of soil nonlinearity on stiffness is to get the vertical distribution of effective moduli using a SHAKE (1972) analysis and to calculate the stiffness at the appropriate frequency using these moduli. The constant stiffnesses calculated in this way are shown also in Figures 3 and 4.

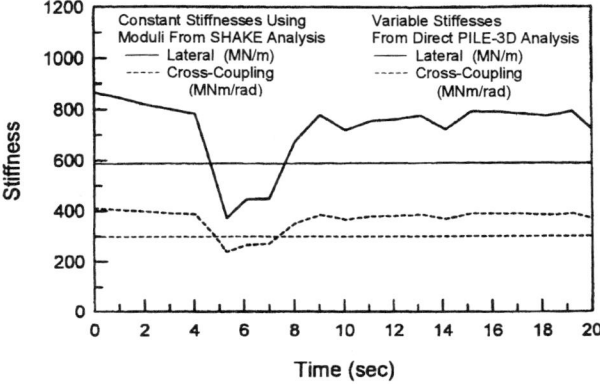

Figure 3 Time history of lateral and cross-coupled stiffness under strong shaking

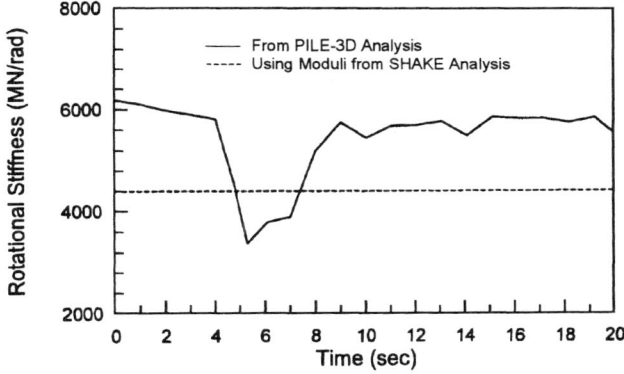

Figure 4 Time history of rotational stiffness under strong shaking

How the period of the system is affected by using the single valued SHAKE springs can be evaluated by doing an eigen value analysis of the bridge using the SHAKE stiffnesses to characterize the springs and comparing the periods with those calculated using the full time histories. The time history of the frequency of the first transverse mode of the bridge is shown in Figure 5. The constant SHAKE frequency is also shown. For most of the excitation, the SHAKE frequency is lower than the PILE-3-D frequency except during the period of strong excitation when the SHAKE frequency is higher. The frequency calculated using the initial small strain moduli is also shown. While it is clearly too high, it is much better than assuming a fixed base which gives a first mode frequency of $f = 5.82$ Hz which is about 50% higher than the lowest frequency from PILE 3-D analysis. Similar results were obtained for the first longitudinal mode of the bridge.

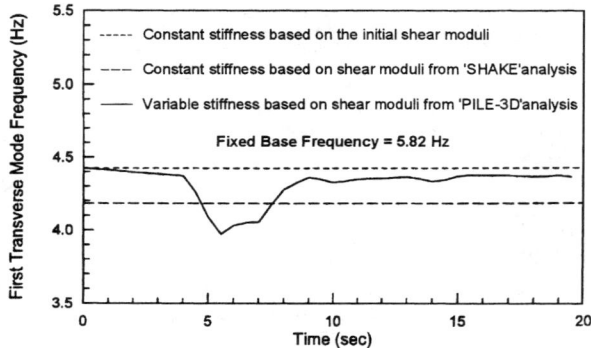

Figure 5 Time history of the frequency of the first transverse mode

Effects of Inertial Interaction

The superstructure is modeled as a single degree of freedom, spring-mass system (SDOF) as shown in Figure 2. The mass is the mass supported by the pile foundation and the supporting column stiffness is chosen so that the period of the SDOF is the same as the fixed base, first mode period of the superstructure. In this case, $M = 370$ Mg, $f = 5.83$ Hz and the stiffness of the column is $K_c = 280$ MN/m. Time histories of lateral stiffness, with and without inertial interaction are shown in Figure 6. The minimum lateral stiffness is now 20% less than minimum stiffness, when inertial interaction is neglected. The effects of inertial interaction are not included in the approximate method using the SHAKE moduli. It requires a coupled analysis of the type described above to include the effects of the additional inertial strains on the moduli used to determine stiffness.

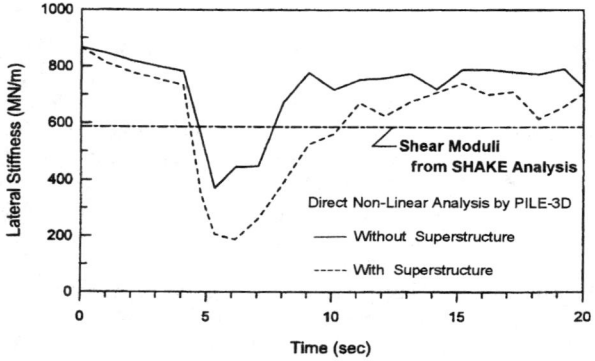

Figure 6 Effects of inertial interaction on lateral pile cap stiffness

Liquefaction and Kinematic Interaction

Many studies have been conducted on the seismic response of large diameter cast-in – place concrete piles for a major research project supported by the Japanese construction company, Anabuki Komuten, who use these piles as foundations in reclaimed land. A typical pile installation is shown in Figure 7 and an idealized model for analysis is shown in Figure 8. Results from these analyses are presented here to show the importance of kinematic interaction on the moments in the pile. The upper 10m are expected to liquefy during the design earthquake. The mass mounted on the pile in Fig. 8 represents the mass equivalent of the reaction force carried by the pile. The purpose in placing the mass on the pile is to model approximately the inertial interaction between the super-structure and the pile foundation. It is mounted on the pile head by a flexible support. The flexibility is selected so that the fixed base period of the mass-support system is 1.4s that is the estimated fundamental period of the prototype structure. In all these analyses, the nonlinearity of the soil and the effects of seismic pore water pressures are taken into account. In general, soil properties are adjusted continuously for current pore water pressures and shear strains. The peak acceleration of the input acceleration record is 0.25g and is amplified to 0.4g at the surface. Fully coupled dynamic effective stress analyses of this system include both inertial and kinematic interactions. Analyses were also conducted without including the mass of the superstructure. These latter analyses include kinematic effects only.

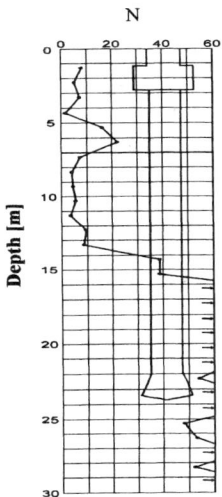
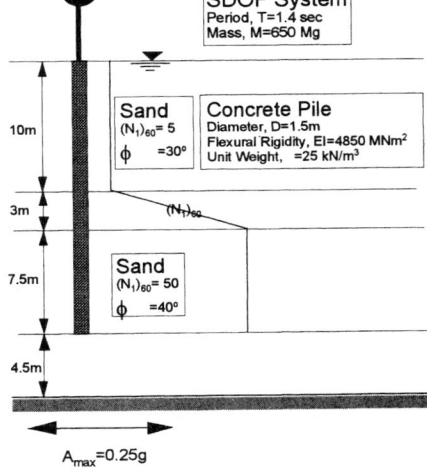

Figure 7 Site in reclaimed land   Figure 8 Model of soil-pile- structure system

Data from these two kinds of analyses are compared to evaluate the significance of kinematic interaction.

Results of Analyses

Pile moments, at the instant of maximum pile head displacement, are shown in Fig. 9 for the case when both inertial and kinematic interactions are taken into account. These moments will be referred to as inertial moments for short. Kinematic moments are shown in Fig. 10. Results are shown for two conditions; the pile head is essentially fixed against rotation and the pile head is essentially free to rotate. When the pile head is fixed against rotation, the maximum moment occurs at the pile head, but very significant moments also occur at the boundary between the softer and stiffer soils. When the pile head is not fixed against rotation, the maximum moment occurs at the boundary between the stiffer and softer soils. The peak kinematic moments are quite significant, being about 60% of the peak total moments.

At some sites a thick surface layer of non-liquefiable soil may lie over the liquefaction zone. Such a layer, 4m thick, is incorporated in the model show in Figure 8.

The inertial moments for this case, at the instant of maximum pile head displacement, are shown in Fig.11 and the kinematic moments in Fig.12. As before, the results are shown for two pile head conditions, no rotation and essentially free to rotate. The stiff upper layer greatly increases the moment demand on the pile during earthquake shaking. This is due to the restraint of the pile by the upper layer and the movement of that layer as a rigid body after liquefaction develops.

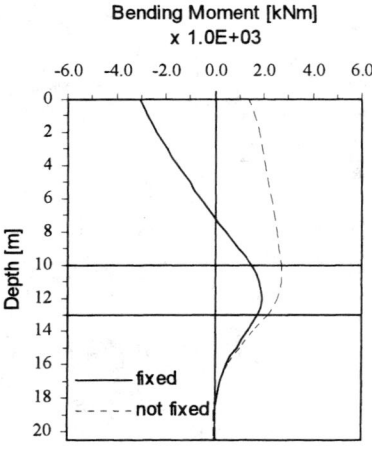
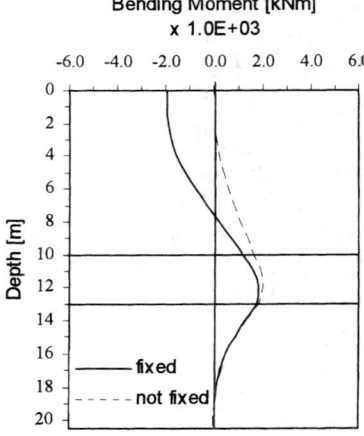

Figure 9 Inertial moments along pile

Figure 10 Kinematic moments along pile

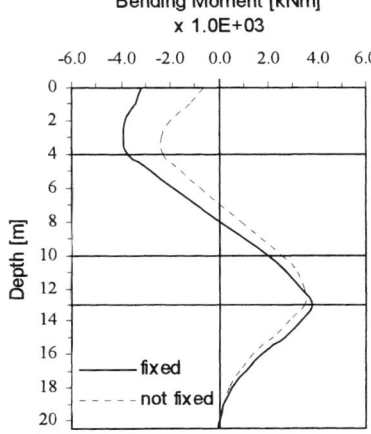 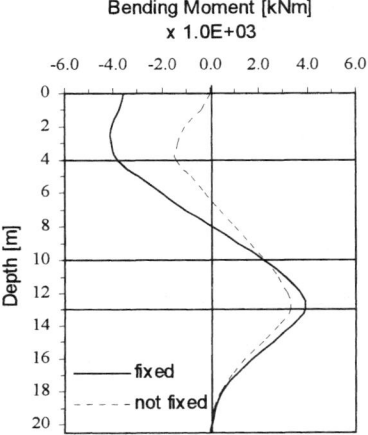

Figure 11 Inertial moments along pile   Figure 12 Kinematic moments along pile

The moments at the pile head and at the interface between the soft and stiff soils has increased by 30%, compared to the case without the upper layer. When the pile is fixed against rotation the moments at the pile head and the interface are about the same. In the case of the stiff upper layer, the kinematic moments are about twice as large as for the case with no such layer and both the total moments and the kinematic moments are about the same magnitude. Clearly analyses that neglect kinematic effects may underestimate significantly design moments and shearing forces and other aspects of dynamic response of piles.

Closing Remarks

This paper explores the reliability of approximate methods for representing the flexibility of pile foundations in the computational model of a superstructure. The study is focused on a 4x4 pile group supporting a bridge pier. The assumptions of the approximate methods were incorporated into a 3-D nonlinear analysis. Most of the approximate methods are based on single pile analysis and further assumptions must be made to establish the group response. The problems in selecting appropriate single valued springs to represent the actions of pile foundations on a superstructure are illustrated by comparing time histories of pile cap stiffnesses during strong earthquake shaking with the single valued spring stiffnesses determined by approximate methods. The consequences of using various approximations to pile cap stiffnesses are investigated by examining their effects on the first modal frequencies of a pile foundation - bridge system. The effects of ignoring inertial or kinematic interactions, as some methods do, are also evaluated. The analysis of a stratified site with a buried liquefied layer shows that, in this case, the kinematic effects are crucially important. The investigation described above is obviously a limited one. Parametric studies are continuing in order to provide a larger data base for a comprehensive evaluation of the many approximate methods in use for the evaluation of pile cap stiffnesses.

Acknowledgements

The research project on the seismic design and analysis of pile foundations is funded by Anabuki Komuten, Takamatsu, Japan. This support is gratefully acknowledged.

References

AASHTO (1983), *Guide specifications for seismic design of highway bridges*, American Assoc. of State Highway and Transportation Officials, Washington, D. C.

API, (1995). *Recommended practice for planning, designing, and constructing fixed offshore platforms*, American Petroleum Institute,

Finn, W. D. Liam and Thavaraj, T. (1999). Pile-3D-EFF: A program for nonlinear dynamic effective stress analysis of pile foundations, Anabuki Chair of Foundation Geodynamics, Kagawa University, Japan

Finn, W. D. Liam, Thavaraj, T., Wilson, D. W., Boulanger, R. W.,and Kutter, B., (1999). Seismic analysis of piles and pile groups in liquefiable sand, *Proceedings, 7th International Symposium on Numerical Models in Geomechanics, NUMOG VI*, Graz, Austria, September, 287-292

Finn, W. D. Liam and Thavaraj, T. (2001). Deep foundations in liquefiable soils: Case histories, centrifuge tests and methods of analysis, CD-ROM *Proceedings, 4th Int. Conf. on Recent Advances in Geotechnical Earthquake Engineering and Soil Dynamics*, San Diego, CA, March 26-31,

Murchison, J. M. and O'Neill, M. W. (1984). An evaluation of p-y relationships in cohesionless soils, *Proceedings of the ASCE Symposium on Analysis and Design of Pile Foundations*, ASCE National Convention, San Francisco, California, Oct 1-5, Edited by J. R. Meyer, 174-191

Schnabel, P. B., Lysmer, J.and Seed, H. B. (1972). SHAKE: A computer program for earthquake response analysis of horizontally layered sites, Report No. EERC72-12, Earthquake Engineering Research Center, University of California, Berkeley, CA

Wu, G. and Finn, W. D. Liam, (1997a). Dynamic elastic analysis of pile foundations using the finite element method in the frequency domain, *Canadian Geotechnical Journal*, (34), 34-43

Wu, G. and Finn, W. D. Liam, (1997b). Dynamic nonlinear analysis of pile foundations using the finite element method in the time domain, *Canadian Geotechnical Journal*, (34), 144-152

LIQUEFACTION AND NON-LIQUEFACTION FROM 1999 CHI-CHI, TAIWAN, EARTHQUAKE

Jonathan P. Stewart[1], Daniel B. Chu[2], Shannon Lee[3], J. S. Tsai[4],
P.S. Lin[5], B.L. Chu[5], Robb E.S. Moss[6], Raymond B. Seed[7],
S. C. Hsu[8], M. S. Yu[9], and Mark C.H. Wang[10]

Abstract:
The 1999 Chi Chi, Taiwan, earthquake provides case histories of ground failure and non-ground failure that are valuable to the ongoing development of liquefaction susceptibility and triggering models because the data occupy sparsely populated parameter spaces (i.e., high cyclic stress ratio and high fines content with low to moderate soil plasticity). In this paper, we synthesize results from several large site investigation programs conducted in Nantou and Wufeng, Taiwan, and compare the data to susceptibility and triggering models. With regard to liquefaction susceptibility, we find components of the well-known Chinese criteria associated with liquid limit (LL) and water content/LL to be reasonably well validated by the Taiwan data, but clay fraction criteria and CPT-based criteria to not be effective. Triggering models are generally validated for ground failure sites, but the data raise important questions regarding non-ground failure sites whose performance is not well predicted.

Introduction

The 1999 $M_w = 7.6$ Chi-Chi Taiwan earthquake triggered numerous significant incidents of liquefaction in inland alluvial areas and in several coastal hydraulic fills (Stewart, 2001). Due to significant interest in the available case histories of liquefaction and non-liquefaction, a series of site investigation programs were undertaken in 2000 by researchers with the National Center for Research in Earthquake Engineering (NCREE) in Taiwan and in 2001-2002 by the authors with funding from the Pacific Earthquake

---
1 Asst. Prof., UCLA Civil Engineering Dept., Los Angeles, CA 90095 (jstewart@seas.ucla.edu)
2 Graduate Student, UCLA Civil Engineering Dept. and Chief Engineer, Ninyo & Moore, Irvine CA
3 Professor and Chair, Civil Engineering Dept., National Chi Nan Univ., Puli, Taiwan
4 Professor, Civil Engineering Dept., National Cheng Kung Univ., Tainan, Taiwan
5 Professor, Civil Engineering Dept., National Chung-Hsing Univ., Taichung, Taiwan
6 Project Engineer, Fugro Inc., Ventura, CA
7 Professor, Civil Engineering Dept., U.C. Berkeley, CA
8 Professor, Civil Engineering Dept., Chao-Yang University, Wufeng, Taiwan
9 Principal, Resources Engineering Services, Inc., Taipei, Taiwan
10 Manager, Moh and Associates, Taipei, Taiwan

Engineering Research Center (PEER). Results of both site investigation programs are synthesized on the web page http://www.cee.ucla.edu/faculty/Taiwanwebpage/Main.htm. The objectives of this paper are to (1) document a number of the particularly significant case histories to emerge from this earthquake, (2) compare these case histories to state-of-practice (i.e., Seed et al., 1985; Robertson and Wride, 1998; Youd et al., 2001) as well as state-of-the-art (Seed et al., 2001; Moss and Seed, 2002) liquefaction triggering procedures, and (3) compare the case histories to established liquefaction susceptibility criteria. As will be shown subsequently, the Taiwan liquefaction data has an important role to play in the ongoing development of empirical liquefaction assessment methodologies for two principal reasons:

- Many of the Taiwan case histories involve high cyclic stress ratios (CSR ≈ 0.4-0.6), where existing data is sparse.

- The Taiwan case histories involve primarily high fines content soils, where the existing data inventory is sparse [i.e., the Youd et al. (2001) triggering model is based on only 13 cases with ≥ 35% fines content, Seed et al. (1985)].

Site Investigation Program

The site investigation programs by NCREE and PEER resulted in a total of 47 Cone Penetration Test (CPT) profiles (of which 18 were seismic CPTs) and 48 soil borings with Standard Penetration Testing (SPT) (typically at 1.0 m spacing). The majority of the NCREE work was performed in the city of Yuanlin, whereas the entirety of the PEER work and some of the NCREE work was performed in the cities of Nantou and Wufeng. In this paper we focus on Nantou and Wufeng, where the CSRs were relatively high (CSR ≈ 0.4-0.6, as compared to CSR ≈ 0.2 in Yuanlin).

Most of the borings/CPTs were limited to depths of 10-15 m. CPT profiling was performed according to standard techniques (ASTM D 5778-95). For SPT sampling, the percentage of the total theoretical energy delivered to the split-spoon sampler, or energy ratio, was controlled by following procedures in ASTM D6066-98 and ASTM D1586. We used a safety hammer with a rope/cathead release mechanism, two turns of the rope around the cathead, standard AW rod, and a 12 cm borehole diameter. Hence, the energy transmitted to the sampler would be assumed to be 60% if no short-rod correction was applied. The actual delivered energy was measured for each blow of the hammer using a rod section instrumented with accelerometers and strain gages (Abou-Matar and Goble 1997). Using the average energy ratio (ER) for each test, we computed the blow-count normalized to 60% of the theoretical energy, N_{60}.

All retrieved soil samples were subjected to a full suite of laboratory index tests per ASTM standards including sieve, hydrometer, liquid limit, plastic limit, and water content. Results are presented on boring logs on the aforementioned web page. These test results were used for liquefaction susceptibility analysis, as discussed below.

Sites selected for subsurface exploration included lateral spread sites, locations of tilted and/or settled buildings, and locations of no apparent ground failure based on post-earthquake reconnaissance. A total of 22 sites in Wufeng and 27 in Nantou were investigated. Locations of the Wufeng sites are overlaid on damage locations in Figure 1. A similar map for Nantou is presented on the aforementioned web page.

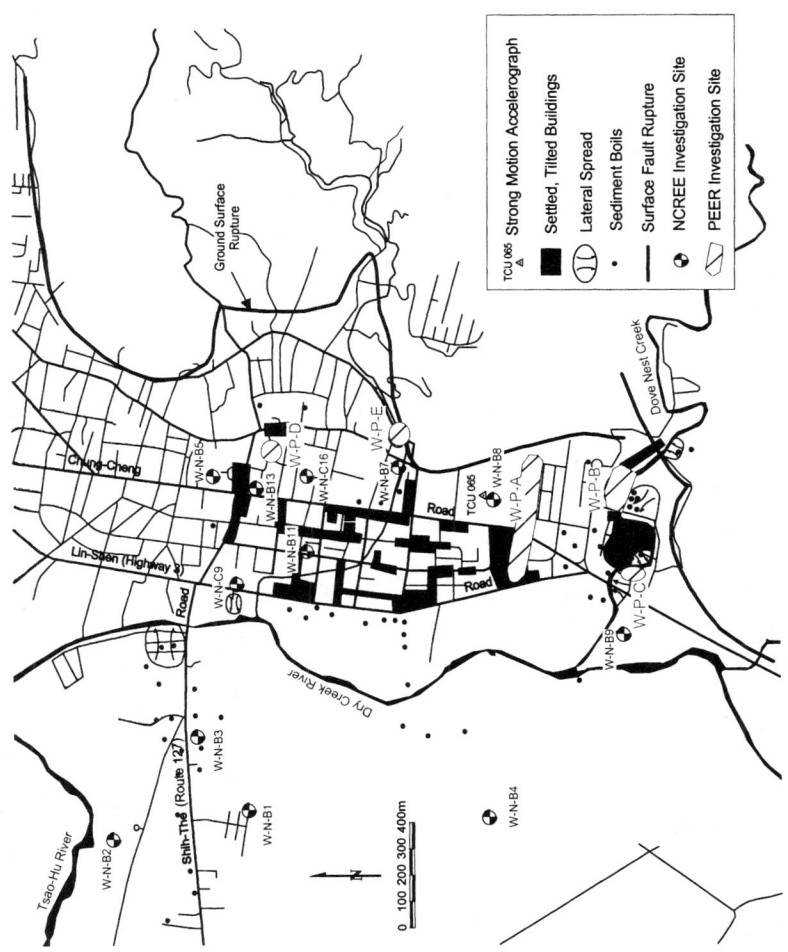

Fig. 1. Map of Wufeng showing ground failure zones and locations of investigated sites

Data Analysis for Liquefaction Susceptibility and Triggering

Seismic Demand in Nantou and Wufeng

Strong motion accelerographs (SMAs) are present in both Nantou (Station TCU076) and Wufeng (TCU065). In both cities, the SMAs are located within about 1 km of the liquefaction/non-liquefaction sites considered in this research, and are on generally similar site conditions (young alluvium). There was no evidence of ground failure in the immediate vicinity of either SMA. Both SMAs are on the footwall side of the ruptured Chelungpu fault, as are the subject liquefaction/non-liquefaction sites. The geometric mean peak horizontal accelerations of the two horizontal components of shaking were PHA = 0.38g in Nantou and 0.67g in Wufeng. These PHA values were used to estimate cyclic stress ratios at the liquefaction/non-liquefaction sites, as discussed below.

Liquefaction Susceptibility Analysis

Most of the soils at the investigated sites in Nantou and Wufeng contain significant fines (i.e. > 35% passing the #200 sieve). Fine-grained soils require analysis to evaluate their liquefaction susceptibility. A well known specification for checking liquefaction susceptibility is the "Chinese criteria," originally presented by Seed and Idriss (1982) and re-stated by Youd et al. (2001). The Chinese criteria specify that liquefaction can only occur if all three of the following conditions are met: (1) weight fraction smaller than 5µm (i.e., "clay fraction," CF) < 15%, (2) liquid limit (LL) < 35%, and (3) natural water content (w_n) > 0.9LL. More recently, Andrews and Martin (2000) stated that silty soils are susceptible to liquefaction if both LL < 32% and the amount finer than 2 µm < 10 %, whereas Sancio et al. (2002) found from Adapazari, Turkey, case histories that the Chinese criteria are effective provided the CF criterion is neglected. As described further below, in this study weight was given to the LL and w_n/LL components of the Chinese criteria during the identification of critical layers at the subject sites.

Liquefaction susceptibility criteria based on CPT test results are not well established, although Robertson and Wride (1998) have proposed that the I_c parameter can distinguish relatively granular soils from potentially plastic soils, with I_c = 2.6 being an approximate boundary between the two. Moss and Seed (2002) found from Bayesian analysis of case history data that I_c correlates poorly with the "clean sand" correction factors needed for CPT-based liquefaction triggering analysis, which suggests that a susceptibility threshold based on I_c = 2.6 may not be reliable. Their results indicate that for soils with an overburden-normalized tip resistance of $q_{c,1}$ > ~ 1 MPa, traditional liquefaction is unlikely to occur if friction ratio R_f > ~ 3%.

Liquefaction Triggering Analysis

The liquefaction triggering analysis procedures used here provide an estimate of cyclic resistance ratio (CRR) based on a measure of penetration resistance (SPT blow count or CPT tip resistance) normalized to 1.0 atm overburden pressure. The current standard of practice for CRR analysis consists of well-known SPT and CPT procedures summarized in Youd et al. (2001). The calculation of seismic demand in terms of a cyclic stress ratio (CSR) in these procedures is performed using ground surface PHA, effective and total stresses at the depth of interest, and stress reduction factors (r_d) by Seed and Idriss (1971) that are a function solely of depth.

New liquefaction triggering procedures for SPT and CPT have been presented by Seed et al. (2001) and Moss and Seed (2002). These procedures differ from the Youd et al. (2001) procedures in that they are based on different data sets (generally larger and more carefully screened) and are fully probabilistic. These procedures also use different r_d models, which are based on statistical interpretation of ground response analysis results. These r_d models are sensitive to depth, depth to groundwater, shear wave velocity in the upper 12 m, and earthquake magnitude.

In the back-analysis of liquefaction/non-liquefaction sites using SPT or CPT procedures, it is necessary to identify a critical layer having the minimum seismic resistance to liquefaction triggering. The identification of this "weakest strata" is ideally performed based on careful study of CPT tip resistance and friction ratio, in conjunction with a boring log with laboratory index testing to evaluate susceptibility. Shown in Figure 2 is a data set used to evaluate the location of the critical layer at an example site that did not show evidence of ground failure. Beginning with the CPT data on the left side of the figure, the critical layer is preliminarily identified as indicated by the dashed lines based on a combination of low q_c (indicating relatively low density) and low R_f (indicating low plasticity). The index properties from the layer (right side of figure) are compared to the LL and w_n/LL components of the Chinese criteria, which in this case suggest the layer is not susceptible to liquefaction. In such cases, additional layers are sought that might be susceptible, and these layers are used for subsequent analysis provided they are not at large depth. In the example, a marginally susceptible zone is identified at about 15 m, but this depth is too great for use with triggering models. Accordingly, since the critical layer is not susceptible, and potentially susceptible layers are deep, data from this site would not be included in data compilations for liquefaction triggering. Identified critical layers that are susceptible to liquefaction are used in the comparisons to liquefaction triggering models described below.

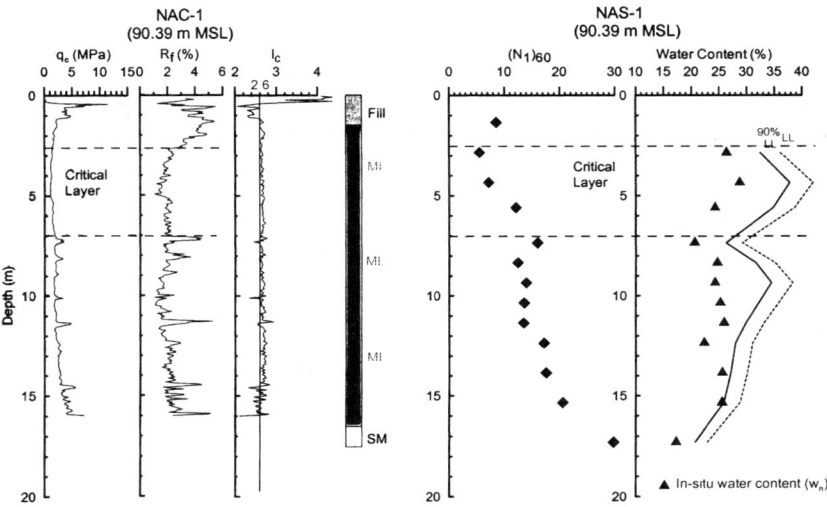

Fig. 2. Example data set from Nantou Site C showing identification of critical layer.

Data Synthesis and Comparison to Susceptibility and Triggering Models

Presented in Table 1 for each investigated site are critical layer depths, mean and standard deviation soil properties within the critical layers, and derived quantities utilized in the triggering analysis models. The statistics on soil properties are evaluated considering both vertical and lateral variability (lateral variability is considered when more than one boring/CPT are available for a site). Also shown are brief descriptions of site performance as observed during post-earthquake reconnaissance. In the following, sites are considered to have "ground failure" when some surface manifestation of ground failure was observed. It should be noted that sites without such surface manifestations of ground failure may have had localized liquefaction that was of little consequence due to the presence of a non-liquefied crust (Ishihara, 1985).

Plotted in Figure 3 are the average soil index properties within the critical layers for the sites in Table 1. The data are plotted as dots at sites with evidence of ground failure, and as open circles for non-ground failure sites. Figure 3(a) shows the data in LL-w_n space along with the boundary curves associated with the Chinese criteria. The results generally support the Chinese criteria (dots in susceptible space, circles in not-susceptible space). Important exceptions are two non-ground failure sites in the susceptible space, which are discussed further below. Figure 3(b) shows the data in LL-CF space. The results shows that some ground failure sites have CF > 15%. Thus, our findings in this regard are consistent with those of Sancio et al. (2002); namely, the CF component of the Chinese criteria appears to be unreliable, whereas the LL and w_n/LL components are generally validated by the Taiwan liquefaction data.

Another important observation relates to the use of CPT-based indices for liquefaction susceptibility evaluations. As shown for example by Site W-N-B8 in Table 1, some soils with $I_c < 2.6$ and $R_f < 3\%$ (indicating relatively granular soils and potentially high liquefaction susceptibility) are *not* susceptible based on the Chinese criteria, and in fact did not show evidence of ground failure. Conversely, Sites W-P-A-W with $I_c > 2.6$ and $R_f > 3\%$ pass the non-CF components of the Chinese criteria and demonstrated evidence of ground failure, although in each case it should be noted that the ground failure involved settlement of rather tall structures (> 4 stories). Accordingly, existing CPT-based indices do not appear to be reliable for evaluating liquefaction susceptibility for such conditions.

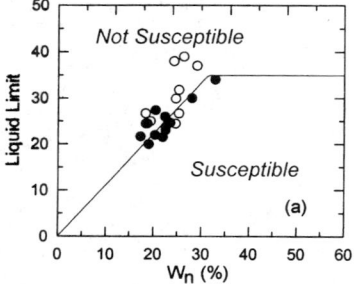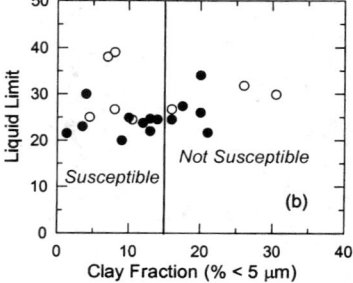

Fig. 3. Ground failure and non-ground failure sites plotted relative to mean soil index properties (dots = ground failure; circles = no ground failure observed)

DISASTER RESPONSE FOR LIFELINE SYSTEMS 1027

Table 1. Inventory of data at selected sites in Nantou and Wufeng (mean ± standard deviation values given for data fields)

										Youd et al. (2001) procedures					Seed et al. (2001)/Moss and Seed (2002)				
Boring/CPT	NO	NO CPT	Z_C (m)	q_c (Mpa)	R_f (%)	$(N_1)_{60}$	FC (%)	LL	w_n/LL	<5 μm (%)	CSR	$(N_1)_{60CS}$	q_{c1N}	I_c	CSR	$(N_1)_{60CS}$	q_{c1} (Mpa)	Δq_c (Mpa)	$q_{c1,mod}$ (Mpa)	Field Observations
W-P-A-E	3	1	3-4.5	0.8±0.2	0.7±0.6	9±4	39.0±9.9	24.5±8.0	1.0±0.1	10.5±0.7	0.65±0.03	16±5	12.8±2.3	2.6±0.1	0.61±0.01	13±5	1.4±0.2	0.2±0.07	1.7±0.7	No ground failure
W-P-A-W	3	1	5.5-9	2.1±2.1	3.0±0.8	18±10	88.0	26.0±5.7	0.9±0.2	20.0	0.69±0.02	15±4	22.7±19.8	2.8±0.2	0.60±0.02	15±7	2.2±2.0	2.8±0.9	5.1±1.8	Building settlement
W-P-B	1	5	2-5.3	2.2±1.3	1.2±5.3	9±5	28.5±8.2			3.5±2.1	0.57±0.03	15±4	29.8±17.3	2.4±0.4	0.52±0.11	12±5	3.3±2.9	0.8±0.56	4.1±8.0	Lateral spread
W-P-C	2	13	0.5-6	2.1±2.1	2.4±2.0	6±2	17.6±8.5	24.5±4.9	1.0±0.3	14.0±12.7	0.59±0.10	6±3	39.2±28.5	2.5±0.4	0.62±0.12	7±3	3.7±2.6	2.1±1.9	5.8±2.5	Lateral spread
W-P-D	1	1	1-3.3			13±4	35.3±0.6			12.7±0.6	0.43±0.00	20±5				16±5				No ground failure
W-P-E																				Building settlement
N-P-A1	1	1	15.7	1.3±0.2	1.9±0.3	8±3	66.7±12.6	39.0±3.0	0.7±0.1	8.0±1.0	0.39±0.02	15±4	19.6±2.3	2.7±0.0	0.40±0.01	14±5	1.8±0.2	1.4±0.3	3.2±0.5	No ground failure
N-P-A2	1	1	3.8-5.5			15±2	100.1±9.9	23.0±1.4	1.0±0.1	3.5±3.5	0.39±0.01	17±1								Building settlement
N-P-A3	1	1	1-2			10±	140	30.0	0.9	4.0	0.28	12				11				Building settlement
N-P-A4	1	1	1.5-2.5			5±	350	20.0	1.0	9.0	0.33	11				8				No ground failure
N-P-B1	1	1	1.6-7.3			11±6	37.4±8.2	26.7±3.2	0.8±0.2	80.±6.7	0.38±0.04	18±7				15±7				Building settlement
N-P-B2	1	1	17-3			8±	50.0	22.0	0.9	13.0	0.34	14				12				No ground failure
N-P-B3	1	1	3-5.5			29±10	40.1±1.0	21.5±3.3	1.0±0.1	13.±0.6	0.38±0.01	29±10				30±10				Building settlement
N-P-B4	1	1	2-6.3			15±1	21.3±9.1	38.0	0.8	7.0±4.7	0.37±0.04	19±4				17±2				No ground failure
N-P-B5	1	1	3.2-6.5			19±9	19.0±0.6	25.0±1.4	0.8±0.1	4.5±2.1	0.37±0.03	23±11				21±11				No ground failure
N-P-C	2	3	15.3-4	4.8±2.9	1.2±1.1	13±0	10.0±0.5			3.8±2.2	0.38±0.02	13±3	73.3±62.8	2.1±0.5	0.36±0.10	11	7.3±6.1	0.6±0.10	7.9±5.7	Lateral spread
W-N-B1	1	1	1.5-2.5			32	10.0			4.0	0.55	7			0.52±0.11					Sand boils
W-N-B2	1	1	0.6-3.1			3	22.0			6.0	0.57	12				4±				No ground failure
W-N-B3	1	1	1-2.5			6	76.0	31.8	0.8	26.0	0.43	11±				11±				No ground failure
W-N-B4	1	1	1.5-4.5			13±6	67.5±31.8	27.4	0.9	17.5±4.9	0.62±0.03	20±7				19±5				Building settlement
W-N-B5	1	1	0.2-4			11±7	65.3±30.5	29.9±10.7	0.9±0.2	30.5±9.1	0.45±0.13	18±8				16±6				No ground failure
W-N-B6	1	1	0.5-5	4.1±3.1	1.6±0.6	11±4	34.5±20.5			7.0±4.2	0.44±0.01	17±2	54.8±35.8	2.3±0.3	0.43±0.02	14±2	5.2±3.3	1.1±0.06	6.4±4.34	Sand boils
W-N-B7	1	1	5.95	5.7±2.4	1.3±0.7	6±2	54.7±17.6	28.7	0.6	16.0±0.7	0.75±0.20	24±9	64.4±26.4	2.2±0.3	0.66±0.03	21±7	6.2±2.5	0.9±0.08	7.1±2.3	No ground failure
W-N-B8	1	1	0-2.3			7	38.0±			9.0	0.43	13				10				Sand boils
W-N-B9	1	1	2-6	1.4±0.9	1.8±1.0	13±1	45.5±33.2	24.5±0.8	0.9±0.3	12.0±11.3	0.65±0.04	19±4	19.3±12.0	2.7±0.2	0.56±0.01	17±2	1.9±1.1	1.4±1.1	3.3±1.7	Sand boils
W-N-B10	1	1	1.6-3.6			10	45.0			9.0	0.64	17				14				Building settlement
W-N-B11	1	1	8-14			17±6	40.5±1.7			16.0±6.0	0.54±0.01	22±5				26±5				Building settlement
W-N-B12	1	1	1-3.5			4±1	34.0±5.7	24.5±0.8	0.8±0.0	9.0±2.8	0.57±0.11	9±2				6±2				Sand boils
W-N-B13	1	1	1-2			15	21.0			7.0	0.43	20				18				Sand boils
W-N-B14																				Building settlement
W-N-C9	1	1	1-2.8	0.8±0.3	0.5±0.5						0.55±0.06		13.1±5.7	2.6±0.2	0.56±0.06		2.1±0.9	0.0±0.5	2.1±0.8	Sand boils
W-N-C15	1	1	2.2-2.8	0.7±0.2	1.9±0.9						0.44±0.01		10.5±3.6	2.9±0.2	0.42±0.01		1.0±0.3	1.4±0.9	2.4±0.9	Sand boils
W-N-C16	1	1	2.5-5.5	2.2±1.6	2.8±1.1						0.56±0.04		30.1±21.7	2.7±0.2	0.54±0.03		2.7±2.5	2.5±1.2	5.2±2.1	No ground failure
N-N-B1	1	1	1-6	1.4±0.4	1.7±0.7	7±1	94.7±1.5	37±8	0.8±0.1	57.7±5.1	0.25±0.00	13±2	18.2±7.7	3.0±0.2	0.25±0.01	14±2	1.7±0.6	3.7±0.6	5.4±1.0	No ground failure
N-N-B2	1	1	1-2.5			10	26.0			6.0	0.25	15			0.25	12				Sand boils
N-N-B3	1	1	3-9.4			16±11	17.8±1.7			8.8±1.5	0.30±0.02	21±11			0.29±0.02	18±12				Sand boils
N-N-B4	1	1	2.4-8	3.9±2.5	0.7±0.5	7±0	33.5±12.0			9.0±2.8	0.25±0.01	13±1	38.4±22.1	2.2±0.3	0.24±0.00	10±1	9.0±2.1	0.2±0.4	3.9±2.1	Sand boils
N-N-B5	1	1	0.5-7.5			8±4	65.0±39.0			15.0±14.7	0.25±0.05	14±4			0.29±0.03	14±6				Lateral spread
N-N-B6	1	1	1-5.11			20±4	255±11.7			6.7±2.3	0.37±0.01	26±5			0.37±0.02	23±4				Lateral spread
N-N-B7	1	1	0.8-11.5	7.1±3.3	1.2±0.9	17±6	40.3±30.5			13.5±19.9	0.43±0.01	24±7	73.3±31.0	2.1±0.3	0.42±0.01	22±7	7.3±3.1	0.7±0.9	7.9±3.1	Building settlement
N-N-B8	1	1	0-2			4	40.0	24.9		10.0	0.27	10			0.27	7				No ground failure
N-N-B9	1	1	7-11	4±	33±35	27.7±16.2			10.3±7.6	0.23±0.00	41±1.39				0.21±0.01	38±38				Sand boils
N-N-B10	1	1	0-5.5			10±7	24.7±12			6.3±0.6	0.28±0.03	15±8			0.27±0.03	12±8				Sand boils
N-N-B11	1	1	1.8-5.8			5±1	92.5±9.2	34±1.3	1.0±0.0	20.0±0.71	0.30±0.02	11±1			0.29±0.02	12±1				Building settlement
N-N-B12	1	1	1.5-4			8±6	87.0±4.2	24.7±2	1.0±0.1	18.0±8.5	0.26±0.00	18±2			0.26±0.02	15±9				Sand boils
N-N-B13	1	1	1.8-5			26	76.0	21.7	0.8	21.0	0.30	36			0.29	38				Sand boils
N-N-B14	1	1	1-8.5			16±4	19.8±12.2			5.0±1.8	0.35±0.07	11±2			0.32±0.05	19±3				Sand boils
N-N-C2	1	1	3-5	4.8±1.3	0.6±0.6						0.24±0.00		53.9±12.0	2.0±0.2			5.4±1.2	0.1±0.05	5.5±1.09	Sand boils
N-N-C3	1	1	1.5-3.5	2.4±2.0	2.7±1.8						0.30±0.03		39.7±13.7	1.9±0.07			3.5±1.2	1.9±0.07	5.4±1.10	Sand boils
N-N-C13	1	1	10-11.5	3.8±1.0	1.6±0.7						0.37±0.00		34.±4.2	2.4±0.1			8.8±0.4	8.6±0.6	7.9±0.06	No ground failure

Plotted in Figure 4 are average CSR-penetration resistance data within the critical layers for selected sites in Table 1. Also shown are the CRR models discussed previously. In the case of the probabilistic models, the CRR curves shown apply for a 20% probability of liquefaction. Only site data that passes the LL and w_n/LL components of the Chinese criteria are plotted in Figure 4. Generally, the upper band of results at CSR ≈ 0.6 are Wufeng sites, whereas the lower band at CSR ≈ 0.4 are Nantou sites. Note that most of the data plotted in Figure 4 are dots (indicating ground failure sites). We actively sought non-ground failure sites during our field work, but many such sites are not susceptible to liquefaction based on soil plasticity.

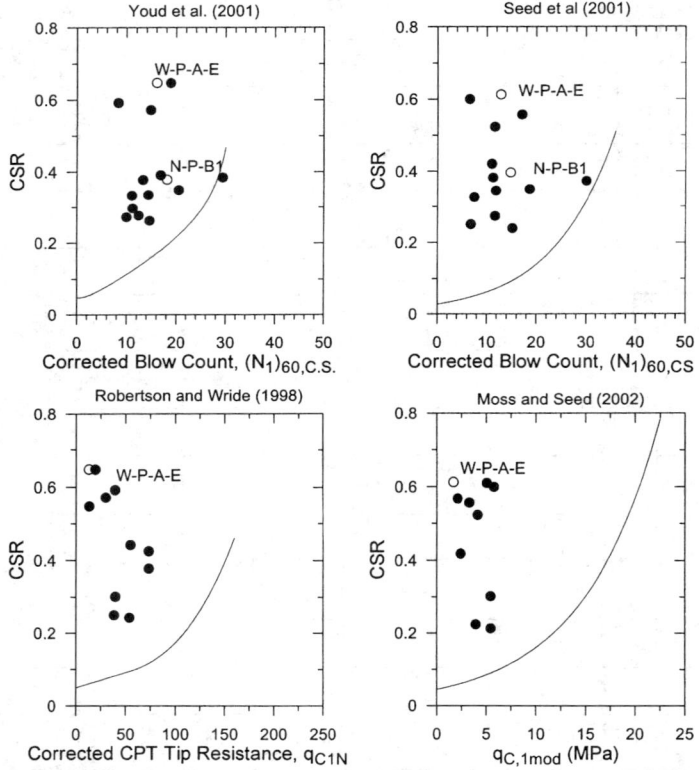

Fig. 4. Ground failure and non-ground failure sites plotted relative to existing liquefaction triggering models (dots = ground failure; circles = no ground failure observed)

The ground failure sites (dots) are generally encompassed by the CRR models, and it is not possible to judge the relative accuracy of these models based on the Taiwan data processed to date. The non-ground failure sites (open circles) that plot to the left of the CRR lines in Figure 4 merit additional discussion (i.e., sites W-P-AE and N-P-B1). These sites also correspond to the open circles within the "susceptible" space in Figure

3. Site W-P-AE consists of a ~2-3 m clay bed overlying a thick deposit of relatively sandy soil. The critical layer in this case was taken as the upper section of sand. Site N-P-B1 consists of ~1.5 m of unsaturated soil overlying ~2 m of silty sand, which was taken as the critical layer. Both sites have essentially "free-field" conditions – namely, the absence of tall structures (local structures at these sites are light, single story buildings). Both sites were observed within 2 weeks of the earthquake by reconnaissance team members, who report no evidence of ground failure in the area.

An appropriate question to ask is whether these sites may have in fact liquefied. Both of the SPT sites would be expected to have surface manifestation of liquefaction based on the relative thicknesses of the surface layer and liquefiable layer per the criteria of Ishihara (1985). Nonetheless, we speculate that the absence of ground failure does not necessarily mean the absence of liquefaction within the critical layers, especially since these layers are overlain by non-liquefiable strata, the sites lacked driving static shear stresses that could mobilize significant ground failure effects through the overlying strata, and the fines contents of the liquefiable soils at depth may be higher than those considered by Ishihara (1985). The driving static shear stresses appear to be particularly important, as most of the ground failure sites in our database *did* have significant static shear stresses. Accordingly, this situation raises fundamental questions about how a site is classified within the "ground failure" or "non-ground failure" categories as a result of observations from post-earthquake reconnaissance.

While there is no question about assigning "dots" to sites with ground failure, a "circle" denoting non-ground failure is strictly only applicable for the site-specific stress conditions, and caution must be exercised in assuming that other sites with similar combinations of CSR and penetration resistance would not liquefy if the static stress conditions were different. Moreover, it is noted that the CRR models presumably apply for a zero static shear stress condition, but for the reasons discussed above, the development of triggering models based on such protocols is likely unconservative for the high fines content materials commonly encountered in Nantou and Wufeng. This issue warrants further research in the development of next-generation triggering models.

Summary of Findings

Case histories of ground failure and non-ground failure from the Chi-Chi earthquake are important for the ongoing development of liquefaction susceptibility and triggering models because the effected soils have large fines contents and marginal plasticity levels, and because the earthquake induced large CSRs in these soils. Prior data sets had a paucity of data for such conditions. Comparisons of the data to models indicate that:

1. The CF component of the Chinese criteria appears to be unreliable, whereas the LL and w_n/LL components are generally validated by the Taiwan data. This is similar to the findings of Sancio et al. (2002) for soils in Adapazari, Turkey.
2. CPT-based indices appear unreliable for evaluating liquefaction susceptibility. It is noted that CPT indices for susceptibility are not formally proposed by Robertson and Wride (1998) or Moss and Seed (2002). Nonetheless, I_c has been applied in practice for this purpose, and this practice is not recommended.
3. Ground failure sites from Wufeng and Nantou are generally encompassed by available CRR models (i.e., the CSR values generally plot above the CRR line).

4. Some non-ground failure sites plot above the CRR lines, which raises questions regarding the nature of the site performance (did the site liquefy in a manner that was not manifest at the surface?) and the degree to which the apparently good site performance may be an adequate predictor of future performance at other sites. These questions remain unanswered, but are important for the ongoing development of robust empirical liquefaction triggering models.

Acknowledgements

This project was sponsored by the Pacific Earthquake Engineering Research Center's Program of Applied Earthquake Engineering Research of Lifeline Systems supported by the State Energy Resources Conservation and Development Commission and the Pacific Gas and Electric Company. This work made use of Earthquake Engineering Research Centers Shared Facilities supported by the National Science Foundation under Award #EEC-9701568. In addition, the support of the California Department of Transportation's PEARL program is acknowledged.

References

Abou-Matar, H., Goble, G. (1997). "SPT dynamic analysis and measurements" *J. Geotech. & Geoenvir. Engrg.*, 123 (10), 921-928.

Andrews, D.C.A. and Martin, G.R. (2000). "Criteria for liquefaction of silty soils," *Proc. 12th World Conf. on Earthquake Engrg.*, New Zealand, Paper No. 0312.

Ishihara, K. (1985). "Stability of natural deposits during earthquakes," *Proc. 11th Int. Conf. Soil Mechanics and Foundation Engineering*, San Francisco, CA, Vol. 1, pp. 321-376.

Moss, R.E.S. and Seed, R.B. (2002). "Probabilistic evaluation of seismic soil liquefaction potential using CPT," *Proc. 8th US-Japan Workshop on Earthquake Resistant Design of Lifeline Facilities and Countermeasures against Liquefaction*, Tokyo, Japan.

Robertson, P.K. and Wride, C.E. (1998). "Evaluation cyclic liquefaction potential using the cone penetration test," *Can. Geotech. J.*, Ottawa, 35(3), 442-459.

Sancio, R.B., Bray, J.D., Stewart, J.P., Youd, T.L., Durgunoğlu, H.T., Önalp, A., Seed, R.B., Christensen, C., Baturay, M.B., and Karadayılar, T. (2002). "Correlation between ground failure and soil conditions in Adapazari, Turkey," *Int. J. Soil Dyn. and Earthquake Engrg.*, 22 (9-12), 1093-1102.

Seed H.B. and Idriss I.M. (1971). "Simplified procedure for evaluating soil liquefaction potential," *J. Soil Mech. & Foundations Div.*, ASCE, 97(9), 1249-1273.

Seed, R.B., Cetin, K.O., Moss, R.E.S., Kammerer, A.M., Wu, J., Pestana, J.M., and Riemer, M.F. (2001). "Recent advances in soil liquefaction engineering and seismic site response evaluation," *Proc. 4th Int. Conf. Recent Adv. Geotech. Eqk. Engrg. Soil Dyn.*, Paper SPL-2.

Seed, H.B. and Idriss, I.M. (1982). "Ground motions and soil liquefaction during earthquakes," Earthquake Engineering Research Institute Monograph, Oakland, CA.

Seed, H.B., Tokimatsu, K., Harder, L.F., and Chung, R.M. (1985). "The influence of SPT procedures in soil liquefaction resistance evaluations," *J. Geotech. Engrg.*, ASCE, 111(12), 1425-1445.

Stewart, J.P.: coordinator (2001). Chapter 4: Soil liquefaction. Chi-Chi, Taiwan Earthquake of September 21, 1999 Reconnaissance Report, J. Uzarski and C. Arnold, eds., *Earthquake Spectra*, Supplement A to Vol. 17, 37-60.

Youd, T.L., Idriss, I.M., Andrus, R.D., Arango, I., Castro, G., Christian, J.T., Dobry, R., Finn, W.D., Harder, L.F., Hynes, M.E., Ishihara, K., Koester, J.P., Liao, S.S.C., Marcuson, W.F., Martin, G.R., Mitchell, J.K., Moriwaki, Y., Power, M.S., Robertson, P.K., Seed, R.B., and Stokoe, K.H. (2001). "Liquefaction resistance of soils: Summary report from the 1996 NCEER and 1998 NCEER/NSF Workshops on evaluation of liquefaction resistance of soils," *J. Geotech. & Geoenvir. Engrg.*, ASCE, 127 (10), 817-833.

Performance of Viscous Damper and its Acceptance Criteria

Li-Hong Sheng[1] and Don Lee[1]

Abstract

After 1989 California Loma Prieta Earthquake, California Department of Transportation, Caltrans, has launched structure retrofit program to improve seismic resistance and performance of its vast bridge inventory. Most bridges can be retrofitted with column jackets, strengthened bent caps, seat extenders, and other mitigation measures. However, a few of bridges need to be retrofitted with seismic isolators and dampers. Although seismic isolation had been used for more than a decade, there were very few cases that viscous dampers were used along with isolation bearings. And there was little performance information available with viscous dampers in other civil engineering fields. As limited maintenance can be performed on bridge with changeable environmental conditions, a rigid quality control and quality assurance procedure has to be established. In this paper, we will examine previous contract specifications in viscous damper requirements and testing data from prototype and proof tests. Finally, performance requirements and acceptance criteria will be discussed and recommended.

Introduction

Energy dissipation is the major principle used in seismic design to protect structures during an earthquake event. Most design practices emphasize on forming plastic hinges in beams (buildings) or columns (bridges) to achieve this goal. For structures which can not be retrofitted through forming plastic hinging to dissipate enough energy to prevent collapse, one of the alternative is seismic isolation and adding dampers. After 1989 Loma Prieta Earthquake, Caltrans launched its structure retrofit program to improve seismic resistance and performance of its vast bridge inventory. Most bridges were retrofitted with column jackets, strengthened bent caps, seat extenders, and other mitigation measures to improve their energy dissipation capacities. A few bridges either were not cost effective or could not be retrofitted

---

[1] Senior Bridge Engineer, Office of Earthquake Engineering, Division of Engineering Services MS 9 2/5I, California Department of Transportation, 1801 30th Street, Sacramento, CA 95816

with these methods were then retrofitted with seismic isolators and viscous dampers. A viscous damper, in theory, does not affect structure's stiffness and thus can not be easily incorporated in design process. A tedious trial and error process in time history analysis is the most common practice for bridge with viscous dampers.

Typical Design

Limited by computer simulation software and manufacturer's capacities, dampers used for energy dissipation were specified to have a linear relationship with velocities in the earlier days. With demand of higher energy absorption and less structure deformation, designer has specified smaller exponentiation related to the velocities. Today, a value of 0.5 is commonly used, while smaller coefficient, such as 0.1, is achievable in computer simulation. In reality, it is very hard if not impossible to make a viscous damper with such performance by any viscous damper manufactures. A coefficient between 0.2 to 0.3 is the best they can achieve nowadays.

As a typical practice, engineers always add redundancy in design for safety reasons. To do this, the common practice is to add extra capacity to cover uncertainties in demand. When this practice is applied to dampers, there is little information to determine how to place the safety factor over stroke or velocity. Some apply it on stroke, while others apply it on velocity or both. Increase stroke can reduce the possibility of hammer action, increase velocity demand may not be necessary since over-strength design is required for internal pressure.

Construction Specification

Since viscous dampers are new to bridge engineers in the beginning of seismic retrofit program, there was no construction standard special provision available. It was then developed from a similar seismic device, isolation bearing. In the earlier stage, the specification detail was primary a collection of information from manufacturers. With limited experience and performance information available, the damper specification is far from complete than most engineers would like. The deficiencies are in the material, testing, and construction inspection.

We used to specify material types and properties in construction specification. However, damper design is proprietary. Information in actual material used is not readily available before contract was bided. Limited information in material used can be specified, for example, stainless steel. There are different grades and types of stainless steel. Each types or grades have its unique material characteristic. Without knowing the need in its performance as part of damper, it is difficult for bridge engineer to decide what will be the best material to use. The most common way to deal with material is to limit the type of material that can not be used instead of the type of material should be used.

Prototype and Proof Tests

All viscous dampers used on bridges are custom designed for specific performance requirements. Prototype dampers are required to be made and tested to verify their performance. Once the designers are satisfied with the test results, approved and accepted, then production units can be manufactured. Prototype tests generally follow the Guide Specifications for Seismic Isolation Design, 2nd Edition, 1999, with minor revisions. The same is true for proof testing. Unfortunately, some designers do not really understand the intention of these tests. Most often, during a retrofit design, the service load demands were ignored or not specified. This situation was caused by either no detail service load analyses were performed or demand by service loads were unknown to designers. Lack of such information may lead to an under-designed viscous damper service life and its redundancies.

Results Interpretation

As recommended by the ASSHTO guide specifications of seismic isolation, all dampers need to be prototype and proof tested. A reasonable acceptance criteria should be specified as part of the construction specification. Data interpretation method needs to be considered at the time the criteria is written. One of the major shortfalls of these criteria we have to date is the consideration of difference in damper performance between tension and compression motion. It is difficult, if not impossible to have a viscous damper to perform similarly in both directions. Either an allowable tolerance relative to average value should be included in the over all variation considered or a specific discussion of different performance in both directions needs to be established. Otherwise a disagreement may be occur when data with significant difference in tension and compression variation was interpreted.

Results from various prototype and proof tests of previous projects have shown:
1. Average test forces in tension or compression motion vs theory values differ from -13.6% to 10.45%.
2. Average Degradation in tension and compression motion based on difference from theory differs from -1.5% to 26.1%.
3. Peak difference between tension and compression motion per cycle is around 9%.
4. The maximum degradation from cycle 1st to 5th cycle can be as high as 37.5% based on deviation from the theoretical values according to the test velocity.
5. The 1st peak force is significantly higher than the force in the following cycles.

Field Performance

There are a total of five bridges in the State of California on which viscous dampers have already been installed. There are two other bridges with viscous dampers yet to be installed at the time this paper was written (Table 1). All seven bridges are located near coastal area with mild climate environment. With limited service time in the field, it is difficult to judge their performance. However, there are several problems we encountered in the last few years and may need further investigation.

First, there is viscous fluid leakage (Figure 1,2) in some dampers. The cause of leakage is unknown at this time and so are the degrees of leakage. There are several potential pathways that fluid can leak out from the internal chamber. The leakage can be in a thin film form or worse. However, based on the previous evaluation of test performance at ETEC during HITEC program, a leaking viscous damper can perform properly as long as the fluid in the internal chamber can produce the needed resistance.

Since bridges will deform and displace during either seismic event or under live loads, spherical bearings are placed at both ends of the viscous damper to accommodate rotations at the connections. However, without a spacer to keep the clevis tang of the dampers positioned at the center of the anchorage bracket, the clevis tang will either move transversely (Figure 3, 4) or rotates (Figure 5) which results in between the tang and the bracket. When the clevis tang and rub against the bracket, it consequently increases stresses on the damper body and fluid seals. Designer should consider such condition and make sure spacers or an over size spherical bearings are in place to ensure the damper position. Un-expected stresses may also caused problems with protective shell over the piston area (Figure 6,7).

Most dampers on our bridges are exposed under direct sun light during daytime. A temperature gradient of a minimum of 30° F has been recorded (Figure 8) recently in mild weather environment. This environmental condition is not addressed in the original design requirements and its effect is not known.

Displacement and deformation on flexible structure often is larger than original estimated. At the Vincent Thomas Bridge in Southern California, the original prototype test called for a wind load of +/- 1 inch movement as the control live load deformation. A latest field study indicated that a daily live load plus temperature variation could reach a total stroke of 4.2 inches (Figure 9).

Conclusion and Recommendation

Viscous damper can absorb energy from seismic excitation to reduce deformation demand. Because of its nonlinear feature (velocity dependent performance), design and analysis procedure are tedious. Demands on deformation and the number of cycle under live loads often are not checked thoroughly. Since such devices need to perform as originally designed during an earthquake, deterioration under service loads needs to be considered.

Construction specification, prototype and proof tests, acceptance criteria, and maintenance procedure need to be well reviewed. Although the 30% rule in performance variation defined in AASHTO guide specification was met in most situations, there were times when such limitation was exceeded. Both testing setup and damper design may be the causes. More information will be needed to determine the true causes.

Information of field environment and damper performance at bridge site is limited. Currently there is no permanent instrumentation installed on the bridges with dampers to monitor their performance under live loads and environmental demand.

Permanent instrumentation at least at one bridge site to collect such information and further study is recommended.

Reference

1. AASHTO (1999). "Guide Specifications for Seismic Isolation Design", 2nd Edition.
2. MCEER (2001). "Recommended LRFD Guidelines for the Seismic Design of Highway Bridges"

Bridge Name	No of Dampers
I-5/SR91	10
East SFOBB (retrofit)	6
Rio Vista Bridge	8
Vincent Thomas Bridge	48
San Diego Coronado	20
West SFOBB	100
Richmond-San Rafael Bridge	16

Table 1 Bridge with Dampers in California

Figure 1 Fluid leakage

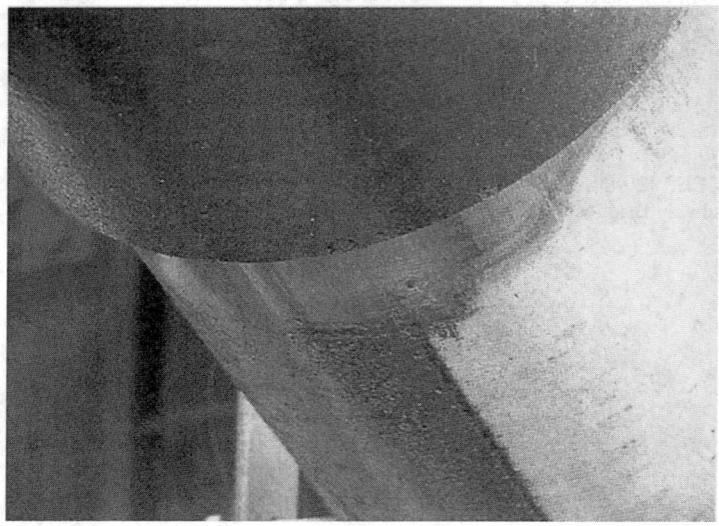

Figure 2 Fluid leakage

Figure 3 Transverse movement and fluid leakage

DISASTER RESPONSE FOR LIFELINE SYSTEMS 1037

Figure 4 Anchorage detail with spacer

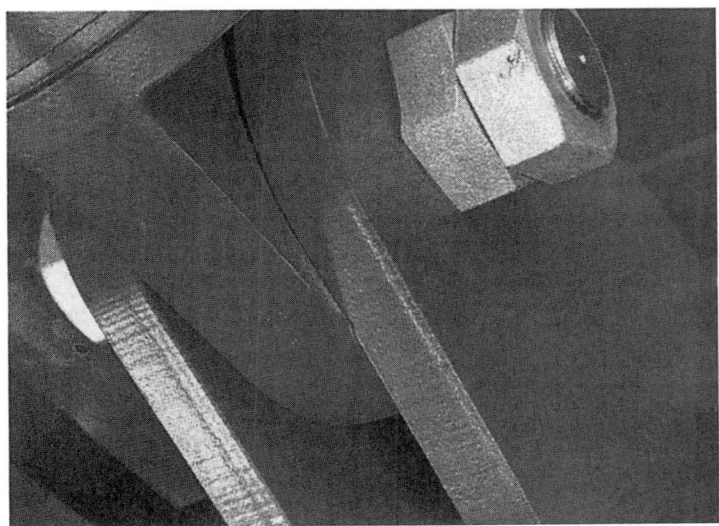

Figure 5 Damper rotation against anchorage

Figure 6 Sheared nuts

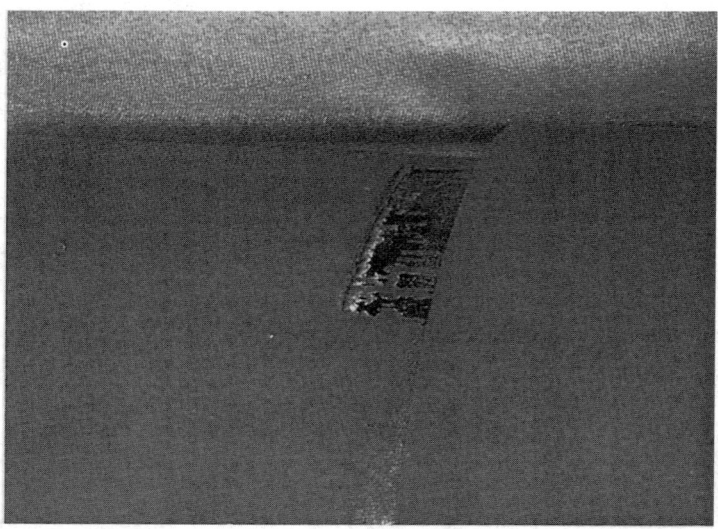

Figure 7 Protection cover

DISASTER RESPONSE FOR LIFELINE SYSTEMS 1039

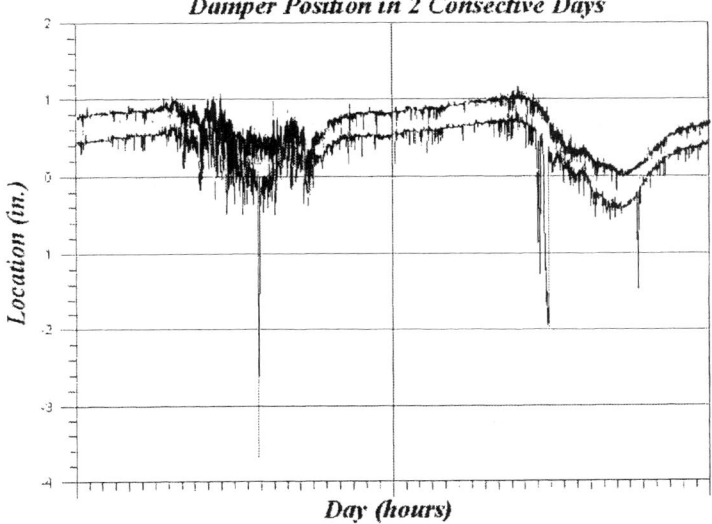

Figure 8 Recorded damper movement

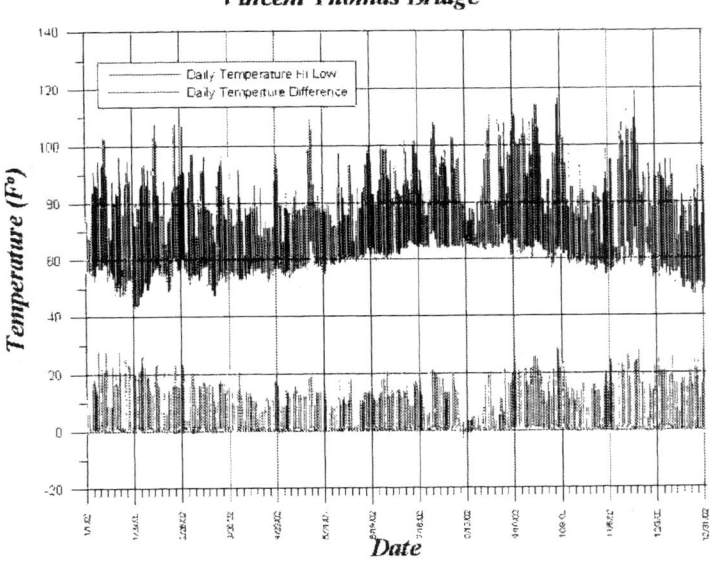

Figure 9 Annual temperature variation at the Vincent Thomas Bridge

Impact of Friction Pendulum Bearings on the Seismic Retrofitting Cost of Typical Bridges with Wall Type Piers in the State of Illinois

Murat Dicleli, Ph.D., P.Eng., M. ASCE[1]., Mouhamad Y. Mansour, Ph.D., Assoc. M. ASCE[2], Anoop Mokha, Ph.D., SE [3], Victor Zayas, Ph.D., SE[4] and Michael C. Constantinou, Ph.D., M. ASCE[5]

Abstract

The economical efficiency of friction pendulum bearings (FPB) for seismic retrofitting of typical Illinois bridges with wall type piers is studied. A typical bridge with heavy wall piers was carefully selected by Illinois Department of Transportation (IDOT) for the purpose of this study. The seismic analysis of the bridge revealed that the bearings, and substructures of the bridge are vulnerable and need to be retrofitted. A conventional retrofitting strategy is developed for the bridge and the cost of retrofit is estimated. Next, the bridge is further studied to develop appropriate techniques for upgrading its seismic capacity using FPB and the seismic analysis is repeated. It is observed that the FPB effectively mitigated the seismic forces and eliminated the need for retrofitting of the substructures of the bridge.

Introduction

The state of Illinois is a region of low to moderate risk of seismic activity. Recognizing the threat of seismic activity, IDOT made a statewide assessment of the seismic vulnerabilities of its highway bridges. The bearings, and substructures of these bridges were found vulnerable. Conventional retrofitting methods based on strengthening and enhancing the ductility of substructure members have been widely used to mitigate the seismic damage risk for such bridges (FHWA, 1995). Yet, most of these methods are expensive and difficult to implement. Thus, seismic retrofitting methods based on response modification may provide a more efficient solution. This may be achieved by replacing the existing bearings by FPB to mitigate the seismic forces and eliminate the need for costly retrofitting of substructures.
   Most of the design and retrofitting applications of FPB to bridges in the past included those with heavy, long-span superstructures and relatively lighter substructures (Zayas and Low, 1999). Since the FPB are placed on top of the bridge

---

[1] Asst. Prof., Dept. of Civil Engrg and Construction, Bradley Univ., Peoria, Illinois 61625, E-mail: mdicleli@bradley.edu, Tel.: (309) 677-3671, Fax.: (309) 677-2867
[2] Structural Engineer, Beirut, Lebanon
[3] Vice President, Earthquake Protection Systems Inc., Richmond, California
[4] President, Earthquake Protection Systems Inc., Richmond, California
[5] Prof. and Chmn., Dept. of Civ., Struct. and Envir. Engrg., Univ. at Buffalo, Buffalo, New York

substructures, they can effectively mitigate the seismic forces transferred from the heavy superstructure of such bridges. However, when applied to bridges with lighter short-span superstructures and relatively heavier substructures, the seismic inertial forces resulting from the larger mass of the substructures may not be mitigated as effectively by the FPB. To assess the economical and structural efficiency of FPB for seismic retrofitting of such bridges, a bridge with a considerably heavier substructure and relatively lighter superstructure was carefully selected by IDOT to represent typical seismically vulnerable bridges of this type in the state of Illinois and studied

Friction Pendulum Bearings

FPB are sliding-based seismic isolators. The main components of FPB are; a stainless steel concave spherical plate, an articulated slider and a housing plate as shown in Fig. 1. FPB use the characteristic of a pendulum motion to lengthen the natural period of the isolated structure, thus, deflect the earthquake input energy to mitigate the seismic forces. A pendulum motion for the supported structure is achieved as the articulated slider rides on the concave surface as illustrated in Fig. 1. The movement of the slider generates friction that results in hysteretic energy dissipation. Fig. 1 also displays the idealized force-displacement hysteresis loop for the FPB.

Description of the Bridge

The bridge is located on Route 14 in Jackson County, Illinois. It has three continuous spans carrying two traffic lanes and is supported by two heavy wall-piers as shown in Fig. 2. Roller steel bearings are provided underneath each girder at both abutments. Steel rocker and fixed bearings are provided at Piers 1 and 2 respectively.

Both abutments are seat type and are nearly identical. Each abutment is supported on a single row of six 356 mm diameter, #7 gauge metal shell piles filled with cast-in-place concrete. Two of the piles are battered at a slope of 1:6. All the piles are embedded 305 mm into the abutments. The length of the piles is 19 m.

The piers are reinforced concrete walls typically used in Illinois bridges. Both piers have identical geometry and are supported by 33 timber piles embedded 152-mm into the pile-cap and driven 8.5-m into the soil. The piles are tapered with a top and bottom end diameters of respectively 305 and 203 mm. The bridge site has stiff soil conditions.

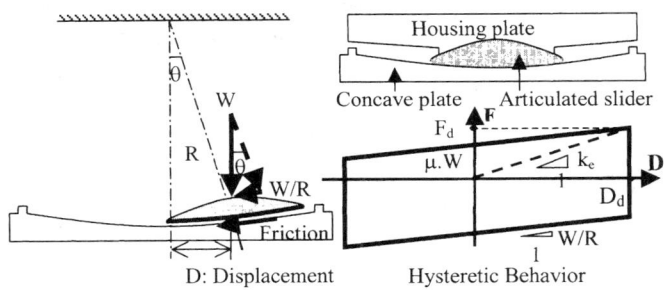

Figure 1. Friction pendulum bearings

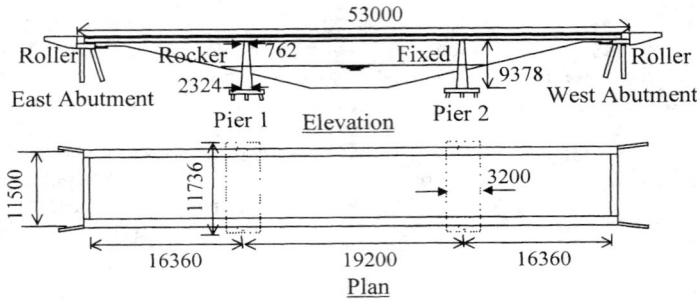

Figure 2. Bridge geometry

Seismic Analysis

Iterative Multi-mode Response Spectrum (MMRS) analyses of the bridge are conducted to simulate (i) sliding at the abutment bearings (ii) piles' nonlinear lateral behavior (iii) flexural yielding of the wingwalls (iv) sliding at the abutment-backfill interface and (v) soil-bridge interaction effects. A detailed description of the iterative analysis procedure is presented elsewhere (Dicleli and Mansour, 2002). The normalized acceleration response spectrum of AASHTO (1998) is used in the analyses. For the bridge site, the zonal acceleration ratio is obtained as 0.12. sec. For the site's stiff soil (AASHTO soil type II), the site coefficient is obtained as 1.20.

Structural Model

A 3-D structural model of the bridge illustrated in Fig. 3 is built and analyzed. The model is capable of simulating the nonlinear behavior of bridge components and soil-bridge interaction effects when used in combination with iterative MMRS analyses. In the model, equivalent linear stiffness (ELS) properties are used for the bridge components exhibiting nonlinear behavior.

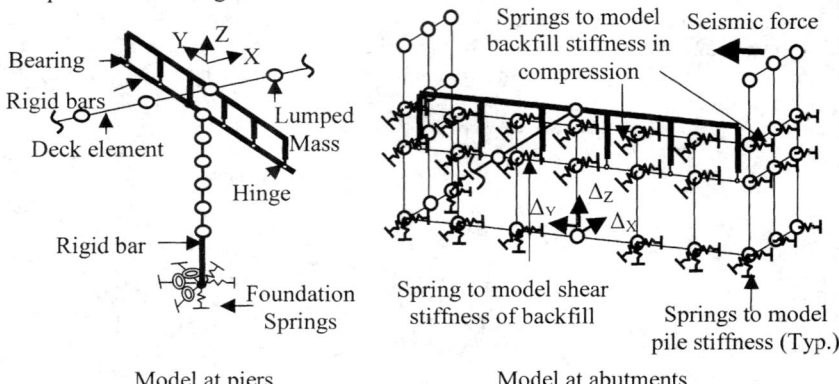

Figure 3. Structural model

Superstructure Modeling. The bridge superstructure is modeled using 3-D beam elements. The 8.49 tons/m superstructure mass is lumped at each nodal point. The bridge deck which has a large in-plane translational stiffness is modeled as a transverse rigid bar at the abutments and piers as shown in Fig. 3

Bearings Modeling. All the bearings are idealized as 3-D beam elements connected between the superstructure and substructures as shown in Fig. 3. Pin connection is assumed at the joints linking the bearings to the substructures. For the steel roller bearings, an ELS is used in combination with iterative MMRS analyses to simulate their sliding behavior in the longitudinal direction. In the transverse direction, the bearing shear stiffness is calculated as 3,736,000 kN/m. For the rocker bearings, the rotation at the bearing top is released about the global Y-axis to obtain a structural mechanism that simulates the rocking motion of the bearings. In the transverse direction, their shear stiffness is calculated as 2,260,000 kN/m. Similarly, the shear stiffness of the fixed steel bearings is calculated as 62,000,000 kN/m.

Substructures modeling. The pier-wall is modeled using a set of 3-D beam elements as shown in Fig 3. Assuming that the wall cross-section remains plane after deformation, the width of the wall is modeled using a horizontal rigid bar at the wall top. This enabled the connection of the bearing elements to the wall. The wall is then connected to a vertical rigid bar representing the footing. The abutments are modeled using a grid of beam elements as shown in Fig. 3. The tributary masses of the abutment and wingwalls are lumped to the nodes.

Modeling pile-soil interaction. At the abutments, three translational springs are connected at each pile location to model the foundation flexibility. At the piers, considering piles' group effect, six translational and rotational springs are connected at the piers' base to model the flexibility of the whole foundation. The horizontal boundary springs connected at the pile-substructure interface nodes are assigned an ELS obtained from the piles' non-linear lateral-force displacement (P-Y) curves as the slope of the line connecting the origin to the point representing pile's seismic lateral force on the curve. The ELS for each abutment pile is obtained as 11,810 kN/m. For the pile group at the piers the ELS is obtained as 78,817 kN/m in the longitudinal and 55,888 kN/m in the transverse directions. Novak's (1977) procedure for floating piles is used to obtain the vertical stiffness of a single pile at the abutments and pile group at the piers as 246,194 and 1,269,075 kN/m respectively. The stiffness of the rotational springs for the rocking and torsional motion of the pier footings is calculated by introducing a unit rotation about the global X, Y and Z axes at the geometric center of the pile group and calculating the moment of the generated pile forces about the geometric center of the pile group. Accordingly, the rotational spring constants about the global X and Y-axis are obtained as 14,440,604 and 1,257,156 kN.m/rad respectively and the torsional spring constant is calculated as 1,269,075 kN.m/rad.

Modeling backfilll-abutment interaction. In the longitudinal direction, translational springs are attached to the nodes of only one of the abutments to simulate backfill-abutment interaction effects as illustrated in Fig 3. In the transverse direction, translational springs are attached to the nodes of only one of the wingwalls at each abutment to simulate backfill-wingwall interaction effects. Translational springs are also attached at each node of the abutment to model the shear stiffness of the backfill. The shear stiffness of the backfill is calculated assuming that only the portion of the backfill between the wing-walls will deform in a shearing mode.

Using the relationship defined by Clough and Duncan (1991) for the variation of the earth pressure coefficient as a function of the abutment movement, the horizontal subgrade constant, k_{sh}, for the backfill is obtained as a function of the depth from the abutment top. The stiffness of the boundary springs connected at the abutment-backfill interface nodes at the abutment and wingwalls are then calculated by multiplying k_{sh} by the area tributary to the node.

Analysis Results

The seismic analysis results are presented in Tables 1 through 3. In the tables two types of results are tabulated. The results of the original bridge with the existing steel bearings (marked by superscript "a") and the results when FPB are used (marked by superscript "b"). The first four vibration periods of the bridge are calculated as 1.191, 0.706, 0.465 and 0.344 s. respectively.

Table 1 Bearing seismic forces and lateral relative displacements

Location	Longitudinal Direction		Transverse Direction	
	Force (kN)	Relative Disp. (mm)	Force (kN)	Relative Disp. (mm)
E. Abutment[a]	25 (100)	41.7	124	0.60
Pier 1[a]	0	2.30	15	0.01
Pier 2[a]	47	0.02	15	0.01
W. Abutment[a]	25	41.8	124	0.54
E. Abutment[b]	100	1.04	8	38.4
Pier 1[b]	17	12.5	25	29.5
Pier 2[b]	17	12.5	25	29.5
W. Abutment[b]	100	1.04	8	38.4

Table 2 Seismic forces in abutment components

Component	Longitudinal Direction		Transverse Direction	
	Shear (kN/m)	Moment (kN-m/m)	Shear (kN/m)	Moment (kN-m/m)
Wingwall[a]	N/A	N/A	71	48
Back-wall[a]	7	9	N/A	N/A
Wingwall[b]	N/A	N/A	19	15
Back-wall[b]	5	6	N/A	N/A

Table 3 Bridge seismic substructure reactions

Substructure		Longitudinal Direction			Transverse Direction		
		Axial (kN)	Shear (kN)	Moment (kN-m)	Axial (kN)	Shear (kN)	Moment (kN-m)
E. Abut. Piles[a]	B	6	65	0	56	158	0
	V	6	33	0	94	158	0
Pier[a]	1	13	589	3700	0	810	2987
	2	20	505	3976	0	796	2925
W. Abut. Piles[a]	B	6	65	0	53	130	0
	V	6	33	0	89	130	0
E. Abut. Piles[b]	B	6	24	0	18	14	0
	V	6	12	0	30	14	0
Pier[b]	1	76	530	1606	0	452	2065
	2	76	530	1606	0	452	2065
W Abut. Piles[b]	B	6	24	0	18	14	0
	V	6	12	0	30	14	0

B: Battered piles, V: Vertical piles

Capacities of Bridge Components

Capacity of steel bearings. The two 19-mm diameter anchor bolts connecting the bearing's base plates to the abutment are found to be the weakest bearing components. The ultimate capacity of the bearings in shear is calculated as 94 kN. Displacement capacity of rocker bearings is calculated as 90 mm.

Capacity of pier-walls. The cracking moment about the weak axis of the pier base is calculated as 13,990 kN.m. The minimum shear capacity of the pier walls is calculated as 11,656 kN. The seismic shear and the moment at the wall base are much smaller than the calculated capacities. Therefore, neither the reinforcement details nor the ultimate shear and flexural capacities of the pier walls need to be assessed.

Capacity of abutment-components. The ultimate flexural capacities of the wingwall and ballast wall are calculated as 48 kN.m/m and 89 kN.m/m, respectively and their shear capacities are calculated as 255 kN/m and 307 kN/m, respectively.

Capacity of piles. For the bridge to remain serviceable, piles' lateral displacements are limited to 38 mm per AASHTO (1998). Based on this displacement, the lateral capacities of the piles at the piers are obtained as 1700 and 1200 kN in the longitudinal and transverse directions respectively and those of the abutment piles are obtained as 207 kN. The axial compression capacity of the piles at the abutments and piers are respectively given as 570 and 355 kN per pile based on the pile driving data.

In seismic regions, AASHTO (1998) requires that the piles be provided with anchoring devices and embedded a minimum of 300 mm into the cap to develop an uplift resistance. For the bridge considered in this study, the connections between the piles and the substructures are not detailed in compliance with AASHTO (1998) to sustain uplift forces. Therefore, the piles are assumed to fail under axial tension.

Capacity of pier footings. Careful examination of the bridge's structural deawings revealed that the footing reinforcement at the piers is not detailed in compliance with AASHTO (1998). Furthermore, no reinforcement is provided at the footing top to resist the moments introduced by the reversed seismic loading. Thus, ductile behavior of the footings is not justified. Based on the provided reinforcement detail, the ultimate shear and flexural capacities of the footings are calculated as 463 kN/m and 195 kN.m/m respectively. The punching shear capacity of the pier footings is found adequate to resist the 355 kN axial compression capacity of the piles.

Capacity / Demand (C/D) Ratios for Vulnerable Components

Bearings. The seismically induced shear forces in the bearings at both abutments of the bridge in the transverse direction are found to be larger than their 94 kN shear capacity. The resulting shear C/D ratio for a typical bearing at both abutments is 0.71.

Abutments. In the longitudinal direction, stability problem exists at the abutments. The single row of piles cannot provide sufficient rotational/overturning resistance at the base of the abutments due to the poor detailing of pile connections. Furthermore, the abutments are not restrained at their top due to the presence of expansion bearings. Consequently, tilting of the abutments is expected. Furthermore, under transverse direction seismic loading, the backfill's passive pressure is found to induce transverse seismic moments in the wingwalls in excess of their ultimate flexural capacities. The flexural C/D ratios for the east and west abutment wingwalls are calculated as 0.53 and 0.64, respectively.

Pier foundations. The seismically induced axial compression forces at both piers' piles are larger than their compression capacities. The axial C/D ratios of the piles at Pier 1 and Pier 2 are calculated as 0.92 and 0.90, respectively. In addition, uplift forces are observed in the piles under longitudinal seismic loading. As the pile-cap connection is not detailed properly to resist such tensile forces, pullout failure is anticipated. Furthermore, no reinforcing steel is provided at the top of the pile cap to resist the seismic moment reversal and flexural failure of the pile cap is anticipated.

Conventional Seismic Retrofitting

Bearings. The abutment bearings are found to be seismically vulnerable. However, as all the bearings are aged, their replacement is recommended. Pot bearings fixed in both orthogonal directions are recommended at the abutments to provide the longitudinal fixity required to ensure the stability of the abutments. On the piers, to reduce the magnitude of the seismic forces transferred from the superstructure and to alleviate the effect of thermal induced loads, pot bearings free in the longitudinal and fixed in the transverse directions are recommended. Two additional analyses of the bridge under thermal and seismic loadings are performed with the new bearing type and configuration. The analyses results revealed that thermal induced forces in the bridge structural components are negligible. The seismic analysis results with the new bearings are used for the seismic retrofitting of the pier foundations.

Wingwalls. The retrofitting of wingwalls at both abutments includes the addition of cleats at the joints between the wingwalls and the abutments to enhance their flexural capacities in the transverse direction.

Pier foundations. Two rows of six steel HP8x36 piles are added to the pier footings to reduce the effect of axial compression loads and to eliminate the uplift loads on the timber piles. With the addition of steel HP piles, the axial compressive forces in the timber piles are reduced to 243 kN and the uplift forces are eliminated. However, the addition of the HP piles requires an enhancement in the flexural capacity of the pile cap. This is achieved by increasing the depth of the footing from 760 to 1000 mm, adding shear dowels at the new and existing concrete interface and adding extra flexural reinforcement at the top of the new footing to resist any seismic moment reversal.

Conventional retrofitting cost. The strengthening of the pier foundations and the wingwalls requires driving piles, concrete removal, placing new concrete, reinforcement and shear dowels, jacking and removing bearings, new bearing installment and excavation of the soil to expose the footings and the wingwalls completely. In addition, retrofitting the pier foundations within the river requires the use of a cofferdam. Mobilization and traffic control costs are not included in the cost estimate. IDOT provided ranges of unit prices for the purpose of cost estimation. Based on this unit prices the total cost of retrofitting the bridge is found to range from $316,324 to $398,623.

Seismic Retrofitting of The Bridge Using FPB

Configuration of bearings. To overcome the abutments' stability problem, guided FPB with a cylindrical concave surface need to be used at the abutments to restrain the displacement of the bridge in the longitudinal direction but allow it in the transverse direction. FPB with a spherical concave surface is recommended over the piers to control and mitigate the seismic forces acting on the piers in both orthogonal directions. It is anticipated that as the massive wall piers vibrate as independent cantilever structures, FPB will dissipate energy due to the friction generated and mitigate the effect of seismic forces acting on the piers.

Structural model with FPB. The structural model used for the detailed seismic analysis of the bridge is slightly modified to incorporate the FPB instead of the existing bearings. The FPB are also modeled using 3-D vertical beam elements. The ELS of the FPB designated as k_e in Fig 1 is used to estimate the stiffness properties of the beam elements. Hysteretic damping of the bearing is also considered in the analyses following the procedures defined by AASHTO (1999). An iterative analysis procedure is performed to calculate the seismic response of the bridge as the ELS and hysteretic damping are functions of the bearing displacement.

Analysis results using FPB. The periods of vibration for the first and second modes of the bridge are 1.34 and 0.572 sec. respectively. In the first mode, the behavior is similar to that of a traditional isolated bridge, where superstructure moves relative to the substructure. However, in the longitudinal direction, the bridge does not exhibit a traditional seismic isolated structure behavior as the superstructure is fixed to the abutments at both ends. In this mode of vibration, the piers vibrate independently resulting in energy dissipation due to friction generated at the bearings.

Table 1 displays the bearing lateral forces and relative displacements when using FPB. In the longitudinal direction, the fixed bearings at the abutments attract the largest seismically induced forces as expected. The bearings at the piers have an out-of-phase displacement of 12.5-mm resulting from the large deflection of the piers in the opposite direction. In the transverse direction, the bearing relative displacements at the abutments and piers are respectively 38.4 and 29.5 mm.

The seismically induced forces in the abutments and wingwalls are displayed in Table 2. Table 3 presents the substructure reactions at the abutments and base of the pier foundations of the bridge retrofitted with FPB. In the longitudinal direction, although the shear forces at the piers of the isolated bridge have magnitudes in the same order as the non-isolated bridge, the pier base moments are reduced considerably. This results from the frictional forces generated by the FPB at the pier top that act opposite to the direction of the inertial forces generated by the large pier mass. As the moment arm of the frictional forces at the pier top is much larger then that of the inertial forces, the pier base moment becomes smaller.

C/D ratios. The C/D ratios for the previously determined vulnerable structural members are now all larger than 1.0 after replacing the existing bearings with FPB. The FPB effectively mitigated the seismically induced forces such that seismic retrofitting of the pier foundations, abutments and wingwalls is no more required. The stability problem for the abutment has been resolved by fixing the FPB at the abutments in the longitudinal direction.

Retrofitting cost. The estimated range of cost for the seismic retrofitting of the bridge using FPB is calculated as $91,670 to $109,650 using the retrofitting cost ranges provided by IDOT. The retrofitting cost includes jacking the bridge and removing the existing bearings, adjusting the elevation of bearing pedestals, and erection and cost of FPB. The average cost of retrofitting using FPB is found to be only 30% of the average conventional retrofitting cost. Thus, FPB may effectively be used for seismic retrofitting of typical bridges with heavy substructures in the state of Illinois or in regions of low to moderate risk of seismic activity.

Conclusions

The economical and structural efficiency of FPB for retrofitting of bridges with relatively lighter superstructures and heavier substructures in the state of Illinois is investigated. The following observations are made:
- FPB resulted in a more uniform distribution of seismic forces among substructures in the transverse direction.

- A considerable reduction in the seismically induced forces in the bridge components is achieved with the FPB in spite of the heavy mass of the substructures. As the massive piers vibrated independently under longitudinal direction seismic excitation, the friction generated at the FPB over the piers resulted in a considerable energy dissipation and reduction in the seismically induced forces at the piers. Thus, FPB effectively mitigated the seismic forces and eliminated the need for costly retrofitting of the bridge substructures.
- An average retrofitting cost using FPB is calculated as only 30% of that using conventional retrofitting method. Retrofitting with FPB requires shorter construction time than that required for conventional retrofitting. Therefore, retrofitting with FPB may become much more economical when the additional cost of construction time and cost of traffic mobilization is considered. Thus, FPB may be effectively used to retrofit typical bridges with lighter superstructures and heavier substructures in the State of Illinois or in regions of low to moderate risk of seismic activity.

References

AASHTO, (1998) LRFD Bridge Design Specifications, Washington, D.C.

AASHTO, (1999) Guide Specifications for Seismic Isolation Design, Washington, D.C.

Clough, G. W. and Duncan, J. M., (1991) Foundation Engineering Handbook, 2nd Edition, Edited by H. Y. Fang, Van Nostrand Reinhold, New York, NY, 223-235.

Dicleli, M. and Mansour, M. Y., (2002) "Seismic Retrofitting of Highway Bridges Using Friction Pendulum Seismic Isolation Bearings", Technical Report: BU-CEC-02-01, Department of Civil Engineering and Construction, Bradley Univ., Peoria IL.

FHWA, (1986) Seismic Design of Highway Bridge Foundations – Volume II: Design Procedures and Guidelines, Publication No. FHWA-RD-94-052, Federal Highway Administration, US Department of Transportation, Washington, D. C.

FHWA, (1995) Seismic Retrofitting Manual for Highway Bridges, Publication No. FHWA-RD-86-102, Federal Highway Administration, US Department of Transportation, Washington, D. C..

Novak, M., (1977) "Vertical vibration of floating piles", ASCE Journal of Engineering Mechanics, 103 (EMI), 153-168.

Zayas, V. A.; Low, S. S. (1999) "Seismic isolation of bridges using friction pendulum bearings", Proceedings of the 1999 Structures Congress 'Structural Engineering in the 21st Century, New Orleans, LA' ASCE, Reston, VA, 99-102.

Development and Analysis of Composite Steel-Concrete Girder for Bridge

Cui Yuping [1], Shi Zhongzhu [2], Sun Shaoping [3]

Abstract

This study aims at the development of new type of bridge girder and analysis of ductile behavior as well as its fatigue strength. In this paper, both experimental results and analytical equations which are affective to resist to earthquake motion are presented.

Introduction

A bridge is a vital link in any road network. If an earthquake occurs, a bridge may be destroyed or damaged. Many old bridges in China are considered to have inadequate seismic resistance and could disrupt transportation lifelines if subjected to severe seismic shaking. So we carry out a study program in order to increase the seismic resistance of bridges. Composite steel-concrete girder of bridge, an economical and practical structure, is able to satisfy the requirement. Thus we focus on its ductile behavior and its fatigue strength. From experiments and analysis, it is shown that the composite steel-concrete girder using welded studs as the means of shear connection has better ductility than other connections, also has good performance of resist-fatigue.

Experiment on Ductility of Composite Girder

In order to study the ductility of composite girder, 12 specimens with simply-supported have been accomplished by two groups. The geometric and mechanical characteristics of the specimens are shown as figure 1 and table 1.

The dimension of composite-site girder specimen is shown in figure 1.

---

[1] Senior Engineer, Beijing Municipal Engineering Research Institute, Beijing, P. R. China cuiyuping@163.net
[2,3] Professor, Beijing Municipal Engineering Research Institute, Beijing, P. R. China

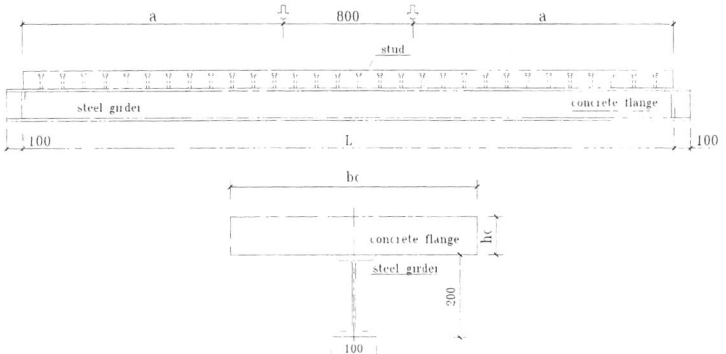

Figure 1 The dimension of composite girder specimen

The main parameters of the specimens   table 1

Specimen	L (mm)	a(mm)	b_c(mm)	h_c (mm)	ρ_{st}(%)	N/N_f	f_y (MPa)	f_{cu} (MPa)	Δ_{yt} (mm)	Δ_{ut} (mm)
SCB-1	3840	1520	500	125	0.60	1.15	310	32	12.6	59.8
SCB-2	3840	1520	800	125	0.60	1.15	310	32	10.5	75.8
SCB-3	3840	1520	800	125	0.60	1.13	310	32	10.2	98.0
SCB-4	3840	1520	800	125	0.60	1.13	310	32	9.8	109.8
SCB-19	4000	1600	700	110	0.45	1.00	320	51	16.4	63.9
SCB-20	4000	1600	700	110	0.77	1.00	320	47	17.4	87.1
SCB-21	4000	1600	700	110	0.58	1.00	320	47	17.0	100.1
SCB-22	4000	1600	700	110	0.58	1.00	320	42	18.7	93.5
SCB-23	4000	1600	700	110	0.58	0.83	320	45	18.1	84.9
SCB-24	4000	1600	700	110	0.58	0.67	320	43	17.8	78.4
SCB-25	4000	1600	700	110	0.58	0.58	320	41	17.4	71.4
SCB-26	4000	1600	700	110	0.58	0.42	320	45	16.9	61.0

Where "L" is the span of the beam, "a" is the shear span, "b_c" is the width of the concrete flange slab, "h_c" is the depth of the concrete slab, "ρ_{st}" is the rate of the tranverse reinforment, "N/N_f" is the coefficient of shear connection degree, "f_y" is the yielding strength of steel beam, "f_{cu}" is the compressive strength of the concrete standard cube. In the last two columns of the table 1, Δ_{yt} is the yielding deflection and Δ_{ut} is the ultimate deflection of mid-span. From the test and analysis, it's shown that the ductility is scale up in proportion to the shear connection degree and the transverse reinforcement rate.

Analysis of Ductility

The quantitative analysis of ductility is mainly depended on the yielding curvature and ultimate curvature. Therefore, it's a very important index to express the deformation and energy dissipation. Its details describe as follows:

- **Curvature of the Elastic Stage**

The yielding curvature φ_y is relative to determined easily. It can be stipulated as equation (1) from the Strength of Materials, as it coincides with the linear elastic theory.

$$\varphi = \frac{M}{E_s I} \tag{1}$$

where $E_s I$ is the conversion section rigidity.

Considering the relative slip between the steel girder and concrete slab, the curvature of the elastic stage can be represented as equation (2):

$$\varphi = \frac{M}{E_s I} + \Delta\varphi = \frac{M}{B} \tag{2}$$

where "$\Delta\varphi$" is the additive curvature which caused by the slip. "B" is the discounting rigidity, which can be determined by equation (3):

$$B = \frac{E_s I}{1+\xi} \tag{3}$$

where "ξ" is the discounting coefficient of the section rigidity. "$E_s I$" and "ξ" can be estimated by equation (4) and (5) respectively:

$$E_s I = E_s (I_0 + A_0 d_c^2) \tag{4}$$

$$\xi = \eta[0.4 - \frac{3}{(\alpha L)^2}] \tag{5}$$

where $\eta = 24 E_c d_c p A_0 / (KhL^2)$; $\alpha = \sqrt{\frac{K}{E_s I_0 A_1 p}}$.

In the η and α expressions:

$$A_0 = \frac{A_s A_c}{n A_s + A_c} \ , \ A_1 = \frac{A_0}{I_0 + A_0 d_c^2} \ , \ I_0 = I_s + \frac{I_c}{n};$$

L   the span of girder;
d_c the centrial distance between the section of steel beam and concrete slab;
p   the space of the studs;
K   the rigidity coefficient of the studs, $K = 0.66 n_s V_u$, where V_u is the shear capacity of the individual stud, n_s is the connection number of the same section, if it is arranged in one line $n_s=1$

In the "I_0" expression:

I_s the moment of inertia of the steel girder;

I_c the moment of inertia of the concrete slab;
n the moduli ratio of the steel and concrete.

In equation (5), αL affects the ξ mainly. The practical condition is $\alpha L \geq 3.0$. If $\alpha L < 3.0$, it can be adopted by $\alpha L = 3.0$. When there is no slip between the interface of the steel girder and concrete slab $K \to \infty$, η is equal to zero, thus ξ is also equal to zero.

In equation (2), replacing the M with M_y, the yielding curvature can be determinded.

● **The Ultimate Curvature**

From the above, it can be known that the equation (1) and (2) are adopted to calculate the curvature of elastic stage. The ultimate curvature and deflection have to be experimentally determined.

The calculating equation of ultimate curvature is based on the equation (2). It can be written in this way:

$$\varphi_u = \left(\frac{N}{N_f}\right)^{3/2} \left(\frac{0.8\varepsilon_u}{x_u} + 1.5\frac{\varepsilon_{su}}{h}\right) \tag{6}$$

where x_u is the equivalent height, ε_u is the ultimate compressive strain. ε_{su} is the ultimate slip strain. ε_u can be derived from the experimental result, it has the following expression:

$$\varepsilon_u = 0.004 - 0.005\xi_1 \tag{7}$$

where ξ_1 is the height of the relative compressive area. It can be described by the equation: $\xi_1 = \frac{x}{h_c} = \frac{x_u}{0.8}\bigg/h_c = 1.25\frac{x_u}{h_c}$.

In equation (6), ε_{su} is experimentally determined in the following

$$\varepsilon_{su} = \varepsilon_{sy}(1 + \frac{3}{0.8 + \xi_1}) \tag{8}$$

where ε_{sy} is the slip strain when the steel beam is starting to yield. The calculating equation of ε_{sy} can be obtained from the literature 5.

$$\varepsilon_{sy}^1 = \frac{\alpha\beta P_y}{2}(1-e^{-\alpha L})/(1+e^{-\alpha L}) \tag{9a}$$

$$\varepsilon_{sy}^2 = \frac{\alpha\beta P_y}{2}(e^{-\alpha b} - \frac{e^{\alpha b} + e^{-\alpha b}}{1+e^{\alpha L}}) \tag{9b}$$

$$\varepsilon_{sy}^3 = \beta q(1 - \sec\frac{\alpha L}{2}) \tag{9c}$$

where the superscript 1, 2 and 3 is the yielding strain correponding to the one-point load, two-point load and uniform load. "P_y" is concentrated load, "q_y" is uniformly distributed load. "b" is the distance from the load point to the mid-span. In equation (9), $\beta = \dfrac{A_1 d_c p}{K}$.

With φ_u confirmed, we can calculate the ultimate deflection by the method of curvature integralation if the length of equivalent plastic-hinge is determined.

- **Confirmation of the Equivalent Plastic-Hinge Length**

In the entire course of bearing the load, the girders which designed by the plastic theory have a forming and developing stage of the plastic-hinge. The upper deformation is nearly coming from the plastic-hinge turnning. So we must confirm the length of equivalent plastic-hinge before calculating the ultimate deflection.

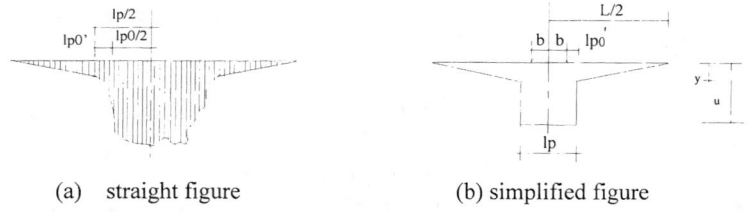

(a) straight figure (b) simplified figure

Figure 2 Calculating module of the equivalent plastic-hinge

The calculating model of the equivalent plastic-hinge is shown in figure 2. In figure 2(a), the section between l_{p0} and l_p is the region from the maximal curvature to yielding curvature, figure 2(b) is the simplified figure, it's demonstrated that the curvature is equal in the region of "l_p".

Let the length of the equivalent plastic-hinge on the side of the varying moment section be "l'_{p0}", it can be experimentally given by

$$l'_{p0} = \lambda h \quad (10)$$

where $\lambda = 0.88ah/(Lh_c)$, "h" is the height of girder and "h_c" is the effective height of girder.

Let "l_p" is the length of equivalent plastic-hinge region, it is represented as follows:

$$l_p = 2\left(b + l'_{p0}\right) \quad (11)$$

DISASTER RESPONSE FOR LIFELINE SYSTEMS 1055

- **Calculating the Ultimate Deflection**

According to the structure machanics, the oblique angle of support θ_A and deflection of mid-span Δ is calculated by the following equation $\theta_A = \int_0^{L/2} \varphi(x)d_x$ and $\Delta = \int_0^{L/2} \varphi(x)xd_x$ respectively. Thus the mid-span deflection can be obtained by integraling the curvature.

Figure 3 Calculating module of the ultimat deflection

When the moment of mid-span $M_c > M_y$, the plastic concertrating aera is formed on the section C (figure 3). It's shown from the figure 2(b) that the curvature is equal in the region of "l_p". Thus the ultimate deflection Δ_{up} can be calculated by equation(12) and its value as table (2).

$$\Delta_{up} = \frac{1}{6}\varphi_y(L-l_p)^2 + \frac{\varphi_u}{8}l_p(L-l_p) \quad (12)$$

- **Ratio of Ductility**

As mentioned above, the ratio of ductility β_φ is denoted by

$$\beta_\varphi = \varphi_u / \varphi_y \quad (13)$$

The yielding and ultimate curvature is determined according to the equation (2) and (6) respectivly. The deflection ratio β_Δ can be represented in the following way

$$\beta_\Delta = \Delta_{up} / \Delta_y \quad (14)$$

where Δ_y is the yielding deflection. It has the following expression

$$\Delta_y = \Delta_e(1+\xi) \quad (15)$$

where Δ_e is the calculating deflection derived from the method of the elastic conversion section. ξ can be obtained from equation (5).

The ratios of ductility are listed in table 2.

The ultimate deflection and the ratio of ductility   table 2

Liter.	specimen	Δ_{ut} (mm)	Δ_{up} (mm)	ξ_1	φ_y (1×10^{-6})	φ_u (1×10^{-6})	$\beta_\Delta = \frac{\Delta_{up}}{\Delta_{ut}}$	$\beta_\varphi = \frac{\varphi_u}{\varphi_y}$	$\frac{\Delta_{ut}}{\Delta_{yt}}$	$\frac{N}{N_f}$	Load mode
This paper	SCB-19	63.9	105.3	0.45	8.6	86.3	1.64	10.0	3.9	1.00	Two-point
	SCB-20	87.1	98.1	0.49	8.7	80.1	1.10	9.2	5.0	1.00	
	SCB-21	100.1	98.2	0.49	8.7	82.3	0.98	9.5	5.9	1.00	

Continue table 2

	Specimen										
This paper	SCB-22	93.5	89.3	0.54	8.7	72.4	0.96	8.3	5.0	1.00	Two-point
	SCB-23	84.9	86.2	0.43	8.9	69.8	1.01	7.8	4.7	0.83	
	SCB-24	78.4	78.3	0.35	9.2	62.9	1.00	6.8	4.4	0.67	
	SCB-25	71.4	71.8	0.31	9.3	57.2	1.00	6.2	4.1	0.58	
	SCB-26	61.0	64.4	0.21	9.5	50.8	1.06	5.3	3.6	0.42	
[5]	SCB-1	58.2	48.2	0.85	8.3	40.4	0.83	4.9	4.6	1.15	Two-point
	SCB-2	75.8	74.1	0.53	7.9	64.2	0.98	8.1	7.3	1.15	
	SCB-3	98.0	89.2	0.57	8.7	68.5	0.91	7.9	9.6	1.13	One-point
	SCB-4	109.8	104.3	0.30	10.0	146.2	0.95	14.6	11.2	1.13	
	B1	28.2	27.7	0.49	12.9	106.7	0.98	8.3	6.6	2.17	
	B2	27.3	27.4	0.50	13.1	105.5	1.00	8.1	7.2	2.25	
	B3	29.1	27.5	0.54	13.6	104.9	0.95	7.7	6.7	1.625	One-point
	B4	32.9	30.2	0.48	13.5	117.4	0.92	8.7	7.4	1.125	
	B6	29.8	29.7	0.44	13.1	115.8	1.00	8.8	6.9	2.00	
	B7	30.7	31.1	0.58	14.9	113.6	1.01	7.6	7.8	1.10	
	BN-1	26.7	23.6	0.81	16.0	90.6	0.88	5.7	6.5	1.00	Two-point
	BN-5	19.5	17.1	1.25	16.6	64.4	0.88	4.0	4.6	1.837	
	BN-6	18.7	21.6	0.89	15.0	82.7	1.15	5.5	5.1	1.00	

- **Comparison of the Ductility with the Ordinary Reinforced Concrete Specimens**

The ductility ratio of β_Δ table 3

Type	Specimen No.	Ductility ratio β_Δ	Loading mode	Type	Specimen No.	Ductility Ratio β_Δ	Loading mode
Comp. girder	SCB-1	4.6	Two-point with positive direction	R.C.	L1-10	2.1	One-point load (compression and bending)
	SCB-2	7.3			L1-11	1.2	
	SCB-3	9.6	One-point with positive direction		L1-12	2.1	
	SCB-4	11.2			L3-1	5.0	
	SCB-5	6.1	One-point with negative direction		L3-2	5.2	
	SCB-6	5.6			L3-3	7.2	
	SCB-7	5.2			L3-4	2.0	
	SCB-8	4.5			L3-5	8.7	
	SCB-19	3.9	Two-point with positive direction		L3-6	2.2	One-point load (bending)
	SCB-20	5.0			L3-7	4.3	
	SCB-21	5.9			L3-8	1.5	
	SCB-22	5.0			L3-10	1.5	
	SCB-23	4.7			L3-12	1.1	
	SCB-24	4.4			L3-13	1.1	
	SCB-25	4.1			L3-14	1.1	
	SCB-26	3.6			L3-15	1.0	
R.C.	L1-1	3.2	One-point load (compression and bending)		L3-16	7.2	
	L1-2	2.9			L3-17	5.1	
	L1-3	2.5			L3-9	2.0	Two-point load
	L1-8	2.0			L3-11	1.5	

DISASTER RESPONSE FOR LIFELINE SYSTEMS 1057

From the table 2 and table 3, we can know that the ductility ratio of composite girders is above 3.6, while the value of reinforcement concrete girder is below 3.0. It's demonstrated that the ductility of composite girders is much higher than that of the reinforcement concrete. Its main cause is that the height of compressive area of composite girder is smaller than that of the reinforcement concrete.

According to analysis and discussion, the main factors which affect the ductility of composite girders include: width of concrete flange; degree of shear connection N/N_f; rate of the transverse reinforcement ρ_{st}; compressive strength of concrete slab; tensile strength of steel girder.

Fatigue Test of Single Stud under Repeated Shearing Force

- **Testing Specimens and Test Results**

To make clear the fatigue strength of studs under repeated shearing forces, five push-out specimens have been tested. The design of testing specimens is adopted the ECCS specification, its dimensions are shown in table 4.

Dimension of push-out specimens   table 4

Item	PF1	PF2	PF3	PF4	PF5
Concrete slab (mm)		a=500	b=460	h=110	
Stud (mm)			$d_s \times h_s = 16 \times 90$		
Transverse reinforcement ρ_{st}(%)			Φ6-67 0.67		
Design strength of concrete (MPa)			40		

The dynamic loads of fatigue testing machine is 400kN, its working frequency is 4Hz. Apply cycles of load at 5×10^6 times, one side of specimen appears cracks, then failure at 7×10^6 times. The failure can be classified into two modes: first is the rod of stud being broken off. Second is the point of the welding part being rupture. If the welded quality is not very well, the second mode is apt to occur.

The test result of push-out specimens   table 5

Serial No.	Upper limit (KN)	Lower limit (KN)	Load amplitude (KN)	Shearing stress Δ τ (MPa)
PUSH_1	140	30	110	136.8
PUSH_2	124	38	96	106.0
PUSH_3	142	41	101	125.6
PUSH_4	140	30	110	137.1
PUSH_5	130	20	110	140.4

Figure 4   The failure mode of studs with the push-out specimen

- **Relation Between the Stud Stress and Repeated Fatigue Times**

We collect experimental values of 15 specimens which include our test and other institute in China. After datum treated, the regression equation is obtained as follows.

$$\log N = 16.205 - 5.13 \log \Delta \tau \quad (16)$$

where $\Delta \tau$ is the shearing stress range, N is repeated times. Its relation curve is shown in figure 5. In order to evaluate the fatigue life, we also put the ECCS's curve in figure 5. From which it can be seen that the fatigue life according the design of ECCS is relative conservation.

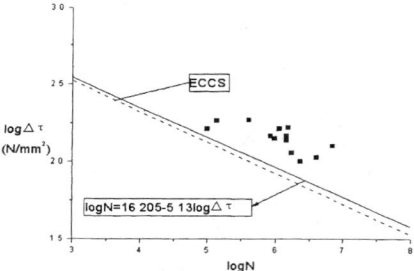

Figure 5   Fatigue life curve

Conclusion

The main results from this study can be summarized as follows:

(1) The method proposed in this paper is able to provide better ductility and energy dissipation ability as well under earthquake action.

(2) The composite girder using studs has good ductility and good performance of resist fatigue.

(3) The quantitative analysis of the ductility index of composite girder mainly

depends on the yielding and ultimate curvatures.

(4) The ductility ratio of composite girders are higher than that of the respective reinforced concrete girders. Based on the experimental data, it is indicated that the ductility ratio of curvature β_φ ($\beta_\varphi = \varphi_u / \varphi_y$) varies between 4~14.6. The ductility ratio of deflection of composite girders β_Δ ($\beta_\Delta = \Delta_{up} / \Delta_y$) varies between 3~11.2.

(5) The slip effect between steel girder and concrete slab of composite girders is favorite to increase the ductility ratio.

(6) The equation of fatigue life presented in this paper is much rational and economic than that of ECCS.

References

1. Viest, I. M., "Review of Research on Composite Steel-concrete Beams", Journal of Structural Division ASCE, June, 1960.
2. R. P. Johnson, "Composite Structural of Steel and Concrete", Volume 1,1975.
3. ECCS, "Composite Structural, 1981.
4. N. Gattesco, E. Giuriani, and A. Gubanan:" Low-cycle fatigue test on stud shear connectors", Journal of Structural Engineering, Febrary 1997, 145-150.
5. Nie Jianguo and Shen Jumin: "The method of discount rigidity with slip-effect", Civil Engineering Transaction, Volume 6, 1995.
6. J. Y. Richard Yen, Yiching Lin, and M. T. Lai: "Composite beams subjected to static and fatigue loads", Journal of Structural Engineering, June 1997, 765-771.
7. Cui Yuping and Nie Jianguo, "The experimental study on fatigue behavior of composite steel-concrete girder", Research paper of Beijing Municipal Engineering Research Institute, Oct. 2002.

Seismic Fragility Curves for Bridges: a Tool for Retrofit Prioritization

Bryant Nielson[1] and Reginald DesRoches[2]

Abstract

An emerging tool in assessing the seismic vulnerability of a highway and bridge network is the use of fragility curves. Fragility curves describe the probability of a structure being damaged beyond a specific damage state for various levels of ground shaking. This, in turn, can be used for prioritizing retrofit, pre-earthquake planning, and a tool for loss estimation. This is particularly useful in regions of moderate seismicity, such as the Central and Southeastern United States, where bridge officials are beginning to develop retrofit programs, as well as conducting pre-earthquake planning. In this paper, fragility curves for the bridges commonly found in the Central and Southeastern United States (CSUS) are presented. Using nonlinear analytical models, and a suite of synthetic ground motion, analytical fragility curves are developed for four bridge types. Comparison of the fragility curves show that the most vulnerable bridge types are the multi-span simply supported and multi-span continuous steel-girder bridges. The fragility curves can be used in connection with functionality information to assist in economic loss estimation.

Keywords: Fragility, Seismic, Bridges, Vulnerability, Retrofit, Moderate Seismic Zones

Introduction

Fragility curves, which describe the probability of a structure being damaged beyond a specific damage state for various levels of ground shaking, are an emerging tool for assessing seismic vulnerability in bridges. This, in turn, can be used for prioritizing retrofit, pre-earthquake planning, evaluation of loss of function of highway systems, and loss estimation tools. This is particularly useful in regions of moderate seismicity, such as the Central and Southeastern United States, where bridge officials are beginning to develop retrofit programs, and conducting pre-earthquake planning. In light of the damage to bridges observed in recent earthquakes (Bruneau, 1996; Buckle, 1994), there is a significant need to perform adequate assessment of the vulnerability of bridges prior to future seismic events.

---

[1]Graduate Research Assistant, Georgia Institute of Technology, Atlanta, GA 30332-0355; PH 404-385-0827; Bryant.nielson@ce.gatech.edu
[2]Assistant Professor, Georgia Institute of Technology, Atlanta, GA 30332-0355; PH 404-385-0826; Reginald.desroches@ce.gatech.edu

Fragility curves can be either empirical or analytical. Empirical fragility curves are usually based on the reported bridge damage from past earthquakes. Basoz et al.(1999) developed empirical fragility curves from the bridge damage resulting from the 1994 Northridge, CA earthquake using logistical regression analysis to account for uncertainties in the damage data. Shinozuka et al. (2000) used the maximum likelihood method to generate empirical fragility curves from the observations of bridge damage in the 1995 Kobe earthquake.

Analytical fragility curves are developed through seismic response data from the analysis of bridges. The fragility analysis generally includes three major parts: (a) the simulation of ground motions, (b) the simulation of bridges to account for uncertainty in bridge properties, and (c) the generation of fragility curves from the seismic response data of the bridges. The seismic response data can be obtained from nonlinear time history analysis (Shinozuka, 2000), elastic spectral analysis (Hwang, 2000), or nonlinear static analysis (Shinozuka, 2000).

Comparisons of empirical and analytical fragility curves have shown good agreement between theory and field observation for the 1994 Northridge, and 1989 Loma Prieta earthquakes (Bazoz, 1999). However, empirical fragility curves which have been developed for bridges in California following the 1989 Loma Prieta earthquake and the 1994 Northridge earthquake are generally not applicable to bridges in the Central and Southeastern United States due to difference in ground motion and bridge characteristics. Therefore to develop fragility curves for bridges in the CSUS, analytical methods must be used. In this study, fragility curves are developed for a class of bridges commonly found in the CSUS to assess the vulnerability of the bridges and assist in the prioritization of retrofit for these bridges. The fragility curves are developed by performing nonlinear time history analyses for six different classes of bridges. For each class of bridge, 10 sample bridges, representing the variability in material properties are developed. A set of 100 ground motion records, with varying magnitudes, distances, and peak ground accelerations, is used in the response history analysis. The fragility curves are developed for various damage states ranging from no damage to complete damage for the bridge columns.

Methodology for Analytical Fragility Curves

The analytical fragility curves developed in this study are based on nonlinear response history analysis. First, a bridge is represented by an analytical model, which includes the inelastic behavior of the appropriate components (i.e., columns, bearings, etc.). Second, earthquake input motion for various characteristic magnitudes, epicentral distances, and local soil conditions are developed. Third, uncertainties in the modeling of seismic source, path attenuation, local soil conditions, and bridge components are quantified to establish a set of earthquake-bridge samples. Fourth, for each earthquake-bridge sample, a nonlinear response history analysis is performed. Using pre-determined damage indices for the various bridge components, an overall damage state is assigned to the bridge. Finally, using a probabilistic seismic demand model obtained by regression analysis on the simulated damage data, the fragility curve can be developed which describes the probability of reaching or exceeding a damage state, as a function of an input parameter (typically peak ground or spectral acceleration).

Bridge Characteristics

Detailed analysis of the bridge inventory in the Central and Southeastern United States, performed in a previous study, shows that approximately 95% of the bridges in the CSUS are multi-span simply supported, multi-span continuous bridges, or single span bridges (Choi, 2002). The majority of these bridges have either pre-stressed concrete girders, or steel girders with a reinforced concrete deck. The substructure in these bridges typically consists of concrete bent cap supported on multi-column bents. A brief description of four of the six bridge types is provided below. Fragility curves for the single span bridges are not developed in this study since they do not have interior column supports.

The multi-span simply supported steel girder (MSSS-SG) bridge consists of 3 simply supported spans supported on multi-column bents, as shown in Figure 1. Each girder is typically supported by a fixed steel bearing and/or expansion steel (rocker/sliding) bearing. The multi-span continuous steel girder (MSC-SG) bridge is similar to the MSSS-SG bridge, except that the girders are continuous across the bent. The multi-span simply supported pre-stressed concrete (MSSS-PSC) girder bridge consists of 3 simply supported spans supported on multi-column bents. The girder connection to the bent cap consists of elastomeric bearing pads with dowel bars projecting 9-inches into the cap and 6 inches up into holes cast in the bottom of the pre-cast girder ends. The multi-span continuous pre-stressed concrete (MSC-PSC) girder bridge is similar to the MSSS-PSC girder bridge, except that the superstructure is made continuous by either casting a parapet between the girders or making the deck continuous.

The multi-span bridges all have similar substructures consisting of multi-column bents. The columns have a one-percent vertical reinforcing ratio with #3 or #4 circular ties spaced at 305 mm vertically. The column vertical bars are spliced at the top of the footing with dowels projecting from the pile cap. The abutment is a stub abutment as designed in the AASHTO specifications with an expansion joint between the deck slab and abutment.

Analytical Modeling of Typical Bridges

Since the six types of bridges discussed above consist of elements that may exhibit highly nonlinear behavior, such as the steel or elastomeric bearings, columns, abutments, and impact, a two-dimensional nonlinear analytical model of the bridges is developed using DRAIN-2DX. Since the superstructure is expected to remain linear under longitudinal earthquake motion, it is modeled using a linear element that represents the stiffness and mass of the composite girder and deck. The columns are modeled using the DRAIN-2DX fiber element. Each fiber has a stress-strain relationship, which can be specified to represent unconfined concrete, confined concrete, and longitudinal steel reinforcement. The steel bearings used in the steel-girder bridges and the elastomeric bearing/dowel combination used in the pre-stressed concrete girder bridges are modeled using nonlinear elements, based on previous tests of similar bearings (Mander, 1996). The behavior of the abutment in both passive and active action, as well as the impact between decks is modeled using nonlinear inelastic elements. The pile foundation is modeled using a combination of linear springs in the horizontal and rotational directions.

Input Ground Motion Simulation

Since few recorded strong ground motion records in the Southeastern and Central United States exists, synthetic ground motions developed by Hwang et al. (2000) are used. The suite consists of 100 ground motion records, distributed from weak to strong probabilistically and includes uncertainty in the soil and seismic characteristics. The peak ground acceleration of the ground motions range from 0.07g to 0.51g. The range of moment magnitude is M_w=6.0-8.0, and the range of epicentral distance is 40-100 km. The simulated ground motion records take into account the uncertainties in seismic source, path attenuation, and local soil conditions. Given a moment magnitude and epicentral distance, the uncertainties in the stress parameter, strong motion duration, cut-off frequency, and quality factor are also taken into account. The variability of soil parameters considered in the site response analysis are the relative density of sand and the undrained shear strength of clay. The uncertainties in dynamic soil parameters, such as the shear modulus and damping ratio are also included in the simulation.

Bridge Model Simulation

To represent the inherent variability in the material properties of steel and concrete, the uncertainties in the bridge strength and stiffness are included in the modeling of the four bridge types. The compressive strength of concrete and the yield strength of steel reinforcing are taken as random variables with a mean and standard deviation as recommended by Ellingwood and Hwang (1985) and Mirza and MacGregor(1979). For each sample set, a cross sectional analysis is performed to obtain the moment curvature relationship for the column sections to be used in the nonlinear analysis. The gap was also included to account for the effect of deck pounding (Choi, 2002).

Using the Latin Hypercube sampling technique (Ayyub, 1989), a set of 10 nominally identical but statistically different bridges were developed for each bridge type. Each bridge sample is matched with a soil profile sample, and 10 earthquake samples, resulting in a total of 100 earthquake-site-bridge samples for each bridge type.

Characterization of Damage

Most studies on fragility analysis of bridges use the column ductility as the primary damage measure. Park and Ang (1985) suggested a damage index based on energy dissipation, and Hwang et al. (2000) used the capacity/demand ratio of the bridge

Table 1. Qualitative and quantitative description of column damage states

Damage States	Description	Ductility Ranges
No Damage	No damage to a bridge	$\mu_\theta < 1.0$
Slight Damage	Minor spalling and cracks at hinges, minor spalling at the column (damage requires no more than cosmetic repair).	$1.0 < \mu_\theta < 2.0$
Moderate Damage	Any column experiencing moderate cracking and spalling (column structurally still sound)	$2.0 < \mu_\theta < 4.0$
Extensive Damage	Any column degrading without collapse (column structurally unsafe).	$4.0 < \mu_\theta < 7.0$
Complete Damage	Any column collapsing.	$7.0 < \mu_\theta$

columns to develop fragility curves. In this study, damage states are defined for column curvature ductility demand. The damage state definitions used are based on recommendations from previous studies (Mander, 1996). The limit states used in this study are presented in Table 1. It should be noted that engineering judgment should always be used when determining the damage states, as these vary depending on the type, age, and condition of the bridge.

Fragility Curves

A fragility curve describes the conditional probability that the structural demand caused by various levels of ground motion exceeds the structural capacity defined by a damage state. In this study, five damage states are quantified in terms of the column curvature ductilities. The probability that the demand on the structure exceeds the structural capacity can be computed as follows:

$$p_f = P\left[\frac{S_d}{S_c} \geq 1\right] \quad (1)$$

Where p_f is the probability of exceeding a specific damage state, S_d is the structural demand, and S_c is the structural capacity.

If the structural capacity and seismic demand are described by a log-normal distribution, the probability of reaching or exceeding a specific damage state will be log-normally distributed, which can be obtained by a log-normal cumulative probability density function as shown in equation 2.

$$p_f = \Phi\left[\frac{\ln(S_d/S_c)}{\sqrt{\beta_d^2 + \beta_c^2}}\right] \quad (2)$$

Where β_c is the logarithmic standard deviation for the capacity, β_d is the logarithmic standard deviation for the demand, and $\Phi[\cdot]$ is the standard normal distribution function. The seismic demand is expressed as

$$Ln(S_d) = a \ln(x) + b \qquad (3)$$

where a and b are unknown regression coefficients, and x is the ground motion intensity parameter (typically PGA or S_a). The composite logarithmic standard deviation $(\beta_c^2 + \beta_d^2)^{1/2}$, known as the dispersion, is taken from values recommended in HAZUS99 (1999).

Results

The fragility curves for the four bridge types are presented in Figure 2. The curves show that the columns for the MSC-PSC girder bridge have a significantly lower probability of exceeding the damage states than do the other three bridge types. The shape of the fragility curves for the other bridge types appear to be relatively close to one another, indicating that their probabilities of exceeding a particular damage state are approximately the same.

To enable a better comparison between the fragilities of the four bridge types, Figure 3 shows the median PGA for the various damage states. This figure shows that the median PGA for the MSC-PSC girder bridge to exceed a slight damage state is 0.79 g. This is a significantly larger PGA than the other three bridge types display, which have PGA's that are in the range of 0.30-0.45g with the MSC-SG bridge being the most vulnerable.

Looking at the moderate damage state, it can be seen that this same trend continues. The MSC-PSC girder bridge has a median PGA of 1.43g while the median PGA for the other bridge types range between 0.48 and 0.79g. The steel girder bridges are the most vulnerable when it comes to column damage, as they have the lowest median PGA's. The MSC-PSC girder bridge, is the least susceptible to damage during an earthquake event, of all the bridge types presented in this study. This supports the common practice of making the pre-stressed concrete girders, continuous to reduce dead and live load moments and to reduce maintenance required at the joints.

It is important to note that all significant vulnerable components, such as bearings, abutments, pile caps, etc., should be evaluated when determining the overall fragility of a bridge. Previous studies have shown that the bearings as well as some of the other components may result in lower median PGA for various damage states (Choi, 2002).

Conclusions and Future Work

The fragility curves in this study have been generated for typical multi-span bridges that are found the Central and Eastern United States. Based on the analytical fragility curves developed, it is found that the multi-span continuous steel girder bridge is the most vulnerable where as the multi-span continuous pre-stressed concrete girder bridge is the least vulnerable. These fragility curves are a good indicator of which bridge types have the highest probability for damage and are useful for assigning retrofit prioritization.

Fragility curves alone, however, are not adequate for determining the potential losses resulting from bridge damage due to an earthquake event. The loss of functionality is critical in understanding the indirect losses resulting from reduced traffic flow and the downtime of bridges needing repair. There is a need to look at the functionality of the bridge (restoration curves) to be able to assess the impact and therefore losses associated with a given earthquake event. Functionality-ground motion relationships for various bridge types and states of retrofit, would provide a much more useful tool for determining the need for retrofit and also what mode of retrofit would give the most desired results.

The Applied Technology Council has attempted to quantify functionality in the form of restoration curves (ATC, 1985). These are done for just two classes of bridges (Major bridges which have spans greater than 500' and conventional bridges that have spans less that 500'). These relationships were developed based on experts who were asked to give an estimated time required to achieve 30%, 60% and 100% recovery of the functionality of a bridge damaged during an earthquake. Since this information was released in 1985, it has been and is currently being used in earthquake loss programs such as HAZUS (1999).

There have been many who have seen a need to incorporate the information from recent earthquakes to better define these relationships. In 1991 as the ATC attempted to expand the methodology in ATC-13 from California to the rest of the country, they also saw the need to look at recent earthquakes to help refine the restoration curves in the ATC-13 (ATC, 1991). The refinement of these relationships should incorporate more functionality information than just a restoration timeline. Additional information that should be obtained includes lane closures, reduced speed limits, weight limits etc. (Werner, 1997). These are all critical in helping to refine network loss estimations.

To adequately address the issue of damage-functionality, a more comprehensive evaluation of the relationship between damage in a bridge and its functionality needs to be performed. This evaluation should take advantage of the lessons that were learned from recent earthquakes. This can be done by conducting a survey experts in the field similar to that by Hwang et al.(2000). This should include experts that represent a broad cross-section of the United States. Actual functionality data from the recent earthquakes should be collected and used to calibrate/validate the results of the survey.

Acknowledgements

This study has been supported by the Earthquake Engineering Research Centers program of the National Science Foundation under Award Number EEC-9701785 (Mid-America Earthquake Center). The authors would like to thank Dr. Howard Hwang for is valuable comments throughout this project.

References

Applied Technology Council (1985). "Earthquake Damage Evaluation Data for California," ATC-13, Redwood City, CA.

Applied Technology Council (1991). "Seismic Vulnerability and Impact of Disruption of Lifelines in the Conterminous United States," ATC-25, Redwood City, CA.

Ayyub, B.M. and Lai, K. (1989). "Structural Reliability Assessment Using Latin Hypercube Sampling," Proceedings of ICOSSAR '89, the 5th International Conference on Structural safety and Reliability, San Francisco, CA, USA.

Basoz, N. I, Kiremidjian, A. S., King, S. A., and Law, K. H. (1999). "Statistical Analysis of Bridge Damage Data from the 1994 Northridge, CA, Earthquake," *Earthquake Spectra*, EERI, Vol. 15, No. 1, February, p.25-53.

Bruneau, M., Wilson, J. C., Tremblay, R (1996). "Performance of Steel Bridges During the 1995 Hyogo-Ken Nanbu (Kobe, Japan) Earthquake," *Canadian Journal of Civil Engineering*, Vol. 23, No. 3, June, p. 678-713.

Buckle, I. G., ed. (1994). "The Northridge California Earthquake of January 17, 1994: Performance of Highway Bridges," *NCEER-94-0008*, National Center for Earthquake Engineering Research, Buffalo, N. Y. March 24, 994.

Choi, E. (2002). "Seismic Analysis and Retrofit of Mid-America Bridges," Ph.D. Thesis, Department of Civil and Environmental Engineering, Georgia Institute of Technology, Atlanta, GA.

Ellingwood, B. and Hwang, H. (1985). "Probabilistic Descriptions of Resistance of Safety-related Structures in Nuclear Power Plant," *Nuclear Engineering and Design* 88, p.167-178.

HAZUS (1999). "Earthquake Loss Estimation Methodology," User's Manual, National Institute of Building for the Federal Emergency Management Agency, Washington, D.C.

Hwang, H., Jernigan, J.B., Billings, S., Werner, S.D. (2000). "Expert Opinion Survey on Bridge Repair Strategy and Traffic Impact," Proceedings of the Post Earthquake Highway Response and Recovery Seminar, St. Louis, MO, September 5-8.

Hwang, H., Jernigan, J.B., and Lin, Y (2000). "Evaluation of Seismic Damage to Memphis Bridges and Highway Systems," *Journal of Bridge Engineering*, Vol. 5, No. 4, ASCE, November, p. 322-330.

Hwang, H., Liu, J., and Chiu, Y. (2000). "Seismic Fragility Analysis of Highway Bridges," Center for Earthquake Research and Information, The University of Memphis, Memphis, TN 38152.

Mander, J. B., Kim, D. K., Chen, S. S., and Premus, G. J. (1996). "Response of Steel Bridge Bearings to the Reversed Cyclic Loading," *Technical Report NCEER 96-0014*, Buffalo, NY.

Mizra, S. and MacGregor, J. (1979). "Variability of Mechanical Properties of Reinforcing Bars," *Journal of Structural Engineering*, ASCE, Vol. 105, No. 5, p.921-937.

Park,Y-J. and Ang, A. H-S. (1985). "Mechanistic seismic damage model for reinforced concrete," *Journal of Structural Engineering*, ASCE, Vol. 3, No.4, p.722-739.

Randall, M. J., Saiidi, M. S., Maragakis E. M., and Isakovic, T. (1999). "Restrainer Design Procedures for Multi-Span Simply Supported Bridges," Technical Report MCEER-99-0011, 1999.

Shinozuka, M., Feng, M. Q., Kim, H., and Kim, S. (2000). "Nonlinear Static Procedure for Fragility Curve Development," *Journal of Engineering Mechanics*, Vol., 126, No. 12, ASCE, December, p.1287-1295.

Shinozuka, M., Feng, M. Q., Lee, J., and Naganuma, T. (2000). "Statistical Analysis of Fragility Curves," *Journal of Engineering Mechanics*, ASCE, Vol. 126, No. 12, December, p.1224-1231.

Werner, S.D., Taylor, C.E., and Moore, J.E. II (1997). "Loss Estimation Due to Seismic Risks to Highway Systems," *Earthquake Spectra*, EERI, Vol. 13, No. 4, November, 1997, p585-604.

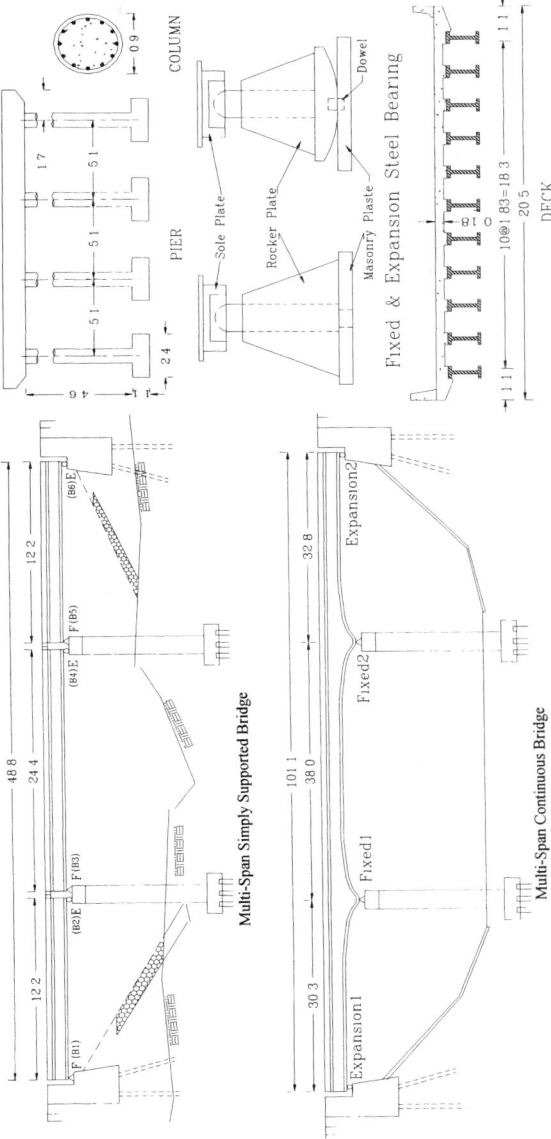

Figure 1. Typical Multi-Span Simply Supported and Multi-Span Continuous Steel Bridge in the Central and Southeastern United States.

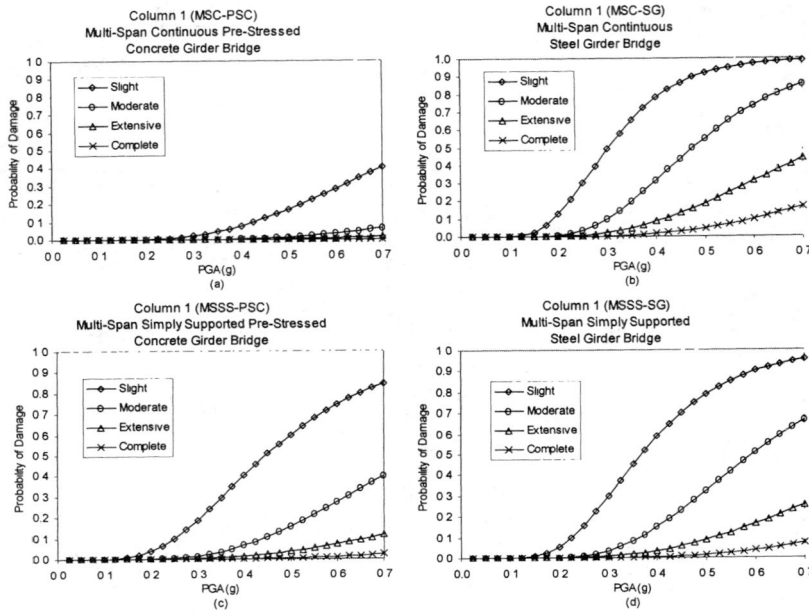

Figure 2. Component Fragility Curves for; (a) MSC-PSC, (b) MSC-SG, (c) MSSS-PSC, (d) MSSS-SG

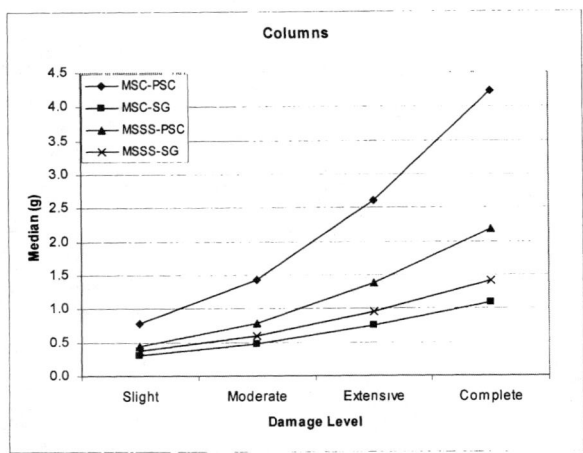

Figure 3. Comparison of the Median Values of PGA for the Typical Bridges

Subject Index

Page number refers to the first page of paper

Alaska, 522, 535, 547, 556, 566, 576, 717
Assessments, 707

Benefit cost ratios, 484, 494
Bents, 153
Bridges, 153, 241, 1060
Bridges, composite, 1050
Bridges, concrete, 906
Bridges, girder, 1050
Bridges, highway, 143, 987
Bridges, piers, 133, 1040
Bridges, railroad, 839
Building codes, 637
Buried pipes, 445, 566, 717, 744, 757, 966

California, 1, 59, 163, 173, 183, 193, 203, 219, 274, 368, 415, 474, 494, 587, 727, 868, 906, 1031
Canada, 274
China, 445
Coastal structures, 385, 395, 415
Codes, 349
Community relations, 19
Computation, 697
Computer software, 484, 494, 878
Connections, 597, 627
Costs, 1040
Coupling, 697
Crossings, 744, 757
Culverts, 294

Dam safety, 1
Damage assessment, 123
Damping, 1031
Disaster relief, 9, 29, 113, 231, 425

Disasters, 19, 103
Displacement, 597
Ductility, 1050
Dynamic response, 946

Earth pressure, 946
Earthquake damage, 39, 224, 284, 308, 405, 484, 494, 868, 878, 906
Earthquakes, 9, 49, 73, 82, 92, 113, 123, 143, 213, 219, 241, 255, 265, 274, 294, 368, 522, 547, 556, 566, 576, 587, 597, 717, 727, 789, 809, 819, 829, 859, 896, 916, 926, 936, 956, 966, 987, 1021, 1031, 1050
Economic factors, 435
El Salvador, 265
Electrical equipment, 607, 617, 627, 647, 667, 677
Emergency services, 9, 29, 219, 231, 576, 926
Equipment, 597, 637
Europe, 339
Experimentation, 667

Facilities, 677
Federal agencies, 359
Field investigations, 213
Friction, 1040

Gas pipelines, 284, 349, 744
Geographic information systems, 39, 926
Government policies, 1
Ground motion, 163, 657, 769
Guidelines, 133, 143, 318, 327, 415, 455, 504
Guideways, 193

Harbors, 425
Highway design, 886
Highways, 92, 849
History, 143, 219
Hospitals, 82, 224

Illinois, 1040
India, 9, 274, 308
Infrastructure, 19, 29, 73, 219, 255, 265, 308, 318, 359, 474, 779, 799, 819, 829, 936, 1001, 1011
International commissions, 213
Iran, 349

Japan, 39, 82, 103, 224, 284, 707, 809, 819

Laws, 1
Liquefaction, 512, 987, 1021

Maps, 113
Methodology, 153, 494, 829
Models, 255, 339, 395, 757, 769, 859, 868, 977, 987
Monitoring, 464

Natural gas, 368
New Zealand, 29

Oil pipelines, 535, 547, 556, 566, 576

Performance evaluation, 1011
Pile foundations, 977, 1011
Pipelines, 378, 522, 736
Pipes, 294
Planning, 59
Ports, 425
Power, 474
Pressures, 1001
Probabilistic models, 707

Rail transportation, 274
Rapid transit systems, 163, 173, 183, 203
Rehabilitation, 39, 405, 799
Remedial action, 59, 474, 484, 512, 779, 987
Research, 143
Restoration, 255
Retaining walls, 946
Retrofitting, 163, 193, 203, 255, 378, 435, 906, 1031, 1040, 1060
Risk analysis, 123, 173, 183, 587, 789, 809, 819, 849, 886, 896
Risk management, 82, 92, 103, 425, 677, 799
Roads, 859, 916

Safety, 368
San Francisco, 163, 173, 183, 193, 203, 868
Sea walls, 405
Security, 19, 359
Seismic analysis, 385, 395, 504, 657, 849, 859, 886
Seismic design, 143, 318, 327, 339, 378, 637, 1001, 1011
Seismic effects, 173, 193, 203, 435, 464, 474, 677, 687
Seismic hazard, 455, 535, 757, 779, 799, 839, 956, 1040
Seismic response, 133, 241, 294, 385, 395, 415, 617, 647, 667, 697, 736, 829, 977, 1060
Seismic stability, 445
Seismic tests, 607, 627
Simulation, 425, 717, 769
Soil deformation, 966
Soil improvement, 987
Soil-structure interaction, 936, 966, 977
Standards, 327
Storage tanks, 327, 339

Structural reliability, 657
Switzerland, 73

Taiwan, 241, 274, 1021
Tennessee, 868
Terrorism, 789
Traffic, 916
Transmission lines, 49, 687, 697
Transmission towers, 687, 697
Transportation networks, 231, 868, 926
Transportation systems, 123, 896
Travel demand, 868
Trenchless technology, 736
Tsunamis, 59
Tubes, 203
Turkey, 224, 274
Two-dimensional models, 956

Underground structures, 597, 1001
United States, 49
Urban areas, 231, 445, 464, 474, 727
Urban planning, 779
Utah, 92
Utilities, 484, 494, 707

Validation, 878

Washington, 405, 512
Water distribution, 435, 464, 504
Water pipelines, 284, 512, 727
Water services, 455
Water supply, 39, 727
Wharves, 385, 395, 415

Author Index

Page number refers to the first page of paper

Abdoun, Tarek, 757
Adib, Hamid R., 789
Antaki, George, 378
Aziz, T. S., 637

Ballantyne, Donald, 512
Ballantyne, Donald B., 494
Barnett, Elson T., 59
Basoz, Nesrin, 896, 926
Beavers, James E., 49
Beckman, Chris J., 294
Billington, S. L., 936
Bortugno, Edward, 484, 494
Brundson, David, 29
Buchanan, James, 789
Buckle, Ian G., 886
Bucknam, Stephen, 484, 494
Buswell, John, 405
Byers, William G., 274

Çagnan, Z., 255
Cameron, Wanda I., 946
Campbell, Kenneth W., 829
Carson, Paul A., 556
Chang, Kuo-Chun, 241
Chang, Stephanie E., 474
Chang, SungHee, 966
Chen, Yanyan, 916
Cho, Sungbin, 868, 878
Chu, B. L., 1021
Chu, Daniel B., 1021
Chu, Dongmei, 977
Cluff, Lloyd S., 535
Constantinou, Michael C., 1040
Cooke, Harry G., 987
Crouse, C. B., 535

Dastous, Jean-Bernard, 597
Davidson, R., 255
Davis, Craig A., 1001
Dawson, E. M., 385, 395
Der Kiureghian, Armen, 617
DesRoches, Reginald, 1060
Dicleli, Murat, 1040
Drobny, John, 789
Duvernay, Blaise, 73

Ebeling, Robert M., 946
Eckhardt, Anne, 73
Edwards, Curt, 213
Edwards, Curtis, P.E., 308
Eguchi, Ronald T., 484, 494, 504, 878, 916
Eidinger, John, 173, 435, 727
Emami, Bardia, 143
Engi, Dennis, 19
Erdman, Craig, 59
Evans, Noel, 29

Fan, Yue Yue, 868, 896
Filiatrault, Andre, 597
Filiatrault, André, 647, 667
Finn, W. D. Liam, 1011
Fok, Eric, 193, 203
Fotinos, George, 203
Fujita, N., 1011

Gadicherla, Vikram, 757
Ger, Jeffrey, 153
Ghosh, Shubharoop, 878
Graf, W. P., 455
Green, Russell A., 946
Greenstein, Jacob, 799
Gregor, N., 163

Habenberger, J., 339
Habibi, Hossein Motevalli, 231
Hackney, David A., 566
Hall, William J., 522, 556
Haque, A., 19
Heubach, William, 512
Higashihara, Hiromichi, 103
Honegger, Douglas, 368
Honegger, Douglas G., 318, 359, 504, 744
Horton, Tom, 173
Hosseini, Mahmood, 231, 349, 779, 839
Hsu, S. C., 1021
Hughes, Bill, 193
Hutson, N. J., 627
Huyck, Charles K., 484, 494, 878

Imanishi, Tatsuhiko, 39
Ivanov, Radan, 717

Johnson, Elden R., 522, 566, 576
Jones, David, 183
Jones, S. L., 936
Jones, Stacey, P.E., 415

Kempner, L., Jr., P.E., 627
Kempner, Leon, Jr., 607
Kempner, Leon, Jr., P.E., 687
Kim, Moon Kyum, 966
Kim, Sang-Hoon, 906
Kim, TaeWook, 966
King, Stephanie A., 789
Kiremidjian, Anne, 896, 926
Kiremidjian, Anne S., 425
Kuo, Kung-Yuan, 241
Kuwata, Yasuko, 82

Laatsch, Edward M., 359
Laatsch, Edwards M., 318
Lang, Kerstin, 73
Lee, Chun-Yu, 123

Lee, Don, 1031
Lee, Shannon, 1021
Leonard, Blaine D., P.E., 92
Li, Hong-Nan, 697
Liang, Jianwen, 464
Lim, Yunmook, 966
Lin, P. S., 1021
Litehiser, J., 163
Loh, Chin-Hsiung, 123
Lu, Chih-Hung, 241
Luna, Ronaldo, 956
Lund, L., 455
Lund, Le Val, P.E., 219, 265, 736

Mageau, Daniel, 405
Mallare, Chip, 193
Mansour, Mouhamad Y., 1040
Marianos, W. N., 143
Marrone, J., 163
Matsuda, Ed, 173, 193
Matt, Howard, 647
Mehrain, M., 385
Metz, Michael C., 566
Meyer, Keith J., 547, 556
Mitchell, James K., 987
Mohaymany, Afshin Shariat, 231
Mokha, Anoop, 1040
Moore, James, 896
Moore, James E., II, 868
Moss, Robb E. S., 1021
Mueller, W. H., III, P.E., 627
Mueller, Wendelin H., III, P.E., 687
Murakami, M., 769
Murono, Y., 769
Murphy, Vivyan, 59

Naecker, Philip A., 113
Nezamabadi, Maryam Firoozi, 839
Nichols, John M., 49
Nielson, Bryant, 1060
Nojima, Nobuoto, 707

Norton, J. David, 522
Nyman, Douglas J., 522, 576, 744

Oberholtzer, Gary, 203
O'Fallon, John, 133, 657
Olgun, C. Guney, 946
Olson, Robert A., 1
O'Rourke, Michael, 757
O'Rourke, T. D., 936
Ostadan, F., 163
Ostrom, Dennis K., 587

Pachakis, Dimitris, 425
Page, Robert A., 535
Perkins, David, 859
Pickett, Mark A., 224
Preuss, Jane, 59
Prospero, Javier, 727

Riley, Michael J., 687
Roach, Christopher H., 576
Roblee, Cliff, 113
Roth, W. H., 385, 395
Rowshanzamir, Farhad, P.E., 405

Sackman, Jerome L., 617
Salmon, Mark, 183
Sato, T., 769
Sayegh, A., 385
Schiff, Anshel J., 213, 607
Schwarz, J., 339
Seed, Raymond B., 1021
Seligson, Hope A., 474, 484, 494, 829
Shaoping, Sun, 445, 1050
Shastid, Tom, 727
Sheckler, Timothy D., 318, 359
Shemirani, Leila Niazi, 779
Shen, J. Jerry, 133, 657
Sheng, Li-Hong, 1031
Shi, Wen-Long, 697
Shimazu, Haruki, 809, 819

Shimazu, Naoki, 809, 819
Shimazu, Takayuki, 809, 819
Shinozuka, Masanobu, 906
Shukla, Vipin C., 9
Shumuta, Yoshiharu, 677
Singh, D. N., 9
Slemmons, D. Burton, 535
Song, Junho, 617
Sorensen, Steve P., 547, 556
Stearns, Christopher, 667
Stewart, H. E., 936
Stewart, Jonathan P., 1021
Sugito, Masata, 707
Swanson, Dave, P.E., 405

Takada, Shiro, 39, 82, 284, 717
Taylor, C. E., 455
Taylor, Craig, 859
Thavaraj, T., 1011
Thenhaus, Paul C., 744
Truman, Kevin Z., 977
Tsai, J. S., 1021
Tseng, Wen, 203
Turner, Fred, 368
Turner, Loren, 113

Uddin, Nasim, P.E., 19
Ueno, Junichi, 284

Vayeghan, Fariborz Yaghoobi, 839
Volz, T., 455

Wald, David J., 113
Wang, James, 183
Wang, Lisa Yunxia, P.E., 327
Wang, Mark C. H., 1021
Wang, Su-Yan, 697
Watkins, Reid M., 484
Weismair, Max, 415
Werner, Stuart D., 849
Wiggins, J. H., 455

Williams, Meredith, 896, 926
Wu, Ching, 173, 183, 203

Yang, Han, 445
Yazlali, Ozgur, 896
Yeh, Chin-Hsun, 123
Yen, Phillip, 153
Yen, W. Phillip, 133, 657
Yin, Peter, 415

Youd, T. Leslie, 294
Youngs, R., 163
Yu, M. S., 1021
Yuping, Cui, 1050

Zayas, Victor, 1040
Zheng, Wei, 956
Zhongzhu, Shi, 1050